Communication Protocol Engineering

Second Edition

Communication Protocol Engineering
Second Edition

Miroslav Popovic

CRC Press
Taylor & Francis Group
Boca Raton London New York

CRC Press is an imprint of the
Taylor & Francis Group, an **Informa** business

CRC Press
Taylor & Francis Group
6000 Broken Sound Parkway NW, Suite 300
Boca Raton, FL 33487-2742

First issued in paperback 2021

© 2018 by Taylor & Francis Group, LLC
CRC Press is an imprint of Taylor & Francis Group, an Informa business

No claim to original U.S. Government works

ISBN-13: 978-1-03-209579-0 (pbk)
ISBN-13: 978-1-138-55812-0 (hbk)

Library of Congress Cataloging-in-Publication Data

Names: Popovic, Miroslav, 1961- author.
Title: Communication protocol engineering / Miroslav Popovic.
Description: Second edition. | Boca Raton : Taylor & Francis, CRC Press, 2018.
Identifiers: LCCN 2017043058| ISBN 9781138558120 (hardback : alk. paper) |
ISBN 9781315151243 (ebook)
Subjects: LCSH: Computer network protocols. | Computer networks--Standards.
Classification: LCC TK5101.55 .P67 2006 | DDC 621.382/12--dc23
LC record available at https://lccn.loc.gov/2017043058

Visit the Taylor & Francis Web site at
http://www.taylorandfrancis.com

and the CRC Press Web site at
http://www.crcpress.com

To my wife, Vlasta, and our sons Marko and Andrej

Contents

Preface to the First Edition

I wrote this book as a textbook for postgraduate students, but it might also be used by people in the industry to update specific knowledge in their life-long learning processes. The book partly covers the actual postgraduate course on computer communications and networks undertaken during the first semester of studies for the M.Sc. degree in computer engineering. Since nowadays we are witnessing the convergence of the Internet and the public telephone network, this book might also be useful to engineers with B.Sc. degrees in telecommunications.

The prerequisite for this book is knowledge of first order logic (predicate calculus), operating systems, and computer network fundamentals. The reader should also be familiar with C++ and Java programming languages.

My approach in writing this book was to provide all the details that the reader may need. I assumed that nothing is obvious. However, if you, the reader, find something obvious while reading the book, you are encouraged to skip ahead. If something is not clear later on, you may always return to what you skipped. Communication protocol engineering is a very interesting combination of abstraction and practice that requires a lot of details. It starts from a vision that gradually materializes in real-world artifacts. This happens through a typical engineering process. This book covers all aspects of communication protocol engineering, including requirements and analysis, design, implementation, and test and verification.

Many people helped me in writing this book. My gratitude goes to all of them. I thank my family for their continuous support; my niece Silvia Likavec for her valuable text corrections; and B.J. Clark, Nora Konopka, and Helena Redshaw, of Taylor & Francis, for their professional support. Special thanks go to my colleagues from the University of Novi Sad; Prof. Vladimir Kovacevic for giving his blessing for this book; Ph.D. student Ivan Velikic for the excellent cooperation (in his M.Sc. thesis we actually developed the FSM Library, one of the anchors of this book); Ph.D. student Ilija Basicevic (for helping me with the preparation of the examples in Sections 3.10.5, 4.5.2, and 5.5.2); Sonja Vukobrat (for helping me with the preparation of the example in Section 3.7); Laslo Benarik and Aleksander Stojicevic (for helping me with the preparation of Chapter 6); Milan Savic; Aleksander Stojicevic; and Cedomir Rebic (for helping me with the preparation of the examples in Sections 3.10.1 and 3.10.2); and Nenad Cetic (for helping me with the preparation of the example in Section 4.5.1). Thank you all!

Miroslav Popovic
Novi Sad

Preface to the Second Edition

The first edition of this book was well accepted by the readers, right from the beginning, back in 2006 when it was printed. Barnes & Noble bestselling rating reports indicated this fact rather well, e.g., the book was the bestseller on Oct. 4th, 2006, in the section "Networking, Telecommunications Protocols, & Standards." From that time to today, *Communication Protocol Engineering* has been a subject on a number of graduate level (M.Sc.) courses at universities worldwide—from the United States (The City College of New York, New York; University Heights Newark, New Jersey; etc.), over Europe (University of Novi Sad, Serbia; Lippe and Hoexter University of Applied Sciences, Germany), to far-east Australia (La Trobe University, Australia), to name just few of the more established points. Nowadays, *Communication Protocol Engineering* sounds like evergreen, similar to its much older predecessors Internet, C, and Linux, which are with us from the 1970s, and it seems that *Communication Protocol Engineering* is here to stay for many years to come, similar to its famous predecessors.

Twelve years after I wrote the first edition, I was glad to see that it was still aligned with the state of the art very well. Still, the book needed to be improved in two important areas, namely, compliance testing based on the standard Testing and Test Control Notation (known as TTCN-3), and model checking based on famous C.A.R. Hoare's process algebra Communicating Sequential Processes (CSP) and its accompanying tool named Process Analysis Toolkit (PAT). Hence, I made this new edition.

Technically, I made appropriate changes in Chapters 3 and 5. In Chapter 3, I have rewritten Sections 3.9 and 3.10 (Examples 1 and 2), and I adapted the TTCN references throughout the book in order to introduce the current standard TTCN-3 instead of the previous standard TTCN-2 (this decision was driven by the fact that TTCN-3 is a superset of the TTCN-2).

In Chapter 5, I revised Section 5.3. The new title of Section 5.3 is "Formal Verification," and it comprises the following two subsections: (i) 5.3.1. Formal Verification Based on Theorem Proving (this is the original Section 5.3), and (ii) 5.3.2 Formal Verification Based on Communicating Sequential Processes (this is the new section based on C.A.R. Hoare's process algebra CSP and the accompanying modeling, simulation, and automatic verification tool PAT).

Many people assisted me during the writing of this second edition. My gratitude goes to all of them. I thank Nora Konopka and Kyra Lindholm, of Taylor & Francis, for their professional support. Thanks again to my family, and my colleagues from the University of Novi Sad for their support throughout all these years.

I would also like to express my special gratitude to Dr. Sun Jun and the PAT Team for providing their PAT Examples in the public domain. I used some of their CSP# models to create the examples in Section 5.3.2.3.

Miroslav Popovic
Novi Sad

Author

Miroslav Popovic, Ph.D., earned all his degrees from the University of Novi Sad, Serbia. He defended his diploma thesis, "An Intelligent System Restart," in 1984; his M.Sc. thesis, "An Efficient Virtual Machine System," in 1988; and his Ph.D. thesis, "A Contribution to Standardization of ISO OSI Presentation Layer," in 1990. He became a full-time professor at the University of Novi Sad in 2002. Currently, he is teaching courses on software tools and real-time systems programming, as well as on intercomputer communications and computer networks. He is a member of IEEE (both the Computer and the Communications Societies) and ACM. He has published approximately 120 papers, and he has supervised many real-world projects for the industry, including telephone exchanges and call centers for Russian, German, Czech, and Serbian telecommunication networks. Taylor & Francis published his book, *Communication Protocol Engineering*, in 2006. He served as Serbian MC Member in EU COST 297 High Altitude Platforms of wireless communications, EU COST IC0703 Traffic Monitoring and Analysis, and EU COST Action IC1001 Transactional Memories (Euro-TM). His current research interests are engineering of computer-based systems, parallel programming, distributed systems, and security.

1

Introduction

Originally, the term **protocol** was related to the customs and regulations dealing with diplomatic formality, precedence, and etiquette. A protocol is actually the original draft, minutes, or record from which a document, especially a treaty, is prepared, e.g., an agreement between states. Today, in the context of computer networks, the term **protocol** is interpreted as a set of rules governing the format of messages that are exchanged between computers. Sometimes, especially if we want to be more specific, we use the term **communication protocol** instead.

The title of this book, *Communication Protocol Engineering*, is used to emphasize the process of developing communication protocols. Like other engineering disciplines, communication protocol engineering typically comprises the following phases (Figure 1.1):

- Requirements and analysis
- Design
- Implementation
- Test and verification

The process described in this book is the union of the UML (Unified Modeling Language)–driven unified development process (Booch et al., 1998) and, Cleanroom engineering (formal system design verification and statistical usage testing), with some elements of Agile programming (particularly unit testing based on JUnit). Of course, each organization should adapt and tune the process to its own needs and goals. For example, one organization may stick to the UML-driven unified development process, another may prefer Cleanroom engineering, yet another may use the combination of both, and so forth.

Because this book is written for the process in which all the existing state-of-the-art methods and techniques in the area are applied, it is independent of any particular engineering process. Therefore, this is as far as we will go in discussions on processes in this book. This book is not about managing processes. Rather, this book is intended for engineers. It provides the knowledge that an engineer needs to work in a modern organization involved in communication protocol engineering.

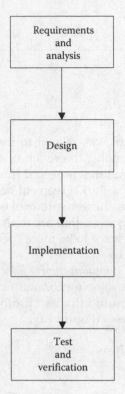

FIGURE 1.1
Typical communication protocol engineering phases.

The chapters are named by typical process phases: requirements and analysis, design, implementation, and test and verification. These chapters are actually used to classify various methods and techniques, and their accompanying tools. As previously stated, the approach taken in this book was to select the best methods and techniques from various methodologies rather than to stick just to a single methodology. The methods and techniques introduced here originate from the following methodologies:

- UML methodology
- ITU-T system specification and description methodology
- Agile unit testing methodology
- Cleanroom engineering methodology

UML methodology is based on various kinds of graphs, also referred to as diagrams. This book covers all of them, namely:

- Use case diagrams (Section 2.1)
- Collaboration diagrams (Section 2.2)

- Class diagrams (Section 3.1)
- Object diagrams (Section 3.2)
- Sequence diagrams (Section 3.3)
- Activity diagrams (Section 3.4)
- Statechart diagrams (Section 3.5)
- Deployment diagrams (Section 3.6)
- Component diagrams (Section 4.1)

ITU-T system specification and description methodology is based on three domain-specific languages, which this book also covers. These languages are

- Specification and description language (SDL) (Section 3.7)
- Message sequence charts (MSC) (Section 3.8)
- Testing and test control notation, ver. 3 (TTCN-3) (Section 3.9)

Agile unit testing methodology assumes writing the test cases before the code. Today, it is supported by the following two open-source packages (both covered in this book):

- JUnit, a package for automated unit testing of Java packages (Section 5.1)
- CppUnit, a library for automated unit testing of C++ modules (Section 5.5.1)

Cleanroom engineering methodology is based heavily on two main methods, both covered in this book. These methods are

- Formal system design verification. Today, more approaches exist to formal system design verification. This book covers formal verification based on automated theorem proving (Section 5.3).
- Statistical usage testing (Section 5.4).

The text of the book is organized as follows. At the end of this chapter, in Section 1.1, we introduce the notion of the communication protocol and related definitions.

Chapter 2 is devoted to the requirements and analysis phase of communication protocol engineering. The first part of that chapter introduces UML use case and collaboration diagrams (Section 2.1 and Section 2.2, respectively). The former is used for capturing both functional and nonfunctional system requirements, whereas the latter is used for making system analysis models. The second part of that chapter presents a real-world example—requirements and analysis of an SIP (Session Initiation Protocol, RFC 3261) Softphone. The example starts with the presentation of the domain-specific information related to SIP, continues with the SIP Softphone requirements model (in the form of the corresponding use case diagram), and ends with the SIP Softphone analysis model (in the form of the corresponding collaboration diagram).

Chapter 3 covers the design phase of communication protocol engineering. In this chapter, we will see that communication protocols are actually modeled as finite state machines (FSMs). The first part of the chapter introduces UML diagrams related to the design phase: class, object, sequence, activity, statechart, and deployment diagrams (Section 3.1, Section 3.2, Section 3.3, Section 3.4, Section 3.5, and Section 3.6, respectively). The second part of Chapter 3 covers domain-specific languages, which originated at ITU-T, namely SDL, MSC, and TTCN-3 (Section 3.7, Section 3.8, and Section 3.9, respectively). The third part consists of design examples. The first three examples are rather academic, while the fourth example shows the design of the sliding window concept. The fifth example is a real-world design example—the design of the SIP INVITE client transaction, which is a part of the SIP protocol stack.

Chapter 4 is devoted to the implementation phase of communication protocol engineering. At the beginning of this chapter, we introduce the UML component diagrams (Section 4.1). The second part of Chapter 4 presents various implementation approaches. Section 4.2 presents three examples of approaches that can be used. The main goal of this study is to provoke dilemmas by studying three different concepts of implementation and to promote creative thinking about a spectrum of possible implementation paradigms before restricting ourselves to a single one. This short overview includes the implementations as nested switch-case statements, the implementation based on the interpretation of protocol messages using a protocol definition data structure, and the implementation based on a class hierarchy and state transition map. The second part of Chapter 4 ends with the introduction of the state design pattern (Section 4.3), with a catalogued FSM implementation approach.

The third part of Chapter 4 (Section 4.4) introduces one concrete, industrial-strength implementation paradigm based on the FSM Library, a library of C++ classes used for modeling communication protocols as FSM. This paradigm has been successfully used on a series of real-world projects, such as SS7, DSS1, V5.2, H.323, SIP, and so on. This part of the book covers FSM Library features and internals as well as the rules for writing FSM Library–based implementations. The last part of Chapter 4 contains two real-world examples of the FSM Library–based implementations. The first is the implementation of the POP3 communication protocol, the TCP/IP Internet protocol for receiving e-mail messages. The second is the SIP INVITE client transaction, a part of the SIP protocol stack.

Chapter 5 deals with the testing and verification phase of communication protocol engineering. The first part starts with the introduction of unit testing based on JUnit, the open-source testing framework for unit testing Java programs, originally developed by Erich Gamma and Kent Beck (Section 5.1). Next, we introduce conformance testing (Section 5.2), actually the first stage of communication protocol acceptance testing. Conformance testing is typically based on the TTCN test suite specification. We then introduce formal verification of both system design and implementation (Section 5.3) based on: (i) automated theorem proving (Section 5.3.1) and (ii) the C.A.R Hoare's process algebra CSP (Section 5.3.2). In this book, we use the theorem prover Theo (in Section 5.3.1) and the modeling, simulation, and automatic verification tool PAT (in Section 5.3.2) as the accompanying tools for this purpose.

The first part of Chapter 5 ends with the introduction of statistical usage testing (Section 5.4) based on product operational profiles. The second part of Chapter 5 consists of two real-world examples. The first example shows the unit testing of the SIP INVITE client transaction based on the usage of the CppUint, the library for unit testing C++ modules. The second example demonstrates the integration testing of the SIP INVITE client transaction.

Chapter 6 is written as a programmer's reference manual for the FSM Library. The first part starts with the introduction of two main classes, *FSMSystem* and *FiniteStateMachine* (Section 6.2). Next, we introduce three main groups of basic functions supported by the FSM Library: time, memory, and message management functions (Section 6.3, Section 6.4, and Section 6.5, respectively). We then introduce two classes that support the communication of FSMs over the TCP/IP Internet (Section 6.6), namely the classes *FSMSystemWithTCP* and *NetFSM*. The first part of Chapter 6 ends with the introduction of global constants, types, and functions (Section 6.7).

The second part of Chapter 6 contains detailed descriptions of the individual FSM Library Application Programming Interface (API) functions (Section 6.8). The third part of Chapter 6 consists of two examples. The first is a simple example with three automata (FSM) instances (Section 6.9), and the second is a simple example with TCP/IP network-aware automata instances (Section 6.10).

1.1 The Notion of the Communication Protocol

What is a communication protocol? A wide range of definitions are available in the literature today, for example: "An established set of conventions by which two computers or communication devices validate the format and content of the messages exchanged;" "A set of defined interfaces that permits the computers to communicate with each other;" "A method by which two computers coordinate their communication;" "Common agreed rules followed in order to interconnect and communicate between computers;" "The rules governing the exchange of information between devices on a data link;" "The set of rules governing how information is exchanged on a network;" and so on.

In this book, we begin with a wider informal definition. A **protocol** is a set of conventions and rules governing their use that regulates the communication of an entity under observation with its environment. Such a definition enables the study of any communication, e.g., an agenda for a technical meeting of representatives of two companies. The subject of this book is one special class of protocols, referred to as **communication protocols**, that regulate the communication of geographically distributed program objects. The communicating program objects are deployed on different processors in the network. We will sometimes use the term **protocol** as an abbreviated form of the phrase **communication protocol** to save space.

A **process**, as generally defined in the theory of operating systems, is a program in execution or prepared for execution. A process may be specialized for data processing, communication, or some other special task (e.g., I/O control or time management). Traditionally, a data processing algorithm is specified by the flowchart. What the flowchart means for the data processing process, the protocol means for the communication process.

The flowchart specifies the program control flow by the use of graphic symbols related to the series of sequential calculations, selection, iteration, procedure/function call, and input/output operations needed to read input data or write output data. On the other hand, the formal specification of a communication protocol is based on messages and consists of the following three parts:

- The message format specification
- The message-processing procedures specification, which is essentially a formal description of process reactions to input stimuli (i.e., messages)
- The error processing specification, which is the formal description of process reactions to exceptional events (i.e., corrupted data or timeouts)

The **message format** completely defines the structure of the message, i.e., it defines the set of fields that constitute the message by defining the width of individual fields (most commonly in bits, bytes, or words), the applied coding scheme (e.g., binary, ASCII, Unicode, ASN.1), and optionally legal values (e.g., constants in binary or some symbolic form or value intervals).

Therefore, a **message** is a series of bits logically divided into various fields. Typically, a message consists of a message header, which most commonly comprises more subfields, and useful data referred to as a payload. The **payload** contains data interpreted by the communicating program objects.

The message header contains data added for supervision and control purposes in accordance with the established conventions.

The **message-processing procedure** (i.e., the process reaction) begins with the message reception and is described as a series of primitive operations that define the rules of the communication, which are the essential parts of a protocol. Typical **primitive operations** include timer-start operations, timer-stop operations, message-send operations, message-receive operations, and message-data processing operations (e.g., cyclic redundancy checking of message data, calculating the expected order number of the next message to be received).

In terms of software implementation, message processing is performed by a message processing routine. Depending on the selected working environment, this routine can be a subroutine that consists of a series of machine instructions in a symbolic form (assembly language) or a function comprising a series of statements in a higher-level programming language, such as C/C++ or Java.

The error-processing specification defines a set of error reactions. An **error reaction** is a special protocol reaction to exceptional events or, in other words, a reaction to unexpected situations, i.e., conditions. Typical examples of unexpected events are the reception of a message that contains corrupted data, the reception of a message that is out of the original order (e.g., after receiving the messages numbered 1, 2, and 3, we receive the message numbered 7 instead of the message numbered 4), timer expiration (e.g., the receiver has not acknowledged the reception of a message to its sender within a certain interval of time, determined by the value of the corresponding timer), and so on.

Note that a protocol can be described informally or formally. The informal description of a protocol is referred to as its **informal specification** and has the following characteristics:

- It frequently has the form of a combination of textual and graphical descriptions of the most common scenarios of communication.
- It may state nothing about the order of the activities to be conducted in the course of the communication.
- It is always incomplete. Most frequently, missing parts are specifications of timers, which determine time limits over individual phases of communication.

Let us forget communication protocols for a moment and use the old example of informal specification of a group of tasks to get a sense of the issues stated above. While leaving the house, a mother says to her daughter:

"Do not forget to finish your homework."

"Have your breakfast when you get hungry."

"Before you go to school, throw the garbage out."

Obviously, this specification does not say anything about the order of the individual tasks. For example, the daughter may complete the tasks in any order without interrupting the individual tasks (e.g., task order may be 1, 2, 3, or 1, 3, 2), or she may complete them in any order and switch between them (e.g., she starts with task 1; then, she switches to task 2 before completing task 1; she completes tasks 2 and 3; and, at the end she finishes task 1). An essential question here is how to organize the task executions within the allocated time. Clearly, a need exists to limit or control the task execution time. What happens if the daughter gets preoccupied with her homework and forgets to have breakfast before it is time to go to school?

The example above might appear to be an exaggeration of the problems we face in reality, but its purpose is to show that informal systems specification alone is insufficient, and that we need a formal systems specification to make a precise and correct system implementation. Formal specification in the area of communication protocols is based on modeling a protocol as a **finite state machine** (FSM). A single FSM is often referred to by the term **automata,** and we will use these two terms interchangeably in this book.

The formal specification of an FSM defines all its states and state transitions, including transitions initiated by expiration of timers, in a unique and detailed way. Today, we may make formal protocol specifications in either UML or ITU-T domain-specific languages. Once we have a formal protocol specification, we can implement it in Java or C++. Finally, we must test and verify it. This procedure is basically what this book is all about.

References

Booch, G., Rumbaugh, J., and Jacobson, I., *The Unified Modeling Language User Guide*, Addison-Wesley, Reading, MA, 1998.

Booch, G., Rumbaugh, J., and Jacobson, I., *The Unified Software Development Process*, Addison-Wesley, Reading, MA, 1998.

2

Requirements and Analysis

At the beginning of any project, engineers face the fundamental question, "What must be done and how do we verify (deliver) the solution (system, device, products, service, hardware or software)?" Answering this question leads to what are called **requirements**. To simplify the matter, the process of answering this question—i.e., the corresponding engineering phase—is also commonly called *requirements*. Although both the working phase and the resulting documents have the same name, the meaning is easily deduced from the context.

The previous question actually consists of the following two questions:

1. What must be done?
2. How can the solution be verified?

Answering the former question leads to a set of functional requirements, most frequently adorned by nonfunctional requirements. **Functional requirements** describe the desired system behavior, while **nonfunctional requirements** can be imagined as the additional attributes to the behavior related to time restrictions, performance, and so on. To answer the latter question, we must quantify the behavior of the system. Normally, we would say, "For this input, the system should produce this output." Such thinking implies the existence of a test setup that enables automated (most preferably automatic) testing, referred to as a **test bed**. A test bed provides a **test harness** by generating the input to the system and capturing its output.

The ordered pair of the given input and the expected output informally stated in the text above is called a **test case**. To verify complex systems, we need many test cases. A set of test cases packed in a suitable form is referred to as a **test suite**. Ideally, we would like the test suite to completely cover the systems behavior (i.e., the functional requirements), which are adorned with their nonfunctional requirements. Typically, one or more test cases will be derived from each functional requirement. Clearly for any nontrivial system, the number of test cases needed to verify the system may be huge.

However, while thinking about the desired behavior of the system and its verification, we inevitably think about the question, "How can we make it?" Actually, we are trying to make a concept of the system or, more precisely, its architecture. This engineering phase is called an **analysis**. Obviously, it is tightly coupled with the requirements. These two phases have a highly interactive relation.

Typically, work on the definition of the system architecture yields the refinement of system functional requirements, and vice versa. This is especially true for communication protocol engineering. Therefore, we think of these two phases, the requirements and the analysis, as one indivisible front-end phase of communication protocol engineering. This is the reason they are covered together in this chapter.

As previously mentioned, the area of communication protocol engineering is very well founded; many standards, recommendations, and well-known experiences exist—hence, this chapter is rather short compared to the others. Unlike other areas of engineering, a vast majority of engineers here will be faced with the task of implementing some already defined standards, such as IETF RFC or ITU-T/ETSI recommendations, and so on. Very few engineers will be in a position to create a completely new protocol, and even then they will have many existing protocols for reference or starting points.

Many existing standards actually represent very detailed designs accompanied by the corresponding test suites, but others are rather informal, bringing nothing more than the message syntax and encoding together with some textual explanations of the message handling procedures. However, most of the standards can be viewed at least as rather good starting functional requirements that must be further formalized and analyzed. This chapter tries to help the reader exactly in this direction. It tries to answer the question, "How can we deal with the requirements in a systematic way?" Or, in other words, "How do we capture the requirements and how do we proceed with forward engineering from there?"

The overall consensus, in both academia and industry today, is that the UML paradigm can help in this respect (Booch et al. 1998). The behavior of the system is described with a set of use cases. Each **use case** captures one functional requirement adorned with its corresponding nonfunctional requirements. The **requirements engineer** models the system are by specifying the individual actors and the corresponding use cases of the system. The result is referred to as a **requirements model** of the system. The means for making such models are **use case diagrams**, which will be introduced in Section 2.1.

The next step in the UML paradigm is to transform the requirements model into the **analysis model**. Typically, a use case is viewed as a collaboration of classifiers. In the analysis model, three different **stereotypes** of classes are used: <<*boundary class*>>, <<*control class*>>, and <<*entity class*>>. The means of specifying the collaborations in UML are **collaboration diagrams**, which will be introduced in a following section.

Sometimes the analysts describe the static structure of the system—in addition to its behavior—with **class diagrams**. This practice can be helpful in really complex systems. In this chapter, we will present the collaboration diagrams sufficient for the examples at hand, therefore the introduction to class diagrams is postponed until the next chapter. Chapter 3 deals with the communication protocol design phase in which class diagrams are essential to show the static relations among classes.

Further on, in accordance with the UML paradigm, the requirements model should be transformed into the **test model** to facilitate the system verification (the test model is actually the test suite needed for the system verification). Essentially, the use cases should be translated into the corresponding test cases described by test scripts of some kind. UML is not specific in that respect. Of course, a few scripting languages are popular today, such as TCL/TK, Perl, and Payton, but being general purpose languages, these might be inappropriate for some of the projects.

To close this gap, we will introduce a domain-specific language known as **testing and test control notation, ver. 3 (TTCN-3)**. The TTCN-3 language is used for specifying the test suites for communication protocols once the software architecture is rather well known. Therefore, we will postpone the introduction to the TTCN-3 language until Chapter 3, which deals with the design phase of communication protocol engineering.

A general problem when transforming use cases to test cases is that the transformation is typically done manually, i.e., it is semiautomatic. Such an approach is both time consuming and prone to error. However, the main conceptual problem is the test coverage of the system behavior. In practice, the number of possible scenarios and all possible combinations of message parameters can be impossible to cover manually. Therefore, testing at least the most frequently used system scenarios and message parameter combinations should somehow be possible.

Clearly, more detailed UML models made during the system design phase (e.g., statecharts, to be introduced in the next chapter) can be used later for the automatic generation of test cases. However, the problem with this approach is that if an error exists in the UML model, it will be propagated into the test suite and the test suite will not be able to detect the error. A well-known principle from mathematical logic is that negation of negation leads to affirmation, so the bug will remain undiscovered. No matter how large test suite we generate, it will not be able to detect the bug.

The former problem can be solved by the application of **statistical usage testing**, also referred to as **behavior testing**. This paradigm is based on the operational profile model of the system, which describes the statistics of the system usage. It enables the practitioners to thoroughly test the system and even estimate the system or software reliability. This practice is recognized as a *de facto* standard by the industry (Broekman and Notenboom, 2003), and it will be covered in detail in Chapter 5 (test and verification phase of communication protocol engineering).

The latter problem can be solved by using one model as a source for the software implementation generated with forward engineering and a completely different model for the system test suite generation. What is also highly desirable is that these two models are made by two separate individuals or teams. For example, the well-known Cleanroom engineering paradigm is conducted by three completely separate teams. The design team makes the design and does its formal verification, the implementation team just does the coding, and the test team makes the operational profile of the system and conducts the statistical usage testing. Cleanroom engineering will be described together with statistical usage testing in Chapter 5.

Before proceeding further to the introduction of the mainstream approach to requirements and analysis, which is based on UML, it is worth mentioning that, until recently, many opponents to this paradigm existed. Some ongoing doubts still exist as to whether this is the correct choice. For example, in his article, "Use-Cases Are Not Requirements," Meyer argues that a better approach to requirements and analysis is transforming the functional requirements into the behavior model that takes the form of a finite state machine (FSM) (Meyer and Apfelbaum 1999). He sees use cases as just walks across the FSM and claims it is possible to generate them automatically rather than writing them manually.

According to the methodology proposed by Meyer, after creating the behavior model, two parallel streams of activities are started. The first stream covers the analysis, the design, and the implementation, and yields the implementation. The second stream covers the operational profile and the performance analysis, as well as the automatic test suite generation. These two streams merge at the automated testing phase.

This approach is very similar to the one used in this book. A slight difference is that the latter promotes separation of concerns between design and implementation and promotes test teams, including the models they make, very much like the Cleanroom engineering model does. Also, it gives more credit to the UML use cases. If we go back to the original ideas of the UML authors (Booch et al., 1998) and try to think of a single use case as a family of closely related collaborations among the same set of objects, clearly a use case really captures a part of the traditional **list of functional requirements**. Use cases help us group simple and closely related functional requirements, as will be illustrated by the examples in this chapter.

As already mentioned, use cases are the starting point of the software development in the unified software development process (Booch et al., 1998). The requirements model, essentially a set of use cases, is used to develop all the models that correspond to the engineering phases of the process, namely, the analysis model (result of the analysis phase), the design and deployment models (results of the design phase), the implementation model (result of the implementation phase), and the test model (result of the test preparation phase). The focus of this chapter is on requirements and analysis modeling.

The rest of the chapter is organized as follows: the use case and collaboration diagrams are introduced in the next two sections. The last section of this chapter illustrates the requirements and analysis phases of communication protocol engineering by presenting the case of the session initiation protocol (SIP), RFC 3261 (Rosenberg et al., 2002). That last section is divided into three subsections: SIP domain specifics, the SIP requirements model, and the SIP analysis model.

2.1 Use Case Diagrams

Use case diagrams are special kinds of graphs whose vertices are connected with arcs. Two types of vertices are found in use case diagrams, namely, actors and use cases (Figure 2.1). The **actors** represent humans, machines, or software components that are the users of the software under development. They are rendered as stick figures. **Use cases** represent possible uses of the software under development and are rendered as ellipses. As already mentioned, we think of use cases as collaborations between the corresponding objects that constitute the part of the software under development. Clearly, they have different roles in the requirements and the analysis phases.

In the requirements phase, we concentrate on the functional requirements and use the use cases to capture them ("What must be done?"). At that time, how these requirements will be fulfilled does not matter. The only important concern is to build a vision of the future system together with the customer. This vision is expressed as a desirable behavior of the system and modeled by drawing the use case diagram and writing down the descriptions of the individual use cases as they are added to the diagram.

In other words, we concentrate on the client's perspective of the system. The requirements engineer tries to define the services that the system under development should provide. They also try to define an interface to these services. Later, the main problems that the requirements engineer must face are

- Structuring the set of use cases by establishing the relationships among them
- Prioritizing the set of use cases by assigning different priorities to the individual use cases (especially important for the evolving systems)

Use cases have another role in the analysis phase. The job of the analyst is to realize the use cases by the corresponding collaborations between objects. The analyst reads the descriptions of the use cases and uses domain-specific knowledge to identify the individual objects (horizontal structuring) and to

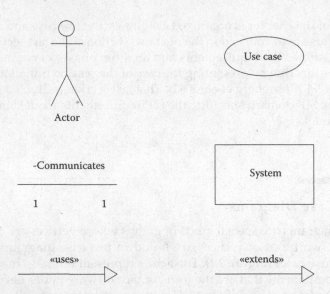

FIGURE 2.1
Basic set of graphical symbols available for rendering use case diagrams.

establish a hierarchy among them (vertical structuring). This process will be described in the next section.

Both actors and use cases are classifiers and, normally, they are connected by associations. The association between the actor and the use case shows the communication between the user and the part of the system modeled by the use case. Using associations enables us to indicate explicitly the points of connection between the users and the system.

Because both actors and use cases are classifiers, we can define general actors and general use cases and then specialize them using the generalization relationship. For example, we may specify the general actor *Client* and its specializations *SIP Client* and *H.323 Client* (Figure 2.2). Or, we can specify the general use case *Make a connection* and its specializations *Make a local connection* and *Make a long distance connection* (Figure 2.3).

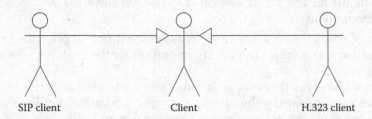

FIGURE 2.2
Example of the generalization and specialization of actors.

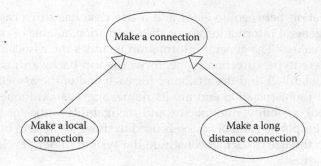

FIGURE 2.3
Example of structuring use cases.

Furthermore, while capturing the individual use cases, it may become obvious that a certain use case extends another use case or that a certain use case includes some other use cases. In such circumstances, the requirements engineer may structure the use cases using <<*extends*>> and <<*includes*>> stereotyped relationships. Especially important things can be indicated by using the sticky notes. Invariants, preconditions, and postconditions can be specified by the corresponding constraints. In more complex use case diagrams, we may need to indicate the packages and the interfaces.

Use case diagrams are normally rendered using the appropriate graphical tools, e.g., Microsoft® Visio. This tool provides the set of graphical symbols that are placed on the working sheet by the drag-and-drop paradigm. The basic set of graphical symbols is shown in Figure 2.1. The requirements engineer must specify the properties for each instance of a symbol in the drawing.

Five categories of actor properties are found: general information, table of attributes, table of operations, table of constraints, and tagged values. The general information includes name, full path, stereotype, visibility (private, protected, or public), and the indicators for *Root*, *Leaf*, and *Abstract* types of actors. The table of attributes includes columns for the attribute name, type, visibility, multiplicity (1, *, 0..1, 0..*, 1..1, or 1..*), and its initial value. The table of operations comprises columns for the operation name, return type, visibility, scope (classifier or instance), and the indicator for the polymorphic operations. The table of constraints consists of four columns: the constraint name, stereotype (precondition, postcondition, or invariant), language type (OCL, text, pseudocode, or code), and body of the constraint. The tagged values include notes for the documentation, location, persistence, responsibility, and semantics.

A use case—being a classifier like an actor—has the same five categories of properties as the actor, as well as the additional sixth category. The sixth category of the use case properties contains the notes about the extension points that are used to describe the <<*extends*>> stereotyped relations.

An association between an actor and a use case has three categories of properties: general information about the association, a table of constraints, and tagged values. The general information includes the association name, full path, stereotype, direction (none, forward, and backward), association end count (default 2), and the attributes for each end of the association. The attributes of the association end are its name, aggregation (none, composite, or shared), visibility, multiplicity, and navigability indicator (navigable or not). The graphical symbol *System* is used to show the system boundaries, i.e., to group the use cases that constitute the system under development. It has no properties.

All the relations between the use cases have three categories of properties: general information, table of constraints, and tagged values. The general information includes the relation name, full path, stereotype (extends, inherits, private, protected, subclass, subtype, or uses), and discriminator. The table of constraints is the same as the table of constraints for the actors and use cases. The tagged values are notes for the documentation.

The additional graphical symbols available for drawing use case diagrams are shown in Figure 2.4. These symbols include notes, general constraints, two-element constraints, OR constraints, packages, and interfaces. The notes have two categories of properties: general properties and tagged values. The general properties include the note name and its stereotype (none or requirement). The tagged values are notes for the documentation.

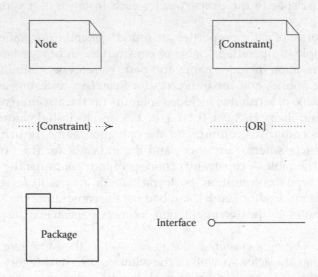

FIGURE 2.4
Additional graphical symbols available for rendering use case diagrams.

All the constraints, including general, two-element, and OR constraints, have the same categories of properties: general properties and tagged values. The general properties include the constraint name, full path, stereotype (precondition, postcondition, or invariant), language type (OCL, code, pseudocode, or text), and constraint body.

Four categories of package properties exist, including general properties, table of events, table of constraints, and tagged values. The general properties are the package name, full path, stereotype (facade, framework, stub, or system), visibility (private, protected, or public), and the indicators for *Root*, *Leaf*, and *Abstract* types of packages. The table of events contains an entry for each event. The attributes of individual events are the event name and event type (call event, signal event, change event, or time event). The table of constraints has the same format as the table of constraints for the actors, and the use cases and tagged values are just the notes for the documentation.

The interface has four categories of properties, actually a subset of the actor properties. These are general properties, table of operations, table of constraints, and tagged values. All of them are the same as the corresponding actor properties.

The requirements engineer renders the use case diagram along as they talk to the customer about the desired behavior of the system to be developed. The use case diagram is intended as a medium to communicate the requirements between the customer and the system provider. Drawing use case diagrams is simple: the right graphical symbol is selected, dragged-and-dropped to the working sheet, the corresponding properties are filled in, and they are connected to the other symbols in the sheet.

As an illustration of a use case diagram, consider a simple program for sending and receiving electronic mail messages over the Internet. The use case diagram for such a program might look like the one shown in Figure 2.5. A single actor is found in this diagram, who is the user of the program (named *User*). On the highest level of abstraction, this program has two main use cases, *Send e-mail* and *Receive e-mail*.

Both of these highest-level use cases make use of the use cases *Use DNS* (Domain Name System) and *Use TCP* (Transmission Control Protocol). The DNS service provides the mapping of the e-mail server domain name into its IP (Internet Protocol) address. The TCP provides reliable data delivery service. Other than that, the use case *Send e-mail* uses the use case *Use SMTP* (Simple Mail Transfer Protocol) and the use case *Receive e-mail* uses the use case *Use POP3* (Post Office Protocol, Version 3). Normally, an e-mail client uses SMTP to send an e-mail message to the e-mail server. Similarly, a user uses POP3 to read the e-mail messages from their mailbox.

The use case *Use DNS* uses the use case *Use IP* to send a DNS request to the DNS server and to receive DNS responses from it. The use case *Use TCP* uses the use case *Use IP* to send and receive segments of data and control

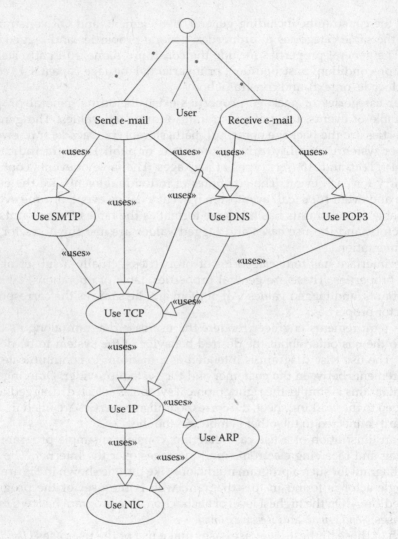

FIGURE 2.5
Use case diagram of the simple program for sending and receiving e-mails.

information over the Internet. The use case *Use IP* uses the use case *Use ARP* (Address Resolution Protocol) to map the IP address of the destination host to its physical (e.g., Ethernet) address. Alternatively, the use case *Use IP* uses the use case *Use NIC* (Network Interface Controller) to send and receive IP datagrams over the Internet. Finally, the use case *Use ARP* uses the use case *Use NIC* to send an ARP request to the ARP server and to receive an ARP response from it.

This hierarchy of use cases actually follows the hierarchy of protocols in the TCP/IP protocol stack. As already mentioned, the concept of layered software architecture, which is traditionally explained by the ISO OSI, was actually invented to enable the separation of functions and the corresponding functional requests, which are referred to as *use cases* in UML.

After creating the skeleton of the use case model, the requirements engineer must fill in the descriptions of the individual use cases. The descriptions in this example are simplified for the sake of clarity. The description of the use case *Send e-mail* in plain text is the following:

Precondition:

```
The user has issued the send mail command.
```

Main flow of events:

```
Extract the recipient's e-mail address from the e-mail message header (defined
by the RFC 822).
Extract the e-mail server domain name from the recipient's e-mail address
(string after the character "@").
Use the use case Use DNS to map the server domain name into its IP address.
Use the use case Use TCP to open the TCP connection.
Use the use case Use SMTP to send the e-mail message to the e-mail server.
Use the use case Use TCP to close the TCP connection.
Prompt the user for the next command.
```

Exceptional flow of events:

```
The user may cancel the use case at any time by issuing the cancel command.
```

Exceptional flow of events:

```
If the use case Use SMTP indicates the problem in the mail delivery, this use
case should report it to the actor User.
```

The use case *Receive e-mail* is identical to the use case *Send e-mail* with the difference being that the former uses the use case Use POP3 instead of the use case *Use SMTP*. The following description of the use case *Use DNS* is rather simple (actually, this is the description of the behavior of the DNS client):

Main flow of events:

```
Send the recursive DNS request by using Use IP.
Receive the DNS response by using Use IP.
```

The use case *Use TCP* is the active (initiator's) side of the TCP. It is defined as follows:

Main flow of events:

```
The procedure to open the TCP connection:
   Send SYN data segment.
   Receive SYN + ACK data segment.
   Send ACK data segment.
   Indicate that the connection is established.
The data transmission procedure:
   Send and receive the data segments using the sliding window.
The procedure to close the TCP connection:
   Send FIN data segment.
   Receive ACK data segment.
   Receive FIN + ACK data segment.
   Send ACK data segment.
   Indicate that the connection is closed both ways.
```

Exceptional flow of events:

```
The use case Send e-mail may close the TCP connection at any time.
```

The use case *Use SMTP* is actually the client side of the SMTP (defined by IETF RFC 821 and RFC 788) and can be described as follows (for simplicity, only one exceptional flow of events is given):

Main flow of events:

```
Receive the message 220 READY FOR MAIL.
Send the message HELLO.
Receive the message 250 OK.
Send the message MAIL FROM: <recipient's e-mail address>.
Receive the message 250 OK.
Send the message RCPT TO: <sender's e-mail address>.
Receive the message 250 OK.
Send the message DATA.
Receive the message 354 START MAIL INPUT.
Send the body of the e-mail message terminated with <CR><LF>.<CR><LF>.
Receive the message 250 OK.
Send the message QUIT.
Receive the message 221.
```

Exceptional flow of events:

```
If a use case receives the message 550 NO SUCH USER HERE, as a reply to its RCPT
TO: message, it indicates the problem to the use case Send e-mail.
```

The use case *Use POP3* is the client side of the POP3 protocol, similar to the use case *Use SMTP*. The use case *Use IP* is actually the IP protocol, which is described as follows:

Main flow of events:

```
The procedure that is used to receive the datagrams:
   Receive a datagram by using the Use NIC.
   Send the received datagram to the use case Use TCP.
The procedure that is used to send the datagrams:
   Decrement the contents of the time-to-live field of the IP datagram.
```

```
Extract the destination IP address from the datagram header.
Extract the destination network id from the destination IP address.
If the destination network is local the network:
  Use the use case Use ARP to determine the physical address.
  Deliver the datagram by using the Use NIC.
Else, route the datagram.
```

Exceptional flow of events:

```
If the datagram has been corrupted during the transmission, drop it.
```

Exceptional flow of events:

```
If the time-to-live field of the datagram counts down to 0, drop it.
```

The use case *Use ARP* is an ARP client and the use case *Use NIC* is a network card driver. The former is defined as follows:

Main flow of events:

```
Send an ARP request by using the use case Use NIC.
Receive the ARP response by using the use case Use NIC.
```

The example above, especially the use cases *Use TCP* and *Use SMTP*, should help the reader understand that a use case is a set of event sequences, not just a single sequence. To keep use cases simple, separating the main and the alternative flows of events is always desirable. Usually, we start by just writing the main flow of events for each use case and refine them later by adding the exceptional flow of events.

After this example, it should be clear that a use case captures the intended behavior of the part of the system (subsystem, class, or interface). Of course, after specifying the intended behavior, we must create a set of classes that work together to implement that behavior. The means of modeling both static and dynamic structures of the society of objects in UML are the collaboration diagrams.

2.2 Collaboration Diagrams

As already mentioned, we think of use cases as collaborations between objects. Actually, in UML we realize a use case as a collaboration of a set of objects. This concept can be explicitly shown in UML by connecting the use case with the corresponding collaboration using the realization relationship.

A **collaboration diagram** is a special kind of graph consisting of a set of vertices interconnected by a set of arcs. Basically, the vertices are the objects and the arcs are the links that carry the messages between the interconnected objects. Additional vertices and arcs are the notes and the constraints (general constraints, two-element constraints, and OR constraints).

Collaboration diagrams are normally rendered using the appropriate graphical tools, e.g., Microsoft® Visio. This tool provides the set of graphical symbols that are placed on the working sheet by the drag-and-drop paradigm. The basic set of graphical symbols is shown in Figure 2.6. The engineer that renders the diagram must specify the properties for each instance of a symbol in the drawing.

Three categories of object properties exist: general properties, table of constraints, and tagged values. The general properties include the object name, full path, classifier name, and multiplicity. The table of constraints and the tagged values contain the same properties as the corresponding categories for the use cases (see the previous section of this chapter).

While adding objects to the collaboration diagram, we are forced to introduce the corresponding classifiers and to specify their properties (at least the classifiers' names, for a start). The classifiers have eight categories of properties, including general properties, table of attributes, table of operations, table of receptions, table of template parameters, list of the components, table of constraints, and tagged values. The general properties, the table of attributes, the table of operations, the table of constraints and tagged values contain the same properties as the corresponding categories for the use cases (see the previous section of this chapter).

The table of receptions has five columns, which contain the reception name, signal name, visibility (private, protected, or public), polymorphic indicator

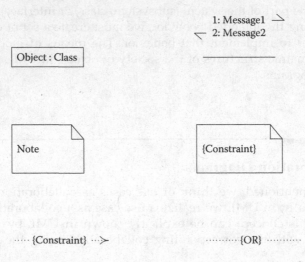

FIGURE 2.6
Set of graphical symbols available for rendering collaboration diagrams.

(false or true), and scope (classifier or instance). The table of template parameters includes the columns for the parameter name and its type. The list of components is just a list of components that implement this class.

The links in collaboration diagrams have four categories of properties, including general properties, table of messages, table of constraints, and tagged values. The general properties are the link name, its full path, and the table of link ends roles, which has two columns, the end name and its stereotype (none, association, global, local, parameter, self). The table of link messages has four columns, including the message name, its direction (forward or backward), flow kind (procedure call, flat, or asynchronous), and sequence expression. The table of constraints contains the same properties as the corresponding category of object (and classifier) properties. The tagged values are just the notes for the documentation. The notes and the constraints have the same properties as in the use case diagrams (see the previous section of this chapter).

Most frequently, we model sequential flow of control with collaboration diagrams. In this case, a message sequence expression takes the simple form of a message sequence number. However, collaboration diagrams allow modeling of more complex flows, such as iteration and branching. Iteration is modeled by prefixing the message sequence number with the iteration expression

*[<control variable> := <start value>..<end value>]

e.g., *[j := 1..m].

Branching is modeled by prefixing the message sequence number with the condition clause [<condition>], e.g., [i > 10]. Alternate paths of the branch have the same message sequence number prefixed by the unique non-overlapping condition, where the set of conditions must cover all the possibilities.

Next, we illustrate the use of collaboration diagrams in the example of a simple program for sending and receiving electronic mail messages over the Internet, which was introduced and modeled in the previous section of this chapter. The use case diagram for this program is shown in Figure 2.5. We start by making the real collaboration between objects that is a realization of the use case model, and continue with the study of virtual collaborations, which correspond to the peer-to-peer protocols present in this example.

To start, imagine that we are provided with the classifier FSM for modeling finite state machines. Clearly a single object of this class could be a realization of a single use case, as shown in Figure 2.5. The assumption that each use case is materialized by a single FSM object leads to a real collaboration between objects, shown in Figure 2.7.

In this class diagram, the object *mailc* (abbreviation for a mail client) is the <<*boundary class*>> object. All other objects are the <<*control class*>>

objects. The e-mail message itself would be the <<*entity object*>>, but it is not shown in Figure 2.7. Obviously, the realization of the individual use cases is as follows:

- The object *sender* is a realization of the use case *Send e-mail*.
- The object *receiver* is a realization of the use case *Receive e-mail*.
- The object *dnsc* (abbreviation for a DNS Client) is a realization of the use case *Use DNS*.
- The object *tcpc* (abbreviation for a TCP Client, i.e., the side that initiates the establishment of the TCP connection) is a realization of the use case *Use TCP*.

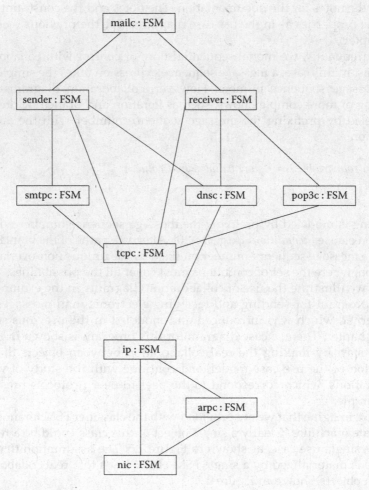

FIGURE 2.7
Collaboration diagram of the simple program for sending and receiving e-mails.

- The object *smtpc* (abbreviation for an SMTP Client) is a realization of the use case *Use SMTP*.
- The object *pop3c* (abbreviation for a POP3 Client) is a realization of the use case *Use POP3*.
- The object *ip* is a realization of the use case *Use IP*.
- The object *arpc* (abbreviation for an ARP Client) is a realization of the use case *Use ARP*.
- The object *nic* is a realization of the use case *Use NIC*.

Figure 2.7 shows general collaboration among the relevant objects, i.e., it just shows the links between objects. Essentially, it shows the software architecture. We may think of it as a family of particular collaborations. For example, the user of the program might select the use case *Send e-mail* and this would lead to a particular collaboration, or the user might select the use case *Receive e-mail* and that would lead to another particular collaboration.

Another important thing to notice and remember is that Figure 2.7 shows only the objects of the system under development. In this case, it is a program that runs on a computer connected to the Internet over its network interface card. If we want the overall picture, we can also add the models of the systems with which our system under development would normally communicate. By adding the models of these external systems, we are modeling end-to-end collaborations.

The system under development communicates with external servers, including the ARP server, the DNS server, and the e-mail server. If we assume that all of these servers run on the same computer, the model of the external environment of the system under development is rather simple (Figure 2.8). The external objects are as follows:

- The object *smtps* is the SMTP server.
- The object *pop3s* is the POP3 server.
- The object *tcps* is the TCP server, i.e., the side that accepts the establishment of the TCP connection.
- The object *dnss* is the DNS server.
- The object *arps* is the ARP server.
- The object *ips* is an instance of IP.
- The object *nics* is an instance of NIC.

The overall collaboration that corresponds to the main flow of events of the use case *Send e-mail*, up to the point when the SMTP client receives the message 220 READY FOR MAIL, is shown in Figure 2.9. The flow of events is as follows:

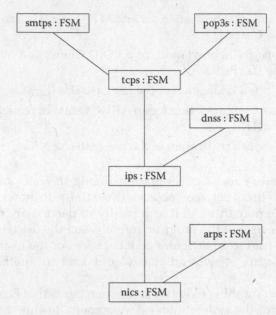

FIGURE 2.8
Collaboration diagram of the e-mail and DNS server.

1: The object *mailc* sends the signal *sendMail(msg)* to the object *sender*. The signal parameter *msg* is the e-mail message itself.

2: The object *sender* sends the signal *domainToIP(domain)* to the object *dnsc*. The signal parameter *domain* is the domain name of the e-mail server.

3: The object *dnsc* sends the signal *dnsReq(domain)* to the object *ip*. The signal *dnsReq* is actually the DNS service request message.

4: The object *ip* sends the signal *data(dnsReq)* to the object *nic*. The general signal *data* is an IP datagram. Together with the parameter *dnsReq*, it represents the datagram carrying the DNS service request message.

5: The object *nic* sends the signal *frame(dnsReq)* to the object *nics*. The general signal *frame* is a data frame from the underlying physical network (e.g., Ethernet). The signal *frame(dnsReq)* is the data frame carrying the datagram that encapsulates the DNS service request message.

6: The object *nics* sends the signal *data(dnsReq)* to the object *ips*.

7: The object *ips* sends the signal *dnsReq(domain)* to the object *dnss*.

8: The object *dnss* sends the signal *dnsRsp(ip)* to the object *ips*. The signal *dnsRsp* is the DNS service response message and its parameter *ip* is the IP address of the target e-mail server.

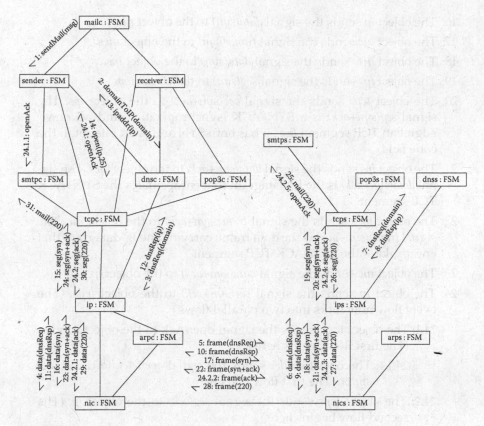

FIGURE 2.9
Overall real collaboration of the simple program for sending and receiving e-mails and its environment.

9: The object *ips* sends the signal *data(dnsRsp)* to the object *nics*.

10: The object *nics* sends the signal *frame(dnsRsp)* to the object *nic*.

11: The object *nic* sends the signal *data(dnsRsp)* to the object *ip*.

12: The object *ip* sends the signal *dnsRsp(ip)* to the object *dnsc*.

13: The object *dnsc* sends the signal *ipaddr(ip)* to the object *sender*.

14: The object *sender* sends the signal *open(ip,25)* to the object *tcpc*. The signal *open* is an active open request to TCP (TCP should send the SYN segment to initiate the TCP connection establishment procedure). Its parameters, *ip* and *25*, are the IP addresses of the target email sever and the well-known TCP port number reserved for the SMTP, respectively.

15: The object *tcpc* sends the signal *seg(syn)* to the object *ip*. The general signal *seg* is a TCP segment. The signal *seg(syn)* is a SYN (synchronization) TCP segment (i.e., it has the SYN bit set in the code field).

16: The object *ip* sends the signal *data(syn)* to the object *nic*.

17: The object *nic* sends the signal *frame(syn)* to the object *nics*.

18: The object *nics* sends the signal *data(syn)* to the object *ips*.

19: The object *ips* sends the signal *seg(syn)* to the object *tcps*.

20: The object *tcps* sends the signal *seg(syn+ack)* to the object *ips*. The signal *seg(syn+ack)* is a SYN+ACK (synchronization and acknowledgment) TCP segment (i.e., it has both SYN and ACK bits set in the code field).

21: The object *ips* sends the signal *data(syn+ack)* to the object *nics*. The signal *data(syn+ack)* is the IP datagram that encapsulates the SYN+ACK TCP segment.

22: The object *nics* sends the signal *frame(syn+ack)* to the object *nic*. The signal *frame(syn+ack)* is the data frame carrying the IP datagram that encapsulates the SYN+ACK TCP segment.

23: The object *nic* sends the signal *data(syn+ack)* to the object *ip*.

24: The object *ip* sends the signal *seg(syn+ack)* to the object *tcpc*. (The event flow now forks into two parallel flows.)

    24.1: The object tcpc sends the signal openAck to the object sender. (The first flow begins here.)

        24.1.1: The object *sender* sends the signal *openAck* to the object *smtpc* (The first flow ends here.)

    24.2: The object *tcpc* sends the signal *seg(ack)* to the object *ip*. (The second flow begins here.)

        24.2.1: The object *ip* sends the signal *data(ack)* to the object *nic*.

        24.2.2: The object *nic* sends the signal *frame(ack)* to the object *nics*.

        24.2.3: The object *nics* sends the signal *data(ack)* to the object *ips*.

        24.2.4: The object *ips* sends the signal *seg(ack)* to the object *tcps*.

        24.2.5: The object *tcps* sends the signal *openAck* to the object *smtps*.

25: The object *smtps* sends the signal *mail(220)* to the object tcps. The general signal *mail* is the SMTP message. The particular signal *mail(220)* is actually the message 220 READY FOR MAIL, where the first three digits are mandatory and the rest of the message is a human-readable comment. (Note: We have restarted the message numbering here for brevity.)

26: The object *tcps* sends the signal *seg(220)* to the object *ips*.

27: The object *ips* sends the signal *data(220)* to the object *nics*.

28: The object *nics* sends the signal *frame(220)* to the object *nic*.

29: The object *nic* sends the signal *data(220)* to the object *ip*.

30: The object *ip* sends the signal *seg(220)* to the object *tcpc*.

31: The object *tcpc* sends the signal *mail(220)* to the object *smtpc*. (The example ends here.)

What we have just described is the real collaboration between objects within the system under development as well as with the relevant objects in its surroundings. The real collaboration for any nontrivial system could be rather complex. This behavior should be clear from the previous example, where we intentionally stopped at the certain point of the event flow, which was selected as a compromise between showing enough complexity and maintaining clarity.

The complete list of events for the use case *Send e-mail* is much longer than the one given above. For modeling the transfer of the rest of the SMTP messages (12 of them), we would need additional 84 (12×7) UML events, almost three times more than already in the list above. This complexity is why we try to break the system down into its parts and analyze them in detail later.

One important aspect of the simplification is the definition of the Application Programming Interfaces (API). For example, we may define the API between the *sender* and the hierarchically lower level objects (*dnsc*, *smtpc*, and *tcpc*), or the API between *tcpc* and *ip*, and so on. Other important items are the virtual collaborations that are governed by the peer-to-peer protocols. Consider for example the virtual collaboration between *dnsc* and *dnss* (Figure 2.10). The corresponding flow comprises only two events, *dnsReq(domain)* and *dnsRsp(ip)*.

The virtual collaboration between *tcpc* and *tcps* is governed by the TCP. It is slightly more complex and comprises the following flow of events (Figure 2.11):

1: The object *tcpc* sends the signal *seg(syn)* to the object *tcps*.

2: The object *tcps* sends the signal *seg(syn+ack)* to the object *tcpc*.

3: The object *tcpc* sends the signal *seg(ack)* to the object *tcps*.

4: The object *tcpc* sends the signal *seg(data)* to the object *tcps*. (Data transmission phase)

5: The object *tcpc* sends the signal *seg(fin)* to the object *tcps*.

6: The object *tcps* sends the signal *seg(ack)* to the object *tcpc*.

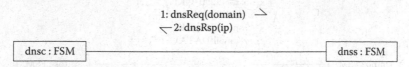

FIGURE 2.10
Virtual collaboration between the DNS client and the DNS server.

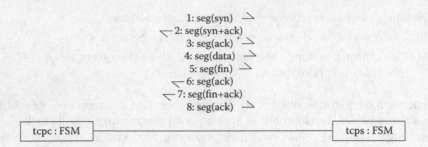

```
        1: seg(syn)  ⟍
      ⟋ 2: seg(syn+ack)
        3: seg(ack)  ⟍
        4: seg(data) ⟍
        5: seg(fin)  ⟍
      ⟋ 6: seg(ack)
      ⟋ 7: seg(fin+ack)
        8: seg(ack)  ⟍
```

| tcpc : FSM | ————————————————————— | tcps : FSM |

FIGURE 2.11
Virtual collaboration between two TCP entities.

7: The object *tcps* sends the signal *seg(fin+ack)* to the object *tcpc*.

8: The object *tcpc* sends the signal *seg(ack)* to the object *tcps*.

Finally, the virtual collaboration between *smtpc* and *smtps* (in accordance with SMTP) is of the same order of complexity (Figure 2.12; note that only the first eight events are shown in the figure). The corresponding flow of events is the following:

1: The object *smtps* sends the signal *mail(220)* to the object *smtpc*.

2: The object *smtpc* sends the signal *mail(HELO)* to the object *smtps*.

3: The object *smtps* sends the signal *mail(250_OK)* to the object *smtpc*.

4: The object *smtpc* sends the signal *mail(MAIL_FROM:)* to the object *smtps*.

5: The object *smtps* sends the signal *mail(250_OK)* to the object *smtpc*.

6: The object *smtpc* sends the signal *mail(RCPT_TO:)* to the object *smtps*.

7: The object *smtps* sends the signal *mail(250_OK)* to the object *smtpc*.

8: The object *smtpc* sends the signal *mail(DATA)* to the object *smtps*.

9: The object *smtps* sends the signal *mail(354_START_MAIL_INPUT)* to the object *smtpc*.

```
      ⟋ 1: mail(220)
        2: mail(HELO)  ⟍
      ⟋ 3: mail(250_OK)
        4: mail(MAIL_FROM:) ⟍
      ⟋ 5: mail(250_OK)
        6: mail(RCPT_TO:) ⟍
      ⟋ 7: mail(250_OK)
        8: mail(DATA) ⟍
```

| smtpc : FSM | ————————————————————— | smtps : FSM |

FIGURE 2.12
Virtual collaboration between the SMTP client and the SMTP server.

10: The object *smtpc* sends the signal *mail(MAIL_BODY)* to the object *smtps*.

11: The object *smtps* sends the signal *mail(250_OK)* to the object *smtpc*.

12: The object *smtpc* sends the signal *mail(QUIT)* to the object *smtps*.

13: The object *smtps* sends the signal *mail(221)* to the object *smtpc*.

2.3 Requirements and Analysis Example

This section of this chapter illustrates the requirements and analysis phases of communication protocol engineering with the example of a simple SIP softphone. Normally, the requirements phase starts by acquiring the relevant domain-specific knowledge and continues by the construction of the corresponding requirements model, which is the input for the analysis phase. As already mentioned, the output of the analysis phase is the corresponding analysis model. Sections 2.3.1 through 2.3.3 cover a short overview of the domain-specific information, the requirements, and the analysis models of a simple SIP softphone.

2.3.1 SIP Domain Specifics

SIP is the application layer protocol used for creating, modifying, and terminating sessions, such as Internet telephone calls and multimedia distribution and conferences, with one or more participants. It has been standardized by the IETF RFC 3261 (Rosenberg et al., 2002) and related series of RFCs (RFC 3262, RFC 3263, RFC 3264, RFC 3265, RFC 3372, RFC 3428, RFC 3485, RFC 3487, and others). In contrast to the ITU-T H.323 family of protocols—which provide the whole protocol stack for multimedia communications—SIP is just the control and signaling component on the top of the multimedia architecture.

Aside from SIP, the multimedia architecture will typically include RTP (Real-Time Transfer Protocol, RFC 1889), RTSP (Real-Time Streaming Protocol, RFC 2326), MEGACO (Media Gateway Control Protocol, RFC 3015), and SDP (Session Description Protocol, RFC 2327). SIP does not provide any service on its own. Instead of full services, it provides primitives for the services that are implemented in the overall architecture. These primitives are based on an HTTP-like (Hyper Text Transport Protocol) request and response transaction model.

The main SIP abstractions are the session, the dialog, and the transaction. A multimedia session is a set of multimedia senders and receivers, as well as data streams flowing from senders to receivers. A dialog is a peer-to-peer relationship between two user agents (end points in the communication) that persists for some time. A transaction is the collaboration between the client and the server, which comprises all the messages from the first request sent from the client to the server up to the final response sent from the server to the client. The requests are processed automatically, meaning that either all requested actions are conducted, if the request has been accepted, or none of the actions are conducted, if the request has not been accepted.

Two main transaction types exist, referred to as invite (officially written in capital letters, i.e., INVITE) and non-invite (or, more formally, non-INVITE) transactions. An invite transaction is a three-way handshake comprising the request, the response, and the acknowledgment. In contrast, a non-invite transaction is the two-way handshake starting with the request and ending with the corresponding response.

Notice that the roles of the user agents (communication end points) are not fixed, and they change on the transaction by transaction bases. The user agent that creates a new request becomes a user agent client (UAC), whereas the user agent that receives the request becomes the user agent server (UAS). Another important detail is that a new transaction (either invite or non-invite) may not be started while an invite transaction is in progress. Alternatively, a new invite transaction may be started while a non-invite transaction is in progress.

Besides user agents, the SIP standard defines three types of SIP servers, namely, the proxy server (stateful or stateless), the registrar, and the redirect server. A proxy server is the mediator that helps end points set up the session. Officially, it is an intermediary entity that acts as both a server and a client for the purpose of making requests on behalf of other clients. A registrar is a server that supports the registration of the user agents by maintaining the corresponding database for the domain it handles. This database is referred to as a location service. These two types of servers are most frequently col-located in the same physical machine. A redirect server can be viewed as a proxy server with limited capabilities. It is only capable of directing the client to contact an alternate set of Uniform Resource Identifications (URI).

Requests and responses between a server and a client are sent as SIP messages. The SIP message comprises the start line, one or more header fields, empty lines (carriage-return line-feed sequences, CRLF), and an optional message body. The start line is different in requests and in responses. In the former case, it is referred to as a request line, and in the latter as a status line. The request line comprises the method name (according to the RFC 3261, six methods are available in SIP: REGISTER, INVITE, ACK, CANCEL, BYE, and OPTIONS), the request URI, and the SIP version (currently "SIP/2.0"). The status line comprises the SIP version, the status code (a three-digit integer result code), and the reason phrase (textual status description).

The SIP protocol stack comprises four layers. Starting from the top and going down the hierarchy, these are the transaction user (TU) layer, the transaction layer, the transport layer, and the syntax and encoding of SIP messages. A transaction user is any SIP entity (client or server) except for the stateless proxy. The transaction layer supports transactions, which are the key component of SIP. The transport layer provides for the transfer of SIP messages across the Internet. SIP may use three types of transport services, including unreliable (UDP), reliable (TCP), and encrypted (Transport Layer Security, TLS) transport service. Most of the SIP message and header field syntax is identical to HTTP/1.1. Although SIP is close to the HTTP philosophy, it is not an extension of HTTP.

As mentioned above, the SIP standard specifies six methods, including REGISTER for registering contact information, INVITE, ACK, CANCEL for setting up sessions, BYE for terminating sessions, and OPTIONS for querying servers regarding their capabilities. Any INVITE after the initial invite to the same destination is called re-INVITE and is used for modifying the session and dialog parameters. The method INVITE starts the invite transaction; all other methods start non-invite transactions. Interestingly enough, six status code types are also found, depending on the value of status code first digit, as follows:

1xx: Provisional (the request has been received and its processing has been started)

2xx: Success (the request has been successfully processed)

3xx: Redirection (further action by the client is needed)

4xx: Client error (the request contains an error or it may not have been fulfilled on this server)

5xx: Server error (the request is valid, but the server failed to fulfill it)

6xx: Global failure (the request cannot be fulfilled on any server)

As an example, consider the typical scenario of the SIP session setup in Figure 2.13. (Note: This figure is actually a UML sequence diagram. Sequence diagrams are intentionally introduced later in Chapter 3. For the moment, it is enough to assume that the rectangular symbols are the communicating entities and that the arrows are the messages they exchange. Time advances downwards.) Two user agents *ua1* and *ua2*, together with their corresponding proxy servers *p1* and *p2*, constitute the SIP trapezoid (imagine the trapezoid by "drawing" the lines that connect *ua1*, *p1*, *p2*, and *ua2*).

Suppose that *ua1* wants to set up a session with *ua2*. It starts by sending an invite request to the proxy server that is responsible for its domain, and that is *p1*. Proxy *p1* locates the proxy server responsible for the destination *ua2*, namely *p2*, and forwards the invite request to it. At the same time, *p1* sends back the response 100 TRYING to *ua1*. Proxy *p2* locates the destination user agent, *ua2*, forwards the invite request to it, and sends back the response 100 TRYING to the proxy *p1*. *ua2* receives the invite request and sends back the response 180 RINGING, which is forwarded by the proxies *p2* and *p1* to *ua1*.

At this point, *ua2* indicates the incoming invite request to its user. The user accepts the request and *ua2* sends back the response 200 OK, which is

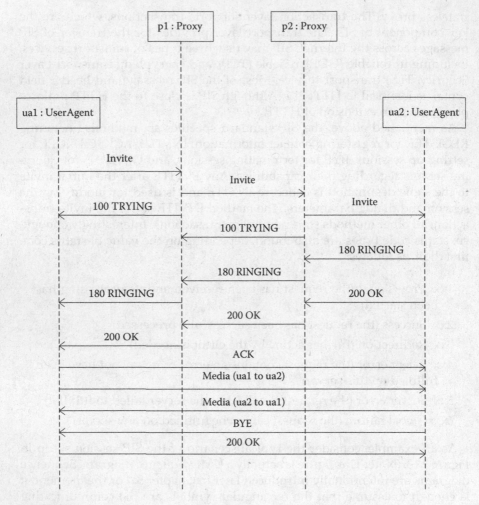

FIGURE 2.13
Example of SIP session setup (with SIP trapezoid).

forwarded by the proxies *p2* and *p1* to *ua1*. The dialog between *ua1* and *ua2* is successfully established. Further on, *ua1* sends the ACK request to *ua2* directly (the end of the three-way handshake). The session is successfully established at this point. The communicating user agents may now exchange the media. In reality, the media is exchanged in the full-duplex mode, i.e., both sides may send data to the other side simultaneously. Unfortunately, in UML sequence diagrams we cannot model the full-duplex communication, because only unidirectional messages may be used. Therefore, we represent the media exchange by the two separate messages, namely by the message Media *(ua1* to *ua2)* and the message Media *(ua2* to *ua1)*.

The session may be terminated by either *ua1* or *ua2*. Suppose that *ua2* wants to terminate the session. It sends the BYE request to *ua1* directly, which in its turn sends back the response 200 OK. The session is successfully closed. This is an example of the non-invite transaction.

This simplified explanation hides one rather important aspect of the invite three-way handshake, and that is the application of the offer-answer procedure. This procedure is used by *ua1* and *ua2* to determine the session parameters in accordance with SDP. The first offer must be carried either by the invite request or by the response 200 OK. If the offer is carried by the invite request (*ua1* makes the first offer), the answer must be included in the response 200 OK. If the offer is carried by the response 200 OK (*ua2* makes the first offer), the answer must be included in the ACK request (the last message in the three-way handshake). The session is successfully established only after the offer-answer procedure is successfully ended.

2.3.2 SIP Softphone Requirements Model

SIP softphone is the application that normally runs on some computer—for example, a desktop PC—and enables its user to set up multimedia sessions and to communicate with other SIP users or entities over the Internet. Such an application would typically use some type of graphical user interface (GUI) and device drivers for the sound card and the web camera, typically provided by the local operating system (out of scope for this book) and, of course, the SIP protocol stack.

This section shows how to construct the requirements model for the SIP protocol stack in a simple SIP softphone. As mentioned previously, the SIP protocol stack comprises the transaction user layer, the transaction layer, and the transport layer. In terms of use cases, the user uses the application (softphone), which in turn uses both the transaction layer and the transport layer. The transaction layer also uses the transport layer. The use case diagram shown in Figure 2.14 is a simple requirements model that captures these relations.

We can refine this simple model by taking into account the details of the individual layers of the SIP protocol stack. To start, the transaction user (TU) layer dynamically creates and uses the user agent clients (UAC) and the user agent servers (UAS) entities to support outgoing and incoming invite requests. Both UAC and UAS use the transaction layer (TAL), as well as the transport layer, which is accessible through the transport layer interface (TLI). TAL and TLI are abbreviations introduced here (they have not been taken from the RFC 3261).

Similar to TU, TAL dynamically creates and uses invite client transactions (INVITE CT), non-invite client transactions (non-INVITE CT), invite server transactions (INVITE ST), and non-invite server transactions (non-INVITE ST). TAL and all transactions use TLI, but they are all also used by

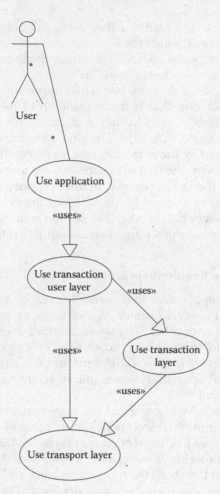

FIGURE 2.14
Use case diagram of the simple SIP softphone.

TU. Finally, TLI uses UDP, TCP, or TLS. The detailed use case diagram of the simple SIP softphone is shown in Figure 2.15.

Before proceeding further, two important points must be emphasized. The first is that the direct relations between TU and TLI are strictly in accordance with the RFC 3261, although this may seem to be an error because it violates the ISO OSI ideal of a strictly layered architecture (no direct communication between layer $i + 1$ and layer i). The second point is that the relations between TU and transactions, and transactions and TLI, are not prescribed by the RFC 3261 but they are also not forbidden. These relations are introduced to minimize the message paths at the expense of the increased relations complexity.

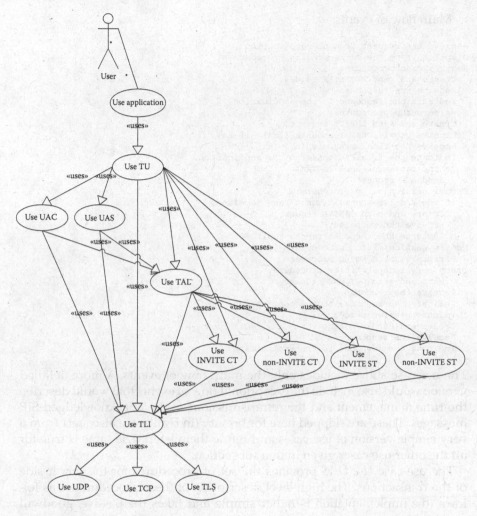

FIGURE 2.15
Detailed use case diagram of the simple SIP softphone.

To complete the requirements model, we need to describe the individual use cases. The use case *Use application* is actually the main program that interacts with the user and makes use of the SIP protocol stack and is out of the scope of this book. The use case *Use TU* is responsible for dispatching TU messages (coming from the application and the lower layers and going to the user agent clients and servers and to the application), as well as for dynamic creation of user agent clients and servers.

The use case *Use UAC* provides a set of procedures for the client side of the transactions. The high-level description of these procedures follows:

Main flow of events:

```
Receive the request from the application.
Dispatch it to the corresponding procedure.
Registration procedure:
  Create and send REGISTER request.
  Receive the response.
  Indicate the response to the application.
Session setup procedure:
  Create and send INVITE request.
  Receive provisional responses (1xx), if any.
  Receive the final response (not 1xx).
  Indicate the final response to the application.
  If the final response is 2xx,
    Send ACK request.
Cancel session setup procedure:
  If the final response has not been received,
    Create and send CANCEL request.
    Receive the response.
    Indicate the response to the application.
Modify session/dialog procedure:
  Perform session setup procedure.
Query server capabilities procedure:
  Create and send OPTIONS request.
  Receive the response.
  Indicate the response to the application.
Terminate session procedure:
  Create and send BYE request.
  Receive the response.
  Indicate the response to the application.
```

The use case above includes only the main flow of events. A more detailed version would also include the exceptional flow of events that would describe the time management and the retransmissions of the unacknowledged SIP messages. These are skipped here for brevity (in reality, we also start from a very simple version of use cases and refine them later). The same is true for all the other use cases given in this subsection.

The use case *Use UAS* provides the set of procedures for the server side of the transactions. The high-level description of these procedures is as follows (the implementation is rather simple and takes the passive, goodwill approach).

Main flow of events:

```
Receive the request from the TU dispatcher (i.e., remote SIP entity).
Dispatch it to the corresponding procedure.
Session setup service procedure:
  Receive the incoming INVITE request.
  Indicate INVITE request to the application.
  Send the provisional response, e.g., 180 RINGING.
  If the user accepts the call,
    Send the final response 200 OK.
    Receive ACK request.
Cancel session setup service procedure:
  Receive CANCEL request.
  Send the final response 200 OK.
  Report the outcome to the application.
```

```
Modify session/dialog service procedure:
  Receive INVITE request.
  Send the final response 200 OK.
  Report the outcome to the application.
Query server capabilities service procedure:
  Receive OPTIONS request.
  Send the final response 200 OK.
  Report the outcome to the application.
Terminate session service procedure:
  Receive BYE request.
  Send the final response 200 OK.
  Report the outcome to the application.
```

The use case Use TAL is responsible for dispatching TAL messages (coming from TU, UAC, UAS, and TLI and going to the TAL transactions), as well as for dynamic creation of TAL transactions. The use case Use INVITE CT is an invite client transaction. Its description is as follows:

Main flow of events:

```
Receive INVITE request from TAL.
Forward INVITE request to TLI.
Receive 1xx response from TAL.
Forward 1xx response to TU.
Receive the final response from TAL.
Forward the final response to TU.
If the final response is 3xx-6xx,
  Send ACK request to TLI.
```

The use case *Use INVITE ST* is an invite server transaction. Its description is as follows:

Main flow of events:

```
Receive INVITE request from TAL.
Forward INVITE request to TU.
Receive 1xx response from TAL.
Forward 1xx response to TLI.
Receive the final response from TAL.
Forward the final response to TLI.
```

The use case *Use non-INVITE CT* is a non-invite client transaction. Its description is as follows:

Main flow of events:

```
Receive the request from TAL.
Forward the request to TLI.
Receive the response from TAL.
Forward the response to TU.
```

The use case *Use non-INVITE ST* is a non-invite server transaction, which is defined as follows:

Main flow of events:

```
Receive the request from TAL.
Forward the request to TU.
Receive the response from TAL.
Forward the response to TLI.
```

The use case *Use TLI* is responsible for dispatching transport messages. It routes the requests from upper layers toward its remote peer in a forward direction, and routes the responses received from its remote peer toward the upper layers in a backward direction (non-ACK responses are sent to TAL, whereas ACK responses are sent to TU). It may use UDP, TCP, or TLS for the communication with its peers over the Internet. The description of this use case is as follows:

Main flow of events:

```
Receive a request from upper layers.
Send the request to the remote peer.
Receive the response from the remote peer.
If the response is ACK,
  Send it to TU,
Else,
  Send it to TAL.
```

Now that we have completed the use case diagram, we can proceed to the next engineering phase. This phase is the analysis, whose main goal is the definition of the software architecture.

2.3.3 SIP Softphone Analysis Model

Generally, the analysis model is constructed by defining the collaboration in a set of objects for each use case in the source requirements model. This process becomes obvious when considering the rough use case diagram shown in Figure 2.14. However, by refining the use cases, we may reach a point when a single class can realize a single use case. Figure 2.15 is an example of exactly such a use case diagram. Each use case is rather simple, so that a single class can realize it. Along this approach, assume the following mapping:

- The instance of the class *FSM* named *app* realizes the use case *Use application*.
- The instance of the class *TUDisp* named *tud* realizes the use case *Use TU*.
- An unnamed instance of the class *UAClient* realizes the use case *Use UAC*.
- An unnamed instance of the class *UAServer* realizes the use case *Use UAS*.

- The instance of the class *TALDisp* named *tald* realizes the use case *Use TAL*.
- An unnamed instance of the class *InClientT* realizes the use case *Use INVITE CT*.
- An unnamed instance of the class *NIClientT* realizes the use case *Use non-INVITE CT*.
- An unnamed instance of the class *InServerT* realizes the use case *Use INVITE ST*.
- An unnamed instance of the class *NIServerT* realizes the use case *Use non-INVITE ST*.
- The instance of the class *TLIDisp* named *tlid* realizes the use case *Use TLI*.
- The instance of the class *FSM* named *udp* realizes the use case *Use UDP*.
- The instance of the class *FSM* named *tcp* realizes the use case *Use TCP*.
- The instance of the class *FSM* named *tls* realizes the use case *Use TLS*.

The mapping above translates the use case diagram (shown in Figure 2.15) into the general collaboration diagram (shown in Figure 2.16). This diagram actually shows the software architecture, which defines the software objects that constitute the software system or product and the associations among them.

The software architecture can be used for the further study of particular object collaborations to check if the architecture is feasible and, if not, to refine the use case or collaboration diagram. An example of a particular collaboration is shown in Figure 2.17. This diagram shows the handling of the invite request initiated by the softphone user. The flow of events is as follows:

1: The object *app* sends the event *inviteReq(adr)* to the object *tud*.
2: The object *tud* sends the event *inviteReq(adr)* to an unnamed instance of the class *UAClient*.
3: The unnamed instance of the class *UAClient* sends the event *req(INVITE)* to the object *tald*.
4: The object *tald* sends the event *req(INVITE)* to an unnamed instance of the class *IClientT*.
5: The unnamed instance of the class *IClientT* sends the event *req(INVITE)* to the object *tlid*.
6: The object *tlid* sends the event *req(INVITE)* to its peer over the object *tcp*.

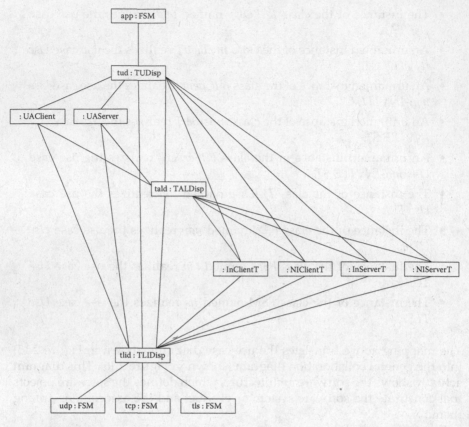

FIGURE 2.16
General collaboration diagram of the simple SIP softphone.

7: The object *tlid* receives the event *rsp(1xx)* from its peer over the object *tcp*.

8: The object *tlid* sends the event *rsp(1xx)* to the object *tald*.

9: The object *tald* sends the event *rsp(1xx)* to an unnamed instance of the class *IClientT*.

10: The unnamed instance of the class *IClientT* sends the even *rsp(1xx)* to the object *tud*.

11: The object *tud* sends the event *rsp(1xx)* to an unnamed instance of the class *UAClient*.

12: The object *tlid* receives the event *rsp(200)* from its peer over the object *tcp*.

13: The object *tlid* sends the event *rsp(200)* to the object *tald*.

14: The object *tald* sends the event *rsp(200)* to an unnamed instance of the class *IClientT*.

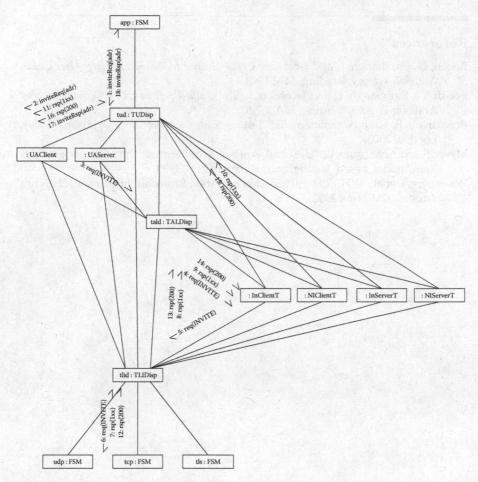

FIGURE 2.17
Collaboration diagram showing the part of the SIP session setup.

15: The unnamed instance of the class *IClientT* sends the event *rsp(200)* to the object *tud*.

16: The object *tud* sends the event *rsp(200)* to an unnamed instance of the class *UAClient*.

17: The unnamed instance of the class *UAClient* sends the event *inviteRsp(adr)* to the object *tud*.

18: The object *tud* sends the event *inviteRsp(adr)* to the object *app*.

Generally, *req()* and *rsp()* designate SIP requests and SIP responses in the flow of events shown above. For example, *req(INVITE)* is the SIP invite request, *rsp(1xx)* is the SIP provisional response, and *rsp(200)* is the SIP final response.

References

Booch, G., Rumbaugh, J., and Jacobson, I., *The Unified Modeling Language User Guide*, Addison-Wesley, Reading, MA, 1998.

Booch, G., Rumbaugh, J., and Jacobson, I., *The Unified Software Development Process*, Addison-Wesley, Reading, MA, 1998.

Broekman, B. and Notenboom, E., *Testing Embedded Software*, Addison-Wesley, London, 2003.

Meyer, S. and Apfelbaum, L., "Use Cases Are Not Requirements," http://www. geocities .com/model_based_testing/online_papers.htm, 1999.

Rosenberg, J. et al., "RFC 3261–SIP: Session Initiation Protocol," http://www.faqs.org /rfcs/rfc3261.html, 2002.

3

Design

System **design** is a phase in engineering work that follows the system requirements and analysis phases. Its main goal is to synthesize a complete solution based on the result of the analysis phase (obtaining the analysis model of the system), which is actually a rough architecture—a skeleton—of the system. We can imagine the system synthesis as a process of creating the body of the system. This body is a reflection of the details related to the system structure and its behavior.

Note that the complete solution of the system mentioned above is not the system itself, but rather a detailed vision of the system that comprises all the details sufficient to construct the system. Technically, we refer to this vision as a **design model**. Therefore, the system synthesis is a process that takes an analysis model as its input and produces the design model as its output.

The design model defines the two most important system aspects:

- System structure
- System behavior

The **system structure** defines the elements of the system and their associations. Sometimes it is referred to as the **static structure** because it defines the static view of the system, i.e., a view without any respect to time. The **system behavior** defines the outputs of the systems as functions of time or their inputs. In the case of a family of communication protocols, which are most frequently modeled as groups of finite state machines (automata), the static structure defines the automata and the links between them whereas the system behavior defines the state transitions for the individual automata and the external messages.

Besides system synthesis, or system design, the communication protocol design phase described in this book includes two additional designs, namely **deployment design** and **test design**, which result in a **deployment model** and a **test model**, respectively. The main goals of the deployment design are identifying network nodes and configurations as well as identifying design subsystems and interfaces. The deployment model is especially important for the complex communication systems comprising many distributed components. For less complex systems, it is not as important, and for very simple systems it may not even be necessary.

Although the system design and deployment models make the complete vision of the system, they do not specify how the system can be verified. Therefore, the engineers conduct the test design by taking the requirements and design models and creating a test model. The test model actually defines the behavior of the testers, who emulate the environment of the system. As already mentioned in the previous chapter, the test model is most frequently referred to as a test suite, which comprises a set of test cases. Each test case specifies a series of test input values (events and messages) to the system and the corresponding output values (events and messages) that are expected at the system output as the results of correct system reactions to the given series.

To summarize, a communication protocol design is a process that takes the requirements and analysis as its input and provides the following models as its output:

- System design model
- System deployment model
- System test model

The means of making these models today are UML diagrams or some domain-specific languages, which are introduced in this chapter. The design engineer starts from the analysis model, essentially a collaboration of <<*boundary*>>, <<*control*>>, and <<*entity*>> classes, described in the corresponding collaboration diagram. The development model is made by mapping each class from the analysis model to a set of new classes in the development model. If the analysis model is well refined, this might even be a one-to-one mapping or close to it. For example, the analysis model of the SIP softphone given at the end of the previous chapter is detailed enough, and the corresponding collaboration diagram is a good base for the refinements that must be made during the system design phase.

The means of defining the static structure of the system in UML are class diagrams and object diagrams. A **class diagram** shows the design classes and the static relations (dependencies, associations, and generalizations) among them without any respect to time. It shows important details about classes, such as their members, fields and functions, types, visibility, and so on. The **object diagram** is similar to the class diagram except that it shows the system frozen at a certain moment of time. Typically, the object diagram will show system objects (class instances) with the characteristic and important values of certain field members.

The means of gathering and refining details about the system behavior are the UML interaction diagrams. Two types of interaction diagrams are found, namely collaboration diagrams (introduced in the previous chapter) and **sequence diagrams**. Collaboration diagrams show the interaction organized by the architecture, meaning that their focus is an architectural view of the system. The architecture is adorned by the flow of events. The sequence of events is shown by adding sequence numbers as prefix labels to the events.

Alternately, sequence diagrams show system interactions from a time progress perspective. The top of the sequence diagram shows the objects of the system without static relations among them. Each object is represented further by a vertical line rendered from its bottom toward the bottom of the diagram. Time advances in the same direction. The interaction itself is shown by the series of events and messages sent among the objects, which are rendered by horizontal arrows from the source object's line to the destination object's line.

The means of specifying complete system behavior are **activity diagrams** and **statechart diagrams** or, more briefly, **statecharts**. An activity diagram shows the action or activity states, starting from the initial and ending in the final state. State transitions can be sequential, branching, or concurrent (through forking and joining). The activity diagram is essentially a flowchart that emphasizes the activity that takes place over time, similar to PERT charts.

Statecharts are the means of specifying finite state machines in UML. They are a type of advanced state transition graphs. A statechart shows simple and composite states starting from the initial and ending in the final state. The composite states are a means to organize states hierarchically. The state transitions can be guarded by conditions and they can indicate firing events and the corresponding actions.

The main goal of the deployment design is the decomposition of the system in two dimensions. Horizontally, the system is partitioned into parts that are deployed onto different network nodes. The term used for nodes by ISO OSI is open systems. Vertically, the system is partitioned into layers. Typical layers recognized by the USDP are the following:

- Application-specific layer
- Application-general layer (e.g., packages common for a set of applications)
- Middleware layer (e.g., Java VM and Java packages)
- System-software layer (e.g., TCP/IP protocol stack)

Furthermore, the system-software layer is generically partitioned by ISO OSI into the following seven layers:

- Application layer
- Presentation layer
- Session layer

- Transport layer
- Network layer
- Data link layer
- Physical layer

Another way to vertically partition is in accordance with the TCP/IP Internet layers, as follows:

- Application layer
- Transport layer
- Network layer
- Network interface layer

In the context of operating systems, we can think of layers as processes. Logically, each process has its own program and the processor that executes it but, in reality, some of the processes may share the program or the processor. The processes sharing the same program are referred to as threads. The processes sharing the same processor constitute the multiprogramming set.

The layers do not exist for themselves—rather, they are typically created to service the requests issued by the upper layers. When the number of requests increases, the engineers face the scalability problem, which can be solved by deploying the same layer on more processors. If the layers are the instances of the same class, we refer to them as replicas. Alternately, on multiprocessor systems with common memory, it might be possible for these layers to share the same program.

The deployment of horizontal system partitions onto different processors or computers is used rather frequently by system designers. Examples include the client–server architecture, the multitier architecture, and others. This convenience is why most engineers think of it in the first place when deployment issues are raised. However, the deployment of a vertical system that partitions onto various processors is also possible. A typical example is the Bluetooth Host Controller Interface (HCI), which is a demarcation line between the host processor that executes the upper layers and the Bluetooth link controller (a microprocessor, a microcontroller, or a digital signal processor) that executes the lower layers.

Horizontal and vertical system partitioning are typically conducted as two interactive activities. The designer typically partitions the horizontal system by rendering the deployment diagram, which shows the network nodes, links between them, and the subsystems deployed on individual nodes. Alternately, vertical partitioning—sometimes referred to as subsystem modeling—results in a class diagram that shows just the subsystems (packages) hierarchically organized in layers, and the dependencies among the subsystems. These two diagrams can be combined in the overall deployment diagram, which shows both the hierarchy and the deployment.

Another important design goal is identifying and providing generic design mechanisms that handle common requirements. The generic design mechanisms can be provided as design classes, collaborations, or subsystems. Examples of the generic design mechanisms in communication protocol engineering are:

- Protocol (finite state machine or automata) state transition management
- Buffer management
- Timer management
- Message management

These mechanisms are common for all communication protocols. Typically, they are designed and implemented once as a separate subsystem that comprises the set of classes, which is then used and refined on a series of projects. In this book, we will use one such subsystem, entitled the FSM library (see Chapter 6). The design and the implementation of such a library is rather specific and rests more in the domain of operating systems. Additionally, such a library frequently already exists and the designers would just use the mechanisms that it provides. Because of these two reasons, we intentionally postpone presenting the FSM library details for Chapter 4.

By accepting this approach, we keep the focus on the activities that are normally conducted during the design phase. We just assume that somebody has written the FSM library that provides all the necessary mechanisms (state transition, buffer, timer, and message management) and concentrate on the design based on these mechanisms. Therefore, for a moment we should simply think of the FSM library as an infrastructure that facilitates the design and implementation of communication protocols.

Going back to the system design itself, this chapter will cover two additional domain-specific languages that have been in use much before UML and are still rather popular today, namely SDL and MSC. The SDL diagrams are semantically equivalent to the UML activity diagrams and statecharts. In principle, establishing a one-to-one mapping between them should not be a problem. The SDL diagram, like the UML activity diagram and statechart, specifies the complete system behavior.

The SDL diagram shows states and state transitions starting from the initial state and ending in the final state. The state transitions are rendered in a style of flowcharts. Each state transition starts with an input message that causes the transition. Typically, a state transition processes the received message and optionally sends the consequent messages.

The MSC chart is semantically equivalent to the UML interaction diagrams, i.e., to both collaboration and sequence diagrams. In fact, the MSC chart can be one-to-one translated into the UML sequence diagram, but the opposite is not the case. By looking at both of them, they make the same impression. Most engineers have the impression that they are almost the same, with the MSC being a little less expressive. Like the UML sequence diagrams, the MSC chart shows the objects that communicate—together with their corresponding vertical lines—and the messages they exchange, which are rendered as horizontal arrows connecting the source and the destination vertical lines.

Finally, this chapter covers the third domain-specific language, TTCN, which is used for making test models more formal than in UML. In contrast to the UML test model, which is rather descriptive and more like a general framework, TTCN is a well-defined language for defining test suites. As already mentioned, it originates from the ISO and has been traditionally used for the conformance testing of communication protocols.

TTCN, much like the higher-level programming language, has built-in types and allows a user to define new types (simple and structured) of variables, constraints, and functions in specialized tables. The essence of the TTCN test case specification is an indented tree of events that is filled in a table, which specifies the behavior of the testers that run the test case and the outcomes of the test case (pass, fail, or inconclusive).

The next sections describe the class diagrams (Section 3.1), the object diagrams (Section 3.2), the sequence diagrams (Section 3.3), the activity diagrams (Section 3.4), the statechart diagrams (Section 3.5), the deployment diagrams (Section 3.6), the SDL diagrams (Section 3.7), the MSC charts (Section 3.8), and the TTCN-3 test suits (Section 3.9). Chapter 3 ends with a series of design examples.

3.1 Class Diagrams

A class diagram is a special type of graph that consists of a set of vertices interconnected by arcs. They are so popular and widely used that most of the newcomers to UML equate the UML and the class diagrams. Normally, we use the class diagrams to model the static design view of the system. More precisely, we typically use them to model the vocabulary of the system, collaborations, or database schemas.

A vocabulary of the system is a set of abstractions that are parts of the system. A collaboration is a group of classes, interfaces, and other elements that cooperate to provide a more complex functionality. A schema is a blueprint that is used for the conceptual design of a database. In communication protocol engineering, we rarely deal with real databases, but we frequently need to design at least a couple of persistent objects that hold the system configuration or similar information.

The basic class diagram vertices are classes, interfaces, and collaborations. These are interconnected with three types of arcs, with dependency, generalization, and association relations. To keep the size of the class diagrams manageable, we typically render smaller collaborations that describe certain aspects of the system. If we want to put those collaborations in a larger context, we can render the surrounding packages or subsystems. Both packages and subsystems enable hierarchical organization of class diagrams. For example, we will render the FSM library as a package that is used by the protocols that are the subjects of design and implementation.

We use packages and subsystems to manage complexity. Alternately, we render class instances (objects) in class diagrams to manage ambiguity, especially when we want to explicitly show the dynamic type of an instance or some other hidden details of the system. A special type of class diagrams are object diagrams, which will be described in the next section of this chapter.

Like use case and collaboration diagrams described in the previous chapter, class diagrams are normally also rendered using some of the commercially available graphical tools, e.g., Microsoft Visio®. The same is true for other UML diagrams described in this chapter. The basic set of graphical symbols available for rendering class diagrams is shown in Figure 3.1. The design engineer must specify properties for each instance of a symbol in the drawing.

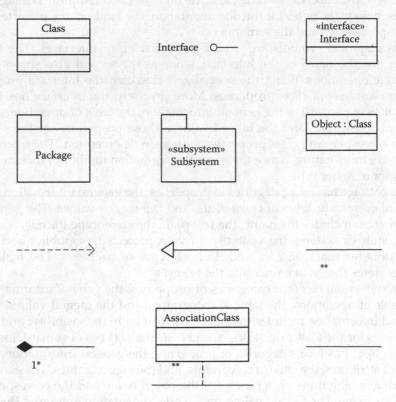

FIGURE 3.1
The basic set of graphical symbols available for rendering class diagrams.

The most frequently used symbol in class diagrams is the class symbol. Eight categories of class properties exist: the general information, the table of attributes, the table of receptions, the table of parameters, the list of components, the table of constraints, and the tagged values. The general information includes the name; the full path; the stereotype (delegate, implementation class, metaclass, structure, type, union, or utility); the visibility (private, protected, or public); and the indicators for the *Root*, *Leaf*, *Abstract*, and *Active* types of classes. The table of attributes comprises columns for the attribute name, the type, the visibility, the multiplicity (1, *, 0..1, 0..*, 1..1, or 1..*), and its initial value. The table of operations comprises columns for the operation name, the return type, the visibility, the scope (classifier or instance), and the indicator for the polymorphic operations. The table of receptions includes columns for the reception name, the corresponding signal name, the visibility, the scope, and the indicator for the polymorphic operations. The table of template parameters stores parameter names and types. The list of components comprises names of the components that implement this class. The table of constraints consists of four columns: the constraint name, the stereotype (precondition, postcondition, or invariant), the language type (OCL, text, pseudocode, or code), and the body of the constraint. The tagged values include the notes for the documentation, the location, the persistence, the responsibility, and the semantics.

Two graphical symbols are available for rendering interfaces. The first shows just the name of the interface, whereas the second also shows the available operations. Being the specialized classifier, the interface properties are a subset of class properties. More precisely, the interface has four categories of properties: the general information, the table of operations, the table of constraints, and the tagged values. Those properties are the same as the corresponding class properties with a single exception. The interface is passive in its nature, hence the general information might not include the indicator of *Active* type.

The package has four categories of properties: the general information, the table of events, the table of constraints, and the tagged values. The general information includes the name; the full path; the stereotype (facade, framework, stub, or system); the visibility (private, protected, or public); and the indicators for the *Root*, *Leaf*, and *Abstract* types of packages. The table of events stores the event names and the types.

The subsystem has four categories of properties: the general information, the table of operations, the table of constraints, and the tagged values. The general information includes the name; the full path; the visibility; and the indicators for the *Root*, *Leaf*, *Abstract*, and *Instantiable* types of subsystems.

The object has four categories of properties: the general information, the table of attributes, the table of constraints, and the tagged values. The general information about the object includes the object name and the corresponding class name. The tagged values are just documentation notes and the tag persistent value.

The dependency relation has three categories of properties: the general information, the table of constraints, and the tagged values. The general information includes the name, the stereotype (becomes, call, copy, derived, friend, import, instance, metaclass, power type, or send), and the description. The tagged values are the notes for the documentation.

The generalization relation has three categories of properties: the general information, the table of constraints, and the tagged values. The general information comprises the name, the full path, the stereotype (extends, inherits, private, protected, subclass, subtype, or uses), and the discriminator. The tagged values are documentation notes.

The association relation has three categories of properties: the general information, the table of constraints, and the tagged values (documentation notes). The general information comprises the name, the full path, the name reading direction (forward or backward), and the information about the association ends, which includes the name, the aggregation (none, composite, or shared), the visibility, the multiplicity, and the indicator *Navigable*. If the end is navigable, it is shown with an arrow symbol, and if not, it is shown without an arrow symbol. Because the composition relation is a specialization of the association relation, it has the same categories of properties (the general information, the table of constraints, and the tagged values), with the exception that the default values for the aggregation and multiplicity (of one of the ends) are composite and 1, respectively.

The association class is a class that models the complex relation; therefore, its set of properties is a union of properties of classes and associations. More precisely, the association class has five categories of properties: the general information, the table of attributes, the table of operations, the table of constraints, and the tagged values. The general information comprises the name, the full path, the information about the association ends (name, aggregation, visibility, multiplicity, and navigability), and the association class details (visibility information and *Root*, *Leaf*, *Abstract*, and *Active* indicators).

The object link has three categories of properties: the general information, the table of constraints, and the tagged values (just documentation notes). The general information includes the name and the information about each of the two link ends. The link end information comprises the name and the stereotype (none, association, global, local, parameter, or self).

This concludes the description of the basic graphical symbols available for rendering class diagrams. The usage of these symbols is illustrated by two examples, as shown in Figures 3.2 and 3.3. The first example is a simple model of the TCP/IP protocol stack, and the second example is a simple model of a finite state machine (automata).

The TCP/IP protocol stack is modeled by the classes that represent its layers: *Application*, *Transport*, *Network*, and *Interface*. The transport layer has a number of ports, which are modeled by the interface *Port*. The application depends on the transport (this fact is modeled by the dependency relation)

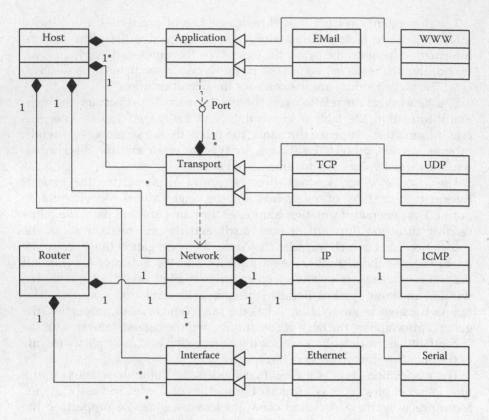

FIGURE 3.2
Example of a simple model of the TCP/IP protocol stack.

and it gets the service it needs through the interface *Port*. Further down, the transport layer depends on the network layer, which in turn is in association with a number of interfaces.

The left side of Figure 3.2 shows the models of the host computers that are connected to the Internet and the routers that interconnect the physical networks that constitute the Internet. The host computer is modeled by the class *Host*. Each host comprises all TCP/IP protocol stack layers. This fact is modeled by the composition relations between the class *Host* and the classes that model the individual layers (*Application*, *Transport*, *Network*, and *Interface*). The router is modeled by the class *Router*. Each router comprises the network and the interface layer. This is modeled by the composition relations between the class *Router* and the classes that model the individual layers.

The right side of Figure 3.2 shows some of the applications and protocols available in the TCP/IP family of protocols. The electronic mail and World Wide Web (WWW) applications—and their corresponding protocols—are modeled by the class *Email* and *WWW*, respectively. These two applications are the examples of particular applications, and this fact is modeled by the

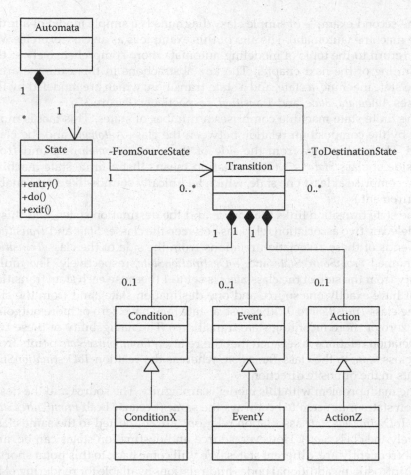

FIGURE 3.3
Example of a simple automata model.

generalization and specialization relations between the class that models
a generic application (*Application*) and the classes that model the particular applications (*Email* and *WWW*). Similarly, TCP and UDP are particular
transport protocols (modeled by the classes *TCP* and *UDP*), and this is modeled by the generalization and specialization relations between the class that
models a generic transport protocol and the class that models TCP and UDP.

Further down the hierarchy, the Internet network layer comprises the IP
and ICMP protocols (modeled as the classes *IP* and *ICMP*). This is modeled
by the composition relations between the classes that model the network
layer and the IP and ICMP protocols. At the bottom of the hierarchy, we
show that various types of interfaces exist, e.g., Ethernet and serial, by generalization and specialization relations between the class *Interface* and the
classes *Ethernet* and *Serial*, which model these particular interfaces.

The second example of simple class diagrams is a simple model of a finite state machine (automata). The aim of this example is as an easy exercise. We will return to the topic of modeling automata more comprehensively at the beginning of the next chapter. The key abstractions in this example are a finite state machine, a state, and a state transition; which are modeled by the classes *Automata*, *State*, and *Transition*, respectively (Figure 3.3).

The finite state machine comprises a number of states. This fact is modeled by the composition relation between the class *Automata* and the class *State*. The multiplicity from the side of the class *Automata* is 1 and from the side of class *State* is *. (This notation means that a finite state machine must comprise at least one state, which technically sounds like a reasonable requirement.)

The state transition links the source and the destination states, and this is modeled by two association relations between the classes *State* and *Transition*. The ends of these association relations from the side of the class *Transition* are named *FromSourceState* and *ToDestinationState*, respectively. The multiplicity from the side of the class *State* is set to 1 (because each state transition must have exactly one source and one destination state), and from the side of the class *Transition* to 0..* (because a state may have zero or more outgoing and zero or more incoming state transitions). The navigability of these two association relations is set such that the relation *FromSourceState* points from the class *State* to the class *Transition*, whereas the relation *ToDestinationState* points in the opposite direction.

The main problem with this model is ambiguity. The source and the destination states may seem to be always the same (because both *FromSourceState* and *ToDestinationState* association relations are connected to the same class, namely the class *State*). However, source and destination states can be, and most frequently are, different states. We will come back to this point shortly, after introducing additional nodes and relations available for rendering class diagrams, to resolve this problem in a less ambiguous way.

The key abstractions related to the transition are the condition that guards the transition, the event that fires the transition, and the action that is taken by the transition, which are modeled by the classes *Condition*, *Event*, and *Action*. Each transition is characterized by these three optional elements, and that is modeled by the composition relations between the class *Transition* and the classes *Condition*, *Event*, and *Action*. The fact that these elements are optional is modeled by setting the multiplicity to 0..1 from the side of the corresponding classes.

Besides actions that are taken during the transitions, we can define state bound actions, such as the action that is taken at the entrance to a certain state, the action that is performed while the system is in a certain state, and the action that is taken at the exit from a certain state (we will encounter these and more in the UML statecharts later in this chapter). These action types are modeled as the state operations *entry()*, *do()*, and *exit()*, which are defined in the table of operations for the class *State*.

Until now, we were modeling a generic finite state machine. To make this model useful for the implementation of a particular finite state machine, first we need to define the concrete conditions, events, and actions. We do so through the specialization of the base classes *Condition*, *Event*, and *Action*. Figure 3.3 shows the examples of the particular condition, event, and action, which are modeled by the classes *ConditionX*, *EventY*, and *ActionZ*, respectively. Finally, to build the particular finite state machine, we need to instantiate the classes.

This concludes the presentation of two simple examples of class diagrams. To make this graphical language more expressive and to reduce the ambiguity of the class diagrams, the graphical tool provides the additional set of graphical symbols, which are shown in Figure 3.4. The first of them is the metaclass, whose instances are classes that are added to the class diagram. We can resolve the problem of ambiguity in the previous example exactly by using the metaclass instead of the class symbol because it is then clear that the source and the destination state may both be the same state or two completely different states. Again, as for the basic set of symbols, the additional symbols have similar categories of properties. The metaclass has the same properties as the class, with the exception that its stereotype (in the general information section) is fixed to metaclass.

Both the signal and the exception symbols have the same four categories of properties, namely, the general information, the table of parameters, the table of constraints, and the tagged values. The general information is the same as for the interfaces (the name, the full path, the visibility, and the indicators *Root*, *Leaf*, and *Abstract*). The table of parameters stores the information about the parameters, which comprise the parameter name, the type, the kind (in, out, or in–out), and the default value.

The data type has five categories of properties. These are the general information, the table of enumeration literals, the table of operations, the table of constraints, and the tagged values. The general information includes the name, the full path, the stereotype (none or enumeration), the visibility, and the indicators *Root*, *Leaf*, and *Abstract*. If the data type is an enumeration, the table of enumeration literals holds the information about the literal names and the corresponding values.

A utility is a special class, therefore it has the same properties as the class with the exception that its stereotype is fixed to utility. Similarly, a parameterized class is a special class that has one or more unbound formal parameters, therefore it has the same categories of properties as the class. Related to the parameterized class is a bind relation, that binds (connects) the designated arguments to the template formal parameters. It has four categories of properties: the general information (just the name and the description), the list of bound arguments, the table of constraints, and the tagged values. The bound element adds the result of binding between the template parameters and their actual values. It has the same categories of properties as the class.

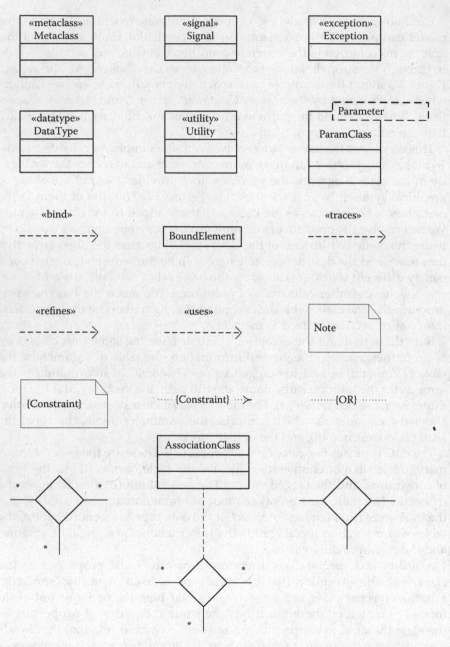

FIGURE 3.4
Additional graphical symbols available for rendering class diagrams.

The next three symbols are the traces, refines, and uses relations. We can think of them as specialized dependency relations. The traces relation connects two model elements from two different models. The refines relation connects a more detailed model element to its previous version. The uses relation indicates the dependency relationship between two model elements where one requires another to fully operate. All these relations have the same categories of information as the dependency relation, with the exception that their stereotype is fixed.

The next four symbols are the note, the constraint note, the constraint shown as arrow, and the OR constraint, which we have already encountered in both use case and collaboration diagrams (described in Chapter 2). The last three symbols are used to describe the relations among more than two model elements. The first is the N-ary association, which models the association among more than two classifiers. Its properties are the same as for the binary association with the additional properties for each association end (the name, the aggregation, the visibility, the multiplicity, and the navigability indicator).

The second symbol is the N-ary association class, which models more complex associations among more than two classifiers. Again, its properties are the same as for the binary association class with additional properties for each association end. The third and the last symbol is the N-ary object link, which interconnects more than two objects. Its properties are the same as the binary object link with additional properties for each end (the name and the stereotype).

At the end of this section, we focus on the domain-specific class diagrams. As already mentioned, the reader should assume and accept that somebody has already prepared the infrastructure for the design and implementation of communication protocols. There is no need to start modeling generic automata every time we start a new project, but rather we do it once and then use it on a number of projects. This practice is what in UML is called providing generic design mechanisms.

In this book, we design and implement communication protocols based on the FSM library. A typical class diagram is shown in Figure 3.5. The FSM library is shown as the package *FSMLibrary* in the diagram and, on most occasions, such representation would be sufficient. It actually comprises a rather ramified hierarchy of C++ classes (we will go into more details in the next chapter). The two most important classes are the *FiniteStateMachine* and *FSMSystem*. The fact that the FSM library contains these classes is modeled by the composition relations between the package *FSMLibrary* and the classes *FiniteStateMachine* and *FSMSystem*. The multiplicity is set to 1 on both sides (one library contains one such class).

The communication protocol is modeled by the class *Automata*. The fact that it is a specific type of finite state machine is modeled by the generalization and specialization relation between the class *Automata* and the class *FiniteStateMachine*. The former inherits all the attributes and operations from

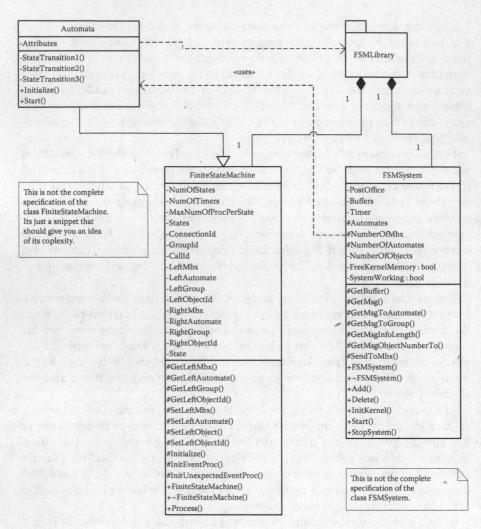

FIGURE 3.5
Typical communication protocol class diagram.

the latter. The list given in Figure 3.5 is not exhaustive, and its purpose is merely to provide the preliminary information about the basic functionality provided by the class *FiniteStateMachine*, and that it is the full set of generic design mechanisms that are needed. Once we have this class, designing a protocol essentially means defining its states and state transitions, and this is basically what we do in this chapter. After the design is finished, implementing the design (in this context) actually means writing the corresponding state transition routines (functions) in C++.

Another important class is the class *FSMSystem*. It actually provides a run-time system for all communication protocols. At the system startup, the main program, here referred to as utility class (not shown in Figure 3.5), registers the given communication protocol by calling the method *Add()* of the class *FSMSystem*, and by giving the reference to the class that models the protocol (*Automata* in this example) as its parameter. Once registered, the protocol can receive, process, and generate events (messages) through the mailboxes provided by the *FSMSystem*.

As we will see in the next chapter, the *FSMSystem* manages all events. It analyzes the event source and destination to locate the destination protocol. Once it is found, the *FSMSystem* looks up its current state, determines the state transition routine based on the event code (type), and calls it. This mechanism is modeled by the Uses relation between the class *FSMSystem* and the class *Automata*.

As we can see, the class *Automata* is a specialization of the class *FiniteStateMachine* and is used by the class *FSMSystem* during the system run-time. More briefly stated, the class *Automata* depends on the package *FSMLibrary*. This fact is also modeled in Figure 3.5 by the corresponding dependency relation between the class and the package.

3.2 Object Diagrams

Object diagrams are a special type of class diagrams that typically show a set of objects (instances of classifiers) and their links. Pure object diagrams contain only objects and their links. However, sometimes we may put some classifiers in the object diagram, especially to clarify the relations between the classes and the objects. We may also use packages or subsystems to deal with complexity.

Object diagrams, like class diagrams, are used to show the static design view of the system. As already mentioned in the previous chapter, the collaboration diagram is used to model the behavior of the system. It also shows the architecture of the system; hence, we say that the collaboration diagram is organized by the architecture. We can think of the object diagram as one snapshot of the collaboration diagram. Imagine that time is frozen. Whatever we can see in the collaboration diagram at that single moment of time is an object diagram.

Later in this chapter, we will introduce deployment diagrams, and in Chapter 4 we will introduce component diagrams. Both deployment and component diagrams can contain only objects and their links. In such cases, they are actually pure object diagrams.

Clearly, the graphical symbols available for rendering object diagrams are the same as the symbols used for class diagrams (sometimes referred to as a static structure). In practice, we use only a very limited subset of those symbols, most frequently only two of them (object and object link). The properties of these symbols are described in Section 3.1.

The usage of object diagrams can reduce the ambiguity of the static structure twofold. First, by rendering instances of classifiers, we can better understand the relations among them. For example, rendering just the classes in the TCP/IP protocol stack model may not give a clear indication of what the network really looks like. Second, by showing the values of the key class attributes, we can recognize reality more easily. For example, by showing the status of the individual protocols, we can comprehend their expectations from other cooperating protocols.

These ideas are illustrated by the following two examples. The first is an object diagram that shows the snapshot from a simple mail transfer protocol (Figure 3.6). The second is an example of a simple finite state machine object diagram (Figure 3.7).

Figure 3.6 shows the software running on two host computers that are connected to two different local area networks, which are interconnected by a router. The host computers clearly require full protocol stacks whereas the router requires only the two lowest level layers (IP and network interface).

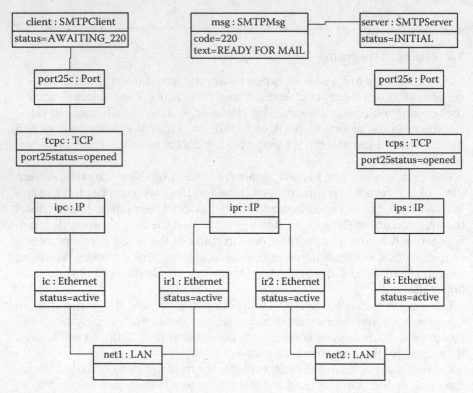

FIGURE 3.6
Snapshot from the simple mail transfer protocol (SMTP).

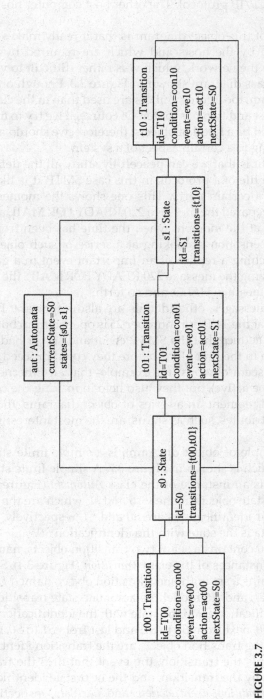

FIGURE 3.7

Example of a simple finite state machine (FSM) object diagram.

One host computer, shown on the left side of the figure, runs the SMTP client on top of the TCP/IP protocol. The other host computer hosts the SMTP server.

The first benefit of this object diagram is that it really makes clear which layers are required by the hosts and which are required by the routers. Graphically, we see the network, which was rather difficult to visualize just by looking at the class diagram shown in Figure 3.3. Enough order is found in this object diagram, too. More symbols are used than in the class diagram, but only five per host and two per router. Of course, if we try to model a large network there would be a flood of objects; therefore, we should always try to restrict our modeling to a certain aspect of a system.

The second benefit is that we can peacefully study all the details of a certain moment in the life of a protocol, in this case SMTP. It is like looking at the photograph of a certain party. This one shows the moment when the SMTP server has prepared the message 220 READY FOR MAIL and its intention was to send it at the moment when the time has been frozen. We can imagine what the sensation of looking at a series of such object diagrams would be, like watching a replica of an important event in a game in slow motion. After receiving the message 220 READY FOR MAIL, the SMTP client would prepare the message HELO, and so forth.

Besides current messages, other details are also important. For example, Figure 3.6 shows that the TCP port number 25 is opened from both sides, and from there we can deduce that the SMTP client and server had to establish the TCP connection in the first place, before they could proceed any further. Some details may seem obvious (for example, that all Ethernet cards and their drivers must be active), but they also help in making the complete picture of the selected moment. In a series of object diagrams, the changes of values of certain attributes, such as status, are the most interesting and most informative parts.

The second example of object diagrams is a simple finite state machine object diagram, which is shown in Figure 3.7. A simple finite state machine object, named *aut*, is an instance of the class *Automata* (Figure 3.4). It comprises a set of two state objects, namely *s0* and *s1*, which are the instances of the class *State*. Their identifications are *S0* and *S1*, respectively. The current state of the automata is the state with the identification *S0*.

The state object *s0* contains a set of two transition objects, namely *t00* and *t01*, which are the instances of the class *Transition* (Figure 3.4). Similarly, the state object *s1* contains a set with one transition object, named *t10*. The transition objects *t00*, *t01*, and *t10* model the automata state transitions from the state with the identification *S0* to the state with the identification *S0*, or more briefly from *S0* to *S0*, next from *S0* to *S1*, and last from *S1* to *S0*, respectively.

The attributes of the transition objects are the transition identification, the condition that guards the transition, the event that fires the transition, the action that is taken by the transition, and the next state identification. Their identifiers are *id*, *condition*, *event*, *action*, and *nextSate*, respectively. *id* and *nextState* would typically be strings or integers. *condition*, *event*, and *action* are the instances of the class *Condition*, *Event*, and *Action*.

An important detail is that the values of these attributes are the instances of classes that are specialized from the classes *Condition*, *Event*, and *Action*. For example, the values of the attribute condition (namely *con00*, *con01*, and *con10*), are the instances of the classes (e.g., *Condition00*, *Condition01*, and *Condition10*), which are actually specializations of the class *Condition*. Such modeling allows us to use polymorphism, the most powerful abstraction of object-oriented design and programming.

3.3 Sequence Diagrams

Two types of UML interaction diagrams are used, namely, sequence diagrams and collaboration diagrams. We have already introduced collaboration diagrams in the previous chapter. They can be used in both the analysis and design phases of communication protocol engineering. Sequence diagrams are just another type of interaction diagrams and are semantically equivalent to collaboration diagrams. This means that a one-to-one mapping exists between these two formalisms that are used for specifying interactions.

An interaction is basically a set of objects and their relationships, together with the messages that are exchanged among the objects. Both sequence and collaboration diagrams show interactions. The major difference between them is that the sequence diagrams emphasize time ordering of messages whereas the collaboration diagrams emphasize the structural organization of a set of objects. The sequence diagrams are particularly useful for visualizing dynamic behavior in the context of the use case scenario. Generally, they are better suited for modeling sequences of events, simple iterations, and branching. Alternately, collaboration diagrams are more useful for modeling complex iterations and branching and for visualizing multiple concurrent flows of control.

Sequence and collaboration diagrams also differ in appearance. As we have already seen in the previous chapter, a collaboration diagram looks like a graph. It consists of objects that are linked together in a certain arrangement. A sequence diagram appears more like a table whose columns are related to individual objects and whose rows are related to the messages that are exchanged among the objects. We can imagine the horizontal axis *x*, at the top of the diagram, pointing from left to right, and the vertical axis *y* that points from top to bottom. The objects that participate in the interaction are arranged across the *x*-axis, starting on the left with the objects that are initiating the interaction and proceeding to the right with more subordinate objects. The messages that are exchanged among the objects are ordered in increasing time along the *y*-axis. (Actually, we have already informally encountered sequence diagrams in Chapter 2. See the example of the SIP session setup in Figure 2.13.)

The sequence diagrams have two key features that distinguish them among other diagrams:

- Object lifeline
- Focus of control

An object lifeline is a dashed vertical line that represents the existence of an object over a period of time. The object lifeline starts with the reception of the message stereotyped as <<*create*>> and ends with the reception of the message stereotyped as <<*destroy*>>. The end of the life of an object is indicated by the mark "X." However, most of the objects will exist throughout the interaction. Such objects are normally placed at the top of the diagram and their lifeline typically goes to the end of the diagram.

The focus of control represents the period of time during which the object executes. It is rendered as a long, thin rectangle. We can model recursion, a call to self-operation, or call-back by placing a new focus of control symbol on top of the current focus of control symbol and slightly to the right, so that both of the symbols are visible. We can explicitly show the part of the focus of control where the actual computation takes place by shading the corresponding region.

We can model the mutation of objects in their state, role, or attribute values in sequence diagrams. Two methods to do this exist: The first is by placing a new copy of the object in the sequence diagram and showing the change by connecting the existing and the new object copy with the transition <<*become*>>. This procedure can be repeated if we want to show a sequence of changes. The second method is by placing a new copy of the object directly on the object's lifeline and showing the change of state, role, or attribute values then and there.

The set of graphical symbols available for rendering sequence diagrams is shown in Figure 3.8. Similar to the diagrams that were previously introduced, each of the symbols has its own properties with the exception of the focus of control, which has no properties on its own (it is a symbol that can exist only on top of the object's lifeline). The designer must fill in the properties after adding the symbol to the diagram.

The object and its lifeline have three categories of properties: the general information, the table of constraints, and the tagged values. The general information includes the name, the full path, the classifier, and the multiplicity. Other categories of properties are already explained in the previous sections.

The message has four categories of properties: These are the general information, the table of arguments, the table of constraints, and the tagged values (documentation notes). The general information includes the name, the

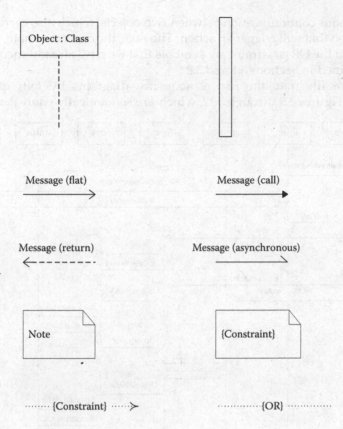

FIGURE 3.8
Set of graphical symbols available for rendering sequence diagrams.

direction (forward or backward), the operation, and the sequence expression. The table of arguments holds information about the arguments, such as the name, the type, the language, and the value.

The following four types of messages are used:

- Flat
- Call
- Return
- Asynchronous

The flat message models the communication between the objects that convey information, which should result in an action. The call message models a synchronous procedure call that should result in some action. The return message models returns from the procedure, which conveys the return value that will cause an action. The asynchronous message models the asynchronous communication between two objects, which also carries some information that will trigger an action. The note, the constraint note, the constraint, and the OR constraint are symbols that we have already encountered and explained in Sections 3.1 and 3.2.

Next, we illustrate the use of sequence diagrams by four examples shown in Figures 3.9 through 3.12, which are semantically equivalent to the

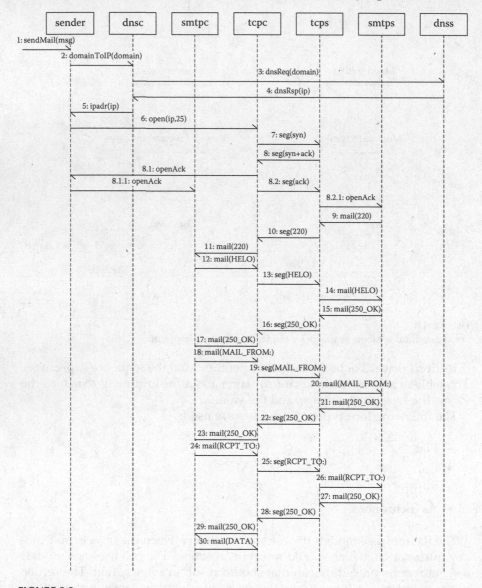

FIGURE 3.9
Sequence diagram showing the interaction between a simple program for sending and receiving e-mails and its environment.

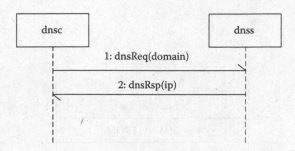

FIGURE 3.10
Sequence diagram showing the interaction between the DNS client and the DNS server.

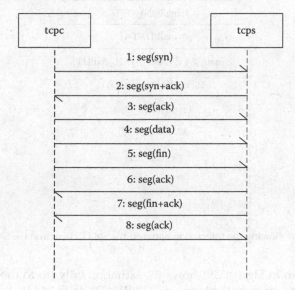

FIGURE 3.11
Sequence diagram showing the interaction between two TCP entities.

collaboration diagrams shown in Figure 2.9 through 2.12, with one exception. Figures 3.9 and 2.9 do relate to the same interaction, but they are not exactly semantically equivalent because of two reasons. First, the former shows fewer objects than the latter, mainly because of the limited diagram width. Second, the latter shows only a part of the interaction shown by the former. Interestingly enough, this seems to be a general rule. The sequence diagrams typically show fewer objects and more messages than collaboration diagrams.

The example shown in Figure 3.9 generally illustrates the same use case *Send e-mail* as the collaboration diagram shown in Figure 2.9. Figure 3.9 shows only the most important subset of objects but, at the same time, it illustrates the interaction long enough to show the moment when the SMTP client sends the SMTP message DATA toward the SMTP server. The collaboration

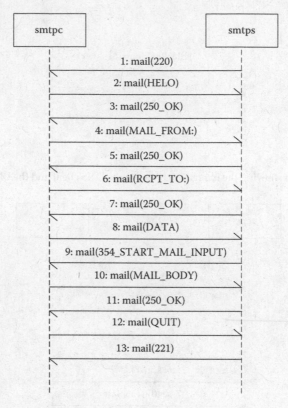

FIGURE 3.12
Sequence diagram showing the interaction between the SMTP client and the SMTP server.

diagram shown in Figure 2.9 shows the situation only up to the point when the SMTP client receives the message 220 READY FOR MAIL, which is actually the very beginning of the SMTP protocol. The names of the objects, messages (signals), and message arguments used in both figures are explained in Chapter 2. The exact flow of events shown in Figure 3.9 is as follows:

1: The object *mailc* (not shown in the diagram) sends the signal *sendMail(msg)* to the object *sender*.
2: The object *sender* sends the signal *domainToIP(domain)* to the object *dnsc*.
3: The object *dnsc* sends the signal *dnsReq(domain)* to the object *dnss*.
4: The object *dnss* sends the signal *dnsRsp(ip)* to the object *dnsc*.
5: The object *dnsc* sends the signal *ipadr(ip)* to the object *sender*.
6: The object *sender* sends the signal *open(ip,25)* to the object *tcpc*.
7: The object *tcpc* sends the signal *seg(syn)* to the object *tcps*.

8: The object *tcps* sends the signal *seg(syn+ack)* to the object *tcpc*. (The event flow now forks into two parallel flows.)

   8.1: The object *tcpc* sends the signal *openAck* to the object *sender*. (The first flow begins here.)

   8.1.1: The object *sender* sends the signal *openAck* to the object *smtpc*. (The first flow ends here.)

   8.2: The object *tcpc* sends the signal *seg(ack)* to the object *tcps*. (The second flow begins here.)

   8.2.1: The object *tcps* sends the signal *openAck* to the object *smtps*.

9: The object *smtps* sends the signal *mail(220)* to the object *tcps*. (Note: We have restarted the message numbering here for brevity. We promoted 8.2.2 to 9.)

10: The object *tcps* sends the signal *seg(220)* to the object *tcpc*.

11: The object *tcpc* sends the signal *mail(220)* to the object *smtpc*.

12: The object *smtpc* sends the signal *mail(HELO)* to the object *tcpc*.

13: The object *tcpc* sends the signal *seg(HELO)* to the object *tcps*.

14: The object *tcps* sends the signal *mail(HELO)* to the object *smtps*.

15: The object *smtps* sends the signal *mail(250_OK)* to the object *tcps*.

16: The object *tcps* sends the signal *seg(250_OK)* to the object *tcpc*.

17: The object *tcpc* sends the signal *mail(250_OK)* to the object *smtpc*.

18: The object *smtpc* sends the signal *mail(MAIL_FROM:)* to the object *tcpc*.

19: The object *tcpc* sends the signal *seg(MAIL_FROM:)* to the object *tcps*.

20: The object *tcps* sends the signal *mail(MAIL_FROM:)* to the object *smtps*.

21: The object *smtps* sends the signal *mail(250_OK)* to the object *tcps*.

22: The object *tcps* sends the signal *seg(250_OK)* to the object *tcpc*.

23: The object *tcpc* sends the signal *mail(250_OK)* to the object *smtpc*.

24: The object *smtpc* sends the signal *mail(RCPT_TO:)* to the object *tcpc*.

25: The object *tcpc* sends the signal *seg(RCPT_TO:)* to the object *tcps*.

26: The object *tcps* sends the signal *mail(RCPT_TO:)* to the object *smtps*.

27: The object *smtps* sends the signal *mail(250_OK)* to the object *tcps*.

28: The object *tcps* sends the signal *seg(250_OK)* to the object *tcpc*.

29: The object *tcpc* sends the signal *mail(250_OK)* to the object *smtpc*.

30: The object *smtpc* sends the signal *mail(DATA)* to the object *tcpc*.

Another practical detail about sequence diagrams is that not only their width but also their height is limited. Because of this, we are normally forced to break the flow of events at a certain point. In the previous example, it was after the object *smtpc* has sent the signal *mail(DATA)* to the object *tcpc*. Typically, we would continue that flow on another sequence diagram. A good practice is to pick the breaking points logically, for example, at the beginning or at the end of certain communication phases.

It is also important to emphasize that the sequence diagram in Figure 3.9 shows only main flows of events. It does not show what happens in the case of errors. The error handling is typically shown in separate sequence diagrams. We can use packages to wrap together all the related sequence diagrams.

Figure 3.9 shows also that the real overall interaction can be fairly complex. To deal with the complexity, we can focus on the individual virtual interactions instead. For example, the sequence diagram showing the interaction between the DNS client and server is a trivial one (Figure 3.10). The overall flow of events is then reduced to only the following two events:

1: The object *dnsc* sends the signal *dnsReq(domain)* to the object *dnss*.

2: The object *dnss* sends the signal *dnsRsp(ip)* to the object *dnsc*.

Similarly, the virtual interaction between two TCP entities, modeled by the objects *tcpc* and *tcps*, is governed by the TCP protocol. It is slightly more complex and comprises the following flow of events (Figure 3.11):

1: The object *tcpc* sends the signal *seg(syn)* to the object *tcps*.

2: The object *tcps* sends the signal *seg(syn+ack)* to the object *tcpc*.

3: The object *tcpc* sends the signal *seg(ack)* to the object *tcps*.

4: The object *tcpc* sends the signal *seg(data)* to the object *tcps*. (This is the data transmission phase.)

5: The object *tcpc* sends the signal *seg(fin)* to the object *tcps*.

6: The object *tcps* sends the signal *seg(ack)* to the object *tcpc*.

7: The object *tcps* sends the signal *seg(fin+ack)* to the object *tcpc*.

8: The object *tcpc* sends the signal *seg(ack)* to the object *tcps*.

Finally, the virtual interaction between the SMTP client and server, modeled by the objects *smtpc* and *smtps*, is of the same order of complexity (Figure 3.12). The interaction is governed by the SMTP protocol. The corresponding flow of events is the following:

1: The object *smtps* sends the signal *mail(220)* to the object *smtpc*.

2: The object *smtpc* sends the signal *mail(HELO)* to the object *smtps*.

3: The object *smtps* sends the signal *mail(250_OK)* to the object *smtpc*.

4: The object *smtpc* sends the signal *mail(MAIL_FROM:)* to the object *smtps*.

5: The object *smtps* sends the signal *mail(250_OK)* to the object *smtpc*.

6: The object *smtpc* sends the signal *mail(RCPT_TO:)* to the object *smtps*.

7: The object *smtps* sends the signal *mail(250_OK)* to the object *smtpc*.

8: The object *smtpc* sends the signal *mail(DATA)* to the object *smtps*.

9: The object *smtps* sends the signal *mail(354_START_MAIL_INPUT)* to the object *smtpc*.

10: The object *smtpc* sends the signal *mail(MAIL_BODY)* to the object *smtps*.

11: The object *smtps* sends the signal *mail(250_OK)* to the object *smtpc*.

12: The object *smtpc* sends the signal *mail(QUIT)* to the object *smtps*.

13: The object *smtps* sends the signal *mail(221)* to the object *smtpc*.

3.4 Activity Diagrams

Up to now, we have introduced three types of diagrams that are used for modeling dynamic aspects of systems. These are the use case, the collaboration, and the sequence diagrams. The use case diagrams are used first for capturing the requirements of the system. Next, they are translated into collaboration diagrams that model the architecture of the system. Then, at the beginning of the design phase, both collaboration and sequence diagrams are used for building up the storyboards of scenarios.

These scenarios describe the interaction among the most interesting objects; hence, we refer to them as interaction diagrams. The interaction itself is shown by the messages that are dispatched among the objects. Generally, interaction (collaboration and sequence) diagrams are similar to Gantt charts. The main difference between the collaboration and sequence diagrams is that the former emphasizes structural relations whereas the latter emphasizes the time ordering of messages.

The storyboards of scenarios are a good place to start the design—therefore, they are a type of design front-end. Although the interaction diagrams make a perfect start of the design, they are seldom used as the final artifacts of the design phase because of two problems:

- The interaction diagrams are most frequently incomplete.
- The interaction diagrams specify the external behavior of individual objects, leaving their internal behavior unknown.

As already mentioned, the interaction diagrams typically cover the main flow of events and, because of the limited space in the diagrams, even the main flow must be partitioned into logical communication phases. Other, less frequent flows (including error handling) are modeled in additional interaction diagrams. All these diagrams can be sorted into packages for easier manipulation. However, no matter how pedantic the engineer is, the set of interaction diagrams remains incomplete by an unwritten rule. Some scenarios are always missing. In the area that is of primary interest for this book, the packages of interaction diagrams are especially vulnerable to the specification of timers and complex, unforeseen error scenarios.

Another problem we encounter while trying to make the packages of inter-action diagrams complete is that they become voluminous and, as a result, hard to comprehend. This behavior is what we should expect when we try to enumerate and describe the cases instead of trying to create the rules that generate these cases. Even a simple program performing some simple arith-metic calculations can produce enormous numbers of execution cases when we take into account the cardinal numbers of sets of values that the com-mon variable types can have. Because of the coverage problems, an implicit engineering rule is that a design based solely on the interaction diagrams is considered as incomplete. This may not be true in the case of simple systems, but generally it is. Therefore, we need the design back-end: the means to end the design.

The secret of how to finish the design is found by turning our attention to the internal behavior of the objects and trying to specify it. This attitude is like turning the interaction diagrams inside out. We want to specify the activities that should take place to provide the desired external behavior and what should be the order (flow) of the activities in the scope of a single object or in the scope of a set of objects that are involved in the interaction. The means to do this in UML are the activity diagrams, which are similar to PERT network charts. The alternative means to specify the behavior of single objects in UML are statecharts, which will be introduced in the next section.

An activity diagram is essentially a flowchart that shows the flow of con-trol from activity to activity. If we model the behavior of a single object, we render the flow of control within that single object. The activity diagrams are even more powerful and they allow us to model the behavior of a group of objects by rendering the flow of control in that larger scope. Additionally, we can model a single flow of control or more concurrent flows of control within both a single object and a group of objects.

In the context of a single object, we typically partition its behavior into a set of its operations and then model the flow of control of these operations indi-vidually. Therefore, the most elementary level of modeling by using activity diagrams is the level of the object's operation. On the opposite side of the scope scale, we can model the workflow of a group of cooperating objects. We will return to that point shortly.

The most elementary activity is an **action state**. It is defined as an atomic (i.e., uninterruptible) program computation. Examples of action sates are the following:

- Create another object
- Destroy another object
- Call an operation on an object
- Return a value
- Send a signal to an object
- Receive a signal from an object
- Evaluate an expression
- Execute a single statement

The action states can be specified in informal text, pseudocode, or a higher-level programming language. Although it is generally assumed that the action state takes a small amount of execution time, that finite amount of time must be taken into account, especially in the models of hard, real-time systems.

By combining more action states, we are building more complex activities, which are referred to as **activity states**. We can think of the activity state as a composite state that is made of other activity states and action states. The activity state can also comprise some special actions, such as entry and exit actions. The former is taken at the entrance to the activity state, and the latter is taken at its exit.

The state transitions in activity diagrams normally take place after completion of the last activity in the originating state. A transition without a guard (condition) immediately passes control to the destination state. Such a transition is referred to as a triggerless, or completion, transition. A transition can branch into two or more guarded transitions, or it can fork into more concurrent transitions. More concurrent transitions can join into a single transition, as we will explain shortly with some simple examples.

An activity diagram is a special type of a graph that comprises a set of vertices that are interconnected by arcs. The basic set of graphical symbols available for rendering activity diagrams is shown in Figure 3.13. Each symbol has a set of properties that must be set by the designer once they add a symbol to the diagram.

The initial state has three categories of properties. These are the general information, the table of constraints, and the tagged values (documentation notes). The general information is just the name and the type (initial). Each activity diagram must start with this symbol.

The final state has the same categories of properties as the initial state symbol, with the exception that its type is final. If the activities specified by the activity diagram go on forever, the diagram will not contain this symbol. Alternately, it can contain one or more such symbols.

The action state has five categories of properties, namely, the general information, the call action, the list of deferred events, the table of constraints, and the tagged values (documentation notes). The general information comprises

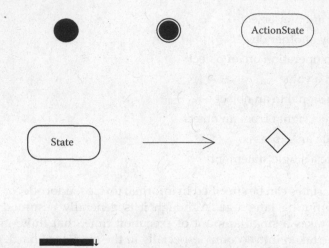

FIGURE 3.13
The basic set of graphical symbols available for rendering activity diagrams.

the name, the stereotype, and the partition. The call action specifies the name of the operation and the table of its arguments, which holds information about the argument name, type, language, and value.

The activity state has six categories of properties. These include the general information, the table of entry actions, the table of exit actions, the table of internal transitions, the table of constraints, and the tagged values. The general information is just the name and the stereotype. Both the table of entry and the table of exit actions store the corresponding action names and their types. The table of internal transitions comprises their properties. Each internal transition is characterized by its name, its stereotype, and the event that triggers the transition.

The control flow transition has four categories of properties, including the general information, the table of actions, the table of constraints, and the tagged values (documentation notes). The general information comprises the name and, optionally, the corresponding event and the guard expression. The table of actions holds action names and their types. The decisions, as well as the fork and join transitions, have three categories of properties, namely, the general information (just the name), the table of constraints, and the tagged values.

We illustrate the usage of these basic symbols by the following four simple examples shown in Figures 3.14 through 3.17. The example in Figure 3.14 shows a simple sequence of interruptible activities (i.e., activity states), namely, *openPort(p)*, *sendData(seg)*, and *closePort(p)*. Normally, these activity states would be modeled by the activity diagrams themselves on the subordinated level of the hierarchy. The control flow transitions between the individual activity states in this example are triggerless, or completion

FIGURE 3.14
An example of a simple sequence of activity states.

transitions, which means that they are not triggered by other events. They also may not be guarded because their sources are not decisions.

The exact semantics of the states in this example are not really important; for example, we can interpret it as open the given port, send the given segment of data, and close the port at the end. Generally, we should think of the activity state as an operation (i.e., procedure or function) which consists of executable statements or calls to other operations, including calls to itself (recursion). Thinking about forward engineering helps make useful activity diagrams. Try to imagine how the model would map to the code. It really does not make any difference how the mapping is made, either automatically with a tool or by hand.

The example in Figure 3.15 is an illustration of activity flow with branching. Actually, it is a simplified implementation of the reliable transport mechanism known as Automatic Repeat Question (ARQ). The whole operation begins by starting the retransmission timer *T1*. This beginning is modeled by the activity state *startTimer(T1)*. The operation then sends the datagram and waits for the answer. These two activities are modeled by the activity state *sendPacket(d)* and *a=waitAnswer()*, respectively.

If the retransmission timer expires, the packet is retransmitted. This mechanism is modeled by the transition guarded by the expression *[T1 expired]*, the activity state *restartTimer(T1)*, and the completion transition back to the activity state *sendPacket(d)*. The reception of the answer is modeled by the

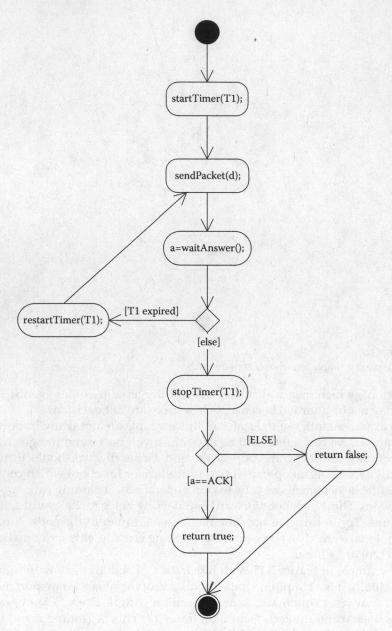

FIGURE 3.15
Example of a simple flow of activities with branching.

transition that covers all the other cases (guard expression *[ELSE]*). The operation proceeds by stopping the retransmission timer, and this action is modeled by the activity state *stopTimer(T1)*. If the answer is the acknowledgment (*ACK*), the operation returns the *value true*; otherwise, it returns the *value false*.

The previous example uses two branches. Each branch has one incoming and two or more outgoing transitions. The outgoing transitions are guarded by the Boolean expressions that are evaluated at the entrance to the branch. The set of guards has two important features:

- The guards must not overlap—this makes the flow of control unambiguous.
- The guards must cover all possibilities—this ensures that the flow of control is not going to freeze.

Precisely these two features force us to make complete models and specifications of activities that describe the behavior of the system. When we render interaction (collaboration and sequence) diagrams, no such enforcements are present. As a result, they remain unfinished. Of course, at the time when we render interaction diagrams, we really do not want to make them final; rather, we want to check the most important aspects and scenarios, and to make our analysis more comprehensive and useful for the finalization later. Therefore, when we start rendering the activity diagram, we already have a good overall vision, but non-overlapping and complete coverage features are the driving forces of the design finalization.

One safe way to provide both of these features is to use only the decisions with two outgoing transitions and to guard one of them by the keyword *ELSE*, as in the example in Figure 3.15. Special attention should be paid to the decisions with more outgoing transitions, which are guarded by explicit expressions (i.e., without the keyword *ELSE*). However, the price that we may pay for safety is ambiguity. For example, if the operation in the previous example returns the *value false*, it might do so because the correct not acknowledge answer (*NAK*) has been received. However, the operation will return the same value if any other message (including corrupted *ACK* or *NAK*) has been received.

The example in Figure 3.16 illustrates the usage of loops in activity diagrams. Imagine that the IP protocol must route a datagram over a physical network, which has the Maximal Transfer Unit (MTU) smaller than the datagram size. Normally, the IP protocol partitions the datagram into fragments (that fit MTU) and routes the resulting fragments individually in such cases. The standard means to model repetitive activities in activity diagrams are loops.

The example in Figure 3.16 starts by setting the control variable i to the value 0. It continues with no operation activity state, followed by the decision

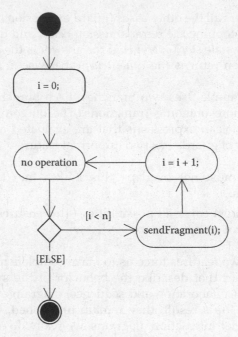

FIGURE 3.16
Example of a loop in an activity diagram.

that checks the loop continuation condition ($i < n$). If the condition is satisfied, the flow enters the loop body (*sendFragment(i)*). The loop body is followed by the activity state that updates the control variable ($i = i + 1$). The example terminates when the loop continuation condition becomes false.

The example in Figure 3.17 shows the usage of concurrent control flows. Imagine that we want to model a simple communication over the TCP connection. First, we must establish the TCP connection by opening a particular TCP port. We model this by the activity state *openPort(p)*. Once the connection is established, the TCP protocol provides simultaneous transfer of data in both directions (full-duplex). To model that, we need to fork a single flow of control into two parallel (concurrent) flows of control. One of them enters the activity state *sendData()*, which models the activity of sending the data to the remote site. The other control flow enters the activity state *receiveData()*, which models the activity of receiving the data from the remote site.

These two activities logically evolve in parallel over time. On a multiprocessor system, they can be deployed on two different processors to maximize the system throughput. In such a case, these two activities would also be parallel in reality. Alternately, single-processor systems create quasi-parallelism using the time-sharing operating system. The activities are then not parallel in reality, but they are still concurrent because they can compete for the same resources. Additionally, the activities can communicate using signals.

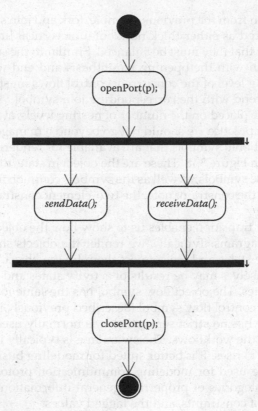

FIGURE 3.17
Example of a simple set of concurrent flows.

Traditionally, such communicating sequential processes are referred to as **coroutines**.

Although the model shown in Figure 3.17 is fairly simple, it may reflect a realistic communication, such as a Telnet session. Imagine that the activity state *sendData()* is a composite state that reads the user keystrokes and sends them to the Telnet server over the TCP connection, in a loop, until the end-of-file key combination is detected. The activity state *receiveData()* in this scenario would be also a composite activity state, which receives the responses from the Telnet server and displays them on the monitor in a loop, until the end-of-communication signal is detected (typically, it would be sent when the end-of-file key combination is detected).

Once one of the parallel activities finishes, it proceeds to the control flow joint synchronization point where it waits for the other parallel activity to finish. When both of the activities are finished, the corresponding parallel control flow joins into a single control flow, which enters the activity state *closePort(p)*; after finishing that activity, it terminates.

As we have seen from the previous example, fork and join synchronization points are rendered as either thick horizontal or vertical lines. It is important to remember that they must be balanced. Similar to the subexpression—which must begin with the opening parenthesis and end with the closing one—each nesting level of the concurrent control flows must begin with the fork symbol and end with the corresponding join symbol. Apart from that, no restrictions are placed on the number of nesting levels, at least not in theory. Of course, in practice we should not go beyond a manageable number.

The set of additional symbols that are available for rendering activity diagrams is shown in Figure 3.18. These are the object in state, the object in flow, and the swim lane symbols, as well as the symbols common for all diagrams, namely, the note, the constraint note, the two-element constraint, and the OR constraint.

The object flow transition enables us to show how the object state changes in the activity diagrams. Typically, we render the objects showing the current and the new states and we connect them by the object flow transition. The objects themselves may be results of activity states and can be used by other activity states. The object flow symbol has the same four categories of properties as the control flow symbol (described previously in this section).

The swim lane has no strict semantics. It is normally used to show individual parties in the workflows. The swim lane is typically implemented as a class or a set of classes. It is better suited for modeling business processes, but it can also be used for modeling communication protocols. The swim lane has three categories of properties: general information (essentially, its name), the table of constraints, and the tagged values.

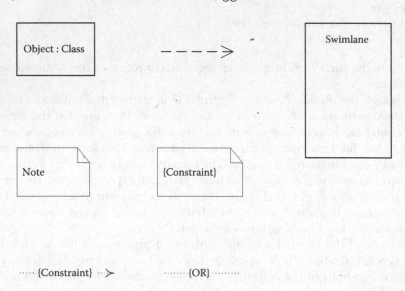

FIGURE 3.18
Additional graphical symbols available for rendering activity diagrams.

The example in Figure 3.19 illustrates the usage of objects, data flow transitions, and swim lanes, with the example of activities initiated by the Domain Name System (DNS) client request for mapping a given domain name onto the corresponding IP address. Figure 3.19 is a type of a workflow conducted by the DNS client and server in their cooperative work of translating a domain name into the IP address. The DNS client is represented by the first swim lane and the DNS server is represented by the second. This activity

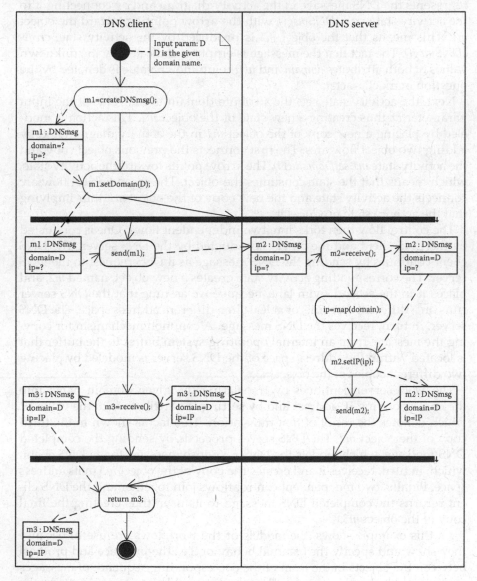

FIGURE 3.19
Workflow between the DNS client and server with the message flow.

diagram shows both the control flow among individual activity states and data flow, which are created by a series of objects that are consumed and produced by the activity states of both DNS client and server.

The given domain name is the input parameter of the DNS client operation that translates the domain name into the corresponding IP address. This operation starts by the activity state *createDNSmsg()*, which creates an empty DNS message. This action is modeled by placing the object *m1* that represents the DNS message in the activity diagram and by connecting it to the activity state *createDNSmsg()*, with the arrow pointing toward the object *m1*. This means that the object *m1* is produced by the activity state *createDNSmsg()*. The fact that the message is empty is indicated by the unknown values of both attributes *domain* and *ip* (the unknown value is denoted by the question mark character, "?").

Next, the activity state sets the attribute domain to the value of the input parameter *D*, thus creating a new state of the object *m1*. This action is modeled by placing a new copy of the object *m1* in the activity diagram and by adding two object flow arcs. The first connects the previous object copy and the activity state *m1.setDomain(D)*. The arrow points toward the activity state, which means that the state consumes the object. The second object flow arc connects the activity state and the new copy of the object *m1*, thus implying that the activity state produces it.

The control flow then forks into two independent flows. One is conducted by the DNS client and the other is conducted by the DNS server. The DNS client continues by sending the DNS message as a DNS request to the DNS server. The corresponding activity state creates a new object, named *m2*, and places it in the second swim lane, because we assume that the DNS server runs on a different machine, or at least in a different address space. The DNS server, in turn, receives the DNS message. A common mechanism for copying the message from an internal operating system buffer to the buffer that is located within the address space of the DNS server is modeled by placing two different copies of the object *m2*.

The DNS server continues by translating the given domain name into the corresponding IP address and by setting the attribute *ip* to the value IP, which denotes the result of that translation. This fact is shown in the third copy of the object *m2*. The DNS server proceeds by sending the completed DNS message, which models the DNS response message, to the DNS client, which, in turn, receives it and creates the copy of the object *m3* in its address space. Finally, two independent control flows join together and the DNS client returns the completed DNS message to its user, thus creating the final copy of the object *m3*.

As this example shows, the models of the workflows are useful because they show and specify the external behavior, i.e., the interface and protocol between the objects in the form of the corresponding sequence of messages exchanged by the objects, as well as the internal behavior of objects in the form of the series of activity states visited by them. The first is created by modeling the data and object flow, and the second is created by modeling the control flow across the objects. Again, by taking care of the complete coverage of possibilities, without any overlaps, we ensure that the model is complete. (This was not the main goal of the last example, at least not to the extent of the previous one, but we should keep that in mind.)

Figure 3.20 shows the activity diagram for one real protocol, TCP, and follows the conventions introduced by the corresponding IETF RFC 793. The user requests are written in capital letters. The user requests are *OPEN*, *SEND*, and *CLOSE*. Two types of *OPEN* requests are used, namely active *OPEN* and passive *OPEN*. The difference between the two depends on which one is taking the initiative in the connection establishment procedure.

The next convention is that the names of the events and actions are written in lowercase letters, with the following abbreviations:

- TCB (Transmission Control Block)
- snd (send)
- rcv (receive)
- SYN (indicates that the synchronization bit of the TCP segment is set)
- ACK (indicates that the acknowledgment bit of the TCP segment is set)
- SYN, ACK (both SYN and ACK bits are set)
- FIN (indicates that the final bit of the TCP segment is set)
- ACK of SYN (denotes the acknowledgment of the SYN segment)
- ACK of FIN (denotes the acknowledgment of the FIN segment)
- MSL (Maximum Segment Lifetime)

The TCP events are actually modeled as guard expressions whereas the TCP activities are modeled as UML action states (a relatively short and uninterruptible series of executable statements). Notice that we could model the TCP activities either by action or by activity states because these activities are essentially interruptible. However, because they can be implemented as rather short routines—which do not involve reception of any signals—modeling them as action states makes more sense than as activity states.

The TCP protocol spends most of the time in one of its stable states waiting for a certain event to occur. The TCP stable states are modeled by the UML activity states. While being in one of its stable states, the TCP protocol just waits for an event (it does not execute any statements). The process that executes the TCP protocol is blocked and it does not compete for the processor's execution time. Therefore, the activity corresponding to the stable state is more than interruptible—it is blocked. Because such an abstraction is missing in the UML activity diagrams, we are forced to model it with an

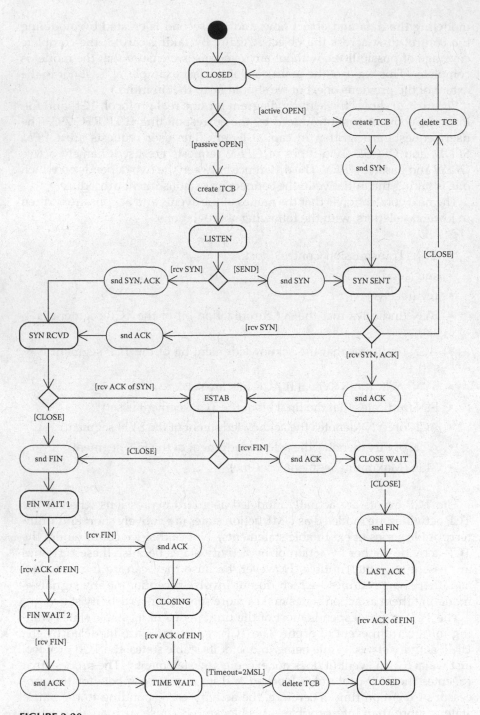

FIGURE 3.20
TCP activity diagram.

abstraction that is the closest to it, and that is the activity state. The model of the TCP protocol shown in Figure 3.20 comprises the following activity states (the names of the states are taken from the RFC 793):

- CLOSED (no connection exists)
- LISTEN (wait for a connection request from any remote TCP and port)
- SYN SENT (wait for a matching connection request after having sent a connection request)
- SYN RCVD (wait for a confirming connection request acknowledgment after having both received and sent a connection request)
- ESTAB (the connection is established, i.e., open)
- FIN WAIT 1 (wait for a connection termination request from the remote TCP, or an acknowledgment of the connection termination request that was previously sent)
- CLOSING (wait for a connection termination request acknowledgment from the remote TCP)
- FIN WAIT 2 (wait for a connection termination request from the remote TCP)
- TIME WAIT (wait for enough time to pass to be sure that the remote TCP has received the acknowledgment of its connection termination request)
- CLOSE WAIT (wait for a connection termination request from the local user)
- LAST ACK (wait for an acknowledgment of the connection termination request previously sent to the remote TCP, which includes an acknowledgment of its connection termination request)

The activity diagram shown in Figure 3.20 is fully compliant with the original TCP standard. Interested readers can refer to IETF RFC 793 for more details.

The last example in this section shows a model of a simplified send e-mail operation. The corresponding activity diagram (Figure 3.21) is a straightforward implementation of a typical SMTP scenario (client side), which has already been introduced in this chapter (Figure 3.12) and in Chapter 2 (Figure 2.12). Although simplified, in the sense that it just follows the successful path of the SMTP scenario, it is a complete specification of a desired behavior because it covers all possibilities in a non-overlapping manner.

Again, like the previous example, the events associated with the reception of the corresponding messages are modeled as guard expressions, while the actions taken by the SMTP client are modeled by the corresponding action states. Additional similarity with the previous example is that the SMTP

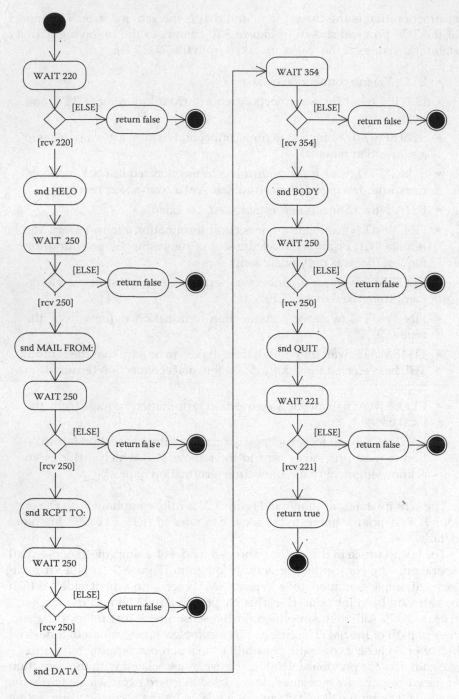

FIGURE 3.21
Simple send e-mail operation activity diagram (SMTP client side).

client, like the TCP protocol, spends most of its time in its stable states, waiting for a message from the SMTP server. If the received message is the one expected, the SMTP client sends the next message, prescribed by the ideal SMTP scenario, and proceeds to the next stable state. If the received message is not the one expected, the SMTP client returns the *value false* and the operation terminates.

The e-mail is successfully sent if all of the prescribed messages between the SMTP client and server are successfully exchanged. In this case, the send e-mail operation returns the *value true* and terminates.

3.5 Statechart Diagrams

In contrast to activity diagrams—which can be used for modeling activities both inside the individual objects and across the workflow of objects—the statechart diagrams are normally used for modeling the lifetime of a single object, typically, an instance of a class or a use case. The activity diagrams emphasize the flow of the action and the activity states, whereas the statecharts emphasize the event-ordered behavior of an object, which is especially suitable for modeling reactive systems.

The common feature of both activity diagrams and statechart diagrams is that they aim at making complete models of behavior, i.e., for use in the design back-end. The driving forces for providing complete behavior specifications are the same, namely, the complete coverage of possibilities without overlaps. The styles differ a bit. By an unwritten rule, the decision symbols are extensively used in activity diagrams and seldom used in statechart diagrams. Therefore, the coverage of possibilities is shown explicitly in activity diagrams and more implicitly in statechart diagrams.

That the activity and statechart diagrams are semantically equivalent is also important to emphasize, i.e., we can use both of them for modeling the same behavior on a comparable level of details. They merely provide two different views of the same behavior. The activity diagrams are better suited for modeling individual operations, whereas the statechart diagrams are better for modeling the behavior of entire stateful objects, especially if the behavior is driven by events (messages).

Statecharts were originally invented for modeling state machines, which makes them a perfect tool for modeling communication protocols because the protocols are essentially state machines. According to the UML terminology, a **state machine** is a sequence of states an object goes through in its lifetime. A **state** is a situation during which an object satisfies a certain condition, performs an activity, or waits for an event. An **event** is an occurrence of a stimulus that triggers the state transition. An **action** is an atomic executable statement (computation). An **activity** is a non-atomic execution composed of actions and other activities. A **transition** is a relation between the source and the target states (these can be different states or the same state) that specifies the actions to be taken when the given event occurs and the given guard condition is satisfied.

The key abstractions in the context of state machines are the object state and the state transition. We can think of the object state as a period of an object's lifetime (it can be just a moment characterized by a certain condition, a period of a certain activity, or an interval of time in which the object waits for a certain event). Alternately, we can think of the state transition as a rather short interval of object's lifetime, which is related to actions caused by a certain event, and is defined by the following five attributes:

- The source state
- The event trigger
- The guard condition
- The actions
- The target state

A statechart diagram is a special type of graph that comprises a set of vertices that are interconnected by arcs. The basic set of graphical symbols available for rendering statechart diagrams is shown in Figure 3.22. Each symbol has a set of properties that must be set by the designer once they add the symbol to a diagram.

The initial state has three categories of properties. These are the general information, the table of constraints, and the tagged values (documentation notes). The general information is just the name and the type (initial). Each statechart diagram must start with this symbol.

The final state has the same categories of properties as the initial state symbol, with the exception that its type is final. If the lifetime specified by the statechart diagram is infinite, the diagram will not contain this symbol. Alternately, it can contain one or more such symbols.

The state has six categories of properties. These include the general information, the table of entry actions, the table of exit actions, the table of internal transitions, the table of constraints, and the tagged values. The general

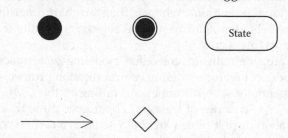

FIGURE 3.22
Basic set of symbols available for rendering statecharts.

information is just the name and the stereotype. Both the table of entry and the table of exit actions store the corresponding action names and their types. The table of internal transitions comprises their properties. Each internal transition is characterized by its name, its stereotype, and the event that triggers the transition.

The following eight common types of actions are used:

- Create an object
- Destroy an object
- Call an operation on another object
- Call an operation on this object (local invocation)
- Send a signal (message) to another (or this) object
- Return a value
- Terminate execution
- Uninterrupted action (other unclassified types of actions)

Four common types of events are also used:

- Signal event: This object has caught (received) the signal (message) that was thrown (sent) by another (or this) object. In UML, we model the signal by the class stereotyped as <<*signal*>>. We can also use a dependency relation, stereotyped as <<*send*>>, between the operation of the class that sends the signal and the class that defines the signal to explicitly show the source of the signal. A signal is an asynchronous event.

- Call event: The object's operation is called by another (or this) object. A call event is a synchronous event. The event name and the parameters are the names and the parameters of the corresponding operations, respectively.

- Change event: The given condition is satisfied. Generally, the condition is related to the state of this object (value of its attributes) or to absolute time. We use the keyword *when* to specify the condition, e.g., *when((time == 17:00)*, or *when(key == pressed)*. A change event is an asynchronous event.

- Time event: The given interval of time has expired. We use the keyword *after* to specify the expression that evaluates to a period of time, e.g., *after(10s)*, or more symbolically *after(T1)*, which means that the timer *T1* has expired. By default, the starting time of such an expression is the time since entering the current state. If we want the starting time to be other than that, we must specify it explicitly. We should note that time events enable implicit timer management, as will be illustrated shortly.

The transition has four categories of properties. These are the general information, the table of actions, the table of constraints, and the tagged values (documentation notes). The general information comprises the name and optionally the corresponding event and the guard expression. The table of actions holds action names and their types. The decision has three categories of properties, namely, the general information (just the name), the table of constraints, and the tagged values (same as the decision in activity diagrams).

Simple examples that illustrate the usage of the basic set of graphical symbols for rendering statechart diagrams seem to be appropriate at this point. The following two examples, shown in Figures 3.23 and 3.24, are semantically equivalent to the simple examples of activity diagrams shown in Figures 3.14 and 3.15, respectively. The activity diagram shown in Figure 3.14 illustrates a sequence of three activity states, namely, *openPort(p)*, *sendData(seg)*, and *closePort(p)*. Figure 3.23 shows three versions of statechart diagrams that model the same behavior. These are the versions A, B, and C.

Version A models the behavior by a sequence of three transient states, namely, *Opening*, *Sending*, and *Closing*. By selecting appropriate names, we can indicate what type of activity is taking place in each of the states. The original activities *openPort(p)*, *sendData(seg)*, and *closePort(p)* are modeled as internal transitions of the states *Opening*, *Sending*, and *Closing*, respectively. We could also use entry or exit actions instead of internal transitions. Alternately, we could model this simple behavior by only one transient state with three internal transitions. Generally, by compressing models we decrease their clarity, and we should seek the compromise appropriate for the project at hand. Of course, defining clarity is tricky because it is essentially a matter of taste.

Version B is the model of the same behavior that employs another way of modeling activities in the statechart diagrams, and that is by actions taken by state transitions. This version of the model comprises three transient states, namely, *Initial*, *Ready*, and *Finished*, which are connected by triggerless transitions. Such transitions take place immediately after their source state is left (finished). The original activities *openPort(p)*, *sendData(seg)*, and *closePort(p)* are modeled here by the actions of the corresponding state transitions.

Finally, version C is the most compressed form of the model with the equivalent semantics. It comprises only one state transition, from the initial to the final state, which conducts a series of actions, namely, *openPort(p)*, *sendData(seg)*, and *closePort(p)*. This extreme shows the power of statechart diagrams. Generally, statecharts are more expressive than activity diagrams when it comes to modeling state machines, therefore we can model the same behavior in less space.

The activity diagram shown in Figure 3.15 is a model of a reliable packet delivery operation, which starts the timer T1, sends a packet, and waits for the answer from the remote site. If the timer T1 expires before the answer is received, the packet is sent again. If the answer is ACK, the operation returns

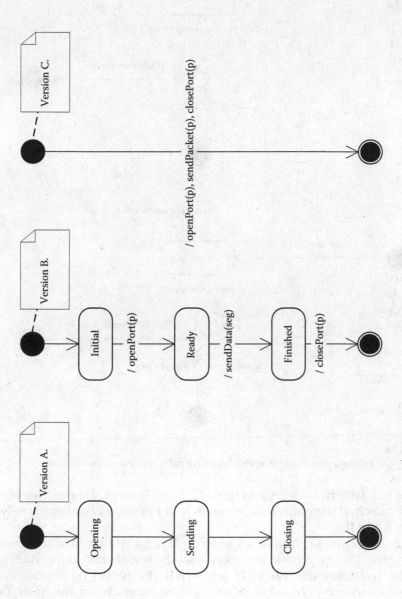

FIGURE 3.23
Example of a simple state machine with a single path of evolution.

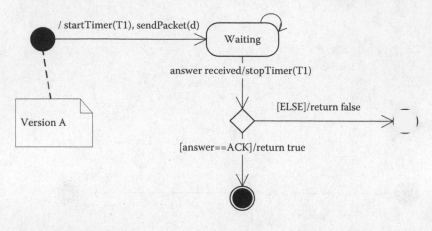

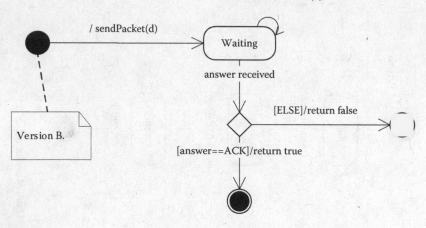

FIGURE 3.24
An example of a simple state machine with alternative paths and loops of evolution.

the *value true*. Otherwise, it returns the *value false*. Figure 3.24 shows two versions of statechart diagrams that are models of the same behavior, namely, versions A and B.

Version A models the given behavior by explicit, rather than implicit, timer management. The triggerless transition from the initial state to the *Waiting* starts the retransmission timer T1 and sends the packet by conducting the actions *startTimer(T1)* and *sendPacket(d)*. The expiration of the timer T1 is modeled here by the signal event *T1 expired*. The corresponding transition restarts the timer T1 and sends the packet again. The reception of the answer from the remote site is modeled by the signal *answer received*. The corresponding transition stops the timer T1 and leads to the decision with two outgoing transitions. The first is taken if the answer is ACK; otherwise, the second is taken. Those who prefer not to use decision symbols in their statechart diagrams should delete it along with the previous transition, and add the event answer received to both transitions that lead to the final state.

Version B, in contrast to version A, models the given behavior by implicit timer management. Here the triggerless state transition from the initial state to the *Waiting* state just sends the packet by conducting the action *sendPacket(d)*. The existence of the state transition triggered by the time event *after: T1* implicitly implies that the timer T1 has started at the entrance of the *Waiting* state. If the timer T1 expires, the packet is sent again by the action *sendPacket(d)* and the timer T1 is restarted at the new entrance to the *Waiting* state. The event *answer received* occurs when this object receives the answer from the remote side. This event triggers the transition that leads to the decision and, later, to the final state. The timer T1 is implicitly stopped at the exit from the *Waiting* state. The result is a more compressed form of a model with more implicit details, which may not be seen at first glance. We can use either one of these two styles, but we should be consistent and stick to one on a certain project.

Now that we have covered the basics of statechart diagrams, we proceed to their more advanced abstractions. First, besides entry and exit actions and internal transitions, a state can perform an ongoing activity that we can specify by using the keyword *do*. Most of the states are stable states, which means that the object is blocked while waiting for an event. Some of the states are transient, which means that they perform certain computations and then finish. Sometimes we need to also model active states, which perform some activities while simultaneously waiting for an event to occur; we do these by using the keyword *do*. Generally, the special *do* transition can name another state machine or a sequence of actions.

Deferred events are the next important abstraction in the context of states. Until now, we were not interested in the events that occur during the state that does not react to them. What happens to these events? They are simply lost. If we want to save them so that they can be processed later in some other states, we must specify that they are to be deferred by using the special action named *defer*. Each event that is associated with this special action will be saved for further processing by the states that explicitly name that event in one of their transitions.

We have already shown how to manage complexity by using hierarchical organization. Statechart diagrams allow us to use that powerful concept in the context of states. Until now, we have dealt with simple states. Actually, a state in UML can also be a composite state, which means that it can comprise simple states and other composite states. This nesting of states can go to an unlimited depth, at least in theory.

A composite state can contain either sequential or concurrent substates. The sequential substates are disjoint, i.e., an object can be in only one of them at a certain point in time. The concurrent substates are orthogonal, which means that an object at a certain point in time is in all of the concurrent substates that are active at that point. We can think of a concurrent state as one aspect (orthogonal axis) of the object's lifetime.

The state transitions until now were transitions between simple states. After the introduction of composite states, the situation becomes more complex in this respect. Besides the transitions between simple states, there exist the transitions between simple states and composite states, as well as the transitions from substates to external states. The transitions from external states to substates of a composite state are not allowed. This asymmetrical relation raises the following question: What happens to the flow of states inside a composite state if a transition from that composite state to another state is triggered?

The answer is that the information about the point of interruption inside the composite state is lost by default. This means that the processing will be restarted from the very beginning when that composite state is reentered once again later. This means that the composite state operates without context saving, which is referred to as a **history** in the UML.

If we want the composite state to operate with the history—which means it is able to restart from the point of interruption at its reentrance—we can use the special history state. The history state is a special type of an initial state that is the target for the transitions from the external states. Once activated, it restarts the operation at the point of interruption. The following two types of history states are used:

- The shallow history state (marked with the symbol H)
- The deep history state (marked with the symbol H*)

The shallow history state ensures context-saving only on the first level of nesting of composite states. Alternately, the deep history state provides context-saving on the innermost state at any depth. If there are more nesting levels, the shallow history remembers the outermost nested state and the deep history remembers the innermost nested state.

Like activity diagrams, statechart diagrams also support modeling concurrency. We model concurrent activities in statechart diagrams by using concurrent sequences of substates inside a certain composite state. Typically, each such sequence begins with the initial state and ends with the final state. The transition from the external state to this composite state forks to concurrent substates, which at the end joins in the transition from this composite state to the external state. The usage of concurrent substates is advisable only if the behavior of one of these concurrent flows is affected by the state of another. Alternately, if the behavior of the concurrent flows is driven by the signals (messages) they exchange, partitioning the object into more active objects is preferable.

The set of additional symbols that are available for rendering statechart diagrams is shown in Figure 3.25. These are the composite state, the shallow history state, the deep history state, the fork or join synchronization point, the note, the constraint note, the constraint, and the OR constraint. These symbols, like others, have their properties. The composite state has the same categories of properties as a simple state, plus two additional indicators (*Concurrent* and *Region*) which determine whether the composite state is concurrent or not and if it is a region or not. Both shallow and deep history states have the same three categories of properties. These are the name, the table of constraints, and the tagged values. The rest of the symbols have already been introduced.

Figure 3.26 shows the simple example of a statechart diagram that uses the shallow history state. Imagine a simple state machine that starts from the state *Idle*. The event *sendCharacter(ch)* triggers its transition to the composite state *Sending Segment*, which starts with the shallow history state to ensure context saving. Because this state comprises only simple states, the application of the deep history state, instead of the shallow history state, would have the same effect because only one level of nesting of composite states is found.

The state machine remains in the substate *Buffering* while it is filling the corresponding buffer with new incoming characters. This status means that the state machine will wait for the additional event *sendCharacter(ch)* until the buffer becomes full, when the state machine will proceed to the state *Sending*. After it sends all the characters from the buffer, the state machine leaves the compound state *Sending Segment* and triggerlessly transits to the state *Idle*.

If the event *break* occurs while the state machine is in the compound state *Sending Segment*, its context will be saved and the state machine will leave it and move to the state *Break*. It will remain in this state until the event *continue*

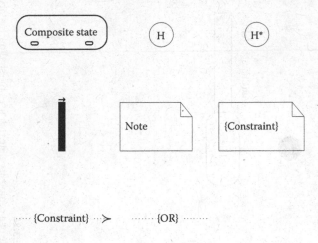

FIGURE 3.25
Additional graphical symbols available for rendering statecharts.

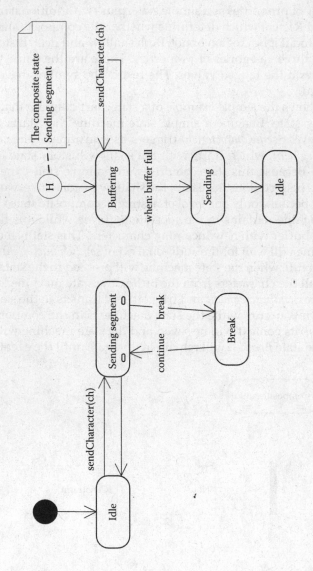

FIGURE 3.26
Example of a composite state that uses the shallow history state.

occurs. Then the state machine will reenter the compound state *Sending Segment*, the context will be restored, and the state machine will resume the processing from the point of interruption.

The example in Figure 3.27 shows simplified DNS client and server statechart diagrams. Both of them have just a single state. Being simple enough, these diagrams make very clear how statechart diagrams are used to make complete designs of communication protocols. Typically, a job performed by the communication protocol is to receive a message, process it, and send one or more messages as the result of this processing. Both DNS client and server go along this simple scheme.

The DNS client starts from the initial state by receiving a call to map the given domain name into the corresponding IP address. This action is modeled by the call event *map(d)* in Figure 3.27. This event triggers the transition of the DNS client from the initial state to the state *Wait DNS Response*. During the course of this transition, the DNS client sends the signal (message) *DNSrequest(d)*, which causes the signal event *receive DNSrequest(d)* at the DNS server side.

The DNS client is simply blocked in the state *Wait DNS Response* while waiting for the signal *DNSresponse(d,ip)*. The signal event *receive DNSresponse(d,ip)* triggers the DNS client transition to its final state. During this transition, the DNS client extracts the IP address from the received signal and returns it as its return value. This is modeled by the return action *return(ip)*.

The DNS server starts with the triggerless transition from its initial state to the state *Wait DNS Request*, where it is blocked while waiting for the signal *DNSrequest(d)*. The signal event *receive DNSrequest(d)* causes the DNS server

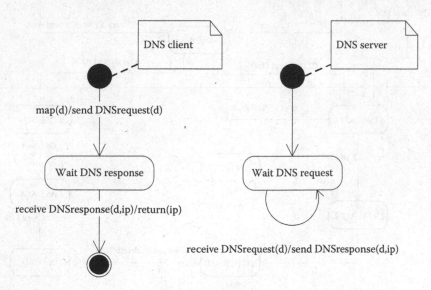

FIGURE 3.27
DNS client and server statechart diagrams.

to map the given domain name to the corresponding IP address, to create the signal (message) *DNSresponse(d,ip)*, and to send it to the DNS client. The DNS server performs all these actions during the transition to the same state, i.e., *Wait DNS Request*. This ensures that after servicing the current request, the DNS server remains available for servicing the next DNS request.

The example in Figure 3.28 shows the statechart diagram for one real protocol, namely TCP. It starts with the triggerless transition from the initial state to the state *CLOSED*, in which it awaits one of the two possible call events. The call event *passive OPEN* causes TCP to create TCB (modeled with the action *create TCB*) and to move to the state *LISTEN*. Alternately, the call

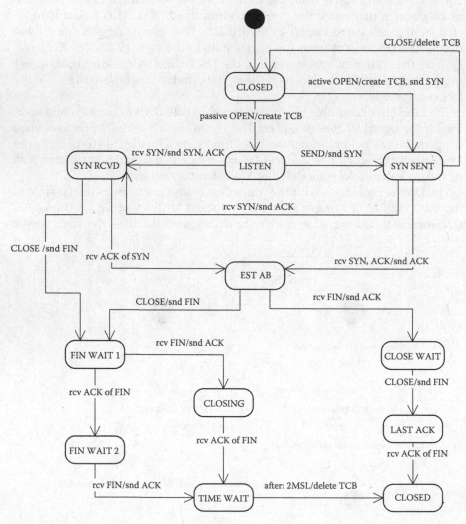

FIGURE 3.28
TCP statechart diagram.

event *active OPEN* causes TCP to additionally send the signal *SYN* (TCP segment with the bit SYN set in the header) to the remote TCP entity. This is modeled with the actions *create TCB* and *snd SYN*.

TCP is blocked in the state *LISTEN* while waiting for one of the two possible events. The signal event *rcv SYN* triggers it to send the signal *SYN, ACK* (TCP segment with both bits SYN and ACK set) to the remote TCP entity and to move to the state *SYN RCVD*. The call signal *SEND* causes TCP to send the signal *SYN* to the remote TCP entity, and to move to the state *SYN SENT*.

While blocked in the state *SYN SENT*, TCP can be triggered by one of three possible events. If the call event *CLOSE* occurs, TCP deletes TCB (modeled with the action *delete TCB*) and returns to the initial state. If the signal event *rcv SYN* occurs, TCP sends the signal *ACK* and moves to the state *SYN RCVD*. If the signal event *rcv SYN, ACK* occurs, TCP sends the signal *ACK* to the remote TCP entity and moves to the state *ESTAB*.

After reaching the state *SYN RCVD*, TCP can react to one of the two possible events. If the call event *CLOSE* occurs, TCP sends the signal *FIN* to the remote TCP entity and moves to the state *FIN WAIT 1*. If the signal event *rcv ACK of SYN*, occurs, TCP moves to the state *ESTAB*.

Two events are recognizable in the state *ESTAB*. If the call event *CLOSE* occurs, TCP sends the signal *FIN* to the remote TCP entity and moves to the state *FIN WAIT 1*. If the signal event *rcv FIN* occurs, TCP sends the signal *ACK* and moves to the state *CLOSE WAIT*.

In the state *FIN WAIT 1*, TCP may receive either *FIN* or *ACK of FIN* signals. In the former case, it sends the signal *ACK* and moves to the state *CLOSING*, whereas in the latter case it just moves to the state *FIN WAIT 2*, where it waits for the signal *FIN* to send the signal *ACK* and move to the state *TIME WAIT*. On the alternative path, TCP moves from the state *CLOSING* to the state *TIME WAIT* after it receives the signal *ACK of FIN*.

Upon the entrance to the state *TIME WAIT*, a timer with the period 2MSL is started. When this period expires, TCP deletes TCB and moves back to its initial state *CLOSED*. After reaching the state *CLOSE WAIT*, TCP waits for the call event *CLOSE* to send the signal *FIN* and move to the state *LAST ACK*, and from there to the initial state *CLOSED* after it receives the signal *ACK of FIN*.

The example in Figure 3.29 shows the statechart diagram of a simple send e-mail operation (SMTP client side). It starts with the triggerless transition from its initial state to the state *WAIT 220*, where it waits for the signal (message) *220* from the SMTP server. When the signal event *rcv 220* occurs, the SMTP client sends the signal *HELO* to the SMTP server and moves to the state *WAIT 250 1*. After receiving the signal *250*, the SMTP client sends the message *MAIL FROM:* to the SMTP server and moves to the state *WAIT 250 2*.

Next, two signals of *250* in succession cause the SMTP client first to send the signal *RCPT TO:*, then to send the signal *DATA* to the SMTP server, and finally to reach the state *WAIT 354*. Upon reception of the signal *354*, the SMTP client sends the body of the e-mail message and moves to the state *WAIT 250 4*. After receiving the signal *250*, it sends the signal *QUIT* to the

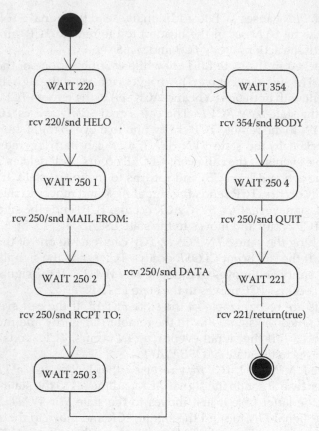

FIGURE 3.29
Simple send e-mail operation statechart diagram (SMTP client side).

SMTP server, and finally, after receiving the signal *250* again, it returns the value *true* and moves to its final state.

The main problem in this oversimplified version of the SMTP client is that it can block indefinitely while waiting for a signal from the SMTP server. The first thing that would be added in a more realistic design is a time limit on waiting for signals, which would be modeled with timers (keyword *after:*). The reaction to the expiration of a timer could be as simple as returning the value *false* and moving to the final state, or it can include some type of a recovery mechanism.

3.6 Deployment Diagrams

Deployment diagrams are used to model the deployment of the compo-nents, the component instances, objects, and packages on nodes and node instances. A **component** is a part of the system that implements a set of interfaces. It typically models a physical package of logical elements, such as classes, interfaces, and collaborations. The common forms of packages are the following:

- Executables
- Libraries
- Tables
- Files
- Documents

A **node** is a physical element that models a computational platform, which comprises a set of resources, such as memory banks, buses, I/O channels, controllers, processors, and so on. The examples of nodes are the following:

- Personal computers
- Mainframes
- Embedded controllers
- Mobile or cellular phones
- Network routers

We use deployment diagrams in the design phase of communication protocol engineering for the following two main purposes:

- To identify network nodes and configurations
- To identify design subsystems and interfaces

The software architecture is closely related to the structure of the physical network. Sometimes the latter can be fixed and, in such a case, it governs the distribution of functionality across the network nodes as well as the selection of active classes. Alternately, both software architecture and network structure can be subjects of design and, in that case, some particular network structures can yield more appropriate software architecture and system solutions.

While trying to identify network nodes and configurations, we typically render network nodes as cubes, interconnect them with association relations, and think how to deploy individual components on these nodes. We show the deployment in the deployment diagrams by adding the component symbols (rectangles with tabs) and by connecting the related nodes and components with the dependency relations. Another way to do this is to adorn the node instances by the names of the components that are deployed on them.

Similarly, while trying to identify the subsystems and interfaces, we typically render the packages with their corresponding interfaces. We try to organize them into hierarchical layers (e.g., application-specific, application-general, middleware, and system-software). Finally, we show which interfaces (services) are provided by which packages or components and also which packages or components are users of the services provided through those interfaces.

Deployment diagrams are a special type of graph that comprise the set of vertices that are interconnected with the corresponding arcs. Figure 3.30 shows the basic set of graphical symbols available for rendering deployment diagrams. These are the node, the node instance, the component, the component instance, the object, the package, the interface, the association relation, the aggregation relation, the dependency relation, the note, the constraint

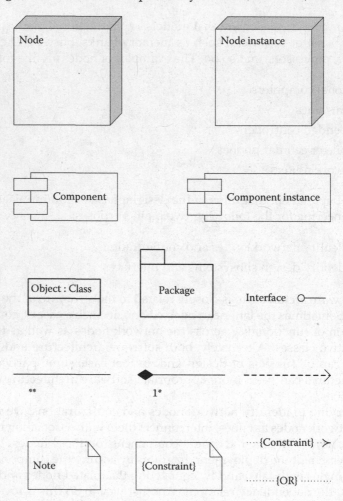

FIGURE 3.30
Basic set of symbols available for deployment diagrams.

note, the two-element constraint, and the OR constraint. Each symbol has a set of properties, which must be set by the designer once they add the symbol to the diagram. The new symbols are the symbols representing the nodes, the components, and their instances. The rest of the symbols are already introduced in the previous sections about class and object diagrams (called together a static structure).

The node has six categories of properties. These are the general information, the table of attributes, the table of operations, the list of components, the table of constraints, and the tagged values. The general information includes the name, the full path, the stereotype, the visibility, and the indicators *Root*, *Leaf*, and *Abstract*. The list of the components comprises the names of the components that are deployed by this node.

The component has seven categories of properties, including the general information, the table of attributes, the table of operations, the list of nodes, the list of classes, the table of constraints, and the tagged values. The general information comprises the name, the full path, the stereotype, the visibility, and the indicators *Root*, *Leaf*, and *Abstract*. The list of nodes holds the names of the nodes that deploy this component. The list of classes stores the names of the classes that are implemented in this component.

The node instance has four categories of properties: These are the general information, the table of attribute values, the table of constraints, and the tagged values (documentation and persistent). The general information comprises the node instance name and the node name. The table of attribute values stores the name, the stereotype, the type, and the value for each attribute. The component instance has the same categories of properties as the node instance, with the exception that its general information differs and it comprises the name of the component instance and the component name.

The deployment diagram in Figure 3.31 shows an example of a network configuration comprised of three personal computers that are connected to the Internet. A personal computer is modeled as the node *PC*. Individual PCs are modeled as node instances, namely *Machine1*, *Machine2*, and *Machine3*. The Internet is modeled as the node instance, named *Network*, of the node type named *Internet*. The real links that connect PCs to the Internet are modeled with the association relations between the node instances *Machine1*, *Machine2*, and *Machine3*, and the node instance *Internet*. The one-to-one nature of these links is modeled by setting the multiplicities on both sides of the associations to 1.

This diagram is what the physical infrastructure of this example looks like. The software components are deployed as follows: The e-mail client executable is deployed to the first PC, the DNS server executable is deployed to the second PC, and the SMTP server is deployed to the third PC. We model the e-mail client executable with the component *EMailClient*, which is stereotyped as the <<*executable*>>, and its particular instance is deployed to the first PC with the component instance *client.exe*. Similarly, the DNS server

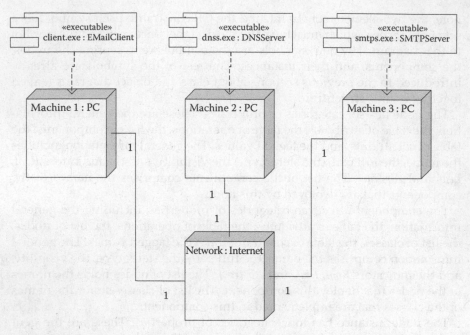

FIGURE 3.31
Example of a network configuration.

executable is modeled with the component *DNSServer* and its particular instance is deployed to the second PC with the component instance *dnss.exe*. Finally, the SMTP server is modeled with the component *SMTPServer* and its particular instance is deployed to the third PC with the component instance *smtps.exe*.

The deployment diagram in Figure 3.32 shows the example of subsystems and interfaces. While thinking about the system shown in the previous example (Figure 3.31), we can identify three application layer packages, two system-software layer packages, and three interfaces. The application layer packages are the packages *EMailClient*, *SMTPServer*, and *DNSServer*, whereas the system-software packages are the packages *TCP/IP* and *OS*.

The package *TCP/IP* provides two service types through the interface *TCPport* and *IPint*, respectively. The services provided through the former interface are used by the package *EMailClient* and *SMTPServer*, whereas the services provided through the latter interface are used by the package *EMailClient* and *DNSServer*. Similarly, the package *OS* provides services through its interface *OSapi*. These services are used by the package *TCP/IP*.

Interested readers can find more information about the UML diagrams in the original books by Booch, Rumbaugh, and Jacobson (Booch et al. 1998). This section concludes the part of this chapter based on UML. The second part of the chapter will be based on domain-specific languages.

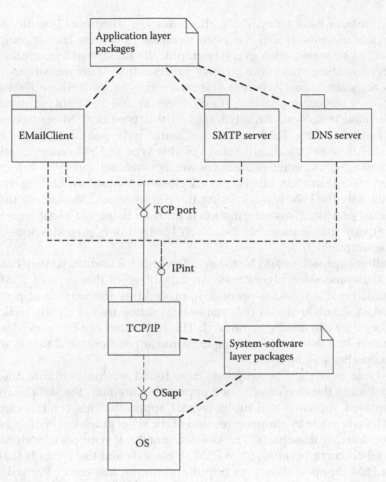

FIGURE 3.32
Example of subsystems and interfaces.

3.7 Specification and Description Language

Software for real communication systems and devices (concentrators, packet switches, gateways, routers, and so on) is very complex and, therefore, hard to understand. Proving that this type of software is correct is very difficult; thus, special attention is paid to its design. Software of this type can be modeled in the form of an individual or a group of finite state machines. Japanese designers were the first to apply this method of specification and description of communication protocols in the 1970s. Not long after its initiation, the CCITT (predecessor of ITU-T) has standardized it in the form of the so-called Specification and Description Language (SDL).

SDL creators have been facing the following dilemma. Traditionally, a finite state machine (FSM) has been modeled by a state transition graph. Typically, a state transition graph is graphically illustrated by circular symbols representing states and arrows representing state transitions. State labels are state names, whereas state transition labels indicate FSM input that causes the corresponding state transition and FSM output produced by the same transition. An advantage of this type of FSM representation is that all the stable FSM states are clearly indicated and can be easily noticed. Alternately, a disadvantage of this type of FSM representation is that message-processing procedures are not defined formally. Informally written state transition labels, placed close to the corresponding arrows, indicate only the FSM input causing the transition and the output that the FSM must produce. This information is far from being sufficient for writing the software that implements the given FSM—it only provides some hints to programmers.

Another approach would be to use a flowchart, a traditional way of specifying data-processing algorithms. An advantage of this type of FSM representation is that message-processing procedures are clearly and precisely defined. A disadvantage is that stable FSM states are not clearly indicated, therefore they can hardly be noticed. The FSM states can be marked as certain points in a flowchart by using informal annotations, and that is simply not comprehensible enough.

The creators of the SDL language have found a solution to this dilemma by combining the abovementioned approaches, namely, the state transition graph-based approach and the flowchart approach. This combination has been cleverly made by simple extension of the set of graphical symbols available for drawing flowcharts. The key new graphical symbols introduced are the symbols corresponding to an FSM stable state and the symbols that represent FSM inputs and outputs (input and output messages). We will fully describe all the SDL graphical symbols later in this chapter.

The protocol designer uses SDL language to specify and describe the corresponding automata instance by listing all its states and state transitions. Although the number of states can be very large, this task is simplified by the fact that in a given state, only a limited number of events can occur. This means that the automata instance can evolve from a given state only for a limited number of new states. For example, consider a telephone call automata instance waiting for the first digit to be dialed (the automata instance enters this state immediately after the user has initiated an outgoing call, i.e., after the so-called "hook-off" event). The telephone call automata instance cannot evolve from this state to any other arbitrary state. More precisely, in this state only the following three events are possible:

- The user ends the call (hook-on event), which causes the automata instance to evolve to its initial idle state.

- The user dials a digit (digit event). This event triggers the state transition from the current state to the state of waiting for the second digit.
- The user does nothing during a certain interval of time. This will cause the expiration of the corresponding timer and a state transition to the state in which the telephone line is blocked.

Communication protocol is by nature a reactive system. Normally, it is blocked in its current state while waiting for one of a few recognizable events to occur. Statistically, it is inactive most of the time. A recognizable event triggers the corresponding state transition to a new state, where the protocol is again blocked while waiting for further events. The state transitions comprise a finite number of primitive operations that are statistically rather short.

An important characteristic of program implementations of the protocols is that they are not trying to monopolize the CPU. This implies that the execution of this type of a program should be organized as a **process with stable states**. In contrast to the conventional time-slicing system, where the task switching is driven by timer interrupts, the switching of processes with stable states is performed at the moment at which the running process reaches its new stable state. Whereas conventional tasks can be interrupted in an arbitrary point of time (determined by the asynchronous occurrence of a timer interrupt signal), a process with stable states is normally not subject to preemption because, unlike conventional tasks, they are not monopolizing the processor. Of course, a process with stable states can be interruptible so that the whole system can react to the urgent events handled by the higher priority tasks.

Enumeration of the possible states and state transitions, as described above, is a logical process that seems to be straightforward for the experts. However, graphical language, such as SDL, is needed to make it possible for design engineers to easily make complete formal specifications of the protocols. The main advantages of graphically oriented languages are as follows:

- Graphical language is easy to read and, because of that, it is easy to check specification completeness and correctness.
- The specification can be easily extended.
- The specification can be directly implemented in software. This means that if the specification is correct, a high probability exists that the software implementation is also correct.

According to ITU-T, the complete software (system) is decomposed into a set of **functional blocks**. Each functional block consists of a set of processes and each process comprises a number of **tasks** (Figure 3.33).

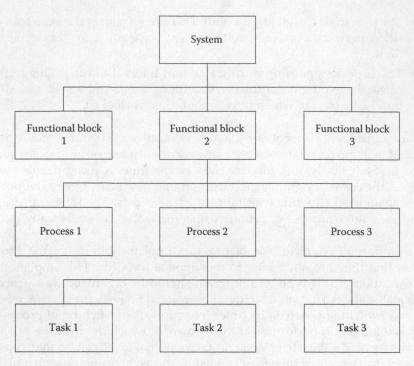

FIGURE 3.33
Structure of the communication software according to ITU-T.

A **process** is essentially an execution of a logical function, which consists of a series of operations applied to message information elements (referred to as tasks) in discrete points of time. Either it is in some of its stable states or it makes its transition from the current to the next state. (In Chapter 4, we refer to the state transition as **unstable states**).

A **signal** is defined as a data stream that delivers information to the receiving process. A data stream among the processes inside the same functional block represents the **internal signal**, whereas a data stream between the processes that are parts of different functional blocks represents the **external signal** to the receiving process. Therefore, from the receiving process point of view, the signal can be classified as internal or external, depending on whether it originates from the same or from a different functional block.

Today, SDL is a standard design language that can be used to specify and describe any system implemented in hardware or software, particularly real-time systems. In this book, we are especially interested in one type of the real-time systems—communications systems.

The basic set of SDL rules is given in ITU-T recommendation Z.100e. Additional explanations are given in a series of subsequent ITU-T recommendations, namely Z.100d1e, Z.100nce, Z.100nfe, Z.100p1e, and Z.100s1e. The main characteristics of the SDL language are as follows:

- It is easy to learn.
- It is easy to extend the specification in case of the new requirements.
- In principle, it can support various methodologies for making the system specifications.

Two forms of SDL language exist, graphical (SDL-GR) and program (SDL-PR). The graphical form has been widely accepted for two reasons. First, it is closer to human understanding because it is easier to understand and follow. Second, in principle, it does not require the support by special, and frequently very expensive, software tools. Of course, a piece of paper and a pencil is hardly sufficient for a professional work. At least a modern graphical editor that supports the SDL set of graphical symbols is needed to enable the making of decent specifications. In this book, we use Microsoft Visio® for that purpose.

The second SDL form, SDL-PR, is practically a higher-level programming language of textual type (similar to C/C++ and Java programming languages). Clearly, this programming language is less synoptic and is harder to follow than the graphical form. It is intended to be used mainly by the accompanying software tools, such as Telelogic® Software Development Tools (SDTs). The goal of using such software tools is not just to make isolated specification and description documents, but rather to make electronic specifications, essentially models of protocols. The software tools can then be used to interpret the models and generate the corresponding program code.

In addition to the tools provided by Telelogic®, other tools exist based on this philosophy that is, as already mentioned, referred to as model integrated computing (MIC). One of them is also already mentioned, GME.

The main SDL applications are the following:

- Call processing in switching systems
- Error supervision and management in telecommunication systems
- Supervision, control, and data acquisition systems
- Telecommunication services
- Data transfer protocols
- Protocols in computer communications

The SDL language basics are as follows: SDL is based on a set of special symbols and the rules for their application. The graphical form (SDL-GR) is based on special graphical symbols whereas the program form (SDL-PR) is based on a set of special keywords. Both SDL forms use the same set of keywords specialized for data representation.

Later, we assume that a system consists of a number of protocols. Also, we refer to a set of hierarchically organized protocols as a **family of protocols** or a **protocol stack**. Typically, each protocol that is a part of the family performs its well-defined task. The family of protocols conducts rather complex tasks by cooperation of its members.

A system is described as a set of interconnected functional blocks. **Channels** are defined as communication links that are used for the interblock communication and for the communication between the blocks and the environment. Each block comprises a number of processes that communicate by exchanging signals. A channel is typically implemented as a FIFO (First-In-First-Out) queue that stores the signals (i.e., messages) to be transferred through the channel. A process is defined as a finite state machine (automata instance) that is described by the given set of states and state transitions.

The next simple example illustrates the notions and terms introduced above. Both graphical and program SDL forms are presented. The only goal of presenting the program form is to provide the intuition for the reader that will help them understand the main differences between the graphical and program forms of the SDL language. The aim of this book is not to fully cover the program form of the SDL language.

The example is a simple game called *Daemongame*. The core of the game is a simple FSM that has only two states, *even* and *odd*. Timing is controlled with a single timer. The expiration of the timer (this event is labeled *none*) causes the FSM to switch from an *even* state to an *odd* state. The player presses a button when they wish (this event is labeled *Probe*), i.e., at arbitrary points of time. If the FSM is in an *even* state, the player gets one negative point (*Lose*). If the FSM is in an *odd* state, the player gets one positive point (*Win*). If the player scores more *Win* than *Lose* points, they win the game.

The first step in describing this simple system is to define input and output signals. Input signals are as follows:

- *Newgame*: The player wants to start the game.
- *Probe*: The player has pressed a button.
- *Result*: The player wants to see the current score.
- *Endgame*: The player wants to quit the game.

Output signals are the following:

- *Gameid*: current game identification
- *Win*: positive point
- *Lose*: negative point
- *Score*: total amount of points (number of *Win* points minus number of *Lose* points)

The specification of the game *Daemongame* in the graphical form of SDL is shown in Figure 3.34. It contains a single functional block labeled *Game*.

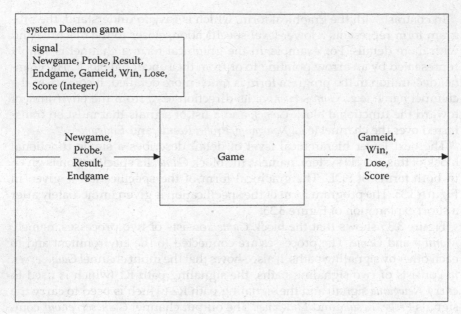

FIGURE 3.34
Structure of the system *Daemongame.*

Input signals are *Newgame, Probe, Result,* and *Endgame.* Output signals are *Gameid, Win, Lose,* and *Score.* Signal declarations are shown in the upper left corner of the figure.

The *Daemongame* system specification in the program form of SDL is as follows:

```
system Daemongame
  signal Newgame, Probe, Result, Endgame, Gameid, Win, Lose, Score(Integer);
  channel Gameserver.in
    from env to Game
    with Newgame, Probe, Result, Endgame;
  endchannel Gameserver.in;
  channel Gameserver.out
    from Game to env
    with Gameid, Win, Lose, Score;
  endchannel Gameserver.out;
block Game referenced;
endsystem Daemongame;
```

Generally, any system SDL program specification starts with the keyword *system* and ends with the keyword *endsystem*. This particular program defines all the required signals (*Newgame, Probe, Result, Endgame, Gameid, Win, Lose,* and *Score*), the input channel *Gameserver.in,* and the output channel *Gameserver.out.*

In contrast with the graphical form, which is easy to understand, the program form represents a lower-level specification, closer to the machine and with more details. For example, in the graphical form a channel is simply represented by an arrow pointing to or from the functional block. The channel declaration in the program form is much more detailed: It comprises the channel name (e.g., *Gameserver.in*), its direction (e.g., from the environment toward the functional block *Game*), and a list of signals that must be transferred over the channel (e.g., *Newgame*, *Probe*, *Result*, and *Endgame*).

The next lower hierarchical level of detail describes a single functional block of this simple system, namely, the block *Game*. Its specification is given in both forms of SDL. The graphical form of the specification is given in Figure 3.35. The program form of the specification is given immediately after a short explanation of Figure 3.35.

Figure 3.35 shows that the block *Game* consists of two processes, namely *Monitor* and *Game*. The processes are connected to the environment and to each other by signaling paths. It also shows that the input channel *Gameserver.in* consists of two signaling paths, the signaling path R1 (which is used to carry *Newgame* signal) and the signaling path R2 (which is used to carry the signals *Probe*, *Result*, and *Endgame*). The output channel *Gameserver.out* comprises the single signaling path R3. A single internal signaling path exists inside the block *Game*, the path R4, which is used to carry the internal signal

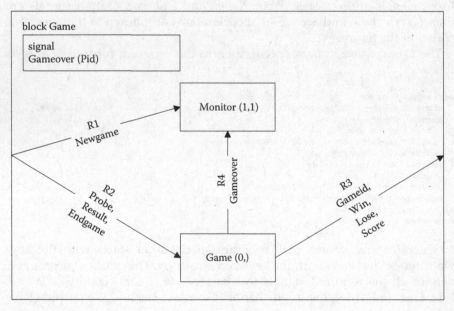

FIGURE 3.35
Structure of the functional block *Game*.

Gameover from the process *Game* to the process *Monitor*. This new signal is declared in the upper left corner of the graphical specification.

The specification of the block *Game* in SDL-PR is as follows:

```
block Game;
    signal Gameover(Pid);
    connect Gameserver.in and R1, R2;
    connect Gameserver.out and R3;
    signalroute R1 from env to Monitor with Newgame;
    signalroute R2 from env to Game with Probe,Result,Endgame;
    signalroute R3 from Game to env with Gameid,Win,Lose,Score;
    signalroute R4 from Game to Monitor with Gameover;
    process Monitor(1,1) referenced;
    process Game(0,) referenced;
endblock Game;
```

The specification given above starts with the keyword *block* and ends with *endblock*. Inside the body of the definition of the block *Game*, we start with the declaration of the internal signal *Gameover* by declaring its name, followed by the list of its parameters enclosed in parenthesis. The signal *Gameover* has a single parameter, the identification of a process (*Pid*) that is sending this signal.

Further on, we connect the channel *Gameserver.in* with the signaling paths R1 and R2. We also connect the channel *Gameserver.out* with the signaling path R3. We proceed with the declarations of signaling paths (keyword *signalroute*). Each declaration indicates the signaling path name, its direction (by using the keywords *from* and *to*), the names of the processes it connects (note that *env* is the special process which represents the environment), and a list of signals it carries (by using the keyword *with*). For example, the first signal path declaration shown in SDL-PR above declares the signaling path R1, which carries the signal *Newgame* from the process *env* (environment) to the process *Monitor*.

We end the definition of the functional block *Game* by declaring the processes it contains. A process in general is declared by the keyword *process*. A process declaration indicates the name of the process followed by the initial and maximal number of process instances that can appear in the system. The maximal number of process instances is an optional parameter, i.e., it can be omitted.

The process *Monitor* is declared as *Monitor(1,1)*, which means that the block *Game* should initially create one instance of this process and, at the same time, it is also the maximal number of *Monitor* instances that can be created in this block. Alternately, the process *Game* is declared as *Game(0,)*, which means that initially there are no *Game* instances, but also that the maximal number of Game instances is not limited, i.e., in theory it is allowed to create an infinite number of process *Game* instances inside the functional block *Game*. Of course, in reality this number is always limited to the available hardware resources.

In this particular example, we have declared two processes, *Monitor* and *Game*, that operate inside the functional block *Game*. The process *Monitor* handles the interaction with a player. It is a mediator between the player and the process *Game*, which is essentially a model of the win–lose game. Due to the fact that the process *Monitor* is trivial and actually insignificant for this example, we will define only the process *Game* on the next hierarchically lower level of abstraction. On this level of detail, the process *Game* is modeled as a finite state machine (automata instance).

As already mentioned, the creators of SDL-GR (the graphical form of SDL) have extended the basic set of traditional flowchart symbols with a set of graphical symbols specialized for modeling finite state machines. The complete set of graphical symbols available for describing processes in SDL-GR is shown in Figure 3.36.

The meaning of the individual graphical symbols shown in Figure 3.36 is as follows:

- *State*: Specifies a stable state in which a process is blocked while waiting for one of the recognizable signals (referred to as *input*).

- *Input*: Specifies the reception of a given input signal (i.e., the occurrence of a certain event).

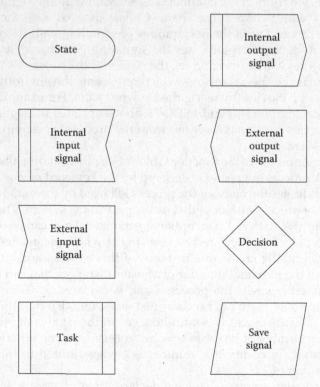

FIGURE 3.36
Set of graphical symbols available in SDL-GR.

- *Output*: Specifies the transmission of a given output signal (normally the output signal generated by a certain process represents an input signal for a process that receives it).

- *Decision*: Specifies an operation that checks if a given condition is true or false and, based on the outcome, selects one of the two possible paths in the current state transition.

- *Task*: Specifies an action in the course of current state transition that is neither *decision* nor *output*.

- *Save signal*: Specifies that recognition (processing) of a given signal should be postponed until it reaches a state where it is recognizable This symbol is used in specifications of signaling systems (e.g., SS7). It is seldom used in other applications, such as call processing.

The specification of a process in SDL-GR is generally made as a combination of the instances of the graphical symbols shown and explained above. An example of this type of specification is shown in Figure 3.37. It specifies and describes the process *Game*, the core of the win–lose game.

The evolution of the process starts from an unnamed state in the upper right corner of the graphical presentation (Figure 3.37). Starting from this state, the process unconditionally transits to its next stable state *even*. During this transition, the process *Game* sends the signal *Gameid* to the player.

While the process *Game* is in its stable state *even*, it awaits one of two possible events: the reception of the signal *Probe* or the expiration of the timer labeled *none*. If the timer expires, the process *Game* receives the corresponding signal *none*, and this causes the process to evolve into the next stable state *odd*. If the process receives the signal *Probe*, it sends the signal *Lose* to the player and updates the player's score, which is stored in the variable *count*, by adding one negative point. The process does not change its stable state, i.e., it remains in its current state (which is denoted with the character "–"), and that is the state *even*.

In its stable state *odd*, the process *Game* recognizes two same possible events, the reception of the signal *Probe* or the expiration of the timer labeled *none*. Actually, the timer *none* determines the time interval the process will spend in either the *even* or *odd* state before switching to the other one. Hence, if the timer *none* expires, the process evolves into the stable state *even*. Alternatively, if the process receives the signal *Probe*, it sends the signal *Win* to the player and updates the player's score (value of the variable *count*) by adding one positive point. The process remains in its current state (i.e., the state *odd*).

The upper left corner of the graphical representation of the process *Game* (Figure 3.37) shows one important example of simplifying SDL-GR diagrams. Because the reception of the input signals *Result* and *Endgame* is possible in both *even* and *odd* states, a straightforward solution would be to mechanically add these inputs and their processing to both states. The result would be a diagram that is much more complex and harder to understand

and follow. A more elegant solution is to draw the description of the processing of the inputs *Result* and *Endgame* in both states as a separate drawing in the diagram, as shown in Figure 3.37.

Generally, it is always useful to try to find identical processing of input signals (state transitions) that repeat in a number of stable states and to

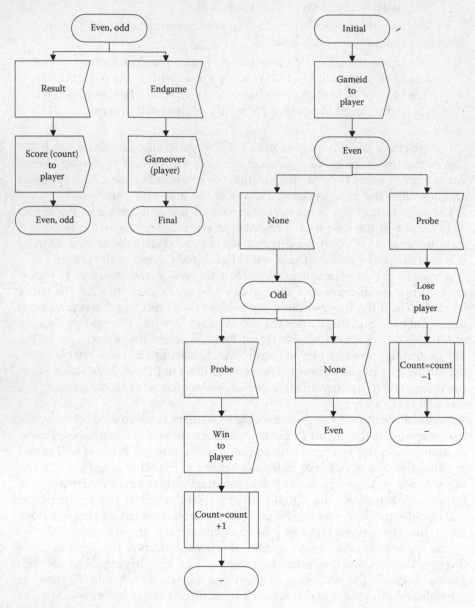

FIGURE 3.37
Process *Game* specification in SDL-GR.

simplify the specification by drawing these parts separately in the diagram. This type of a model reduction is really easy. We just draw an oval state symbol and write a list of the states (the list comprises the state names separated by commas) that share the common inputs inside the state symbol. Then we can copy and paste common state transitions. At the end, we can just remove the redundant state transitions. Of course, in the simple diagrams such as in the example at hand, we can see this in advance and draw accordingly, as we did for the processing of the inputs *Result* and *Endgame* in the states *even* and *odd*.

If the process *Game* receives the signal *Result*, which comes from the environment, i.e., from the player, the process sends the signal *Score(count)* to the environment (actually to the player) and it remains in its current state (*even* or *odd*). Alternately, if the process *Game* receives the signal *Endgame*, it sends the signal *Gameover* to the process *Monitor* and the game ends, i.e., the functional block deletes the process *Game*.

The specification of the process *Game* in SDL-PR (SDL program form) is as follows:

```
process Game(0,); fpar player Pid;
  dcl count Integer := 0; /* the counter that contains the result */
  start;
    output Gameid to player;
      nextstate even;
  state even;
    input none;
      nextstate odd;
    input Probe;
      output Lose to player;
      task count:=count-1;
      nextstate -;
  state odd;
    input Probe;
      output Win to player;
      task count:=count+1;
      nextstate -;
    input none;
      nextstate even;
  state even,odd;
    input Result;
      output Score(count) to player;
      nextstate -;
    input Endgame;
      output Gameover(player);
      stop;
endprocess Game;
```

The definition of the process starts with the keyword *process* and ends with the keyword *endprocess*. As already mentioned, initially no instances of the process *Game* are used, and the maximal number of its instances is unlimited. The process declaration is followed by the construct *fpar player Pid*, which defines the formal process parameter *player* that is assigned the value *Pid*. At the beginning of the game, the run-time environment creates an instance of the process, and assigns a unique *Pid* number to it.

Next, we declare the integer variable *count* (using the keyword *Integer*), which contains the current total value of points that the player has scored so far. After the label *start*, we define a series of statements that are executed by the process at its startup. In this example, the process *Game* at its startup sends the signal *Gameid* to the player and enters its initial stable state *even* (next state of the process is defined by the keyword *nextstate*).

For each stable state (keyword *state*) of the process, we define all the recognizable input signals (using the keyword *input*) and on the next level of indentation, we define the corresponding state transition as a series of statements that ends with the *nextstate* statement. For example, the recognizable input signals in the stable state *even* are the signal *none*, which relates to the expiration of the corresponding timer, and the signal *Probe*, generated by the player's stroke of the pushbutton. In the case the timer *none* expires, the process evolves to its next stable state *odd*. Alternatively, if the process receives the signal *Probe*, it sends the signal *Lose* to the player (using the keyword *output*), performs the task of decrementing the score by 1 (using the keyword *task*), and remains in its current state (the statement *nextstate -;*).

The stable state *odd* is defined in a similar manner. The input signals recognized by the process in its stable state *odd* are the signal *Probe* and the expiration of the timer *none*. If the process receives the signal *Probe*, it sends the signal *Win* to the player, increments the score by 1, and remains in its current stable state as *odd*. Alternatively, if the timer *none* expires, the process evolves into its stable state *even*. Finally, we define the state transitions initiated by the reception of the input signals *Result* and *Endgame* in either the state *even* or *odd*.

Understanding the principals of SDL-PR helps in more easily understanding the communications protocol software implementation in the state-of-the-art, higher-level programming languages such as C/C++ or Java. Although SDL-PR can resemble a pseudolanguage when compared to these programming languages, in reality it is a specialized language of higher level abstraction and it is feasible to construct a compiler for it. However, the study of the compilers is out of the scope of this book. The primary goal of this book in this respect is to provide an insight into the manual coding of SDL graphical diagrams in some of the abovementioned programming languages (C/C++ or Java).

The example under study can help in this respect. Obviously, two levels of nesting are included in it. The first level of nesting corresponds to the current stable state, in which the process is blocked while waiting for the next input signal, i.e., *start*, *even*, and *odd*. The second level of nesting corresponds to the type of input signal, i.e., *Probe*, *none*, *Result*, or *Endgame*.

The simplest method to implement this selection construction with two levels of nesting in the C/C++ or Java programming language is to use nested *switch-case* statements. The first *switch-case* statement is used to locate the current state. Then in each *case* clause of the first *switch-case* statement, another *switch-case* statement is used to locate the state transition statements that correspond to the given input signal. This type of protocol implementation will be covered in detail in the next chapter.

3.7.1 Telephone Call Processing Example

The second example of the system specification made in SDL-GR is the specification of the telephone call processing system. The description of this system is given in the separate ITU-T recommendation Q.71. The Q.71 compliant program system consists of six mutually interconnected functional entities (referred to as functional blocks), namely FE1, FE2, FE3, FE4, FE5, and FE6 (Figure 3.38). The aim of this example is just to illustrate SDL-GR applicability, and the details of the recommendation Q.71 (such as the concrete names of the entities, their types and links, i.e., relations) are not really significant for the comprehension of the usage of SDL-GR. The reader that is more interested in Q.71 details can refer to the corresponding ITU-T recommendation. We use the hypothetical telephone call processing system *CallProcessor* to make further illustrations more concrete, without diving into the bulk of details of Q.71 recommendation. Comparing it to the real Q.71-compliant system, the *CallProcessor* is a very simplified academic example that consists of a single functional block, namely *TelephoneLine* (Figure 3.39). This functional block is linked with the environment by one input channel, named *input*, and one output channel, named *output*. So far, this example is very similar to the previous example *Daemongame*, which also comprises the single functional block *Game* that is interconnected with the environment with one input and one output channel.

The functional block *TelephoneLine* is shown in Figure 3.40. This simple functional block consists of the single process *FE1*. Two lists of signals are

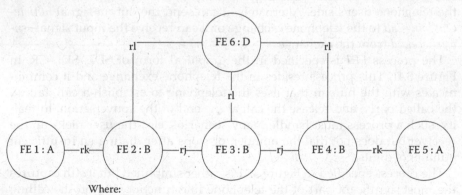

Where:
A, B, and D are the types of functional entities
FE1, FE2, FE3, FE4, FE5, and FE6 are the names of the functional entities
rk, rj, and rl are the relations between the functional entity types

FIGURE 3.38
Functional model of the telephone call processing system.

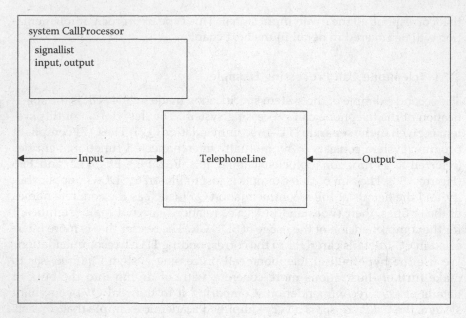

FIGURE 3.39
Hypothetical system *CallProcessor*.

declared (using the keyword *signallist*) in the upper left corner of Figure 3.39, namely, *input* and *output*. The process *FE1* is connected both to the telephone user (shown by the arrows placed at the right of *FE1*) and to the telephone exchange (indicated by the arrows placed at the bottom of *FE1*). It can receive one of the three possible input signals (*hookOff*, *dialDigit*, and *hookOn*) from the telephone user's side. Alternately, it can send the output signal *initiateOutgoingCall* to the telephone exchange or it can receive the input signal *asnwerReceived* from the exchange.

The process *FE1* is specified in the graphical form of SDL, SDL-GR, in Figure 3.41. This process resides in the telephone exchange and it communicates with the human that uses the telephone to establish a call, talk to the called party, and release the call at the end of the conversation. In reality, such a process must handle many scenarios, e.g., the user picks up the receiver but does not dial the number, or stops after dialing an insufficient number of digits.

The process specified in Figure 3.41 is rather simplified but it still captures the most significant part of the telephone line functionality on the calling party side. The telephone line in this context is a processor that hosts *FE1*, together with the interfacing hardware that connects it to both the calling party user's telephone and switching unit of the telephone exchange. For brevity, we refer to the former simply as a user and to the latter as a telephone exchange, or just an exchange.

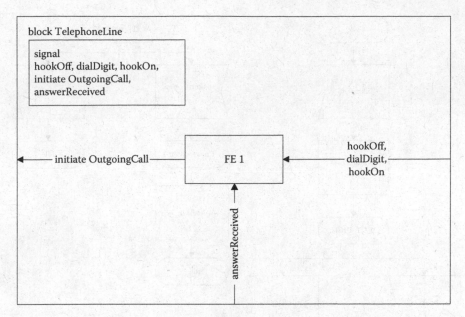

FIGURE 3.40
Structure of the functional block *TelephoneLine*.

The process *FE1* has four stable states, namely, *IDLE, WAIT_DIGIT, WAIT_ ANSWER*, and *CONVERSATION*. The evolution of the process starts from the state *IDLE*. The single recognizable input signal in this state is the signal *hookOff*. If the process *FE1* receives the signal *hookOff*, it performs the task *prepareForDialing* and moves to its next stable state *WAIT_DIGIT*. While performing the task *prepareForDialing*, the process connects the free-to-dial tone to the calling party user. This tone serves as the indication to the user that they can start dialing the number of another user to which they wish to talk.

Two recognizable input signals are used in the stable state *WAIT_DIGIT*, i.e., the process can either receive the input signal *hookOn* or the input signal *dialDigit*. In this simplified example, we assume that the telephone number of the called party consists of a single digit. However, in real ISDN telephone networks, a so-called *enblock* dialing mode exists in which the ISDN terminal sends the complete telephone number to the telephone exchange in a single *SETUP* message. Therefore, this simplified example is not so far from reality. If the process *FE1* receives the input signal *hookOn*, it evolves into its initial state *IDLE*. If it receives the input signal *dialDigit*, it sends the output signal *initiateOutgoingCall* to the telephone exchange and it moves to the stable state *WAIT_ANSWER*.

In the stable state *WAIT_ANSWER*, two events are again possible—the reception of the input signal *hookOn* or the reception of the input signal *answerReceived*. In the former case, the process goes back to its initial state *IDLE*,

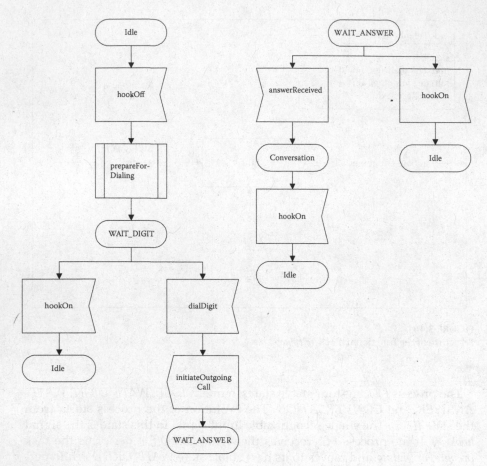

FIGURE 3.41
Simplified model of the Q.71 FE1 in SDL-GR.

whereas in the latter it evolves into its next stable state *CONVERSATION*. The input signal *asnwerReceived* is actually the result of the series of events that start with the input signal *hookOff* at the called party side. The telephone line entity at the called party translates it to the signal *answerIncomingCall* and sends it to the exchange at the called party side, which in turn sends it to the exchange at the calling party side. Finally, the exchange at the calling party side translates it to *answerReceived* and sends it to *FE1*.

In the final stable state *CONVERSATION*, only a single event is possible. The process *FE1* can receive the input signal *hookOff*, and if it does, that is the end of the conversation phase of the call and the process will return to its initial stable state *IDLE*. This closes the circle and the process is ready to process a new call originating from the same telephone line. Clearly, an instance of the process *FE1* is assigned to each telephone line in the telephone exchange.

In this example, we described the process *FE1* that is assigned to the calling party telephone line without going into a detailed specification of the operations performed by the telephone exchanges and the called party telephone line involved in the call. Obvious from this example should be that SDL diagrams are self-documented formal specifications and that no need really exists for any additional textual descriptions.

The SDL diagram shows the possible evolution paths of a process (a call processing in the example above). It defines unambiguously all telephone stable states, as well as all possible input signals for each state. The functional specification is based on the logical advance of a call, expressed in terms of telephony events. This makes it completely independent of both the hardware structure of the hosting system and the selected programming language and framework.

The SDL diagram is drawn based on the observations of a single telephone call without thinking about other calls, which are processed simultaneously (quasi-parallel by a single CPU or genuinely parallel by a multi-CPU system). This approach greatly simplifies software design. Finally, the existing SDL diagram can be easily extended by adding new states and input signals without the need to start drawing a new diagram from the very beginning. This possibility also enables the easy removal of revealed design errors.

3.8 Message Sequence Charts

An alternative method of specifying communication systems is by drawing message sequence charts that show the sequences of messages (signals) exchanged by the communicating entities. The ITU-T has developed a special language for this purpose, briefly referred to as MSC (Message Sequence Charts), and has standardized it in the Z.120 series of ITU-T recommendations.

MSC is based on the idea of following a single evolution path of a process. We start from a certain, most frequently initial, state of the process (e.g., the state *IDLE* in the previous example). After that, we select one of the possible input signals and follow the evolution path to which it points. In the previous example, a single input signal can be received in the state *IDLE*: signal *hookOff*, which causes the transition to the state *WAIT_DIGIT*.

In the newly reached stable state, we select again one of the recognizable events (the input signals that may be received in the stable state *WAIT_DIGIT* are *hookOff* or *dialDigit*; let us assume that we have selected *dialDigit*) and we follow the process evolution along the corresponding path (in the case of the input signal *dialDigit*, the process moves to the state *WAIT_ANSWER*). At the same time, as we mentally follow the evolution path of the process, we draw the messages that are exchanged between the process and its environment on the paper or, even better, the corresponding graphical editor.

The messages are represented by the graphical arrow symbols that are labeled by the message names. This is how we get the MSC charts.

Clearly, an MSC chart represents a single trace over the corresponding path, through the SDL diagram, or some other form of specifying finite state machines. We can see intuitively that for the real automata that we come across in practical applications, a finite number of paths exist that cover the SDL diagram. The set of the MSC charts that are obtained by visiting these paths represents the specification that is in a logical sense equivalent to the SDL diagram.

However, an obvious disadvantage of this type of a specification, in a form of a set of the MSC charts, is that it is much less evident than the SDL diagram. Therefore, when communication protocol designers refer to the formal specification, they really assume the SDL diagram. This disadvantage becomes obvious if, instead of dealing with a single automaton, we try to follow the evolution of a group of automata, which communicate between themselves, as well as with the environment, e.g., the group of automata defined in the abovementioned recommendation Q.71. The number of evolution traces of such systems can be extraordinarily large.

Not only must we select the initial state of a single automata, we must do it for all the automata from the group we want to analyze. Furthermore, in the case of simple and loosely coupled automata, an increase in the number of possible path combinations is not so high, but in the case of complex or tightly coupled automata, it is clear that the number of evolutions of the system can be huge.

The discussion above naturally raises the following questions: For what purpose are MSC charts useful? Do we need them at all? Practical experience shows that making the MSC charts can be useful at the beginning of the design process, when the designers talk rather freely about possible communication scenarios. These scenarios of message exchange most frequently represent the so-called main branches, i.e., main paths, through the protocol. Typically, they go from the beginning (the initial state) to the end (logically, the last state in the chain of states), e.g., from the state *IDLE* to the state *CONVERSATION*, such as in the previous example, without any errors or other exceptional events. Later, after finishing the analysis of the main paths, the paths of minor importance are analyzed. These are related to various less frequent cases, such as handling timer expirations, error recovery procedures, and so on.

All these scenarios, in the form of MSC charts, would be very useful in the later stages. Actually, these charts will be used as individual test cases during the implementation phase to partially check the functionality of the individual software modules (this is the so-called unit testing). They are also used during the final phase of the software verification as test cases for the compliance testing. The goal of **compliance testing** is to check if the software is compliant with the specification.

In most cases, the number of manually written MSC charts is finite and not too large (on the order of a few hundred at most). Later, during the testing and verification phase, automatically generating a much larger number of test cases would be an ideal way (logically equivalent to MSC charts) to check the system much more thoroughly. This testing most frequently takes the form of statistical usage testing, which enables quality engineers to estimate the software reliability without any previous knowledge about the system under examination.

As already mentioned, the MSC language—similar to the SDL language—has both the graphical and program form. The graphical form of the MSC language is more interesting than the program form for developing communications software. The next example illustrates the message exchange among the functional entities *FE1*, *FE2*, *FE3*, *FE4*, and *FE5*, in the case of the successful establishment and successful release of the ISDN connection between two subscribers. From this example, MSC is obviously useful for tracing the message exchange between more processes, which is not so easy and clear by looking at the set of corresponding SDL diagrams.

We start drawing the MSC chart by placing the rectangle graphical symbols that represent the communicating entities (i.e., processes) at the top of the chart sheet. The names of the entities are used to label these rectangular symbols. Next, we draw a vertical line from each rectangular symbol to the bottom of the sheet. After that, we enter a series of messages exchanged by the processes shown on the top of the chart. Each message (i.e., signal) is represented by the arrow symbol labeled with the message name. The arrow starts from the vertical line that represents the process sending the message and ends at the vertical line that represents the process receiving the message. The time advances in the direction from top to bottom of the sheet, i.e., the messages that appear on the top of the chart are exchanged before the messages that appear at the bottom of the chart.

An example of the MSC chart is shown in Figure 3.42. This example illustrates the scenario of successful establishment and release of the ISDN connection. The functional entities *FE1* and *FE5* are assigned to the calling and called party user, respectively. Initially, the functional entity *FE1* receives the signal *SETUP_req* from the environment (in reality, this signal is generated by the signaling system DSS1). After receiving the signal *SETUP_req*, *FE1* translates it to the signal *SETUP_req_ind* and sends this new signal to *FE2*. *FE2* forwards this signal to *FE3*, *FE3* forwards it to *FE4*, and finally *FE4* forwards it to *FE5*.

After receiving the signal *SETUP_req_ind*, the functional entity *FE5* immediately sends two signals, the signal *SETUP_ind* to its environment and the signal *REPORT_req_ind* back to *FE4*. The latter signal is forwarded from *FE4* to *FE3*, then from *FE3* to *FE2*, and finally from *FE2* to *FE1*. *FE1* translates this signal to *REPORT_ind* and sends the latter to its environment.

The acceptance of the call by the calling party is signaled to *FE5* by the signal *SETUP_resp*. *FE5* translates this signal to the signal *SETUP_resp_conf* and sends the latter over the chain of FEs back to *FE1*. *FE1* in its turn translates it to *SETUP_conf* and sends the latter to its environment. This is the final step

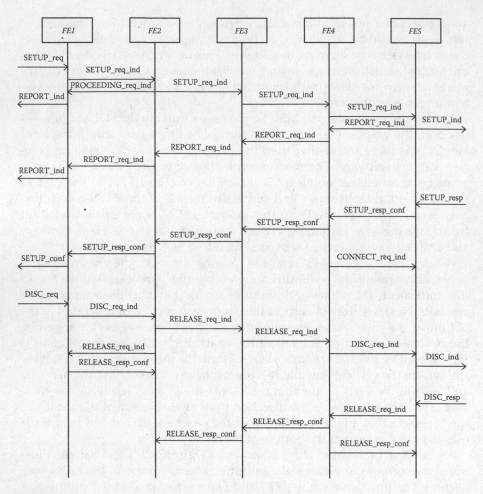

FIGURE 3.42
Example of the MSC chart: Successful ISDN call establishment and release.

of the connection establishment procedure. The next communication phase is a conversation.

At the end of the conversation, the calling party user initiates the call release procedure by sending the signal *DISC_req* to the functional entity *FE1*, which in turn translates it to *DISC_req_ind* and sends the latter to *FE2*. The functional entity *FE2* translates this signal to the signal *RELEASE_req_ind* and sends the latter to both *FE1* and *FE3*. From there, we have two parallel flows of messages. *FE1* replies to the signal *RELEASE_req_ind* by the signal *RELEASE_req_conf*. Alternately, *FE3* forwards the signal *RELEASE_req_ind* to *FE4*, which translates it to *DISC_req_ind* and sends the latter to *FE5*. *FE5* indicates the reception of that signal by sending the signal *DISC_ind* to its environment.

The environment answers with the signal *DISC_resp*, which is then translated to *RELEASE_req_ind* and sent to *FE4*. The functional entity *FE4* translates that to the signal *RELEASE_resp_conf* and sends the latter to both *FE3* and *FE5*. Finally, *FE3* forwards that final signal to *FE2*. This is the final step of the call release procedure.

This real-world example shows the main advantage of using MSC charts—instead of speculatively analyzing the parallel work of five finite state machines (*FE1*, *FE2*, *FE3*, *FE4*, and *FE5*) by looking at their SDL diagrams, here on a single chart we see how the system evolves through the procedures of call establishment and release. At this level of abstraction, we are not interested in the individual work of the individual automata. We just follow the interaction based on the message exchange between the automata in a given group.

3.9 Tree and Tabular Combined Notation Version 3

In this section, we cover the Testing and Test Control Notation Version 3 (TTCN-3), a language that was originally standardized by the European Telecommunication Standardization Institute (ETSI) by extending the previous language Version 2 (TTCN-2). A group of designers may employ TTCN-3 to make a formal specification of test procedures that are used to check if the implementation behaves in conformance with the system's formal specification. The type of testing that is conducted in accordance with such test procedures is referred to as **conformance testing**. The object of the testing is typically called an **implementation under test** (IUT) or a **system under test** (SUT). Since an IUT might be a part of a SUT, in this section we use the term SUT as a more general term. The system that is used to test a SUT is called the **test system** (TS), see Figure 3.43.

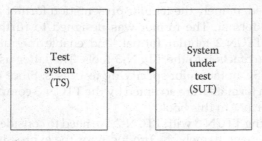

FIGURE 3.43
Standard test configuration.

Since TTCN-3 is a rather complex standard, here we cover the TTCN-3 basic features, which are sufficient for making simple test suites, and we leave the more advanced TTCN-3 features (such as multicomponent TTCN-3 and procedure-based communication) to the reader as an option for further study (see Willock, 2011). Therefore, this section is organized in the following subsections:

- TTCN-3 Language, Test Suite, and Test Systems
- Basic TTCN-3 Constructs and Statements
- Single Component TTCN-3 Test Suites

3.9.1 TTCN-3 Language, Test Suite, and Test Systems

TTCN-3 is an internationally standardized language specially designed for testing. Besides reusing many basic constructs and statements from conventional programming languages, TTCN-3 introduces testing-oriented extensions and more advanced concepts, including (1) the type system with native types for lists, test verdicts, and test components; (2) direct support for timers, message-based and procedure-based communication; and (3) built-in data matching, distributed test system architecture, and test components concurrent execution.

TTCN-3 standards provide clear and precise language definitions, so test cases written in TTCN-3 are unambiguous, and their execution on any TTCN-3 compliant testing system must have the same behavior. This independence from testing tool vendors enables easy test suite migration to other testing tools and their reuse. Although in this book, we primarily use TTCN-3 for conformance testing, actually TTCN-3 may be used across the whole product development cycle.

The TTCN-3 core notation is an intuitive textual format for defining test cases that is quite similar to conventional programming languages. Additionally, TTCN-3 supports specifying test cases using other presentation formats. These presentation formats may be converted into the core notation with the same semantics. Initially, two presentation formats have been standardized, namely the tabular presentation format and the graphical presentation format. The former was designed to further the support of the existing TTCN-2 tabular format, and enable migration of existing TTCN-2 legacy artifacts into the TTCN-3 tools. The latter uses an extended version of the MSC notation for specifying test cases. Since neither of these two presentation formats were accepted by the TTCN-3 community, they are not described further in this book.

When comparing TTCN-3 with TTCN-2, we need to consider the four major areas of improvement, namely the productivity, the expressiveness, the flexibility, and the extensibility. The core TTCN-3 notation has been developed as a textual language resembling conventional programming languages, with the intention to enable productivity. TTCN-3's better expressiveness and flexibility is based on various language extensions, such as (1) support for the testing IP based systems and text based protocols like HTTP and SIP, and (2) support for testing systems based on remote procedure calls, like CORBA and web services. In order to support extensibility, TTCN-3 has explicit hooks and mechanisms that allow new language features and notations to be easily integrated.

Generally, there are two kinds of new features: the self-contained and the multifaceted features. Some examples of self-contained features are the integration of IDL and XML type definitions and the definition of a common set of documentation tags, whereas some examples of multifaceted features are behavior types, type parameterization, and test deployment support. The self-contained features have been defined by the new parts in the TTCN-3 standards, whereas the multifaceted features have been defined by separate extension packages, including the necessary modifications to the core language, the operational semantics, and the parts of the TTCN-3 standards.

Next, we introduce the TTCN-3 Test Suite. The TTCN-3 test suite is a collection of modules. A **module** comprises definitions of data types, values, and test cases, as well as a control part that specifies how different test cases are to be executed. A module may import necessary definitions from other modules, which is key for test suite modularization.

Obviously, a test case running on a test system must be able to communicate with a SUT in order to test it. Generally, TTCN-3 supports two types of communication among test cases and SUTs, namely message-based communication and procedure-based communication. Since message-based communication is still dominantly used, we focus on it in this book. Normally, we use the TTCN-3 type **record** to define needed message types. A record is an ordered structured type, which is a collection of basic type elements (such as **integer** and **charstring**) and other structured type elements that correspond to individual message fields. Many message types comprise a field that defines the kind of the message, and such a field is typically defined using the TTCN-3 type **enumerated**.

To define a message type, we normally first define the types of individual message fields, and then we define the message type itself. For example, let's define the message type *Msg*, which comprises the four fields, namely *ID*, *Kind*, *Question*, and *Answer*. Assume that the types of these fields are 16-bit integer, enumerated, charstring, and charstring, respectively. We would then use the following definitions to define the message type *Msg*:

type integer *ID* (0..65535);
type enumerated *Kind* (e_Question, e_Answer);
type charstring *Question*;
type charstring *Answer*;
type record *Msg* {

ID id;
Kind kind;
Question question;
Answer answer;
}

The communication messages are the actual instances of message types, and these instances are called **templates**. In TTCN-3, templates are used to send particular messages or to test whether received messages are in the set of expected messages. We may specify a set of expected messages using ranges, lists, and matching attributes (we will illustrate this later on). It is important to remember that a template for a type must specify a value or a matching expression for each field of this type. If the value of some field type is unknown, i.e., it may contain any value, we encode such a value with the character "?." Furthermore, if some field of a message type is optional, and if this field should not be the part of the template we are creating, then we have to assign the special value **omit** to this field.

The template definition resembles the definition of a function. We specify the type (**template**), the name, and the list of its formal parameters. Instantiating a template (i.e., creating the concrete message) resembles a function call—we specify the template name and the list of its real parameters, which are used to replace the formal parameters.

For example, using the message type *Msg*, we may define the parametrized template *t_request* as follows:

template *Msg t_request(ID p_id, Question p_question) :=* {
  id := p_id,
  kind := e_Question,
  question := p_question,
  answer := **omit**
}

Similarly, we may define the parametrized template *t_response*:

template *Msg t_response(ID p_id, Answer p_answer) :=* {
  id := p_id,
  kind := e_Answer,
  question := ?,
  answer := p_answer
}

Next, we introduce test **components** and communication **ports**. Each test case **runs on** the test component it has been assigned to. A test suite may use a single or multiple test components, which may communicate with each other, and/or with the SUT over communication ports. A port is theoretically an infinite first-in-first-out (FIFO) queue oriented in the receive direction, which stores messages in message-based communication (or calls in procedure-based communication) until they are processed by the test component that is the owner of that port. Many simple test suites use a single test component to execute test cases, and a single communication port to communicate with the SUT. Here is an example definition of the test component named *ComponentS*, which uses the single port *pt* of the type *PortS*:

type component *ComponentS* {
  port *PortS pt*
}

Each port has a type, which defines the type of the communication (message-based or procedure-based) and the types of messages that may be communicated over that port. Within the definition of a port, each message type is given the attribute **in/out/inout** that determines whether a message of that type may be received (**in**), sent (**out**), or received and sent (**inout**) from the test component owning that port. For example, the port type *PortS* that may be used to both send and receive the message type *Msg* is defined as follows:

type port *PortS* {
  inout *Msg*
}

A test case may send and receive messages (i.e., templates) on a port using the method **send** and the method **receive**, respectively. For example, a test case sends the message *t_requestMsg* on the **out/inout** port *pt* using the following statement:

pt.**send**(*t_requestMsg*(12345, "SUT what is your name?"));

Similarly, a test case receives the message *t_responseMsg* on the **in/inout** port *pt* using the following statement:

pt.**receive**(*t_responseMsg*(12345, "My name is SUT XY."));

Although the syntax of statements for sending and receiving messages is very similar (only the method name is different), there is a fundamental difference in their semantics, i.e., in the way they operate. The method **send** is a nonblocking method and it always successfully sends the outgoing message, whereas the method **receive** is a blocking method and returns when the specified incoming message appears at the head of the input FIFO queue. More precisely, if the input FIFO queue is empty, the method **receive** blocks until the specified message arrives into the input FIFO queue. If some other message is at the head of the input FIFO queue, the method **receive** remains blocked forever (we normally use timers to recover from such situations, as will be shown later).

A typical test case resembles a single- or multiphase interview. In each phase, the test case asks a question by sending a request message to the SUT, and then it checks the SUT's answer by matching the SUT's response message with the expected message (i.e., particular template). If the response matches the expected message, SUT passed that phase, and the test case proceeds to the next phase. If SUT passes all the phases, the test case sets the final verdict by using the keyword **setverdict** to the value **pass**. An example of the body of a simple single-phase test case is the following:

pt.**send**(*t_requestMsg*(12345, "SUT what is your name?"));
pt.**receive**(*t_responseMsg*(12345, "My name is SUT XY."));
setverdict(pass);
stop;

In this simple example above, the test case sends the messages *t_request-Msg* over the port *pt*, matches the SUT's response form the same port *pt* to the message *t_replyMsg*, and if the test case receives that message, it sets the verdict to **pass**, and stops its execution using the keyword **stop**.

Besides the verdict **pass**, the test verdict may be **none**, **inconc** (i.e., inconclusive), **fail**, or **error**. The meaning of these verdicts are as follows (we will return to more detailed technical treatment of test verdicts later in the text that follows):

- The verdict **none** is the default verdict and it is implicitly set by the test system before the test case starts executing.
- The verdict **pass** means that the test case has been completed successfully.
- The verdict **inconc** means that there is not enough evidence to proclaim that the SUT is conformant to the specification.
- The verdict **fail** indicates that the SUT is not compliant with the specification.
- The verdict **error** indicates that the test case terminated with a runtime error, e.g., divide by zero.

In order to complete the simple test case above, we have to give it a name and define the test component that will execute it. If we give it the name *tc_simple1* and if we assume that it will run on the component *ComponentS*, the complete test case would be the following:

testcase *tc_simple1*() **runs on** *ComponentS* {
  pt.**send**(*t_requestMsg*(12345, "SUT what is your name?"));
  pt.**receive**(*t_responseMsg*(12345, "My name is SUT XY."));

```
setverdict(pass);
stop;
}
```

If we want to activate this test case, we have to declare that it should execute, by using the keyword **execute**, within the control part of the test module, which is declared by the keyword **control**, as follows:

```
control {
execute( tc_simple1() )
}
```

Although our test case above looks simple and elegant, it will be blocked forever in two cases. The first case is when the SUT sends the unexpected response message, i.e., when it sends any other message not equal to *t_response Msg*(12345, "My name is SUT XY."). The second case is when the SUT, for some reason, does not send any response message at all. So, besides the successful case when the SUT returns the expected response message within some reasonable amount of time, we have two failure cases.

Generally, in TTCN-3 we use the statement **alt** to specify alternative SUT behaviors at a given point of a test case. The statement **alt** blocks until any of its alternatives matches. The alternatives are checked, starting from the first and towards the last, until the first matching alternative is found. The way to receive any message is to use the method **receive** without parameters, which will match with any message at the head of the input FIFO queue.

We may fix the initial test case above by introducing the statement **alt** with three alternatives, which correspond to three possible SUT behaviors (i.e., the one successful and the two failure cases). Additionally, we must introduce a timer that will limit the time interval for awaiting a response message from SUT. Let's give this timer the name *responseTimer*. This timer should be started before the statement **alt**, and it should be stopped when any response message from SUT is received. Of course, there is no need to stop the expired timer.

The fixed test case operates as follows. It sends the request message to the SUT, starts the timer *responseTimer*, and checks the alternatives. If the response message is the expected one, it sets the verdict to **pass**. If the response message is any other (unexpected) message, it stops the time *responseTimer*, and sets the verdict to **fail**. If the timer *responseTimer* expires, it just sets the verdict to **fail**. The complete code of the fixed test case is as follows:

```
testcase tc_simple1() runs on ComponentS {
timer responseTimer;
pt.send(t_requestMsg(12345, "SUT what is your name?"));
responseTimer.start(5.0);
alt {
[]pt.receive(t_responseMsg(12345, "My name is SUT XY.")){
```

```
    responseTimer.stop;
    setverdict(pass);
  }
[]pt.receive {
  responseTimer.stop;
  setverdict(fail);
  }
[]responseTimer.timeout {
  setverdict(fail);
  }
}
stop;
}
```

Obviously, dealing with unexpected or untimely SUT behavior may lead to considerable code duplication. If we needed to add two additional cases for every receive statement in order to catch incorrect or missing responses, then our test cases would become very long and verbose, thus hard to comprehend and maintain. Because of this, TTCN-3 offers a so-called **default behavior** construct, which allows us to handle unexpected situations implicitly. Instead of writing code to handle unexpected situations explicitly, we write the default behavior handler in a single place and define that this handler should be used implicitly when none of the explicitly available alternatives match (we will come to this later in this section).

Sometimes, test cases that require access to more than one interface can be better structured by having one dedicated test component per interface. These interfaces need to be described within the so-called **test system interface** (TSI), which defines the common interface towards the SUT that different test components share in order to test the SUT. One of these components is called the **Main Test Component** (MTC), which is typically responsible for creating other test components, collecting their individual verdicts, and calculating the final verdict for the whole test case.

Next, we introduce TTCN-3 Test Systems. So far, we learned the TTCN-3 language elements for writing the so-called **abstract test suites**, which do not provide any system specific information, such as message encoding or practical communication setup. In the abstract test cases shown in this section, we send and receive messages without being concerned with the details how these messages are sent in the physical world. However, in order to create real test suite, we have to commute from an abstract to the real world. An abstract test suite is not directly executable, so we have to provide a TTCN-3 compiler or interpreter for it. Additionally, outside of an abstract test suite, we have to provide the following parts:

- **Message Codecs**, which are able to encode messages that are sent to the SUT and decode messages that are received from the SUT.

- An **SUT Adapter** that maps the TTCN-3 port to the real port used by the SUT, and the TTCN-3 communication mechanism to the real communication mechanism used by the SUT.
- A **Platform Adapter**, which typically provides the real implementation of timers and the mechanism for calling external platform specific functions.
- **Test Management** provides support for creating test campaigns, or for customizing log formats and handling log records. It is especially important for a dynamic test environment, where test cases and/or order of their execution are frequently changed. In this case, we need advanced test management support in order to avoid unnecessary, time-consuming test suite recompilations.

The additional parts, mentioned above, communicate with the abstract test suite using the two standard interfaces, namely the **TTCN-3 Runtime Interface** (TRI) and the **TTCN-3 Control Interface** (TCI). The TRI specifies operations for the SUT adapter and the platform adapter, whereas the TCI specifies operations for the test management, the component handling, the logging, and the encoders and the decoders (i.e., codecs). Figure 3.44 shows the block diagram of the complete TTCN-3 Test System architecture.

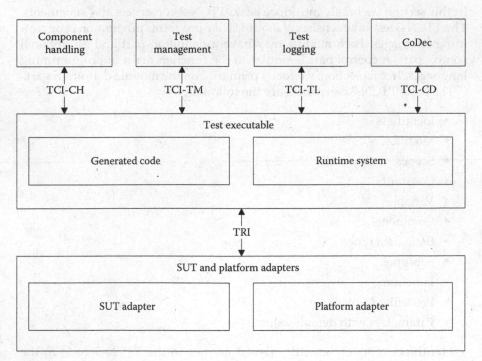

FIGURE 3.44
Architecture of the TTCN-3 Test System.

The source abstract test suite is compiled into the module **Generated Code**, shown in the center of the Figure 3.44. The generated code is executed on the module **Runtime System**, which implements the TTCN-3 operational semantics. These two modules together are called the **TTCN-3 Executable** (TE). The test executable uses the interface TRI to call functions provided by the **SUT Adapter** and the **Platform Adapter**, shown in Figure 3.44. These adapters map common operational abstractions, like communication ports and timers, to real mechanisms available on particular test system platforms.

On the other hand, the interface TCI connects the test executable with the rest of the modules shown in Figure 3.44, namely the **component handling** (CH), the **test management** (TM), the **test logging** (TL), and the **codec** (CD). Since the interface TCI is rather complex, it has been partitioned into the four sub-interfaces: the interface TCI-CH, the interface TCI-TM, the interface TCI-TL, and the interface TCI-CD. The roles of these four modules (and their corresponding sub-interfaces) are the following: The module CH is used to specify how test components are created and implemented when the test system is actually deployed, the module TM is used to control test case creation and execution, the module TL used to create execution logs, and the module CD is used to specify external codecs.

3.9.2 Basic TTCN-3 Constructs and Statements

In this section we briefly introduce basic TTCN-3 constructs and statements. The TTCN-3 test suite consists of modules like programs in common programming languages. Each module may have a definition part and an optional control part. A control part is similar to the function **main** in programming languages. In this section, we focus primarily on the module definition part.

The basic TTCN-3 constructs are the following:

- Identifiers
- Modules
- Scopes
- Constants
- Variables
- Comments
- Basic data types
- Subtypes
- Functions
- Predefined functions
- Parameters with default values

Identifiers uniquely identify named entities in the TTCN-3 code in the same way that identifiers in programming languages do. They consist of alphanumeric characters and underscores, must start with a letter, and are case sensitive. TTCN-3 has its own naming convention for identifiers, which is rather similar to naming conventions in programming languages. So, we skip its formal specification here, and instead use it consistently in the code snippets in this section, so readers will become familiar with it.

Modules are defined using the keyword **module** followed by the module name and the module body, which is enclosed in the curly brackets. The module body consists of a definition part and an optional control part. The control part is defined using the keyword **control** followed by the control part body that is enclosed in the curly brackets. The body of the control part defines how the defined test cases are to be executed. The syntax for defining modules is the following:

```
module module_name {
  // Here goes the definition part, which defines data types and constants

  control {
  // Here goes the control part that executes the test cases
  }
}
```

Scopes are defined by code blocks enclosed in the curly brackets. The code blocks may contain code statements and nested code blocks. The outermost scope is the current module. The purpose of TTCN-3 scopes is the same as in programming languages, and they follow the same rules. Definitions made in the current scope are only visible within that scope and in the nested scopes. In TTCN-3, it is not possible to reuse the identifiers that were introduced in the outer scopes. The following are the nine basic scope units:

- Module definitions part
- Control part of a module
- Component types
- Functions
- Altsteps
- Test cases
- Statement blocks
- Templates
- User-defined named types

All the identifiers must be declared before they are used, except the module identifiers, which may be declared and referred to in any order.

Constants are defined using the keyword **const** in any scope. A constant is assigned the value within its declaration, which has the following syntax:

```
const const_type const_name := const_value;
```

where *const_type* is the type of the constant, *const_name* is the identifier of the constant, and *const_value* is the value assigned to the constant. Generally, *const_value* may be an expression with constants, but references to other constants must be made without creating cycles. By using constants, we create test suites that are easier to understand and maintain.

Variables are declared using the keyword **var** in any scope, except at the top module level, because in TTCN-3 there are no global variables. Global variables are not allowed in TTCN-3 because of data races that would otherwise occur when distributed test components would try to update them. Like in programming languages, variables are used to save temporary values during program execution. A variable may be assigned the initial value within its declaration, or later in a separate assignment statement. However, using a variable before it is assigned a value results in a run-time error.

Besides simple variables, we can declare arrays the same way we do in other programming languages, after the array name we define its size enclosed by square brackets. Arrays are indexed starting from 0, and any attempt to access a value outside of the permitted range would lead to an error.

Comments in TTCN-3 are classified as block comments, line comments, or documentation comments. The block comment starts with characters "/*", may span several lines, and ends with characters "*/", whereas the line comment starts with characters "//" and extends to the end of the line. Documentation comments are defined in standard ETSI ES 201 873-10 (see Part 10: TTCN-3 Documentation Comment Specification). Like in some other programming languages, such as Java, an external documenting tool processes these documentation comments to automatically generate the up-to-date test suite user documentation.

Basic data types, also known as built-in data types, are a constitutive part of the TTCN-3 language. The TTCN-3 may be classified as a strongly typed language with a very rich type system. Here we will introduce only the most frequently used simple data types and the subtyping mechanisms for introducing user-defined types. The most frequently used basic data types are **integer**, **Boolean**, and **charstring**. Possible values of the type integer are positive and negative whole numbers, including zero, possible values of the type **Boolean** are **true** and **false**, whereas possible values of the type **charstring** are strings of ASCII characters that are enclosed by double quotes. However, unlike in other programming languages, nonprintable control characters, such as new line or tab, cannot be expressed using escape sequences.

Subtypes in TTCN-3 may be defined using two available subtyping mechanisms. The first subtyping mechanism restricts the set of possible values of a given ordered type to a particular range of values. For example, the type **integer** may be subtyped to a range of its values, by specifying a lower and an upper bound of that range. According to the first subtyping mechanism, a new subtype is defined by the type declaration of the following syntax:

type *parent_type new_type new_type_range*

Here, *parent_type* is the parent type, *new_type* is the name for the newly defined type, and *new_type_range* is the new subtype's restricted range of values. We already saw the following example of the first subtyping mechanism in the previous section:

type integer *ID* (0..65535);

A constant or a variable of the given subtype must obey the subtype restrictions. An assignment outside of the allowed range of values would cause a compile time or run time error. For example, assigning the value -1 to a variable of the type *ID* would cause such an error.

The second subtyping mechanism restricts the set of possible values of a given ordered type to a particular list of values. According to the second subtyping mechanism, a new subtype is defined by the type declaration of the following syntax:

type *parent_type new_type new_type_list*

where *parent_type* is the parent type, *new_type* is the name for the newly defined type, and *new_type_list* is the new subtype's restricted list of values. For example, we define the new type *SomeNumbers* by listing the list of its possible values 1, 3, 5, and 8:

type integer *SomeNumbers* (1, 3, 5, 8);

While introducing subtyping, we already touch upon compatibility restrictions. TTCN-3 enforces type compatibility of values in assignments, instantiations, expressions, and comparisons. We already mentioned that assigning the value to the given variable that is outside of its set of possible values causes a compile time or a run time error. In principle, a variable can be assigned a value of another type if they have the same root type and the value conforms to the associated subtype constraints of that variable.

Functions are defined in the module definitions part by the keyword **function**, followed by a function name, an optional parameter list, an optional return value, and the function body enclosed by curly brackets. A function body typically contains definitions of local constants and variables, and statements that define dynamic behavior. Functions may be called from the module control part, from test cases, or from other functions.

The function's return value may be a value like in common programming languages or a template. The return value is defined by the keyword **return** after the parameter list in the function header, followed by the return type. In this case, the function body must contain at least one return statement followed by a value or template, which must be compatible with the specified type in the function header.

Function parameters are declared with an optional passing mode, their type, and their name. There are the three parameter passing modes, namely the passing mode **in** (this is the default mode), the passing mode **out**, and the passing mode **inout**. In case of the passing mode **in**, function parameters are passed by value, i.e., the actual parameters are copied into the formal parameters before the function body is executed. In cases of the passing modes **out** and **inout**, function parameters are passed by reference. In particular, in the case of the passing mode **out**, the formal parameters are copied into actual parameters, whereas in the case of the passing mode **inout**, parameter passing is performed in both directions. Obviously, an actual parameter cannot be a constant if it is to be passed in the modes **out** and **inout**.

TTCN-3 introduces a term **instantiating** a function, which corresponds to a function call in other programming languages. We may instantiate a function by specifying the function name and its actual parameters. There are two possible ways to specify the actual parameters—with or without the parameter names. If actual parameters are specified without their names, they must be specified in the same order as the corresponding formal parameters that are specified in the function header. If the actual parameters are specified by referring to the names of formal parameters, they may be specified in any order.

As in other programming languages, functions may be defined externally, i.e., outside of the current module. We use the keyword **external** in front of the function prototype to declare such a function.

Predefined functions are functions prepared in advance that are already available for use, much like built-in (or basic) data types. These functions enable productive work—without them the user would need to write everything from scratch. The most important predefined functions are as follows:

- Value conversion functions, e.g., integer to a character
- String handling functions
- Length and size functions
- Presence checking functions
- Codec functions

Parameters with default values enable smooth evolution of test suite libraries by adding parameters without breaking previous releases. Once a formal parameter with a default value is defined in the list of formal parameters, it may be omitted in an actual parameter list. Obviously, an **in** parameter may have a default value, whereas **out** and **inout** parameters may not have a default value.

The parameter default value is used when no actual parameter is provided for a formal one. Typically, when the trailing formal parameters in a parameter list have default values, they can all be omitted in an actual parameter list. In case of other parameters that follow a parameter with a default value, the parameter with the default value could be omitted in the actual parameter list by using the character dash '-' instead of a value.

The alternative convention of providing actual parameters is to assign actual parameters to the formal parameter names explicitly. This alternative convention may be used for all the parameter passing modes (**in**, **out**, and **inout**). However, it is not allowed to mix the conventional and the assignment conventions. Also, the assignment convention may not be used for the parameter with a default value.

Here we conclude our brief introduction to basic TTCN-3 constructs, and we switch to basic TTCN-3 statements. The basic TTCN-3 statements are

- Operators
- Expressions
- Assignments
- Conditional statements
- Loops
- Labels and goto statements
- Log statements
- Control part
- Preprocessing macros

Operators are classified into five categories: arithmetic operators (+, -, *, /, mod, rem), relational operators (==, <, >, != , >=, <=), logical operators (not and, or, xor), binary string operators (not4b, and4b, xor4b, or4b), and string operators (&, <<, >>, <@, @>). Operator precedence (i.e., operator priorities) is defined similar to other programming languages, e.g., / has higher priority than +, etc.

We construct **expressions** by applying operators to operands, which may be literals, constants, and variables. Expressions are evaluated according to operator priorities, or from left to right when operators have the same priority. If in doubt, we may group subexpressions by parentheses. Of course, operands of arithmetic, logical, and string concatenation operators must have the same root type. In TTCN-3, all variables must be initialized before the expression is evaluated (unlike common programming languages where this is not required).

Assignments are used to update variables. The expression on the right-hand side and the variable of the left-hand side must be of compatible types and the expression must evaluate to a value. If these conditions are met, the value of the expression is stored into the variable.

Conditional statements, like in other programming languages, are used to organize control flow within the dynamic parts of test suites. There are two kinds of conditional statements: the statement **if–else** and the statement

select–case–else. These statements may be nested and mutually nested. The syntax of the statement **if–else** is as follows:

```
if (condition_expression)
  statement_true
else
  statement_false
```

where *condition_expression* is a Boolean expression, *statement_true* is a statement that is executed if the expression evaluates to the value **true**, and *statement_false* is a statement that is executed otherwise. Most frequently, these statements are some block statements wherein some processing is performed:

```
if (condition_expression) {
  // Do something if the condition is true
}
else {
  // Do something else if the condition is not true
}
```

The syntax of the statement **select–case–else** is

```
select (control_variable) {
  case (values_1)
    statement_1
  ...
  case (values_n)
    statement_n
  ...
  else
    statement_else
}
```

where *control_variable* is the name of the control variable that governs the selection of possible cases, *values_1* to *values_n* are the specifications of possible values, *statement_1* to *statement_n* are the corresponding statements, and *statement_else* is the statement that is executed if none of the cases was selected. Here is a simple example:

```
integer v_int;
  // assume that a value has been assigned to v_int
  select (v_int) {
    case (0 .. 9) {
      log(v_int, " is a one digit positive integer");
    } case (10 .. 19) {
      log(v_int, " is a two digits positive integer");
```

```
}else case{
  log( v_int, " is not a one digit or a two digits positive integer" );
}
}
```

Loops are used to specify repetitive behavior. There are the three kinds of loops in TTCN-3: the statement **for**, the statement **do–while**, and the statement **while**. Within a loop, the statement **break** may be used to exit the loop, whereas the statement **continue** may be used to skip the current iteration. The syntax of the statement **for** is as follows:

for (*initial_stmt; condition_exp; next_stmt*)
  body_statement

where *initial_stmt* is the initial statement typically used to declare a control variable and to assign it an initial value, *condition_exp* is the condition expression that is checked before the next loop iteration starts (if the expression is not **true**, the loop terminates), *next_stmt* is the statement that is executed after each iteration, and *body_statement* is the loop's body. The following simple example, with the typical control variable *i*, looks familiar:

for (**integer** $i := 0$; $i < n$; $i := i + 1$)
  // Do something that depends on the value of *i*

The syntax of the statement **do–while** is as follows:

do
  body_statement
while (*condition_exp*)

The syntax of the statement **while** is

while (*condition_exp*)
  body_statement

Labels and goto statements provide a mechanism to jump from one part of a program to another. Although they provide compatibility with TTCN-2, their usage in TTCN-3 is strongly discouraged. The statement **label** defines a label within a logical block statement (e.g., function or control part), whereas the statement **goto** is a control flow statement that transfers control to the specified label within the same block statement. So, it is not possible to jump out of (or into) the functions, test cases, and the control part; it is not possible to jump into both the loop and conditional statements.

Log statements are used for writing relevant information on the test system's logging interface. The particular format of logged values depends on the logging interface implementation. We may log the variables, arrays (whole arrays by specifying their name), constants, function parameters, function instances (that have the statement **return**), test component references, templates, timers, and related operations.

The **control part** of the module is the entry point for execution of a test suite, which is similar to the function **main** in other programing languages. The control part specifies the dynamic behavior of the test system. It may contain control statements and function calls. The main role of the control part is to execute test cases. The control part is not allowed to directly communicate with the SUT, to set a verdict, or to create dynamic configurations. These operations must be performed only within test cases.

Preprocessing macros are used in definition or control parts to locate the position of the macro call. TTCN-3 compiler replaces these macros with their **charstring** or **integer** values. More precisely, these values are inserted in the program source code instead of the macro calls. By the convention, the macro's names are enclosed by underscores. Currently, TTCN-3 offers the following preprocessing macros: _MODULE_, _FILE_, _BFILE_, _LINE_, and _SCOPE_.

The value of the macro _MODULE_ is the name of TTCN-3 module in which the macro was called.

The value of the macro _FILE_ is the full pathname (ending with the basic file name) of the file in which the macro was called.

The value of the macro _BFILE_ is the basic file name (without its path) of the file in which the macro was called.

The value of the macro _LINE_ is the number of the source code (i.e., file) line in which the macro was called.

The value of the macro _SCOPE_ depends on whether the corresponding scope is named or unnamed. The following basic scopes are named: the module, the control part (has a special name "Control"), the function, the component, the test case, the altstep, the template, and the user-defined type. If the corresponding scope is named, the value of the macro _SCOPE_ is its name; otherwise, the value is the name of the next higher basic scope.

3.9.3 Single Component TTCN-3 Test Suites

Although TTCN-3 resembles a common programming language, it's a domain-specific language for developing test cases, which defines the interaction between the test system and the SUT. In this section, we study the message-based communication with the SUT and test cases executed on a single test component (i.e., nonconcurrent TTCN-3 test suites).

We introduce the concepts for message-based communication and single component test suites through examples for testing the Address Resolution Protocol (ARP) server. So, the test setup is such that test cases executing on a test system (also called the tester) imitate an ARP client, whereas the SUT is the real ARP server under testing, see Figure 3.45.

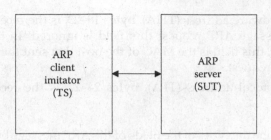

FIGURE 3.45
Test configuration for testing the ARP Server.

The main task of the ARP is to map a given network address, such as the Internet Protocol version 4 (IPv4) address, into the corresponding physical (or hardware) address, such as Ethernet address, which is also known as the Media Access Control (MAC) address. The ARP is a simple client–server protocol, which uses a simple message format containing one address resolution request or response. The size of the ARP messages depends on the size of the particular network and physical addresses. For example, the size of the IPv4 address is 32 bits (4 bytes), the size of the MAC address is 48 bits (6 bytes), and the size of ARP messages used to map IPv4 to the MAC addresses is 28 bytes.

The fields of the ARP message used for mapping IPv4 to MAC addresses are as follows (we refer to individual bytes, also called octets, of the message by using their index, which starts from 0):

- Hardware type (HTYPE), bytes 0–1, specifies the type of the physical address (for Ethernet, HTYPE is equal to 1).
- Protocol type (PTYPE), bytes 2–3, specifies the network protocol (for IPv4, PTYPE is equal to 0x0800).
- Hardware address length (HLEN), byte 4, is the length of the hardware address (for Ethernet, HLEN is equal to 6).
- Protocol address length (PLEN), byte 5, is the length of the network address (for IPv4, PLEN is equal to 4).
- Operation (OPER), bytes 6–7, specifies the operation that the sender is performing (1 for request, 2 for reply).
- Sender hardware address (SHA), bytes 8–13, is the sender's MAC. In the message ARP request, this field is the MAC of the host sending the request. In the message ARP reply, this field is the MAC of the host that the request was looking for, i.e., the result of the request mapping.
- Sender protocol address (SPA), bytes 14–17, is the sender's IPv4 address.

- Target hardware address (THA), bytes 18–23, is the receiver's MAC. In the message ARP request this field is ignored. In the message ARP reply, this field is the MAC of the host that sent the initial message ARP request.
- Target protocol address (TPA), bytes 24–27, is the receiver's IPv4 address.

We may describe the types of the fields of the ARP message by the following supplementary types (note that generally we may specify hexadecimal numbers using the construct '*h_num*'H, where *h_num* is a hexadecimal number):

type integer *Int8* (0..'FF'H)
type integer *Int16* (0..'FFFF'H)

where *Int8* corresponds to a single byte field and *Int16* corresponds to a double byte field. Then we may define possible values of the field OPER using the following enumerated type (note that generally we may explicitly assign a value to an enumeration element by writing the particular value enclosed in the parenthesis after the particular enumeration element name):

type enumerated *ARPOperation* (
  e_ARPRequest(1),
  e_ARPReplay(2)
);

Using these supplementary types, we may describe the ARP message by the following record type:

type record *ARPMessage* {
  Int16 htype,
  Int16 ptype,
  Int8 hlen,
  Int8 plen,
  Int16 oper,
  charstring *sha,*
  charstring *spa,*
  charstring *tha,*
  charstring *tpa*
}

Finally, we may construct individual ARP messages using the following parametrized template:

template *ARPMessage t_ARPMessage(*
  Int16 p_oper, Int8 p_sha, Int32 p_spa, Int48 p_tha, Int32 p_tpa

```
):= {
    htype := 1,
    ptype := 0x0800,
    hlen := 6,
    plen := 4,
    oper := p_oper,
    sha := p_sha,
    spa := p_spa,
    tha := p_tha,
    tpa := p_tpa
}
```

The ARP operates as follows: Assume that the router R has to deliver an IPv4 datagram to the host H, which for example has the IPv4 address 192.168.0.48 and the MAC address 00:EB:24:B2:05:C8. First, R will have a look in its own local routing table for the entry corresponding to H's IPv4 address. If R finds it there, then R reads the H's MAC address from that entry and uses it to perform direct datagram delivery to H.

If R does not find the entry for the IPv4 address 192.168.0.48, then R broadcasts the message ARP request for this IPv4 address by sending the Ethernet frame to the destination MAC address FF:FF:FF:FF:FF:FF. The ARP server S receives this message, finds the mapping in its local table, creates the corresponding message ARP reply, and sends it to R, which, in turn, performs direct datagram delivery to H, and updates its local routing table accordingly.

The tester (i.e., test system) may test ARP by executing a simple test case, which first sends the message ARP request (with SPA set to its IPv4 address, SHA set to its MAC address, and TPA set to the IPv4 address 192.168.0.48; THA is ignored), and then receives the message ARP reply with the required mapping (with SPA set to the IPv4 address 192.168.0.48 and SHA set to the MAC address 00:EB:24:B2:05:C8), see the MSC in the Figure 3.46. If the

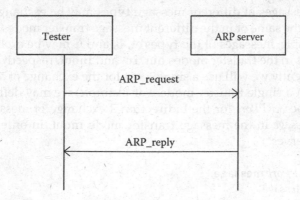

FIGURE 3.46
MSC for mapping the IPv4 address into the corresponding MAC address.

received message ARP reply contains the correct mapping, the tester would set the test verdict to **pass**; otherwise, it would set the test verdict to **fail**.

Here, we introduce the following concepts for message-based communication and single component test suites:

- Ports
- Components
- Test Cases
- Templates
- Message-Based Communication
- Timers
- Alt Statement
- Altsteps
- Default Altsteps
- Functions

Ports are used for exchanging messages. The messages sent to a port are immediately delivered to the related receiver, whereas the messages received from a port are stored in the unbounded FIFO queue, which is implicitly assigned to a port. Although the queue is theoretically unbounded, i.e., of infinite length, TTCN-3 implementations may introduce some practical limits.

Directions of messages exchanged over ports are defined from the test system point of view. There are the three possible message transfer modes for exchanging messages over ports, namely the mode **out**, the mode **in**, and the mode **inout**. The mode **out** is used for sending messages from the test system to the SUT, the mode **in** is used for receiving messages sent from the SUT to the test system, whereas the mode **inout** is used for the bidirectional exchange of messages between the test system and the SUT.

Generally, a single port may be used for exchanging more message types. Moreover, messages of different message types may be exchanged over the same port in the same or in the different message transfer modes. In the most general example, messages of the types *A*, *B*, and *C* may be exchanged over the same port in the transfer modes **out**, **in**, and **inout**, respectively.

Most frequently we will use a single port for the exchange of a single type of message in a single transfer mode. For example, we may define the message port type *ARPPort* for the bidirectional exchange of messages, or the type *ARPMessage* in the message transfer mode **inout**, in order to test the target ARP server:

```
type port ARPPort message {
 inout ARPMessage
};
```

However, sometimes we will need to define different message types to be exchanged over the same port type in various message transfer modes. For example, imagine that we want to test the Email server. Since Email clients use SMTP protocol for sending email messages to the Email server, and POP3 protocol for receiving email messages from the Email server, we would define one message port type with two different message types, for example, as follows:

type port *MailPort* **message** {
  inout *SMTPMessage*;
  inout *POP3Message*
}

Components are used for executing test cases. A component may have its local state that comprises its constants, variables, and timers. The component's interface is defined by its ports. In order to define a component type, we have to provide the list of particular port instances used by that component type, where each item in that list indicates the type of the port and the name of the port instance. It is not necessary that all the ports have different types. Some of the ports may have the same type, but their names must be different, i.e., unique.

For example, we may define the component type *ARPTester*, which uses a single port instance of the type *ARPPort* with the name *serverPort*, as follows:

type component *ARPTester* {
  port *ARPPort serverPort*
}

Optionally, we may define component's constants, variables, and timers, within the component's type definition. As shown in the previous section, constants are defined by their type, name, and value; variables are defined by their type and name and optional initial value; and timers are defined by their name and optional default duration of the type **float**. It is important to notice that each instance of a component type has its own instances of the ports, variables, and timers (i.e., they are analogous to nonstatic class attributes in programming languages).

For example, we may extend the previous definition of the component type *ARPTesterS* by introducing the constant *c_maxRequests* (the max number of ARP requests that an *ARPTester* may send in a burst, i.e., without waiting for a reply before issuing the next request), the variable *v_noRequests* (the number of requests sent to the SUT), and the timer *t_inactive* (that may bound the time interval for waiting the reply from the SUT), with the default duration of 0.5 s. The extended type *ARPTesterS* is as follows:

```
type component ARPTesterS {
  const integer c_maxRequests := 1000;
  var integer v_noRequests;
  timer t_inactive := 0.5;
  port ARPPort serverPort
}
```

Test cases are used to describe the expected behavior of the SUT, and to set the test verdict depending on the real behavior of the SUT. More precisely, test cases define the behavior of the main test component within a given test configuration that may generally have more test components. As its name suggests, a single component TTCN-3 test suite's test configuration has a single test component, which must be the main test component.

The Test System Interface (TSI) is the interface between the TS and the SUT. In case of a single component TTCN-3 test suite, TSI is completely defined by the set of ports of the main test component, thus TSI is defined implicitly and there is no need to define it separately.

When writing a test case, we use the clause **runs on** to specify the component type that will execute that test case. Most frequently, a test case will not have parameters, and in such a case we simply omit the list of formal parameters by writing the empty pair of parenthesis after the test case name. For example, the following empty test case *tc_nop* (which does not perform any operation) is designed to be executed on the component type *ARPTester*, which has no parameters:

testcase *tc_nop*() **runs on** *ARPTester* {};

We have already introduced possible test verdicts (**none, pass, inconc, fail,** and **error**) without going too much into detail. Actually, each test component has its own local verdict, which is a variable of the type **verdicttype** that we may set or get using the test component's operations **setverdict** or **getverdict**, respectively. The exception is the verdict **error**, which can be set only by the runtime execution system (within the error handling routine) and cannot be set by a test case. Like any other variable, we may log the local verdict current value by the statement **log**.

As already mentioned, the initial value of the local verdict (i.e., its default value) is the verdict **none**. For example, since the test case *tc_nop* performs no operation, its final verdict is the verdict **none**, too.

Unlike simple variables, the possible values of the local verdict are not just elements of a conventional enumeration. Instead, verdicts are assigned different strengths, such that all the verdicts are ordered by their strength, from the weakest (**none**) to the strongest (**error**), according to the following list: <**none, pass, inconc, fail, error**>. Assignment of a value to the local verdict is governed by the following important rule: The current value of the local verdict can be assigned the next value only when the next value is stronger than the current value.

For example, the value of the local verdict can be changed from **none** to **pass** or **fail**, but it cannot be changed, for example, from **fail** to **pass**. Therefore, in the test case *tc_remains_fail*, as shown below, the final test verdict remains **fail**, because the assignment of the verdict **pass** after the assignment of the verdict **fail** is not possible (and thus the runtime execution system just ignores it):

```
testcase tc_remains_fail() runs on ARPTester {
  var verdicttype current_verdict;
  setverdict(fail);
  ...
  // later in the code...
  setverdict(pass);
  current_verdict = getverdict; // verdict remains fail
};
```

Besides the local verdict, a test component also has the implicit variable of the type **charstring**, which may be used to describe the reason for the particular verdict assignment. This variable is assigned by the **setverdict** operation and the reason string is passed as one or more optional parameters at the end of the **setverdict** operation's parameter list (these parameters are specified the same way as those for the **log** statement). For example, in the test case *tc_always_pass*, we describe the reason for setting the test verdict to **pass**:

```
testcase tc_always_pass() runs on ARPTester {
  // Check the SUT behavior
  setverdict(pass, 'The SUT behavior was as expected.')
};
```

We should note that in the case of a single component TTCN-3 test suite, the overall test verdict is equal to the local verdict of the main test component (whereas in the case of the multi component test suite, it is evaluated based on the local verdicts of individual test components).

As we have already seen, a test case is executed from the control part by the statement **execute**, which returns the overall test case's verdict. The verdict returned by the statement may be stored in the variable of the type **verdicttype** for further processing, or it may be ignored if it is not needed. Note that assignments to the user-defined variable of the type **verdicttype** are not governed by the assignment rule for the test component's local verdict, because it is a simple variable, so its value can be changed freely.

The second parameter of the operation execute is optional, and when it is supplied it defines the upper bound on the test case execution time. Under the hood, the runtime execution system starts the corresponding timer, and if the timer expires it terminates the test case with the verdict **error**. We should note that even if some of the test cases have the overall test verdict **error**, other test cases defined within the control part will be executed as requested.

The following example illustrates the control part that executes the three previously introduced test cases. The execution time for all the test cases is bounded to the time interval of 5s and the return verdict is stored into the user-defined variable *result* for all the executions:

```
control {
  var verdicttype result;
  result := execute(tc_nop(), 5.0);
  result := execute(tc_remains_fail(), 5.0);
  result := execute(tc_always_pass(), 5.0);
};
```

Usually, a control part, such as the one shown above, is just a list of execute statements, but when needed, we may use conditional statements and loops (introduced in the previous section) within a more complex control part.

Like functions, test cases may have **in**, **out**, and **inout** parameters. The **in** parameters are passed by a value, whereas the **out** and **inout** parameters are passed by a reference. In the latter case, changes of parameters within the test case cause updates of real parameters in the control part. But, if the test case verdict is **error**, the values of **out** parameters are undefined.

The function's restrictions of its real parameters (which we have already seen) apply to test case parameters, too. The real **inout** parameter cannot be uninitialized, and the real **out**, as well as the real **inout** parameter, cannot be a constant expression. For example, the following test case *tc_counting* has the **inout** parameter *p_count*, which may be used for counting the number of test case executions:

```
testcase tc_counting(inout p_count) runs on ARPTester {
  p_count := p_count + 1;
  setverdict(pass);
};
```

Within the control part, we may define the initialized **inout** variable *v_count* in order to count the number of test case executions:

```
control {
  var integer v_count := 1;
  // execute tc_counting 10 times
  for ( integer i := 0; i < 10; i := i + 1 ) {
  log("v_count = ", v_count);
  execute( tc_counting(v_count) );
  }
}
```

The local constants, variables, and timers of the test component which the test case **runs on**, are in the scope of this test case. These constants, variables, and timers may be used the same way as ordinary test case's local variables. This concept is similar to the concept of inheriting attributes of a supper class in a subclass in programming languages.

In TTCN-3, a function may also inherit local constants, variables, and timers of the test component on which it runs. Note that for all the test cases and functions running on the same test component, these inherited local constants, variables, and timers appear as global entities, and we should use them carefully (the same way we use global variables in other programming languages).

In the following test case *tc_using_comp_vars*, we set the local variable of test component *ARPTesterS* the same way as we set the test case's local variable *v_current*:

```
testcase tc_using_comp_vars() runs on ARPTesterS {
  var integer v_current := 1
  ...
  v_noRequests := 10;
  v_current := 1;
  ...
}
```

A test case implicitly terminates with its last statement. We may explicitly terminate a test case using the operation **stop** or the operation **testcase. stop**. We use the operation **stop** to terminate an error-free test case execution and the operation **testcase.stop** to terminate an erroneous test case execution. The operation **stop** returns the overall test verdict to the control part (analogously to the statement **return** that returns the return value of the called function to the calling function in other programming languages). On the other hand, the operation **testcase.stop** sets the test verdict to **error** and terminates the test case. We may use the operation's optional arguments to indicate the reason for termination (the same way as we use the optional arguments of the operation **setverdict**).

Templates are used to define messages exchanged between the test system and the SUT. When we want to send a particular message from the test system to the SUT, we use the template instance that defines a single value of the corresponding type, i.e., that particular message. But, when we want to receive a reply from the SUT, we would more frequently use matching expressions with template instances specifying more possible reply messages.

Generally, a template defines a set of values of a given type. This set may contain just a single value, more values, or even all the values of the given type (we specify all the values using the character '?'). In the example below, we use the nonparametrized template *t_fixedARPRequest* to define the fixed ARP request message from the test system to the SUT, with SPA set to "192.168.0.40" (this is the test system's IPv4 address), SHA set to "00:EB:24:B2:05:C0" (this is the test system's MAC address), TPA set to "192.168.0.48" (this is the IPv4 address that has to be resolved), and THA set to 0 (actually, it could be any value, because ARP protocol ignores THA field in the ARP request message):

```
template ARPMessage t_fixedARPRequest () := {
  htype := 1,
  ptype := 0x0800,
  hlen := 6,
  plen := 4,
  oper := 1,
  sha := "00:EB:24:B2:05:C0",
  spa := "192.168.0.40",
  tha := 0,
  tpa := "192.168.0.48"
}
```

On the other hand, the previously introduced parametrized template *t_ARPMessage* defines a subset of all the possible values of the record type *ARPMessage*, with the first four fields fixed to the values 1, 0x0800, 6, and 4, respectively.

Although templates are used to specify values, they are not values. Even a single-valued template is not a value. Thus, a template cannot be directly used in an expression.

However, templates can be passed as **in** parameters to functions and test cases. Such a parameter must be defined with the additional keyword **template** in order to distinguish it from the simple value. For example, the following test case has the template as its input parameter:

```
testcase tc_withParam(
 in template t_ARPMessage p_msg
) runs on ARPTester {
  // some statements that depend on p_msg
};
```

Message-based communication between the test system and the SUT is conducted over TSI ports in order to effectively test the SUT. The type **port** supports the three main operations, namely **send**, **receive**, and **check**. The operation **send** sends the specified message to the SUT. The operation **receive** compares the received message with the specified template, and if they match, it receives the message from the port's queue; otherwise it blocks. The operation **check** is similar to the operation **receive**, but it does not remove the received message from the port's queue. Besides receiving a message from a single port, it is also possible to receive a message from **any port**. In the following paragraphs, we study these operations in more detail.

The port's operation **send** sends the particular message (single value template instance) over the specified port. For example, the following test case creates the ARP request message *req_msg*, with SPA set to "192.168.0.40" (the test system's IPv4 address), SHA set to "00:EB:24:B2:05:C0" (the test system's MAC address), TPA set to "192.168.0.48" (the IPv4 address that has to be resolved), and THA set to 0 (actually, it could be any value), and sends this message over the port *serverPort* to the SUT:

```
testcase tc_resolve_part_1() runs on ARPTester {
  // create the ARP request message
  ARPMessage req_msg := t_ARPMessage(
    1,                        // ARP operation: 1 – request
    "00:EB:24:B2:05:C0",      // test system's MAC address
    "192.168.0.40",           // test system's IPv4 address
    0,                        // this field is ignored by ARP
    "192.168.0.48"            // target IPv4 address to be resolved
  );
  // send the ARP request message
  serverPort.send(req_msg);

  // part 2 - to be finished later
};
```

The state of the SUT cannot influence the execution of the operation **send**, which is executed by the test system. Once the message is delivered over the specified port, the operation **send** is finished, and the test case proceeds to the next statement following it, no matter whether SUT really received the message or not.

When we define a template using a simple type rather than a record, the particular template instance might not be distinguished from the ordinary value of the corresponding type. In such a case, the value must be preceded by a type name. For example, assume that we defined the template *t_MyIPAddresses* using the type charstring, and assume that "128.0.0.0" is a member of *t_MyIPAddresses*. In order to **send** the value "128.0.0.0" as one of the *t_MyIPAddresses* instances, we must explicitly write the template name before the particular value:

```
type charstring t_MyIPAddresses {"128.0.0.0", …};
somePort.send(t_MyIPAddresses: "128.0.0.0");
```

The port's operation **receive** is generally used for receiving messages from the SUT. Unlike the operation **send**, its argument is a template that may specify more possible SUT replies rather than just one particular SUT reply (which is allowed as a special case). Also, the operation **receive** is a blocking operation, whereas the operation **send** is a nonblocking operation.

The operation **receive** performs two steps. In the first step, it compares the message at the head of the port's queue with the specified template. If this message is a member of the set of messages specified by the template, we say that the message matches the template. More precisely, if the template specifies a single message, the message at the head of the queue must be that message. If the template specifies a subset of messages of the message type that may be received over the specified port, the message at the head of the queue must be a member of that subset. Finally, if the template specifies any message of the corresponding message type, then the message at the head of the queue must be of that type.

In the second step of the operation **receive** there are two possible cases. If the message at the head of the queue matches the template, this message is dequeued from the head of the queue and delivered to the receiving process, which proceeds to the next statement that follows the operation **receive**. If the message at the head of the queue does not match the template, and if there are no alternatives, then the receiving process blocks within the operation **receive** (we introduce alternatives later in the following text).

The message at the head of the queue may mismatch the template in two possible cases. The first case is when the queue is empty. No message mismatches any template, and consequently the receiving process blocks. The second case is when there is some message at the head of the queue that mismatches the template, so the receiving process again blocks. However, there is a fundamental difference between these two cases. In the latter case, the receiving process blocks forever (even if the right message is received later, because it will still not be positioned at the head of the queue), whereas in the former case, the receiving process blocks temporarily. If the right message is received later, the receiving process would be unblocked.

The following test case *tc_resolve* tests the whole ARP. Its first part is the same as in the previous test case *tc_resolve_part_1*. In its second part, the test case *tc_resolve* creates the expected ARP reply message *rpy_msg*, with SPA set to"192.168.0.48" (the IPv4 address that has to be resolved); SHA set to "00:EB:24:B2:05:C8" (the expected MAC address that should be the result of the ARP resolution); TPA set to"192.168.0.40" (test system's IPv4 address); and THA set to "00:EB:24:B2:05:C0" (test system's MAC address), which receives this message over the port *serverPort*, and sets the test verdict to **pass**.

```
testcase tc_resolve() runs on ARPTester {
// create the ARP request message
ARPMessage req_msg := t_ARPMessage(
  1,                      // ARP operation: 1 - request
  "00:EB:24:B2:05:C0",    // test system's MAC address
  "192.168.0.40",         // test system's IPv4 address
  0,                      // this field is ignored by ARP
  "192.168.0.48"          // target IPv4 address to be resolved
);
// send the ARP request message
```

```
serverPort.send(req_msg);

// part 2 – create ARP reply, receive it, and set test verdict
// create the ARP reply message
ARPMessage rpy_msg := t_ARPMessage(
2,                        // ARP operation: 2 - reply
"00:EB:24:B2:05:C8",      // target MAC address – expected value
"192.168.0.48",           // target IPv4 address to be resolved
"00:EB:24:B2:05:C0",      // test system's MAC address
"192.168.0.40"            // test system's IPv4 address
);
// receive the ARP reply message
serverPort.receive(rpy_msg);
// set test verdict to pass
setverdict(pass);
};
```

In the previous test case, the operation **receive** may block the receiving process temporarily if the test system still did not receive a reply from the SUT. Alternatively, the operation receive may block forever if the test system received the message that mismatched the expected message *rpy_msg*. The receiving process will not block, or will be unblocked, if the test system receives the expected message *rpy_msg*. Once this expected message is received, the test case will set the test verdict to **pass** and it will successfully terminate.

The operation **receive** also offers an option to save the received message into the specified variable of the corresponding type (e.g., the type that is used in the definition of the template). The syntax of the statement using this option is as follows:

port.**receive**(*template*) -> **value** *variable*

where *port* is the name of the port over which the message is to be received, *template* is the name of the template that the received message should match, and *variable* is the name of the variable where the received message should be stored.

Alternatively, by using the operation **receive** without the argument, we may receive any message over the specified port, but we cannot save that message. Of course, the received message must be of the correct type. For example, the following statement will receive any message of the type *ARPMessage*:

serverPort.**receive**;

Like in the case of the operation **send**, if the type of the operation's argument could not be uniquely determined, it must be specified explicitly as follows:

port.**receive**(*type: template*)

where *port* is the name of the receiving port, *type* is the name of the message type, and the *template* is the name of the template.

The port's operation **check** receives the message from the specified port, but it does not remove it from the port's queue. The receiving process will block if the queue is empty or if the message at the head of the queue mismatches the specified template. Alternatively, if the message at the head of the queue matches the specified template, the operation check successfully finishes, and the receiving process proceeds to the next statement following it.

The operation **check** is the operation on the specified port whose argument is the operation **receive** with its argument. For example, the following statement checks any message on the port *serverPort*:

serverPort.**check**(**receive**);

Alternatively, the following statement checks the particular ARP reply *rpy_msg* on the port *serverPort*:

serverPort.**check**(**receive**(*rpy_msg*));

Like the operation **receive**, the operation **check** offers the option for saving the checked message into the specified variable. The syntax of the statement for using this option is the same as for the operation **receive**. For example, the following statement saves the checked message *rpy_msg* into the variable *v_msg* of the type *ARPMessage*:

ARPMessage v_msg;
serverPort.**check**(**receive**(*rpy_msg*)) -> **value** *v_msg*;

Besides receiving and checking messages on the particular port, we may receive or check messages on **any port**. We may want to do this in order to receive or check the unexpected messages and we may do this simply by using the keyword **any port** as the port name in the corresponding statements. Of course, sending some message on any port would be an ambiguous operation, thus this option is not supported.

For the sake of illustration, assume that *SysTester* is the test component with two ports, namely *serverPort* and *serverPort2*. Further assume that the port *serverPort* connects the test system with the primary ARP server, and the port *serverPort2* connects the test system with the secondary ARP server (which is a backup in case of the primary ARP server failure).

Generally, we may receive any message on **any port** by using the operation receive on **any port** and without a template, as follows:

any port.receive;

If this statement is executed on the test component *SysTester*, it would block until there is a message in at least one of the two message queues.

Alternatively, if both queues contain messages, this statement would randomly select one of the two queues, and it would dequeue the message from the head of the selected queue. However, in this case, there are no means to determine from which port the message was dequeued.

Alternatively, we may receive the specified message(s) on any port. For example, if the following statement is executed on the test component *SysTester*, it would receive the message *rpy_msg* either from the port *serverPort* or the port *serverPort2*:

any port.receive(*rpy_msg*);

Again, it would not be possible to determine whether the message *rpy_msg* was received from the port *serverPort* or the port *serverPort2*. In this particular example, this would mean that the system as a whole (primary plus secondary ARP servers) reacted as expected. However, in some other protocols, receiving excepted messages from **any port** might not be what we are really looking for. Receiving unexpected messages from any port is the intended usage of the keyword **any port**.

Like in the case of the ordinary receipt of the specified port, we may save the message received on any port into the specified variable. The syntax is the same. For example, if the following statement is executed on the test component *SysTester*, it would save the received message *rpy_msg* (received from either of two available ports) into the variable *v_msg*:

any port.receive(*rpy_msg*) -> **value** *v_msg*;

Timers are used to describe the protocol's timing properties. The moment in time is represented by the nonnegative floating point number (**float**). The type **timer** supports the five main operations: **start**, **stop**, **timeout**, **read**, and **running**. The operation **start** starts the specified timer, the operation **stop** stops the specified timer, the operation **timeout** waits for the specified timer to expire, the operation **read** returns the duration since the specified timer was started, and the operation **running** returns the Boolean indicator indicating whether the specified timer is running (the indicator has the value **true** if the timer running; otherwise it has the value **false**).

We may declare a timer within the test component, the test case, the control part of a module, the function, or the altstep. Each timer exists only within the scope in which it was declared. Once the timer's scope is left, the timer is destroyed, and thereafter becomes unavailable. We may declare a timer without explicitly specifying its default duration. For example, the following declaration declares the timer *t_T1* without the explicit default duration:

timer *t_T1*;

Alternatively, we may declare a timer with the explicit default duration. For example, the following declaration declares the timer t_T2 with the default duration of 1s:

timer t_T2 := 1.0;

We start the specified timer by the operation **start**, which has the timer duration as an optional argument. If we use this optional argument, and if the timer was declared with the explicit default duration, the value of the optional argument will overwrite the default value. For example, the following statement starts the timer t_T2 for the duration of 2s:

t_T2.**start**(2.0);

We typically use the operation **timeout** to simulate the desired rhythm of messages that are sent towards the SUT. For example, imagine that we want to send the ten *req_msg* messages towards the SUT over the port *serverPort*, with the 1s time interval between two adjacent messages. We may do this by the following snippet of code:

```
for ( integer i := 0; i < 10; i := i + 1 ) {
serverPort.send(req_msg);
t_T2.start;
t_T2.timeout;
}
```

We may stop the running timer by the operation **stop**. It is important to remember that the timer's states stopped and expired are two different states. Note that the operation **timeout** on the previously stopped timer would block forever, because this timer would remain in the state stopped and would never go (back) to the state expired. Another important detail to remember is that starting the running, or expired, timer is equivalent to first stopping and then restarting the timer.

The operation **running** and the operation **read** are typically combined. In the following example, we start the timer t_T1 with the duration of 10s and then while it is running, we use the timer t_T2 to report, every 1s, the time that elapsed from the moment when the timer t_T1 was started:

```
t_T1.start(10.0);
while(t_T1.running) {
 log(t_T1.read, " seconds elapsed since t_T1 was started...");
 t_T2.start(1.0);
 t_T2.timeout;
}
log("t_T1 expired.");
```

We may pass timers as **inout** arguments to altsteps or functions, but we cannot pass them to test cases. A timer does not need to be in the running state in order to be passed as an argument.

Alt Statements are used to combine several blocking operations as possible alternatives to continue process execution, in order to avoid unbounded blocking of individual blocking operations. The statement **alt** executes the first blocking operation that is ready to proceed.

For example, as already mentioned, the standalone operation **receive** will block forever if no message, or some unexpected message, is received. The usual way to overcome this situation is to guard this blocking operation **receive** by using a timer. We do this by starting a timer and using the **alt** statement with two alternatives, namely the operation **receive** on the specified port, with the template specifying the expected message(s), and the operation **timeout** on the running timer.

However, this solution with these two alternatives does not eliminate possible indefinite blocking in case when some unexpected message is received on the specified port. Therefore, if we want to completely eliminate indefinite blocking we must use the statement **alt** with the three alternatives in the order listed below:

- The operation **receive** on the specified port with the template specifying the expected message(s)
- The operation **receive** on the specified port without any template, which is used to receive the unexpected messages.
- The operation **timeout** on the running timer, which is used to terminate indefinite blocking in case when no messages are received in some reasonable interval of time (which is equal to the duration of the timer)

This order of alternatives in the statement **alt** is important, because the alternatives are evaluated from top to bottom, and the first one that is ready to proceed will be executed. So, the position of the alternative may be seen as its priority, because if two alternatives are ready to proceed, the one that is closer to the top of the list of alternatives will get executed.

So, how should we order the alternatives? Generally, we put the alternatives for the expected messages on the top of the list, and then we proceed to various kinds of unexpected messages and errors going down the list.

The following test case uses this strategy to test the ARP:

```
testcase tc_resolve_guarded() runs on ARPTester {
  timer t_T1;
  // create the ARP request message
  ARPMessage req_msg := t_ARPMessage(
    1, "00:EB:24:B2:05:C0", "192.168.0.40", 0, "192.168.0.48"
```

```
);
// send the ARP request message
serverPort.send(req_msg);
// part 2
// create the ARP reply message
ARPMessage rpy_msg := t_ARPMessage(
  2, "00:EB:24:B2:05:C8", "192.168.0.48",
  "00:EB:24:B2:05:C0", "192.168.0.40"
);
// start the timer t_T1 with duration 1s
t_T1.start(1.0);
// use the statement alt with 3 alternatives
alt {
  []serverPort.receive(rpy_msg) { // rpy_msg received
    t_T1.stop;
    setverdict(pass);
  };
  []serverPort.receive { // unexpected message received
    t_T1.stop;
    setverdict(fail);
  };
  []t_T1.timeout { // timer expired
    setverdict(fail)
  }
}
};
```

What happens if a message arrives on some port, or some timer expires, while the other alternative is evaluated? Obviously, immediate and continuous reevaluation of all the alternatives would lead to race conditions. Therefore, the statement **alt** uses the concept of the **snapshot** in order to keep the top-down order of evaluation and to avoid race conditions. More precisely, the statement **alt** performs the following steps in a loop until it breaks from it:

- Take a snapshot of the current state of the test component.
- Evaluate all the alternatives from the top to the bottom of the list.
- When the first alternative that is ready to proceed is found, break this loop and execute that alternative.

Furthermore, the statement **alt** offers the option to specify **Boolean guards** for its alternatives, which we did not use so far. Actually, the empty square brackets that we used to mark the beginning of an alternative are the placeholder for an optional Boolean guard. The Boolean guard is the Boolean expression, which evaluates to the values **true** or **false**.

The statement **alt** considers only the alternatives whose Boolean guards evaluate the value **true**, and skips the alternatives whose Boolean guards evaluate the value **false**. The special guard **else** is used to mark the default alternative at the end of the list of alternatives, which will be selected if none of the previous alternatives were selected.

In the following example, we use two Boolean guards to guard the reception of the corresponding messages, and we also use the default guard **else**:

```
alt {
  [select_msg == 1] pt.receive(t_msg1) { setverdict(pass); };
  [select_msg == 2] pt.receive(t_msg2) { setverdict(pass); };
  [else] { setverdict(fail); }
}
```

In the previous example, the test verdict would be set to **pass** if the first Boolean guard evaluates the value **true** and the message *t_msg1* is received over the port *pt*, or if the second Boolean guard evaluates to the value **true** and the message *t_msg2* is received over the port *pt*. Otherwise, the test verdict would be set to **fail**.

Generally, the Boolean guards in the list of alternatives do not have to be orthogonal and complete, i.e., more or none of them may evaluate the value **true**. If more Boolean guards evaluate the value true, the corresponding alternatives are evaluated top-down until the first alternative ready to proceed is selected. If none of the Boolean guards evaluate the value **true** and we do not use the default guard **else**, there are two possible cases: (1) the Boolean guards are independent of the snapshot and (2) the Boolean guards are dependent on the snapshot.

If the first case, the statement **alt** would block forever, which is considered to be a test case design error. In the second case, there is a chance that the statement **alt** will not block forever, because it will continue taking snapshots in a loop, and for some future snapshot some Boolean guards may evaluate the value true. However, there is the risk that this does not happen, because of a design error, so we would be better off by avoiding such designs.

Motivated by these concerns, TTCN-3 standard forbids using operations in Boolean guards whose results may change in repeated evaluations, such as checking whether a timer is running or not. Also, functions that are called from Boolean guards must not change the current snapshot. The examples of forbidden operations, within such functions, are the operation **receive** on a port; the operations **start**, **stop**, and **timeout** on a timer; and operations that update the test component's local variables.

As discussed so far, the statement **alt** may be seen as a selection of alternatives—once the alternative with the highest priority that is ready to proceed is selected, it is executed, and the execution continues with the next statement following the statement **alt**. But, sometimes we would like to repeat the whole selection from the beginning. A traditional way to do it would be to introduce a loop with a break indicator around the statement **alt**. The more elegant way to do it is to use the statement **repeat**.

The statement **repeat** repeats the whole enclosing statement **alt** from the very beginning—the Boolean guards and the alternatives are evaluated again and the next alternative is selected. We may use the statement **repeat** only within the alternatives of the statement **alt** (typically, as the last statement in the alternative) or within the alternatives of an **altstep**. The way the statement **repeat** operates is somewhat similar to the tail recursion in functional programming languages.

As an example, we may use the statement **repeat** to construct a simple ARP server robustness test. Sometimes, the SUT may return the correct reply to the single request, but when the same request is repeated more times, the SUT may become overloaded or some internal synchronization error may lead to a failure, which may cause incorrect replies from the SUT or absence of replies. To test robustness of the SUT, we adapt the test case *tc_resolve_guarded* such that we send the burst of the same ten ARP requests (by using a simple for loop) and then we expect to receive the same ten ARP replies (by using the statement **repeat**):

```
testcase tc_resolve_robustness() runs on ARPTester{
timer t_T10;
// create the ARP request message
ARPMessage req_msg := t_ARPMessage(
 1, "00:EB:24:B2:05:C0", "192.168.0.40", 0, "192.168.0.48"
);
// send the burst of 10 ARP request messages
for( integer i := 0; i < 10; i := i + 1 ){
serverPort.send(req_msg);
}
// part 2
// create the ARP reply message
ARPMessage rpy_msg := t_ARPMessage(
 2,"00:EB:24:B2:05:C8","192.168.0.48",
 "00:EB:24:B2:05:C0","192.168.0.40"
);
// start the timer t_T1 with duration 10s
t_T10.start(10.0);
// use the statement alt and repeat to receive 10 ARP replies
alt{
[] serverPort.receive(rpy_msg){ // rpy_msg received
   setverdict(pass);
   repeat; // repeat in order to receive the next reply
 };
[] serverPort.receive{ // unexpected message received
   t_T10.stop;
```

```
     setverdict(fail);
   };
  [] t_T10.timeout{ // timer expired
     setverdict(fail)
  }
 }
};
```

So far, we have seen only the statements **alt** with more alternative blocking operations. Since the statement **alt** with a single alternative behaves as a single alternative without the enclosing statement **alt**, we would naturally write the single alternative as a stand-alone operation (without the enclosing statement **alt**).

Interestingly enough, and for the reason that would become apparent later on when we introduce **default altsteps**, according to the TTCN-3 standard, a stand-alone blocking operation will be treated by implicitly wrapping it into the enclosing statement **alt**. For example, the stand-alone blocking statement:

serverPort.**receive**(*rpy_msg*);

 is implicitly expanded to:
```
alt {
[] serverPort.receive(rpy_msg) {}
}
```

Altsteps are named groups of alternatives, which may be referred to within the statement **alt**. Like functions, they may have parameters, but unlike functions they may use the Boolean guards and the operations **receive** and **timeout**. The following typical altstep has the timer *p_timer* as its parameter, and it checks the timeout condition on this timer:

```
altstep alt_timeout(inout timer p_timer) {
 [] p_timer.timeout { setverdict(fail) }
};
```

We may now use the altstep *alt_timeout* within the statement **alt**, for example, in order to bound time interval for waiting the message *rpy_msg*:

```
t_T1.start(1.0);
alt {
[] serverPort.receive(rpy_msg){ // rpy_msg received
     t_T1.stop;
     setverdict(pass);
   };
[] serverPort.receive{ // unexpected message received
```

```
  t_T1.stop;
  setverdict(fail);
  };
[] alt_timeout(t_T1) // timer expired
};
```

The altstep *alt_timeout* has the single alternative. If an altstep has more alternatives, they are evaluated the same way as in the statement **alt**. Once the first alternative that may proceed is selected, individual statements in this alternative are executed until the last statement in this alternative is completed, or the explicit statement **return** is encountered. The statement **return** cannot specify the return value, and it transfers control back to the statement following the altstep call within the enclosing statement **alt**.

An altstep may also have local variables, which are typically used for saving received messages and some intermediate results. Like a test case, an altstep may also use the clause **runs on** to inherit ports, timers, variables, and constants of the corresponding test component. The following altstep *alt_receive_10* uses the variable *v_count* to count the number of received *rpy_msg* messages, and also uses the clause **runs on** to inherit the port *serverPort* from the test component *ARPTester*:

```
altstep alt_receive_10(in ARPMessage rpy_msg) runs on ARPTester {
var integer v_count := 0;
alt{
[] serverPort.receive(rpy_msg){ // expected message received
   v_count := v_count + 1;
   if(v_count == 10){ // expected number of replies
    setverdict(pass)
   }
   else if(v_count > 10){ // unexpected number of replies
    setverdict(fail)
   }
   else {
    repeat // repeat in order to receive the next message
   }
  };
[] serverPort.receive{ // unexpected message received
    setverdict(fail)
  }
};
```

It is important to remember that an altstep must not change the current snapshot by the initialization of its local variables. The restrictions on operations that may be used for initializing the altstep's local variables are actually the same as the restrictions for the Boolean guards of the statement **alt**, which we have already discussed previously. An example of the initialization that does not change the current snapshot is the statement for saving the received message into the altstep's local variable.

The altstep call has an optional block statement following it, which is executed after the altstep if any of the alternatives within the altstep are triggered. This block statement may be, for example, used to stop a running timer.

We may now use the altsteps *alt_timeout* and *alt_receive_10* to construct the statement **alt** for receiving the ten *rpy_msg* messages with the guard against unexpected messages and within the time interval bounded by the timer *t_T10*; we also use the optional statement block after the altstep *alt_receive_10* call to stop the timer *t_T10*:

```
// create the ARP reply message
ARPMessage rpy_msg := t_ARPMessage(
  2, "00:EB:24:B2:05:C8","192.168.0.48","00:EB:24:B2:05:C0", "192.168.0.40"
);
t_T10.start(10.0);
alt {
[] alt_receive_10(rpy_msg){ // receive 10 rpy_msg
  t_T10.stop
  };
[] alt_timeout(t_T10) // timer expired
};
```

The reception of the ten *rpy_msg* messages is performed by the altstep *alt_receive_10*, which contains the statement **repeat**. Note that the inner of the two nested **alt** statements would be repeated. More precisely, the statement **alt** defined within the altstep *alt_receive_10* would be repeated.

We may use the operation **return** to end the execution of altstep at the desired point. The operation **return** returns the control to the enclosing statement **alt**, and then the optional block statement following the altstep call is executed. Alternatively, we may use the operation **break** to end the execution of the altstep at some point. The operation **break** returns control to the statement following the enclosing statement **alt**. Note that the optional block statement following the altstep call would not be executed in this case.

So, we should remember that the operation **return** leaves the enclosing altstep, whereas the operation **break** leaves the enclosing statement **alt** from which the altstep was called.

Normally, some more simple altsteps appear in many **alt** statements. The altstep *alt_timeout*, which we introduced earlier, is a typical example of such an altstep. Another typical example is the following altstep named *alt_receive_any*, which is typically used to catch unexpected messages:

```
altstep alt_receive_any() runs on ARPTester {
[] any port.receive {
```

```
setverdict(fail);
};
};
```

We may avoid adding such frequently used altstep to all the **alt** statements in our test suite by using them as **default altsteps**. Although they have a special name, we define the default altsteps exactly the same way we define the nondefault altsteps that we have used so far, such as the altstep *alt_receive_ any* we have defined above.

The **default altstep** is an altstep that has been activated by the operation **activate**, and it remains the default altstep until it is deactivated by the operation **deactivate**. The operation **activate** adds the default altstep at the head of the list of default altsteps. This list of default altsteps is implicitly added at the end of each **alt** statement in the test suite. The operation **deactivate** removes the specified default altstep from the list of default altsteps.

Since the list of the default altsteps is evaluated from head to tail, the default altstep Y that has been activated after the default altstep X has a higher priority than the altstep X. In other words, if the altstep Z has been activated last and the altstep A has been activated first, Z would have the highest priority and A would have the lowest priority. In practice, we use this rule such that we activate the more general default altsteps before the more specific default altsteps, thus the latter would have a higher priority.

The parameter of the operation **activate** is the altstep together with its arguments, and the return value of the operation **activate** is the reference to the activated default altstep, which is of the type **default**. The parameter of the operation **deactivate** is the reference to the default altstep that should be deactivated.

There is one important rule related to the default altstep's call by reference parameters, i.e., **out** and **inout** parameters. Values and templates cannot be **out** or **inout** parameters of default altsteps. The reason for introducing this rule is that a value or a template passed by a reference to the default altstep might not exist at the time when the default altstep has to be executed.

Another important rule is that timers and ports may be passed as **inout** parameters to the default altsteps. Alternatively, a default altstep may inherit timers and ports of the test component that it **runs on**. In other words, in order to provide access to timers and ports within the default altstep, we may either pass them as **inout** parameters or we may provide access to the test component's timers and ports by using the clause **runs on**.

In the following two examples, we adapt the previously introduced test case *tc_resolve_guraded* by using the default altsteps *alt_timeout* and *alt_ receive_any*. We do the adaptation in two steps: In the first step we just introduce the default altsteps and then in the second step we use the convention of the implicit expansion of stand-alone blocking statements into the corresponding statement **alt**, but in the reverse order, to further shorten the final test case. The result of the first step of adaptation is the following test case:

```
testcase tc_resolve_default1() runs on ARPTester {
  timer t_T1;
  var default v_ref1, v_ref2;
  // activate the default altsteps - more general first
  v_ref1 = activate(alt_timeout(t_T1));
  v_ref2 = activate(alt_receive_any());
  // create the ARP request message
  ARPMessage req_msg := t_ARPMessage(
    1,"00:EB:24:B2:05:C0","192.168.0.40",0,"192.168.0.48"
  );
  // send the ARP request message
  serverPort.send(req_msg);
  // part 2
  // create the ARP reply message
  ARPMessage rpy_msg := t_ARPMessage(
    2,"00:EB:24:B2:05:C8","192.168.0.48",
    "00:EB:24:B2:05:C0","192.168.0.40"
  );
  // start the timer t_T1 with duration 1s
  t_T1.start(1.0);
  // use the statement alt with 3 alternatives (2 are implicit)
  alt {
  [] serverPort.receive(rpy_msg){ // rpy_msg received
    t_T1.stop;
    setverdict(pass);
  };
  // alt_receive_any is implicitly considered first
  // alt_timeout is implicitly considered second
  }
  // deactivate the default altsteps
  deactivate(v_ref1);
  deactivate(v_ref2);
};
```

Remember that we intentionally activate the more specific default altsteps later than the more general, so that the former have a higher priority. In this example, we activated the default altstep *alt_receive_any* after the default altstep *alt_timeout*, so that the unexpected message may be cached before the timer *t_T1* expires.

Next, we transform the statement **alt** with a single alternative into the corresponding stand-alone blocking statement. The resulting test case is the following:

```
testcase tc_resolve_default2() runs on ARPTester {
  timer t_T1;
  var default v_ref1, v_ref2;
```

```
// activate the default altsteps – more general first
v_ref1 = activate(alt_timeout(t_T1));
v_ref2 = activate(alt_receive_any());
// create the ARP request message
ARPMessage req_msg := t_ARPMessage(
  1, "00:EB:24:B2:05:C0","192.168.0.40",0,"192.168.0.48"
);
// send the ARP request message
serverPort.send(req_msg);
// part 2
// create the ARP reply message
ARPMessage rpy_msg := t_ARPMessage(
  2,"00:EB:24:B2:05:C8","192.168.0.48",
  "00:EB:24:B2:05:C0","192.168.0.40"
);
// start the timer t_T1 with duration 1s
t_T1.start(1.0);
// this stand-alone blocking statement is implicitly expanded
// into the corresponding single-alternative alt statement
serverPort.receive(rpy_msg){ // rpy_msg received
  t_T1.stop;
  setverdict(pass);
};
// deactivate the default altsteps
deactivate(v_ref1);
deactivate(v_ref2);
};
```

Obviously, by using the default altsteps we may get rather compact code. However, the disadvantage of using the default altsteps is that we may forget which default altsteps are currently active and their order of activation, especially if we often activate and deactivate them. The code using the default altsteps may be hard to understand and maintain, so we should use the default altsteps carefully.

Functions in TTCN-3 may be also used to specify communication behavior, and they may contain all the kinds of statements that we have introduced so far. Unlike the altsteps that must start with the statement **alt** at the topmost level, the functions may start with any statement, including, for example, the statement **send**.

In the following example, we define the function *sendReqBurst* whose parameter is the number of ARP requests to be sent (*p_noReqs*) and that runs on the test component *ARPTester*:

```
function sendReqBurst(in integer p_noReqs) runs on ARPTester {
  // create the ARP request message
```

```
ARPMessage req_msg := t_ARPMessage(
  1, "00:EB:24:B2:05:C0", "192.168.0.40", 0, "192.168.0.48"
);
for(var integer i := 0; i < p_noReqs; i++) {
  serverPort.send(req_msg)
};
return;
};
```

The function *sendReqBurst* has access to the port *serverPort* because it **runs on** the test component *ARPTester*, which comprises this port. Alternatively, we may pass the port as an **inout** parameter of a function. The function *sendReqBurst* does not have a return value, and we use the statement **return** at the end of the function to explicitly indicate the end of the function.

Next, we adapt the previously introduced test case *tc_resolve_robustness* to use the newly introduced function *sendReqBurst*:

```
testcase tc_resolve_robustness_fun() runs on ARPTester {
timer t_T10;
// call the function to send the burst of 10 requests
sendReqBurst (10);

// part 2
// create the ARP reply message
ARPMessage rpy_msg := t_ARPMessage(
  2,"00:EB:24:B2:05:C8","192.168.0.48",
  "00:EB:24:B2:05:C0","192.168.0.40"
);
// start the timer t_T1 with duration 10s
t_T10.start(10.0);
// use the statement alt and repeat to receive 10 ARP replies
alt {
[] serverPort.receive(rpy_msg){ // rpy_msg received
  setverdict(pass);
  repeat; // repeat in order to receive the next reply
  };
[] serverPort.receive{ // unexpected message received
  t_T10.stop;
  setverdict(fail);
  };
[] t_T10.timeout{ // timer expired
  setverdict(fail)
  }
 }
};
```

TTCN-3 makes no distinction between the ordinary value-computing functions and the communication-behavioral functions. We may call the former functions from the latter and vice versa. Both simple and recursive function calls are allowed.

The function defined using the clause **runs on** naturally can be executed on the instance of the specified component type, but it can also be executed on the instance of the component type that is the extension of the specified component type. The extended component type must have all the timers, ports, constants, and variables of the original type and it may have additional timers, ports, constants, and variables. For example, the function *send-ReqBurst* can be also executed on the test component type *ARPTesterS*, which is the extension of the component type *ARPTester*.

Analogously, the altstep defined using the clause **runs on** can be executed both on the instance of the specified component type and on the instance of the component type that is the extension of the specified component type. For example, the altstep *alt_timeout* can be executed both on the component types *ARPTester* and *ARPTesterS*.

The next restriction applies to both functions and altsteps. A function or an altstep that is defined without the clause **runs on**, cannot be called with the clause **runs on**.

Finally, we summarize similarities and differences between the functions and the altsteps. The similarities between the functions and the altsteps are as follows:

- Both may define communication behavior.
- Both may have parameters.
- Both may be defined using the clause **runs on**.
- Both may call functions and altsteps.

The differences between the functions and the altsteps are as follows:

- Altsteps can be used at the top level of the statement alt, whereas functions can only be used in the statements within alternatives or the Boolean guards.
- Altsteps without values and template parameters passed by a reference can be activated as the default altsteps, whereas functions cannot be used to specify the default behavior.
- Altsteps must start with the statement **alt**, whereas functions may start with any statement.
- Altsteps cannot use initializations of local variables that change the current snapshot, whereas functions can use local variables without any restrictions.
- Altsteps cannot have return value, whereas functions can have return value.

3.10 Examples

This section contains some examples that are related to the communication protocol design. These should help the reader to consolidate their understanding of the concepts and techniques introduced so far.

3.10.1 Example 1

This example demonstrates the procedures for connection establishment and release that are performed by two communicating processes, namely *TE1* and *TE2*. The processes *TE1* and *TE2* are specified by their statechart diagrams shown in Figures 3.47 and 3.48, respectively. The semantically equivalent SDL diagrams are shown in Figures 3.49 and 3.50, respectively.

The process *TE1* has four stable states, labeled *TE1_IDLE*, *TE1_CONNECTING*, *TE1_CONNECTED*, and *TE1_DISCONNECTING*. While the process *TE1* is in the state *TE1_IDLE*, it can receive only the message *CONNECT_req* from the user and after receiving that message, the process *TE1* sends the message *CONNECT_ind* to the process *TE2*, and evolves to its next stable state *TE1_CONNECTING*. In that state, the process may receive one of two possible input messages, namely *CONNECT_conf* or *CONNECT_reject*. In the former case, the process moves to the stable state *TE1_CONNECTED*, whereas in the latter case, it evolves to its initial stable state *TE1_IDLE*.

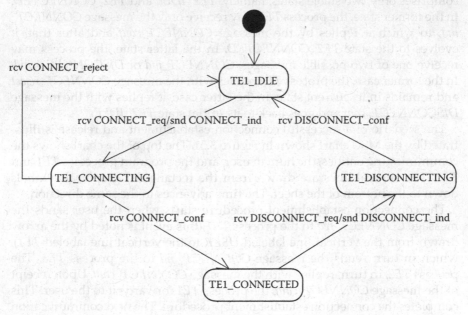

FIGURE 3.47
Statechart diagram of the process *TE1*.

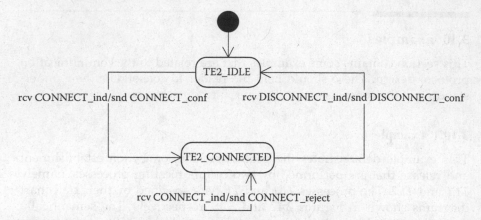

FIGURE 3.48
Statechart diagram of the process *TE2*.

In its stable state *TE1_CONNECTED*, the process *TE1* may receive the message *DISCONNECT_req* from the user. In that case, it sends the message *DISCONNECT_ind* to the process *TE2* and evolves to the stable state *TE1_DISCONNECTING*. From that stable state, it returns to its initial stable state *TE1_IDLE* after receiving the message *DISCONNECT_conf* from its peer process *TE2*.

The SDL diagram specification of the process *TE2* is much simpler because it comprises only two stable states, namely, *TE2_IDLE* and *TE2_CONNECTED*. In the former state, the process *TE2* may receive only the message *CONNECT_ind*, to which it replies by the message *CONNECT_conf* and after that, it evolves to the state *TE2_CONNECTED*. In the latter state, the process may receive one of two possible messages, *CONNECT_ind* or *DISCONNECT_ind*. In the former case, the process *TE2* replies with the message *CONNECT_reject* and remains in its current state. In the latter case, it replies with the message *DISCONNECT_conf* and goes back to its initial state *TE2_IDLE*.

The scenario of a successful connection establishment and release is illustrated by the MSC chart shown in Figure 3.51. The top of the chart shows the communicating entities, the human user, and the program processes *TE1* and *TE2*. The vertical lines are drawn from the rectangular graphical symbols down to the bottom of the sheet. The time advances in the same direction.

The connection establishment procedure starts when the user sends the message *CONNECT_req* to the process *TE1* (this event is noted by the arrow drawn from the vertical line labeled *USER* to the vertical line labeled *TE1*), which in turn sends the message *CONNECT_ind* to the process *TE2*. The process *TE2*, in turn, replies with the message *CONNECT_conf*. Upon receipt of the message *CONNECT_conf*, the process *TE1* forwards it to the user. This completes the connection establishment procedure. The next communication phase is normally used for the desired data transfer. Because of that, it is most frequently referred to as a data transfer phase.

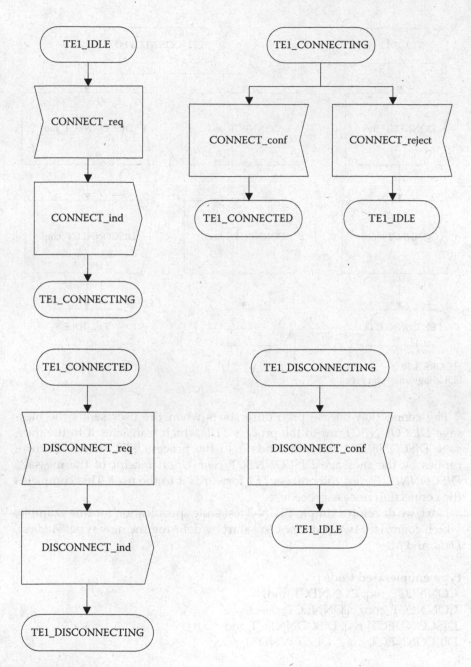

FIGURE 3.49
SDL diagram of the process *TE1*.

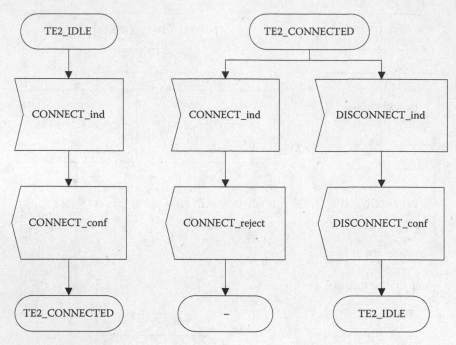

FIGURE 3.50
SDL diagram of the process *TE2*.

The connection release procedure starts when the user sends the message *DISCONNECT_req* to the process *TE1*, which translates it to the message *DISCONNECT_ind* and sends it to the process *TE2*, which, in turn, replies by the message *DISCONNECT_conf*. Upon receipt of the message *DISCONNECT_conf*, the process *TE1* forwards it to the user. This completes the connection release procedure.

Next, we develop a simple TTCN-3 test suite specification for this example, which comprises two test cases. We start by defining the new types *Address*, *Data*, and *Msg*:

```
type enumerated Code (
 CONNECT_req, CONNECT_ind,
 CONNECT_conf, CONNECT_reject,
 DISCONNECT_req, DISCONNECT_ind,
 DISCONNECT_conf, DISCONNECT_reject
);
type integer Adress;
type integer Data;
type record Msg {
 Code code;
 Address source_address;
```

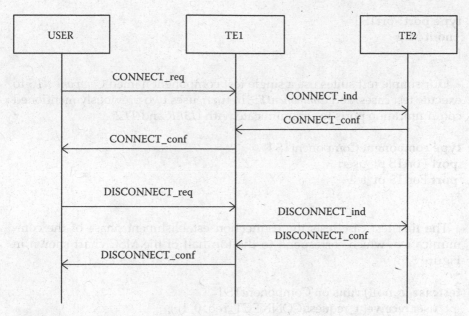

FIGURE 3.51
Successful connection establishment and release MSC.

```
 Address destination_address;
 Data user_data;
}
```
Then by using the message type *Msg,* we define the suitable parametrized templates *t_request* and *t_indication*:
```
template Msg t_request(Code p_code, Address p_src, Address p_dst) := {
code := p_code,
source_address := p_src,
destination_address:= p_dst,
user_data := ?
}

template Msg t_indication(Code p_code, Address p_src, Address p_dst) := {
code := p_code,
source_address := p_src,
destination_address:= p_dst,
user_data := ?
}
```

Let's assume that *USER, TE1,* and *TE2,* are assigned the addresses 0, 1, and 2, respectively. Let's also assume that the test system plays the role of *TE1,* and that it communicates with *USER* and *TE2* over the ports *pt_user* and *pt_te2,* respectively. Finally, we assume that both of these ports are of type *PortTS,* which are defined as follows:

```
type port PortTS {
 inout Msg
}
```

Our simple test suites use a single test component named *ComponentTS* to execute test cases, and *ComponentTS*, in turn, uses two previously mentioned communication ports to communicate with *USER* and *TE2*:

```
type component ComponentTS {
 port PortTS pt_user;
 port PortTS pt_te2
}
```

The first test case tests the connection establishment phase of the communication, which correspond to the top half of the MSC chart shown in Figure 3.51:

```
testcase tc_no1() runs on ComponentTS {
 pt_user.receive( t_request(CONNECT_req, 0, 1) );
 pt_te2.send( t_indication(CONNECT_ind, 1, 2) );
 alt {
 [] pt_te2.receive( t_indication(CONNECT_conf, 2, 1) ) {
 pt_user.send( t_indication(CONNECT_conf, 1, 0) );
 setverdict( pass );
 }
 [] pt_te2.receive( t_indication(CONNECT_reject, 2, 1) ) {
 setverdict( inconc );
 }
 }
 stop;
}
```

The second test case tests both the connection establishment phase and the connection release phase of the communication, which correspond to the complete MSC chart shown in Figure 3.51:

```
testcase tc_no2() runs on ComponentTS {
 // check the connection establishment phase
 pt_user.receive( t_request(CONNECT_req,0,1) );
 pt_te2.send( t_indication(CONNECT_ind,1,2) );
 alt {
 [] pt_te2.receive( t_indication(CONNECT_conf,2,1) ) {
 pt_user.send( t_indication(CONNECT_conf,1,0) );
 // the connection is successfully established
 }
```

```
[] pt_te2.receive( t_indication(CONNECT_reject,2,1)) {
    setverdict(inconc);
    stop
    }
}
// check the connection release phase
pt_user.receive( t_request(DISCONNECT_req, 0, 1) );
pt_te2.send( t_indication(DISCONNECT_ind, 1, 2) );
alt {
[] pt_te2.receive( t_indication(DISCONNECT_conf, 2, 1) ) {
    pt_user.send( t_indication(DISCONNECT_conf, 1, 0) );
    // the connection is successfully released
    setverdict( pass );
    }
[] pt_te2.receive {
    // receive any other message
    setverdict( fail );
    }
}
stop;
}
```

We may execute both of these test cases by using the following control part:

```
control {
execute( tc_no1() )
execute( tc_no2() )
}
```

The reader is encouraged to play more with this simple example. For example, we can change the previous example so that before the existing connection is established, the process *User* checks if the process *TE1* is ready for the communication. The MSC chart that specifies a new connection establishment procedure is shown in Figure 3.52.

3.10.2 Example 2

Figure 3.53 shows a hypothetical computer network with a star topology. Three terminal nodes (N1, N2, and N3) are connected to one transit node (TN). The routing table residing in TN is shown in Figure 3.53 to the right of TN. Terminal nodes generate messages for other terminal nodes in the network. Depending on the value of the message parameter (1, 2, or 3), a transit node delivers the message to its destination by sending it to the corresponding port (A, B, or C).

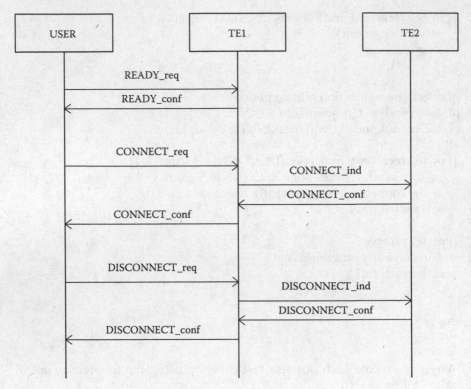

FIGURE 3.52
New connection establishment procedure MSC.

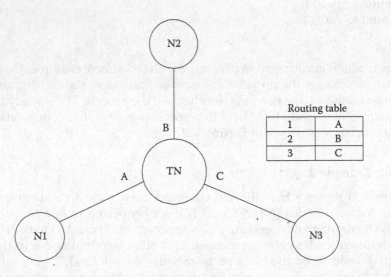

FIGURE 3.53
Hypothetical star network with one transit and three terminal nodes.

The communication process that resides in the terminal node of the network is specified by the statechart diagram shown in Figure 3.54. The process that executes in the transit node is described by the statechart diagram shown in Figure 3.55. The semantically equivalent SDL diagrams are shown in Figures 3.56 and 3.57, respectively.

The process that runs in the terminal node of the network has two stable states, *N123_IDLE* and *N123_MSG_SENT*. The state transition is initiated by the user message *MSG_req*. The process returns to its initial state after the reception of one of three possible messages, namely, *MSG_conf*, *MSG*, or *MSG_reject*. The process that resides in the transit node of the network has a single state, *TN_IDLE*. This process routes the input message toward its destination.

Figure 3.58 shows the scenario of a successful message delivery. The node N1 sends the correct message to the node N3 over the node TN. The user is informed about the successful delivery by the message *MSG_conf*. Figure 3.59 shows the scenario of an unsuccessful message delivery. The node N1 has

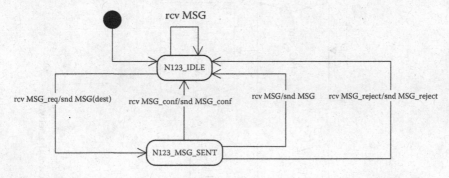

FIGURE 3.54
Statechart diagram of the process that runs in a terminal node of the network.

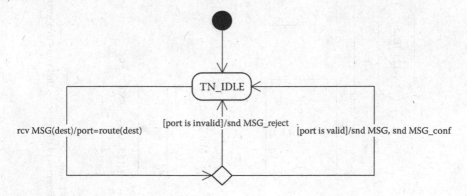

FIGURE 3.55
Statechart diagram of the process that resides in the transit node of the network.

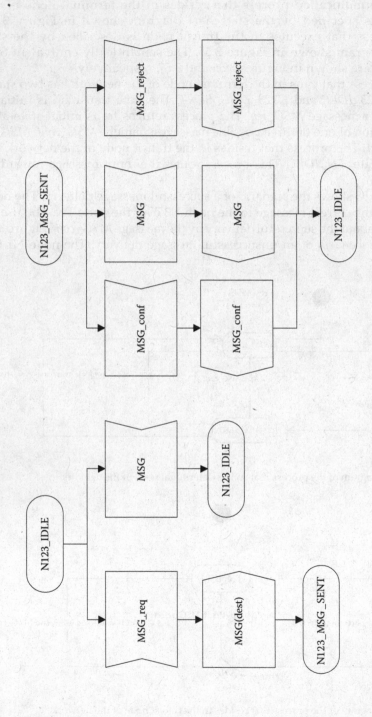

FIGURE 3.56
SDL diagram of the process that runs in a terminal node of the network.

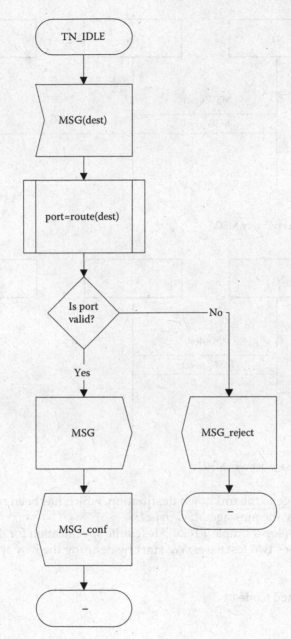

FIGURE 3.57
SDL diagram of the process that resides in the transit node of the network.

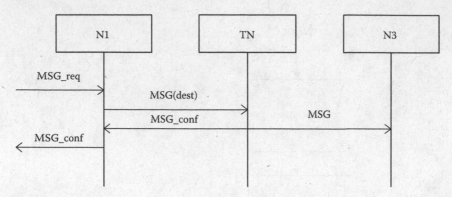

FIGURE 3.58
Successful message delivery MSC.

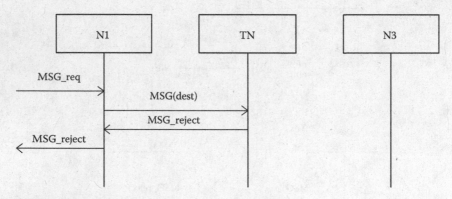

FIGURE 3.59
Unsuccessful message delivery MSC.

sent the message to the unknown destination, which has been rejected from the node TN by the message *MSG_reject*.

Next, we develop a simple TTCN-3 test suite specification for this example, which comprises two test cases. We start by defining the new types *Address*, *Data*, and *Msg*:

```
type enumerated Code (
 MSG_req,
 MSG_conf,
 MSG_reject
);
type integer Adress;
type integer Data;
type record Msg {
 Code code;
```

```
Address destination_address;
Data user_data;
}
```

Then by using the message type *Msg* we define the suitable parametrized templates *t_request* and *t_response*:

```
template Msg t_request(Code p_code, Address p_dst) := {
code := p_code,
destination_address:= p_dst,
user_data := ?
}
```

```
template Msg t_response(Code p_code, Address p_dst) := {
code := p_code,
destination_address:= p_dst,
user_data := ?
}
```

Let's assume that the test system plays the role of N1 in Figures 3.58 and 3.59, and that it communicates with USER and TN over the ports *pt_user* and *pt_tn*, respectively. We assume that both of these ports are of type *PortN*, which is defined as following:

```
type port PortN {
inout Msg
}
```

Our simple test suites use a single test component named *ComponentN* to execute test cases, and *ComponentN*, in turn, uses two previously mentioned communication ports to communicate with USER and TN:

```
type component ComponentTS {
port PortN pt_user;
port PortN pt_tn
}
```

The first test case tests the successful delivery of the correct message from N1 to N3, in accordance with the MSC chart shown in Figure 3.58:

```
testcase tc_no1() runs on ComponentN {
pt_user.receive( t_request(MSG_req, 3) );
pt_n1.send( t_request(MSG_req, 3) );
alt {
[] pt_n1.receive( t_response(MSG_conf, 3) ) {
  pt_user.send( t_response(MSG_conf, 3) );
```

```
        setverdict( pass );
        }
 [] pt_n1.receive( t_response(MSG_reject, 3) ) {
      pt_user.send( t_response(MSG_reject, 3) );
      setverdict( fail );
      }
 }
 stop;
 }
```

The second test case tests the successful drop of the incorrect message from N1 to non-existing N4, in accordance with the MSC chart shown in Figure 3.59:

```
testcase tc_no2() runs on ComponentN {
 pt_user.receive( t_request(MSG_req, 4));
 pt_n1.send( t_request(MSG_req, 4));
 alt {
 [] pt_n1.receive( t_response(MSG_conf, 4)) {
      pt_user.send( t_response(MSG_conf, 4));
      setverdict( fail );
      }
 [] pt_n1.receive( t_response(MSG_reject, 4) ) {
      pt_user.send( t_response(MSG_reject, 4) );
      setverdict( pass );
      }
 }
 stop;
 }
```

We may execute both of these test cases by using the following control part:

```
control {
 execute( tc_no1() )
 execute( tc_no2() )
 }
```

The reader is encouraged to play more with this example. One interesting direction of generalization would be to consider a more complex network, such as the one shown in Figure 3.60.

3.10.3 Example 3

This example illustrates reliable packet delivery based on message acknowledgment. Each communication process expects the acknowledgment of the

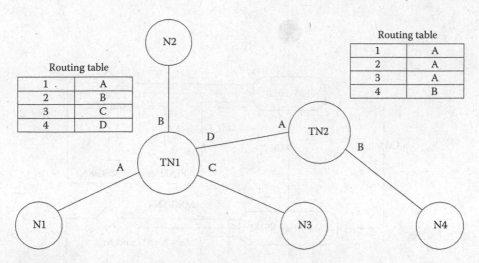

FIGURE 3.60
Topology of a more complex hypothetical network.

message that it has previously sent. If the acknowledgment is not received within the limited period of time, the corresponding timer will expire, the process will assume that the message or its acknowledgment have been lost, and the process will retransmit the message once again.

The statechart diagram and the SDL diagram of the process are shown in Figures 3.61 and 3.62, respectively. The process has two stable states, *FSM_IDLE* and *FSM_MSG_SENT*. In its initial state, the process starts the timer *T1*, sends the message with the sequence number *SN*, and evolves into its next stable state *FSM_MSG_SENT*. In that state, the process either receives the acknowledgment, stops the timer *T1*, and returns to its initial state, or the timer *T1* expires and, in turn, the process retransmits the message.

In any state (*FSM_IDLE* or *FSM_MSG_SENT*), the process can receive a message from its peer process. The process acknowledges the message if the sequence number of the message is valid (in communication protocols, the process would normally maintain the counter of the next expected message in a sequence by incrementing its contents for each received message—a validity check in this context would be to compare the sequence number in the received message with the contents of this counter). If the sequence number, *RN*, of the message is invalid, the process throws the message away.

Figure 3.63 illustrates two scenarios of the communication between two peer processes. The MSC on the left in Figure 3.63 shows a successful message delivery. The process *FSM1* sends the message *M1* to the process *FSM2*, which in turn sends the acknowledgment *ACK* to the process *FSM1*.

The MSC on the right in Figure 3.63 shows a more complex scenario of successful message retransmission after the unsuccessful first message delivery attempt. The process *FSM1* sends the message *M1*, the process *FSM2* receives it and sends its acknowledgment *ACK*, but gets lost. The timer *T1*

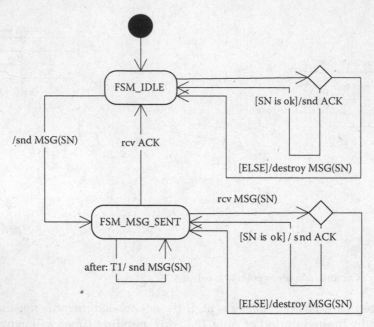

FIGURE 3.61
Statechart diagram of the communicating process that provides the reliable message delivery based on the retransmission scheme.

expires and the process *FSM1* retransmits the message *M1*. The process *FSM2* receives it and sends its acknowledgment *ACK*, which is successfully received by *FSM1*.

3.10.4 Example 4

This example illustrates the sliding window concept, which provides a reliable and efficient transport service. Voluminous literature can be found that addresses this topic (Halsall, 1988). The design shown here is based on the *Go-back-N* retransmission mechanism. It also supports the robust frame acknowledgment procedure (one ACK may acknowledge more than one frame).

The collaboration diagram in Figure 3.64 shows two distributed applications that communicate with the help of two communication objects, which are deployed at the local and remote side. The application *a1* sends the data packed into messages (*M*) to the object *p* (primary), which, in turn, encapsulates the messages into *I* (information) frames, together with its sequence number V(*s*), and sends them to the object *s* (secondary). The object *s* checks the frame *I* sequence number against the number it expects V(*r*), and if they match, it accepts the frame *I* and acknowledges it by sending the message *ACK* to the object *p*. If these numbers do not match, the object *s* rejects the received *I* frame and sends the corresponding message *NAK*. We assume

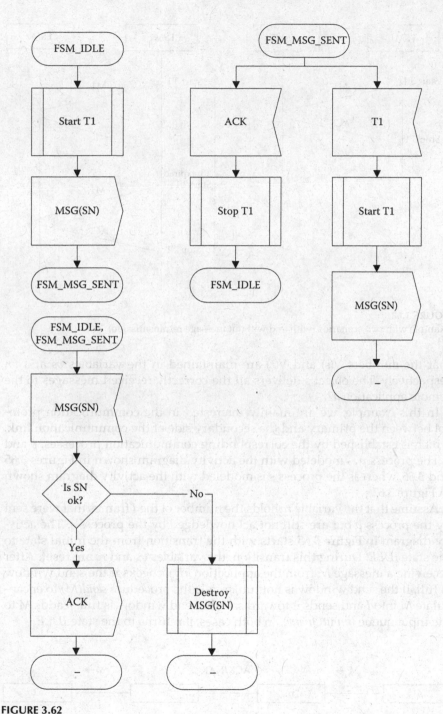

FIGURE 3.62

SDL diagram of the communicating process that provides the reliable message delivery, based on the retransmission scheme.

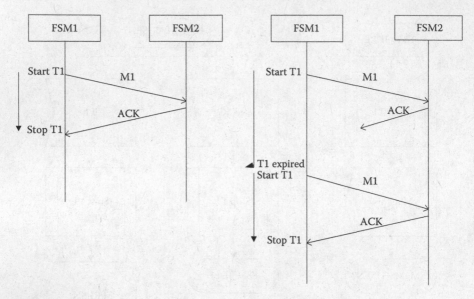

FIGURE 3.63
Example with two scenarios (with and without message retransmission).

that the numbers V(*s*) and V(*r*) are maintained in the variables *vs* and *vr*, respectively. The object *s* delivers all the correctly received messages to the remote application *a2*.

In this example, we are mainly interested in the communication protocol between the primary and the secondary side of the communication link, which is established by the corresponding communication processes, *p* and *s*. The process *p* is modeled with the activity diagram shown in Figures 3.65 and 3.66, whereas the process *s* is modeled with the activity diagram shown in Figure 3.67.

Assume that the variable *rc* holds the number of the *I* frames that were sent by the process *p* but are still not acknowledged by the process *s*. The activity diagram in Figure 3.65 starts with the transition from the initial state to the state *IDLE*. During this transition, the variables *vs* and *rc* are reset. After receiving a message *M* from the application *a1*, *p* checks if the send window is full. If the send window is not full, *p* calls the procedure *send(M)* to encapsulate *M* into *I* and sends it toward *s*. If the send window is full, *p* adds *M* to the input queue (*inputQueue*). In both cases, it returns to the state *IDLE*.

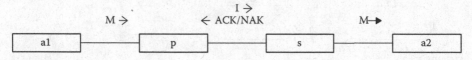

FIGURE 3.64
Example 4 collaboration diagram.

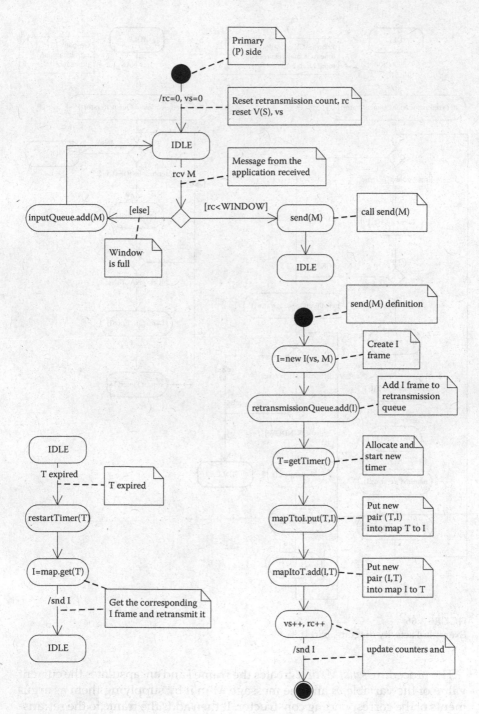

FIGURE 3.65
Example 4 activity diagram, part I.

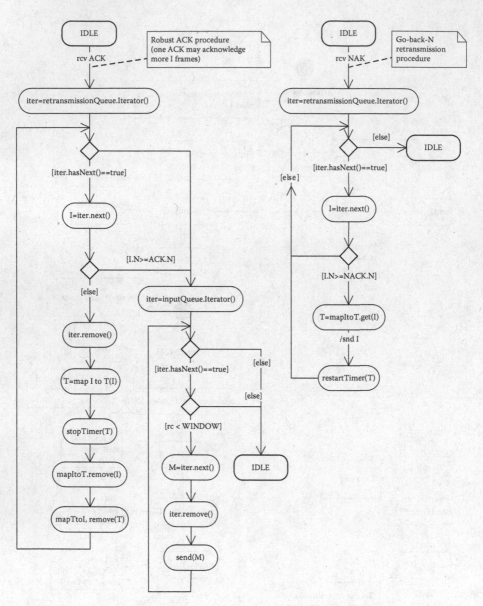

FIGURE 3.66
Example 4 activity diagram, part II.

The procedure *send(M)* first creates the frame *I* and encapsulates the current value of the variable *vs* and the message *M* in it by supplying them as arguments of the corresponding constructor. It then adds the frame to the retransmission queue (*retransmissionQueue*), allocates and starts a new timer (*T*), adds the pair (*T,I*) to the map *mapTtoI*, adds the pair (*I,T*) to the map *mapItoT*,

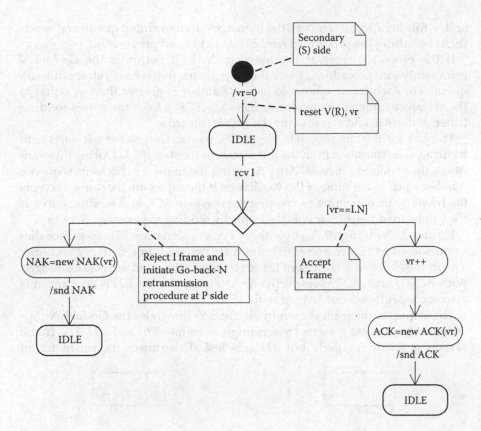

FIGURE 3.67
Example 4 activity diagram, part III.

increments *vs* and *rc*, and sends the frame *I* toward *s*. The map *mapTtoI* is used to search for the frame *I* that corresponds to the given timer *T*, whereas the map *mapItoT* is used to search for the timer *T* that corresponds to the given frame *I*. Notice that the procedure *send(M)* assigns a timer to each frame it sends. When the timer expires, *p* restarts the timer (*restartTimer(T)*), finds the corresponding frame by using the map *mapTtoI*, and retransmits the frame toward *s*.

When *p* receives the message *ACK* from *s*, it provides the iterator on the list *retransmissionQueue* and starts iterating through this list. For all the frames whose sequence number is smaller than the sequence number in the received *ACK* message, *p* finds the corresponding timer (by using the map *mapItoT*), stops it, and removes both the pair (*T,I*) from the map *mapTtoI* and the pair (*I,T*) from the map *mapItoT*.

Because some of the slots (or at least one of them) should be free after the previous iteration, *p* provides the iterator on the list *inputQueue* and starts iterating through it. It iterates while empty slots exist in the send window, and, while iterating, it removes the messages from the input queue and sends them by calling the procedure *send(M)*, as explained previously.

If the process *p* receives the message *NAK*, it performs the *Go-back-N* retransmission procedure. Essentially, *p* scans the whole retransmission queue. For each frame whose sequence number is greater than or equal to the sequence number in the receive message *ACK*, *p* finds the corresponding timer, restarts it, and retransmits the frame toward *s*.

The activity diagram shown in Figure 3.67 models the process *s*. It starts with the triggerless transition from the initial state to the state *IDLE*. During this transition, the variable *vr* is reset. After receiving the frame *I*, *s* checks its sequence number equal to the value of the variable *vr*. If the values are the same, *s* accepts the frame by incrementing *vs*, creating the message *ACK*, and sending it to *p*. If the values are different, *s* rejects the frame by sending the message *NAK* to *p*.

Figures 3.68 through 3.70 show three typical scenarios. The sequence diagram shown in Figure 3.68 illustrates a successful frame delivery scenario. The frames *I(0)* and *I(1)* are sent through the window and are acknowledged with *ACK(1)* and *ACK(2)*, respectively. After some delay, *I(2)* is sent and it is also successfully acknowledged with *ACK(3)*.

The sequence diagram shown in Figure 3.69 illustrates the *Go-back-N* procedure. The process *p* starts by sending the frames *I(0)* and *I(1)*. The frame arrives at *s* side regularly but *I(1)* gets lost. This causes the mismatch of

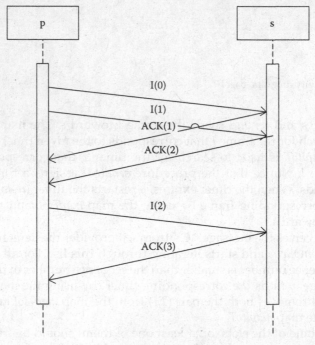

FIGURE 3.68
Example 4 MSC diagram: Successful frame delivery.

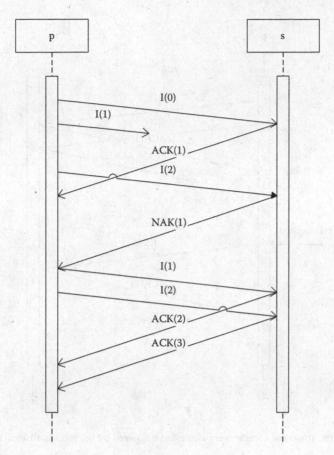

FIGURE 3.69
Example 4 MSC diagram: *Go-back-N* retransmission.

sequence numbers at the secondary side when it successfully receives *I(2)*, because the value of the variable *vr* is 1 (which indicates that *s* is awaiting *I(1)* instead of *I(2)*). Because the sequence number of the frame and the value of the variable are not the same, *s* rejects the frame by sending the message *NAK(1)*. The process *p*, in turn, retransmits both *I(1)* and *I(2)*.

The sequence diagram shown in Figure 3.70 illustrates the frame retransmission triggered by the retransmission timer. The process *p* starts again by sending *I(0)* and *I(1)* in succession. The process *s* in its turn acknowledges them by *ACK(1)* and *ACK(2)*, respectively. The message *ACK(1)* arrives successfully at the primary side, but the message *ACK(2)* gets lost. This causes the corresponding timer to expire after a while. Triggered by that event, *p* restarts the timer and retransmits the frame *I(1)*. During the second time, both *I(1)* and the corresponding *ACK(2)* are successfully transferred over the communication link. After receiving *ACK(2)*, *p* stops the timer and removes *I(1)* from the retransmission queue.

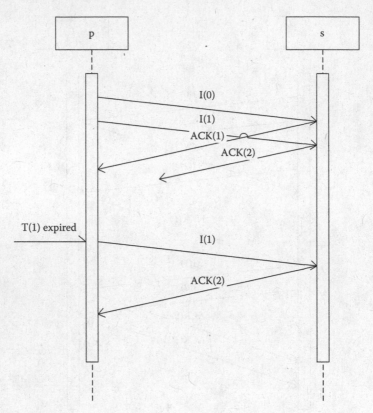

FIGURE 3.70
Example 4 MSC diagram: *I* frame retransmission triggered by the retransmission timer.

3.10.5 Example 5

In this example, we design the SIP INVITE client transaction in accordance with RFC 3261, Section 17.11. First, let us return to the requirements and analysis of a SIP Softphone, introduced as an example at the end of Chapter 2. In that example, we constructed the use case diagram and transformed it into the corresponding general collaboration diagram. At the very end of that example, we showed the one particular collaboration related to the successful session establishment.

Now, let us zoom in on the general collaboration diagram of a SIP Softphone with the focus on the SIP INVITE client transaction and the surrounding objects with which it directly communicates. The resulting general collaboration diagram is shown in Figure 3.71. The SIP INVITE client transaction is modeled as an unnamed object of the class *InClientT* because this object

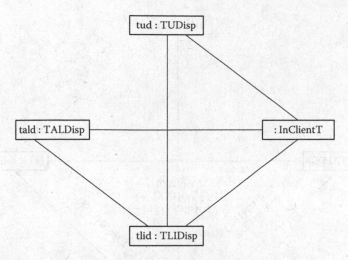

FIGURE 3.71
SIP INVITE client transaction collaboration diagram.

is dynamically created upon user request. It collaborates with the following three objects:

- *tud*, which represents the transaction user dispatcher
- *tald*, which represents the transaction layer dispatcher
- *tlid*, which represents the transport layer dispatcher

Similarly, we can zoom in on the particular collaboration diagram that illustrates a successful session establishment scenario (Figure 2.17) to provide the corresponding particular collaboration of the SIP INVITE client transaction with its surrounding objects (Figure 3.72). As already mentioned in Chapter 2, *req()* and *rsp()* designate requests and responses, respectively. More precisely, *req(INVITE)* is the SIP invite request, *rsp(1xx)* is the SIP provisional response, and *rsp(200)* is the SIP final response. Note that the first message 1:*req(INVITE)* sent from the object *tald* to the SIP INVITE client transaction object in Figure 3.72 corresponds to the fourth message 4:*req(INVITE)* sent from the object *tald* to the SIP INVITE client transaction object in Figure 2.17. Note also that Figure 3.72 shows only the messages exchanged among the objects shown in this figure, and that the sequence numbers of these messages are assigned accordingly.

Another particular collaboration that corresponds to an unsuccessful session establishment scenario is shown in Figure 3.73. This scenario is the same as the previous one up to the step number 6, when instead of the successful final response *rsp(200)*, the unsuccessful final response *rsp(300-699)* is

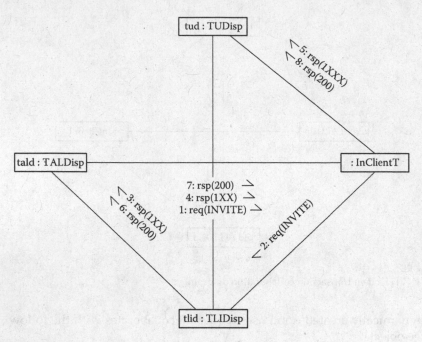

FIGURE 3.72
Successful session establishment collaboration diagram.

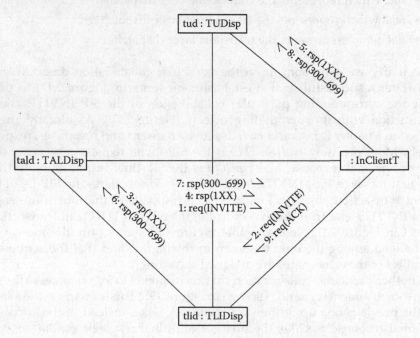

FIGURE 3.73
Unsuccessful session establishment collaboration diagram.

received. In step 7, *tald* forwards *rsp(300-699)* to the SIP INVITE client transaction, which in accordance with RFC 3261, forwards it toward the upper layer and sends the message *ACK* to the remote site. These two actions are performed in steps 8 and 9, respectively. Semantically equivalent sequence diagrams are shown in Figures 3.74 and 3.75. Figure 3.74 illustrates a successful session establishment, whereas Figure 3.75 shows an unsuccessful session establishment scenario.

Based on the SIP INVITE client transaction state transition graph (RFC 3261, page 127) we can construct the corresponding statechart diagram (Figure 3.76). This statechart diagram starts with the transition from the initial state to the state *Calling*, which is triggered by the reception of the signal (message) *req(INVITE)* from the transaction user (TU). The signal *req(INVITE)* models the original request SIP INVITE. During this transition, the SIP INVITE client transaction forwards the message *req(INVITE)* to the transport layer.

At the entrance to the state *Calling*, two timers are started, timer A (TA) and timer B (TB). The former corresponds to the time interval that must elapse before the response to the request INVITE can be received, whereas the latter limits the time interval during which the SIP INVITE client transaction waits for the response to the request INVITE. Initially, TA is set to the value T1 (estimated round-trip time, RTT, which is by default 500 ms) and TB is set to $64 \times T1$.

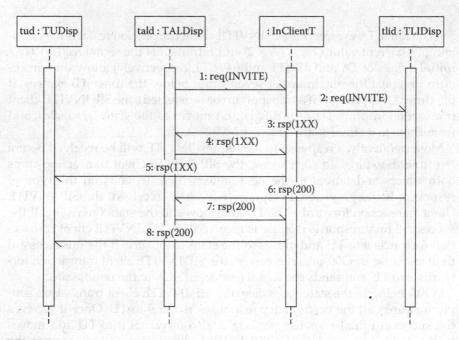

FIGURE 3.74
Successful session establishment sequence diagram.

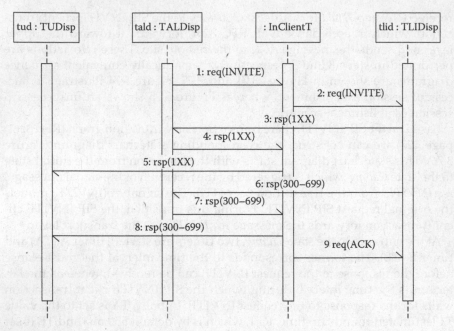

FIGURE 3.75
Unsuccessful session establishment sequence diagram.

If the timer TA expires, the SIP INVITE client transaction restarts it by doubling its current value (TA = TA × 2) and retransmits the signal *req(INVITE)*. Initial values of TA and TB (T1 and 64 × T1, respectively) allow this procedure to repeat the maximum of seven times before the timer TB expires. If the timer TB expires (or if a transport error is detected), the SIP INVITE client transaction informs TU accordingly and moves to the state *Terminated*, and from there to its final state.

Most frequently, a response to the request INVITE will be received before the timer B expires. In such a case, the SIP INVITE client transaction stops both timers and moves to the next state, which depends on the type of response. If the provisional response *rsp(1xx)* is received, the SIP INVITE client transaction forwards it to TU and moves to the state *Proceeding*. If the successful final response *rsp(2xx)* is received, the SIP INVITE client transaction forwards it to TU and moves to the state *Terminated*. If the unsuccessful final response *rsp(300-699)* is received, the SIP INVITE client transaction forwards it to TU and sends the signal (message) *ACK* to the remote site.

While being in the state *Proceeding*, the SIP INVITE client transaction simply forwards all the preliminary responses *rsp(1xx)* to TU. Once it receives the successful final response *rsp(2xx)*, it also forwards it to TU and moves to the state *Terminated*. If the SIP INVITE client transaction receives the

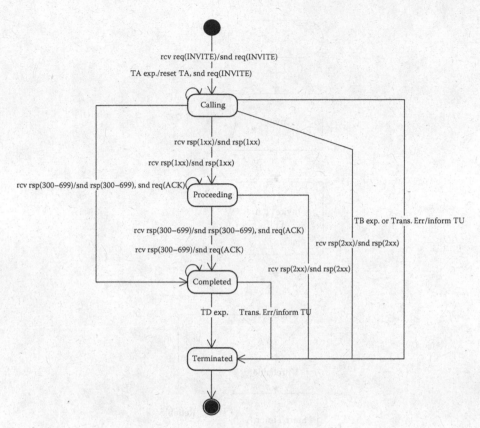

FIGURE 3.76
Statechart diagram of the SIP INVITE client transaction.

unsuccessful final response *rsp(300–699)* in the state *Proceeding*, it forwards that response to TU, sends the signal *req(ACK)* to the remote site, and moves to the state *Completed*.

At the entrance to the state *Completed*, the third timer, the timer D (TD), is started. While being in the state *Completed*, the SIP INVITE client transaction just confirms any unsuccessful final responses *rsp(300-699)* by sending the SIP message *ACK* to the remote site. If the SIP INVITE client transaction detects a transport error, it informs TU accordingly and moves to the state *Terminated*. Finally, when the timer D expires, the SIP INVITE client transaction finishes simply by moving to the state *Terminated*.

We finalize this example with the semantically equivalent SDL diagram, which, due to its size, is shown in the next four figures (in these figures, TPL stands for the transport layer and TU stands for the transaction user). Figures 3.77 through 3.80 illustrate the processing of events in the states *Calling, Proceeding, Completed,* and *Terminated,* respectively.

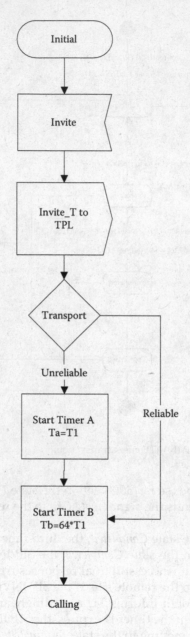

FIGURE 3.77
SDL diagram of the SIP INVITE client transaction, part I.

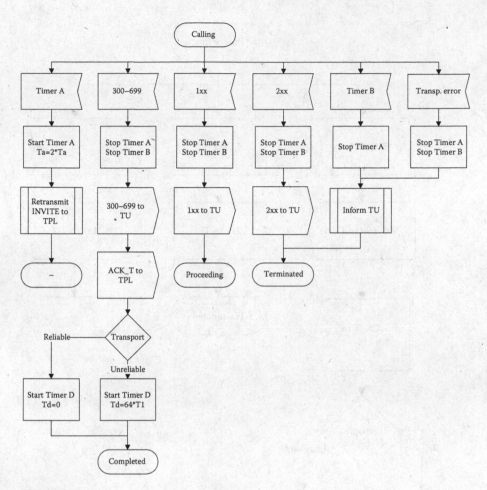

FIGURE 3.78
SDL diagram of the SIP INVITE client transaction, part II.

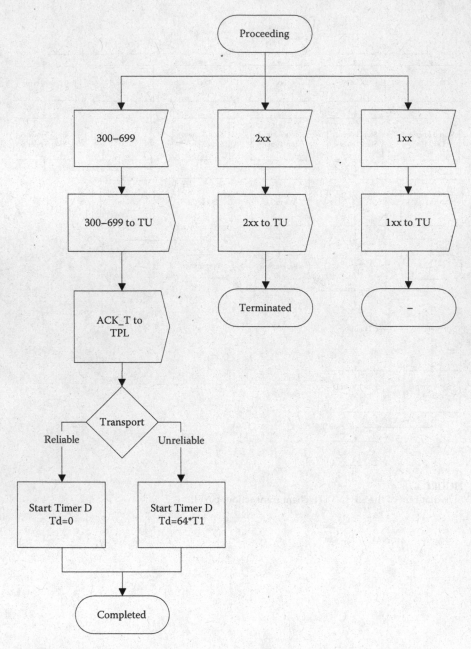

FIGURE 3.79
SDL diagram of the SIP INVITE client transaction, part III.

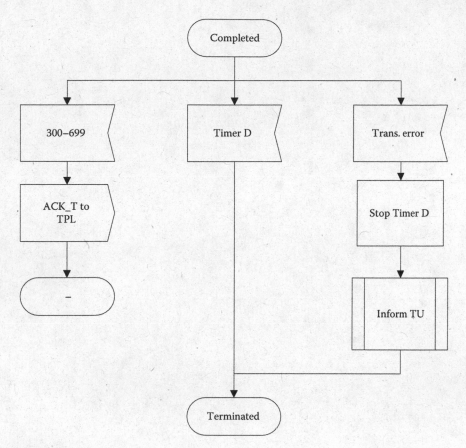

FIGURE 3.80
SDL diagram of the SIP INVITE client transaction, part IV.

References

Booch, G., Rumbaugh, J., and Jacobson, I., *The Unified Software Development Process*, Addison-Wesley, Reading, MA, 1998.

Willock, C., Deiß, T., Tobies, S., Keil, S., Engler, F., Schulz, S., *An Introduction to TTCN-3, Second Edition*, John Wiley & Sons, Chichester, West Sussex, UK, 2011.

Halsall, F., *Data Communications, Computer Networks and OSI*, Addison-Wesley, Reading, MA, 1988.

4

Implementation

The system **implementation** is a phase in engineering work that follows the system design phase. This phase consists of the following two steps:

- Transform a design model into the implementation model
- Transform the implementation model into a higher-level programming language code

A design model is given in the form of the corresponding UML (Booch et al., 1998) or SDL diagrams, which are the results of the previous phases of communication protocol engineering, i.e., requirements, analysis, and design. The implementation model takes the form of the corresponding UML component diagram. The output of the implementation phase is a set of source code modules, today most frequently in C/C++ or Java, which is also referred to as the implementation. This may sound confusing, but in reality, the correct meaning of the term is easily deduced from its context.

Logically, *implementation as a phase* of the production process is a well-defined mapping of a design model into a higher-level programming language source code. *Implementation as a product* is a result of this mapping. The attribute *well-defined* reflects the assumption that both detailed procedures and adequate tools are provided for transforming models into program source code. This well-defined mapping of a model into the program source code is referred to as **forward engineering** in UML terminology. Likewise, the reverse mapping of a program source code into the model is referred to as **backward engineering**.

In a mathematical sense, both the mapping of a program into the program source code and the result of that mapping (i.e., the implementation in both of its meanings) are not unique. Therefore, logically, more than one *correct implementation* exists for a given model of the communication protocol. Under the correct implementation, we assume an implementation that for given input produces expected outputs within the expected time frame, which is defined with the corresponding timers. We say for such implementation that it is compliant (conformant) with (to) the given model. The terms *compliant* and *conformant* are synonyms in this context. If the model has been standardized (e.g., by IETF or ITU-T), we say that the implementation is compliant with the standard.

The concept of forward and backward engineering is an intriguing one. Proponents of the model-based software development and various initiatives in Model-Driven Architecture (MDA) strongly believe that forward and backward engineering is possible, and they are putting forth tremendous efforts to make it real. Quite a number of commercially available tools are made with this goal in mind. The agile programming community is strongly opposed to it because their members believe that only the program source code is the complete specification of the system. From their point of view, only the set of test cases that successfully pass are proof that the implementation is correct.

Other groups also exist between these two extremes that are trying to close the gap between software modeling and programming (also called coding). For example, the creators of the StateWORKS® tool and the corresponding approach claim that although UML tool vendors made serious attempts to generate code from models, they are facing major difficulties, and that these tools can so far produce only header files or code skeletons. As an alternative, they introduced the notion of the totally complete models in an attempt to completely eliminate programming. The models in StateWORKS® are sets of virtual finite state machines (VFSMs) that run on top of the VFSM Executor, which is essentially an interpreter.

This book has a similar but different approach. We try to shrink the gap between communication protocol modeling and programming, both by making detailed models and by providing the FSM library, which forces programmers to transform models into code in a uniform way. This methodology makes forward engineering well defined. As already mentioned in the previous chapter, the FSM library provides two main classes, namely *FiniteStateMachine* and *FSMSystem*. The former is used to model and implement individual FSMs and the latter is used as their execution platform, which comprises common services and an event (message) interpreter.

When it comes to programming interpreters and FSM-related libraries, a broad spectrum of possible implementations exists, starting with the traditional structural or procedural solution, continuing with a series of mixed solutions, and ending with the object-oriented solutions of both static and dynamic type. This situation is justified by the fact that the implementation style depends highly on the type of target architecture. For example, if we consider a microcontroller as the target architecture, we are naturally forced to select a structural solution in the C/C++ programming language. If we consider more powerful architectures, in terms of resources, we may also take into consideration the object-oriented approaches supported by the C++ and Java programming languages.

In Section 4.1, we introduce the component diagrams, which are the means of making implementation models. We then illustrate a spectrum of possible finite state machine implementations, including the catalogued state design pattern (Gamma et al. 1995), which is explained in Section 4.3. After that, we cover the concepts and, most importantly, the design and implementation details of the FSM library (its reference manual is given in Chapter 6). We conclude this chapter with two implementation examples.

4.1 Component Diagrams

In Chapter 3 we were dealing with abstractions in the conceptual world. The design phase typically starts with exploration in the realm of interaction diagrams, where we try to get a better feeling of the system. We finish the design phase by defining the static structure and the complete behavior of the system in the corresponding class and activity, or statechart diagrams, respectively. At the end of the design phase, we also specify the deployment of individual software components by rendering the corresponding deployment diagrams.

In the implementation phase, we are materializing the design abstractions (such as classes, interfaces, and collaborations) into the components that live in the physical world. As already mentioned, a component is a physical and replaceable part of the system that realizes the given set of interfaces. What we actually do at the beginning of the implementation phase is pack the design abstractions into packages with well-defined interfaces, referred to as components. Examples of such packages are traditional binary object libraries, dynamically linkable libraries (DLLs), and executables; as well as tables, files, and documents.

The components and classes are very much alike. Both can:

- Realize a set of interfaces
- Participate in relations (dependencies, generalizations, and associations)
- Be nested
- Have instances
- Participate in interactions

The differences between the components and the classes are as follows:

- The former represents physical entities, whereas the latter is a conceptual abstraction, so they exist on different levels of abstraction.
- The former only has operations that are accessible through its interfaces, whereas the latter may have both operations and attributes.

The most important feature of the component is that it is replaceable. This means that we can substitute a component with another one without any influence on the system as a whole. This replacement is completely transparent to the users of the replaced component. A new component provides the same or perhaps even better services through the exact same interfaces.

We distinguish the following three types of components:

- The deployment components (already introduced in the context of deployment diagrams) are the parts of the executable system, such as executables and DLLs.
- The work product components are the artifacts of the development process (such as project settings or the source code) and data files that are used to build the deployment components.
- The executable components are the parts of the run-time system, e.g., DCOM and CORBA components.

We make the implementation models by rendering the component diagrams. The set of graphical symbols that are available for rendering component diagrams is shown in Figure 4.1. As usual, we select a symbol from the set of available symbols, drag and drop it onto the working sheet, and fill in the data related to its properties. The set of symbols available for rendering component diagrams is obviously a subset of the set of symbols available for rendering deployment diagrams. The properties of these symbols are explained in Chapter 3 (see Section 3.6).

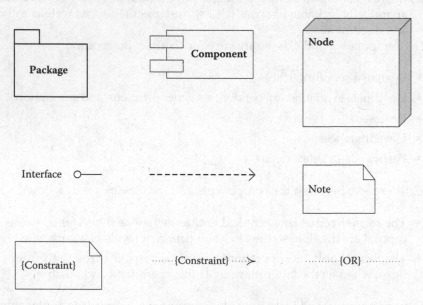

FIGURE 4.1
Set of symbols available for rendering component diagrams.

In communication protocol engineering, we are mainly using component diagrams for:

- Modeling APIs
- Modeling executables and libraries
- Modeling source code

Well-defined application programming interfaces (APIs) are some of the most important features of the well-structured software system. An API is an interface that is realized by one or more components. Being an interface, it actually defines a set of services. It represents a clear demarcation line between the service users and the service providers. The former receives the service without caring who is providing it. The same also holds true in the opposite direction: the latter provides the service without caring who receives it.

We may think of APIs as the programmatic seams of the system. We use them to connect more components together to create more complex systems. Each component is replaceable. We can replace it with another component whenever there is a need. The developers of the component that use some APIs do not care who or how it will be provided. They only care about how to fulfill the requirements for the component they are working on currently. Alternately, the system integrator must care that all of the needed components are provided and that they are compliant with their APIs.

Figure 4.2 illustrates the modeling of APIs by means of a very simple example. Imagine that we have been provided with the TCP/IP protocol stack packed as a dynamically linkable library, named *tcpipstack.dll*. It defines the API that comprises three interfaces, namely, *TCPSockets*, *UDPSockets*, and *IPInterface*. The first provides communication services over TCP ports, the second over UDP ports, and the third directly over IP.

Provided with such a component, we are now able to create a new component that uses it. For example, we can create the DLL *sip.dll* (Figure 4.3). This

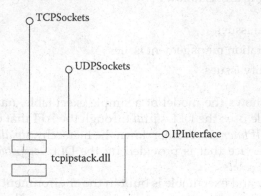

FIGURE 4.2
Example of a simple API.

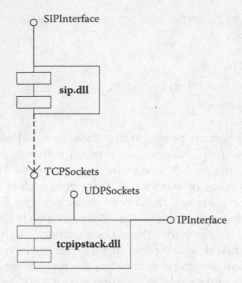

FIGURE 4.3
Example of a simple API user.

new component provides the SIP services through the interface *SIPInterface*.
The fact that *sip.dll* uses services provided through the interface *TCPSockets*
is modeled by connecting these two with the dependency relation.

Besides modeling APIs, we can use component diagrams to model executables and libraries. Generally, if the system under development comprises
more executables and associated object libraries, it may be wise to make a
model that illustrates their relationships. This is especially important if we
want to keep versioning and configuration management during the system
lifetime under control.

Modeling of executables and libraries can help in making the decision
regarding the physical partitioning of the system. The issues that affect this
decision-making are as follows:

- Technical issues
- Configuration management issues
- Reusability issues

Figure 4.4 shows the model of a simple executable, named *softphone.exe*.
This executable uses the DLL *sip.dll* through the API that comprises the single interface *SIPInterface*. Farther down the hierarchy, *sip.dll* receives the communication service that is provided by the DLL *tcpipstack.dll* through the
interface *TCPSockets*.

Each library and executable is built in the environment of a separate software project. Generally, a software project comprises the project configuration (settings) files, the source code files, and the object libraries. The source

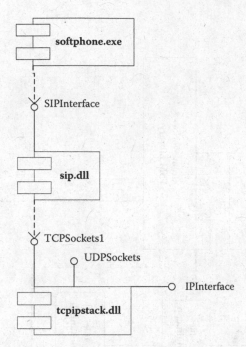

FIGURE 4.4
Model of a simple executable.

code files typically include the module declaration (header) and the module definition files. The developers try to logically organize these files into a file system structure by placing the related files into the same directory (folder).

In the case of complex projects, the corresponding directory tree can get rather ramified, and sometimes it may not be clear where to put new software modules. This can be especially confusing for the new members of the development team. Things get even worse when we must manage splitting and merging of groups of files as development paths fork and join.

In such cases, it is advisable to make a model of the software project, also referred to as the source code model. An example of such a model is shown in Figure 4.5. The executable *Main.exe* is built in accordance with the project definition file *Main.dsw*. Because the project comprises all the module headers and module definition files, the file *Main.dsw* has a dependency relation with all of them. (For clarity, only some of these dependencies are shown in Figure 4.5.)

Farther down the hierarchy, the source code files *AutomataA.cpp* and *AutomataB.cpp* use the header files *AutomataA.h* and *AutomataB.h*, respectively. Both of these header files use the header file *Constants.h*. Finally, all of the header and source code files, except *Constants.h*, use the framework *FSMLibrary*.

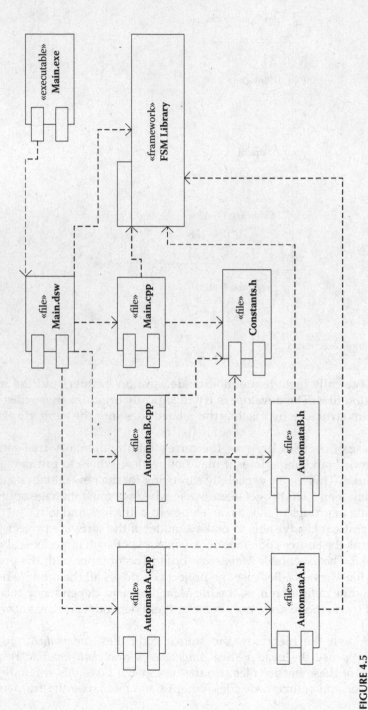

FIGURE 4.5
Model of a simple project.

4.2 Spectrum of FSM Implementations

As mentioned in Chapter 3, we model communication protocols as finite state machines (FSMs). A broad spectrum of various solutions exists for the implementation of FSMs. This section contains a short overview of only three, perhaps the most representative approaches to the implementation of FSMs. The complete treatment of all methodologies and corresponding tools is outside the scope of this book, and as an alternative we simply want to develop ideas by exploring different implementations of a simple FSM (counter by modulo 2). The goal is to familiarize the reader with this subject by showing what the problems are and how they can be tackled.

The three approaches to FSM implementation are illustrated by simple implementations of modulo 2 counters in the Java programming language. As already mentioned, communication protocol developers today mainly use C/C++ and Java, and the selection of the programming languages for certain projects mainly depends on the target platform. By mixing examples in Java and C/C++, we want to show that all these languages are applicable in the area of communication protocol engineering, and that the selection of a programming language is not the highest priority issue. Actually, we start with Java in Sections 4.2 and 4.3, switching to C++ later.

The state design pattern is a particular FSM implementation type that is special because it was catalogued by Gamma et al. in 1995. Because of that, it receives its own separate section. However, none of these four approaches are used later in this book. Instead, we introduce the FSM Library-based implementation paradigm, which is more like the state-of-the-art paradigm. In other words, first we show what is possible, and perhaps what is next, and then we turn to the current practice in communication protocol engineering.

Let us turn our attention to the subject of the implementation, a communication protocol. As already mentioned in Chapter 1, the communication protocol is defined with the syntax of its messages, the set of procedures (actions) that process the messages, and the set of reactions to exceptional events (timer and error management). In the programming world, they are modeled as finite state machines, also referred to as automata. Mathematically, the abstract automata are defined as

$$A = (X, Y, S, t, o, S_0)$$

where
$X = \{X_1, X_2, \ldots X_n\}$ is a set of input signals (input alphabet)
$Y = \{Y_1, Y_2, \ldots Y_m\}$ is a set of output signals (output alphabet)
$S = \{S_1, S_2, \ldots S_k\}$ is a set of states (state alphabet)
S_0 is the initial state
t is the transition function that maps the Cartesian product of $S \times X$ to S
o is an output function that maps the Cartesian product of $S \times X$ to Y

Abstract automata are typically illustrated in the form of a state transition graph. The example of the state transition graph in Figure 4.6 illustrates the counter by modulo 2, which is actually the example of a finite state machine we want to implement in Java. It is formally defined as follows:

$$C = (X, Y, S, t, o, S_0)$$

where
  $X = \{0, 1\}$
  $Y = \{0, 1, 2\}$
  $S = \{S1, S2, S3\}$
  $S_0 = S1$

The functions t and o are defined in Table 4.1.

The input and output alphabets comprise the signals $\{0, 1\}$ and the signals $\{0, 1, 2\}$, respectively. The automata can take one of the three possible states, namely, $S1$, $S2$, and $S3$. The initial state of the automata (S_0) is the state

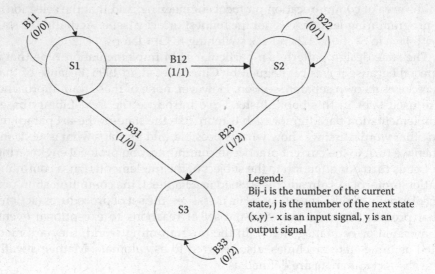

Legend:
Bij-i is the number of the current
state, j is the number of the next state
(x,y) - x is an input signal, y is an
output signal

FIGURE 4.6
Counter by modulo 2 state transition graph.

TABLE 4.1

The Counter by Modulo 2 Transition Table

Next State//Output Signal	Input Signal 0	Input Signal 1
State *S1*	1/0	2/1
State *S2*	2/1	3/2
State *S3*	3/2	1/0

S1. Both transition and output functions are defined in Table 4.1. The rows of this table correspond to the automata states (*S1*, *S2*, and *S3*), whereas the columns correspond to the input signals (0 and 1). The elements of Table 4.1 have the format s/y, where s corresponds to the next state number and y corresponds to the output signal.

The same information about the next state and the output signal is shown differently in the state transition graph (Figure 4.6). The arcs of the state transition graph are labeled as $B_{ij}(x/y)$, where i is the number of the current state, j is the number of the next state, x is the input signal that triggers the transition, and y is the output signal generated by the transition. The corresponding statechart diagram is shown in Figure 4.7.

The simplest but perhaps still the most frequently used FSM implementation is based on the structural or procedural approach. This implementation is made in the form of nested selection statements in higher-level programming languages. In the programming languages C/C++ and Java, we typically use *switch–case* statements for this purpose, because the control flow structures made with *if* and *else–if* statements are less readable.

Typically, the outermost *switch–case* statement selects a case that corresponds to the current state of automata. In the code paragraph that defines the processing of the current state, normally we use the second, nested *switch–case* statement, which selects the case that corresponds to the input signal. The program paragraph that corresponds to that input signal effectively performs the transition by creating the corresponding output signals and evolving to the next state. This evolution is made simply by updating the content of a variable that holds the identification of the current state (most frequently, this is just the index of the state).

Actually, the structure of the resulting program code is very similar to the program representation of SDL (SDL-PR), which was introduced in Chapter 3, and this fact was also mentioned there. Generally, communication

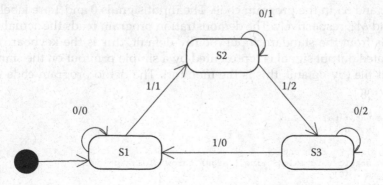

FIGURE 4.7
Counter by modulo 2 statechart diagram.

protocol implementation based on nested *switch–case* statements looks like the following:

```
switch(state) {
 case STATE_1:
  switch(message_code) {
   case MESSAGE_CODE_1:
   // processing of the message code 1 in the state 1
   break;
   case MESSAGE_CODE_2:
   // processing of the message code 2 in the state 1
   break;
   case MESSAGE_CODE_3:
   // processing of the message code 3 in the state 1
   break;
   ...
   default:
    // processing of the unexpected message in the state 1
    break;
  }
 case STATE_2:
  switch(message_code) {
   case MESSAGE_CODE_1:
   // processing of the message code 1 in the state 2
   break;
   case MESSAGE_CODE_2:
   // processing of the message code 2 in the state 2
   break;
   ase MESSAGE_CODE_3:
   // processing of the message code 3 in the state 2
   break;
   ...
   default:
   // processing of the unexpected message in the state 2
   break;
  }
 ...
 case STATE_N:
 ...
}
```

We illustrate this general scheme by applying it to the implementation of the counter by modulo 2 in Java. The three states of the counter are labeled as *S1*, *S2*, and *S3* in the program code. The input signals 0 and 1 are labeled as *M1* and *M2*, respectively. The demonstration program reads the actual input signals from the standard input file (by default, this is the keyboard). The generated output signal is represented by a simple printout on the standard output file (by default, this is the monitor). The demo program code is the following:

```
package automata;
import java.util.*;
import java.io.*;
public class Environment1 {
 public static void main(String[] args) throws IOException {
  char ch = '0';
  Automata1 a1 = new Automata1();
  System.out.println("This is the example of counter by modulo 2.");
  System.out.println("Automata evolution has started...");
```

```
while(true) {
  System.out.print("Enter input signal (0/1 and <ENTER>):");
  ch = (char)System.in.read();
  System.in.skip(2);
  if(((ch!='0') && (ch!='1'))) break;
  a1.processMsg(ch);
  }
 }
}
```

The demo program initially creates the object *a1*, an instance of the class *Automata1*, which is the structural and procedural implementation of the counter by modulo 2. After printing two welcome messages, it falls into an infinite *while* loop in which it prompts the user for the input signal and reads it. If the input signal is neither 0 nor 1, the demo program breaks the loop and terminates. Otherwise, it performs one step of the automata evolution by calling the procedure *processMsg()* of the object *a1*.

The Java code for the class *Automata1* is the following:

```
package automata;
public class Automata1 {
  private static final int S1 = 0;
  private static final int S2 = 1;
  private static final int S3 = 2;
  private static final char M1 = '0';
  private static final char M2 = '1';
  private int state=S1;
  public void processMsg(char msg) {
    switch(state) {
    case S1:
      switch(msg) {
      case M1:
        System.out.println("Output signal: 0");
        break;
      case M2:
        System.out.println("Output signal: 1");
        state = S2;
        break;
      default:
        break;
      }
     break;
   case S2:
    switch(msg) {
    case M1:
      System.out.println("Output signal: 1");
      break;
    case M2:
      System.out.println("Output signal: 2");
      state = S3;
      break;
    default:
      break;
    }
   break;
   case S3:
    switch(msg) {
    case M1:
      System.out.println("Output signal: 2");
      break;
```

```
 case M2:
  System.out.println("Output signal: 0");
  state = S1;
  break;
 default:
  break;
 }
 break;
default:
 break;
 }
}
}
```

The implementation above starts with the definition of the symbolic constants that correspond to the possible automata states (namely *S1*, *S2*, and *S3*) and valid input signals *M1* and *M2* (input signals 0 and 1). Next, we define the variable state that holds the current automata state and we set it to the value *S1* (the automata initial state).

The method *processMsg* starts with the *switch–case* statement that selects the further execution path depending on the content of the variable state (i.e., the current automata state). Three possible cases are found that are defined by the corresponding case clauses. Each of these clauses contains a further *switch–case* statement that distinguishes between two valid input signals, namely *M1* and *M2*. The nested *case* clause that corresponds to the particular input signal prints the message, which corresponds to the output signal, and updates the variable *state*, if the current state of the automata changes.

This example demonstrates the main advantage of the structural or procedural approach: simplicity, which yields greater performance in terms of execution speed. Another advantage is that we can easily construct a compiler or a code generator that generates such implementations (a good example that justifies this claim is SDL-PR). The main disadvantage of this approach is its bad scalability, which becomes evident in the case of large-scale implementations, i.e., implementations of automata that have a large number of states and state transitions.

The code size for such program implementations increases linearly with the number of states and the number of state transitions. Another disadvantage of this approach is that it is monolithic which implies that it is static regarding the possible need to change the automata, either by adding new, or deleting the existing states, or by adding or deleting state transitions.

In this type of implementation, the structure of the automata (its vertex and arcs) is built into the machine code of the implementation (hard-coded). We say that the input signal processing flow is governed by the structure of the machine code. If we want to add or delete a state or a state transition, we must change the program code, recompile it, and install the new version on the target platform. Most frequently, the installation procedure requires the system to be restarted at its end. Restarting the system means that effectively it will not be operational for a certain short interval of time. The problem is that some types of systems, such as nonstop systems, may not tolerate restarts no matter how short the time interval is.

Some systems try to make restarts allowable by providing processor tandem configurations. Typically, in such a system, one of the processors continues the normal operation while the other restarts after an update. In that case, we have a synchronization problem, which of course can be solved but could become rather complex. Generally, system restarts are problematic and should be handled with special care.

On the other end of the spectrum of FSM implementations, we have the diametrical approach to FSM implementation in which the structure of the automata is not defined by the program control flow, but rather with the corresponding data structure. The simple interpreter uses this data structure to process the incoming events (messages), therefore it is referred to as an event interpreter. The data structure implementations in assembler and C programming languages are built from lists and lookup tables.

The automata evolution is driven by the incoming events. Each input event triggers one step of the evolution. The event interpreter carries out the evolution step by traversing the data structure to determine the current state and the state transition that corresponds to the input event type. In contrast to this common part of the message processing flow—which is directed by the data structure—program parts that correspond to particular reaction tasks are dedicated routines that perform specific functions, which cannot be generalized.

Figure 4.8 illustrates the FSM implementation based on the event interpreter and the data structure that defines the FSM structure (essentially, the state transition graph). New, incoming events (messages) are added at the end of the message queue (see the top left corner of Figure 4.8). The interpreter takes the messages from the head of the message queue and processes them by using the data structure, which comprises

- An automata control table
- An automata state table
- A list of valid events (one such list exists for each automata state)

The automata control table is assigned to automata to store its current state and optionally some of its additional attributes. The automata state table is a lookup table that maps the state index onto the address of the corresponding list of valid events in that state. The elements of this list contain the complete information necessary and sufficient to perform the state transition from the current state to the next state, which is determined by the event type. This information is stored in the following fields:

- *event ID*: holds the event type to which this element corresponds
- *task address*: contains the pointer to the corresponding routine (procedure)

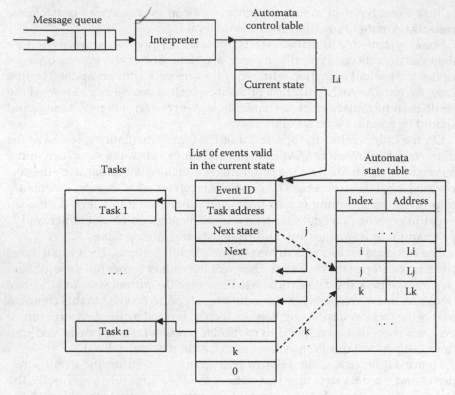

FIGURE 4.8
Event interpreter and the data structure that defines the FSM structure.

- *next state*: stores the index of the next state
- *next*: contains the pointer to the next element in the list

The event interpreter processes the message through the following steps:

- Get the message from the head of the message queue.
- Locate the automata control table by examining the content of the message header (the message destination field, in particular).
- Read the current state and locate the corresponding list of valid events by looking up the automata state table.
- Determine the event type by examining the content of the message header (the message code field, in particular) and locate the corresponding element in the list of valid events (ignore the event if such an element does not exist).
- Perform the task by calling the corresponding task routine as a subroutine (procedure).

- Read the index of the next state from the field *next state*.
- Update the field *current state* by storing the index of the next state into the field *current state*.

The advantage of this approach is that we can construct a compiler that transforms the design FSM model into the corresponding data structure and the set of task routines. The automatic translation performed by the compiler increases the probability that the implementation is compliant with the design model and, therefore, that it is correct. Moreover, the routine performed by the event (described above) is fairly simple and short. The price that is paid for the correctness and simplicity is poor performance. The decrease in the processing throughput is proportional to the number of memory accesses to the corresponding elements of the data structure.

Two characteristics of this approach are not obvious from Figure 4.8 and require further explanation. The first characteristic is universality. Since the FSM structure is built into the corresponding data structure, the event interpreter routine is completely independent from it. The event interpreter always repeats the same routine. This is the same for all FSMs. Therefore, this routine is universal in contrast to the implementation with nested **switch–case** statements, which implement just one particular FSM. This characteristic is especially important from the point of software maintenance. If we want to change the FSM structure by adding or deleting states or state transitions, we must update the data structure. There is no need to change the simple interpreter routine at all.

The second characteristic of the event interpreter-based approach is that it enables sharing of common tasks between more state transitions. In principle, this is also possible in the nested **switch–case**-based approach by introducing common functions, which are called from the corresponding case program clauses, but this is seldom used by their practitioners. In the event interpreter-based approach, this possibility becomes more apparent and is, therefore, implemented because tasks are already specified as procedures (subroutines) rather than *case* program clauses.

Because of task sharing, the number of tasks may generally be smaller than the number of state transitions. We can also organize tasks hierarchically, such that higher-level tasks call their subordinate tasks. This makes it possible to implement more complex tasks by using simple primitives. Such organization has the following advantages:

- Better performance in terms of code size
- Enables dynamic mutation of tasks

By exploiting these characteristics in environments with dynamic loaders, such as Java, we can implement dynamically reconfigurable automata. The automata in such environments change during normal system operation, and those changes do not demand any system restarts. In such environments, it is desirable to use the object-oriented approach and to define the FSM structure with the set of objects rather than with a data structure, such as the one previously described. The event interpreters in such implementations interact with the objects that materialize the FSM structure instead of using the traditional data structures.

The following code illustrates FSM structure modeling with the group of classes written in Java:

```java
package automata2;
import java.util.*;
import java.io.*;

class Task {
 public int id;
 public Task(int ident) {id=ident;}
 public void processMsg() {System.out.println(id);}
}

class Branch {
 private String msgcode;
 private Task task;
 private String nextstateid;

 public Branch(String msg, Task tsk, String nextsts) {
  msgcode=msg;
  task=tsk;
  nextstateid=nextsts;
 }
 public String getMsgCode() {return msgcode;}
 public Task getTask() {return task;}
 public String getNextStateId() {return nextstateid;}
}

class State {
 private String stateid;
 public Set setofbranches;

 public State(String id,Set branches) {
  stateid=id;
  setofbranches=branches;
 }
 public String getStateId() {return stateid;}
 public Set getSetOfBranches() {return setofbranches;}
}

class AStructure {
 private String automataid;
 private Set setofstates;

 public AStructure(String id,Set states) {
  automataid=id;
  setofstates=states;
 }
 public String getAutomataId() {return automataid;}
 public Set getSetOfStates() {return setofstates;}
}

class Automata {
 protected AStructure structure;
 protected String stateId;
```

```
protected State initial;

public Automata(AStructure str,String id,State s) {
  structure = str;
  stateId=id;
  initial=s;
}
public void processMsg(String msg) {
  State currentS = initial;
  Iterator iterA =
structure.getSetOfStates().iterator(); while(iterA.hasNext()) {
    State eachS = (State)iterA.next();
    if(eachS.getStateId().equals(stateId)) {
      currentS=eachS;
      break;
    }
  }
  Iterator iterS =
currentS.getSetOfBranches().iterator(); while(iterS.hasNext()) {
    Branch eachB = (Branch)iterS.next();
    if(eachB.getMsgCode().equals(msg)) {
      Task t=eachB.getTask();
      t.processMsg();
      stateId=eachB.getNextStateId();
      break;
    }
  }
}
}
```

The class *Task* models the task that is performed during the transition from the current state to the next state. The task identification is stored in the class field *id*. The user of the class *Task* specifies the particular task identification as the parameter of the class constructor. The default message processing function, named *processMsg()*, just prints the task identification to the standard output file.

The class *Branch* models the arc of the state transition graph. The attributes of the state transition are the message code that triggers the state transition, the task that is performed during the state transition, and the identification of the next stable state. The corresponding fields are named *msgcode, task,* and *nextstateid*, respectively. These fields are set by the class constructor. The current content of these fields is returned by the functions *getMsgCode(), get-Task(),* and *getNextStateId()*, respectively.

The class *State* models a single FSM state. The state attributes are the state identification and the set of the outgoing state transitions (the target state is irrelevant; it can be this state or some other state). The corresponding class fields are named *id* and *branches*, respectively. Their content is set by the class constructor and returned by the functions *getStateId()* and *getSetOfBranches()*, respectively.

The class *AStructure* models the FSM structure. Its attributes are the automata identification and the corresponding set of states. The corresponding class fields are *automataid* and *setofstates*. The class constructor gets particular values for these fields through its parameters. The functions *getAutomataId()* and *getSetOfStates()* return the current values of these fields.

Finally, the class *Automata* models the complete FSM. Its attributes are the FSM structure (essentially the set of sets of state transitions), the current state identification, and the initial state identification. The corresponding class fields are named *structure*, *stateId*, and *initial*, respectively. These fields are set by the class constructor.

The function *processMsg(String msg)* is the event interpreter. The input argument *msg* is the message, which triggered the state transition. The interpretation starts with the iteration through the set of states to locate the object that corresponds to the FSM current state (its identification is stored in the field *stateId*). This is a typical object-oriented approach, which avoids the unpopular *switch–case* and similar selection statements. Principally, this first iteration is really not needed and can be easily eliminated by saving the current state object instead of the current state identification. However, the first iteration is intentionally kept to make the example more informative by showing how we can use two subsequent iterations to search through the set of sets of state transitions.

The second iteration searches through the set of state transitions that correspond to the current state to locate the state transition that corresponds to the input message *msg*. After locating the state transition, it gets the object that corresponds to the state transition task and calls its *processMsg()* functions, which, in turn, prints the task identification to the standard output file.

From the program code given above, the classes *Task*, *Branch*, *AStructure*, and *Automata* are obviously generic and can be used for the construction of any FSM. Besides that, this solution enables the design and implementation of dynamically reconfigurable FSMs, because sets in Java can be dynamically updated with the corresponding task object dynamically loaded and unloaded.

We illustrate the applicability of this set of classes with the following implementation of the counter by modulo 2 in Java (the corresponding overall class architecture is shown in Figure 4.9):

```
class Task0 extends Task {
 public Task0(int ident) {super(ident);}
 public void processMsg() {System.out.println("0");}
}

class Task1 extends Task {
 public Task1(int ident) {super(ident);}
 public void processMsg() {System.out.println("1");}
}

class Task2 extends Task {
 public Task2(int ident) {super(ident);}
 public void processMsg() {System.out.println("2");}
}

class Automata2 {
 public static void main(String[]args) throws IOException {
  Automata a2 = makeAutomata();
  char ch;
  String msg;
  System.out.println("This is the example of counter by modulo 2.");
```

```
  System.out.println("The automata evolution has started...");
  while(true) {
    System.out.print("Enter input signal (0/1 and <ENTER>): ");
    ch = (char)System.in.read();
    System.in.skip(2);
    if(((ch!='0') && (ch!='1'))) break;
    if(ch=='0') msg="0"; else msg="1";
    a2.processMsg(msg);
  }
}
private static Automata makeAutomata() {
  Branch b11 = new Branch("0",new Task0(0),"0");
  Branch b12 = new Branch("1",new Task1(1),"1");
  Set s1 = new HashSet();
  s1.add(b11); s1.add(b12);
  state S1 = new State("0",s1);

  Branch b22 = new Branch("0",new Task1(1),"1");
  Branch b23 = new Branch("1",new Task2(2),"2");
  Set s2 = new HashSet();
  s2.add(b22); s2.add(b23);
  State S2 = new State("1",s2);

  Branch b33 = new Branch("0",new Task2(2),"2");
  Branch b31 = new Branch("1",new Task0(0),"0");
  Set s3 = new HashSet();
  s3.add(b33); s3.add(b31);
  State S3 = new State("2",s3);

  Set a = new HashSet();
  a.add(S1); a.add(S2); a.add(S3);
  AStructure as = new AStructure("0",a);

  Automata au = new Automata(as,"0",S1);
  return au;
}
}
```

At the beginning of this example, we define the application-specific tasks, namely, *Task0*, *Task1*, and *Task2*, which are responsible for printing the counter by modulo 2 outputs (0, 1, and 2, respectively). Note that the number of tasks (three) is smaller than the number of state transitions (six) in this particular example. The application-specific *processMsg()* functions are defined by overriding the default functions.

The definitions of the classes *Task0*, *Task1*, and *Task2* are followed by the definition of the class *Automata2*, which comprises two public functions: *main()* and *makeAutomata()*. The function *main()* starts by calling the function *makeAutomata()*, which, in turn, returns the counter by the modulo 2 object, named *a2*. After that, it falls into an infinite *while* loop in which it reads the standard input file. If the input character is neither "0" nor "1," it breaks the loop and the program terminates. Otherwise, it converts an input character into the corresponding string ("0" and "1," respectively) and passes it as an input event to the event interpreter.

The function *makeAutomata()* constructs individual state transitions (instances of the class *Branch*), individual states (instances of the class *State*),

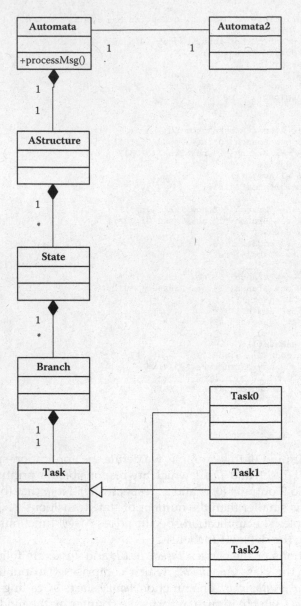

FIGURE 4.9
Static structure used in the second approach to the FSM implementation.

the counter by modulo 2 structure (an instance of the class *AStructure*), and the counter by modulo 2 itself (an instance of the class *Automata*). It first constructs the state transition *b11*, which for the input "0" moves the FSM from the state *S1* to the same state, and during that transition it performs the task *Task0*. Similarly, it constructs the state transition *b12*, which for the input "1" moves the FSM from the state *S1* to the state *S2*, and during that transition it performs the task *Task1*. Next, it constructs the set of state transitions *s1* and the state *S1*.

Likewise, this function constructs the state transitions *b22* and *b23* and the state *S2*, as well as the state transitions *b33* and *b31* and the state *S3*. Finally, it constructs the structure of the counter by modulo 2, named *as*, and the counter by modulo 2, named *au*.

The third approach to FSM implementation, from the broad spectrum of implementations, is illustrated next. In this approach, we define the FSM structure with the corresponding class hierarchy and the set of lookup tables that map FSM inputs into the corresponding state transitions. This approach also uses message interpretation and is therefore universal, like the previous one, but it yields much better performance that is comparable with the performance of the first approach (nested **switch–case** statements).

The first idea behind this concept is to model each FSM stable state with the class that is derived from the basic class *State*. The second idea is to consider a state transition (represented with the corresponding arc of the state transition graph) as a transient (i.e., unstable) state. Each state transition is modeled with a class that is derived from the class that represents its originating stable state.

These two ideas lead to a class hierarchy with two hierarchical levels. The root of the class hierarchy is the basic class *State*. The first level of hierarchy defines the FSM stable states, whereas the second level of hierarchy defines its unstable states, i.e., state transitions.

We illustrate this approach with the example of counter by modulo 2. The corresponding class hierarchy is shown in Figure 4.10. The first hierarchy level defines the FSM stable states *S1*, *S2*, and *S3*. All of these are derived from the basic class *State*. The second level defines FSM state transitions *B11*, *B12*, *B22*, *B23*, *B33*, and *B31*. Notice that *B11* and *B12* are derived from their originating state *S1*. Similarly, *B22* and *B23* are derived from *S2*, and *B33* and *B31* are derived from *S3*.

The third idea behind this approach is that FSM evolution takes place by traversing the class hierarchy tree and by using polymorphism, one of the most powerful abstractions of object-oriented programming. Concretely, the event interpreter performs the following steps:

- Use the FSM input message (signal) and the lookup table (map), which are associated with the FSM current state, to determine the corresponding unstable state (state transition).
- Perform the application-specific task by calling the message processing function defined within the class that models the corresponding unstable state.
- Move the FSM into its next stable state.

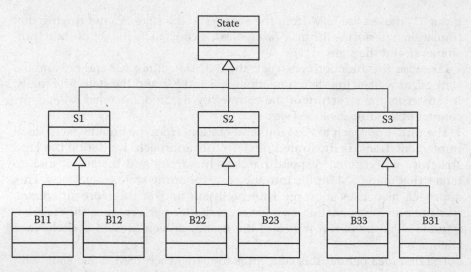

FIGURE 4.10
Counter by modulo 2 state class hierarchy.

The class hierarchy for the counter by modulo 2 is defined with the following Java module:

```java
package automata;
import java.util.*;

class State {
 public State msgToBranch(String msg) {return new State();}
 public State processMsg() {return new State();}
}

class S1 extends State {
 public State msgToBranch(String msg) {
  return Structure3.getBranch("0",msg);
 }
}
class S2 extends State {
 public State msgToBranch(String msg) {
  return Structure3.getBranch("1",msg);
 }
}
class S3 extends State {
 public State msgToBranch(String msg) {
  return Structure3.getBranch("2",msg);
 }
}

class B11 extends S1 {
 public State processMsg() {
  System.out.println("Output: 0");
  return new S1();
 }
}
class B12 extends S1 {
```

```
public State processMsg() {
System.out.println("Output: 1");
return new S2();
}
}

class B22 extends S2 {
public State processMsg() {
System.out.println("Output: 1");
return new S2();
}
}
class B23 extends S2 {
public State processMsg() {
          System.out.println("Output: 2");
          return new S3();
}
}

class B33 extends S3 {
public State processMsg() {
System.out.println("Output: 2");
return new S3();
}
}
class B31 extends S3 {
public State processMsg() {
System.out.println("Output: 0");
return new S1();
}
}

public class Automata3 {
private State state;

public Automata3() {
state = new S1();
}
public void processMsg (char chmsg) {
String msg;
if(chmsg=='0') msg="0"; else msg="1";
state = state.msgToBranch(msg);
state = state.processMsg();
}
}
```

The basic class *State* has two default functions, *msgToBranch()* and *processMsg()*. Both functions return an instance of the class *State*. The fact that the instance of the class derived from the class *State* is also considered to be the instance of the class *State* that enables the event interpreter to employ polymorphism. We will return to this point shortly.

The function *msgToBranch()* is responsible for mapping the FSM input message into the corresponding state transition object. The input message in this simple example is a one-character string ("0" or "1"). The function can return any instance of the basic class *State*, but normally in this example, it should return the instance of the class *B11*, *B12*, *B22*, *B23*, *B33*, or *B31*.

The function *processMsg()* carries out the application-specific task for the given input message. It returns the FSM's next stable state. The idea is that the FSM dynamically changes its behavior. The FSM is in a certain state, either stable or unstable, at any point in time, but it is always represented by a single object. That object is actually returned by one of these two functions, which are called in the course of FSM evolution.

Next, we define the classes that model the FSM stable states, namely, *S1, S2,* and *S3*. Each of these classes extends the basic class *State* and overrides the default function *msgToBranch()* with the application-specific one. These particular functions actually delegate their responsibility to the function *getBranch()* of the class *Structure3* by passing their identification ("0," "1," and "2" for *S1, S2,* and *S3*, respectively) and the input message to it. More precisely, these simple functions just return the unstable state object that is provided by the function *getBranch()* to their caller, and that is the event interpreter.

The stable state classes are followed by the classes that model the FSM unstable states, namely, *B11, B12, B22, B23, B33,* and *B31*. Each of these classes extends the corresponding stable state class and overrides the default function *processMsg()*, which each individual class inherits from the basic class *State*, with the application-specific one. These particular functions perform the application-specific tasks and return the corresponding next stable state object (*S1* for *B11* and *B31*, *S2* for *B12* and *B22*, and *S3* for *B23* and *B33*). The application-specific tasks in this simple example are implemented as the corresponding print statements to the standard output file.

The FSM is modeled with the class *Automata3*. This class has a single attribute named *state*, which is set by the class constructor to the FSM initial stable state, namely *S1*. Later, during the FSM evolution, it changes and can become any FSM state, either stable or unstable.

The class *Automata3* has a single function, named *processMsg()*, that is the FSM event interpreter. This function performs one state transition in two steps. In the first step, it calls the function *msgToBranch()* of the FSM current stable state object. This effectively starts the state transition by moving the FSM from its current stable state to the unstable state that corresponds to the input message. In the second step, the event interpreter calls the function *processMsg()* of the FSM unstable state, which performs the application-specific task and returns the next FSM stable state object. This effectively completes the state transition. Interestingly, the state class hierarchy in this approach is completely application-specific, whereas the event interpreter is very simple and generic and therefore can be reused in the implementations of other FSMs.

The following utility classes support the mapping of input messages to the corresponding state transitions (unstable state objects):

```
package automata;
import java.util.*;

class MapContainer {
 private String identification;
 private Map map;

 public MapContainer(String id,Map m){
  identification = id;
  map = m;
```

```
  }
  public String getId() {return identification;}
  public Map getMap() {return map;}
}

public class Structure3 {
  private static Set maps;

  public void setMaps(Set m) {
   maps = m;
  }
  public static State getBranch(String id,String msg) {
   Map m = new HashMap();
   Iterator iter = maps.iterator();
   while(iter.hasNext()) {
    MapContainer each = (MapContainer)iter.next();
    if(each.getId().equals(id)) {
     m = each.getMap();
     break;
    }
   }
   return (State)m.get(msg);
  }
}
```

The class *MapContainer* stores the map identification and the map itself in the attributes *identification* and *map*, respectively. These attributes are set by the class constructor. Their current content is available through the corresponding *get* functions.

The class *Structure3* contains a set of maps for all FSM stable states. This set is established by the function *setMaps()* and is searched by the function *getBranch()*. The input parameters of the function *getBranch()* are the map (i.e., stable state) identification and the input message. The function *getBranch()* iterates through the set of map containers, locates the one with the given identification, uses the located map to get the state transition that corresponds to the input message, and returns it to its caller.

An important feature of this approach is that it is based on Java sets and maps, which makes it an ideal environment for making dynamically reconfigurable FSMs as Java sets and maps can be dynamically updated. For example, if we want to add a new state transition *B21*, it would be sufficient to write, compile, and dynamically load a new class *B21* that represents it, and add the corresponding entry in the map that is associated to the FSM stable state *S2*.

Because the current Java version does not support a map of maps, the solution for mapping input events to the corresponding state transitions presented here is based on the usage of a set of maps. It is worth mentioning that an environment with a map of maps would enable top performance implementations based on two connected mappings. The key for the first mapping would be the FSM current stable state, whereas the key for the second mapping would be the input message. The performance of such implementations would be even better than the performance of the implementations based on nested **switch–case** statements.

The class *Environment3* uses the previously defined classes and demonstrates their usability. The corresponding Java code is the following (the overall class architecture is shown in Figure 4.11):

```
package automata;
import java.util.*;
import java.io.*;

public class Environment3 {
 public static void main(String[] args) throws IOException {
  char ch = '0';
  Automata3 a3 = new Automata3();

  Map m1 = new HashMap();
  m1.put("0",new B11()); m1.put("1",new B12());
  MapContainer M1 = new MapContainer("0",m1);

  Map m2 = new HashMap();
  m2.put("0",new B22()); m2.put("1",new B23());
  MapContainer M2 = new MapContainer("1",m2);

  Map m3 = new HashMap();
  m3.put("0",new B33()); m3.put("1",new B31());
  MapContainer M3 = new MapContainer("2",m3);

  Set maps = new HashSet();
  maps.add(M1); maps.add(M2); maps.add(M3);

  Structure3 st3 = new Structure3();
  st3.setMaps(maps);

  System.out.println("This is the example of counter by modulo 2.");
  System.out.println("The automata evolution has started...");
  while(true) {
   System.out.print("Enter input signal (0/1 and <ENTER>): ");
   ch = (char)System.in.read();
   System.in.skip(2);
   if(((ch!='0') && (ch!='1'))) break;
   a3.processMsg(ch);
  }
 }
}
```

The function *main* starts by creating the object *a3*, an instance of the counter by modulo 2. It then creates all the necessary maps and map containers, the set of maps named *maps*, the object *st3*, and an instance of the class *Structure3*. After this, it sets the set of maps by calling the function *setMaps()* and falls into an infinite *while* loop in which it reads FSM input messages and calls the event (message) interpreter until the user enters a signal that is neither "0" nor "1."

The keys for searching Java maps in this simple example are just simple strings ("0" and "1"). This Java map is a rather powerful abstraction because its key may be any class whose instances are comparable. This makes it possible to model real communication protocol messages with such classes and to build Java maps for them. Once we model the messages by the corresponding objects, FSM objects can interact with them in an object-oriented fashion.

If we want to provide a full object-oriented treatment of communication protocol messages, we must provide the corresponding serialization functions. Two

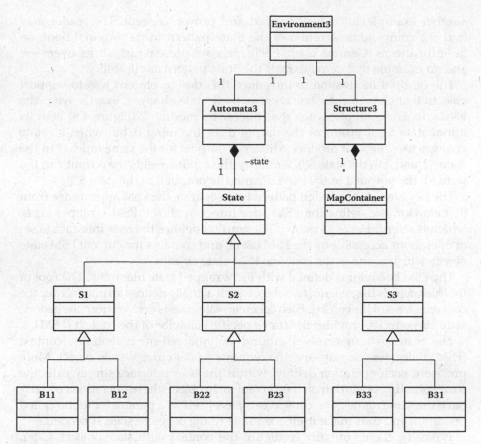

FIGURE 4.11
The static structure used in the third approach to the FSM implementation.

types of these functions are actually used. The first type is used for converting an object into a series of octets that can be transported over the communication line. The second type performs the reverse operation by converting the received series of octets into the corresponding object. If we do not provide these serialization functions, we are forced to operate directly on numbers and use **switch–case** and similar statements unpopular in the object-oriented world.

4.3 State Design Pattern

The State design pattern is one of the approaches to FSM implementation. As previously mentioned, the State pattern is shown in a separate section because it was catalogued by Gamma et al., and therefore it is not just another example, but a well-defined and proven concept. The reader may find the complete description of the State pattern in the original book on design patterns (Gamma et al., 1995). Here we present just a brief overview and an example that demonstrates the State pattern applicability.

The original motivation to introduce this design pattern was to support objects that change their behavior as their state changes, exactly what the FSMs do. For example, when the counter by modulo 2 (Figure 4.6) is in its initial state *S1*, it produces the output 0 for the input 0, but when its state changes to *S2* or *S3*, it produces different outputs for the same input (1 in the state *S2*, and 2 in the state *S3*). Similarly, the input 1 yields the output 1 in the state *S1*, the output 2 in the state *S2*, and the output 0 in the state *S3*.

The key idea of this design pattern is to separate the FSM appearance from its behavior. We define the FSM appearance with the FSM wrapper class, which is referred to as a context. The **context** defines the user interface (a set of operations accessible by the FSM users) and contains the current FSM state object, which is one of the concrete FSM state objects.

The FSM behavior is defined with the wrapped state hierarchy. The root of this hierarchy is the generic state class, which actually defines an interface for the concrete states of the context. Each concrete state class is derived from the generic state class, and it provides the state-specific behavior of the context (FSM).

The State pattern revolves around polymorphism. Essentially, context (FSM) delegates the state-specific requests to the current state object. More precisely, each operation defined within the user interface simply calls the corresponding operation on the current state object (these operations usually have the same name). The context can pass itself as a parameter to the called operation and thus make itself accessible to the concrete state, if needed.

Typically, clients initially configure the context with state objects. Later, during the normal system operation, clients do not deal with state objects directly. Notice that either the context class or the concrete state subclass can change the context current state. Therefore, the FSM transition logic can be centralized, distributed, or hybrid.

According to the authors, the State pattern consequences are the following:

- It localizes state-specific behavior.
- It makes state transitions explicit.
- State objects can be shared.

At the end of this short overview of the State pattern, we illustrate its applicability with the simple example of a State pattern-based implementation of the counter by modulo 2. The corresponding class diagram is shown in Figure 4.12. The context in this example is the class *Automata4*. The attribute *state* holds the current FSM state object. The key function *processMsg()* delegates message processing to the current FSM state object by calling its function *processMsg()*.

The generic state class *State* defines a simple interface, which comprises a single function, *processMsg()*. Generally, such a function would define the

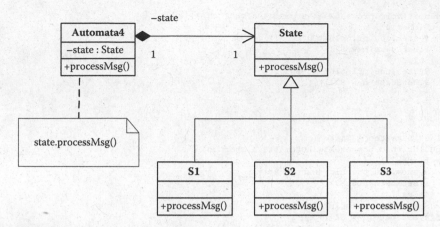

FIGURE 4.12
Static structure used by the State design pattern.

default FSM behavior, which can then be overridden in the concrete substate classes. In this simple example, as we will shortly see, no such behavior is allowed, and therefore the corresponding operation is simply empty.

The concrete substate classes *S1*, *S2*, and *S3* are derived from the generic state class *State*. Each of these classes provides a state-specific behavior by overriding the function *processMsg()* with its own particular definition. The corresponding code in Java is the following:

```java
package automata4;
import java.util.*;

public class Automata4 {
 private State state;
 public Automata4() {state = new S1();}
 public void setState(State s) {state = s;}
 public void processMsg(char msg) {
  state.processMsg(this,msg);
 }
}

class State {
 public void processMsg(Automata4 a,char ch) {
 }
}

class S1 extends State {
 public void processMsg(Automata4 a,char ch) {
  if(ch=='0') {
   System.out.println("Output 0");
   a.setState(new S1());
  } else {
   System.out.println("Output 1");
   a.setState(new S2());
  }
 }
}

class S2 extends State {
```

```
public void processMsg(Automata4 a,char ch) {
 if(ch=='0') {
  System.out.println("Output 1");
  a.setState(new S2());
 } else {
  System.out.println("Output 2");
  a.setState(new S3());
 }
}
}

class S3 extends State {
 public void processMsg(Automata4 a,char ch) {
  if(ch=='0') {
   System.out.println("Output 2");
   a.setState(new S3());
  } else {
   System.out.println("Output 0");
   a.setState(new S1());
  }
 }
}
```

The definition of the class *Automat4* begins with the definition of the field *state*, which is used to store the FSM current state object. The class constructor sets this field to the FSM initial state object, which is an instance of the class *S1*. The function *setState()* is used by the FSM concrete state objects to change the FSM state (an example of distributed transit logic). The function *processMsg()* simply calls the corresponding function on the FSM current state object.

The class *State* defines a simple state interface with just one function—*processMsg( )*—which is empty because this example has no default behavior. The class *S1* is an example of a concrete substate class. It defines the *S1*-specific FSM behavior by overriding the function *processMsg()* that it inherits from the base class *State*. This function checks whether the input signal is 0 or 1, prints the corresponding output signal, and changes the FSM state by calling the function *setState()*. We made the context accessible by passing it as a parameter to the function *processMsg()*.

The following Java code creates the working environment for this example (given without the comments because a similar code is already explained in a previous section):

```
package automata4;
import java.util.*;
import java.io.*;

public class Environment4 {
 public static void main(String[] args) throws IOException {
  char ch = '0';
  Automata4 a4 = new Automata4();
  System.out.println("This is the example of counter by modulo 2.");
  System.out.println("The automata evolution has started...");
  while(true) {
   System.out.print("Enter input signal (0/1 and <ENTER>): ");
   ch = (char)System.in.read();
   System.in.skip(2);
```

```
   if(((ch!='0') && (ch!='1'4))) break;
   a4.processMsg(ch);
  }
 }
}
```

4.4 Implementation Based on the FSM Library

In the previous two sections, we have explored various approaches to the FSM implementations by means of simple examples. The reader should be much more familiar with FSM implementation by now, but for serious communication protocol engineering we need much more. We need a well-established working environment that will enable productive and repeatable development processes that yield maintainable products (communication protocols) of high quality.

The main measure (metrics) of quality in the context of communication protocols is their reliability, which is considered to be proportional to the number of remaining software bugs. Another important quality measure is the product performance measure with its throughput (the number of messages processed in the given interval of time) and hardware resources needed to achieve that throughput (RAM and ROM size and processor speed measured in MIPS or MHz). Generally, one of the key factors to successful software quality assurance is the quality of the software tools used in the development process. Communication protocol engineering is by no means an exception in this respect.

In this section, we present an example of the state-of-the-art working environment for the productive development of communication protocols. The environment is effectively created by an integrated development environment, which includes a C++ compiler and the domain-specific C++ library, named FSM Library. As already mentioned, the FSM Library includes two fundamental classes, *FSMSystem* and *FiniteStateMachine*. The former creates the execution platform for a group of FSMs whereas the latter is the base class for implementing individual FSMs.

The FSM Library API comprises two interfaces, which are defined by the class *FSMSystem* and *FiniteStateMachine*. The complete FSM Library programmer reference manual is given in Chapter 6. The reference manual also includes two representative implementation examples. In this section, we focus on the FSM Library concepts and internals.

The key concept behind the FSM Library is to enable productive implementations of FSMs in a uniform way. The main task of the FSM Library user is to implement the FSM state transition functions. The user does this by translating the design artifacts (statechart diagram, activity diagram, or SDL diagram) into the corresponding C++ class function members. This translation can be done manually or with a software tool (typically used if the product performance is not critical).

The process of translation is both productive and uniform because the FSM Library provides all the functions needed to effectively construct an FSM state transition. These functions can be classified into the following function groups:

- Message handling functions (both message header and message payload handling functions). These functions support both message coding and decoding (i.e., message synthesis and analysis).

- Message sending functions.

- Timer handling functions (essentially start, stop, and restart timer).

The reader may be puzzled by the fact that the list given above does not include any message receiving functions. The FSM Library is specific in this respect. The developer does not need to explicitly call a function that receives a message (signal). Rather, the FSM execution platform (provided by the class *FSMSystem*) routes all sent messages toward their destination automata, locates the state transition function that corresponds to the message type (determined by the content of the corresponding message header field), and calls it as its subroutine. We will see shortly that the function that performs the message routing and processing (named *Start*) is actually the event interpreter.

Therefore, the FSM Library completely supports the message handling style present in the design artifacts (statecharts, activity diagrams, and SDL diagrams), which just name the input event (message) without taking care of how that event is effectively recognized (received). The FSM Library provides the class *FSMSystem* to support the straightforward implementations of design artifacts. Once provided with the class *FSMSystem*, the developers do not care how the message is received; they simply write the C++ function that performs the state transition when the message is received.

Other FSM Library specifics are the following:

- The FSM implementation is independent from the underlying real-time kernel.

- The FSM Library provides the mechanism to send messages to the dynamically allocated automata instances, which are referred to as unknown automata instances.

- The FSM Library provides public mailboxes, which can be used as message queues with different priorities.

- The FSM Library separates the message handling functions from the real-time kernel. This feature is referred to as the encapsulation of the message handling functions.

- The FSM Library treats timers as special messages, which are distinguished from the communication protocol messages by the code that determines the message type.

- The logging system provided by the FSM Library is based on the test version of the real-time kernel, which is derived from the target (final) real-time kernel.

- The FSM implementation is independent from the concrete formats of the communication protocol messages.

- The FSM Library provides automatic message buffer reallocation in cases where current buffer capacity becomes insufficient for storing additional message parameters.

The following paragraphs provide short comments on each of these FSM Library specifics. We proceed through the list of specifics from its beginning toward its end.

An important design decision was to make the FSM Library independent from the underlying run-time kernel. This decision is important because it enables easy porting of the FSM implementations to various target platforms (bare machine, UNIX, Windows NT). The internal class *KernelAPI* facilitates this independence. It represents a clean interface between the FSM implementation and the run-time system. The kernel developer must derive a new class from the class *KernelAPI* and write its real member functions by taking into account the details of the particular target platform. An example of such implementation is shown later in this section.

The second FSM Library-specific feature is related to the beginning of the communication between two FSMs, namely, FSM A and FSM B, where the former has the active role and the latter is passive. The problem is simple if A always communicates with the same B, but it becomes more complex if B is not known in advance (B is an unknown FSM). Consider a pool of FSMs, where each is capable of performing the same task. FSM A is principally interested in engaging with any instance from the pool that is free.

The FSM Library facilitates the communication with the unknown automata by placing all relevant data into the header of the message that is sent to it. The message destination is set to the special code, named *UNKNOWN_ AUTOMATA*. The function member *Start* of the class *FSMSystem* recognizes this code and dynamically allocates an automata instance, which will be the message destination and therefore involved in the further communication with the message originator. In the case when there are no free automata instances available in the pool, the function *Start* calls the special function *NoFreeInstances*, which is responsible for the recovery procedure. Typically, this function informs the message originator about the automata instance outage by sending it an appropriate signal, such as *NAK, DISCONNECT*, and so on.

The third FSM Library-specific feature is the provision of general purpose mailboxes, which can be used both as public mailboxes and private mailboxes. The former are actually FIFO message queues that contain messages for various destinations, whereas the latter contain messages for a single destination, which is an FSM that owns the private mailbox. Generally, we can use only a single public mailbox to enable the communication between all FSMs present in the system. Such a solution can suffice in the case of simple systems with a small number of FSMs and soft real-time requirements. However, a single public mailbox may not be sufficient in the case of more complex systems because the FSM Library mailbox is just a FIFO message queue without any support for message prioritization.

The absence of message prioritization can lead to a case where an FSM processes an outdated message instead of processing the corresponding timeout message, just because the outdated message is ahead of the timeout message in the public mailbox. Such cases can lead to dysfunctional behaviors that are not caused by design oversights but, rather, inappropriate implementation.

The regular method of supporting message prioritization in the FSM Library-based implementations is to use more public mailboxes that are assigned different priorities. For example, we can use three public mailboxes for three different priorities. These three public mailboxes are effectively treated as three FIFO message queues with different priorities (e.g., high, medium, and low). We can select a strategy of using private mailboxes instead. We can also mix public and private mailboxes if we wish. Actually, the function *Start* (the member of the class *FSMSystem*) treats them equally. In its loop, it searches all the mailboxes for messages. The effective mailbox priority is determined by the order of that search (i.e., it starts from the mailbox index 0).

The fourth FSM Library–specific feature is the encapsulation of the message handling functions. Generally, real-time kernels can store the message source and destination information in the message header or in the separate data structure. By separating the message handling functions into a group that handles the message header and a group that handles the message payload, the FSM Library provides complete FSM implementation independence from the message source and destination information location.

An additional enhancement related to the message destination provided by the FSM Library is the support for sending messages to the *left* or to the *right* FSM. The abstraction of the *left* and *right* FSM originally comes from SDL. If the SDL symbol for sending a message points to the left, we say that the message is sent to the *left* FSM. Similarly, if the symbol points to the right, we say that the message is sent to the *right* FSM.

The internal class *KernelAPI* provides the functions *SendMessageLeft* and *SendMessageRight*, which are inherited by the class *FiniteStateMachine*, to support this abstraction. These two functions enable the direct coding of corresponding parts of SDL diagrams, and the resulting C++ code has a great similarity with the original SDL diagrams. For example, consider the following snippet of C++ code that corresponds to a state transition:

```
StopTimer(FE4_TIMER1);
DisconnectRingTone();
PrepareNewMessage(0x00,r2_SetupRespConf);
SendMessageLeft();
StartChargingIncoming();
Connect();
SetState(FE4_ACTIVE);
```

The call of the function *SendMessageLeft()* above is a direct encoding of the corresponding left-pointing SDL graphical symbol. This snippet of code is a typical state transition implementation based on the FSM Library, which is rather short and easy to read and map to the original design model. These are two key implementation features that ensure productivity and quality.

The fifth FSM Library–specific feature is that it treats timers as special messages, distinguished from the communication protocol messages by the code that determines the message type. Some of the message header parameters are meaningless for timers. The corresponding message header fields are used by the FSM Library API functions related to timers to store the data specific for individual timers, such as timer duration.

All timers used by a certain FSM type must be initialized in the FSM class function member *Initialize()* by calling the function *InitTimerBlock()* (see Section 6.8.74). The parameters of this function are the timer identification, the timer duration, and the identification of the message to be sent when the timer expires. In response to a series of *InitTimerBlock()* calls, the system creates the corresponding array of timers. The identification of a timer effectively becomes the index of this array.

Once initialized, the timer can be started by the function *StartTimer()*, stopped by the function *StopTimer()*, restarted by the function *RestartTimer()*, or checked by the function *IsTimerRunning()*. All these functions have a single parameter, the identification of the timer. Therefore, the resulting C++ code resembles the original design model to a great extent. Moreover, when the timer expires, the corresponding message is automatically sent to the FSM that started it, which processes this message in the same fashion as all other messages. This feature also contributes to the similarity of the resulting C++ code and the original design model.

The sixth FSM Library–specific feature is that the logging subsystem provided by the FSM Library is based on the test version of the real-time kernel, which is derived from the target (final) real-time kernel. The logging subsystem is important in communication protocol engineering because certain design oversights or implementation errors become evident only in complex circumstances, which can happen only after long run-time periods. Typically, such circumstances are difficult to repeat and therefore developers normally use log files to backtrack the sources of errors once they occur.

The FSM Library provides a complete logging subsystem that is used both during system testing and normal system exploitation. The internal class *LogAutomata* defines the necessary set of functions. FSM tracing is based on the interception of all relevant internal functions, such as FSM state updating, message processing, timer management functions, and so on. Automatic logging of various events makes the resulting log file outlook uniform, and thus easy to read by any member of the development team. All logging events are prioritized, which helps developers to easily define exactly which events they want to trace.

Traditionally, log files are located on mass storage devices such as hard disks or flash memory. The FSM Library introduces an enhancement in this respect. The internal class *LogInterface* defines the interface between the system implementation and the concrete logging media, such as the conventional log file, the TCP/IP connection to the logging server, and so on. Logging to the concrete media is provided by a subclass that is derived from the base class *LogInterface*. Examples of such classes are the classes *LogFile* and *LogTCP*.

The seventh FSM Library–specific feature is that the FSM implementation is independent of the concrete formats of communication protocol messages. The feature is facilitated by the internal class *MessageHandler*, which provides a set of generic functions for manipulating message parameters. Basically, two families of these functions exist, namely, *get* and *add*. The former returns the value of the given parameter, whereas the latter adds the given message parameter to the message. The parameter is specified with its identification (code) and its value.

The class *MessageHandler* uses the class *MessageInterface*, which is an abstract class that defines the interface for the abstract message format. Normally, the developer derives a class from the class *MessageInterface* for each concrete message format and writes its function in accordance with the format-specific details. An example of such a class is the class *StandardMessage*, which models a message that comprises a sequence of octets (characters). Such an approach centralizes message handling functionality. This centralization eliminates code redundancy and increases code coverage. Additionally, development team productivity is increased because message handling functions and FSMs can be developed in parallel.

The eighth and last FSM Library-specific feature is that it provides automatic message buffer reallocation in cases where the current buffer capacity becomes insufficient for storing additional message parameters. Although this functionality is rather easily implemented, it is important because it makes the process of message creation completely transparent. The programmer just adds parameters to the new message as needed, without having to take care about the size of the free space in the corresponding buffer. This detail is completely hidden by the message handling functions.

4.4.1 Using the FSM Library

Using the FSM Library is rather easy. It helps a lot in both the design and implementation phases of the development process. The author's experience shows that both students and engineers working in the industry can start using it only after a couple of days of training. Actually, it does not take more than writing one example based on the FSM Library to start using it. Besides that, it is a well-established working environment that has been used in a series of the real-world projects for the industry.

When it comes to design, the FSM Library greatly simplifies matters by providing two fundamental classes, *FSMSystem* and *FiniteStateMachine*. The existence of these two classes makes the system static structure well known from the start (Figure 3.5). Each protocol is modeled by the subclass derived from the base class *FiniteStateMachine*. The resulting FSM is executed by the event interpreter, which is hidden inside the class *FSMSystem*. These two classes practically encapsulate all domain-specific design patterns needed for designing a communication protocol.

The overall result is that the class diagram is almost not needed at all, at least not for realistic communication systems that comprise less than a dozen communication protocols. Even for very complex communication systems based on the FSM Library, the class diagram can be used more as an accompanying document. The most informative part of such a class diagram would be the one that specifies the mailboxes present in the system, as well as the timers used by individual FSM types.

The real valuable design artifacts for the paradigm based on the FSM Library are the complete models of the system behavior in the form of the activity, statechart, or SDL diagrams. This is the case because the FSM Library *de facto* specifies the skeleton of the system static structure, but it does not (and cannot) specify the complete system behavior. It provides only primitive behavior from which we can build more complex behavior, in particular, the state transitions.

Once we have finalized the detailed design diagrams (activity, statechart, or SDL diagrams), we are ready to proceed to the implementation phase of the development process. The main task of implementing FSMs by using the FSM Library, besides writing the initialization function and a couple of simple auxiliary functions, is the encoding of state transitions by using the set of primitives provided within the FSM Library application programming interface (see Section 6.8). A good thing about these primitives is that they provide mapping of SDL steps in almost a one-to-one manner. The names of the primitives are almost self-documenting, at least after the short experience you get by using them. The code resembles the original design artifacts (especially SDL diagrams). All these attributes help any member of the development team to read, understand, and continue the work that was done by some other member of the development team, especially if they have the design artifact at their disposal.

It is also worth mentioning that besides forward engineering, the FSM Library helps backward engineering too. This is especially true if the backward engineering is done by hand. Using software tools for that purpose is also possible if the development team strictly obeys certain coding guidelines. The key for successful forward and backward engineering is a well-defined API (see Section 6.8).

We demonstrate the usage of the FSM Library API by the examples at the end of this chapter, as well as with the examples at the end of Chapter 6.

4.4.2 FSM Library Internals

This section describes the FSM Library internals. The main FSM Library components are the following:

- The class *FSMSystem*
- The class *FiniteStateMachine*
- The real-time kernel

The class *FSMSystem* provides the following functionalities:

- Initialization of the FSM objects: The result is a set of the corresponding transition tables, which determine which state transitions are triggered by the individual events (messages).
- Routing of messages: This component locates the message destination FSM, looks up its state transition table to find the state transition that corresponds to the message type, and calls the corresponding function as its subroutine.
- Public mailbox prioritization: The public mailbox priority decreases as its identification increases. The identification is actually the index of the corresponding mailbox array. The public mailbox with the identification 0 has the highest priority.
- Allocation of FSMs from the pool of FSMs: If the message destination is an unknown object of a certain type, a free FSM from the corresponding pool is allocated to process that message.

The class *FiniteStateMachine* provides the following functionalities:

- Maintaining the current state variable (the field member of this class)
- Maintaining the state transition table
- FSM evolution support by providing the address of the state transition function that corresponds to the incoming message type
- Message handling (message checking, parsing, and creation)
- Message exchange (the message send operation is explicit whereas the message receive operation is implicit)
- Memory management (supports requesting and releasing buffers for messages)
- Timer management (supports starting, stopping, restarting, and testing timers)

The functionalities provided by the real-time kernel are inherited by the class *FiniteStateMachine* (message exchange, buffer, and timer management). The following subsections describe the internals of these three components.

4.4.2.1 FSMSystem Internals

As already mentioned, the class *FSMSystem* provides the execution platform for all FSMs present in the system. The list of concrete functionalities provided by this class is already given in the previous section. The heart of the class *FSMSystem* is the function *Start*, which actually provides all the listed functionalities. Essentially, it is the event (message) interpreter. Its program code in C++ is as follows:

```
void FSMSystem::Start(){
 SystemWorking = true;
 while(SystemWorking) {
  Sleep(1);
  for(uint8 i=0; i<NumberOfMbx; i++) {
   uint8 *msg = GetMsg(i);
   if(msg == NULL){
    continue;
   }
   uint8 automataType = GetMsgToAutomata(msg);
   if(((automataType > NumberOfAutomata) ||
     (NumberOfObjects[automataType] == 0))){
    // Error handling
    DiscardMsg(msg);
    continue;
   }
   uint32 objNum = GetMsgObjectNumberTo(msg);
   if(objNum == UNKNOWN_AUTOMATA){
    ptrFiniteStateMachine object =
     FreeAutomata[automataType].Get();
    if(object != 0) object->Process(msg);
    else
     (Automata[automataType][0])->NoFreeObjectProcedure(msg);
    continue;
   }
   else if(objNum > NumberOfObjects[automataType]) {
    // Error handling
    DiscardMsg(msg);
    continue;
   }
   else {
    (Automata[automataType][objNum])->Process(msg);
   }
  }
 }
}
```

The function *Start* initially sets its field member *SystemWorking* to the value *true* and enters the loop, which is executed while *SystemWorking* has the value *true*. Once this variable is set to the value *false* (this is exactly what the API function *StopSystem()* does), the function *Start* exits the loop and terminates. Because this function is the FSM event interpreter, once it stops, the whole system stops.

Inside the *while* loop, this function enters the nested *for* loop in which it checks all mailboxes for messages. This *for* loop starts from the mailbox with the identification (index) 0, thus making it the highest priority mailbox. As it proceeds toward the identification *NumberOfMbx*, the priority of the corresponding mailboxes decreases.

Once it finds a message in the mailbox, it exits the nested *for* loop and continues with determining the destination automata (FSM) type identification by calling the function *GetMsgToAutomata()*. If the identification is invalid (greater than the configuration parameter *NumberOfAutomata*) or if no instances of that type are found, the function discards the message by calling the function *DiscardMsg()* and continues the main loop.

If the automata type identification is valid and at least one instance of that type is found, the function *Start* determines the destination object identification by calling the function *GetMsgObjectNumberTo()*. If this identification is equal to *UNKNOWN_AUTOMATA*, the function *Start* tries to allocate an object from the pool of objects of the given type by calling the function *Get()* on the object of that type.

If at least one free object is found in the pool (actually an array of objects of the given type), the function *Get()* will return the identification (array index) of the first one and, in turn, the function *Start* will call its function *ProcessMsg()*. Behind the scenes, the function *ProcessMsg()* locates the state transition that corresponds to the message type, calls it its subroutine, and continues the main loop. If no free objects are in the pool, the function *Start* discards the message and continues the main loop.

Finally, if the message destination is a known object (its identification is not equal to *UNKNOWN_AUTOMATA*), the function *Start* checks if its identification is valid (not greater than the configuration parameter *NumberOfObjects[automataType]*). If the object identification is valid, the function Start calls object function *ProcessMsg()* and continues the main loop.

4.4.2.2 *FiniteStateMachine Internals*

The class *FiniteStateMachine* is at the top of the FSM Library class hierarchy (Figure 4.13). It hides the details of the FSM Library internal static structure from its user. The class *FiniteStateMachine* inherits logging-related functionality from the class *LogAutomata* (shown as the left branch of the class hierarchy in Figure 4.13). Alternately, the class *FiniteStateMachine* inherits the buffer, timer, and message management functionality from the class *KernelAPI* (shown as the right branch of the class hierarchy in Figure 4.13). Both *FiniteStateMachine* and *KernelAPI* inherit the message management functionality from the class *MessageHandler*.

The class *LogAutomata* conceptually uses the logging services provided through the interface created by the class *LogInterface*. The logging services are provided in run-time reality by the object that is an instance of a subclass, which is derived from the base class *LogInterface*. Figure 4.13 shows two examples of such classes, namely, *LogFile* and *LogTCP*. The former provides the recording of log events into the file located on some mass storage device. The latter uses the TCP/IP network to send log events packed into messages to the logging server, which, in turn, writes the log events to a file, perhaps located on its hard disk.

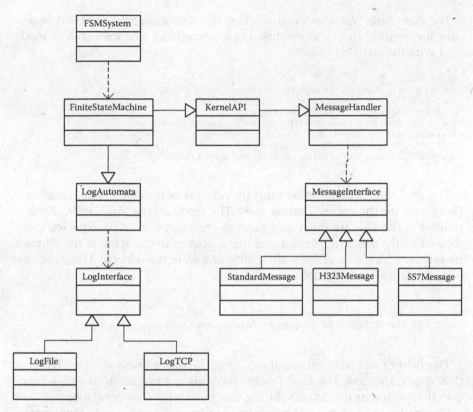

FIGURE 4.13
Internal FSM Library static structure.

Similarly, the class *MessageHandler* uses services of the abstract interface provided by the class *MessageInterface*. The real providers of the message handling services are subclasses derived from the base class *MessageInterface*. Figure 4.13 shows three examples of such classes, namely, *StandardMessage*, *H323Message*, and *SS7Message*. In the examples in this book, we use the class *StandardMessage*, which creates an abstraction of the message comprising a series of octets (characters) that can be partitioned into an arbitrary number of message fields (carrying message parameters) of arbitrary size (given as a number of octets).

In the text that follows, we cover the most important details of the class *FiniteStateMachine*, *KernelAPI*, and *MessageHandler*. The effect of this top-down approach is that we introduce first the functionality solely provided by the class *FiniteStateMachine*, then the functionality that the class *FiniteStateMachine* inherits from the class *KernelAPI*, and finally the functionality that the class *FiniteSateMachine* inherits from the class *MessageHandler*.

The class *FiniteStateMachine* comprises all attributes and operations necessary for the definition and evolution of a single FSM. The FSM state is modeled with the structure *SState*:

```
struct SState {
 SState(uint16 maxNumOfProceduresPerState);
 ~SState();
 bool StateValid; // if true, data are valid
 unsigned short NumOfBranches; // number of branches in a state
 // procedure for processing unexpected message
 PROC_FUN_PTR UnexpectedEventProcPtr;
 SBranch* PBranch; // pointer on data for each branch
};
```

The field *NumOfBranches* contains the number of outgoing state transitions (branches) for the corresponding state. The field *UnexpectedEventProcPtr* is a pointer to the C++ function that handles the reception of unexpected messages. Finally, the field *PBranch* contains a pointer to the array of the *SBranch* instances, which model individual outgoing state transitions. The structure *SBranch* definition is the following:

```
struct SBranch {
 uint16 EventCode; // message code
 PROC_FUN_PTR ProcPtr; // message processing function
};
```

The field *EventCode* contains the code of the event (message) that triggers this state transition. The field *ProcPtr* contains the pointer to the C++ function that performs the actions during this particular state transition.

Generally, an FSM can use a number of timers. Each timer is represented with an instance of the structure *TimerBlock*:

```
struct TimerBlock {
 TimerBlock(uint16 v, uint16 s) :
  Count(v), SignalId(s), Valid(false), TimerBuffer(0){}
 TimerBlock() :
  Count(INVALID_32), SignalId(INVALID_16), Valid(false),
  TimerBuffer(0) {};
 uint32 Count; // in time slices
 uint16 SignalId; // message code
 bool Valid; // if true, data is valid
 ptrBuff TimerBuffer; // Ptr to timer buffer
};
```

The field *Count* defines the timer duration, the field *SignalId* defines the code of the message (signal) that is generated when the timer expires, the field *Valid* is set if the timer is running, and the field *TimerBuffer* contains the pointer to the buffer used by the timer expiration message.

The main private field members of the class *FiniteStateMachine* are as follows:

```
class FiniteStateMachine : public KernelAPI, LogAutomate {…
 private:
```

```
uint16 NumOfStates; // Number of FSM states
uint16 NumOfTimers; // Number of timers
uint16 MaxNumOfProcPerState; // Max. no. of branches
SState *States[MAX_STATE_NO]; // State data
uint32 ConnectionId; // Current connection
uint32 CallId; // Current call
uint8 State; // Current state
```

The fields *NumOfStates, NumOfTimers,* and *MaxNumOfProcPerState* are the dimensions of the corresponding arrays. They define the number of FSM states, the number of timers it uses, and the maximum number of branches, respectively. The field *States* is an array of pointers to the instances of the structure *SState* that contains pointers to arrays of instances of the structure *SBranch*. This data structure corresponds to the FSM state transition table.

The field *ConnectionId* carries the domain-specific name but actually contains the FSM object identification that is unique within the scope of objects of the same type. During the system initialization, the class *FSMSystem* creates the array of FSM objects of the same type. The index of the object in that array is written into this field at that time. This identification can be used as appropriate for the application at hand. The FSM Library user can take advantage of the fact that all message sending functions automatically copy the content of this field into the object identification field of the message header.

The field *CallId* carries another domain-specific name but it can be used for various purposes in various applications. In contrast to the field *ConnectionId* whose uniqueness is limited to the scope of a single FSM type, the value of the field *CallId* is unique in the scope of the whole system. Traditionally, it has been used to identify a single call, but generally it can be used to identify any communication process of interest. Like the field *ConnectionId*, this field is also copied by the message sending functions to the message header automatically.

Finally, the field *State* is the FSM current state identification, which is the value of the index of array defined in the field *States*. This field defines the context of the FSM.

As already mentioned, the FSM Library supports the abstraction of the *left* and *right* FSM. The message sending functions, namely *SendLeftAutomata()* and *SendRightAutomata()*—originally defined in the class *KernelAPI*—require data about the *left* and *right* FSM. Relevant *FiniteStateMachine* attributes are as follows:

```
// Left automata data
uint8 LeftMbx; // left mbx id
uint8 LeftAutomata; // left automata
uint8 LeftGroup; // left group
uint32 LeftObjectId; // left object
// Right automata data
uint8 RightMbx; // right mbx id
uint8 RightAutomata; // right automata
uint8 RightGroup; // right group
uint32 RightObjectId; // right object
```

We finish the overview of the *FiniteStateMachine* internals with its initialization and control functions:

```
FiniteStateMachine(
 uint16 numOfTimers = DEFAULT_TIMER_NO,
 uint16 numOfState = DEFAULT_STATE_NO,
 uint16 maxNumOfProceduresPerState = DEFAULT_PROCEDURE_NO_PER_STATE);
virtual void Initialize(void) = 0;
void InitEventProc(uint8 state, uint16 event, PROC_FUN_PTR fun);
void InitUnexpectedEventProc(uint8 state, PROC_FUN_PTR fun);
PROC_FUN_PTR GetProcedure(uint16 event);
virtual void NoFreeInstances() = 0;
virtual void Process(uint8 *msg);
void FreeFSM();
```

The class constructor first sets the number of timers, the number of states, and the maximal number of branches per state. It then calls the function *Initialize()*, provided by the user. This function typically uses a series of calls to functions *InitEventProc()* and *InitUnexpectedEventProc()*. The former defines the state transition function for the given state and message type whereas the latter defines the unexpected message handler for the given state.

The function *GetProcedure()* is a control function that returns the address of the state transition function for the given message type in the current state. The function *NoFreeInstances()* is a recovery function that is called in cases where no more free objects of this type are found. The function *Process()* is the prototype of the state transition function. The function *FreeFSM()* releases the FSM object by returning it to the pool of objects of this type.

The class *KernelAPI* provides the following groups of functions:

- Initialization functions
- Memory management functions
- Message management functions
- Timer management functions

The initialization functions provided by the class *KernelAPI* are its constructors (see Section 6.8) and the function *setKernelObjects*, whose prototype is as follows:

```
void setKernelObjects(TPostOffice *o, TBuffers *b, CTimer *t);
```

The parameters of this function are the pointers to the objects that comprise the system mailboxes, buffers, and timers. These objects will be described in the next section.

The memory management functions provided by the class *KernelAPI* are the following:

```
uint8 *GetBuffer(uint32 length);
void RetBuffer(uint8 *buff);
```

```
bool IsBufferSmall(uint8 *buff, uint32 length);
uint32 GetBufferLength(uint8 *buff);
```

The function *GetBuffer()* returns the pointer to the buffer of the sufficient size (not less than specified by its parameter). The function *RetBuffer()* releases the given buffer. The function *IsBufferSmall()* checks the size of the given buffer. The function *GetBufferLength()* returns the size of the given buffer.

The message management functions provided by the class *KernelAPI* are the following:

```
void Discard(uint8* buff);
void SetMessageFromData();
void SendMessage(uint8 mbxId);
void SendMessage(uint8 mbxId, uint8 *msg);
void SendMessageLeft();
void SendMessageRight();
void ReturnMsg(uint8 mbxId);
```

The function *Discard()* releases the given message. The function *SetMessageFromData()* copies the data about this FSM (type, group, and instance identifications) to the corresponding fields of the new message header. According to the FSM Library terminology, the current message is the one that has been received and processed, whereas the new message is the message that is currently under construction (and will be subsequently sent).

The function *SendMessage(uint8 mbxId)* sends the new message to the given mailbox. The function *SendMessage(uint8 mbxId, unit8 *msg)* sends the given message to the given mailbox. The functions *SendMessageLeft()* and *SendMessageRight()* send the new message to the left and right automata, respectively. The function *ReturnMsg()* sends the current message to the given mailbox.

The timer management functions provided by the class *KernelAPI* are as follows:

```
uint8 *StartTimer(uint16 code, uint32 count, uint8 *info=0);
void StopTimer(uint8 *timer);
bool IsTimerRunning(uint8 *timer);
```

The function *StartTimer()* starts the given timer by setting its duration and the corresponding message buffer. The function *StopTimer()* stops the given timer. The function *IsTimerRunning()* checks if the given timer is running.

The interface defined by the class *MessageHandler* comprises the following two parts:

- Message header handling
- Message payload handling

The message header handling part provides getting and setting functions for the individual message header fields. The main message header fields are as follows:

- *MSG_FROM_AUTOMATA*: the identification of the originating FSM type
- *MSG_TO_AUTOMATA*: the identification of the destination FSM type
- *MSG_CODE*: the identification of the message type
- *MSG_OBJECT_ID_FROM*: the identification of the originating FSM object
- *MSG_OBJECT_ID_TO*: the identification of the destination FSM object
- *MSG_CALL_ID*: the identification of the application-specific communication process
- *MSG_INFO_CODING*: the identification of the message format type
- *MSG_LENGTH*: the message payload length in octets

The timer message is a special message. If the timer expires, it is sent to the same FSM that created it. Because of this, the message header fields *MSG_FROM_AUTOMATA* and *MSG_OBJECT_ID_FROM* are not needed, and thus can be used to hold information about the timer duration and the destination mailbox identification.

The class *MessageInterface* defines the set of abstract functions that handle the message payload. The key idea behind the abstraction introduced by the class *MessageInterface* is the generic message parameter definition, which is independent from the particular message format. Each message parameter is uniquely defined by the following data:

- The message parameter identification
- The message parameter length (size)
- The message parameter value (content)

Depending on the message format type, the first and the second items listed may be implicit or explicit. Some of the messages carry the message parameter identification and length, and some do not. However, all three items must be known to the message handling functions.

Another important fact related to the message format is that particular message formats can be disassembled to a series of primitive elements of the following types:

- Byte (1 byte)
- Word (2 bytes)

- DWord (4 bytes)
- Sequence of bytes (*n* bytes)

Therefore, the class *MessageInterface* includes the functions that provide access to these primitive types of information. These functions can be partitioned into the following two groups:

- Current message handing functions
- New message handling functions

The current message handling functions are as follows:

```
uint8 *GetParam(uint8 paramCode);
bool GetParamByte(uint8 paramCode, BYTE &param);
bool GetParamWord(uint8 paramCode, WORD &param);
bool GetParamDWord(uint8 paramCode, DWORD &param);
```

The first function returns a pointer to the parameter (sequence of octets) whose identification (*paramCode*) is given. The next three functions return the requested parameter of the size *Byte, Word,* and *DWord,* respectively. The new message handling functions are as follows:

```
uint8 *AddParam(uint8 paramCode, uint8 paramLength, uint8 *param);
uint8 *AddParamByte(uint8 paramCode, BYTE param);
uint8 *AddParamWord(uint8 paramCode, WORD param);
uint8 *AddParamDWord(uint8 paramCode, DWORD param);
bool RemoveParam(uint8 paramCode);
```

The first four functions add the given sequence of octets, *Byte, Word,* and *DWord* parameter, respectively, to the new message. The function *RemoveParam()* removes the parameter—whose identification is given—from the message.

Each message handling function consists of two parts, a preparation part and an operation part. The preparation part of the current message handling functions includes preparing temporary data and message parsing. In case of message syntax errors, message handling functions report an error by returning the *value false.* The preparation part of the new message handling functions includes allocation of the message buffer and initialization of the message header fields *MSG_CODE, MSG_INFO_CODING,* and *MSG_LENGTH* (initially set to 0).

4.4.2.3 Kernel Internals

As already mentioned, the class *FiniteStateMachine* is independent of the particular real-time kernel with the introduction of the API defined by the class *KernelAPI.* Generally, the class *FiniteStateMachine* can use services provided by any real-time kernel that is a subclass of the class *KernelAPI.* In this section, we cover the internals of one such kernel (a default one), which is simply referred to as *Kernel.*

Figure 4.14 shows the static structure of *Kernel*. The root of the structure is the class *KernelAPI*, which acts as the wrapper of *Kernel*. This class contains pointers to the following three main parts of *Kernel*:

- Memory manager
- Message manager
- Time manager

The interfaces to these three resource managers are defined by the classes *TBuffers*, *TPostOffice*, and *CTimer*, respectively. The memory manager comprises the class *TBuffers* and a set of instances of the class *TBufferQueue*. The message manager consists of the class *TPostOffice* and a set of instances of the class *TMailBox*. The time manager is implemented by the class *CTimer* itself.

The class *TBuffers* creates an abstraction of a set of buffer pools. The size of the buffers in the pool is the same, but these sizes are different between the pools.

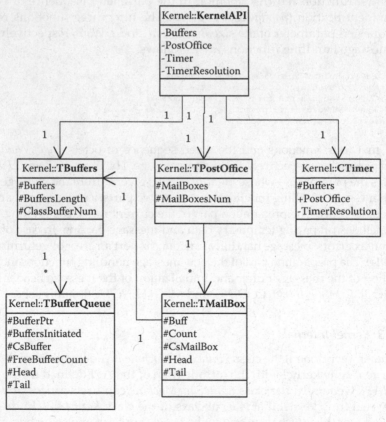

FIGURE 4.14
Internal *Kernel* static structure.

For example, we can have three pools with three different sizes, namely, small, medium, and large. The class *TBufferQueue* models one such a pool.

The constructor of the class *TBufferQueue* initially allocates an array of bytes (uint8), which is the actual memory space that accommodates the memory pool:

```
// calculate memory size for all buffers and get memory for them
memSize = bufferLength + BUFF_HEADER_LENGTH;
memSize *= buffersNo;
BufferPtr = new uint8[memSize];
```

This memory space is then partitioned into individual memory buffers that are added to the list of free buffers that actually represent the buffer pool. A buffer consists of the buffer header and the space for useful data. The buffer header comprises the pointers to the previous and to the next element in the list and the buffer code that indicates buffer size. Each buffer pool is defined with the pointer to the list of free buffers and the size of the buffers in that list. The class *TBuffers* holds the array of pointers to the instances of the class *TBufferQueue* (in the field member *Buffers*), as well as the array of the corresponding buffer sizes (in the field member *BuffersLength*).

The function *GetBuffer()* provided by the class *KernelAPI* first searches the field *BuffersLength* to find the pool of buffers of the sufficient size. It then gets the buffer from the head of the list of free buffers and returns the pointer to it. The function *RetBuffer()* uses buffer code from its header to return the buffer by adding it to the end of the corresponding list.

The class *TPostOffice* stores the array of pointers to the corresponding mailboxes. A mailbox is implemented as an instance of the class *TMailBox*. Actually, the class *TMailBox* is very similar to the class *TBufferQueue*. The main difference between them is that the former provides atomic (uninterruptible) access to the list of messages. This feature is needed because the list of messages is a resource shared by two concurrent processes, namely the event interpreter and the time interrupt routine.

The atomic mailbox access is ensured by two virtual functions, namely *MbxLock()* and *MbxUnlock()*. The former function locks the mailbox and the latter unlocks it. These functions ensure the FSM Library's portability. They can be implemented by the use of semaphores provided by the local operating systems. (The FSM Library supports OS Linux® and Windows® NT at the moment.)

The class *CTimer* is the most target-platform-dependent part of *Kernel*. It consists of two parts, a platform-dependent part and a platform-independent part. The platform-dependent part comprises the time-driven routine that is periodically called by the local operating system and the routines that provide access to shared data. The platform-independent part consists of the list of running timers and routines that maintain that list. The list of running timers is implemented as a traditional delta list (the timer at the head of the list contains the absolute time interval whereas all other timers contain the time interval relative to the previous timer in the list).

To simplify timer maintenance, the function *StopTimer()* does not analyze the current status of the given timer (already expired or still running)—it simply marks the timer as expired. If the timer was still running, it will remain in the list of running timers. When it expires, it is forwarded to the given mailbox and from there it is discarded by the function member *Get()* of the class *TMailBox*.

4.4.3 Writing FSM Library–Based Implementations

Normally, we start by deriving subclasses from the base class *FiniteStateMachine*. For each such subclass, we must define the following functions (see Section 6.8 for more details):

- *GetMessageInterface()*: This function returns the pointer to the particular message interface object.
- *SetDefaultHeader()*: This function sets the default message header parameters.
- *GetMbxId()*: This function returns the identification of the mailbox associated to this FSM type.
- *GetAutomata()*: This function returns the identification of this FSM type.
- *SetDefaultFSMData()*: This function sets default FSM data.
- *NoFreeInstances()*: This recovery function is called when the pool of objects is exhausted.
- *Initialize()*: This function initializes FSM-related data, including the state transition table.

We then write the main program, which typically follows these steps:

- Create an instance of the class *FSMSystem*.
- Initialize the real-time kernel.
- Set the system parameters.
- Register (add) all FSM objects with the instance of the class *FSMSystem*.
- Start the system by calling the function *Start()* (defined within the class *FSMSystem*).

4.5 Examples

This section includes two representative examples of FSM Library–based implementations. The first example is the implementation of an application for reading Internet electronic mail. The second example shows an implementation of the SIP invite client transaction.

4.5.1 Example 1

This example demonstrates how an application for reading Internet electronic mail can be constructed. The application is actually an e-mail client that comprises the following three objects (see the general collaboration diagram in Figure 4.15):

- *user*: a user interface
- *pop3*: the implementation of the POP3 protocol (refer to the original RFC 1939, freely available on the Internet at www.ietf.org/rfc /rfc1939.txt)
- *channel*: responsible for the direct communication with the e-mail server over the TCP protocol

As shown in Figure 4.15, the objects *user*, *pop3*, and *channel* are the instances of the classes *UserAuto*, *ClAuto*, and *ChAuto*, respectively. The object *pop3* is the central object. On its left side is the object *user*, and on its right side is the object *channel*. The interaction between these objects is illustrated with three typical scenarios that are shown in Figures 4.16 through 4.18. Figure 4.16 shows a successful session during which all pending e-mails are received and saved as files on a mass storage device. The flow of events from the point of view of the object *pop3* is as follows:

- Triggered by the reception of the message *User_Check_Mail* from the left object, it sends the message *Cl_Connection_Request* to the right object.
- Upon the reception of the message *Cl_Connection_Accept* from the right object, it sends the message *User_Connected* to the left object. The connection with the e-mail server is successfully established at this point.
- After receiving the username and password carried by the message *User_Name_Password* from the left object, it first sends the username in the message *MSG(USER name)* to the right object, which is acknowledged with the message *MSG(+OK)* from the right object, and it then sends the password in the message *MSG(PASS password)* to the right object, which is also acknowledged with the message *MSG(+OK)* from the right object. The user authentication procedure is successfully finished at this point.

FIGURE 4.15
Receive e-mail application collaboration diagram.

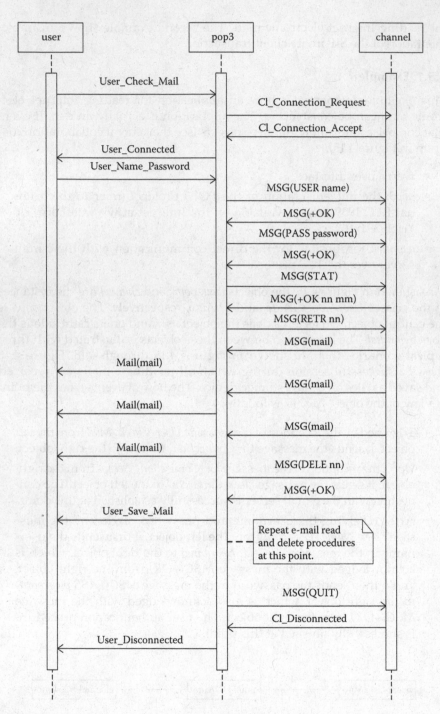

FIGURE 4.16
Successful receive e-mail session establishment scenario.

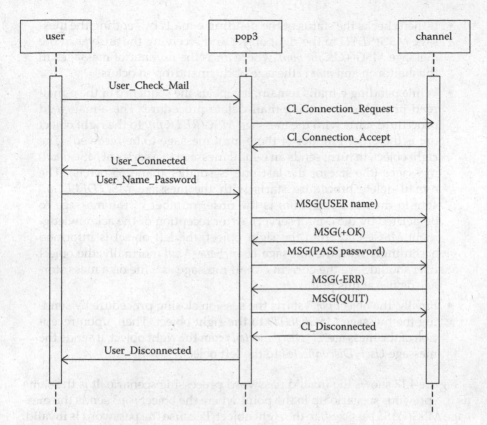

FIGURE 4.17
Invalid e-mail password processing scenario.

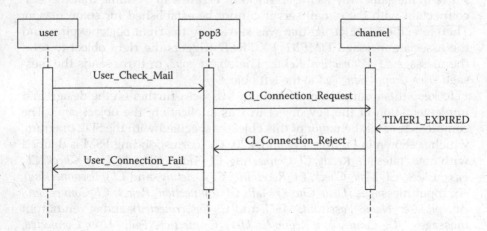

FIGURE 4.18
Unsuccessful receive e-mail session establishment scenario.

- It then checks the status of the pending e-mails by sending the message *MSG(STAT)* to the right object and receiving the answer in the message *MSG(+OK nn mm)*, where *nn* is the number of messages in the maildrop and *mm* is the size of the maildrop in octets.

- While pending e-mails remain, it repeats the sequence of the e-mail read procedure and the e-mail delete procedure. The e-mail read procedure starts with the message *MSG(RETR nn)* to the right object (*nn* is the order number of the e-mail message to be received). The right object, in turn, sends an e-mail message in a series of *MSG(mail)* messages (the size of the last one is smaller than 255 octets). The e-mail delete procedure starts with the message *MSG(DELE nn)* sent to the right object (*nn* is the order number of the message to be deleted by the e-mail server). After reception of the acknowledgment *MSG(+OK)* from the right object, the left object is informed accordingly with the message *User_Save_Mail* (normally, the object user should save the current e-mail message as a file on a mass storage device at this point).

- Finally, the object *pop3* starts the session closing procedure by sending the message *MSG(QUIT)* to the right object. Then, upon reception of the message *Cl_Disconnected* from the right object, it sends the message *User_Disconnected* to the left object.

Figure 4.17 shows the invalid password processing scenario. It is the same as the previous scenario up to the point where the object *pop3* sends the message *MSG(PASS password)* to the right object. Because the password is invalid, the right object responds with the message *MSG(-ERR)* and the object *pop3* immediately proceeds to the session closing procedure.

Figure 4.18 shows the unsuccessful session establishment scenario. It starts in the same way as the scenario in Figure 4.16. Assume that the TCP connection with the e-mail server cannot be established for some reason. Therefore, the *TIMER1_ID* that was started by the right object expires and the associate message *TIMER1_EXPIRED* triggers the right object to send the message *Cl_Connection_Reject*. The object *pop3*, in turn, sends the message *User_Connection_Fail* to the left object.

To keep this example simple enough, we focus further on the design and implementation of the key object in this application, the object *pop3*. The complete dynamic behavior of this object is specified with the SDL diagram, which is shown in Figures 4.19 and 4.20. The corresponding FSM is defined with nine states (*Cl_Ready, Cl_Connecting, Cl_Authorizing, Cl_User_Check, Cl_Pass_Check, Cl_Mail_Check, Cl_Receiving, Cl_Deleting,* and *Cl_Disconnecting*), six input messages (*User_Check_Mail, Cl_Connection_Reject, Cl_Connection_Accept, User_Name_Password, MSG,* and *Cl_Disconnected*), and seven output messages (*Cl_Connection_Request, User_Connection_Fail, User_Connected, MSG, Mail, User_Save_Mail,* and *User_Disconnected*).

FIGURE 4.19
POP3 client SDL diagram, part I.

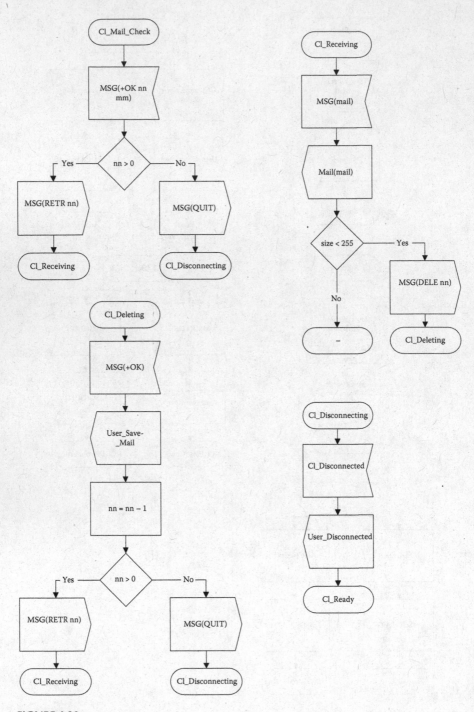

FIGURE 4.20
POP3 client SDL diagram, part II.

By convention, the names of all messages (except *Mail*) exchanged between the object *pop3* and the left object begin with the prefix *User_*. The names of the control messages exchanged between the object *pop3* and the right object begin with the prefix *Cl_*. The names of the POP3-related messages exchanged between the object *pop3* and the right object are named *MSG*. Two types of *MSG* messages are used—commands directed to the e-mail server and responses received from it.

The *MSG* commands are as follows:

- *MSG(USER name)* corresponds to the original POP3 command for specifying the name of the user mailbox.
- *MSG(PASS password)* corresponds to the original POP3 command for specifying the password for the previously specified mailbox.
- *MSG(STAT)* corresponds to the original POP3 command for inquiring about the mailbox status.
- *MSG(RETR nn)* corresponds to the original POP3 command for reading the pending e-mail message number *nn*.
- *MSG(DELE nn)* corresponds to the original POP3 command for deleting the pending e-mail message number *nn*.
- *MSG(QUIT)* corresponds to the original POP3 command for closing the current session.

The *MSG* responses are the following:

- *MSG(+OK)* corresponds to the original POP3 acknowledgment message.
- *MSG(ERR)* corresponds to the original POP3 error message.
- *MSG(mail)* corresponds to the actual e-mail message that was received from the e-mail server.

Figure 4.19 shows valid state transitions for the states *Cl_Ready*, *Cl_Connecting*, *Cl_Authorizing*, *Cl_User_Check*, and *Cl_Pass_Check*. The eight state transitions are shown in Figure 4.19, as follows:

- From *Cl_Ready* to *Cl_Connecting*, triggered by *User_Check_Mail*
- From *Cl_Connecting* to *Cl_Ready*, triggered by *Cl_Connection_Reject*
- From *Cl_Connecting* to *Cl_Authorizing*, triggered by *Cl_Connection_ Accepted*
- From *Cl_Authorizing* to *Cl_User_Check*, triggered by *User_Name_ Password*
- From *Cl_User_Check* to *Cl_Pass_check*, triggered by *MSG(+OK)*
- From *Cl_User_Check* to *Cl_Disconnecting*, triggered by *MSG(ERR)*

- From *Cl_Pass_Check* to *Cl_Mail_check*, triggered by *MSG(+OK)*
- From *Cl_Pass_Check* to *Cl_Disconnecting*, triggered by the *MSG(ERR)*

Figure 4.20 shows valid state transitions for the states *Cl_Mail_Check*, *Cl_Receiving*, *Cl_Deleting*, and *Cl_Disconnecting*. The seven state transitions are shown in Figure 4.20, as follows:

- From *Cl_Mail_Check* to *Cl_Receiving*, triggered by *MSG(+OK)* and guarded by the condition $nn > 0$
- From *Cl_Mail_Check* to *Cl_Disconnecting*, triggered by *MSG(+OK)* and guarded by the condition $!(nn > 0)$
- From *Cl_Receiving* to *Cl_Deleting*, triggered by *MSG(mail)* and guarded by the condition *mail(size)* < 255
- From *Cl_Receiving* to *Cl_Receiving*, triggered by *MSG(mail)* and guarded by the condition !(*mail(size)* < 255)
- From *Cl_Deleting* to *Cl_Receiving*, triggered by *MSG(+OK)* and guarded by the condition $nn > 0$
- From *Cl_Deleting* to *Cl_Disconnecting*, triggered by *MSG(+OK)* and guarded by the condition $!(nn > 0)$
- From *Cl_Disconnecting* to *Cl_Ready*, triggered by *Cl_Disconnected*

Next, we proceed to the implementation in C++ based on the FSM Library. First, we define symbolic constants specific for this project in a header file, which is typically named *const.h*. The content of this file is as follows:

```
#ifndef _CONST_H_
#define _CONST_H_
#include <fsm.h>
const uint8 CH_AUTOMATA_TYPE_ID = 0x00;
const uint8 CL_AUTOMATA_TYPE_ID = 0x01;
const uint8 USER_AUTOMATA_TYPE_ID = 0x02;

const uint8 CH_AUTOMATA_MBX_ID = 0x00;
const uint8 CL_AUTOMATA_MBX_ID = 0x01;
const uint8 USER_AUTOMATA_MBX_ID = 0x02;

// channel messages
const uint16 MSG_Connection_Request = 0x0001;
const uint16 MSG_Sock_Connection_Reject = 0x0002;
const uint16 MSG_Sock_Connection_Accept = 0x0003;
const uint16 MSG_Cl_MSG = 0x0004;
const uint16 MSG_Sock_MSG = 0x0005;
const uint16 MSG_Disconnect_Request = 0x0006;
const uint16 MSG_Sock_Disconnected = 0x0007;
const uint16 MSG_Sock_Disconnecting_Conf = 0x0008;

// pop3 client messages
const uint16 MSG_User_Check_Mail = 0x0009;
const uint16 MSG_Cl_Connection_Reject = 0x000a;
const uint16 MSG_Cl_Connection_Accept = 0x000b;
const uint16 MSG_User_Name_Password = 0x000c;
```

```
const uint16 MSG_MSG = 0x000d;
const uint16 MSG_Cl_Disconnected = 0x000f;

// user messages
const uint16 MSG_Set_All = 0x0010;
const uint16 MSG_User_Connected = 0x0011;
const uint16 MSG_User_Connection_Fail = 0x0012;
const uint16 MSG_Mail = 0x0013;
const uint16 MSG_User_Save_Mail = 0x0015;
const uint16 MSG_User_Disconnected = 0x0014;

#define ADRESS "krtlab8"
#define PORT 110

#define TIMER1_ID 1
#define TIMER1_COUNT 10
#define TIMER1_EXPIRED 0x20

#define PARAM_DATA 0x01
#define PARAM_Name 0x02
#define PARAM_Pass 0x03
#endif // _CONST_H_
```

The file *const.h* starts with the definitions of automata types and their private mailbox identifications. The identifications assigned to the classes *ChAuto*, *ClAuto*, and *UserAuto* are *CH_AUTOMATA_TYPE_ID*, *CL_AUTOMATA_TYPE_ID*, and *USER_AUTOMATA_TYPE_ID*, respectively. The identifications of their private mailboxes are *CH_AUTOMATA_MBX_ID*, *CL_AUTOMATA_MBX_ID*, and *USER_AUTOMATA_MBX_ID*, respectively. Next, we define the symbols that correspond to the codes of the messages recognized by the classes *ChAuto*, *ClAuto*, and *UserAuto*, respectively. By convention, these symbols are provided by prefixing the names of the messages from the SDL diagram (Figures 4.19 and 4.20) with the prefix *MSG_*.

At the end of the file *const.h*, we define the domain name and the number of the port, which are used to establish the TCP connection with the e-mail server (symbols *ADDRESS* and *PORT*), channel timer-related constants (symbols *TIMER1_ID*, *TIMER1_COUNT*, and *TIMER1_EXPIRED*), and the identifications of the message parameters (symbols *PARAM_DATA*, *PARAM_Name*, and *PARAM_Pass*).

Next, we write the header file *ClAuto.h*. Its content is as follows:

```
#ifndef _Cl_AUTO_H_
#define _Cl_AUTO_H_
#include <NetFSM.h>
#include <fsmsystem.h>
#include "const.h"
class ClAuto : public FiniteStateMachine {
 // for FSM
 StandardMessage StandardMsgCoding;
 MessageInterface *GetMessageInterface(uint32 id);
 void SetDefaultHeader(uint8 infoCoding);
 void SetDefaultFSMData();
 void NoFreeInstances();
 void Reset();
 uint8 GetMbxId();
```

```
uint8 GetAutomata();
uint32 GetObject();
void ResetData();
// FSM States
enum ClStates {
 FSM_Cl_Ready,
 FSM_Cl_Connecting,
 FSM_Cl_Authorizing,
 FSM_Cl_User_Check,
 FSM_Cl_Pass_Check,
 FSM_Cl_Mail_Check,
 FSM_Cl_Receiving,
 FSM_Cl_Deleting,
 FSM_Cl_Disconnecting
 };
public:
 ClAuto();
 ~ClAuto();
 void Initialize();
 void FSM_Cl_Ready_User_Check_Mail();
 void FSM_Cl_Connecting_Cl_Connection_Reject();
 void FSM_Cl_Connecting_Cl_Connection_Accept();
 void FSM_Cl_Authorizing_User_Name_Password();
 void FSM_Cl_User_Check_MSG();
 void FSM_Cl_Pass_Check_MSG();
 void FSM_Cl_Mail_Check_MSG();
 void FSM_Cl_Receiving_MSG();
 void FSM_Cl_Deleting_MSG();
 void FSM_Cl_Disconnecting_Cl_Disconnected();
protected:
 int m_MessageCount;
 char m_UserName[20];
 char m_Password[20];
 };
#endif /* _Cl_AUTO_H */
```

After listing all necessary header files, we declare the class *ClAuto*, which is derived from the base class *FiniteStateMachine*. The declaration of the class *ClAuto* starts with the declaration of field and function members that are mandatory for any class that is derived from the class *FiniteStateMachine* (as explained previously in this chapter). It continues with the declaration of FSM state names and state transition function prototypes.

By convention, FSM state names are the names from the SDL diagram with the prefix *FSM_* (e.g., the initial state *Cl_Ready* is named *FSM_Cl_Ready* in the C++ code). The state transition function is named by concatenating the state name and the input message name and by prefixing this composite name with *FSM_* (e.g., the state transition function performed when the FSM in state *Cl_Ready* receives the message *User_Check_Mail* is named *FSM_Cl_Ready_User_Check_Mail*). As previously mentioned, *ClAuto* FSM has nine states and fourteen state transitions.

The reader may be puzzled with the fact that there are fourteen valid FSM state transitions and only ten state transition functions declared in the header file *ClAuto.h*. This circumstance is because some of the state transitions are triggered with the same message type but different message content—e.g., *MSG(+OK)* and *MSG(-ERR)*—or they are guarded with the complementary conditions—e.g., ($nn > 0$) and !($nn > 0$). To clearly understand these matters, remember that *FiniteStateMachine* derivatives react to various message types in various FSM states. This is how we calculate the number of state transitions.

If we apply the principle stated above to the class *ClAuto*, we have the situation where all the states react to a single message with the exception of the state *Cl_Connecting*, which reacts to two valid messages, *Cl_Connection_Reject* and *Cl_Connection_Accept*. Because of this, we have (8×1) + (1×2) state transition functions, which resolves to ten state transition functions, as mentioned above.

Finally, we write the class *ClAuto* definition file, named *ClAuto.cpp*. The content of this file is as follows:

```
#include <stdio.h>
#include "const.h"
#include "ClAuto.h"
#define StandardMessageCoding 0x00

ClAuto::ClAuto() : FiniteStateMachine(0, 9, 2) {}
ClAuto::~ClAuto() {}

uint8 ClAuto::GetAutomate() {
 return CL_AUTOMATA_TYPE_ID;
}

uint8 ClAuto::GetMbxId() {
 return CL_AUTOMATA_MBX_ID;
}

uint32 ClAuto::GetObject() {
 return GetObjectId();
}

MessageInterface *ClAuto::GetMessageInterface(uint32 id) {
 return &StandardMsgCoding;
}

void ClAuto::SetDefaultHeader(uint8 infoCoding) {
 SetMsgInfoCoding(infoCoding);
 SetMessageFromData();
}

void ClAuto::SetDefaultFSMData() {
 SetDefaultHeader(StandardMessageCoding);
}

void ClAuto::NoFreeInstances() {
 printf("[%d] ClAuto::NoFreeInstances()\n", GetObjectId());
}

void ClAuto::Reset() {
 printf("[%d] ClAuto::Reset()\n", GetObjectId());
}

void ClAuto::Initialize() {
 SetState(FSM_Cl_Ready);

 // set message handlers
 InitEventProc(FSM_Cl_Ready, MSG_User_Check_Mail,
(PROC_FUN_PTR)&ClAuto::FSM_Cl_Ready_User_Check_Mail));
```

```
 InitEventProc(FSM_Cl_Connecting,  MSG_Cl_Connection_Reject,

(PROC_FUN_PTR)&ClAuto::FSM_Cl_Connecting_Cl_Connection_Reject));

 InitEventProc(FSM_Cl_Connecting,  MSG_Cl_Connection_Accept,

(PROC_FUN_PTR)&ClAuto::FSM_Cl_Connecting_Cl_Connection_Accept));

 InitEventProc(FSM_Cl_Authorizing,  MSG_User_Name_Password,

(PROC_FUN_PTR)&ClAuto::FSM_Cl_Authorizing_User_Name_Password));

 InitEventProc(FSM_Cl_User_Check, MSG_MSG,
   (PROC_FUN_PTR)&ClAuto::FSM_Cl_User_Check_MSG));

 InitEventProc(FSM_Cl_Pass_Check, MSG_MSG,
   (PROC_FUN_PTR)&ClAuto::FSM_Cl_Pass_Check_MSG));

 InitEventProc(FSM_Cl_Mail_Check, MSG_MSG,
   (PROC_FUN_PTR)&ClAuto::FSM_Cl_Mail_Check_MSG));

 InitEventProc(FSM_Cl_Receiving, MSG_MSG,
   (PROC_FUN_PTR)&ClAuto::FSM_Cl_Receiving_MSG));

 InitEventProc(FSM_Cl_Deleting, MSG_MSG,
   (PROC_FUN_PTR)&ClAuto::FSM_Cl_Deleting_MSG));

 InitEventProc(FSM_Cl_Disconnecting,  MSG_Cl_Disconnected,
(PROC_FUN_PTR)&ClAuto::FSM_Cl_Disconnecting_Cl_Disconnected));
}

void ClAuto::FSM_Cl_Ready_User_Check_Mail(){
 PrepareNewMessage(0x00, MSG_Connection_Request);
 SetMsgToAutomata(CH_AUTOMATA_TYPE_ID);
 SetMsgObjectNumberTo(0);
 SendMessage(CH_AUTOMATA_MBX_ID);
 SetState(FSM_Cl_Connecting);
}

void ClAuto::FSM_Cl_Connecting_Cl_Connection_Reject(){
 PrepareNewMessage(0x00, MSG_User_Connection_Fail);
 SetMsgToAutomata(USER_AUTOMATA_TYPE_ID);
 SetMsgObjectNumberTo(0);
 SendMessage(USER_AUTOMATA_MBX_ID);
 SetState(FSM_Cl_Ready);
}

void ClAuto::FSM_Cl_Connecting_Cl_Connection_Accept(){
 PrepareNewMessage(0x00, MSG_User_Connected);
 SetMsgToAutomata(USER_AUTOMATE_TYPA_ID);
 SetMsgObjectNumberTo(0);
 SendMessage(USER_AUTOMATA_MBX_ID);
 SetState(FSM_Cl_Authorizing);
}

void ClAuto::FSM_Cl_Authorizing_User_Name_Password(){
 char* name = new char[20];
 char* pass = new char[20];
 uint8* buffer = GetParam(PARAM_Name);

 memcpy(m_UserName,buffer+2,buffer[1]);
 m_UserName[buffer[1]] = 0;      // terminate string
 buffer = GetParam(PARAM_Pass);
```

```
 memcpy(m_Password,buffer+2,buffer[1]);
 m_Password[buffer[1]] = 0;      // terminate string
 char l_Command[20] = "user";
 strcpy(l_Command+5,m_UserName);
 strcpy(l_Command+5+strlen(m_UserName),"\r\n");

 PrepareNewMessage(0x00, MSG_Cl_MSG);
 SetMsgToAutomata(CH_AUTOMATA_TYPE_ID);
 SetMsgObjectNumberTo(0);
AddParam(PARAM_DATA,strlen(l_Command),(uint8*)l_Command);
 SendMessage(CH_AUTOMATA_MBX_ID);
 SetState(FSM_Cl_User_Check);
}

void ClAuto::FSM_Cl_User_Check_MSG(){
 char* data = new char[255];
 uint8* buffer = GetParam(PARAM_DATA);
 uint16 size = buffer[1];

 memcpy(data,buffer + 2,size);
 data[size]=0;
 printf("%s",data);
 if((data[0] == '+')) {
  char l_Command[20] = "pass ";
  strcpy(l_Command+5,m_Password);
  strcpy(l_Command+5+strlen(m_Password),"\r\n");
  PrepareNewMessage(0x00, MSG_Cl_MSG);
  SetMsgToAutomata(CH_AUTOMATA_TYPE_ID);
  SetMsgObjectNumberTo(0);
AddParam(PARAM_DATA,strlen(l_Command),(uint8*)l_Command);
  SendMessage(CH_AUTOMATA_MBX_ID);
  SetState(FSM_Cl_Pass_Check);
  else {
  char l_Command[20] = "quit\r\n";
  PrepareNewMessage(0x00, MSG_Cl_MSG);
  SetMsgToAutomata(CH_AUTOMATA_TYPE_ID);
  SetMsgObjectNumberTo(0);
  AddParam(PARAM_DATA,6,(uint8*)l_Command);
  SendMessage(CH_AUTOMATA_MBX_ID);
  SetState(FSM_Cl_Disconnecting);
 }
}

void ClAuto::FSM_Cl_Pass_Check_MSG(){
 char* data = new char[255];
 uint8* buffer = GetParam(PARAM_DATA);
 uint16 size = buffer[1];

 memcpy(data,buffer + 2,size);
 data[size]=0;
 printf("%s",data);
 if((data[0] == '+')) {
  char l_Command[20] = "stat\r\n";
  PrepareNewMessage(0x00, MSG_Cl_MSG);
  SetMsgToAutomata(CH_AUTOMATA_TYPE_ID);
  SetMsgObjectNumberTo(0);
  AddParam(PARAM_DATA,6,(uint8*)l_Command);
  SendMessage(CH_AUTOMATA_MBX_ID);
  SetState(FSM_Cl_Mail_Check);
  else {
  char l_Command[20] = "quit\r\n";
  PrepareNewMessage(0x00, MSG_Cl_MSG);
  SetMsgToAutomata(CH_AUTOMATA_TYPE_ID);
  SetMsgObjectNumberTo(0);
```

```
  AddParam(PARAM_DATA,6,(uint8*)l_Command);
  SendMessage(CH_AUTOMATA_MBX_ID);
  SetState(FSM_Cl_Disconnecting);
 }
}

void ClAuto::FSM_Cl_Mail_Check_MSG(){
 char* data = new char[255];
 uint8* buffer = GetParam(PARAM_DATA);
 uint16 size = buffer[1];

 memcpy(data,buffer+2,size);
 data[size]=0;
 printf("%s",data);
 int l_nDigit = 1;
 while(buffer[l_nDigit+6] != ' ') l_nDigit++;
 memcpy(data,buffer +6,l_nDigit);
 data[l_nDigit]=0;
 m_MessageCount = atoi(data);

 if((m_MessageCount == 0) {
   char l_Command[20] = "quit\r\n";
   PrepareNewMessage(0x00, MSG_Cl_MSG);
   SetMsgToAutomata(CH_AUTOMATA_TYPE_ID);
   SetMsgObjectNumberTo(0);
   AddParam(PARAM_DATA,6,(uint8*)l_Command);
   SendMessage(CH_AUTOMATA_MBX_ID);
   SetState(FSM_Cl_Disconnecting);
   else {
   char l_Command[20] = "retr ";
   strcpy(l_Command+5,data);
   strcpy(l_Command+5+l_nDigit,"\r\n");
   PrepareNewMessage(0x00, MSG_Cl_MSG);
   SetMsgToAutomata(CH_AUTOMATA_TYPE_ID);
  SetMsgObjectNumberTo(0);

AddParam(PARAM_DATA,5+l_nDigit+2,(uint8*)l_Command);
  SendMessage(CH_AUTOMATA_MBX_ID);
  SetState(FSM_Cl_Receiving);
 }
}

void ClAuto::FSM_Cl_Receiving_MSG(){
 char* data = new char[255];
 uint8* buffer = GetParam(PARAM_DATA);
 uint16 size = buffer[1];

 memcpy(data,buffer + 2,size);
 char temp[4];
 memcpy(temp,data,3); temp[3] = 0;
 if((strcmp(temp,"+OK") != 0) {
   PrepareNewMessage(0x00, MSG_Mail);
   SetMsgToAutomata(USER_AUTOMATA_TYPE_ID);
   SetMsgObjectNumberTo(0);
   AddParam(PARAM_DATA,size,(uint8*)data);
   SendMessage(USER_AUTOMATA_MBX_ID);
   if((size < 255) {
     char l_Command[20] = "dele ";
     itoa(m_MessageCount,data,10);
     strcpy(l_Command+5,data);
     strcpy(l_Command+5+strlen(data),"\r\n");
     PrepareNewMessage(0x00, MSG_Cl_MSG);
     SetMsgToAutomata(CH_AUTOMATA_TYPE_ID);
     SetMsgObjectNumberTo(0);
```

```
AddParam(PARAM_DATA,5+strlen(data)+2,(uint8*)l_Command);
    SendMessage(CH_AUTOMATA_MBX_ID);
    SetState(FSM_Cl_Deleting);
  }
 }
}

void ClAuto::FSM_Cl_Deleting_MSG(){
 PrepareNewMessage(0x00, MSG_User_Save_Mail);
SetMsgToAutomata(USER_AUTOMATA_TYPE_ID);
 SetMsgObjectNumberTo(0);
 SendMessage(USER_AUTOMATA_MBX_ID);
 m_MessageCount—;
 if(m_MessageCount > 0) {
  char data[5];
  char l_Command[20] = "retr ";
  itoa(m_MessageCount,data,10);
  strcpy(l_Command+5,data);
  strcpy(l_Command+5+strlen(data),"\r\n");
  PrepareNewMessage(0x00, MSG_Cl_MSG);
  SetMsgToAutomata(CH_AUTOMATA_TYPE_ID);
  SetMsgObjectNumberTo(0);

AddParam(PARAM_DATA,5+strlen(data)+2,(uint8*)l_Command);
  SendMessage(CH_AUTOMATA_MBX_ID);
  SetState(FSM_Cl_Receiving);
  else {
  char l_Command[20] = "quit\r\n";
  PrepareNewMessage(0x00, MSG_Cl_MSG);
  SetMsgToAutomata(CH_AUTOMATA_TYPE_ID);
  SetMsgObjectNumberTo(0);
  AddParam(PARAM_DATA,6,(uint8*)l_Command);
  SendMessage(CH_AUTOMATA_MBX_ID);
  SetState(FSM_Cl_Disconnecting);
 }
}

void ClAuto::FSM_Cl_Disconnecting_Cl_Disconnected(){
 PrepareNewMessage(0x00, MSG_User_Disconnected);
 SetMsgToAutomata(USER_AUTOMATA_TYPE_ID);
 SetMsgObjectNumberTo(0);
 SendMessage(USER_AUTOMATA_MBX_ID);
 SetState(FSM_Cl_Ready);
}
```

The file *ClAuto.cpp* starts with a list of all necessary header files (*stdio.h,*
const.h, and *ClAuto.h*), followed by the definition of the symbolic constant
StandardMessageCoding and the set of mandatory function definitions: class
constructor, class destructor, and functions *GetAutomata(), GetMbxId(),*
GetObject(), GetMessageInterface(), SetDefaultHeader(), SetDefaultFSMData(),
NoFreeInstances(), Reset(), and *Initialize().*

The class constructor *ClAuto()* calls the constructor of the class
FiniteStateMachine with a list of parameters, which specifies that the
ClAuto FSM has no timers, nine states, and the maximum of two state
transitions per state (see the FSM Library API specification in Section 6.8,
particularly, Section 6.8.11). The class destructor performs no particular
operation.

The mandatory functions provide the following functionalities:

- The function *GetAutomata()* returns the *ClAutomata* type identification (the constant *CL_AUTOMATA_TYPE_ID*). See also Section 6.8.24.

- The function *GetMbxId()* returns the associated mailbox identification (the constant *CL_AUTOMATA_MBX_ID*). See also Section 6.8.38.

- The function *GetObject()* returns the object identification. Actually, it returns the value returned by the FSM Library function *GetObject Id()*. See also Section 6.8.60.

- The function *GetMessageInterface()* returns the pointer to the message coding object (an instance of the class *StandardMessage*). See also Section 6.8.39.

- The function *SetDefaultHeader()* sets default data in the new message header by calling two FSM Library functions, *SetMsgInfoCoding()* and *SetMessageFromData()*. See also Section 6.8.97, Section 6.8.117, and Section 6.8.108.

- The function *SetDefaultFSMData()* sets the new message header default values by calling the function *SetDefaultHeader()* and specifying the constant *StandardMessageCoding* as its parameter.

- The function *NoFreeInstances()* just prints the information message to the standard output file. See also Section 6.8.78.

- The function *Reset()* also just prints the information message to the standard output file. See also Section 6.8.85.

The most important mandatory function is the function *Initialize()*. It starts by setting the FSM initial state, *Cl_Ready* (denoted with the constant *FSM_Cl_Ready*). It continues by setting the state transition functions (also referred to as message handlers). Each message handler is set by a single call to the FSM Library function *InitEventProc()*. The first parameter of this function is the state name, the second is the input message name, and the third is the address of the corresponding *ClAuto* function member (see also Section 6.8.73).

The set of mandatory functions is followed by the set of state transition functions. As already mentioned, ten such functions are used. Each of these functions processes a single message type in a single state, as follows:

- The function *FSM_Cl_Ready_User_Check_Mail()* processes the message *User_Check_Mail* in the state *Cl_Ready*.

- The function *FSM_Cl_Connecting_Cl_Connection_Reject()* processes the message *Cl_Connection_Reject* in the state *Cl_Connecting*.

- The function *FSM_Cl_Connecting_Cl_Connection_Accept()* processes the message *Cl_Connection_Accept* in the state *Cl_Connecting*.

- The function *FSM_Cl_Authorizing_User_Name_Password()* processes the message *User_Name_Password* in the state *Cl_Authorizing*.
- The function *FSM_Cl_User_Check_MSG()* processes the message *MSG* in the state *Cl_User_Check*.
- The function *FSM_Cl_Pass_Check_MSG()* processes the message *MSG* in the state *Cl_Pass_Check*.
- The function *FSM_Cl_Mail_Check_MSG()* processes the message *MSG* in the state *Cl_Mail_Check*.
- The function *FSM_Cl_Receiving_MSG()* processes the message *MSG* in the state *Cl_Receiving*.
- The function *FSM_Cl_Deleting_MSG()* processes the message *MSG* in the state *Cl_Deleting*.
- The function *FSM_Cl_Disconnecting_Cl_Disconnected()* processes the message *Cl_Disconnected* in the state *Cl_Disconnecting*.

The function *FSM_Cl_Ready_User_Check_Mail()* is a typical simple state transition function. First, it creates a new message by calling the function *PrepareNewMessage()*. (Its first parameter is the message length and the second is the message type; the third parameter is optional and is not used in this example. See also Section 6.8.81.) It then sets the destination FSM type and object identification by calling the function *SetMsgToAutomata()* (its parameter is the FSM type identification; see also Section 6.8.125) and the function *SetMsgObjectNumberTo()* (its parameter is the FSM object identification; see also Section 6.8.123), respectively. Next, it sends the new message to the destination mailbox by calling the function *SendMessage()* (its parameter is the mailbox identification; see also Section 6.8.106). Finally, it sets the new FSM state by calling the function *SetState* (its parameter is the state identification; see also Section 6.8.137).

The next two functions, *FSM_Cl_Connecting_Cl_Connection_Reject()* and *FSM_Cl_Connecting_Cl_Connection_Accept()*, are very similar to the one previously described (only the message type and the new state name are different). But the fourth state transition function, *FSM_Cl_Authorizing_User_Name_Password()*, is more complex. It demonstrates well how a state transition function can get a parameter from the current message and how it can add a parameter to the new message. This concrete state transition function gets two parameters (username and password) from the current message by calling the function *GetParam()* (its parameter is the identification of the parameter type; see also Section 6.8.61). It also adds one parameter (username) to the new message by calling the function *AddParam()* (its parameters are the message parameter type, length, and pointer; see also Section 6.8.12).

The fifth state transition function, *FSM_Cl_User_Check_MSG()*, is even more complex because it involves branching depending on the value of the current message parameter. By making a branch, the state transition function actually selects one of two possible paths of the FSM evolution, which yields two different output (new) messages and two different destination FSM states. The sixth state transition function is very similar to the fifth one.

The seventh state transition function, *FSM_Cl_Mail_Check_MSG()*, brings one new important detail. It shows how a state transition function can save some data (in this example, the number of pending e-mail messages, which is stored in the class field member *m_MessageCount*) so that it can be shared or used by other state transitions—in this example, by the ninth state transition function, *FSM_Cl_Deleting_MSG()*.

The rest of the state transition functions do not bring anything essentially new. However, the reader is advised to study them in detail as an additional exercise.

4.5.2 Example 2

The aim of this example is to implement the SIP invite client transaction design, which is given in Section 3.10.5 (Chapter 3, Example 5). Briefly, in that section we examined the general collaboration diagram of the SIP Softphone (see Section 2.3.3, Figure 2.16) with the focus on the invite client transaction. The result is the general collaboration diagram shown in Figure 3.69. We then made two particular collaboration diagrams and their semantically equivalent sequence diagrams for the cases of successful and unsuccessful SIP session establishment (Figures 3.70 through 3.73). Finally, we devised the complete dynamic behavior specification in the form of the statechart diagram (Figure 3.74) and semantically equivalent SDL diagram (Figures 3.75 through 3.78).

We start the implementation of this design by defining the symbolic constants, such as the FSM type names (e.g., the name of the invite client FSM type is *InviteClienteTE_FSM*), mailbox names (e.g., the name of the invite client mailbox is *InviteClienteTE_FSM_MBX*), names of the FSM Library related message types, timer names (e.g., *TIMER_A*, *TIMER_B*, and *TIMER_D*), names of the SIP messages (e.g., *INVITE, OPTIONS, CANCEL, ACK, BYE,* and *RESISTER*), names of the response codes (e.g., *_180_RINGING, _200_OK, _302_MOVED_TEMPORARILY, _401_UNAUTHORIZED, _403_FORBIDDEN,* and *_404_NOT_FOUND*), and names of situations (e.g., *URI_IN_TO_UNRECOGNIZED* and *NOT_TO_CURRENT_USER*). Traditionally, we write definitions of all these constants in the file *constants.h*.

Next, we write the class that represents an SIP message, simply named *Message*. The most important field member of this class is the last (also referred to as the current) SIP message (its type is the C++ type string). Other field members hold the relevant SIP session related information. The function members support SIP message analysis and synthesis (parsing and creation). Actually, the class *Message* that is used in this example is a simple wrapper around the OpenSIP SIP message parser. (OpenSIP is freely available on the Internet at https://www.opensips.org/.)

We skip the content of the file *constants.h* and the source code of the class *Message* intentionally to keep this example short enough and easily comprehendible, and we proceed with the introduction of the supplementary class *TALE*. The declaration of the class *TALE* is as follows:

```
#ifndef _TALE_FSM_
#define _TALE_FSM_
#include "../kernel/fsm.h"
#include "../message/message.h"
#include "../constants.h"

class TALE : public FiniteStateMachine {
 uint8 MessageCopy[MAX_LENGTH_MESSAGE];
 uint32 IndexTLI;
 BOOL IndexTLISet;
public:
 void SetIndexTLI(uint32 newIndexTLI);
 uint32 GetIndexTLI();
 BOOL IsTransportReliable();
 void SendMessageToTU();
 void SendMessageToTPL();
 void SendErrorMessageToTU();
 void MakeLocalCopyOfMsg();
 void SendCopiedMessageToTPL();

public:
 TALE(uint16 numOfTimers, uint16 numOfState, uint16 maxNumOfPrPerSt);
 ~TALE();
};
```

The class *TALE* is a good example of how we can make our implementations more compact. As we can see from the previous example, sending a single message requires a series of FSM Library function calls. For example, forwarding the current message would require a series of calls to the function *CopyMessage()*, *SetMsgToAutomata()*, *SetMsgToGroup()*, *SetMsgObjectNumberTo()*, and function *SendMessage()*—five function calls. In the case of simple designs, we can tolerate repetition of this series of function calls, but in cases requiring more complex design or platforms with limited resources, this repetition may not be tolerated.

Consider the SIP invite client transaction FSM. It has thirteen state transitions, and most of them require sending a message to either the TPL (transport layer) or TU (transaction user). We would need to repeat the same series of function calls about ten times. Consider now the whole SIP Softphone, which supports four types of transactions (invite, non-invite, client, and server transactions). In such situations, replacing this series of function calls with a single function call (which, in turn, performs the original sequence of function calls) makes sense.

This replacement is exactly the reason why the class *TALE* has been introduced in the first place. This class inherits all field and function members from the class *FiniteStateMachine*, from which it is derived. It also adds some new field and function class members. All classes that implement SIP transactions are derived from the class *TALE*. The most important field member of the class *TALE* is the field *MessageCopy*, which holds the copy of the last sent message. Actually, this field is the retransmission buffer (remember that SIP invite client in the state *Calling* must retransmit the message *INVITE* in case the timer A expires).

The two most important function members are the functions *SendMessageToTU()* and *SendMessageToTPL()*. The former sends the current message to TU and the latter to TPL. They are very similar; therefore, it is sufficient to study just one of them. Here is the source code of the former function:

```
void TALE::SendMessageToTU() {
 CopyMessage();
 SetMsgToAutomata(UA_Disp_FSM);
 SetMsgToGroup(INVALID_08);
 SetMsgObjectNumberTo(0);
 SendMessage(UA_Disp_FSM_MBX);
}
```

This is the most elegant way to forward a message in FSM Library-based implementations. The function *CopyMessage()* copies the current (last received) message to the new (output) message. The symbolic constant *UA_Disp_FSM* is the name of the UA (user agent) FSM type, and the constant *UA_Disp_FSM_MBX* is the name of its mailbox. As we will shortly see, the use of the functions *SendMessageToTU()* and *SendMessageToTPL()* significantly compresses the source code. They make one-to-one mapping of SDL diagrams to C++ code possible.

Next, we proceed to the implementation of the invite client transaction FSM. We implement it by writing the class *InviteClientTE*. Note that in Figures 3.69 through 3.73, we used the abbreviation *InClientT* for this name. The declaration of the class *InviteClientTE* is as follows:

```
#ifndef _InviteClientTE_FSM_
#define _InviteClientTE_FSM_
#include "TALE.h"

class InviteClientTE : public TALE {
 Message SIPMsg;
 uint32 cseq_number;
 uint32 TimerADuration;

public:
 enum States {
  STATE_INITIAL,
  STATE_CALLING,
  STATE_PROCEEDING,
  STATE_COMPLETED
 };
 // state Initial message handlers
 void Evt_Init_INVITE();
 // state Calling message handlers
 void Evt_Calng_TIMER_A_EXP();
 void Evt_Calng_RESPONSE_1XX();
 void Evt_Calng_RESPONSE_2XX();
 void Evt_Calng_TIMER_B_EXP();
 void Evt_Calng_RESPONSE_3_6XX();
 void Evt_Calng_TRANSPORT_ERR();
 // state Proceeding message handlers
```

```
void Evt_Proc_RESPONSE_1XX();
void Evt_Proc_RESPONSE_2XX();
void Evt_Proc_RESPONSE_3_6XX();
// state Completed message handlers
void Evt_Comptd_TIMER_D_EXP();
void Evt_Comptd_RESPONSE_3_6XX();
void Evt_Comptd_TRANSPORT_ERR();
// unexpected messages message handler
void Event_UNEXPECTED();
// problem specific functions
void RetransmitInvite();
BOOL SendAckMessageToTPL();
// FiniteStateMachine abstract functions
StandardMessage StandardMsgCoding;
MessageInterface *GetMessageInterface(uint32 id);
void SetDefaultHeader(uint8 infoCoding);
void SetDefaultFSMData();
void NoFreeInstances();
void Reset();
uint8 GetMbxId();
uint8 GetAutomate();
uint32 GetObject();
void ResetData();
public:
```

The class *InviteClientTE* is derived from the class *TALE*. The meaning of its field members is as follows:

- The field *SIPMsg* is the SIP message parser (an instance of the class *Message*).

- The field *cseq_number* holds the value of the SIP message header field *CSeq*, which is used to identify and order transactions (see RFC 3261, Subsection 8.1.1.5).

- The field *TimerADuration* contains the current value of the timer A (remember, the value of the timer A is doubled each time it expires).

Next, we enumerate the names of the FSM states. There are altogether four FSM states: *STATE_INITIAL*, *STATE_CALLING*, *STATE_PROCEEDING*, and *STATE_COMPLETED*. A short explanation is needed at this point. According to the original specification (RFC 3261, Figure 5, page 128), the invite client transaction FSM also has four explicitly rendered states, namely, *Calling*, *Proceeding*, *Completed*, and *Terminated*. The initial state is omitted in the original specification. In our implementation, we create a pool of *InviteClientTE* objects, which are dynamically allocated on demand by the TU. These objects are never really terminated. Once they play their simple role, they are returned to the pool of free *InviteClientTE* objects, and from there they are dynamically assigned to play the same role again. Therefore, we renamed the state *Terminated* to *Initial*. We also made this state the source of the initial state transition (triggered with the message *INVITE* from TU), thus making the FSM a never-terminating one.

We then list the state transition function prototypes for each state individually. The naming convention is the same as in the previous example:

The name of the state transition function is constructed by concatenating the state name and the message name and by prefixing that name with a certain prefix. The naming convention is applied more freely in this example by shortening the state names. This practice is frequently done to keep the name lengths acceptable (short enough, but providing code readability at the same time). Thirteen valid state transitions and their corresponding state transition functions (message handlers) are used. The fourteenth message handler, named *Event_UNEXPECTED()*, handles all unexpected messages in all states.

Finally, we list the function prototypes of the problem-specific functions and mandatory *FiniteStateMachine* abstract functions. These functions—except the function *RetransmitInvite()*—are intentionally skipped in the text that follows to keep the presentation of this example short.

We finish the implementation by writing the class *InviteClientTE* definition file, named *InvClientTE.cpp*. The content of this file is as follows:

```
#include <stdio.h>
#include "InvClientTE.h"
#include "../Message/message.h"
#include "timer_values.h"
#define StandardMessageCoding 0x00

InviteClientTE::InviteClientTE() : TALE(10, 10, 10) {}
InviteClientTE::~InviteClientTE() {}

void InviteClientTE::Initialize() {
 SetState(STATE_INITIAL);
 // define timers
 InitTimerBlock(TIMER_A,1,TIMER_A_EXPIRED);
 InitTimerBlock(TIMER_B,1,TIMER_B_EXPIRED);
 InitTimerBlock(TIMER_D,1,TIMER_D_EXPIRED);
 // state STATE_INITIAL message handlers
 InitEventProc(STATE_INITIAL, INVITE,
   (PROC_FUN_PTR)&InviteClientTE::Evt_Init_INVITE);
 // state STATE_CALLING message handlers
InitEventProc(STATE_CALLING, TIMER_A_EXPIRED,
  (PROC_FUN_PTR)&InviteClientTE::Evt_Calng_TIMER_A_EXP);

InitEventProc(STATE_CALLING, RESPONSE_1XX_T,
  (PRO_FUN_PTR)&InviteClientTE::Evt_Calng_RESPONSE_1XX);

InitEventProc(STATE_CALLING, RESPONSE_2XX_T,
  (PROC_FUN_PTR)&InviteClientTE::Evt_Calng_RESPONSE_2XX);

InitEventProc(STATE_CALLING, TIMER_B_EXPIRED,
  (PROC_FUN_PTR)&InviteClientTE::Evt_Calng_TIMER_B_EXP);

InitEventProc(STATE_CALLING, RESPONSE_3XX_T,
  (PROC_FUN_PTR)&InviteClientTE::Evt_Calng_RESPONSE_3_6XX);

InitEventProc(STATE_CALLING, RESPONSE_4XX_T,
  (PROC_FUN_PTR)&InviteClientTE::Evt_Calng_RESPONSE_3_6XX);

InitEventProc(STATE_CALLING, RESPONSE_5XX_T,
  (PROC_FUN_PTR)&InviteClientTE::Evt_Calng_RESPONSE_3_6XX);

InitEventProc(STATE_CALLING, RESPONSE_6XX_T,
  (PROC_FUN_PTR)&InviteClientTE::Ev_Calng_RESPONSE_3_6XX);
```

```
InitEventProc(STATE_CALLING, TRANSPORT_ERR,
  (PROC_FUN_PTR)&InviteClientTE::Evt_Calng_TRANSPORT_ERR);

  // state STATE_PROCEEDING message handlers
InitEventProc(STATE_PROCEEDING, RESPONSE_1XX_T,
  (PROC_FUN_PTR)&InviteClientTE::Evt_Proc_RESPONSE_1XX);

InitEventProc(STATE_PROCEEDING, RESPONSE_2XX_T,
  (PROC_FUN_PTR)&InviteClientTE::Evt_Proc_RESPONSE_2XX);

InitEventProc(STATE_PROCEEDING, RESPONSE_3XX_T,
  (PROC_FUN_PTR)&InviteClientTE::Evt_Proc_RESPONSE_3_6XX);

InitEventProc(STATE_PROCEEDING, RESPONSE_4XX_T,
  (PROC_FUN_PTR)&InviteClientTE::Evt_Proc_RESPONSE_3_6XX);

InitEventProc(STATE_PROCEEDING, RESPONSE_5XX_T,
  (PROC_FUN_PTR)&InviteClientTE::Evt_Proc_RESPONSE_3_6XX);

InitEventProc(STATE_PROCEEDING, RESPONSE_6XX_T,
  (PROC_FUN_PTR)&InviteClientTE::Evt_Proc_RESPONSE_3_6XX);

  // state STATE_COMPLETED message handlers
InitEventProc(STATE_COMPLETED, TIMER_D_EXPIRED,
  (PROC_FUN_PTR)&InviteClientTE::Evt_Comptd_TIMER_D_EXP);

InitEventProc(STATE_COMPLETED, RESPONSE_3XX_T,
  (PROC_FUN_PTR)&InviteClientTE::Evt_Comptd_RESPONSE_3_6XX);

InitEventProc(STATE_COMPLETED, RESPONSE_4XX_T,
  (PROC_FUN_PTR)&InviteClientTE::Evt_Comptd_RESPONSE_3_6XX);

InitEventProc(STATE_COMPLETED, RESPONSE_5XX_T,
  (PROC_FUN_PTR)&InviteClientTE::Evt_Comptd_RESPONSE_3_6XX);
InitEventProc(STATE_COMPLETED, RESPONSE_6XX_T,
  (PROC_FUN_PTR)&InviteClientTE::Evt_Comptd_RESPONSE_3_6XX);

InitEventProc(STATE_COMPLETED, TRANSPORT_ERR,
  (PROC_FUN_PTR)&InviteClientTE::Evt_Comptd_TRANSPORT_ERR);

  // unexpected messages message handler
InitUnexpectedEventProc(STATE_INITIAL,
  (PROC_FUN_PTR)&InviteClientTE::Event_UNEXPECTED);

InitUnexpectedEventProc(STATE_CALLING,
  (PROC_FUN_PTR)&InviteClientTE::Event_UNEXPECTED);

InitUnexpectedEventProc(STATE_PROCEEDING,
  (PROC_FUN_PTR)&InviteClientTE::Event_UNEXPECTED);

InitUnexpectedEventProc(STATE_COMPLETED,
  (PROC_FUN_PTR)&InviteClientTE::Event_UNEXPECTED);
}

void InviteClientTE::Evt_Init_INVITE() {
 SendMessageToTPL();
 if (!IsTransportReliable()){
  TimerADuration = GetT1();
  setTimerCount(TIMER_A, TimerADuration);
  StartTimer(TIMER_A);
 }
 setTimerCount(TIMER_B, 64*GetT1());
 StartTimer(TIMER_B);
 MakeLocalCopyOfMsg();
```

```
 SetState(STATE_CALLING);
}

void InviteClientTE::Evt_Calng_TIMER_A_EXP(){
 TimerADuration = 2 * TimerADuration;
 setTimerCount(TIMER_A, TimerADuration);
 RestartTimer(TIMER_A);
 RetransmitInvite();
}

void InviteClientTE::Evt_Calng_RESPONSE_1XX(){
 uint16 val;
 StopTimer(TIMER_A);
 StopTimer(TIMER_B);
 SendMessageToTU();
 GetParamWord(INDEX_TLI_PARAM, val);
 SetIndexTLI(val);
 SetState(STATE_PROCEEDING);
}

void InviteClientTE::Evt_Calng_RESPONSE_2XX(){
 StopTimer(TIMER_A);
 StopTimer(TIMER_B);
 SendMessageToTU();
 SetState(STATE_INITIAL);
}

void InviteClientTE::Evt_Calng_TIMER_B_EXP(){
 StopTimer(TIMER_A);
 SendErrorMessageToTU();
 SetState(STATE_INITIAL);
}

void InviteClientTE::Evt_Calng_TRANSPORT_ERR(){
 StopTimer(TIMER_A);
 StopTimer(TIMER_B);
 SendErrorMessageToTU();
 SetState(STATE_INITIAL);
}

void InviteClientTE::Evt_Calng_RESPONSE_3_6XX(){
 uint16 val;
 StopTimer(TIMER_A);
 StopTimer(TIMER_B);
 SendMessageToTU();
 GetParamWord(INDEX_TLI_PARAM, val);
 SetIndexTLI(val);
 SendAckMessageToTPL();
 if (IsTransportReliable())
  setTimerCount(TIMER_D, ZERO_TIMER_VAL_APPROX);
 else
  setTimerCount(TIMER_D, 64*GetT1());//64T1
 StartTimer(TIMER_D);
 SetState(STATE_COMPLETED);
}

void InviteClientTE::Evt_Proc_RESPONSE_1XX(){
 SendMessageToTU();
}

void InviteClientTE::Evt_Proc_RESPONSE_2XX(){
 SendMessageToTU();
 SetState(STATE_INITIAL);
}
```

```
void InviteClientTE::Evt_Proc_RESPONSE_3_6XX(){
 SendMessageToTU();
 SendAckMessageToTPL();
 if (IsTransportReliable())
  setTimerCount(TIMER_D, ZERO_TIMER_VAL_APPROX);
 else
  setTimerCount(TIMER_D, 64*GetT1()); //64T1
 StartTimer(TIMER_D);
 SetState(STATE_COMPLETED);
}

void InviteClientTE::Evt_Comptd_TIMER_D_EXP(){
 SetState(STATE_INITIAL);
}

void InviteClientTE::Evt_Comptd_RESPONSE_3_6XX(){
 SendAckMessageToTPL();
}

void InviteClientTE::Evt_Comptd_TRANSPORT_ERR(){
 StopTimer(TIMER_D);
 SendErrorMessageToTU();
 SetState(STATE_INITIAL);
}

void InviteClientTE::Event_UNEXPECTED() {
}

void InviteClientTE::RetransmitInvite(){
 SendCopiedMessageToTPL();
}
```

The mandatory function *Initialize()* starts by setting the FSM initial state *STATE_INITIAL*. It then initializes the timers A, B, and D by calling the FSM Library function *InitTimerBlock()* (its parameters are the timer identification, the timer interval duration, and the identification of the associated message; see also Section 6.8.74). The function *Initialize()* finishes by setting the FSM state transition functions. These functions process various message types in different states, as follows:

- The function *Evt_Init_INVITE()* processes the message *INVITE* in the state *STATE_INITIAL*.
- The function *Evt_Calng_TIMER_A_EXP()* processes the message *TIMER_A_EXPIRED* in the state *STATE_CALLING*.
- The function *Evt_Calng_RÈSPONSE_1XX()* processes the message *RESPONSE_1XX_T* in the state *STATE_CALLING*.
- The function *Evt_Calng_ RESPONSE_2XX()* processes the message *RESPONSE_2XX_T* in the state *STATE_CALLING*.
- The function *Evt_Calng_TIMER_B_EXP()* processes the message *TIMER_B_EXPIRED* in the state *STATE_CALLING*.
- The function *Evt_Calng_RESPONSE_3_6XX()* processes the messages *RESPONSE_3XX_T, RESPONSE_4XX_T, RESPONSE_5XX_T,* and *RESPONSE_6XX_T* in the state *STATE_CALLING*.

- The function *Evt_Calng_TRANSPORT_ERR()* processes the message *TRANSPORT_ERR* in the state *STATE_CALLING*.

- The function *Evt_Proc_RESPONSE_1XX()* processes the message *RESPONSE_1XX_T* in the state *STATE_PROCEEDING*.

- The function *Evt_Proc_RESPONSE_2XX()* processes the message *RESPONSE_2XX_T* in the state *STATE_PROCEEDING*.

- The function *Evt_Proc_RESPONSE_3_6XX()* processes the messages *RESPONSE_3XX_T, RESPONSE_4XX_T, RESPONSE_5XX_T,* and *RESPONSE_6XX_T* in the state *STATE_PROCEEDING*.

- The function *Evt_Comptd_TIMER_D_EXP()* processes the message *TIMER_D_EXPIRED* in the state *STATE_COMPLETED*.

- The function *Evt_Comptd_RESPONSE_3_6XX()* processes the messages *RESPONSE_3XX_T, RESPONSE_4XX_T, RESPONSE_5XX_T,* and *RESPONSE_6XX_T* in the state *STATE_COMPLETED*.

- The function *Evt_Comptd_TRANSPORT_ERR()* processes the message *TRANSPORT_ERR* in the state *STATE_COMPLETED*.

- The function *Event_UNEXPECTED()* processes all unexpected messages in all states.

As we can see from the source code above, the state transition functions (message handlers) are short and easily readable because each program statement is easily traceable back to the original statechart and SDL diagrams. For example, consider the first state transition function *Evt_Init_INVITE()*. The original SDL specification of this state transition starts with the reception of the message *INVITE* (Figure 3.75). This step is provided by the class *FSMSystem*. The next step in the SDL diagram says: "Invite_T to TPL." This step is implemented with a single program statement, namely, the function call to the function *SendMessageToTPL()*.

The next step in the SDL diagram is the question, "Is transport reliable?" We implement it also with a single function call to the function *IsTransportReliable()*. We continue the SDL coding in this manner. If the transport is reliable, the initial value of the timer A is provided by calling the function *GetT1()*—a way to parameterize the software. Next, we set the timer A duration by calling the function *setTimerCount()*—this is the undocumented FSM Library function at the moment, to be included in the next official release—and start the timer A by calling the function *StartTimer()* (the parameter of this function is the timer identification; see also Section 6.8.138).

At the end of this function, we set the duration of the timer B and start it, make the local copy of the last sent message by calling the function *MakeLocalCopy()*—remember that it is needed for the possible retransmission—and set the new state by calling the function *SetState()* (its parameter is the state identification; see also Section 6.8.137).

Next, the state transmission function *EvtCalng_TIMER_A_EXP()* performs the reaction to the timer A expiration (see the corresponding SDL specification in Figure 3.75) with only four program statements. The first one doubles the timer A duration, the second sets this new duration, the third restarts the timer A by calling the FSM Library function *RestartTimer()* (see Section 6.8.87), and the fourth retransmits the message *INVITE* by calling the function *RetransmitInvite()*. Also, all the other state transition functions are made in this spirit of one-to-one mapping from the original SDL diagram. The reader is advised to study them as an additional exercise.

References

Booch, G., Rumbaugh, J., and Jacobson, I., *The Unified Modeling Language User Guide*, Addison-Wesley, Reading, MA, 1998.

Gamma, E., Helm, R., Johnson, R., and Vlissides, J., *Design Patterns: Elements of Reusable Object-Oriented Software*, Addison-Wesley, Reading, MA, 1995.

5

Test and Verification

The test and verification phase is a phase of communication protocol engineering work that follows the implementation phase. The primary goal of this phase is to verify that the implementation in the higher-level programming language is correct. The implementation is correct if it meets its original requirements, which are modeled in the form of use cases (see Chapter 2).

The correctness of the implementation is checked with the test suite, which is typically designed in TTCN-3 (see Section 3.9). The test suite itself is implemented in a higher-level programming language, e.g., Java or C++. But how do we verify the correctness of the test suite implementation? The answer is that we do not check the correctness of the test suite independently. We always check the correctness of the implementation under the test and test suite simultaneously. Theoretically, a bug in a test suite can cover a bug in the implementation; we should be aware of this, but such cases seldom happen in practice.

Typical testing activities conducted in the communication protocol engineering test and verification phase are the following:

- Unit testing
- Integration testing
- Conformance testing
- Load testing
- In-field testing
- Formal verification
- Statistical usage testing

The first four types of activities (unit testing, conformance testing, load testing, and in-field testing) stem from traditional software engineering, whereas the last two (formal verification and statistical usage testing) originate from Cleanroom engineering. Today, communication protocol engineers tend to complement software engineering with Cleanroom engineering testing approaches, therefore we cover all the above listed activities in this chapter.

As its name suggests, unit testing is used for testing individual software units before their integration into the product. Typically, a software unit is a single class written in a separate Java compilation unit or C++ module. This class most commonly implements a simple communication protocol or part of a more complex communication protocol. In the case of the FSM Library–based paradigm, such a unit would be a C++ module that defines the class derived from the class *FiniteStateMachine*.

Unit testing of communication protocols is relatively straightforward. Typically, we construct a set of test cases that check individual FSM state transitions, as well as more complex FSM transactions (series of FSM state transitions). We will use JUnit and CppUinit testing frameworks for unit testing of communication protocols in this book. Details of unit testing are given in Section 5.1 (Unit Testing) and Section 5.5.1 (Example 1).

The next phase is integration testing. The philosophy of integration testing starts from the fact that some of the units have successfully undergone unit testing and that they are available for further testing, whereas the rest of them are not. For the purpose of integration testing, we introduce replacements for the units that are not available, which are referred to as the imitators (or simulators).

There are two kinds of imitators, namely drivers and stubs. A driver is an active imitator that generates input messages for the real objects (units) under test. A stub is a passive imitator that accepts the output messages generated by the objects under test. Stubs can also send replays that are expected from the objects they are imitating. Of course, we can construct more complex imitators that act as both drivers and stubs. In this book, we will call the collaborations of real objects, *drivers,* and stubs simply *integration test collaborations.*

Generally, communication protocols are well suited for integration testing because families of communication protocols are hierarchically organized in layers with well-defined interfaces. The communication between individual protocols is based on messages, which are traditionally exchanged through the mailboxes (as in implementations based on the FSM Library). Simulating the environment of a real object under test in such a situation is easy. Drivers and stubs simply exchange messages with objects under test. Actually, they act on behalf of the units that will communicate with the units under test in the final product.

Normally, protocol stacks are implemented in the bottom-up fashion, starting from the lowest layer of the protocol stack and building the next layer on top of the previous one. Drivers and stubs in such an approach simulate only a part of the environment, the higher layer of the protocol stack in particular. An example of simple integration test collaboration is given in Section 5.5.2 (Example 2).

When all software units have undergone unit and integration testing, the final product is integrated and ready for acceptance testing, which comprises conformance testing (also referred to as compliance testing), load testing, and in-field testing. Preliminary acceptance testing can be organized solely by the production organization and conducted on its premises. However, final acceptance testing is organized and conducted by the organization that has the legal authority to issue acceptance certificates.

As suggested by its name, the aim of conformance testing is to prove that the product (implementation) under test conforms to the original requirements. In the area of communication protocol engineering, these requirements would normally be standards issued by the IETF, ISO, ITU-T, ETSI, and similar organizations. The newer standards made by ITU-T and ETSI most frequently include the conformance test suite specification in TTCN-3.

Conformance testing is a kind of functional testing (also referred to as black box testing). The testers are not interested in the structure of the product and its internal behavior. They only ensure that the external behavior of the product meets the original specification. Typically, this behavior is specified with the set of scenarios described in TTCN-3. We will return to the subject of conformance testing in Section 5.2.

The load testing typically involves exposing the implementation under test to the conditions of the real exploitation. Conceptually, this means that the implementation under test must service the requests coming from more independent sources simultaneously. While conformance testing focuses on the correctness of services given to the minimal number of request sources, load testing checks the correctness of services driven by the requests coming from independent sources, preferably in an interleaved fashion.

Normally, load testing is conducted in a laboratory-simulated environment. Typically, we would construct, purchase, or lease the specialized equipment referred to as a load generator. A load generator is normally a programmable device that offers a selection of predefined scenarios and their parameters (such as number of request sources, duration of individual communication phases, and so on) as well as definitions of completely new scenarios.

The name *load generator* may be misleading because it suggests that the device generates only the requests—which it does—but it also receives the responses from the implementation under test and checks if it operates correctly. For example, after the connection is successfully established, it sends and receives test tones to check that the connection is really usable. During load testing, we primarily check declared traffic capabilities of the product. A typical requirement would be that the number of lost requests must not exceed the given limit after the given number of requests has been issued in accordance with the given request arrival distribution.

We also normally check the behavior of the implementation under test for both lower and higher rates of request arrivals. With an extremely low rate of requests, we want to check the sustainability of long-lasting connections, whereas with an extremely high rate, we want to make sure that the overload protection mechanisms are in place and that they function correctly. After successful load testing, the implementation under test is integrated into the target network for in-field testing. In-field testing is essentially the experimental exploitation of the product for the given interval of time (e.g., three months).

The aim of in-field testing is to detect, locate, and eliminate bugs that are exposed by the real-world scenario (also referred to as a traffic case) that could not be simulated in the laboratory. During this last phase of acceptance testing, log files always prove to be extremely useful. Today, the log files can be collected over the Internet and analyzed remotely. Also, installing software upgrades can be done by uploading new software patches over the Internet.

Detecting bugs through the analysis of the log files can be augmented by adding program hooks for certain, really infrequent traffic cases. Defining state transition preconditions, postconditions, and invariants and checking them at run-time is also extremely useful for detecting bugs during in-field testing, and later during normal system exploitation. Although communication protocol maintenance is an integral part of communication protocol engineering, it is out of scope of this book (see directions for further reading in Section 5.6).

Traditional software engineering comprises a number of development phases, such as requirements, analysis, design, implementation, unit test, integration, integration test, verification, and maintenance. These phases can be cascaded in the case of the waterfall process model or revisited in the case of the spiral-incremental process model. The number of remaining bugs is the main software quality metric. Another important metric used in software engineering is test coverage (measured as the percentage of tested software paths, variable usages, and so on).

Cleanroom engineering, in contrast to traditional software engineering, is organized as a sequence of the following development activities:

- Formal model development.
- Formal verification of the formal model.
- Handing formal model to the implementation team, which implements it in a higher-level programming language.
- Operational profile modeling.
- Automatic test suite generation, which is based on the given operational profile model.
- Statistical usage testing and software reliability estimation: If at least one test case from the automatically generated test suite fails, the implementation under test is thrown away and the complete development cycle is repeated from the very beginning (starting with the formal model development).

The complete treatment of formal modeling and verification is out of the scope of this book (see directions for further reading in Section 5.6). As a means of introduction to the area of formal methods, formal modeling and verification based on theorem proving and model checking is covered in Section 5.3, which is divided into Subsections 5.3.1 and 5.3.2. The paradigm described in the subsection 5.3.1 is based on the application of the theorem prover named THEO, whereas the paradigm described in the subsection 5.3.2 is based on the model checker PAT.

Operational profile modeling, automatic test suite generation, statistical usage testing, and software reliability estimation are described in Section 5.4. The paradigm described in that section is based on the application of the software tool, which is named generic test case generator (GTCG).

5.1 Unit Testing

The aim of unit testing is to check the correctness of an individual software unit (Java compilation unit or C/C++ module). A generally accepted belief, especially among proponents of agile methods such as extreme programming, is that unit testing should be conducted by the programmer who is implementing the target software unit, because it greatly improves programmer's productivity. In principle, unit tests should be written before, or at least during, the implementation of the target software unit.

Of course, the programmer must clearly distinguish between the roles of an implementer and a tester (the author of extreme programming, Kent Beck, uses the metaphor: "by changing hats" to explain this paradigm). The programmer, as unit tester, concentrates on the unit interface. By thinking about the interface and by writing unit tests, the programmer gets a clearer picture about the services that the target software unit must provide. The programmer should also try to make test cases that cover boundary conditions, as well as situations that would be potentially hard to manage for the target software unit.

The programmer, as the unit implementer, concentrates on the implementation of the original unit design. They should forget about unit tests and concentrate on mapping the design to code. This should be a straightforward task if a proper framework (such as the FSM Library) is provided.

Unit testing helps programmers produce software units of better quality in shorter time intervals and this has been proven in practice. First, by creating unit tests, the programmer becomes even more familiar with the implementation at hand. Second, the programmer gets immediate feedback. If there is a bug, it is easy to detect in the scope of a particular test case. If the test case passes, the programmer gets immediate satisfaction that they have done their job well.

Unit test cases should be executed frequently during the target unit's implementation. As time passes, new test cases are added and old cases are run again. Even if no new test cases are used, we should rerun all existing unit tests every time we add new functionality. Testing that is conducted by running an unchanged test suite to check if the new software functionality has not affected existing functionalities is referred to as regression testing.

Regression testing is the key point of this paradigm. It enables a dramatic increase of productivity because it builds the programmer's confidence that everything is in good order and under control; therefore, the programmer can work more relaxed. Regression testing also encourages experimenting. In situations when alternative paths may be used in the course of implementation, the programmer may try out a way that seems most appropriate. If one or more test cases fail in the regression testing that is subsequently conducted, the programmer may decide to reset to the starting point by retrieving the previous version from the installed version of the control system database.

Unit testing (including regression testing) definitively has a positive impact on a programmer's psychology. It is estimated to be the key factor for increases in the programmer's productivity. The next question is "to what extent should we go with the unit testing?" The answer is not easy. Certainly, any amount of unit testing is better than none. Alternately, an attempt at exhaustive unit testing might be counterproductive.

The right choice is somewhere between these two extremes. We do not need to test trivial things, such as class function members that set or get the value of a certain private field member. Rather, we should concentrate on the boundary conditions and parts of code where it becomes more complex. Although generally unpopular among professionals, copy–paste practice may be tolerated for generating a set of similar test cases.

Three principal preconditions exist for a successful unit testing practice:

- A proper unit testing framework must be provided.
- Test cases should not involve any human intervention.
- The implementation under test must not be changed.

A proper unit testing framework must provide three main functions:

- Test case registration: This function enables registering new test cases within the given test suite hierarchy. On each level of the hierarchy, a set of individual test cases may be found, as well as other hierarchically subordinated test suites (very similar to the file system structure).
- Test case execution: This function provides automatic execution of all test cases defined within the given test suite hierarchy. It must not require more than a single push button to be started. Otherwise, the framework is simply not usable.
- Test case reporting: This function must provide a general report on the outcome of the execution of all test cases, as well as individual reports for all test cases that failed or caused errors.

The second precondition is that test suite execution should not involve any human intervention. This is the essential precondition to make unit testing completely automatic. If we want to eliminate human intervention, we must secure two conditions: First, the input data required by a test case must be defined as symbolic constants in its source code or in other external files. Second, the results of the test case must be automatically checked by a test case itself. The unit testing framework must provide adequate functions for this purpose.

A typical function for checking test case results is the function *assert(condition)*, where *condition* is a Boolean expression that evaluates to either the value *true* or *false*. The test case continues (*pass*) in the former case and breaks (*fail*) in the latter case. If the test case execution successfully reaches the end of the test case, it is considered successful (qualified with the verdict *pass*). Otherwise, it is considered unsuccessful (qualified with the verdict *fail*). If the test case execution breaks because of some error (most typically, an exception such as "divide by zero"), it is qualified with the verdict *error*.

Another typical function for checking test case results is the function *assertEquals(p1,p2)*. This function call is semantically equivalent to the function call *assert(p1==p2)*. This means that if the parameters *p1* and *p2* are equal (of course, they must be comparable), the test case execution continues; otherwise, it breaks. Typically, one of the parameters is a constant and another is a program variable.

Although these two functions are semantically equivalent, the function *assertEquals()* is advantageous when it comes to test case reporting. If the function *assert()* breaks the test case execution, the unit testing framework reports only that the condition evaluated to the value *false*, which is not a very informative report. Alternately, if the function *assertEquals()* breaks, the framework provides the report "expected *C* but was *V*," where *C* is the value of the constant (e.g., *p1*) and *V* is the real value of the variable (e.g., *p2*).

We can further improve the readability of the test case execution reports by using the optional text string parameter of the function *assertEquals()*. Generally, the function call format for this function is *assertEquals(text, condition)*, where *text* is the text string that explains the meaning of this assertion point in more detail. The string *text* is used as a prefix of the test report shown above. For example, if the value of the variable *ch* should be 'A' but it turns out to be 'B' instead, the function call *assertEquals("Check ch:," 'A', ch)* would produce the report, "Check ch: expected 'A' but was 'B.'"

Besides the functions *assert()* and *assertEquals()*, unit testing framework typically provides two additional functions for writing test cases: *setUp()* and *tearDown()*. The former sets up the test fixture whereas the latter destroys it. A test fixture is a set of objects that act as samples for testing. Normally, the test fixture comprises the instance of the unit under test (e.g., the instance of the class that is derived from the class *FiniteStateMachine*) and also other supplementary objects, which are required for effective unit testing.

Typically, the unit testing framework offers the base class for writing test cases, which provides the functions *assert()*, *assertEquals()*, *setUp()*, and *tearDown()*. The programmer normally derives his tester class from this base class, fills in *setUp()* and *tearDown()* functions, and starts writing individual test cases. Each function member of the tester class—whose name follows the given naming convention—is a single test case.

Remember that concrete *setUp()* and *tearDown()* implementations are shared by all test cases defined within a single tester class. Actually, these two functions are implemented as null (empty) methods on test cases. The execution of each test case starts with the call to the function *setUp()*, proceeds with a call to the user-defined function that implements a single test case, and ends with the call to the function *tearDown()*. Normally, we put the test case initialization and cleanup code in the functions *setUp()* and *tearDown()*, respectively.

The third unit testing postulate is that the unit under test must not be touched at all. We are only allowed to write new classes that are derived from the base class, which is provided by the unit testing framework. Changing the source code of the unit under test for the purpose of its testing is strictly forbidden, even by adding a simple print statement to the standard output file. Because of that, the only proper way to do the unit testing is to drive the unit under test with various messages, capture its responses, and check the correctness of the unit's external behavior.

This kind of controlled execution of the implementation under test is referred to as the test harness. The key request is that it must be fully automatic. The programmer should provide the mechanisms that support the test harness while he plays the role of the implementer (what we refer to as the *design for testability*). Otherwise, providing a test harness can be a very hard task. For example, consider a simple program that reads its input from the keyboard and writes its output to the monitor by using the operating system services, which cannot be replaced. Because we are not allowed to change the source code of the implementation under test, providing a test harness in this case is hardly achievable.

An example of the unit testing framework is JUnit, an open-source testing framework for unit testing Java programs that was originally developed by Erich Gamma and Kent Beck. Based on this framework, the open-source community came up with CppUnit, a semantically equivalent testing framework for unit testing C++ modules. These frameworks are very simple but powerful enough to enable industrial-strength unit testing of individual software units. Because JUnit and CppUnit are semantically equivalent, we will treat them as two implementations of the same framework.

The framework comprises the interface *Test* and two fundamental classes, the classes *TestSuite* and *TestCase*, as in Figure 5.1. As shown in the figure, the test suite (an instance of the class *TestSuite*) can contain an arbitrary number of test cases (instances of the class *TestCase*), as well as an arbitrary number of other hierarchically subordinated test suites. This arrangement allows

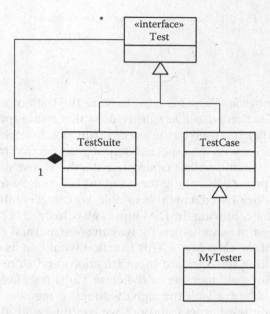

FIGURE 5.1
Structure of the JUnit testing framework.

programmers (playing the role of unit testers) to organize test cases into a hierarchy of test suites to their convenience.

Any concrete tester class (such as the class *MyTester* in Figure 5.1) must be derived from the base class *TestCase*, which, among others, provides the four fundamental functions described above, namely, *setUp()*, *tearDown()*, *assert()*, and *assertEquals()*. By convention, an individual test case is written as the function member of the tester class, whose name starts with the word "test," for example, *test1*, *test2*, and so on.

Next, we illustrate JUnit's usability on a concrete example. In the example that follows, we demonstrate the unit testing paradigm for the case where the implementation under test is counter by modulo 2. The particular implementation we are interested in is the one based on the State design pattern. This implementation was presented in Section 4.3.

As already mentioned in Section 4.3, the function *processMsg()*, which processes FSM input (message), prints its results by calling the function member *println()* of the class *MyIO*, rather than by calling the standard I/O function *System.out.println()*. This is a good example of how we can provide support for the test harness in our design and implementation. Here is the source code of the class *MyIO*:

```
package automata4;
import java.util.*;

public class MyIO {
private static String lastOutput;
```

```
public static String getLastOutput() { return lastOutput; }
public static void println(String s) {
lastOutput = s;
System.out.println(s);
}
}
```

The field member *lastOutput* is used to store the last output generated by the FSM. The function *getLastOutput()* returns this last output generated by the FSM to its caller. It is used by the test case function to retrieve the last FSM output to compare it with the expected output (also referred to as the "golden output"). The function *println()* is simple enough—it just stores the output of the FSM and prints it by calling the standard function *System.out.println()*.

Although we do not need it in this example, we can generally use an analogous approach for capturing the FSM inputs also. Instead of calling the standard function *System.in.read()* directly, we can construct and call the function member *read()* of the class *MyIO*. This function would, in its own turn, read the input by calling the standard input functions and store that input into the corresponding field member of the class *MyIO* (e.g., *lastInput*). The last FSM input would be available through the function member *getLastInput()*.

After providing test harness support, we continue with the definition of the tester class, which is named *Automata4Tester* in this example. The source code of this class is as follows:

```
/*
 * Automata4 tester
 *
 */

package automata4;
import junit.framework.*;

public class Automata4Tester extends TestCase {
  protected Automata4 a4;
  public Automata4Tester(String name) {
  super(name);
  }

  protected void setUp() {
   // setup code
   a4 = new Automata4();
  }

  protected void tearDown() {
   // cleanup code
  }

  // test case 1
  public void test1() {
    a4.processMsg('0');
    assertEquals(MyIO.getLastOutput(),"Output 0");
    a4.processMsg('0');
    assertTrue(MyIO.getLastOutput() == "Output 0");
  }
```

```
// test case 2
public void test2() {
  for(int i=0;i<100;i++) {
  a4.processMsg('0');
  assertEquals(MyIO.getLastOutput(),"Output 0");
  }
}

// test case 3
public void test3() {
  a4.processMsg('0');
  assertEquals(MyIO.getLastOutput(),"Output 0");
  a4.processMsg('1');
  assertEquals(MyIO.getLastOutput(),"Output 1");
  a4.processMsg('0');
  assertEquals(MyIO.getLastOutput(),"Output 1");
  a4.processMsg('1');
  assertEquals(MyIO.getLastOutput(),"Output 2");
  a4.processMsg('0');
  assertEquals(MyIO.getLastOutput(),"Output 2");
  a4.processMsg('1');
  assertEquals(MyIO.getLastOutput(),"Output 0");
}
// test case 4
public void test4() {
  a4.processMsg('1');
  assertEquals(MyIO.getLastOutput(),"Output 1");
  a4.processMsg('1');
  assertEquals(MyIO.getLastOutput(),"Output 2");
  a4.processMsg('1');
  assertEquals(MyIO.getLastOutput(),"Output 0");
}

// test case 5
public void test5() {
  for(int i=0;i<1000;i++) {
    test3();
    test4();
  }
}
public static TestSuite suite() {
  return new TestSuite(Automata4Tester.class);
}

public static void main(String[] args) {
  junit.textui.TestRunner.run(suite());
}
}
```

The tester class *Automata4Tester* is derived from the class *TestCase*. Its field member *a4* is an instance of the implementation under test, namely, the class *Automata4*. The constructor of the class *Automata4* simply calls the constructor of its super class (the class *TestCase*) and passes its input parameter (*String name*).

The function *setUp()* creates an instance of implementation under test by instantiating the class *Automata4*, and storing its instance into the field member *a4*. The function *tearDown()* is empty in this example because the Java garbage collector takes care of unused objects. The garbage collector destroys the object that is stored in the field member *a4* at the end of the test case.

The function *test1()* is the first test case defined within the tester class *Automata4Tester*. Basically, it tests the FSM state transition from the state *S0* to the state *S0*, which is driven by the input value *0*. It does the same operation twice. It supplies input *0* to the implementation under test (stored in the field member *a4*) each time by calling its function *processMsg()* and passing it the parameter, '0'.

Assuming that the implementation under test was in its initial state and that it reacted correctly to the given input, its last output should be the text, "Output 0". The test case function *test1()* checks that assumption by calling the function *assertEquals()*. The first real parameter of that function call is the value of the last output, which is returned by the function member *getLast Output()* of the class *MyIO*, whereas the second parameter is the expected string, "Output 0".

Second, the test case function *test1()* again supplies input *0* to the implementation under test (stored in the field member *a4*) by calling its function *processMsg()* and passing the parameter '0' to it. Assuming that the implementation under test has reacted properly in the first place, it would be in the initial state at the time the second call to the function *processMsg()* happens. Driven with the input '0', it should again produce the output string "Output 0". The test case function *test1()* checks this assumption again, only this time it does so by calling the function *assert()*. The real parameter of this function call is the condition *MyIO.getLastOutput()* == "Output 0".

The function *test2()* is the second test case defined within the tester class *Automata4Tester*. This test case is slightly more complex than the previous one. The previous test case checks if the implementation under test reacts correctly when it is driven twice with the same input value '0' in the same current state (*S0*). We did this on purpose—first, to demonstrate the usage of both *assert()* and *assertEquals()*, and second, the implementation under test may not always react correctly if it is driven with a certain input value in the given state, at least not in theory.

This practice may seem paranoid but, in reality, various types of time- and FSM evolution-dependent bugs are hidden at the beginning and become evident only later during the FSM evolution. Returning to the problem at hand, we ask ourselves: Will this FSM react correctly many times, for example, 100 times? With JUnit at our disposal, we can easily construct a test case that resolves such dilemmas.

This is exactly what the test case function *test2()* does. It does so by executing the body of the *for* loop 100 times. Inside the body of the loop, it drives the implementation under test with input value '0' by calling its function *processMsg()*. After each of these calls, it checks if the last output was the string "Output 0" by calling the function *assertEquals()*.

The function *test3()* is the third test case defined within the tester class *Automata4Tester*. This is a typical FSM-related test case, characterized with complete coverage of the FSM state transition graph. The flow of the state transitions checked by this test case is the following:

- From S0 to S0, driven with the input 0 (expected output 0)
- From S0 to S1, driven with the input 1 (expected output 1)
- From S1 to S1, driven with the input 0 (expected output 1)
- From S1 to S2, driven with the input 1 (expected output 2)
- From S2 to S2, driven with the input 0 (expected output 2)
- From S2 to S0, driven with the input 1 (expected output 0)

The function *test4()* is the fourth test case defined within the tester class *Automata4Tester*. This is another typical FSM-related test case, characterized by its progressive nature. The counter is always driven with the input "1" so that its content is incremented every time. This test case does not provide the full state transition graph coverage, but it is valid, and we can think of many partial graph coverage test cases. The flow of the state transitions checked by this test case is as follows:

- From S0 to S1, driven with the input 1 (expected output 1)
- From S1 to S2, driven with the input 1 (expected output 2)
- From S2 to S0, driven with the input 1 (expected output 0)

The function *test5()* is the fifth, and the last, test case defined within the tester class *Automata4Tester*. It is a fairly simple, yet rather intensive, test case that is based on the combination of the previous two test cases. The test case function *test5()* repeats the body of the *for* loop 1,000 times. Inside the body of the loop, it just calls the functions *test3()* and *test4()* in succession.

The function *suite()* returns the test suite, which it creates by calling the constructor of the class *TestSuite*. The real parameter of this function call is the name of the implementation under test class file (*Automata4Tester.class*). The constructor of the class *TestSuite* finds all the functions whose names start with the word "test" defined within the class *Automata4Tester* and automatically adds them to the test suite it creates.

The function *main()* runs the test suite defined by the previous function *suite()*. It does that by calling the function *run()* of the class *TestRunner*, which is an integral part of the JUnit testing framework. The real parameter of this function call is the test suite that is created by the function *suite()*. This test suite contains all test cases defined within the class *Automata4Tester*.

In the case of more complex implementations, we may decide to create more tester classes rather than define all test cases within a single tester class, such as the class *Automata4Tester*. In such a situation, we would need to create a hierarchy of test suites and an overall tester class that would automatically run all test cases in all test suites. The source code of such a tester class is the following:

```
/*
 * Tester
 *
 */

package automata4;
import junit.framework.*;

/*
 * TestSuite that runs all test suites
 *
 */
public class AllTests {
  public static void main (String[] args) {
    junit.textui.TestRunner.run(suite());
  }
  public static TestSuite suite() {
    TestSuite suite = new TestSuite("All Tests");
    suite.addTest(Automata4Tester.suite());
    // add other test suites here
    return suite;
  }
}
```

The class *AllTests* comprises two function members, namely, the functions *suite()* and *main()*. The former function creates and returns the test suite that is in the root of the test suite hierarchy. This means that it contains all other hierarchically subordinated test suites. The latter function executes the root test suite, i.e., it executes all test suites that were added to it.

The function *suite()* creates the root test suite simply by calling the constructor of the class *TestSuite*. The real parameter of this function call is the name of that test suite (the string "All Tests"). It then adds the test suite that contains the test cases defined within the tester class *Automata4Tester* to the root test suite. It does this by calling the function member *addTests()* of the root test suite object *suite*. Generally, in the case when we have multiple tester classes, we would repeat the call to the function *addTests()* for each tester class.

The function *main()* runs the test suite defined by the previous function *suite()*. It does this by calling the function member *run()* of the class *TestRunner*. The real parameter of this function call is the test suite created by the function member *suite()* of the class *AllTests*. This test suite contains a single, hierarchically subordinated test suite, which in turn contains all test cases defined within the class *Automata4Tester*.

We start the automatic execution of all test cases defined within the class *Automata4Tester* by running the file *Automata4Tester.class*. Similarly, we start the automatic execution of all test cases defined within all tester classes (in this simple example, we have just one of them: the class *Automata4Tester*) by running the file *AllTests.class*. In both cases, we should get the same result. Each test case function will print its own outputs to the standard output file. At the end, the test runner will print out the final report, which should look like this:

```
Time: 1,783
OK (5 tests)
Press any key to continue...
```

The number 1783 corresponds to the number of seconds that were needed to execute all test cases, whereas the number 5 in parenthesis corresponds to the total number of test cases that were executed.

5.2 Conformance Testing

As already mentioned at the beginning of this chapter, conformance testing is the first step of acceptance testing (followed by load testing and in-field testing). The aim of conformance testing is to check the functional correctness of external behavior of the implementation under test without checking its inner workings. Essentially, conformance testing is functional testing that is based on the "black box" approach.

The main goal of conformance testing is to separately check the correctness of each individual function of the implementation under test (IUT). The sample test case for a simple SIP softphone (IUT) is: "Initiate session setup. Check if IUT sends the message INVITE to the outbound proxy server (imitated by the testing framework). Make the testing framework replay with the message 404 (not found). Check if IUT replays with the message ACK" (see sequence diagram in Figure 5.2). We are intentionally making test cases

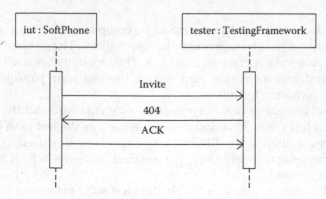

FIGURE 5.2
Example of the conformance testing test case.

as simple as possible so we can easily interpret their outcomes. Of course, some of the test cases are inevitably complex and we cannot do anything about this, but we should never make them more complex than they need to be.

More precisely, we do not try to check more functions simultaneously by interleaving the corresponding scenarios. For example, consider the SIP proxy server as the implementation under test. In the case of conformance testing, we are only interested if it can support a single session establishment at a time. Normally, we would not be interested in checking if it can support multiple session establishments simultaneously. Actually, that is exactly the purpose of load testing.

When it comes to specifying official conformance test suites for real-world protocols (like SIP), this is a really serious business conducted by the international standardization institutions, such as IEEE, ISO, IETF, ITU-T, ETSI, and others. The results are rather voluminous specifications that most frequently use TTCN language. The most recent version of TTCN at the time of this writing is the TTCN-3 (see Section 3.9), which enables both tabular and program formats of specifications.

For a better understanding of the scope of conformance testing, consider the documents currently available from ETSI (you can download them from the Internet; see http://www.etsi.org) that are related to conformance testing of SIP (IETF RFC 3261). These documents are the following:

- Conformance test specification for SIP, Part 1: Protocol implementation conformance statement proforma (ETSI TS 102 027-1)
- Conformance test specification for SIP, Part 2: Test suite structure and test purposes (ETSI TS 102 027-2)
- Conformance test specification for SIP, Part 3: Abstract test suite and partial protocol implementation of extra information for testing (ETSI TS 102 027-3)

The first document is the proforma to be completed by the vendor of the implementation to claim implementation capabilities. The guidance for completing the proforma is given in Section 5. This document is used both during static conformance review and during the test suite parameterization phase of conformance testing.

The second document describes the test suite structure and the purposes of individual test cases. This document was used as the test plan before the test suite was written in the TTCN-3 language. Now it is used as the reference document for understanding the abstract test suite, which is given in the third document.

The third document specifies the abstract test suite to be used for SIP conformance testing. Actually, it is composed of two files, the archive (ZIP file) that contains SIP test suite in TTCN-3 program format, and the SIP test suite overview file (PDF file). The SIP test suite in TTCN-3 program format can be executed using a commercially available TTCN-3 tool.

The SIP conformance test suite specification by ETSI (the three documents listed above) considers four types of implementations under test. The implementations are as follows (see IETF RFC 3261 for their definitions):

- User agent that behaves as client or server
- Registrar
- Proxy server (both outbound and simple proxy server)
- Redirect server

The present version of the specification considers the following three types of sessions:

- Sessions that are established using a proxy server
- Sessions that are established directly (without proxy)
- Sessions that are established using the redirect server

The way the SIP conformance test suite is structured is a good example of typical conformance test suite structuring. All test cases are classified into the following four main groups (which correspond to the main SIP functionalities):

- Registration
- Call control
- Querying for capabilities
- Messaging

The test cases in the main groups are further classified according to the role that should be checked. The roles for the main group *registration* are the *registrant* and the *registrar*. The roles for the main group *call control* are *originating endpoint, terminating endpoint, proxy*, and *redirect server*. The roles for the main group *querying for capabilities* are *originating endpoint, terminating endpoint*, and *proxy*. The roles for the main group *messaging* are *registrant, registrar, originating endpoint, terminating endpoint, proxy*, and *redirect server*.

Some of the role subgroups are further divided into functional subgroups. For example, the role subgroup *originating endpoint* of the main group *call control* is divided into three functional subgroups, namely, *call establishment, call release*, and *session modification*. Finally, functional subgroups of test cases can be divided into three test groups: *valid behavior* (V), *invalid behavior* (I), and *inopportune behavior* (O).

Notice that official conformance testing can be conducted only by authorized organizations (national certification centers, telecom operators, and so on) that use special tools that themselves were certified for such usage. These tools are professional equipment, most frequently referred to as testers, e.g., a SIP tester. A tester typically comprises the framework that supports test suite administration, execution (most frequently based on interpretation), and associated reporting. Such a framework is referred to as the testing framework.

The testers may be rather sophisticated. Most of them support most of— if not all—the state-of-the-art protocols. Alternately, almost unique testers are also used that support ultramodern protocols that have not become part of the mainstream protocols. Both of these types of testers can be rather expensive. Most frequently, competent and efficient operating of protocol testers requires special training.

Because of that, most of the small- and even middle-scale organizations involved in protocol development cannot afford purchasing testers and employing full-time employees (confusingly enough, also called testers) for the purpose of conformance testing. Rather, they rent the equipment or the person who can operate it for the purpose of the unofficial and preliminary conformance testing at the client location. The goals of this preliminary conformance testing are to reduce the overall cost and to minimize the risk of failing the official conformance testing.

Some organizations use open source test suites to reduce the cost of the preliminary conformance testing. An example of such a test suite is the SIP Forum Basic UA Test Suite created by Nils Ohlmeier, freely available on the Internet at https://github.com/nils-ohlmeier/sipsak (in accordance with the GPL license). This test suite is comprised of the following two parts:

- SIP Forum Testing Framework (SFTF)
- Basic UA tests

SFTF provides regular functions of test suite administration (e.g., adding new test cases, simply referred to as "the tests"), test suite execution control (executing all tests, selected groups of tests, or individual tests), and test suite execution reporting (both by printouts in the interactive window and in the log files, with five possible levels of logging details). The testing framework contains the logic required to execute the test, parse incoming messages, and create replies.

The second part (listed above) is simply a subdirectory that contains all basic user agent tests (i.e., test cases). The tests and SFTF itself are written in Python. The goal of these tests is not to provide complete conformance testing of SIP implementations, as the ETSI specification does. Rather, the goal is to check the well-known SIP interoperability problems, which frequently occur in immature SIP User Agent (UA) implementations, such as the simple SIP softphone.

Additionally, these tests can discover the implementation under test behavior that conforms to the original SIP specification but is considered a suboptimal implementation solution. Such cases are reported as *warnings* (W). The developer should consider revising the implementation in the case of warnings to make it more robust.

Many tests in this test suite are adopted from the IETF's SIP torture tests Internet draft (available on the Internet under the name *draft-ietf-sipping-torture-tests-02*). The rest of the tests are the contributions from the SIP Forum members. Original IETF SIP torture tests focus on areas that have caused problems in the past or have particularly unfavorable characteristics if handled improperly. Some of them test only the parser and others test both the parser and the application above it. Some use valid and some use invalid SIP messages to check target functionality.

The SIP Forum tests are classified into the following eight test groups: protocol tortures (26 tests), authentication (4 tests), registration (1 test), dialog and transaction processing (19 tests), DNS (2 tests), NAT capabilities (2 tests), services (2 tests), and warnings about obsolete features (5 tests). All tests are defined in one spreadsheet (XLS file). The test attributes (spreadsheet columns) are the following: number, title, tested device, expected behavior, typical failures, notes, call flow, source (the corresponding section in RFC 3261), and comment.

For example, the test number 201 entitled "A Short Tortuous Request" tests the SIP user agent server behavior. The expected behavior is, "Server considers the request valid and generates a proper response". The call flow is illustrated with the sequence diagram shown in Figure 5.3.

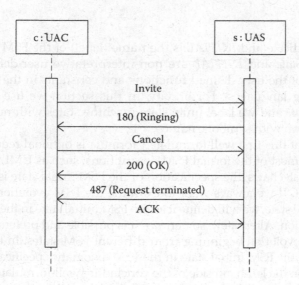

FIGURE 5.3
Example of the SIP protocol torture test.

5.3 Formal Verification

5.3.1 Formal Verification Based on Theorem Proving

This section covers the formal verification of communication protocols based on automated theorem proving. The reader will learn how to use automated theorem proving for formal verification of both communication protocol specification and its implementation. Normally, the communication protocol is modeled as the finite state machine. Basic knowledge of predicate calculus (first-order logic) is assumed for easy and complete understanding of this section.

The outline of this section is the following:

- Axiomatic specification of finite state machines
- Theoretic specification of test cases
- Formal verification of the specification
- Directions for generating test cases
- Formal verification of the implementation
- Software development process based on the formal verification
- A realistic example

The axiomatic specification of the finite state machine is the model of the FSM in the predicate calculus. This model is the set of well-formulated formulas. The first well-formulated formula in the model is optional and it defines the initial state of the FSM. Its general format is the following:

```
State(INITIAL).
```

State is a predicate and *INITIAL* is the name (label) of the FSM initial state. The names *State* and *INITIAL* are non-interpretative user-defined names (like names of the user-defined functions and constants in the higher-level programming languages). For brevity, in this section we use the name *S* instead of *State* and we label finite state machine states with numbers (*0, 1, 2...*) rather than with symbolic names.

The fact that this first well-formulated formula is optional requires a short comment. In most of the formal FSM descriptions, such as UML activity diagrams and statecharts, the specification of the FSM initial state is mandatory. Here, it is not. If we always want to examine the FSM evolution, beginning from the same state, we will define it as the FSM initial state in the FSM axiomatic specification. Alternately, sometimes it is possible and preferable to examine the FSM evolution beginning from different FSM states. In that case, we do not define the FSM initial state in the FSM axiomatic specification; instead we define it on the left-hand side of the concluding well-formulated formula.

The rest of the well-formulated formulas in the FSM axiomatic specification are obligatory. Each of the mandatory well-formulated formulas models a single FSM state transition (also referred to as a FSM branch). The format of the well-formulated formula that models the time invariant FSM state transition from the state X to the state Y triggered with the input T and generating the output R is as follows:

$$\{State(X)\&Input(T)\} => \{State(Y)\&Output(R)\}$$

State, *Input*, and *Output* are predicates. X, Y, T, and R are constants that label the source FSM state, the destination FSM state, the particular FSM input, and the particular FSM output, respectively. Most frequently, we use abbreviated names I and S instead of *Input* and *Output*, respectively. In the case that the state transition generates more, say N, output signals (messages), the corresponding well-formulated formula has the following format:

$$\{State(X)\&Input(T)\} => \{State(Y)\&Output(R\_1)\& \\ Output(R\_2)\&...\&Output(R\_N)\}$$

where $R\_1, R\_2...R\_N$ are the labels of particular output signals.

Next, we introduce the concept of control predicates. As their name suggests, the control predicates are used to control the FSM activity. A global control predicate is used to enable or disable the complete FSM activity. Usually we name it $A(N\_I)$, where A stands for *Automata* and N_I labels the particular FSM.

Besides the global control predicate, state transition control predicates also exist, one for each FSM state transition. A state transition control predicate enables or disables the associated state transition. We typically name it $T(M\_I)$, where T stands for *Transition* and M_I labels the particular FSM state transition. The state transition well-formulated formula that includes control predicates has the following format:

$$\{Automata(I)\&Transition(J)\&State(X)\&Input(T)\}=>\{State(Y)\&Output(R)\}$$

I is the label of the particular FSM and J is the label of the particular state transition modeled with this formula. If we include both *Automata(I)* and *Transition(J)*, the state transition is enabled. If we skip *Automata(I)*, the FSM (i.e., all its state transitions) are disabled. If we skip *Transition(J)*, this individual state transition is disabled. This concludes the presentation of the axiomatic specification of a single FSM.

A theoretical test case for a single FSM is the theorem about the particular FSM evolution path, which states that for a given series of inputs $(I_1, I_2...I_n)$, FSM performs a series of state transitions $(S_1, S_2...S_n)$, which will produce a series of particular output values $(O_1, O_2...O_n)$. The corresponding well-formulated formula has the following format:

$$\{Automata(N)\&Transition(M)\&Input(I_1)\&...\&Input(I_n)\} => \\ \{Output(O_1)\&...\&Output(O_n)\&State(S_1)\&...\&State(S_n)\}$$

Most frequently, we only want to check that FSM produces the expected series of outputs and that at the end it reaches the expected final state S_n. The corresponding theorem has a very similar, but simpler format:

$$\{Automata(N)\&Transition(M)\&Input(I_1)\&...\&Input(I_n)\} =>$$
$$\{Output(O_1)\&...\&Output(O_n)\&State(S_n)\}$$

Before proceeding to modeling the groups of communicating FSMs, let us look at a simple example. The following shows the axiomatic specification of the counter by modulo 2 (see the statechart diagram in Figure 5.4) and a sample theorem about its expected behavior. The FSM axiomatic specification is as follows:

$S(0)$
$\{A(0)\&T(0)\&S(0)\&I(0)\} => \{S(0)\&O(0)\}$
$\{A(0)\&T(1)\&S(0)\&I(1)\} => \{S(1)\&O(1)\}$
$\{A(0)\&T(2)\&S(1)\&I(0)\} => \{S(1)\&O(1)\}$
$\{A(0)\&T(3)\&S(1)\&I(1)\} => \{S(2)\&O(2)\}$
$\{A(0)\&T(4)\&S(2)\&I(0)\} => \{S(2)\&O(2)\}$
$\{A(0)\&T(5)\&S(2)\&I(1)\} => \{S(0)\&O(0)\}$

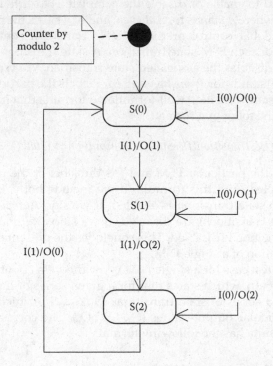

FIGURE 5.4
Counter by modulo two statechart.

The first well-formulated formula defines the state S(0) as the FSM initial state. Next, six well-formulated formulas define six FSM state transitions— from the state $S(0)$ to $S(0)$, from $S(0)$ to $S(1)$, from $S(1)$ to $S(1)$, from $S(1)$ to $S(2)$, from $S(2)$ to $S(2)$, and from $S(2)$ to $S(0)$, respectively. $A(0)$ is the global control predicate. $T(0)$, $T(1)...T(5)$ are the individual state transition control predicates. The sample theorem is as follows:

$$\{A(0)\&T(0)\&I(0)\&T(1)\&I(1)\} => \{O(0)\&O(1)\&S(1)\}$$

It may be interpreted as follows: The FSM is globally enabled by including the general control predicate $A(0)$ on the left-hand side of the concluding well-defined formula. The first FSM state transition is enabled by including the state transition predicate $T(0)$. The FSM is stimulated with the input $I(0)$, which should result in the output $O(0)$. The second FSM state transition is enabled by including the state transition control predicate $T(1)$. The FSM is stimulated with the input $I(1)$, and the FSM should generate $O(1)$ at its output. Finally, the FSM should reach the state $S(1)$.

We can prove this theorem with the automated theorem prover THEO developed by Monty Newborn (2001). To do that, we must write the theorem in a text file, compile it using the program Compile (*cc.exe*), and prove it by running the program THEO (*teo.exe*). The final result looks like this:

```
Predicates: S A T I O
Functions: 0 1 . 2 3 4 5 :
EQ:
ESAF:
ESAP:
 0 <BC: 19 NC: 6 AC: 3 U: 0>
 1 {T0 N1 R1 F0 C9 H0 h0 U11} *
.Proof Found!
```

Of course, realistic FSMs never operate in isolation. Rather, they normally operate in groups of cooperating finite state machines. For example, according to ITU-T, the system consists of functional blocks interconnected with communication channels (see Section 3.7, SDL). Each functional block comprises finite state machines (processes) interconnected with signaling paths (routes). A communication channel may comprise one or more signaling paths. Finite state machines communicate by exchanging signals (events, messages) over signaling paths.

We can use such a kind of traditional system decomposition for our convenience, but it is not required. In the opposite extreme, we can have a chaotic system in which each FSM talks to all other FSMs (like stations in wireless networks). We can even connect more FSMs in signaling networks with all kinds of topologies, such as start, bus, or a network that connects an arbitrary number of FSMs. The means to model all these abstractions in the first-order logic are predicates and their compositions.

To start, we can introduce the notation *Signal(SIG_N)* that represents the act of signaling the particular signal, where *Signal* is a predicate and

SIG_N is the label of a particular signal. We then can introduce the notation *SignalOverPath(SIG_N,PATH_M)* that represents the act of signaling the particular signal over the particular signaling path, and so on. The well-formulated formulas that model state transitions do not change much. For example, the state transition from the state *X* to the state *Y* is triggered with the signal *P* and generates the signal *Q*, and looks like this:

$$\{State(X)\&Signal(P)\} => \{State(Y)\&Signal(Q)\}$$

In the formula above, *Signal(P)* is received and *Signal(Q)* is sent out of any signaling path, channel, or network. In the case where the former signal is transferred over path *M* and the latter signal is sent over the path *N*, the formula would look like this:

$$\{State(X)\&SignalOverPath(P,M)\} => \{State(Y)\&SignalOverPath(Q,N)\}$$

After introducing the concept of signaling between finite state machines in a group of cooperating FSMs, we can proceed to the axiomatic specification of the group of FSMs. As shown above, each FSM in a group is specified with a set of well-formulated formulas (one optional for the initial state and one mandatory for each individual state transition). Consequently, the specification of a group of FSMs is the union of sets of well-formulated formulas for individual FSMs that constitute that group.

The theoretical test case for the group of FSMs is just a generalization of the theoretical test case for the individual FSM. The left-hand side of the corresponding well-formulated formula consists of control predicates, if any, and staring signals whereas the right-hand side of the formula lists the resulting signals and final states of individual FSMs. The format of the typical theorem about the evolution of the group of FSMs is as follows (assume the system with two FSMs):

$$\{Signal(A)\} => \{Signal(B)\&Signal(C)\&Signal(D)\&State(X)\&State(Y)\}$$

In the sample theorem above, *Signal(A)* triggers the evolution of the system. As the result of the evolution, the system generates three signals: *Signal(B)*, *Signal(C)*, and *Signal(D)*. At the end of the evolution, the FSMs reach their final states, namely, *State(X)* and *State(Y)*.

We now illustrate the concepts introduced above by the means of a simple example. Consider a simple system with three FSMs (see their statechart diagrams in Figure 5.5). The first FSM waits for the signal *E(0)* in its state *S(0)*. After receiving that signal, it sends the signal *E(10)* and goes to the state *S(1)*, where it waits for the signal *E(1)*. Once it receives the signal *E(1)*, it sends the signal *E(20)* and goes to the state *S(2)*. The second and the third FSMs are very much alike. The former waits for the signal *E(10)* and, after receiving that signal, it sends the signal *E(11)*. The latter waits for *E(20)* and sends *E(21)*.

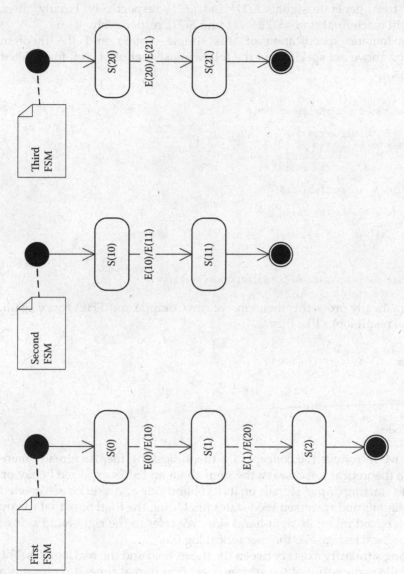

FIGURE 5.5
Statecharts of three communicating FSMs.

Next, we construct the theorem about the expected behavior of this simple system. This theorem says that if we supply signals *E(0)* and *E(1)* to this system, the first FSM will start evolving and will generate the signals *E(10)* and *E(20)*. These two signals will trigger the second and the third FSMs, which will, in turn, generate signals *E(11)* and *E(21)*, respectively. Finally, these FSMs will reach final states *S(2)*, *S(11)*, and *S(21)*, respectively.

The axiomatic specification of this simple system and the theorem explained above are specified in the following sequence of well-formulated formulas: .

```
; Simple system with 3 FSMs
; Axiomatic spec. of the first FSM
S(0).
{S(0)&E(0)} => {S(1)&E(10)}.
{S(1)&E(1)} => {S(2)&E(20)}.

; Axiomatic spec. of the second FSM
S(10).
{S(10)&E(10)} => {S(11)&E(11)}.

; Axiomatic spec. of the third FSM
S(20).
{S(20)&E(20)} => {S(21)&E(21)}.

; Theorem
conclusion
{E(0)&E(1)} => {S(2)&S(11)&S(21)&E(10)&E(20)&E(11)&E(21)}.
```

To automatically prove this theorem, we run Compile and THEO once again. The final result looks like this:

```
Predicates: S E
Functions: 0 1 10 2 20 11 21 : .
EQ:
ESAF:
ESAP:
 0 <BC: 14 NC: 3 AC: 3 U: 0>
 1 {T0 N1 R1 F0 C1 H0 h0 U14} *
.Proof Found!
```

Next, we introduce the concept of a theoretical log file. As already mentioned, a theoretical test case is a theorem about an FSM's expected behavior. It defines starting (input) signals on its left-hand side and a series of expected output signals and traversed FSM states (including the final ones that we are most interested in) on its right-hand side. We refer to the right-hand side of the theoretical test case as the theoretical log file.

A strong similarity exists between the theoretical and the real log files. The real log file is the result of the system execution in real time. It represents a particular path of the system evolution. The theoretical log file is the result of the virtual (speculative) system execution. It shows the expected outcomes, such as generated signals and traversed states (including the final states).

However, one principal difference between the two of them is that the logs in the real log file usually have a time stamp. The value of the time stamp is usually unique (with the exception of the logs in multiprocessor systems). Alternately, logs in the theoretical log files are individual predicates that correspond to signals and states, and they do not have any time stamp at all.

Furthermore, we can write logs in the theoretical log file in any order, because the operator "&" is a commutative one. The easiest way to think about it is that the theoretical test case is true forever. Hence, it really does not matter in which order we name the logs. Another way to think about it is that all of them have happened at the same moment of time. Therefore, all logs have the same "time stamp," which may be omitted because it does not provide any meaningful information, and then again the order of logs does not matter.

Actually, when we look at the FSM axiomatic specification, and the theoretical test case more closely, we notice that no explicit notion of time exists at all. The only notion of time present there is an implicit one, and it is made through control predicates. Although the absence of an explicit notion of time may seem confusing and disadvantageous, it is the main source of the power of proving theorems.

To understand why, imagine that we made a system that reacts in certain ways when it receives two different messages, but we are not sure what will happen if these two signals arrive at exactly the same time. If the probability of this event is very low, it can take a long period of time before the event happens and we face a system failure. With the theorem-proving approach, we check such situations immediately. Imagine the enormous amounts of test time that are saved this way.

Another powerful characteristic of this approach is that each theoretical test case actually represents a family of test cases. For example, let us return to the counter by modulo 2. Consider the theorem:

$$\{A(0)\&T(0)\&I(0)\} => \{O(0)\&S(0)\}$$

Because in first-order logic, $I(0) <=> I(0)\&I(0)$, we can rewrite the theorem as follows:

$$\{A(0)\&T(0)\&I(0)\&I(0)\&I(0)\&I(0)\&I(0)\} => \{O(0)\&S(0)\}$$

We may interpret this theorem as follows: If we apply the same signal $I(0)$ many times (even up to infinity), we will always get the signal $O(0)$ at the FSM output and it will remain in the state $S(0)$. Therefore, by proving individual theoretical test cases, most frequently we are actually checking the families of test cases. This concludes the presentation of axiomatic specification and theoretical test cases related to FSMs.

Now let us see how we can use this in communication protocol engineering. We start with the formal verification of the specification. The concept is rather simple, although it can prove to be difficult to realize in practice. Ideally, two independent teams must be present (or at least a person who is "changing hats"), namely, the design and testing teams. The former writes the axiomatic specification of the family of communication protocols that is modeled as a group of FSMs. The latter writes and proves the theoretical test cases.

If a theoretical test case fails (the proof of the theorem cannot be found), at least one error is generated in either axiomatic specification or in the theorem. It may be the case that two or even more errors occur in both of them. Most frequently, the errors are trivial oversights made by theorem writers because they are not so familiar with the system at hand. If not, the errors are typically caused by rather nontrivial oversights in the system design.

Finding these errors is not a trivial task at all. Typically, we would try to shorten the theorem or the axiomatic specification and see what happens. Of course, with an automated theorem prover, such as THEO, at our disposal, this is much easier than doing it by hand. Control predicates may help, also—with them, we can sequence the events to our convenience. The need for them is typically a clue that we have synchronization problems.

We can also use an automated theorem prover for automatic test case generation. To do that, we assume that axiomatic specification of the system is errorless. We start by selecting one of the possible input signals on the left-hand side of the theorem. We then check various output signals at the right-hand side of the theorem by trying to prove the theorem. If the proof is found, our assumption was correct and we keep that signal at the right-hand side. If not, we continue by checking other signals.

Of course, some input signals can just cause internal state transitions and no signals at the output of the system. The right-hand side will remain empty in that case. By continuing this process, we can generate theoretical test cases of arbitrary length:

$$\{I(A)\&I(B)\&I(C)\} => \{O(X)\&O(Y)\&O(Z)\}$$

Similarly, we can make guesses about transient or final states of the system, for example:

$$\{I(A)\&I(B)\&I(C)\} => \{O(X)\&O(Y)\&O(Z)\&S(P)\&S(Q)\}$$

The real benefit of such automatically generated test cases is that they can be translated into executable test cases and used for automatic testing of the system implementation. Generating test cases in the previously described fashion is not very efficient, and neither it is well coordinated. We can generate test cases more cleverly by respecting the structure of the FSM axiomatic specification rather than viewing it as a black box. Actually, the FSM axiomatic specification introduced in this section is yet another means of modeling the FSM state transition graph.

Generating test cases by traversing the FSM state transition graph is possible with the goal to achieve its complete coverage. Three possible types of

FSM state transition coverage exist, namely, node, branch (arc), and path coverage. That the path coverage cannot be achieved if the graph is cyclic is well known. Alternately, branch coverage subsumes node coverage and, because of that, seems to be the best selection.

Sometimes we may have the opposite problem. The test suite (a set of test cases) may already be available, such as the SIP conformance test suite available from ETSI in TTCN-3 language (see Section 5.3). In such a situation, we can use a tool to translate TTCN-3 test cases into theorems, and then we can use the automated theorem prover to formally verify conformance of the system axiomatic specification with the standard.

Yet another application of the automated theorem prover is the formal verification of the system implementation. To do this, we assume that a conformance test suite is already available and use the reverse engineering tool to extract the axiomatic specification of the system from the implementation source code and, optionally, from log files if some are available. The reverse engineering tool normally relies on conventions that govern the structure of the source code and log files.

For example, the reverse engineering tool for the FSM Library-based implementations relies on the specification of the FSM Library API (see Section 6.8). This tool simply searches the source code for specific library functions and their real parameters to retrieve the well-formulated formulas that constitute system axiomatic specification. More precisely, the tool extracts the elements of the left-hand side of the state transition well-formulated formula by searching for library functions *InitEventProc()* and *InitUnexpectedEventProc()*.

The real parameters of the function *InitEventProc()* are the source state, the triggering signal (event, message), and the state transition function. The first two parameters (state and signal) are exactly the elements of the left-hand side of the corresponding well-formulated formula. The real parameters of the function *InitUnexpectedEventProc()* are the source state and the state transition function. The state is the first element of the left-hand side of the well-formulated formula. The second element is any signal that is not valid for the given state.

The reverse engineering tool proceeds by examining an individual state transition function. It creates one well-formulated formula (they all have the same left-hand side) for each state transition function execution path. For example, a state transition function with a simple sequence of statements yields a single formula, whereas a state transition function that has a switch with three cases yields three formulas.

The right-hand side of the state transition well-formulated formula is constructed by the analysis of the state transition function. The tool first searches for the functions *PrepareNewMessage()* and *SendMessage()* to extract symbolic names of the signals that are generated by that execution path of the state transition function. It then searches for the function *SetState()*, whose real parameter is the name of the destination state. If this function is not found, the tool assumes that the FSM state should not be changed and copies the state name from the left-hand side to the right-hand side of the formula.

This procedure is repeated for all state transition functions. Finally, the tool provides complete axiomatic specification of the system in ASCII format, which is readable by the automated theorem prover. We then use already available test cases to formally verify the system implementation source code.

Although most frequently we assume that the tools and other components we use are bug-free (in this particular case, these tools are the reverse engineering tool, compiler, linker, loader, and operating system), sometimes they are not. No matter how low the probability of such a failure is, it can happen and when it does, it compromises the formal verification of the source code. In such a case, we can use the reverse engineering tool that extracts the axiomatic system specification from log files. The example of the particular log file that was created by the FSM Library-based implementation is given in Section 5.5.1. Principally, the axiomatic specification that is provided from the log file is usually incomplete (except when it contains traces of all possible system execution paths), but even as such, it is sufficient to locate and eliminate the problem at hand.

When it comes to the application of formal verification methods, software development processes can be classified into three different categories. The Cleanroom engineering is a typical representative of the first category. It uses formal verification methods to formally verify the system design. The second category uses formal methods to formally verify the system implementation, whereas the third uses it to formally verify both the system design and implementation.

We will end this section with a more realistic example—the axiomatic specification of the FSM that implements both ITU-T Q.71 FE1 and FE5 call control functional entities (see Figure 3.38, Section 3.7.1) and a sample theoretical test case. The former functional entity models the functionality of the calling party (also referred to as subscriber A) whereas the latter models the functionality of the called party (also referred to as subscriber B). The following is the axiomatic specification of the FSM, named FE1FE5 (ITU-T Q.71 FE1 and FE5 merged together):

```
;
;   FE1FE5 definition
;
;   Initial state definition:
S(FE1FE5_ON_HOOK).

{S(FE1FE5_ON_HOOK)&E(r3_DisconnectReqInd)} =>
{S(FE1FE5_ON_HOOK)&E(r3_DisconnectRespConf)}.

{S(FE1FE5_ON_HOOK)&E(r3_SetupReqInd)} =>
{S(FE1FE5_WAIT_OFF_HOOK)&E(r3_ReportReqInd)}.

{S(FE1FE5_ACTIV)&E(r3_SetupReqInd)} =>
{S(FE1FE5_ACTIV)&E(r3_DisconnectReqInd)}.

{S(FE1FE5_ACTIV)&E(r3_DisconnectReqInd)} =>
{S(FE1FE5_WAIT_ON_HOOK)&E(r3_DisconnectRespConf)}.
```

```
{S(FE1FE5_ACTIV)&E(User_ON_HOOK)} =>
{S(FE1FE5_ON_HOOK?)&E(r3_DisconnectReqInd)}.

{S(FE1FE5_WAIT_ON_HOOK)&E(User_ON_HOOK)} =>
{S(FE1FE5_ON_HOOK)}.

{S(FE1FE5_WAIT_ON_HOOK)&E(r3_DisconnectReqInd)} =>
{S(FE1FE5_WAIT_ON_HOOK)&E(r3_DisconnectRespConf)}.

{S(FE1FE5_WAIT_ON_HOOK)&E(r3_SetupReqInd)} =>
{S(FE1FE5_WAIT_ON_HOOK)&E(r3_DisconnectReqInd)}.

{S(FE1FE5_WAIT_OFF_HOOK)&E(User_OFF_HOOK)} =>
{S(FE1FE5_ACTIV)&E(r3_SetupRespConf)}.

{S(FE1FE5_WAIT_OFF_HOOK)&E(r3_DisconnectReqInd)} =>
{S(FE1FE5_ON_HOOK)&E(r3_DisconnectRespConf)}.

{S(FE1FE5_WAIT_OFF_HOOK)&E(r3_SetupReqInd)} =>
{S(FE1FE5_WAIT_OFF_HOOK)&E(r3_DisconnectReqInd)}.

conclusion
; {S(FE1FE5_ON_HOOK)&E(User_OFF_HOOK)} =>
; {S(FE1FE5_UNKNOWN_FE2)&E(r1_SetupReqInd)}.

; {S(FE1FE5_UNKNOWN_FE2)&E(User_ON_HOOK)} =>
; {S(FE1FE5_DISCONNECTING_FE2)}.

{S(FE1FE5_ON_HOOK)&E(User_OFF_HOOK)&E(User_ON_HOOK)} =>
{S(FE1FE5_DISCONNECTING_FE2)&E(r1_SetupReqInd)}.
```

Actually, this file contains three theorems (starting after the keyword *conclusion*). The first two are commented out (the semicolon character ";" at the beginning of the line means that the line is a comment) leaving only the third open as a subject to prove by the automated theorem prover. The first commented theorem claims that if the FSM FE1FE5 is stimulated with the input signal *User_OFF_HOOK* in its initial state *FE1FE5_ON_HOOK*, it will generate the output signal *r1_SetupReqInd* and move to the state *FE1FE5_UNKNOWN_FE2*. The second commented theorem claims that if the FSM FE1FE5 is further stimulated with the signal *User_ON_HOOK* in the state *FE1FE5_UNKNOWN_FE2*, it will just move to the state *FE1FE5_DISCONNECTING_FE2*.

Finally, the third theorem—which is actually the subject of automated theorem proving—is a simple composition of the previous two theorems. It states that if the FSM FE1FE5 is stimulated by the sequence of the input signals *User_OFF_HOOK* and *User_ON_HOOK* in its initial state *FE1FE5_ON_HOOK*, it will generate the output signal *r1_SetupReqInd* and finish in the state *FE1FE5_DISCONNECTING_FE2*. To automatically prove this theorem, we run Compile and THEO once again. The final result looks like this:

```
Predicates: S E
Functions: FE1FE5_ON_HOOK User_OFF_HOOK r1_SetupReqInd User_ON_HOOK
FE1FE5_DISCONNECTING_FE2 . r1_DisconnectRespConf FE1FE5_UNKNOWN_FE2
r1_DisconnectReqInd User_DIGIT r1_ProceedingReqInd
FE1FE5_WAIT_FOR_DIGITS r1_ADDL_AddrReqInd r3_DisconnectReqInd
FE1FE5_WAIT_ON_HOOK r1_SetupRespConf FE1FE5_ACTIV r1_ReportReqInd
```

```
r3_DisconnectRespConf r3_SetupReqInd FE1FE5_WAIT_OFF_HOOK
r3_ReportReqInd FE1FE5_ON_HOOK? r3_SetupRespConf :
EQ:
ESAF:
ESAP:
0 <BC: 56 NC: 4 AC: 4 U: 0>
1 {T1 N1 R1 F0 C49 H1 h0 U8} *
.Proof Found!
```

5.3.2 Formal Verification Based on Communicating Sequential Processes

This section covers formal verification of communication protocols based on the process algebra named Communicating Sequential Processes (CSP) and aided by the toolkit named Process Analysis Toolkit (PAT). PAT supports a rich modeling language named CSP#, which is essentially the CSP extended with elements of the programming language C#. PAT also supports the First-Order Logic (FOL) and Linear Temporal Logic (LTL) formulas. Actually, PAT is a powerful toolkit comprised of many modules, including the module CSP#, the module Real-Time Systems (RTS), the module Probability CSP (PCSP), the module Probability RTS (PRTS), the module Labeled Transition Systems (LTS), the module Timed Automata (TA), the module NesC (targeting sensor networks), the module Orc (targeting Service Oriented Architecture), the module Stateflow (MDL), the module Security, the module Web Services (WS), and the module UML to PAT (for translating UML state machines to CSP#). We focus on the module CSP# in this book, because it is the most commonly used module.

In this section, the reader will learn from examples how to model protocols in CSP# and how to formally verify them by checking their desired properties, which are normally specified in the form of the corresponding FOL and/or LTL formulas. We will start with some more simple, classical examples (such as alternating bit protocol and two-phase commit protocol), we will continue with various leader election protocols (in complete graphs, in rings, and in rooted trees), and we will end with an example of a real-world communication protocol for providing telecomm services (such as basic call establishment and release, unconditional call forwarding, etc.).

The outline of this section is the following:

- Brief overview of CSP in Section 5.3.2.1
- Brief overview of PAT and CSP# in Section 5.3.2.2
- Examples of formal verification based on CSP and PAT in Section 5.3.2.3

5.3.2.1 Brief Overview of CSP

Process algebra is a formal method that uses an algebraic approach to study the communications of concurrent systems. Three well-known process algebras are Calculus of Communicating Systems (CCS), Communicating Sequential Processes (CSP), and Algebra of Communicating Processes (ACP). This section serves as a brief introduction to CSP (Hoare, 1985), which was initially proposed by C.A.R. Hoare in 1978, and since then it has been developed into one of the most mature formal methods that are based on process algebras. CSP is specialized in modeling the inter-action between concurrent systems using mathematical theories. Due to its powerful expressiveness, CSP is widely used in many different fields, such as real-time systems, web services, security, etc. CSP processes are composed of primitive processes and actions, which are connected by operators.

Here are the most important notions related to CSP processes:

- $\alpha P = \alpha(P)$ is the alphabet of the process P, i.e., the set of actions that P can engage in.
- αc is the set of messages that are communicable on channel c.
- $a \rightarrow P$ means that the process first performs the action a and then behaves as the process P. We read it as a then P.
- $(a \rightarrow P) \mid (b \rightarrow Q)$ means the choice between $(a \rightarrow P)$ and $(b \rightarrow Q)$, where b is the second action and Q is the second process.
- $(x : A \rightarrow P(x))$ means the choice of x from A then $P(x)$.
- $\mu\, X : A \bullet F(X)$ means the process X with the alphabet A such that $X = F(X)$.
- $P\, /\, s$ means P after engaging in events of trace s.
- $P \parallel Q$ means P in parallel with Q (i.e., the parallel execution of P and Q).
- $P\, [\,|X|\,]\, Q$ means that P and Q perform concurrent events on a set of channels X.
- $l : P$ means P with the name l.
- $L : P$ means P with names from the set L.
- $P \sqcap Q$ means the nondeterministic choice between P and Q. We read it as P or Q.
- $P \square Q$ means the deterministic choice between P and Q. We read it as P choice Q.
- $P \setminus C$ means P without the elements of the set C.
- $P \mid\mid\mid Q$ means the interleaving of P and Q. We read it P interleaves Q.
- $P \gg Q$ means P is chained to Q.
- $P\, //\, Q$ means P is subordinate to Q.
- $P\, ;\, Q$ means that P is successfully followed by Q. We read it P followed by Q.

- $P \lhd b \rhd Q$ means P if b (is true), else Q.
- $*P$ means repeat P (more precisely, P repeats an arbitrary number of times).
- $b * P$ means while b (is true) repeat P.
- $x := e$ means x becomes (the value of) e.
- $b!e$ means on (the channel) b output (the value of) e.
- $b?x$ means from (the channel) b input to x.
- $l!e?x$ means the call of the shared subroutine named l with the value parameter e and the results to x.
- P sat S means that the process P satisfies the specification S.
- *tr* is an arbitrary trace of the specified process, e.g., $\langle x, y \rangle$ where x and y are the elements of the alphabet of the specified process.
- *ref* is an arbitrary refusal of the specified process. The *refusal* of the process is the set of actions (events) in which the process cannot engage.
- $x^{\sqrt{}}$ means the final value of x produced by the specified process.
- *var(P)* is the set of variables assignable by the process P.
- *acc(P)* is the set of variables accessible by the process P.
- *Skip* is a process which does nothing but terminates successfully.
- *Stop* is a process which is in the state of deadlock and does nothing.

The most important notions related to CSP special events are the following:

- $\sqrt{}$ means success (successful termination of the specified process).
- *l.a* means participation in event a by a process named l.
- *c.v* means communication of the value v on the channel c.
- *l.c* is the channel c of a process named l.
- *l.c.v* means communication of the message v on the channel $l.c$.
- *acquire* means acquisition of a resource.
- *release* means release of a resource.

Formal system verification in CSP is based on the trace model of a process, which is a set of traces, where each trace represents a sequence of events that the process may perform. The most important notions related to process traces are as follows:

- $\langle \rangle$ is the empty trace.
- $\langle a \rangle$ is the trace containing only a (the singleton sequence).
- $\langle a, b, c \rangle$ is the trace with three symbols, a then b, then c.

- "+ is the trace catenation operator in this book. For example, $\langle a, b, c \rangle = \langle a, b \rangle$ "+ $\langle \rangle$ "+ $\langle c \rangle$.
- s^n is the trace s repeated n times. For example, $\langle a, b \rangle^2 = \langle a, b, a, b \rangle$.
- $s \uparrow A$ means s restricted to A (in this book $\uparrow$ is used as a restriction operator). For example, $\langle b, c, d, a \rangle \uparrow \langle a, c \rangle = \langle c, a \rangle$.
- $s \leq t$ means s is a prefix of t. For example, $\langle a, b \rangle \leq \langle a, b, c \rangle$.
- $s \leq^n t$ means s is like t with up to n symbols removed. For example, $\langle a, b \rangle \leq^2 \langle a, b, c, d \rangle$.
- s in t means trace s is in the trace t (i.e., s is the subtrace of the trace t). For example, $\langle b, c \rangle$ in $\langle a, b, c, d \rangle$.
- #s means the length of trace s. For example, #$\langle b, c, b, a \rangle = 4$.
- $s \downarrow b$ means the count of b in s. For example, $\langle b, c, b, a \rangle \downarrow b = 2$.
- $s \downarrow c$ means the communications on the channel c recorded in s. For example, $\langle c.1, a.4, c.3, d.1 \rangle \downarrow c = \langle 1, 3 \rangle$.
- $s ; t$ means the trace s successfully followed by the trace t. For example, $(s$ "+ $\langle \sqrt{} \rangle) ; t = s$ "+ t.
- A^* means the set of sequences with elements in A, or more formally $A^* = \{s \mid s \uparrow A = s\}$.
- s_0 means the head of s. For example, $\langle a, b, c \rangle_0 = a$.
- s' means the tail of s. For example, $\langle a, b, c \rangle' = \langle b, c \rangle$
- $s[i]$ means the ith element of s. For example, $\langle a, b, c \rangle[1] = b$.
- $f^*(s)$ means apply f on each element of s; we read it as f star of s. For example, $square^*(\langle 1, 5, 3 \rangle) = \langle 1, 25, 9 \rangle$.

The syntax of CSP core language is defined as follows:

$P, Q ::= Skip \mid Stop \mid a \to P \mid P ; Q \mid c?x \to P \mid$
$c!x \to P \mid P \parallel Q \mid P [|X|] Q \mid P \lhd b \rhd Q$

At the end of this section, let's have a look in some of the evergreen examples of CSP processes from Hoare (1985).

Example 1: The process *COPY*, which immediately copies every message it has input from the channel named *left* by outputting it to the channel named *right*.

$\alpha left(COPY) = \alpha right(COPY)$
$COPY = \mu X \bullet (left?x \to right!x \to X)$

The process *COPY* satisfies the following specification:

COPY sat *right* $\leq^1$ *left*

Example 2: The process *DOUBLE* is like process COPY, except that every input number is doubled before it is output.

$\alpha left(DOUBLE) = \alpha right(DOUBLE) = N$
$DOUBLE = \mu X \bullet (left?x \rightarrow right!(x + x) \rightarrow X)$

The process *DOUBLE* satisfies the following specification:

$DOUBLE$ sat $right \leq^1 double^*(left)$

5.3.2.2 Brief Overview of PAT and CSP#

PAT is an extensible and modularized framework for automatic system analysis based on CSP, which is freely available for noncommercial research at http:// sav.sutd.edu.sg/PAT/. This self-contained framework supports modeling, simulating, and verifying concurrent real-time systems including communication protocols. PAT supports various model checking techniques targeting different properties, such as deadlock-freeness, divergence-freeness, reachability, LTL properties with fairness assumptions, refinement checking, and probabilistic model checking. Moreover, the PAT development team implemented advanced optimization techniques, including partial order reduction, symmetry reduction, process counter abstraction, and parallel model checking, in order to achieve good performance from the user point of view.

The main PAT facilities are as follows:

- Multidocument and multilanguage editor for creating models
- Simulator for visual and interactive simulation of system behaviors, including random simulation, step-by-step simulation, complete state graph generation, trace playback, and counterexample visualization
- Verifiers for deadlock-free analysis, reachability analysis, state/event LTL checking (with and without fairness assumptions), and refinement checking
- Documentation and many examples (we focus on some of them in the next section)

The PAT framework has been developed by J. Sun, Y. Liu, J.S. Dong, and their colleagues at the National University of Singapore since 2007 (Sun 2009). The first time that PAT was successfully demonstrated internationally was at the 30th International Conference on Software Engineering in 2008. Over the last decade, many other researches worldwide have been using PAT to model and verify various systems, ranging from recently proposed distributed algorithms and security protocols to real-world systems like multilifts and pacemakers. However, in this book, we focus on communication protocols.

We continue with a brief overview of CSP#. CSP# (pronounced "CSP sharp") is a super-language for CSP, which combines high-level operators (mostly coming from CSP) such as conditional and nondeterministic choices, interrupt, parallel composition, interleaving, hiding, and asynchronous message passing, with low-level C# programming language constructs such as variables, arrays, and control flow statements like if–then–else and while and for loops. CSP# supports both general models of communication among processes, namely *shared memory* and *message passing*. The former model of communication is supported by the means of global variables, whereas the second model is supported by the channels for asynchronous message passing or by the CSP multiparty barrier synchronization. The main CSP# design principle is to keep the original CSP as a core sublanguage and additionally to provide access to data states and executable data operators from C#.

The CSP# language constructs may be divided into the following four groups:

- The core subset of CSP operators, including event-prefixing, internal and external choices, alphabetized lock-step synchronization, conditional branching, recursion, etc.
- The language constructs that are regarded as a syntactic sugar to CSP, including global variables and asynchronous channels: Although, the original CSP supports modeling of shared variables and asynchronous channels as processes, the dedicated language constructs offer better usability and improved verification efficiency.
- The set of event annotations: Since CSP supports only the notion of safety, the event annotations provide additional means for modeling fairness using event-based compositional language.
- The language constructs for stating assertions that may be automatically verified using PAT built-in verifiers.

The language syntax structures are classified as follows:

- Global definitions
- Process definitions
- Assertions

5.3.2.2.1 CSP# Global Definitions

CSP# global definitions include:

- Model names
- Global constants
- Global variables and arrays

- Asynchronous channels
- Macros

Model names are given using a declaration *//@@ model_name @@*.

Constants are specified using the C macro directive **#define** or the key word **enum**. A constant value may be either integer or Boolean Examples:

```
#define N 10; // N == 10
enum {zero, one, two}; // zero == 0, one == 1, two == 2
```

Variables and arrays are specified using the keyword **var**. Since CSP# is a weekly typed language, no typing information is required. However, casting between incompatible types leads to run-time exceptions. PAT supports multidimensional arrays by converting them into one-dimensional arrays. The index range of an array dimension may be specified explicitly by giving the lower bound or the upper bound or both. Examples are as follows:

```
var x = 0; // variable x set to 0
var ba = [1, 2, 3, 4]; // array ba with 4 elements
var leader[3]; // array leader with 3 elements set to 0
var knight : {0..} = 0; // array knight with specified lower index 0
```

Elements of an array may be set using event-prefixing, e.g.,

```
P() = a {m[1][9] = 1} -> Skip
```

User-defined type may be specified using the declaration **var**<*type*> *var_name*; e.g.,

```
var<MyType> x; // default constructor MyType is called
var<MyType> x = new MyType(100); // constructor with one parameter
```

Channels and channel arrays are specified using the keyword **channel**, which has two parameters, namely the channel name and the corresponding buffer size. Examples are as follows:

```
channel c 10; // channel c with buffer size 10
channel c[3] 10; // channel array c comprising 3 channels
```

Macros may be defined using the C macro directive **#define**, which has two parameters, namely the macro (instruction) name and its definition. An example is

```
#define condition x==0; // the name is condition, the definition is x==0
```

Model inclusion: a submodel may be included in the current model using the directive **#include**. For example:

```
#include "c:\submodel.csp";
```

5.3.2.2.2 *CSP# Process Definitions*

CSP# process definitions include:

- Stop
- Skip
- Event prefixing
- Statement block inside events
- Channel input/output
- Sequential composition
- External/internal choice
- Conditional choice
- Case
- Guarded process
- Interleaving
- Parallel composition
- Interrupt
- Hiding
- Atomic sequence
- Recursion
- Assert

Processes may be defined using the equations of the following format:

$$P(x_1, x_2, \ldots, x_n) = exp$$

where P is a process name, x_1, ..., x_n is an optional list of formal parameters, and exp is a process expression. A process P may be referenced by the expression:

$$P(y_1, y_2, \ldots, y_n)$$

where y_1, ..., y_n are the real parameters (or arguments). Self-recursion and mutual-recursion among processes is normally allowed.

Stop is the deadlock process, whereas **Skip** is a process that immediately terminates.

Even prefixing $e \rightarrow P$ describes a process which performs an event e first and then behaves as process P. An event may be in a simple form (just the event name) or in a compound form, such as $event\_name.e_1.e_2$ where e_1 and e_2 are expressions composed from variables, including process parameters, channel input variables, and global variables. Examples are as follows:

```
VM() = coin -> chocolate -> VM(); // chocolate vending machine
Phil(i) = get.i.(i + 1)%N -> Rest(); // i is a process parameter
```

Event name is an arbitrary (user-defined) string. It may also be a channel name. It cannot be the name of a global variable/constant, a process, a process parameter, or a proposition.

Statement blocks inside events (a.k.a. **data operations**): A statement block {*statements*} may be attached to an event simply using the expression *event_name*{*statements*}, where *statements* may include declarations of local variables and arrays, control flow constructs made using the keywords like **if–then–else** and **while**, references to global variables, C# functions, etc., e.g.,

```
P() = incx{x = x + 1;} -> Stop // increment global variable x
```

The attached statement block is executed atomically (i.e., without interleaving with other processes). On the other hand, an event with an attached statement block may be viewed as a labeled piece of code, which is also sometimes used for constructing counterexamples. A reader should note that here are no per process local variables in CSP#, so processes need to use global variables instead.

Invisible events may be specified using the keyword **tau**, e.g., **tau**{$x = x + 1$;} is equivalent with {$x = x + 1$;}.

Channel input/output is written similar to simple event prefixing. Simple examples are as follows (let c be the channel name and P be a process expression; imagine channel as a FIFO buffer):

```
c!a.b -> P  // output values of expressions a and b
c?x.y -> P // input values of local free variables x and y
c!10 -> P // output constant 10
c?[x > y]x.y -> P // if x > y input values of x and y
```

We may use an arbitrary number of variables/expressions in channel input/output by separating them with dots ('.'), but we cannot use global variables in channel input expressions. Here is an example of two processes involved in an asynchronous communication over the channel c:

```
channel c 1;
P(i) = c!i -> P(i)
Q() = c?x -> a.x -> Q()
System() = P(3) ||| Q()
```

In the example above, communication over channel c is asynchronous because the size of the channel is nonzero (it is 1). We may turn this communication into synchronous communication by setting the size of the channel *c* to zero (**channel** c 0). Furthermore, we may attach statement blocks to asynchronous/synchronous channel input/output, e.g.,

```
channel c 1; // or channel c 0;
var x = 0;
P() = c!x{x = 1} -> P()
Q() = c?y{x = y} -> Q()
System() = P() ||| Q()
```

The execution sequence in the example above is $c!x$, $(x = 1)$, $c?y$, $(x = y)$. Note that the scope of the channel's input variable (such as x and y above) is after the channel input event and within the enclosing process. Such variables may be referenced in the scope, but cannot be updated.

We should also remember that if channel input expressions evaluate to constants, the process can receive only the matching channel outputs. For example, process $P(i) = (c?i.(i + 1) \rightarrow Skip)$ can receive only the sequence of values i, $(i + 1)$ from the channel c. Furthermore, local free variables used in channel input can be reused again in the next channel inputs.

CSP# also supports channel arrays, for example:

```
channel c[2] 1;
S(i) = c[i]!i - S(i);
R() = c[0]?x -> a.x -> R() [] c[1]?x -> a.x -> R();
System() = (|||i:{0..2}@S(i)) ||| R()
```

Channel operations may be used to query the buffer information of an asynchronous channel. A channel operation is invoked by the static method call: **call**(*channel_operation, channel_name*). There are five channel operations:

- **cfull** is a Boolean function that tests weather the buffer if full or not.
- **cepmty** is a Boolean function that tests weather the buffer if empty or not.
- **ccount** is an integer function that returns the number of elements in the buffer.
- **csize** is an integer function that returns the buffer size.
- **cpeek** returns the first element (the head) of the buffer.

Sequential composition P; Q means that P and Q execute sequentially. There are three kinds of choices in CSP#:

- The **general choice** P [] Q, which means that either P or Q may execute. If P performs an event first, it takes control, otherwise Q takes control.
- The **external choice** P [*] Q is resolved by the environment through observation of a visible event (not a tau event). If the first event of both P and Q are visible, P [] Q and P [*] Q have the same result.

- The **internal choice** $P <> Q$ means that either P or Q may execute and the choice is made internally and nondeterministically. Although nondeterminism is normally undesirable, it may be useful in the modeling phase for hiding irrelevant (or unknown) details.

The generalized forms of general/external/internal choices are as follows:

- $[] \, x : \{1..n\} @ P(x)$ is the generalized form for $P(1) \, [] \, ... \, [] \, P(n)$
- $[*] \, x : \{1..n\} @ P(x)$ is the generalized form for $P(1) \, [*] \, ... \, [*] \, P(n)$
- $<> x : \{1..n\} @ P(x)$ is the generalized form for $P(1) <> ... <> P(n)$

Conditional choices are also supported in CSP#. Besides the **traditional conditional choices** used in programming languages which are based on the keywords **if, else,** and **else if,** CSP# introduces more specialized conditional choices, such as the **atomic conditional choice (ifa)** and the **blocking conditional choice (ifb)**. The formats of conditional choices are as follows:

- **if** (*condition1*) P **else if** (*condition2*) Q **else** R. Of course, the shorter **if** and **if–then–else** formats are also allowed.
- **ifa** (*condition*) P **else** Q. The atomic conditional choice (**ifa**) performs condition checking and the first event of P or Q atomically.
- **ifb** (*condition*) P. The blocking conditional choice (**ifb**) is similar to the guarded process, but unlike the guarded process, in **ifb** condition checking and process execution are separated. There is no **else** in **ifb**, and side effects are not allowed.

The **case** construct in CSP# is somewhat similar to the **switch–case** construct in say, C#:

```
case {
    condition1: P
    condition2: Q
    ...
    default: R
}
```

The **guarded process** executes when its guarded condition is satisfied (i.e., the *condition* is **true**), otherwise the whole process waits:

$$[condition] \; P$$

Interleaving $P \; ||| \; Q$ means that P and Q execute concurrently without barrier synchronization, except during termination events (termination events must be executed jointly by all the interleaving processes). Of course,

P and *Q* may communicate over shared variables and channels. The generalized format of interleaving is as follows:

$$||| \; x : \{0..n\} @ P(x)$$

We may specify groups of interleaving processes by using looping variables with a finite, or even infinite, range. We may do the same with the parallel composition and the internal/external choices. Examples are as follows:

```
||| {50} @ P(); // interleaving  of 50 P()
||| {..} @ Q(); // interleaving  of infinite number of Q()
||| {} @ P(); // this is equivalent to Skip
[] x : {0..1} @ ( (||| {x} @ P()) ||| (||| {x} @ Q()) )
// <=> (Skip|||Skip) [] (||| {1} @ P()) ||| (||| {1} @ Q())
```

A looping variable *x* may also be used as a process parameter within the process, e.g.:

```
||| x : {0..n} @ (a.x -> Skip)
```

Generally, the symbols used to define a looping variable's range (like *n* in the example above) can be global constants or process parameters, but they cannot be global variables.

Parallel composition *P* || *Q* means that *P* and *Q* execute concurrently with possible **lock-step synchronization**, a.k.a., **barrier synchronization**. Lock-step synchronization means that *P* and *Q* simultaneously perform the same event. In the following example, *P* and *Q* are lock-step synchronized by the event *b*:

```
P() = a -> b -> Stop;
Q() = b -> Stop;
System() = P() || Q()
```

The execution sequence for the example above is *a, b, Stop,* because *P* performs *a* first, then both *P* and *Q* perform *b*, and, finally, both *P* and *Q* perform *Stop*. Obviously, lock-step synchronization assumes that the alphabets of parallel processes are known. It is well-known that determining the alphabet of a process automatically is generally not trivial (because of process self/mutual referencing and the usage of process parameters), and, in fact, sometimes it is not even possible (for example, in the case of a nonterminating processes). However, PAT provides a best-effort automatic procedure for determining the **default alphabet** of a given process. When the default alphabet is not as expected, we may manually modify it. For example, if we use data operations (statement blocks attached to events), PAT will not cover them when determining the default alphabet, and we would have to manually modify the default alphabet.

Alternatively, we may decide to manually specify the alphabet of a process using the directive **#alphabet** *P* {*events*}, where *P* is the process name, and *events* is a comma-separated list of event expressions. The event expressions may also contain variables, for example, the alphabet of a process *P(x)* may be specified as **#alphabet** *P* {*a.x*};.

The important principle of CSP# (and PAT) is that the process signature comprises both the process expression and the process alphabet, i.e., processes with the same process expression, but with different alphabets, which are seen as different processes. To cope with this, we sometimes need to introduce supplementary processes. For example, if the process *P* has different alphabets in two different parts of a model, we may introduce supplementary processes *Q* and *R*:

```
Q() = P();
#alphabet Q = {x};
R() = P();
#alphabet R = {y};
```

Generalized parallel composition has the following format:

$$|| \; x : \{0..n\} \; @ \; P(x);$$

We may also use indexed event lists within alphabets. For example:

```
#alphabet P {x:{0..N}; y:{0..N} @ e.x.y};
```

The **interrupt** composition *P* **interrupt** *Q* means that *P* executes until the first visible event of *Q* is engaged, and then control is switched to *Q* (the first visible event may occur at any point of *P*). The corresponding execution trace is a trace of *P*, followed by a trace of *Q*. The main purpose of interrupt abstraction is modeling the interrupt processing behavior.

Hiding may be used to define a process with a reduced alphabet, e.g., *P* \ *A* is a process whose alphabet is **#alphabet**(*P*) \ *A*, where *A* is any subset of **#alphabet**(*P*). We use hiding to hide unimportant events from a process alphabet (for example, to prevent unwanted synchronization in parallel composition) or in order to introduce nondeterminism. For example *Phil* specifies a philosopher who gets forks, eats, and puts the forks away when finished, whereas *dashPhil* hides the events related to forks:

```
Phil() = getfork.1 -> getfork.2 -> eat ->
         putfork.1 -> putfork.2 -> Phil();
dashPhil() = Phil() \ { getfork.1, getfork.2, putfork.1, putfork.2};
```

For our convenience, we may use indexed event lists for defining a set of events with the same prefix, for example:

```
dashPhil() = Phil() \ { x:{1..2}@getfork.x, y:{1..2}@putfork.y};
```

Atomic process *P* is declared using the declaration **atomic**{*P*}, which means that its events should be executed atomically. Also, if a statement block is prefixed with the keyword atomic, this block should be executed as one superstep without interleaving with other processes. Such a statement block may contain any process statements and may be nondeterministic. Generally, an atomic process has a higher priority than a non-atomic process, i.e., if an atomic process has an enabled event, that event will execute before the events of non-atomic processes. However, if multiple atomic processes are enabled, they interleave each other.

We may use atomic processes and atomic statement blocks to reduce the model state space and thus speedup model checking, especially when the model comprises parallel process compositions. The state space may be sometimes reduced exponentially. Using **atomic** is actually similar to manual partial order reduction. An important rule is that local events that are invisible to the verifying property and independent of other events will get the higher priority.

Recursion is constructed by the self or mutual process referencing. The following example illustrates a system with mutual recursion:

```
P(i) = a.i - Q(i);
Q(i) = b.i - P(i);
System() = P(1) || Q(2);
```

A parameter of the recursive process may be any valid expression, e.g., $P(x + y)$, $P(\mathbf{new}\ List())$, etc. However, when a parameter is a user-defined type, the user must take special care to pass the correct value type, because CSP# is not a typed language, so PAT does not support compile time type checking. The user should be also very conscious of possible negative side effects, which may, for example, appear within constructs with choices (e.g., *exp1* [] *exp2*—a side effect in *exp1* may remain even when *exp2* is selected).

We may also use recursion to implement common loops. For example, the behavior of the while loop **while** (*condition*) {*P()*} is equivalent to the behavior of the following process:

```
Q() = if(condition) {P(); Q()};
```

Assert is used to add an assertion in the program. PAT simulator and verifiers check the assertion in run-time, and, if the assertion fails, the corresponding run-time exception is thrown and the system evaluation is stopped. For example:

```
var x = 0;
P() = assert(x = 0); a{x = x + 1;} -> P();
```

5.3.2.2.3 CSP# Assertions

Assertions are queries about system behavior. CSP# assertions include:

- Deadlock-freeness
- Divergence-free

- Reachability analysis
- Linear Temporal Logic (LTL)
- Refinement/Equivalence

The deadlock-freeness assertion claims that a process *P* is deadlock-free:

#assert *P*() **deadlockfree**;

PAT uses a depth-first-search or a breadth-first-search algorithm to search the process's state space. A **deadlock state** is a state with no further movement, except a **successfully terminated** state. A process is **deadlock-free** if it does not have any deadlock states.

A **divergence-free** assertion claims that a process *P* is divergence free:

#assert *P*() **divergencefree**;

A process is **divergent** if it performs internal transitions forever without engaging in any useful events, e.g., $P = (a \rightarrow P) \setminus \{a\}$; a **divergent-free** process is a process that is not divergent.

A **deterministic** assertion claims that a process *P* is deterministic:

#assert P() **deterministic**;

A process is **deterministic** if it does not have a state with more than one outgoing transitions driven with the same event. Otherwise it is **nondeterministic**, e.g., $P = a \rightarrow Skip \ [] \ a \rightarrow Stop$.

The **nonterminating** assertion claims that a process *P* is nonterminating:

#assert *P*() **nonterminating**;

A process is **nonterminating** if it does not have a terminating state (either a successfully terminated state or a deadlock state), e.g., such as $P = a \rightarrow P$.

The **reachability** assertion claims that a process *P* may reach a state satisfying a given *condition* (where a condition is a proposition defined as a global definition):

#assert *P*() **reaches** *condition*;

In the following example, the reachability assertion claims that the process *P* reaches a state satisfying the condition $(x < 0)$:

```
#define goal x < 0;
var x = 0;
P() = a{x = x + 1;} -> P();
#assert P() reaches goal; // this will not be satisfied
```

The **optimized reachability** allows, for example, minimizing a given *cost* function during the reachability search:

#assert *P*() **reaches** *goal* with min(*cost*);

For example:

```
#define goal x = 14;
var cost = 0;
var x = 0;
P() = if (x <= 14) {{some arithmetic with x} -> P()};
#assert P() reaches goal with min(cost);
```

LTL assertion claims that a process *P* satisfies a given LTL formula *F*:

#assert *P*() |= *F*;

An LTL formula is evaluated on an infinite sequence of truth evaluations over a path traversing the process state space, and a specified position on that path. The syntax of an LTL formula *F* is as follows:

$$F = event \mid proposition \mid [] F \mid <> F \mid X F \mid F1\ \mathbf{U}\ F2 \mid F1\ \mathbf{R}\ F2$$

where [] (or 'G') reads as **always** , <> (or 'F') reads as **eventually**, **X** reads as **next**, **U** reads as **until**, and **R** (or 'V') reads as **release** (note that in PAT [], <>, **R** may also be written as 'G', 'F', and 'V', respectively).

The semantic of unary modal operators is as follows:

- **X** φ, neXt, φ holds in the next state: • → • φ - - - → • → •
- **G** φ, Globally, φ holds on the entire subsequent path: • φ → • φ - - - → • φ → • φ
- **F** φ, Finally, φ eventually has to hold: • → • - - - → • φ → • (holds somewhere on the subsequent path)

The semantic of binary modal operators is as follows:

- **U** φ, Until, φ holds at the current or future position, and ψ has to hold until that position; at that position ψ does not have to hold anymore: • ψ → • ψ - - - → • ψ → • φ
- **R** φ, Release, φ is true until the first position in which ψ becomes true, or φ is true forever if such position does not exist: • φ → • φ - - - → • ψ → • ψ, or • φ → • φ - - - → • φ → • φ

The LTL assertion is true if and only if the given formula *F* is satisfied for all the possible paths corresponding to all the possible system executions.

Internally, PAT first constructs a Buchi automaton equivalent to the negation of *F* and a Buchi automaton of the process *P*, and then uses these automatons to check the LTL assertion. For example, the following LTL assertion claims that the philosopher can always eventually eat, i.e., the **nonstarvation property**:

```
#assert Phil() |= [] <> eat;
```

Events in LTL formulas may also be component events like *eat*.0, and channel events like "c!3.8" and "c?19" (here we must use "" because '!' and '?' are special characters). In case of synchronous channels, PAT automatically converts channel input/output operators ('!' and '?') to dots in the events, e.g., the channel event "c!3.8" is converted to c.3.8.

Refinement/Equivalence is the FDR (Failures-Divergences Refinement) approach for checking whether an implementation meets its specification. In contrast to an LTL assertion, a **refinement assertion** compares the complete behaviors of two processes, for example, whether one is a subset of another. CSP# supports the following notions of a refinement relationship:

- **#assert** *P*() **refines** *Q*(): *P*() refines *Q*() in the trace semantics
- **#assert** *P*() **refines** <F> *Q*(): *P*() refines *Q*() in the stable failures semantics
- **#assert** *P*() **refines** <FD> *Q*(): *P*() refines *Q*() in the failures· divergence semantics

When it comes to verifying CSP# assertions, PAT supports the following options of admissible behavior (the process-level options are enabled only for the systems with interleaving or parallel composition):

- All (or No Special Fairness) is a default option that allows all behaviors to occur. We choose this option to give all the next states (that have the same previous state) the same fairness, i.e., the same possibility to happen, which also means that there is no special fairness for each process.
- Event-level Weak Fair Only means that for every event in the system, if the event is eventually always enabled, then the event always eventually occurs.
- Event-level Strong Fair Only means that for every process in the system, if the process is always eventually enabled, then the event always eventually occurs.
- Process-level Weak Fair Only means that for every process in the system, if the process is eventually always enabled, then the process is always eventually engaged.

- Process-level Strong Fair Only means that for every process in the system, if the process is always eventually enabled, then the process is always eventually engaged.
- Global Fair Only (or Strong Global Fairness) means that for every transition in the system, if the transition can always eventually be taken, then the transition is actually always eventually taken.

A detailed discussion of the above listed options is outside the scope of this book. Also, in order to save the space in the following examples, we sometimes present verification results for only some of the options. Shorter counterexamples are provided without comment so that the reader may analyze and think about them, while longer counterexamples are skipped to save space (of course, an interested reader may repeat the presented experiments using the freely available PAT and reproduce all the counterexamples on their own).

5.3.2.3 Examples of Formal Verification Based on CSP# and PAT

In this section, we study the following examples:

- Alternating bit protocol
- Two-phase commit protocol
- Various leader election protocols in the complete graphs, the rings, and the rooted directed trees
- Telecomm service system

5.3.2.3.1 Alternating Bit Protocol

Alternating Bit Protocol (ABP) is a data link layer protocol that retransmits lost or corrupted messages. Actually, it is a special case of a sliding window protocol where a timer regulates the order of messages to provide reliable message transmission over a data link, using the 1-bit window. Transmitter **A** sends messages to receiver **B** (initially the channel from **A** to **B** is empty). Each message contains data and a 1-bit **sequence number** (SN) whose value is 0 or 1. **B** acknowledges the successfully received messages by sending the appropriate ACK: ACK0 for a message with SN 0 or ACK1 for a message with SN 1.

A resends a message continuously with the same sequence number until it receives an ACK with the same sequence number, then **A** toggles (complements) the sequence number and starts sending the next message. Symmetrically, when **B** receives an uncorrupted message with SN 0, **B** resends ACK0 continuously until it receives an uncorrupted message with SN 1, then it switches to ACK1, etc. Therefore, **A** may still receive ACK0 when it has already switched to resending a message with SN 1, and vice versa. **A** treats such ACKs as negative-ACKs (NAKs) by simply ignoring them.

The ABP is initialized by sending a bogus message and ACKs with SN 1, so the first real message is a message with SN 0.

We illustrate the ABP in Figure 5.6. **A** starts by sending the information message I1 with the data bit 1, and it keeps resending it until it receives ACK1. Once **B** receives the message, it acknowledges it by the ACK1, and it keeps resending ACK1 until it receives the message I0. In order to keep the figure readable, we show the message I1 and ACK1 as quarter-length arrows. Later on, when **A** receives ACK1 it starts sending I0, when **B** receives I0 it starts sending ACK0, and so on. In order to keep the figure readable, we do not show other messages that were resent.

The parametrized ABP model in CSP# was created by Dr. Sun Jun. The following is ABP model for the parameter *ChannelSize* set to 1 (in the model that has the constant *CHANNELSIZE*):

```
#define CHANNELSIZE 1;
channel c CHANNELSIZE;
channel d CHANNELSIZE;
channel tmr 0;

Sender(alterbit) =
(c!alterbit -> Skip [] lost -> Skip);
tmr!1 -> Wait4Response(alterbit);

Wait4Response(alterbit) =
(d?x -> ifa (x == alterbit) {
        tmr!0 -> Sender(1 - alterbit)
      } else {
        Wait4Response(alterbit)
      })
[] tmr?2 -> Sender(alterbit);

Receiver(alterbit) =
c?x -> ifa (x == alterbit) {
      d!alterbit -> Receiver(1 - alterbit)
    } else {
      Receiver(alterbit)
    };

Timer = tmr?1 -> (tmr?0 -> Timer [] tmr!2 -> Timer);

ABP = Sender(0) ||| Receiver(0) ||| Timer;

#assert ABP deadlockfree;
#assert ABP |= []<> lost;
```

In the model above we model the sender (i.e., transmitter), the receiver, and the sender's timer by the processes *Sender* (and *Wait4Response*), *Receiver*, and *Timer*, respectively. For simplicity, messages only have the sequence number and no data. Messages from *Sender* to *Receiver* are transferred over the channel *c*, whereas messages from *Receiver* to *Sender* are transferred over the channel *d*. Both channels *c* and *d* are ordinary CSP# channels, but in this

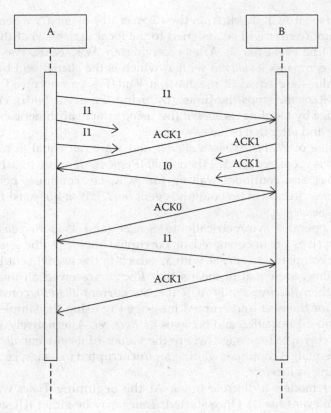

FIGURE 5.6
ABP sequence diagram.

model, we call them the unreliable channel *c* and the perfect (or reliable) channel *d*, because the former models an unreliable channel and the latter models a reliable channel. The third channel in this model is a synchronous channel *tmr* between *Sender* and *Timer*, which is used to model timer-related events, namely start timer (the event *tmr.1*), stop timer (the event *tmr.0*), and timeout (the event *tmr.2*).

As mentioned above, the sender is modeled by the processes *Sender* and *Wait4Response*. Within *Sender*, we use external choice [] to model the unreliability of channel *c*. In particular, the construct (*c!alterbit -> Skip* [] *lost -> Skip*) means that *Sender* will either successfully send the message *alterbit*, or it will skip sending it, which is equivalent to losing the message on an unreliable channel—in the model, this case corresponds to the event *lost*. Next, *Sender* starts the timer using the channel output event *tmr!1*, and, further on, behaves as *Wait4Response*.

Within *Wait4Response* we use external choice [] to model a possible timer expiration event (i.e., the timeout behavior). In particular, *Wait4Response* will either receive an ACK/NAK from the *Receiver* (the event $d?x$ when the *alterbit* sent from *Receiver* will be assigned to the local variable x) or the timeout will occur (the event $tmr?2$). After receiving an ACK/NAK, *Wait4Response* atomically compares its *alterbit* with x, which is the *alterbit* sent by *Receiver*. If these values are equal (it means that *Wait4Response* received an ACK), then *Wait4Response* stops the timer (by using the event $tmr!1$), creates the new message by toggling its *alterbit* (by using simple arithmetic expression: $1 - alterbit$), and after that behaves as *Sender*.

If the value of *Wait4Response*'s *alterbit* and x are not equal (it means that *Wait4Response* received NAK), then *Wait4Response* ignores that NAK (just does noting), and continues waiting for ACK, i.e., continues behaving as *Wait4Response*. In case of timeout (the event $tmr?2$), *Wait4Response* further on behaves as *Sender*.

Receiver operates symmetrically to *Sender* (and *Wait4Response*). When *Receiver* receives an uncorrupted or corrupted message (the event $c?x$), it atomically compares its *alterbit* with x, which is the *alterbit* sent by *Sender*. If these values are equal (it means that *Receiver* received an uncorrupted message), then *Receiver* sends ACK (i.e., its current *alterbit*), constructs the new ACK for the next uncorrupted message (by using the simple arithmetic expression: $1 - alterbit$) and behaves as *Receiver*. Alternatively, if *Receiver* received a corrupted message (where the values of its current *alterbit* and x where not equal), it continues waiting an uncorrupted message, i.e., it continues behaving as *Receiver*.

The *timer* models a discrete timer. At the beginning *Timer* waits to be started (the event $tmr?1$). Once started, *Timer* may be either ([]) stopped by *Wait4Response* (the event $tmr?0$) or it may expire and generate a timeout signal (the event $tmr!2$) towards *Wait4Response*. In both cases, after engaging in a prefix event ($tmr?0$ or $tmr!2$) it continues behaving as *Timer*.

The complete system is modeled as a parallel composition of *Sender*, *Receiver*, and *Timer*. Initially, both *Sender* and *Receiver* set their local variables *alterbit* to 0.

There are two assertions at the end of the model. The first assertion claims that ABP is deadlock-free. As a result of verifying this assertion, PAT produces the following positive report:

```
The Assertion (ABP() deadlockfree) is VALID.
```

The second assertion claims that always, at some point, a message from *Sender* to *Receiver* will be lost (the event *lost* will happen). The verification result for this assertion depends on the admissible behavior option that we select. As expected, if we select the options "Event-level Strong Fair Only" or "Global Fair Only", the verification result is positive:

```
The Assertion (ABP() |= []<> lost) is VALID.
```

But, if we select the option "All", the result is negative with the following counterexample:

```
The Assertion (ABP() |= []<> lost) is NOT valid.
A counterexample is presented as follows.
<init -> c!0 -> (c?0 -> d!0 -> tmr.1 -> d?0 -> tmr.0 -> c!1 -> c?1
-> d!1 -> tmr.1 -> d?1 -> tmr.0 -> c!0)*>
```

Similarly, if we select the options "Event-level Weak Fair Only," "Process-level Weak Fair Only," or "Process-level Strong Fair Only," the result is also negative, with rather lengthy counterexamples that a reader may reproduce on their own.

5.3.2.3.2 *Two-Phase Commit Protocol*

Two-phase commit protocol (2PC) is one of the most widely used atomic commitment protocols (ACPs). It coordinates all the processes participating in a distributed atomic transaction on whether they should **commit** or **abort** (or **rollback**) the transaction. In the theory of distributed computing, 2PC is viewed as a specialized **consensus protocol**. The 2PC advantages are simplicity and resilience to many temporary system failures, such as process, network node, or communication failures. However, in some rare cases, system administrators must perform manual failure recovery procedures. To enable failure recovery, which is automatic in most of the cases, participating processes must maintain logs of the protocol's states. Many existing 2PC variants use different logging strategies and recovery procedures.

The protocol relies on the following three assumptions:

- One node is the coordinator, whereas the rest of the nodes are participants (the coordinator may be selected using a leader election protocol).
- Each node has a stable storage for storing a write-ahead log, which is never lost or corrupted in a node crash.
- No node crashes forever.
- Any two nodes can (directly or indirectly) communicate with each other.

During **normal operation** (i.e., when there are no failures) the protocol consists of the following two phases:

- The **commit request phase** (or **voting phase**), in which the process **coordinator** requests from all the processes participating in the transaction (or **participants**, **cohorts**, **workers**, or **pages**) to prepare to commit/abort the transaction by performing all the necessary steps locally, and to vote "yes" (**commit**) if the local preparation was successful or "no" (**abort**) if some problem during local preparation was detected.

- The **commit phase**, in which the coordinator decides whether to commit (if all the participants voted "yes") or abort the transaction (if at least one participant voted "no"), and notifies the decision result to all the participants. The participants, in turn, perform all necessary local actions (effectively realizing commit) on their local resources (or **recoverable resources**) and their portions in the transaction's output (if any).

The commit request phase consists of the following steps:

1. The coordinator sends the message **query to commit** to all the participants and waits until it receives replies from all of them.
2. The participants execute the transaction locally (and update their logs) to the point where they will be asked to commit/abort.
3. Each participant replies to the coordinator with the message **agreement**, which carries its vote—**yes** (commit) if its actions were successful, or **no** (abort) if otherwise.

The completion phase in case of success (commit) consists of the following steps:

1. The coordinator sends the message **commit** to all the participants and waits for their **ACK**s.
2. Each participant completes the transaction locally and releases all locks and resources.
3. Each participant sends the message **ACK** to the coordinator.
4. The coordinator completes the transaction once it receives all the **ACK**s.

The transaction will fail if any of the participants votes **no**, or the coordinator's timer expires (and signals a timeout). The completion phase in case of failure (abort) consists of the following steps:

1. The coordinator sends the message **rollback** to all the participants and waits for their **ACK**s.
2. Each participant undoes the transaction locally, and then releases all locks and resources.
3. Each participant sends the message **ACK** to the coordinator.
4. The coordinator completes the transaction once it receives all the **ACK**s.

We illustrate the 2PC by the sequence diagram in Figure 5.7. As shown, the protocol consists of two phases. In the commit request phase, the coordinator

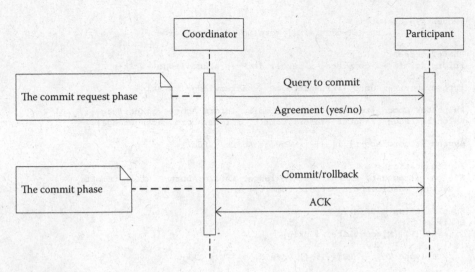

FIGURE 5.7
2PC sequence diagram.

sends the message **query to commit** to each participant, and all the participants vote yes or no by using the corresponding message **agreement**. In the commit phase, depending on the results of the previous phase, the coordinator sends either the message **commit** or the message **rollback** to each participant, and all the participants reply with the message **ACK**.

The parametrized 2PC model in CSP# was created by Dr. Sun Jun slightly different than the one explained above. The following is the simplified 2PC model for the parameter *Page* set to 2 (in the model that is the constant *N*):

```
#define N 2;
enum {Yes, No, Commit, Abort};
channel vote 0;
var hasNo = false;

//The following models the coordinator
Coord(decC) =
(||||{N}@ request -> Skip);

(||||{N}@ vote?vo -> atomic{tau{if (vo == No) {hasNo = true;}} -> Skip});

decide -> (
  ([hasNo == false] (||||{N}@inform.Commit -> Skip);
    CoordPhaseTwo(Commit))
  []
  ([hasNo == true] (||||{N}@inform.Abort -> Skip);
    CoordPhaseTwo(Abort))
);

CoordPhaseTwo(decC) = ||||{N}@acknowledge -> Skip;

//The following models a page
Page(decP, stable) =
```

```
request -> execute ->
  (vote!Yes -> PhaseTwo(decP) [] vote!No -> PhaseTwo(decP));

PhaseTwo(decP) =
inform.Commit -> complete -> result.decP -> acknowledge -> Skip
[]
inform.Abort -> undo -> result.decP -> acknowledge -> Skip;

#alphabet Coord {request, inform.Commit, inform.Abort, acknowledge};
#alphabet Page {request, inform.Commit, inform.Abort, acknowledge};

System = Coord(Abort) || (||||{N}@Page(Abort, true));

Implementation =
System \ {request, execute, acknowledge, inform.Abort, inform.Commit,
  decide, result.Abort, result.Commit};

Specification = PC(N);
PC(i) =
[i == 0](||||{N}@complete -> Skip)
[]
[i > 0](vote.Yes -> PC(i-1) [] vote.No -> PU(i-1));

PU(i) =
[i == 0](||||{N}@undo -> Skip)
[]
[i > 0](vote.Yes -> PU(i-1) [] vote.No -> PU(i-1));

#assert System deadlockfree;
#define has hasNo == 1;
#assert System |= [](has -> <> undo);
#assert System |= [](request -> <> undo);

#assert Specification deadlockfree;
#assert Implementation refines Specification;
```

In the model above, we model the coordinator and the participant (page) by the processes *Coord* (and *CoordPhaseTwo*) and *Page* (and *PhaseTwo*), respectively. For simplicity, only messages carrying a *Page*'s vote (*Yes/No*) are sent over the channel *vote* to *Cord*. The rest of the communication is modeled as a barrier synchronization, mostly using component events like *inform. Commit*, where *inform* corresponds to a channel and *Commit* corresponds to a message. In the two special cases, simple events are used rather than component events, namely the event *request* models the exchange of a **query to commit** message, whereas the event *acknowledge* models the exchange of an **ACK** message. This mapping of message names (given in the informal protocol specification at the beginning of this section) to the corresponding events used in the CSP# model was done with a good choice of event names, so that the reader would not have any difficulties in recognizing the correspondences.

At the beginning of the model, we define the global constants *N, Yes, No, Commit,* and *Abort*; the channel *vote* with the (FIFO buffer) size 0 (which implies synchronous communication); and the Boolean variable *hasNo* with the initial value **false** (assuming final success, i.e., commit).

The process *Coord* initially executes the event *request* once per each *Page* in the system (here twice, because *N*==2): (||||{*N*}@ *request* –> *Skip*); and then *Coord* waits for the votes from all *Pages*: ||||{*N*}@ *vote?vo*. After receiving a vote from a *Page*, *Coord* atomically (see the keyword **atomic**) checks whether the vote (in the variable *vo*) is *No*, and if it is, *Coord* sets *hasNo* to **true** (effectively changing the final result to failure, i.e., abort); otherwise it ignores the vote (*Skip*).

Next, *Coord* decides the final outcome (success/failure) based on the contents of the variable *hasNo* and informs the pages accordingly. In particular, it first executes the observable event *decide* and then executes either the event *inform.Commit* (if *hasNo* is **true**), or *inform.Abort* (otherwise), once per each *Page* in the system: ||||{*N*}@*inform.Commit* or ||||{*N*}@*inform.Abort*. Further on, *Cord* behaves as *CoordPhaseTwo* wherein it simply ignores the event *acknowledge* (the reader should note that *CoordPhaseTwo* corresponds only to the points 3 and 4 in the informal specification of the second phase of 2PC, given at the beginning of this section).

Page initially synchronizes with *Coord* using the event *request* and executes the externally observable event *execute* (which models local transaction processing). The local page's actions may be either successful or unsuccessful, and we model this possibility using the external choice operator [] (remember, we used [] similarly in the model of ABP in the previous section). Next, *Page* sends its vote to *Coord* – *Yes* (if local processing was successful) or *No* (otherwise). Further on, *Page* behaves as *PhaseTwo* (which corresponds to the page's side of the second phase of 2PC). It is important to notice that *Page* passes the value *Commit/Abort* (if its vote was *Yes/No*) to *PhaseTwo* using the process parameter *decP* (decision of a *Page*).

PhaseTwo's actions depend on the notification from *Coord*. In case the notification was *inform.Commit*, *PhaseTwo* sequentially executes the events *complete* (which models successful commit), *result.decP* (which is in this case equal to *result.Commit*), and *acknowledge*. Similarly, in the case that the notification was *inform.Abort*, *PhaseTwo* sequentially executes the events *undo* (which models abort), *result.decP* (which is, in this case, equal to *result.Abort*), and *acknowledge*.

Next, we define alphabets of *Coord* and *Page* (they are equal) by listing the events that are used for barrier synchronization between them, namely, *request, inform.Commit, inform.Abort,* and *acknowledge*. The complete *System* is defined as a parallel composition of *Coord* and interleaving of *Pages*:

System = *Coord(Abort)* || (||||{*N*}@*Page(Abort, true)*);

In this example, we also demonstrate usage of a refinement assertion. Therefore, we firstly define the process *Implementation* as *System* without all the events related to internal operation of *System*. More precisely, *Implementation* inherits only the events *complete, undo, vote.Yes,* and *vote.No* from *System*.

Second, we define the process *Specification* as the process *PC(N)*, where the process *PC(i)*, in turn, is defined as a mutual recursion of itself and the process *PU(i)*. A reader may easily see that *PC(i)* and *PU(i)* are essentially countdown processes, where PC counts down *Yes* votes, whereas PU counts down both *Yes* and *No* votes starting with the first *No* vote. If the parameter *i* during counting down of votes reaches the value $i==0$ within the process *PC(i)*, *PC(i)* will execute the *N* instances of the event *complete*; otherwise *PU(i)* will finally execute the *N* instances of the event *undo*.

At the end of the model, we define five assertions—three of them are related to *System*, one is related to *Specification*, and the fifth is a refinement assertion. The *System*-related assertions are the following:

- *System* is deadlock free.
- *System* satisfies that always after the point when the condition *has* holds (i.e., *has* is a macro which is defined as $hasNo == 1$, i.e. **true**), the event *undo* will be eventually executed.
- *System* satisfies that always after the point when the event *request* was executed, the event *undo* will be eventually executed.

When these assertions are verified by PAT, as expected, PAT reports that the first two are valid, whereas the third is invalid. The third assertion is invalid because after the initial execution of the event *request*, the resulting event may be either *undo* or *complete*, and not always *undo* as claimed. Here is the counterexample produced by PAT:

```
The Assertion (System() |= [] ( request-><> undo)) is NOT valid.
A counterexample is presented as follows.
<init -> request -> request -> execute -> vote.Yes -> τ ->
execute -> vote.Yes -> τ -> decide -> inform.2 -> inform.2 ->
complete -> result.Abort -> acknowledge -> complete ->
result.Abort -> acknowledge -> terminate>
```

The *Implementation*-related assertion claims that it is deadlock free. The fifth, and the last assertion in this example claims that *Implementation* **refines** *Specification*. When these two assertions are verified by PAT, as expected, PAT reports that both are valid.

5.3.2.3.3 *Leader Election in Complete Graphs*

Generally, **leader election** is a fundamental problem in distributed systems, because many hard-distributed problems are easy to solve once a central coordinator is available. An attractive approach to solve the leader election is by using **self-stabilizing algorithms**, which do not require initialization in order to operate correctly, and which can recover from transient faults that may destroy the system state information. Also, among many models, a **network of finite-state anonymous agents** is a rather interesting one, because it models many distributed systems of identical, simple computational nodes, such as wireless sensor networks, etc. It is well-known that the self-stabilizing leader election is impossible without a **failure detector**, which is a kind of oracle that provides some information to the system that it is unable to compute on its own.

Therefore, Fischer and Jiang (2006) introduced the **eventual leader detector** Ω?. We may imagine Ω? as a black box that provides global status information about the protocol, in particular, whether or not there is a leader in the system. This detector is **weak** in the sense that it does not respond to status changes immediately, but with some indeterminate delay, and it does not report its findings to all the processes (agents) simultaneously. (So some agents may discover status changes sooner than the other processes.) Formally, Ω? provides a Boolean input to each process at each step, such that the following conditions are satisfied by every execution E:

- If all, except finitely many, configurations of E lack a leader, then all processes receive **false** in all, except finitely many, steps.
- If all, except finitely many, configurations of E have one or more leaders, then all processes receive **true** in all, except finitely many, steps.

Thanks to its weakness, Ω? may be simply implemented using timeouts. Each leader periodically sends a keep-alive message, whereas each agent restarts its timer after receiving such a message, and sets the leader detector flag to **true** (indicating that leader is present). On timeout, the process sets the leader detector flag to **false** (indicating that leader is absent). Of course, in an adverse environment, Ω? may temporarily produce incorrect information. However, eventually after the environment stabilizes, Ω? will produce correct information.

Further on in this section, we introduce, model, and analyze the self-stabilizing leader election algorithm for complete graphs (see the example of the complete graph with five nodes in the Figure 5.8) using Ω? (Fischer, 2006), which works under either local or global fairness condition. According to this algorithm, each node has a memory slot that can hold either a leader mark "x" or nothing "-" for a total of two states. Each node receives its current input **true** (T) or **false** (F) from Ω?. A nonleader becomes a leader, when Ω? signals the absence of a leader, and the responder is not a leader. When two leaders interact, the responder becomes a nonleader. Otherwise, no state change occurs.

The algorithm can be formally described by the three pattern rules, which are matched against the state and the input of the initiator and the responder, respectively. If the match succeeds, the states of the two interacting nodes are replaced by the respective states on the right-hand side of the rule. According to the **star convention**, "*" is a symbol that always matches the slot or the input. On the rule's right-hand side, "*" specifies that the contents of the corresponding slot do not change. If no explicit rules match, neither node changes state (i.e., a null transition takes place).

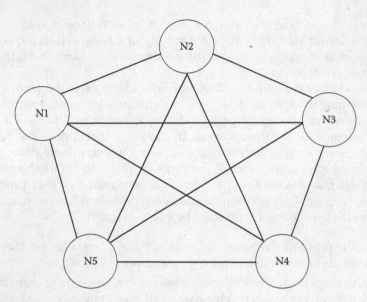

FIGURE 5.8
Example of the complete graph with five nodes.

The three pattern rules are as follows:

Rule 1: ((x, *), (x, *)) -> ((x), (-))
Rule 2: ((-, F), (-, *)) -> ((x), (-))
Rule 3: ((-, T), (-, *)) -> ((-), (-))

The parametrized model of a leader election algorithm for complete graphs in CSP# was created by the PAT Team. The following is the particular model for the parameter *Number of Processes* set to 3 (in the model that is the constant *N*):

```
#define N 3;
var dok = 0;  // detector correct (ok)
var detector = false;
var leader[N];

/*Rule 1*/
Rule1(i, r) =
[leader[i] == 1 && leader[r] == 1]
  (rule1.i.r{leader[r] = 0;} -> Rule1(i, r));

/*Rule 2*/
Rule2(i, r) =
[leader[i] == 0 && leader[r] == 0 && !(( dok == 0 && detector) ||
(dok != 0 && ( leader[0] + leader[1] + leader[2] > 0)))]
  (rule2.i.r{leader[i] = 1;} -> Rule2(i, r));
```

```
/*Rule 3*/
Rule3(i, r) =
[leader[i] == 0 && leader[r] == 0 && ((dok == 0 && detector) ||
(dok != 0 && (leader[0] + leader[1] + leader[2] > 0)))]
  (rule3.i.r -> Rule3(i, r));

// eventual leader detector
DetectorCorrect() =
[dok == 0]
  (progress{dok = 1;} -> DetectorCorrect());
// detector
RandomDetector() =
[dok == 0]
  ((random1{detector = false;} -> RandomDetector()) []
   (random2{detector = true;} -> RandomDetector()));

Initialization() =
((tau{leader[0] = 0;} -> Skip) [] (tau{leader[0] = 1;} -> Skip));
((tau{leader[1] = 0;} -> Skip) [] (tau{leader[1] = 1;} -> Skip));
((tau{leader[2] = 0;} -> Skip) [] (tau{leader[2] = 1;} -> Skip));

LeaderElection() =
Initialization();
(DetectorCorrect() ||| RandomDetector() |||

Rule1(0,1)|||Rule1(1,0)|||Rule1(0,2)|||Rule1(2,0)|||
Rule1(1,2)|||Rule1(2,1)|||

Rule2(0,1)|||Rule2(1,0)|||Rule2(0,2)|||Rule2(2,0)|||
Rule2(1,2)|||Rule2(2,1)|||

Rule3(0,1)|||Rule3(1,0)|||Rule3(0,2)|||Rule3(2,0)|||
Rule3(1,2)|||Rule3(2,1));

// The Property
#define oneLeader (leader[0] + leader[1] + leader[2] == 1);
#assert LeaderElection() |= <>[]oneLeader;
```

In the model above, we modeled the three rules: Rule 1, Rule 2, and Rule 3, by the processes *Rule1*, *Rule2*, and *Rule3*, respectively. Each of these processes has two parameters, namely *i* and *r*, where *i* is the index of the initiator node and *r* is the index of the responder node. Next, we model the eventual leader detector by the processes *DetectorCorrect* and *RandomDetector*, the random setup of the node's initial states by the process *Initialization*, and the complete system by the process *LeaderElection*.

At the beginning of the model, we define global constants and variables. The global constant *N* is equal to the number of processes (i.e., 3). The value of the global integer variable *dok* determines whether the information provided by the leader detector is correct (value 1) or not (value 0). Initially, *dok* is set to 0, indicating the presence of transient errors in the environment, and later on it is set to 1, indicating that the environment becomes stable and well-behaved. The value of the Boolean variable *detector* represents the output of the leader detector (**true** if there is a leader in the system), which may be erroneous. The integer array *leader* (of size *N*) corresponds to memory slots at each node, which hold the current state of the node (the value 0 means that the node is not a leader, whereas the value 1 means that the node is a leader).

According to Rule 1, the process *Rule1* checks if both *leader*[*i*] and *leader*[*r*] are set to 1 (i.e., if both initiator and responder are leaders), and if they are, it then sets *leader*[*r*] to 0 (i.e., the responder becomes a nonleader). Further on, it behaves again as *Rule1*. Obviously, the process *Rule1* is a straightforward encoding of Rule 1.

Similarly, *Rule2* and *Rule3* are rather straightforward encodings of Rule 2 and Rule 3, respectively. However, there is one important difference; unlike the process *Rule1*, the processes *Rule2* and *Rule3* behave differently in the first phase of the system evolution, when the environment is unstable (i.e., when the value of the variable *dok* is 0), and in the second phase when the environment becomes stable (i.e., when *dok* is set to 1). In the first phase, the processes *Rule2* and *Rule3* behave as specified by the rules (i.e., Rule 2 and Rule 3, respectively), whereas in the second phase (when *dok* is not equal to 0), *Rule2* and *Rule3* execute if there is at least one leader in the system (when the sum of the elements of the array *leader* is greater than 0).

The process *DetectorCorrect* models the system transition from an unstable to a stable state. Initially, the variable *dok* is set to 0 (indicating an unstable environment). Once the process *DetectorCorrect* sees that the variable *dok* is set to 0, it simply sets it to 1 (indicating a stable environment). Similarly, the process *RandomDetector* models possibly erroneous readings from the leader detector. As long as the variable *dok* is set to 0 (indicating an unstable environment), the process *RandomDetector* randomly sets the variable *detector* to either **true** or **false**, but once *dok* is set to 1, it stops writing to the variable *detector* (so its value stabilizes).

The process *Initialization* is a sequence of three sub processes, where each of the sub processes randomly sets the initial state of the corresponding node (i.e., the corresponding element of the array *leader*) to either 0 (nonleader) or 1 (leader), and then terminates (*Skip*). So, the system may start from any possible combination of node states.

The process *LeaderElection* is the sequential composition of the process *Initialization* and the process that is the interleaving of the processes *DetectorCorrect*, *RandomDetector*, and the process instances of the processes *Rule1*, *Rule2*, and *Rule3*, for all possible combinations of values of their parameters *i* and *r*, i.e., (0, 1), (1, 0), (0, 2), (2, 0), (1, 2), and (2, 1).

At the end of the model, we define the macro *oneLeader* and the one *LeaderElection*-related assertion. The macro *oneLeader* is defined as the equation of the sum of the elements of the array *leader* (which corresponds to the number of leaders currently present in the system) and the constant 1. Of course, the goal of any leader election protocol is that this number of leaders is finally equal to 1, which means that there is exactly one leader in the system. Therefore, the assertion at the end of the model claims that the process *LeaderElection* eventually always satisfies the goal (i.e., the equation) *oneLeader*.

PAT verification reports are as expected. If we select the admissible behavior to be Global Fair Only or Event-level Strong/Weak Fair Only, the assertion is found to be valid. Alternatively, if we select the admissible behavior option All, the assertion is found to be invalid. Here is the counterexample produced by PAT:

```
The Assertion (LeaderElection() |= <>[] oneLeader) is NOT valid.
A counterexample is presented as follows.
<init -> τ -> τ -> τ -> rule1.2.1 -> rule1.2.0 -> (rule2.1.0 ->
rule1.2.1 -> rule2.1.0)*>
```

5.3.2.3.4 *Leader Election in Rings*

In this section we introduce, model, and analyze the uniform, self-stabilizing leader election algorithm for rings using Ω? (Fischer, 2006), which requires global fairness (and under local fairness is not feasible). See the example of the ring graph with five nodes in the Figure 5.9.

The algorithm is based on the following assumptions: The ring is directed such that each node has a sense of forward (clockwise) and backward (counterclockwise) directions and every interaction takes place between the initiator and its forward neighbor. Each node can store zero or one of each of three kinds of tokens: a bullet "o", a leader mark "x", and a shield "|", for a total of eight possible states. Corresponding to each kind of token is a slot which is empty if the corresponding token is not present, and full if it is present. An empty slot is denoted by "-" whereas a full slot is denoted by the token. The slots in each node are ordered with the bullet first, leader mark second, and shield third. Extending this to a clockwise ordering of all slots in the ring,

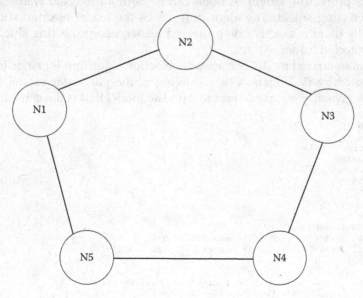

FIGURE 5.9
Example of the ring graph with five nodes.

the shield slot of one node is followed by the bullet slot of the next node in the clockwise order.

The algorithm can be formally described by the five pattern rules:

Rule 1. ((* * *, F), (* * *, *)) -> ((o x |), (* * *))

Rule 2. ((* - |, T), (* * *, *)) -> ((* - -), (- * |))

Rule 3. ((* x |, T), (* * *, *)) -> ((o x -), (- * |))

Rule 4. ((* x -, T), (- * *, *)) -> ((o x -), (- * *))

Rule 5. ((* * -, T), (o * *, *)) -> ((o - -), (- * *))

When two nodes interact and the initiator's input is false (F), a leader and shield are created, and at the same time, a bullet is fired (Rule 1). This is the only way for leaders and shields to be created. When the initiator's input is true (T), the following rules apply: Shields move forward around the ring (Rules 2 and 3), and bullets move backward (Rule 5). Bullets are absorbed by any shield they encounter (Rules 2 and 3) but kill any leaders along the way (Rule 5). If a bullet moves into a node already containing a bullet, the two bullets merge into one. Similarly, when two shields meet, they merge into one. A leader fires a bullet whenever it is the initiator of an interaction (Rules 3 and 4).

In a configuration in which the node i has a leader mark, the node j has a shield, and all of the slots between i's leader mark and j's shield in clockwise order are empty, the node i is called the **protected leader**, and the node j is called its **protecting shield**. A node can be both a protected leader and its own protecting shield. The algorithm solves the leader election such that eventually there is exactly one protected leader, one protecting shield, and no unprotected leader.

The parametrized model of the leader election algorithm for rings in CSP# was created by PAT Team. The following is the particular model for the parameter *Number of Processes* set to 3 (in the model that is the constant N):

```
#define N 3;
var dok = 0;
var detector = false;
var leader[N];
var bullet[N];
var shield[N];

// Processes
Process(i) =
[!((dok==0 && detector) ||
(dok!=0 && leader[0]+leader[1]+leader[2] > 0))]
  rule1.i.(i+1)%N{bullet[i]=1; leader[i]=1; shield[i]=1;} ->
  Process(i)
[]
[leader[i] == 0 && shield[i] == 1 && ((dok==0 && detector) ||
(dok!=0 && leader[0]+leader[1]+leader[2] > 0))]
  rule2.i.(i+1)%N{leader[i]=0; shield[i]=0; bullet[(i+1)%N] = 0;
  shield[(i+1)%N] = 1;} -> Process(i)
```

```
[]
[leader[i] == 1 && shield[i] == 1 && ((dok ==0 && detector) ||
(dok!=0 && leader[0]+leader[1]+leader[2] > 0))]
  rule3.i.(i+1)%N{ bullet[i] = 1; leader[i] = 1; shield[i] = 0;
  bullet[(i+1)%N] = 0; shield[(i+1)%N] = 1;} -> Process(i)
[]
[leader[i] == 1 && shield[i] == 0 && bullet[(i + 1) % N] == 0
&& ((dok==0 && detector) ||
(dok!=0 && leader[0]+leader[1]+leader[2] > 0))]
  rule4.i.(i+1)%N{ bullet[i] = 1; leader[i] = 1; shield[i]=0;
  bullet[(i+1) % N] = 0;} -> Process(i)
[]
[shield[i] == 0 && bullet[(i+1)% N] == 1 && ((dok==0 && detector) ||
(dok!=0 && leader[0]+leader[1]+leader[2] > 0))]
  rule5.i.(i+1)%N{bullet[i] = 1; leader[i] = 0; shield[i] = 0;
  bullet[(i+1)%N] = 0;} -> Process(i);

// eventual leader detector
DetectorCorrect() =
[dok == 0](progress{ dok = 1;} -> DetectorCorrect());
//detector
RandomDetector() =
[dok == 0]
  ((guess1{detector = false;} -> RandomDetector())
  []
  (guess2{detector = true;} -> RandomDetector()));

Initialization() =
((tau{leader[0] = 0;} -> Skip) [] (tau{leader[0] = 1;} -> Skip));
((tau{leader[1] = 0;} -> Skip) [] (tau{leader[1] = 1;} -> Skip));
((tau{leader[2] = 0;} -> Skip) [] (tau{leader[2] = 1;} -> Skip));
((tau{bullet[0] = 0;} -> Skip) [] (tau{bullet[0] = 1;} -> Skip));
((tau{bullet[1] = 0;} -> Skip) [] (tau{bullet[1] = 1;} -> Skip));
((tau{bullet[2] = 0;} -> Skip) [] (tau{bullet[2] = 1;} -> Skip));
((tau{shield[0] = 0;} -> Skip) [] (tau{shield[0] = 1;} -> Skip));
((tau{shield[1] = 0;} -> Skip) [] (tau{shield[1] = 1;} -> Skip));
((tau{shield[2] = 0;} -> Skip) [] (tau{shield[2] = 1;} -> Skip));

LeaderElection() =
Initialization();
(DetectorCorrect() ||| RandomDetector() |||
  Process(0)|||Process(1)|||Process(2));

// The Property
#define oneLeader (leader[0] + leader[1] + leader[2] == 1);
#assert LeaderElection() |= <>[]oneLeader;
```

This model is rather similar to the model in the previous section. Actually, some processes are identical or almost identical. The main differences are as follows: In this model, the integer arrays *leader*, *bullet*, and *shield* (each of size N), correspond to memory slots at each node, which hold the current state of the node (the element value 1 means that the node holds the corresponding token, whereas the element value 0 means that the node does not hold the corresponding token). All the five pattern rules are encoded within a single process, namely *Process*, rather than being defined as separate processes (as was done in the model in Section 3.2.3.2). *Process* corresponds to a single node within a ring, and its parameter i is simply the index of the node in the ring. This index is used to access the corresponding elements of arrays *leader*, *bullet*, and *shield*. Obviously, in order to define the process *LeaderElection* in this model, we need to make the three *Process*'s instances, namely *Process*(0), *Process*(1), and *Process*(2). Apart from these differences, the models are analogous, and thus the reader should have no difficulties in analyzing the model above, and so we leave it as an individual reader's exercise.

The PAT verification reports are as expected. If we select the admissible behavior option All, the assertion is found to be valid. Alternatively, if we select the admissible behavior to be Global Fair Only or Event-level Strong/Weak Fair Only, the assertion is found to be invalid. Here is the counterexample produced by the PAT for the option Global Fair Only (the other two counterexamples are longer, so we skip them, and leave the reader to reproduce them as an exercise):

```
The Assertion (LeaderElection() |= <>[] oneLeader) is NOT valid.
A counterexample is presented as follows.
<init -> τ -> τ -> τ -> τ -> τ -> τ -> τ -> τ -> τ -> guess2 -> rule3.0.1 ->
rule3.2.0 -> guess1 -> rule1.1.2 -> rule1.0.1 -> guess2 -> rule5.2.0 ->
rule3.1.2 -> rule3.0.1 -> rule2.2.0 -> guess1 -> rule1.1.2 -> progress ->
rule3.0.1 -> (rule5.2.0 -> rule4.0.1 -> rule5.2.0)*>
```

5.3.2.3.5 *Leader Election in Trees*

In this section, we introduce, model, and analyze the deterministic, uniform, and self-stabilizing leader election algorithm for rooted directed trees using Ω? (Canepa, 2008), which requires global fairness (like the algorithm in the previous example). See the example of the tree graph with five nodes in Figure 5.10. This algorithm is space optimal because it requires only one memory bit per agent.

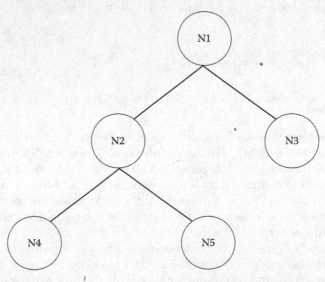

FIGURE 5.10
Example of the tree graph with five nodes.

The algorithm is based on the following assumptions: The root of a rooted directed tree is the only node of in-degree 0, and for each node in the tree, there is a directed path from the root to that node. Each node has a memory slot that can hold either a leader mark "x" or nothing "-", for a total of two states per node, and each node receives its current input true (T) or false (F) from Ω?.

The algorithm can be formally described by the three pattern rules:

Rule 1. $((x, *), (x, *)) \rightarrow ((x), (-))$

Rule 2. $((-, F), (-, *)) \rightarrow ((x), (-))$

Rule 3. $((-, *), (x, *)) \rightarrow ((x), (-))$

Intuitively the algorithm works as follows: A clean agent (i.e., an agent without a leader mark) becomes leader mark holder, when Ω? signals the absence of leader marks, and the responder does not hold a leader mark (Rule 2). When two agents holding a leader mark each interact, the responder becomes clean (Rule 1). If the responder has a leader mark and the initiator is a clean agent, the latter becomes a leader mark holder and the former becomes clean (Rule 3). Otherwise, no state change occurs.

The parametrized model of the leader election algorithm for rings in CSP# was created by the PAT Team. The following is the particular model for the parameter *Number of Processes* set to 3 (in the model that is the constant N):

```
#define N 3;
var dok = 0;
var detector = false;
var leader[N];

/*Rule 1*/
Rule1(i, r) =
[leader[i] == 1 && leader[r] == 1]
  (rule1.i.r{leader[r] = 0;} -> Rule1(i, r));

/*Rule 2*/
Rule2(i, r) =
[leader[i] == 0 && leader[r] == 0 && !(( dok == 0 && detector) ||
(dok != 0 && ( leader[0] + leader[1] + leader[2] > 0)))]
  (rule2.i.r{leader[i] = 1;} -> Rule2(i, r));

/*Rule 3*/
Rule3(i, r) =
[leader[i] == 0 && leader[r] == 1]
  (rule3.i.r{leader[i] = 1; leader[r] = 0;} -> Rule3(i, r));

// eventual leader detector
DetectorCorrect() =
[dok == 0]
  (progress{dok = 1;} -> DetectorCorrect());
// detector
RandomDetector() =
[detectorcorrect == 0]
  ((random1{detector = false;} -> RandomDetector()) []
   (random2{detector = true;} -> RandomDetector()));
```

```
Initialization() =
((tau{leader[0] = 0;} -> Skip) [] (tau{leader[0] = 1;} -> Skip));
((tau{leader[1] = 0;} -> Skip) [] (tau{leader[1] = 1;} -> Skip));
((tau{leader[2] = 0;} -> Skip) [] (tau{leader[2] = 1;} -> Skip));

// The topology is a rooted tree
LeaderElection() =
Initialization();
(DetectorCorrect() ||| RandomDetector() |||

Rule1(0,1)|||Rule1(0,2)|||
Rule2(0,1)|||Rule2(0,2)|||
Rule3(0,1)|||Rule3(0,2));

// The Property
#define oneLeader (leader[0] + leader[1] + leader[2] == 1);
#assert LeaderElection() |= <>[]oneLeader;
```

This CSP# code is completely analogous to the CSP# code given in Section 5.3.2.3.3 (Leader Election in Complete Graphs), so the reader should have no difficulties understanding it. Actually, it uses the same variables and conventions for encoding the pattern rules. Moreover, this example and the example in Section 5.3.2.3.3 use the processes *DetectorCorrect*, *RandomDetector*, and *Initialization*.

The main difference between these two examples is the way how their respective *LeaderElection* processes instantiate the processes *Rule1*, *Rule2*, and *Rule3*. In this section, we introduce the convention that the node with index 0 is the root of the tree, whereas nodes with indexes 1 and 2 are the leaves of the tree. We also assume that the first parameter (i) of the processes *Rule1*, *Rule2*, and *Rule3* is the index of the root, whereas the second parameter (r) is the index of a leaf. Thus, we create two process instances of each of the processes *Rule1*, *Rule2*, and *Rule3*, for two possible combinations of values for their parameters i and r, i.e., (0, 1), and (0, 2).

The PAT verification reports are as expected. If we select the admissible behavior to be Global Fair Only or Event-level Strong/Weak Fair Only, the assertion is found to be valid. Alternatively, if we select the admissible behavior option All, the assertion is found to be invalid. Here is the counterexample produced by PAT:

```
The Assertion (LeaderElection() |= <>[] oneLeader) is NOT valid.
A counterexample is presented as follows.
<init -> τ -> τ -> τ -> rule1.0.2 -> random2 -> (random2 -> random2)*>
```

5.3.2.3.6 Telecomm Service System

The Telecomm Service System (TSS) presented in this section is a simplified model of a telephone exchange (a.k.a., Private Branch eXchange, PBX) that supports local calls only. The main PBX call-processing functions are as follows:

- Establishing connections (circuits) between the telephone sets of the two users

- Maintaining such connections as long as the users require them
- Disconnecting those connections as per the user's request
- Providing information for accounting purposes, i.e., metering calls (not modeled in TSS)

Besides these basic functions, PBXs offer many other calling features and capabilities, also known as supplementary services or add-ons. Modeling all of them would result in a rather complex model, thus TSS supports only some of the most frequently used supplementary services, namely the following:

- Call Forward Unconditional (CFU)
- Call Forward when Busy (CFB)
- Originating Call Screening (OCS)
- Originating Dial Screening (ODS)
- Terminating Call Screening (TCS)
- Ring Back When Free (RBWF)

Normally, systems like TSS are designed (and implemented) incrementally. We start with the basic functions, and then add individual, supplementary services incrementally. However, the complex interaction between users and incremental system extensions may lead to unpredictable and undesirable results. Some new features may conflict with each other, or hinder the basic services. Thus, the model of TSS must reflect the high-level design of the system, both the interaction between the users and the compatibility of new services. We also need a comprehensive set of properties covering basic call-processing and the new services in order to verify if all the system requirements are satisfied.

Since the model of TSS is rather complex and there are many properties to verify, we first analyze only the model, and then we introduce and discuss the system properties separately. The parametrized model of TSS in CSP# was created by the PAT Team. The following is the particular model for the parameter *Number of Users*, set to 2 (in the model that is the constant *NoOfUsers*):

```
#define NoOfUsers        2;
#define NoOfChannels     4; // = NoOfUsers+2
#define NIL              3; // = NoOfUsers+1
#define INVALID_USER     2; // = NoOfUsers

// Model variables
var partner=[NIL(NoOfChannels)];
var chan=[NIL(NoOfChannels)];
var connect=[0(NoOfChannels)];
var dev=[1(NoOfChannels)];
var CFU:{0..NIL}=[NIL(NoOfChannels)];
var CFB:{0..NIL}=[NIL(NoOfChannels)];
var RBWF:{0..NIL}=[0(NoOfChannels)];
var lastCall:{0..NIL}=[NIL(NoOfChannels)];
var OCS:{0..NIL}=[NIL(NoOfChannels)];
```

```
var ODS:{0..NIL}=[NIL(NoOfChannels)];
var TCS:{0..NIL}=[NIL(NoOfChannels)];

// Verification variables
var dialNum=[NIL(NoOfUsers)];
var justDial=[0(NoOfUsers)];

System() = (User());
// All users start from Idle
User() = ||id:{0..NoOfUsers-1}@(Idle(id));
#alphabet Idle
  {keepTalking.0.1,stopTalking.0.1,keepTalking.1.0,stopTalking.1.0};

// Processes corresponding to individual call-processing states
Idle(id) =
  idle.id{chan[id]=NIL; partner[id]=NIL; dev[id]=1; connect[id]=0;
  dialNum[id]=NIL; } ->
  ([chan[id]== NIL] dialing.id{dev[id]=0;} -> Dialing(id)
  []
  [chan[id]>= 0 && chan[id] != NIL]
    answerCall.id{partner[id]=chan[id]; dev[id]=0;} -> Answer(id)
  []
  [RBWF[id]==1 && lastCall[id]!=NIL]
  // RBWF feature, reply last call
    if (lastCall[id]!=ODS[id]) {
      ringback.id.lastCall[id]{partner[id]=lastCall[id];
      lastCall[id]=NIL;} -> Calling(id)
    } else {forbidCall.id -> Idle(id)}
  );

Answer(id) =
case{
  partner[id]==TCS[id] :
    ignoreCall.id{if (chan[partner[id]]==id) chan[partner[id]]=NIL;}
    -> Idle(id)

  chan[partner[id]]==id:
    //partner[id] is waiting
    t_alert.id.partner[id] -> T_Alert(id)

  default:
    //chan[partner[id]]!= id -> partner[id] has changed dial number
    partnerChanged.id.partner[id] -> Idle(id)
};

Dialing(id) =
noCall.id -> Idle(id)
[]
([] callId:{0..NoOfUsers}@(
  if (callId!=ODS[id]) {
    dial.id.callId{justDial[id]=1;} ->
    makeCall.id{partner[id]=callId; dialNum[id]=callId;} ->
    Calling(id)
  } else {
    forbidCall.id -> Idle(id)
  }
));

Calling(id) =
if (partner[id]!=OCS[id]) {
  doCall.id.partner[id]{justDial[id]=0;} ->
  case {
    partner[id]==id:
    //id call to itself
```

```
    selfCall.id -> CallBusy(id)

    partner[id]==INVALID_USER:
    //id call to unobtainable user
    invalidCall.id -> CallUnobtainable(id)

CFU[partner[id]] != NIL:
    //partner[id] has CFU add-on
    forwardCall.partner[id].CFU[partner[id]]{
      partner[id]=CFU[partner[id]];} -> Calling(id)

chan[partner[id]]==NIL:
    //partner channel is free
    ringPartner.id{
      chan[partner[id]]=id; chan[id]=partner[id];} -> O_Alert(id)

    chan[partner[id]]!=NIL && CFB[partner[id]]==NIL:
    //partner channel is busy, no CFB add-on
    if (RBWF[partner[id]]==0) {
      //partner does not have RBWF add-on
      busyPartner.id -> CallBusy(id)
    } else { //partner has RBWF add-on
      busyPartner.id{lastCall[partner[id]]=id;} -> CallBusy(id)
                                 }

    chan[partner[id]]!=NIL && CFB[partner[id]]!=NIL:
    //partner channel is busy, has CFB add-on
      forwardCall.partner[id].CFB[partner[id]]{
        partner[id]=CFB[partner[id]];} -> Calling(id)

    default:
    errorCall.id -> ErrorState(id)
    }
} else {
  forbidCall.id{justDial[id]=0;} -> Idle(id)
};

CallBusy(id) =
//id receives busy signal
soundBusy.id -> Idle(id);

CallUnobtainable(id) =
//id calls to unobtainable user
soundInvalid.id -> Idle(id);

O_Alert(id) =
[chan[id]==partner[id] && connect[id]==1]
  callConnect.id.partner[id] -> O_Connected(id)
[]
[chan[id]==partner[id] && connect[id]==0]
  ringOut.id{if (chan[partner[id]]==id) chan[partner[id]]=NIL;
  chan[id]=NIL;} -> Idle(id)
[]
[chan[id]!= partner[id]]
  callStopped.id.partner[id] -> Idle(id)
[]
[chan[id]==partner[id] && connect[id]!= 1 && connect[id] != 0]
  alertError.id -> ErrorState(id);

O_Connected(id) =
[connect[id]==1 && connect[partner[id]]==1]
  keepTalking.id.partner[id] -> O_Connected(id)
[]
tau{connect[id]=0;connect[partner[id]]=0;} ->
```

```
    stopTalking.id.partner[id] -> Idle(id)
[]
stopTalking.id.partner[id] -> Idle(id);

T_Alert(id) =
case {
    chan[partner[id]]!=id:
    //partner calls others before id can establish connection
    partnerBusy.id.partner[id] -> Idle(id)

    chan[partner[id]]==id:
    //partner is still calling id -> pickup
    partnerReady.id.partner[id] -> T_Pickup(id)

    default:
    // errors
    errorT_Alert.id -> ErrorState(id)
};

T_Pickup(id) =
[chan[partner[id]]==id]
    pickup.id.partner[id]{
        dev[id]=0; connect[partner[id]]=1; connect[id]=1;
    } -> T_Connected(id)
[]
[chan[partner[id]]==NIL || chan[partner[id]]!=id]
    Idle(id);

T_Connected(id) =
[connect[id]==1 && connect[partner[id]]==1]
    keepTalking.partner[id].id -> T_Connected(id)
[]
stopTalking.partner[id].id -> Idle(id)
[]
tau{connect[id]=0;connect[partner[id]]=0;} ->
    stopTalking.partner[id].id -> Idle(id);

ErrorState(id) =
error -> Stop; // an error happened
```

At the beginning of the model, we define global constants and variables. The global constant *NoOfChannels* is equal to the number of users plus 2 (i.e., 4), so that each user has its own local channel and there are two additional channels, which are left for future work on this model (e.g., one outgoing trunk and one incoming trunk). The constant *NIL* is equal to the number of users plus 1 (i.e., 3), and it designates the inactive channel. When an element of the array *chan* (a shorthand for channel) is assigned the value *NIL*, it means that the corresponding channel is inactive. The constant *INVALID_USER* is equal to the number of users (i.e., 2), and it represents the upper bound on the variable *id*, which holds an index of a user (*id* must be less than this constant).

Next, we define global variables, which we classify as the model variables and the verification variables. The model variables include the arrays *partner*, *chan*, *connect*, *dev*, *CFU*, *CFB*, *RBWF*, *lastCall*, *OCS*, *ODS*, and *TCS*, which are of size *NoOfChannels*. The verification variables are the arrays *dailNum* and *justDail*, which are of size *NoOfUsers*. The conventions for the possible values of these variables are as follows:

The value *partner*[i] = k means that the user i is connecting with the user k, whereas the value *partner*[i] = NIL means that the user i is free (i.e., in the state *Idle*). The value *chan*[i] = NIL means no incoming call for the user i, whereas the value *chan*[i] = k means the user k is calling the user i. The value *connect*[i] = 1 means that the user i is connected to other side, whereas the value *connect*[i] = 0 means that it is not connected. The value *dev*[i] = 0 means that a device of the user i is busy, whereas the value *dev*[i] = 1 means that this device is ready.

The value *CFU*[i] = NIL means that the user i not subscribed to the *CFU* service, whereas the value *CFU*[i] = k means that a call to the user i shall be unconditionally forwarded to the user k. The value *CFB*[i] = NIL means that the user i has not subscribed to the *CFB* service, whereas the value *CFB*[i] = k means that a call to the user i shall be forwarded to the user k, if the user i is busy. The value *RBWF*[i] = 0 means that the user i has not subscribed to *RBWF* service, whereas the value *RBWF*[i] = 1 means that the user i shall ring back the user *lastCall*[i] when the user i becomes free. The value *lastCall*[i] = NIL means there is no last call for the user i, whereas the value *lastCall*[i] = k means the last call to the user i (when the user i was busy) was from the user k.

Generally, a screen (block) list can be implemented using a hash table. For simplicity, here we use a list of size one, i.e., just one screened (blocked) number per user, which is quite sufficient for modeling and verification purposes. Thus, the arrays *OCS*, *ODS*, and *TCS*, contain these minimal one-element lists for each user. The value *OCS*[i] = NIL means that the user i is not subscribed to the *OCS* service, whereas the value *OCS*[i] = k means that the user k is screened. The conventions for *ODS*[i] and *TCS*[i] are the same as for *OCS*[i]. The difference between these three services is the moment when the screening takes place (i.e., in which call-processing state).

The conventions for the verification variables are as follows: The value *dialNum*[i] = k means that the user i dialed the number k (originally), whereas the value *dialNum*[i] = NIL means there is no such number. The value *justDial*[i] = 0 means that the user i did not just dial a number, whereas the value *justDial*[i] = 1 means the user i did just dial a number.

Next, we define the processes in the model. The process *System* behaves as the process *User*, which, in turn, is defined as a concurrent execution of *NoOfUser* (i.e., 2) instances of the process *Idle*. In fact, the process *Idle* models the initial state of each user. Further on in the model, we define an individual process for each possible call-processing state of the user. Besides *Idle* these processes are *Answer*, *Dialing*, *Calling*, *Callbusy*, *CallUnobtainable*, *O_Alert*, *O_Connected*, *T_Alert*, *T_Pickup*, *T_Connected*, and *ErrorState*. The single parameter of all these processes is the user identification (*id*), where *id* is an element of the set {0..*NoOfUsers*–1}, i.e., {0, 1}. In the following text, we briefly describe each process in turn.

The process *Idle* first initializes model variables according to the conventions introduced above, in particular it sets *chan*[*id*] to *NIL*, *partner*[*id*] to *NIL*, *dev*[*id*] to 1, *connect*[*id*] to 0, and *dialNum*[*id*] to *NIL*. Further on, *Idle* nondeterministically selects one of the three possible activities (by using the external choice operator []). The guard for the first activity is that there is no incoming call to the user *id* (*chan*[*id*] == *NIL*), and in this case, *Idle* initiates the outgoing call (by setting *dev*[*id*] to 0), and transforms into the process *Dialing*. The guard for the second activity is that there is an incoming call to the user *id*, and, in this case, *Idle* accepts this incoming call (by setting *partner*[*id*] to *chan*[*id*] and *dev*[*id*] to 0), and transforms into the process *Alert*. The guard for the third activity is that the user id is subscribed to RBWF service and that there was an incoming call to the user *id* while it was busy (*RBWF*[*id*] == 1 && *lastCall*[*id*] != *NIL*), and, in this case, *Idle* checks whether the initiator of that incoming call is in the ODS screen list. If that initiator is not in the ODS list (*lastCall*[*id*] != *ODS*[*id*]), *Idle* initiates the ring back (by setting *partner*[*id*] to *lastCall*[*id*] and *lastCall*[*id*] to *NIL*) and transforms into the process *Calling*; otherwise it ignores this situation and continues to behave as the same process *Idle*.

The process *Answer* performs one of three possible cases. If the calling user (*partner*[*id*]) is in the TCS list of the user *id* (*partner*[*id*] == *TCS*[*id*]), *Answer* ignores this incoming call, and if the element *chan*[*partner*[*id*]] is set to *id*, it resets it to *NIL*, and ultimately transforms into the process *Idle*. Otherwise, if the calling user is still waiting for the user *id* to answer (*chan*[*partner*[*id*]] == *id*), *Answer* transforms into the process *T_Alert*. Otherwise (in the third case), *Answer* transforms into the process *Idle*.

The process *Dialing* nondeterministically selects one of the two possible activities. In the first activity, *Dialing* stops the outgoing call of the user *id* and transforms into the process *Idle* (this activity corresponds to the case when the user *id*, for some reason, quits the call). In the second activity, *Dialing* nondeterministically selects the called user (the variable *callId*). If this user is in the ODS list of the user *id* (*callId* == *ODS*[*id*]), *Dialing* forbids the call and transforms it to the process *Idle*. Otherwise (if the call is not screened), *Dialing* sets *partner*[*id*] to *callId* and *dialNum*[*id*] to *callId*, and transforms into the process *Calling*.

The process *Calling* first checks whether the called user (*partner*[*id*]) is in the ODS list of the user *id*, and if it is in this list, then it forbids the call and transforms into the process *Idle*. Otherwise (if the call is not screened), *Calling* performs one of the seven possible cases. The first case is when the user *id* calls itself, then *Calling* ignores the call and transforms into the process *CallBusy*. The second case is when the called user is invalid, then *Calling* ignores the call and transforms into the process *CallUnobtainable*. The third case is when the called user is subscribed to CFU service, then *Calling* sets *partner*[*id*] to *CFU*[*partner*[*id*]] and continues to behave as the process *Calling*. The fourth case is when the channel of the called user is free, then *Calling* sets *chan*[*partner*[*id*]] to *id* and *chan*[*id*] to *partner*[*id*], and transforms into the process *O_Alert*. The fifth case is when the called user is busy and is not subscribed to CFB service, then if the called user is subscribed to RBWF service, *Calling* sets *lastCall[partner[id]]* to *id* and transforms into the process *CallBusy*, else (if the called user is not subscribed to RBWF service), *Calling* just transforms into the process *CallBusy*. The sixth case is when the called user is busy and it is subscribed to CFB service, then *Calling* sets *partner[id]* to *CFB[partner[id]]* and continues to behave as the process *Calling*. The seventh case is the default case (none of the previous cases, i.e., some error occurred), then *Calling* transforms into the process *ErrorState*.

The process *CallBusy* notifies the user *id* that called user is busy (by the event *soundBusy.id*) and transforms into the process *Idle*. Similarly, the process *CallUnobtainable* notifies the user *id* that the called user is invalid (by the event *soundInvalid.id*) and transforms into the process *Idle*.

The process *O_Alert* nondeterministically selects one of the four possible activities. The guard for the first activity is that the user *id* was still calling the same partner (*chan[id] == partner[id]*) and that partner answered the call (*connect[id] == 1*), and in this case *O_Alert* connects the call (by the event *callConnect.id.partner[id]*) and transforms into the process *O_Connected*. The guard for the second activity is that the user *id* was still calling the same partner (*chan[id] == partner[id]*), and that the partner did not answer the call (*connect[id] == 0*); in this case *O_Alert* quits the call (by the event *ringOut.id*), sets *chan[partner[id]]* to *NIL* and *chan[id]* to *NIL*, and transforms into the process *Idle*. The guard for the third activity is that the user *id* quit the call (*chan[id] != partner[id]*), and in this case *O_Alert* indicates that the call was stopped and transformed into the process *Idle*. The guard for the fourth case is that error occurred (*connect[id]* is neither 0 nor 1), and in this case *O_Alert* indicates that an error occurred and transformed the call into the process *ErrorState*.

The process *O_Connected* nondeterministically selects one of the three possible activities. The guard for the first activity is that both calling and called users are still connected, and in this case *O_Connected* indicates that the conversation phase is ongoing (*keepTalking.id.partner[id]*), and continues to behave as the process *O_Connected*. In the second activity, *O_Connected* disconnects both users (by setting *connect[id]* to 0 and *connect[partner[id]]* to 0), indicates the end of the conversation phase (*stopTalking.id.partner[id]*) and transforms into the process *Idle*—this activity corresponds to the case when the calling user *id* ends the call first. The guard for the third activity is at the end of the conversation phase (*stopTalking.id.partner[id]*), which has been indicated by the called user, and, in this case, *O_Connected* transforms into the process *Idle*.

The process *T_Alert* performs one of three possible cases. The first case is when the called user is busy, then *T_Alert* indicates that partner is busy (*partnerBusy.id.partner[id]*) and transforms into the process *Idle*. The second case is when the user *id* is still calling the same *partner[id]* and the called user is not busy, then *T_Alert* indicates that partner is ready (*partnerReady.id.partner[id]*) and transforms into the process *T_Pickup*. The third case is the default case when some error occurs, then *T_Alert* indicates the error and transforms into the process *ErrorState*.

The process *T_Pickup* nondeterministically selects one of the two possible activities. The guard for the first activity is that the user *id* is still calling the same *partner[id]*, and in this case *T_Pickup* indicates the partner's answer, sets *dev[id]* to 0, *connect[partner[id]]* to 1, and *connect[id]* to 1, and transforms into the process *T_Connected*. The guard for the second activity is that either no incoming call is present at the called side (*chan[partner[id]]* == *NIL*) or that the incoming call at the called side is not from the user *id* (*chan[partner[id]]* != *id*), and, in this case, *T_Pickup* transforms into the process *Idle*.

The process *T_Connected* nondeterministically selects one of the three possible activities. The guard for the first activity is that both users are still connected, then *T_Connected* indicates that the conversation phase is still ongoing (*keepTalking.partner[id].id*), and continues to behave as the process *T_Connected*. The guard for the second activity is the calling user *id* has disconnected the call (*stopTalking.partner[id].id*), then *T_Connected* transforms into the process *Idle*. The third activity is when the called user decides to disconnect the call, then *T_Connected* sets *connect[id]* to 0 and *connect[partner[id]]* to 0, indicates the end of the conversation phase (*stopTalking.partner[id].id*), and transforms into the process *Idle*.

The process *ErrorSate* just indicates that an error occurred and stops execution by transforming into the process *Stop*. Next, we define various system properties.

Property no. 1 states that a connection between two users is possible. We use the event *pickup.1.0* (user 1 picks up the phone dialed by user 0) to check this property. The assertion below claims that user 1 will never pick up the phone dialed by user 0. PAT finds this assertion to be invalid and produces a lengthy witness trace, demonstrating that user 1 will pick up the phone dialed by user 0, i.e., that connection between 0 and 1 is possible.

```
#assert System |= [] (!pickup.1.0);
```

Property no. 2 states that if a user dials itself, then it will receive the engaged tone before it returns to the idle state. We use the events *dial.0.0* (the user 0 dials themselves), *soundBusy.0* (user 0 hears a busy tone, i.e., is engaged), and *idle.0* (user 0 goes back to the *Idle* state) to check this property. The assertion below claims that if user 0 dials itself, then they must hear an engaged tone before returning to idle. As expected, the PAT finds this assertion to be valid.

```
#assert System |= [] ( dial.0.0 -> (soundBusy.0 R (!idle.0) ) ) ;
```

Property no. 3 states that either a busy tone or a ringing tone will directly follow calling. We use the events *makeCall.0* (user 0 makes an outgoing call), *soundBusy.0* (user 0 hears busy tone), *ringPartner.0* (user 0 waits for a partner to answer the call, thus hearing the ringing tone), and *soundInvalid.0* (user 0 dials an invalid number) to check this property. The assertion below claims that if user 0 makes a call, it must hear either a busy or ringing tone before any next major event (*idle.0, callConnect.0, ringOut.0*). Note that the attribute directly in the property specification is not equal to X (next) in LTL logic, because there are other events in between. Thus, we encode the attribute directly as before any next major event. As expected, the PAT finds this assertion to be valid.

```
#assert System |= [](makeCall.0 -> (soundBusy.0 || ringPartner.0 ||
soundInvalid.0) R (!idle.0 && !callConnect.0 && !ringOut.0) );
```

Property no. 4 states that the dialed number is the same as the number of the connection attempt. We use the condition *dialed01* (user 0 has dialed user 1) and the event *doCall.0.1* (user 0 attempts to call user 1) to check this property. The assertion below claims that in all traces, if user 0 attempts to call user 1, then user 0 must have dialed user 1. As expected, the PAT finds this assertion to be valid.

```
#define dialed01 (dialNum[0]==1);
#assert System |= []( doCall.0.1 -> dialed01 );
```

Property no. 5 states that if the user dials a busy number, then either the busy line is cleared before a call is attempted, or the user will hear the engaged (busy) tone before returning to the idle state.

We use the event *dial.0.1* (user 0 dials user 1) and the condition *rcvReady* (user 1 is ready to receive a call), as well as the events *makeCall.0* (user 0 starts a call), *soundBusy.0* (user 0 hears the busy tone), and *idle.0* (user 0 goes back to *Idle* state) to check this property. The assertion below claims that if user 0 dials user 1, then either user 1 is ready to receive a call before user 0 starts making a call, or user 0 will hear busy signal before it goes back to *Idle* state. As expected, the PAT finds this assertion to be valid.

```
#define rcvReady (chan[partner[0]] == NIL);
#assert System |= [] ( dial.0.1 -> ( (rcvReady R (!makeCall.0) )
   || (soundBusy.0 R (!idle.0) ) ) );
```

Property no. 6 states that a user cannot make a call without having just dialed a number. Recall that the flag *justDial[id]* is set to 1 immediately after event *dial.id.**, and is cleared just after the next *makeCall.id* event. We use the conditions *justDialed0* (user 0 just dialed a number) and *justDialed1* (user 1 just dialed a number), as well as the events *makeCall.0* and *makeCall.1* to check this property. The assertion below claims that if user 0 starts making a call, then the event *dial.0.** has just happened, as indicated by the condition *justDialed0*, and the same holds for the user 1. As expected, the PAT finds this assertion to be valid.

```
#define justDialed0 justDial[0]==1;
#define justDialed1 justDial[1]==1;
#assert System |= []( (makeCall.0 -> justDialed0) &&
   (makeCall.1 -> justDialed1) );
```

Property no. 7 describes the CFU (Call Forward Unconditional) service. To verify this property, we initialize *systemCFU* by setting *CFU*[1]=2 and perform verification on that system.

The first assertion below claims that if *CFU*[1]==2, then in all traces, if user 0 dials user 1 (the event *dial.0.1*), then user 0 will call user 2 (the event *docall.0.2*) before user 0 can go back to the state *Idle* (the event *idle.0*), see the sequence diagram in the Figure 5.11. The second assertion below claims that in all the traces, user 0 will not connect to user 1. As expected, the PAT finds both assertions to be valid.

```
SystemCFU = initCFU{CFU[1]=2;} -> System();
#assert SystemCFU |= [] (dial.0.1 -> (doCall.0.2 R (!idle.0)) );
#assert SystemCFU |= ([] !callConnect.0.1);
```

Property no. 8 describes the CFB (Call Forward when Busy) service. To verify this property, we initialize a system with *CFB*[1]=2. The assertion below claims that in all the traces, if user 0 dials user 1 (the event *dial.0.1*) and user 1 is busy (the condition *Busy1*) then user 0 will not go back to the state idle before it has dialed user 2. Since the system is symmetric, similar assertions hold for other users. As expected, the PAT finds this assertion to be valid.

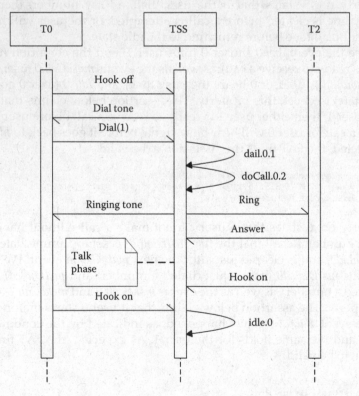

FIGURE 5.11
Sequence diagram for property no. 7.

```
SystemCFB = initCFB{CFB[1]=2;} -> System();
#define Busy1 chan[1]!=NIL && chan[1]!=0;
#assert SystemCFB |= [] ( (dial.0.1 && Busy1)->
  (doCall.0.2 R (!idle.0)) );
```

Property no. 9 describes the OCS (Originating Call Screening) service. To verify this property, we initialize a system with *OCS*[0]=1. The assertion below claims that, in such a system, the event *doCall.0.1* (user 0 calls user 1) will never happen, see the sequence diagram in the Figure 5.12. Since the system is symmetric, similar assertions hold for other users. As expected, the PAT finds this assertion to be valid.

```
SystemOCS = initOCS{OCS[0]=1;} -> System();
#assert SystemOCS |= []!doCall.0.1;
```

Property no. 10 describes the ODS (Originating Dial Screening) service. To verify this property, we initialize a system with *ODS*[0]=1. The assertion below claims that, in such a system, the event *dial.0.1* (user 0 dials user 1) will never happen. As expected, the PAT finds this assertion to be valid.

```
SystemODS = initODS{ODS[0]=1;} -> System();
#assert SystemODS |= []!dial.0.1;
```

Property no. 11 describes the TCS (Terminating Call Screening) service. To verify this property, we initialize a system with *TCS*[0]=1. The assertion below claims that in such a system, the event *t_alert.0.1* (user 0 responses to a call

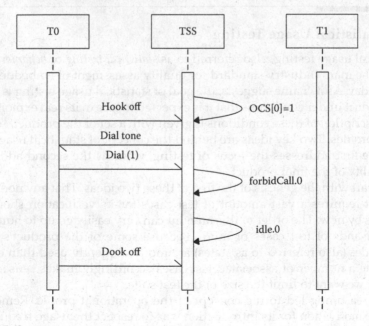

FIGURE 5.12
Sequence diagram for property no. 9.

from user 1) will never happen. Since the system is symmetric, similar assertions hold for other users. As expected, the PAT finds this assertion to be valid.

```
SystemTCS = initTCS{TCS[0]=1;} -> System();
#assert SystemTCS |= []!t_alert.0.1;
```

Property no. 12 describes the RBWF (Ring Back When Free) service. To
verify this property, we initialize a system with *RBWF*[1]=1. Next, we define
the supplementary condition *dial10*, meaning that user 1 must have dialed
user 0. The first assertion below claims that, in such a system, the event *ring-
back.1.0* (user 1 rings back user 0) will never happen. As expected, the PAT
finds that this assertion is not valid, and produces a lengthy counterexample. The second assertion below claims that whenever *dial.0.1* happens, if
user 1 calls user 0, then user 1 must have dialed user 0 (the condition *dial10*).
As expected, the PAT finds that this assertion is not valid, and produces a
lengthy counterexample. The second assertion is invalid because user 1 may
make a simple call to user 0 (the event *dial.0.1*) and not because user 1 is using
the ring back service (the event *ringback.1.0*).

```
SystemRBWF = initRBWF{RBWF[1]=1;} -> System();
#define dialed10 dialNum[1]==0;
#assert SystemRBWF |= [] !ringback.1.0;
#assert SystemRBWF |= [](dial.0.1 -> [](callConnect.1.0 -> dialed10));
```

5.4 Statistical Usage Testing

Statistical usage testing, also referred to as *statistical testing* or *behavioral testing*, is the main industry standard for quality assessment of embedded systems today. As its name suggests, the goal of statistical usage testing is to test
the product under conditions that it is expected to face in its real exploitation.
The description of these conditions is given with a set of the product's operational profiles. Two key ideas are behind the concept of statistical usage testing. The first addresses the focus of testing, whereas the second addresses
the quality of the final product.

We start with the genesis of the first of these two ideas. That any nontrivial
product requires a vast amount of test cases for its verification should be
obvious by now. The order of this amount can very easily go up to hundreds
of thousands of test cases or more. Because some of the product's working modes (also referred to as states) are more frequently used than others,
selecting a number of associated test cases accordingly makes sense, especially if we want to limit the size of the test suite.

This reasoning led to the concept of the operational profile. Remember
that the motivation for its introduction was to respect the usage frequencies
of individual operational states. Actually, because product state transitions

are triggered by corresponding events (signals, messages), the state usage frequencies are equal to the frequencies of these events. Furthermore, if we want to make our considerations independent of the total number of usages (tests), introducing the probabilities of events is convenient. (In this context, we define probability as the number of real occurrences of the event divided by the total number of its possible occurrences.)

Mathematically, the operational profile is a Markov process. It can be modeled as a special kind of graph whose vertices are product states, and whose arcs are state transitions triggered by the corresponding events of the given probability. The operational profile is essentially an FSM with given probabilities of its state transitions. Of course, the sum of probabilities of all outgoing state transitions for a single state must be equal to 1 (100%).

The second idea behind the concept of statistical usage testing is to use the product's reliability as the main measure of its quality. The genesis of this idea is that the traditional software engineering measures of product quality are the *number of remaining bugs* and the *test coverage* of the implementation under test that was achieved through its testing. However, achieving good results with respect to these two measures is not sufficient for assuring the high quality of the product.

For example, consider the following paradox: Imagine a software product that has a single bug that causes a system crash every time the software is started. Although the product has the excellent value of the metric *number of remaining bugs* (only 1 bug remaining), it is completely unreliable and therefore practically unusable. In real life, we are not interested in how good the product is with respect to the number of remaining bugs and test coverage. Rather, we are primarily interested in its reliability.

Of course, we cannot measure the product reliability directly, but we can estimate this from the number of test cases that it has successfully passed. More precisely, in real engineering practice we have the opposite problem. We want to calculate the number of test cases needed for the desired product reliability, and for the given level of risk we are ready to accept. We can do this by solving the following equation:

$$B = R^N$$

where

 B is an upper bound on the probability that the model assertions are erroneous.

 R is a lower bound on the estimate of product reliability.

 N is the number of random test cases that the product must successfully pass.

For example, achieving even moderate reliability of $R = 0.999$ with $B = 0.007$ would require the successful pass of $N = 5,000$ random test cases. Similarly, achieving $R = 0.9999$ with $B = 0.007$ requires $N = 50,000$ random test cases, and achieving $R = 0.99999$ with $B = 0.007$ requires $N = 500,000$ random test cases. Alternatively, we can run a smaller number of test cases on more product samples in parallel. For example, instead of running $N = 500,000$ random test cases on a single sample, we can run $N = 50,000$ random test cases on 10 product samples simultaneously.

By considering these examples, we can deduce two conclusions. The first is that conducting statistical usage testing of the final product may require a significant amount of time. The order of magnitude of this amount is calendar weeks or even months, depending on the characteristics of the concrete product. The second conclusion is that we definitely need tools that automatically generate and execute test suites of that size. We simply cannot do this by hand.

An example of the automated working environment for generating statistical test suites is described by Popovic and Velikic (2005). This working environment consists of two parts, namely, the front-end and the back-end (Figure 5.13). The front-end is the Generic Modeling Environment (GME) developed at the Institute for Software Integrated Systems at Vanderbilt University. GME is a configurable toolkit for creating domain-specific modeling and program synthesis environments.

Generally, we configure GME by creating metamodels that specify the *modeling language*, and therefore the *modeling paradigm*, of the application domain. Once we create a metamodel, we must interpret and register it by GME to create a new working environment for making domain-specific models. We normally use such working environments for building domain-specific models and for storing them in a model database. The domain-specific models are essentially graphs, and we render them by dragging and dropping the graphical symbols on the working sheet that is maintained by the GME graphical user interface (GUI). The symbols in GME have their attributes, preferences, and properties.

The particular metamodel that specifies the language (and the paradigm) for modeling operational profiles is represented with the metaclass *OperationalProfile* in Figure 5.13. Each concrete operational profile model (represented with the class *OpProfile* in Figure 5.13) is created by using the operational profile modeling paradigm (the class *OpProfile* is derived from the class *OperationalProfile*). Creating operational profile models by using this paradigm is quite easy.

The modeling language for rendering operational profile models has a single symbol, *State*. This symbol has a single attribute, which is the name of the state. Normally, we just drag and drop the state symbol icon to the working sheet, click on the name field, and type in its name. Each of the state symbols we place on the working sheet represents a single working state (mode) of the product that we want to test.

Rendering state transitions requires a little more work. To render a state transition, we select a connecting tool (symbolized by the operator "+"), click on the source state, and click on the destination state. When the state transition is in place, we enter the particular data for its attributes. A state transition has the following three attributes:

- *EventClass* specifies the class of events that trigger the state transition.
- *Output* specifies the expected output of the state transition.
- *Probability* specifies the probability of the state transition (in percent).

The most frequently used format of the attribute *EventClass* definition is as follows:

```
E(a,b,c...);->a := A1/A2/...; b := B1/B2/...; c := C1/C2/...
```

The event class definition above consists of two parts. The first one is on the left-hand side of the substring "->" and is referred to as the event class. The event class $E(a,b,c...)$; is a string with an arbitrary number of parameters (substrings), labeled here as a, b, c, and so on. The second part of the definition is on the right-hand side of the substring "->". It provides definitions of possible replacements (which are also strings) for each event class parameter. As indicated above, the parameter a may be replaced with the string A_1 or A_2 and so on.

A particular event (also referred to as the *constant event*) is an event class without parameters. We may also think about it as the event class with a single member. Particular events are generated from the event class by substituting each event class parameter with the randomly selected replacement from the list of possible replacements. All replacements have equal selection probabilities. Examples of particular events for the event class definition given above are $E(A_1,B_1,C_1...)$, $E(A_1,B_1,C_2...)$, $E(A_1, B_2, C_1...)$, $E(A_2, B_1, C_1...)$, and so on.

The event class format shown above is feasible as far as the number of the possible values of event class parameters is relatively small. But when the number of the possible values is large, writing them explicitly becomes impractical, if not impossible. For example, consider the integer parameter whose possible values are from the interval [0,10000). Writing all 10,000 of its possible values would be really annoying. To make it easier for the user, the working environment supports the following two intrinsic functions:

- *randInt<i,j>* randomly selects an integer number from the interval $[i,j)$.
- *randFloat<x,y>* randomly selects a float number from the interval $[x,y)$.

When we place and name all state symbols, interconnect them with state transitions, and enter the data for attributes of all state transitions, the operational profile model is finished, and we can store it in a file (or a database).

This is exactly the main purpose of the working environment front-end (Figure 5.13). Of course, later we may modify the model by adding or deleting states or state transitions, as well as by changing the data for attributes of state transitions, and store it again. All these manipulations are supported by the GME's GUI.

The working environment back-end consists of two parts. The first is the operational profile model interpreter (represented by the class *ModelInterpreter* in Figure 5.13), which is registered to GME. The second part of the back-end is a separate program written in Java, which is named Generic Test Case Generator (GTCG). The main task of the model interpreter is to transform the operational profile model to the operational profile specification, a simple text file of the well-defined format (represented with the class *OpProfileSpec* in Figure 5.13). Alternately, the main task of GTCG is to automatically generate the test suite to be used for statistical usage testing and the corresponding statistical report (represented with the classes *TestSuite* and *Statistics* in Figure 5.13).

The operational profile model interpreter is a Java package that is registered to GME with the program JavaCompRegister. The package comprises the following three classes:

- *OPBONComponent*: the interface between GME and the model interpreter
- *OPState*: the state interpreter
- *OPTransition*: the state transition interpreter

The model interpreter behaves similarly to traditional plug-in components of GUIs. We activate it by a click on the corresponding model interpreter icon. As the result of this activation, GME calls the model interpreter interface function *invokeEx*, which, in turn, creates temporary container objects for state names, event classes, state transition probabilities, event class definitions, and next state definitions.

Next, the model interpretation is performed by traversing the multigraph architecture of the model in focus. While visiting individual states and state transitions, GME calls the function *traverseChildren* of the classes *OPState* and *OPTransition*, respectively. These two functions effectively interpret the model by reading the data of the attributes and filling the above-mentioned container objects. At the end of the interpretation, the content of these container objects is saved into the operational profile specification file named *opspec.txt*.

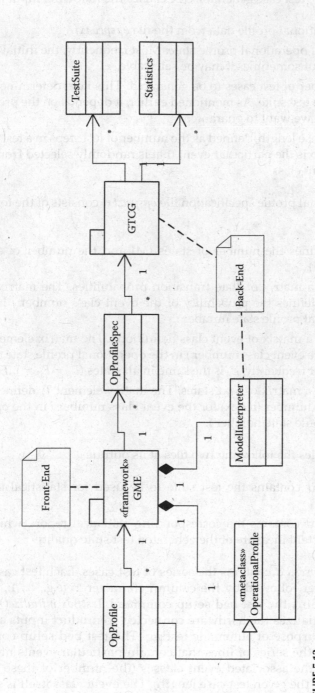

FIGURE 5.13
Working environment for generating statistical test suites.

The automatic test case generator GTCG uses the following input items:

- The operational profile data from the file *opspec.txt*.
- The initial operational profile state: Most frequently, the initial state is fixed, but sometimes it may be selectable.
- The number of test cases to be generated: This item determines the size of the test suite. As mentioned earlier, it depends on the product reliability we want to guarantee.
- The test case length, defined as the number of test steps in a test case. A test step is the particular event that is randomly selected from the given event class.

The operational profile specification file *opspec.txt* consists of the following four parts:

- Part I defines the number of states (M) and the number of event classes (N).
- Part II is a matrix of state transition probabilities. The matrix element P_{ij} defines the probability of the event class number j in the operational profile state number i.
- Part III is a matrix of event class definitions. The matrix element E_{ij} defines the event class number j in the operational profile state number i. Most frequently, E_{ij} is the same in all states ($E_{i1} = E_{i2} = \ldots E_{iM}$).
- Part IV is a matrix of next states. The matrix element T_{ij} defines the next state number (index) for the event class number j in the operational profile state number i.

GTCG provides the following two files at its output:

- *testcases.txt* contains the test suite to be used for statistical usage testing.
- *statistics.txt* contains the corresponding statistical report, which is the important measure of the generated test suite quality.

The file *testcases.txt* contains the series of test cases. Each test case starts with its number followed by the column character ':' (e.g., 0:, 1:, 2:). The next line contains the test bed setup command *TestBox.initialize()*, which essentially initializes the hardware connected to product inputs and outputs for the purpose of automatic testing. The test bed setup command is followed by the series of lines that contain particular events randomly selected from the associated event classes (the number of these lines is determined by the given test case length). The event class itself is selected randomly from the distribution defined by the operational profile data (*opspec.txt*, Part II).

The file *statistics.txt* consists of two parts. The first part contains a series of lines, one per operational profile state. Each of these lines indicates the number of occurrences of the corresponding operational profile state (c_i), the discrepancy between the observed and expected frequency of state occurrence (d_i), and the significance level (SL_i). The significance level is actually the probability that the discrepancies as large as those observed would occur with random variation. The second part of the statistical report shows the mean value of the discrepancy and the mean value of the significance value.

A detailed explanation of the statistical measures mentioned above is outside the scope of this book but can be found elsewhere (e.g., Woit, 1994). Practically, it is enough to remember the following guides:

- A significance level greater than or equal to 20% is considered large. This result means that the test suite is of sufficient quality and we may use it for statistical usage testing.

- A significance level less than or equal to 1% is considered small. This result means that the test suite quality is poor and it should not be used for statistical usage testing.

The statistical usage testing methodology governs the usage of tools that create the working environment. This methodology subsumes the following steps:

- Make the operational profile model of the product (implementation under test).
- Interpret the model.
- Determine the desired level of reliability.
- Calculate the required size of the test suite (the number of test cases).
- Generate the test suite.
- Check the test suite quality. If the quality is not acceptable, return to the previous step.
- Execute the test suite. If all test cases successfully pass, the final verdict is *pass*. In that case, we can claim that product reliability is at least at the level of the desired reliability. If at least one test case fails, the final verdict is *fail* and the product is considered not usable, at least not at the desired level of reliability.

This methodology can be used for testing both parts of the products and their complete forms. We will illustrate such applications by the following two examples. The implementation under test in the first example is the SIP invite client transaction. We start with modeling its operational profile in accordance with the methodology outlined above (Figure 5.14).

The operational profile shown in Figure 5.14 has five working states, namely, *Initial, Calling, Proceeding, Completed,* and *Terminated.* At the same time, it has nine event classes that are intentionally labeled with names that resemble the original specification (see RFC 3261, Figure 5). The definitions of the event classes (not shown in Figure 5.14) are the following:

- The event class labeled *INVITE* is defined as *INVITE* (this class has a single member).
- The event class labeled *300–699* is defined as `M3->M3:=randInt<300,700>;`
- The event class labeled *TA* is defined as TA (original RFC 3261 label: Timer A fires).
- The event class labeled *1XX* is defined as `M1->M1:=randInt<100,200>;`
- The event class labeled *TB* or *TransportERR* is defined as `E->E:=TB/TransportERR;`
- The event class labeled *2XX* is defined as `M2->M2:=randInt<200,300>;`
- The event class labeled *TD* is defined as TD (original RFC 3261 label: Timer D fires).
- The event class labeled *TransportERR* is defined as TransportERR (constant event).
- The event class labeled *End* is defined as End (added because the sum of outgoing state transition probabilities for each state must be equal to 100%).

The probabilities of individual state transitions are shown in Figure 5.14. Note that outgoing state transition probabilities add up to 100% for each state (an essential request for a Markov process). Generally, we set the state transition probabilities according to what we expect the product will face in its real exploitation. Of course, we should use statistical data available for some similar product or the previous version of the same product whenever we can.

Next, we start the model interpreter, which transforms the model into the operational profile specification file *opspec.txt.* When writing GME model interpreters, we should make no assumptions about the order in which the model is traversed. For example, assuming that individual states and state transitions are going to be visited in the same order in which they were originally entered would be a mistake because this is not going to happen. The best assumption we can make in this respect is to assume a completely random visiting order.

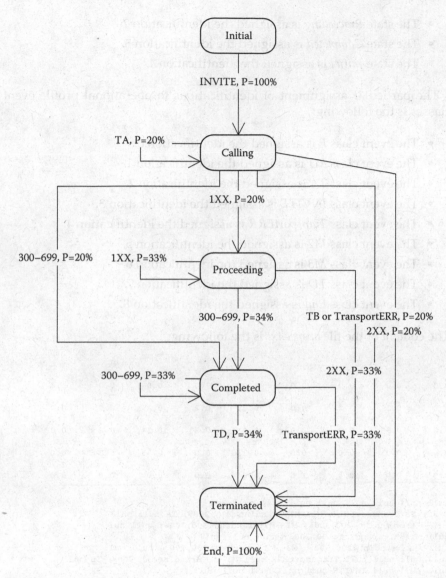

FIGURE 5.14
SIP INVITE client transaction operational profile.

Based on this assumption, the model interpreter simply assigns identifications to states and state transitions according to the order they are visited. The particular assignment of identifications to operational profile states in this example is the following:

- The state *Terminated* is assigned the identification 0.
- The state *Calling* is assigned the identification 1.

- The state *Proceeding* is assigned the identification 2.
- The state *Completed* is assigned the identification 3.
- The state *Initial* is assigned the identification 4.

The particular assignment of identifications to operational profile event classes is the following:

- The event class *E* is assigned the identification 0.
- The event class *M1* is assigned the identification 1.
- The event class *TA* is assigned the identification 2.
- The event class *INVITE* is assigned the identification 3.
- The event class *TransportERR* is assigned the identification 4.
- The event class *M2* is assigned the identification 5.
- The event class *M3* is assigned the identification 6.
- The event class *TD* is assigned the identification 7.
- The event class *End* is assigned the identification 8.

The content of the file *opspec.txt* is the following:

```
5       9

0.0     0.0     0.0     0.0     0.0     0.0     0.0     0.0     1.0

0.2     0.2     0.2     0.0     0.0     0.2     0.2     0.0     0.0

0.0     0.33    0.0     0.0     0.0     0.33    0.34    0.0     0.0

0.0     0.0     0.0     0.0     0.33    0.0     0.33    0.34    0.0

0.0     0.0     0.0     1.0     0.0     0.0     0.0     0.0     0.0
```

```
null null null null null null null null End
E->E:=TB/TransportERR; M1->M1:=randInt<100,200>; TA null null
  M2->M2:=randInt<200,300>; M3->M3:=randInt<300,700>; null null
null M1->M1:=randInt<100,200>; null null null
  M2->M2:=randInt<200,300>; M3->M3:=randInt<300,700>; null null
null null null null TransportERR null M3->M3:=randInt<300,700>; TD null
null null null INVITE null null null null null
```

```
0    0    0    0    0    0    0    0    0
0    2    1    0    0    0    3    0    0
0    2    0    0    0    0    3    0    0
0    0    0    0    0    0    3    0    0
0    0    0    1    0    0    0    0    0
```

NOTE: The specifications of event classes for the states 1, 2, and 3 (*Calling, Proceeding,* and *Completed*) were too long to fit into a single line. Therefore, definitions of event classes for each of these states spans across two lines (the second starts at the next level of indentation).

Next, we activate GTCG with the script that specifies the starting state identification 4 (*Initial*), the number of test cases that is equal to 1,000, and the test case length that is equal to 4 (this means 4 steps, i.e., particular events, per test case). Selection of this particular test case length requires a short comment. This value is exactly the length of the shortest path across all five states, starting from the state *Initial* (path *Initial–Calling–Proceeding–Completed–Terminated*, with five states and four state transitions). Of course, other paths of length 4 are possible and will be generated.

As already mentioned, the GTCG creates two output files, *testcases.txt* and *statistics.txt*. According to the methodology outlined above, we first check the quality of the generated test suite by inspecting the file *statistics.txt*. Its content is the following:

```
Calculating statistics

i=0    ci=1104          di=0.0                    SLi=1.0
i=1    ci=1237          di=2.470493128536783      SLi=0.0
i=2    ci=291           di=0.7498208280500565     SLi=0.7014229616104999
i=3    ci=368           di=0.1864198248469353     SLi-0.910066579962014
i=4    ci=1000          di=0.0                    SLi=1.0
Mean   d=0.6813467562867549
Mean   SL=0.7222979083145027
```

The average significance level *SL* is equal to 72% (0.72). Because this number is greater than the required 20%, we conclude that the quality of the generated test suite is sufficient, and that we can use it for statistical usage testing.

Next, we look more closely at a couple of test cases from the beginning of the file *testcases.txt* to get a better feeling of the nature of statistical test cases. The relevant comments are interleaved with the test cases:

```
0:
TestBox.initialize();
INVITE
443
TransportERR
End
```

Test case number 0: After the initial *INVITE*, GTCG randomly selects the event class labeled *300–699* and the particular event 443 from that class. This action causes the state to transition to the state *Completed* (Figure 5.14). Next, GTCG randomly selects the event *TransportERR*, thus causing the state to transition to the state *Terminated*. *End* is the only possible event in that state.

```
1:
TestBox.initialize();
INVITE
TA
586
TD
```

Test case number 1: After the initial *INVITE*, GTCG randomly selects the event class *TA* (Timer A fires). The current state remains in the state *Calling* (Figure 5.14). Next, GTCG randomly selects the event *586*, thus causing the state transition to the state *Completed*. Finally, GTCG randomly selects the event *TD* (Timer D fires), which causes the state transition to the state *Terminated*.

```
2:
TestBox.initialize();
INVITE
190
267
End
```

. Test case number 2: After the initial *INVITE*, GTCG randomly selects the event class *1XX* and the particular event *190*. This causes the state transition to the state *Proceeding* (Figure 5.14). Next, GTCG randomly selects the event *267*, thus causing the state transition to the state *Terminated*. The next event must be the event *End*.

```
3:
TestBox.initialize();
INVITE
494
TD
End
```

Test case number 3: After the initial *INVITE*, GTCG randomly selects the event class *300–699* and the particular event *494*. This causes the state transition to the state *Completed* (Figure 5.14). Next, GTCG randomly selects the event *TD*, thus causing the state transition to the state *Terminated*. The next event must be the event *End*.

In the short descriptions of the generated test cases given above, we used the construct, "GTCG randomly selects the event class X and the particular event Y," for brevity. One should remember that the selection of the event class is always in accordance with the given operational profile probability distribution, whereas the selection of the particular event from the given class is really random.

The previous example shows how we can use statistical usage testing for testing a part of the product. As already mentioned, we can employ statistical usage testing for testing whole products, too. The next example shows such an application—statistical usage testing of the simple SIP softphone.

The operational profile of the SIP softphone is shown in Figure 5.15. It has 8 states and 13 event classes. The states are *Connecting*, *Terminating*, *Disconnecting*, *Connected*, *Calling*, *Initial*, *Proceeding*, and *Ringing* (listed here in the ascending order of their identification). The event classes are *RELEASE*, *200*, *ACK*, *180*, *ERR*, *END*, *ANSWER*, *100*, *INVITE*, *SETUP*, *BYE*, *TH*, and *TB* (also listed in the ascending order of their identification).

All event classes have just one member, and their definition is equal to the label shown in Figure 5.15, with the exception of the event class that is labeled *ERR*, which is defined as follows:

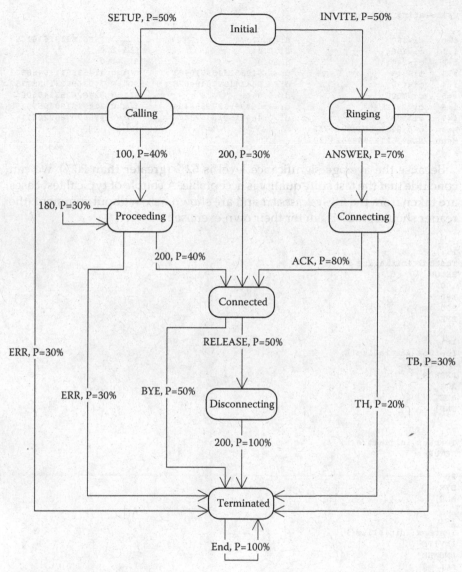

FIGURE 5.15
SIP softphone operational profile.

```
M3->M3:=randInt<300,381>/randInt<400,494>/randInt<500,514>/randInt<600,607>;
```

This definition is a good example of how we can specify a random value that may be selected from more disjointed intervals of values. Next, we generate 1,000 test cases with five test steps each. The content of the file *statistics.txt* is the following:

```
Calculating statistics

i=0    ci=360              di=0.625              SLi=0.4686783191616166
i=1    ci=1564             di=0.0                SLi=1.0
i=2    ci=244              di=0.0                SLi=1.0
i=3    ci=546              di=1.6483516483516483 SLi=0.21453651135488572
i=4    ci=496              di=0.5843413978494628 SLi=0.7503695231083775
i=5    ci=1000             di=0.064              SLi=0.8248262531456066
i=6    ci=286              di=3.0879953379953404 SLi=0.21451818049555796
i=7    ci=504              di=0.4897959183673477 SLi=0.4966702889206116
Mean   d=0.8124355378204748
Mean   SL=0.6211998845233321
```

Because the average significance level is 62% (greater than 20%), we can conclude that the test suite quality is acceptable. A couple of typical test cases are taken from the file *testcases.txt* and are shown here without comment (the reader should study them for their own exercise):

```
15:
TestBox.initialize();
SETUP
100
180
200
BYE

16:
TestBox.initialize();
INVITE
ANSWER
ACK
BYE
END

17:
TestBox.initialize();
SETUP
100
200
BYE
END

18:
TestBox.initialize();
INVITE
ANSWER
ACK
RELEASE
200
```

5.5 Examples

This section includes two examples and two related problems. The first example demonstrates unit testing of the FSM Library–based implementations. The second example illustrates integration testing of FSM Library–based products.

5.5.1 Example 1

This example demonstrates unit testing of the SIP invite client transaction implementation, which is described in Section 4.5.2 (Example 2). The SIP invite client transaction implementation is based on the requirements and analysis made in Section 2.3.3 (Figure 2.16) and the design presented in Section 3.10.5 (Example 5).

Because the implementation under test (SIP invite client transaction) is implemented in C++, we use CppUnit implementation of the unit testing framework, introduced in Section 5.1. In this simple example, we will construct just one test case to keep it short enough. Also, we will skip some SIP message-specific message handling, which is really not essential for this example.

We start this example by constructing two classes: *ExampleTestCase* and *ExampleMessageFactory*. The former is the tester class, which comprises one sample test case, whereas the latter is the supplementary class, which provides the functions for message management. The content of the class *ExampleTestCase* declaration file, named *ExampleTestCase.h*, is the following:

```
#ifndef CPP_UNIT_EXAMPLETESTCASE_H
#define CPP_UNIT_EXAMPLETESTCASE_H
// CppUnit helper macros
#include <cppunit/extensions/HelperMacros.h>
// Problem specific headers
#include "../kernel/fsmsystem.h"
#include "../kernel/logfile.h"
#include "../NewSIP/InvClientTE.h"
#include "ExampleMessageFactory.h"
/*
 * A sample test case
 *
 */
class ExampleTestCase : public CPPUNIT_NS::TestFixture {
  CPPUNIT_TEST_SUITE(ExampleTestCase);
  CPPUNIT_TEST(example);
  CPPUNIT_TEST_SUITE_END();

protected:
  FSMSystemWithTCP *pSys;
  LogFile *lf;
  InviteClientTE* pInviteCltTE[NUMBER_OF_TES];
  ExampleMessageFactory* pEMF;
  uint8 *msg;
  uint16 msgcode;
public:
  void setUp();
protected:
  void example();
};
#endif
```

The declaration file above includes the CppUnit helper macros header file (*HelperMacros.h*) and the problem-specific header files (*fsmsystem.h*, *logfile.h*, *InvClientTE.h*, and *ExampleMessageFactory.h*). The class *ExampleTestCase* is derived from the class that is defined by the macro instruction

CPPUNIT_NS::TestFixture. The definition of the test suite starts with the macro instruction *CPPUNIT_TEST_SUITE()* and ends with the macro instruction *CPPUNIT_TEST_SUITE_END()*. The parameter of the former macro instruction is the name of the test suite (*ExampleTestCase*, in this example).

Generally, we use the macro instruction *CPPUNIT_TEST()* to define individual test cases inside the body of test suite definition. The parameter of this macro instruction is the name of the test case function that is defined within the tester class and that we want to add to the test suite. In this particular example, we add a single test case function, named *example()*, with a single macro instruction, *CPPUNIT_TEST()*, whose real parameter is the string "example".

Next, we define the test case fixture. In this example, it comprises the following:

- The pointer to the instance of the class *FSMSystemWithTCP* (see Section 6.8.9)
- The pointer to the instance of the class LogFile (which is the interface to the log file)
- The array of pointers to the instances of the class InviteClientTE (which is actually the implementation under test)
- The pointer to the instance of the class ExampleMessageFactory (which is the supplementary tester class)
- The pointer to the message
- The code of the message

At the end of this file we declare the function *setUp()* and the test case function *example()*. The content of the class *ExampleTestCase* definition file, named *ExampleTestCase.cpp*, is as follows:

```
#include "ExampleTestCase.h"
#include "../kernel/fsmsystem.h"
#include "../kernel/logfile.h"
#include "../NewSIP/InvClientTE.h"
#include "ExampleMessageFactory.h"

CPPUNIT_TES_SUITE_REGISTRATION(ExampleTestCase);
void ExampleTestCase::setUp() {
  pSys = new FSMSystemWithTCP(11,11);
  pEMF = new ExampleMessageFactory();
  for (int i = 0; i < NUMBER_OF_TES; i++) {
    pInviteCltTE[i] = new InviteClientTE();
  }

  uint8 buffClassNo = 4;
  uint32 buffsCount[4] = {50, 50, 50, 50};
  uint32 buffsLength[4] = {1025, 1025, 1025, 1025};
  pSys->InitKernel(buffClassNo, buffsCount, buffsLength, 1);
```

```
lf = new LogFile("log.log", "log.ini");
LogAutomateNew::SetLogInterface(lf);

pSys->Add(pInviteCltTE[0], InviteClientTE_FSM, 10, true);
for (i = 1; i < NUMBER_OF_TES; i++){
  pSys->Add(pInviteCltTE[i], InviteClientTE_FSM);
 }
}

void ExampleTestCase::example() {
 msg = pEMF->MakeInviteToTALMsg();
 pInviteCltTE[0]->Process(msg);
 msgcode = pEMF->GetMsgCodeFromMBX(TLI_Test_FSM_MBX);
 CPPUNIT_ASSERT_EQUAL(msgcode,(uint16)INVITE);

 msg = pEMF->Make1XXToTAL();
 pInviteCltTE[0]->Process(msg);
 msgcode = pEMF->GetMsgCodeFromMBX(UA_Disp_FSM_MBX);
 CPPUNIT_ASSERT_EQUAL(msgcode,(uint16)RESPONSE_1XX);

 msg = pEMF->Make2XXToTAL();
 pInviteCltTE[0]->Process(msg);
 msgcode = pEMF->GetMsgCodeFromMBX(UA_Disp_FSM_MBX);
 CPPUNIT_ASSERT_EQUAL(msgcode,(uint16)RESPONSE_2XX);
}
```

At the beginning of this file, we register the test suite with the macro instruction *CPPUNIT_TEST_SUITE_REGISTRATION()*. The real parameter of this macro instruction is the name of the test suite. Next, we define the function *setup()* and the test case function *example()*.

The function *setup()* starts by creating an instance of the class *FSMSystemWithTCP*, an instance of the class *ExampleMessageFactory*, and the given number (*NUMBER_OF_TES*) of instances of the implementation under test (the class *InviteClientTE*). After that, it defines the types of buffers to be used by the FSM Library kernel, initializes the kernel by calling the function *InitKernel()* (see Section 6.8.4), creates the log file by calling the function *LogFile()*, and sets the log interface by calling the function *SetLogInterface()* (see Section 6.8.105). At the end, it adds the given number (*NUMBER_OF_TES*) of instances of the implementation under test to the FSM system by calling its function *Add()* (see Section 6.8.2 and Section 6.8.3).

The function *example()* performs the test case by checking state transitions of the implementation under test in the following three steps:

- Check the state transition from the state *STATE_IDLE* (see Section 4.5.2) to the state *STATE_CALLING*, driven by the message *INVITE*.
- Check the state transition from the state *STATE_CALLING* to the state *STATE_PROCEEDING*, driven by the message *1XX*.
- Check the state transition from the state *STATE_PROCEEDING* to the state *STATE_INITIAL*, driven by the message *2XX*.

Each of these three steps consists of the following four substeps:

- Create the message (*INVITE, 1XX*, or *2XX*).
- Send the message to the implementation under test by calling its function member *Process()* (see Section 6.8.82).
- Get the message code of the output message by calling the function member *GetMsgCodeFromMBX()* of the class *ExampleMessageFactory*. The output message is retrieved from the destination FSM Library mailbox. The destination mailbox is either the mailbox of the transport layer (TPL) or the mailbox of the transaction user (TU).
- Check the retrieved message code against the expected one (message code of the message *INVITE, 1XX*, or *2XX*) by calling the macro *CPPUNIT_ASSERT_EQUAL()*.

The particular substeps of the first step are the following:

- Create the message *INVITE* by calling the function member *MakeInviteToTALMsg()* of the class *ExampleMessageFactory*.
- Send the message to the implementation under test.
- Get the message code of the message that is retrieved from the TPL mailbox.
- Check it against the code of the message *INVITE*.

The particular substeps of the second step are the following:

- Create the message *1XX* by calling the function member *Make1XXToTAL()* of the class *ExampleMessageFactory*.
- Send the message to the implementation under test.
- Get the message code of the message that is retrieved from the TU mailbox.
- Check the message code against the code of the message *1XX*.

The particular substeps of the third step are the following:

- Create the message *2XX* by calling the function member *Make2XXToTAL()* of the class *ExampleMessageFactory*.
- Send the message to the implementation under test.
- Get the message code of the message that is retrieved from the TU mailbox.
- Check the message code against the code of the message *2XX*.

Next, we construct the supplementary class *ExampleMessageFactory*. The content of its declaration file, named *ExampleMessageFactory.h*, is as follows:

```
#ifndef _ExampleMessageFactory_FSM_
#define _ExampleMessageFactory_FSM_
#include "../constants.h"
#include "../kernel/fsm.h"
#include "../message/message.h"
class ExampleMessageFactory : public FiniteStateMachine {
  int cseq_number;
  Message SIPMsg;
  sip_t *mes;
  stringresponseBody;
public:
  uint8* MakeInviteToTALMsg();
  uint16 GetMsgCodeFromMBX(uint8 mbx);
  uint8* Make1XXToTAL();
  uint8* Make2XXToTAL();

  // FiniteStateMachine abstract functions
  StandardMessage StandardMsgCoding;
  MessageInterface *GetMessageInterface(uint32 id);
  void SetDefaultHeader(uint8 infoCoding);
  void SetDefaultFSMData();
  void NoFreeInstances();
  void Reset();
  uint8 GetMbxId();
  uint8 GetAutomate();
  uint32 GetObject();
  void ResetData();
public:
  ExampleMessageFactory();
  ~ExampleMessageFactory();
  void Initialize();
};
#endif
```

The content of the class *ExampleMessageFactory* definition file, named *ExampleMessageFactory.cpp*, is as follows (the parts that are not essential for this example are omitted to keep the example short):

```
#include "ExampleMessageFactory.h"
#include "../parser/smsgtypes.h"
#include "../parser/smsg.h"
#define SipMessageCoding 0x00
extern char* IPString(unsigned int addr, char* buf, int len);

ExampleMessageFactory::ExampleMessageFactory() : FiniteStateMachine(16, 2, 3) {}

ExampleMessageFactory::~ExampleMessageFactory() {}

void ExampleMessageFactory::Initialize() {}

uint8* ExampleMessageFactory::MakeInviteToTALMsg(){
  char temp[10];
  char szHostName[255];
  hostent* HostData;
  uint8* recmsg;
  uint8* msg;
  ...
  PrepareNewMessage(0x00,INVITE);
```

```
    SetMsgToAutomate(InviteClientTE_FSM);
    SetMsgToGroup(INVALID_08);
    SetMsgObjectNumberTo(0);
    AddParam(SIP_RAW_MESSAGE, SIPMsg.getLastMessage().length(),
            (uint8*) SIPMsg.getLastMessage().c_str());
    AddParamDWord(SIP_PARSED_MESSAGE, (unsigned long) mes);
    SendMessage(InviteClientTE_FSM_MBX);
    msg = GetMsg(InviteClientTE_FSM_MBX);
    return msg;
}
uint16 ExampleMessageFactory::GetMsgCodeFromMBX(uint8 mbx) {
    uint8* msg;
    uint16 msgCode;
    msg = GetMsg(mbx);
    msgCode = GetUint16((uint8*)(msg+MSG_CODE));
    return msgCode;
}

uint8* ExampleMessageFactory::Make1XXToTAL(){
    uint8* msg;
    ...
    PrepareNewMessage(0x00,RESPONSE_1XX_T);
    SetMsgToAutomate(TAL_Disp_FSM);
    SetMsgToGroup(INVALID_08);
    SetMsgObjectNumberTo(0);
    AddParamDWord(SIP_PARSED_MESSAGE, (unsigned long) mes);
    SendMessage(InviteClientTE_FSM_MBX);
    msg = GetMsg(InviteClientTE_FSM_MBX);
    return msg;
}

uint8* ExampleMessageFactory::Make2XXToTAL(){
    uint8* msg;
    SIPMsg.makeResponse("200","OK",responseBody,0);
    PrepareNewMessage(0x00,RESPONSE_2XX_T);
    SetMsgToAutomate(TAL_Disp_FSM);
    SetMsgToGroup(INVALID_08);
            SetMsgObjectNumberTo(0);
    AddParamDWord(SIP_PARSED_MESSAGE, (unsigned long) mes);
    SendMessage(InviteClientTE_FSM_MBX);
    msg = GetMsg(InviteClientTE_FSM_MBX);
    return msg;
}
...
```

The main reason we must introduce the supplementary class *ExampleMessageFactory* is because most of the functions defined in the FSM Library API are protected, which means that they cannot be used in the tester class directly. Alternately, as defined at the moment, CppUnit does not allow us to use multiple inheritance when we are defining tester classes. Rather, a tester class may be derived only from the class that is defined by the macro instruction *CPPUNIT_NS::TestFixture*.

The source code from the file *ExampleMessageFactory.cpp* should be obvious by now. The only detail that deserves a short explanation is the method by which we create messages. We use typical snippets of code, which start with the *PrepareNewMessage()* function call and are followed with the series of *SetXX()* and *AddParamXX()* function calls. The way we end these code snippets may seem odd. First, we send a new message by calling the function *SendMessage()* and, immediately after that, we read that message from the same destination mailbox by calling the function *GetMsg()*. Although it may seem odd, this is the most effective method of creating the complete message in the format that is expected by the function *Process()*.

Finally, we write the main module, named *Main.cpp*. This module creates the collaboration of objects necessary to automatically execute the test suite and report the results of its execution (Figure 5.16). The function *main()* performs the following steps:

- Create the event manager and the test controller.
- Add a listener that collects test results.
- Add a listener that prints dots as test cases are executed (one dot per test case).
- Add the top suite to the test runner.
- Print the test results in a compiler-compatible format.

The source code of the module *Main.cpp* follows:

```
#include <cppunit/BriefTestProgressListener.h>
#include <cppunit/CompilerOutputter.h>
#include <cppunit/extensions/TestFactoryRegistry.h>
#include <cppunit/TestResult.h>
#include <cppunit/TestResultCollector.h>
#include <cppunit/TestRunner.h>

int main(int argc,char* argv[]) {
  CPPUNIT_NS::TestResult controller;
  CPPUNIT_NS::TestResultCollector result;
  controller.addListener(&result);
  CPPUNIT_NS::BriefTestProgressListener progress;
```

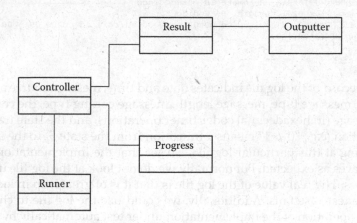

FIGURE 5.16
Collaboration of objects necessary for the automatic execution of the CppUnit test suite.

```
controller.addListener(&progress);
CPPUNIT_NS::TestRunner runner;
    runner.addTest(CPPUNIT_NS::TestFactoryRegistry::getRegistry().makeTest());
runner.run(controller);
CPPUNIT_NS::CompilerOutputter    outputter(&result,std::cerr);
outputter.write();
return result.wasSuccessful() ? 0 : 1;
}
```

As a result of automatic test suite execution, we get the following report on the monitor:

```
ExampleTestCase::example : OK
OK(1)
Press any key to continue...
```

Additionally, we will get the log file with the following content:

```
Fri Sep 16 19:32:50 2005
Msg To: UNKNOWN (0x02), Automate ID: 0x00000000
MsgFrom: UNKNOWN (0x0f), Automate ID: 0xcdcdcdcd
Received Msg: (0x0000), Length: 502  Coding type: 0
0f cd 02 ff | 00 00 cd cd | cd cd 00 00 | 00 00 cd cd | cd cd 00 f6 |
...
Start Timer:  (2)
State: 0 -> 1
-------------------------------------------------------
Fri Sep 16 19:32:50 2005
Msg To: UNKNOWN (0x02), Automate ID: 0x00000000
MsgFrom: UNKNOWN (0x0f), Automate ID: 0xcdcdcdcd
Received Msg: (0x0029), Length: 9  Coding type: 0
0f cd 06 ff | 29 00 cd cd | cd cd 00 00 | 00 00 cd cd | cd cd 00 09 | 00 01 00
04 00 | 50 9c 4c 00 | 00
Stop Timer:  (2)
State: 1 -> 2
-------------------------------------------------------
Fri Sep 16 19:32:50 2005
Msg To: UNKNOWN (0x02), Automate ID: 0x00000000
MsgFrom: UNKNOWN (0x0f), Automate ID: 0xcdcdcdcd
Received Msg: (0x002a), Length: 9  Coding type: 0
0f cd 06 ff | 2a 00 cd cd | cd cd 00 00 | 00 00 cd cd | cd cd 00 09 | 00 01 00
04 00 | 50 9c 4c 00 | 00
State: 2 -> 0
-------------------------------------------------------
```

Each record of the log file indicates date and time, message source and destination, message type, message length, message coding type, the content of the message (in hexadecimal code), timer operations, and the state transition information (e.g., "0 -> 1" means a transition from the state S_0 to the state S_1). By looking at this particular log file, we see that the implementation under test behaves as expected. But normally we do not look at the log file if all test cases pass. The real value of the log file is that it is of great help in localizing bugs if a test case fails. Additionally, we could use the log file to check the internal operation of the implementation under test automatically by the tester class. We skipped that step to keep the example simple enough.

5.5.2 Example 2

This example illustrates one of the steps in integration testing of an SIP-based softphone. Imagine that the SIP invite client transaction and the transaction layer dispatcher have undergone complete unit testing. The next normal step would be to integrate them into the final product. Furthermore, imagine that TU and TPL are not yet developed. The only thing we can do is to replace TU and TPL with their imitator classes, named *UA_Test* and *TLI_Test* (TLI stands for Transport Layer Interface), respectively (see the collaboration diagram in Figure 5.17).

The aim of this simple example is to check one particular interaction, illustrated with the collaboration diagram in Figure 5.17. To achieve that goal, we construct the class *UA_Test* that acts as a simple test driver, and the class *TLI_Test* that acts as a simple test stub. Both classes are derived from the class *FiniteStateMachine*. The former class has a single state and a single state transition, whereas the latter has two states and two state transitions.

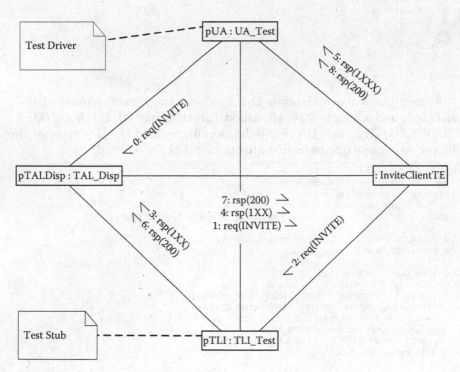

FIGURE 5.17
Example of integration testing collaboration.

The class *UA_Test* declaration file, named *UA_Test.h*, has the following content:

```
#ifndef _UA_Test_FSM_
#define _UA_Test_FSM_
#include "../constants.h"
#include "../kernel/fsm.h"
#include "../message/message.h"

class UA_Test : public FiniteStateMachine {
 int cseq_number;
 Message SIPMsg;
 void SendInviteToTAL();
public:
 enum States { STATE_INITIAL };
 void Evt_Init_TIMER_TINV_EXP();
 void Event_UNEXPECTED();
 // FiniteStateMachine abstract functions
 StandardMessage StandardMsgCoding;
 MessageInterface *GetMessageInterface(uint32 id);
 void SetDefaultHeader(uint8 infoCoding);
 void SetDefaultFSMData();
 void NoFreeInstances();
 void Reset();
 uint8 GetMbxId();
 uint8 GetAutomate();
 uint32 GetObject();
 void ResetData();
public:
 UA_Test();
 ~UA_Test();
 void Initialize();
};
#endif
```

As mentioned above, the class *UA_Test* has a single state, named *STATE_INITIAL*, and a single state transition function, named *Evt_Init_TIMER_TINV_EXP()*. The class *UA_Test* definition file, named *UA_Test.cpp*, has the following content (the parts that are not essential are omitted):

```
#include "UA_Test.h"
#include "../parser/smsgtypes.h"
#include "../parser/smsg.h"
#define SipMessageCoding 0x00
extern char* IPString(unsigned int addr, char* buf, int len);

UA_Test::UA_Test() : FiniteStateMachine(16, 2, 3) {}
UA_Test::~UA_Test() {}

void UA_Test::Initialize() {
 SetState(STATE_INITIAL);
 InitTimerBlock(TIMER_TINV,1,TIMER_TINV_EXPIRED);
 InitEventProc(STATE_INITIAL,TIMER_TINV_EXPIRED,
    (PROC_FUN_PTR)&UA_Test::Evt_Init_TIME_TINV_EXP);
 InitUnexpectedEventProc(STATE_INITIAL,
    (PROC_FUN_PTR)&UA_Test::Event_UNEXPECTED);
 StartTimer(TIMER_TINV);
}

void UA_Test::Evt_Ini_TIMER_TINV_EXP() {
 SendInviteToTAL();
}
```

```
void UA_Test::SendInviteToTAL(){
  char temp[10];
  char szHostName[255];
  hostent* HostData;
  uint8* recmsg;
  sip_t *mes;
  ...
  PrepareNewMessage(0x00,INVITE);
  SetMsgToAutomate(TAL_Disp_FSM);
  SetMsgToGroup(INVALID_08);
  SetMsgObjectNumberTo(0);
  AddParam((SIP_RAW_MESSAGE, SIPMsg.getLastMessage().length(),
    (uint8*) SIPMsg.getLastMessage().c_str());
  AddParamDWord((SIP_PARSED_MESSAGE, (unsigned long) mes);
  SendMessage(TAL_Disp_FSM_MBX);
}
...
```

The function *Initialize()* sets the FSM initial state, initializes the timer *TIMER_TINV* to a 1-s delay, sets the state transition functions, and starts the timer *TIMER_TINV*. When the timer expires, the state transition function *Evt_Init_TIMER_TINV_EXP()* is called. This function sends the *INVITE* message to the transaction layer dispatcher (*TAL_Disp*) by calling the function *SendInviteToTAL()*, which is very similar to the one given in the Example 1 (see Section 5.5.1). Further on, the *INVITE* message is routed toward the test stub class *TLI_Test*.

The class *TLI_Test* declaration file, named *TLI_Test.h*, has the following content (the parts that are not essential are omitted):

```
#ifndef _TLI_Test_FSM_
#define _TLI_Test_FSM_
#include "../constants.h"
#include "../kernel/fsm.h"
#include "../message/message.h"

class TLI_Test : public FiniteStateMachine {
  ...
  Message SIPMsg;
  sip_t *mes;
  // Message management functions
  void Send1XXToTAL();
  void Send2XXToTAL();
public:
  enum States {
    STATE_INITIAL,
    STATE_1XX_SENT
};
  void Evt_Init_INVITE_T();
  void Evt_1XXSent_TIMER_T2XX_EXP();
  void Event_UNEXPECTED();
// FiniteStateMachine abstract functions
  ...
public:
  TLI_Test();
  ~TLI_Test();
  void Initialize();
};
#endif
```

As mentioned above, the class *TLI_Test* has two states, named *STATE_INITIAL* and *STATE_1XX_SENT*, and two state transition functions, named *Evt_Init_INVITE_T()* and *Evt_1XXSent_TIMER_T2XX_EXP()*. The class *TLI_Test* definition file, named *TLI_Test.cpp*, has the following content (the parts that are not essential are omitted):

```
#include "TLI_Test.h"
#define SipMessageCoding 0x00
extern char* IPString(unsigned int addr, char* buf, int len);
TLI_Test::TLI_Test() : FiniteStateMachine(16, 2, 3) {}
TLI_Test::~TLI_Test() {}

void TLI_Test::Initialize() {
 char szHostName[255];
 hostent* HostData;
 SetState(STATE_INITIAL);
 InitTimerBlock(TIMER_T2XX,2,TIMER_T2XX_EXPIRED);
 InitEventProc(STATE_INITIAL,INVITE,
 (PROC_FUN_PTR)&TLI_Test::Evt_Init_INVITE_T);
 InitEventProc(STATE_1XX_SENT,TIMER_T2XX_EXPIRED,

(PROC_FUN_PTR)&TLI_Test::Evt_1XXSent_TIME_T2XX_EXP);
 InitUnexpectedEventProc(STATE_INITIAL,
 (PROC_FUN_PTR)&TLI_Test::Event_UNEXPECTED);
 // Problem specific part
 ...
}

void TLI_Test::Evt_Init_INVITE_T() {
 Send1XXToTAL();
 StartTimer(TIMER_T2XX);
 SetState(STATE_1XX_SENT);
}

void TLI_Test::Evt_1XXSent_TIMER_T2XX_EXP() {
 Send2XXToTAL();
}

void TLI_Test::Send1XXToTAL(){
 uint8* recmsg;
 recmsg = GetParam(SIP_RAW_MESSAGE);
 ...
 SIPMsg.makeResponse("100","Trying",responseBody,0);
 PrepareNewMessage(0x00,RESPONSE_1XX_T);
 SetMsgToAutomate(TAL_Disp_FSM);
 SetMsgToGroup(INVALID_08);
 SetMsgObjectNumberTo(0);
 AddParamDWord((SIP_PARSED_MESSAGE, (unsigned long) mes);
 SendMessage(TAL_Disp_FSM_MBX);
}

void TLI_Test::Send2XXToTAL(){
 SIPMsg.makeResponse("200","OK",responseBody,0);
 PrepareNewMessage(0x00,RESPONSE_2XX_T);
 SetMsgToAutomate(TAL_Disp_FSM);
 SetMsgToGroup(INVALID_08);
 SetMsgObjectNumberTo(0);
 AddParamDWord((SIP_PARSED_MESSAGE, (unsigned long) mes);
  SendMessage(TAL_Disp_FSM_MBX);
}
...
```

The function *Initialize()* sets the initial state, initializes the timer *TIMER_T2XX* to a 2-s delay, sets the state transition functions, and finishes with some problem-specific initializations. The state transition function *Evt_Init_INVITE_T()*, triggered by the reception of the message *INVITE*, sends the preliminary response *100 (Trying)* by calling the function *Send1XXToTAL()*, starts the timer *TIMER_T2XX*, and changes its state to *STATE_1XX_SENT*. The state transition function *Evt_1XXSent_TIMER_T2XX_EXP()*, triggered with the expiration of the timer *TIMER_T2XX*, sends the final response *200 (OK)* by calling the function *Send2XXToTAL()*.

The content of the main module, named *test_main.cpp*, is as follows (the parts that are not essential are omitted):

```
#include <conio.h>
#include "kernel/fsmsystem.h"
#include "kernel/logfile.h"
#include "NewSIP/TAL_Disp.h"
#include "Test/UA_Test.h"
#include "Test/TLI_Test.h"
#include "NewSIP/InvClientTE.h"
 FSMSystemWithTCP *pSys;
 LogFile *lf;
 TAL_Disp* pTALDisp;
 TLI_Test* pTLI;
 UA_Test* pUA;
 InviteClientTE* pInviteCltTE[NUMBER_OF_TES];
 DWORD thread_id;
 HANDLE thread_handle;
...
DWORD WINAPI SystemThread(void *data){
 FSMSystem *sysAutomate = (FSMSystem *)data;
 sysAutomate->Start();
 return 0;
}
int init(){
 pSys = new FSMSystemWithTCP(11,11);
 pTALDisp = new TAL_Disp();
 pTLI = new TLI_Test();
 pUA = new UA_Test();
 for (int i = 0; i < NUMBER_OF_TES; i++){
 pInviteCltTE[i]= new InviteClientTE();
 }
 uint8 buffClassNo = 4;
 uint32 buffsCount[4] = { 50, 50, 50, 50 };
 uint32 buffsLength[4] = { 1025, 1025, 1025, 1025};
 pSys->InitKernel(buffClassNo, buffsCount, buffsLength, 1);
   lf = new LogFile("log.log", "log.ini");
 LogAutomateNew::SetLogInterface(lf);
 pSys->Add(pTALDisp, TAL_Disp_FSM, 1, false);
 pSys->Add(pInviteCltTE[0], InviteClientTE_FSM, 10, true);
 pSys->Add(pTLI, TLI_Test_FSM, 1, false);
 pSys->Add(pUA, UA_Test_FSM, 1, false);
 for (i = 1; i < NUMBER_OF_TES; i++){
   pSys->Add(pInviteCltTE[i], InviteClientTE_FSM);
 }
 thread_handle = CreateThread(NULL, 0, SystemThread, pSys,
 THREAD_PRIORITY_ABOVE_NORMAL, &thread_id);
 return 1;
}
...
```

```
void main (void) {
  parser_init();
  init();
while(!kbhit());
  exit_app();
}
```

As a result of the execution of the main module, we get the log file with nine records that correspond to the messages that are exchanged between implementations under test (the transaction layer dispatcher and SIP invite client transaction), test driver (*UA_Test*), and test stub (*TLI_Test*). This file is very similar to the one given in Example 1 (see Section 5.5.1) but three times longer, and hence is not included here.

Test automation of integration tests based on log files is possible for simple collaborations like the one shown in this example, although it may be cumbersome. However, if we must deal with more complex collaborations that evolve concurrently, this approach is hardly applicable. Using log files in such situations would normally require human intervention for checking the results of the integration tests. Generally, we should try to use the style of unit testing based on the automatic checking of results (see Section 5.1), even for the integration of the parts of the system.

5.6 Further Reading

The reader can find more information related to this chapter in the references. The research by Berard et al. (2001) contains comprehensive coverage of the state-of-the-art model-checking techniques and tools. Newborn (2001) provides detailed information on the theorem prover THEO used in Section 5.3. Hoare (1985) wrote the famous book on CSP (nowadays, it is also available online for free as a PDF). The study of Sun (2009) is about PAT. Fischer et al. (2006) and Canepa et al. (2008) are the original papers on leader election algorithms, which appeared in the examples in Sections 5.3.2.3.3, 5.3.2.3.4, and 5.3.2.3.5. Popovic et al. (2001) provide a software maintenance case study in the area of communication protocol engineering. The research by Popovic and Velikic (2005) contains more information on the generic test case generator used in Section 5.5. Woit (1994; Chapter 3 and Section 3.1.1, in particular) provides more information on the reliability estimation model used in Section 5.5.

References

Berard, B., Bidoit, M., Finkel, A., Laroussinie, F., Petit, A., Petrucci, L., Schnoebelen, Ph., and McKenzie, P., *Systems and Software Verification: Model-Checking Techniques and Tools*, Springer-Verlag, Berlin, 2001.

Canepa, D. and Gradinariu Potop-Butucaru, M., Stabilizing token schemes for population protocols, arXiv:0806.3471v1 [cs.DC], 2008. Available online at https://arxiv.org/abs/0806.3471v1 (accessed June 28, 2017).

Fischer, M.J. and Jiang, H., "Self-stabilizing leader election in networks of finite-state anonymous agents," *Proceedings of the 10th Conference on Principles of Distributed Systems*, Vol. 4305 of LNCS, Springer, pp. 395–409, 2006.

Hoare, C.A.R., *Communicating Sequential Processes*, Prentice-Hall, 1985. Available online at http://usingcsp.com/cspbook.pdf (accessed June 28, 2017).

Newborn, M., *Automated Theorem Proving*, Springer-Verlag, New York, 2001.

Popovic, M., Atlagic., B., and Kovacevic, V., "Case study: A maintenance practice used with real-time telecommunication software," *Journal of Software Maintenance and Evolution: Research and Practice*, John Wiley & Sons, West Sussex, No. 13, pp. 97, 2001.

Popovic, M. and Velikic, I., "A generic model-based test case generator," *Proc. IEEE International Conference and Workshop on Engineering of Computer Based Systems*, Greenbelt, MD, April 4–7, 2005.

Sun, J., Liu, Y., Dong, J.S., Pang, J., "PAT: Towards Flexible Verification under Fairness," *Proceedings of the 21th International Conference on Computer Aided Verification (CAV'09)*, Vol. 5643 of LNCS, Springer, pp. 709–714, 2009.

Woit, D.M., "Operational profile specification, test case generation, and reliability estimation for modules," Ph.D. thesis, Queens University Kingstone, Ontario, Canada, February 1994.

6

FSM Library

The purpose of this chapter is to familiarize the reader with an example of a real-world library for making families of communication protocols. Although it is not perfect, it is in use and evolving. The main argument against it may be that there are too few C++ classes with too many function members. Alternately, this disadvantage is a tradeoff for a rather simple API, which is quite easy to learn and use.

6.1 Introduction

The FSM Library described in this book was created to be used as a working environment for the implementation of groups of communication protocols. The programmer has two basic classes at his or her disposal, namely, *FSMSystem* and *FiniteStateMachine*. The class *FSMSystem* models a platform for a group of communication processes (otherwise called finite state machines or automata). An instance of this class interconnects individual communication processes by handling all of the resources needed for the operation of individual finite state machines.

The class *FiniteStateMachine* models a generic communication process (i.e., communication protocol). Each individual communication protocol is represented by an instance of this class. The implementation of a particular communication protocol is narrowed down to writing state transition functions in C++. The transition function comprises procedures that process the message received in a given FSM state. This processing results in a transition to a new FSM state and the optional generation of corresponding outgoing messages. All state transition functions must be defined for all the finite state machines registered to a single FSM system (an instance of the class *FSMSystem*). Additionally, all the FSM system run-time elements must be initialized properly before they can be successfully started.

The relationship between the classes *FSMSystem* and *FiniteStateMachine* is symbiosis—one cannot operate without the other. The FSM system clearly represents an infrastructure, or an unused platform. In reality, an FSM system is always used so that at least a couple of finite state machines are registered to it, together representing a group of finite state machines. Because of this, and in order to achieve simplicity and brevity, we frequently use the term "FSM system" as a synonym for the group of automata, assuming that some individual automata are actually registered to it, and vice versa. Although an instance of the class *FiniteStateMachine* cannot operate on its own, we simply refer to it as a "finite state machine."

6.2   Basic FSM System Components

The FSMSystem Library is written in C++ using an object-oriented approach. The basic components are written as C++ classes that provide functionality of both individual finite state machines and a group of finite state machines. These classes are the following:

- *FiniteStateMachine*
- *FSMSystem*

A class can inherit the functionality of a single finite state machine by specializing the base class *FiniteStateMachine*. The programmer implements this class by writing the real functions for those declared as virtual, by adding new problem-specific functions (e.g., state transition functions), and by optionally overriding the inherited functions to redefine the functionality of the base class.

A class can inherit the functionality of a group of finite state machines by specializing the class *FSMSystem*. Normally, this class is simply instantiated as an oracle of a group of finite state machines.

6.2.1   Class *FSMSystem*

An instance of the class *FSMSystem* is an object representing a finite state machine system, i.e., a group of finite state machines (a group of automata). The protected attributes of this class represent the resources available for all the automata included in a group of automata. The basic task of this class is the initialization and management of FSMs, buffers (memory zones), messages, and timers. During a normal lifecycle of an instance of the class *FSMSystem*, its user typically performs the following steps or operations:

- Create FSM system
- Initialize FSM system
- Start FSM system
- Stop FSM system

In the list above, the idiom "FSM system" represents an instance of the class *FSMSystem*.

6.2.1.1 FSM System Initialization

The initialization of the FSM system consists of the following steps:

- Create the FSM system—see the constructor *FSMSystem()*.
- Create and initialize individual finite state machines—see the constructor *FiniteStateMachine()*.
- Add individual finite state machines to the FSM system.
- Initialize the FSM system.
- Start FSM system logging.

The constructor *FSMSystem()* requires two parameters:

- The number of types of finite state machines
- The number of mailboxes

Individual instances of the class *FiniteStateMachine* can be added to the FSM system by using one of two the possible functions:

```
void Add(ptrFiniteStateMachine object, // Automata instance address
 uint8 automataType, // Automata type
 uint32 numOfObjects, // Number of instances
 bool useFreeList = false); // List of free automata

void Add(ptrFiniteStateMachine object, // Automata instance address
 uint8 automataType); // Automata type
```

The first of the overloaded functions above is used to add the first finite state machine of each type. The other instances of the same type are added using the second function.

The initialization of the FSM system kernel is performed by calling the following function:

```
void InitKernel(uint8 buffClassNo, // Number of different types
 uint32 *buffersCount, // Number of buffers per type
 uint32 *buffersLength, // Buffer lengths per type
 uint8 numOfMbxs=0, // Number of mailboxes
 TimerResolutionEnum timerRes = Timer1s); // Timer  resolution in ms
```

The parameters of the function *InitKernel* specify the number of buffer types, the numbers of the instances of different types, their sizes, the number of mailboxes to be used by the automata in a group, and the basic timer resolution. The default number of mailboxes is 0. The default basic timer resolution is 1 sec (just as an example, it can be much smaller, e.g., 10 ms).

The FSM system logging functionality provides message content recording in a sequence resulting from the evolution of the FSM system. These messages are recorded automatically into a file created at the FSM system startup. The file *log.ini* is optional and is used to define textual titles (names) of the messages exchanged among the finite state machines included in the corresponding FSM system. If *log.ini* file is defined, the message binary codes are substituted by the corresponding message names, thus making the log files human readable. On Windows® machines, the *log.ini* file must be placed in the system folder (*c:\winnt* or *c:\windows*). The format of this file is as follows:

```
[AUTOMATA]
1=AUTOMATA1_FSM
2=AUTOMATA2_FSM
SequenceNumber=AUTOMATA_TYPE
[MESSAGES]
0=0xe000,MSG_1,0
1=0xe002,MSG_2,0
SequenceNumber=MSG_CODE,TEXT_TITLE,0
```

A typical example is as follows:

```
#define NO_BUFFERS 3
#define NO_AUTOMATA_1 5
#define NO_AUTOMATA_2 9
...

// Definition of buffers: three types, where number of buffers per type
// is 50, 30, and 20, and their lengths are 128, 256, and 512 bytes,
// respectively.
uint8 buffClassNo = NO_BUFFERS;
uint32 buffersCount[NO_BUFFERS] = {50,30,20};
uint32 buffersLength[NO_BUFFERS] = {128,256,512};

// Create FSM system that has two automata types and uses
// two mailboxes (one mailbox per each automata type)
FSMSystem *fsmSystem = new FSMSystem(2,2);

// Create individual automata
Automata1 *automata1 = new Automata1[NO_AUTOMATA_1];
Automata2 *automata2 = new Automata2[NO_AUTOMATA_2];

// Add individual automata to FSM system and implicitly initialize each
// automata instance by calling its function Initialize(). This call is
// made from the function Add.
fsmSystem->Add(&automata1[0],AUTOMATA1_FSM,NO_AUTOMATA_1,false);
for((i=1; i<NO_AUTOMATA_1; i++))
 fsmSystem->Add(&automata1[i],AUTOMATA1_FSM);

fsmSystem->Add(&automata2[0],AUTOMATA2_FSM,NO_AUTOMATA_2,true);
for((i=1; i<NO_AUTOMATA_2; i++))
 fsmSystem->Add((&automata2[i],AUTOMATA2_FSM);

// Initialize kernel
fsmSystem->InitKernel(buffClassNo,buffersCount,buffersLength,2);
// Create and set logging system (log file name, message definition file)
lf = new LogFile("log.log", "log.ini");
LogAutomataNew::SetLogInterface(lf);
...
```

The example above starts with the definition of the number of buffer types. In this example, three buffer types are defined (i.e., small, medium, and large buffers) by setting the symbolic constant *NO_BUFFERS* value to 3. Next, we define the number of instances of two automata types by setting the values of symbolic constants *NO_AUTOMATA_1* to 5 and *NO_AUTOMATA_2* to 9. This means that five instances of the first automata type and nine instances of the second automata type will exist in the group of automata we are going to create.

Next, the program paragraph defines the number of buffers, as well as their size, for each buffer type. Fifty small buffers of size 128 bytes, thirty medium buffers of size 256 bytes, and twenty large buffers of size 512 bytes would be used. The number of buffer types is stored in the variable *buffClassNo*. The number of buffers of each type and their lengths are stored in the arrays *buffersCount* and *buffersLength*.

We then create the FSM system by calling the constructor of the class *FSMSystem*. This constructor has two parameters: the number of automata types and the number of mailboxes to be used by the system for its own purposes. Next, we create two groups of automata of two different types. In this program, these groups are represented as arrays of instances of classes, namely, the classes *Automata1* and *Automata2*. In this example, we assume that these classes have already been defined by extending the base class *FiniteStateMachine*.

After creating two groups of automata of different types, all the automata are added to the already created FSM system. The first instance of each automata type is added by calling the overloaded function *Add* with the first type of signature, which specifies the instance address, the instance type, the total number of instances of this type, and the indicator specifying if a list of free automata of this type exists or not. The rest of the instances are added by calling the overloaded function *Add* with the second type of signature, specifying just the instance address and its type.

The first automata type in this example does not have a list of free automata, whereas the second type does have a list of free automata. This means that the instance of the second automata type can be viewed as a pool of resources of the same type. They may be dynamically allocated to be engaged in a certain communication scenario. When a programmer decides to use this opportunity, they must provide the function *NoFreeInstance*, which is called when the dynamic allocation request cannot be satisfied, because no more free automata instances of that type are found.

The FSM system is initialized by simply calling its function *InitKernel*. The parameters of this function specify the number of buffer types, the number of buffers of each type, their sizes, and the number of mailboxes to be used for FSMs. Normally, we use one mailbox per automata type. This is not a restriction imposed by the class *FSMSystem*, it is simply a convention. Other arrangements are also allowed; for example, we can create more mailboxes for messages of different priorities, or we can create additional mailboxes dedicated to communication between the given groups of automata types. Most generally, we can use mailboxes as queues of any kinds of messages. Because the last parameter of the function *InitKernel* is omitted, the timer resolution is set to its default value (1 sec, in this example).

At the end of this example, we create and set the logging system by call-ing its constructor *LogFile* and the function *SetLogInterface*, respectively. The parameters of the constructor specify the name of the log file (*log.log*) as well as the name of the file containing the textual names of the messages (*log.ini*). The parameter of the function *SetLogInterface* specifies the logging system interface, which generally is a file. In this example, the disk file is named *log.log* but it could be any file, including special files representing devices handled by the corresponding device drivers, such as */dev/lpt* or */dev/com1*.

6.2.1.2 FSM System Startup

The FSM system is started by calling its function *Start*. Most frequently, this function is called by the thread assigned to the FSM system. Here is an example:

```
DWORD WINAPI FsmSystemThreadFunc((void* param)){
  try {
    fsmSystem->Start();
  }
  catch(...){
    OutputDebugString('Exception — terminating FSM system\n');
    return 0;
  }
  OutputDebugString('FSM system terminated\n');
  return 0;
}
...

// Somewhere in the main function
DWORD fsmSystemThreadId;
CreateThread(NULL,0,FsmSystemThreadFunc,0,0,fsmSystemThreadId);
...
```

In the example above, we start the FSM system by calling its function *Start* from the thread function *FsmSystemThreadFunction*. We assume that thread has already been created and that its identification is stored in the variable *fsmSystemThreadId*.

6.2.2 Class *FiniteStateMachine*

All the automata added to the FSM system are implemented by extending the base class *FiniteStateMachine*. This class defines a set of virtual functions that must be defined by the programmer. These functions are as follows:

```
MessageInterface *GetMessageInterface(uint32 id);
void SetDefaultHeader(uint8 infoCoding);
uint8 GetMbxId();
uint8 GetAutomata();
void SetDefaultFSMData();
void NoFreeInstances();
void Initialize();
```

The following example illustrates the most frequently used definitions of *FiniteStateMachine* functions. A detailed description of all the functions is given in Section 6.8.

```
// This function returns the message interface for the given interface ID.
// It is assumed that standardMsgCoding is defined as:
// StandardMessage standardMsgCoding;
MessageInterface *Automata::GetMessageInterface(uint32 id){
 switch(id){
  case 0x00:
  return &standardMsgCoding;

  // Other definitions
  // case 0x01:
  // case 0x02:
 }
 throw TErrorObject(__LINE__,__FILE__,0x01010400);
}

// This function fills in the message header.
void Automata::SetDefaultHeader(uint8 infoCoding){
 SetMsgInfoCoding(infoCoding);
 SetMessageFromData();

}
// This function defines the mailbox number (ID) to be used as default
// by the automata of the type defined by this class.
uint8 Automata::GetMbxId(){
 return AUTOMATA_MB_ID;
}

// This function returns the number (ID) which identifies the automata
// type defined by this class.
uint8 Automata::GetAutomate(){
 return AUTOMATA_TYPE_ID;
}

// This function sets the values of the instance attributes.
void Automata::SetDefaultFSMData(){
 attribut1 = VALUE_1;
 attribut2 = VALUE_2;
}

// This function is called if there are no more free automata of this
// type. It may be used if the instances of this class have been added to
// the FSM system with the parameter useFreeList set to value true.
void Automata::NoFreeInstances(){
 // The activity if there are no free automata of this  type.
}

 // This function defines state transition functions and timers to be used
 // by the automata of this type. It is called by the function Add, which
 // is used to add an automata instance to the given FSM system.
 // It is assumed that state transition functions are declared and defined
 // elsewhere.
void Automata::Initialize(){
 // Here we place a series of initializations:
 // InitEventProc(uint8 state, uint16 event,  PROC_FUN_PTR fun);
 // InitUnexpectedEventProc(uint8 state, PROC_FUN_PTR  fun);
 // InitTimerBlock(uint16 timerId, uint32 timerCount,  uint16 signalId);
```

```
InitEventProc(IDLE, MSG_SEND,  (PROC_FUN_PTR)  &Automata::Idle_MsgSend);
InitEventProc(IDLE, MSG_RCV,   (PROC_FUN_PTR)  &Automata::Idle_MsgReceive);

InitEventProc(SEND, MSG_NEW,   (PROC_FUN_PTR)  &Automata::Send_MsgNew);
InitEventProc(SEND, MSG_END,   (PROC_FUN_PTR)  &Automata::Send_MsgEnd);
InitEventProc(IDLE, T200_CODE, (PROC_FUN_PTR)  &Automata::T200Expired);

InitUnexpectedEventProc(IDLE,  (PROC_FUN_PTR)  &Automata::Idle_Unexpected);
InitUnexpectedEventProc(SEND,  (PROC_FUN_PTR)  &Automata::Send_Unexpected);

InitTimerBlock(T200,T200_VALUE,T200_CODE);
}
```

In the example above, we would like to create the class *Automata* that models one type of finite state machines (automata). The definition of the class comprises the definitions of its function members. The function member *GetMessageInterface* returns the object that embodies the coding of messages to be used by the instances of the class *Automata*. In this example, it is an instance of the class *StandardMessage*.

The member function *SetDefaultHeader* is used to automatically fill in the message header defaults. Normally, these are the data about the automata instance that has created the message to send to some other automata instance. In this example, it uses the function *SetMsgInfoCoding* to specify the type of coding to be applied. It also uses the function *SetMessageFromData* to specify the type of originating automata instance, the identification of the group to which the automata instance belongs, and the identification of the originating automata instance.

The member function *GetMbxId* returns the identification of the mailbox used by the automata instance of this type. In this example, it is the value of the symbolic constant *AUTOMATA_MBX_ID*. The member function *GetAutomata* returns the identification of the automata type. It is the value of the symbolic constant *AUTOMATA_TYPE_ID*. The member function *SetDefaultFSMData* is used by the automata instance to set its specific data before it commences its normal operation. In this example, *attribute1* is set to the value *VALUE_1* and *attribute2* is set to the value *VALUE_2*.

The member function *NoFreeInstances* can be used to specify the action to be performed, if no more free automata instances of this type are found, e.g., to make a small system restart, allocate some additional automata instances, and so on. This mechanism is available to the programmer if the instances of automata have been added (function *Add*) to the FSM system with the parameter *useFreeList*, set to the value *true*.

The member function *Initialize* is used to define automata state transition functions and timers (referred to as timer blocks throughout the FSM Library documentation) to be used by the automata. The FSM Library distinguishes two types of events, expected and unexpected, and allows the programmer to specify the corresponding event handlers, which are just specialized C++ functions. These handlers are defined by calling the registration functions, namely, the function *InitEventProc* for expected events and the function *InitUnexpectedEventProc* for unexpected events. The parameters of both

TABLE 6.1

Example of a State Transition Table

	MSG_RCV	MSG_SEND	T200_CODE	MSG_NEW	MSG_END	?
Idle	Idle_MsgReceive	Idle_MsgSend	T200Expired			Idle_Unexpected
Send				Send_MsgNew	Send_MsgEnd	Send_Unexpected

of these functions specify the state code, the event (message) code, and the pointer to the event handler.

In this example, we have defined seven automata state transition functions altogether, five of them triggered by expected events and two triggered by unexpected events. The part of the automata shown in the example has two states, *IDLE* and *SEND*. The expected events in the state *IDLE* are *MSG_SEND, MSG_RCV*, and *T200_CODE*. The corresponding event handlers are *Idle_MsgSend*, *Idle_MsgReceive*, and *T200Expired*, respectively. Two legible events exist in the state *SEND*, *MSG_NEW* and *MSG_END*. The corresponding handlers are *Send_MsgNew* and *Send_MsgEnd*. The unexpected event handler for the state *IDLE* is *Idle_Unexpected* whereas for the state *SEND* it is *Send_Unexpected*. The corresponding state transition table is shown in Table 6.1.

The timers are initialized by calling the function *InitTimerBlock*. The parameters of this function specify the unique timer identification, its duration (as the number of basic timer resolution units), and the code of the message sent when the timer expires. In the example above, these are the symbolic constants *T200, T200_VALUE*, and *T200_CODE*.

To sum, automata states and attributes are defined in accordance with the problem at hand. The state transition function, referred to as the event handler, is called upon the reception of a given message in a given state, as defined by the function *Initialize*. Each event handler is defined as a class member function responsible for handling a given event.

The timers to be used by the automata are also defined by the function *Initialize*. This is done by calling the function *InitTimerBlock*, which, in turn, creates the internal kernel timer block (essentially a program object) and fills in its identification, duration, and corresponding timer message code.

6.3 Time Management

In Section 6.2, automata timers are initialized during the FSM system startup by the function *Initialize*. The automata type that uses timers in its regular operation manages them through the corresponding FSM Library API functions, which maintain the internal kernel object behind the scenes. The API functions are the following:

```
void InitTimerBlock(uint16 tmrId,uint32 count,uint16 signalId);
void StartTimer(uint16 tmrId);
void StopTimer(uint16 tmrId);
void RestartTimer(uint16 tmrId)
bool IsTimerRunning(uint16 tmrId);
```

The function *InitTimerBlock* is used to define (initialize) the timer. Its parameters specify the unique timer identification, its duration as a multiple of the basic timer resolution unit, and the code of the message sent to the automata mailbox when the timer expires. This is explained in the previous section. Notice that each timer has the unique identification *tmrId* used as a parameter of all the API functions to identify the timer.

Each API function represents a primitive timer operation. The function *StartTimer* is used to start the timer, the function *StopTimer* stops the timer, the function *RestartTimer* restarts the timer, and the function *IsTimerRunning* is used to check if the timer is running or not.

The following example illustrates the usage of these primitives:

```
if(!IsTimerRunning(T200)){
 StartTimer(T200);
}
else
 StopTimer(T200);
...
```

A normal timer life cycle has the following phases:

- Define, i.e., initialize, the timer.
- Use the timer by alternative application of the following primitives:
  - *Start* (applicable if the timer is not running, meaning it was either newly defined or previously stopped)
  - *Stop* (applicable if the timer is running)
  - *Restart* (logically equivalent to *Stop* plus *Start*)
  - *IsTimerRunning* (returns true if it does; otherwise, it returns false)

6.4 Memory Management

Because the main application of the FSM Library is in real-time systems, efficient memory allocation must be provided. The FSM Library does not rely on a hosting operating system because some of the operating systems suffer from a memory fragmentation problem. Furthermore, in some applications on bare machines, the operating system may not even be available. Because of that, memory management is one of the main functions of the FSM Library.

The working memory is partitioned into certain zones referred to as **buffers**. The programmer defines the number of different buffer types, the number of buffers of each type, and their sizes. The programmer specifies this data as parameters of the function *InitKernel* (see Section 6.8.4) and the FSM Library kernel, in turn, creates them as its own internal objects.

The buffers are most frequently used indirectly through message management (message create, send, receive, and similar operations) and timer operations (timer definition and usage operations). Besides this indirect buffer usage, the buffers can be managed directly, if needed, through the following API functions:

```
uint8  *GetBuffer(uint32 length);
void   RetBuffer(uint8 *buff);
bool   IsBufferSmall(uint8 *buff,uint32 length);
uint32 GetBufferLength(uint8 *buff);
```

The programmer requests a buffer by calling the function *GetBuffer*. The parameter of this function is the minimal size of the desired buffer. All the buffers provided by the kernel must be returned by calling the function *RetBuffer*. Untidy memory management can cause buffer loss, commonly referred to as **memory leak**, which may cause irregular kernel operation and a system crash.

Besides memory allocation (*malloc*) and *free* primitives, two additional primitives provide the information about the buffer already allocated to the finite state machine. The function *IsBufferSmall* checks if the buffer size is smaller than the value of its parameter. If yes, it returns *true*, otherwise, it returns *false*. Another function, named *GetBufferLength*, returns the buffer size in octets (bytes).

The following example illustrates the usage of the buffer management primitives:

```
// We define two buffer types, small and large.
// There are ten small buffers and fifteen large buffers.
// The small buffer size is 128 bytes. The large buffer size is
// 256 bytes.
uint8 buffClassNo = 2;
uint32 buffersCount[2] = {10,15};
uint32 buffersLength[2] = {128,256};
...

// Kernel initialization (noMBX is irrelevant in this example)
fsmSystem->InitKernel(buffClassNo,buffersCount,buffersLength,noMBX);
...

uint32 bufferLength;
uint8 *pointer = GetBuffer(100);
if((IsBufferSmall(pointer,129)){
 RetBuffer(pointer);
 pointer = GetBuffer(129);
}
if((pointer != NULL))
 bufferLength = GetBufferLength(pointer);
...
```

In the example above, we first define two buffer types—small and large—by calling the function *InitKernel*. Its fourth parameter (*noMBX*, the number of the mailboxes) is not relevant for this example. The rest of the program illustrates the usage of the FSM Library's buffer management functions. First, the program asks for a buffer not smaller than 100 bytes, then it checks if this buffer is smaller than 129 bytes. If yes, it returns the allocated buffer and requests a new one not smaller than 129 bytes (in this example, it will get one large buffer of size 256 bytes). At the end, the program checks if the pointer is defined, which also means that it points to a certain buffer. If it is defined, the program asks for its size by calling the function *GetBufferLength*.

6.5 Message Management

The main communication among individual automata included in the FSM system is achieved through the messages exchanged through the mailboxes typically assigned to individual automata. The message sent from the originating automata instance towards the destination automata instance is placed temporarily in the mailbox assigned to the destination automata instance. There, it waits to be taken over and subsequently processed by the destination automata instance (process).

As already mentioned, a **mailbox** is a message queue that can contain messages for any automata type, thus it does not need to be assigned to some particular automata type. In contrast to a typical paradigm, it can be used as a general message queue shared by more destination automata. Essentially, in such a paradigm, the source automata instance can put the message in any mailbox hosted by the FSM system, and it will eventually be delivered to its proper destination.

This message routing and delivery is performed automatically by the FSM system and is hidden from the automata, which are just service users. The FSM system has an abstraction of the mailbox from which it takes messages, one at a time (mailbox abstraction provides buffering functionality by employing the FIFO memory type). Upon the reception of each individual message, the FSM system consults the message header to determine the destination automata instance and passes the message to it. The destination automata instance looks up the message code and, based on the current automata state, calls the appropriate automata state transition function.

Message reception is completely transparent for the programmer writing the program code for the finite state machine. The above mechanism is absolutely hidden from them. The programmer must simply accept that the message reception and its classification are done automatically by the system. They just write the message processing functions that are called automatically by the system upon the reception of the corresponding message.

The API functions can be partitioned into two groups:

- The functions that work with the received message.
- The functions that work with the new message that must be prepared and sent.

The functions in the first group are used to provide the information about the originating automata instance. The source of this information is the message header and the values of the message parameters. The functions in the second group provide primitives needed to make and send a message:

- Buffer allocation (indirect call to *GetBuffer* primitive)
- Filling the message header with the data about the originating automata instance
- Adding the message parameters and setting them to the given values
- Sending the message to the mailbox assigned to the destination automata instance

The messages may be sent only from a finite state machine or a FSM system. Note that during normal system operation, a FSM system does not send any messages. In this context, a finite state machine is an instance of the class *FiniteStateMachine*, or a class derived from it, and an FSM system is an instance of the class *FSMSystem*.

Example 1:

```
// Get parameter of type PARAM_1 from the received message.
// The size of PARAM_1 is WORD.
WORD word;
GetParamWord(PARAM_1,word);

// Get parameter of arbitrary size. Maximum size for StandardMessage is
// 256 bytes. If that is not sufficient, a programmer must derive a new
// class and redefine its functions.
uint8    *pointer;
uint8    text[300];
uint8    msgLength;

pointer = GetParam(TEXT);
if(pointer != NULL){
// StandardMessage format: bytes 1 and 2 contain parameter name,
// byte 3 contains parameter length in bytes,
// byte 4 and further contain the parameter itself.
memcpy(text,pointer+3,*((pointer+2)));

// Make a string by placing null at the end of  character array.
memset(text+(*((pointer+2))),0x00,1);
}
```

The example above shows how the programmer can get a parameter from the *current* message. A current message is the last message received by the automata instance, i.e., it is the last message taken from the mailbox and assigned to the automata instance for processing. The parameter size is *WORD* (2 bytes). First, the programmer declares the variable *word* in which he wants to store the parameter value.

The message can contain many parameters, and therefore the programmer must specify the unique identifier of the parameter they want to get. In this example, the identifier is the value of the symbolic constant *PARAM_1*. Finally, a copy of the desired parameter is provided by calling the API function *GetParam*. The first parameter of this function is the parameter identifier (*PARAM_1*) and the second is the variable (*word*) in which the desired parameter is to be copied.

The second part of the example above demonstrates how the programmer may handle textual parameters of arbitrary size. The *StandardMessage* format prescribes that the first 2 bytes of such a parameter are reserved for the parameter name, the next byte is used for the parameter length (in bytes), and the rest of the bytes in the parameter represent its value. The example shows how a copy of such a parameter can be provided and how a null terminated string can be constructed by adding the NULL character at its end.

Example 2:

```
...
// PrepareNewMessage parameters: buffer size and message type.
PrepareNewMessage(0xAA,MSG_NAME);

// Fill in the message header:
// destination automata type, its ID, and optionally its group ID.
SetMsgToAutomata(AUTOMATA_TYPE);
SetMsgObjectNumberTo(automataId);
SetMsgToGroup(INVALID_08);

// Add parameters: see also other AddParam functions.
AddParamByte(PARAM_1,byte);
AddParamWord(PARAM_2,word);
AddParam(PARAM_3,parameterLength,parameterPointer);

// Send message to the specified mailbox.
SendMessage(AUTOMATA_MBX_ID);
```

The example above shows a common way to construct and send a message. The first step is to call the function *PrepareNewMessage*. The parameters of this function specify the expected buffer size (*0xAA* in this example) and the message name, which also specifies the message type (*MSG_NAME*).

Next, we fill in the message header by calling the following functions:

- *SetMsgToAutomata*: set the destination automata instance type (*AUTOMATA_TYPE*)
- *SetMsgObjectNumberTo*: set the destination automata instance identification (*automataId*)
- *SetMsgToGroup*: set the automata instance group identification (*INVALID_08*)

We then add three message parameters by calling members of the *AddParam* family of functions. The first function shown in the example is *AddParamByte*. Its parameters specify the unique parameter identifier (*PARAM_1*) and the variable containing the value of the parameter to be copied to the corresponding field of the message (*byte*). The second function is *AddParamWord*. Similarly, its parameters specify the parameter identification (*PARAM_2*) and the variable holding its value (*word*). The last function is *AddParam*. The parameters of this function specify the parameter identification (*PARAM_3*), its length (*parameterLength*), and a pointer to it (*parameterPointer*).

At the end of the example above, we send the message by calling the function *SendMessage*. The parameter of this function specifies the destination mailbox identification (*AUTOMATA_MBX_ID*).

Example 3:

```
// Send a message from the FSM system.
uint8 *msg = GetBuffer(messageInfoLength+MSG_HEADER_LENGTH);

// infoBuffer must be properly formatted.
memcpy(msg+MSG_HEADER_LENGTH,infoBuffer,infoBufferLength);

SetMsgFromAutomata(AUTOMATA_TYPE_FROM_ID,msg);
SetMsgFromGroup(INVALID_08,msg);
SetMsgObjectNumberFrom(automataFromId,msg);

SetMsgToAutomata(AUTOMATA_TYPE_TO_ID,msg);
SetMsgToGroup(INVALID_08,msg);
SetMsgObjectNumberTo(automataToId,msg);

SetMsgInfoCoding(0,msg); // 0 = StandardMessage
SetMsgCode(MSG_FROM_SYSTEM_AUTOMATA,msg);
SetMsgInfoLength(infoBufferLength,msg);
SendMessage(AUTOMATA_TO_MBX_ID,msg);
...
```

The example above shows how a message can be created and sent within the FSM system. This process is done through the following steps:

- Allocate a buffer by calling the function *GetBuffer*.
- Copy the information payload.
- Fill in the data about the originating automata instance by calling the function *SetMsgFromAutomata* fill in the originating automata instance type identification (AUTOMATA_TYPE_FROM_ID); by calling the function *SetMsgFromGroup*, fill in the originating automata instance group identification (*INVALID_08*); and by calling the function *SetMsgObjectNumberFrom*, fill in the automata instance identification (*automataFromId*).
- Fill in the data about the destination automata instance. The function *SetMsgToAutomata* sets the destination automata instance type identification (*AUTOMATA_TYPE_TO_ID*), the function

SetMsgToGroup sets the destination automata instance group identification (*INVALID_08*), and the function *SetMsgObjectNumberTo* sets the destination automata instance identification (*automataToId*).

- Finalize the message. The function *SetMsgInfoCoding* sets the type of coding (*StandardMessage*), the function *SetMsgCode* sets the message code (*MSG_FROM_SYSTEM_AUTOMATA*), and the function *SetMsgInfoLength* sets the payload length (*infoBufferLength*).

- Send message by calling the function *SendMessage* with the second type of the signature. The parameters of this function specify the destination mailbox identification (*AUTOMATA_TO_MBX_ID*) and the pointer to the message to be sent (*msg*).

6.6   TCP/IP Support

One of the primary design goals of creating the FSM Library was to support the design of scalable applications based on distributed processing. The FSM Library enables both single-processor and multiprocessor applications. In the former case, all groups of automata execute in a single processor. They share processor resources, such as its processing unit, operating memory, flash, and so on. The automata communicate over the mailboxes placed in the common operating memory.

In the latter case, various groups of automata are deployed on more processors, which can be logically viewed as a multiprocessor system. The groups of automata execute on different processors in parallel and use the mailboxes physically located in separate operating memories. The FSM Library transparently uses the network infrastructure to pass messages among the communicating automata. Most frequently, the communication infrastructure is the TCP/IP technology.

In both cases, the communicating automata are unaware of the real physical infrastructure because the physical details are hidden from them. This is accomplished by providing a unique API. An individual automata instance manages just its timers, buffers, and messages (new and current, i.e., last received). The rest is handled by the FSM Library kernel behind the scenes. This means that the FSM Library inherently provides implicit support for TCP/IP. For example, if an automata instance wishes to send a message to some other automata instance physically located on a different machine, it just prepares the message and calls the API function *SendMessage*. The class *FSMSystem* takes care of transporting the message over the TCP/IP network and placing it in the local mailbox assigned to the destination automata.

Since individual automata based on the FSM Library only need to communicate among themselves, implicit TCP/IP support is sufficient. The need to communicate with other program components that are not based on the

FSM Library, and that use TCP/IP sockets, directly leads to the requirement for explicit TCP/IP support. To fulfill that requirement, the FSM Library also provides explicit (in addition to implicit) TCP/IP support in a form of traditional TCP/IP socket abstraction. Of course, the automata instance that uses these additional API features must be aware and capable of handling details of TCP/IP communication (IP addresses and port numbers).

Explicit TCP/IP support is provided by two additional classes, namely, *FSMSystemWithTCP* and *NetFSM*. These two classes enable the FSM Library–based automata to directly communicate over the TCP/IP protocol stack with other FSM Library–based automata, or with other TCP/IP program components, e.g., a Web server or SIP client. As their names suggest, the class *FSMSystemWithTCP* is used instead of the class *FSMSystem*, and the class *NetFSM* is a logical counterpart of the class *FiniteStateMachine*.

6.6.1 Class *FSMSystemWithTCP*

The class *FSMSystemWithTCP* is derived from the class *FSMSystem* by extending it with support for communication over the TCP/IP family of protocols. It inherits the basic functionality of the base class, which has been described previously (see Section 6.2.1 describing the class *FSMSystem*).

In contrast to single-processor applications, distributed applications comprise parts (i.e., groups of automata) that are started independently. Because of this, two groups of automata executing on different processors must establish a TCP/IP connection at their startup. The connection establishing procedure is symmetric: This means that either side of the party—or both—must start their local TCP servers by calling the function *InitTCPServer*. The opposite side establishes the connection by calling the function *establishConnection*.

Example:

```
// In processor 1 (server)
//
// Initialize kernel.
fsmSystem1->InitKernel(buffClassNo,buffersCount,buffersLength,2);

// Initialize TCP/IP server on port number 5000.
// NetFSM_Automata1 is derived from NetFSM.
fsmSystem1->InitTCPServer(5000,NetFSM_Automata1);

// In processor 2 (client)
//
// Set server TCP/IP parameters (port, IP address).
// Establish the connection.
fsmSystem2.setPort(5000);
fsmSystem2.setIP("192.168.77.77");
fsmSystem2.establishConnection();
...
```

This example shows the code excerpts for the TCP/IP server and client machines, named processor 1 and processor 2. At startup, the server initializes the FSM Library kernel by calling the function *InitKernel* (its parameters are the number of buffer types, their count, length, and the number of the mailboxes to be used). Next, it calls the function *InitTCPServer* to start the TCP/IP server. We assumed in this example that the class *NetFSM_Automata1* is derived from the class *NetFSM*.

Alternately, the client sets the TCP port number (5000) by calling the function *setPort* and the IP address of the TCP server (192.168.77.77) by calling the function *setIP*, and establishes the connection with the server by calling the function *establishConnection*.

6.6.2 Class *NetFSM*

The class *NetFSM* is derived from the base class *FiniteStateMachine* by extending its basic functionality with support for the communication over the TCP/IP infrastructure. The inherited basic functionality has been described previously (see Section 6.2.2 describing the class *FiniteStateMachine*). The basic functionality is extended with the abstraction enabling TCP/IP communication by adding three new function members. The new functions are the following:

```
virtual void convertFSMToNetMessage()=0;
virtual uint16 convertNetToFSMMessage()=0;
virtual uint8 getProtocolInfoCoding()=0;
```

These functions are used to convert the internal message format (abbreviated as *FSM*) into an external, or network message format (abbreviated as *Net*), and vice versa. Normally, automata executing in the same processor exchange internal messages coded in internal message format. However, this message format is not suitable for transmission over the network. Most commonly, the message must be serialized, i.e., transformed from the data object and structure form into an external message in accordance with a given external message format. This is a series of bits, sometimes grouped in octets or words, that are transmitted over the communication line.

The functions listed above are virtual functions and therefore the programmer must define them while they write a class that is derived from the class *NetFSM*. The message format conversion functions naturally read a message from some input buffer, convert it into a requested format, and write the output to an output buffer.

The function *convertFSMToNetMessage* is not intended to be used directly by the communicating automata, but rather to be called internally by the FSM Library kernel to convert an internal message into the external one before it can be sent over the network. Therefore, the input of this function is the internal message, and its output is the corresponding output message. The parameters of this function specify the pointer to the internal message *fsmMessageS*, its length *fsmMessageLength*, the pointer to the output, the external message *protocolMessageS*, and its length *sendMsgLength*. The programmer must specify the mapping algorithm by writing this function.

Symmetrically, the function *convertNetToFSM* is intended to be used by the FSM library kernel to convert an external message received over the network into an internal message representation, which must be delivered to the local mailbox and processed further by the corresponding local automata. The input of this function is the external message and the output is the internal message. The parameters of this function specify the pointer to the external message *protocolMessageR*, its length *receivedMessageLength*, the pointer to the output, internal message *fsmMessageR*, and its length *fsmMessageRLength*.

The function *getProtocolInfoCoding* returns the code of the type of external information coding. An instance of the class *NetFSM*, referred to as *net automata*, initiates the transmission of the message across the TCP/IP network by calling the function *sentToTCP*. This function may throw an exception in the case of an error, e.g., when net automata wants to send a message after the TCP connection has been closed.

Example:

```
// PrepareNewMessage parameters: buffer size and message type
PrepareNewMessage(0xAA,MESSAGE_NAME);

// Fill in message header:
// destination automata type, its ID, and its group ID (if relevant)
SetMsgToAutomata(AUTOMATA_TYPE);
SetMsgObjectNumberTo(automataId);
SetMsgToGroup(INVALID_08);
// Add parameters.
AddParamByte(PARAM_1,byte);
AddParamWord(PARAM_2,word);
AddParam(PARAM_3,parameterLength,parameterPointer);

// Send message to local mailbox:
// SendMessage(AUTOMATA_MBX_ID);
// or send it over TCP/IP network:
sendToTCP();
```

The example above demonstrates how automata can prepare a message and send it over a TCP/IP network. The message is prepared like any other message. The function *PrepareNewMessage* is used to allocate a buffer for the message and to specify a message name. A series of already described functions is then used to fill in the message header and add the message parameters (see the second example in Section 6.5 describing message management). At the end, instead of sending the message to the local mailbox by calling the function *SendMessage*, the message is sent over the TCP/IP network by calling the function *sendToTCP*.

A net finite state machine receives the messages equally as simple automata (instances of the class *FiniteStateMachine*) do, just by reading its local mailbox.

6.7 Global Constants, Types, and Functions

The file *kernelConsts.h* defines the global constants, types, and functions used by the FSM Library kernel. The constants and their values are as follows:

```
MSG_FROM_AUTOMATA = 0; // Source automata ID (BYTE)
MSG_FROM_GROUP = 1; // Source automata group ID (BYTE)
MSG_TO_AUTOMATA = 2; // Destination automata ID (BYTE)
MSG_TO_GROUP = 3; // Destination automata group ID (BYTE)
MSG_CODE = 4; // Message code(WORD)
MSG_OBJECT_ID_FROM = 6; // Source automata instance ID (DWORD)
MSG_OBJECT_ID_TO = 10; // Destination automata ID (DWORD)
CALL_ID = 14; // Call (process) ID
MSG_INFO_CODING = 18; // Info coding type, 0 = StandardMessage
MSG_LENGTH = 19; // Message payload length
MSG_INFO = 21; // Message payload offset
MSG_HEADER_END = MSG_INFO; // End of message header

INVALID_08 = 0xff; // Mask for 8 bits
INVALID_16 = 0xfff; // Mask for 16 bits
INVALID_32 = 0xffffffff; // Mask for 32 bits
```

The global data types are as follows:

```
int8, uint8 // BYTE
int16, uint16 // WORD
int32, uint32 // DWORD
```

The utility functions provided for the load-store manipulation with various data types are as follows:

```
void SetUint16(uint8 *addr,uint16 value);
void SetUint32(uint8 *addr,uint32 value);
uint16 GetUint16(uint8 *addr);
uint32 GetUint32(uint8 *addr);
```

The utility functions are provided to avoid cast operators in C/C++ programs because some microcontrollers do not allow word or double-word memory access to odd memory addresses.

6.8 API Functions

The FSM Library API functions are grouped into the following eight groups:

- *FSMSystem* constructor (Table 6.2)
- *FSMSystem* member functions (Table 6.3)
- *FSMSystemWithTCP* constructor (Table 6.4)

TABLE 6.2

FSMSystem Constructor Summary

FSMSystem(uint8 numOfAutomata, uint8 numberOfMbx)
The constructor initializes the object that represents the FSM system, along with the data structures needed for its proper operation.

TABLE 6.3

FSMSystem Member Functions Summary

Type	Member Function
Void	Add (ptrFiniteStateMachine object, uint8 automataType, uint32 numOfObjects, bool useFreeList=false) This function adds the first instance of each automata type to the FSM system.
Void	Add(ptrFiniteStateMachine object, uint8 automataType) This function adds all the automata instances of the given type to the FSM system, except for the first instance.
Void	InitKernel(uint8 buffClassNo, uint32 *buffersCount, uint32 *buffersLength, uint8 numOfMbxs=0, TimerResolutionEnum timerRes=Timer1s) This function initializes the elements of the kernel responsible for time, buffer, and message management.
Void	Remove(uint8 automataType) This function removes all the instances of the given automata type from the FSM system.
ptrFiniteStateMachine	Remove(uint8 automataType, uint32 object) This function removes the given instance of the given automata type.
Virtual void	Start() This function starts the FSM system.
Void	StopSystem() This function stops the FSM system.

TABLE 6.4

FSMSystemWithTCP Constructor Summary

FSMSystemWithTCP(uint8 numOfAutomata, uint8 numberOfMbx)
The constructor initializes the object that represents the FSM system supporting communication over the TCP/IP network, along with the data structures needed for its proper operation.

- *FSMSystemWithTCP* member functions (Table 6.5)
- *FiniteStateMachine* constructor (Table 6.6)
- *FiniteStateMachine* member functions (Table 6.7)
- *NetFSM* constructor (Table 6.8)
- *NetFSM* member functions (Table 6.9)

The following sections contain a detailed description of *FSMSystem* library API functions.

TABLE 6.5

FSMSystemWithTCP Member Functions Summary

Type	Member Function
int	InitTCPServer(uint16 port, uint8 automataType, char *ipAddress=0, unsigned char *parm=0, int length=0) This function initializes the TCP server. Once initialized, the server waits for a request to establish the TCP connection with a remote client.

TABLE 6.6

FiniteStateMachine Constructor Summary

FiniteStateMachine(uint16 numOfTimers=DEFAULT_TIMER_NO, uint16 numOfState=DEFAULT_STATE_NO, uint16 maxNumOfProceduresPerState=DEFAULT_PROCEDURE_NO_PE_STATE, bool getMemory=true)
This constructor initializes the object that represents the instance of a given automata type, along with the data structures needed for its proper operation.

TABLE 6.7

FiniteStateMachine Member Functions Summary

Type	Member Function
uint8*	AddParam(uint16 paramCode, uint32 paramLength, uint8 *param) This function is used to add a given parameter of the given length to the new message.
uint8*	AddParamByte(uint16 paramCode, BYTE param) This function is used to add the given parameter of length 1 byte to the new message.
uint8*	AddParamDWord(uint16 paramCode, DWORD param) This function is used to add the given parameter of length 4 bytes to the new message.
uint8*	AddParamWord(uint16 paramCode, WORD param) This function is used to add the given parameter of length 2 bytes to the new message.
virtual void	CheckBufferSize(uint32 paramLength) This function provides a new message buffer with a size sufficient enough to accept a parameter of the given length.

(Continued)

TABLE 6.7 (CONTINUED)

FiniteStateMachine Member Functions Summary

Type	Member Function
virtual void	ClearMessage() This function returns the buffer allocated for the current message to the kernel and assigns value *NULL* to the internal pointer to the current message. The current message is the last message received by the automata instance.
virtual void	CopyMessage() This function makes a copy of the current message and assigns that copy to the new message.
virtual void	CopyMessage(uint8 *msg) This function makes a copy of the given message and assigns that copy to the new message.
virtual void	CopyMessageInfo(uint8 infoCoding, uint16 lengthCorrection=0) This function copies the part of the message containing useful information, referred to as a payload (a message without its header), from the current to the new message.
virtual void	Discard(uint8* buff) This function deletes the message placed in the given buffer and returns the buffer to the kernel.
void	DoNothing() This function performs no operation. It is called when the automata receives an unexpected message, unless a new function is provided to handle unexpected messages.
void	Free FSM() This function reports to the FSM system that the automata instance has finished its current assignment and is free for further assignments.
virtual uint8	GetAutomata()=0 This function returns the identification of the automata type for this automata instance.
uint8	GetBitParamByteBasic(uint32 offset, uint32 mask=MASK_32_BIT) This function returns the value of the current message parameter of length 1 byte masked with the given mask.
uint16	GetBitParamWordBasic(uint32 offset, uint32 mask=MASK_32_BIT) This function returns the value of the current message parameter of length 2 bytes masked with the given mask.
uint32	GetBitParamDWordBasic(uint32 offset, uint32 mask=MASK_32_BIT) This function returns the value of the current message parameter of length 4 bytes masked with the given mask.
virtual uint8*	GetBuffer(uint32 length) This function returns the buffer whose size is not less than the size given by the value of its parameter.
uint32	GetBufferLength(uint8 *buff) This function returns the size of the given buffer in bytes.

(Continued)

TABLE 6.7 (CONTINUED)

FiniteStateMachine Member Functions Summary

Type	Member Function
`virtual inline uint32`	`GetCallId()` This function returns the identification of the communication process in which this instance is currently involved, e.g., the call ID.
`uint32`	`GetCount(uint8 mbx)` This function returns the current number of messages in the given mailbox.
`virtual uint8`	`GetGroup()` This function returns the identification of the group of automata to which this instance belongs.
`virtual uint8`	`GetInitialState()` This function returns the identification of the initial state of this automata type.
`virtual inline uint8`	`GetLeftMbx()` This function returns the identification of the mailbox assigned to the automata instance that is logically to the left of this automata instance.
`virtual inline uint8`	`GetLeftAutomata()` This function returns the identification of the automata type that is logically to the left of this automata instance.
`virtual inline uint8`	`GetLeftGroup()` This function returns the identification of the group of automata that is logically to the left of this automata instance.
`virtual inline uint32`	`GetLeftObjectId()` This function returns the identification of the automata instance that is logically to the left of this automata instance.
`virtual uint8`	`GetMbxId()` This function returns the identification of the mailbox assigned to this automata instance.
`virtual MessageInterface*`	`GetMessageInterface(uint32 id)` This function returns the object that governs the coding of messages used by this automata instance. The returned object is an instance of the class derived from the class *Message Interface*.
`uint8*`	`GetMsg()` This function returns the first unread message from the mailbox assigned to this automata instance.
`static uint8*`	`GetMsg(uint8 mbx)` This function returns the first unread message from the mailbox identified by the value of its parameter.
`inline uint32`	`GetMsgCallId()` This function returns the identification of the communication process (e.g., call ID) from the current message.
`inline uint16`	`GetMsgCode()` This function returns the message code from the current message header.

(Continued)

TABLE 6.7 (CONTINUED)

FiniteStateMachine Member Functions Summary

Type	Member Function
inline uint8	GetMsgFromAutomata() This function returns the identification of the originating automata type from the current message.
inline uint8	GetMsgFromGroup() This function returns the identification of the group of the originating automata instance for the current message.
inline uint8	GetMsgInfoCoding() This function returns the identification of the information coding scheme used for the current message.
inline uint16	GetMsgInfoLength() This function returns the payload length of the current message in bytes.
inline uint16	GetMsgInfoLength(uint8 *msg) This function returns the payload length of the given message in bytes. The message is specified by its pointer.
inline uint32	GetMsgObjectNumberFrom() This function returns the identification of the originating automata instance from the current message.
inline uint32	GetMsgObjectNumberTo() This function returns the identification of the destination automata instance from the current message.
inline uint8	GetMsgToAutomata() This function returns the identification of the destination automata type from the current message.
inline uint8	GetMsgToGroup() This function returns the identification of the type of group of the destination automata from the current message.
inline uint8*	GetNewMessage() This function returns the address of the buffer that contains the new message.
inline uint8	GetNewMsgInfoCoding() This function returns the identification of the information coding scheme used for the new message.
inline uint16	GetNewMsgInfoLength() This function returns the payload length of the new message in bytes.
uint8*	GetNextParam(uint16 paramCode) This function returns the address of the next instance of the given type of message parameter within the current message.
bool	GetNextParamByte(uint16 paramCode, BYTE &param) This function searches for the next instance of the given type of the single-byte parameter in the current message. If the instance is found, the function copies it into its parameter specified by the reference and returns the value *true*; otherwise, it returns the value *false*.

(Continued)

TABLE 6.7 (CONTINUED)

FiniteStateMachine Member Functions Summary

Type	Member Function
bool	GetNextParamDWord(uint16 paramCode, DWORD &param) This function searches for the next instance of the given type of the 4-byte parameter in the current message. If the instance is found, the function copies it into its parameter specified by the reference and returns the value *true*; otherwise, it returns the value *false*.
bool	GetNextParamWord(uint16 paramCode, WORD &param) This function searches for the next instance of the given type of the 2-byte parameter in the current message. If the instance is found, the function copies it into its parameter specified by the reference and returns the value *true*; otherwise, it returns the value *false*.
virtual uint32	GetObjectId() This function returns the unique identification of this automata instance.
uint8*	GetParam(uint16 paramCode) This function returns the address of the first instance of the given type of the message parameter within the current message.
bool	GetParamByte(uint16 paramCode, BYTE &param) This function searches for the first instance of the given type of single-byte parameter in the current message. If the instance is found, the function copies it into its parameter specified by the reference and returns the value *true*; otherwise, it returns the value *false*.
bool	GetParamDWord(uint16 paramCode, DWORD &param) This function searches for the first instance of the given type of 4-byte parameter in the current message. If the instance is found, the function copies it into its parameter specified by the reference and returns the value *true*; otherwise, it returns the value *false*.
bool	GetParamWord(uint16 paramCode, WORD &param) This function searches for the first instance of the given type of 2-byte parameter in the current message. If the instance is found, the function copies it into its parameter specified by the reference and returns the value *true*; otherwise, it returns the value *false*.
PROC_FUN_PTR	GetProcedure(uint16 event) This function returns the pointer to the event handler for the given event identifier and the current state of automata.
virtual inline uint8	GetRightMbx() This function returns the identification of the mailbox assigned to the automata instance that is logically to the right of this automata instance.
virtual inline uint8	GetRightAutomata() This function returns the identification of the automata type that is logically to the right of this automata instance.

(Continued)

TABLE 6.7 (CONTINUED)

FiniteStateMachine Member Functions Summary

Type	Member Function
virtual inline uint8	GetRightGroup() This function returns the identification of the type of the group of automata that is logically to the right of this automata instance.
virtual inline uint32	GetRightObjectId(); This function returns the identification of the automata instance that is logically to the right of this automata instance.
virtual inline uint8	GetState() This function returns the identification of the current state of this automata instance.
virtual bool	IsBufferSmall(uint8 *buff, uint32 length) This function returns the value *true* if the size of the given buffer is not greater than the given size specified as the value of its second parameter; otherwise, it returns the value *false*.
virtual void	Initialize() This function defines the automata state transition event handlers and timers used by this automata type.
void	InitEventProc(uint8 state, uint16 event, PROC_FUN_PTR fun) This function defines the given state transition event handler for the given automata state and the given event (message code).
void	InitTimerBlock(uint16 tmrId, uint32 count, uint16 signalId) This function initializes the given timer by the given duration and the timer expiration message code.
void	InitUnexpectedEventProc(uint8 state, PROC_FUN_PTR fun) This function defines the given state transition event handler for unexpected events in the given automata state.
bool	IsTimerRunning(uint16 id) This function returns the value *true* if the given timer is active (running); otherwise, it returns the value *false*.
void	NoFreeObjectProcedure(uint8 *msg) This function defines the behavior of this automata type if the list of free automata of this type is used and if it is empty at the moment when a free instance is requested.
virtual void	NoFreeInstances() This function defines the behavior of the FSM system if a list of free automata is used and if it is empty at the moment when a free instance is requested.
virtual bool	ParseMessage(uint8 *msg) This function checks if the given message is coded properly and, if it is, it becomes the current message (its pointer is assigned to the internal variable *CurrentMessage*).

(Continued)

TABLE 6.7 (CONTINUED)

FiniteStateMachine Member Functions Summary

Type	Member Function
virtual void	`PrepareNewMessage(uint8 *msg)` This function defines the given buffer as the new message buffer by assigning the given pointer to the internal variable *NewMessage*. The buffer is used as a working area for the construction of the new message.

virtual void	`PrepareNewMessage(uint32 length, uint16 code,`
	`uint8 infoCode = LOCAL_PARAM_CODING)`
	This function creates a new message of the given length with the given message code and the given type of information coding.

virtual void	`Process(uint8 *msg)` This function performs the preparations for the message processing and selects the state transition event handler based on the message code and current state of this automata instance.
void	`PurgeMailBox()` This function purges all the messages from the mailbox assigned to this automata type and releases all the buffers assigned to the messages.
bool	`RemoveParam(uint16 paramCode)` This function removes the given type of message parameter from the new message.
virtual void	`Reset()` This function resets this automata instance by returning it to its initial state and by stopping all its active timers.
void	`ResetTimer(uint16 id)` This function resets the internal timer block object and returns the buffer allocated by the *StartTimer* primitive to the FSM Library kernel.
void	`RestartTimer(uint16 tmrId)` This function restarts the given timer. It is logically equivalent to a sequence of *StopTimer* and *StartTimer* primitives.
virtual void	`RetBuffer(uint8 *buff)` This function returns the given buffer to the FSM Library kernel. Normally, each memory buffer is returned at the end of its life cycle. Failure to do so leads to a memory leak problem.
void	`ReturnMsg(uint8 mbxId)` This function makes a copy of the current message and sends it to the given mailbox. This primitive is used frequently for message forwarding. On many occasions, the communication process must react in this simple way.

void	`SetBitParamByteBasic(BYTE param, uint32`
	`offset, uint32 mask=MASK_32_BIT)`
	This function sets the given single byte parameter of the new message to the result of the bit-wise inclusive OR operation applied to the given parameter and its previous value masked (bit-wise AND operation) with the given bit-mask.

(Continued)

TABLE 6.7 (CONTINUED)

FiniteStateMachine Member Functions Summary

Type	Member Function
void	`SetBitParamDWordBasic(DWORD param, uint32 offset, uint32 mask=MASK_32_BIT)` This function sets the given 4-byte parameter of the new message to the result of the bit-wise inclusive OR operation applied to the given parameter and its previous value masked (bit-wise AND operation) with the given bit-mask.
void	`SetBitParamWord(WORD param, uint32 offset, uint32 mask=MASK_32_BIT)` This function sets the given 2-byte parameter of the new message to the result of the bit-wise inclusive OR operation applied to the given parameter and its previous value masked (bit-wise AND operation) with the given bit-mask.
inline void	`SetCallId()` This function sets the default value of the attribute *CallId* of this automata instance.
inline void	`SetCallId(uint32 id)` This function sets the given value of the attribute *CallId* of this automata instance.
inline void	`SetCallIdFromMsg()` This function sets the attribute *CallId* of this automata instance to the value of the parameter *CallId* of the current message. This primitive is used to store the reference number specific to the communication protocol.
virtual void	`SetDefaultFSMData()` This function sets the automata-specific data to their default values. It is typically used before the normal operation phase.
virtual void	`SetDefaultHeader(uint8 infoCoding)` This function sets the default header field values for the given type of the message information coding.
inline void	`SetGroup(uint8 id)` This function sets the identification of the group of automata for this automata type to the given value. This primitive is used to declare the group membership.
virtual void	`SetInitialState()` This function sets the current state of this automata instance to its initial state.
static void	`SetKernelObjects(TPostOffice *postOffice, TBuffers *buffers, CTimer *timer)` This function sets the *FSMSystem* library kernel objects (post office, buffers, and timers), which are common for all of the automata in the FSM system.
inline void	`SetLeftMbx(uint8 mbx)` This function sets the identification of the mailbox assigned to the automata instance that is logically to the left of this automata instance.
inline void	`SetLeftAutomata(uint8 automata)` This function sets the identification of the automata type that is logically to the left of this automata instance.

(Continued)

TABLE 6.7 (CONTINUED)

FiniteStateMachine Member Functions Summary

Type	Member Function
inline void	SetLeftObject(uint8 group) This function sets the identification of the type of the group of automata that is logically to the left of this automata instance.
inline void	SetLeftObjectId(uint32 id) This function sets the identification of the automata instance that is logically to the left of this automata instance.
static void	SetLogInterface(LogInterface *logingObject) This function defines the object responsible for message logging. The object is an instance of a class derived from the class *LogInterface*.
inline void	SendMessage(uint8 mbxId) This function sends a new message to the given mailbox. The mailbox is specified by its identification.
inline void	SendMessage(uint8 mbxId, uint8 *msg) This function sends the given message to the given mailbox.
void	SetMessageFromData() This function sets the header fields of the new message related to the originating automata instance to the values specific to this automata instance.
inline void	SetMsgCallId(uint32 id) This function sets the call ID parameter of the new message to the given value.
inline void	SetMsgCallId(uint32 id, uint8 *msg) This function sets the call ID parameter of the given message to the given value.
inline void	SetMsgCode(uint16 code) This function sets the message code parameter of the new message to the given value.
inline void	SetMsgCode(uint16 code, uint8 *msg) This function sets the message code parameter of the given message to the given value.
inline void	SetMsgFromAutomata(uint8 from) This function sets the type of the originating automata parameter of the new message to the given value.
inline void	SetMsgFromAutomata(uint8 from, uint8 *msg) This function sets the type of the originating automata parameter of the given message to the given value.
inline void	SetMsgFromGroup(uint8 from) This function sets the type of the originating group of automata parameters of the new message to the given value.
inline void	SetMsgFromGroup(uint8 from, uint8 *msg) This function sets the type of the originating group of automata parameters of the given message to the given value.
inline void	SetMsgInfoCoding(uint8 codingType) This function sets the message information coding parameter of the new message to the given value.

(Continued)

TABLE 6.7 (CONTINUED)

FiniteStateMachine Member Functions Summary

Type	Member Function
inline void	`SetMsgInfoCoding(uint8 codingType, uint8 *msg)` This function sets the message information coding parameter of the given message to the given value.
inline void	`SetMsgInfoLength(uint16 length)` This function sets the message payload (useful information) length parameter of the new message.
inline void	`SetMsgInfoLength(uint16 length, uint8 *msg)` This function sets the message payload (useful information) length parameter of the given message.
inline void	`SetMsgObjectNumberFrom(uint32 from)` This function sets the originating automata instance identification parameter of the new message to the given value.
inline void	`SetMsgObjectNumberFrom(uint32 from, uint8 *msg)` This function sets the originating automata instance identification parameter of the given message to the given value.
inline void	`SetMsgObjectNumberTo(uint32 to)` This function sets the destination automata instance identification parameter of the new message to the given value.
inline void	`SetMsgObjectNumberTo(uint32 to, uint8 *msg)` This function sets the destination automata instance identification parameter of the given message to the given value.
inline void	`SetMsgToAutomata(uint8 to)` This function sets the destination automata type identification parameter of the new message to the given value.
inline void	`SetMsgToAutomata(uint8 to, uint8 *msg)` This function sets the destination automata type identification parameter of the given message to the given value.
inline void	`SetMsgToGroup(uint8 to)` This function sets the destination automata group identification parameter of the new message to the given value.
inline void	`SetMsgToGroup(uint8 to, uint8 *msg)` This function sets the destination automata group identification parameter of the given message to the given value.
void	`SendMessageLeft()` This function sends the new message to the mailbox assigned to the automata instance that is logically to the left of this automata instance.
void	`SendMessageRight()` This function sends the new message to the mailbox assigned to the automata instance that is logically to the right of this automata instance.
inline void	`SetNewMessage(uint8 *msg)` This function sets the new message to the given message by assigning the given message pointer to the internal pointer to the new message.
inline void	`SetObjectId(uint32 id)` This function sets the identification of this automata instance to the given value.

(Continued)

TABLE 6.7 (CONTINUED)

FiniteStateMachine Member Functions Summary

Type	Member Function
inline void	SetRightMbx(uint8 mbx) This function sets the identification of the mailbox assigned to the automata instance that is logically to the right of this automata instance.
inline void	SetRightAutomata(uint8 automata) This function sets the identification of the automata type that is logically to the right of this automata instance.
inline void	SetRightObject(uint8 group) This function sets the identification of the type of the group of automata that is logically to the right of this automata instance.
inline void	SetRightObjectId(uint32 id) This function sets the identification of the automata instance that is logically to the right of this automata instance.
inline void	SetState(uint8 state) This function sets the identification of the current state of this automata instance.
void	StartTimer(uint16 tmrId) This function starts the given timer. The timer is specified by its identification.
void	StopTimer(uint16 tmrId) This function stops the given timer. The timer is specified by its identification.
static void	SysClearLogFlag() This function stops the logging of the messages exchanged by the automata.
static void	SysStartAll() This function starts the logging of the messages exchanged by the automata.

TABLE 6.8

NetFSM Constructor Summary

NetFSM(uint16 numOfTimers=DEFAULT_TIMER_NO, uint16
numOfState=DEFAULT_STATE_NO, uint16 maxNumOfProceduresPerState=DEFA
ULT_PROCEDURE_NO_PER_STATE, bool getMemory=true)
    The constructor initializes the object that represents an instance of the given automata type, along with the data structures needed for its proper operation.

TABLE 6.9

NetFSM Member Functions Summary

Type	Member Function
virtual void	convertFSMToNetMessage()
	This function converts the internal message format into the external message format appropriate for the transmission over the TCP/IP network.
virtual uint16	convertNetToFSMMessage()
	This function converts the external message format into the internal message format appropriate for communication within the FSM system.
void	establishConnection()
	This function establishes the TCP connection between two geographically distributed FSM systems.
virtual uint8	getProtocolInfoCoding()
	This function returns the identification of the type of the external message coding.
void	sendToTCP()
	This function sends the new message to the remote FSM system over the previously established TCP connection.

6.8.1 *FSMSystem*

Function prototype:

```
FSMSystem(
 uint8 numOfAutomata,
 uint8 numberOfMbx)
```

Function description: This constructor initializes the object that represents the FSM system together with the data structures needed for its proper operation.

Parameters:

numOfAutomata: the number of various automata types to be added to the FSM system

numberOfMbx: the number of mailboxes to be used by the FSM system

Note: Typically, a single mailbox is assigned to each automata type, but other arrangements are also allowed. Normally, an automata type corresponds to a protocol. For example, the IP protocol may be implemented as one automata type, and the TCP protocol may be implemented as another automata type. A typical arrangement would be to assign one mailbox to IP and one to TCP. Another arrangement would be to assign two mailboxes to each protocol. For example, in this arrangement, IP would use the first mailbox to receive the messages from network interfaces (drivers) and the second mailbox to receive the messages from TCP. Yet another arrangement would be to assign a single mailbox to all the protocols. Finally, a set of mailboxes can be used to prioritize the messages. For example, three mailboxes may be used to distinguish high, middle, and low priority messages.

6.8.2   Add(*ptrFiniteStateMachine, uint8, uint32, bool*)

Function prototype:

```
void Add(
 ptrFiniteStateMachine object,
 uint8 automataType,
 uint32 numberOfObjects,
 bool useFreeList = false)
```

Function description: This function adds the first instance of each automata type to the FSM system. At the same time, this function defines the unique identification of this automata type and the number of instances of this automata type that will be subsequently added to the FSM system. It also declares a group of instances of this automata type as either a set of resources to be used individually or as a pool of resources of the same type available for dynamic allocation.

Function parameters:

object: the pointer to the first instance of this automata type to be added to the FSM system

automataType: the unique identification of this type of automata

numberOfObjects: the total number of instances of this type to be added to the FSM system

useFreeList: the indicator selecting the mode of usage of individual instances of this type

Note: Typically, the FSM system is created at system startup, and then groups of various automata types are added to it. As a rule, the first instance of the given automata type is added by this function. Its parameters specify, in order from left to right, the pointer to the first object of this type, the identification of this automata type, the total number of instances that will be added to the FSM system, and the mode of individual instance allocation. This last parameter has a default value *false*, which means that each automata instance represents an individual resource. If this default is over-ridden by the value *true*, the group of instances of this automata type represents a pool of resources of the same type. The individual instances from this pool are allocated dynamically and on-demand, based on the use of the internal FSMSystem library kernel list of resources of the given type. (This is the origin of the name of the last parameter of this function, *useFreeList*.) This dynamic allocation is requested by sending a message to an unknown automata, which is identified by the instance identification set to the value −1 (see function *SetMsgObjectNumberTo*).

6.8.3 Add(*ptrFiniteStateMachine, uint8*)

Function prototype:

```
void Add(
 ptrFiniteStateMachine object,
 uint8 automataType)
```

Function description: This function adds all the automata instances except the first instance of the given type to the FSM system. It assumes that the first instance of this automata type has been added previously to the FSM system by calling the overloaded function *Add* with four parameters in its signature.

Function parameters:

object: the pointer to the instance of this automata type to be added to the FSM system

automataType: the unique identification of this automata type

Note: As already mentioned, after the FSM system is created at system startup, the groups of various automata types are added to it. As a rule, the first instance of the given automata type is added by the overloaded function *Add* with four parameters in its signature (see the previous section for more details on its parameters). All the other instances of the given automata type are added to the FSM system by this overloaded function *Add*. An advantage of differentiating these two functions becomes obvious in a dynamic environment where objects are created on-demand and added to the FSM system. If the given automata type already exists, and a need arises for another instance of it, this overloaded *Add* function is sufficient.

6.8.4 *InitKernel*

Function prototype:

```
void InitKernel(
 uint8 buffClassNo,
 uint32 *buffersCount,
 uint32 *buffersLength,
 uint8 numOfMbxs=0,
 TimerResolutionEnum timerRes = Timer1s)
```

Function description: This function initializes the elements of the kernel responsible for time, buffer, and message management. The parameters of this function specify the number of buffer types, the number of instances per buffer type and their lengths, the number of mailboxes to be used by the automata added to the FSM system, and the basic timer resolution. The default value of the basic timer resolution is 1 sec, which is defined by the symbolic constant *Timer1s*.

Function parameters:

buffClassNo: the number of buffer types

buffersCount: the pointer to the array of the numbers of instances of the corresponding buffer types

buffersLength: the pointer to the array of the sizes of the corresponding buffer types

numOfMbxs: the number of the mailboxes

timerRes: the basic timer resolution

Note: This function essentially initializes the *FSMSystem* library kernel. It must be called after the FSM system has been created and before it can be started. It also assumes that the arrays of the cardinal numbers and the sizes of individual buffer types were already created and filled by the programmer. Because the specification of the buffers to be provided by the kernel may look cumbersome, we provide the following example. Suppose that a need arises for three buffer types, namely, small, medium, and large. The programmer should set the first parameter of this function to the number 3. Next, suppose that the programmer needs 300 small buffers, 200 medium buffers, and 100 large buffers, and that their sizes should be 64, 128, and 256 bytes, respectively. Before calling this function, the programmer should create the following two arrays:

- Array of cardinal numbers = [300, 200, 100]
- Array of sizes = [64, 128, 256]

Finally, the programmer should specify the pointers to these two arrays as the second and the third parameter of this function.

6.8.5   *Remove(uint8)*

Function prototype:

```
void Remove(unit8 automataType)
```

Function description: This function removes all instances of the given automata type from the FSM system.

Function parameters:

automataType: the type of automata to be removed from the system

Note: First, the FSM system removes all instances of the given automata type from the FSM system. Next, the kernel frees all the memory zones occupied by the internal data structures used by the automata of this type.

6.8.6   *Remove(uint8, uint32)*

Function prototype:

```
ptrFiniteStateMachine Remove(
 uint8 automataType
 uint32 object)
```

Function description: This function removes the given instance of the given automata type. The parameters of this function specify the identification of the automata type and the identification of the automata instance.

Function parameters:

automataType: the identification of the automata type

object: the identification of the instance of the given automata type

Function returns: This function returns the pointer to the automata instance removed from the FSM system.

6.8.7   *Start*

Function prototype:

```
virtual void Start()
```

Function description: This function starts the FSM system and is the main function of the FSM system. In this function, the FSM system thread enters a loop in which it reads the kernel mailboxes and distributes the messages to the destination automata.

Note: The FSM system thread remains in the loop while the internal attribute *SystemWorking* is set to the value *true*. A typical implementation of the FSM system thread is shown in the example in Section 6.2.1.2.

6.8.8   *StopSystem*

Function prototype:

```
void StopSystem()
```

Function description: This function stops the FSM system. It sets the internal attribute *SystemWorking* to the value *false*, thus causing the FSM system thread to exit its loop and stop the FSM system.

Note: If the function *Start* has been called from the separate operating system thread, the call to the function *StopSystem* will cause the termination of that thread.

6.8.9   *FSMSystemWithTCP*

Function prototype:

```
FSMSystemWithTCP(
 uint8 numOfAutomata,
 uint8 numberOfMbx)
```

Function description: This constructor initializes the object that represents the FSM system supporting communication over TCP/IP network, along with the data structures needed for its proper operation. Its parameters specify the number of automata types to be added to the FSM system and the number of mailboxes.

Function parameters:

numOfAutomata: the number of automata types that will be added to the FSM system

numberOfMbx: the number of mailboxes that will be used by the automata added to the FSM system

Note: Typically, a single mailbox is assigned to each automata type included in the FSM system, but other arrangements are also allowed. For example, a single mailbox may be assigned to all the automata types included in the FSM system. Also allowed is to assign an arbitrary number of mailboxes to each automata type, e.g., to enable message prioritization.

6.8.10   *InitTCPServer*

Function prototype:

```
int InitTCPServer(
 uint16 port,
 unit8 automataType,
 char *ipAddress = 0,
 unsigned char *parm = 0,
 int length = 0)
```

Function description: This function initializes the TCP server. Once initialized, the server waits for a request to establish the TCP connection with a remote client. The parameters of this function specify the number of the TCP port on which the server awaits the connection request, the automata type included in the FSM system engaged in the communication, the server IP address, the pointer to the area where the connection parameters should be passed to the specified automata type, and the parameter lengths in bytes. After reception of the request, the server allocates an instance of the given automata type and passes the connection together with the received parameters to the allocated automata instance. Further communication continues directly between the remote client and the allocated automata instance, i.e., the server is completely isolated from it.

Function parameters:

Port: the number of the TCP port on which the server awaits a connection request

automataType: the automata type included in the FSM system that is engaged in the communication. This automata type must be derived from the class *NetFSM*. After the connection has been initially established, the server transfers it to the allocated instance of this automata type.

ipAddress: the pointer to the server IP address

parm: the pointer to the area where the parameters received while establishing the connection should be passed and subsequently taken by to the specified automata type

length: the parameter lengths specified by the previous pointer, in bytes

Function returns: If the TCP server awaiting a request from a remote client is successfully started, this function returns the value 0. Otherwise, it returns the value –1.

Note: This function should be called only once, just initially to start the TCP server.

6.8.11 *FiniteStateMachine*

Function prototype:

```
FiniteStateMachine(
 unit16 numOfTimers = DEFAULT_TIMER_NO,
 uint16 numOfState = DEFAULT_STATE_NO,
 uint16 maxNumOfProceduresPerState = DEFAULT_PROCEDURE_NO_PER_STATE,
 bool getMemory = true)
```

Function description: This constructor initializes the object that represents the instance of a given automata type together with the data structures needed for its proper operation. Its parameters specify the number of timers to be used by this automata type, the number of the states that this automata type has, the maximal number of state transitions per state, and the indicator specifying whether this constructor should reserve the memory for the objects that represent the states and state transitions of this automata type or not. The default value of this indicator is *true*, which means that this constructor is responsible for memory allocation.

Function parameters:

numOfTimers: the number of the timers to be used by this automata type

numOfState: the number of the states that this automata type has

maxNumOfProceduresPerState: the maximal number of state transitions per state

getMemory: the memory allocation indicator (by default, its value is *true*)

Note: This constructor may be called either with some or without any of the parameters. If the parameter is not specified, the constructor will use its default value. The indicator *getMemory* may be set to the value *false* when the programmer wants to do manual memory allocation to optimize overall memory consumption.

6.8.12   *AddParam*

Function prototype:

```
uint8 *AddParam(
 uint16 paramCode,
 uint32 paramLength,
 uint8 *param)
```

Function description: This function is used to add a given parameter of a given length to the new message. The parameters of this function specify the unique identification of the parameter type, the parameter length in bytes, and the pointer to the parameter itself. If the parameter to be added to the message is too large to fit in the buffer that is assigned to the new message, this function will get a bigger buffer, copy the new message into it, add the parameter, and release the old buffer.

Function parameters:

paramCode: the parameter type

paramLength: the parameter length, in bytes

param: the pointer to the parameter

Function returns: This function returns the pointer to the buffer that contains the new message.

Note: This function enables the programmer to add a parameter of an arbitrary size to the new message with the limitation that it must not exceed the maximal parameter length specified for the given type of message coding (e.g., for the type *StandardMessage*, the maximal parameter length is 256 bytes). The message parameters in *StandardMessage* are sorted by ascending order of their corresponding type identifiers.

6.8.13 *AddParamByte*

Function prototype:

```
uint8 *AddParamByte(
 uint16 paramCode,
 BYTE param)
```

Function description: This function is used to add the given parameter of length 1 byte to the new message. The parameters of this function specify the unique identification of the parameter type and the parameter value.

Function parameters:

paramCode: the parameter type

param: the parameter value

Function returns: This function returns the pointer to the buffer that contains the new message.

Note: The total message length must not exceed the limit specified for the given type of message coding. In any case, it must not exceed 8G bytes.

6.8.14 *AddParamDWord*

Function prototype:

```
uint8 *AddParamDWord(
 uint16 paramCode,
 DWORD param)
```

Function description: This function is used to add the given parameter of length 4 bytes to the new message. The parameters of this function specify the unique identification of the parameter type and the parameter value.

Function parameters:

paramCode: the parameter type

param: the parameter value

Function returns: This function returns the pointer to the buffer that contains the new message.

Note: The total message length must not exceed the limit specified for the given type of message coding. In any case, it must not exceed 232 bytes.

6.8.15 *AddParamWord*

Function prototype:

```
uint8 *AddParamDWord(
 uint16 paramCode,
 WORD param)
```

Function description: This function is used to add the given parameter of length 2 bytes to the new message. The parameters of this function specify the unique identification of the parameter type and the parameter value.

Function parameters:

paramCode: the parameter type

param: the parameter value

Function returns: This function returns the pointer to the buffer that contains the new message.

Note: The total message length must not exceed the limit specified for the given type of message coding. In any case, it must not exceed 8G bytes.

6.8.16 *CheckBufferSize*

Function prototype:

```
uint8 *CheckBufferSize(uint32 paramLength)
```

Function description: This function provides a new message buffer with the size sufficient enough to accept the parameter of the given length. The parameter of this function specifies the parameter length in bytes.

Function parameters:

paramLength: the parameter length

Function returns: This function returns the pointer to the new message.

Note: This function is obsolete. In the previous version of the FSM Library, this function ensured the new message buffer management was transparent for the programmer. Typically, the programmer would call this function before calling some of the *AddParam* functions to ensure that the new message is stored in a buffer of sufficient size. This means that the buffer is large enough to accept a new parameter of the given size, in addition to the current content of the new message. Behind the scenes, this function checked the current size of the new message. If it was not sufficient, the function allocated a new, larger buffer; copied the current new message into it; released the old buffer; and returned the pointer to the newly allocated buffer containing the new message. In the current version of the FSM Library, all the *AddParam* functions call this function internally at their very beginning, and the programmer no longer needs to call it explicitly.

6.8.17 *ClearMessage*

Function prototype:

```
virtual void ClearMessage()
```

Function description: This function returns the buffer allocated for the current message to the kernel and assigns the value *NULL* to the internal pointer to the current message. The current message is the last message received by the automata instance.

Note: If the *FSMSystem* library has been compiled for the debug mode, this function will additionally verify that the return value of the function is *NULL*.

6.8.18 *CopyMessage( )*

Function prototype:

```
virtual void CopyMessage()
```

Function description: This function makes a copy of the current message and assigns that copy to the new message. By definition, a current message is the last received message, and a new message is the message under construction to be subsequently sent. The value of the pointer to the current message copy is assigned to the internal pointer to the new message.

Note: This function first checks if the new message already exists by checking the internal pointer to the new message. If the new message has already been defined or is under construction (the internal pointer is not equal to the value *NULL*), the function releases the buffer that contains the new message and assigns the value *NULL* to the internal pointer. Next, the function makes a copy of the current message and assigns its address to the pointer to the new message. This function is typically used for message forwarding. The protocol *A* sends a message to the protocol *B*, which, in turn, forwards the copy of the same message to the protocol *C*.

6.8.19 *CopyMessage(uint*)*

Function prototype:

```
virtual void CopyMessage(uint8 *msg)
```

Function description: This function makes a copy of the given message and assigns that copy to the new message. The parameter of this function specifies the pointer to the original message.

Function parameters:

msg: the pointer to the original message

Note: This function assumes that the new message does not exist, i.e., the internal pointer to the new message should contain the value *NULL* before this function is called. However, if the new message already exists, this function will return its buffer and get a fresh buffer for the new message before copying the given message into it.

6.8.20   *CopyMessageInfo*

Function prototype:

```
virtual void CopyMessageInfo(
 uint8 infoCoding,
 uint16 lengthCorrection = 0)
```

Function description: This function copies the part of the message containing the useful information, referred to as a payload (message without its header), from the current message into the new message stored in a newly allocated buffer. The parameters of this function specify the type of information coding that governs the formatting and length correction of the message.

Function parameters:

infoCoding: the identification of the type of information coding

lengthCorrection: the message length correction

Note: The message length correction depends on the type of applied information coding. If the new message buffer does not exist, this function will get a buffer, assign it to the new message, and make the required copy.

6.8.21   *Discard*

Function prototype:

```
virtual void Discard(uint8* buff)
```

Function description: This function deletes the message placed in the given buffer and returns the buffer to the kernel. The parameter of this function specifies the buffer to be cleared and released.

Function parameters:

buff: the pointer to the buffer

6.8.22 *DoNothing*

Function prototype:

```
void DoNothing()
```

Function description: This function performs no operation. It is called when the automata receives an unexpected message, unless a new function to handle unexpected messages is defined. By definition, an unexpected message is any type of message that has not been defined as a legal type of message in the current automata state.

Note: This function may be redefined by calling the function *Init UnexpectedEventProc*, if a need exists for concrete functionality handling unexpected messages.

6.8.23 *FreeFSM*

Function prototype:

```
void FreeFSM()
```

Function description: This function reports to the FSM system that an automata instance has finished its current assignment and is free for further assignments. If the first instance of this automata type has been added to the FSM system with the parameter *useFreeList* set to the value *true*, the group of instances of this automata type is viewed as a pool of resources. In that case, this function returns the resource to the corresponding pool by queuing it to the internal list of the resources of the same type.

Note: If a group of instances of this automata type is used as a set of individual resources, rather than as a pool of resources (the parameter *useFree List* has been set to the value *false* when the first automata instance has been added to the FSM system), this function has no effect.

6.8.24 *GetAutomata*

Function prototype:

```
virtual uint8 GetAutomata() = 0
```

Function description: This function returns the identification of the automata type for this automata instance.

Function returns: This function returns the unique ID of the automata type.

Note: This function is a pure virtual function, which means that it must be defined in the class that models some concrete automata type. Typically, this function returns the constant value that represents the required identification. It finds this constant by looking up the table of identifications created by reading the file of all the known automata types at the FSM system startup time.

6.8.25   *GetBitParamByteBasic*

Function prototype:

```
unit8 GetBitParamByteBasic(
 uint32 offset,
 uint32 mask=MASK_32_BIT)
```

Function description: This function returns the value of the current message parameter of length 1 byte masked with the given mask. The parameters of this function specify the offset of the original parameter of the message and the value of the mask.

Function parameters:

offset: the offset of the original parameter of the message

mask: the value of the mask

Function returns: This function returns the result of the bit-wise AND operation between the value of the message parameter at the given message *offset* and the given value of the parameter *mask*.

Note: Normally, depending on the value of the parameter mask, testing the value of a single bit, or of a group of bits simultaneously, is possible in the parameter of size 1 byte that is at a given distance from the beginning of the message.

6.8.26   *GetBitParamWordBasic*

Function prototype:

```
unit8 GetBitParamWordBasic(
 uint32 offset,
 uint32 mask=MASK_32_BIT)
```

Function description: This function returns the value of the current message parameter of length 2 bytes masked with the given mask. The parameters of this function specify the offset of the original parameter of the message and the value of the mask.

Function parameters:

offset: the offset of the original parameter of the message

mask: the value of the mask

Function returns: This function returns the result of the bit-wise AND operation between the value of the message parameter at the given message *offset* and the given value of the parameter *mask*.

Note: Normally, depending on the value of the parameter mask, testing the value of a single bit, or a group of bits simultaneously, is possible in the parameter of size 2 bytes that is at a given distance from the beginning of the message.

6.8.27   *GetBitParamDWordBasic*

Function prototype:

```
unit8 GetBitParamDWordBasic(
  uint32 offset,
  uint32 mask=MASK_32_BIT)
```

Function description: This function returns the value of the current message parameter of length 4 bytes masked with the given mask. The parameters of this function specify the offset of the original parameter of the message and the value of the mask.

Function parameters:
offset: the offset of the original parameter of the message
mask: the value of the mask

Function returns: This function returns the result of the bit-wise AND operation between the value of the message parameter at the given message *offset* and the given value of the parameter *mask*.

Note: Normally, depending on the value of the parameter mask, testing the value of a single bit, or of a group of bits simultaneously, is possible in the parameter of size 4 bytes that is at a given distance from the beginning of the message.

6.8.28   *GetBuffer*

Function prototype:

```
virtual uint8 *GetBuffer(uint32 length)
```

Function description: This function returns a buffer whose size is not less than the size given by the value of its parameter. The parameter of this message specifies the minimal buffer length in bytes.

Function parameters:
length: the buffer length

Function returns: This function returns the pointer to a newly allocated buffer.

Note: The *FSMSystem* library kernel handles a limited number of buffer types with a limited number of instances per each type defined during the kernel initialization by calling the function *InitKernel*. By definition, this function first searches for the buffer types of the size that ideally match the desired buffer. If such a type does not exist, the function searches for the next size buffer type (in the increasing order of size). This allocation policy may yield a buffer of a size much bigger than needed, and the frequent occurrence of this type of allocation may lead to inefficient memory usage. For example, suppose that the programmer has mistakenly defined only two buffer sizes, small and large, such that not a single protocol message can fit into the small buffer. In this case, only the large buffers will be consumed, and the small buffers will not be used at all. Therefore, special care must be taken when defining the buffers before calling the function *InitKernel*.

Now, let us go back to the buffer allocation algorithm. When this function finds a buffer type of a sufficient size, it checks for a free buffer of that type. If no such type is found, the system is badly designed and a new buffer type must be added to the system. If such a buffer type exists, but no free buffers of that type are available, the function will look for the next size buffer. If all the buffers of the sufficient size are already allocated, the FSM system experiences a memory exhaustion problem. In the academic examples, the system is allowed to crash under these circumstances. However, industrial-strength applications require implementation of additional mechanisms, such as overload protection and intelligent automatic restarts.

6.8.29 *GetBufferLength*

Function prototype:

```
uint32 GetBufferLength(uint8 *buff)
```

Function description: This function returns the size of the given buffer in bytes. The parameter of this function specifies the pointer to the buffer.

Function parameters:
buff: the address of the buffer

Function returns: This function returns the specified buffer length in bytes.

6.8.30 *GetCallId*

Function prototype:

```
virtual inline uint32 GetCallId()
```

Function description: This function returns the identification of the communication process that this instance is currently involved in, e.g., the call ID. The actual meaning of this identification is application specific.

Function returns: This function returns the value of the attribute *CallId*.

Note: Historically, the attribute *CallId* is tied to call processing (e.g., Q.71) and signaling (e.g., SS7, DSS1) protocols, but it has also proved to be useful in modern multimedia protocols (e.g., H.323 and SIP). Generally, this attribute may be used as an identification of the process or transaction that engages more cooperative automata. If a single attribute is not sufficient, the programmer may introduce additional attributes in classes derived from the base class *FiniteStateMachine*.

6.8.31   *GetCount*

Function prototype:

```
uint32 GetCount(uint8 mbx)
```

Function description: This function returns the current number of messages in the given mailbox. The parameter of this message specifies the identification of the mailbox.

Function parameters:

mbx: the mailbox identification

Function returns: This function returns the number of unread messages contained in the mailbox of interest.

6.8.32   *GetGroup*

Function prototype:

```
virtual uint8 GetGroup()
```

Function description: This function returns the identification of the group of automata to which this instance belongs.

Function returns: This function returns a number that uniquely identifies the group of automata which, besides other members, includes this automata instance.

6.8.33   *GetInitialState*

Function prototype:

```
virtual uint8 GetInitialState()
```

Function description: This function returns the identification of the initial state of this automata type.

Function returns: This function returns the number that uniquely identifies the initial state of this automata type.

Note: The default value of the initial state is 0.

6.8.34   *GetLeftMbx*

Function prototype:

```
virtual inline uint8 GetLeftMbx()
```

Function description: This function returns the identification of the default mailbox assigned to the automata instance that is logically to the left of this automata instance.

Function returns: This function returns the number that uniquely identifies the default mailbox assigned to the left automata instance.

Note: Historically, the terms *left* and *right* automata instance originate from SDL, where an automata instance typically communicates with its left and right neighbors. These neighbors might have their own mailboxes, sometimes briefly called left and right mailboxes.

6.8.35   *GetLeftAutomata*

Function prototype:

```
virtual inline uint8 GetLeftAutomata()
```

Function description: This function returns the identification of the automata type that is logically to the left of this automata instance.

Function returns: This function returns the number that uniquely identifies the left automata type.

Note: By definition, left automata are logically placed to the left of the currently observed automata instance.

6.8.36   *GetLeftGroup*

Function prototype:

```
virtual linline uint8 GetLeftGroup()
```

Function description: This function returns the identification of the group of automata that is logically to the left of this automata instance.

Function returns: This function returns the number that uniquely identifies the left group of automata.

Note: By definition, a left group of automata is a group that contains left automata.

6.8.37 *GetLeftObjectId*

Function prototype:

```
virtual inline uint32 GetLeftObjectId()
```

Function description: This function returns the identification of the automata instance that is logically to the left of this automata instance.

Function returns: This function returns the number that uniquely identifies the left automata instance.

Note: By definition, left automata are logically placed to the left of the currently observed automata instance. This function returns the identification of the particular left automata instance with which the currently observed automata instance communicates.

6.8.38 *GetMbxId*

Function prototype:

```
virtual uint8 GetMbxId()
```

Function description: This function returns the identification of the default mailbox assigned to this automata type. Note that an instance of a given automata type may receive its messages through any mailbox, i.e., through the default mailbox as well as through other mailboxes. Alternately, a single mailbox may be assigned to more than one automata type.

Function returns: This function returns the number that uniquely identifies the default mailbox assigned to this automata instance.

Note: This function is a pure virtual function, which means that it must be defined by the programmer when they write a class derived from the class *FiniteStateMachine*. Typically, this function returns the constant value that represents the required mailbox identification (the content of the corresponding class field). This constant can be initially determined by looking up the table of identifications, and set by calling the function *SetMbxId*. The table of identifications can be created by reading the file containing all the known automata types at the FSM system startup time. A mailbox ID is typically a record field that describes a single automata type.

6.8.39 *GetMessageInterface*

Function prototype:

```
virtual MessageInterface *GetMessageInterface(uint32 id) = 0
```

Function description: This function returns the object that governs the coding of messages used by this automata instance. The parameter of this function specifies the identification of the information coding scheme. The returned object is an instance of the class derived from the class *MessageInterface*.

Function parameters:

id: the information coding scheme

Function returns: This function returns the pointer to the object responsible for parsing and coding the messages used by this automata instance.

Note: This function is a virtual function, which means that it must be defined when the programmer writes a class derived from the class *FiniteStateMachine*. The identification with the value 0 is reserved for the information coding used by the format of the class *StandardMessage*, which is a basic type of message supported by the *FSMSystem* library.

6.8.40 *GetMsg()*

Function prototype:

```
uint8* GetMsg()
```

Function description: This function returns the first unread message from the mailbox assigned to this automata instance.

Function returns: This function returns a pointer to the buffer that has been removed from the head of the list, which is hidden by the abstraction of the mailbox assigned to this automata instance. If no such buffer exists, i.e., if the list is empty, the function returns the value *NULL*.

6.8.41 *GetMsg(uint8)*

Function prototype:

```
static uint8* GetMsg(uint8 mbx)
```

Function description: This function returns the first unread message from the given mailbox. The parameter of this function specifies the identification of the mailbox.

Function parameters:

mbx: the mailbox ID

Function returns: This function returns the pointer to the buffer that has been removed from the head of the list, which is hidden by the abstraction of the given mailbox. If no such buffer exists, i.e., if the list is empty, the function returns the value *NULL*.

Note: Although this function is defined as a static function, a call to this function is not allowed before the kernel initialization and the FSM system startup. The call to this function made before that may cause unpredictable behavior.

6.8.42   *GetMsgCallId*

Function prototype:

```
inline uint32 GetMsgCallId()
```

Function description: This function returns the identification of the communication process (e.g., call ID) from the current message.

Function returns: This function returns the value of the attribute *CallId*.

Note: The attribute *CallId* is application specific. It can be used to indicate a process or a transaction in which more cooperating automata are involved. The size of *CallId* is 32 bits. It is considered large enough for most of the applications. To increase the size of *CallId*, the programmer would need to modify the base class *FiniteStateMachine*.

6.8.43   *GetMsgCode*

Function prototype:

```
inline uint16 GetMsgCode()
```

Function description: This function returns the message code from the current message header.

Function returns: This function returns the value of the message code from the header of the current (last received) message.

6.8.44   *GetMsgFromAutomata*

Function prototype:

```
inline uint8 GetMsgFromAutomata()
```

Function description: This function returns the identification of the originating automata type from the current message. This value is provided from the header of the current message.

Function returns: This function returns the value of the identification of the automata type that has created and sent the current message to this automata instance.

6.8.45   *GetMsgFromGroup*

Function prototype:

```
inline uint8 GetMsgFromGroup()
```

Function description: This function returns the identification of the group of the originating automata instance for the current message. This value is provided from the header of the current message.

Function returns: This function returns the value of the identification of the group of automata instance that has created and sent the current message to this automata instance.

6.8.46   *GetMsgInfoCoding*

Function prototype:

```
inline uint8 GetMsgInfoCoding()
```

Function description: This function returns the identification of the information coding scheme used for the current message.

Function returns: This function returns the value that identifies the type of information coding that has been used to create the current message.

Note: This information is provided from the header of the current message.

6.8.47   *GetMsgInfoLength()*

Function prototype:

```
inline uint16 GetMsgInfoLength()
```

Function description: This function returns the payload length of the current message in bytes.

Function returns: This function returns the value of the current message payload size in bytes.

Note: The length of the message header is not included in the length returned by this message. By definition, the total message length is the sum of the length of the message header and the length of the message payload.

6.8.48   *GetMsgInfoLength(uint8*)*

Function prototype:

```
inline uint16 GetMsgInfoLength(uint8 *msg)
```

Function description: This function returns the payload length of the given message in bytes. The parameter of this function specifies the pointer to the message.

Function parameters:

msg: the pointer to the message

Function returns: This function returns the value of the size of the given message payload in bytes.

Note: The length of the message header is not included in the length returned by this message. By definition, the total message length is the sum of the length of the message header and the length of the message payload.

6.8.49 *GetMsgObjectNumberFrom*

Function prototype:

```
inline uint32 GetMsgObjectNumberFrom()
```

Function description: This function returns the identification of the originating automata instance from the current message.

Function returns: This function returns the value that identifies the automata instance that has created and sent the message.

Note: This value is provided from the header of the current (last received) message.

6.8.50 *GetMsgObjectNumberTo*

Function prototype:

```
inline uint32 GetMsgObjectNumberTo()
```

Function description: This function returns the identification of the destination automata instance from the current message. This value is actually this automata instance.

Function returns: This function returns the value that identifies the automata instance that has received the message and that must process it.

Note: This value is provided from the header of the current (last received) message.

6.8.51 *GetMsgToAutomata*

Function prototype:

```
inline uint8 GetMsgToAutomata()
```

Function description: This function returns the identification of the destination automata type from the current message. This value is actually this automata type.

Function returns: This function returns the value that identifies the automata type that should receive the message and that should process it.

Note: This value is provided from the header of the current (last received) message.

6.8.52 *GetMsgToGroup*

Function prototype:

```
inline uint8 GetMsgToGroup()
```

Function description: This function returns the identification of the type of the group of the destination automata from the current message. This value is actually the group to which this automata type belongs.

Function returns: This function returns the value that identifies the group of automata that has received the message and that must process it.

Note: This value is provided from the header of the current (last received) message.

6.8.53 *GetNewMessage*

Function prototype:

```
inline uint8 *GetNewMessage()
```

Function description: This function returns the address of the buffer that contains the new message.

Function returns: This function returns the pointer to the already defined new message or the message under construction.

Note: If the new message does not exist, this function returns the value *NULL*. This function assumes that the programmer has already allocated a buffer for the new message by previously calling the function *PrepareNewMessage* or calling the function *GetBuffer*.

6.8.54 *GetNewMsgInfoCoding*

Function prototype:

```
inline uint8 GetNewMsgInfoCoding()
```

Function description: This function returns the identification of the information coding scheme used for the new message.

Function returns: This function returns the value that uniquely identifies the type of information coding.

Note: This value is provided from the header of the new message.

6.8.55 *GetNewMsgInfoLength*

Function prototype:

```
inline uint16 GetNewMsgInfoLength()
```

Function description: This function returns the payload length of the new message in bytes.

Function returns: This function returns the value of the new message payload size in bytes.

Note: The length of the message header is not included in the length returned by this message. By definition, the total message length is the sum of the length of the message header and the length of the message payload.

6.8.56 *GetNextParam*

Function prototype:

```
uint8 *GetNextParam(uint16 paramCode)
```

Function description: This function returns the address of the next instance of the given parameter type within the current message. The parameter of this function specifies the type of message parameter.

Function parameters:

paramCode: the identification of the type of message parameter

Function returns: The function returns the pointer to the next instance of the message parameter. If it does not exist, the function returns the value *NULL*.

Note: This function cannot be used by the programmer to get the first instance of the message parameter of a given type. It assumes that the first instance has already been provided by calling the function *GetParam*. Typically, the function *GetParam* is called once to provide the first instance of the parameter and then called iteratively to provide the next instances of the parameter.

6.8.57 *GetNextParamByte*

Function prototype:

```
bool GetNextParamByte(
 uint16 paramCode,
 BYTE &param)
```

Function description: This function searches for the next instance of the given type of the single-byte parameter in the current message. If the instance is found, the function copies it into its parameter specified by the reference and returns the value *true*; otherwise, it returns the value *false*. The parameters of this function specify the identification of the type of message parameter and the pointer to the memory area, where this function should store the next instance of the message parameter.

Function parameters:

paramCode: the identification of the type of the message parameter

param: the pointer to the memory area reserved by the programmer for the next instance of the message parameter

Function returns: This function returns the value *true* if the next instance of the message parameter is found. If the instance is not found, this function returns the value *false*.

Note: The programmer cannot use this function to get the first instance of the message parameter of the given type. This function assumes that the first instance has already been provided by calling the function *GetParamByte*. Typically, the function *GetParamByte* is called once to provide the first instance of the parameter and then called iteratively to provide the next instances of the parameter.

6.8.58   *GetNextParamDWord*

Function prototype:

```
bool GetNextParamDWord(
 uint16 paramCode,
 DWORD &param)
```

Function description: This function searches for the next instance of the given type of parameter 4 bytes in the current message. If the instance is found, the function copies it into its parameter specified by the reference and returns the value *true*; otherwise, it returns the value *false*. The parameters of this function specify the identification of the type of the message parameter and the pointer to the memory area, where this function should store the next instance of the message parameter.

Function parameters:

paramCode: the identification of the type of message parameter

param: the pointer to the memory area reserved by the programmer for the next instance of the message parameter

Function returns: This function returns the value *true* if the next instance of the message parameter is found. If the instance is not found, this function returns the value *false*.

Note: The programmer cannot use this function to get the first instance of the message parameter of the given type. This function assumes that the first instance has already been provided by calling the function *GetParamDWord*. Typically, the function *GetParamDWord* is called once to provide the first instance of the parameter and then called iteratively to provide the next instances of the parameter.

6.8.59   *GetNextParamWord*

Function prototype:

```
bool GetNextParamWord(
 uint16 paramCode,
 WORD &param)
```

Function description: This function searches for the next instance of the given type of parameter 2 bytes in the current message. If the instance is found, the function copies it into its parameter specified by the reference and returns the value *true*; otherwise, it returns the value *false*. The parameters of this function specify the identification of the type of the message parameter and the pointer to the memory area, where this function should store the next instance of the message parameter.

Function parameters:

paramCode: the identification of the type of message parameter

param: the pointer to the memory area reserved by the programmer for the next instance of the message parameter

Function returns: This function returns the value *true* if the next instance of the message parameter is found. If the instance is not found, this function returns the value *false*.

Note: The programmer cannot use this function to get the first instance of the message parameter of the given type. This function assumes that the first instance has already been provided by the call to the function *GetParamWord*. Typically, the function *GetParamWord* is called once to provide the first instance of the parameter and then called iteratively to provide the next instances of the parameter.

6.8.60   *GetObjectId*

Function prototype:

```
virtual uint32 GetObjectId()
```

Function description: This function returns the unique identification of this automata instance.

Function returns: This function returns the value that uniquely identifies this particular automata instance.

Note: This value has been automatically assigned to this automata instance by the function *Add*, which is called to add this automata instance to the FSM system.

6.8.61   *GetParam*

Function prototype:

```
uint8 *GetParam(uint16 paramCode)
```

Function description: This function returns the address of the first instance of the given type of message parameter within the current message. The parameter of this function specifies the identification of the parameter type.

Function parameters:

paramCode: the identification of the parameter type

Function returns: This function returns the pointer to the first instance of the message parameter within the current message. If no message parameters of the given type are found, this function returns the value *NULL*.

Note: This function returns the pointer to the beginning of the message parameter. The format of the message parameter is governed by the selected type of message information coding. For example, the parameter of the message *StandardMessage* consists of three fields. These fields are the parameter type (stored in 2 bytes), the parameter length (stored in 1 byte), and the information part of the parameter (stored in the number of bytes determined by the content of the previous field of the parameter).

6.8.62   *GetParamByte*

Function prototype:

```
bool GetParamByte(
 uint16 paramCode,
 BYTE &param)
```

Function description: This function searches for the first instance of the given type of single-byte parameter in the current message. If the instance is found, the function copies it into its parameter specified by the reference and returns the value *true*; otherwise, it returns the value *false*. The parameters of this function specify the identification of the type of message parameter and the pointer to the memory area, where this function should store the first instance of the message parameter.

Function parameters:

paramCode: the identification of the type of message parameter

param: the pointer to the memory area reserved by the programmer for the next instance of the message parameter

Function returns: This function returns the value *true* if the first instance of the message parameter is found. If the instance is not found, this function returns the value *false*.

Note: The programmer must use this function to get the first instance of the message parameter of the given type. Typically, this function is called once to provide the first instance of the parameter, and then the function *GetNextParamByte* is called iteratively to provide the next instances of the parameter.

6.8.63 *GetParamDWord*

Function prototype:

```
bool GetParamDWord(
 uint16 paramCode,
 DWORD &param)
```

Function description: This function searches for the first instance of the given type of parameter 4 bytes in the current message. If the instance is found, the function copies it into its parameter specified by the reference and returns the value *true*; otherwise, it returns the value *false*. The parameters of this function specify the identification of the type of message parameter and the pointer to the memory area, where this function should store the first instance of the message parameter.

Function parameters:

paramCode: the identification of the type of message parameter

param: the pointer to the memory area reserved by the programmer for the next instance of the message parameter

Function returns: This function returns the value *true* if the first instance of the message parameter is found. If the instance is not found, this function returns the value *false*.

Note: The programmer must use this function to get the first instance of the message parameter of the given type. Typically, this function is called once to provide the first instance of the parameter, and then the function *GetNextParamDWord* is called iteratively to provide the next instances of the parameter.

6.8.64 *GetParamWord*

Function prototype:

```
bool GetParamWord(
 uint16 paramCode,
 BYTE &param)
```

Function description: This function searches for the first instance of the given type of parameter 2 bytes in the current message. If the instance is found, the function copies it into its parameter specified by the reference and returns the value *true*; otherwise, it returns the value *false*. The parameters of this function specify the identification of the type of message parameter and the pointer to the memory area, where this function should store the first instance of the message parameter.

Function parameters:

paramCode: the identification of the type of message parameter

param: the pointer to the memory area reserved by the programmer for the next instance of the message parameter

Function returns: This function returns the value *true* if the first instance of the message parameter is found. If the instance is not found, this function returns the value *false*.

Note: The programmer must use this function to get the first instance of the message parameter of the given type. Typically, this function is called once to provide the first instance of the parameter, and then the function *GetNextParamWord* is called iteratively to provide the next instances of the parameter.

6.8.65 *GetProcedure*

Function prototype:

```
PROC_FUN_PTR GetProcedure(uint16 event)
```

Function description: This function returns the pointer to the event handler for the given event identifier and the current state of automata. The parameter of this function specifies the identification of the event type.

Function parameters:

event: the identification of the event type (message code)

Function returns: This function returns the pointer to the event handler. Essentially, the event handler is a C++ class function member that handles the given event type in the current state.

Note: The FSM system internal data structures contain all the necessary information about the automata states, the sets of recognizable events (messages) for all automata states, and the corresponding event handlers. This information must be defined for each automata type after it has been added to the FSM system by the function *Add*. The programmer specifies this information in the parameters of the function *Initialize*. If the event handler has not been specified by the function *Initialize* for the given event type in the current automata state, this function returns the pointer to the function *DoNothing*, which performs the default processing of the unexpected events (messages).

6.8.66   *GetRightMbx*

Function prototype:

```
virtual inline uint8 GetRightMbx()
```

Function description: This function returns the identification of the default mailbox assigned to the automata instance that is logically to the right of this automata instance.

Function returns: This function returns the number that uniquely identifies the default mailbox for the right automata instance.

Note: Historically, the terms *left* and *right* automata instance originate from SDL, where an automata instance typically communicates with its left and right neighbors. These neighbors have their own mailboxes, sometimes briefly called left and right mailboxes.

6.8.67   *GetRightAutomata*

Function prototype:

```
virtual inline uint8 GetRightAutomata()
```

Function description: This function returns the identification of the automata type that is logically to the right of this automata instance.

Function returns: This function returns the number that uniquely identifies the right automata type.

Note: By definition, right automata are logically placed to the right of the currently observed automata instance.

6.8.68   *GetRightGroup*

Function prototype:

```
virtual linline uint8 GetRightGroup()
```

Function description: This function returns the identification of the group of automata that is logically to the right of this automata instance.

Function returns: This function returns the number that uniquely identifies the right group of automata.

Note: By definition, a right group of automata is a group that contains right automata.

6.8.69   *GetRightObjectId*

Function prototype:

```
virtual inline uint32 GetRightObjectId()
```

Function description: This function returns the identification of the automata instance that is logically to the right of this automata instance.

Function returns: This function returns the number that uniquely identifies the right automata instance.

Note: By definition, right automata are logically placed to the right of the currently observed automata instance. This function returns the identification of the particular right automata instance with which the currently observed automata instance communicates.

6.8.70   *GetState*

Function prototype:

```
virtual inline uint8 GetState()
```

Function description: This function returns the identification of the current state of this automata instance.

Function returns: This function returns the value that uniquely identifies the current state of this automata instance.

6.8.71   *IsBufferSmall*

Function prototype:

```
virtual bool IsBuferSmall(
 uint8 *buff,
 uint32 length)
```

Function description: This function returns the value *true* if the size of the given buffer is not greater than the given size specified as the value of its second parameter; otherwise, it returns the value *false*. The parameters of this function specify the buffer whose size is to be checked and the size to be used as a measuring unit.

Function parameters:

buff: the pointer to the buffer whose size is to be checked

length: the value of the measuring unit

Function returns: This function returns the value *true* if the size of the given buffer is less than or equal to the given size. If the buffer size is greater than the given size, the function returns the value *false*.

6.8.72   *Initialize*

Function prototype:

```
virtual void Initialize() = 0
```

Function description: This function defines the automata state transition event handlers and timers used by this automata type. State transition event handlers are essentially the C++ functions defined by the programmer, which process events (messages). Timers are primitive time mechanisms used to restrict the duration of certain communication phases.

Note: While writing the function *Initialize*, the programmer normally defines the functions that process the expected events (messages) by calling the function *InitEventProc*, the functions that process the unexpected events by calling the function *InitUnexpectedEventProc*, and the timers by calling the function *InitTimerBlock*.

6.8.73   *InitEventProc*

Function prototype:

```
void InitEventProc(
 uint8 state,
 uint16 event,
 PROC_FUN_PTR fun)
```

Function description: This function defines the given state transition event handler for the given automata state and the given event (message code). The parameters of this function specify the identification of the state of this automata type, the identification of the event type, and the pointer to the event handler.

Function parameters:

state: the identification of the state of this automata type

event: the identification of the event type

fun: the pointer to the event handler

Note: This function may be used only within the definition of the function *Initialize*. A sequence of calls to this function fills in the internal state table for this automata type. This table is used by the FSM system and this automata type during its normal operation to locate the event handler that corresponds to the given pair (state, event).

6.8.74 *InitTimerBlock*

Function prototype:

```
void InitTimerBlock (
 uint16 tmrId,
 uint32 count,
 uint16 signalId)
```

Function description: This function initializes the given timer by the given duration and the timer expiration message code. The parameters of this function specify the timer identification, the timer duration, and the identification of the message to be sent to this automata type when the specified timer expires.

Function parameters:

tmrId: the timer identification

count: the timer duration (in timer ticks)

signalId: the identification of the message (signal) to be sent by the specified timer

Note: The timer identification is a value selected by the programmer. This value uniquely identifies the timer to the automata type that uses it in all the timer-related primitives, namely, *InitTimerBlock, ResetTimer, RestartTimer, StartTimer*, and *StopTimer*. Uniqueness of identifiers is limited to the scope of a single automata type. If the timer expires, it sends a special message (referred to as a *signal*) to the automata instance that has started that timer. The code of this message is set to the value of the parameter *SignalId*. The kernel calculates the absolute timer duration in seconds by dividing the time resolution specified for automata type with the time resolution of the FSM system and by multiplying this result with the basic timer resolution specified as the parameter of the function *InitKernel*.

6.8.75 *InitUnexpectedEventProc*

Function prototype:

```
void InitUnexpectedEventProc(
 uint8 state,
 PROC_FUN_PTR fun)
```

Function description: This function defines the given state transition event handler for unexpected events in the given automata state. The parameters of the function specify the automata state and the unexpected event handler, which is essentially a C++ function that handles unexpected events (messages).

Function parameters:

state: the value that uniquely identifies the automata state

fun: the pointer to the unexpected event handler

Note: If the unexpected event (message) handler does not exist because it has not been defined by this function, the FSM system and this automata type will use the function *DoNothing* to handle unexpected messages for all the states in which the unexpected message is not defined.

6.8.76   *IsTimerRunning*

Function prototype:

```
bool IsTimerRunning(uint16 id)
```

Function description: This function returns the value *true* if a given timer is active (running); otherwise, it returns the value *false*. The parameter of this function specifies the timer identification.

Function parameters:

id: the timer identification

Function returns: This function returns the value *true* if the timer is running. If the timer is not active, this function returns the value *false*.

Note: The timer may not be active because it has not been started at all, or it has been started but has expired in the meantime.

6.8.77   *NoFreeObjectProcedure*

Function prototype:

```
void NoFreeObjectProcedure(uint8 *msg)
```

Function description: This function defines the behavior of this automata type if the list of free automata of this type is used, and if it is empty at the moment when a free instance is requested. The parameter of this function specifies the pending event (message).

Function parameters:

msg: the pointer to the pending message

Note: This function is used if a group of automata of this type is used as a pool of resources of the same type. This function is called if the message related to this automata type appears and no available automata instances (resources) of this type are available. The programmer should write their own function to handle this situation in an application-specific way. This situation is additionally handled at the level of the FSM system by the function *NoFreeInstances*.

6.8.78 NoFreeInstances

Function prototype:

```
virtual void NoFreeInstances() = 0
```

Function description: This function defines the behavior of the FSM system if a list of free automata is used, and if it is empty at the moment when a free instance is requested.

Note: This function is used if a group of automata of this type is used as a pool of resources of the same type within the FSM system. This function is called if the message related to this automata type appears and no available automata instances (resources) of this type are available. The programmer should write their own function to handle this situation in an application-specific way. This situation is additionally handled at the level of this automata type by the function *NoFreeObjectProcedure*.

6.8.79 ParseMessage

Function prototype:

```
virtual bool ParseMessage(uint8 *msg)
```

Function description: This function checks if the given message is coded properly and, if it is, it becomes the current message (its pointer is assigned to the internal variable *CurrentMessage*). The parameter of this function specifies the message to be parsed.

Function parameters:

msg: the pointer to the message to be parsed

Function returns: This function returns the value *true* if the message syntax is correct; otherwise, it returns the value *false*.

Note: This function is called internally for each received message. Normally, this function is called after the reception of the message to check its syntax. If the message syntax is correct, further message processing functions are called. Otherwise, the FSM system reports an error and discards the syntactically incorrect message.

6.8.80   *PrepareNewMessage(uint8*)*

Function prototype:

```
virtual void PrepareNewMessage(uint8 *msg)
```

Function description: This function defines the given buffer as the new message buffer by assigning the given pointer to the internal variable *NewMessage*. The buffer is used by this automata instance as a working area for the construction of the new message. The parameter of this function specifies the buffer.

Function parameters:
 msg: the pointer to the buffer

Note: If the programmer wants to create a new message, they would normally call the function *GetBuffer* to obtain the buffer for the construction of the message. Next, the programmer would call this function to declare the buffer provided by the kernel as the buffer that will contain the new message. After this declaration, the programmer may use all the functions from the family of functions that operate on the new message to construct the new message. Basically, these are the *AddParamX* functions.

6.8.81   *PrepareNewMessage(uint32, uint16, uint8)*

Function prototype:

```
virtual void PrepareNewMessage(
 uint32 length,
 uint16 code,
 uint8 infoCode = LOCAL_PARAM_CODING)
```

Function description: This function creates the new message of the given length with the given message code and the given type of information coding. The parameters of this function specify the message length, the message code, and the identification of the type of message information coding.

Function parameters:
 length: the message length
 code: the value of the message code
 infoCode: the identification of the type of message information coding

Note: Dealing with static messages of fixed and known sizes is easy. In this case, the programmer normally knows the size of the message they must create. The programmer creates the new message by calling this function and specifying the size as the value of the function parameter *length*. However, dealing with dynamic messages is more complicated, because the message length might not be known in advance. In this case, the programmer may specify the value 0 as the value of the parameter *length*. This function, in turn, will create the empty message that has its header, but has no payload. Further on, the programmer typically uses functions *AddParamX* to dynamically add new parameters to the message. Whenever not enough room exists for the new parameter in the existing new message buffer, the function *AddParamX* transparently allocates a bigger buffer, moves the content of the new message into it, and releases the smaller buffer. Of course, the price paid for this flexibility is the processing overhead for transparent buffer management.

6.8.82 *Process*

Function prototype:

```
virtual void Process(uint8 *msg)
```

Function description: This function performs the preparations for the message processing and selects the state transition event handler based on the message code and current state of this automata instance. After completion of the message processing, this function releases the buffer used by the message. The parameter of this function specifies the message to be processed.

Function parameters:

msg: the pointer to the message to be processed

Note: This function is called internally by this automata type. Because this function is virtual, the programmer may define the message handling procedure in accordance with the application-specific requirements.

6.8.83 *PurgeMailBox*

Function prototype:

```
void PurgeMailBox()
```

Function description: This function purges all the messages from the mailbox assigned to this automata type and releases all the buffers assigned to the messages.

Note: Notice that the mailbox is assigned to an automata type rather than to an individual instance of this type. This means that the mailbox may contain the messages addressed to different instances of this type. This function does not differentiate the messages. Instead, it simply purges all of them.

6.8.84 *RemoveParam*

Function prototype:

```
bool RemoveParam(uint16 paramCode)
```

Function description: This function removes the given type of message parameter from the new message. The parameter of this function specifies the identification of the type of message parameter.

Function parameters:

paramCode: the value that uniquely identifies the type of message parameter

Function returns: This function returns the value *true* if the given type of message parameter is successfully found and removed. If the new message does not contain the given type, this function returns the value *false*.

Note: Removing the type of message parameter with identification 0 is not recommended because it marks the end of the parameters in the message. The *FSMSystem* library debug version will report an error in that case and stop the program execution.

6.8.85 *Reset*

Function prototype:

```
virtual void Reset()
```

Function description: This function resets this automata instance by returning it to its initial state and stopping all its active timers.

Note: If the programmer wants to specify some additional actions to be undertaken during the restart operation, they may redefine this default behavior by writing the corresponding function member of a class derived from the class *FiniteStateMachine*.

6.8.86 *ResetTimer*

Function prototype:

```
void ResetTimer(uint16 id)
```

Function description: This function resets the internal timer block object and returns the buffer allocated by the *StartTimer* primitive to the FSM Library kernel. The parameter of this function specifies the identification of the timer.

Function parameters:
id: the value that uniquely identifies the timer

6.8.87   *RestartTimer*

Function prototype:

```
void RestartTimer(uint16 tmrId)
```

Function description: This function restarts the given timer. It is logically equivalent to a sequence of *StopTimer* and *StartTimer* primitives. The parameter of this function specifies the identification of the timer.

Function parameters:
tmrId: the value that uniquely identifies the timer

6.8.88   *RetBuffer*

Function prototype:

```
virtual void RetBuffer(uint8 *buff)
```

Function description: This function returns the given buffer to the FSM Library kernel. Normally, each memory buffer is returned at the end of its life cycle. Failure to do so leads to a memory leak problem. The parameter of this function specifies the buffer to be released.

Function parameters:
buff: the pointer to the buffer to be released

Note: The programmer must pay special attention to releasing the buffers when they are not needed anymore because the *FSMSystem* library does not include a garbage collector. Memory outage causes the exception that will stop the program execution.

6.8.89   *ReturnMsg*

Function prototype:

```
void ReturnMsg(uint8 mbxId)
```

Function description: This function makes a copy of the current message and sends it to the given mailbox. This primitive is used frequently for message forwarding. On many occasions, the communication process must react in this simple way. The parameter of this function specifies the identification of the mailbox.

Function parameters:

mbxId: the value that uniquely identifies the mailbox

6.8.90   *SetBitParamByteBasic*

Function prototype:

```
void SetBitParamByteBasic(
 BYTE param,
 uint32 offset,
 uint32 mask = MASK_32_BIT)
```

Function description: This function sets the given single-byte parameter of the new message to the result of the bit-wise inclusive OR operation applied to the given parameter and its previous value masked (bit-wise AND operation) with the given bit-mask. The parameters of this function specify the value of the single-byte parameter, the offset of the target parameter of the new message, and the value of the bit-mask.

Function parameters:

param: the value of the single-byte parameter

offset: the target parameter of the new message

mask: the value of the bit-mask

6.8.91   *SetBitParamDWordBasic*

Function prototype:

```
void SetBitParamDWordBasic(
 DWORD param,
 uint32 offset,
 uint32 mask = MASK_32_BIT)
```

Function description: This function sets the given 4-byte parameter of the new message to the result of the bit-wise inclusive OR operation applied to the given parameter and its previous value masked (bit-wise AND operation) with the given bit-mask. The parameters of this function specify the value of the 4-byte parameter, the offset of the target parameter of the new message, and the value of the bit-mask.

Function parameters:

param: the value of the 4-byte parameter

offset: the target parameter of the new message

mask: the value of the bit-mask

6.8.92   SetBitParamWordBasic

Function prototype:

```
void SetBitParamWordBasic(
 WORD param,
 uint32 offset,
 uint32 mask = MASK_32_BIT)
```

Function description: This function sets the given 2-byte parameter of the new message to the result of the bit-wise inclusive OR operation applied to the given parameter and its previous value masked (bit-wise AND operation) with the given bit-mask. The parameters of this function specify the value of the 2-byte parameter, the offset of the target parameter of the new message, and the value of the bit-mask.

Function parameters:

param: the value of the 2-byte parameter

offset: the target parameter of the new message

mask: the value of the bit-mask

6.8.93   *SetCallId()*

Function prototype:

```
inline void SetCallId()
```

Function description: This function sets the default value of the attribute *CallId* of this automata instance.

Note: This function automatically allocates the first available identification and assigns it to the protected class attribute *CallId*, completely transparent to the programmer.

6.8.94   *SetCallId(uint32)*

Function prototype:

```
inline void SetCallId(uint32 id)
```

Function description: This function sets the given value of the attribute *CallId* of this automata instance. The parameter of this function specifies the value to be assigned to the attribute *CallId*.

Function parameters:

  id: the value to be assigned to the attribute *CallId*

Note: In contrast to an overloaded function without any parameters in its signature, this function enables the programmer to manually assign the value to the attribute *CallId*. However, this value must be unique. The programmer must pay special attention to the assignment of these numbers, especially if they mix this function call with function calls to the overloaded function that assigns the default values.

6.8.95   *SetCallIdFromMsg*

Function prototype:

```
inline void SetCallIdFromMsg()
```

Function description: This function sets the attribute *CallId* of this automata instance to the value of the parameter *CallId* of the current message. This primitive is used to store the reference number specific to the communication protocol.

6.8.96   *SetDefaultFSMData*

Function prototype:

```
virtual void SetDefaultFSMData() = 0
```

Function description: This function sets the automata-specific data to their default values. It is typically used before the normal operation phase.

Note: The programmer must define this virtual function for a class derived from the class *FiniteStateMachine*. They do so by writing a C++ function that initializes the problem-specific data.

6.8.97   *SetDefaultHeader*

Function prototype:

```
virtual void SetDefaultHeader(uint8 infoCoding = 0)
```

Function description: This function sets the default header field values for the given type of message information coding. The parameter of this function specifies the identification of the type of message information coding.

Function parameters:

infoCoding: the type of message information coding

Note: The programmer must define this virtual function for a class derived from the class *FiniteStateMachine*. They do so by writing a C++ function that fills in the protocol-specific data in the new message header.

6.8.98   *SetGroup*

Function prototype:

```
inline void SetGroup(uint8 id)
```

Function description: This function sets the identification of the group of automata for this automata type to the given value. This primitive is used to declare group membership. The parameter of this function specifies the value to be assigned to the corresponding class attribute.

Function parameters:

id: the value that uniquely identifies the group of automata

6.8.99   *SetInitialState*

Function prototype:

```
virtual void SetInitialState()
```

Function description: This function sets the current state of this automata instance to its initial state.

Note: The programmer must obey the rule that the value of the identification of the initial automata state is 0.

6.8.100   *SetKernelObjects*

Function prototype:

```
static void SetKernelObjects(
 TPostOffice *postOffice,
 TBuffers *buffers,
 CTimer *timer)
```

Function description: This function sets the *FSMSystem* library kernel objects (post office, buffers, and timers), which are common for all the automata in the FSM system. The parameters of this function specify the post office object, the buffers object, and the timers object.

Function parameters:

postOffice: the pointer to the post office object

buffers: the pointer to the buffers object

timer: the pointer to the timers object

Note: This function is called internally by the function *InitKernel*. Remember that this function defines the kernel objects that are common for all automata types and all their instances. An accidental call to this function may cause unpredictable behavior in the FSM system.

6.8.101 *SetLeftMbx*

Function prototype:

```
inline void SetLeftMbx(uint8 mbx)
```

Function description: This function sets the default identification of the mailbox assigned to the automata instance that is logically to the left of this automata instance. The parameter of this function specifies the identification of the mailbox.

Function parameters:

mbx: the value that uniquely identifies the mailbox

6.8.102 *SetLeftAutomata*

Function prototype:

```
inline void SetLeftAutomata(uint8 automata)
```

Function description: This function sets the identification of the automata type that is logically to the left of this automata instance. The parameter of this function specifies the identification of the automata type.

Function parameters:

automata: the value that uniquely identifies the automata type

6.8.103 *SetLeftObject*

Function prototype:

```
inline void SetLeftObject(uint8 group)
```

Function description: This function sets the identification of the type of the group of automata that is logically to the left of this automata instance. The parameter of this function specifies the identification of the group of automata.

Function parameters:

group: the value that uniquely identifies the group of automata

6.8.104   *SetLeftObjectId*

Function prototype:

```
inline void SetLeftObjectId(uint32 id)
```

Function description: This function sets the identification of the automata instance that is logically to the left of this automata instance. The parameter of this function specifies the identification of the automata instance.

Function parameters:

id: the identification of the automata instance

6.8.105   *SetLogInterface*

Function prototype:

```
static void SetLogInterface(LogInterface *logingObject)
```

Function description: This function defines the object responsible for message logging. The object is an instance of a class derived from the class *LogInterface*. The parameter of this function specifies the message logging object.

Function parameters:

logingObject: the pointer to the message logging object

Note: The programmer must not call this function before the initialization of all the automata included in the FSM system has been finished. The logging object may log data to the file on the local mass memory unit (e.g., flash memory) or to the network file server. The log file is essential for debugging and test and verification purposes.

6.8.106   *SendMessage(uint8)*

Function prototype:

```
inline void SendMessage(uint8 mbxId)
```

Function description: This function sends the new message to the given mailbox. The parameter of this function specifies the identification of the mailbox.

Function parameters:

mbxId: the value that uniquely specifies the mailbox

Note: By definition, the internal pointer *NewMessage* points to the buffer that contains the new message. The programmer initializes this pointer by calling the function *PrepareNewMessage*.

6.8.107   SendMessage(uint8, uint8*)

Function prototype:

```
inline void SendMessage(
 uint8 mbxId,
 uint8 *msg)
```

Function description: This function sends the given message to the given mailbox. The parameters of this function specify the identification of the mailbox and the message to be sent to that mailbox.

Function parameters:

mbxId: the value that uniquely identifies the mailbox

msg: the pointer to the message

6.8.108   SetMessageFromData

Function prototype:

```
void SetMessageFromData()
```

Function description: This function sets the header fields of the new message related to the originating automata instance to the values specific to this automata instance. The data specifying the originating automata instance are its type, group, and identification.

Note: This function is automatically called from the function *SendMessage*.

6.8.109   SetMsgCallId(uint32)

Function prototype:

```
inline void SetMsgCallId(uint32 id)
```

Function description: This function sets the call ID parameter of the new message to the given value. The parameter of this function specifies the value of the call ID.

Function parameters:

id: the value of the call ID

Note: The call ID parameter has been traditionally used to identify a single telephone call. In general, it may be used to uniquely identify a communication process or a transaction that engages a group of automata that participates in its processing.

6.8.110   *SetMsgCallId(unit32, unit8*)*

Function prototype:

```
inline void SetMsgCallId(
 uint32 id,
 uint8 *msg)
```

Function description: This function sets the call ID parameter of the given message to the given value. The parameters of this function specify the value of the call ID and the target message.

Function parameters:

id: the value of the call ID

msg: the pointer to the buffer that contains the target message

Note: The value of the call ID parameter is the same for all the messages involved in a transaction or a process, e.g., a single telephone call.

6.8.111   *SetMsgCode(uint16)*

Function prototype:

```
inline void SetMsgCode(uint16 code)
```

Function description: This function sets the message code parameter of the new message to the given value. The parameter of this message specifies the message code.

Function parameters:

code: the message code

6.8.112   *SetMsgCode(uint16, uint8*)*

Function prototype:

```
inline void SetMsgCode(
 uint16 code,
 uint8 *msg)
```

Function description: This function sets the message code parameter of the given message to the given value. The parameters of this function specify the message code and the target message.

Function parameters:

code: the message code

msg: the pointer to the buffer that contains the target message

6.8.113   SetMsgFromAutomata(uint8)

Function prototype:

```
inline void SetMsgFromAutomata(uint8 from)
```

Function description: This function sets the type of the originating automata parameter of the new message to the given value. The parameter of this function specifies the identification of the automata type that is the message source.

Function parameters:

from: the identification of the automata type

Note: This function is automatically called by the function *SetMessage FromData*.

6.8.114   SetMsgFromAutomata(uint8, uint8*)

Function prototype:

```
inline void SetMsgFromAutomata(
 uint8 from,
 uint8 *msg)
```

Function description: This function sets the type of the originating automata parameter of the given message to the given value. The parameters of this function specify the type of automata that is the message source and the target message.

Function parameters:

from: the automata type that is the message source

msg: the pointer to the buffer that contains the target message

6.8.115   SetMsgFromGroup(uint8)

Function prototype:

```
inline void SetMsgFromGroup(uint8 from)
```

Function description: This function sets the type of the originating group of automata parameter of the new message to the given value. The parameter of this message specifies the identification of the group of automata that is the message source.

Function parameters:

from: the identification of the group of automata that is the message source

Note: This function is automatically called by the function *SetMessage FromData*.

6.8.116   *SetMsgFromGroup(uint8, uint8*)*

Function prototype:

```
inline void SetMsgFromGroup(
 uint8 from,
 uint8 *msg)
```

Function description: This function sets the type of the originating group of automata parameter of the given message to the given value. The parameters of this function specify the identification of the group of automata that is the message source and the target message.

Function parameters:

from: the identification of the group of automata that is the message source

msg: the pointer to the buffer that contains the target message

6.8.117   *SetMsgInfoCoding(uint8)*

Function prototype:

```
inline void SetMsgInfoCoding(uint8 codingType)
```

Function description: This function sets the message information coding parameter of the new message to the given value. The parameter of this message specifies the identification of the information coding scheme.

Function parameters:

codingType: the value that uniquely specifies the information coding scheme

Note: This function is automatically called by the function *PrepareNewMessage*.

6.8.118 SetMsgInfoCoding(uint8, uint8*)

Function prototype:

```
inline void SetMsgInfoCoding(
  uint8 codingType,
  uint8 *msg)
```

Function description: This function sets the message information coding parameter of the given message to the given value. The parameters of this function specify the identification of the information coding scheme and the target message.

Function parameters:
codingType: the identification of the information coding scheme
msg: the pointer to the target message

6.8.119 SetMsgInfoLength(uint16)

Function prototype:

```
inline void SetMsgInfoLength(uint16 length)
```

Function description: This function sets the message payload (useful information) length parameter of the new message. The parameter of this function specifies the value of the payload length.

Function parameters:
length: the payload length in octets (bytes)

Note: All the *AddParamX* functions—which are responsible for adding parameters to the new message—call this function automatically to update the length of the message payload.

6.8.120 SetMsgInfoLength(uint16, uint8*)

Function prototype:

```
inline void SetMsgInfoLength(
  uint16 length,
  uint8 *msg)
```

Function description: This function sets the message payload (useful information) length parameter of the given message. The parameters of this function specify the value of the payload length and the target message.

Function parameters:

length: the payload length in octets (bytes) ·

msg: the pointer to the buffer that contains the target message

6.8.121   *SetMsgObjectNumberFrom(uint32)*

Function prototype:

```
inline void SetMsgObjectNumberFrom(uint32 from)
```

Function description: This function sets the originating automata instance identification parameter of the new message to the given value. The parameter of this function specifies the identification of the automata instance that is the message source.

Function parameters:

from: the identification of the automata instance that is the message source

Note: This function is automatically called by the function *SetMessage FromData*.

6.8.122   *SetMsgObjectNumberFrom(uint32, uint8*)*

Function prototype:

```
inline void SetMsgObjectNumberFrom(
 uint32 from,
 uint8 *msg)
```

Function description: This function sets the originating automata instance identification parameter of the given message to the given value. The parameters of this message specify the identification of the automata instance that is the message source and the target message.

Function parameters:

from: the identification of the automata instance that is the message source

msg: the pointer to the buffer that contains the target message

6.8.123   *SetMsgObjectNumberTo(uint32)*

Function prototype:

```
inline void SetMsgObjectNumberTo(uint32 to)
```

Function description: This function sets the destination automata instance identification parameter of the new message to the given value. The parameter of this function specifies the automata instance that is the message destination.

Function parameters:

to: the automata instance that is the message destination

6.8.124 *SetMsgObjectNumberTo(uint32, uint8*)*

Function prototype:

```
inline void SetMsgObjectNumberTo(uint32 to,uint8 *msg)
```

Function description: This function sets the destination automata instance identification parameter of the given message to the given value. The parameters of this function specify the automata instance that is the message destination and the target message.

Function parameters:

to: the automata instance that is the message destination

msg: the pointer to the buffer that contains the target message

6.8.125 *SetMsgToAutomata(uint8)*

Function prototype:

```
inline void SetMsgToAutomata(uint8 to)
```

Function description: This function sets the destination automata type identification parameter of the new message to the given value. The parameter of this function specifies the automata type that is the message destination.

Function parameters:

to: the automata type that is the message destination

6.8.126 *SetMsgToAutomata(uint8, uint8*)*

Function prototype:

```
inline void SetMsgToAutomata(
 uint8 to,
 uint8 *msg)
```

Function description: This function sets the destination automata type identification parameter of the given message to the given value. The parameters of this function specify the identification of the automata type that is the message destination and the target message.

Function parameters:

to: the identification of the automata type that is the message destination

msg: the pointer to the buffer that contains the target message

6.8.127   *SetMsgToGroup(uint8)*

Function prototype:

```
inline void SetMsgToGroup(uint8 to)
```

Function description: This function sets the destination automata group identification parameter of the new message to the given value. The parameter of this function specifies the identification of the group of automata that is the message destination.

Function parameters:

to: the identification of the group of automata that is the message destination

6.8.128   *SetMsgToGroup(uint8, uint8*)*

Function prototype:

```
inline void SetMsgToGroup(
 uint8 to,
 uint8 *msg)
```

Function description: This function sets the destination automata group identification parameter of the given message to the given value. The parameters of this function specify the identification of the group of automata that is the message destination and the target message.

Function parameters:

to: the identification of the group of automata that is the message destination

msg: the pointer to the buffer that contains the target message

6.8.129   *SendMessageLeft*

Function prototype:

```
void SendMessageLeft()
```

Function description: This function sends the new message to the mailbox assigned to the automata instance that is logically to the left of this automata instance.

Note: The programmer may use this function if they have already defined the left automata instance for the currently observed automata instance. This definition includes the definition of the mailbox assigned to the left automata instance. If the left automata instance and its mailbox are defined, this function automatically fills in all the data related to both the source (originating) and destination automata instances within the new message and sends the new message to the left mailbox.

6.8.130 *SendMessageRight*

Function prototype:

```
void SendMessageLeft()
```

Function description: This function sends the new message to the mailbox assigned to the automata instance that is logically to the right of this automata instance.

Note: The programmer may use this function if they have already defined the right automata instance for the currently observed automata instance. This definition includes the definition of the mailbox assigned to the right automata instance. If the right automata instance and its mailbox are defined, this function automatically fills in all the data related to both the source (originating) and destination automata instances within the new message and sends the new message to the right mailbox.

6.8.131 *SetNewMessage*

Function prototype:

```
inline void SetNewMessage(uint8 *msg)
```

Function description: This function sets the new message to the given message by assigning the given message pointer to the internal pointer to the new message. The parameter of this function specifies the target message.

Function parameters:

msg: the pointer to the buffer that contains the target message

6.8.132 *SetObjectId*

Function prototype:

```
inline void SetObjectId(uint32 id)
```

Function description: This function sets the identification of this automata instance to the given value. The parameter of this function specifies the identification of this automata instance.

Function parameters:

id: the value that uniquely identifies this automata instance

6.8.133   *SetRightMbx*

Function prototype:

```
inline void SetRightMbx(uint8 mbx)
```

Function description: This function sets the identification of the mailbox assigned to the automata instance that is logically to the right of this automata instance. The parameter of this message specifies the identification of the right mailbox for this automata instance.

Function parameters:

mbx: the identification of the right mailbox for this automata instance

6.8.134   *SetRightAutomata*

Function prototype:

```
inline void SetRightAutomata(uint8 automata)
```

Function description: This function sets the identification of the automata type that is logically to the right of this automata instance. The parameter of this function specifies the automata type that is to the right of this automata instance.

Function parameters:

automata: the identification of the automata type

6.8.135   *SetRightObject*

Function prototype:

```
inline void SetRightObject(uint8 group)
```

Function description: This function sets the identification of the type of the group of automata that is logically to the right of this automata instance. The parameter of this function specifies the type of the group of automata that is to the right of this automata instance.

Function parameters:

group: the identification of the group of automata

6.8.136 *SetRightObjectId*

Function prototype:

```
inline void SetRightObjectId(uint32 id)
```

Function description: This function sets the identification of the automata instance that is logically to the right of this automata instance. The parameter of this function specifies the identification of the automata instance that is to the right of this automata instance.

Function parameters:

id: the identification of the automata instance

6.8.137 *SetState*

Function prototype:

```
inline void SetState(uint8 state)
```

Function description: This function sets the identification of the current state of this automata instance. The parameter of this function specifies the identification of the state.

Function parameters:

state: the value that uniquely identifies the particular state of automata

6.8.138 *StartTimer*

Function prototype:

```
void StartTimer(uint16 tmrId)
```

Function description: This function starts the given timer. The parameter of this function specifies the identification of the timer.

Function parameters:

tmrId: the value that uniquely identifies the particular timer

Note: Uniqueness of the timer identifier is limited to the scope of a single automata type that uses it.

6.8.139 *StopTimer*

Function prototype:

```
void StopTimer(uint16 tmrId)
```

Function description: This function stops the given timer. The parameter of this function specifies the identification of the timer.

Function parameters:
tmrId: the value that uniquely identifies the particular timer

Note: Uniqueness of the timer identifier is limited to the scope of a single automata type that uses it.

6.8.140　*SysClearLogFlag*

Function prototype:

```
static void SysClearLogFlag()
```

Function description: This function stops the logging of the messages exchanged by the automata.

6.8.141　*SysStartAll*

Function prototype:

```
Static void SysStartAll()
```

Function description: This function starts the logging of the messages exchanged by the automata.

Note: Normally, the programmer should start the logging of messages before they start the individual automata included in the FSM system.

6.8.142　*NetFSM*

Function prototype:

```
NetFSM(
 uint16 numOfTimers = DEFAULT_TIMER_NO,
 uint16 numOfState = DEFAULT_STATE_NO,
 uint16 maxNumOfProceduresPerState = DEFAULT_PROCEDURE_NO_PER_STATE,
 bool getMemory = true)
```

Function description: This constructor initializes the object that represents an instance of the given automata type together with the data structures needed for its proper operation. The parameters of this function specify the number of timers to be used by this automata type, the total number of states for this automata type, the maximal number of state transitions per state for this automata type, and the memory allocation indicator. All the parameters have their default values as shown in the function prototype declaration above.

Function parameters:

numOfTimers: the number of timers to be used by this automata type

numOfState: the total number of states for this automata type

maxNumOfProceduresPerState: the maximal number of state transitions per state

getMemory: the memory allocation indicator

Note: The programmer may call this a constructor without parameters. In this case, the parameters will be set to their corresponding default values. The value of the fourth parameter *getMemory* regulates memory allocation. By default, this indicator is set to the value *true*, which means that the constructor will take care of memory allocation. Default memory allocation is not optimal because it is based on the maximal number of transitions per state. This compromise has been made intentionally because it leads to a very simple FSM definition API. If the programmer wants to optimize memory allocation, they may build the data structure describing the FSM by allocating necessary memory blocks from the memory heap, linking them together, and storing the pointer to this data structure in the protected class field member *States* before this function is called. In that case, the programmer would set the fourth parameter *getMemory* to the value *false*.

6.8.143   *convertFSMToNetMessage*

Function prototype:

```
virtual void convertFSMToNetMessage() = 0
```

Function description: This function converts the internal message format into the external message format appropriate for the transmission over the TCP/IP network.

Note: The programmer must define this virtual function by writing the corresponding function member of a class derived from the class *NetFSM*.

6.8.144   *convertNetToFSMMessage*

Function prototype:

```
virtual uint16 convertNetToFSMMessage() = 0
```

Function description: This function converts the external message format into the internal message format appropriate for the communication within the FSM system.

Function returns: This function returns the code of the received message.

Note: The programmer must define this virtual function by writing the corresponding function member of a class derived from the class *NetFSM*.

6.8.145 *establishConnection*

Function prototype:

```
void establishConnection()
```

Function description: This function establishes the TCP connection between two geographically distributed FSM systems.

Note: The programmer must call this function before they can call the function *sendToTCP* to send the message to the remote FSM system.

6.8.146 *getProtocolInfoCoding*

Function prototype:

```
virtual uint8 getProtocolInfoCoding() = 0
```

Function description: This function returns the identification of the type of external message coding.

Function returns: This function returns the value that uniquely identifies the type of coding of the external message.

6.8.147 *sendToTCP*

Function prototype:

```
void sendToTCP()
```

Function description: This function sends the new message to the remote FSM system over the previously established TCP connection.

Note: The programmer must call the function *establishConnection* before they can call this function.

6.9 A Simple Example with Three Automata Instances

This section shows how the programmer can construct the FSM system and how they can add individual automata instances to it. To keep the example simple, we include only one use case, *Show Simple Demo* (Figure 6.1). The realization of this use case is a simple collaboration that comprises three instances (*instance_1*, *instance_2*, and *instance_3*) of the same automata type

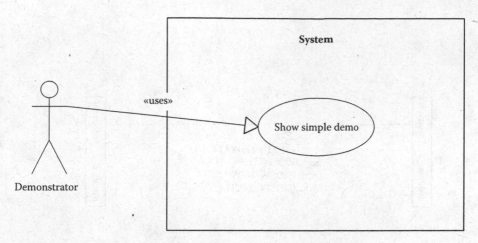

FIGURE 6.1
Simple use case diagram for the example with three automata instances.

(*Automata*), which are added to the FSM system (Figure 6.2). These three automata instances have the trivial task of exchanging the given number of messages in a "round robin" fashion.

At the beginning, the main thread calls the function *StartDemo* of *instance_1*, which, in turn, asynchronously sends itself the message *IDLE_START*. Upon reception of this message, *instance_1* sends the message *IDLE_MSG* to *instance_2*, which increments the message sequence number and forwards the message to *instance_3*; the latter translates it to the message *MSG_MSG* and sends it back to *instance_1*. This message then makes two full circles around the collaborating objects. Finally, *instance_1* translates it to the message *MSG_STOP* and sends it to *instance_2*, which, in turn, forwards it to *instance_3*. The corresponding sequence diagram is shown in Figure 6.3. The conditions A, B, and C regulate the already mentioned translations of the messages.

The statechart diagram that describes the behavior of a single automata instance is organized into two hierarchical levels. The top level comprises two simple states (*IDLE* and *MESSAGE*) and four composite states (*Automata_IDLE_START*, *Automata_IDLE_MSG*, *Automata_MSG_MSG*, and *Automata_MSG_STOP*) (Figure 6.4). The symbolic constant *MAX_MSG_NUM* is defined to have the value 10 in this example. The variable *msgno* is the message sequence number, whose values are shown in parentheses in Figures 6.2 and 6.3. Later in the program text, this short variable name suitable for figures is replaced with the longer self-documenting name *msgNumber*.

The individual composite states *Automata_IDLE_START*, *Automata_IDLE_MSG*, *Automata_MSG_MSG*, and *Automata_MSG_STOP* are shown in

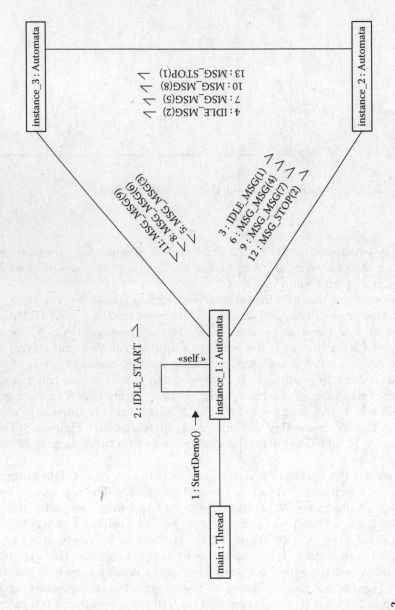

FIGURE 6.2
Collaboration diagram for the example with three automata instances.

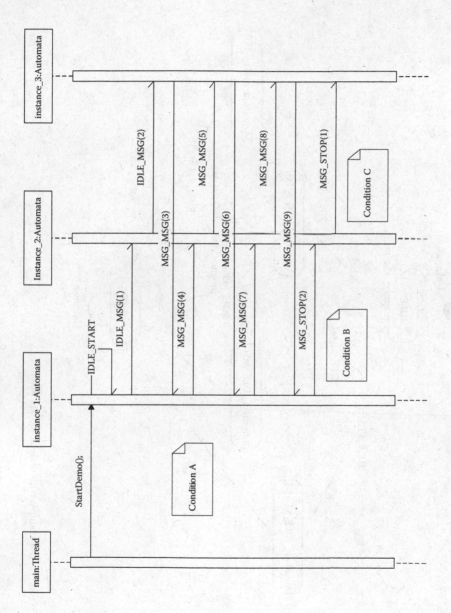

FIGURE 6.3
Sequence diagram for the example with three automata instances.

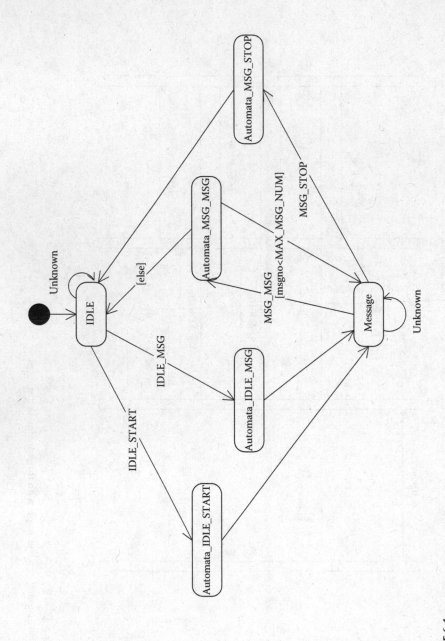

FIGURE 6.4
Statechart diagram for the example with three automata instances.

Figures 6.5 through 6.8, respectively. These have been made rather detailed to show how to provide the mapping from the UML model to the corresponding program code by the application of forward engineering. Essentially, the state transition actions are sequences of calls to functions provided by the FSM Library, such as *PrepareNewMessage, AddParamDWord, SendMessage,* and so on.

Each of the composite states can be modeled as an operation by the corresponding activity diagram. The activity diagrams for the operations *Automata_IDLE_START, Automata_IDLE_MSG, Automata_MSG_MSG,* and *Automata_MSG_STOP* are shown in Figures 6.9 through 6.12, respectively. Again, these diagrams have been made by applying forward engineering, but on a slightly higher abstraction level, using informal text statements instead of explicit functions calls. Essentially, composite statechart and activity diagrams have the same semantics in this example.

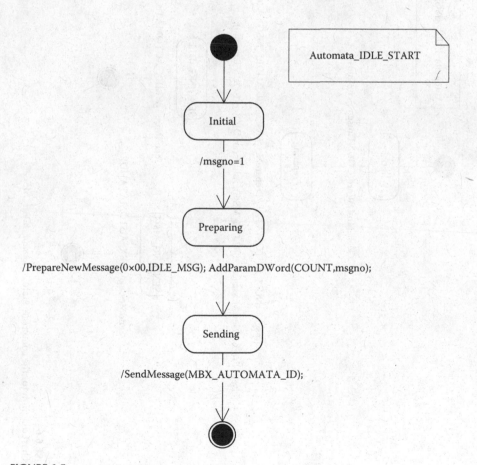

FIGURE 6.5
Statechart diagram for the composite state *Automata_IDLE_START.*

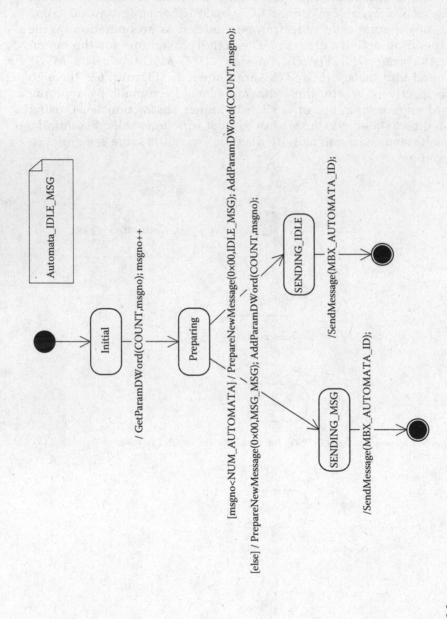

FIGURE 6.6
Statechart diagram for the composite state *Automata_IDLE_MSG*.

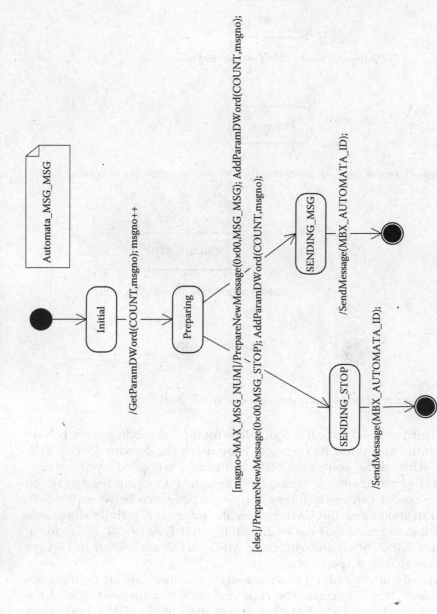

FIGURE 6.7
Statechart diagram for the composite state *Automata_MSG_MSG*.

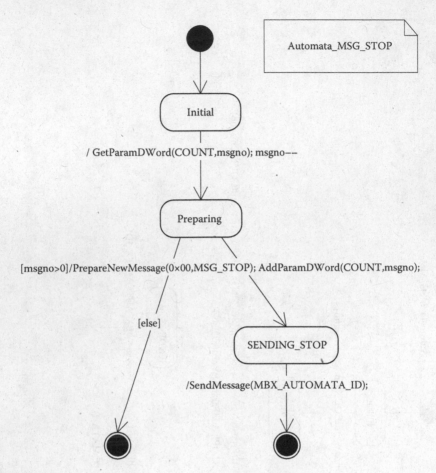

FIGURE 6.8
Statechart diagram for the composite state *Automata_MSG_STOP*.

The third and semantically equivalent method of modeling the behavior of individual automata instances is by using the domain-specific SDL model. This model comprises state transitions triggered by the reception of the corresponding messages. The same names are used again so that the reader can easily follow the correspondence between the SDL state transitions and the UML composite states and activity diagrams. The SDL state transitions *Automata_IDLE_START*, *Automata_IDLE_MSG*, *Automata_MSG_MSG*, and *Automata_MSG_STOP* are shown in Figures 6.13 through 6.16, respectively.

As already mentioned, all three automata instances in this example are of the same type, i.e., class. The class *Automata* is a specialization of the FSM Library class *FiniteStateMachine* and is used by the FSM Library class *FSMSystem* (see the corresponding UML class diagram in Figure 6.17). The

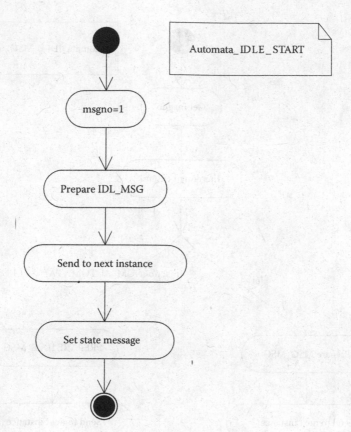

FIGURE 6.9
Activity diagram for the operation *Automata_IDLE_START*.

class *Automata* inherits all the members from its parent class and adds some field members (such as *msgno*) and function members (such as *Automata_ IDLE_START*, *Automata_IDLE_MSG*, *Automata_MSG_MSG*, *Automata_ MSG_STOP*, *Initialize*, and *StartDemo*). The first four correspond to composite states from the previous UML statechart model.

An object diagram, such as the one shown in Figure 6.18, helps us to better understand the structural relationships among objects. A collaboration diagram (Figure 6.2) shows the logical communication of automata instances over their virtual, peer-to-peer connections. On a more detailed level of abstraction, we see that the real communication is governed by the FSM system, which is the owner of the mailboxes (not shown in the figure) used for storing the messages, e.g., *StandardMessage* (shown in Figure 6.18). This particular message shown in one snapshot of object collaboration is the first message sent from *instance_1* to *instance_2*. The message code is *IDLE_MSG*, and the value of the message sequence parameter is 1.

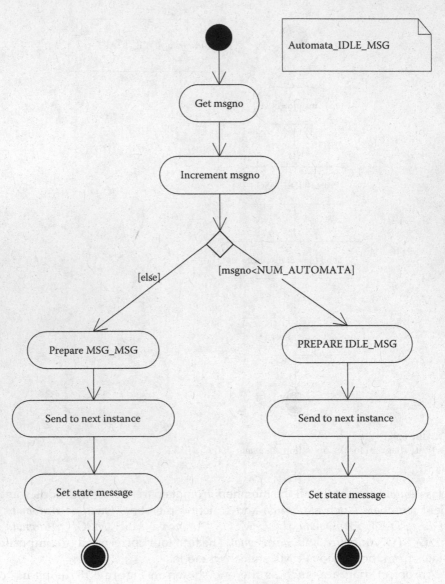

FIGURE 6.10
Activity diagram for the operation *Automata_IDLE_MSG*.

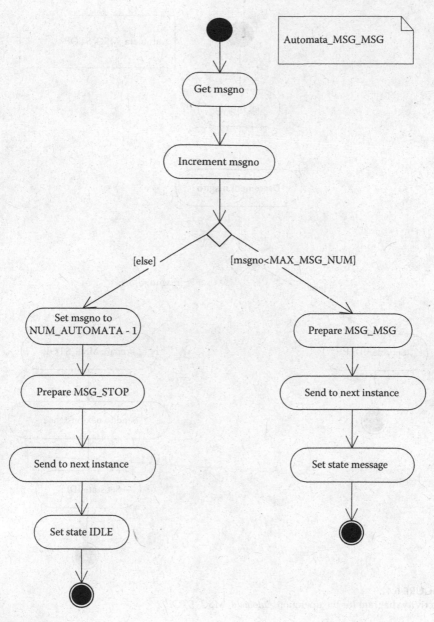

FIGURE 6.11
Activity diagram for the operation *Automata_MSG_MSG*.

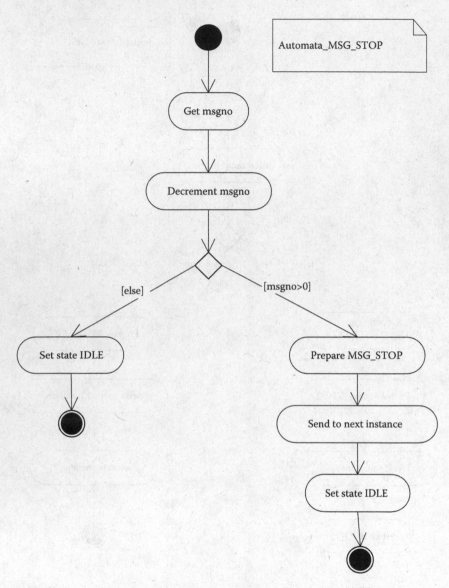

FIGURE 6.12
Activity diagram for the operation *Automata_MSG_STOP*.

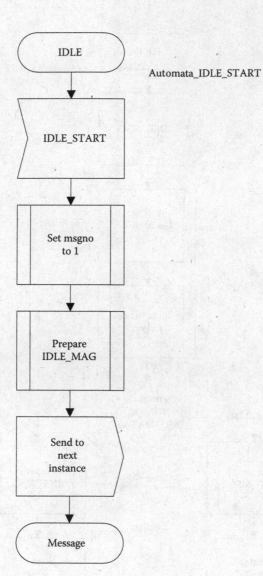

FIGURE 6.13
SDL diagram for the transition *Automata_IDLE_START.*

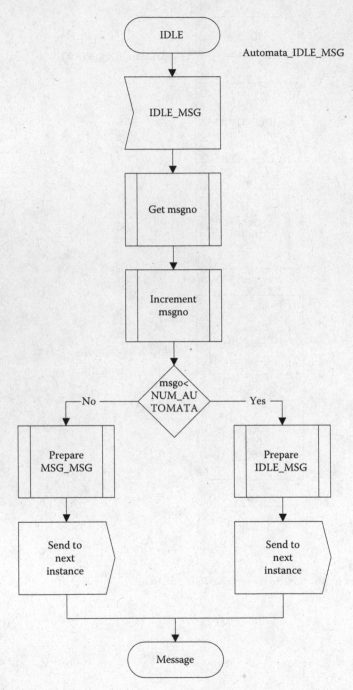

FIGURE 6.14
SDL diagram for the transition *Automata_IDLE_MSG*.

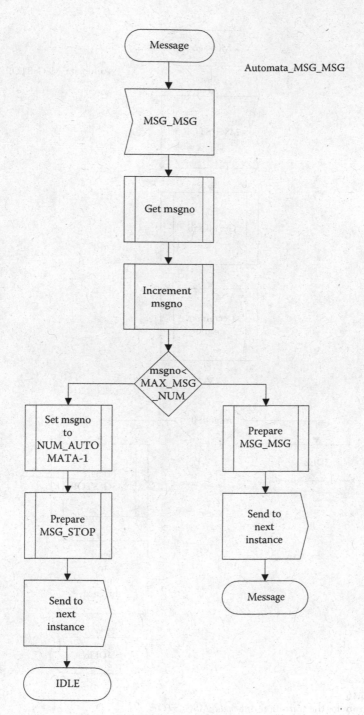

FIGURE 6.15
SDL diagram for the transition *Automata_MSG_MSG*.

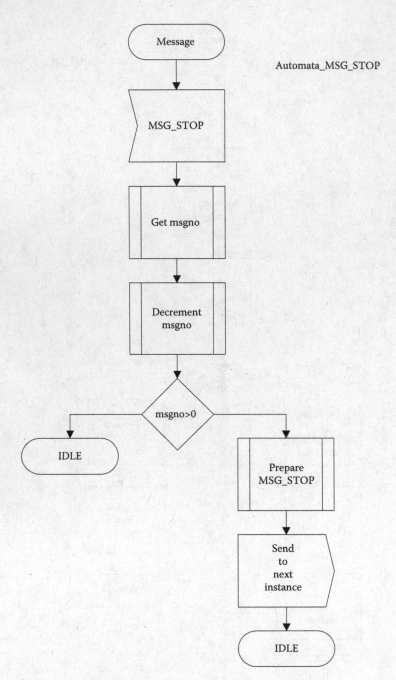

FIGURE 6.16
SDL diagram for the transition *Automata_MSG_STOP*.

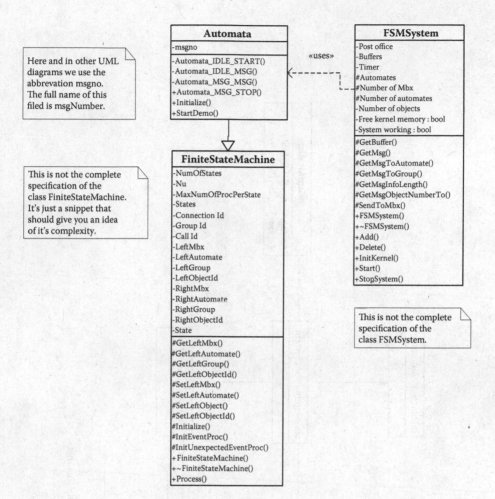

FIGURE 6.17
Class diagram for the example with three automata instances.

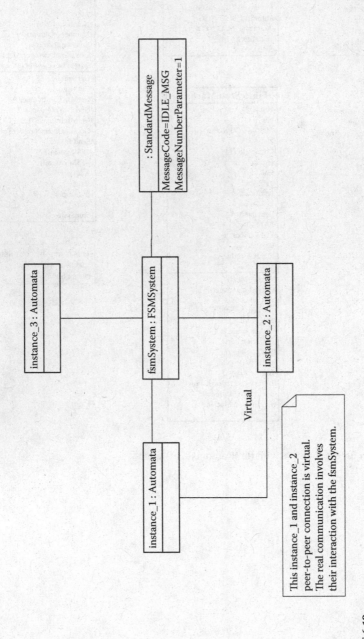

FIGURE 6.18
Object diagram for the example with three automata instances.

The program project in this example comprises the files *Automata.h*, *Automata.cpp*, *Constants.h*, *Main.cpp*, and the FSM Library (see the corresponding component diagram in Figure 6.19). Building this project in Microsoft® Visual Studio 6.0 yields a single executable, which is executed on a single PC machine (see the corresponding deployment diagram in Figure 6.20).

The rest of this section is devoted to the program implementation of the previous models. The content of the corresponding program files is as follows.

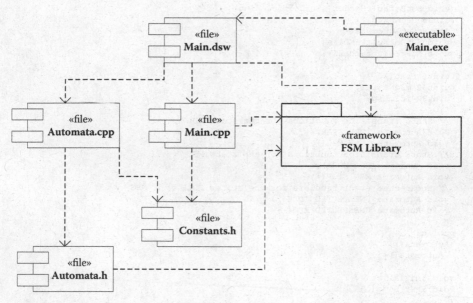

FIGURE 6.19
Component diagram for the example with three automata instances.

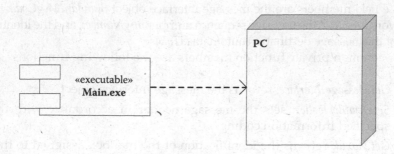

FIGURE 6.20
Deployment diagram for the example with three automata instances.

File *Automata.h*:

```
#ifndef __AUTOMATA__
#define __AUTOMATA__
#include <stdio.h>
#include "stdlib.h"
#include "kernel\fsm.h"
#include "kernel\errorObject.h"
#include "Constants.h"

class Automata: public FiniteStateMachine {
 private:
  StandardMessage StandardMsgCoding;
  MessageInterface *GetMessageInterface(uint32 id);

  void SetDefaultHeader(uint8 infoCoding);
  uint8 GetMbxId();
  uint8 GetAutomata();
  void SetDefaultFSMData();
  void NoFreeInstances();

  uint8 text[20];
  uint32 msgNumber;
  uint32 idToMsg;

  // State transition functions for the state IDLE
  void Automata_IDLE_START();
  void Automata_IDLE_MSG();
  // State transition functions for the state MSG
  void Automata_MSG_MSG();
  void Automata_MSG_STOP();
  // Unexpected event handlers for the states IDLE and  MSG
  void Automata_UNEXPECTED_IDLE();
  void Automata_UNEXPECTED_MSG();

 public:
  Automata();
  ~Automata(){};

 void Initialize();
 void StartDemo();
};
#endif
```

The file *Automata.h* contains a declaration of the class *Automata* derived from the class *FiniteStateMachine*. This declaration has its private and public parts. The private field members are the message interface object *StandardMsgCoding*, the text work area *text*, the message sequence number *msgNumber*, and the identification of the message destination automata *idToMsg*.

The common private function members are the following functions:

- *GetMessageInterface*: returns the message interface object
- *SetDefaultHeader*: sets the message header in accordance with the specified information coding
- *GetMbxId*: returns the identification of the mailbox assigned to this automata type

- *GetAutomata*: returns the identification of this automata type
- *SetDefaultFSMData*: sets the data specific for this automata type (*msgNumber* and *idToMsg*)
- *NoFreeInstances*: handles the situation when no more free instances of this type are found

The application-specific private function members are the following state transition functions:

- *Automata_IDLE_START*: handles the message *IDLE_START* in the state *IDLE*
- *Automata_IDLE_MSG*: handles the message *IDLE_MSG* in the state *IDLE*
- *Automata_MSG_MSG*: handles the message *MSG_MSG* in the state *MESSAGE*
- Automata_*MSG_STOP*: handles the message *MSG_STOP* in the state *MESSAGE*
- *Automata_UNEXPECTED_IDLE*: handles unexpected messages in the state *IDLE*
- Automata_*UNEXPECTED_MSG*: handles unexpected messages in the state *MESSAGE*

The public function members are the class constructor, the class destructor, the initialization function *Initialize*, and the startup function *StartDemo*.

File *Automata.cpp*:

```
#include "kernel/LogFile.h"
#include "Automata.h"

Automata::Automata() : FiniteStateMachine(
 0, // uint16 numOfTimers = DEFAULT_TIMER_NO,
 2, // uint16 numOfState = DEFAULT_STATE_NO,
 3) // uint16 maxNumOfProceduresPerState = DEFAULT_PROCEDURE_NO_PER_STATE
 {
 SetDefaultFSMData();
 }

// This function returns the pointer to the object that governs the
// message information coding (the pointer to the message interface).
// This automata instance works only with the standard messages
// (ID 0x00). If the caller specifies another type of coding,
// this function throws the exception TErrorObject. The message
// interface is defined in Automata.h
MessageInterface *Automata::GetMessageInterface(uint32 id){
 switch(id){
  case 0x00:
   return &StandardMsgCoding;
 }
```

```
 throw TErrorObject(__LINE__,__FILE__,0x01010400);
}

// This function fills in the message header.
void Automata::SetDefaultHeader(uint8 infoCoding){
 SetMsgInfoCoding(infoCoding);
 SetMessageFromData();
}

// This function returns the identification of the mailbox that is
// assigned to this automata type.
uint8 Automata::GetMbxId(){
 return MBX_AUTOMATA_ID;
}

// This function returns the identification of this automata type.
uint8 Automata::GetAutomata(){
 return FSM_TYPE_AUTOMATA;
}

// This function initializes the data specific to individual
// instance of this automata type.
void Automata::SetDefaultFSMData(){
 msgNumber = 0;
 idToMsg  = INVALID_32;
}
// This function is called if there are no free instances of this
// automata type. If the programmer wants to use this option, they must
// add the first automata instance of this type to the parameter
// useFreeList of the function Add set to true. In this example, it
// is empty. In real applications, the programmer should provide
// some recovery mechanism, such as overload protection or restart.
void Automata::NoFreeInstances(){
}

// This function initializes the state transition functions and the
// timers that are used by this automata type. This function is
// called implicitly by the function Add, which is responsible for
// adding individual automata instances to the FSM system.
// Each state transition function is separately declared and defined.
void Automata::Initialize(){
 // Here the programmer does the following initializations:
 // InitEventProc(uint8 state, uint16 event, PROC_FUN_PTR fun);
 // InitUnexpectedEventProc(uint8 state, PROC_FUN_PTR fun);
 // InitTimerBlock(uint16 timerId, uint32 timerCount, uint16 signalId);
 InitEventProc(IDLE,IDLE_START,(PROC_FUN_PTR)
  &Automata::Automata_IDLE_START);
 InitEventProc(IDLE,IDLE_MSG,(PROC_FUN_PTR)
  &Automata::Automata_IDLE_MSG);

 InitEventProc(MESSAGE,MSG_MSG,(PROC_FUN_PTR)
  &Automata::Automata_MSG_MSG);
 InitEventProc(MESSAGE,MSG_STOP,(PROC_FUN_PTR)
  &Automata::Automata_MSG_STOP);

 InitUnexpectedEventProc(IDLE,(PROC_FUN_PTR)
  &Automata::Automata_UNEXPECTED_IDLE);
 InitUnexpectedEventProc(MESSAGE,(PROC_FUN_PTR)
  &Automata::Automata_UNEXPECTED_MSG);
}

// State transition functions for the state IDLE.
void Automata::Automata_IDLE_START(){
 msgNumber = 1;
```

```
 idToMsg  = GetObjectId()+1;

 // Round Robin message transfer among automata instances 0-2
 if(idToMsg == 3)
  idToMsg = 0;

 // The automata instance prepares and sends the message,
 // and changes its state to MESSAGE.
 PrepareNewMessage(0x00,IDLE_MSG);

 char text[] = "THIS IS THE FIRST MESSAGE";
 AddParam(PARAM_TEXT,strlen(text),(unsigned char *)text);
 AddParamDWord(COUNT,msgNumber);

 SetMsgToAutomata(FSM_TYPE_AUTOMATA);
 SetMsgToGroup(INVALID_08);
 SetMsgObjectNumberTo(idToMsg);
 SendMessage(MBX_AUTOMATA_ID);
 SetState(MESSAGE);
}

void Automata::Automata_IDLE_MSG(){
 idToMsg = GetObjectId()+1;

 // Round Robin message transfer among automata instances 0-2
 if((idToMsg == 3)
  idToMsg = 0;
 // Get parameters from the message
 unsigned char *tmp;
 tmp = GetParam(PARAM_TEXT);
 assert(tmp);
 memcpy(text,tmp+2,*(tmp+1));
 memset(text+(*(tmp+1)),0x00,1); // make the string
 GetParamDWord(COUNT,msgNumber);

 // Round Robin - this instance receives the message from the previous one
 uint32 idFromMsg = GetObjectId()-1;
 if(idFromMsg == -1)
  idFromMsg = 2;

 printf("Text received: %s\n from automata:%u \n",text,idFromMsg);

 // If the message sequence number is less than NUM_AUTOMATA,
 // send IDLE_MSG. If not, send MSG_MSG.
 msgNumber++;
 if(msgNumber < NUM_AUTOMATA){
  // Prepare and send the message.
  // Change automata state to MESSAGE.
  PrepareNewMessage(0x00,IDLE_MSG);

  char text[] = "THIS IS THE SECOND MESSAGE";
  AddParam(PARAM_TEXT,strlen(text),(unsigned char *)text);
  AddParamDWord(COUNT,msgNumber);

  SetMsgToAutomata(FSM_TYPE_AUTOMATA);
  SetMsgToGroup(INVALID_08);
  SetMsgObjectNumberTo(idToMsg);
  SendMessage(MBX_AUTOMATA_ID);
 }
 else {
  // Prepare and send the message.
  // Change automata state to MESSAGE.
  PrepareNewMessage(0x00,MSG_MSG);
  AddParamDWord(COUNT,msgNumber);
```

```
 SetMsgToAutomata(FSM_TYPE_AUTOMATA);
 SetMsgToGroup(INVALID_08);
 SetMsgObjectNumberTo(idToMsg);
 SendMessage(MBX_AUTOMATA_ID);
 }
 SetState(MESSAGE);
}

void Automata::Automata_MSG_MSG(){
 GetParamDWord(COUNT,msgNumber);
 msgNumber++;
 if(msgNumber < MAX_MSG_NUM){
  // Forward the message to the next automata    instance.
  PrepareNewMessage(0x00,MSG_MSG);
  AddParamDWord(COUNT,msgNumber);
  SetMsgToAutomata(FSM_TYPE_AUTOMATA);
  SetMsgToGroup(INVALID_08);
  SetMsgObjectNumberTo(idToMsg);
  SendMessage(MBX_AUTOMAT_ID);
 }
 else {
  printf("Stop automata:%with message:%u\n",GetObjectId(),msgNumber);

  // Prepare and send the message.
  // Change automata state to IDLE.
  PrepareNewMessage(0x00,MSG_STOP);
  AddParamDWord(COUNT,NUM_AUTOMATA-1);
  SetMsgToAutomata(FSM_TYPE_AUTOMATA);
  SetMsgToGroup(INVALID_08);
  SetMsgObjectNumberTo(idToMsg);
  SendMessage(MBX_AUTOMATA_ID);
  SetState(IDLE);
 }
}

void Automata::Automata_MSG_STOP(){
 printf("Stop automata instance: %u\n",GetObjectId());

 GetParamDWord(COUNT,msgNumber);
 msgNumber--;
 if(msgNumber > 0){
  // Prepare and send the message.
  // Change automata state to IDLE.
  PrepareNewMessage(0x00,MSG_STOP);
  AddParamDWord(COUNT,msgNumber);
  SetMsgToAutomata(FSM_TYPE_AUTOMATA);
  SetMsgToGroup(INVALID_08);
  SetMsgObjectNumberTo(idToMsg);
  SendMessage(MBX_AUTOMATA_ID);
 }
 SetState(IDLE);
}

void Automata::Automata_UNEXPECTED_IDLE(){
 printf("Unexpected message in the state IDLE \n");
}

void Automata::Automata_UNEXPECTED_MSG(){
 printf("Unexpected message in the state MESSAGE \n");
}

void Automata::StartDemo(){
 uint8 *msg = GetBuffer(MSG_HEADER_LENGTH);
```

```
SetMsgFromAutomata(FSM_TYPE_AUTOMATA,msg);
SetMsgFromGroup(INVALID_08,msg);
SetMsgObjectNumberFrom(0,msg);

SetMsgToAutomata(FSM_TYPE_AUTOMATA,msg);
SetMsgToGroup(INVALID_08,msg);
SetMsgObjectNumberTo(0,msg);

SetMsgInfoCoding(0,msg); // 0 = StandardMessage
SetMsgCode(IDLE_START,msg);
SetMsgInfoLength(0,msg);
SendMessage(MBX_AUTOMATA_ID,msg);
}
```

The file *Automata.cpp* contains the definition of the class *Automata*. This definition starts with the class constructor that first calls the base class constructor specifying no timers, two states, and the maximum of three state transitions per state for this automata type. After that, the constructor calls the function *SetDefaultFSMData*, which sets the data specific for this automata type.

The function *GetMessageInterface* returns the pointer to the message interface object for the given type of information coding. This class operates with only standard messages (the corresponding ID is 0x00). If the caller of this function specifies the identification of the standard message as its parameter, the function returns the pointer to the object *StandardMsgCoding*. If the caller specifies some other message type, this function throws the exception *TErrorObject*.

The function *SetDefaultHeader* sets the message information coding by calling the function *SetMsgInfoCoding* and the automata specific data by calling the function *SetMessageFromData*. The function *GetMbxId* returns the value *MBX_AUTOMATA_ID* as the identification of the mailbox assigned to this automata type. The function *GetAutomata* returns the value *FSM_TYPE_AUTOMATA* as the identification of this automata type. The function *SetDefaultFSMData* sets the field *msgNumber* to the value 0 and the field *idToMsg* to the value *INVALID_32*. The function *NoFreeInstances* is empty in this simple example. In real-world projects, it would be used to trigger some higher-level protection or recovery mechanism.

The function *Initialize* defines the event handlers by calling the function *InitEventProc* and the unexpected event handlers by calling the function *InitUnexpectedEventProc*. More precisely, this function defines the event handlers for the messages *IDLE_START* and *IDLE_MSG* in the state *IDLE*, and for the messages *MSG_MSG* and *MSG_STOP* in the state *MESSAGE*. It also defines the handlers for unexpected messages in both states.

The function *Automata_IDLE_START* handles the message *IDLE_START* in the state *IDLE*. First, it sets the message sequence number *msgNumber* to the value 1. It then determines the identification of the destination automata instance by incrementing its own identification by modulo 3. (This means that the destination of the messages created and sent by *instance_0* is *instance_1*, the destination for *instance_1* is *instance_2*, and the destination for *instance_2* is *instance_0*.) Next, this function prepares and sends the message, "THIS IS THE FIRST MESSAGE". At the end, it performs the state transition from

IDLE to *MESSAGE* by calling the function *SetState* and specifying the value *MESSAGE* as its parameter.

The function *Automata_IDLE_MSG* handles the message *IDLE_MSG* in the state *IDLE*. First, it determines the identifications of the source and destination automata instances for the received message and prints them to the monitor. It then increments the message sequence numbers and checks if they are less than the number of communicating automata instances *NUM_AUTOMATA* (value 3). If yes, the function prepares and sends the message *IDLE_MSG* with the text, "THIS IS THE SECOND MESSAGE". If not, the function prepares and sends the message *MSG_MSG* without any text. In both cases, it sets the current state of the automata instance to the value *MESSAGE*.

The function *Automata_MSG_MSG* handles the message *MSG_MSG* in the state *MESSAGE*. First, it gets the message sequence number from the received message and increments that number. It then checks if the new value of the message sequence number has reached the given limit. If not, this function prepares and sends the message *MSG_MSG* to the next automata instance in the chain. If it has, this function prepares and sends the message *MSG_STOP* to the next automata instance in the chain, and sets the current state of this automata instance to *IDLE*.

The function *Automata_MSG_STOP* handles the message *MSG_STOP* in the state *MESSAGE*. First, it decrements the message sequence number and checks its new value. If the value is positive, the function prepares and sends the message *MSG_STOP* to the next automata instance in the chain, and sets the current state of this automata instance to *IDLE*.

The unexpected event handlers in this example just print the warning messages. In real applications, these functions would trigger some higher-level recovery mechanisms. The function *StartDemo* creates the first message in the system. It fills in its header as if the automata instance with the identification 0 had sent that message to itself, and sends the message to the mailbox assigned to this automata type.

File *Constants.h*:

```
// FSM
#define FSM_TYPE_AUTOMATA 0

// MBX
#define MBX_AUTOMATA_ID 0

#define MAX_MSG_NUM 10
#define NUM_AUTOMATA 3
#define COUNT 1
#define PARAM_TEXT 2

enum AutomataStates{
  IDLE = 0,
  MESSAGE,
};
```

```
enum Messages{
 IDLE_START = 0,
 IDLE_MSG,
 MSG_MSG,
 MSG_STOP
};
```

The file *Constants.h* first defines general symbolic constants. The identification of the automata type *FSM_TYPE_AUTOMATA* is assigned the value 0, the identification of the mailbox related to the automata type *MBX_AUTOMATA_ID* is assigned the value 0, the maximal message sequence number *MAX_MSG_NUM* is assigned the value 10, the number of automata instances of this type *NUM_AUTOMATA* is assigned the value 3, the identification of the message parameter that contains the messages sequence number *COUNT* is assigned the value 1, and the identification of the message parameter that contains the text *PARAM_TEXT* is assigned the value 2.

Next, the identifications of the individual states of this automata type are enumerated. The identification of the state *IDLE* is assigned the value 0 and the identification of the state *MESSAGES* is assigned the value 1. Finally, the identifications of various message types (message codes) are enumerated. The message types are named as *IDLE_START*, *IDLE_MSG*, *MSG_MSG*, and *MSG_STOP*. These symbols are assigned the values 0, 1, 2, and 3, respectively.

File *Main.cpp*:

```
#include "conio.h"
#include "Kernel/fsmsystem.h"
#include "Kernel/LogFile.h"
#include "Automata.h"

// Assume the following.
// The FSM system hosts a single automata type.
// The FSM system uses a single mailbox for the message exchange.
// Create the FSM system.
FSMSystem fsmSystem(1,1);

// Create three instances of the class Automata.
Automata instance_1, instance_2, instance_3;

// FSM system thread
DWORD WINAPI ThreadFunction(void* dummy){
 uint32 buffersCount[3] = {5,3,2};
 uint32 buffersLength[3] = {128,256,512};
 uint8 buffClassNo = 3;
 // Initialize the FSM system.
 printf("Initialize the FSM system... \n");
 fsmSystem.Add(&instance_1,FSM_TYPE_AUTOMATA,3,false);
 fsmSystem.Add(&instance_2,FSM_TYPE_AUTOMATA);
 fsmSystem.Add(&instance_3,FSM_TYPE_AUTOMATA);

 fsmSystem.InitKernel(buffClassNo,buffersCount,buffersLength,1);
```

```
LogFile lf("log.log", "log.ini");
LogAutomataNew::SetLogInterface(&lf);

// Start the FSM system.
printf("Start the FSM system... \n");
try {
   fsmSystem.Start();
}
catch(...) {
   OutputDebugString("Exception - stop the FSM     system...\n");
   return 0;
}
OutputDebugString("The end of the operation.\n");
return 0;
}

void main(int argc,char* argv[]){
DWORD threadID;
bool end = false;
char ret;

// Start the FSM system thread.
HANDLE hTemp = CreateThread(NULL,0,ThreadFunction,NULL,0,&threadID);
Sleep(100);

// Program works until the character 'Q' or 'q' is        pressed.
while(!end) {
  if(_kbhit()) {
   ret = _getch();
   switch(ret) {
    case 'Q':
    case 'q':
     fsmSystem.StopSystem();
     end = true;
     Sleep(100);
     break;
    case 'S':
    case 's':
     instance_1.StartDemo();
     break;
    default:
     break;
   }
  }
}
CloseHandle(hTemp);
printf("The end. \n");
}
```

The file *Main.cpp* starts with the instantiation of the class *FSMSystem* by calling its constructor. The parameters used in this call specify that an instance of the *FSMSystem*, named *fsmSystem*, will include a single automata type, and this automata type will use a single mailbox. Next, three instances of the class *Automata* are made, namely, *instance_1*, *instance_2*, and *instance_3*. Additionally, this file contains the definitions of the FSM system thread function *ThreadFunction* and the function *main*.

The function *ThreadFunction* first prepares the data needed to define three buffer types. The sizes and quantities of these buffers are five at 128 bytes, three at 256 bytes, and two at 512 bytes. Next, three automata instances are added to *fsmSystem*. Note that the fourth parameter of the first call to the function *Add* is set to the value *false*, which means that these three instances are to be used as three distinctive instances, rather than as a pool of instances of the same type. After that, this function initializes the kernel by calling the function *InitKernel*, defines and sets the logging interface by calling the function *SetLogInterface*, and starts the *fsmSystem* by calling its function *Start*.

The function *main* starts the FSM system thread (which executes the function *ThreadFunction*) and suspends itself for 100 ms. After that, it just waits for the character 'Q' or 'q' to be pressed and to subsequently terminate the program.

6.10 A Simple Example with Network-Aware Automata Instances

This section shows how the programmer can construct FSM systems with TCP support that is able to communicate over the TCP/IP network, and how they can add individual, network-aware automata instances to it. Normally, the programmer creates the FSM system with TCP support by instantiating the class *FSMSystemWithTCP*. Alternately, network-aware automata types are normally derived from the base class *NetFSM*. Of course, network-aware automata instances of a given type are then created simply by instantiating that automata type.

This example is very similar to the previous one. Actually, it has been created from it with a few rather simple modifications. Only one instance of the given automata type is added to the FSM system (now with TCP/IP support). This automata instance has a trivial task of exchanging the given number of messages with its peers in the remote FSM system. The main difference is that the whole program is instantiated twice. These program instances run as two separate processes that communicate over the TCP/IP protocol stack (see the corresponding collaboration diagram in Figure 6.21).

At the beginning, as in the previous example, the main thread calls the function *StartDemo* of *instance_1*, which, in turn, sends itself asynchronously the message *IDLE_START*. Upon reception of this message, *instance_1* sends the message *IDLE_MSG* to its peer *instance_1* that resides at the remote FSM system. These two automata instances, local and remote, then exchange nine *MSG_MSG* messages (the last *MSG_MSG* message is not shown in the figure). At the end of the communication, the local instance sends the message *MSG_STOP* to the remote instance (not shown in the figure). The corresponding sequence diagram is shown in Figure 6.22. This diagram shows all the messages.

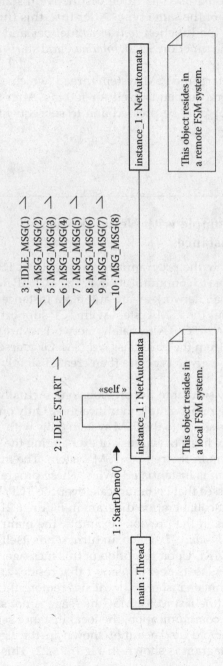

FIGURE 6.21

Collaboration diagram for the example with network-aware automata.

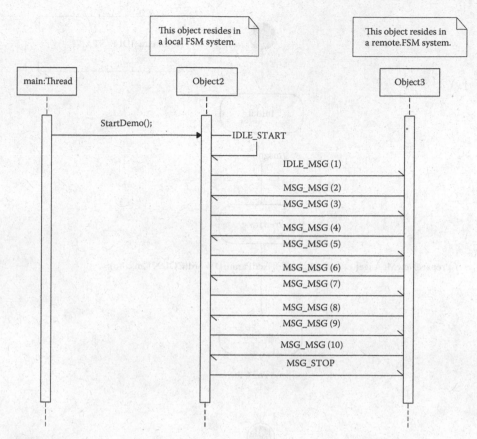

FIGURE 6.22
Sequence diagram for the example with network-aware automata.

The statechart diagram that describes the behavior of an individual automata instance is again organized into two hierarchical levels. The top level is exactly the same as the one shown in Figure 6.4. The composite states *Automata_IDLE_START*, *Automata_IDLE_MSG*, *Automata_MSG_MSG*, and *Automata_MSG_STOP* are a little simpler in this example and are shown in Figures 6.23 through 6.26, respectively.

The program code given in this example assumes that both processes run on the same machine whose IP address is 192.168.0.57. To get this code running on another machine, the reader should change this parameter accordingly. If the reader wants to experiment on two different machines, they must set this parameter to the IP addresses of those machines (see the corresponding deployment diagram shown in Figure 6.27).

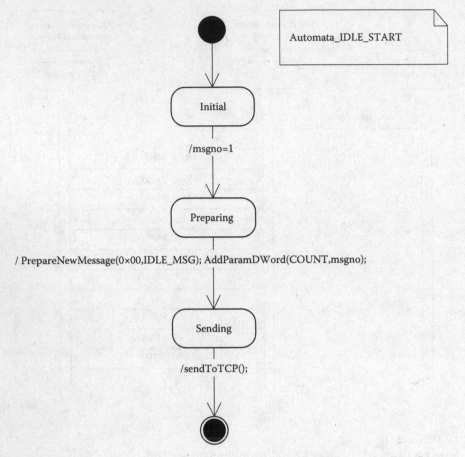

FIGURE 6.23
Composite state *Automata_IDLE_START*.

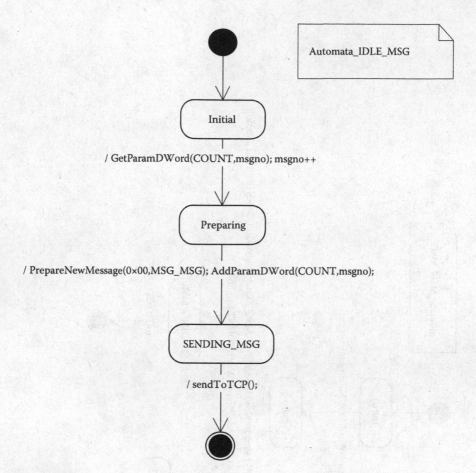

FIGURE 6.24
Composite state *Automata_IDLE_MSG*.

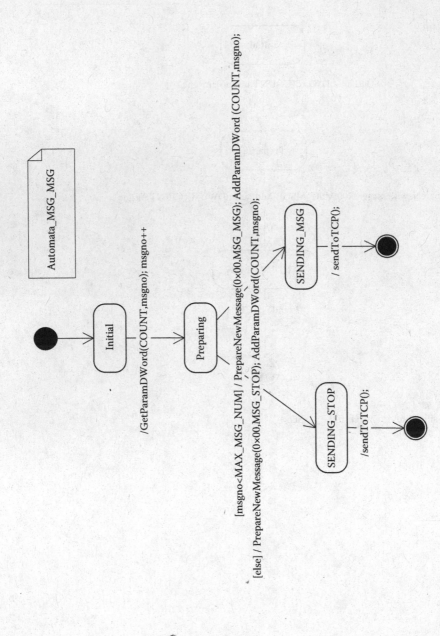

FIGURE 6.25

Composite state *Automata_MSG_MSG*.

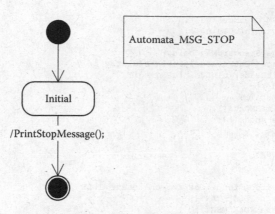

FIGURE 6.26
Composite state *Automata_MSG_STOP*.

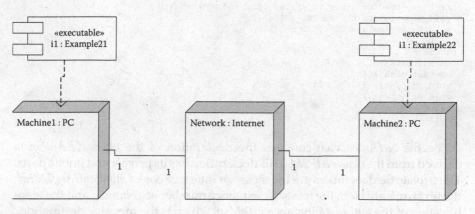

FIGURE 6.27
Deployment diagram for the example with network-aware automata.

Before proceeding further, studying the previous example first is strongly recommended. The content of the program files are as follows:

File *NetAutomata.h*:

```
#ifndef __NET_AUTOMATA__
#define __NET_AUTOMATA__
#include <stdio.h>
#include "stdlib.h"
#include "kernel\NetFSM.h"
#include "kernel\errorObject.h"
#include "Constants.h"

class NetAutomata: public NetFSM {
 private:
  // NetFSM
  uint16 convertNetToFSMMessage();
```

```
     void convertFSMToNetMessage();
     uint8 getProtocolInfoCoding();
     // FSM
     StandardMessage StandardMsgCoding;
     MessageInterface *GetMessageInterface(uint32 id);
     void SetDefaultHeader(uint8 infoCoding);
     uint8 GetMbxId();
     uint8 GetAutomata();
     void SetDefaultFSMData();
     void NoFreeInstances();

     uint8 text[20];
     uint32 msgNumber;
     uint32 idToMsg;

     // State transition functions for the state IDLE
     void NetAutomata_IDLE_START();
     void NetAutomata_IDLE_MSG();
     // State MSG
     void NetAutomata_MSG_MSG();
     void NetAutomata_MSG_STOP();
     // Unexpected messages in states IDLE and MSG
     void NetAutomata_UNEXPECTED_IDLE();
     void NetAutomata_UNEXPECTED_MSG();

public:
   NetAutomata();
   ~NetAutomata(){};
   void Initialize();
   void StartDemo();
   };
#endif
```

The file *NetAutomata.h* contains the declaration of the class *NetAutomata* derived from the class *NetFSM*. This declaration has its private and public parts. The private field members are the message interface object *StandardMsgCoding*, the text work area *text*, the message sequence number *msgNumber*, and the identification of the automata instance *idToMsg*, which is the message destination.

The private function members specific to the class *NetFSM* are the following functions:

- *convertNetToFSMMessage*: converts the external message format into the internal message format appropriate for communication within the FSM system
- *convertFSMToNetMessage*: converts the internal message format into the external message format appropriate for the transmission over the TCP/IP network
- *getProtocolInfoCoding*: returns the identification of the type of the external message coding

The private function members specific to the class *FinteStateMachine* are the following functions:

- *GetMessageInterface*: returns the message interface object

- *SetDefaultHeader*: sets the message header according to the specified information coding
- *GetMbxId*: returns the identification of the mailbox that is assigned to this automata type
- *GetAutomata*: returns the identification of this automata type
- *SetDefaultFSMData*: sets the data specific for this automata type (*msgNumber* and *idToMsg*)
- *NoFreeInstances*: handles the situation when no more free instances of this type are found

The application-specific private function members are the following state transition functions:

- *Automata_IDLE_START*: handles the message *IDLE_START* in the state *IDLE*
- *Automata_IDLE_MSG*: handles the message *IDLE_MSG* in the state *IDLE*
- *Automata_MSG_MSG*: handles the message *MSG_MSG* in the state *MESSAGE*
- *Automata_MSG_STOP*: handles the message *MSG_STOP* in the state *MESSAGE*
- *Automata_UNEXPECTED_IDLE*: handles unexpected messages in the state *IDLE*
- *Automata_UNEXPECTED_MSG*: handles unexpected messages in the state *MESSAGE*

The public function members are the class constructor, the class destructor, the initialization function *Initialize*, and the startup function *StartDemo*.

File *NetAutomata.cpp*:

```cpp
#include "kernel/LogFile.h"
#include "NetAutomata.h"

NetAutomata::NetAutomata() : NetFSM(
 0, // uint16 numOfTimers = DEFAULT_TIMER_NO,
 2, // uint16 numOfState = DEFAULT_STATE_NO,
 3) // uint16 maxNumOfProceduresPerState = DEFAULT_PROCEDURE_NO_PER_STATE
{
 SetDefaultFSMData();
}

// This function returns the pointer to the object that governs the
// message information coding (the pointer to the message interface).
// This automata instance works only with the standard messages
// (ID 0x00). If the caller specifies another type of coding,
// this function throws the exception TErrorObject.
// The message interface is defined in NetAutomata.h
MessageInterface *NetAutomata::GetMessageInterface(uint32 id){
```

```
switch(id) {
 case 0x00:
   return &StandardMsgCoding;
 }
 throw TErrorObject(__LINE__,__FILE__,0x01010400);
}
// This function fills in the message header.
void NetAutomata::SetDefaultHeader(uint8 infoCoding){
 SetMsgInfoCoding(infoCoding);
 SetMessageFromData();
}

// This function returns the identification of the mailbox that is
// assigned to this automata type.
uint8 NetAutomata::GetMbxId(){
 return MBX_AUTOMATA_ID;
}

// This function returns the identification of this automata type.
uint8 NetAutomata::GetAutomata(){
 return FSM_TYPE_AUTOMATA;
}

// This function initializes the data specific for individual
// instance of this automata type.
void NetAutomata::SetDefaultFSMData(){
 msgNumber = 0;
 idToMsg = INVALID_32;
}

// This function is called if there are no free instances of this
// automata type. If the programmer wants to use this option they must
// add the first automata instance of this type with the parameter
// useFreeList of the function Add set to true. In this example it is
// empty. In real applications the programmer should provide some
// recovery mechanism, such as overload protection or restart.
void NetAutomata::NoFreeInstances(){}

// This function initializes the state transition functions and the
// timers that are used by this automata type. This function is called
// implicitly by the function Add responsible for adding individual
// automata instances to the FSM system.
// Each state transition function is separately declared and defined.
void NetAutomata::Initialize(){
 // Here the programmer does the following  initializations:
 // InitEventProc(uint8 state, uint16 event,  PROC_FUN_PTR fun);
 // InitUnexpectedEventProc(uint8 state, PROC_FUN_PTR  fun);
 // InitTimerBlock(uint16 timerId, uint32 timerCount,  uint16 signalId);

 InitEventProc(IDLE,IDLE_START,(PROC_FUN_PTR)
  &NetAutomata::NetAutomata_IDLE_START);
 InitEventProc(IDLE,IDLE_MSG,(PROC_FUN_PTR)
  &NetAutomata::NetAutomata_IDLE_MSG);

 InitEventProc(MESSAGE,MSG_MSG,(PROC_FUN_PTR)
  &NetAutomata::NetAutomata_MSG_MSG);
 InitEventProc(MESSAGE,MSG_STOP,(PROC_FUN_PTR)
  &NetAutomata::NetAutomata_MSG_STOP);

 InitUnexpectedEventProc(IDLE,(PROC_FUN_PTR)
  &NetAutomata::NetAutomata_UNEXPECTED_IDLE);
 InitUnexpectedEventProc(MESSAGE,(PROC_FUN_PTR)
  &NetAutomata::NetAutomata_UNEXPECTED_MSG);
}
```

```
// State transition functions for the state IDLE.
void NetAutomata::NetAutomata_IDLE_START(){
 msgNumber = 1;
 idToMsg = 0;

 // The automata instance prepares and sends the  message,
 // and changes its state to MESSAGE.
 PrepareNewMessage(0x00,IDLE_MSG);

 char text[] = "THIS IS THE FIRST MESSAGE";
 AddParam(PARAM_TEXT,strlen(text),(unsigned char *)text);
 AddParamDWord(COUNT,msgNumber);

 SetMsgToAutomata(FSM_TYPE_AUTOMATA);
 SetMsgToGroup(INVALID_08);
 SetMsgObjectNumberTo(idToMsg);
 sendToTCP();
 SetState(MESSAGE);
}

void NetAutomata::NetAutomata_IDLE_MSG(){
 idToMsg = 0;

 // Get parameters from the message
 unsigned char *tmp;
 tmp = GetParam(PARAM_TEXT);
 assert(tmp);
 memcpy(text,tmp+2,*(tmp+1));
 memset(text+(*(tmp+1)),0x00,1); // make the string

 GetParamDWord(COUNT,msgNumber);
 printf("Text received: %s\n",text);

 // If the message sequence number is less than given  limit,
 // continue message counting. If not stop the program.
 msgNumber++;

 // Prepare and send the message.
 // Change automata state to MESSAGE.
 PrepareNewMessage(0x00,MSG_MSG);
 AddParamDWord(COUNT,msgNumber);
 SetMsgToAutomata(FSM_TYPE_AUTOMATA);
 SetMsgToGroup(INVALID_08);
 SetMsgObjectNumberTo(idToMsg);
 sendToTCP();
 SetState(MESSAGE);
}

void NetAutomata::NetAutomata_MSG_MSG(){
 GetParamDWord(COUNT,msgNumber);
 msgNumber++;
 if(msgNumber < MAX_MSG_NUM){
  // Forward the message.
  PrepareNewMessage(0x00,MSG_MSG);
  AddParamDWord(COUNT,msgNumber);
  SetMsgToAutomata(FSM_TYPE_AUTOMATA);
  SetMsgToGroup(INVALID_08);
  SetMsgObjectNumberTo(idToMsg);
  sendToTCP();
 }
 else {
  printf("Stop automata: %u\n",GetObjectId());
```

```
   // Prepare and send the message.
   // Change automata state to IDLE.
   PrepareNewMessage(0x00,MSG_STOP);
   SetMsgToAutomata(FSM_TYPE_AUTOMATA);
   SetMsgToGroup(INVALID_08);
   SetMsgObjectNumberTo(idToMsg);
   sendToTCP();
   SetState(IDLE);
  }
}

void NetAutomata::NetAutomata_MSG_STOP(){
 printf("Stop automata: %u\n",GetObjectId());
 SetState(IDLE);
}
void NetAutomata::NetAutomata_UNEXPECTED_IDLE(){
 printf("Unexpected message in the state IDLE \n");
}

void NetAutomata::NetAutomata_UNEXPECTED_MSG(){
 printf("Unexpected message in the state MESSAGE \n");
}

void NetAutomata::StartDemo(){
 uint8 *msg = GetBuffer(MSG_HEADER_LENGTH);
 SetMsgFromAutomata(FSM_TYPE_AUTOMATA,msg);
 SetMsgFromGroup(INVALID_08,msg);
 SetMsgObjectNumberFrom(0,msg);

 SetMsgToAutomata(FSM_TYPE_AUTOMATA,msg);
 SetMsgToGroup(INVALID_08,msg);
 SetMsgObjectNumberTo(0,msg);

 SetMsgInfoCoding(0,msg); // 0 = StandardMessage
 SetMsgCode(IDLE_START,msg);
 SetMsgInfoLength(0,msg);
 SendMessage(MBX_AUTOMATA_ID,msg);
}

uint16 NetAutomata::convertNetToFSMMessage(){
 // Manipulate only data because automata sends the new
 // message to itself.
 int length = receivedMessageLength-MSG_HEADER_LENGTH;
 memcpy(fsmMessageR,  protocolMessageR+MSG_HEADER_LENGTH, length);
 fsmMessageRLength=length; // mandatory - used by  workWhenReceive()

 // Rotate bytes
 uint16 msgCode =  GetUint16((uint8*)(protocolMessageR+MSG_CODE));

 switch((msgCode)){
  case IDLE_START:
   msgCode = IDLE_START;
   break;
  case IDLE_MSG:
   msgCode = IDLE_MSG;
   break;
  case MSG_MSG:
   msgCode = MSG_MSG;
   break;
  case MSG_STOP:
   msgCode = MSG_STOP;
   break;
  default:
   msgCode = 0xffff;
```

```
}
  return msgCode;
}

void NetAutomata::convertFSMToNetMessage(){
  // Here we send the whole message.
    memcpy(protocolMessageS,fsmMessageS,fsmMessageSLength);
  sendMsgLength = fsmMessageSLength;
}

uint8 NetAutomata::getProtocolInfoCoding(){
  // Standard msg info coding
  return 0;
}
```

The file *NetAutomata.cpp* contains the definition of the class *NetAutomata*. This definition starts with the class constructor that first calls the base class constructor specifying no timers, two states, and the maximum of three state transitions per state for this automata type. After this, the constructor calls the function *SetDefaultFSMData*, which sets the data specific for this automata type.

The functions *GetMessageInterface*, *SetDefaultHeader*, *GetMbxId*, *GetAutomata*, *SetDefaultFSMData*, *NoFreeInstances*, and *Initialize* are the same as in the previous example. The only difference is that the name of the class *Automata* has been renamed to *NetAutomata*.

The function *NetAutomata_IDLE_START* handles the message *IDLE_START* in the state *IDLE*. First, it sets the message sequence number *msgNumber* to the value 1 and the identification of the destination automata instance *idToMsg* to the value 0. Next, this function prepares and sends the message, "THIS IS THE FIRST MESSAGE," to its peer in the remote FSM system by calling the function *SendToTCP*. At the end, it performs the state transition from *IDLE* to *MESSAGE* by calling the function *SetState* and specifying the value *MESSAGE* as its parameter.

The function *NetAutomata_IDLE_MSG* handles the message *IDLE_MSG* in the state *IDLE*. First, it prints the received message to the monitor. It then prepares and sends the message with the code *MSG_MSG* to its peer by calling the function *SendToTCP*, and sets the current state of this automata instance to the value *MESSAGE*.

The function *NetAutomata_MSG_MSG* handles the message *MSG_MSG* in the state *MESSAGE*. First, it gets the message sequence number from the received message and increments this value. It then checks if the new value of the message sequence number has reached the given limit. If not, this function prepares and sends the message *MSG_MSG* to its peer at the remote FSM system by calling the function *SendToTCP*. If it has reached the limit, this function prepares and sends the message *MSG_STOP* to its peer at the remote FSM system, and sets the current state of this automata instance to *IDLE*.

The function *NetAutomata_MSG_STOP* handles the message *MSG_STOP* in the state *MESSAGE*. It is fairly simple and just sets the current state of this automata instance to *IDLE*. The unexpected event handlers in this example just print the warning messages. In real-world applications, these functions would trigger some higher-level recovery mechanisms. The function *StartDemo* creates the first message in the system. It fills in its header as if the automata instance with the identification 0 had sent that message to itself and sends this message to the mailbox assigned to this automata type.

The function *convertNetToFSMMessage* just copies the payload of the external message received from the remote FSM system to the current FSM system internal message (the last received message), because in this simple example, the two communicating instances have the same IDs and no need exists for any mappings between them. The pointer *fsmMessageR* points to the current internal message, the pointer *protocolMessageR* points to the current external message, and the variable *fsmMessageRLength* is equal to the payload size of the current external message. At the end, this function determines the message code and returns it as its return value.

The function *convertFSMToNetMessage* copies the whole new internal message to the new external message and sets the value of its length. The pointer *fsmMessageS* points to the new internal message, the pointer *protocolMessageS* points to the new external message, and the variables *fsmMessageSLength* and *sendMsgLength* contain their lengths.

The function *getProtocolInfoCoding* returns the code of the standard message coding (code 0x00) used for coding external messages. Note that in this simple example, both internal and external messages are actually standard messages.

File *Constants.h*:

```
// FSM
#define FSM_TYPE_AUTOMATA 0

// MBX
#define MBX_AUTOMATA_ID 0
#define MAX_MSG_NUM 10
#define COUNT 1
#define PARAM_TEXT 2
#define IP_ADDRESS "192.168.0.57"
#define PORT_1 7000
#define PORT_2 8000

enum AutomataStates {
 IDLE = 0,
 MESSAGE,
};

enum Messages {
 IDLE_START = 0,
 IDLE_MSG,
 MSG_MSG,
 MSG_STOP
};
```

The file *Constants.h* first defines general symbolic constants. It is very similar to the file with the same name in the previous example. The identification of this automata type *FSM_TYPE_AUTOMATA* is assigned the value 0, the identification of the mailbox related to this automata type

MBX_AUTOMATA_ID is assigned the value 0, the maximal message sequence number *MAX_MSG_NUM* is assigned the value 10, the identification of the message parameter that contains the message sequence number *COUNT* is assigned the value 1, and the identification of the message parameter that contains the text *PARAM_TEXT* is assigned the value 2.

The main difference with the previous example is the definition of the symbolic constants related to the communication over the TCP/IP infrastructure. The IP address *IP_ADDRESS* is assigned the value 192.168.0.57, the TCP port number for the first server *PORT_1* is assigned the value 7000, and the TCP port number for the second server *PORT_2* is assigned the value 8000. Next, the identifications of the individual states of this automata type, as well as possible message codes, are enumerated. This part of the file is the same as in the previous example.

File *Main.cpp*:

```
#include "conio.h"
#include "Kernel/fsmsystem.h"
#include "Kernel/LogFile.h"
#include "NetAutomata.h"

// If the following line is not commented out we get the code for the
// server listening to the port number PORT_1.
// If the following line is commented out we get the code for the
// server listening to the port number PORT_2.
#define AUTOMATA1

// Assume the following.
// The FSM system hosts a single automata type.
// The FSM system uses a single mailbox for the message exchange.
// Create the FSM system.
FSMSystemWithTCP fsmSystem(1,1);

// Create the instance of the class NetAutomata.
NetAutomata instance_1;

DWORD WINAPI ThreadFunction(void* dummy){
  uint32 buffersCount[3] = {5,3,2};
  uint32 buffersLength[3] = {128,256,512};
  uint8 buffClassNo = 3;

  // Initialize the FSM system.
  printf("Initialize the FSMSystemWithTCP... \n");
  fsmSystem.Add(&instance_1,FSM_TYPE_AUTOMATA,1,true);
    fsmSystem.InitKernel(buffClassNo,buffersCount,buffersLength,1);
  LogFile lf("log.log", "log.ini");
  LogAutomataNew::SetLogInterface(&lf);

  // Server in machine number 1 will listen to the port  number PORT_1.
  // Server in machine number 2 will listen to the port  number PORT_2.
  // It does not matter which instance will establish  the TCP
  // connection by calling the function  establishConection().
#ifdef AUTOMATA1
  printf("Start server...on port:%u\n",PORT_1);
  fsmSystem.InitTCPServer(PORT_1,FSM_TYPE_AUTOMATA);
#else
```

```
 printf("Start server...on port:%u\n",PORT_2);
 fsmSystem.InitTCPServer(PORT_2,FSM_TYPE_AUTOMATA);

#endif
 // Start the FSM system.
 printf("Start the FSM system...\n");
 try {
  fsmSystem.Start();
 }
 catch(...) {
  OutputDebugString("Exception - stop the FSM system...\n");
  return 0;
 }
 OutputDebugString("The end of the operation.\n");
 return 0;
}

void main(int argc,char* argv[]){
 DWORD threadID;
 bool end = false;
 char ret;

 // Start the FSM system thread.
 HANDLE hTemp = CreateThread(NULL,0,ThreadFunction,NULL,0,&threadID);
 Sleep(100);

 // Program works until the character 'Q' or 'q' is pressed.
 while((!end)) '{
  if(_kbhit()) {
   ret = _getch();
   switch((ret)) {
    case 'Q':
    case 'q':
     fsmSystem.StopSystem();
     end = true;
     Sleep(100);
     break;
    case 'S':
    case 's':
     instance_1.StartDemo();
     break;
    case 'E':
    case 'e':
// Press 'e' to establish the connection with the remote server.
// This will enable the communication with the remote system.
#ifdef AUTOMATA1
     instance_1.setPort(PORT_2);
     instance_1.setIP((IP_ADDRESS));
     printf("establishConection on port:%u",PORT_2);
     instance_1.establishConnection();
#else
     instance_1.setPort(PORT_1);
     instance_1.setIP(IP_ADDRESS);
     printf("establishConection on port:%u",PORT_1);
     instance_1.establishConnection();
#endif
    default:
     break;
   }
  }
 }
 CloseHandle(hTemp);
 printf("The end. \n");
}
```

The file *Main.cpp* starts with a list of the necessary included files and the definition of the symbolic constant *AUTOMATA1*. This constant should be defined for the local process and not for the remote process (this is done by commenting out the source code line that defines the symbol *AUTOMATA1*).

Next, the instantiation of the class *FSMSystemWithTCP* is performed by a call to its constructor. The parameters used in this call specify that an instance of the *FSMSystemWithTCP*, named *fsmSystem*, will include a single automata type and this automata type will use a single mailbox. After that, a single instance of the class *NetAutomata* is made, *instance_1*. Additionally, this file contains the definitions of the FSM system thread function *ThreadFunction* and the function *main*.

The function *ThreadFunction* first prepares the data needed to define three buffer types. The sizes and quantities of these buffers are five at 128 bytes, three at 256 bytes, and two at 512 bytes. Next, the three automata instances are added to *fsmSystem*. Note that the fourth parameter of the first call to the function *Add* is set to the value *true*, which means that the instances are to be used as a pool of instances of the same type. After that, this function initializes the kernel by calling the function *InitKernel*, defines and sets the logging interface by calling the function *SetLogInterface*, starts the TCP server by calling the function *InitTCPServer*, and starts the *fsmSystem* by calling its function *Start*.

The function *main* starts the FSM system thread (which executes the function *ThreadFunction*) and suspends itself for 100 ms. After this, it waits for the user command. If the user presses the character 'E' or 'e', it establishes the TCP connection with the remote TCP server by calling the function *establishConnection*. If the user presses the character 'Q' or 'q', it terminates the program.

Index

Printed in the United States
by Baker & Taylor Publisher Services

Springer Series in
Computational
Mathematics

8

Springer
Berlin
Heidelberg
New York
Barcelona
Hong Kong
London
Milan
Paris
Tokyo

E. Hairer
S. P. Nørsett
G. Wanner

Solving Ordinary Differential Equations I

Nonstiff Problems

Second Revised Edition
With 135 Figures

Springer

Ernst Hairer
Gerhard Wanner

Université de Genève, Section des Mathématiques
C.P. 240, 1211 Genève 24, Switzerland
e-mail: Ernst.Hairer|Gerhard.Wanner@math.unige.ch

Syvert Paul Nørsett

University of Trondheim, NTH
Institute of Numerical Mathematics
7034 Trondheim, Norway
e-mail: norsett@math.ntnu.no

Library of Congress Cataloging-in-Publication Data

Hairer, E. (Ernst)
Solving ordinary differential equations / E. Hairer, S. P. Nørsett, G. Wanner. p. cm.
(Springer series in computational mathematics; 8)
Includes bibliographical references and index.
Contents: 1. Nonstiff problems.
ISBN 3-540-56670-8 (v.1)
1. Differential equations - Numerical solutions.
I. Nørsett, S. P. (Syvert, Paul), 1944 -. II. Wanner, Gerhard. III. Title. IV. Series.
QA372.H16 1993 515'.352-dc20 93-7847 CIP

Second, Corrected Printing 2000

Mathematics Subject Classification (2000): 65Lxx, 34A50

ISSN 0179-3632
ISBN 978-3-540-56670-0 2nd Ed. Springer-Verlag Berlin Heidelberg New York
ISBN 978-3-540-17145-4 1st Ed. Springer-Verlag Berlin Heidelberg New York

Springer-Verlag Berlin Heidelberg New York
a member of Springer Science+Business Media

© Springer-Verlag Berlin Heidelberg 1987, 1993

Camera-ready copy by the authors
Printed on acid-free paper   SPIN 11517863   46/3111ck - 5 4 3 2 1

This edition is dedicated to
Professor John Butcher
on the occasion of his 60th birthday

His unforgettable lectures on Runge-Kutta methods, given in June 1970 at the University of Innsbruck, introduced us to this subject which, since then, we have never ceased to love and to develop with all our humble abilities.

From the Preface to the First Edition

> So far as I remember, I have never seen an Author's Preface
> which had any purpose but one — to furnish reasons for the
> publication of the Book.   (Mark Twain)

> Gauss' dictum, "when a building is completed no one should
> be able to see any trace of the scaffolding," is often used
> by mathematicians as an excuse for neglecting the motiva-
> tion behind their own work and the history of their field.
> Fortunately, the opposite sentiment is gaining strength, and
> numerous asides in this Essay show to which side go my
> sympathies.   (B.B. Mandelbrot 1982)

> This gives us a good occasion to work out most of the book
> until the next year.   (the
> Authors in a letter, dated Oct. 29, 1980, to Springer-Verlag)

There are two volumes, one on non-stiff equations, ..., the second
on stiff equations, ... The first volume has three chapters, one on
classical mathematical theory, one on Runge-Kutta and extrapolation
methods, and one on multistep methods. There is an Appendix containing some Fortran codes which we have written for our numerical
examples.

Each chapter is divided into sections. Numbers of formulas, theorems, tables and figures are consecutive in each section and indicate,
in addition, the section number, but not the chapter number. Cross
references to other chapters are rare and are stated explicitly. ...
References to the Bibliography are by "Author" plus "year" in parentheses. The Bibliography makes no attempt at being complete; we
have listed mainly the papers which are discussed in the text.

Finally, we want to thank all those who have helped and encouraged us to prepare this book. The marvellous "Minisymposium" which
G. Dahlquist organized in Stockholm in 1979 gave us the first impulse
for writing this book. J. Steinig and Chr. Lubich have read the whole
manuscript very carefully and have made extremely valuable mathematical and linguistical suggestions. We also thank J.P. Eckmann
for his troff software with the help of which the whole manuscript
has been printed. For preliminary versions we had used textprocess-
ing programs written by R. Menk. Thanks also to the staff of the
Geneva computing center for their help. All computer plots have
been done on their beautiful HP plotter. Last but not least, we would
like to acknowledge the agreable collaboration with the planning and
production group of Springer-Verlag.

October 29, 1986   The Authors

Preface to the Second Edition

The preparation of the second edition has presented a welcome opportunity to improve the first edition by rewriting many sections and by eliminating errors and misprints. In particular we have included new material on

- Hamiltonian systems (I.14) and symplectic Runge-Kutta methods (II.16);

- dense output for Runge-Kutta (II.6) and extrapolation methods (II.9);

- a new Dormand & Prince method of order 8 with dense output (II.5);

- parallel Runge-Kutta methods (II.11);

- numerical tests for first- and second order systems (II.10 and III.7).

Our sincere thanks go to many persons who have helped us with our work:

- all readers who kindly drew our attention to several errors and misprints in the first edition;

- those who read preliminary versions of the new parts of this edition for their invaluable suggestions: D.J. Higham, L. Jay, P. Kaps, Chr. Lubich, B. Moesli, A. Ostermann, D. Pfenniger, P.J. Prince, and J.M. Sanz-Serna.

- our colleague J. Steinig, who read the entire manuscript, for his numerous mathematical suggestions and corrections of English (and Latin!) grammar;

- our colleague J.P. Eckmann for his great skill in manipulating Apollo workstations, font tables, and the like;

- the staff of the Geneva computing center and of the mathematics library for their constant help;

- the planning and production group of Springer-Verlag for numerous suggestions on presentation and style.

This second edition now also benefits, as did Volume II, from the marvels of TEXnology. All figures have been recomputed and printed, together with the text, in Postscript. Nearly all computations and text processings were done on the Apollo DN4000 workstation of the Mathematics Department of the University of Geneva; for some long-time and high-precision runs we used a VAX 8700 computer and a Sun IPX workstation.

November 29, 1992   The Authors

Contents

Chapter I. Classical Mathematical Theory

Chapter II. Runge-Kutta and Extrapolation Methods

Chapter III. Multistep Methods and General Linear Methods

Chapter I. Classical Mathematical Theory

> ... halte ich es immer für besser, nicht mit dem Anfang anzufan-
> gen, der immer das Schwerste ist.
> (B. Riemann copied this from F. Schiller into his notebook)

This first chapter contains the classical theory of differential equations, which we judge useful and important for a profound understanding of numerical processes and phenomena. It will also be the occasion of presenting interesting examples of differential equations and their properties.

We first retrace in Sections I.2-I.6 the historical development of classical integration methods by series expansions, quadrature and elementary functions, from the beginning (Newton and Leibniz) to the era of Euler, Lagrange and Hamilton. The next part (Sections I.7-I.14) deals with theoretical properties of the solutions (existence, uniqueness, stability and differentiability with respect to initial values and parameters) and the corresponding flow (increase of volume, preservation of symplectic structure). This theory was initiated by Cauchy in 1824 and then brought to perfection mainly during the next 100 years. We close with a brief account of boundary value problems, periodic solutions, limit cycles and strange attractors (Sections I.15 and I.16).

I.1 Terminology

A *differential equation of first order* is an equation of the form

$$y' = f(x, y) \tag{1.1}$$

with a given function $f(x, y)$. A function $y(x)$ is called a *solution* of this equation if for all x,

$$y'(x) = f(x, y(x)). \tag{1.2}$$

It was observed very early by Newton, Leibniz and Euler that the solution usually contains a free parameter, so that it is uniquely determined only when an *initial value*

$$y(x_0) = y_0 \tag{1.3}$$

is prescribed. Cauchy's existence and uniqueness proof of this fact will be discussed in Section I.7. Differential equations arise in many applications. We shall see the first examples of such equations in Section I.2, and in Section I.3 how some of them can be solved explicitly.

A *differential equation of second order* for y is of the form

$$y'' = f(x, y, y'). \tag{1.4}$$

Here, the solution usually contains *two* parameters and is only uniquely determined by *two* initial values

$$y(x_0) = y_0, \qquad y'(x_0) = y_0'. \tag{1.5}$$

Equations of second order can rarely be solved explicitly (see I.3). For their numerical solution, as well as for theoretical investigations, one usually sets $y_1(x) := y(x)$, $y_2(x) := y'(x)$, so that equation (1.4) becomes

$$
\begin{aligned}
y_1' &= y_2 & y_1(x_0) &= y_0 \\
y_2' &= f(x, y_1, y_2) & y_2(x_0) &= y_0'.
\end{aligned}
\tag{1.4'}
$$

This is an example of a *first order system of differential equations*, of dimension n (see Sections I.6 and I.9),

$$
\begin{aligned}
y_1' &= f_1(x, y_1, \ldots, y_n) & y_1(x_0) &= y_{10} \\
&\cdots & &\cdots \\
y_n' &= f_n(x, y_1, \ldots, y_n) & y_n(x_0) &= y_{n0}.
\end{aligned}
\tag{1.6}
$$

Most of the theory of this book is devoted to the solution of the initial value problem for the system (1.6). At the end of the 19th century (Peano 1890) it became customary to introduce the vector notation

$$y = (y_1, \ldots, y_n)^T, \qquad f = (f_1, \ldots, f_n)^T$$

so that (1.6) becomes $y' = f(x,y)$, which is again the same as (1.1), but now with y and f interpreted as vectors.

Another possibility for the second order equation (1.4), instead of transforming it into a system (1.4'), is to develop *methods specially adapted to second order equations (Nyström methods)*. This will be done in special sections of this book (Sections II.13 and III.10). Nothing prevents us, of course, from considering (1.4) as a second order system of dimension n.

If, however, the initial conditions (1.5) are replaced by something like $y(x_0) = a$, $y(x_1) = b$, i.e., if the conditions determining the particular solution are not all specified at the same point x_0, we speak of a *boundary value problem*. The theory of the existence of a solution and of its numerical computation is here much more complicated. We give some examples in Section I.15.

Finally, a problem of the type

$$\frac{\partial u}{\partial t} = f\left(t, u, \frac{\partial u}{\partial x}, \frac{\partial^2 u}{\partial x^2}\right) \tag{1.7}$$

for an unknown function $u(t,x)$ of *two independent variables* will be called a *partial differential equation*. We can also deal with partial differential equations of higher order, with problems in three or four independent variables, or with systems of partial differential equations. Very often, initial value problems for partial differential equations can conveniently be transformed into a system of ordinary differential equations, for example with finite difference or finite element approximations in the variable x. In this way, the equation

$$\frac{\partial u}{\partial t} = a^2 \frac{\partial^2 u}{\partial x^2}$$

would become

$$\frac{du_i}{dt} = \frac{a^2}{\Delta x^2}\left(u_{i+1} - 2u_i + u_{i-1}\right),$$

where $u_i(t) \approx u(t, x_i)$. This procedure is called the "method of lines" or "method of discretization in space" (Berezin & Zhidkov 1965). We shall see in Section I.6 that this connection, the other way round, was historically the origin of partial differential equations (d'Alembert, Lagrange, Fourier). A similar idea is the "method of discretization in time" (Rothe 1930).

I.2 The Oldest Differential Equations

Newton

Differential equations are as old as differential calculus. Newton considered them in his treatise on differential calculus (Newton 1671) and discussed their solution by series expansion. One of the first examples of a first order equation treated by Newton (see Newton (1671), Problema II, Solutio Casus II, Ex. I) was

$$y' = 1 - 3x + y + x^2 + xy. \tag{2.1}$$

For each value x and y, such an equation prescribes the derivative y' of the solutions. We thus obtain a *vector field*, which, for this particular equation, is sketched in Fig. 2.1a. (So, contrary to the belief of many people, vector fields existed long before Van Gogh). The solutions are the curves which respect these prescribed directions everywhere (Fig. 2.1b).

Newton discusses the solution of this equation by means of infinite series, whose terms he obtains recursively ("... & ils se jettent sur les series, oú M. Newton m'a precedé sans difficulté; mais ...", Leibniz). The first term

$$y = 0 + \ldots$$

is the initial value for $x = 0$. Inserting this into the differential equation (2.1) he obtains

$$y' = 1 + \ldots$$

which, integrated, gives

$$y = x + \ldots.$$

Again, from (2.1), we now have

$$y' = 1 - 3x + x + \ldots = 1 - 2x + \ldots$$

and by integration

$$y = x - x^2 + \ldots.$$

E x e m p l. I.

Sit Æquatio $\dfrac{\dot{y}}{x} = 1 - 3x + y + xx + xy$, cujus Terminos:

$1 - 3x + xx$ non affectos *Relatâ* Quantitate difpofitos vides in lateralem Seriem primo loco, & reliquos y & xy in finiftrâ Columnâ.

	$+ 1 - 3x + xx$
$+ y$	$* + x - xx + \dfrac{1}{3}x^3 - \dfrac{1}{6}x^4 + \dfrac{1}{30}x^5$; &c.
$+ xy$	$* \quad x + xx - x^3 + \dfrac{1}{3}x^4 - \dfrac{1}{6}x^5 + \dfrac{1}{30}x^6$; &c.
Aggreg.	$+ 1 - 2x + xx - \dfrac{2}{3}x^3 + \dfrac{1}{6}x^4 - \dfrac{4}{30}x^5$; &c.
$y =$	$+ x - xx + \dfrac{1}{3}x^3 - \dfrac{1}{6}x^4 + \dfrac{1}{30}x^5 - \dfrac{1}{45}x^6$; &c.

Nunc:

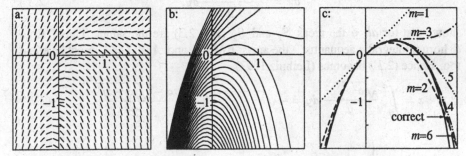

Fig. 2.1. a) vector field, b) various solution curves of equation (2.1),
c) Correct solution vs. approximate solution

The next round gives

$$y' = 1 - 2x + x^2 + \dots, \qquad y = x - x^2 + \frac{x^3}{3} + \dots.$$

Continuing this process, he finally arrives at

$$y = x - xx + \frac{1}{3}x^3 - \frac{1}{6}x^4 + \frac{1}{30}x^5 - \frac{1}{45}x^6; \text{ \&c.} \tag{2.2}$$

These approximations, term after term, are plotted in Fig. 2.1c together with the correct solution. It can be seen that these approximations are closer and closer to the true solution for small values of x. For more examples see Exercises 1-3. Convergence will be discussed in Section I.8.

Leibniz and the Bernoulli Brothers

A second access to differential equations is the consideration of geometrical problems such as *inverse tangent problems* (Debeaune 1638 in a letter to Descartes). A particular example describes the path of a silver pocket watch ("horologio portabili suae thecae argentae") and was proposed around 1674 by "Claudius Perraltus Medicus Parisinus" to Leibniz: a curve $y(x)$ is required whose tangent AB is given, say everywhere of constant length a (Fig. 2.2). This leads to

$$y' = -\frac{y}{\sqrt{a^2 - y^2}}, \tag{2.3}$$

a first order differential equation. Despite the efforts of the "plus célèbres mathématiciens de Paris et de Toulouse" (from a letter of Descartes 1645, "Toulouse" means "Fermat") the solution of these problems had to wait until Leibniz (1684) and above all until the famous paper of Jacob Bernoulli (1690). Bernoulli's idea applied to equation (2.3) is as follows: let the curve BM in Fig. 2.3 be such that LM is equal to $\sqrt{a^2 - y^2}/y$. Then (2.3), written as

$$dx = -\frac{\sqrt{a^2 - y^2}}{y} \, dy, \tag{2.3'}$$

shows that for *all* y the areas S_1 and S_2 (Fig. 2.3) are the same. Thus ("Ergo & horum integralia aequantur") the areas $BMLB$ and $A_1 A_2 C_2 C_1$ must be equal too. Hence (2.3') becomes (Leibniz 1693)

$$x = \int_y^a \frac{\sqrt{a^2 - y^2}}{y} \, dy = -\sqrt{a^2 - y^2} - a \cdot \log \frac{a - \sqrt{a^2 - y^2}}{y}. \tag{2.3''}$$

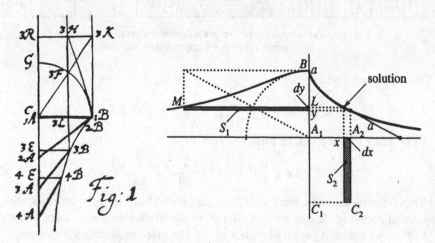

Fig. 2.2. Illustration from   Fig. 2.3. Jac. Bernoulli's
        Leibniz (1693)   Solution of (2.3)

Variational Calculus

In 1696 Johann Bernoulli invited the brightest mathematicians of the world ("Profundioris in primis Mathesos cultori, Salutem!") to solve the *brachystochrone* (shortest time) problem, mainly in order to fault his brother Jacob, from whom he expected a wrong solution. The problem is to find a curve $y(x)$ connecting two points P_0, P_1, such that a point gliding on this curve under gravitation reaches P_1 in the shortest time possible. In order to solve his problem, Joh. Bernoulli (1697b) imagined thin layers of homogeneous media and knew from optics (Fermat's principle) that a light ray with speed v obeying the law of Snellius

$$\sin \alpha = Kv$$

passes through in the shortest time. Since the speed is known to be proportional to the square root of the fallen height, he obtains, by passing to thinner and thinner layers,

$$\sin \alpha = \frac{1}{\sqrt{1 + y'^2}} = K\sqrt{2g(y - h)}, \tag{2.4}$$

a differential equation of the first order.

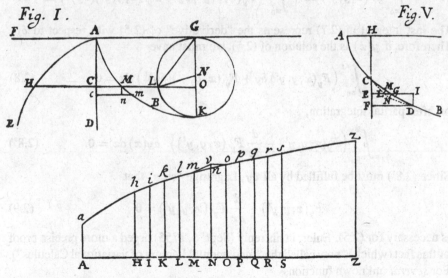

Fig. 2.4. Solutions of the variational problem (Joh. Bernoulli, Jac. Bernoulli, Euler)

The solutions of (2.4) can be shown to be cycloids (see Exercise 6 of Section I.3). Jacob, in his reply, also furnished a solution, much less elegant but unfortunately correct. Jacob's method (see Fig. 2.4) was something like today's (inverse) "finite element" method and more general than Johann's and led to the famous work of Euler (1744), which gives the general solution of the problem

$$\int_{x_0}^{x_1} F(x, y, y') \, dx = \min \tag{2.5}$$

with the help of the differential equation of the second order

$$F_y(x, y, y') - \frac{d}{dx} \left(F_{y'}(x, y, y') \right) = F_y - F_{y'y'} y'' - F_{y'y} y' - F_{y'x} = 0, \tag{2.6}$$

and treated 100 variational problems in detail. Equation (2.6), in the special case where F does not depend on x, can be integrated to give

$$F - F_{y'} y' = K. \tag{2.6'}$$

Euler's original proof used polygons in order to establish equation (2.6). Only the ideas of Lagrange, in 1755 at the age of 19, led to the proof which is today the usual one (letter of Aug. 12, 1755; Oeuvres vol. 14, p. 138): add an arbitrary "variation" $\delta y(x)$ to $y(x)$ and linearize (2.5).

$$\int_{x_0}^{x_1} F\left(x, y + \delta y, y' + (\delta y)'\right) dx \tag{2.7}$$

$$= \int_{x_0}^{x_1} F(x, y, y') \, dx + \int_{x_0}^{x_1} \left(F_y(x, y, y') \, \delta y + F_{y'}(x, y, y')(\delta y)' \right) dx + \ldots$$

The last integral in (2.7) represents the "derivative" of (2.5) with respect to δy. Therefore, if $y(x)$ is the solution of (2.5), we must have

$$\int_{x_0}^{x_1} \left(F_y(x, y, y') \, \delta y + F_{y'}(x, y, y')(\delta y)' \right) dx = 0 \tag{2.8}$$

or, after partial integration,

$$\int_{x_0}^{x_1} \left(F_y(x, y, y') - \frac{d}{dx} F_{y'}(x, y, y') \right) \cdot \delta y(x) \, dx = 0. \tag{2.8'}$$

Since (2.8') must be fulfilled by all δy, Lagrange "sees" that

$$F_y(x, y, y') - \frac{d}{dx} F_{y'}(x, y, y') = 0 \tag{2.9}$$

is necessary for (2.5). Euler, in his reply (Sept. 6, 1755) urged a more precise proof of this fact (which is now called the "fundamental Lemma of variational Calculus"). For *several* unknown functions

$$\int F(x, y_1, y_1', \ldots, y_n, y_n') \, dx = \min \tag{2.10}$$

the same proof leads to the equations

$$F_{y_i}(x, y_1, y_1', \ldots, y_n, y_n') - \frac{d}{dx} F_{y_i'}(x, y_1, y_1', \ldots, y_n, y_n') = 0 \tag{2.11}$$

for $i = 1, \ldots, n$. Euler (1756) then gave, in honour of Lagrange, the name "Variational calculus" to the whole subject ("... tamen gloria primae inventionis acutissimo Geometrae Taurinensi La Grange erat reservata").

Clairaut

A class of equations with interesting properties was found by Clairaut (see Clairaut (1734), Problème III). He was motivated by the movement of a rectangular wedge (see Fig. 2.5), which led him to differential equations of the form

$$y - xy' + f(y') = 0. \tag{2.12}$$

This was the first *implicit* differential equation and possesses the particularity that not only the lines $y = Cx - f(C)$ are solutions, but also their enveloping curves (see Exercise 5). An example is shown in Fig. 2.6 with $f(C) = 5(C^3 - C)/2$.

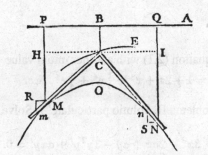

Fig. 2.5. Illustration from Clairaut (1734)

Since the equation is of the third degree in y', a given initial value may allow up to three different solution lines. Furthermore, where a line touches an enveloping curve, the solution may be continued either along the line or along the curve. There is thus a huge variety of different possible solution curves. This phenomenon attracted much interest in the classical literature (see e.g., Exercises 4 and 6). Today we explain this curiosity by the fact that at these points no Lipschitz condition is satisfied (see also Ince (1944), p. 538–539).

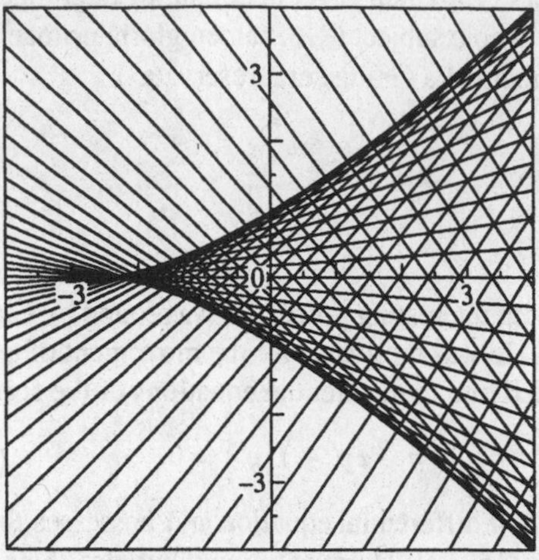

Fig. 2.6. Solutions of a Clairaut differential equation

Exercises

1. (Newton). Solve equation (2.1) with another initial value $y(0) = 1$.
   Newton's result: $y = 1 + 2x + x^3 + \frac{1}{4}x^4 + \frac{1}{4}x^5$, &c.

2. (Newton 1671, "Problema II, Solutio particulare"). Solve the total differential equation
   $$3x^2 - 2ax + ay - 3y^2 y' + axy' = 0.$$
   Solution given by Newton: $x^3 - ax^2 + axy - y^3 = 0$. Observe that he missed the arbitrary integration constant C.

3. (Newton 1671). Solve the equations

   a) $\quad y' = 1 + \dfrac{y}{a} + \dfrac{xy}{a^2} + \dfrac{x^2 y}{a^3} + \dfrac{x^3 y}{a^4}$, &c.

   b) $\quad y' = -3x + 3xy + y^2 - xy^2 + y^3 - xy^3 + y^4 - xy^4 + 6x^2 y$
   $$- 6x^2 + 8x^3 y - 8x^3 + 10x^4 y - 10x^4, \text{ \&c.}$$

   Results given by Newton:

   a) $\quad y = x + \dfrac{x^2}{2a} + \dfrac{x^3}{2a^2} + \dfrac{x^4}{2a^3} + \dfrac{x^5}{2a^4} + \dfrac{x^6}{2a^5}$, &c.

   b) $\quad y = -\dfrac{3}{2}x^2 - 2x^3 - \dfrac{25}{8}x^4 - \dfrac{91}{20}x^5 - \dfrac{111}{16}x^6 - \dfrac{367}{35}x^7$, &c.

4. Show that the differential equation

$$x + yy' = y'\sqrt{x^2 + y^2 - 1}$$

possesses the solutions $2ay = a^2 + 1 - x^2$ for all a. Sketch these curves and find yet another solution of the equation (from Lagrange (1774), p. 7, which was written to explain the "Clairaut phenomenon").

5. Verify that the envelope of the solutions $y = Cx - f(C)$ of the Clairaut equation (2.12) is given in parametric representation by

$$x(p) = f'(p)$$
$$y(p) = pf'(p) - f(p) .$$

Show that this envelope is also a solution of (2.12) and calculate it for $f(C) = 5(C^3 - C)/2$ (cf. Fig. 2.6).

6. (Cauchy 1824). Show that the family $y = C(x + C)^2$ satisfies the differential equation $(y')^3 = 8y^2 - 4xyy'$. Find yet another solution which is not included in this family (see Fig. 2.7).

Answer: $y = -\frac{4}{27}x^3$.

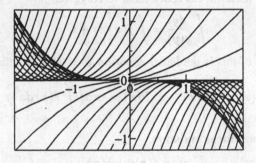

Fig. 2.7. Solution family of Cauchy's example in Exercise 6

I.3 Elementary Integration Methods

We now discuss some of the simplest types of equations, which can be solved by the computation of integrals.

First Order Equations

The equation with separable variables.

$$y' = f(x)g(y).$$

(3.1)

Extending the idea of Jacob Bernoulli (see (2.3')), we divide by $g(y)$, integrate and obtain the solution (Leibniz 1691, in a letter to Huygens)

$$\int \frac{dy}{g(y)} = \int f(x)\,dx + C.$$

A special example of this is the *linear equation* $y' = f(x)y$, which possesses the solution

$$y(x) = C R(x), \qquad R(x) = \exp\left(\int f(x)\,dx\right).$$

The inhomogeneous linear equation.

$$y' = f(x)y + g(x).$$

(3.2)

Here, the substitution $y(x) = c(x)R(x)$ leads to $c'(x) = g(x)/R(x)$ (Joh. Bernoulli 1697). One thus obtains the solution

$$y(x) = R(x)\left(\int_{x_0}^{x} \frac{g(s)}{R(s)}\,ds + C\right).$$

(3.3)

Total differential equations. An equation of the form

$$P(x,y) + Q(x,y)y' = 0$$

(3.4)

is found to be immediately solvable if

$$\frac{\partial P}{\partial y} = \frac{\partial Q}{\partial x}.$$

(3.5)

One can then find by integration a potential function $U(x, y)$ such that

$$\frac{\partial U}{\partial x} = P, \qquad \frac{\partial U}{\partial y} = Q.$$

Therefore (3.4) becomes $\frac{d}{dx}U(x, y(x)) = 0$, so that the solutions can be expressed by $U(x, y(x)) = C$. For the case when (3.5) is not satisfied, Clairaut and Euler investigated the possibility of multiplying (3.4) by a suitable factor $M(x, y)$, which sometimes allows the equation $MP + MQy' = 0$ to satisfy (3.5).

Second Order Equations

Even more than for first order equations, the solution of *second* order equations by integration is very seldom possible. Besides linear equations with constant coefficients, whose solutions for the second order case were already known to Newton, several tricks of reduction are possible, as for example the following:

For a *linear equation*

$$y'' = a(x)y' + b(x)y$$

we make the substitution (Riccati 1723, Euler 1728)

$$y = \exp\left(\int p(x)\, dx\right). \tag{3.6}$$

The derivatives of this function contain only derivatives of p of lower order

$$y' = p \cdot \exp\left(\int p(x)\, dx\right), \qquad y'' = (p^2 + p') \cdot \exp\left(\int p(x)\, dx\right)$$

so that inserting this into the differential equation, after division by y, leads to a *lower order* equation

$$p^2 + p' = a(x)p + b(x) \tag{3.7}$$

which, however, is nonlinear.

If the equation is *independent* of y, $y'' = f(x, y')$, it is natural to put $y' = v$ which gives $v' = f(x, v)$.

An important case is that of *equations independent of x*:

$$y'' = f(y, y').$$

Here we consider y' as function of y: $y' = p(y)$. Then the chain rule gives $y'' = p'p = f(y, p)$, which is a first order equation. When the function $p(y)$ has been found, it remains to integrate $y' = p(y)$, which is an equation of type (3.1) (Riccati (1712): "Per liberare la premessa formula dalle seconde differenze,..., chiamo p la sunnormale BF ... ", see also Euler (1769), Problema 96, p. 33).

The investigation of all possible differential equations which can be integrated by analytical methods was begun by Euler. His results have been collected, in more than 800 pages, in Volumes XXII and XXIII of Euler's Opera Omnia. For a more recent discussion see Ince (1944), p. 16-61. An irreplaceable document on this subject is the book of Kamke (1942). It contains, besides a description of the solution methods and general properties of the solutions, a systematically ordered list of more than 1500 differential equations with their solutions and references to the literature.

The computations, even for very simple looking equations, soon become very complicated and one quickly began to understand that elementary solutions would not always be possible. It was Liouville (1841) who gave the first *proof* of the fact that certain equations, such as $y' = x^2 + y^2$, cannot be solved in terms of elementary functions. Therefore, in the 19th century mathematicians became more and more interested in general existence theorems and in numerical methods for the computation of the solutions.

Exercises

1. Solve Newton's equation (2.1) by quadrature.

2. Solve Leibniz' equation (2.3) in terms of elementary functions.
   Hint. The integral for y might cause trouble. Use the substitution $a^2 - y^2 = u^2$, $-y\,dy = u\,du$.

3. Solve and draw the solutions of $y' = f(y)$ where $f(y) = \sqrt{|y|}$.

4. Solve the master-and-dog problem: a dog runs with speed w in the direction of his master, who walks with speed v along the y-axis. This leads to the differential equation
   $$xy'' = \frac{v}{w}\sqrt{1 + (y')^2}.$$

5. Solve the equation $my'' = -k/y^2$, which describes a body falling according to Newton's law of gravitation.

6. Verify that the cycloid
   $$x - x_0 = R(\tau - \sin\tau), \qquad y - h = R(1 - \cos\tau), \qquad R = \frac{1}{4gK^2}$$
   satisfies the differential equation (2.4) for the brachystochrone problem. Solving (2.4) in a forward manner, one arrives after some simplifications at the integral
   $$\int \sqrt{\frac{y}{1-y}}\, dy,$$
   which is computed by the substitution $y = (\sin t)^2$.

7. Reduce the "Bernoulli equation" (Jac. Bernoulli 1695)

$$y' + f(x)y = g(x)y^n$$

with the help of the coordinate transformation $z(x) = (y(x))^q$ and a suitable choice of q, to a linear equation (Leibniz, Acta Erud. 1696, p. 145, Joh. Bernoulli, Acta Erud. 1697, p. 113).

8. Compute the "Linea Catenaria" of the hanging rope. The solution was given by Joh. Bernoulli (1691) and Leibniz (1691) (see Fig. 3.2) without any hint.

   Hint. (Joh. Bernoulli, "Lectiones ... in usum Ill. Marchionis Hospitalii" 1691/92). Let H resp. V be the horizontal resp. vertical component of the tension in the rope (Fig. 3.1). Then $H = a$ is a constant and $V = q \cdot s$ is proportional to the arc length. This leads to $Cp = s$ or $Cdp = ds$ i.e., $Cdp = \sqrt{1 + p^2}\,dx$, where $p = y'$, a differential equation.

   Result. $y = K + C \cosh\left(\dfrac{x - x_0}{C}\right)$.

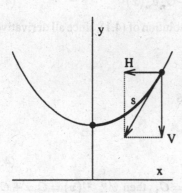

Fig. 3.1. Solution of the Catenary problem

Fig. 3.2. "Linea Catenaria" ▶ drawn by Leibniz (1691)

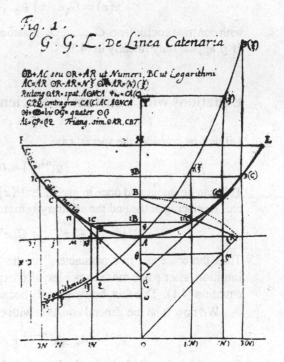

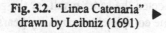

I.4 Linear Differential Equations

Following in the footsteps of Euler (1743), we want to understand the general solution of nth order linear differential equations. We say that the equation

$$\mathcal{L}(y) := a_n(x)y^{(n)} + a_{n-1}(x)y^{(n-1)} + \ldots + a_0(x)y = 0 \qquad (4.1)$$

with given functions $a_0(x), \ldots, a_n(x)$ is *homogeneous*. If n solutions $u_1(x)$, $\ldots, u_n(x)$ of (4.1) are known, then any linear combination

$$y(x) = C_1 u_1(x) + \ldots + C_n u_n(x) \qquad (4.2)$$

with constant coefficients $C_1, \ldots, C_n$ is also a solution of (4.1), since all derivatives of y appear only linearly in (4.1).

Equations with Constant Coefficients

Let us first consider the special case

$$y^{(n)}(x) = 0. \qquad (4.3)$$

This can be integrated once to give $y^{(n-1)}(x) = C_1$, then $y^{(n-2)}(x) = C_1 x + C_2$, etc. Replacing at the end the arbitrary constants C_i by new ones, we finally obtain

$$y(x) = C_1 x^{n-1} + C_2 x^{n-2} + \ldots + C_n.$$

Thus there are n "free parameters" in the "general solution" of (4.3). Euler's intuition, after some more examples, also expected the same result for the general equation (4.1). This fact, however, only became completely clear many years later.

We now treat the general equation with constant coefficients,

$$y^{(n)} + A_{n-1}y^{(n-1)} + \ldots + A_0 y = 0. \qquad (4.4)$$

Our problem is to find a basis of n linearly independent solutions $u_1(x), \ldots, u_n(x)$. To this end, Euler's inspiration was guided by the transformation (3.6), (3.7) above: if $a(x)$ and $b(x)$ are constants, we assume p constant in (3.7) so that p' vanishes, and we obtain the quadratic equation $p^2 = ap + b$. For any root of this equation,

(3.6) then becomes $y = e^{pz}$. In the general case we thus assume $y = e^{pz}$ with an unknown constant p, so that (4.4) leads to the *characteristic equation*

$$p^n + A_{n-1}p^{n-1} + \ldots + A_0 = 0. \qquad (4.5)$$

If the roots $p_1, \ldots, p_n$ of equation (4.5) are distinct, all solutions of (4.4) are given by

$$y(x) = C_1 e^{p_1 x} + \ldots + C_n e^{p_n x}. \qquad (4.6)$$

It is curious to see that the "brightest mathematicians of the world" struggled for many decades to find this solution, which appears so trivial to today's students.

A difficulty arises with the solution (4.6) when (4.5) does not possess n distinct roots. Consider, with Euler, the example

$$y'' - 2qy' + q^2 y = 0. \qquad (4.7)$$

Here $p = q$ is a double root of the corresponding characteristic equation. If we set

$$y = e^{qx}u, \qquad (4.8)$$

(4.7) becomes $u'' = 0$, which brings us back to (4.3). So the general solution of (4.7) is given by $y(x) = e^{qx}(C_1 x + C_2)$ (see also Exercise 5 below). After some more examples of this type, one sees that the transformation (4.8) effects a *shift* of the characteristic polynomial, so that if q is a root of multiplicity k, we obtain for u an equation ending with $\ldots + Bu^{(k+1)} + Cu^{(k)} = 0$. Therefore

$$e^{qx}(C_1 x^{k-1} + \ldots + C_k)$$

gives us k independent solutions.

Finally, for a pair of complex roots $p = \alpha \pm i\beta$ the solutions $e^{(\alpha+i\beta)x}$, $e^{(\alpha-i\beta)x}$ can be replaced by the real functions

$$e^{\alpha x}(C_1 \cos \beta x + C_2 \sin \beta x).$$

The study of the *inhomogeneous* equation

$$\mathcal{L}(y) = f(x) \qquad (4.9)$$

was begun in Euler (1750), p. 13. We mention from this work the case where $f(x)$ is a polynomial, say for example the equation

$$Ay'' + By' + Cy = ax^2 + bx + c. \qquad (4.10)$$

Here Euler puts $y(x) = Ex^2 + Fx + G + v(x)$. Inserting this into (4.10) and eliminating all possible powers of x, one obtains

$$CE = a, \qquad CF + 2BE = b, \qquad CG + BF + 2AE = c,$$

$$Av'' + Bv' + Cv = 0.$$

This allows us, when C is different from zero, to compute E, F and G and we observe that *the general solution of the inhomogeneous equation is the sum of a particular solution of it and of the general solution of the corresponding homogeneous equation.* This is also true in the general case and can be verified by trivial linear algebra.

The above method of searching for a particular solution with the help of unknown coefficients works similarly if $f(x)$ is composed of exponential, sine, or cosine functions and is often called the "fast method". We see with pleasure that it was historically the first method to be discovered.

Variation of Constants

The *general treatment of the inhomogeneous equation*

$$a_n(x)y^{(n)} + \ldots + a_0(x)y = f(x) \tag{4.11}$$

is due to Lagrange (1775) ("... par une nouvelle méthode aussi simple qu'on puisse le désirer", see also Lagrange (1788), seconde partie, Sec. V.) We assume known n independent solutions $u_1(x), \ldots, u_n(x)$ of the *homogeneous* equation. We then set, in extension of the method employed for (3.2), instead of (4.2)

$$y(x) = c_1(x)u_1(x) + \ldots + c_n(x)u_n(x) \tag{4.12}$$

with unknown functions $c_i(x)$ ("method of variation of constants"). We have to insert (4.12) into (4.11) and thus compute the first derivative

$$y' = \sum_{i=1}^{n} c_i' u_i + \sum_{i=1}^{n} c_i u_i'.$$

If we continue blindly to differentiate in this way, we soon obtain complicated and useless formulas. Therefore Lagrange astutely requires the first term to vanish and puts

$$\sum_{i=1}^{n} c_i' u_i^{(j)} = 0 \qquad j = 0, \quad \text{then also for } j = 1, \ldots, n-2. \tag{4.13}$$

Then repeated differentiation of y, with continued elimination of the undesired terms (4.13), gives

$$y' = \sum_{i=1}^{n} c_i u_i', \qquad \ldots \qquad y^{(n-1)} = \sum_{i=1}^{n} c_i u_i^{(n-1)},$$

$$y^{(n)} = \sum_{i=1}^{n} c_i' u_i^{(n-1)} + \sum_{i=1}^{n} c_i u_i^{(n)}.$$

If we insert this into (4.11), we observe wonderful cancellations due to the fact that the $u_i(x)$ satisfy the homogeneous equation, and finally obtain, together with (4.13),

$$
\begin{pmatrix}
u_1 & \cdots & u_n \\
u_1' & \cdots & u_n' \\
\vdots & & \vdots \\
u_1^{(n-1)} & \cdots & u_n^{(n-1)}
\end{pmatrix}
\begin{pmatrix}
c_1' \\
c_2' \\
\vdots \\
c_n'
\end{pmatrix}
=
\begin{pmatrix}
0 \\
\vdots \\
0 \\
f(x)/a_n(x)
\end{pmatrix}.
\tag{4.14}
$$

This is a linear system, whose determinant is called the "Wronskian" and whose solution yields $c_1'(x), \ldots, c_n'(x)$ and after integration $c_1(x), \ldots, c_n(x)$.

Much more insight into this formula will be possible in Section I.11.

Exercises

1. Find the solution "huius aequationis differentialis quarti gradus" $a^4 y^{(4)} + y = 0$, $a^4 y^{(4)} - y = 0$; solve the equation "septimi gradus" $y^{(7)} + y^{(5)} + y^{(4)} + y^{(3)} + y^{(2)} + y = 0$. (Euler 1743, Ex. 4, 5, 6).

2. Solve by Euler's technique $y'' - 3y' - 4y = \cos x$ and $y'' + y = \cos x$.

   Hint. In the first case the particular solution can be searched for in the form $E \cos x + F \sin x$. In the second case (resonance) one puts $Ex \cos x + Fx \sin x$ just as in the solution of (4.7).

3. Find the solution of
$$
y'' - 3y' - 4y = g(x), \qquad g(x) = \begin{cases} \cos(x) & 0 \le x \le \pi/2 \\ 0 & \pi/2 \le x \end{cases}
$$

   such that $y(0) = y'(0) = 0$,

   a) by using the solution of Exercise 2,

   b) by the method of Lagrange (variation of constants).

4. (Reduction of the order if one solution is known). Suppose that a nonzero solution $u_1(x)$ of $y'' + a_1(x)y' + a_0(x)y = 0$ is known. Show that a second independent solution can be found by putting $u_2(x) = c(x)u_1(x)$.

5. Treat the case of multiple characteristic values (4.7) by considering them as a limiting case $p_2 \to p_1$ and using the solutions
$$
u_1(x) = e^{p_1 x}, \qquad u_2(x) = \lim_{p_2 \to p_1} \frac{e^{p_2 x} - e^{p_1 x}}{p_2 - p_1} = \frac{\partial e^{p_1 x}}{\partial p_1}, \text{ etc.}
$$

   (d'Alembert (1748), p. 284: "Enfin, si les valeurs de p & de p' sont égales, au lieu de les supposer telles, on supposera $p = a + \alpha$, $p' = a - \alpha$, α étant quantité infiniment petite ...").

I.5 Equations with Weak Singularities

Der Mathematiker weiss sich ohnedies beim Auftreten von singu-
lären Stellen gegebenenfalls leicht zu helfen.   (K. Heun 1900)

Many equations occurring in applications possess *singularities*, i.e., points at which
the function $f(x,y)$ of the differential equation becomes infinite. We study in some
detail the classical treatment of such equations, since numerical methods, which
will be discussed later in this book, often fail at the singular point, at least if they
are not applied carefully.

Linear Equations

As a first example, consider the equation

$$y' = \frac{q + bx}{x}\, y, \qquad q \neq 0 \tag{5.1}$$

with a singularity at $x = 0$. Its solution, using the method of separation of variables
(3.1), is

$$y(x) = Cx^q e^{bx} = C(x^q + bx^{q+1} + \ldots). \tag{5.2}$$

These solutions are plotted in Fig. 5.1 for different values of q and show the
fundamental difference in the behaviour of the solutions in dependence of q.

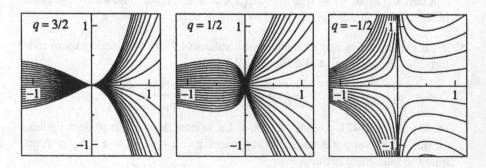

Fig. 5.1. Solutions of (5.1) for $b = 1$

Euler started a systematic study of equations with singularities. He asked which type of equation of the second order can conveniently be solved by a series as in (5.2) (Euler 1769, Problema 122, p. 177, "... quas commode per series resolvere licet..."). He found the equation

$$\mathcal{L}y := x^2(a+bx)y'' + x(c+ex)y' + (f+gx)y = 0. \tag{5.3}$$

Let us put $y = x^q(A_0 + A_1x + A_2x^2 + \ldots)$ with $A_0 \neq 0$ and insert this into (5.3). We observe that the powers x^2 and x which are multiplied by y'' and y', respectively, just re-establish what has been lost by the differentiations and obtain by comparing equal powers of x

$$\Big(q(q-1)a + qc + f\Big)A_0 = 0 \tag{5.4a}$$

$$\Big((q+i)(q+i-1)a + (q+i)c + f\Big)A_i \tag{5.4b}$$

$$= -\Big((q+i-1)(q+i-2)b + (q+i-1)e + g\Big)A_{i-1}$$

for $i = 1, 2, 3, \ldots$. In order to get $A_0 \neq 0$, q has to be a root of the *index equation*

$$\chi(q) := q(q-1)a + qc + f = 0. \tag{5.5}$$

For $a \neq 0$ there are two characteristic roots q_1 and q_2 of (5.5). Since the left-hand side of (5.4b) is of the form $\chi(q+i)A_i = \ldots$, this relation allows us to compute $A_1, A_2, A_3, \ldots$ at least for q_1 (if the roots are ordered such that $\operatorname{Re} q_1 \geq \operatorname{Re} q_2$). Thus we have obtained a first non-zero solution of (5.3). A second linearly independent solution for $q = q_2$ is obtained in the same way if $q_1 - q_2$ is not an integer.

Case of double roots. Euler found a second solution in this case with the inspiration of some acrobatic heuristics (Euler 1769, p. 150: "... quod $\frac{x^0}{0}$ aequivaleat ipsi $\ell x \, x \ldots$"). Fuchs (1866, 1868) then wrote a monumental paper on the form of all solutions for the general equation of order n, based on complicated calculations. A very elegant idea was then found by Frobenius (1873): fix A_0, say as $A_0(q) = 1$, completely ignore the index equation, choose q arbitrarily and consider the coefficients of the recursion (5.4b) as functions of q to obtain the series

$$y(x,q) = x^q \sum_{i=0}^{\infty} A_i(q)x^i, \tag{5.6}$$

whose convergence is discussed in Exercise 8 below. Since all conditions (5.4b) are satisfied, with the exception of (5.4a), we have

$$\mathcal{L}y(x,q) = \chi(q)x^q. \tag{5.7}$$

A second independent solution is now found simply by differentiating (5.7) with respect to q:

$$\mathcal{L}\Big(\frac{\partial y}{\partial q}(x,q)\Big) = \chi(q)\cdot\log x\cdot x^q + \chi'(q)\cdot x^q. \tag{5.8}$$

If we set $q = q_1$

$$\frac{\partial y}{\partial q}(x, q_1) = \log x \cdot y(x, q_1) + x^{q_1} \sum_{i=0}^{\infty} A_i'(q_1) x^i, \tag{5.9}$$

we obtain the desired second solution since $\chi(q_1) = \chi'(q_1) = 0$ (remember that q_1 is a double root of χ).

The case $q_1 - q_2 = m \in \mathbf{Z}$, $m \geq 1$. In this case we define a function $z(x)$ by satisfying $A_0(q) = 1$ and the recursion (5.4b) for all i with the exception of $i = m$. Then

$$\mathcal{L}z = \chi(q)x^q + Cx^{q+m} \tag{5.10}$$

where C is some constant. For $q = q_2$ the first term in (5.10) vanishes and a comparison with (5.8) shows that

$$\chi'(q_1)z(x) - C\frac{\partial y}{\partial q}(x, q_1) \tag{5.11}$$

is the required second solution of (5.3).

Euler (1778) later remarked that the formulas obtained become particularly elegant, if one starts from the differential equation

$$x(1 - x)y'' + (c - (a + b + 1)x)y' - aby = 0 \tag{5.12}$$

instead of from (5.3). Here, the above method leads to

$$q(q - 1) + cq = 0, \qquad q_1 = 0, \qquad q_2 = 1 - c, \tag{5.13}$$

$$A_{i+1} = \frac{(a + i)(b + i)}{(c + i)(1 + i)} A_i \qquad \text{for } q_1 = 0. \tag{5.14}$$

The resulting solutions, later named *hypergeometric functions*, became particularly famous throughout the 19th century with the work of Gauss (1812).

More generally, the above method works in the case of a differential equation

$$x^2 y'' + xa(x)y' + b(x)y = 0 \tag{5.15}$$

where $a(x)$ and $b(x)$ are regular analytic functions. One then says that 0 is a *regular singular point*. Similarly, we say that the equation $(x - x_0)^2 y'' + (x - x_0)a(x)y' + b(x)y = 0$ possesses the regular singular point x_0. In this case solutions can be obtained by the use of algebraic singularities $(x - x_0)^q$.

Finally, we also want to study the behaviour at *infinity* for an equation of the form

$$a(x)y'' + b(x)y' + c(x)y = 0. \tag{5.16a}$$

For this, we use the coordinate transformation $t = 1/x$, $z(t) = y(x)$ which yields

$$t^4 a\left(\frac{1}{t}\right)z'' + \left(2t^3 a\left(\frac{1}{t}\right) - t^2 b\left(\frac{1}{t}\right)\right)z' + c\left(\frac{1}{t}\right)z = 0. \tag{5.16b}$$

∞ is called a regular singular point of (5.16a) if 0 is a regular singular point of (5.16b). For examples see Exercise 9.

Nonlinear Equations

For nonlinear equations also, the above method sometimes allows one to obtain, if not the complete series of the solution, at least a couple of terms.

EXEMPLUM. Let us see what happens if we try to solve the classical brachystochrone problem (2.4) by a series. We suppose $h = 0$ and the initial value $y(0) = 0$. We write the equation as

$$(y')^2 = \frac{L}{y} - 1 \qquad \text{or} \qquad y(y')^2 + y = L. \tag{5.17}$$

At the initial point $y(0) = 0$, y' becomes infinite and most numerical methods would fail. We search for a solution of the form $y = A_0 x^q$. This gives in (5.17) $q^2 A_0^3 x^{3q-2} + A_0 x^q = L$. Due to the initial value we have that $y(x)$ becomes negligible for small values of x. We thus set the first term equal to L and obtain $3q - 2 = 0$ and $q^2 A_0^3 = L$. So

$$u(x) = \left(\frac{9Lx^2}{4}\right)^{1/3} \tag{5.18}$$

is a first approximate solution. The idea is now to use (5.18) just to escape from the initial point with a small x, and then to continue the solution with any numerical step-by-step procedure from the later chapters.

A more refined approximation could be tried in the form $y = A_0 x^q + A_1 x^{q+r}$. This gives with (5.17)

$$q^2 A_0^3 x^{3q-2} + q(3q + 2r) A_0^2 A_1 x^{3q+r-2} + A_0 x^q + \ldots = L.$$

We use the second term to neutralize the third one, which gives $3q + r - 2 = q$ or $r = q = 2/3$ and $5q^2 A_0 A_1 = -1$. Therefore

$$v(x) = \left(\frac{9Lx^2}{4}\right)^{1/3} - \left(\frac{9^2 x^4}{4^2 L5^3}\right)^{1/3} \tag{5.19}$$

is a better approximation. The following numerical results illustrate the utility of the approximations (5.18) and (5.19) compared with the correct solution $y(x)$ from I.3, Exercise 6, with $L = 2$:

$x = 0.10$	$y(x) = 0.342839$	$u(x) = 0.355689$	$v(x) = 0.343038$
$x = 0.01$	$y(x) = 0.076042$	$u(x) = 0.076631$	$v(x) = 0.076044.$

Exercises

1. Compute the general solution of the equation $x^2 y'' + xy' + gx^n y = 0$ with g constant (Euler 1769, Problema 123, Exemplum 1).

2. Apply the technique of Euler to the *Bessel equation*

$$x^2 y'' + xy' + (x^2 - g^2)y = 0.$$

Sketch the solutions obtained for $g = 2/3$ and $g = 10/3$.

3. Compute the solutions of the equations $x^2 y'' - 2xy' + y = 0$ and $x^2 y'' - 3xy' + 4y = 0$. Equations of this type are often called Euler's or even Cauchy's equation. Its solution, however, was already known to Joh. Bernoulli.

4. (Euler 1769, Probl. 123, Exempl. 2). Let

$$y(x) = \int_0^{2\pi} \sqrt{\sin^2 s + x^2 \cos^2 s}\, ds$$

be the perimeter of the ellipse with axes 1 and $x < 1$.

a) Verify that $y(x)$ satisfies the differential equation

$$x(1 - x^2)y'' - (1 + x^2)y' + xy = 0. \tag{5.20}$$

b) Compute the solutions of this equation.

c) Show that the coordinate change $x^2 = t$, $y(x) = z(t)$ transforms (5.20) to a hypergeometric equation (5.12).

Hint. The computations for a) lead to the integral

$$\int_0^{2\pi} \frac{1 - 2\cos^2 s + q^2 \cos^4 s}{(1 - q^2 \cos^2 s)^{3/2}}\, ds, \qquad q^2 = 1 - x^2$$

which must be shown to be zero. Develop this into a power series in q^2.

5. Try to solve the equation

$$x^2 y'' + (3x - 1)y' + y = 0$$

with the help of a series (5.6) and study its convergence.

6. Find a series of the type

$$y = A_0 x^q + A_1 x^{q+s} + A_2 x^{q+2s} + \ldots$$

which solves the nonlinear "Emden-Fowler equation" of astrophysics $(x^2 y')' + y^2 x^{-1/2} = 0$ in the neighbourhood of $x = 0$.

7. Approximate the solution of Leibniz's equation (2.3) in the neighbourhood of the singular initial value $y(0) = a$ by a function of the type $y(x) = a - Cx^q$. Compare the result with the correct solution of Exercise 2 of I.3.

8. Show that the radius of convergence of series (5.6) is given by

$$\text{i)} \quad r = |a/b| \qquad\qquad \text{ii)} \quad r = 1$$

for the coefficients given by (5.4) and (5.14), respectively.

9. Show that the point ∞ is a regular singular point for the hypergeometric equation (5.12), but not for the Bessel equation of Exercise 2.

10. Consider the initial value problem

$$y' = \frac{\lambda}{x} y + g(x), \qquad y(0) = 0. \tag{5.21}$$

a) Prove that if $\lambda \leq 0$, the problem (5.21) possesses a unique solution for $x \geq 0$;

b) If $g(x)$ is k-times differentiable and $\lambda \leq 0$, then the solution $y(x)$ is $(k+1)$-times differentiable for $x \geq 0$ and we have

$$y^{(j)}(0) = \left(1 - \frac{\lambda}{j}\right)^{-1} g^{(j-1)}(0), \qquad j = 1, 2, \ldots.$$

I.6 Systems of Equations

Newton (1687) distilled from the known solutions of planetary motion (the Kepler laws) his "Lex secunda" together with the universal law of gravitation. It was mainly the "Dynamique" of d'Alembert (1743) which introduced, the other way round, second order differential equations as a general tool for computing mechanical motion. Thus, Euler (1747) studied the movement of planets via the equations in 3-space

$$m\frac{d^2x}{dt^2} = X, \qquad m\frac{d^2y}{dt^2} = Y, \qquad m\frac{d^2z}{dt^2} = Z, \qquad (6.1)$$

where X, Y, Z are the forces in the three directions. ("... & par ce moyen j'evite quantité de recherches penibles").

The Vibrating String and Propagation of Sound

Suppose a string is represented by a sequence of identical and equidistant mass points and denote by $y_1(t)$, $y_2(t), \dots$ the deviation of these mass points from the equilibrium position (Fig. 6.1a). If the deviations are supposed small ("fort petites"), the repelling force for the i-th mass point is proportional to $-y_{i-1} + 2y_i - y_{i+1}$ (Brook Taylor 1715, Johann Bernoulli 1727). Therefore equations (6.1) become

$$y_1'' = K^2(-2y_1 + y_2)$$
$$y_2'' = K^2(y_1 - 2y_2 + y_3)$$
$$\dots \qquad\qquad\qquad (6.2)$$
$$y_n'' = K^2(y_{n-1} - 2y_n).$$

This is a system of n linear differential equations. Since the finite differences $y_{i-1} - 2y_i + y_{i+1} \approx c^2 \frac{\partial^2 y}{\partial x^2}$, equation (6.2) becomes, by the "inverse" method of lines, the famous partial differential equation (d'Alembert 1747)

$$\frac{\partial^2 u}{\partial t^2} = a^2 \frac{\partial^2 u}{\partial x^2}$$

for the vibrating string.

The *propagation of sound* is modelled similarly (Lagrange 1759): we suppose the medium to be a sequence of mass points and denote by $y_1(t)$, $y_2(t)$, ... their longitudinal displacements from the equilibrium position (see Fig. 6.1b). Then by Hooke's law of elasticity the repelling forces are proportional to the differences of displacements $(y_{i-1} - y_i) - (y_i - y_{i+1})$. This leads to equations (6.2) again ("En examinant les équations,... je me suis bientôt aperçu qu'elles ne différaient nullement de celles qui appartiennent au problème *de chordis vibrantibus*... ").

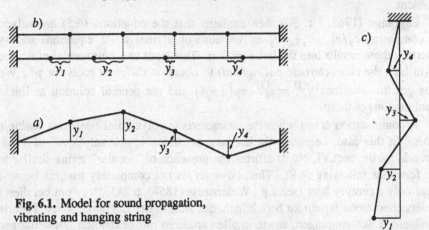

Fig. 6.1. Model for sound propagation, vibrating and hanging string

Another example, treated by Daniel Bernoulli (1732) and by Lagrange (1762, Nr. 36), is that of mass points attached to a *hanging* string (Fig. 6.1c). Here the tension in the string becomes greater in the upper part of the string and we have the following equations of movement

$$y_1'' = K^2(-y_1 + y_2)$$
$$y_2'' = K^2(y_1 - 3y_2 + 2y_3)$$
$$y_3'' = K^2(2y_2 - 5y_3 + 3y_4)$$

...

$$y_n'' = K^2\big((n-1)y_{n-1} - (2n-1)y_n\big). \tag{6.3}$$

In all these examples, of course, the deviations y_i are supposed to be "infinitely" small, so that linear models are realistic.

Using a notation which came into use only a century later, we write these equations in the form

$$y_i'' = \sum_{j=1}^{n} a_{ij}y_j, \qquad i = 1, \ldots, n, \tag{6.4}$$

which is a *system of 2nd order linear equations with constant coefficients*. Lagrange solves system (6.4) by putting $y_i = c_i e^{pt}$, which leads to

$$p^2 c_i = \sum_{j=1}^{n} a_{ij} c_j, \qquad i = 1, \ldots, n \qquad (6.5)$$

so that p^2 must be an *eigenvalue* of the matrix $A = (a_{ij})$ and $c = (c_1, \ldots, c_n)^T$ a corresponding *eigenvector*. We see here the first appearance of an eigenvalue problem.

Lagrange (1762, Nr. 30) then explains that the equations (6.5) are solved by computing $c_2/c_1, \ldots, c_n/c_1$ as functions of p from $n-1$ equations and by inserting these results into the last equation. This leads to a polynomial of degree n (in fact, the *characteristic polynomial*) to obtain n different roots for p^2. We thus get $2n$ solutions $y_i^{(j)} = c_i^{(j)} \exp(\pm p_j t)$ and the general solution as linear combinations of these.

A complication arises when the characteristic polynomial possesses *multiple roots*. In this case, Lagrange (in his famous "Mécanique Analytique" of 1788, seconde partie, sect. VI, No.7) affirms the presence of "secular" terms similar to the formulas following (4.8). This, however, is not completely true, as became clear only a century later (see e.g., Weierstrass (1858), p.243: "... um bei dieser Gelegenheit einen Irrtum zu berichtigen, der sich in der Lagrange'schen Theorie der kleinen Schwingungen, sowie in allen späteren mir bekannten Darstellungen derselben, findet."). We therefore postpone this subject to Section I.12.

We solve equations (6.2) in detail, since the results obtained are of particular importance (Lagrange 1759). The corresponding eigenvalue problem (6.5) becomes in this case $p^2 c_1 = K^2(-2c_1 + c_2)$, $p^2 c_i = K^2(c_{i-1} - 2c_i + c_{i+1})$ for $i = 2, \ldots, n-1$ and $p^2 c_n = K^2(c_{n-1} - 2c_n)$. We introduce $p^2/K^2 + 2 = q$, so that

$$c_{j+1} - q c_j + c_{j-1} = 0, \qquad c_0 = 0, \qquad c_{n+1} = 0. \qquad (6.6)$$

This means that the c_i are the solutions of a *difference equation* and therefore $c_j = A a^j + B b^j$ where a and b are the roots of the corresponding characteristic equation $z^2 - qz + 1 = 0$, hence

$$a + b = q, \qquad\qquad ab = 1.$$

The condition $c_0 = 0$ of (6.6), which means that $A + B = 0$, shows that $c_j = A(a^j - b^j)$ with $A \neq 0$. The second condition $c_{n+1} = 0$, or equivalently $(a/b)^{n+1} = 1$, implies together with $ab = 1$ that

$$a = \exp\left(\frac{k\pi i}{n+1}\right), \quad b = \exp\left(\frac{-k\pi i}{n+1}\right)$$

for some $k = 1, \ldots, n$. Thus we obtain

$$q_k = 2\cos\frac{\pi k}{n+1}, \qquad k = 1, \ldots, n, \qquad (6.7a)$$

$$p_k^2 = 2K^2\left(\cos\frac{\pi k}{n+1} - 1\right) = -4K^2\left(\sin\frac{\pi k}{2n+2}\right)^2. \qquad (6.7\text{b})$$

Finally, Euler's formula from 1740, $e^{iz} - e^{-iz} = 2i\sin x$ ("... si familière aujourd'hui aux Géomètres") gives for the eigenvectors (with $A = -i/2$)

$$c_j^{(k)} = \sin\frac{jk\pi}{n+1}, \qquad j, k = 1, \ldots, n. \qquad (6.8)$$

Since the p_k are purely imaginary, we also use for $\exp(\pm p_k t)$ the "familière" formula and obtain the general solution

$$y_j(t) = \sum_{k=1}^n \sin\frac{jk\pi}{n+1}(a_k\cos r_k t + b_k\sin r_k t), \qquad r_k = 2K\sin\frac{\pi k}{2n+2}. \quad (6.9)$$

Lagrange then observed after some lengthy calculations, which are today seen by using the orthogonality relations

$$\sum_{\ell=1}^n \sin\frac{\ell j\pi}{n+1}\sin\frac{\ell k\pi}{n+1} = \begin{cases} 0 & j \neq k \\ \frac{n+1}{2} & j = k \end{cases} \qquad j, k = 1, \ldots, n$$

that

$$a_k = \frac{2}{n+1}\sum_{j=1}^n \sin\frac{kj\pi}{n+1}\, y_j(0), \qquad b_k = \frac{1}{r_k}\frac{2}{n+1}\sum_{j=1}^n \sin\frac{kj\pi}{n+1}\, y_j'(0)$$

are determined by the initial positions and velocities of the mass points. He also studied the case where n, the number of mass points, tends to infinity (so that, in the formula for r_k, $\sin x$ can be replaced by x) and stood, 50 years before Fourier, at the portal of Fourier series theory. "Mit welcher Gewandtheit, mit welchem Aufwande analytischer Kunstgriffe er auch den ersten Theil dieser Untersuchung durchführte, so liess der Uebergang vom Endlichen zum Unendlichen doch viel zu wünschen übrig..." (Riemann 1854).

Fourier

> J'ajouterai que le livre de Fourier a une importance capitale dans l'histoire des mathématiques.   (H. Poincaré 1893)

The first *first order systems* were motivated by the problem of heat conduction (Biot 1804, Fourier 1807). Fourier imagined a rod to be a sequence of molecules, whose temperatures we denote by y_i, and deduced from a law of Newton that the energy which a particle passes to its neighbours is proportional to the difference of their temperatures, i.e., $y_{i-1} - y_i$ to the left and $y_{i+1} - y_i$ to the right ("Lorsque deux molécules d'un même solide sont extrêmement voisines et ont des températures inégales, la molécule plus échauffée communique à celle qui l'est moins une quantité de chaleur exactement exprimée par le produit formé de la durée de l'instant, de la différence extrêmement petite des températures, et d'une certaine fonction de la distance des molécules"). This long sentence means, in formulas, that the total gain of energy of the ith molecule is expressed by

$$y_i' = K^2(y_{i-1} - 2y_i + y_{i+1}), \tag{6.10}$$

or, in general by

$$y_i' = \sum_{j=1}^{n} a_{ij} y_i, \qquad i = 1, \ldots, n, \tag{6.11}$$

a first order system with constant coefficients.

By putting $y_i = c_i e^{pt}$, we now obtain the eigenvalue problem

$$p c_i = \sum_{j=1}^{n} a_{ij} c_j, \qquad i = 1, \ldots, n. \tag{6.12}$$

If we suppose the rod cooled to zero at both ends ($y_0 = y_{n+1} = 0$), we can use Lagrange eigenvectors from above and obtain the solution

$$y_j(t) = \sum_{k=1}^{n} a_k \sin \frac{jk\pi}{n+1} \exp(-w_k t), \qquad w_k = 4K^2 \left(\sin \frac{\pi k}{2n+2} \right)^2. \tag{6.13}$$

By taking n larger and larger, Fourier arrived from (6.10) (again the inverse "method of lines") at his famous heat equation

$$\frac{\partial u}{\partial t} = a^2 \frac{\partial^2 u}{\partial x^2} \tag{6.14}$$

which was the origin of Fourier series theory.

Lagrangian Mechanics

> Dies ist der kühne Weg, den *Lagrange* ..., freilich ohne ihn gehörig zu rechtfertigen, eingeschlagen hat.
>
> (Jacobi 1842/43, Vorl. Dynamik, p. 13)

This combines d'Alembert's dynamics, the "principle of least action" of Leibniz–Maupertuis and the variational calculus; published in the monumental treatise "Mécanique Analytique" (1788). It furnishes an excellent means for obtaining the differential equations of motion for complicated mechanical systems (arbitrary coordinate systems, constraints, etc.).

If we define (with Poisson 1809) the "Lagrange function"

$$\mathcal{L} = T - U \tag{6.15}$$

where

$$T = m \frac{\dot{x}^2 + \dot{y}^2 + \dot{z}^2}{2} \qquad \text{(kinetic energy)} \tag{6.16}$$

and U is the "potential energy" satisfying

$$\frac{\partial U}{\partial x} = -X, \qquad \frac{\partial U}{\partial y} = -Y, \qquad \frac{\partial U}{\partial z} = -Z \tag{6.17}$$

then the equations of motion (6.1) are *identical* to Euler's equations (2.11) for the variational problem

$$\int_{t_0}^{t_1} \mathcal{L} \, dt = \min \tag{6.18}$$

(this, mainly through a misunderstanding of Jacobi, is often called "Hamilton's principle"). The important idea is now to forget (6.16) and (6.17) and to apply (6.15) and (6.18) to *arbitrary mass points* and *arbitrary coordinate systems*.

Example. The *spherical pendulum* (Lagrange 1788, Seconde partie, Section VIII, Chap. II, §I). Let $\ell = 1$ and

$$x = \sin\theta \cos\varphi$$
$$y = \sin\theta \sin\varphi$$
$$z = -\cos\theta.$$

We set $m = g = 1$ and have

$$T = \frac{1}{2}(\dot{x}^2 + \dot{y}^2 + \dot{z}^2) = \frac{1}{2}(\dot{\theta}^2 + \sin^2\theta \cdot \dot{\varphi}^2)$$
$$U = z = -\cos\theta \tag{6.19}$$

so that (2.11) becomes

$$\mathcal{L}_\theta - \frac{d}{dt}(\mathcal{L}_{\dot\theta}) = -\sin\theta + \sin\theta\cos\theta \cdot \dot\varphi^2 - \ddot\theta = 0$$
$$\mathcal{L}_\varphi - \frac{d}{dt}(\mathcal{L}_{\dot\varphi}) = -\sin^2\theta \cdot \ddot\varphi - 2\sin\theta\cos\theta \cdot \dot\varphi \cdot \dot\theta = 0. \tag{6.20}$$

We have thus obtained, by simple calculus, the equations of motion for the problem. These equations cannot be solved analytically. A solution, computed numerically by a Runge-Kutta method (see Chapter II) is shown in Fig. 6.2.

In general, suppose that the mechanical system in question is described by n coordinates $q_1, q_2, \ldots, q_n$ and that $\mathcal{L} = T - U$ depends on $q_1, q_2, \ldots, q_n$, $\dot{q}_1, \dot{q}_2, \ldots, \dot{q}_n$. Then the equations of motion are

$$\frac{d}{dt}\mathcal{L}_{\dot{q}_i} = \sum_{k=1}^{n} \mathcal{L}_{\dot{q}_i \dot{q}_k} \ddot{q}_k + \sum_{k=1}^{n} \mathcal{L}_{\dot{q}_i q_k} \dot{q}_k = \mathcal{L}_{q_i}, \qquad i = 1, \ldots, n. \tag{6.21}$$

These equations allow several generalizations to time-dependent systems and non-conventional forces.

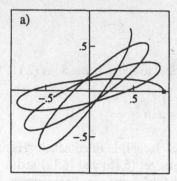 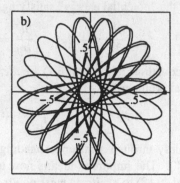

Fig. 6.2. Solution of the spherical pendulum, a) $0 \le x \le 20$, b) $0 \le x \le 100$
($\varphi_0 = 0$, $\dot{\varphi}_0 = 0.17$, $\theta_0 = 1$, $\dot{\theta}_0 = 0$)

Hamiltonian Mechanics

> Nach dem Erscheinen der ersten Ausgabe der Mécanique analytique wurde der wichtigste Fortschritt in der Umformung der Differentialgleichungen der Bewegung von *Poisson* ... gemacht ... im 15ten Hefte des polytechnischen Journals ... Hier führt *Poisson* die Grössen $p = \partial T / \partial q'$ ... ein.
>
> (Jacobi 1842/43, Vorl. Dynamik, p. 67)

Hamilton, having worked for many years with variational principles (Fermat's principle) in his researches on optics, discovered at once that his ideas, after introducing a "principal function", allowed very elegant solutions for Kepler's motion of a planet (Hamilton 1833). He then undertook in several papers (Hamilton 1834, 1835) to revolutionize mechanics. After many pages of computation he thereby discovered that it was "more convenient in many respects" (Hamilton 1834, Math. Papers II, p. 161) to work with the momentum coordinates (idea of Poisson)

$$p_i = \frac{\partial \mathcal{L}}{\partial \dot{q}_i} \tag{6.22}$$

instead of $\dot{q}_i$, and with the function

$$H = \sum_{k=1}^{n} \dot{q}_k p_k - \mathcal{L} \tag{6.23}$$

considered as function of $q_1, \dots, q_n, p_1, \dots, p_n$. This idea, to let derivatives $\partial \mathcal{L} / \partial \dot{q}_i$ and independent variables p_i interchange their parts in order to simplify differential equations, is due to Legendre (1787). Differentiating (6.23) by the chain rule, we obtain

$$\frac{\partial H}{\partial p_i} = \sum_{k=1}^{n} \frac{\partial \dot{q}_k}{\partial p_i} \cdot p_k + \dot{q}_i - \sum_{k=1}^{n} \frac{\partial \mathcal{L}}{\partial \dot{q}_k} \frac{\partial \dot{q}_k}{\partial p_i}$$

and

$$\frac{\partial H}{\partial q_i} = \sum_{k=1}^{n} \frac{\partial \dot{q}_k}{\partial q_i} \cdot p_k - \frac{\partial \mathcal{L}}{\partial q_i} - \sum_{k=1}^{n} \frac{\partial \mathcal{L}}{\partial \dot{q}_k} \frac{\partial \dot{q}_k}{\partial q_i}.$$

By (6.22) and (6.21) both formulas simplify to

$$\dot{q}_i = \frac{\partial H}{\partial p_i}, \qquad \dot{p}_i = -\frac{\partial H}{\partial q_i}, \qquad i = 1, \dots, n. \tag{6.24}$$

These equations are marvellously symmetric "... and to integrate these differential equations of motion... is the chief and perhaps ultimately the only problem of mathematical dynamics" (Hamilton 1835). Jacobi (1843) called them *canonical* differential equations.

Remark. If the kinetic energy T is a quadratic function of the velocities $\dot{q}_i$, Euler's identity (Euler 1755, Caput VII, § 224, "... si V fuerit functio homogenea...") states that

$$2T = \sum_{k=1}^{n} \dot{q}_k \frac{\partial T}{\partial \dot{q}_k}. \tag{6.25}$$

If we further assume that the potential energy U is independent of $\dot{q}_i$, we obtain

$$H = \sum_{k=1}^{n} \dot{q}_k p_k - \mathcal{L} = \sum_{k=1}^{n} \dot{q}_k \frac{\partial T}{\partial \dot{q}_k} - \mathcal{L} = 2T - \mathcal{L} = T + U. \tag{6.26}$$

This is the *total* energy of the system.

Example. The spherical pendulum again. From (6.19) we have

$$p_\theta = \frac{\partial T}{\partial \dot{\theta}} = \dot{\theta}, \qquad p_\varphi = \frac{\partial T}{\partial \dot{\varphi}} = \sin^2 \theta \cdot \dot{\varphi} \tag{6.27}$$

and, by eliminating the undesired variables $\dot{\theta}$ and $\dot{\varphi}$,

$$H = T + U = \frac{1}{2}\left(p_\theta^2 + \frac{p_\varphi^2}{\sin^2 \theta}\right) - \cos \theta. \tag{6.28}$$

Therefore (6.26) becomes

$$\dot{p}_\theta = p_\varphi^2 \cdot \frac{\cos \theta}{\sin^3 \theta} - \sin \theta \qquad \dot{p}_\varphi = 0$$
$$\dot{\theta} = p_\theta \qquad\qquad\qquad \dot{\varphi} = \frac{p_\varphi}{\sin^2 \theta}. \tag{6.29}$$

These equations appear to be a little simpler than Lagrange's formulas (6.20). For example, we immediately see that $p_\varphi = Const$ (Kepler's second law).

Exercises

1. Verify that, if $u(x)$ is sufficiently differentiable,

$$\frac{u(x-\delta) - 2u(x) + u(x+\delta)}{\delta^2} = u''(x) + \frac{\delta^2}{12}\, u^{(4)}(x) + \mathcal{O}(\delta^4).$$

   Hint. Use Taylor series expansions for $u(x+\delta)$ and $u(x-\delta)$. This relation establishes the connection between (6.10) and (6.14) as well as between (6.2) and the wave equation.

2. Solve equation (6.3) for $n = 2$ and $n = 3$ by using the device of Lagrange described above (1762) and discover naturally the characteristic polynomial of the matrix.

3. Solve the first order system (6.11) with initial values $y_i(0) = (-1)^i$, where the matrix A is the same as in Exercise 2, and draw the solutions. Physically, this equation would represent a string with weights hanging, say, in honey.

4. Find the first terms of the development at the singular point $x = 0$ of the following system of nonlinear equations

$$\begin{aligned} x^2 y'' + 2xy' = 2yz^2 + \lambda x^2 y(y^2 - 1), \qquad & y(0) = 0 \\ x^2 z'' = z(z^2 - 1) + x^2 y^2 z, \qquad & z(0) = 1 \end{aligned} \tag{6.30}$$

   where λ is a constant parameter. Equations (6.30) are the Euler equations for the variational problem

$$I = \int_0^\infty \left((z')^2 + \frac{x^2(y')^2}{2} + \frac{(z^2-1)^2}{2x^2} + y^2 z^2 + \frac{\lambda}{4} x^2 (y^2 - 1)^2 \right) dx,$$

$$y(\infty) = 1, \qquad z(\infty) = 0$$

   which gives the mass of a "monopole" in nuclear physics (see 't Hooft 1974).

5. Prove that the Hamiltonian function $H(q_1, \dots, q_n, p_1, \dots, p_n)$ is a first integral for the system (6.24), i.e., every solution satisfies

$$H(q_1(t), \dots, q_n(t), p_1(t), \dots, p_n(t)) = Const.$$

I.7 A General Existence Theorem

> M. Cauchy annonce, que, pour se conformer au voeu du Conseil,
> il ne s'attachera plus à donner, comme il a fait jusqu'à présent, des
> démonstrations parfaitement rigoureuses.
> (Conseil d'instruction de l'Ecole polytechnique, 24 nov. 1825)
>
> You have all professional deformation of your minds; *convergence*
> does not matter here ...   (P. Henrici 1985)

We now enter a new era for our subject, more theoretical than the preceding
one. It was inaugurated by the work of Cauchy, who was not as fascinated by
long numerical calculations as was, say, Euler, but merely a fanatic for perfect
mathematical rigor and exactness. He criticized in the work of his predecessors
the use of infinite series and other infinite processes without taking much account
of error estimates or convergence results. He therefore established around 1820 a
convergence theorem for the polygon method of Euler and, some 15 years later,
for the power series method of Newton (see Section I.8). Beyond the estimation
of errors, these results also allow the statement of *general existence theorems*
for the solutions of arbitrary differential equations ("d'une équation différentielle
quelconque"), whose solutions were only known before in a very few cases. A
second important consequence is to provide results about the *uniqueness* of the
solution, which allow one to conclude that the computed solution (numerically or
analytically) is the only one with the same initial value and that there are no others.
Only then we are allowed to speak of *the* solution of the problem.

His very first proof has recently been discovered on fragmentary notes (Cauchy
1824), which were never published in Cauchy's lifetime (did his notes not satisfy
the Minister of education?: "... mais que le second professeur, M. Cauchy, n'a
présenté que des feuilles qui n'ont pu satisfaire la commission, et qu'il a été jusqu'à
présent impossible de l'amener à se rendre au voeu du Conseil et à exécuter la
décision du Ministre").

Convergence of Euler's Method

Let us now, with bared head and trembling knees, follow the ideas of this historical
proof. We formulate it in a way which generalizes directly to higher dimensional
systems.

Starting with the one-dimensional differential equation

$$y' = f(x,y), \qquad y(x_0) = y_0, \qquad y(X) = ? \tag{7.1}$$

we make use of the method explained by Euler (1768) in the last section of his
"Institutiones Calculi Integralis" (Caput VII, p. 424), i.e., we consider a subdivision

of the interval of integration

$$x_0, x_1, \ldots, x_{n-1}, x_n = X \qquad (7.2)$$

and replace in each subinterval the solution by the first term of its Taylor series

$$y_1 - y_0 = (x_1 - x_0)f(x_0, y_0)$$
$$y_2 - y_1 = (x_2 - x_1)f(x_1, y_1)$$
$$\ldots \qquad (7.3)$$
$$y_n - y_{n-1} = (x_n - x_{n-1})f(x_{n-1}, y_{n-1}).$$

For the subdivision above we also use the notation

$$h = (h_0, h_1, \ldots, h_{n-1})$$

where $h_i = x_{i+1} - x_i$. If we connect y_0 and y_1, y_1 and $y_2, \ldots$ etc by straight lines we obtain the *Euler polygon*

$$y_h(x) = y_i + (x - x_i)f(x_i, y_i) \qquad \text{for} \quad x_i \leq x \leq x_{i+1}. \qquad (7.3a)$$

Lemma 7.1. *Assume that $|f|$ is bounded by A on*

$$D = \Big\{ (x,y) \mid x_0 \leq x \leq X, \; |y - y_0| \leq b \Big\}.$$

If $X - x_0 \leq b/A$ then the numerical solution (x_i, y_i) given by (7.3), remains in D for every subdivision (7.2) and we have

$$|y_h(x) - y_0| \leq A \cdot |x - x_0|, \qquad (7.4)$$

$$\Big| y_h(x) - \Big(y_0 + (x - x_0)f(x_0, y_0) \Big) \Big| \leq \varepsilon \cdot |x - x_0| \qquad (7.5)$$

if $|f(x,y) - f(x_0, y_0)| \leq \varepsilon$ on D.

Proof. Both inequalities are obtained by adding up the lines of (7.3) and using the triangle inequality. Formula (7.4) then shows immediately that for $A(x - x_0) \leq b$ the polygon remains in D. □

Our next problem is to obtain an estimate for the change of $y_h(x)$, when the initial value y_0 is changed: let z_0 be another initial value and compute

$$z_1 - z_0 = (x_1 - x_0)f(x_0, z_0). \qquad (7.6)$$

We need an estimate for $|z_1 - y_1|$. Subtracting (7.6) from the first line of (7.3) we obtain

$$z_1 - y_1 = z_0 - y_0 + (x_1 - x_0)\Big(f(x_0, z_0) - f(x_0, y_0) \Big).$$

This shows that we need an estimate for $f(x_0, z_0) - f(x_0, y_0)$. If we suppose

$$|f(x,z) - f(x,y)| \leq L|z - y| \qquad (7.7)$$

we obtain

$$|z_1 - y_1| \le (1 + (x_1 - x_0)L)|z_0 - y_0|. \tag{7.8}$$

Lemma 7.2. *For a fixed subdivision h let $y_h(x)$ and $z_h(x)$ be the Euler polygons corresponding to the initial values y_0 and z_0, respectively. If*

$$\left|\frac{\partial f}{\partial y}(x,y)\right| \le L \tag{7.9}$$

in a convex region which contains $(x, y_h(x))$ and $(x, z_h(x))$ for all $x_0 \le x \le X$, then

$$|z_h(x) - y_h(x)| \le e^{L(x-x_0)}|z_0 - y_0|. \tag{7.10}$$

Proof. (7.9) implies (7.7), (7.7) implies (7.8), (7.8) implies

$$|z_1 - y_1| \le e^{L(x_1-x_0)}|z_0 - y_0|.$$

If we repeat the same argument for $z_2 - y_2$, $z_3 - y_3$, and so on, we finally obtain (7.10). □

Remark. Condition (7.7) is called a "Lipschitz condition". It was Lipschitz (1876) who rediscovered the theory (footnote in the paper of Lipschitz: "L'auteur ne connaît pas évidemment les travaux de Cauchy ...") and advocated the use of (7.7) instead of the more stringent hypothesis (7.9). Lipschitz's proof is also explained in the classical work of Picard (1891-96), Vol. II, Chap. XI, Sec. I.

If the subdivision (7.2) is refined more and more, so that

$$|h| := \max_{i=0,\dots,n-1} h_i \to 0,$$

we expect that the Euler polygons converge to a solution of (7.1). Indeed, we have

Theorem 7.3. *Let $f(x,y)$ be continuous, and $|f|$ be bounded by A and satisfy the Lipschitz condition (7.7) on*

$$D = \Big\{(x,y) \mid x_0 \le x \le X, \ |y - y_0| \le b\Big\}.$$

If $X - x_0 \le b/A$, then we have:

a) *For $|h| \to 0$ the Euler polygons $y_h(x)$ converge uniformly to a continuous function $\varphi(x)$.*

b) *$\varphi(x)$ is continuously differentiable and solution of (7.1) on $x_0 \le x \le X$.*

c) *There exists no other solution of (7.1) on $x_0 \le x \le X$.*

Proof. a) Take an $\epsilon > 0$. Since f is uniformly continuous on the compact set D, there exists a $\delta > 0$ such that

$$|u_1 - u_2| \le \delta \qquad \text{and} \qquad |v_1 - v_2| \le A \cdot \delta$$

imply

$$|f(u_1,v_1) - f(u_2,v_2)| \le \varepsilon. \qquad (7.11)$$

Suppose now that the subdivision (7.3) satisfies

$$|x_{i+1} - x_i| \le \delta, \qquad \text{i.e.,} \qquad |h| \le \delta. \qquad (7.12)$$

We first study the effect of adding new mesh-points. In a first step, we consider a subdivision $h(1)$, which is obtained by adding new points only to the *first* subinterval (see Fig. 7.1). It follows from (7.5) (applied to this first subinterval) that for the new refined solution $y_{h(1)}(x_1)$ we have the estimate $|y_{h(1)}(x_1) - y_h(x_1)| \le \varepsilon|x_1 - x_0|$. Since the subdivisions h and $h(1)$ are identical on $x_1 \le x \le X$ we can apply Lemma 7.2 to obtain

$$|y_{h(1)}(x) - y_h(x)| \le e^{L(x-x_1)}(x_1 - x_0)\varepsilon \qquad \text{for} \quad x_1 \le x \le X.$$

We next add further points to the subinterval (x_1, x_2) and denote the new subdivision by $h(2)$. In the same way as above this leads to $|y_{h(2)}(x_2) - y_{h(1)}(x_2)| \le \varepsilon|x_2 - x_1|$ and

$$|y_{h(2)}(x) - y_{h(1)}(x)| \le e^{L(x-x_2)}(x_2 - x_1)\varepsilon \qquad \text{for} \quad x_2 \le x \le X.$$

The entire situation is sketched in Fig. 7.1. If we denote by $\widehat{h}$ the final refinement, we obtain for $x_i < x \le x_{i+1}$

$$|y_{\widehat{h}}(x) - y_h(x)| \qquad (7.13)$$

$$\le \varepsilon\Big(e^{L(x-x_1)}(x_1 - x_0) + \ldots + e^{L(x-x_i)}(x_i - x_{i-1})\Big) + \varepsilon(x - x_i)$$

$$\le \varepsilon \int_{x_0}^{x} e^{L(x-s)}\, ds = \frac{\varepsilon}{L}\Big(e^{L(x-x_0)} - 1\Big).$$

If we now have two different subdivisions h and $\widetilde{h}$, which both satisfy (7.12), we introduce a *third* subdivision $\widehat{h}$ which is a refinement of both subdivisions (just as is usually done in proving the existence of Riemann's integral), and apply (7.13) twice. We then obtain from (7.13) by the triangle inequality

$$|y_h(x) - y_{\widetilde{h}}(x)| \le 2\frac{\varepsilon}{L}\Big(e^{L(x-x_0)} - 1\Big).$$

For $\varepsilon > 0$ small enough, this becomes arbitrarily small and shows the uniform convergence of the Euler polygons to a continuous function $\varphi(x)$.

  b) Let

$$\varepsilon(\delta) := \sup\Big\{|f(u_1,v_1) - f(u_2,v_2)|\, ;\, |u_1 - u_2| \le \delta,\, |v_1 - v_2| \le A\delta,\, (u_i,v_i) \in D\Big\}$$

be the modulus of continuity. If x belongs to the subdivision h then we obtain from (7.5) (replace (x_0, y_0) by $(x, y_h(x))$ and x by $x + \delta$)

$$|y_h(x+\delta) - y_h(x) - \delta f(x, y_h(x))| \le \varepsilon(\delta)\delta. \qquad (7.14)$$

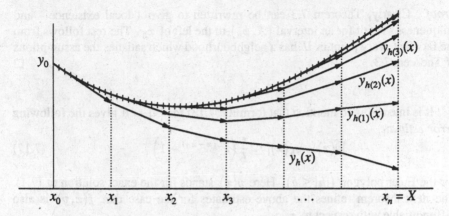

Fig. 7.1. Lady Windermere's Fan (O. Wilde 1892)

Taking the limit $|h| \to 0$ we get

$$|\varphi(x + \delta) - \varphi(x) - \delta f(x, \varphi(x))| \le \varepsilon(\delta)\delta. \tag{7.15}$$

Since $\varepsilon(\delta) \to 0$ for $\delta \to 0$, this proves the differentiability of $\varphi(x)$ and $\varphi'(x) = f(x, \varphi(x))$.

c) Let $\psi(x)$ be a second solution of (7.1) and suppose that the subdivision h satisfies (7.12). We then denote by $y_h^{(i)}(x)$ the Euler polygon to the initial value $(x_i, \psi(x_i))$ (it is defined for $x_i \le x \le X$). It follows from

$$\psi(x) = \psi(x_i) + \int_{x_i}^{x} f(s, \psi(s))\, ds$$

and (7.11) that

$$|\psi(x) - y_h^{(i)}(x)| \le \varepsilon|x - x_i| \qquad \text{for} \quad x_i \le x \le x_{i+1}.$$

Using Lemma 7.2 we deduce in the same way as in part a) that

$$|\psi(x) - y_h(x)| \le \frac{\varepsilon}{L}\Big(e^{L(x-x_0)} - 1\Big). \tag{7.16}$$

Taking the limits $|h| \to 0$ and $\varepsilon \to 0$ we obtain $|\psi(x) - \varphi(x)| \le 0$, proving uniqueness. □

Theorem 7.3 is a *local* existence - and uniqueness - result. However, if we interpret the endpoint of the solution as a new initial value, we can apply Theorem 7.3 again and continue the solution. Repeating this procedure we obtain

Theorem 7.4. *Assume U to be an open set in $\mathbf{R}^2$ and let f and $\partial f/\partial y$ be continuous on U. Then, for every $(x_0, y_0) \in U$, there exists a unique solution of (7.1), which can be continued up to the boundary of U (in both directions).*

Proof. Clearly, Theorem 7.3 can be rewritten to give a local existence - and uniqueness - result for an interval (X, x_0) to the left of x_0. The rest follows from the fact that every point in U has a neighbourhood which satisfies the assumptions of Theorem 7.3.   □

It is interesting to mention that formula (7.13) for $|\widehat{h}| \to 0$ gives the following *error estimate*

$$|y(x) - y_h(x)| \leq \frac{\varepsilon}{L}\left(e^{L(x-x_0)} - 1\right) \tag{7.17}$$

for the Euler polygon ($|h| \leq \delta$). Here $y(x)$ stands for the exact solution of (7.1). The next theorem refines the above estimates for the case that $f(x,y)$ is also differentiable with respect to x.

Theorem 7.5. *Suppose that in a neighbourhood of the solution*

$$|f| \leq A, \qquad \left|\frac{\partial f}{\partial y}\right| \leq L, \qquad \left|\frac{\partial f}{\partial x}\right| \leq M.$$

We then have the following error estimate for the Euler polygons:

$$|y(x) - y_h(x)| \leq \frac{M + AL}{L}\left(e^{L(x-x_0)} - 1\right) \cdot |h|, \tag{7.18}$$

provided that $|h|$ *is sufficiently small.*

Proof. For $|u_1 - u_2| \leq |h|$ and $|v_1 - v_2| \leq A|h|$ we obtain, due to the differentiability of f, the estimate

$$|f(u_1, v_1) - f(u_2, v_2)| \leq (M + AL)|h|$$

instead of (7.11). When we insert this amount for ε into (7.16), we obtain the stated result.   □

The estimate (7.18) shows that the global error of Euler's method is proportional to the maximal step size $|h|$. Thus, for an accuracy of, say, three decimal digits, we would need about a thousand steps; a precision of six digits will normally require a million steps etc. We see thus that the present method is not recommended for computations of high precision. In fact, the main subject of Chapter II will be to find methods which converge faster.

Existence Theorem of Peano

> Si a est un complexe d'ordre n, et b un nombre réel, alors on peut déterminer b' et f, où b' est une quantité plus grande que b, et f est un signe de fonction qui à chaque nombre de l'intervalle de b à b' fait correspondre un complexe (en d'autres mots, ft est un complexe fonction de la variable réelle t, définie pour toutes les valeurs de l'intervalle (b, b')); la valeur de ft pour $t = b$ est a; et dans tout l'intervalle (b, b') cette fonction ft satisfait à l'équation différentielle donnée.   (Original version of Peano's Theorem)

The Lipschitz condition (7.7) is a crucial tool in the proof of (7.10) and finally of the Convergence Theorem. If we completely abandon condition (7.7) and only require that $f(x, y)$ be continuous, the convergence of the Euler polygons is no longer guaranteed.

An example, plotted in Fig. 7.2, is given by the equation

$$y' = 4\left(\text{sign}\,(y)\sqrt{|y|} + \max\left(0, x - \frac{|y|}{x}\right)\cdot\cos\left(\frac{\pi \log x}{\log 2}\right)\right) \qquad (7.19)$$

with $y(0) = 0$. It has been constructed such that

$$f(h, 0) = 4(-1)^i h \qquad\qquad \text{for } h = 2^{-i},$$
$$f(x, y) = 4\,\text{sign}(y)\cdot\sqrt{|y|} \qquad \text{for } |y| \geq x^2.$$

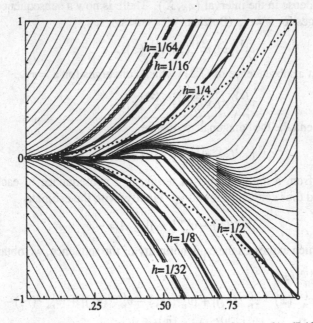

Fig. 7.2. Solution curves and Euler polygons for equation (7.19)

There is an infinity of solutions for this initial value, some of which are plotted in Fig. 7.2. The Euler polygons converge for $h = 2^{-i}$ and even i to the maximal solution $y = 4x^2$, and for odd i to $y = -4x^2$. For other sequences of h all intermediate solutions can be obtained as well.

Theorem 7.6 (Peano 1890). *Let $f(x,y)$ be continuous and $|f|$ be bounded by A on*

$$D = \left\{ (x,y) \mid x_0 \leq x \leq X, \ |y - y_0| \leq b \right\}.$$

If $X - x_0 \leq b/A$, then there is a subsequence of the sequence of the Euler polygons which converges to a solution of the differential equation.

The original proof of Peano is, in its crucial part on the convergence result, very brief and not clear to unexperienced readers such as us. Arzelà (1895), who took up the subject again, explains his ideas in more detail and emphasizes the need for an *equicontinuity* of the sequence. The proof usually given nowadays (for what has become the theorem of Arzelà-Ascoli), was only introduced later (see e.g. Perron (1918), Hahn (1921), p. 303) and is sketched as follows:

Proof. Let

$$v_1(x), v_2(x), v_3(x), \ldots \tag{7.20}$$

be a sequence of Euler polygons for decreasing step sizes. It follows from (7.4) that for fixed x this sequence is bounded. We choose a sequence of numbers $r_1, r_2, r_3, \ldots$ dense in the interval (x_0, X). There is now a subsequence of (7.20) which converges for $x = r_1$ (Bolzano-Weierstrass), say

$$v_1^{(1)}(x), v_2^{(1)}(x), v_3^{(1)}(x), \ldots \tag{7.21}$$

We next select a subsequence of (7.21) which converges for $x = r_2$

$$v_1^{(2)}(x), v_2^{(2)}(x), v_3^{(2)}(x), \ldots \tag{7.22}$$

and so on. Then take the "diagonal" sequence

$$v_1^{(1)}(x), v_2^{(2)}(x), v_3^{(3)}(x), \ldots \tag{7.23}$$

which, apart from a finite number of terms, is a subsequence of each of these sequences, and thus converges for all r_i. Finally, with the estimate

$$|v_n^{(n)}(x) - v_m^{(m)}(r_j)| \leq A|x - r_j|$$

(see (7.4)), which expresses the equicontinuity of the sequence, we obtain

$$|v_n^{(n)}(x) - v_m^{(m)}(x)|$$
$$\leq |v_n^{(n)}(x) - v_n^{(n)}(r_j)| + |v_n^{(n)}(r_j) - v_m^{(m)}(r_j)| + |v_m^{(m)}(r_j) - v_m^{(m)}(x)|$$
$$\leq 2A|x - r_j| + |v_n^{(n)}(r_j) - v_m^{(m)}(r_j)|.$$

For fixed $\varepsilon > 0$ we then choose a finite subset R of $\{r_1, r_2, \ldots\}$ satisfying $\min\{|x - r_j| ; \; r_j \in R, \; x_0 \le x \le X\} \le \varepsilon/A$ and secondly we choose N such that

$$|v_n^{(n)}(r_j) - v_m^{(m)}(r_j)| \le \varepsilon \qquad \text{for} \quad n, m \ge N \quad \text{and} \quad r_j \in R.$$

This shows the uniform convergence of (7.23). In the same way as in part b) of the proof of Theorem 7.3 it follows that the limit function is a solution of (7.1). One only has to add an $\mathcal{O}(|h|)$-term in (7.14), if x is not a subdivision point.   □

Exercises

1. Apply Euler's method with constant step size $x_{i+1} - x_i = 1/n$ to the differential equation $y' = ky$, $y(0) = 1$ and obtain a classical approximation for the solution $y(1) = e^k$. Give an estimate of the error.

2. Apply Euler's method with constant step size to

   a) $y' = y^2$, $y(0) = 1$, $y(1/2) = ?$

   b) $y' = x^2 + y^2$, $y(0) = 0$, $y(1/2) = ?$

   Make rigorous error estimates using Theorem 7.4 and compare these estimates with the actual errors. The main difficulty is to find a suitable region in which the estimates of Theorem 7.4 hold, without making the constants A, L, M too large and, at the same time, ensuring that the solution curves remain inside this region (see also I.8, Exercise 3).

3. Prove the result: if the differential equation $y' = f(x, y)$, $y(x_0) = y_0$ with f continuous, possesses a unique solution, then the Euler polygons converge to this solution.

4. "There is an elementary proof of Peano's existence theorem" (Walter 1971). Suppose that A is a bound for $|f|$. Then the sequence

   $$y_{i+1} = y_i + h \cdot \max\{f(x, y) | x_i \le x \le x_{i+1}, y_i - 3Ah \le y \le y_i + Ah\}$$

   converges for all continuous f to a (the maximal) solution. Try to prove this. Unfortunately, this proof does not extend to systems of equations, unless they are "quasimonotone" (see Section I.10, Exercise 3).

I.8 Existence Theory using Iteration Methods and Taylor Series

A second approach to existence theory is possible with the help of an iterative refinement of approximate solutions. The first appearances of the idea are very old. For instance many examples of this type can be found in the work of Lagrange, above all in his astronomical calculations. Let us consider here the following illustrative example of a Riccati equation

$$y' = x^2 + y + 0.1y^2, \qquad y(0) = 0. \tag{8.1}$$

Because of the quadratic term, there is no elementary solution. A very natural idea is therefore to neglect this term, which is in fact very small at the beginning, and to solve for the moment

$$y_1' = x^2 + y_1, \qquad y_1(0) = 0. \tag{8.2}$$

This gives, with formula (3.3), a first approximation

$$y_1(x) = 2e^x - (x^2 + 2x + 2). \tag{8.3}$$

With the help of this solution, we now know more about the initially neglected term $0.1y^2$; it will be close to $0.1y_1^2$. So the idea lies at hand to reintroduce this solution into (8.1) and solve now the differential equation

$$y_2' = x^2 + y_2 + 0.1 \cdot \left(y_1(x)\right)^2, \qquad y_2(0) = 0. \tag{8.4}$$

We can use formula (3.3) again and obtain after some calculations

$$y_2(x) = y_1(x) + \frac{2}{5}e^{2x} - \frac{2}{15}e^x(x^3 + 3x^2 + 6x - 54)$$
$$- \frac{1}{10}(x^4 + 8x^3 + 32x^2 + 72x + 76).$$

This is already much closer to the correct solution, as can be seen from the following comparison of the errors $e_1 = y(x) - y_1(x)$ and $e_2 = y(x) - y_2(x)$:

$$x = 0.2 \qquad e_1 = 0.228 \times 10^{-07} \qquad e_2 = 0.233 \times 10^{-12}$$
$$x = 0.4 \qquad e_1 = 0.327 \times 10^{-05} \qquad e_2 = 0.566 \times 10^{-09}$$
$$x = 0.8 \qquad e_1 = 0.534 \times 10^{-03} \qquad e_2 = 0.165 \times 10^{-05}.$$

It looks promising to continue this process, but the computations soon become very tedious.

Picard-Lindelöf Iteration

The general formulation of the method is the following: we try, if possible, to split up the function $f(x, y)$ of the differential equation

$$y' = f(x, y) = f_1(x, y) + f_2(x, y), \qquad y(x_0) = y_0 \tag{8.5}$$

so that any differential equation of the form $y' = f_1(x, y) + g(x)$ can be solved analytically and so that $f_2(x, y)$ is small. Then we start with a first approximation $y_0(x)$ and compute successively $y_1(x)$, $y_2(x), \ldots$ by solving

$$y'_{i+1} = f_1(x, y_{i+1}) + f_2(x, y_i(x)), \qquad y_{i+1}(x_0) = y_0. \tag{8.6}$$

The most primitive form of this process is obtained by choosing $f_1 = 0$, $f_2 = f$, in which case (8.6) is immediately integrated and becomes

$$y_{i+1}(x) = y_0 + \int_{x_0}^{x} f(s, y_i(s)) \, ds. \tag{8.7}$$

This is called the *Picard-Lindelöf iteration method*. It appeared several times in the literature, e.g., in Liouville (1838), Cauchy, Peano (1888), Lindelöf (1894), Bendixson (1893). Picard (1890) considered it merely as a by-product of a similar idea for partial differential equations and analyzed it thoroughly in his famous treatise Picard (1891-96), Vol. II, Chap. XI, Sect. III.

The fast *convergence* of the method, for $|x - x_0|$ small, is readily seen: if we subtract formula (8.7) from the same with i replaced by $i - 1$, we have

$$y_{i+1}(x) - y_i(x) = \int_{x_0}^{x} \Big(f(s, y_i(s)) - f(s, y_{i-1}(s)) \Big) \, ds. \tag{8.8}$$

We now apply the Lipschitz condition (7.7) and the triangle inequality to obtain

$$|y_{i+1}(x) - y_i(x)| \leq L \int_{x_0}^{x} |y_i(s) - y_{i-1}(s)| \, ds. \tag{8.9}$$

When we assume $y_0(x) \equiv y_0$, the triangle inequality applied to (8.7) with $i = 0$ yields the estimate

$$|y_1(x) - y_0(x)| \leq A|x - x_0|$$

where A is a bound for $|f|$ as in Section I.7. We next insert this into the right hand side of (8.9) repeatedly to obtain finally the estimate (Lindelöf 1894)

$$|y_i(x) - y_{i-1}(x)| \leq AL^{i-1} \frac{|x - x_0|^i}{i!}. \tag{8.10}$$

The right-hand side is a term of the Taylor series for $e^{L|x-x_0|}$, which converges for all x; we therefore conclude that $|y_{i+k} - y_i|$ becomes arbitrarily small when i is large. The error is bounded by the remainder of the above exponential series. So the sequence $y_i(x)$ converges uniformly to the solution $y(x)$. For example, if $L|x - x_0| \leq 1/10$ and the constant A is moderate, 10 iterations would provide a numerical solution with about 17 correct digits.

The main practical drawback of the method is the need for repeated computation of integrals, which is usually not very convenient, if at all analytically possible, and soon becomes very tedious. However, its fast convergence and new machine architectures (parallelism) coupled with numerical evaluations of the integrals have made the approach interesting for large problems (see Nevanlinna 1989).

Taylor Series

> Après avoir montré l'insuffisance des méthodes d'intégration fondées sur le développement en séries, il me reste à dire en peu de mots ce qu'on peut leur substituer.   (Cauchy)

A third existence proof can be based on a study of the convergence of the Taylor series of the solutions. This was mentioned in a footnote of Liouville (1836, p. 255), and brought to perfection by Cauchy (1839-42).

We have already seen the recursive computation of the Taylor coefficients in the work of Newton (see Section I.2). Euler (1768) then formulated the general procedure for the higher derivatives of the solution of

$$y' = f(x,y), \quad y(x_0) = y_0 \tag{8.11}$$

which, by successive differentiation, are obtained as

$$\begin{aligned} y'' &= f_x + f_y y' = f_x + f_y f \\ y''' &= f_{xx} + 2f_{xy}f + f_{yy}f^2 + f_y(f_x + f_y f) \end{aligned} \tag{8.12}$$

etc. Then the solution is

$$y(x_0 + h) = y(x_0) + y'(x_0)h + y''(x_0)\frac{h^2}{2!} + \dots . \tag{8.13}$$

The formulas (8.12) for higher derivatives soon become very complicated. Euler therefore proposed to use only a few terms of this series with h sufficiently small and to repeat the computations from the point $x_1 = x_0 + h$ ("analytic continuation").

We shall now outline the main ideas of Cauchy's *convergence proof* for the series (8.13). We suppose that $f(x,y)$ is *analytic* in the neighbourhood of the initial value x_0, y_0, which for simplicity of notation we assume located at the origin $x_0 = y_0 = 0$:

$$f(x,y) = \sum_{i,j \geq 0} a_{ij}x^i y^j, \tag{8.14}$$

where the a_{ij} are multiples of the partial derivatives occurring in (8.12). If the series (8.14) is assumed to converge for $|x| \leq r$, $|y| \leq r$, then the Cauchy inequalities from classical complex analysis give

$$|a_{ij}| \leq \frac{M}{r^{i+j}}, \qquad \text{where} \qquad M = \max_{|x| \leq r, |y| \leq r} |f(x,y)|. \qquad (8.15)$$

The idea is now the following: since all signs in (8.12) are positive, we obtain the worst possible result if we replace in (8.14) all a_{ij} by the largest possible values (8.15) ("method of majorants"):

$$f(x,y) \rightarrow \sum_{i,j \geq 0} M \frac{x^i y^j}{r^{i+j}} = \frac{M}{(1 - x/r)(1 - y/r)}.$$

However, the majorizing differential equation

$$y' = \frac{M}{(1 - x/r)(1 - y/r)}, \qquad y(0) = 0$$

is readily integrated by separation of variables (see Section I.3) and has the solution

$$y = r \left(1 - \sqrt{1 + 2M \log\left(1 - \frac{x}{r}\right)} \right). \qquad (8.16)$$

This solution has a power series expansion which converges for all x such that $|2M \log(1 - x/r)| < 1$. Therefore, the series (8.13) also converges at least for all $|h| < r(1 - \exp(-1/2M))$. $\qquad\qquad\square$

Recursive Computation of Taylor Coefficients

> ... dieses Verfahren praktisch nicht in Frage kommen kann.
> (Runge & König 1924)

> The exact opposite is true, if we use the right approach ...
> (R.E. Moore 1979)

The "right approach" is, in fact, an extension of Newton's approach and has been rediscovered several times (e.g., Steffensen 1956) and implemented into computer programs by Gibbons (1960) and Moore (1966). For a more extensive bibliography see the references in Wanner (1969), p. 10-20.

The idea is the following: let

$$Y_i = \frac{1}{i!} y^{(i)}(x_0), \qquad F_i = \frac{1}{i!} \Big(f(x, y(x)) \Big)^{(i)} \Big|_{x=x_0} \qquad (8.17)$$

be the Taylor coefficients of $y(x)$ and of $f(x, y(x))$, so that (8.13) becomes

$$y(x_0 + h) = \sum_{i=0}^{\infty} h^i Y_i.$$

Then, from (8.11),

$$Y_{i+1} = \frac{1}{i+1} F_i. \tag{8.18}$$

Now suppose that $f(x, y)$ is the composition of a sequence of algebraic operations and elementary functions. This leads to a sequence of items,

$$x, y, p, q, r, \ldots, \text{ and finally } f. \tag{8.19}$$

For each of these items we find formulas for generating the ith Taylor coefficient from the preceding ones as follows:

a) $r = p \pm q$:

$$R_i = P_i \pm Q_i, \qquad i = 0, 1, \ldots \tag{8.20a}$$

b) $r = pq$: the Cauchy product yields

$$R_i = \sum_{j=0}^{i} P_j Q_{i-j}, \qquad i = 0, 1, \ldots \tag{8.20b}$$

c) $r = p/q$: write $p = rq$, use formula b) and solve for R_i:

$$R_i = \frac{1}{Q_0} \left(P_i - \sum_{j=0}^{i-1} R_j Q_{i-j} \right), \qquad i = 0, 1, \ldots \tag{8.20c}$$

There also exist formulas for many elementary functions (in fact, because these functions are themselves solutions of rational differential equations).

d) $r = \exp(p)$: use $r' = p' \cdot r$ and apply (8.20b). This gives for $i = 1, 2, \ldots$

$$R_0 = \exp(P_0), \qquad R_i = \frac{1}{i} \sum_{j=0}^{i-1} (i-j) R_j P_{i-j}. \tag{8.20d}$$

e) $r = \log(p)$: use $p = \exp(r)$ and rearrange formula d). This gives

$$R_0 = \log(P_0), \qquad R_i = \frac{1}{P_0} \left(P_i - \frac{1}{i} \sum_{j=1}^{i-1} (i-j) P_j R_{i-j} \right). \tag{8.20e}$$

f) $r = p^c, c \neq 1$ constant. Use $pr' = crp'$ and apply (8.20b):

$$R_0 = P_0^c, \qquad R_i = \frac{1}{iP_0} \left(\sum_{j=0}^{i-1} (ci - (c+1)j) R_j P_{i-j} \right). \tag{8.20f}$$

g) $r = \cos(p)$, $s = \sin(p)$: as in d) we have

$$R_0 = \cos P_0, \qquad R_i = -\frac{1}{i} \sum_{j=0}^{i-1} (i-j) S_j P_{i-j},$$

$$\hspace{8cm} (8.20\text{g})$$

$$S_0 = \sin P_0, \qquad S_i = \frac{1}{i} \sum_{j=0}^{i-1} (i-j) R_j P_{i-j}.$$

The alternating use of (8.20) and (8.18) then allows us to compute the Taylor coefficients for (8.17) to any wanted order in a very economical way. It is not difficult to write subroutines for the above formulas, which have to be called in the same order as the differential equation (8.11) is composed of elementary operations. There also exist computer programs which "compile" Fortran statements for $f(x,y)$ into this list of subroutine calls. One has been written by T. Szymanski and J.H. Gray (see Knapp & Wanner 1969).

Example. The differential equation $y' = x^2 + y^2$ leads to the recursion

$$Y_0 = y(0), \qquad Y_{i+1} = \frac{1}{i+1}\left(P_i + \sum_{j=0}^{i} Y_j Y_{i-j}\right), \qquad i = 0, 1, \dots$$

where $P_i = 1$ for $i = 2$ and $P_i = 0$ for $i \neq 2$ are the coefficients for x^2. One can imagine how much easier this is than formulas (8.12).

An important property of this approach is that it can be executed in *interval analysis* and thus allows us to obtain *reliable error bounds* by the use of Lagrange's error formula for Taylor series. We refer to the books by R.E. Moore (1966) and (1979) for more details.

Exercises

1. Obtain from (8.10) the estimate

$$|y_i(x) - y_0| \leq \frac{A}{L}\left(e^{L(x-x_0)} - 1\right)$$

and explain the similarity of this result with (7.16).

2. Apply the method of Picard to the problem $y' = Ky$, $y(0) = 1$.

3. Compute three Picard iterations for the problem $y' = x^2 + y^2$, $y(0) = 0$, $y(1/2) = ?$ and make a rigorous error estimate. Compare the result with the correct solution $y(1/2) = 0.041791146154681863220768806849179$.

4. Compute with an iteration method the solution of

$$y' = \sqrt{x} + \sqrt{y}, \qquad y(0) = 0$$

and observe that the method can work well for equations which pose serious problems with other methods. An even greater difference occurs for the equations

$$y' = \sqrt{x} + y^2, \quad y(0) = 0 \qquad \text{and} \qquad y' = \frac{1}{\sqrt{x}} + y^2, \quad y(0) = 0.$$

5. Define $f(x,y)$ by

$$f(x,y) = \begin{cases} 0 & \text{for } x \leq 0 \\ 2x & \text{for } x > 0, \ y < 0 \\ 2x - \dfrac{4y}{x} & \text{for } 0 \leq y \leq x^2 \\ -2x & \text{for } x > 0, \ x^2 < y. \end{cases}$$

a) Show that $f(x,y)$ is continuous, but not Lipschitz.

b) Show that for the problem $y' = f(x,y)$, $y(0) = 0$ the Picard iteration method does not converge.

c) Show that there is a unique solution and that the Euler polygons converge.

6. Use the method of Picard iteration to prove: if $f(x,y)$ is continuous and satisfies a Lipschitz condition (7.7) on the infinite strip $D = \{(x,y)\, ; \, x_0 \leq x \leq X\}$, then the initial value problem $y' = f(x,y)$, $y(x_0) = y_0$ possesses a unique solution on $x_0 \leq x \leq X$.

Compare this global result with Theorem 7.3.

7. Define a function $y(x)$ (the "inverse error function") by the relation

$$x = \frac{2}{\sqrt{\pi}} \int_0^y e^{-t^2} dt$$

and show that it satisfies the differential equation

$$y' = \frac{\sqrt{\pi}}{2} e^{y^2}, \qquad y(0) = 0.$$

Obtain recursion formulas for its Taylor coefficients.

I.9 Existence Theory for Systems of Equations

The first treatment of an existence theory for simultaneous systems of differential equations was undertaken in the last existing pages (p. 123-136) of Cauchy (1824). We write the equations as

$$y_1' = f_1(x, y_1, \ldots, y_n), \qquad y_1(x_0) = y_{10}, \qquad y_1(X) = ?$$
$$\cdots \qquad\qquad\qquad \cdots \qquad\qquad\qquad \cdots \qquad (9.1)$$
$$y_n' = f_n(x, y_1, \ldots, y_n), \qquad y_n(x_0) = y_{n0}, \qquad y_n(X) = ?$$

and ask for the existence of the n solutions $y_1(x), \ldots, y_n(x)$. It is again natural to consider, in analogy to (7.3), the method of Euler

$$y_{k,i+1} = y_{ki} + (x_{i+1} - x_i) \cdot f_k(x_i, y_{1i}, \ldots, y_{ni}) \qquad (9.2)$$

(for $k = 1, \ldots, n$ and $i = 0, 1, 2, \ldots$). Here y_{ki} is intended to approximate $y_k(x_i)$, where $x_0 < x_1 < x_2 \ldots$ is a subdivision of the interval of integration as in (7.2).

We now try to carry over everything we have done in Section I.7 to the new situation. Although we have no problem in extending (7.4) to the estimate

$$|y_{ki} - y_{k0}| \le A_k |x_i - x_0| \qquad \text{if} \qquad |f_k(x, y_1, \ldots, y_n)| \le A_k, \qquad (9.3)$$

things become a little more complicated for (7.7): we have to estimate

$$f_k(x, z_1, \ldots, z_n) - f_k(x, y_1, \ldots, y_n) = \frac{\partial f_k}{\partial y_1} \cdot (z_1 - y_1) + \ldots + \frac{\partial f_k}{\partial y_n} \cdot (z_n - y_n),$$
$$(9.4)$$

where the derivatives $\partial f_k / \partial y_i$ are taken at suitable intermediate points. Here Cauchy uses the inequality now called the "Cauchy-Schwarz inequality" ("Enfin, il résulte de la formule (13) de la 11e leçon du calcul différentiel ...") to obtain

$$|f_k(x, z_1, \ldots, z_n) - f_k(x, y_1, \ldots, y_n)| \qquad (9.5)$$
$$\le \sqrt{\left(\frac{\partial f_k}{\partial y_1}\right)^2 + \ldots + \left(\frac{\partial f_k}{\partial y_n}\right)^2} \cdot \sqrt{(z_1 - y_1)^2 + \ldots + (z_n - y_n)^2}.$$

At this stage, we begin to feel that further development is advisable only after the introduction of vector notation.

Vector Notation

This was promoted in our subject by the papers of Peano, (1888) and (1890), who was influenced, as he says, by the famous "Ausdehnungslehre" of Grassmann and the work of Hamilton, Cayley, and Sylvester. We introduce the vectors (Peano called them "complexes")

$$y = (y_1,\ldots,y_n)^T, \quad y_i = (y_{1i},\ldots,y_{ni})^T, \quad z = (z_1,\ldots,z_n)^T \quad \text{etc},$$

and hope that the reader will not confuse the components y_i of a vector y with vectors with indices. We consider the "vector function"

$$f(x,y) = (f_1(x,y),\ldots,f_n(x,y))^T,$$

so that equations (9.1) become

$$y' = f(x,y), \qquad y(x_0) = y_0, \qquad y(X) =?, \tag{9.1'}$$

Euler's method (9.2) is

$$y_{i+1} = y_i + (x_{i+1} - x_i)f(x_i,y_i), \qquad i = 0,1,2,\ldots \tag{9.2'}$$

and the Euler polygon is given by

$$y_h(x) = y_i + (x - x_i)f(x_i,y_i) \qquad \text{for} \qquad x_i \le x \le x_{i+1}.$$

There is no longer any difference in notation with the one-dimensional cases (7.1), (7.3) and (7.3a).

In view of estimate (9.5), we introduce for a vector $y = (y_1,\ldots,y_n)^T$ the *norm* (originally "modulus")

$$\|y\| = \sqrt{y_1^2 + \ldots + y_n^2} \tag{9.6}$$

which satisfies all the usual properties of a norm, for example the triangle inequality

$$\|y + z\| \le \|y\| + \|z\|, \qquad \left\|\sum_{i=1}^n y_i\right\| \le \sum_{i=1}^n \|y_i\|. \tag{9.7}$$

The Euclidean norm (9.6) is not the only one possible, we also use ("on pourrait aussi définir par mx la plus grande des valeurs absolues des élements de x; alors les propriétes des modules sont presqu'évidentes.", Peano)

$$\|y\| = \max(|y_1|,\ldots,|y_n|), \tag{9.6'}$$
$$\|y\| = |y_1| + \ldots + |y_n|. \tag{9.6''}$$

We are now able to formulate estimate (9.3) as follows, in perfect analogy with (7.4): if for some norm $\|f(x,y)\| \le A$ on $D = \{(x,y)\,|\,x_0 \le x \le X, \|y - y_0\| \le b\}$ and if $X - x_0 \le b/A$ then the numerical solution (x_i,y_i), given by (9.2'), remains in D and we have

$$\|y_h(x) - y_0\| \le A \cdot |x - x_0|. \tag{9.8}$$

The analogue of estimate (7.5) can be obtained similarly.

In order to prove the implication "(7.9) $\Rightarrow$ (7.7)" for vector-valued functions it is convenient to work with norms of matrices.

Subordinate Matrix Norms

The relation (9.4) shows that the difference $f(x, z) - f(x, y)$ can be written as the product of a matrix with the vector $z - y$. It is therefore of interest to estimate $\|Qv\|$ and to find the best possible estimate of the form $\|Qv\| \leq \beta \|v\|$.

Definition 9.1. Let Q be a matrix (n columns, m rows) and $\|\dots\|$ be one of the norms defined in (9.6), (9.6') or (9.6"). The *subordinate matrix norm* of Q is then defined by

$$\|Q\| = \sup_{v \neq 0} \frac{\|Qv\|}{\|v\|} = \sup_{\|u\|=1} \|Qu\|. \tag{9.9}$$

By definition, $\|Q\|$ is the smallest number such that

$$\|Qv\| \leq \|Q\| \cdot \|v\| \qquad \text{for all } v \tag{9.10}$$

holds. The following theorem gives explicit formulas for the computation of (9.9).

Theorem 9.2. *The norm of a matrix Q is given by the following formulas: for the Euclidean norm (9.6),*

$$\|Q\| = \sqrt{\text{largest eigenvalue of } Q^T Q}; \tag{9.11}$$

for the max-norm (9.6'),

$$\|Q\| = \max_{k=1,\dots,m} \left(\sum_{i=1}^{n} |q_{ki}| \right); \tag{9.11'}$$

for the norm (9.6"),

$$\|Q\| = \max_{i=1,\dots,n} \left(\sum_{k=1}^{m} |q_{ki}| \right). \tag{9.11"}$$

Proof. Formula (9.11) can be seen from $\|Qv\|^2 = v^T Q^T Q v$ with the help of an orthogonal transformation of $Q^T Q$ to diagonal form.

Formula (9.11') is obtained as follows (we denote (9.6') by $\|\dots\|_\infty$):

$$\|Qv\|_\infty = \max_{k=1,\dots,m} \left| \sum_{i=1}^{n} q_{ki} v_i \right| \leq \left(\max_{k=1,\dots,m} \sum_{i=1}^{n} |q_{ki}| \right) \cdot \|v\|_\infty \tag{9.12}$$

shows that $\|Q\| \leq \max_k \sum_i |q_{ki}|$. The equality in (9.11') is then seen by choosing a vector of the form $v = (\pm 1, \pm 1, \dots, \pm 1)^T$ for which equality holds in (9.12). The formula (9.11") is proved along the same lines. $\qquad \square$

All these formulas remain valid for *complex matrices*. Q^T has only to be replaced by Q^* (transposed and complex conjugate). See e.g., Wilkinson (1965), p. 55-61, Bakhvalov (1976), Chap. VI, Par. 3. With these preparations it is possible to formulate the desired estimate.

Theorem 9.3. *If $f(x,y)$ is differentiable with respect to y in an open convex region U and if*

$$\left\| \frac{\partial f}{\partial y}(x,y) \right\| \leq L \qquad \text{for} \qquad (x,y) \in U \tag{9.13}$$

then

$$\|f(x,z) - f(x,y)\| \leq L \|z - y\| \qquad \text{for} \qquad (x,y), (x,z) \in U. \tag{9.14}$$

(Obviously, the matrix norm in (9.13) is subordinate to the norm used in (9.14).)

Proof. This is the "mean value theorem" and its proof can be found in every textbook on calculus. In the case where $\partial f/\partial y$ is continuous, the following simple proof is possible. We consider $\varphi(t) = f\big(x, y + t(z-y)\big)$ and integrate its derivative (componentwise) from 0 to 1

$$f(x,z) - f(x,y) = \varphi(1) - \varphi(0) = \int_0^1 \varphi'(t)\, dt$$
$$= \int_0^1 \frac{\partial f}{\partial y}\big(x, y + t(z-y)\big) \cdot (z-y)\, dt. \tag{9.15}$$

Taking the norm of (9.15), using

$$\left\| \int_0^1 g(t)\, dt \right\| \leq \int_0^1 \|g(t)\|\, dt, \tag{9.16}$$

and applying (9.10) and (9.13) yields the estimate (9.14). The relation (9.16) is proved by applying the triangle inequality (9.7) to the finite Riemann sums which define the two integrals. $\square$

We thus have obtained the analogue of (7.7). All that remains to do is, *Da capo al fine,* to read Sections I.7 and I.8 again: *Lemma 7.2, Theorems 7.3, 7.4, 7.5, and 7.6 together with their proofs and the estimates (7.10), (7.13), (7.15), (7.16), (7.17), and (7.18) carry over to the more general case with the only changes that some absolute values are to be replaced by norms.*

The *Picard-Lindelöf iteration* also carries over to systems of equations when in (8.7) we interpret $y_{i+1}(x), y_0$ and $f(s, y_i(s))$ as vectors, integrated componentwise. The convergence result with the estimate (8.10) also remains the same; for its proof we have to use, between (8.8) and (8.9), the inequality (9.16).

The Taylor series method, its convergence proof, and the recursive generation of the Taylor coefficients also generalize in a straightforward manner to systems of equations.

Exercises

1. Solve the system

$$y_1' = -y_2, \qquad y_1(0) = 1$$
$$y_2' = +y_1, \qquad y_2(0) = 0$$

by the methods of Euler and Picard, establish rigorous error estimates for all three norms mentioned. Verify the results using the correct solution $y_1(x) = \cos x$, $y_2(x) = \sin x$.

2. Consider the differential equations

$$y_1' = -100y_1 + y_2, \qquad y_1(0) = 1, \qquad y_1(1) = ?$$
$$y_2' = y_1 - 100y_2, \qquad y_2(0) = 0, \qquad y_2(1) = ?$$

a) Compute the exact solution $y(x)$ by the method explained in Section I.6.

b) Compute the error bound for $\|z(x) - y(x)\|$, where $z(x) = 0$, obtained from (7.10).

c) Apply the method of Euler to this equation with $h = 1/10$.

d) Apply Picard's iteration method.

3. Compute the Taylor series solution of the system with constant coefficients $y' = Ay$, $y(0) = y_0$. Prove that this series converges for all x. Apply this series to the equation of Exercise 1.

Result.

$$y(x) = \sum_{i=0}^{\infty} \frac{x^i}{i!} A^i y_0 =: e^{Ax} y_0.$$

I.10 Differential Inequalities

Differential inequalities are an elegant instrument for gaining a better understanding of equations (7.10), (7.17) and much new insight. This subject was inaugurated in the paper, once again, Peano (1890) and further developed by Perron (1915), Müller (1926), Kamke (1930). A classical treatise on the subject is the book of Walter (1970).

Introduction

The basic idea is the following: let $v(x)$ denote the Euler polygon defined in (7.3) or (9.2), so that

$$v'(x) = f(x_i, y_i) \qquad \text{for} \quad x_i < x < x_{i+1}. \tag{10.1}$$

For any chosen norm, we investigate the *error*

$$m(x) = \|v(x) - y(x)\| \tag{10.2}$$

as a function of x and we naturally try to estimate its growth.

Unfortunately, $m(x)$ is not necessarily differentiable, due firstly to the corners of the Euler polygons and secondly, to corners originating from the norms, especially the norms (9.6') and (9.6"). Therefore we consider the so-called *Dini derivatives* defined by

$$D^+ m(x) = \limsup_{h \to 0, h > 0} \frac{m(x+h) - m(x)}{h},$$

$$D_+ m(x) = \liminf_{h \to 0, h > 0} \frac{m(x+h) - m(x)}{h},$$

(see e.g., Scheeffer (1884), Hobson (1921), Chap. V, §260, §280). The property

$$\|w(x+h)\| - \|w(x)\| \le \|w(x+h) - w(x)\| \tag{10.3}$$

is a simple consequence of the triangle inequality (9.7). If we divide (10.3) by $h > 0$, we obtain the estimates

$$D_+ \|w(x)\| \le \|w'(x+0)\|, \qquad D^+ \|w(x)\| \le \|w'(x+0)\|, \tag{10.4}$$

where $w'(x + 0)$ is the right derivative of the vector function $w(x)$. If we apply this to $m(x)$ of (10.2), we obtain

$$D_+ m(x) \leq \|v'(x + 0) - y'(x)\|$$
$$= \|v'(x + 0) - f(x, v(x)) + f(x, v(x)) - f(x, y(x))\|$$

and, using the triangle inequality and the Lipschitz condition (9.14),

$$D_+ m(x) \leq \delta(x) + L \cdot m(x). \tag{10.5}$$

Here, we have introduced

$$\delta(x) = \|v'(x + 0) - f(x, v(x))\| \tag{10.6}$$

which is called the *defect* of the approximate solution $v(x)$. This fundamental quantity measures the extent to which the function $v(x)$ does *not* satisfy the imposed differential equation. (7.11) together with (10.1) tell us that $\delta(x) \leq \varepsilon$, so that (10.5) can be further estimated to become

$$D_+ m(x) \leq L \cdot m(x) + \varepsilon, \qquad m(x_0) = 0. \tag{10.7}$$

Formula (10.7) (or (10.5)) is what one calls a *differential inequality*. The question is: are we allowed to replace "$\leq$" by "$=$", i.e., to solve instead of (10.7) the equation

$$u' = Lu + \varepsilon, \qquad u(x_0) = 0 \tag{10.8}$$

and to conclude that $m(x) \leq u(x)$? This would mean, by the formulas of Section I.3 or I.5, that

$$m(x) \leq \frac{\varepsilon}{L}\left(e^{L(x - x_0)} - 1\right). \tag{10.9}$$

We would thus have obtained (7.17) in a natural way and have furthermore discovered an elegant and powerful tool for many kinds of new estimates.

The Fundamental Theorems

A general theorem of the type

$$\left.\begin{array}{l} D_+ m(x) \leq g(x, m(x)) \\ D_+ u(x) \geq g(x, u(x)) \\ m(x_0) \leq u(x_0) \end{array}\right\} \quad\Longrightarrow\quad m(x) \leq u(x) \quad \text{for} \quad x_0 \leq x \tag{10.10}$$

cannot be true. Counter-examples are provided by any differential equation with non-unique solutions, such as

$$g(x, y) = \sqrt{y}, \qquad m(x) = \frac{x^2}{4}, \qquad u(x) = 0. \tag{10.11}$$

The important observation, due to Peano and Perron, which allows us to overcome this difficulty, is that *one* of the first two inequalities must be replaced by a *strict* inequality (see Peano (1890), §3, Lemme 1):

Theorem 10.1. *Suppose that the functions* $m(x)$ *and* $u(x)$ *are continuous and satisfy for* $x_0 \leq x < X$

$$
\begin{aligned}
&a)\quad D_+ m(x) \leq g(x, m(x)) \\
&b)\quad D_+ u(x) > g(x, u(x)) \\
&c)\quad m(x_0) \leq u(x_0).
\end{aligned}
\tag{10.12}
$$

Then

$$
m(x) \leq u(x) \qquad for \qquad x_0 \leq x \leq X.
\tag{10.13}
$$

The same conclusion is true if both D_+ *are replaced by* D^+.

Proof. In order to be able to compare the derivatives $D_+ m$ and $D_+ u$ in (10.12), we consider points at which $m(x) = u(x)$. This is the main idea.

If (10.13) were not true, we could choose a point x_2 with $m(x_2) > u(x_2)$ and look for the first point x_1 to the left of x_2 with $m(x_1) = u(x_1)$. Then for small $h > 0$ we would have

$$
\frac{m(x_1 + h) - m(x_1)}{h} > \frac{u(x_1 + h) - u(x_1)}{h}
$$

and, by taking limits, $D_+ m(x_1) \geq D_+ u(x_1)$. This, however, contradicts (a) and (b), which give

$$
D_+ m(x_1) \leq g(x_1, m(x_1)) = g(x_1, u(x_1)) < D_+ u(x_1). \qquad \square
$$

Many variant forms of this theorem are possible, for example by using left Dini derivates (Walter 1970, Chap. II, §8, Theorem V).

Theorem 10.2 (The "fundamental lemma"). *Suppose that* $y(x)$ *is a solution of the system of differential equations* $y' = f(x, y)$, $y(x_0) = y_0$, *and that* $v(x)$ *is an approximate solution. If*

$$
\begin{aligned}
&a)\quad \|v(x_0) - y(x_0)\| \leq \varrho \\
&b)\quad \|v'(x+0) - f(x, v(x))\| \leq \varepsilon \\
&c)\quad \|f(x, v) - f(x, y)\| \leq L\|v - y\|,
\end{aligned}
$$

then, for $x \geq x_0$, *we have the error estimate*

$$
\|y(x) - v(x)\| \leq \varrho e^{L(x - x_0)} + \frac{\varepsilon}{L}\left(e^{L(x - x_0)} - 1\right).
\tag{10.14}
$$

Remark. The two terms in (10.14) express, respectively, the influence of the error ϱ in the initial values and the influence of the defect ε to the error of the approximate solution. It implies that the error depends continuously on both, and that for $\varrho = \varepsilon = 0$ we have $y(x) = v(x)$, i.e., uniqueness of the solution.

Proof. We put $m(x) = \|y(x) - v(x)\|$ and obtain, as in (10.7),

$$D_+ m(x) \le L \cdot m(x) + \varepsilon, \qquad m(x_0) \le \varrho.$$

We shall try to compare this with the differential equation

$$u' = Lu + \varepsilon, \qquad u(x_0) = \varrho. \tag{10.15}$$

Theorem 10.1 is not directly applicable. We therefore replace in (10.15) ε by $\varepsilon + \eta, \eta > 0$ and solve instead

$$u' \doteq Lu + \varepsilon + \eta > Lu + \varepsilon, \qquad u(x_0) = \varrho.$$

Now Theorem 10.1 gives the estimate (10.14) with ε replaced by $\varepsilon + \eta$. Since this estimate is true for *all* $\eta > 0$, it is also true for $\eta = 0$. $\square$

Variant form of Theorem 10.2. *The conditions*

$$a) \quad \|v(x_0) - y(x_0)\| \le \varrho$$

$$b) \quad \|v'(x+0) - f(x, v(x))\| \le \delta(x)$$

$$c) \quad \|f(x, v) - f(x, y)\| \le \ell(x)\|v - y\|$$

imply for $x \ge x_0$

$$\|y(x) - v(x)\| \le e^{L(x)} \left(\varrho + \int_{x_0}^{x} e^{-L(s)} \delta(s)\, ds \right), \qquad L(x) = \int_{x_0}^{x} \ell(s)\, ds.$$

Proof. This is simply formula (3.3). $\square$

Theorem 10.3. *If the function* $g(x, y)$ *is continuous and satisfies a Lipschitz condition, then the implication (10.10) is true for continuous functions* $m(x)$ *and* $u(x)$.

Proof. Define functions $w_n(x)$, $v_n(x)$ by

$$w_n'(x) = g(x, w_n(x)) + 1/n, \qquad w_n(x_0) = m(x_0),$$
$$v_n'(x) = g(x, v_n(x)) - 1/n, \qquad v_n(x_0) = u(x_0),$$

so that from Theorem 10.1

$$m(x) \le w_n(x), \qquad v_n(x) \le u(x) \qquad \text{for} \qquad x_0 \le x \le X. \tag{10.16}$$

It follows from Theorem 10.2 that the functions $w_n(x)$ and $v_n(x)$ converge for $n \to \infty$ to the solutions of

$$w'(x) = g(x, w(x)), \qquad w(x_0) = m(x_0),$$
$$v'(x) = g(x, v(x)), \qquad v(x_0) = u(x_0),$$

since the defect is $\pm 1/n$. Finally, because of $m(x_0) \leq u(x_0)$ and uniqueness we have $w(x) \leq v(x)$. Taking the limit $n \to \infty$ in (10.16) thus gives $m(x) \leq u(x)$.

$\square$

A further generalization of Theorem 10.2 is possible if the Lipschitz condition (c) is replaced by something nonlinear such as

$$\|f(x,v) - f(x,y)\| \leq \omega(x, \|v - y\|).$$

Then the differential inequality for the error $m(x)$ is to be compared with the solution of

$$u' = \omega(x,u) + \delta(x) + \eta, \qquad u(x_0) = \varrho, \qquad \eta > 0.$$

See Walter (1970), Chap. II, §11 for more details.

Estimates Using One-Sided Lipschitz Conditions

As we already observed in Exercise 2 of I.9, and as has been known for a long time, much information about the errors can be lost by the use of positive Lipschitz constants L (e.g (9.11), (9.11'), or (9.11")) in the estimates (7.16), (7.17), or (7.18). The estimates all grow exponentially with x, even if the solutions and errors decay. Therefore many efforts have been made to obtain better error estimates, as for example the papers Eltermann (1955), Uhlmann (1957), Dahlquist (1959), and the references therein. We follow with great pleasure the particularly clear presentation of Dahlquist.

Let us estimate the derivative of $m(x) = \|v(x) - y(x)\|$ with more care than we did in (10.5): for $h > 0$ we have

$$\begin{aligned}
m(x+h) &= \|v(x+h) - y(x+h)\| \\
&= \|v(x) - y(x) + h(v'(x+0) - y'(x))\| + \mathcal{O}(h^2) \\
&\leq \left\|v(x) - y(x) + h\Big(f(x,v(x)) - f(x,y(x))\Big)\right\| + h\delta(x) + \mathcal{O}(h^2)
\end{aligned}$$

(10.17)

by the use of (10.6) and (9.7). Here, we apply the mean value theorem to the function $y + hf(x,y)$ and obtain

$$m(x+h) \leq \left(\max_{\eta \in [y(x),v(x)]} \left\|I + h\frac{\partial f}{\partial y}(x,\eta)\right\|\right) \cdot m(x) + h\delta(x) + \mathcal{O}(h^2)$$

and finally for $h > 0$,

$$\frac{m(x+h) - m(x)}{h} \leq \max_{\eta \in [y(x),v(x)]} \frac{\|I + h\frac{\partial f}{\partial y}(x,\eta)\| - 1}{h} m(x) + \delta(x) + \mathcal{O}(h).$$

(10.18)

The expression on the right hand side of (10.18) leads us to the following definition:

Definition 10.4. Let Q be a square matrix, then we call

$$\mu(Q) = \lim_{h \to 0, h > 0} \frac{\|I + hQ\| - 1}{h} \tag{10.19}$$

the *logarithmic norm* of Q.

Here are formulas for its computation (Dahlquist (1959), p. 11, Eltermann (1955), p. 498, 499):

Theorem 10.5. *The logarithmic norm (10.19) is obtained by the following formulas: for the Euclidean norm (9.6),*

$$\mu(Q) = \lambda_{\max} = \text{largest eigenvalue of } \frac{1}{2}(Q^T + Q); \tag{10.20}$$

for the max-norm (9.6'),

$$\mu(Q) = \max_{k=1,\dots,n} \left(q_{kk} + \sum_{i \neq k} |q_{ki}| \right); \tag{10.20'}$$

for the norm (9.6"),

$$\mu(Q) = \max_{i=1,\dots,n} \left(q_{ii} + \sum_{k \neq i} |q_{ki}| \right). \tag{10.20''}$$

Proofs. Formulas (10.20') and (10.20") follow quite trivially from (9.11') and (9.11") and the definition (10.19). The point is that the presence of I suppresses, for h sufficiently small, the absolute values for the diagonal elements. (10.20) is seen from the fact that the eigenvalues of

$$(I + hQ)^T (I + hQ) = I + h(Q^T + Q) + h^2 Q^T Q,$$

for $h \to 0$, converge to $1 + h\lambda_i$, where λ_i are the eigenvalues of $Q^T + Q$.   □

Remark. For complex-valued matrices the above formulas remain valid if one replaces Q by Q^* and q_{kk}, q_{ii} by $\text{Re} q_{kk}$, $\text{Re} q_{ii}$.

We now obtain from (10.18) the following improvement of Theorem 10.3.

Theorem 10.6. *Suppose that we have the estimates*

$$\mu\left(\frac{\partial f}{\partial y}(x, \eta) \right) \leq \ell(x) \qquad \text{for} \qquad \eta \in [y(x), v(x)] \qquad \text{and} \tag{10.21}$$

$$\|v'(x + 0) - f(x, v(x))\| \leq \delta(x), \qquad \|v(x_0) - y(x_0)\| \leq \varrho.$$

Then for $x > x_0$ we have

$$\|y(x) - v(x)\| \leq e^{L(x)} \left(\varrho + \int_{x_0}^{x} e^{-L(s)} \delta(s) \, ds \right), \tag{10.22}$$

with $L(x) = \int_{x_0}^{x} \ell(s) \, ds$.

Proof. Since, for a fixed x, the segment $[v(x), y(x)]$ is compact,

$$K = \max_i \max_{[v(x),y(x)]} \left| \frac{\partial f_i}{\partial y_i} \right|$$

is finite. Then (see the proof of Theorem 10.5)

$$\frac{\|I + h\frac{\partial f}{\partial y}(x,\eta)\| - 1}{h} = \mu\left(\frac{\partial f}{\partial y}(x,\eta)\right) + \mathcal{O}(h)$$

where the $\mathcal{O}(h)$-term is uniformly bounded in η. (For the norms (9.6') and (9.6")) this term is in fact zero for $h < 1/K$). Thus the condition (10.21) inserted into (10.18) gives

$$D_+m(x) \leq \ell(x)m(x) + \delta(x).$$

Now the estimate (10.22) follows in the same way as that of Theorem 10.3.

$\square$

Exercises

1. Apply Theorem 10.5 to the example of Exercise 2 of I.9. Observe the substantial improvement of the estimates.

2. Prove the following (a variant form of the famous "Gronwall lemma", Gronwall 1919): suppose that a positive function $m(x)$ satisfies

$$m(x) \leq \varrho + \varepsilon(x - x_0) + L \int_{x_0}^x m(s)\, ds =: w(x) \qquad (10.23)$$

then

$$m(x) \leq \varrho e^{L(x-x_0)} + \frac{\varepsilon}{L}\left(e^{L(x-x_0)} - 1\right); \qquad (10.24)$$

a) directly, by subtracting from (10.23)

$$u(x) = \varrho + \varepsilon(x - x_0) + L \int_{x_0}^x u(s)\, ds;$$

b) by differentiating $w(x)$ in (10.23) and using Theorem 10.1.

c) Prove Theorem 10.2 with the help of the above lemma of Gronwall. The same interrelations are, of course, also valid in more general situations.

3. Consider the problem $y' = \lambda y$, $y(0) = 1$ with $\lambda \geq 0$ and apply Euler's method with constant step size $h = 1/n$. Prove that

$$\frac{\lambda}{1 + \lambda/n}\, y_h(x) \leq D_+ y_h(x) \leq \lambda y_h(x)$$

and derive the estimate

$$\left(1 + \frac{\lambda}{n}\right)^n \leq e^\lambda \leq \left(1 + \frac{\lambda}{n}\right)^{n+\lambda} \qquad \text{for} \quad \lambda \geq 0.$$

4. Prove the following properties of the logarithmic norm:

 a) $\mu(\alpha Q) = \alpha \mu(Q)$ for $\alpha \geq 0$

 b) $-\|Q\| \leq \mu(Q) \leq \|Q\|$

 c) $\mu(Q + P) \leq \mu(Q) + \mu(P)$, $\mu\left(\int Q(t)\,dt\right) \leq \int \mu(Q(t))\,dt$

 d) $|\mu(Q) - \mu(P)| \leq \|Q - P\|$.

5. For the Euclidean norm (10.20), $\mu(Q)$ is the smallest number satisfying

$$\langle v, Qv \rangle \leq \mu(Q)\|v\|^2.$$

This property is valid for all norms associated with a scalar product. Prove this.

6. Show that for the Euclidean norm the condition (10.21) is equivalent to

$$\langle y - z, f(x,y) - f(x,z) \rangle \leq \ell(x)\|y - z\|^2.$$

7. Observe, using an example of the form

$$y_1' = y_2, \qquad y_2' = -y_1,$$

that a generalization of Theorem 10.1 to *systems* of first order differential equations, with inequalities interpreted component-wise, is not true in general (Müller 1926).

However, it is possible to prove such a generalization of Theorem 10.1 under the additional hypothesis that the functions $g_i(x, y_1, \ldots, y_n)$ are *quasimonotone*, i.e., that

$$g_i(x, y_1, \ldots, y_j, \ldots, y_n) \leq g_i(x, y_1, \ldots, z_j, \ldots, y_n)$$
$$\text{if} \quad y_j < z_j \quad \text{for all} \quad j \neq i.$$

Try to prove this.

An important fact is that many systems from parabolic differential equations, such as equation (6.10), *are* quasimonotone. This allows many interesting applications of the ideas of this section (see Walter (1970), Chap. IV).

I.11 Systems of Linear Differential Equations

[Wronski] ... beschäftigte sich mit Mathematik, Mechanik und Physik, Himmelsmechanik und Astronomie, Statistik und politischer Ökonomie, mit Geschichte, Politik und Philosophie, ... er versuchte seine Kräfte in mehreren mechanischen und technischen Erfindungen.   (S. Dickstein, III. Math. Kongr. 1904, p. 515)

With more knowledge about existence and uniqueness, and with more skill in linear algebra, we shall now, as did the mathematicians of the 19th century, better understand many points which had been left somewhat obscure in Sections I.4 and I.6 about linear differential equations of higher order.

Equation (4.9) divided by $a_n(x)$ (which is $\neq 0$ away from singular points) becomes

$$y^{(n)} + b_{n-1}(x)y^{(n-1)} + \ldots + b_0(x)y = g(x), \qquad b_i(x) = a_i(x)/a_n(x). \quad (11.1)$$

with $g(x) = f(x)/a_n(x)$. Introducing $y = y_1$, $y' = y_2, \ldots, y^{(n-1)} = y_n$ we arrive at

$$\begin{pmatrix} y_1' \\ y_2' \\ \vdots \\ y_n' \end{pmatrix} = \begin{pmatrix} 0 & 1 & & \\ 0 & 0 & \ddots & \\ \vdots & \vdots & \ldots & 1 \\ -b_0(x) & -b_1(x) & \ldots & -b_{n-1}(x) \end{pmatrix} \begin{pmatrix} y_1 \\ y_2 \\ \vdots \\ y_n \end{pmatrix} + \begin{pmatrix} 0 \\ \vdots \\ 0 \\ g(x) \end{pmatrix}. \quad (11.1')$$

We again denote by y the vector $(y_1, \ldots, y_n)^T$ and by $f(x)$ the inhomogeneity, so that (11.1') becomes a special case of the following *system of linear differential equations*

$$y' = A(x)y + f(x), \tag{11.2}$$

$$A(x) = \big(a_{ij}(x)\big), \qquad f(x) = \big(f_i(x)\big), \qquad i,j = 1, \ldots, n.$$

Here, the theorems of Section I.9 and I.10 apply without difficulty. Since the partial derivatives of the right hand side of (11.2) with respect to y_i are given by $a_{ki}(x)$, we have the Lipschitz estimate (see condition (c) of the variant form of Theorem 10.2), where $\ell(x) = \|A(x)\|$ in any subordinate matrix norm (9.11, 11', 11''). We apply Theorem 7.4, and the variant form of Theorem 10.2 with $v(x) = 0$ as "approximate solution". We may also take $\ell(x) = \mu(A(x))$ (see (10.20, 20', 20'')) and apply Theorem 10.6.

Theorem 11.1. *Suppose that $A(x)$ is continuous on an interval $[x_0, X]$. Then for any initial values $y_0 = (y_{10}, \ldots, y_{n0})^T$ there exists for all $x_0 \leq x \leq X$ a unique* *solution of (11.2) satisfying*

$$\|y(x)\| \le e^{L(x)} \left(\|y_0\| + \int_{x_0}^{x} e^{-L(s)} \|f(s)\| \, ds \right) \tag{11.3}$$

$$L(x) = \int_{x_0}^{x} \ell(s) \, ds, \qquad \ell(x) = \|A(x)\| \qquad or \qquad \ell(x) = \mu(A(x)).$$

For $f(x) \equiv 0$, $y(x)$ depends linearly on the initial values, i.e., there is a matrix $R(x, x_0)$ (the "resolvent"), such that

$$y(x) = R(x, x_0) y_0. \tag{11.4}$$

Proof. Since $\ell(x)$ is continuous and therefore bounded on any compact interval $[x_0, X]$, the estimate (11.3) shows that the solutions can be continued until the end. The linear dependence follows from the fact that, for $f \equiv 0$, linear combinations of solutions are again solutions, and from uniqueness.   □

Resolvent and Wronskian

From uniqueness we have that the solutions with initial values y_0 at x_0 and $y_1 = R(x_1, x_0) y_0$ at x_1 (see (11.4)) must be the same. Hence we have

$$R(x_2, x_0) = R(x_2, x_1) R(x_1, x_0) \tag{11.5}$$

for $x_0 \le x_1 \le x_2$. Finally by integrating backward from x_1, y_1, i.e., by the coordinate transformation $x = x_1 - t$, $0 \le t \le x_1 - x_0$, we must arrive, again by uniqueness, at the starting values. Hence

$$R(x_0, x_1) = \left(R(x_1, x_0) \right)^{-1} \tag{11.6}$$

and (11.5) is true without any restriction on x_0, x_1, x_2.

Let $y_i(x) = (y_{1i}(x), \ldots, y_{ni}(x))^T$ (for $i = 1, \ldots, n$) be a set of n solutions of the homogeneous differential equation

$$y' = A(x) y \tag{11.7}$$

which are *linearly independent* at $x = x_0$ (i.e., they form a *fundamental system*). We form the *Wronskian matrix* (Wronski 1810)

$$W(x) = \begin{pmatrix} y_{11}(x) & \cdots & y_{1n}(x) \\ \vdots & & \vdots \\ y_{n1}(x) & \cdots & y_{nn}(x) \end{pmatrix},$$

so that

$$W'(x) = A(x)W(x)$$

and all solutions can be written as

$$c_1 y_1(x) + \ldots + c_n y_n(x) = W(x) \cdot c \qquad \text{where} \qquad c = (c_1, \ldots, c_n)^T. \quad (11.8)$$

If this solution must satisfy the initial conditions $y(x_0) = y_0$, we obtain $c = W^{-1}(x_0)y_0$ and we have the formula

$$R(x, x_0) = W(x)W^{-1}(x_0). \quad (11.9)$$

Therefore all solutions are known if one has found n linearly independent solutions.

Inhomogeneous Linear Equations

Extending the idea of Joh. Bernoulli for (3.2) and Lagrange for (4.9), we now compute the solutions of the *inhomogeneous* equation (11.2) by letting c be "variable" in the "general solution" (11.8): $y(x) = W(x)c(x)$ (Liouville 1838). Exactly as in Section I.3 for (3.2) we obtain from (11.2) and (11.7) by differentiation

$$y' = W'c + Wc' = AWc + Wc' = AWc + f.$$

Hence $c' = W^{-1}f$. If we integrate this with integration constants c, we obtain

$$y(x) = W(x) \int_{x_0}^{x} W^{-1}(s)f(s)\, ds + W(x) \cdot c.$$

The initial conditions $y(x_0) = y_0$ imply $c = W^{-1}(x_0)y_0$ and we obtain:

Theorem 11.2 ("Variation of constants formula"). *Let $A(x)$ and $f(x)$ be continuous. Then the solution of the inhomogeneous equation $y' = A(x)y + f(x)$ satisfying the initial conditions $y(x_0) = y_0$ is given by*

$$
\begin{aligned}
y(x) &= W(x)\left(W^{-1}(x_0)\, y_0 + \int_{x_0}^{x} W^{-1}(s)f(s)\, ds \right) \\
&= R(x, x_0)\, y_0 + \int_{x_0}^{x} R(x, s)f(s)\, ds.
\end{aligned}
\qquad (11.10)
$$

The Abel-Liouville-Jacobi-Ostrogradskii Identity

We already know from (11.6) that $W(x)$ remains regular for all x. We now show that the *determinant* of $W(x)$ can be given explicitly as follows (Abel 1827, Liouville 1838, Jacobi 1845, §17):

$$\det\big(W(x)\big) = \det\big(W(x_0)\big) \cdot \exp\left(\int_{x_0}^{x} Tr\big(A(s)\big)\, ds\right), \tag{11.11}$$

$$Tr\big(A(x)\big) = a_{11}(x) + a_{22}(x) + \ldots + a_{nn}(x)$$

which connects the determinant of $W(x)$ to the *trace* of $A(x)$.

For the *proof* of (11.11) (see also Exercise 2) we compute the derivative $\frac{d}{dx}\det\big(W(x)\big)$. Since $\det\big(W(x)\big)$ is multilinear, this derivative (by the Leibniz rule) is a sum of n terms, whose first is

$$T_1 = \det \begin{pmatrix} y'_{11} & y'_{12} & \cdots & y'_{1n} \\ y_{21} & y_{22} & \cdots & y_{2n} \\ \vdots & \vdots & & \vdots \\ y_{n1} & y_{n2} & \cdots & y_{nn} \end{pmatrix}.$$

We insert $y'_{1i} = a_{11}(x)y_{1i} + \ldots + a_{1n}(x)y_{ni}$ from (11.7). All terms $a_{12}(x)y_{2i}$, $\ldots, a_{1n}(x)y_{ni}$ disappear by subtracting multiples of lines 2 to n, so that $T_1 = a_{11}(x)\det\big(W(x)\big)$. Summing all these terms we obtain finally

$$\frac{d}{dx}\det\big(W(x)\big) = \big(a_{11}(x) + \ldots + a_{nn}(x)\big) \cdot \det\big(W(x)\big) \tag{11.12}$$

and (11.11) follows by integration.   □

Exercises

1. Compute the resolvent matrix $R(x, x_0)$ for the two systems

$$\begin{aligned} y'_1 &= y_1 & \qquad y'_1 &= y_2 \\ y'_2 &= 3y_2 & \qquad y'_2 &= -y_1 \end{aligned}$$

and check the validity of (11.5), (11.6) as well as (11.11).

2. Reconstruct Abel's original proof for (11.11), which was for the case

$$y''_1 + py'_1 + qy_1 = 0, \qquad y''_2 + py'_2 + qy_2 = 0.$$

Multiply the equations by y_2 and y_1 respectively and subtract to eliminate q. Then integrate.

Use the result to obtain an identity for the two integrals

$$y_1(a) = \int_0^\infty e^{ax - x^2} x^{\alpha - 1}\, dx, \qquad y_2(a) = \int_0^\infty e^{-ax - x^2} x^{\alpha - 1}\, dx,$$

which both satisfy

$$\frac{d^2 y_i}{da^2} - \frac{a}{2} \cdot \frac{dy_i}{da} - \frac{\alpha}{2} y_i = 0. \tag{11.13}$$

Hint. To verify (11.13), integrate from 0 to infinity the expression for $\frac{d}{dx}(\exp(ax - x^2)x^\alpha)$ (Abel 1827, case IV).

3. (Kummer 1839). Show that the general solution of the equation

$$y^{(n)}(x) = x^m y(x) \tag{11.14}$$

can be obtained by quadrature.

Hint. Differentiate (11.14) to obtain

$$y^{(n+1)} = x^m y' + m x^{m-1} y. \tag{11.15}$$

Suppose by recursion that the general solution of

$$\psi^{(n+1)} = x^{m-1}\psi, \quad \text{i.e.,} \quad \frac{d^{n+1}}{dx^{n+1}}\psi(xu) = x^{m-1}u^{m+n}\psi(xu) \tag{11.16}$$

is already known. Show that then

$$y(x) = \int_0^\infty u^{m-1}\exp\left(-\frac{u^{m+n}}{m+n}\right)\psi(xu)\,dx$$

is the general solution of (11.15), and, under some conditions on the parameters, also of (11.14). To simplify the computations, consider the function

$$g(u) = u^m \exp\left(-\frac{u^{m+n}}{m+n}\right)\psi(xu),$$

compute its derivative with respect to u, multiply by x^{m-1}, and integrate from 0 to infinity.

4. (Weak singularities for systems). Show that the linear system

$$y' = \frac{1}{x}\left(A_0 + A_1 x + A_2 x^2 + \ldots\right)y \tag{11.17}$$

possesses solutions of the form

$$y(x) = x^q\left(v_0 + v_1 x + v_2 x^2 + \ldots\right) \tag{11.18}$$

where $v_0, v_1, \ldots$ are vectors. Determine first q and v_0, then recursively v_1, v_2, etc. Observe that there exist n independent solutions of the form (11.18) if the eigenvalues of A_0 satisfy $\lambda_i \neq \lambda_j \bmod (\mathbf{Z})$ (Fuchs 1866).

5. Find the general solution of the weakly singular systems

$$y' = \frac{1}{x}\begin{pmatrix} \frac{3}{4} & 1 \\ \frac{1}{4} & -\frac{1}{4} \end{pmatrix}y \quad \text{and} \quad y' = \frac{1}{x}\begin{pmatrix} \frac{3}{4} & 1 \\ -\frac{1}{4} & -\frac{1}{4} \end{pmatrix}y. \tag{11.19}$$

Hint. While the first is easy from Exercise 4, the second needs an additional idea (see formula (5.9)). A second possibility is to use the transformation $x = e^t$, $y(x) = z(t)$, and apply the methods of Section I.12.

I.12 Systems with Constant Coefficients

Die Technik der Integration der linearen Differentialgleichungen
mit constanten Coeffizienten wird hier auf das Höchste entwickelt.

(F. Klein in Routh 1898)

Linearization

Systems of linear differential equations with constant coefficients form a class of
equations for which the resolvent $R(x, x_0)$ can be computed explicitly. They generally occur by *linearization* of time-independent (i.e., *autonomous* or *permanent*)
nonlinear differential equations

$$y_i' = f_i(y_1, \ldots, y_n) \qquad \text{or} \qquad y_i'' = f_i(y_1, \ldots, y_n) \qquad (12.1)$$

in the neighbourhood of a stationary point (Lagrange (1788), see also Routh (1860),
Chap. IX, Thomson & Tait 1879). We choose the coordinates so that the stationary
point under consideration is the origin, i.e., $f_i(0, \ldots, 0) = 0$. We then expand f_i
in its Taylor series and neglect all nonlinear terms:

$$y_i' = \sum_{k=1}^{n} \frac{\partial f_i}{\partial y_k}(0) y_k \qquad \text{or} \qquad y_i'' = \sum_{k=1}^{n} \frac{\partial f_i}{\partial y_k}(0) y_k. \qquad (12.1')$$

This is a system of equations with constant coefficients, as introduced in Section
I.6 (see (6.4), (6.11)),

$$y' = Ay \qquad \text{or} \qquad y'' = Ay. \qquad (12.1'')$$

Autonomous systems are invariant under a *shift* $x \to x + C$. We may therefore
always assume that $x_0 = 0$. For arbitrary x_0 the resolvent is given by

$$R(x, x_0) = R(x - x_0, 0). \qquad (12.2)$$

Diagonalization

We have seen in Section I.6 that the assumption $y(x) = v \cdot e^{\lambda x}$ leads to

$$Av = \lambda v \qquad \text{or} \qquad Av = \lambda^2 v, \qquad (12.3)$$

hence $v \neq 0$ must be an *eigenvector* of A and λ the corresponding *eigenvalue* (in
the first case; a square root of the eigenvalue in the second case, which we do not

consider any longer). From (12.3) we obtain by subtraction that there exists such a $v \neq 0$ if and only if the determinant

$$\chi_A(\lambda) := \det(\lambda I - A) = (\lambda - \lambda_1)(\lambda - \lambda_2)\ldots(\lambda - \lambda_n) = 0. \tag{12.4}$$

This determinant is called the *characteristic polynomial of A*.

Suppose now that for the n eigenvalues λ_i the n eigenvectors v_i can be chosen linearly independent. We then have from (12.3)

$$A\left(v_1, v_2, \ldots, v_n\right) = \left(v_1, v_2, \ldots, v_n\right) \operatorname{diag}\left(\lambda_1, \lambda_2, \ldots, \lambda_n\right),$$

or, if T is the matrix *whose columns are the eigenvectors of A*,

$$T^{-1}AT = \operatorname{diag}\left(\lambda_1, \lambda_2, \ldots, \lambda_n\right). \tag{12.5}$$

On comparing (12.5) with (12.1''), we see that the differential equation simplifies considerably if we use the coordinate transformation

$$y(x) = Tz(x), \qquad y'(x) = Tz'(x) \tag{12.6}$$

which leads to

$$z'(x) = \operatorname{diag}\left(\lambda_1, \lambda_2, \ldots, \lambda_n\right) z(x). \tag{12.7}$$

Thus the original system of differential equations decomposes into n single equations which are readily integrated to give

$$z(x) = \operatorname{diag}\left(\exp(\lambda_1 x), \exp(\lambda_2 x), \ldots, \exp(\lambda_n x)\right) z_0,$$

from which (12.6), yields

$$y(x) = T \operatorname{diag}\left(\exp(\lambda_1 x), \exp(\lambda_2 x), \ldots, \exp(\lambda_n x)\right) T^{-1} y_0. \tag{12.8}$$

The Schur Decomposition

> Der Beweis ist leicht zu erbringen.   (Schur 1909)

The foregoing theory, beautiful as it may appear, has several drawbacks:

a) Not all $n \times n$ matrices have a set of n linearly independent eigenvectors;

b) Even if it is invertible, the matrix T can behave very badly (see Exercise 1).

However, for *symmetric* matrices a classical theory tells that A can always be diagonalized by orthogonal transformations. Let us therefore, with Schur (1909), extend this classical theory to non-symmetric matrices. A real matrix Q is called *orthogonal* if its column vectors are mutually orthogonal and of norm 1, i.e., if $Q^T Q = I$ or $Q^T = Q^{-1}$. A complex matrix Q is called *unitary* if $Q^* Q = I$ or $Q^* = Q^{-1}$, where Q^* is the *adjoint* matrix of Q, i.e., transposed and complex conjugate.

Theorem 12.1. a) (Schur 1909). *For each complex matrix A there exists a unitary matrix Q such that*

$$Q^* A Q = \begin{pmatrix} \lambda_1 & \times & \times & \cdots & \times \\ & \lambda_2 & \times & \cdots & \times \\ & & \ddots & & \vdots \\ & & & & \lambda_n \end{pmatrix}; \tag{12.9}$$

b) (Wintner & Murnaghan 1931). *For a real matrix A the matrix Q can be chosen real and orthogonal, if for each pair of conjugate eigenvalues $\lambda, \overline{\lambda} = \alpha \pm i\beta$ one allows the block*

$$\begin{pmatrix} \lambda & \times \\ & \overline{\lambda} \end{pmatrix} \qquad \text{to be replaced by} \qquad \begin{pmatrix} \times & \times \\ \times & \times \end{pmatrix}.$$

Proof. a) The matrix A has at least one eigenvector with eigenvalue λ_1. We use this (normalized) vector as the first column of a matrix Q_1. Its other columns are then chosen by arbitrarily completing the first one to an orthonormal basis. Then

$$A Q_1 = Q_1 \left(\begin{array}{c|ccc} \lambda_1 & \times & \cdots & \times \\ \hline 0 & & A_2 & \end{array} \right). \tag{12.10}$$

We then apply the same argument to the $(n-1)$-dimensional matrix A_2. This leads to

$$A_2 \widetilde{Q}_2 = \widetilde{Q}_2 \left(\begin{array}{c|ccc} \lambda_2 & \times & \cdots & \times \\ \hline 0 & & A_3 & \end{array} \right).$$

With the unitary matrix

$$Q_2 = \left(\begin{array}{c|c} 1 & 0 \\ \hline 0 & \widetilde{Q}_2 \end{array} \right)$$

we obtain

$$Q_1^* A Q_1 Q_2 = Q_2 \left(\begin{array}{cc|ccc} \lambda_1 & \times & \times & \cdots & \times \\ & \lambda_2 & \times & \cdots & \times \\ \hline & 0 & & A_3 & \end{array} \right).$$

A continuation of this process leads finally to a triangular matrix as in (12.9) with $Q = Q_1 Q_2 \cdots Q_{n-1}$.

b) Suppose A to be a real matrix. If λ_1 is real, Q_1 can be chosen real and orthogonal. Now let $\lambda_1 = \alpha + i\beta$ ($\beta \neq 0$) be a *non-real* eigenvalue with a corresponding eigenvector $u + iv$, i.e.,

$$A(u \pm iv) = (\alpha \pm i\beta)(u \pm iv) \tag{12.11}$$

or

$$Au = \alpha u - \beta v, \qquad Av = \beta u + \alpha v. \tag{12.11'}$$

Since $\beta \neq 0$, u and v are linearly independent. We choose an orthogonal basis $\hat{u}$, $\hat{v}$ of the subspace spanned by u and v and take $\hat{u}$, $\hat{v}$ as the first two columns of the orthogonal matrix Q_1. We then have from (12.11')

$$AQ_1 = Q_1 \left(\begin{array}{cc|ccc} \times & \times & \times & \cdots & \times \\ \times & \times & \times & \cdots & \times \\ \hline 0 & & A_3 & & \end{array} \right).$$

$\square$

Schur himself was not very proud of "his" decomposition, he just derived it as a tool for proving interesting properties of eigenvalues (see e.g., Exercise 2).

Clearly, if A is real and symmetric, $Q^T A Q$ will also be symmetric, and therefore diagonal (see also Exercise 3).

Numerical Computations

The above theoretical proof is still not of much practical use. It requires that one know the eigenvalues, but the computation of eigenvalues from the characteristic polynomial is one of the best-known stupidities of numerical analysis. Good numerical analysis turns it the other way round: the real matrix A is directly reduced, first to Hessenberg form, and then by a sequence of orthogonal transformations to the real Schur form of Wintner & Murnaghan ("QR-algorithm" of Francis, coded by Martin, Peters & Wilkinson, contribution II/14 in Wilkinson & Reinsch 1970). The eigenvalues then drop out. However, the produced code, called "HQR2", does *not* give the Schur form of A, since it continues for the eigenvectors of A. Some manipulations must therefore be done to interrupt the code at the right moment (in the FORTRAN translation HQR2 of Eispack (1974), for example, the "340" of statement labelled "60" has to be replaced by "1001"). Happy "Matlab"-users just call "SCHUR".

Whenever the Schur form has been obtained, the transformation $y(x) = Qz(x)$, $y'(x) = Qz'(x)$ (see (12.6)) leads to

$$\begin{pmatrix} z_1' \\ \vdots \\ z_{n-1}' \\ z_n' \end{pmatrix} = \begin{pmatrix} \lambda_1 & b_{12} & \cdots & b_{1,n-1} & b_{1n} \\ & \ddots & & \vdots & \vdots \\ & & & \lambda_{n-1} & b_{n-1,n} \\ & & & & \lambda_n \end{pmatrix} \begin{pmatrix} z_1 \\ \vdots \\ z_{n-1} \\ z_n \end{pmatrix}. \tag{12.12}$$

The last equation here is $z_n' = \lambda_n z_n$ and can be integrated to give $z_n = \exp(\lambda_n x)z_{n0}$. Next, the equation for z_{n-1} is

$$z_{n-1}' = \lambda_{n-1} z_{n-1} + b_{n-1,n} z_n \tag{12.12'}$$

with z_n known. This is a linear equation (inhomogeneous, if $b_{n-1,n} \neq 0$) which can be solved by Euler's technique (Section I.4). Two different cases arise:

a) If $\lambda_{n-1} \neq \lambda_n$ we put $z_{n-1} = E \exp(\lambda_{n-1}x) + F \exp(\lambda_n x)$, insert into (12.12') and compare coefficients. This gives $F = b_{n-1,n} z_{n0}/(\lambda_n - \lambda_{n-1})$ and $E = z_{n-1,0} - F$.

b) If $\lambda_{n-1} = \lambda_n$ we set $z_{n-1} = (E + Fx)\exp(\lambda_n x)$ and obtain $F = b_{n-1,n} z_{n0}$ and $E = z_{n-1,0}$.

The next stage, following the same ideas, gives z_{n-2}, etc. Simple recursive formulas for the elements of the resolvent, which work in the case $\lambda_i \neq \lambda_j$, are obtained as follows (Parlett 1976): we assume

$$z_i(x) = \sum_{j=i}^n E_{ij} \exp(\lambda_j x)$$

and insert this into (12.12). After comparing coefficients, we obtain for $i = n$, $n-1$, $n-2$, etc.

$$E_{ik} = \frac{1}{\lambda_k - \lambda_i}\Big(\sum_{j=i+1}^k b_{ij} E_{jk}\Big), \qquad k = i+1, i+2, \ldots$$

$$(12.13)$$

$$E_{ii} = z_{i0} - \sum_{j=i+1}^n E_{ij}.$$

The Jordan Canonical Form

Simpler Than You Thought
(Amer. Math. Monthly 1980)

Whenever one is not afraid of badly conditioned matrices (see Exercise 1), and many mathematicians are not, the Schur form obtained above can be further transformed into the famous *Jordan canonical form*:

Theorem 12.2 (Jordan 1870, Livre deuxième, §5 and 6). *For every matrix A there exists a non-singular matrix T such that*

$$T^{-1}AT = \text{diag}\left\{\begin{pmatrix} \lambda_1 & 1 & \\ & \ddots & 1 \\ & & \lambda_1 \end{pmatrix}, \begin{pmatrix} \lambda_2 & 1 & \\ & \ddots & 1 \\ & & \lambda_2 \end{pmatrix}, \ldots\right\}. \quad (12.14)$$

(The dimensions (≥ 1) of the blocks may vary and the λ_i are not necessarily distinct).

Proof. We may suppose that the matrix is already in the Schur form. This is of course possible in such a way that identical eigenvalues are grouped together on the principal diagonal.

The next step (see Fletcher & Sorensen 1983) is to remove all nonzero elements outside the upper-triangular blocks containing identical eigenvalues. We let

$$A = \begin{pmatrix} B & C \\ 0 & D \end{pmatrix}$$

where B and D are upper-triangular. The diagonal elements of B are all equal to λ_1, whereas those of D are $\lambda_2, \lambda_3, \ldots$ and all different from λ_1. We search for a matrix S such that

$$\begin{pmatrix} B & C \\ 0 & D \end{pmatrix} \begin{pmatrix} I & S \\ 0 & I \end{pmatrix} = \begin{pmatrix} I & S \\ 0 & I \end{pmatrix} \begin{pmatrix} B & 0 \\ 0 & D \end{pmatrix}$$

or, equivalently,

$$BS + C = SD. \tag{12.15}$$

From this relation the matrix S can be computed column-wise as follows: the first column of (12.15) is $BS_1 + C_1 = \lambda_2 S_1$ (here S_j and C_j denote the jth column of S and C, respectively) which yields S_1 because λ_2 is not an eigenvalue of B. The second column of (12.15) yields $BS_2 + C_2 = \lambda_3 S_2 + d_{12} S_1$ and allows us to compute S_2, etc.

In the following steps we treat each of the remaining blocks separately: we thus assume that all diagonal elements are equal to λ and transform the block recursively to the form stated in the theorem. Since $(A - \lambda I)^n = 0$ (n is the dimension of the matrix A) there exists an integer k ($1 \leq k \leq n$) such that

$$(A - \lambda I)^k = 0, \qquad (A - \lambda I)^{k-1} \neq 0. \tag{12.16}$$

We fix a vector v such that $(A - \lambda I)^{k-1} v \neq 0$ and put

$$v_j = (A - \lambda I)^{k-j} v, \qquad j = 1, \ldots, k$$

so that

$$A v_1 = \lambda v_1, \qquad A v_j = \lambda v_j + v_{j-1} \qquad \text{for} \qquad j = 2, \ldots, k.$$

The vectors $v_1, \ldots, v_k$ are linearly independent, because a multiplication of the expression $\sum_{j=1}^{k} c_j v_j = 0$ with $(A - \lambda I)^{k-1}$ yields $c_k = 0$, then a multiplication with $(A - \lambda I)^{k-2}$ yields $c_{k-1} = 0$, etc. As in the proof of the Schur decomposition (Theorem 12.1) we complete $v_1, \ldots, v_k$ to a basis of $\mathbf{C}^n$ in such a way that (with $V = (v_1, \ldots, v_n)$)

$$AV = V \begin{pmatrix} J & C \\ 0 & D \end{pmatrix}, \qquad J = \left. \begin{pmatrix} \lambda & 1 & & \\ & \ddots & \ddots & \\ & & & 1 \\ & & & \lambda \end{pmatrix} \right\} k \tag{12.17}$$

where D is upper-triangular with λ on its diagonal.

Our next aim is to eliminate the nonzero elements of C in (12.17). In analogy to (12.15) it is natural to search for a matrix S such that $JS + C = SD$. Unfortunately, such an S does not always exist because the eigenvalues of J and of D are the same. However, it is possible to find S such that all elements of C are removed with the exception of its last line, i.e.,

$$\begin{pmatrix} J & C \\ 0 & D \end{pmatrix} \begin{pmatrix} I & S \\ 0 & I \end{pmatrix} = \begin{pmatrix} I & S \\ 0 & I \end{pmatrix} \begin{pmatrix} J & e_k c^T \\ 0 & D \end{pmatrix} \tag{12.18}$$

or equivalently

$$JS + C = e_k c^T + SD,$$

where $e_k = (0, \ldots, 0, 1)^T$ and $c^T = (c_1, \ldots, c_{n-k})$. This can be seen as follows: the first column of this relation becomes $(J - \lambda I) S_1 + C_1 = c_1 e_k$. Its last component yields c_1 and the other components determine the 2nd to kth elements of S_1. The first element of S_1 can arbitrarily be put equal to zero. Then we compute S_2 from $(J - \lambda I) S_2 + C_2 = c_2 e_k + d_{12} S_1$, etc. We thus obtain a matrix S (with vanishing first line) such that (12.18) holds.

We finally show that the assumption $(A - \lambda I)^k = 0$ implies $c = 0$ in (12.18). Indeed, a simple calculation yields

$$\begin{pmatrix} J - \lambda I & e_k c^T \\ 0 & D - \lambda I \end{pmatrix}^k = \begin{pmatrix} 0 & \widehat{C} \\ 0 & 0 \end{pmatrix}$$

where the first row of $\widehat{C}$ is equal to the row-vector c^T.

We have thus transformed A to block-diagonal form with blocks J of (12.17) and D. The procedure can now be repeated with the lower-dimensional matrix D. The product of all the occurring transformation matrices is then the matrix T in (12.14). $\qquad\square$

Corollary 12.3. *For every matrix A and for every number $\varepsilon \neq 0$ there exists a non-singular matrix T (depending on ε) such that*

$$T^{-1} A T = \mathrm{diag} \left\{ \begin{pmatrix} \lambda_1 & \varepsilon & \\ & \ddots & \varepsilon \\ & & \lambda_1 \end{pmatrix}, \begin{pmatrix} \lambda_2 & \varepsilon & \\ & \ddots & \varepsilon \\ & & \lambda_2 \end{pmatrix}, \ldots \right\}. \tag{12.14'}$$

Proof. Multiply equation (12.14) from the right by $D = \mathrm{diag}(1, \varepsilon, \varepsilon^2, \varepsilon^3, \ldots)$ and from the left by D^{-1}. $\qquad\square$

Numerical difficulties in determining the Jordan canonical form are described in Golub & Wilkinson (1976). There exist also several computer programs, for example the one described in Kågström & Ruhe (1980).

When the matrix A has been transformed to Jordan canonical form (12.14), the solutions of the differential equation $y' = Ay$ can be calculated by the method explained in (12.12'), case b):

$$y(x) = T D T^{-1} y_0 \tag{12.19}$$

where D is a block-diagonal matrix with blocks of the form

$$\begin{pmatrix} e^{\lambda x} & x e^{\lambda x} & \cdots & \dfrac{x^k}{k!} e^{\lambda x} \\ & e^{\lambda x} & & \vdots \\ & & \ddots & x e^{\lambda x} \\ & & & e^{\lambda x} \end{pmatrix}$$

This is an extension of formula (12.8).

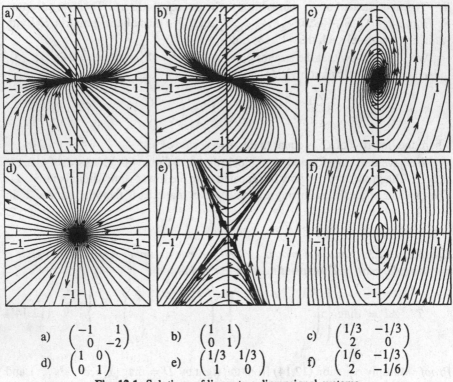

a) $\begin{pmatrix} -1 & 1 \\ 0 & -2 \end{pmatrix}$ b) $\begin{pmatrix} 1 & 1 \\ 0 & 1 \end{pmatrix}$ c) $\begin{pmatrix} 1/3 & -1/3 \\ 2 & 0 \end{pmatrix}$

d) $\begin{pmatrix} 1 & 0 \\ 0 & 1 \end{pmatrix}$ e) $\begin{pmatrix} 1/3 & 1/3 \\ 1 & 0 \end{pmatrix}$ f) $\begin{pmatrix} 1/6 & -1/3 \\ 2 & -1/6 \end{pmatrix}$

Fig. 12.1. Solutions of linear two dimensional systems

Geometric Representation

The geometric shapes of the solution curves of $y' = Ay$ are presented in Fig. 12.1 for dimension $n = 2$. They are plotted as paths in the phase-space (y_1, y_2). The cases a), b), c) and e) are the linearized equations of (12.20) at the four critical points (see Fig. 12.2).

Much of this structure remains valid also for *nonlinear* systems (12.1) in the *neighbourhood of equilibrium points*. Exceptions may be "structurally unstable" cases such as complex eigenvalues with $\alpha = \mathrm{Re}\,(\lambda) = 0$. This has been the subject of many papers discussing "critical points" or "singularities" (see e.g., the famous treatise of Poincaré (1881, 82, 85)).

In Fig. 12.2 we show solutions of the quadratic system

$$y_1' = \frac{1}{3}(y_1 - y_2)(1 - y_1 - y_2)$$
$$y_2' = y_1(2 - y_2) \tag{12.20}$$

which possesses four critical points of all four possible structurally stable types (Exercise 4).

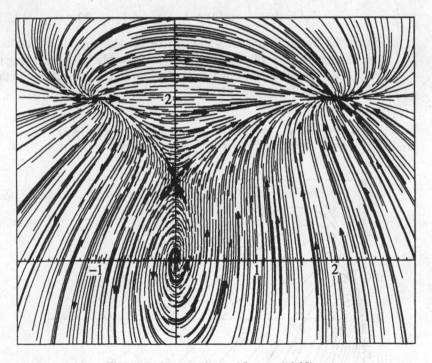

Fig. 12.2. Solution flow of System (12.20)

Exercises

1. a) Compute the eigenvectors of the matrix

$$A = \begin{pmatrix} -1 & 20 & & & & \\ & -2 & 20 & & & \\ & & -3 & 20 & & \\ & & & \ddots & \ddots & \\ & & & & -19 & 20 \\ & & & & & -20 \end{pmatrix} \tag{12.21}$$

by solving $(A - \lambda_i I)v_i = 0$.

Result. $v_1 = (1, 0, \ldots)^T$, $v_2 = (1, -1/20, 0, \ldots)^T$, $v_3 = (1, -2/20, 2/400, 0, \ldots)^T$, $v_4 = (1, -3/20, 6/400, -6/8000, 0, \ldots)^T$, etc.

b) Compute numerically the inverse of $T = (v_1, v_2, \ldots, v_n)$ and determine its largest element (answer: 4.5×10^{12}). The matrix T is thus very badly conditioned.

c) Compute numerically or analytically from (12.13) the solutions of

$$y' = Ay, \qquad y_i(0) = 1, \qquad i = 1, \ldots, 20. \tag{12.22}$$

Observe the "hump" (Moler & Van Loan 1978): although all eigenvalues of A are negative, the solutions first grow enormously before decaying to zero. This is typical of non-symmetric matrices and is connected with the bad condition of T (see Fig. 12.3).

Result.

$$y_1 = -\frac{20^{19}}{19!} e^{-20x} + \frac{(1+20)20^{18}}{18!} e^{-19x} - \frac{(1+20+20^2/2!)20^{17}}{17!} e^{-18x} \pm \ldots$$

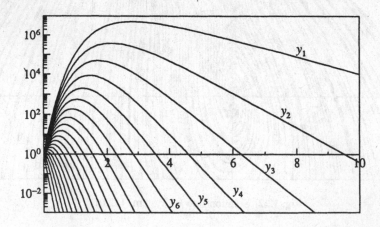

Fig. 12.3. Solutions of equation (12.22) with matrix (12.21)

2. (Schur). Prove that the eigenvalues of a matrix A satisfy the estimate

$$\sum_{i=1}^{n} |\lambda_i|^2 \le \sum_{i,j=1}^{n} |a_{ij}|^2$$

and that equality holds iff A is orthogonally diagonizable (see also Exercise 3).

Hint. $\sum_{i,j} |a_{ij}|^2$ is the trace of A^*A and thus invariant under unitary transformations Q^*AQ.

3. Show that the Schur decomposition $S = Q^*AQ$ is diagonal iff $A^*A = AA^*$. Such matrices are called *normal.* Examples are symmetric and skew-symmetric matrices.

   Hint. The condition is equivalent to $S^*S = SS^*$.

4. Compute the four critical points of System (12.20), and for each of these points the eigenvalues and eigenvectors of the matrix $\partial f / \partial y$. Compare the results with Figs. 12.2 and 12.1.

5. Compute a Schur decomposition and the Jordan canonical form of the matrix

$$A = \frac{1}{9} \begin{pmatrix} 14 & 4 & 2 \\ -2 & 20 & 1 \\ -4 & 4 & 20 \end{pmatrix}.$$

   Result. The Jordan canonical form is

$$\begin{pmatrix} 2 & 1 & \\ & 2 & \\ & & 2 \end{pmatrix}.$$

6. Reduce the matrices

$$A = \begin{pmatrix} \lambda & 1 & b & c \\ & \lambda & 1 & d \\ & & \lambda & 1 \\ & & & \lambda \end{pmatrix}, \qquad A = \begin{pmatrix} \lambda & 1 & b & c \\ & \lambda & 0 & d \\ & & \lambda & 1 \\ & & & \lambda \end{pmatrix}$$

   to Jordan canonical form. In the second case distinguish the possibilities $b + d = 0$ and $b + d \ne 0$.

I.13 Stability

The Examiners give notice that the following is the subject of the Prize to be adjudged in 1877: *The Criterion of Dynamical Stability.*

(S.G. Phear

(Vice-Chancellor), J. Challis, G.G. Stokes, J. Clerk Maxwell)

Introduction

"To illustrate the meaning of the question imagine a particle to slide down inside a smooth inclined cylinder along the lowest generating line, or to slide down outside along the highest generating line. In the former case a slight derangement of the motion would merely cause the particle to oscillate about the generating line, while in the latter case the particle would depart from the generating line altogether. The motion in the former case would be, in the sense of the question, stable, in the latter unstable ... what is desired is, a corresponding condition enabling us to decide when a dynamically possible motion of a system is such, that *if slightly deranged the motion shall continue to be only slightly departed from.*" ("The Examiners" in Routh 1877).

Whenever no analytical solution of a problem is known, numerical solutions can only be obtained for specified initial values. But often one needs information about the stability behaviour of the solutions for all initial values in the neighbourhood of a certain equilibrium point. We again transfer the equilibrium point to the origin and define:

Definition 13.1. Let

$$y_i' = f_i(y_1, \ldots, y_n), \qquad i = 1, \ldots, n \tag{13.1}$$

be a system with $f_i(0, \ldots, 0) = 0$, $i = 1, \ldots, n$. Then the origin is called *stable in the sense of Liapunov* if for any $\varepsilon > 0$ there is a $\delta > 0$ such that for the solutions, $\|y(x_0)\| < \delta$ implies $\|y(x)\| < \varepsilon$ for all $x > x_0$.

The first step, taken by Routh in his famous Adams Prize essay (Routh 1877), was to study the *linearized equation*

$$y_i' = \sum_{j=1}^{n} a_{ij} y_j, \qquad a_{ij} = \frac{\partial f_i}{\partial y_j}(0). \tag{13.2}$$

("The quantities $x, y, z, \ldots$ etc are said to be *small* when their squares can be neglected.") From the general solution of (13.2) obtained in Section I.12, we immediately have

Theorem 13.1. *The linearized equation (13.2) is stable (in the sense of Liapunov) iff all roots of the characteristic equation*

$$\det(\lambda I - A) = a_0\lambda^n + a_1\lambda^{n-1} + \ldots + a_{n-1}\lambda + a_n = 0 \qquad (13.3)$$

satisfy $\mathrm{Re}\,(\lambda) \leq 0$, *and the multiple roots, which give rise to Jordan chains, satisfy the strict inequality* $\mathrm{Re}\,(\lambda) < 0$.

Proof. See (12.12) and (12.19). For Jordan chains the "secular" term (e.g., $E + Fx$ in the solution of (12.12), case (b)) which tends to infinity for increasing x, must be "killed" by an exponential with strictly negative exponent. $\qquad\square$

The Routh-Hurwitz Criterion

The next task, which leads to the famous Routh-Hurwitz criterion, was the verification of the conditions $\mathrm{Re}\,(\lambda) < 0$ directly from the coefficients of (13.3), without computing the roots. To solve this problem, Routh combined two known ideas: the first was Cauchy's *argument principle,* saying that the number of roots of a polynomial $p(z) = u(z) + iv(z)$ inside a closed contour is equal to the number of (positive) rotations of the vector $(u(z), v(z))$, as z travels along the boundary in the positive sense (see e.g., Henrici (1974), p. 276). An example is presented in Fig. 13.1 for the polynomial

$$z^6 + 6z^5 + 16z^4 + 25z^3 + 24z^2 + 14z + 4$$
$$= (z+1)(z+2)(z^2 + z + 1)(z^2 + 2z + 2). \qquad (13.4)$$

On the half-circle $z = Re^{i\theta}$ ($\pi/2 \leq \theta \leq 3\pi/2$, R very large) the argument of $p(z)$, due to the dominant term z^n, makes $n/2$ positive rotations. In order to have all zeros of p in the negative half plane, we therefore need an additional $n/2$ positive rotations along the imaginary axis:

Lemma 13.2. *Let $p(z)$ be a polynomial of degree n and suppose that $p(iy) \neq 0$ for $y \in \mathrm{R}$. Then all roots of $p(z)$ are in the negative half-plane iff, along the imaginary axis, $\arg(p(iy))$ makes $n/2$ positive rotations for y from $-\infty$ to $+\infty$.* $\qquad\square$

The second idea was the use of Sturm's theorem (Sturm 1829) which had its origin in Euclid's algorithm for polynomials. Sturm made the discovery that in the division of the polynomial $p_{i-1}(y)$ by $p_i(y)$ it is better to take the remainder $p_{i+1}(y)$ with negative sign

$$p_{i-1}(y) = p_i(y)q_i(y) - p_{i+1}(y). \qquad (13.5)$$

Then, due to the "Sturm sequence property"

$$\mathrm{sign}\,(p_{i+1}(y)) \neq \mathrm{sign}\,(p_{i-1}(y)) \qquad \text{if} \qquad p_i(y) = 0, \qquad (13.6)$$

Fig. 13.1. Vector field of arg $(p(z))$ for the polynomial $p(z)$ of (13.4)

the number of *sign changes*

$$w(y) = \text{No. of sign changes of } \big(p_0(y), p_1(y), \dots, p_m(y)\big) \qquad (13.7)$$

does not vary at the zeros of $p_1(y), \dots, p_{m-1}(y)$. A consequence is the following

Lemma 13.3. *Suppose that a sequence $p_0(y), p_1(y), \dots, p_m(y)$ of real polynomials satisfies*

  i) $\deg(p_0) > \deg(p_1)$,

  ii) $p_0(y)$ *and* $p_1(y)$ *not simultaneously zero*,

  iii) $p_m(y) \neq 0$ *for all* $y \in \mathbf{R}$,

  iv) *and the Sturm sequence property (13.6).*

Then

$$\frac{w(\infty) - w(-\infty)}{2} \qquad (13.8)$$

is equal to the number of rotations, measured in the positive direction, of the vector $(p_0(y), p_1(y))$ as y tends from $-\infty$ to $+\infty$.

Proof. Due to the Sturm sequence property, $w(y)$ does not change at zeros of $p_1(y), \dots, p_{m-1}(y)$. By assumption (iii) also $p_m(y)$ has no influence. Therefore $w(y)$ can change only at zeros of $p_0(y)$. If $w(y)$ increases by one at $\widehat{y}$, either $p_0(y)$

changes from $+$ to $-$ and $p_1(\widehat{y}) > 0$ or it changes from $-$ to $+$ and $p_1(\widehat{y}) < 0$ ($p_1(\widehat{y}) = 0$ is impossible by (ii)). In both situations the vector $(p_0(y), p_1(y))$ crosses the imaginary axis in the positive direction (see Fig. 13.2). If $w(y)$ decreases by one, $(p_0(y), p_1(y))$ crosses the imaginary axis in the negative direction. The result now follows from (i), since the vector $(p_0(y), p_1(y))$ is horizontal for $y \to -\infty$ and for $y \to +\infty$.   □

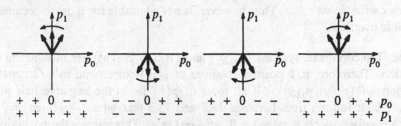

$$
\begin{array}{cccc}
+ \; + \; 0 \; - \; - & - \; - \; 0 \; + \; + & + \; + \; 0 \; - \; - & - \; - \; 0 \; + \; + & p_0 \\
+ \; + \; + \; + \; + & - \; - \; - \; - \; - & + \; + \; 0 \; - \; - & + \; + \; + \; + \; + & p_1
\end{array}
$$

Fig. 13.2. Rotations of $(p_0(y), p_1(y))$ compared to $w(y)$

The two preceding lemmas together give us the desired criterion for stability: let the characteristic polynomial (13.3)

$$
p(z) = a_0 z^n + a_1 z^{n-1} + \ldots + a_n = 0, \qquad a_0 > 0
$$

be given. We divide $p(iy)$ by i^n and separate real and imaginary parts,

$$
\begin{aligned}
p_0(y) &= \operatorname{Re} \frac{p(iy)}{i^n} = a_0 y^n - a_2 y^{n-2} + a_4 y^{n-4} \pm \ldots \\
p_1(y) &= -\operatorname{Im} \frac{p(iy)}{i^n} = a_1 y^{n-1} - a_3 y^{n-3} + a_5 y^{n-5} \pm \ldots .
\end{aligned}
\tag{13.9}
$$

Due to the special structure of these polynomials, the Euclidean algorithm (13.5) is here particularly simple: we write

$$
p_i(y) = c_{i0} y^{n-i} + c_{i1} y^{n-i-2} + c_{i2} y^{n-i-4} + \ldots ,
\tag{13.10}
$$

and have for the quotient in (13.5) $q_i(y) = (c_{i-1,0}/c_{i0})y$, provided that $c_{i0} \neq 0$. Now (13.10) inserted into (13.5) gives the following recursive formulas for the computation of the coefficients c_{ij}:

$$
c_{i+1,j} = c_{i,j+1} \cdot \frac{c_{i-1,0}}{c_{i0}} - c_{i-1,j+1} = \frac{1}{c_{i0}} \det \begin{pmatrix} c_{i-1,0} & c_{i-1,j+1} \\ c_{i,0} & c_{i,j+1} \end{pmatrix} .
\tag{13.11}
$$

If $c_{i0} = 0$ for some i, the quotient $q_i(y)$ is a higher degree polynomial and the Euclidean algorithm stops at $p_m(y)$ with $m < n$.

The sequence $(p_i(y))$ obtained in this way obviously satisfies conditions (i) and (iv) of Lemma 13.3. Condition (ii) is equivalent to $p(iy) \neq 0$ for $y \in \mathbf{R}$, and (iii) is a consequence of (ii) since $p_m(y)$ is the *greatest common divisor* of $p_0(y)$ and $p_1(y)$.

Theorem 13.4 (Routh 1877). *All roots of the real polynomial (13.3) with $a_0 > 0$ lie in the negative half plane* $\mathrm{Re}\,\lambda < 0$ *if and only if*

$$c_{i0} > 0 \qquad for \qquad i = 0, 1, 2, \ldots, n. \tag{13.12}$$

Remark. Due to the condition $c_{i0} > 0$, the division by c_{i0} in formula (13.11) can be omitted (common positive factor of $p_{i+1}(y)$), which leads to the same theorem (Routh (1877), p. 27: "... so that by remembering this simple cross-multiplication we may write down ..."). This, however, is not advisable for n large because of possible overflow.

Proof. The coordinate systems (p_0, p_1) and $(\mathrm{Re}\,(p), \mathrm{Im}\,(p))$ are of *opposite* orientation. Therefore, $n/2$ positive rotations of $p(iy)$ correspond to $n/2$ negative rotations of $(p_0(y), p_1(y))$. If all roots of $p(\lambda)$ lie in the negative half plane $\mathrm{Re}\,\lambda < 0$, it follows from Lemmas 13.2 and 13.3 that $w(\infty) - w(-\infty) = -n$, which is only possible if $w(\infty) = 0$, $w(-\infty) = n$. This implies the positivity of all leading coefficients of $p_i(y)$.

On the other hand, if (13.12) is satisfied, we see that $p_n(y) \equiv c_{n0}$. Hence the polynomials $p_0(y)$ and $p_1(y)$ cannot have a common factor and $p(\lambda) \neq 0$ on the imaginary axis. We can now apply Lemmas 13.2 and 13.3 again to obtain the result.

□

Table 13.1.
Routh tableau for (13.4)

	$j = 0$	$j = 1$	$j = 2$	$j = 3$
$i = 0$	1	-16	24	-4
$i = 1$	6	-25	14	
$i = 2$	11.83	-21.67	4	
$i = 3$	14.01	-11.97		
$i = 4$	11.56	-4		
$i = 5$	7.12			
$i = 6$	4			

Table 13.2.
Routh tableau for (13.13)

	$j = 0$	$j = 1$	$j = 2$
$i = 0$	1	$-q$	s
$i = 1$	p	$-r$	
$i = 2$	$pq - r$	$-ps$	
$i = 3$	$(pq - r)r - p^2 s$		
$i = 4$	$((pq - r)r - p^2 s)ps$		

Example 1. The Routh tableau (13.11) for equation (13.4) is given in Table 13.1. It clearly satisfies the conditions for stability.

Example 2 (Routh 1877, p. 27). Express the stability conditions for the biquadratic

$$z^4 + pz^3 + qz^2 + rz + s = 0. \tag{13.13}$$

The c_{ij} values (without division) are given in Table 13.2. We have stability iff

$$p > 0, \qquad pq - r > 0, \qquad (pq - r)r - p^2 s > 0, \qquad s > 0.$$

Computational Considerations

The actual computational use of Routh's criterion, in spite of its high historical importance and mathematical elegance, has two drawbacks for higher dimensions:

1) It is not easy to compute the characteristic polynomial for higher order matrices;
2) The use of the characteristic polynomial is very dangerous in the presence of rounding errors.

So, whenever one is not working with exact algebra or high precision, it is advisable to avoid the characteristic polynomial and use numerically stable algorithms for the eigenvalue problem (e.g., Eispack 1974).

Numerical experiments. 1. The $2n \times 2n$ dimensional matrix

$$
A = \left(
\begin{array}{ccc|ccc}
-.05 & & & -1 & & \\
& \ddots & & & \ddots & \\
& & -.05 & & & -n \\
\hline
1 & & & -.05 & & \\
& \ddots & & & \ddots & \\
& & n & & & -.05
\end{array}
\right)
$$

has the characteristic polynomial

$$
p(z) = \prod_{j=1}^{n} (z^2 + 0.1z + j^2 + 0.0025).
$$

We computed the coefficients of p using double precision, and then applied the Routh algorithm in single precision (machine precision $= 1.5 \times 10^{-8}$). The results indicated stability for $n \leq 15$, but not for $n \geq 16$, although the matrix always has its eigenvalues $-0.05 \pm ki$ in the negative half plane. On the other hand, a direct computation of the eigenvalues of A with the use of Eispack subroutines gave no problem for any n.

2. We also tested the Routh algorithm at the (scaled) *numerators of the diagonal Padé approximations to* $\exp(z)$

$$
1 + \frac{n}{2n}(nz) + \frac{n(n-1)}{(2n)(2n-1)} \frac{(nz)^2}{2!} + \frac{n(n-1)(n-2)}{(2n)(2n-1)(2n-2)} \frac{(nz)^3}{3!} + \dots, \quad (13.14)
$$

which are also known to possess all zeros in $\mathbf{C}^-$. Here, the results were correct only for $n \leq 21$, and wrong for larger n due to rounding errors.

Liapunov Functions

We now consider the question whether the stability of the nonlinear system (13.1) "can really be determined by examination of the terms of the first order only" (Routh 1877, Chapt. VII). This theory, initiated by Routh and Poincaré, was brought to perfection in the famous work of Liapunov (1892). As a general reference to the enormous theory that has developed in the meantime we mention Rouche, Habets & Laloy (1977) and W. Hahn (1967).

Liapunov's (and Routh's) main tools are the so-called *Liapunov functions* $V(y_1, \ldots, y_n)$, which should satisfy

$$V(y_1, \ldots, y_n) \geq 0,$$

$$V(y_1, \ldots, y_n) = 0 \qquad \text{iff} \qquad y_1 = \ldots = y_n = 0 \qquad (13.15)$$

and along the solutions of (13.1)

$$\frac{d}{dx} V\big(y_1(x), \ldots, y_n(x)\big) \leq 0. \qquad (13.16)$$

Usually $V(y)$ behaves quadratically for small y and condition (13.15) means that

$$c\|y\|^2 \leq V(y) \leq C\|y\|^2, \qquad C \geq c > 0. \qquad (13.17)$$

The existence of such a Liapunov function is then a sufficient condition for stability of the origin.

We start with the *construction of a Liapunov function* for the linear case

$$y' = Ay. \qquad (13.18)$$

This is best done in the basis which is naturally given by the *eigenvectors* (or Jordan chains) of A. We therefore introduce $y = Tz$, $z = T^{-1}y$, so that A is transformed to Jordan canonical form (12.14') $J = T^{-1}AT$ and (13.18) becomes

$$z' = Jz. \qquad (13.19)$$

If we put

$$V_0(z) = \|z\|^2 \qquad \text{and} \qquad V(y) = V_0(T^{-1}y) = V_0(z), \qquad (13.20)$$

the derivative of $V(y(x))$ becomes

$$\frac{d}{dx} V\big(y(x)\big) = \frac{d}{dx} V_0\big(z(x)\big) = 2\text{Re}\,\langle z(x), z'(x)\rangle \qquad (13.21)$$

$$= 2\text{Re}\,\langle z(x), Jz(x)\rangle \leq 2\mu(J)V\big(y(x)\big).$$

By (10.20) the logarithmic norm is given by

$$2\mu(J) = \text{ largest eigenvalue of } (J + J^*).$$

The matrix $J + J^*$ is block-diagonal with tridiagonal blocks

$$\begin{pmatrix} 2\,\mathrm{Re}\,\lambda_i & \varepsilon & & \\ \varepsilon & 2\,\mathrm{Re}\,\lambda_i & \ddots & \\ & \ddots & \ddots & \varepsilon \\ & & \varepsilon & 2\,\mathrm{Re}\,\lambda_i \end{pmatrix}. \tag{13.22}$$

Subtracting the diagonal and using formula (6.7a), we see that the eigenvalues of the m-dimensional matrix (13.22) are given by

$$2\left(\mathrm{Re}\,\lambda_i + \varepsilon\cos\frac{\pi k}{m+1}\right), \qquad k = 1,\ldots,m. \tag{13.23}$$

As a consequence of this formula or by the use of Exercise 4 we have:

Lemma 13.5. *If all eigenvalues of A satisfy $\mathrm{Re}\,\lambda_i < -\varrho < 0$, then there exists a (quadratic) Liapunov function for equation (13.18) which satisfies*

$$\frac{d}{dx}V(y(x)) \le -\varrho\,V(y(x)). \tag{13.24}$$

$\square$

This last differential inequality implies that (Theorem 10.1)

$$V(y(x)) \le V(y_0)\cdot\exp(-\varrho(x - x_0))$$

and ensures that $\lim_{x\to\infty}\|y(x)\| = 0$, i.e., *asymptotic stability.*

Stability of Nonlinear Systems

It is now easy to extend the same ideas to *nonlinear* equations. The following theorem is an example of such a result.

Theorem 13.6. *Let the nonlinear system*

$$y' = Ay + g(x,y) \tag{13.25}$$

be given with $\mathrm{Re}\,\lambda_i < -\varrho < 0$ for all eigenvalues of A. Further suppose that for each $\varepsilon > 0$ there is a $\delta > 0$ such that

$$\|g(x,y)\| \le \varepsilon\|y\| \qquad for \qquad \|y\| < \delta,\ x \ge x_0. \tag{13.26}$$

Then the origin is (asymptotically) stable in the sense of Liapunov.

Proof. We use the Liapunov function $V(y)$ constructed for Lemma 13.5 and obtain from (13.25)

$$\frac{d}{dx}V(y(x)) \le -\varrho\,V(y(x)) + 2\,\mathrm{Re}\,\langle T^{-1}y(x),\ T^{-1}g(x,y(x))\rangle. \tag{13.27}$$

Cauchy's inequality together with (13.26) yields

$$\frac{d}{dx}V\big(y(x)\big) \le \big(-\varrho + \|T\| \cdot \|T^{-1}\|\varepsilon\big) \cdot V\big(y(x)\big). \tag{13.28}$$

For sufficiently small ε the right hand side is negative and we obtain asymptotic stability. $\square$

We see that, for nonlinear systems, stability *is only assured in a neighbourhood* of the origin. This can also be observed in Fig. 12.2. Another difference is that the *stability for eigenvalues on the imaginary axis can be destroyed.* An example for this (Routh 1877, pp. 95-96) is the system

$$y_1' = y_2, \qquad y_2' = -y_1 + y_2^3. \tag{13.29}$$

Here, with the Liapunov function $V = (y_1^2 + y_2^2)/2$, we obtain $V' = y_2^4$ which is > 0 for $y_2 \ne 0$. Therefore all solutions with initial value $\ne 0$ increase. A survey of this question ("the centre problem") together with its connection to limit cycles is given in Wanner (1983).

Stability of Non-Autonomous Systems

When the coefficients are not constant,

$$y' = A(x)y, \tag{13.30}$$

it is *not* a sufficient test of stability that the eigenvalues of A satisfy the conditions of stability for each instantaneous value of x.

Examples. 1. (Routh 1877, p. 96).

$$y_1' = y_2, \qquad y_2' = -\frac{1}{4x^2}y_1 \tag{13.31}$$

which is satisfied by $y_1(x) = a\sqrt{x}$.

2. An example with eigenvalues strictly negative: we start with

$$B = \begin{pmatrix} -1 & 0 \\ 4 & -1 \end{pmatrix}, \qquad y' = By.$$

An inspection of the derivative of $V = (y_1^2 + y_2^2)/2$ shows that V *increases* in the sector $2 - \sqrt{3} < y_2/y_1 < 2 + \sqrt{3}$. The idea is to take the initial value in this region and, for x increasing, to rotate the coordinate system with the same speed as the solution rotates:

$$y' = T(x)BT(-x)y = A(x)y, \qquad T(x) = \begin{pmatrix} \cos ax & -\sin ax \\ \sin ax & \cos ax \end{pmatrix}. \tag{13.32}$$

For $y(0) = (1,1)^T$, the good choice for a is $a = 2$ and (13.32) possesses the solution

$$y(x) = \Big((\cos 2x - \sin 2x)e^x, (\cos 2x + \sin 2x)e^x\Big)^T. \qquad (13.33)$$

This solution is clearly unstable, while -1 remains for all x the double eigenvalue of $A(x)$. For more examples see Exercises 6 and 7 below.

We observe that stability theory for non-autonomous systems is more complicated. Among the cases in which stability can be shown are the following:
1) $a_{ii}(x) < 0$ and $A(x)$ is diagonally dominant; then $\mu(A(x)) \le 0$ such that stability follows from Theorem 10.6.
2) $A(x) = B + C(x)$, with B constant and satisfying $\text{Re } \lambda_i < -\varrho < 0$ for its eigenvalues, and $\|C(x)\| < \varepsilon$ with ε so small that the proof of Theorem 13.6 can be applied.

Exercises

1. Express the stability conditions for the polynomials $z^2 + pz + q = 0$ and $z^3 + pz^2 + qz + r = 0$.

   Result. a) $p > 0$ and $q > 0$; b) $p > 0$, $r > 0$ and $pq - r > 0$.

2. (Hurwitz 1895). Verify that condition (13.12) is equivalent to the positivity of the principal minors of the matrix

   $$H = \begin{pmatrix} a_1 & a_3 & a_5 & \cdots \\ a_0 & a_2 & a_4 & \cdots \\ & a_1 & a_3 & \cdots \\ & a_0 & a_2 & \cdots \\ & & \cdots & \cdots \end{pmatrix} = \Big(a_{2j-i}\Big)_{i,j=1}^n$$

   ($a_k = 0$ for $k < 0$ and $k > n$). Understand that Routh's algorithm (13.11) is identical to a sort of Gaussian elimination transforming H to triangular form.

3. The polynomial

   $$\frac{5 \cdot 4 \cdot 3 \cdot 2 \cdot 1}{10 \cdot 9 \cdot 8 \cdot 7 \cdot 6} \frac{z^5}{5!} + \frac{5 \cdot 4 \cdot 3 \cdot 2}{10 \cdot 9 \cdot 8 \cdot 7} \frac{z^4}{4!} + \frac{5 \cdot 4 \cdot 3}{10 \cdot 9 \cdot 8} \frac{z^3}{3!} + \frac{5 \cdot 4}{10 \cdot 9} \frac{z^2}{2!} + \frac{5}{10} z + 1$$

   is the numerator of the $(5,5)$-Padé approximation to $\exp(z)$. Verify that all its roots satisfy $\text{Re } z < 0$. Try to establish the result for general n (see e.g., Birkhoff & Varga (1965), Lemma 7).

4. (Gerschgorin). Prove that the eigenvalues of a matrix $A = (a_{ij})$ lie in the union of the discs

   $$\Big\{ z \, ; \, |z - a_{ii}| \le \sum_{j \ne i} |a_{ij}| \Big\}.$$

Hint. Write the formula $A\boldsymbol{x} = \lambda\boldsymbol{x}$ in coordinates $\sum_j a_{ij}x_j = \lambda x_i$, put the diagonal elements on the right hand side and choose i such that $|x_i|$ is maximal.

5. Determine the stability of the origin for the system

$$y_1' = -y_2 - y_1^2 - y_1 y_2 \,,$$
$$y_2' = y_1 + 2y_1 y_2 \,.$$

Hint. Find a Liapunov function of degree 4 starting with $V = (y_1^2 + y_2^2)/2 + \ldots$ such that $V' = K(y_1^2 + y_2^2)^2 + \ldots$ and determine the sign of K.

6. (J. Lambert 1987). Consider the system

$$y' = A(x) \cdot y \qquad \text{where} \qquad A(x) = \begin{pmatrix} -1/4x & 1/x^2 \\ -1/4 & -1/4x \end{pmatrix}. \tag{13.34}$$

a) Show that both eigenvalues of $A(x)$ satisfy $\operatorname{Re}\lambda < 0$ for all $x > 0$.

b) Compute $\mu(A)$ (from (10.20)) and show that

$$\mu(A) \le 0 \qquad \text{iff} \qquad \sqrt{5} - 1 \le x \le \sqrt{5} + 1.$$

c) Compute the general solution of (13.34).

Hint. Introduce the new functions $z_2(x) = y_2(x)$, $z_1(x) = xy_1(x)$ which leads to the second equation of (11.19) (Exercise 5 of Section I.11). The solution is

$$y_1(x) = x^{-3/4}\left(a + b\log x\right), \qquad y_2(x) = x^{1/4}\left(-\frac{a}{2} + b\left(1 - \frac{1}{2}\log x\right)\right).$$
$$\tag{13.35}$$

d) Determine a and b such that $\|y(x)\|_2^2$ is *increasing* for $0 < x < \sqrt{5} - 1$.

e) Determine a and b such that $\|y(x)\|_2^2$ is *increasing* for $\sqrt{5} + 1 < x < \infty$.

Results. $b = 1.8116035 \cdot a$ for (d) and $b = 0.2462015 \cdot a$ for (e).

7. Find a counter-example for Fatou's conjecture

If $\ddot{y} + A(t)y = 0$ and $\forall t \quad 0 < C_1 \le A(t) \le C_2$ then $y(t)$ is stable

(C.R. 189 (1929), p.967-969; for a solution see Perron (1930)).

8. Help James Watt (see original drawing from 1788 in Fig. 13.3) to solve the stability problem for his steam engine governor: if ω is the rotation speed of the engine, its acceleration is influenced by the steam supply and exterior work as follows:

$$\omega' = k\cos(\varphi + \alpha) - F, \qquad k, F > 0.$$

Here φ and α are the angles shown in Fig. 13.3. The acceleration of φ is determined by centrifugal force, weight, and friction as

$$\varphi'' = \omega^2 \sin\varphi \cos\varphi - g\sin\varphi - b\varphi', \qquad g, b > 0.$$

Compute the equilibrium point $\varphi'' = \varphi' = \omega' = 0$ and determine under which conditions it is stable (the solution is easier for $\alpha = 0$).

Correct solutions should be sent to: James Watt, famous inventor of the steam engine, Westminster Abbey, 6HQ 1FX London.

Remark. Hurwitz' paper (1895) was motivated by a similar practical problem, namely "... die Regulirung von Turbinen des Badeortes Davos".

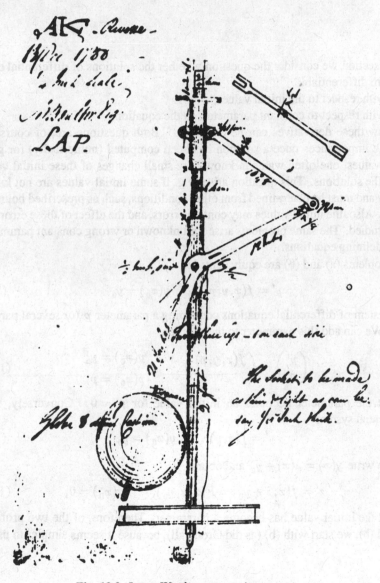

Fig. 13.3. James Watt's steam engine governor

I.14 Derivatives with Respect to Parameters and Initial Values

For a single equation, Dr. Ritt has solved the problem indicated in the title by a very simple and direct method ... Dr. Ritt's proof cannot be extended immediately to a system of equations.

(T.H. Gronwall 1919)

In this section we consider the question whether the solutions of differential equations are differentiable

   a) with respect to the initial values;

   b) with respect to constant parameters in the equation;

and how these derivatives can be computed. Both questions are, of course, of extreme importance: once a solution has been computed (numerically) for given initial values, one often wants to know how small changes of these initial values affect the solutions. This question arises e.g. if some initial values are not known exactly and must be determined from other conditions, such as prescribed boundary values. Also, the initial values may contain errors, and the effect of these errors has to be studied. The same problems arise for unknown or wrong constant parameters in the defining equations.

   Problems (a) and (b) are equivalent: let

$$y' = f(x,y,p), \qquad y(x_0) = y_0 \tag{14.1}$$

be a system of differential equations containing a parameter p (or several parameters). We can add this parameter to the solutions

$$\begin{pmatrix} y' \\ p' \end{pmatrix} = \begin{pmatrix} f(x,y,p) \\ 0 \end{pmatrix}, \qquad \begin{aligned} y(x_0) &= y_0 \\ p(x_0) &= p, \end{aligned} \tag{14.1'}$$

so that the parameter becomes an initial value for $p' = 0$. Conversely, for a differential system

$$y' = f(x,y), \qquad y(x_0) = y_0 \tag{14.2}$$

we can write $y(x) = z(x) + y_0$ and obtain

$$z' = f(x, z + y_0) = F(x,z,y_0), \qquad z(x_0) = 0, \tag{14.2'}$$

so that the initial value has become a parameter. Therefore, of the two problems (a) and (b), we start with (b) (as did Gronwall), because it seems simpler to us.

The Derivative with Respect to a Parameter

Usually, a given problem contains *several* parameters. But since we are interested in partial derivatives, we can treat one parameter after another while keeping the remaining ones fixed. It is therefore sufficient in the following theory to suppose that $f(x, y, p)$ depends only on *one* scalar parameter p.

When we replace the parameter p in (14.1) by q we obtain another solution, which we denote by $z(x)$:

$$z' = f(x, z, q), \qquad z(x_0) = y_0. \tag{14.3}$$

It is then natural to subtract (14.1) from (14.3) and to linearize

$$
\begin{aligned}
z' - y' &= f(x, z, q) - f(x, y, p) \\
&= \frac{\partial f}{\partial y}(x, y, p)(z - y) + \frac{\partial f}{\partial p}(x, y, p)(q - p) + \varrho_1 \cdot (z - y) + \varrho_2 \cdot (q - p).
\end{aligned}
\tag{14.4}
$$

If we put $(z(x) - y(x))/(q - p) = \psi(x)$ and drop the error terms, we obtain

$$\psi' = \frac{\partial f}{\partial y}(x, y(x), p)\psi + \frac{\partial f}{\partial p}(x, y(x), p), \qquad \psi(x_0) = 0. \tag{14.5}$$

This equation is the key to the problem.

Theorem 14.1 (Gronwall 1919). *Suppose that for $x_0 \le x \le X$ the partial derivatives $\partial f/\partial y$ and $\partial f/\partial p$ exist and are continuous in the neighbourhood of the solution $y(x)$. Then the partial derivatives*

$$\frac{\partial y(x)}{\partial p} = \psi(x)$$

exist, are continuous, and satisfy the differential equation (14.5).

Proof. This theorem was the origin of the famous Gronwall lemma (see I.10, Exercise 2). We prove it here by the equivalent Theorem 10.2. Set

$$L = \max \left\| \frac{\partial f}{\partial y} \right\|, \qquad A = \max \left\| \frac{\partial f}{\partial p} \right\| \tag{14.6}$$

where the max is taken over the domain under consideration. When we consider $z(x)$ as an approximate solution for (14.1) we have for the defect

$$\| z'(x) - f(x, z(x), p) \| = \| f(x, z(x), q) - f(x, z(x), p) \| \le A|q - p|,$$

therefore from Theorem 10.2

$$\| z(x) - y(x) \| \le \frac{A}{L} |q - p| (e^{L(x - x_0)} - 1). \tag{14.7}$$

So for $|q - p|$ sufficiently small and $x_0 \le x \le X$, we can have $\| z(x) - y(x) \|$ arbitrarily small. By definition of differentiability and by (14.7), for each $\varepsilon > 0$

there is a δ such that the error terms in (14.4) satisfy

$$\|\varrho_1 \cdot (z - y) + \varrho_2 \cdot (q - p)\| \le \varepsilon |q - p| \qquad \text{if} \qquad |q - p| \le \delta. \qquad (14.8)$$

(The situation is, in fact, a little more complicated: the δ for the bounds $\|\varrho_1\| < \varepsilon$ and $\|\varrho_2\| < \varepsilon$ may depend on x. But due to compactness and continuity, it can then be replaced by a uniform bound. Another possibility to overcome this little obstacle would be a bound on the second derivatives. But why should we worry about this detail? Gronwall himself did not mention it).

We now consider $(z(x) - y(x))/(q - p)$ as an approximate solution for (14.5) and apply Theorem 10.2 a second time. Its defect is by (14.8) and (14.4) bounded by ε and the linear differential equation (14.5) *also* has L as a Lipschitz constant (see (11.2)). Therefore from (10.14) we obtain

$$\left\| \frac{z(x) - y(x)}{q - p} - \psi(x) \right\| \le \frac{\varepsilon}{L} \left(e^{L(x - x_0)} - 1 \right)$$

which becomes arbitrarily small; this proves that $\psi(x)$ is the derivative of $y(x)$ with respect to p.

Continuity. The partial derivatives $\partial y / \partial p = \psi(x)$ are solutions of the differential equation (14.5), which we write as $\psi' = g(x, \psi, p)$, where by hypothesis g depends continuously on p. Therefore the continuous dependence of ψ on p follows again from Theorem 10.2. $\qquad \Box$

Theorem 14.2. *Let $y(x)$ be the solution of equation (14.1) and consider the Jacobian*

$$A(x) = \frac{\partial f}{\partial y}(x, y(x), p). \qquad (14.9)$$

Let $R(x, x_0)$ be the resolvent of the equation $y' = A(x)y$ (see (11.4)). Then the solution $z(x)$ of (14.3) with a slightly perturbed parameter q is given by

$$z(x) = y(x) + (q - p) \int_{x_0}^{x} R(x, s) \frac{\partial f}{\partial p}(s, y(s), p) \, ds + o(|q - p|) \qquad (14.10)$$

Proof. This is the variation of constants formula (11.10) applied to (14.5). $\qquad \Box$

It can be seen that the sensitivity of the solutions to changes of parameters is influenced firstly by the partial derivatives $\partial f / \partial p$ (which is natural), and secondly by the size of $R(x, s)$, i.e., by the stability of the differential equation with matrix (14.9).

Derivatives with Respect to Initial Values

Notation. We denote by $y(x, x_0, y_0)$ the solution $y(x)$ at the point x satisfying the initial values $y(x_0) = y_0$, and hope that no confusion arises from the use of the same letter y for two different functions.

The following identities are trivial by definition or follow from uniqueness arguments as for (11.6):

$$\frac{\partial y(x, x_0, y_0)}{\partial x} = f(x, y(x, x_0, y_0)) \tag{14.11}$$

$$y(x_0, x_0, y_0) = y_0 \tag{14.12}$$

$$y(x_2, x_1, y(x_1, x_0, y_0)) = y(x_2, x_0, y_0). \tag{14.13}$$

Theorem 14.3. *Suppose that the partial derivative of f with respect to y exists and is continuous. Then the solution $y(x, x_0, y_0)$ is differentiable with respect to y_0 and the derivative is given by the matrix*

$$\frac{\partial y(x, x_0, y_0)}{\partial y_0} = \Psi(x) \tag{14.14}$$

where $\Psi(x)$ is the resolvent of the so-called "variational equation"

$$\Psi'(x) = \frac{\partial f}{\partial y}(x, y(x, x_0, y_0)) \cdot \Psi(x),$$
$$\Psi(x_0) = I. \tag{14.15}$$

Proof. We know from (14.2) and (14.2') that $\partial F/\partial z$ and $\partial F/\partial y_0$ are both equal to $\partial f/\partial y$, so the derivatives are known to *exist* by Theorem 14.1. In order to obtain formula (14.15), we just have to differentiate (14.11) and (14.12) with respect to y_0. $\qquad\square$

We finally compute the derivative of $y(x, x_0, y_0)$ with respect to x_0.

Theorem 14.4. *Under the same hypothesis as in Theorem 14.3, the solutions are also differentiable with respect to x_0 and the derivative is given by*

$$\frac{\partial y(x, x_0, y_0)}{\partial x_0} = -\frac{\partial y(x, x_0, y_0)}{\partial y_0} \cdot f(x_0, y_0). \tag{14.16}$$

Proof. Differentiate the identity

$$y(x_1, x_0, y(x_0, x_1, y_1)) = y_1,$$

which follows from (14.13), with respect to x_0 and apply (14.11) (see Exercise 1). $\qquad\square$

The Nonlinear Variation-of-Constants Formula

The following theorem is an extension of Theorem 11.2 to systems of non-linear differential equations.

Theorem 14.5 (Alekseev 1961, Gröbner 1960). *Denote by y and z the solutions of*

$$y' = f(x,y), \qquad\qquad y(x_0) = y_0, \qquad\qquad (14.17a)$$
$$z' = f(x,z) + g(x,z), \qquad z(x_0) = y_0, \qquad\qquad (14.17b)$$

respectively and suppose that $\partial f/\partial y$ exists and is continuous. Then the solutions of (14.17a) and of the "perturbed" equation (14.17b) are connected by

$$z(x) = y(x) + \int_{x_0}^{x} \frac{\partial y}{\partial y_0}(x,s,z(s)) \cdot g(s,z(s))\, ds. \qquad\qquad (14.18)$$

Proof. We choose a subdivision $x_0 = s_0 < s_1 < s_2 < \ldots < s_N = x$ (see Fig. 14.1). The descending curves represent the solutions of the unperturbed equation (14.17a) with initial values s_i, $z(s_i)$. The differences d_i are, due to the different slopes of $z(s)$ and $y(s)$ ((14.17b) minus (14.17a)), equal to $d_i = g(s_i, z(s_i)) \cdot \Delta s_i + o(\Delta s_i)$. This "error" at s_i is then "transported" to the final value x by the amount given in Theorem 14.3, to give

$$D_i = \frac{\partial y}{\partial y_0}(x,s_i,z(s_i)) \cdot g(s_i, z(s_i)) \cdot \Delta s_i + o(\Delta s_i). \qquad\qquad (14.19)$$

Since $z(x) - y(x) = \sum_{i=1}^{N} D_i$, we obtain the integral in (14.18) after insertion of (14.19) and passing to the limit $\Delta s_i \to 0$. □

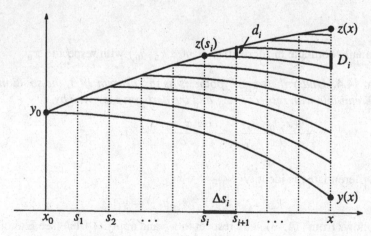

Fig. 14.1. Lady Windermere's fan, Act 2

If we also want to take into account a possible difference in the initial values, we may formulate:

Corollary 14.6. *Let $y(x)$ and $z(x)$ be the solutions of*

$$y' = f(x, y), \qquad\qquad y(x_0) = y_0,$$
$$z' = f(x, z) + g(x, z), \qquad z(x_0) = z_0,$$

then

$$z(x) = y(x) + \int_0^1 \frac{\partial y}{\partial y_0} \Big(x, x_0, y_0 + s(z_0 - y_0) \Big) \cdot (z_0 - y_0)\, ds$$
$$+ \int_{x_0}^x \frac{\partial y}{\partial y_0} \Big(x, s, z(s) \Big) \cdot g\big(s, z(s)\big)\, ds. \qquad\qquad \square$$

(14.20)

These two theorems allow many estimates of the stability of general nonlinear systems. For *linear* systems, $\partial y / \partial y_0 (x, s, z)$ is independent of z, and formulas (14.20) and (14.18) become the variation-of-constants formula (11.10). Also, by majorizing the integrals in (14.20) in a trivial way, one obtains the fundamental lemma (10.14) and also the variant form of Theorem 10.2.

Flows and Volume-Preserving Flows

> Considérons des molécules fluides dont l'ensemble forme à l'origine des temps une certaine figure F_0 ; quand ces molécules se déplaceront, leur ensemble formera une nouvelle figure qui ira en se déformant d'une manière continue, et à l'instant t l'ensemble des molécules envisagées formera une nouvelle figure F.
>
> (H. Poincaré, Mécanique Céleste 1899, Tome III, p.2)

We now turn our attention to a new interpretation of the Abel-Liouville-Jacobi-Ostrogradskii formula (11.11). Liouville and above all Jacobi (in his "Dynamik" 1843) used this formula extensively to obtain "first integrals", i.e., relations between the solutions, so that the dimension of the system could be decreased and the analytic integration of the differential equations of mechanics becomes a little less hopeless. Poincaré then (see the quotation) introduced a much more geometric point of view: for an autonomous system of differential equations [1]

$$\frac{dy}{dt} = f(y) \tag{14.21}$$

we define the *flow* $\varphi_t : \mathbf{R}^n \to \mathbf{R}^n$ to be the function which associates, for a given t, to the initial value $y^0 \in \mathbf{R}^n$ the corresponding solution value at time t

$$\varphi_t(y^0) := y(t, 0, y^0). \tag{14.22}$$

---

[1] Due to the origin of these topics in Mechanics and Astronomy, we here use t for the independent variable.

For sets A of initial values we also study its behaviour under the action of the flow and write

$$\varphi_t(A) = \{y \mid y = y(t, 0, y^0),\ y^0 \in A\}. \tag{14.22'}$$

We can imagine, with Poincaré, sets of "molecules" moving (and being deformed) with the flow.

Example 14.7. Fig. 14.2 shows, for the two-dimensional system (12.20) (see Fig. 12.2), the transformations which three sets A, B, C [2] undergo when t passes from 0 to $0.2, 0.4$ and (for C) 0.6. It can be observed that these sets quickly lose very much of their beauty.

Fig. 14.2. Transformation of three sets under a flow

Now divide A into "infinitely small" cubes I of sides $dy_1^0, \ldots, dy_n^0$. The image $\varphi_t(I)$ of such a cube is an infinitely small parallelepiped. It is created by the columns of $\partial y / \partial y^0(t, 0, y^0)$ scaled by dy_i^0, and its volume is $\det(\partial y / \partial y^0(t, 0, y^0)) \cdot dy_1^0 \ldots dy_n^0$. Adding up all these volumes (over A) or, more precisely, using the transformation formula for multiple integrals (Euler 1769b, Jacobi 1841), we

---

[2] The resemblance of these sets with a certain feline animal is not entirely accidental; we chose it in honour of V.I. Arnol'd.

obtain

$$\text{Vol}\left(\varphi_t(A)\right) = \int_{\varphi_t(A)} dy = \int_A \left| \det\left(\frac{\partial y}{\partial y^0}(t,0,y^0)\right) \right| dy^0.$$

Next we use formula (11.11) together with (11.9) and (14.15)

$$\det\left(\frac{\partial y}{\partial y^0}(t,0,y^0)\right) = \exp\left(\int_0^t \text{Tr}(f'(y(s,0,y^0)))\, ds\right) \tag{14.23}$$

and we obtain

Theorem 14.8. *Consider the system (14.21) with continuously differentiable function $f(y)$.*

a) *For a set $A \subset \mathbb{R}^n$ the total volume of $\varphi_t(A)$ satisfies*

$$\text{Vol}\left(\varphi_t(A)\right) = \int_A \exp\left(\int_0^t \text{Tr}(f'(y(s,0,y^0)))\, ds\right) dy^0. \tag{14.24}$$

b) *If $\text{Tr}(f'(y)) = 0$ along the solution, the flow is volume-preserving, i.e.,*
   $\text{Vol}\left(\varphi_t(A)\right) = \text{Vol}(A)$.   □

Example 14.9. For the system (12.20) we have

$$f'(y) = \begin{pmatrix} (1-2y_1)/3 & (2y_2-1)/3 \\ 2-y_2 & -y_1 \end{pmatrix} \qquad \text{and} \qquad \text{Tr}(f'(y)) = (1-5y_1)/3.$$

The trace of $f'(y)$ changes sign at the line $y_1 = 1/5$. To its left the volume increases, to the right we have decreasing volumes. This can clearly be seen in Fig. 14.2.

Example 14.10. For the mathematical pendulum (with y_1 the angle of deviation from the vertical)

$$\begin{aligned} \dot{y}_1 &= y_2 \\ \dot{y}_2 &= -\sin y_1 \end{aligned} \qquad f'(y) = \begin{pmatrix} 0 & 1 \\ -\cos y_1 & 0 \end{pmatrix} \tag{14.25}$$

we have $\text{Tr}(f'(y)) = 0$. Therefore the flow, although treating the cats quite badly, at least preserves their areas (Fig. 14.3).

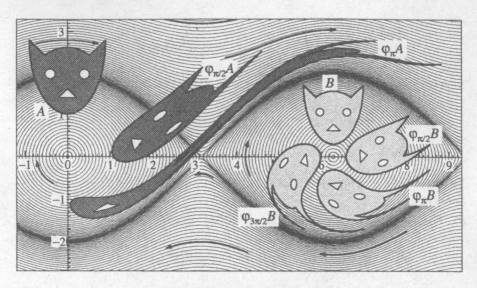

Fig. 14.3. Cats, beware of pendulums!

Canonical Equations and Symplectic Mappings

Let $H(p_1, \ldots, p_n, q_1, \ldots, q_n)$ be a twice continuously differentiable function of $2n$ variables and (see (6.26))

$$\dot{p}_i = -\frac{\partial H}{\partial q_i}(p, q), \qquad \dot{q}_i = \frac{\partial H}{\partial p_i}(p, q) \tag{14.26}$$

the corresponding canonical system of differential equations. Small variations of the initial values lead to variations $\delta p_i(t), \delta q_i(t)$ of the solution of (14.26). By Theorem 14.3 (variational equation) these satisfy

$$\dot{\delta p_i} = -\sum_{j=1}^{n} \frac{\partial^2 H}{\partial p_j \partial q_i}(p, q) \cdot \delta p_j - \sum_{j=1}^{n} \frac{\partial^2 H}{\partial q_j \partial q_i}(p, q) \cdot \delta q_j$$

$$\dot{\delta q_i} = \sum_{j=1}^{n} \frac{\partial^2 H}{\partial p_j \partial p_i}(p, q) \cdot \delta p_j + \sum_{j=1}^{n} \frac{\partial^2 H}{\partial q_j \partial p_i}(p, q) \cdot \delta q_j. \tag{14.27}$$

The upper left block of the Jacobian matrix is the negative transposed of the lower right block. As a consequence, the trace of the Jacobian of (14.27) is identically zero and *the corresponding flow is volume-preserving* ("Theorem of Liouville").

But there is much more than that (Poincaré 1899, vol. III, p. 43): consider a two-dimensional manifold A in the $2n$-dimensional flow. We represent it as a (differentiable) map of a compact set $K \subset \mathbb{R}^2$ into $\mathbb{R}^{2n}$ (Fig. 14.4)

$$\Phi : \quad \begin{array}{ccc} K & \longrightarrow & A \subset \mathbb{R}^{2n} \\ (u, v) & \longmapsto & (p^0(u, v), q^0(u, v)) \end{array} \tag{14.28}$$

We let $\pi_i(A)$ be the projection of A onto the (p_i, q_i)-coordinate plane and consider the *sum of the oriented areas of* $\pi_i(A)$. We shall see that this is also an invariant.

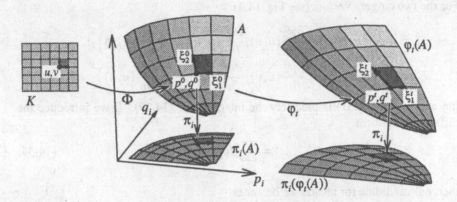

Fig. 14.4. Two-dimensional manifold in the flow

The oriented area of $\pi_i(A)$ is a surface integral over A which is defined, with the transformation formula in mind, as

$$\text{or.area}(\pi_i(A)) = \iint_K \det \begin{pmatrix} \dfrac{\partial p_i^0}{\partial u} & \dfrac{\partial p_i^0}{\partial v} \\ \dfrac{\partial q_i^0}{\partial u} & \dfrac{\partial q_i^0}{\partial v} \end{pmatrix} du\, dv \,. \tag{14.29}$$

For the computation of the area of $\pi_i\big(\varphi_t(A)\big)$, after the action of the flow, we use the composition $\varphi_t \circ \Phi$ as coordinate map (Fig. 14.4). This produces, with p_i^t, q_i^t being the ith respectively $(n+i)$th component of this map,

$$\text{or.area}(\pi_i(\varphi_t(A))) = \iint_K \det \begin{pmatrix} \dfrac{\partial p_i^t}{\partial u} & \dfrac{\partial p_i^t}{\partial v} \\ \dfrac{\partial q_i^t}{\partial u} & \dfrac{\partial q_i^t}{\partial v} \end{pmatrix} du\, dv \,. \tag{14.30}$$

There is no theoretical difficulty in differentiating this expression with respect to t and summing for $i = 1, \ldots, n$. This will give zero and the invariance is established.

The proof, however, becomes more elegant if we introduce *exterior differential forms* (E. Cartan 1899). These, originally "expressions purement symboliques", are today understood as *multilinear maps* on the *tangent space* (for more details see "Chapter 7" of Arnol'd 1974). In our case the one-forms dp_i, respectively dq_i, map a tangent vector ξ to its ith, respectively $(n+i)$th, component. The *exterior product* $dp_i \wedge dq_i$ is a bilinear map acting on a pair of vectors

$$\begin{aligned} (dp_i \wedge dq_i)(\xi_1, \xi_2) &= \det \begin{pmatrix} dp_i(\xi_1) & dp_i(\xi_2) \\ dq_i(\xi_1) & dq_i(\xi_2) \end{pmatrix} . \\ &= dp_i(\xi_1)dq_i(\xi_2) - dp_i(\xi_2)dq_i(\xi_1) \end{aligned} \tag{14.31}$$

and satisfies Grassmann's rules for exterior multiplication

$$dp_i \wedge dp_j = -dp_j \wedge dp_i , \qquad dp_i \wedge dp_i = 0 . \tag{14.32}$$

For the two tangent vectors (see Fig. 14.4)

$$
\begin{aligned}
\xi_1^0 &= \left(\frac{\partial p_1^0}{\partial u}(u,v), \ldots, \frac{\partial p_n^0}{\partial u}(u,v), \frac{\partial q_1^0}{\partial u}(u,v), \ldots, \frac{\partial q_n^0}{\partial u}(u,v) \right)^T \\
\xi_2^0 &= \left(\frac{\partial p_1^0}{\partial v}(u,v), \ldots, \frac{\partial p_n^0}{\partial v}(u,v), \frac{\partial q_1^0}{\partial v}(u,v), \ldots, \frac{\partial q_n^0}{\partial v}(u,v) \right)^T
\end{aligned}
\tag{14.33}
$$

the expression (14.31) is precisely the integrand of (14.29). If we introduce the differential 2-form

$$\omega^2 = \sum_{i=1}^{n} dp_i \wedge dq_i \tag{14.34}$$

then our candidate for invariance becomes

$$\sum_{i=1}^{n} \text{or.area}\big(\pi_i(A)\big) = \iint_K \omega^2(\xi_1^0, \xi_2^0) \, du \, dv.$$

After the action of the flow we have the tangent vectors

$$\xi_1^t = \varphi_t'(p^0, q^0) \cdot \xi_1^0 , \qquad \xi_2^t = \varphi_t'(p^0, q^0) \cdot \xi_2^0$$

and

$$\sum_{i=1}^{n} \text{or.area}\big(\pi_i(\varphi_t(A))\big) = \iint_K \omega^2(\xi_1^t, \xi_2^t) \, du \, dv$$

(see (14.30)). We shall see that $\omega^2(\xi_1^t, \xi_2^t) = \omega^2(\xi_1^0, \xi_2^0)$.

Definition 14.11. For a differentiable function $g : \mathbb{R}^{2n} \to \mathbb{R}^{2n}$ we define the differential form $g^* \omega^2$ by

$$(g^* \omega^2)(\xi_1, \xi_2) := \omega^2\big(g'(p,q)\xi_1, g'(p,q)\xi_2\big) . \tag{14.35}$$

Such a function g is called *symplectic* (a name suggested by H. Weyl 1939, p. 165) if

$$g^* \omega^2 = \omega^2, \tag{14.36}$$

i.e., if the 2-form ω^2 is invariant under g.

Theorem 14.12. *The flow of a canonical system (14.26) is symplectic, i.e.,*

$$(\varphi_t)^* \omega^2 = \omega^2 \qquad \text{for all } t. \tag{14.37}$$

Proof. We compute the derivative of $\omega^2(\xi_1^t, \xi_2^t)$ (see (14.35)) with respect to t by the Leibniz rule. This gives

$$\frac{d}{dt}\left(\sum_{i=1}^{n}(dp_i \wedge dq_i)(\xi_1^t, \xi_2^t)\right) = \sum_{i=1}^{n}(dp_i \wedge dq_i)(\dot{\xi}_1^t, \xi_2^t) + \sum_{i=1}^{n}(dp_i \wedge dq_i)(\xi_1^t, \dot{\xi}_2^t) .$$

(14.38)

Since the vectors ξ_1^t and ξ_2^t satisfy the variational equation (14.27), we have

$$\frac{d}{dt}\omega^2(\xi_1^t, \xi_2^t) = \sum_{i,j=1}^{n}\left(-\frac{\partial^2 H}{\partial p_j \partial q_i}\,dp_j \wedge dq_i - \frac{\partial^2 H}{\partial q_j \partial q_i}\,dq_j \wedge dq_i \right.$$

(14.39)

$$\left. + \frac{\partial^2 H}{\partial p_j \partial p_i}\,dp_i \wedge dp_j + \frac{\partial^2 H}{\partial q_j \partial p_i}\,dp_i \wedge dq_j\right)(\xi_1^t, \xi_2^t).$$

The first and last terms in this formula cancel by symmetry of the partial derivatives. Further, the properties (14.32) imply that

$$\sum_{i,j=1}^{n}\frac{\partial^2 H}{\partial p_i \partial p_j}(p,q)\,dp_i \wedge dp_j = \sum_{i<j}\left(\frac{\partial^2 H}{\partial p_i \partial p_j}(p,q) - \frac{\partial^2 H}{\partial p_j \partial p_i}(p,q)\right)dp_i \wedge dp_j$$

vanishes. Since the last remaining term cancels in the same way, the derivative (14.38) vanishes identically.   □

Example 14.13. We use the spherical pendulum in canonical form (6.28)

$$\dot{p}_1 = p_2^2\,\frac{\cos q_1}{\sin^3 q_1} - \sin q_1 \qquad \dot{p}_2 = 0$$

$$\dot{q}_1 = p_1 \qquad\qquad\qquad \dot{q}_2 = \frac{p_2}{\sin^2 q_1}$$

(14.40)

and for A the familiar two-dimensional cat placed in $\mathbb{R}^4$ such that its projection to (p_1, q_1) is a line; i.e., with zero area. It can be seen that with increasing t the area in (p_1, q_1) increases and the area in (p_2, q_2) decreases. Their sum remains constant. Observe that for larger t the left ear in (p_1, q_1) is twisted, i.e., surrounded in the negative sense, so that this part counts for negative area (Fig. 14.5). If time proceeded in the negative sense, *both* areas would increase, but the first area would be oriented negatively.

Between the two-dimensional invariant of Theorem 14.12 and the $2n$-dimensional of Liouville's theorem, there are many others; e.g., the differential 4-form

$$\omega^4 = \sum_{i<j} dp_i \wedge dp_j \wedge dq_i \wedge dq_j.$$

(14.41)

These invariants, however, are not really new, because (14.41) is proportional to the exterior square of ω^2, $\omega^2 \wedge \omega^2 = -2\omega^4$.

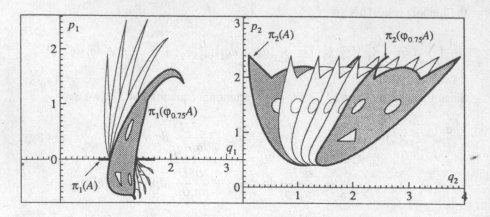

Fig. 14.5. Invariance of $\omega^2 = \sum_{i=1}^n dp_i \wedge dq_i$ for the spherical pendulum

Writing (14.31) in matrix notation

$$\omega^2(\xi_1, \xi_2) = \xi_1^T J \xi_2 \qquad \text{with} \qquad J = \begin{pmatrix} 0 & I \\ -I & 0 \end{pmatrix} \qquad (14.42)$$

we obtain the following criterion:

Theorem 14.14. *A differentiable transformation* $g : \mathbb{R}^{2n} \to \mathbb{R}^{2n}$ *is symplectic if and only if its Jacobian* $R = g'(p, q)$ *satisfies*

$$R^T J R = J \qquad (14.43)$$

with J *given in (14.42).*

Proof. This follows at once from (see (14.35))

$$(g^*\omega^2)(\xi_1, \xi_2) = (R\xi_1)^T J (R\xi_2) = \xi_1^T R^T J R \xi_2. \qquad \square$$

Exercises

1. Prove the following lemma from elementary calculus which is used in the proof of Theorem 14.4: if for a function $F(x, y)$, $\partial F/\partial y$ exists and $y(x)$ is differentiable and such that $F(x, y(x)) = Const$, then $\partial F/\partial x$ exists at $(x, y(x))$ and is equal to

$$\frac{\partial F}{\partial x}\big(x, y(x)\big) = -\frac{\partial F}{\partial y}\big(x, y(x)\big) \cdot y'(x).$$

Hint. Use the identity

$$F\big(x_1, y(x_1)\big) - F\big(x_0, y(x_1)\big) = F\big(x_0, y(x_0)\big) - F\big(x_0, y(x_1)\big).$$

I.15 Boundary Value and Eigenvalue Problems

Although our book is mainly concerned with initial value problems, we want to include in this first chapter some properties of boundary and eigenvalue problems.

Boundary Value Problems

They arise in systems of differential equations, say

$$y_1' = f_1(x, y_1, y_2),$$
$$y_2' = f_2(x, y_1, y_2),$$

(15.1)

when there is *no* initial point x_0 at which $y_1(x_0)$ and $y_2(x_0)$ are known simultaneously. Questions of existence and uniqueness then become much more complicated.

Example 1. Consider the differential equation

$$y'' = \exp(y) \qquad \text{or} \qquad y_1' = y_2, \quad y_2' = \exp(y_1) \qquad (15.2a)$$

with the *boundary conditions*

$$y_1(0) = a, \qquad y_1(1) = b. \qquad (15.2b)$$

In order to apply our existence theorems or to do numerical computations (say by Euler's method (7.3)), we can proceed as follows: guess the missing initial value y_{20}. We can then compute the solution and check whether the computed value for $y_1(1)$ is equal to b or not. So our problem is, whether the function of the single variable y_{20}

$$F(y_{20}) := y_1(1) - b \qquad (15.3)$$

possesses a zero or not.

Equation (15.2a) is *quasimonotone*, which implies that $F(y_{20})$ depends monotonically on y_{20} (Fig. 15.1a, see Exercise 7 of I.10). Also, for y_{20} very small or very large, $y_1(1)$ is arbitrarily small or large, or even infinite. Therefore, (15.2) possesses for all a, b a unique solution (see Fig. 15.1b).

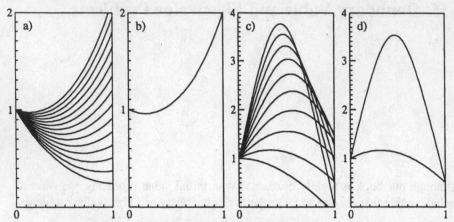

Fig. 15.1. a) Solutions of (15.2a) for different initial values $y_{20} = -1.7, \ldots, -0.4$
b) Unique solution of (15.2a) for $a = 1, b = 2, y_{20} = -0.476984656$
c) Solutions of (15.4a) for $y(0) = 1$ and $y_{20} = 0, 1, 2, \ldots, 9$
d) The two solutions of (15.4a), $y(0) = 1$, $y(1) = 0.5$, $y_{20} = 7.93719$, $y_{20} = 0.97084$

The root of $F(y_{20}) = 0$ can be computed by an iterative method, (bisection, regula falsi, $\ldots$; if the derivative of $y_1(1)$ with respect to y_{20} is used from Theorem 14.3 or numerically from finite differences, also by Newton's method). The initial value problem is then computed several times. Small problems, such as the above example, can be done by a simple dialogue with the computer. Harder problems with more unknown initial values need more programmational sophistication. This method is one of the most commonly used and is called *the shooting method.*

Example 2. For the differential equation

$$y'' = -\exp(y) \qquad \text{or} \qquad y_1' = y_2, \quad y_2' = -\exp(y_1) \qquad (15.4a)$$

with the *boundary conditions*

$$y_1(0) = a, \qquad y_1(1) = b \qquad (15.4b)$$

the monotonicity of $F(y_{20})$ is lost and things become more complicated: solutions for different initial values y_{20} are sketched for $a = 1$ in Fig. 15.1c. It can be seen that for b above a certain value (which is 1.499719998) there exists *no* solution of the problem at all, and for b below this value there exist *two* solutions (Fig. 15.1d).

Example 3.

$$y_1' = y_2, \qquad y_2' = y_1^3, \qquad y_1(0) = 1, \qquad y_1(100) = 2. \qquad (15.5)$$

This equation is similar to (15.2) and the same statement of existence and uniqueness holds as above. However, if one tries to compute the solutions by the shooting method, one gets into trouble because of the length of the interval: *the solution nearly never exists on the whole interval;* in fact, the correct solution is $y_{20} = -0.70710616655$. But already for $y_{20} = -0.7071061$, $y_1(x)$ tends to $+\infty$ for $x \to 98.2$. On the other side, for $y_{20} = -0.70711$, we have $y_1(94.1) = -\infty$. So the domain where $F(y_{20})$ of (15.3) *exists* is of length less than 4×10^{-6}.

In a case like this, one can use the *multiple shooting technique:* the interval is split up into several subintervals, on each of which the problem is solved with well-chosen initial values. At the endpoints of the subintervals, the solutions are then matched together. Equation (15.3) thereby becomes a system of higher dimension to be solved. Another possibility is to apply *global methods* (finite differences, collocation). Instead of integrating a sequence of initial value problems, a global representation of the approximate solution is sought. There exists an extensive literature on methods for boundary value problems. As a general reference we give Ascher, Mattheij & Russel (1988) and Deuflhard (1980).

Sturm-Liouville Eigenvalue Problems

This subject originated with a remarkable paper of Sturm (Sturm 1836) in Liouville's newly founded Journal. This paper was followed by a series of papers by Liouville and Sturm published in the following volumes. It is today considered as the starting point of the "geometric theory", where the main effort is not to try to integrate the equation, but merely to obtain geometric properties of the solution, such as its form, oscillations, sign changes, zeros, existence of maxima or minima and so on, *directly from the differential equation* ("Or on peut arriver à ce but par la seule considération des équations différentielles en elles-mêmes, sans qu'on ait besoin de leur intégration.")

The physical origin was, as in Section I.6, the study of heat and small oscillations of elastic media. Let us consider the heat equation with non-constant conductivity

$$\frac{\partial u}{\partial t} = \frac{\partial}{\partial x}\left(k(x)\frac{\partial u}{\partial x}\right) - \ell(x)u, \qquad k(x) > 0, \tag{15.6}$$

which was studied extensively in Poisson's "Théorie de la chaleur". Poisson (1835) assumes $u(x,t) = y(x)e^{-\lambda t}$, so that (15.6) becomes

$$\frac{d}{dx}\left(k(x)\frac{dy}{dx}\right) - \ell(x)y = -\lambda y. \tag{15.7}$$

We write (15.7) in the form

$$(k(x)y')' + G(x)y = 0 \tag{15.8}$$

and state the following comparison theorem of Sturm:

Theorem 15.1. *Consider, with (15.8), the differential equation*

$$(\widehat{k}(x)\widehat{y}')' + \widehat{G}(x)\widehat{y} = 0, \tag{15.9}$$

and assume k, $\widehat{k}$ *differentiable*, G, $\widehat{G}$ *continuous*,

$$0 < \widehat{k}(x) \le k(x), \qquad \widehat{G}(x) \ge G(x) \tag{15.10}$$

for all x *and let* $y(x)$, $\widehat{y}(x)$ *be linearly independent solutions of (15.8) and (15.9), respectively. Then, between any two zeros of* $y(x)$ *there is at least one zero of* $\widehat{y}(x)$, *i.e., if* $y(x_1) = y(x_2) = 0$ *with* $x_1 < x_2$ *then there exists* x_3 *in the open interval* (x_1, x_2) *such that* $\widehat{y}(x_3) = 0$.

Proof. The original proof of Sturm is based on the quotient

$$q(x) = \frac{y(x)}{k(x)y'(x)}$$

which is the slope of the line connecting the origin with the solution point in the (ky', y)-plane and satisfies a first-order differential equation. In order to avoid the singularities caused by the zeros of $y'(x)$, we prefer the use of polar coordinates (Prüfer 1926)

$$k(x)y'(x) = \varrho(x)\cos\varphi(x), \qquad y(x) = \varrho(x)\sin\varphi(x). \tag{15.11}$$

Differentiation of (15.11) yields the following differential equations for φ and ϱ:

$$\varphi' = \frac{1}{k(x)}\cos^2\varphi + G(x)\sin^2\varphi \tag{15.12}$$

$$\varrho' = \left(\frac{1}{k(x)} - G(x)\right) \cdot \sin\varphi \cdot \cos\varphi \cdot \varrho. \tag{15.13}$$

In the same way we also introduce functions $\widehat{\varrho}(x)$ and $\widehat{\varphi}(x)$ for the second differential equation (15.9). They satisfy analogous relations with $k(x)$ and $G(x)$ replaced by $\widehat{k}(x)$ and $\widehat{G}(x)$.

Suppose now that x_1, x_2 are two consecutive zeros of $y(x)$. Then $\varphi(x_1)$ and $\varphi(x_2)$ must be multiples of π, since $\varrho(x)$ is always different from zero (uniqueness of the initial value problem). By (15.12) $\varphi'(x)$ is positive at x_1 and at x_2. Therefore we may assume that

$$\varphi(x_1) = 0, \qquad \varphi(x_2) = \pi, \qquad \widehat{\varphi}(x_1) \in [0, \pi). \tag{15.14}$$

The fact that equation (15.12) is first-order and the inequalities (15.10) allow the application of Theorem 10.3 to give

$$\widehat{\varphi}(x) \ge \varphi(x) \qquad \text{for} \qquad x_1 \le x \le x_2.$$

It is impossible that $\widehat{\varphi}(x) = \varphi(x)$ everywhere, since this would imply $\widehat{G}(x) = G(x)$, $\cos\widehat{\varphi}(x)/\widehat{k}(x) = \cos\varphi(x)/k(x)$ by (15.12) and (15.10). As a consequence of (15.13) we would have $\widehat{\varrho}(x) = C \cdot \varrho(x)$ and the solutions $y(x)$, $\widehat{y}(x)$ would be linearly dependent. Therefore, there exists $x_0 \in (x_1, x_2)$ such that $\widehat{\varphi}(x_0) > \varphi(x_0)$. In this situation $\widehat{\varphi}(x) > \varphi(x)$ for all $x \geq x_0$ and the existence of $x_3 \in (x_1, x_2)$ with $\widehat{\varphi}(x_3) = \pi$ is assured. □

The next theorem shows that our eigenvalue problem possesses an *infinity* of solutions. We add to (15.7) the boundary conditions

$$y(x_0) = y(x_1) = 0. \tag{15.15}$$

Theorem 15.2. *The eigenvalue problem (15.7), (15.15) possesses an infinite sequence of eigenvalues $\lambda_1 < \lambda_2 < \lambda_3 < \ldots$ whose corresponding solutions $y_i(x)$ ("eigenfunctions") possess respectively $0, 1, 2, \ldots$ zeros in the interval (x_0, x_1). The zeros of $y_{j+1}(x)$ separate those of $y_j(x)$. If $0 < K_1 \leq k(x) \leq K_2$ and $L_1 \leq \ell(x) \leq L_2$, then*

$$L_1 + K_1 \frac{j^2 \pi^2}{(x_1 - x_0)^2} \leq \lambda_j \leq L_2 + K_2 \frac{j^2 \pi^2}{(x_1 - x_0)^2}. \tag{15.16}$$

Proof. Let $y(x, \lambda)$ be the solution of (15.7) with initial values $y(x_0) = 0$, $y'(x_0) = 1$. Theorem 15.1 (with $\widehat{k}(x) = k(x)$, $\widehat{G}(x) = G(x) + \Delta\lambda$) implies that for increasing λ the zeros of $y(x, \lambda)$ move towards x_0, so that the number of zeros in (x_0, x_1) is a non-decreasing function of λ.

Comparing next (15.7) with the solution ($\lambda > L_1$)

$$\sin\left(\sqrt{(\lambda - L_1)/K_1} \cdot (x - x_0)\right)$$

of $K_1 y'' + (\lambda - L_1)y = 0$ we see that for $\lambda < L_1 + K_1 j^2 \pi^2/(x_1 - x_0)^2$, $y(x, \lambda)$ has at most $j - 1$ zeros in $(x_0, x_1]$. Similarly, a comparison with

$$\sin\left(\sqrt{(\lambda - L_2)/K_2} \cdot (x - x_0)\right)$$

which is a solution of $K_2 y'' + (\lambda - L_2)y = 0$, shows that $y(x, \lambda)$ possesses at least j zeros in (x_0, x_1), if $\lambda > L_2 + K_2 j^2 \pi^2/(x_1 - x_0)^2$. The statements of the theorem are now simple consequences of these three properties. □

Example. Fig. 15.2 shows the first 5 solutions of the problem

$$((1 - 0.8 \sin^2 x)y')' - (x - \lambda)y = 0, \qquad y(0) = y(\pi) = 0. \tag{15.17}$$

The first eigenvalues are $2.1224, 3.6078, 6.0016, 9.3773, 13.7298, 19.053, 25.347, 32.609, 40.841, 50.041,$ etc.

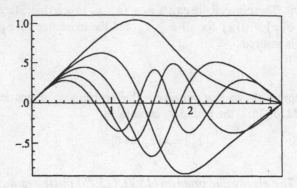

Fig. 15.2. Solutions of the Sturm-Liouville eigenvalue problem (15.17)

For more details about this theory, which is a very important page of history, we refer to the book of Reid (1980).

Exercises

1. Consider the equation

$$L(x)y'' + M(x)y' + N(x)y = 0.$$

Multiply it with a suitable function $\varphi(x)$, so that the ensuing equation is of the form (15.8) (Sturm 1836, p. 108).

2. Prove that two solutions of (15.7), (15.15) satisfy the orthogonality relations

$$\int_{x_0}^{x_1} y_j(x)y_k(x)dx = 0 \qquad \text{for} \qquad \lambda_j \neq \lambda_k.$$

Hint. Multiply this by λ_j, replace $\lambda_j y_j(x)$ from (15.7) and do partial integration (Liouville 1836, p. 257).

3. Solve the problem (15.5) by elementary functions. Explain why the given value for y_{20} is so close to $-\sqrt{2}/2$.

4. Show that the boundary value problem (see Collatz 1967)

$$y'' = -y^3, \qquad y(0) = 0, \qquad y(A) = B \tag{15.18}$$

possesses infinitely many solutions for each pair (A, B) with $A \neq 0$.

Hint. Draw the solution $y(x)$ of (15.18) with $y(0) = 0$, $y'(0) = 1$. Show that for each constant a, $z(x) = ay(ax)$ is also a solution.

I.16 Periodic Solutions, Limit Cycles, Strange Attractors

The phenomenon of limit cycles was first described theoretically by Poincaré (1882) and Bendixson (1901), and has since then found many applications in Physics, Chemistry and Biology. In higher dimensions things can become much more chaotic and attractors may look fairly "strange".

Van der Pol's Equation

The first practical examples were studied by Rayleigh (1883) and later by Van der Pol (1920-1926) in a series of papers on nonlinear oscillations: the solutions of

$$y'' + \alpha y' + y = 0$$

are *damped* for $\alpha > 0$, and *unstable* for $\alpha < 0$. The idea is to change α (with the help of a triode, for example) so that $\alpha < 0$ for small y and $\alpha > 0$ for large y. The simplest expression, which describes the physical situation in a somewhat idealized form, would be $\alpha = \varepsilon(y^2 - 1)$, $\varepsilon > 0$. Then the above equation becomes

$$y'' + \varepsilon(y^2 - 1)y' + y = 0, \tag{16.1}$$

or, written as a system,

$$\begin{aligned} y_1' &= y_2 \\ y_2' &= \varepsilon(1 - y_1^2)y_2 - y_1, \quad \varepsilon > 0. \end{aligned} \tag{16.2}$$

In this equation, small oscillations are amplified and large oscillations are damped. We therefore expect the existence of a stable periodic solution to which all other solutions converge. We call this a *limit cycle* (Poincaré 1882, "Chap. VI"). The original illustrations of the paper of Van der Pol are reproduced in Fig. 16.1.

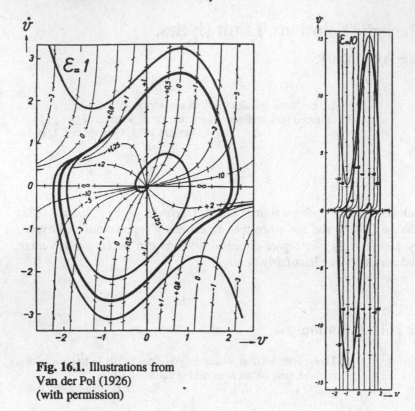

Fig. 16.1. Illustrations from
Van der Pol (1926)
(with permission)

Existence proof. The existence of limit cycles is studied by the method of *Poincaré sections* (Poincaré 1882, "Chap. V, Théorie des conséquents"). The idea is to cut the solutions transversally by a hyperplane Π and, for an initial value $y_0 \in \Pi$, to study the first point $\Phi(y_0)$ where the solution again crosses the plane Π in the same direction.

For our example (16.2), we choose for Π the half-line $y_2 = 0$, $y_1 > 0$. We then examine the signs of y_1' and y_2' in (16.2). The sign of y_2' changes at the curve

$$y_2 = \frac{y_1}{\epsilon(1 - y_1^2)}, \tag{16.3}$$

which is drawn as a broken line in Fig. 16.2. It follows (see Fig. 16.2) that $\Phi(y_0)$ exists for all $y_0 \in \Pi$. Since two different solutions *cannot intersect* (due to uniqueness), the map Φ is *monotone*. Further, Φ is bounded (e.g., by every solution starting on the curve (16.3)), so $\Phi(y_0) < y_0$ for y_0 large. Finally, since the origin is unstable, $\Phi(y_0) > y_0$ for y_0 small. Hence there must be a fixed point of $\Phi(y_0)$, i.e., a limit cycle.   □

The limit cycle is, in fact, *unique*. The proof for this is more complicated and is indicated in Exercise 8 below (Liénard 1928).

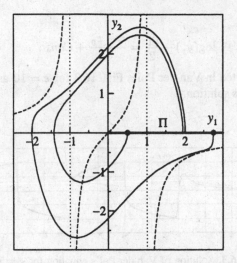

Fig. 16.2. The Poincaré map for Van der Pol's equation, $\varepsilon = 1$

With similar ideas one proves the following general result:

Theorem 16.1 (Poincaré 1882, Bendixson 1901). *Each bounded solution of a two-dimensional system*

$$y_1' = f_1(y_1, y_2), \qquad y_2' = f_2(y_1, y_2) \tag{16.4}$$

must

  i) *tend to a critical point $f_1 = f_2 = 0$ for an infinity of points $x_i \to \infty$; or*

 ii) *be periodic; or*

iii) *tend to a limit cycle.* □

Remark. Exercise 1 below explains why the possibility (i) is written in a form somewhat more complicated than seems necessary.

Steady-state approximations for ε large. An important tool for simplifying complicated nonlinear systems is that of steady-state approximations. Consider (16.2) with ε very large. Then, in the neighbourhood of $f_2(y_1, y_2) = 0$ for $|y_1| > 1$, the derivative of $y_2' = f_2$ with respect to y_2 is very large negative. Therefore the solution will very rapidly approach an equilibrium state in the neighbourhood of $y_2' = f_2(y_1, y_2) = 0$, i.e., in our example, $y_2 = y_1/(\varepsilon(1 - y_1^2))$. This can be inserted into (16.2) and leads to

$$y_1' = \frac{y_1}{\varepsilon(1 - y_1^2)}, \tag{16.5}$$

an equation of lower dimension. Using the formulas of Section I.3, (16.5) is easily solved to give

$$\log(y_1) - \frac{y_1^2}{2} = \frac{x - x_0}{\varepsilon} + Const.$$

These curves are dotted in Van der Pol's Fig. 16.3 for $\varepsilon = 10$ and show the good approximation of this solution.

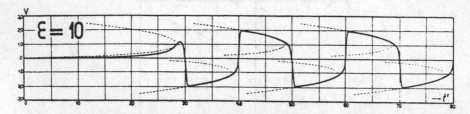

Fig. 16.3. Solution of Van der Pol's equation for $\varepsilon = 10$ compared with steady state approximations

Asymptotic solutions for ε small. The computation of periodic solutions for *small* parameters was initiated by astronomers such as Newcomb and Lindstedt and brought to perfection by Poincaré (1893). We demonstrate the method for the Van der Pol equation (16.1). The idea is to develop the solution as a series in powers of ε. Since the period will change too, we also introduce a coordinate change

$$t = x(1 + \gamma_1\varepsilon + \gamma_2\varepsilon^2 + \ldots) \tag{16.6}$$

and put

$$y(x) = z(t) = z_0(t) + \varepsilon z_1(t) + \varepsilon^2 z_2(t) + \ldots. \tag{16.7}$$

Inserting now $y'(x) = z'(t)(1 + \gamma_1\varepsilon + \ldots)$, $y''(x) = z''(t)(1 + \gamma_1\varepsilon + \ldots)^2$ into (16.1) we obtain

$$\begin{aligned}
&\left(z_0'' + \varepsilon z_1'' + \varepsilon^2 z_2'' + \ldots\right)\left(1 + 2\gamma_1\varepsilon + (2\gamma_2 + \gamma_1^2)\varepsilon^2 + \ldots\right) \\
&+ \varepsilon\left((z_0 + \varepsilon z_1 + \ldots)^2 - 1\right)(z_0' + \varepsilon z_1' + \ldots)(1 + \gamma_1\varepsilon + \ldots) \\
&+ (z_0 + \varepsilon z_1 + \varepsilon^2 z_2 + \ldots) = 0.
\end{aligned} \tag{16.8}$$

We first compare the coefficients of ε^0 and obtain

$$z_0'' + z_0 = 0. \tag{16.8;0}$$

We fix the initial value on the Poincaré section P, i.e., $z'(0) = 0$, so that $z_0 = A\cos t$ with A, for the moment, a free parameter. Next, the coefficients of ε yield

$$\begin{aligned}
z_1'' + z_1 &= -2\gamma_1 z_0'' - (z_0^2 - 1)z_0' \\
&= 2\gamma_1 A\cos t + \left(\frac{A^3}{4} - A\right)\sin t + \frac{A^3}{4}\sin 3t.
\end{aligned} \tag{16.8;1}$$

Here, the crucial idea is that we are looking for *periodic* solutions, hence the terms in $\cos t$ and $\sin t$ on the right-hand side of (16.8;1) must disappear, in order to avoid that $z_1(t)$ contain terms of the form $t \cdot \cos t$ and $t \cdot \sin t$ ("... et de faire disparaître ainsi les termes dits *séculaires* ..."). We thus obtain $\gamma_1 = 0$ and $A = 2$. Then (16.8;1) can be solved and gives, together with $z_1'(0) = 0$,

$$z_1 = B \cos t + \frac{3}{4} \sin t - \frac{1}{4} \sin 3t. \tag{16.9}$$

The continuation of this process is now clear: the terms in ε^2 in (16.8) lead to, after insertion of (16.9) and simplification,

$$z_2'' + z_2 = \left(4\gamma_2 + \frac{1}{4}\right) \cos t + 2B \sin t + 3B \sin 3t - \frac{3}{2} \cos 3t + \frac{5}{4} \cos 5t. \tag{16.8;2}$$

Secular terms are avoided if we set $B = 0$ and $\gamma_2 = -1/16$. Then

$$z_2 = C \cos t + \frac{3}{16} \cos 3t - \frac{5}{96} \cos 5t.$$

The next round will give $C = -1/8$ and $\gamma_3 = 0$, so that we have: *the periodic orbit of the Van der Pol equation (16.1) for ε small is given by*

$$y(x) = z(t), \qquad t = x(1 - \varepsilon^2/16 + \ldots),$$

$$z(t) = 2 \cos t + \varepsilon \left(\frac{3}{4} \sin t - \frac{1}{4} \sin 3t\right)$$

$$+ \varepsilon^2 \left(-\frac{1}{8} \cos t + \frac{3}{16} \cos 3t - \frac{5}{96} \cos 5t\right) + \ldots \tag{16.10}$$

and is of period $2\pi(1 + \varepsilon^2/16 + \ldots)$.

Chemical Reactions

The laws of chemical kinetics give rise to differential equations which, for multi-molecular reactions, become nonlinear and have interesting properties. Some of them possess periodic solutions (e.g. the Zhabotinski-Belousov reaction) and have important applications to the interpretation of biological phenomena (e.g. Prigogine, Lefever).

Let us examine in detail the model of Lefever and Nicolis (1971), the so-called "Brusselator": suppose that six substances A, B, D, E, X, Y undergo the following reactions:

$$
\begin{aligned}
A &\xrightarrow{k_1} X \\
B + X &\xrightarrow{k_2} Y + D \quad \text{(bimolecular reaction)} \\
2X + Y &\xrightarrow{k_3} 3X \quad \text{(autocatalytic trimol. reaction)} \\
X &\xrightarrow{k_4} E
\end{aligned} \tag{16.11}
$$

If we denote by $A(x), B(x), \ldots$ the *concentrations* of $A, B, \ldots$ as functions of the time x, the reactions (16.11) become by the mass action law the following differential equations

$$A' = -k_1 A$$
$$B' = -k_2 BX$$
$$D' = k_2 BX$$
$$E' = k_4 X$$
$$X' = k_1 A - k_2 BX + k_3 X^2 Y - k_4 X$$
$$Y' = k_2 BX - k_3 X^2 Y.$$

This system is now simplified as follows: the equations for D and E are left out, because they do not influence the others; A and B are supposed to be maintained constant (positive) and all reaction rates k_i are set equal to 1. We further set $y_1(x) := X(x)$, $y_2(x) := Y(x)$ and obtain

$$y_1' = A + y_1^2 y_2 - (B+1)y_1$$
$$y_2' = By_1 - y_1^2 y_2. \tag{16.12}$$

The resulting system has one critical point $y_1' = y_2' = 0$ at $y_1 = A$, $y_2 = B/A$. The linearized equation in the neighbourhood of this point is unstable iff $B > A^2 + 1$. Further, a study of the domains where y_1', y_2', or $(y_1 + y_2)'$ is positive or negative leads to the result that all solutions remain bounded. Thus, for $B > A^2 + 1$ there must be a limit cycle which, by numerical calculations, is seen to be unique (Fig. 16.4).

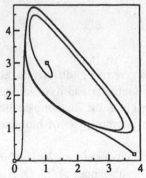

Fig. 16.4. Solutions of the Brusselator, $A = 1$, $B = 3$

An interesting phenomenon (Hopf bifurcation, see below) occurs, when B approaches $A^2 + 1$. Then the limit cycle becomes smaller and smaller and finally disappears in the critical point. Another example of this type is given in Exercise 2.

Limit Cycles in Higher Dimensions, Hopf Bifurcation

The Theorem of Poincaré-Bendixson is apparently true only in two dimensions. Higher dimensional counter-examples are given by nearly every mechanical movement without friction, as for example the spherical pendulum (6.20), see Fig. 6.2. Therefore, in higher dimensions limit cycles are usually found by numerical studies of the Poincaré section map Φ defined above.

There is, however, one situation where limit cycles occur quite naturally (Hopf 1942): namely when at a critical point of $y' = f(y, \alpha)$, $y, f \in \mathbb{R}^n$, all eigenvalues of $(\partial f / \partial y)(y_0, \alpha)$ have strictly negative real part with the exception of *one* pair which, by varying α, crosses the imaginary axis. The eigenspace of the stable eigenvalues then continues into an analytic two dimensional manifold, inside which a limit cycle appears. This phenomenon is called "Hopf bifurcation". The proof of this fact is similar to Poincaré's parameter expansion method (16.7) (see Exercises 6 and 7 below), so that Hopf even hesitated to publish it ("... ich glaube kaum, dass an dem obigen Satz etwas wesentlich Neues ist ...").

As an example, we consider the "full Brusselator" (16.11): we no longer suppose that B is kept constant, but that B is constantly added to the mixture with

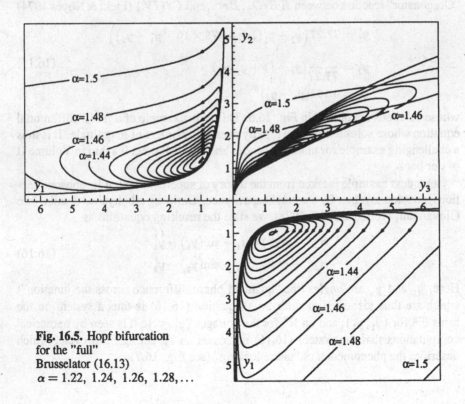

Fig. 16.5. Hopf bifurcation for the "full" Brusselator (16.13) $\alpha = 1.22, 1.24, 1.26, 1.28, \ldots$

rate α. When we set $y_3(x) := B(x)$, we obtain instead of (16.12) (with $A = 1$)

$$y_1' = 1 + y_1^2 y_2 - (y_3 + 1)y_1$$
$$y_2' = y_1 y_3 - y_1^2 y_2 \tag{16.13}$$
$$y_3' = -y_1 y_3 + \alpha.$$

This system possesses a critical point at $y_1 = 1$, $y_2 = y_3 = \alpha$ with derivative

$$\frac{\partial f}{\partial y} = \begin{pmatrix} \alpha - 1 & 1 & -1 \\ -\alpha & -1 & 1 \\ -\alpha & 0 & -1 \end{pmatrix}. \tag{16.14}$$

This matrix has $\lambda^3 + (3 - \alpha)\lambda^2 + (3 - 2\alpha)\lambda + 1$ as characteristic polynomial and satisfies the condition for stability iff $\alpha < (9 - \sqrt{17})/4 = 1.21922$ (see I.13, Exercise 1). Thus when α increases beyond this value, there arises a limit cycle which exists for all values of α up to approximately 1.5 (see Fig. 16.5). When α continues to grow, the limit cycle "explodes" and $y_1 \to 0$ while y_2 and $y_3 \to \infty$. So the system (16.13) has a behaviour completely different from the simplified model (16.12).

A famous chemical reaction with a limit cycle in three dimensions is the "Oregonator" reaction between $HBrO_2, Br^-$, and $Ce\,(IV)$ (Field & Noyes 1974)

$$y_1' = 77.27\left(y_2 + y_1(1 - 8.375 \times 10^{-6}y_1 - y_2)\right)$$
$$y_2' = \frac{1}{77.27}(y_3 - (1 + y_1)y_2) \tag{16.15}$$
$$y_3' = 0.161(y_1 - y_3)$$

whose solutions are plotted in Fig. 16.6. This is an example of a "stiff" differential equation whose solutions change rapidly over many orders of magnitude. It is thus a challenging example for numerical codes and we shall meet it again in Volume II of our book.

Our next example is taken from the theory of superconducting Josephson junctions, coupled together by a mutual capacitance. Omitting all physical details, (see Giovannini, Weiss & Ulrich 1978), we state the resulting equations as

$$c(y_1'' - \alpha y_2'') = i_1 - \sin(y_1) - y_1'$$
$$c(y_2'' - \alpha y_1'') = i_2 - \sin(y_2) - y_2'. \tag{16.16}$$

Here, y_1 and y_2 are *angles* (the "quantum phase difference across the junction") which are thus identified modulo 2π. Equation (16.16) is thus a system on the torus T^2 for (y_1, y_2), and on $\mathbf{R}^2$ for the voltages (y_1', y_2'). It is seen by numerical computations that the system (16.16) possesses an attracting limit cycle, which describes the phenomenon of "phase locking" (see Fig. 16.7).

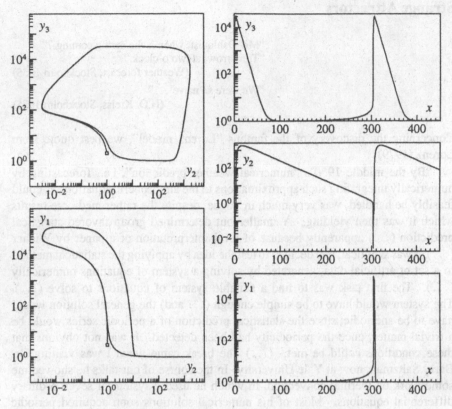

Fig. 16.6. Limit cycle of the Oregonator

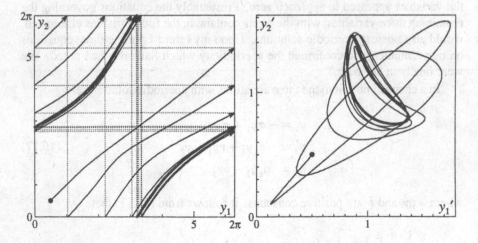

Fig. 16.7. Josephson junctions (16.16) for $c = 2$, $\alpha = 0.5$, $i_1 = 1.11$, $i_2 = 1.08$

Strange Attractors

> "Mr. Dahlquist, when is the spring coming ?"
> "Tomorrow, at two o'clock."
> (Weather forecast, Stockholm 1955)
> "We were so naïve ..."
> (H.O. Kreiss, Stockholm 1985)

Concerning the discovery of the famous "Lorenz model", we best quote from Lorenz (1979):

"By the middle 1950's "numerical weather prediction", i.e., forecasting by numerically integrating such approximations to the atmospheric equations as could feasibly be handled, was very much in vogue, despite the rather mediocre results which it was then yielding. A smaller but determined group favored statistical prediction (...) apparently because of a misinterpretation of a paper by Wiener (...). I was skeptical, and decided to test the idea by applying the statistical method to a set of artificial data, generated by solving a system of equations numerically (...). The first task was to find a suitable system of equations to solve (...). The system would have to be simple enough (... and) the general solution would have to be aperiodic, since the statistical prediction of a periodic series would be a trivial matter, once the periodicity had been detected. It was not obvious that these conditions could be met. (...) The break came when I was visiting Dr. Barry Saltzman, now at Yale University. In the course of our talks he showed me some work on thermal convection, in which he used a system of seven ordinary differential equations. Most of his numerical solutions soon acquired periodic behavior, but one solution refused to settle down. Moreover, in this solution four of the variables appeared to approach zero. Presumably the equations governing the remaining three variables, with the terms containing the four variables eliminated, would also possess aperiodic solutions. Upon my return I put the three equations on our computer, and confirmed the aperiodicity which Saltzman had noted. We were finally in business."

In a changed notation, the three equations with aperiodic solutions are

$$
\begin{aligned}
y_1' &= -\sigma y_1 + \sigma y_2 \\
y_2' &= -y_1 y_3 + r y_1 - y_2 \\
y_3' &= y_1 y_2 - b y_3
\end{aligned}
\tag{16.17}
$$

where σ, r and b are positive constants. It follows from (16.17) that

$$
\begin{aligned}
\frac{1}{2}\frac{d}{dx}\left(y_1^2 + y_2^2 + (y_3 - \sigma - r)^2\right) \\
= -\left(\sigma y_1^2 + y_2^2 + b(y_3 - \frac{\sigma}{2} - \frac{r}{2})^2\right) + b\left(\frac{\sigma}{2} + \frac{r}{2}\right)^2.
\end{aligned}
\tag{16.18}
$$

Therefore the ball

$$R_0 = \left\{ (y_1, y_2, y_3) \mid y_1^2 + y_2^2 + (y_3 - \sigma - r)^2 \le c^2 \right\} \tag{16.19}$$

is mapped by the flow φ_1 (see (14.22)) into itself, provided that c is sufficiently large so that R_0 wholly contains the ellipsoid defined by equating the right side of (16.18) to zero. Hence, if x assumes the increasing values $1, 2, 3, \ldots$, R_0 is carried into regions $R_1 = \varphi_1(R_0)$, $R_2 = \varphi_2(R_0)$ etc., which satisfy $R_0 \supset R_1 \supset R_2 \supset R_3 \supset \cdots$ (applying φ_1 to the inclusion $R_0 \supset R_1$ gives $R_1 \supset R_2$ and so on).

Since the trace of $\partial f / \partial y$ for the system (16.17) is the negative constant $-(\sigma + b + 1)$, the *volumes* of R_k tend exponentially to zero (see Theorem 14.8). Every orbit is thus ultimately trapped in a set $R_\infty = R_0 \cap R_1 \cap R_2 \ldots$ of zero volume.

System (16.17) possesses an obvious critical point $y_1 = y_2 = y_3 = 0$; this becomes unstable when $r > 1$. In this case there are two additional critical points C and C' respectively given by

$$y_1 = y_2 = \pm\sqrt{b(r-1)}, \qquad y_3 = r - 1. \tag{16.20}$$

These become unstable (e.g. by the Routh criterion, Exercise 1 of Section I.13) when $\sigma > b + 1$ and

$$r \ge r_c = \frac{\sigma(\sigma + b + 3)}{\sigma - b - 1}. \tag{16.21}$$

In the first example we shall use Saltzman's values $b = 8/3$, $\sigma = 10$, and $r = 28$. ("Here we note another lucky break: Saltzman used $\sigma = 10$ as a crude approximation to the Prandtl number (about 6) for water. Had he chosen to study air, he would probably have let $\sigma = 1$, and the aperiodicity would not have been discovered", Lorenz 1979). In Fig. 16.8 we have plotted the solution curve of (16.17) with the initial value $y_1 = -8$, $y_2 = 8$, $y_3 = r - 1$, which, indeed, looks pretty chaotic.

For a clearer understanding of the phenomenon, we choose the plane $y_3 = r - 1$, especially the square region between the critical points C and C', as Poincaré section Π. The critical point $y_1 = y_2 = y_3 = 0$ possesses (since $r > 1$) one unstable eigenvalue $\lambda_1 = (-1 - \sigma + \sqrt{(1 - \sigma)^2 + 4r\sigma})/2$ and two stable eigenvalues $\lambda_2 = -b$, $\lambda_3 = (-1 - \sigma - \sqrt{(1 - \sigma)^2 + 4r\sigma})/2$. The eigenspace of the stable eigenvalues continues into a two-dimensional manifold of initial values, whose solutions tend to 0 for $x \to \infty$. This "stable manifold" cuts Π in a curve Σ (see Fig. 16.9). The one-dimensional *unstable* manifold (created by the unstable eigenvalue λ_1) cuts Π in the points D and D' (Fig. 16.9).

All solutions starting in Π_u *above* Σ (the dark cat) surround the above critical point C and are, at the first return, mapped to a narrow stripe S_u, while the solutions starting in Π_d *below* Σ surround C' and go to the left stripe S_d. At the *second* return, the two stripes are mapped into two very narrow stripes *inside* S_u and S_d. After the third return, we have 8 stripes closer and closer together, and so on. The

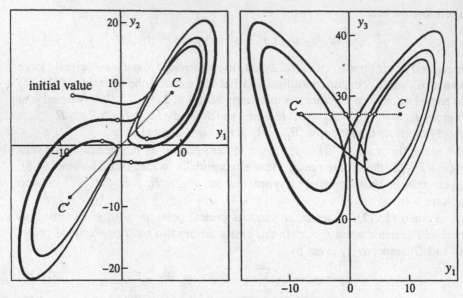

Fig. 16.8. Two views of a solution of (16.17)
(small circles indicate intersection of solution with plane $y_3 = r - 1$)

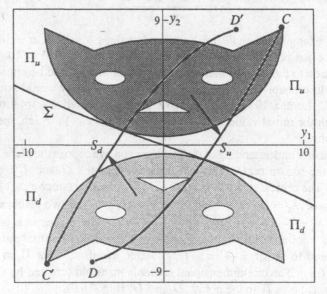

Fig. 16.9. Poincaré map for (16.17)

intersection of all these stripes is a Cantor-like set and, continued into 3-space by the flow, forms the *strange attractor* ("An attractor of the type just described can therefore not be thrown away as non-generic pathology", Ruelle & Takens 1971).

The Ups and Downs of the Lorenz Model

> "Mr. Laurel and Mr. Hardy have many ups and downs — Mr. Hardy takes charge of the upping, and Mr. Laurel does most of the downing — "
> (from "Another Fine Mess", Hal Roach 1930)

If one watches the solution $y_1(x)$ of the Lorenz equation being calculated, one wonders who decides for the solution to go up or down in an apparently unpredictable fashion. Fig. 16.9 shows that Σ cuts both stripes S_d and S_u. Therefore the *inverse image* of Σ (see Fig. 16.10) consists of *two* lines Σ_0 and Σ_1 which cut, together with Σ, the plane Π into *four* sets Π_{uu}, Π_{ud}, Π_{du}, Π_{dd}. If the initial value is in one of these, the corresponding solution goes up-up, up-down, down-up, down-down. Further, the inverse images of Σ_0 and Σ_1 lead to four lines Σ_{00}, Σ_{01}, Σ_{10}, Σ_{11}. The plane Π is then cut into 8 stripes and we now know the fate of the first three ups and downs. The more inverse images of these curves we compute, the finer the plane Π is cut into stripes and all the future ups and downs are coded in the position of the initial value with respect to these stripes (see Fig. 16.10). It appears that a *very small* change in the initial value gives rise, after a couple of rotations, to a *totally different* solution curve. This phenomenon, discovered merely by accident by Lorenz (see Lorenz 1979), is highly interesting

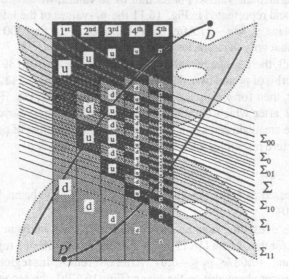

Fig. 16.10. Stripes deciding for the ups and downs

and explains why the theorem of uniqueness (Theorem 7.4), of whose philosophical consequences Laplace was so proud, has its practical limits.

Remark. It appears in Fig. 16.10 that not all stripes have the same width. The sequences of "u"'s and "d"'s which repeat u or d a couple of times (but not too often) are more probable than the others. More than 25 consecutive "ups" or "downs" are (for the chosen constants and except for the initial phase) never possible. This has to do with the position of D and D', the outermost frontiers of the attractor, in the stripes of Fig. 16.10.

Feigenbaum Cascades

However nicely the beginning of Lorenz' (1979) paper is written, the affirmations of his last section are only partly true. As Lorenz did, we now vary the parameter b in (16.17), letting at the same time $r = r_c$ (see (16.21)) and

$$\sigma = b + 1 + \sqrt{2(b+1)(b+2)}. \tag{16.22}$$

This is the value of σ for which r_c is minimized. Numerical integration shows that for b very small (say $b \leq 0.139$), the solutions of (16.17) evidently converge to a stable limit cycle, which cuts the Poincaré section $y_3 = r - 1$ twice at two different locations and surrounds both critical points C and C'. Further, for b large (for example $b = 8/3$) the coefficients are not far from those studied above and we have a strange attractor. But what happens in between? We have computed the solutions of the Lorenz model (16.17) for b varying from 0.1385 to 0.1475 with 1530 intermediate values. For each of these values, we have computed 1500 Poincaré cuts and represented in Fig. 16.11 the y_1-values of the intersections with the Poincaré plane $y_3 = r - 1$. After each change of b, the first 300 iterations were not drawn so that only the attractor becomes visible.

For b small, there is one periodic orbit; then, at $b = b_1 = 0.13972$, it suddenly splits into an orbit of period two, this then splits for $b = b_2 = 0.14327$ into an orbit of period four, then for $b = b_3 = 0.14400$ into period eight, etc. There is a point $b_\infty = 0.14422$ after which the movement becomes chaotic. Beyond this value, however, there are again and again intervals of stable attractors of periods 5, 3, etc. The whole picture resembles what is obtained by the recursion

$$x_{n+1} = a(x_n - x_n^2) \tag{16.23}$$

which is discussed in many papers (e.g. May 1976, Feigenbaum 1978, Collet & Eckmann 1980).

But where does this resemblance come from? We study in Fig. 16.12 the Poincaré map for the system (16.17) with b chosen as 0.146 of a region $-0.095 \leq y_1 \leq -0.078$ and $-0.087 \leq y_2 \leq -0.07$. After one return, this region is compressed to a thin line somewhere else on the plane (Fig. 16.12b), the second return bends this line to U-shape and maps it into the original region (Fig. 16.12c).

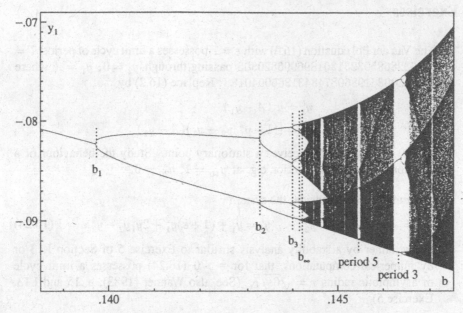

Fig. 16.11. Poincaré cuts y_1 for (16.17) as function of b

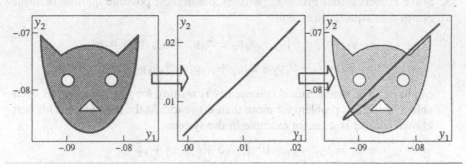

Fig. 16.12. Poincaré map for system (16.17) with $b = 0.146$

Therefore, the Poincaré map is essentially a map of the interval $[0, 1]$ to itself similar to (16.23). It is a great discovery of Feigenbaum that for *all* maps of a similar shape, the phenomena are always the same, in particular that

$$\lim_{i \to \infty} \frac{b_i - b_{i-1}}{b_{i+1} - b_i} = 4.6692016091029906715\ldots$$

is a universal constant, the *Feigenbaum number*. The repeated doublings of the periods at $b_1, b_2, b_3, \ldots$ are called *Feigenbaum cascades*.

Exercises

1. The Van der Pol equation (16.2) with $\varepsilon = 1$ possesses a limit cycle of period $T = 6.6632868593231301896996820305$ passing through $y_2 = 0$, $y_1 = A$ where $A = 2.00861986087484313650940188$. Replace (16.2) by

$$y_1' = y_2(A - y_1)$$
$$y_2' = ((1 - y_1^2)y_2 - y_1)(A - y_1)$$

so that the limit cycle receives a stationary point. Study the behaviour of a solution starting in the interior, e.g. at $y_{10} = 1$, $y_{20} = 0$.

2. (Frommer 1934). Consider the system

$$y_1' = -y_2 + 2y_1 y_2 - y_2^2, \qquad y_2' = y_1 + (1+\varepsilon)y_1^2 + 2y_1 y_2 - y_2^2. \qquad (16.24)$$

Show, either by a stability analysis similar to Exercise 5 of Section I.13 or by numerical computations, that for $\varepsilon > 0$ (16.24) possesses a limit cycle of asymptotic radius $r = \sqrt{6\varepsilon/7}$. (See also Wanner (1983), p. 15 and I.13, Exercise 5).

3. Solve Hilbert's 16th Problem: what is the highest possible number of limit cycles that a quadratic system

$$y_1' = \alpha_0 + \alpha_1 y_1 + \alpha_2 y_2 + \alpha_3 y_1^2 + \alpha_4 y_1 y_2 + \alpha_5 y_2^2$$
$$y_2' = \beta_0 + \beta_1 y_1 + \beta_2 y_2 + \beta_3 y_1^2 + \beta_4 y_1 y_2 + \beta_5 y_2^2$$

can have? The mathematical community is waiting for *you*: nobody has been able to solve this problem for more than 80 years. At the moment, the highest known number is 4, as for example in the system

$$y_1' = \lambda y_1 - y_2 - 10y_1^2 + (5+\delta)y_1 y_2 + y_2^2$$
$$y_2' = y_1 + y_1^2 + (-25 + 8\varepsilon - 9\delta)y_1 y_2,$$
$$\delta = -10^{-13}, \qquad \varepsilon = -10^{-52}, \qquad \lambda = -10^{-200}$$

(see Shi Songling 1980, Wanner 1983, Perko 1984).

4. Find a change of coordinates such that the equation

$$my'' + (-A + B(y')^2)y' + ky = 0$$

becomes the Van der Pol equation (16.2) (see Kryloff & Bogoliuboff (1947), p. 5).

5. Treat the pendulum equation

$$y'' + \sin y = y'' + y - \frac{y^3}{6} + \frac{y^5}{120} \pm \ldots = 0, \qquad y(0) = \varepsilon, \quad y'(0) = 0,$$

by the method of asymptotic expansions (16.6) and (16.7) and study the period as a function of ε.

Result. The period is $2\pi(1+\varepsilon^2/16+\ldots)$.

6. Compute the limit cycle (Hopf bifurcation) for

$$y'' + y = \varepsilon^2 y' - (y')^3$$

for ε small by the method of Poincaré (16.6), (16.7) with $z'(0) = 0$.

7. Treat in a similar way as in Exercise 6 the Brusselator (16.12) with $A = 1$ and $B = 2+\varepsilon^2$.

Hint. With the new variable $y = y_1 + y_2 - 3$ the differential equation (16.12) becomes equivalent to $y' = 1 - y_1$ and

$$y'' + y = -\varepsilon^2(y' - 1) - (y')^2(y + y') + 2yy'.$$

Result. $z(t) = \varepsilon(2/\sqrt{3})\cos t + \ldots$, $t = x(1 - \varepsilon^2/18 + \ldots)$, so that the period is asymptotically $2\pi(1 + \varepsilon^2/18 + \ldots)$.

8. (Liénard 1928). Prove that the limit cycle of the Van der Pol equation (16.1) *is unique* for every $\varepsilon > 0$.

Hint. The identity

$$y'' + \varepsilon(y^2 - 1)y' = \frac{d}{dx}\left(y' + \varepsilon\left(\frac{y^3}{3} - y\right)\right)$$

suggests the use of the coordinate system $y_1(x) = y(x)$, $y_2(x) = y' + \varepsilon(y^3/3 - y)$. Write the resulting first order system, study the signs of y_1', y_2' and the increase of the "energy" function $V(x) = (y_1^2 + y_2^2)/2$.

Also generalize the result to equations of the form $y'' + f(y)y' + g(y) = 0$. For more details see e.g. Simmons (1972), p. 349.

9. (Rayleigh 1883). Compute the periodic solution of

$$y'' + \kappa y' + \lambda(y')^3 + n^2 y = 0$$

for κ and λ small.

Result. $y = A\sin(nx) + (\lambda n A^3/32)\cos(3nx) + \ldots$ where A is given by $\kappa + (3/4)\lambda n^2 A^2 = 0$.

10. (Bendixson 1901). If in a certain region Ω of the plane the expression

$$\frac{\partial f_1}{\partial y_1} + \frac{\partial f_2}{\partial y_2}$$

is always negative or always positive, then the system (16.4) cannot have closed solutions in Ω.

Hint. Apply Green's formula

$$\int\int \left(\frac{\partial f_1}{\partial y_1} + \frac{\partial f_2}{\partial y_2}\right) dy_1 dy_2 = \int \left(f_1 \, dy_2 - f_2 \, dy_1\right).$$

Chapter II. Runge-Kutta and Extrapolation Methods

Numerical methods for ordinary differential equations fall naturally into two classes: those which use *one* starting value at each step ("one-step methods") and those which are based on *several* values of the solution ("multistep methods" or "multi-value methods"). The present chapter is devoted to the study of one-step methods, while multistep methods are the subject of Chapter III. Both chapters can, to a large extent, be read independently of each other.

We start with the theory of Runge-Kutta methods: the derivation of order conditions with the help of labelled trees, error estimates, convergence proofs, implementation, methods of higher order, dense output. Section II.7 introduces implicit Runge-Kutta methods. More attention will be drawn to these methods in Volume II on stiff differential equations. Two sections then discuss the elegant idea of *extrapolation* (Richardson, Romberg, etc) and its use in obtaining high order codes. The methods presented are then tested and compared on a series of problems. The potential of parallelism is discussed in a separate section. We then turn our attention to an algebraic theory of the composition of methods. This will be the basis for the study of order properties for many general classes of methods in the following chapter. The chapter ends with special methods for second order differential equations $y'' = f(x, y)$, for Hamiltonian systems (symplectic methods) and for problems with delay.

We illustrate the methods of this chapter with an example from Astronomy, the restricted three body problem. One considers two bodies of masses $1 - \mu$ and μ in circular rotation in a plane and a third body of negligible mass moving around in the same plane. The equations are (see e.g., the classical textbook Szebehely 1967)

$$y_1'' = y_1 + 2y_2' - \mu' \frac{y_1 + \mu}{D_1} - \mu \frac{y_1 - \mu'}{D_2},$$

$$y_2'' = y_2 - 2y_1' - \mu' \frac{y_2}{D_1} - \mu \frac{y_2}{D_2}, \tag{0.1}$$

$$D_1 = ((y_1 + \mu)^2 + y_2^2)^{3/2}, \qquad D_2 = ((y_1 - \mu')^2 + y_2^2)^{3/2},$$

$$\mu = 0.012277471, \qquad \mu' = 1 - \mu.$$

There exist initial values, for example

$$y_1(0) = 0.994, \qquad y_1'(0) = 0, \qquad y_2(0) = 0,$$
$$y_2'(0) = -2.00158510637908252240537862224,$$
$$x_{end} = 17.0652165601579625588917206249,$$

(0.2)

such that the solution is periodic with period x_{end}. Such periodic solutions have fascinated astronomers and mathematicians for many decades (Poincaré; extensive numerical calculations are due to Sir George Darwin (1898)) and are now often called "Arenstorf orbits" (see Arenstorf (1963) who did numerical computations "on high speed electronic computers"). The problem is C^∞ with the exception of the two singular points $y_1 = -\mu$ and $y_1 = 1 - \mu$, $y_2 = 0$, therefore the Euler polygons of Section I.7 are known to converge to the solution. But are they really numerically useful here? We have chosen 24000 steps of step length $h = x_{end}/24000$ and plotted the result in Figure 0.1. The result is not very striking.

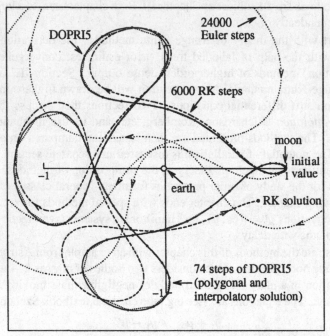

Fig. 0.1. An Arenstorf orbit computed by equidistant Euler, equidistant Runge-Kutta and variable step size Dormand & Prince

The performance of the Runge-Kutta method (left tableau of Table 1.2) is already much better and converges faster to the solution. We have used 6000 steps of step size $x_{end}/6000$, so that the numerical work becomes equivalent. Clearly, most accuracy is lost in those parts of the orbit which are close to a singularity. Therefore, codes with automatic step size selection, described in Section II.4, perform much better and the code DOPRI5 (Table 5.2) computes the orbit with a precision of 10^{-3} in 98 steps (74 accepted and 24 rejected). The step size becomes very large in some regions and the graphical representation as polygons connecting the solution points becomes unsatisfactory. The solid line is the interpolatory solution (Section II.6), which is also precise for all intermediate values and useful for many other questions such as delay differential equations, event location or discontinuities in the differential equation.

For still higher precision one needs methods of higher order. For example, the code DOP853 (Section II.5) computes the orbit faster than DOPRI5 for more stringent tolerances, say smaller than about 10^{-6}. The highest possible order is obtained by extrapolation methods (Section II.9) and the code ODEX (with $K_{max} = 15$) obtains the orbit with a precision of 10^{-30} with about 25000 function evaluations, precisely the same amount of work as for the above Euler solution.

II.1 The First Runge-Kutta Methods

> Die numerische Berechnung irgend einer Lösung einer gegebenen
> Differentialgleichung, deren analytische Lösung man nicht kennt,
> hat, wie es scheint, die Aufmerksamkeit der Mathematiker bisher
> wenig in Anspruch genommen ...
> (C. Runge 1895)

The Euler method for solving the initial value problem

$$y' = f(x, y), \qquad y(x_0) = y_0 \tag{1.1}$$

was described by Euler (1768) in his "Institutiones Calculi Integralis" (Sectio Secunda, Caput VII). The method is easy to understand and to implement. We have studied its convergence extensively in Section I.7 and have seen that the global error behaves like Ch, where C is a constant depending on the problem and h is the maximal step size. If one wants a precision of, say, 6 decimals, one would thus need about a million steps, which is not very satisfactory. On the other hand, one knows since the time of Newton that much more accurate methods can be found, if f in (1.1) is independent of y, i.e., if we have a quadrature problem

$$y' = f(x), \qquad y(x_0) = y_0 \tag{1.1'}$$

with solution

$$y(X) = y_0 + \int_{x_0}^{X} f(x)\,dx. \tag{1.2}$$

As an example consider the midpoint rule (or first Gauss formula)

$$y(x_0 + h_0) \approx y_1 = y_0 + h_0 f\left(x_0 + \frac{h_0}{2}\right)$$

$$y(x_1 + h_1) \approx y_2 = y_1 + h_1 f\left(x_1 + \frac{h_1}{2}\right) \tag{1.3'}$$

$$\cdots$$

$$y(X) \approx Y = y_{n-1} + h_{n-1} f\left(x_{n-1} + \frac{h_{n-1}}{2}\right),$$

where $h_i = x_{i+1} - x_i$ and $x_0, x_1, \ldots, x_{n-1}, x_n = X$ is a subdivision of the integration interval. Its global errror $y(X) - Y$ is known to be bounded by Ch^2. Thus for a desired precision of 6 decimals, a thousand steps will usually do, i.e., the method here is a thousand times faster. Therefore Runge (1895) asked whether it would also be possible to extend method (1.3') to problem (1.1). The first step with $h = h_0$ would read

$$y(x_0 + h) \approx y_0 + hf\left(x_0 + \frac{h}{2}, y\left(x_0 + \frac{h}{2}\right)\right), \tag{1.3}$$

but which value should we take for $y(x_0 + h/2)$? In the absence of something better, it is natural to use one small Euler step with step size $h/2$ and obtain from (1.3) [1]

$$k_1 = f(x_0, y_0)$$

$$k_2 = f\left(x_0 + \frac{h}{2}, y_0 + \frac{h}{2}k_1\right) \tag{1.4}$$

$$y_1 = y_0 + hk_2.$$

One might of course be surprised that we propose an Euler step for the computation of k_2, just half a page after preaching its inefficiency. The crucial point is, however, that k_2 is multiplied by h in the third expression and therefore its error becomes less important. To be more precise, we compute the Taylor expansion of y_1 in (1.4) as a function of h,

$$
\begin{aligned}
y_1 &= y_0 + hf\left(x_0 + \frac{h}{2}, y_0 + \frac{h}{2}f_0\right) \\
&= y_0 + hf(x_0, y_0) + \frac{h^2}{2}\left(f_x + f_y f\right)(x_0, y_0) \\
&\quad + \frac{h^3}{8}\left(f_{xx} + 2f_{xy}f + f_{yy}f^2\right)(x_0, y_0) + \dots.
\end{aligned} \tag{1.5}
$$

This can be compared with the Taylor series of the exact solution, which is obtained from (1.1) by repeated differentiation and replacing y' by f every time it appears (Euler (1768), Problema 86, §656, see also (8.12) of Chap. I)

$$
\begin{aligned}
y(x_0 + h) &= y_0 + hf(x_0, y_0) + \frac{h^2}{2}\left(f_x + f_y f\right)(x_0, y_0) \\
&\quad + \frac{h^3}{6}\left(f_{xx} + 2f_{xy}f + f_{yy}f^2 + f_y f_x + f_y^2 f\right)(x_0, y_0) + \dots.
\end{aligned} \tag{1.6}
$$

Subtracting these two equations, we obtain for the error of the first step

$$
y(x_0 + h) - y_1 = \frac{h^3}{24}\left(f_{xx} + 2f_{xy}f + f_{yy}f^2 + 4(f_y f_x + f_y^2 f)\right)(x_0, y_0) + \dots. \tag{1.7}
$$

When all second partial derivatives of f are bounded, we thus obtain $\|y(x_0 + h) - y_1\| \le Kh^3$.

In order to obtain an approximation of the solution of (1.1) at the endpoint X, we apply formula (1.4) successively to the intervals (x_0, x_1), (x_1, x_2),..., (x_{n-1}, X), very similarly to the application of Euler's method in Section I.7. Again similarly to the convergence proof of Section I.7, it will be shown in Section II.3 that, as in the case (1.1'), the error of the numerical solution is bounded by Ch^2 (h the maximal step size). Method (1.4) is thus an improvement on the Euler method. For high precision computations we need to find still better methods; this will be the main task of what follows.

---

[1] The analogous extension of the *trapezoidal rule* has been given in an early publication by Coriolis in 1837; see Chapter II.4.2 of the thesis of D. Tournès, Paris VII, 1996.

General Formulation of Runge-Kutta Methods

Runge (1895) and Heun (1900) constructed methods by including additional Euler steps in (1.4). It was Kutta (1901) who then formulated the general scheme of what is now called a Runge-Kutta method:

Definition 1.1. Let s be an integer (the "number of stages") and $a_{21}, a_{31}, a_{32}, \ldots, a_{s1}, a_{s2}, \ldots, a_{s,s-1}, b_1, \ldots, b_s, c_2, \ldots, c_s$ be real coefficients. Then the method

$$
\begin{aligned}
k_1 &= f(x_0, y_0) \\
k_2 &= f(x_0 + c_2 h, y_0 + h a_{21} k_1) \\
k_3 &= f(x_0 + c_3 h, y_0 + h(a_{31} k_1 + a_{32} k_2)) \\
&\cdots \\
k_s &= f(x_0 + c_s h, y_0 + h(a_{s1} k_1 + \ldots + a_{s,s-1} k_{s-1})) \\
y_1 &= y_0 + h(b_1 k_1 + \ldots + b_s k_s)
\end{aligned}
\tag{1.8}
$$

is called an s-*stage explicit Runge-Kutta method* (ERK) for (1.1).

Usually, the c_i satisfy the conditions

$$
c_2 = a_{21}, \quad c_3 = a_{31} + a_{32}, \quad \ldots \quad c_s = a_{s1} + \ldots + a_{s,s-1},
\tag{1.9}
$$

or briefly,

$$
c_i = \sum_{j=1}^{i-1} a_{ij}.
\tag{1.9'}
$$

These conditions, already assumed by Kutta, express that all points where f is evaluated are first order approximations to the solution. They greatly simplify the derivation of order conditions for high order methods. For low orders, however, these assumptions are not necessary (see Exercise 6).

Definition 1.2. A Runge-Kutta method (1.8) has *order p* if for sufficiently smooth problems (1.1),

$$
\|y(x_0 + h) - y_1\| \le K h^{p+1},
\tag{1.10}
$$

i.e., if the Taylor series for the exact solution $y(x_0 + h)$ and for y_1 coincide up to (and including) the term h^p.

With the paper of Butcher (1964b) it became customary to symbolize method (1.8) by the tableau (1.8').

$$
\begin{array}{c|ccccc}
0 \\
c_2 & a_{21} \\
c_3 & a_{31} & a_{32} \\
\vdots & \vdots & \vdots & \ddots \\
c_s & a_{s1} & a_{s2} & \cdots & a_{s,s-1} \\
\hline
 & b_1 & b_2 & \cdots & b_{s-1} & b_s
\end{array}
\tag{1.8'}
$$

Examples. The above method of Runge as well as methods of Runge and Heun of order 3 are given in Table 1.1.

Table 1.1. Low order Runge-Kutta methods

$$
\begin{array}{c|cc}
0 \\
1/2 & 1/2 \\
\hline
 & 0 & 1
\end{array}
\qquad
\begin{array}{c|ccc}
0 \\
1/2 & 1/2 \\
1 & 0 & 1 \\
\hline
 & 1/6 & 2/3 & 0 & 1/6
\end{array}
\qquad
\begin{array}{c|ccc}
0 \\
1/3 & 1/3 \\
2/3 & 0 & 2/3 \\
\hline
 & 1/4 & 0 & 3/4
\end{array}
$$

Runge, order 2   Runge, order 3   Heun, order 3

Discussion of Methods of Order 4

> Von den neueren Verfahren halte ich das folgende von Herrn Kutta angegebene für das beste.
> (C. Runge 1905)

Our task is now to determine the coefficients of 4-stage Runge-Kutta methods (1.8) in order that they be of order 4. We have seen above what we must do: compute the derivatives of $y_1 = y_1(h)$ for $h = 0$ and compare them with those of the true solution for orders 1, 2, 3, and 4. In theory, with the known rules of differential calculus, this is a completely trivial task and, by the use of (1.9), results in the following conditions:

$$\sum_i b_i = b_1 + b_2 + b_3 + b_4 = 1 \tag{1.11a}$$

$$\sum_i b_i c_i = b_2 c_2 + b_3 c_3 + b_4 c_4 = 1/2 \tag{1.11b}$$

$$\sum_i b_i c_i^2 = b_2 c_2^2 + b_3 c_3^2 + b_4 c_4^2 = 1/3 \tag{1.11c}$$

$$\sum_{i,j} b_i a_{ij} c_j = b_3 a_{32} c_2 + b_4 (a_{42} c_2 + a_{43} c_3) = 1/6 \tag{1.11d}$$

$$\sum_i b_i c_i^3 = b_2 c_2^3 + b_3 c_3^3 + b_4 c_4^3 = 1/4 \tag{1.11e}$$

$$\sum_{i,j} b_i c_i a_{ij} c_j = b_3 c_3 a_{32} c_2 + b_4 c_4 (a_{42} c_2 + a_{43} c_3) = 1/8 \tag{1.11f}$$

$$\sum_{i,j} b_i a_{ij} c_j^2 = b_3 a_{32} c_2^2 + b_4 \left(a_{42} c_2^2 + a_{43} c_3^2 \right) = 1/12 \qquad (1.11g)$$

$$\sum_{i,j,k} b_i a_{ij} a_{jk} c_k = b_4 a_{43} a_{32} c_2 = 1/24. \qquad (1.11h)$$

These computations, which are not reproduced in Kutta's paper (they are, however, in Heun 1900), are very tedious. And they grow enormously with higher orders. We shall see in Section II.2 that by using an appropriate notation, they can become very elegant.

Kutta gave the general solution of (1.11) without comment. A clear derivation of the solutions is given in Runge & König (1924), p. 291. We shall follow here the ideas of J.C. Butcher, which make clear the role of the so-called *simplifying assumptions*, and will also apply to higher order cases.

Lemma 1.3. *If*

$$\sum_{i=j+1}^{s} b_i a_{ij} = b_j (1 - c_j), \qquad j = 1, \ldots, s, \qquad (1.12)$$

then the equations (d), (g), and (h) in (1.11) follow from the others.

Proof. We demonstrate this for (g):

$$\sum_{i,j} b_i a_{ij} c_j^2 = \sum_j b_j c_j^2 - \sum_j b_j c_j^3 = \frac{1}{3} - \frac{1}{4} = \frac{1}{12}$$

by (c) and (e). Equations (d) and (h) are derived similarly.   □

We shall now show that (1.12) is also *necessary* in our case:

Lemma 1.4. *For $s = 4$, the equations (1.11) and (1.9) imply (1.12).*

The proof of this lemma will be based on the following:

Lemma 1.5. *Let U and V be 3×3 matrices such that*

$$UV = \begin{pmatrix} a & b & 0 \\ c & d & 0 \\ 0 & 0 & 0 \end{pmatrix}, \qquad \det \begin{pmatrix} a & b \\ c & d \end{pmatrix} \neq 0. \qquad (1.13)$$

Then either $V e_3 = 0$ or $U^T e_3 = 0$ where $e_3 = (0, 0, 1)^T$.

Proof of Lemma 1.5. If $\det U \neq 0$, then $UV e_3 = 0$ implies $V e_3 = 0$. If $\det U = 0$, there exists $x = (x_1, x_2, x_3)^T \neq 0$ such that $U^T x = 0$, and therefore $V^T U^T x = 0$. But (1.13) implies that x must be a multiple of e_3.   □

Proof of Lemma 1.4. Define

$$d_j = \sum_i b_i a_{ij} - b_j(1 - c_j) \qquad \text{for} \qquad j = 1, \ldots, 4,$$

so that we have to prove $d_j = 0$. We now introduce the matrices

$$U = \begin{pmatrix} b_2 & b_3 & b_4 \\ b_2 c_2 & b_3 c_3 & b_4 c_4 \\ d_2 & d_3 & d_4 \end{pmatrix}, \qquad V = \begin{pmatrix} c_2 & c_2^2 & \sum_j a_{2j} c_j - c_2^2/2 \\ c_3 & c_3^2 & \sum_j a_{3j} c_j - c_3^2/2 \\ c_4 & c_4^2 & \sum_j a_{4j} c_j - c_4^2/2 \end{pmatrix}. \quad (1.14)$$

Multiplication of these two matrices, using the conditions of (1.11), gives

$$UV = \begin{pmatrix} 1/2 & 1/3 & 0 \\ 1/3 & 1/4 & 0 \\ 0 & 0 & 0 \end{pmatrix} \qquad \text{with} \qquad \det \begin{pmatrix} 1/2 & 1/3 \\ 1/3 & 1/4 \end{pmatrix} \neq 0.$$

Now the last column of V cannot be zero, since $c_1 = 0$ implies

$$\sum_j a_{2j} c_j - c_2^2/2 = -c_2^2/2 \neq 0$$

by condition (h). Thus $d_2 = d_3 = d_4 = 0$ follows from Lemma 1.5. The last identity $d_1 = 0$ follows from $d_1 + d_2 + d_3 + d_4 = 0$, which is a consequence of (1.11a,b) and (1.9). □

From Lemmas 1.3 and 1.4 we obtain

Theorem 1.6. *Under the assumption (1.9) the equations (1.11) are equivalent to*

$$b_1 + b_2 + b_3 + b_4 = 1 \tag{1.15a}$$

$$b_2 c_2 + b_3 c_3 + b_4 c_4 = 1/2 \tag{1.15b}$$

$$b_2 c_2^2 + b_3 c_3^2 + b_4 c_4^2 = 1/3 \tag{1.15c}$$

$$b_2 c_2^3 + b_3 c_3^3 + b_4 c_4^3 = 1/4 \tag{1.15e}$$

$$b_3 c_3 a_{32} c_2 + b_4 c_4 (a_{42} c_2 + a_{43} c_3) = 1/8 \tag{1.15f}$$

$$b_3 a_{32} + b_4 a_{42} = b_2(1 - c_2) \tag{1.15i}$$

$$b_4 a_{43} = b_3(1 - c_3) \tag{1.15j}$$

$$0 = b_4(1 - c_4). \tag{1.15k}$$

□

It follows from (1.15j) and (1.11h) that

$$b_3 b_4 c_2(1 - c_3) \neq 0. \tag{1.16}$$

In particular this implies $c_4 = 1$ by (1.15k).

Solution of equations (1.15). Equations (a)-(e) and (k) just state that b_i and c_i are the coefficients of a fourth order quadrature formula with $c_1 = 0$ and $c_4 = 1$. We distinguish four cases for this:

1) $c_2 = u$, $c_3 = v$ and $0, u, v, 1$ are all distinct;   (1.17)

then (a)-(e) form a regular linear system for b_1, b_2, b_3, b_4. This system has the solution

$$b_1 = \frac{1 - 2(u + v) + 6uv}{12uv}, \qquad b_2 = \frac{2v - 1}{12u(1 - u)(v - u)},$$

$$b_3 = \frac{1 - 2u}{12v(1 - v)(v - u)}, \qquad b_4 = \frac{3 - 4(u + v) + 6uv}{12(1 - u)(1 - v)}.$$

Due to (1.16) we have to assume that u, v are such that $b_3 \neq 0$ and $b_4 \neq 0$. The three other cases with double nodes are built upon the Simpson rule:

2) $c_3 = 0$, $c_2 = 1/2$, $b_3 = w \neq 0$, $b_1 = 1/6 - w$, $b_2 = 4/6$, $b_4 = 1/6$;

3) $c_2 = c_3 = 1/2$, $b_1 = 1/6$, $b_3 = w \neq 0$, $b_2 = 4/6 - w$, $b_4 = 1/6$;

4) $c_2 = 1$, $c_3 = 1/2$, $b_4 = w \neq 0$, $b_2 = 1/6 - w$, $b_1 = 1/6$, $b_3 = 4/6$.

Once b_i and c_i are chosen, we obtain a_{43} from (j), and then (f) and (i) form a linear system of two equations for a_{32} and a_{42}. The determinant of this system is

$$\det \begin{pmatrix} b_3 & b_4 \\ b_3 c_3 c_2 & b_4 c_4 c_2 \end{pmatrix} = b_3 b_4 c_2 (c_4 - c_3)$$

which is $\neq 0$ by (1.16). Finally we obtain a_{21}, a_{31}, and a_{41} from (1.9).

Two particular choices of Kutta (1901) have become especially popular: case (3) with $w = 2/6$ and case (1) with $u = 1/3$, $v = 2/3$. They are given in Table 1.2. Both methods generalize classical quadrature rules in keeping the same order. The first is more popular, the second is more precise ("Wir werden diese Näherung als im allgemeinen beste betrachten ...", Kutta).

Table 1.2. Kutta's methods

0					0				
1/2	1/2				1/3	1/3			
1/2	0	1/2			2/3	−1/3	1		
1	0	0	1		1	1	−1	1	
	1/6	2/6	2/6	1/6		1/8	3/8	3/8	1/8
	"The" Runge-Kutta method					3/8–Rule			

"Optimal" Formulas

Much research has been undertaken, in order to choose the "best" possibilities from the variety of possible 4th order RK-formulas.

The first attempt in this direction was the very popular method of Gill (1951), with the aim of reducing the need for computer storage ("registers") as much as possible. The first computers in the fifties largely used this method which is therefore of historical interest. Gill observed that most computer storage is needed for the computation of k_3, where "registers are required to store in some form"

$$y_0 + a_{31}hk_1 + a_{32}hk_2, \quad y_0 + a_{41}hk_1 + a_{42}hk_2, \quad y_0 + b_1hk_1 + b_2hk_2, \quad hk_3.$$

"Clearly, three registers will suffice for the third stage if the quantities to be stored are linearly dependent, i.e., if"

$$\det \begin{pmatrix} 1 & a_{31} & a_{32} \\ 1 & a_{41} & a_{42} \\ 1 & b_1 & b_2 \end{pmatrix} = 0.$$

Gill observed that this condition is satisfied for the methods of type (3) if $w = (1 + \sqrt{0.5})/3$. The resulting method can then be reformulated as follows ("As each quantity is calculated it is stored in the register formerly holding the corresponding quantity of the previous stage, which is no longer required"):

$$y := \text{initial value}, \quad k := hf(y), \quad y := y + 0.5k, \quad q := k,$$
$$k := hf(y), \quad y := y + (1 - \sqrt{0.5})(k - q),$$
$$q := (2 - \sqrt{2})k + (-2 + 3\sqrt{0.5})q,$$
$$k := hf(y), \quad y := y + (1 + \sqrt{0.5})(k - q), \tag{1.18}$$
$$q := (2 + \sqrt{2})k + (-2 - 3\sqrt{0.5})q,$$
$$k := hf(y), \quad y := y + \frac{k}{6} - \frac{q}{3}, \quad (\rightarrow \text{compute next step}).$$

Today, in large high-speed computers, this method is no longer used, but could still be of interest for very high dimensional equations.

Other attempts have been made to choose u and v in (1.17), case (1), such that the *error terms* (terms in h^5, see Section II.3) become as small as possible. We shall discuss this question in Section II.3.

Numerical Example

> Zu grosses Gewicht darf man natürlich solchen Beispielen nicht
> beilegen ...
> (W. Kutta 1901)

We compare five different choices of 4th order methods on the Van der Pol equation
(I.16.2) with $\varepsilon = 1$. As initial values we take $y_1(0) = A$, $y_2(0) = 0$ on the limit
cycle and we integrate over one period T (the values of A and T are given in
Exercise I.16.1). For a comparison of these methods with lower order ones we have
also included the explicit Euler method, Runge's method of order 2 and Heun's
method of order 3 (see Table 1.1).

We have applied the methods with several fixed step sizes. The errors of both
components and the number of function evaluations (fe) are displayed in logarithmic
scales in Fig. 1.1. Whenever the error behaves like $C \cdot h^p = C_1 \cdot (fe)^{-p}$, the curves
appear as straight lines with slope $1/p$. We have chosen the scales such that the
theoretical slope of the 4th order methods appears to be $45°$.

These tests clearly show up the importance of higher order methods. Among
the various 4th order methods there is usually no big difference. It is interesting to
note that in our example the method with the smallest error in y_1 has the biggest
error in y_2 and vice versa.

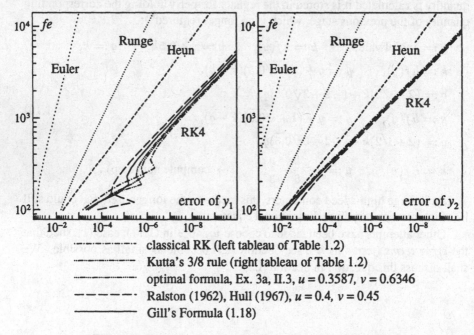

⎯·⎯·⎯·⎯·⎯	classical RK (left tableau of Table 1.2)
⎯⎯⎯⎯⎯	Kutta's 3/8 rule (right tableau of Table 1.2)
⎯⎯·⎯⎯·⎯⎯	optimal formula, Ex. 3a, II.3, $u = 0.3587$, $v = 0.6346$
⎯ ⎯ ⎯ ⎯	Ralston (1962), Hull (1967), $u = 0.4$, $v = 0.45$
⎯⎯⎯⎯⎯	Gill's Formula (1.18)

Fig. 1.1. Global errors versus number of function evaluations

Exercises

1. Show that every s-stage explicit RK method of order s, when applied to the problem $y' = \lambda y$ (λ a complex constant), gives

$$y_1 = \left(\sum_{j=0}^{s} \frac{z^j}{j!}\right) y_0, \qquad z = h\lambda.$$

Hint. Show first that y_1/y_0 must be a polynomial in z of degree s and then determine its coefficients by comparing the derivatives of y_1, with respect to h, to those of the true solution.

2. (Runge 1895, p. 175; see also the introduction to Adams methods in Chap. III.1). The theoretical form of drops of fluids is determined by the differential equation of Laplace (1805)

$$-z = \alpha^2 \frac{(K_1 + K_2)}{2} \tag{1.21}$$

where α is a constant, $(K_1 + K_2)/2$ the mean curvature, and z the height (see Fig. 1.2). If we insert $1/K_1 = r/\sin\varphi$ and $K_2 = d\varphi/ds$, the curvature of the meridian curve, we obtain

$$-2z = \alpha^2 \left(\frac{\sin\varphi}{r} + \frac{d\varphi}{ds}\right), \tag{1.22}$$

where we put $\alpha = 1$. Add

$$\frac{dr}{ds} = \cos\varphi, \qquad \frac{dz}{ds} = -\sin\varphi, \tag{1.22'}$$

to obtain a system of three differential equations for $\varphi(s)$, $r(s)$, $z(s)$, s being the arc length. Compute and plot different solution curves by the method of Runge (1.4) with initial values $\varphi(0) = 0$, $r(0) = 0$ and $z(0) = z_0$ ($z_0 < 0$ for lying drops; compute also hanging drops with appropriate sign changes in (1.22)). Use different step sizes and compare the results.

Hint. Be careful at the singularity in the beginning: from (1.22) and (1.22') we have for small s that $r = s$, $\varphi = \zeta s$ with $\zeta = -z_0$, hence $(\sin\varphi)/r \to -z_0$. A more precise analysis gives for small s the expansions ($\zeta = -z_0$)

$$\varphi = \zeta s + \frac{\zeta}{4}s^3 + \left(\frac{\zeta}{48} - \frac{\zeta^3}{120}\right)s^5 + \dots$$

$$r = s - \frac{\zeta^2}{6}s^3 + \left(-\frac{\zeta^2}{20} + \frac{\zeta^4}{120}\right)s^5 + \dots$$

$$z = -\zeta - \frac{\zeta}{2}s^2 + \left(-\frac{\zeta}{16} + \frac{\zeta^3}{24}\right)s^4 + \left(-\frac{\zeta}{288} + \frac{\zeta^3}{45} - \frac{\zeta^5}{720}\right)s^6 + \dots .$$

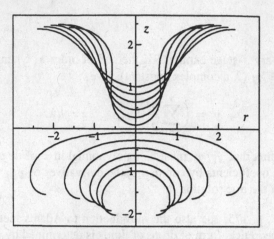

Fig. 1.2. Drops

3. Find the conditions for a 2-stage explicit RK-method to be of order two and determine all such methods ("... wozu eine weitere Erörterung nicht mehr nötig ist", Kutta).

4. Find all methods of order three with three stages (i.e., solve (1.11;a-d) with $b_4 = 0$).

   Result. $c_2 = u$, $c_3 = v$, $a_{32} = v(v-u)/(u(2-3u))$, $b_2 = (2-3v)/(6u(u-v))$, $b_3 = (2-3u)/(6v(v-u))$, $b_1 = 1 - b_2 - b_3$, $a_{31} = c_3 - a_{32}$, $a_{21} = c_2$ (Kutta 1901, p. 438).

5. Construct all methods of order 2 of the form

$$
\begin{array}{c|ccc}
0 & & & \\
c_2 & c_2 & & \\
c_3 & 0 & c_3 & \\
\hline
& 0 & 0 & 1
\end{array}
$$

   Such methods "have the property that the corresponding Runge-Kutta process requires relatively less storage in a computer" (Van der Houwen (1977), §2.7.2). Apply them to $y' = \lambda y$ and compare with Exercise 1.

6. Determine the conditions for order two of the RK methods with two stages which do *not* satisfy the conditions (1.9):

$$
k_1 = f(x_0 + c_1 h, y_0)
$$
$$
k_2 = f(x_0 + c_2 h, y_0 + a_{21} h k_1)
$$
$$
y_1 = y_0 + h (b_1 k_1 + b_2 k_2).
$$

   Discuss the use of this extra freedom for c_1 and c_2 (Oliver 1975).

II.2 Order Conditions for Runge-Kutta Methods

... I heard a lecture by Merson ...
(J. Butcher's first contact with RK methods)

In this section we shall derive the general structure of the order conditions (Merson 1957, Butcher 1963). The proof has evolved very much in the meantime, mainly under the influence of Butcher's later work, many personal discussions with him, the proof of "Theorem 6" in Hairer & Wanner (1974), and our teaching experience. We shall see in Section II.11 that exactly the same ideas of proof lead to a general theorem of composition of methods (= B-series), which gives access to order conditions for a much larger class of methods.

A big advantage is obtained by transforming (1.1) to *autonomous* form by appending x to the dependent variables as

$$\begin{pmatrix} x \\ y \end{pmatrix}' = \begin{pmatrix} 1 \\ f(x,y) \end{pmatrix}. \tag{2.1}$$

The main difficulty in the derivation of the order conditions is to understand the correspondence of the formulas to certain rooted labelled trees; this comes out most naturally if we use well-chosen indices and tensor notation (as in Gill (1951), Henrici (1962), p. 118, Gear (1971), p. 32). As is usual in tensor notation, we denote (in this section) the components of vectors by *superscript* indices which, in order to avoid confusion, we choose as *capitals*. Then (2.1) can be written as

$$(y^J)' = f^J(y^1, \ldots, y^n), \qquad J = 1, \ldots, n. \tag{2.2}$$

We next rewrite the method (1.8) for the autonomous differential equation (2.2). In order to get a better symmetry in all formulas of (1.8), we replace k_i by the argument g_i such that $k_i = f(g_i)$. Then (1.8) becomes

$$g_i^J = y_0^J + \sum_{j=1}^{i-1} a_{ij} h f^J(g_j^1, \ldots, g_j^n), \qquad i = 1, \ldots, s$$

$$y_1^J = y_0^J + \sum_{j=1}^{s} b_j h f^J(g_j^1, \ldots, g_j^n). \tag{2.3}$$

If the system (2.2) originates from (2.1), then, for $J = 1$,

$$g_i^1 = y_0^1 + \sum_{j=1}^{i-1} a_{ij} h = x_0 + c_i h$$

by (1.9). We see that (1.9) becomes a natural condition. If it is satisfied, then for the derivation of order conditions only the autonomous equation (2.2) has to be considered.

As indicated in Section II.1 we have to compare the Taylor series of y_1^J with that of the exact solution. Therefore we compute the derivatives of y_1^J and g_i^J with respect to h for $h = 0$. Due to the similarity of the two formulas, it is sufficient to do this for g_i^J. On the right hand side of (2.3) there appear expressions of the form $h\varphi(h)$, so we make use of Leibniz' formula

$$\left(h\varphi(h)\right)^{(q)}\big|_{h=0} = q\cdot\left(\varphi(h)\right)^{(q-1)}\big|_{h=0}. \tag{2.4}$$

The reader is now asked to take a deep breath, take five sheets of reversed computer paper, remember the basic rules of differential calculus, and begin the following computations:

$q = 0$: from (2.3)

$$\left(g_i^J\right)^{(0)}\big|_{h=0} = y_0^J. \tag{2.5;0}$$

$q = 1$: from (2.3) and (2.4)

$$\left(g_i^J\right)^{(1)}\big|_{h=0} = \sum_j a_{ij} f^J\big|_{y=y_0}. \tag{2.5;1}$$

$q = 2$: because of (2.4) we shall need the first derivative of $f^J(g_j)$

$$\left(f^J(g_j)\right)^{(1)} = \sum_K f_K^J(g_j)\cdot(g_j^K)^{(1)}, \tag{2.6;1}$$

where, as usual, f_K^J denotes $\partial f^J/\partial y^K$. Inserting formula (2.5;1) (with i, j, J replaced by j, k, K) into (2.6;1) we obtain with (2.4)

$$\left(g_i^J\right)^{(2)}\big|_{h=0} = 2\sum_{j,k} a_{ij} a_{jk} \sum_K f_K^J f^K\big|_{y=y_0}. \tag{2.5;2}$$

$q = 3$: we differentiate (2.6;1) to obtain

$$\left(f^J(g_j)\right)^{(2)} = \sum_{K,L} f_{KL}^J(g_j)\cdot(g_j^K)^{(1)}(g_j^L)^{(1)} + \sum_K f_K^J(g_j)(g_j^K)^{(2)}. \tag{2.6;2}$$

The derivatives $(g_j^K)^{(1)}$ and $(g_j^K)^{(2)}$ at $h = 0$ are already available in (2.5;1) and (2.5;2). So we have from (2.3) and (2.4)

$$\left(g_i^J\right)^{(3)}\big|_{h=0} = 3\sum_{j,k,l} a_{ij} a_{jk} a_{jl} \sum_{K,L} f_{KL}^J f^K f^L\big|_{y=y_0}$$

$$+ 3\cdot 2\sum_{j,k,l} a_{ij} a_{jk} a_{kl} \sum_{K,L} f_K^J f_L^K f^L\big|_{y=y_0}. \tag{2.5;3}$$

The same formula holds for $(y_1^J)^{(3)}\big|_{h=0}$ with a_{ij} replaced by b_j.

The Derivatives of the True Solution

The derivatives of the correct solution are obtained much more easily just by differentiating equation (2.2): first

$$(y^J)^{(1)} = f^J(y). \tag{2.7;1}$$

Differentiating (2.2) and inserting (2.2) again for the derivatives we get

$$(y^J)^{(2)} = \sum_K f_K^J(y) \cdot (y^K)^{(1)} = \sum_K f_K^J(y) f^K(y). \tag{2.7;2}$$

Differentiating (2.7;2) again we obtain

$$(y^J)^{(3)} = \sum_{K,L} f_{KL}^J(y) f^K(y) f^L(y) + \sum_{K,L} f_K^J(y) f_L^K(y) f^L(y). \tag{2.7;3}$$

Conditions for Order 3

For order 3, the derivatives (2.5;1-3), (with a_{ij} replaced by b_j) must be equal to the derivatives (2.7;1-3), and this for every differential equation. Thus, comparing the corresponding expressions, we obtain:

Theorem 2.1. *The RK method (2.3) (and thus (1.8)) is of order 3 iff*

$$\sum_j b_j = 1, \qquad\qquad 2\sum_{j,k} b_j a_{jk} = 1,$$

$$3\sum_{j,k,l} b_j a_{jk} a_{jl} = 1, \qquad 6\sum_{j,k,l} b_j a_{jk} a_{kl} = 1. \tag{2.8}$$

$\square$

Inserting $\sum_k a_{jk} = c_j$ from (1.9), we can simplify these expressions still further and obtain formulas (a)-(d) of (1.11).

Trees and Elementary Differentials

> But without a more convenient notation, it would be difficult to find the corresponding expressions ... This, however, can be at once effected by means of the analytical forms called trees ...
>
> (A. Cayley 1857)

The continuation of this process, although theoretically clear, soon leads to very complicated formulas. It is therefore advantageous to use a graphical representation:

indeed, the indices j, k, l and J, K, L in the terms of (2.5;3) are linked together as pairs of indices in $a_{jk}, a_{jl}, \ldots$ in exactly the same way as upper and lower indices in the expressions f_{KL}^J, f_K^J, namely

$$t_{31} = \quad \searrow_{j} \nearrow^{k \; l} \qquad \text{and} \qquad t_{32} = \quad \searrow_{j}^{l} \; k \qquad (2.9)$$

for the first and second term respectively. We call these objects *labelled trees*, because they are connected graphs (trees) whose vertices are labelled with summation indices. They can also be represented as *mappings*, e.g.,

$$l \mapsto j, \quad k \mapsto j \qquad \text{and} \qquad l \mapsto k, \quad k \mapsto j \qquad (2.9')$$

for the above trees. This mapping indicates to which lower letter the corresponding vertices are attached.

Definition 2.2. Let A be an ordered chain of indices $A = \{j < k < l < m < \ldots\}$ and denote by A_q the subset consisting of the first q indices. A *(rooted) labelled tree of order* q $(q \geq 1)$ is a mapping (the son-father mapping)

$$t : A_q \setminus \{j\} \to A_q$$

such that $t(z) < z$ for all $z \in A_q \setminus \{j\}$. The *set of all labelled trees of order* q is denoted by LT_q. We call "z" the *son* of "$t(z)$" and "$t(z)$" the *father* of "z". The vertex "j", the forefather of the whole dynasty, is called the *root* of t. The *order* q of a labelled tree is equal to the number of its vertices and is usually denoted by $q = \varrho(t)$.

Definition 2.3. For a labelled tree $t \in LT_q$ we call

$$F^J(t)(y) = \sum_{K,L,\ldots} f_{K,\ldots}^J(y) f_{..}^K(y) f_{..}^L(y) \cdot \ldots$$

the corresponding *elementary differential*. The summation is over $q-1$ indices $K, L, \ldots$ (which correspond to $A_q \setminus \{j\}$) and the summand is a product of q f's, where the upper index runs through all vertices of t and the lower indices are the corresponding sons. We also denote by $F(t)(y)$ the vector $\left(F^1(t)(y), \ldots, F^n(t)(y) \right)$.

If the set A_q is written as

$$A_q = \{j_1 < j_2 < \ldots < j_q\}, \qquad (2.10)$$

then we can write the definition of $F(t)$ as follows:

$$F^{J_1}(t) = \sum_{J_2,\ldots,J_q} \prod_{i=1}^{q} f_{t^{-1}(J_i)}^{J_i}, \qquad (2.11)$$

since the sons of an index are its inverse images under the map t.

Examples of elementary differentials are

$$\sum_{K,L} f_{KL}^J f^K f^L \quad \text{and} \quad \sum_{K,L} f_K^J f_L^K f^L$$

for the labelled trees t_{31} and t_{32} above. These expressions appear in formulas (2.5;3) and (2.7;3).

The three labelled trees

$$(2.12)$$

all look topologically alike, moreover the corresponding elementary differentials

$$\sum_{K,L,M} f_{KM}^J f^M f_L^K f^L, \quad \sum_{K,L,M} f_{KL}^J f^L f_M^K f^M, \quad \sum_{K,L,M} f_{LK}^J f^K f_M^L f^M \quad (2.12')$$

are the same, because they just differ by an exchange of the summation indices. Thus we give

Definition 2.4. Two labelled trees t and u are *equivalent*, if they have the same order, say q, and if there exists a permutation $\sigma : A_q \to A_q$, such that $\sigma(j) = j$ and $t\sigma = \sigma u$ on $A_q \setminus \{j\}$.

This clearly defines an equivalence relation.

Definition 2.5. An equivalence class of qth order labelled trees is called a *(rooted) tree* of *order q*. The set of all trees of order q is denoted by T_q. The *order* of a tree is defined as the order of a representative and is again denoted by $\varrho(t)$. Furthermore we denote by $\alpha(t)$ (for $t \in T_q$) the number of elements in the equivalence class t; i.e., the number of possible different monotonic labellings of t.

Geometrically, a tree is distinguished from a labelled tree by omitting the labels. Often it is advantageous to include $\emptyset$, the empty tree, as the only tree of order 0. The only tree of order 1 is denoted by τ. The number of trees of orders $1, 2, \ldots, 10$ are given in Table 2.1. Representatives of all trees of order ≤ 5 are shown in Table 2.2.

Table 2.1. Number of trees up to order 10

q	1	2	3	4	5	6	7	8	9	10
card(T_q)	1	1	2	4	9	20	48	115	286	719

Table 2.2. Trees and elementary differentials up to order 5

q	t	graph	$\gamma(t)$	$\alpha(t)$	$F^J(t)(y)$	$\Phi_j(t)$
0	$\emptyset$	$\emptyset$	1	1	y^J	
1	τ	$\bullet j$	1	1	f^J	1
2	t_{21}		2	1	$\sum_K f_K^J f^K$	$\sum_k a_{jk}$
3	t_{31}		3	1	$\sum_{K,L} f_{KL}^J f^K f^L$	$\sum_{k,l} a_{jk} a_{jl}$
	t_{32}		6	1	$\sum_{K,L} f_K^J f_L^K f^L$	$\sum_{k,l} a_{jk} a_{kl}$
4	t_{41}		4	1	$\sum_{K,L,M} f_{KLM}^J f^K f^L f^M$	$\sum_{k,l,m} a_{jk} a_{jl} a_{jm}$
	t_{42}		8	3	$\sum_{K,L,M} f_{KM}^J f_L^K f^L f^M$	$\sum_{k,l,m} a_{jk} a_{kl} a_{jm}$
	t_{43}		12	1	$\sum_{K,L,M} f_K^J f_{LM}^K f^L f^M$	$\sum_{k,l,m} a_{jk} a_{kl} a_{km}$
	t_{44}		24	1	$\sum_{K,L,M} f_K^J f_L^K f_M^L f^M$	$\sum_{k,l,m} a_{jk} a_{kl} a_{lm}$
5	t_{51}		5	1	$\sum f_{KLMP}^J f^K f^L f^M f^P$	$\sum a_{jk} a_{jl} a_{jm} a_{jp}$
	t_{52}		10	6	$\sum f_{KMP}^J f_L^K f^L f^M f^P$	$\sum a_{jk} a_{kl} a_{jm} a_{jp}$
	t_{53}		15	4	$\sum f_{KP}^J f_{ML}^K f^L f^M f^P$	$\sum a_{jk} a_{kl} a_{km} a_{jp}$
	t_{54}		30	4	$\sum f_{KP}^J f_L^K f_M^L f^M f^P$	$\sum a_{jk} a_{kl} a_{lm} a_{jp}$
	t_{55}		20	3	$\sum f_{KM}^J f_L^K f^L f_P^M f^P$	$\sum a_{jk} a_{kl} a_{jm} a_{mp}$
	t_{56}		20	1	$\sum f_K^J f_{LMP}^K f^L f^M f^P$	$\sum a_{jk} a_{kl} a_{km} a_{kp}$
	t_{57}		40	3	$\sum f_K^J f_{LP}^K f_M^L f^M f^P$	$\sum a_{jk} a_{kl} a_{lm} a_{kp}$
	t_{58}		60	1	$\sum f_K^J f_L^K f_{MP}^L f^M f^P$	$\sum a_{jk} a_{kl} a_{lm} a_{lp}$
	t_{59}		120	1	$\sum f_K^J f_L^K f_M^L f_P^M f^P$	$\sum a_{jk} a_{kl} a_{lm} a_{mp}$

The Taylor Expansion of the True Solution

We can now state the general result for the qth derivative of the true solution:

Theorem 2.6. *The exact solution of (2.2) satisfies*

$$(y)^{(q)}(x_0) = \sum_{t \in LT_q} F(t)(y_0) = \sum_{t \in T_q} \alpha(t) F(t)(y_0). \qquad (2.7;q)$$

Proof. The theorem is true for $q = 1, 2, 3$ (see (2.7;1-3) above). For the computation of, say, the 4th derivative, we have to differentiate (2.7;3). This consists of two terms (corresponding to the two trees of (2.9)), each of which contains three factors $f_{\cdots}$ (corresponding to the three nodes of these trees). The differentiation of these by Leibniz' rule and insertion of (2.2) for the derivatives is geometrically just the addition of a new branch with a new summation letter to *each* vertex (Fig. 2.1).

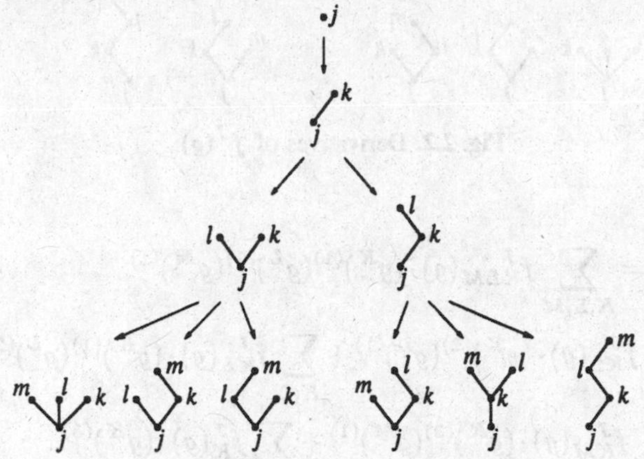

$$(2.7;1)$$

$$(2.7;2)$$

$$(2.7;3)$$

$$(2.7;4)$$

Fig. 2.1. Derivatives of exact solution

It is clear that by this process *all* labelled trees of order q appear for the qth derivative, each of them *exactly once*.

If we group together the terms with identical elementary differentials, we obtain the second expression of (2.7;q).   □

Faà di Bruno's Formula

Our next goal will be the computation of the qth derivative of the numerical solution y_1 and of the g_j. For this, we have first to generalize the formulas (2.6;1) (the chain rule) and (2.6;2) for the qth derivative of the composition of two functions. We represent these two formulas graphically in Fig. 2.2.

Formula (2.6;2) consists of two terms; the first term contains three factors, the second contains only two. Here the node "l" is a "dummy" node, not really present in the formula, and just indicates that we have to take the second derivative. The derivation of (2.6;2) will thus lead to *five* terms which we write down for the convenience of the reader (but not for the convenience of the printer ...)

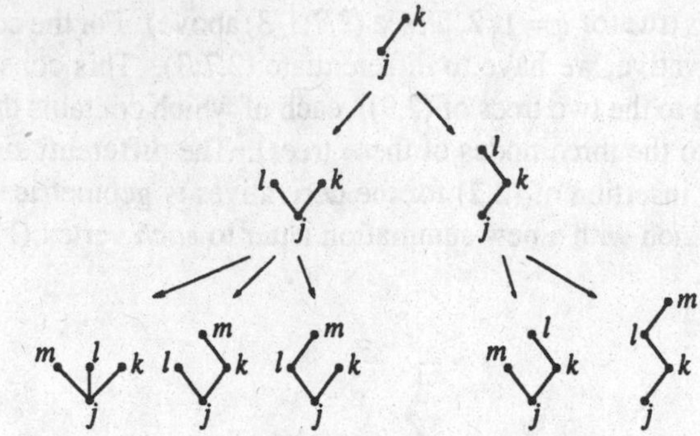

(2.6;1)

(2.6;2)

(2.6;3)

Fig. 2.2. Derivatives of $f^J(g)$

$$(f^J(g))^{(3)} = \sum_{K,L,M} f^J_{KLM}(g) \cdot (g^K)^{(1)}(g^L)^{(1)}(g^M)^{(1)}$$

$$+ \sum_{K,L} f^J_{KL}(g) \cdot (g^K)^{(2)}(g^L)^{(1)} + \sum_{K,L} f^J_{KL}(g) \cdot (g^K)^{(1)}(g^L)^{(2)} \qquad (2.6;3)$$

$$+ \sum_{K,M} f^J_{KM}(g) \cdot (g^K)^{(2)}(g^M)^{(1)} + \sum_{K} f^J_K(g) \cdot (g^K)^{(3)}.$$

The corresponding trees are represented in the third line of Fig. 2.2. Each time we differentiate, we have to
i) differentiate the first factor $f^J_{K\ldots}$; i.e., we add a new branch to the root j;
ii) increase the derivative numbers of each of the g's by 1; we represent this by lengthening the corresponding branch.
Each time we add a new label. *All* trees which are obtained in this way are those "special" trees which have no ramifications except at the root.

Definition 2.7. We denote by LS_q the set of *special labelled trees of order* q which have no ramifications except at the root.

Lemma 2.8 (Faà di Bruno's formula). *For $q \geq 1$ we have*

$$(f^J(g))^{(q-1)} = \sum_{u \in LS_q} \sum_{K_1,\ldots,K_m} f^J_{K_1,\ldots,K_m}(g) \cdot (g^{K_1})^{(\delta_1)} \ldots (g^{K_m})^{(\delta_m)} \quad (2.6;q-1)$$

Here, for $u \in LS_q$, m is the number of branches leaving the root and $\delta_1, \ldots, \delta_m$ are the numbers of nodes in each of these branches, such that $q = 1 + \delta_1 + \ldots + \delta_m$.
□

Remark. The usual multinomial coefficients are absent here, as we use labelled trees.

The Derivatives of the Numerical Solution

> It is difficult to keep a cool head when discussing the various derivatives ...
> (S. Gill 1956)

In order to generalize (2.5;1-3), we need the following definitions:

Definition 2.9. Let t be a labelled tree with root j; we denote by

$$\Phi_j(t) = \sum_{k,l,\ldots} a_{jk} a_{\ldots} \cdots$$

the sum over the $q-1$ remaining indices $k, l, \ldots$ (as in Definition 2.3). The summand is a product of $q-1$ a's, where all fathers stand two by two with their sons as indices. If the set A_q is written as in (2.10), we have

$$\Phi_{j_1}(t) = \sum_{j_2,\ldots,j_q} a_{t(j_2),j_2} \cdots a_{t(j_q),j_q}. \tag{2.13}$$

Definition 2.10. For $t \in LT_q$ let $\gamma(t)$ be the product of $\varrho(t)$ and all orders of the trees which appear, if the roots, one after another, are removed from t. (See Fig. 2.3 or formula (2.17)).

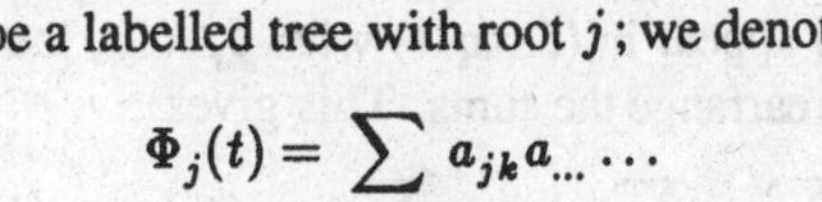

$$\gamma(t) = 9 \qquad \cdot 2 \cdot 6 \qquad \cdot 4 \qquad = 432$$

Fig. 2.3. Example for the definition of $\gamma(t)$

The above expressions are of course independent of the labellings, so $\Phi_j(t)$ as well as $\gamma(t)$ also make sense in T_q. Examples are given in Table 2.2.

Theorem 2.11. *The derivatives of g_i satisfy*

$$g_i^{(q)}\big|_{h=0} = \sum_{t \in LT_q} \gamma(t) \sum_j a_{ij} \Phi_j(t) F(t)(y_0). \tag{2.5;q}$$

The numerical solution y_1 of (2.3) satisfies

$$y_1^{(q)}\big|_{h=0} = \sum_{t \in LT_q} \gamma(t) \sum_j b_j \Phi_j(t) F(t)(y_0)$$

$$= \sum_{t \in T_q} \alpha(t)\gamma(t) \sum_j b_j \Phi_j(t) F(t)(y_0). \tag{2.14}$$

Proof. Because of the similarity of y_1 and g_i (see (2.3)) we only have to prove the first equation. We do this by induction on q, in exactly the same way as we obtained (2.5;1-3): we first apply Leibniz' formula (2.4) to obtain

$$(g_i^J)^{(q)}\big|_{h=0} = q \sum_j a_{ij} \big(f^J(g_j)\big)^{(q-1)}\big|_{y=y_0}. \tag{2.15}$$

Next we use Faà di Bruno's formula (Lemma 2.8). Finally we insert for the derivatives $(g_j^{K_*})^{(\delta_*)}$, which appear in (2.6;q-1) with $\delta_* < q$, the induction hypothesis (2.5;1) - (2.5;q-1) and rearrange the sums. This gives

$$(g_i^J)^{(q)}\big|_{h=0} = q \sum_{u \in LS_q} \sum_{t_1 \in LT\delta_1} \cdots \sum_{t_m \in LT\delta_m} \gamma(t_1)\ldots\gamma(t_m)\cdot$$
$$\sum_j a_{ij} \sum_{k_1} a_{jk_1} \Phi_{k_1}(t_1) \ldots \sum_{k_m} a_{jk_m} \Phi_{k_m}(t_m)\cdot \tag{2.16}$$
$$\sum_{K_1,\ldots,K_m} f_{K_1,\ldots,K_m}^J(y_0) F^{K_1}(t_1)(y_0)\ldots F^{K_m}(t_m)(y_0).$$

The main difficulty is now to understand that to each tuple

$$(u, t_1, \ldots, t_m) \qquad \text{with} \qquad u \in LS_q, \ t_s \in LT_{\delta_s}$$

there corresponds a labelled tree $t \in LT_q$ such that

$$\gamma(t) = q \cdot \gamma(t_1)\ldots\gamma(t_m) \tag{2.17}$$
$$F^J(t)(y) = \sum_{K_1,\ldots,K_m} f_{K_1,\ldots,K_m}^J(y) F^{K_1}(t_1)(y)\ldots F^{K_m}(t_m)(y) \tag{2.18}$$
$$\Phi_j(t) = \sum_{k_1,\ldots,k_m} a_{jk_1} \ldots a_{jk_m} \Phi_{k_1}(t_1)\ldots\Phi_{k_m}(t_m). \tag{2.19}$$

This labelled tree t is obtained if the branches of u are replaced by the trees $t_1,\ldots,t_m$ and the corresponding labels are taken over in a natural way, i.e., in the same order (see Fig. 2.4 for some examples).

In this way, *all* trees $t \in LT_q$ appear exactly *once*. Thus (2.16) becomes (2.5;q) after inserting (2.17), (2.18) and (2.19).   □

The above construction of t can also be used for a recursive definition of trees. We first observe that the equivalence class of t (in Fig. 2.4) depends only on the equivalence classes of $t_1,\ldots,t_m$.

Definition 2.12. We denote by

$$t = [t_1,\ldots,t_m] \tag{2.20}$$

the tree, which leaves over the trees $t_1,\ldots,t_m$ when its root and the adjacent branches are chopped off (Fig. 2.5).

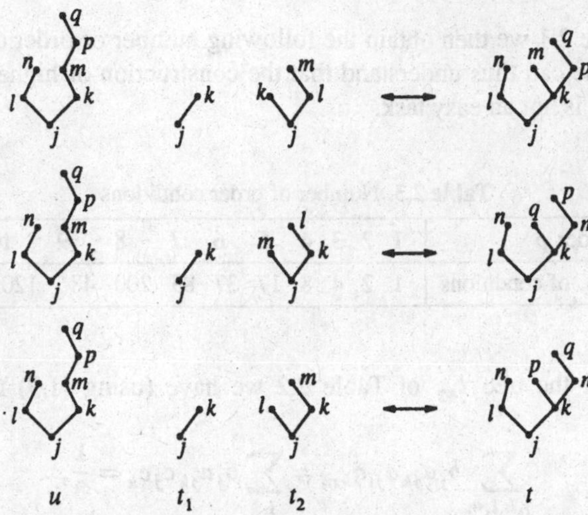

Fig. 2.4. Example for the bijection $(u, t_1, \ldots, t_m) \leftrightarrow t$

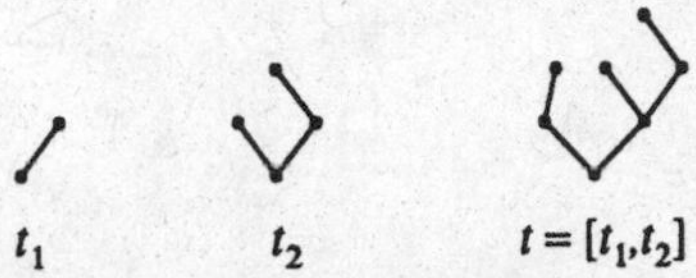

Fig. 2.5. Recursive definition of trees

With (2.20) all trees can be expressed in terms of τ; e.g., $t_{21} = [\tau]$, $t_{31} = [\tau, \tau]$, $t_{32} = [[\tau]]$, ..., etc.

The Order Conditions

Comparing Theorems 2.6 and 2.11 we now obtain:

Theorem 2.13. *A Runge-Kutta method (1.8) is of order p iff*

$$\sum_{j=1}^{s} b_j \Phi_j(t) = \frac{1}{\gamma(t)} \tag{2.21}$$

for all trees of order $\leq p$.

Proof. While the "if" part is clear from the preceding discussion, the "only if" part needs the fact that the elementary differentials for different trees are actually *independent*. See Exercises 3 and 4 below. □

From Table 2.1 we then obtain the following number of order conditions (see Table 2.3). One can thus understand that the construction of higher order Runge Kutta formulas is not an easy task.

Table 2.3. Number of order conditions

order p	1	2	3	4	5	6	7	8	9	10
no. of conditions	1	2	4	8	17	37	85	200	486	1205

Example. For the tree t_{42} of Table 2.2 we have (using (1.9) for the second expression)

$$\sum_{j,k,l,m} b_j a_{jk} a_{jl} a_{km} = \sum_{j,k} b_j a_{jk} c_j c_k = \frac{1}{8},$$

which is (1.11;f). All remaining conditions of (1.11) correspond to the other trees of order ≤ 4.

Exercises

1. Find all trees of order 6 and order 7.

   Hint. Search for all representations of $p-1$ as a sum of positive integers, and then insert all known trees of lower order for each term in the sum. You may also use a computer for general p.

2. (A. Cayley 1857). Denote the number of trees of order q by a_q. Prove that

   $$a_1 + a_2 x + a_3 x^2 + a_4 x^3 + \ldots = (1-x)^{-a_1}(1-x^2)^{-a_2}(1-x^3)^{-a_3}\ldots.$$

   Compare the result with Table 2.1.

3. Compute the elementary differentials of Table 2.2 for the case of the scalar non-autonomous equation (2.1), i.e., $f^1 = 1$, $f^2 = f(x,y)$. One imagines the complications met by the first authors (Kutta, Nyström, Huťa) in looking for higher order conditions. Observe also that in this case the expressions for t_{54} and t_{57} are the same, so that here Theorem 2.13 is sufficient, but not necessary for order 5.

   Hint. For, say, t_{54} we have non-zero derivatives only if $K = L = 2$. Letting M and P run from 1 to 2 we then obtain

   $$F^2(t) = (f_x + ff_y)(f_{yx} + ff_{yy})f_y$$

(see also Butcher 1963a).

4. Show that for every $t \in T_q$ there is a system of differential equations such that $F^1(t)(y_0) = 1$ and $F^1(u)(y_0) = 0$ for all other trees u.

   Hint. For t_{54} this system would be

   $$y_1' = y_2 y_5, \quad y_2' = y_3, \quad y_3' = y_4, \quad y_4' = 1, \quad y_5' = 1$$

   with all initial values $= 0$. Understand this and the general formula

   $$y_{\text{father}}' = \prod y_{\text{sons}}.$$

5. Kutta (1901) claimed that the scheme given in Table 2.4 is of order 5. Was he correct in his statement? Try to correct these values.

   Result. The values for $a_{6j} (j = 1, \ldots, 5)$ should read $(6, 36, 10, 8, 0)/75$; the correct values for b_j are $(23, 0, 125, 0, -81, 125)/192$ (Nyström 1925).

Table 2.4. A method of Kutta

0					
$\dfrac{1}{3}$	$\dfrac{1}{3}$				
$\dfrac{2}{5}$	$\dfrac{4}{25}$	$\dfrac{6}{25}$			
1	$\dfrac{1}{4}$	-3	$\dfrac{15}{4}$		
$\dfrac{2}{3}$	$\dfrac{6}{81}$	$\dfrac{90}{81}$	$-\dfrac{50}{81}$	$\dfrac{8}{81}$	
$\dfrac{4}{5}$	$\dfrac{7}{30}$	$\dfrac{18}{30}$	$-\dfrac{5}{30}$	$\dfrac{4}{30}$	0
	$\dfrac{48}{192}$	0	$\dfrac{125}{192}$	0	$-\dfrac{81}{192}$	$\dfrac{100}{192}$

6. Verify
   $$\sum_{\varrho(t)=p} \alpha(t) = (p-1)!$$

7. Prove that a Runge-Kutta method, when applied to a linear system

   $$y' = A(x)y + g(x), \tag{2.22}$$

   is of order p iff

   $$\sum_j b_j c_j^{q-1} = 1/q \quad \text{for } q \le p$$

   $$\sum_{j,k} b_j c_j^{q-1} a_{jk} c_k^{r-1} = 1/((q+r)r) \quad \text{for } q + r \le p$$

   $$\sum_{j,k,l} b_j c_j^{q-1} a_{jk} c_k^{r-1} a_{kl} c_l^{s-1} = 1/((q+r+s)(r+s)s) \quad \text{for } q + r + s \le p$$

   ... etc (write (2.22) in autonomous form and investigate which elementary differentials vanish identically; see also Crouzeix 1975).

II.3 Error Estimation and Convergence for RK Methods

> Es fehlt indessen noch der Beweis dass diese Näherungs-Ver-
> fahren convergent sind oder, was practisch wichtiger ist, es fehlt
> ein Kriterium, um zu ermitteln, wie klein die Schritte gemacht
> werden müssen, um eine vorgeschriebene Genauigkeit zu erre-
> ichen.
> (Runge 1905)

Since the work of Lagrange (1797) and, above all, of Cauchy, a numerically established result should be accompanied by a reliable error estimation ("... l'erreur commise sera inférieure à ..."). Lagrange gave the well-known error bounds for the Taylor polynomials and Cauchy derived bounds for the error of the Euler polygons (see Section I.7). A couple of years after the first success of the Runge-Kutta methods, Runge (1905) also required error estimates for these methods.

Rigorous Error Bounds

Runge's device for obtaining bounds for the error in one step ("local error") can be described in a few lines (free translation):

"For a method of order p consider the local error

$$e(h) = y(x_0 + h) - y_1 \tag{3.1}$$

and use its Taylor expansion

$$e(h) = e(0) + he'(0) + \ldots + \frac{h^p}{p!} e^{(p)}(\theta h) \tag{3.2}$$

with $0 < \theta < 1$ and $e(0) = e'(0) = \ldots = e^{(p)}(0) = 0$. Now compute explicitly $e^{(p)}(h)$, which will be of the form

$$e^{(p)}(h) = E_1(h) + h E_2(h), \tag{3.3}$$

where $E_1(h)$ and $E_2(h)$ contain partial derivatives of f up to order $p-1$ and p respectively. Further, because of $e^{(p)}(0) = 0$, we have $E_1(0) = 0$. Thus, if all partial derivatives of f up to order p are bounded, we have $E_1(h) = \mathcal{O}(h)$ and $E_2(h) = \mathcal{O}(1)$. So there is a constant C such that $|e^{(p)}(h)| \leq Ch$ and

$$|e(h)| \leq C \frac{h^{p+1}}{p!}. \text{ "} \tag{3.4}$$

A slightly different approach is adopted by Bieberbach (1923, 1. Abschn., Kap. II, §7), explained in more detail in Bieberbach (1951): we write

$$e(h) = y(x_0 + h) - y_1 = y(x_0 + h) - y_0 - h \sum_{i=1}^{s} b_i k_i \qquad (3.5)$$

and use the Taylor expansions

$$y(x_0 + h) = y_0 + y'(x_0)h + y''(x_0)\frac{h^2}{2!} + \ldots + y^{(p+1)}(x_0 + \theta h)\frac{h^{p+1}}{(p+1)!}$$

$$k_i(h) = k_i(0) + k_i'(0)h + \ldots + k_i^{(p)}(\theta_i h)\frac{h^p}{p!}, \qquad (3.6)$$

where, for vector valued functions, the formula is valid componentwise with possibly different θ's. The first terms in the h expansion of (3.5) vanish because of the order conditions. Thus we obtain

Theorem 3.1. *If the Runge-Kutta method (1.8) is of order p and if all partial derivatives of $f(x, y)$ up to order p exist (and are continuous), then the local error of (1.8) admits the rigorous bound*

$$\|y(x_0 + h) - y_1\| \le h^{p+1} \Big(\frac{1}{(p+1)!} \max_{t \in [0,1]} \|y^{(p+1)}(x_0 + th)\| $$
$$+ \frac{1}{p!} \sum_{i=1}^{s} |b_i| \max_{t \in [0,1]} \|k_i^{(p)}(th)\| \Big) \qquad (3.7)$$

and hence also

$$\|y(x_0 + h) - y_1\| \le C h^{p+1}. \qquad (3.8)$$

$\square$

Let us demonstrate this result on Runge's first method (1.4), which is of order $p = 2$, applied to a scalar differential equation. Differentiating (1.1) we obtain

$$y^{(3)}(x) = \Big(f_{xx} + 2f_{xy}f + f_{yy}f^2 + f_y(f_x + f_y f) \Big)(x, y(x)) \qquad (3.9)$$

while the second derivative of $k_2(h) = f(x_0 + \frac{h}{2}, y_0 + \frac{h}{2}f_0)$ is given by

$$k_2^{(2)}(h) = \frac{1}{4}\Big(f_{xx}(x_0 + \frac{h}{2}, y_0 + \frac{h}{2}f_0) + 2f_{xy}(...)f_0 + f_{yy}(...)f_0^2 \Big) \qquad (3.10)$$

(f_0 stands for $f(x_0, y_0)$). Under the assumptions of Theorem 3.1 we see that the expressions (3.9) and (3.10) are bounded by a constant independent of h, which gives (3.8).

The Principal Error Term

For higher order methods rigorous error bounds, like (3.7), become very unpractical. It is therefore much more realistic to consider the first non-zero term in the Taylor expansion of the error. For autonomous systems of equations (2.2), the error term is best obtained by subtracting the Taylor series and using (2.14) and (2.7;q).

Theorem 3.2. *If the Runge-Kutta method is of order p and if f is $(p+1)$-times continuously differentiable, we have*

$$y^J(x_0+h)-y_1^J = \frac{h^{p+1}}{(p+1)!}\sum_{t\in T_{p+1}} \alpha(t)e(t)F^J(t)(y_0)+\mathcal{O}(h^{p+2}) \qquad (3.11)$$

where

$$e(t) = 1-\gamma(t)\sum_{j=1}^{s}b_j\Phi_j(t). \qquad (3.12)$$

$\square$

$\gamma(t)$ and $\Phi_j(t)$ are given in Definitions 2.9 and 2.10; see also formulas (2.17) and (2.19). The expressions $e(t)$ are called the *error coefficients*.

Example 3.3. For the two-parameter family of 4th order RK methods (1.17) the error coefficients for the 9 trees of Table 2.2 are ($c_2=u$, $c_3=v$):

$$e(t_{51}) = -\frac{1}{4}+\frac{5}{12}(u+v)-\frac{5}{6}uv, \qquad e(t_{52}) = \frac{5}{12}v-\frac{1}{4},$$

$$e(t_{53}) = \frac{5}{8}u-\frac{1}{4}, \qquad e(t_{54}) = -\frac{1}{4},$$

$$e(t_{55}) = 1-\frac{5(b_4+b_3(3-4v)^2)}{144b_3b_4(1-v)^2}, \qquad (3.13)$$

$$e(t_{56}) = -4e(t_{51}), \qquad e(t_{57}) = -4e(t_{52}),$$

$$e(t_{58}) = -4e(t_{53}), \qquad e(t_{59}) = -4e(t_{54}).$$

Proof. The last four formulas follow from (1.12). $e(t_{59})$ is trivial, $e(t_{58})$ and $e(t_{57})$ follow from (1.11h). Further

$$e(t_{51}) = 5\int_0^1 t(t-1)(t-u)(t-v)\,dt$$

expresses the quadrature error. For $e(t_{55})$ one best introduces $c_i' = \sum_j a_{ij}c_j$ such that $e(t_{55}) = 1-20\sum_i b_ic_i'c_i'$. Then from (1.11d,f) one obtains

$$c_1' = c_2' = 0, \qquad b_3c_3' = \frac{1}{24(1-v)}, \qquad b_4c_4' = \frac{3-4v}{24(1-v)}.$$

$\square$

For the classical 4th order method (Table 1.2a) these error coefficients are given by Kutta (1901), p. 448 (see also Lotkin 1951) as follows

$$\left(-\frac{1}{24}, -\frac{1}{24}, \frac{1}{16}, -\frac{1}{4}, -\frac{2}{3}, \frac{1}{6}, \frac{1}{6}, -\frac{1}{4}, 1\right)$$

Kutta remarked that for the second method (Table 1.2b) ("Als besser noch erweist sich ...") the error coefficients become

$$\left(-\frac{1}{54}, \frac{1}{36}, -\frac{1}{24}, -\frac{1}{4}, -\frac{1}{9}, \frac{2}{27}, -\frac{1}{9}, \frac{1}{6}, 1\right)$$

which, with the exception of the 4th and 9th term, are all smaller than for the above method. A tedious calculation was undertaken by Ralston (1962) (and by many others) to determine optimal coefficients of (1.17). For solutions which minimize the constants (3.13), see Exercise 3 below.

Estimation of the Global Error

Das war auch eine aufregende Zeit ...   (P. Henrici 1983)

The global error is the error of the computed solution after *several* steps. Suppose that we have a one-step method which, given an initial value (x_0, y_0) and a step size h, computes a numerical solution y_1 approximating $y(x_0 + h)$. We shall denote this process by Henrici's notation

$$y_1 = y_0 + h\Phi(x_0, y_0, h) \tag{3.14}$$

and call Φ the *increment function* of the method.

The numerical solution for a point $X > x_0$ is then obtained by a step-by-step procedure

$$y_{i+1} = y_i + h_i \Phi(x_i, y_i, h_i), \qquad h_i = x_{i+1} - x_i, \qquad x_N = X \tag{3.15}$$

and our task is to estimate the *global error*

$$E = y(X) - y_N. \tag{3.16}$$

This estimate is found in a simple way, very similar to Cauchy's convergence proof for Theorem 7.3 of Chapter I: *the local errors are transported to the final point x_N and then added up*. This "error transport" can be done in two different ways:

a) either along the exact solution curves (see Fig. 3.1); this method can yield sharp results when sharp estimates of error propagation for the exact solutions are known, e.g., from Theorem 10.6 of Chapter I based on the logarithmic norm $\mu(\partial f/\partial y)$.

b) or along $N - i$ steps of the numerical method (see Fig. 3.2); this is the method used in the proofs of Cauchy (1824) and Runge (1905), it generalizes easily to multistep methods (see Chapter III) and will be an important tool for the existence of asymptotic expansions (see II.8).

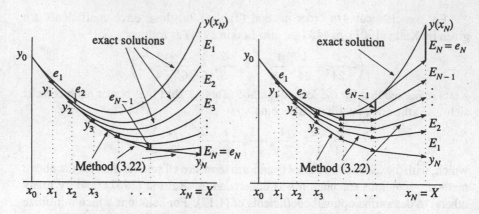

Fig. 3.1. Global error estimation, method (a)

Fig. 3.2. Global error estimation, method (b)

In both cases we first estimate the local errors e_i with the help of Theorem 3.1 to obtain

$$\|e_i\| \le C \cdot h_{i-1}^{p+1}. \tag{3.17}$$

Warning. The e_i of Fig. 3.1 and Fig. 3.2, for $i \ne 1$, are *not* the same, but they allow similar estimates.

We then estimate the transported errors E_i: for method (a) we use the known results from Chapter I, especially Theorem I.10.6, Theorem I.10.2, or formula (I.7.17). The result is

Theorem 3.4. *Let U be a neighbourhood of $\{(x, y(x)) | x_0 \le x \le X\}$ where $y(x)$ is the exact solution of (1.1). Suppose that in U*

$$\left\| \frac{\partial f}{\partial y} \right\| \le L \qquad or \qquad \mu\left(\frac{\partial f}{\partial y}\right) \le L, \tag{3.18}$$

and that the local error estimates (3.17) are valid in U. Then the global error (3.16) can be estimated by

$$\|E\| \le h^p \frac{C'}{L} \Big(\exp\big(L(X - x_0)\big) - 1 \Big) \tag{3.19}$$

where $h = \max h_i$,

$$C' = \begin{cases} C & L \ge 0 \\ C \exp(-Lh) & L < 0, \end{cases}$$

and h is small enough for the numerical solution to remain in U.

Remark. For $L \to 0$ the estimate (3.19) tends to $h^p C (x_N - x_0)$.

Proof. From Theorem I.10.2 (with $\varepsilon = 0$) or Theorem I.10.6 (with $\delta = 0$) we obtain

$$\|E_i\| \le \exp\big(L(x_N - x_i)\big)\|e_i\|. \tag{3.20}$$

We then insert this together with (3.17) into

$$\|E\| \le \sum_{i=1}^{N} \|E_i\|.$$

Using $h_{i-1}^{p+1} \le h^p \cdot h_{i-1}$ this leads to

$$\|E\| \le h^p C \Big(h_0 \exp\big(L(x_N - x_1)\big) + h_1 \exp\big(L(x_N - x_2)\big) + \dots\Big).$$

The expression in large brackets can be bounded by

$$\int_{x_0}^{x_N} \exp(L(x_N - x))dx \qquad \text{for} \qquad L \ge 0 \tag{3.21}$$

$$\int_{x_0}^{x_N} \exp(L(x_N - h - x))dx \qquad \text{for} \qquad L < 0 \tag{3.22}$$

(see Fig. 3.3). This gives (3.19). □

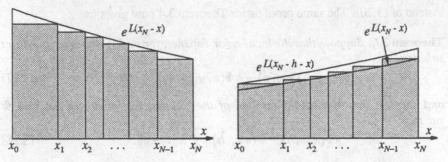

Fig. 3.3. Estimation of Riemann sums

For the second method (b) we need an estimate for $\|z_{i+1} - y_{i+1}\|$ in terms of $\|z_i - y_i\|$, where, besides (3.15),

$$z_{i+1} = z_i + h_i \Phi(x_i, z_i, h_i)$$

is a second pair of numerical solutions. For RK-methods z_{i+1} is defined by

$$\ell_1 = f(x_i, z_i),$$
$$\ell_2 = f(x_i + c_2 h_i, z_i + h_i a_{21} \ell_1), \quad \text{etc.}$$

We now subtract formulas (1.8) from this and obtain

$$\|\ell_1 - k_1\| \le L\|z_i - y_i\|,$$
$$\|\ell_2 - k_2\| \le L(1 + |a_{21}|h_i L)\|z_i - y_i\|, \quad \text{etc.}$$

This leads to the following

Lemma 3.5. *Let L be a Lipschitz constant for f and let $h_i \le h$. Then the increment function Φ of method (1.8) satisfies*

$$\|\Phi(x_i, z_i, h_i) - \Phi(x_i, y_i, h_i)\| \le \Lambda\|z_i - y_i\| \tag{3.23}$$

where

$$\Lambda = L\left(\sum_i |b_i| + hL\sum_{i,j} |b_i a_{ij}| + h^2 L^2 \sum_{i,j,k} |b_i a_{ij} a_{jk}| + \ldots\right). \tag{3.24}$$

$\square$

From (3.23) we obtain

$$\|z_{i+1} - y_{i+1}\| \le (1 + h_i \Lambda)\|z_i - y_i\| \le \exp(h_i \Lambda)\|z_i - y_i\| \tag{3.25}$$

and for the errors in Fig. 3.2,

$$\|E_i\| \le \exp\big(\Lambda(x_N - x_i)\big)\|e_i\| \tag{3.26}$$

instead of (3.20). The same proof as for Theorem 3.4 now gives us

Theorem 3.6. *Suppose that the local error satisfies, for initial values on the exact solution,*

$$\|y(x + h) - y(x) - h\Phi(x, y(x), h)\| \le Ch^{p+1}, \tag{3.27}$$

and suppose that in a neighbourhood of the solution the increment function Φ satisfies

$$\|\Phi(x, z, h) - \Phi(x, y, h)\| \le \Lambda\|z - y\|. \tag{3.28}$$

Then the global error (3.16) can be estimated by

$$\|E\| \le h^p \frac{C}{\Lambda}\Big(\exp\big(\Lambda(x_N - x_0)\big) - 1\Big) \tag{3.29}$$

where $h = \max h_i$.

$\square$

Exercises

1. (Runge 1905). Show that for explicit Runge Kutta methods with $b_i \geq 0$, $a_{ij} \geq 0$ (all i, j) of order s the Lipschitz constant Λ for Φ satisfies
$$1 + h\Lambda < \exp(hL)$$
   and that (3.29) is valid with Λ replaced by L.

2. Show that $e(t_{55})$ of (3.13) becomes
$$e(t_{55}) = 1 - 5\,\frac{(4v^2 - 15v + 9) - u(6v^2 - 42v + 27) - u^2(26v - 18)}{12(1 - 2u)(6uv - 4(u + v) + 3)}$$
   after inserting (1.17).

3. Determine u and v in (1.17) such that in (3.13)

   a) $\max_{i=5,6,7,8} |e(t_{5i})| = \min$ b) $\sum_{i=1}^{9} |e(t_{5i})| = \min$

   c) $\max_{i=5,6,7,8} \alpha(t_{5i}) |e(t_{5i})| = \min$ d) $\sum_{i=1}^{9} \alpha(t_{5i}) |e(t_{5i})| = \min$

   Results.

a)	$u = 0.3587$,	$v = 0.6346$,	$\min = 0.1033$;
b)	$u = 0.3995$,	$v = 0.6$,	$\min = 1.55$;
c)	$u = 0.3501$,	$v = 0.5839$,	$\min = 0.1248$;
d)	$u = 0.3716$,	$v = 0.6$,	$\min = 2.53$.

   Such optimal formulas were first studied by Ralston (1962), Hull & Johnston (1964), and Hull (1967).

4. Apply an explicit Runge-Kutta method to the problem $y' = f(x, y)$, $y(0) = 0$, where
$$f(x, y) = \begin{cases} \dfrac{\lambda}{x} y + g(x) & \text{if } x > 0 \\[2mm] (1 - \lambda)^{-1} g(0) & \text{if } x = 0, \end{cases}$$
   $\lambda \leq 0$ and $g(x)$ is sufficiently differentiable (see Exercise 10 of Section I.5).

   a) Show that the error after the first step is given by
$$y(h) - y_1 = C_2 h^2 g'(0) + \mathcal{O}(h^3)$$
   where C_2 is a constant depending on λ and on the coefficients of the method. Also for high order methods we have in general $C_2 \neq 0$.

   b) Compute C_2 for the classical 4th order method (Table 1.2).

II.4 Practical Error Estimation and Step Size Selection

Even the simplified error estimates of Section II.3, which are content with the leading error term, are of little practical interest, because they require the computation and majorization of several partial derivatives of high orders. But the main advantage of Runge-Kutta methods, compared with Taylor series, is precisely that the computation of derivatives should be no longer necessary. However, since practical error estimates are necessary (on the one hand to ensure that the step sizes h_i are chosen sufficiently small to yield the required precision of the computed results, and on the other hand to ensure that the step sizes are sufficiently large to avoid unnecessary computational work), we shall now discuss alternative methods for error estimates.

The oldest device, used by Runge in his numerical examples, is to repeat the computations with *halved* step sizes and to compare the results: those digits which haven't changed are assumed to be correct ("... woraus ich schliessen zu dürfen glaube ...").

Richardson Extrapolation

The idea of Richardson, announced in his classical paper Richardson (1910) which treats mainly partial differential equations, and explained in full detail in Richardson (1927), is to use more carefully the known behaviour of the error as a function of h.

Suppose that, with a given initial value (x_0, y_0) and step size h, we compute *two* steps, using a fixed Runge-Kutta method of order p, and obtain the numerical results y_1 and y_2. We then compute, starting from (x_0, y_0), *one big step* with step size $2h$ to obtain the solution w. The error of y_1 is known to be (Theorem 3.2)

$$e_1 = y(x_0 + h) - y_1 = C \cdot h^{p+1} + \mathcal{O}(h^{p+2}) \tag{4.1}$$

where C contains the error coefficients of the method and the elementary differentials $F^J(t)(y_0)$ of order $p+1$. The error of y_2 is composed of two parts: the transported error of the first step, which is

$$\left(I + h\frac{\partial f}{\partial y} + \mathcal{O}(h^2)\right)e_1,$$

and the local error of the second step, which is the same as (4.1), but with the elementary differentials evaluated at $y_1 = y_0 + \mathcal{O}(h)$. Thus we obtain

$$e_2 = y(x_0 + 2h) - y_2 = \left(I + \mathcal{O}(h)\right)Ch^{p+1} + \left(C + \mathcal{O}(h)\right)h^{p+1} + \mathcal{O}(h^{p+2})$$
$$= 2Ch^{p+1} + \mathcal{O}(h^{p+2}). \tag{4.2}$$

Similarly to (4.1), we have for the big step

$$y(x_0 + 2h) - w = C(2h)^{p+1} + \mathcal{O}(h^{p+2}). \tag{4.3}$$

Neglecting the terms $\mathcal{O}(h^{p+2})$, formulas (4.2) and (4.3) allow us to eliminate the unknown constant C and to "extrapolate" a better value $\widehat{y}_2$ for $y(x_0 + 2h)$, for which we obtain:

Theorem 4.1. *Suppose that y_2 is the numerical result of two steps with step size h of a Runge-Kutta method of order p, and w is the result of one big step with step size $2h$. Then the error of y_2 can be extrapolated as*

$$y(x_0 + 2h) - y_2 = \frac{y_2 - w}{2^p - 1} + \mathcal{O}(h^{p+2}) \tag{4.4}$$

and

$$\widehat{y}_2 = y_2 + \frac{y_2 - w}{2^p - 1} \tag{4.5}$$

is an approximation of order $p+1$ to $y(x_0 + 2h)$. ☐

Formula (4.4) is a very simple device to estimate the error of y_2 and formula (4.5) allows one to increase the precision by one additional order ("... The better theory of the following sections is complicated, and tends thereby to suggest that the practice may also be complicated; whereas it is really simple." Richardson).

Embedded Runge-Kutta Formulas

> Scraton is right in his criticism of Merson's process, although Merson did not claim as much for his process as some people expect.
> (R. England 1969)

The idea is, rather than using Richardson extrapolation, to construct Runge-Kutta formulas which themselves contain, besides the numerical approximation y_1, a second approximation $\widehat{y}_1$. The difference then yields an estimate of the local error for the less precise result and can be used for step size control (see below). Since it is at our disposal at every step, this gives more flexibility to the code and makes step rejections less expensive.

We consider two Runge-Kutta methods (one for y_1 and one for $\widehat{y}_1$) such that both use the *same* function values. We thus have to find a scheme of coefficients (see (1.8')),

$$
\begin{array}{c|ccccc}
0 & & & & & \\
c_2 & a_{21} & & & & \\
c_3 & a_{32} & a_{32} & & & \\
\vdots & \vdots & & \ddots & & \\
c_s & a_{s1} & a_{s2} & \cdots & a_{s,s-1} & \\
\hline
 & b_1 & b_2 & \cdots & b_{s-1} & b_s \\
 & \widehat{b}_1 & \widehat{b}_2 & \cdots & \widehat{b}_{s-1} & \widehat{b}_s
\end{array}
\tag{4.6}
$$

such that

$$
y_1 = y_0 + h(b_1 k_1 + \ldots + b_s k_s)
\tag{4.7}
$$

is of order p, and

$$
\widehat{y}_1 = y_0 + h(\widehat{b}_1 k_1 + \ldots + \widehat{b}_s k_s)
\tag{4.7'}
$$

is of order $\widehat{p}$ (usually $\widehat{p} = p-1$ or $\widehat{p} = p+1$). The approximation y_1 is used to continue the integration.

From Theorem 2.13, we have to satisfy the conditions

$$
\sum_{j=1}^{s} b_j \Phi_j(t) = \frac{1}{\gamma(t)} \qquad \text{for all trees of order} \leq p ,
\tag{4.8}
$$

$$
\sum_{j=1}^{s} \widehat{b}_j \Phi_j(t) = \frac{1}{\gamma(t)} \qquad \text{for all trees of order} \leq \widehat{p} .
\tag{4.8'}
$$

The first methods of this type were proposed by Merson (1957), Ceschino (1962), and Zonneveld (1963). Those of Merson and Zonneveld are given in Tables 4.1 and 4.2. Here, "name $p(\widehat{p})$" means that the order of y_1 is p and the order of the error estimator $\widehat{y}_1$ is $\widehat{p}$. Merson's $\widehat{y}_1$ is of order 5 only for *linear* equations with constant coefficients; for nonlinear problems it is of order 3. This method works quite well and has been used very often, especially by NAG users. Further embedded methods were then derived by Sarafyan (1966), England (1969), and Fehlberg (1964, 1968, 1969). Let us start with the construction of some low order embedded methods.

Methods of order 3(2). It is a simple task to construct embedded formulas of order 3(2) with $s = 3$ stages. Just take a 3-stage method of order 3 (Exercise II.1.4) and put $\widehat{b}_3 = 0$, $\widehat{b}_2 = 1/2c_2$, $\widehat{b}_1 = 1 - 1/2c_2$.

Table 4.1. Merson 4("5")

0					
$\dfrac{1}{3}$	$\dfrac{1}{3}$				
$\dfrac{1}{3}$	$\dfrac{1}{6}$	$\dfrac{1}{6}$			
$\dfrac{1}{2}$	$\dfrac{1}{8}$	0	$\dfrac{3}{8}$		
1	$\dfrac{1}{2}$	0	$-\dfrac{3}{2}$	2	
y_1	$\dfrac{1}{6}$	0	0	$\dfrac{2}{3}$	$\dfrac{1}{6}$
$\widehat{y}_1$	$\dfrac{1}{10}$	0	$\dfrac{3}{10}$	$\dfrac{2}{5}$	$\dfrac{1}{5}$

Table 4.2. Zonneveld 4(3)

0					
$\dfrac{1}{2}$	$\dfrac{1}{2}$				
$\dfrac{1}{2}$	0	$\dfrac{1}{2}$			
1	0	0	1		
$\dfrac{3}{4}$	$\dfrac{5}{32}$	$\dfrac{7}{32}$	$\dfrac{13}{32}$	$-\dfrac{1}{32}$	
y_1	$\dfrac{1}{6}$	$\dfrac{1}{3}$	$\dfrac{1}{3}$	$\dfrac{1}{6}$	
$\widehat{y}_1$	$-\dfrac{1}{2}$	$\dfrac{7}{3}$	$\dfrac{7}{3}$	$\dfrac{13}{6}$	$-\dfrac{16}{3}$

Methods of order 4(3). With $s = 4$ it is impossible to find a pair of order 4(3) (see Exercise 2). The idea is to add y_1 as 5th stage of the process (i.e., $a_{5i} = b_i$ for $i = 1, \ldots, 4$) and to search for a third order method which uses all five function values. Whenever the step is accepted this represents no extra work, because $f(x_0 + h, y_1)$ has to be computed anyway for the following step. This idea is called FSAL (First Same As Last). Then the order conditions (4.8') with $\widehat{p} = 3$ represent 4 linear equations for the five unknowns $\widehat{b}_1, \ldots, \widehat{b}_5$. One can arbitrarily fix $\widehat{b}_5 \neq 0$ and solve the system for the remaining parameters. With $\widehat{b}_5$ chosen such that $\widehat{b}_4 = 0$ the result is

$$\widehat{b}_1 = 2b_1 - 1/6, \qquad \widehat{b}_2 = 2(1 - c_2)b_2,$$
$$\widehat{b}_3 = 2(1 - c_3)b_3, \qquad \widehat{b}_4 = 0, \qquad \widehat{b}_5 = 1/6. \tag{4.9}$$

Automatic Step Size Control

> D'ordinaire, on se contente de multiplier ou de diviser par 2 la
> valeur du pas ...
> (Ceschino 1961)

We now want to write a code which automatically adjusts the step size in order to achieve a prescribed tolerance of the local error.

Whenever a starting step size h has been chosen, the program computes two approximations to the solution, y_1 and $\widehat{y}_1$. Then an estimate of the error for the less precise result is $y_1 - \widehat{y}_1$. We want this error to satisfy componentwise

$$|y_{1i} - \widehat{y}_{1i}| \leq sc_i, \qquad sc_i = Atol_i + \max(|y_{0i}|, |y_{1i}|) \cdot Rtol_i \tag{4.10}$$

where $Atol_i$ and $Rtol_i$ are the desired tolerances prescribed by the user (relative errors are considered for $Atol_i = 0$, absolute errors for $Rtol_i = 0$; usually both tolerances are different from zero; they may depend on the component of the solution). As a measure of the error we take

$$err = \sqrt{\frac{1}{n} \sum_{i=1}^{n} \left(\frac{y_{1i} - \widehat{y}_{1i}}{sc_i}\right)^2};$$ (4.11)

other norms, such as the max norm, are also of frequent use. Then err is compared to 1 in order to find an optimal step size. From the error behaviour $err \approx C \cdot h^{q+1}$ and from $1 \approx C \cdot h_{\text{opt}}^{q+1}$ (where $q = \min(p, \widehat{p})$) the optimal step size is obtained as ("... le procédé connu", Ceschino 1961)

$$h_{\text{opt}} = h \cdot (1/err)^{1/(q+1)}.$$ (4.12)

Some care is now necessary for a good code: we multiply (4.12) by a safety factor fac, usually $fac = 0.8, 0.9, (0.25)^{1/(q+1)}$, or $(0.38)^{1/(q+1)}$, so that the error will be acceptable the next time with high probability. Further, h is not allowed to increase nor to decrease too fast. For example, we may put

$$h_{\text{new}} = h \cdot \min\left(facmax, \max\left(facmin, fac \cdot (1/err)^{1/(q+1)}\right)\right)$$ (4.13)

for the new step size. Then, if $err \le 1$, the computed step is *accepted* and the solution is advanced with y_1 and a new step is tried with h_{new} as step size. Else, the step is *rejected* and the computations are repeated with the new step size h_{new}. The maximal step size increase $facmax$, usually chosen between 1.5 and 5, prevents the code from too large step increases and contributes to its safety. It is clear that, when chosen too small, it may also unnecessarily increase the computational work. It is also advisable to put $facmax = 1$ in the steps right after a step-rejection (Shampine & Watts 1979).

Whenever y_1 is of lower order than $\widehat{y}_1$, then the difference $y_1 - \widehat{y}_1$ is (at least asymptotically) an estimate of the local error and the above algorithm keeps this estimate below the given tolerance. But isn't it more natural to continue the integration with the higher order approximation? Then the concept of "error estimation" is abandoned and the difference $y_1 - \widehat{y}_1$ is only used for the purpose of step size selection. This is justified by the fact that, due to unknown stability and instability properties of the differential system, the local errors have in general very little in common with the global errors. The procedure of continuing the integration with the higher order result is called "local extrapolation".

A modification of the above procedure (PI step size control), which is particularly interesting when applied to mildly stiff problems, is described in Section IV.2 (Volume II).

Starting Step Size

> If anything has been made foolproof, a better fool will be developed.
>
> (Heard from Dr. Pirkl, Baden)

For many years, the starting step size had to be supplied to a code. Users were assumed to have a rough idea of a good step size from mathematical knowledge or previous experience. Anyhow, a bad starting choice for h was quickly repaired by the step size control. Nevertheless, when this happens too often and when the choices are too bad, much computing time can be wasted. Therefore, several people (e.g., Watts 1983, Hindmarsh 1980) developed ideas to let the computer do this choice. We take up an idea of Gladwell, Shampine & Brankin (1987) which is based on the hypothesis that

$$\text{local error} \approx Ch^{p+1}y^{(p+1)}(x_0).$$

Since $y^{(p+1)}(x_0)$ is unknown we shall replace it by approximations of the first and second derivative of the solution. The resulting algorithm is the following one:

a) Do one function evaluation $f(x_0, y_0)$ at the initial point. It is in any case needed for the first RK step. Then put $d_0 = \|y_0\|$ and $d_1 = \|f(x_0, y_0)\|$, where the norm is that of (4.11) with $sc_i = Atol_i + |y_{0i}| \cdot Rtol_i$.

b) As a first guess for the step size let

$$h_0 = 0.01 \cdot (d_0/d_1)$$

so that the increment of an explicit Euler step is small compared to the size of the initial value. If either d_0 or d_1 is smaller than 10^{-5} we put $h_0 = 10^{-6}$.

c) Perform one explicit Euler step, $y_1 = y_0 + h_0 f(x_0, y_0)$, and compute $f(x_0 + h_0, y_1)$.

d) Compute $d_2 = \|f(x_0 + h_0, y_1) - f(x_0, y_0)\|/h_0$ as an estimate of the second derivative of the solution; the norm being the same as in (a).

e) Compute a step size h_1 from the relation

$$h_1^{p+1} \cdot \max(d_1, d_2) = 0.01.$$

If $\max(d_1, d_2) \leq 10^{-15}$ we put $h_1 = \max(10^{-6}, h_0 \cdot 10^{-3})$.

f) Finally we propose as starting step size

$$h = \min(100 \cdot h_0, h_1). \tag{4.14}$$

An algorithm like the one above, or a similar one, usually gives a good guess for the initial step size (or at least avoids a very bad choice). Sometimes, more information about h is known, e.g., from previous experience or computations of similar problems.

Numerical Experiments

As a representative of 4-stage 4th order methods we consider the "3/8 Rule" of Table 1.2. We equipped it with the embedded formula (4.9) of order 3.

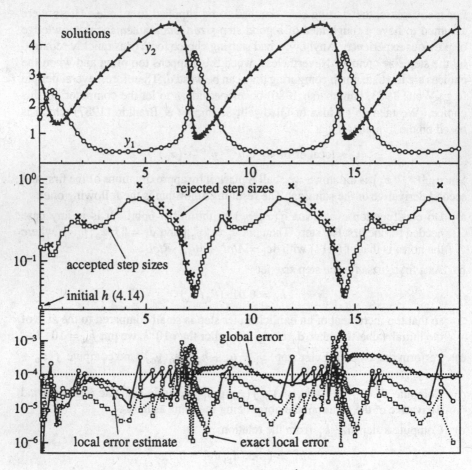

Fig. 4.1. Step size control, $Rtol = Atol = 10^{-4}$, 96 steps + 32 rejected

Step control mechanism. Fig. 4.1 presents the results of the step control mechanism (4.13) described above. As an example we choose the Brusselator (see Section I.16).

$$y_1' = 1 + y_1^2 y_2 - 4y_1$$
$$y_2' = 3y_1 - y_1^2 y_2$$

(4.15)

with initial values $y_1(0) = 1.5$, $y_2(0) = 3$, integration interval $0 \leq x \leq 20$ and $Atol = Rtol = 10^{-4}$. The following results are plotted in this figure:

i) At the top, the solutions $y_1(x)$ and $y_2(x)$ with all accepted integration steps;

ii) then all step sizes used; the accepted ones are connected by a polygon; the rejected ones are indicated by $\times$;

iii) the third graph shows the local error estimate err, the exact local error and the global error; the desired tolerance is indicated by a broken horizontal line.

It can be seen that, due to the instabilities of the solutions with respect to the initial values, quite large global errors occur during the integration with small local tolerances everywhere. Further many step rejections can be observed in regions where the step size has to be decreased. This cannot easily be prevented, because right after an accepted step, the step size proposed by formula (4.13) is (apart from the safety factor) always increasing.

Numerical comparison. We are now curious to see the behaviour of the variable step size code, when compared to a fixed step size implementation. We applied both implementations to the Brusselator problem (4.15) with the initial values used there. The tolerances $(Atol = Rtol)$ are chosen between 10^{-2} and 10^{-10} with ratio $\sqrt[3]{10}$. The results are then plotted in Fig. 4.2. There, the abscissa is the global error at the endpoint of integration (the "precision"), and the ordinate is the number of function evaluations (the "work"). We observe that for this problem the variable step size code is about twice as fast as the fixed step size code. There are, of course, problems (such as equation (0.1)) where variable step sizes are *much* more important than here.

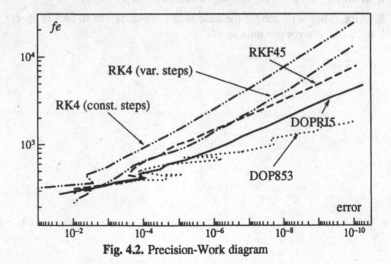

Fig. 4.2. Precision-Work diagram

In this comparison we have included some higher order methods, which will be dicussed in Section II.5. The code RKF45 (written by H.A. Watts and L.F. Shampine) is based on an embedded method of order 5(4) due to Fehlberg. The codes DOPRI5 (order 5(4)) and DOP853 (order 8(5,3)) are based on methods of

Dormand & Prince. They will be discussed in the following section. It can clearly be seen that higher order methods are, especially for higher precision, more efficient than lower order methods. We shall also understand why the 5th order method of Dormand & Prince is clearly superior to RKF45.

Exercises

1. Show that Runge's method (1.4) can be interpreted as two Euler steps (with step size $h/2$), followed by a Richardson extrapolation.

2. Prove that no 4-stage Runge-Kutta method of order 4 admits an embedded formula of order 3.

   Hint. Replace d_j by $\widehat{b}_j - b_j$ in the proof of Lemma 1.4 and deduce that $\widehat{b}_j = b_j$ for all j, which is a contradiction.

3. Show that the step size strategy (4.13) is invariant with respect to a rescaling of the independent variable. This means that it produces equivalent step size sequences when applied to the two problems

$$y' = f(x, y), \qquad y(0) = y_0, \qquad y(x_{\text{end}}) = ?$$
$$z' = \sigma \cdot f(\sigma t, z), \qquad z(0) = y_0, \qquad z(x_{\text{end}}/\sigma) = ?$$

   with initial step sizes h_0 and h_0/σ, respectively.

   Remark. This is no longer the case if one replaces *err* in (4.13) by err/h and q by $q - 1$ ("error per unit step").

II.5 Explicit Runge-Kutta Methods of Higher Order

Gehen wir endlich zu Näherungen von der fünften Ordnung über,
so werden die Verhältnisse etwas andere.   (W. Kutta 1901)

This section describes the construction of Runge-Kutta methods of higher orders, particularly of orders $p = 5$ and $p = 8$. As can be seen from Table 2.3, the complexity and number of the order conditions to be solved increases rapidly with p. An increasingly skilful use of simplifying assumptions will be the main tool for this task.

The Butcher Barriers

For methods of order 5 there are 17 order conditions to be satisfied (see Table 2.2). If we choose $s = 5$ we have 15 free parameters. Already Kutta raised the question whether there might nevertheless exist a solution ("Nun wäre es zwar möglich ..."), but he had no hope for this and turned straight away to the case $s = 6$ (see II.2, Exercise 5). Kutta's question remained open for more than 60 years and was answered around 1963 by three authors independently (Ceschino & Kuntzmann 1963, p. 89, Shanks 1966, Butcher 1964b, 1965b). Butcher's work is the farthest reaching and we shall mainly follow his ideas in the following:

Theorem 5.1. *For $p \geq 5$ no explicit Runge-Kutta method exists of order p with $s = p$ stages.*

Proof. We first treat the case $s = p = 5$: define the matrices U and V by

$$U = \begin{pmatrix} \sum_i b_i a_{i2} & \sum_i b_i a_{i3} & \sum_i b_i a_{i4} \\ \sum_i b_i a_{i2} c_2 & \sum_i b_i a_{i3} c_3 & \sum_i b_i a_{i4} c_4 \\ g_2 & g_3 & g_4 \end{pmatrix}, \quad V = \begin{pmatrix} c_2 & c_2^2 & \sum_j a_{2j} c_j - c_2^2/2 \\ c_3 & c_3^2 & \sum_j a_{3j} c_j - c_3^2/2 \\ c_4 & c_4^2 & \sum_j a_{4j} c_j - c_4^2/2 \end{pmatrix} \tag{5.1}$$

where

$$g_k = \sum_{i,j} b_i a_{ij} a_{jk} - \frac{1}{2} \sum_i b_i a_{ik}(1 - c_k). \tag{5.2}$$

Then the order conditions for order 5 imply

$$UV = \begin{pmatrix} 1/6 & 1/12 & 0 \\ 1/12 & 1/20 & 0 \\ 0 & 0 & 0 \end{pmatrix}. \tag{5.3}$$

Lemma 1.5 gives $g_4 = 0$ and consequently $c_4 = 1$ as in Lemma 1.4. Next we put in (5.1)

$$g_j = \left(\sum_i b_i a_{ij} - b_j(1 - c_j) \right)(c_j - c_5). \tag{5.4}$$

Again it can be verified by trivial computations that UV is the same as above. This time it follows that $c_4 = c_5$, hence $c_5 = 1$. Consequently, the expression

$$\sum_{i,j,k} b_i(1 - c_i)a_{ij}a_{jk}c_k \tag{5.5}$$

must be zero (because of $2 \le k < j < i$). However, by multiplying out and using two fifth-order conditions, the expression in (5.5) should be $1/120$, a contradiction.

The case $p = s = 6$ is treated by considering all "one-leg trees", i.e., the trees which consist of one leg above the root and the 5th order trees grafted on. The corresponding order conditions have the form

$$\sum_{i,j,\ldots} b_i a_{ij}(a_{j\ldots} \cdots \text{ expressions for order 5}) = \frac{1}{\gamma(t)}.$$

If we let $b'_j = \sum_i b_i a_{ij}$ we are back in the 5th order 5-stage business and can follow the above ideas again. However, the $\gamma(t)$ values are not the same as before; as a consequence, the product UV in (5.3) now becomes

$$UV = \begin{pmatrix} \dfrac{1!}{(s-2)!} & \dfrac{2!}{(s-1)!} & 0 \\ \dfrac{2!}{(s-1)!} & \dfrac{3!}{s!} & 0 \\ 0 & 0 & 0 \end{pmatrix} \quad (s = 6). \tag{5.3'}$$

Further, for $p = s = 7$ we use the "stork-trees" with order conditions

$$\sum_{i,j,\ldots} b_i a_{ij} a_{jk}(a_{k\ldots} \cdots \text{ expressions for order 5}) = \frac{1}{\gamma(t)}$$

and let $b''_k = \sum_{i,j} b_i a_{ij} a_{jk}$ and so on. The general case $p = s \ge 5$ is now clear.
□

6-Stage, 5th Order Processes

We now demonstrate the construction of 5th order processes with 6 stages in full detail following the ideas which allowed Butcher (1964b) to construct 7-stage, 6th order formulas.

"In searching for such processes we are guided by the analysis of the previous section to make the following assumptions:"

$$\sum_{i=1}^{6} b_i a_{ij} = b_j(1 - c_j) \qquad j = 1,\ldots,6, \tag{5.6}$$

$$\sum_{j=1}^{i-1} a_{ij}c_j = \frac{c_i^2}{2} \qquad\qquad i = 3,\ldots,6, \tag{5.7}$$

$$b_2 = 0. \tag{5.8}$$

The advantage of condition (5.6) is known to us already from Section II.1 (see Lemma 1.3): we can disregard all one-leg trees other than t_{21}.

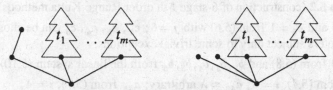

Fig. 5.1. Use of simplifying assumptions

Condition (5.7) together with (5.8) has a similar effect: for $[[\tau], t_2,\ldots,t_m]$ and $[\tau, \tau, t_2,\ldots,t_m]$ of Fig. 5.1 (with identical but arbitrary subtrees $t_2,\ldots,t_m$) the order conditions read

$$\sum_{i,j} b_i a_{ij} c_j \Phi_i = \frac{1}{r \cdot 2} \qquad \text{and} \qquad \sum_i b_i c_i^2 \Phi_i = \frac{1}{r} \tag{5.9}$$

with known values for Φ_i and r. Since $b_2 = 0$ by (5.8) it follows from (5.7) that both conditions of (5.9) are equivalent (the condition $b_2 = 0$ is necessary for this reduction, because (5.7) cannot be satisfied for $i = 2$; otherwise we would have $c_2 = 0$ and the method would be equivalent to one of fewer stages).

The only trees left after the above reduction are the quadrature conditions

$$\sum_{i=1}^{6} b_i c_i^{q-1} = \frac{1}{q} \qquad q = 1,2,3,4,5 \tag{5.10}$$

and the two equations

$$\sum_{i,j,k} b_i c_i a_{ij} a_{jk} c_k = \frac{1}{5 \cdot 3 \cdot 2}, \tag{5.11}$$

$$\sum_{i,j} b_i c_i a_{ij} c_j^2 = \frac{1}{5 \cdot 3}. \tag{5.12}$$

We multiply (5.12) by $1/2$ and then subtract both equations to obtain

$$\sum_{i,j} b_i c_i a_{ij} \left(\sum_k a_{jk} c_k - c_j^2/2 \right) = 0.$$

From (5.7) the parenthesis is zero except when $j = 2$, and therefore

$$\sum_{i=3}^{6} b_i c_i a_{i2} = 0 \tag{5.13}$$

replaces (5.11). Our last simplification is to subtract other order conditions from (5.12) to obtain

$$\sum_{i,j} b_i (1 - c_i) a_{ij} c_j (c_j - c_3) = \frac{1}{60} - \frac{c_3}{24}, \tag{5.14}$$

which has fewer terms than before, in particular because $c_6 = 1$ by (5.6) with $j = 6$. The resulting *reduced system* (5.6)-(5.8), (5.10), (5.13), (5.14) can easily be solved as follows:

Algorithm 5.2 (construction of 6-stage 5th order Runge-Kutta methods).

a) $c_1 = 0$ and $c_6 = 1$ from (5.6) with $j = 6$; c_2, c_3, c_4, c_5 can be chosen as free parameters subject only to some trivial exceptions;

b) $b_2 = 0$ from (5.8) and b_1, b_3, b_4, b_5, b_6 from the linear system (5.10);

c) a_{32} from (5.7), $i = 3$; $a_{42} = \lambda$ arbitrary; a_{43} from (5.7), $i = 4$;

d) a_{52} and a_{62} from the two linear equations (5.13) and (5.6), $j = 2$;

e) a_{54} from (5.14) and a_{53} from (5.7), $i = 5$;

f) a_{63}, a_{64}, a_{65} from (5.6), $j = 3, 4, 5$;

g) finally a_{i1} ($i = 2, \ldots, 6$) from (1.9).

Condition (5.6) for $j = 1$ and (5.7) for $i = 6$ are automatically satisfied. This follows as in the proof of Lemma 1.4.

Embedded Formulas of Order 5

Methods of Fehlberg. The methods obtained from Algorithm 5.2 do not all possess an embedded formula of order 4. Fehlberg, interested in the construction of Runge-Kutta pairs of order 4(5), looked mainly for simplifying assumptions which depend only on c_i and a_{ij}, but not on the weights b_i. In this case the simplifying assumptions are useful for the embedded method too. Therefore Fehlberg (1969) considered (5.7), (5.8) and replaced (5.6) by

$$\sum_{j=1}^{i-1} a_{ij} c_j^2 = \frac{c_i^3}{3}, \qquad i = 3, \ldots, 6. \tag{5.15}$$

Table 5.1. Fehlberg 4(5)

0						
$\dfrac{1}{4}$	$\dfrac{1}{4}$					
$\dfrac{3}{8}$	$\dfrac{3}{32}$	$\dfrac{9}{32}$				
$\dfrac{12}{13}$	$\dfrac{1932}{2197}$	$-\dfrac{7200}{2197}$	$\dfrac{7296}{2197}$			
1	$\dfrac{439}{216}$	-8	$\dfrac{3680}{513}$	$-\dfrac{845}{4104}$		
$\dfrac{1}{2}$	$-\dfrac{8}{27}$	2	$-\dfrac{3544}{2565}$	$\dfrac{1859}{4104}$	$-\dfrac{11}{40}$	
y_1	$\dfrac{25}{216}$	0	$\dfrac{1408}{2565}$	$\dfrac{2197}{4104}$	$-\dfrac{1}{5}$	0
$\widehat{y}_1$	$\dfrac{16}{135}$	0	$\dfrac{6656}{12825}$	$\dfrac{28561}{56430}$	$-\dfrac{9}{50}$	$\dfrac{2}{55}$

As with (5.9) this allows us to disregard all trees of the form $[[\tau,\tau],t_2,\ldots,t_m]$. In order that the reduction process of Fig. 5.1 also work on a higher level, we suppose, in addition to $b_2 = 0$, that

$$\sum_i b_i a_{i2} = 0, \qquad \sum_i b_i c_i a_{i2} = 0, \qquad \sum_{i,j} b_i a_{ij} a_{j2} = 0. \qquad (5.16)$$

Then the last equations to be satisfied are

$$\sum_{i,j} b_i a_{ij} c_j^3 = \frac{1}{20} \qquad (5.17)$$

and the quadrature conditions (5.10). We remark that the equations (5.7) and (5.15) for $i = 3$ imply

$$c_3 = \frac{3}{2} c_2. \qquad (5.18)$$

We now want the method to possess an embedded formula of order 4. Analogously to (5.8) we set $\widehat{b}_2 = 0$. Then conditions (5.7) and (5.15) simplify the conditions of order 4 to 5 linear equations (the 4 quadrature conditions and $\sum_i \widehat{b}_i a_{i2} = 0$) for the 5 unknowns $\widehat{b}_1, \widehat{b}_3, \widehat{b}_4, \widehat{b}_5, \widehat{b}_6$. This system has a second solution (other than the b_i) only if it is singular, which is the case if (see Exercise 1 below)

$$c_4 = \frac{3c_2}{4 - 24c_2 + 45c_2^2}. \qquad (5.19)$$

With c_2, c_5, c_6 as free parameters, the above system can be solved and yields an embedded formula of order 4(5). The coefficients of a very popular method, constructed by Fehlberg (1969), are given in Table 5.1.

All of the methods of Fehlberg are of the type $p(\widehat{p})$ with $p < \widehat{p}$. Hence, the lower order approximation is intended to be used as initial value for the next step. In order to make his methods optimal, Fehlberg tried to minimize the error coefficients for the lower order result y_1. This has the disadvantage that the local extrapolation mode (continue the integration with the higher order result) does not make sense and the estimated "error" can become substantially smaller than the true error.

> It is possible to do a lot better than the pair of Fehlberg currently regarded as "best."
>
> (L.F. Shampine 1986)

Table 5.2. Dormand-Prince 5(4) (DOPRI5)

0						
$\dfrac{1}{5}$	$\dfrac{1}{5}$					
$\dfrac{3}{10}$	$\dfrac{3}{40}$	$\dfrac{9}{40}$				
$\dfrac{4}{5}$	$\dfrac{44}{45}$	$-\dfrac{56}{15}$	$\dfrac{32}{9}$			
$\dfrac{8}{9}$	$\dfrac{19372}{6561}$	$-\dfrac{25360}{2187}$	$\dfrac{64448}{6561}$	$-\dfrac{212}{729}$		
1	$\dfrac{9017}{3168}$	$-\dfrac{355}{33}$	$\dfrac{46732}{5247}$	$\dfrac{49}{176}$	$-\dfrac{5103}{18656}$	
1	$\dfrac{35}{384}$	0	$\dfrac{500}{1113}$	$\dfrac{125}{192}$	$-\dfrac{2187}{6784}$	$\dfrac{11}{84}$
y_1	$\dfrac{35}{384}$	0	$\dfrac{500}{1113}$	$\dfrac{125}{192}$	$-\dfrac{2187}{6784}$	$\dfrac{11}{84}$	0
$\widehat{y}_1$	$\dfrac{5179}{57600}$	0	$\dfrac{7571}{16695}$	$\dfrac{393}{640}$	$-\dfrac{92097}{339200}$	$\dfrac{187}{2100}$	$\dfrac{1}{40}$

Dormand & Prince pairs. The first efforts at minimizing the error coefficients of *the higher order result*, which is then used as numerical solution, were undertaken by Dormand & Prince (1980). Their methods of order 5 are constructed with the help of Algorithm 5.2 under the additional hypothesis (5.15). This condition is achieved by fixing the parameters c_3 and a_{42} in such a way that (5.15) holds for $i = 3$ and $i = 4$. The remaining two relations ($i = 5, 6$) are then automatically satisfied. To see this, multiply the difference $e_i = \sum_{j=1}^{i-1} a_{ij} c_j^2 - c_i^3/3$ by b_i and $b_i c_i$, respectively, sum up and deduce that all e_i must vanish.

In order to equip the method with an embedded formula, Dormand & Prince propose to use the FSAL idea (i.e., add y_1 as 7th stage). In this way the restriction (5.19) for c_4 is no longer necessary. We fix arbitrarily $\widehat{b}_7 \neq 0$, put $\widehat{b}_2 = 0$ (as in (5.8)) and compute the remaining $\widehat{b}_i$, as above for the Fehlberg case from the 4 quadrature conditions and from $\sum_i \widehat{b}_i a_{i2} = 0$.

We have thus obtained a family of 5th order Runge-Kutta methods with 4th order embedded solution with c_2, c_4, c_5 as free parameters. Dormand & Prince (1980) have undertaken an extensive search to determine these parameters in order to minimize the error coefficients for y_1 and found that $c_2 = 1/5$, $c_4 = 4/5$ and $c_5 = 8/9$ was a close rational approximation to an optimal choice. Table 5.2 presents the coefficients of this method. The corresponding code of the Appendix is called DOPRI5.

Higher Order Processes

Order 6. By Theorem 5.1 at least 7 stages are necessary for order 6. A. Huťa (1956) constructed 6th order processes with 8 stages. Finally, methods with $s = 7$, the optimal number, were derived by Butcher (1964b) along similar lines as above. He arrived at an algorithm where c_2, c_3, c_5, c_6 are free parameters.

Order 7. The existence of such a method with 8 stages is impossible by the following barrier:

Theorem 5.3 (Butcher 1965b). *For $p \geq 7$ no explicit Runge-Kutta method exists of order p with $s = p + 1$ stages.*

Since the proof of this theorem is much more complicated than that of Theorem 5.1, we do not reproduce it here.

This raises the question, whether 7th order methods with 9 stages exist. Such methods, announced by Butcher (1965b), do exist; see Verner (1971).

Order 8. As to methods of order 8, Curtis (1970) and Cooper & Verner (1972) have constructed such processes with $s = 11$. It was for a long time an open question whether there exist methods with 10 stages. John Butcher's dream of settling this difficult question before his 50th birthday did not become true. But he finally succeeded in proving the non-existence for Dahlquist's 60th birthday:

Theorem 5.4 (Butcher 1985b). *For $p \geq 8$ no explicit Runge-Kutta method exists of order p with $s = p + 2$ stages.*

For the proof, which is still more complicated, we again refer to Butcher's original paper.

Order 10. These are the highest order explicitly constructed explicit Runge-Kutta methods. Curtis (1975) constructed an 18-stage method of order 10. His construction was based solely on simplifying assumptions of the type (5.7), (5.8) and their extensions. Hairer (1978) then constructed a 17-stage method by using the complete arsenal of simplifying ideas. For more details, see the first edition, p. 189.

Embedded Formulas of High Order

It was mainly the formula manipulation genius Fehlberg who first derived high order embedded formulas. His greatest success was his 7th order formula with 8th order error estimate (Fehlberg 1968) which is of frequent use in all high precision computations, e.g., in astronomy. The coefficients are reproduced in Table 5.3.

Table 5.3. Fehlberg 7(8)

0													
$\dfrac{2}{27}$	$\dfrac{2}{27}$												
$\dfrac{1}{9}$	$\dfrac{1}{36}$	$\dfrac{1}{12}$											
$\dfrac{1}{6}$	$\dfrac{1}{24}$	0	$\dfrac{1}{8}$										
$\dfrac{5}{12}$	$\dfrac{5}{12}$	0	$-\dfrac{25}{16}$	$\dfrac{25}{16}$									
$\dfrac{1}{2}$	$\dfrac{1}{20}$	0	0	$\dfrac{1}{4}$	$\dfrac{1}{5}$								
$\dfrac{5}{6}$	$-\dfrac{25}{108}$	0	0	$\dfrac{125}{108}$	$-\dfrac{65}{27}$	$\dfrac{125}{54}$							
$\dfrac{1}{6}$	$\dfrac{31}{300}$	0	0	0	$\dfrac{61}{225}$	$-\dfrac{2}{9}$	$\dfrac{13}{900}$						
$\dfrac{2}{3}$	2	0	0	$-\dfrac{53}{6}$	$\dfrac{704}{45}$	$-\dfrac{107}{9}$	$\dfrac{67}{90}$	3					
$\dfrac{1}{3}$	$-\dfrac{91}{108}$	0	0	$\dfrac{23}{108}$	$-\dfrac{976}{135}$	$\dfrac{311}{54}$	$-\dfrac{19}{60}$	$\dfrac{17}{6}$	$-\dfrac{1}{12}$				
1	$\dfrac{2383}{4100}$	0	0	$-\dfrac{341}{164}$	$\dfrac{4496}{1025}$	$-\dfrac{301}{82}$	$\dfrac{2133}{4100}$	$\dfrac{45}{82}$	$\dfrac{45}{164}$	$\dfrac{18}{41}$			
0	$\dfrac{3}{205}$	0	0	0	0	$-\dfrac{6}{41}$	$-\dfrac{3}{205}$	$-\dfrac{3}{41}$	$\dfrac{3}{41}$	$\dfrac{6}{41}$	0		
1	$-\dfrac{1777}{4100}$	0	0	$-\dfrac{341}{164}$	$\dfrac{4496}{1025}$	$\dfrac{289}{82}$	$\dfrac{2193}{4100}$	$\dfrac{51}{82}$	$\dfrac{33}{164}$	$\dfrac{12}{41}$	0	1	
y_1	$\dfrac{41}{840}$	0	0	0	0	$\dfrac{34}{105}$	$\dfrac{9}{35}$	$\dfrac{9}{35}$	$\dfrac{9}{280}$	$\dfrac{9}{280}$	$\dfrac{41}{840}$	0	0
$\widehat{y}_1$	0	0	0	0	0	$\dfrac{34}{105}$	$\dfrac{9}{35}$	$\dfrac{9}{35}$	$\dfrac{9}{280}$	$\dfrac{9}{280}$	0	$\dfrac{41}{840}$	$\dfrac{41}{840}$

Fehlberg's methods suffer from the fact that they give identically zero error estimates for quadrature problems $y' = f(x)$. The first high order embedded formulas which avoid this drawback were constructed by Verner (1978). One of Verner's methods (see Table 5.4) has been implemented by T.E. Hull, W.H. Enright and K.R. Jackson as DVERK and is widely used.

Table 5.4. Verner's method of order 6(5) (DVERK)

0								
$\dfrac{1}{6}$	$\dfrac{1}{6}$							
$\dfrac{4}{15}$	$\dfrac{4}{75}$	$\dfrac{16}{75}$						
$\dfrac{2}{3}$	$\dfrac{5}{6}$	$-\dfrac{8}{3}$	$\dfrac{5}{2}$					
$\dfrac{5}{6}$	$-\dfrac{165}{64}$	$\dfrac{55}{6}$	$-\dfrac{425}{64}$	$\dfrac{85}{96}$				
1	$\dfrac{12}{5}$	-8	$\dfrac{4015}{612}$	$-\dfrac{11}{36}$	$\dfrac{88}{255}$			
$\dfrac{1}{15}$	$-\dfrac{8263}{15000}$	$\dfrac{124}{75}$	$-\dfrac{643}{680}$	$-\dfrac{81}{250}$	$\dfrac{2484}{10625}$	0		
1	$\dfrac{3501}{1720}$	$-\dfrac{300}{43}$	$\dfrac{297275}{52632}$	$-\dfrac{319}{2322}$	$\dfrac{24068}{84065}$	0	$\dfrac{3850}{26703}$	
y_1	$\dfrac{3}{40}$	0	$\dfrac{875}{2244}$	$\dfrac{23}{72}$	$\dfrac{264}{1955}$	0	$\dfrac{125}{11592}$	$\dfrac{43}{616}$
$\widehat{y}_1$	$\dfrac{13}{160}$	0	$\dfrac{2375}{5984}$	$\dfrac{5}{16}$	$\dfrac{12}{85}$	$\dfrac{3}{44}$	0	0

An 8th Order Embedded Method

The first high order methods with small error constants of the *higher* order solution were constructed by Prince & Dormand (1981, Code DOPRI8 of the first edition). In the following we describe the construction of a new Dormand & Prince pair of order 8(6) which will also allow a cheap and accurate dense output (see Section II.6). This method has been announced, but not published, in Dormand & Prince (1989, p. 983). We are grateful to P. Prince for mailing us the coefficients and for his help in recovering their construction.

The essential difficulty for the construction of a high order Runge-Kutta method is to set up a "good" reduced system which implies all order conditions of Theorem 2.13. At the same time it should be simple enough to be easily solved. In extending the ideas for the construction of a 5th order process (see above), Dormand & Prince proceed as follows:

Reduced system. Suppose $s = 12$ and consider for the coefficients c_i, b_i and a_{ij} the equations:

$$\sum_{i=1}^{s} b_i c_i^{q-1} = 1/q, \qquad q = 1,\dots,8 \tag{5.20a}$$

$$\sum_{j=1}^{i-1} a_{ij} = c_i, \qquad i = 1,\dots,s \tag{5.20b}$$

$$\sum_{j=1}^{i-1} a_{ij} c_j = c_i^2/2, \qquad i = 3,\dots,s \tag{5.20c}$$

$$\sum_{j=1}^{i-1} a_{ij}c_j^2 = c_i^3/3, \qquad i = 3, \ldots, s \tag{5.20d}$$

$$\sum_{j=1}^{i-1} a_{ij}c_j^3 = c_i^4/4, \qquad i = 6, \ldots, s \tag{5.20e}$$

$$\sum_{j=1}^{i-1} a_{ij}c_j^4 = c_i^5/5, \qquad i = 6, \ldots, s \tag{5.20f}$$

$$b_2 = b_3 = b_4 = b_5 = 0 \tag{5.20g}$$

$$a_{i2} = 0 \quad \text{for} \quad i \geq 4, \qquad a_{i3} = 0 \quad \text{for} \quad i \geq 6 \tag{5.20h}$$

$$\sum_{i=j+1}^{s} b_i a_{ij} = b_j(1 - c_j), \qquad j = 4, 5, 10, 11, 12 \tag{5.20i}$$

$$\sum_{i=j+1}^{s} b_i c_i a_{ij} = 0, \qquad j = 4, 5 \tag{5.20j}$$

$$\sum_{i=j+1}^{s} b_i c_i^2 a_{ij} = 0, \qquad j = 4, 5 \tag{5.20k}$$

$$\sum_{i=k+2}^{s} b_i c_i \sum_{j=k+1}^{i-1} a_{ij}a_{jk} = 0, \qquad k = 4, 5 \tag{5.20l}$$

$$\sum_{i=1}^{s} b_i c_i \sum_{j=1}^{i-1} a_{ij}c_j^5 = 1/48. \tag{5.20m}$$

Verification of the order conditions. The equations (5.20a) are the order conditions for the bushy trees $[\tau, \ldots, \tau]$ and (5.20m) is that for the tree $[\tau, [\tau, \tau, \tau, \tau, \tau]]$. For the verification of further order conditions we shall show that the reduced system implies

$$\sum_{i=j+1}^{s} b_i a_{ij} = b_j(1 - c_j) \qquad \text{for all } j. \tag{5.21}$$

If we denote the difference by $d_j = \sum_{i=j+1}^{s} b_i a_{ij} - b_j(1 - c_j)$ then $d_2 = d_3 = 0$ by (5.20g,h) and $d_4 = d_5 = d_{10} = d_{11} = d_{12} = 0$ by (5.20i). The conditions (5.20a-g) imply

$$\sum_{j=1}^{s} d_j c_j^{q-1} = 0 \qquad \text{for} \quad q = 1, \ldots, 5. \tag{5.22}$$

Hence, the remaining 5 values must also vanish if c_1, c_6, c_7, c_8, c_9 are distinct. The significance of condition (5.21) is already known from Lemma 1.3 and from formula (6.6). It implies that all one-leg trees $t = [t_1]$ can be disregarded.

Fig. 5.2. Use of simplifying assumptions

Conditions (5.20c-f) are an extension of (5.6) and (5.15). Their importance will be, once more, demonstrated on an example. Consider the two trees of Fig. 5.2 and suppose that their encircled parts are identical. Then the corresponding order conditions are

$$\sum_{i,j=1}^{s} \Theta_i a_{ij} c_j^3 = \frac{1}{r \cdot 5 \cdot 4} \quad \text{and} \quad \sum_{i=1}^{s} \Theta_i c_i^4 = \frac{1}{r \cdot 5} \tag{5.23}$$

with known values for Θ_i and r. If (5.20e) is satisfied and if

$$\Theta_2 = \Theta_3 = \Theta_4 = \Theta_5 = 0 \tag{5.24}$$

then both conditions are equivalent so that the left-hand tree can be neglected. The conditions (5.20g,i-l) correspond to (5.24) for certain trees. Finally the assumption (5.20h) together with (5.20g,i-k) implies that for arbitrary Φ_i, Ψ_j and for $q \in \{1,2,3\}$,

$$\sum_i b_i \Phi_i a_{i2} = 0$$
$$\sum_{i,j} b_i \Phi_i a_{ij} \Psi_j a_{j2} = 0 \qquad \text{and} \qquad \begin{aligned} &\sum_i b_i \Phi_i a_{i3} = 0 \\ &\sum_{i,j} b_i c_i^{q-1} a_{ij} \Phi_j a_{j3} = 0 \end{aligned}$$
$$\sum_{i,j,k} b_i c_i^{q-1} a_{ij} \Phi_j a_{jk} \Psi_k a_{k2} = 0$$

which are again conditions of type (5.24). Using these relations the verification of the order conditions (order 8) is straightforward; all trees are reduced to those corresponding to (5.20a) and (5.20m).

Solving the reduced system. Compared to the original 200 order conditions of Theorem 2.13 for the 78 coefficients b_i, a_{ij} (the c_i are defined by (5.20b)), the 74 conditions of the reduced system present a considerable simplification. We can hope for a solution with 4 degrees of freedom.

We start by expressing the coefficients b_i, a_{ij} in terms of the c_i. Because of (5.20g), condition (5.20a) represents a linear system for $b_1, b_6, \ldots, b_{12}$, which has a unique solution if $c_1, c_6, \ldots, c_{12}$ are distinct. For a fixed i ($1 \le i \le 8$) conditions (5.20b-f) represent a linear system for $a_{i1}, \ldots, a_{i,i-1}$. Since there are sometimes less unknowns than equations (mainly due to (5.20h)) restrictions have to be imposed on the c_i. One verifies (similarly to (5.18)) that the relations

$$c_1 = 0, \qquad c_2 = \frac{2}{3} c_3, \qquad c_3 = \frac{2}{3} c_4,$$
$$c_4 = \frac{6 - \sqrt{6}}{10} c_6, \qquad c_5 = \frac{6 + \sqrt{6}}{10} c_6, \qquad c_6 = \frac{4}{3} c_7 \tag{5.25a}$$

allow the computation of the a_{ij} with $i \le 8$ (Step 1 in Fig. 5.3).

If $b_{12} \ne 0$ (which will be assumed in our construction), condition (5.20i) for $j = 12$ implies

$$c_{12} = 1, \tag{5.25b}$$

and for $j = 11$ it yields the value for $a_{12,11}$. We next compute the expressions

$$e_j = \sum_{i=j+1}^{s} b_i c_i a_{ij} - \frac{b_j}{2} (1 - c_j^2), \qquad j = 1, \ldots, s. \tag{5.26}$$

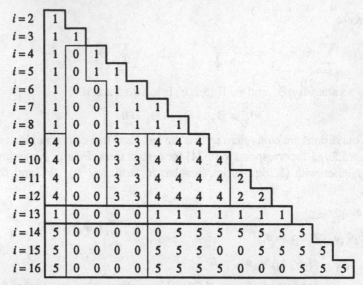

Fig. 5.3. Steps in the construction of an 8th order RK method;
the entries 0 indicate vanishing coefficients;
the stages $i = 14, 15, 16$ will be used for dense output, see II.6.

We have $e_{12} = 0$ by (5.25b), $e_{11} = b_{12}a_{12,11} - b_{11}(1 - c_{11}^2)/2$ is known and $e_2 = e_3 = e_4 = e_5 = 0$ by (5.20g,h,j). The remaining 6 values are determined by the system

$$\sum_{j=1}^{s} e_j c_j^{q-1} = 0, \qquad q = 1, \ldots, 6 \tag{5.27}$$

which follows from (5.20a-f,m). The conditions (5.20i) and (5.26) for $j = 10$ then yield $a_{12,10}$ and $a_{11,10}$ (Step 2 in Fig. 5.3).

We next compute a_{ij} ($i = 9, 10, 11, 12$; $j = 4, 5$) from the remaining 8 equations of (5.20i-l). This is indicated as Step 3 in Fig. 5.3. Finally, we use the conditions (5.20b-f) with $i \geq 9$ for the computation of the remaining coefficients (Step 4). A difficulty still arises from the case $i = 9$, where only 4 parameters for five equations are at our disposal. A tedious computation shows that this system has a solution if (see Exercise 6 below)

$$2c_9 = \frac{3\sigma_1 - 28\sigma_2 + 189\sigma_3 + 14\sigma_1\sigma_2 - 168\sigma_1\sigma_3 + 98\sigma_2\sigma_3}{6 - 21\sigma_1 + 35\sigma_2 - 42\sigma_3 + 21\sigma_1^2 + 98\sigma_2^2 + 735\sigma_3^2 - 84\sigma_1\sigma_2 + 168\sigma_1\sigma_3 - 490\sigma_2\sigma_3} \tag{5.25c}$$

where

$$\sigma_1 = c_6 + c_7 + c_8, \qquad \sigma_2 = c_6c_7 + c_6c_8 + c_7c_8, \qquad \sigma_3 = c_6c_7c_8. \tag{5.28}$$

The reduced system (5.20) leaves c_7, c_8, c_{10}, c_{11} as free parameters. Dormand

& Prince propose the following numerical values:

$$c_7 = 1/4, \qquad c_8 = 4/13, \qquad c_{10} = 3/5, \qquad c_{11} = 6/7 .$$

All remaining coefficients are then determined by the above procedure. Since c_4 and c_5 (see (5.25a)) are not rational, there is no easy way to present the coefficients in a tableau.

Embedded method. We look for a second method with the same c_i, a_{ij} but with different weights, say $\widehat{b}_i$. If we require that

$$\sum_{i=1}^{s} \widehat{b}_i c_i^{q-1} = 1/q, \qquad q = 1, \dots, 6 \qquad (5.29a)$$

$$\widehat{b}_2 = \widehat{b}_3 = \widehat{b}_4 = \widehat{b}_5 = 0 \qquad (5.29b)$$

$$\sum_{i=j+1}^{s} \widehat{b}_i a_{ij} = 0, \qquad j = 4, 5 \qquad (5.29c)$$

then one can verify (similarly as above for the 8th order method) that the corresponding Runge-Kutta method is of order 6. The system (5.29) consists of 12 linear equations for 12 unknowns. A comparison with (5.20) shows that $b_1, \dots, b_{12}$ is a solution of (5.29). Furthermore, the corresponding homogeneous system has the nontrivial solution $e_1, \dots, e_{12}$ (see (5.27) and (5.20l)). Therefore

$$\widehat{b}_i = b_i + \alpha e_i \qquad (5.30)$$

is a solution of (5.29) for all values of α. Dormand & Prince suggest taking α in such a way that $\widehat{b}_6 = 2$.

A program based on this method (with a different error estimator, see Section II.10) has been written and is called DOP853. It is documented in the Appendix. The performance of this code, compared to methods of lower order, is impressive. See for example the results for the Brusselator in Fig. 4.2.

Exercises

1. Consider a Runge-Kutta method with s stages that satisfies (5.7)-(5.8), (5.15), (5.17) and the first two relations of (5.16).

   a) If the relation (5.19) holds, then the method possesses an embedded formula of order 4.

   b) The condition (5.19) implies that the last relation of (5.16) is automatically satisfied.

   Hint. The order conditions for the embedded method constitute a linear system for the $\widehat{b}_i$ which has to be singular. This implies that

$$a_{i2} = \alpha c_i + \beta c_i^2 + \gamma c_i^3 \qquad \text{for} \qquad i \neq 2. \qquad (5.31)$$

Multiplying (5.31) with b_i and $b_i c_i$ and summing up, yields two relations for $\alpha, \beta, J\gamma$. These together with (5.31) for $i = 3, 4$ yield (5.19).

2. Construct a 6-stage 5th order formula with $c_3 = 1/3$, $c_4 = 1/2$, $c_5 = 2/3$ possessing an embedded formula of order 4.

3. (Butcher). Show that for any Runge-Kutta method of order 5,

$$\sum_i b_i \left(\sum_j a_{ij} c_j - \frac{c_i^2}{2} \right)^2 = 0.$$

Consequently, there exists no explicit Runge-Kutta method of order 5 with all $b_i > 0$.

Hint. Multiply out and use order conditions.

4. Write a code with a high order Runge-Kutta method (or take one) and solve numerically the Arenstorf orbit of the restricted three body problem (0.1) (see the introduction) with initial values

$$y_1(0) = 0.994, \qquad y_1'(0) = 0, \qquad y_2(0) = 0,$$
$$y_2'(0) = -2.00317326295573368357302057924,$$

Compute the solutions for

$$x_{\mathrm{end}} = 11.124340337266085134999734047.$$

The initial values are chosen such that the solution is periodic to this precision. The plotted solution curve has one loop less than that of the introduction.

5. (Shampine 1979). Show that the storage requirement of a Runge-Kutta method can be substantially decreased if s is large.

Hint. Suppose, for example, that $s = 15$. After computing (see (1.8)) $k_1, k_2, \ldots, k_9$, compute the sums

$$\sum_{j=1}^{9} a_{ij} k_j \quad \text{for } i = 10, 11, 12, 13, 14, 15, \qquad \sum_{j=1}^{9} b_j k_j, \qquad \sum_{j=1}^{9} \widehat{b}_j k_j;$$

then the memories occupied by $k_2, k_3, \ldots, k_9$ are not needed any longer. Another possibility for reducing the memory requirement is offered by the zero-pattern of the coefficients.

6. Show that the reduced system (5.20) implies (5.25c).

Hint. The equations (5.20b-f) imply that for $i \in \{1, 6, 7, 8, 9\}$

$$\alpha a_{i4} + \beta a_{i5} = \sigma_3 \frac{c_i^2}{2} - \sigma_2 \frac{c_i^3}{3} + \sigma_1 \frac{c_i^4}{4} - \frac{c_i^5}{5} \tag{5.32}$$

with σ_j given by (5.28). The constants α and β are not important. Further, for the same values of i one has

$$0 = c_i(c_i - c_6)(c_i - c_7)(c_i - c_8)(c_i - c_9) \qquad (5.33)$$
$$= \sigma_3 c_9 c_i - (\sigma_3 + c_9 \sigma_2)c_i^2 + (\sigma_2 + c_9 \sigma_1)c_i^3 - (\sigma_1 + c_9)c_i^4 + c_i^5.$$

Multiplying (5.32) and (5.33) by $e_i, b_i, b_i c_i, b_i c_i^2$, summing up from $i = 1$ to s and using (5.20) gives the relation

$$\begin{pmatrix} \times & \times & \times \\ \times & \times & \times \\ 0 & 0 & b_{12}^{-1} \end{pmatrix} \begin{pmatrix} e_{10} & b_{10} & b_{10}c_{10} & b_{10}c_{10}^2 \\ e_{11} & b_{11} & b_{11}c_{11} & b_{11}c_{11}^2 \\ 0 & b_{12} & b_{12} & b_{12} \end{pmatrix} = \begin{pmatrix} 0 & \gamma_1 & \gamma_2 & \gamma_3 \\ 0 & \delta_1 & \delta_2 & \delta_3 \\ 0 & 1 & 1 & 1 \end{pmatrix}$$
$$(5.34)$$

where

$$\gamma_j = \frac{\sigma_3}{2 \cdot (j+2)} - \frac{\sigma_2}{3 \cdot (j+3)} + \frac{\sigma_1}{4 \cdot (j+4)} - \frac{1}{5 \cdot (j+5)}$$
$$\delta_j = \frac{\sigma_3 c_9}{j+1} - \frac{\sigma_3 + c_9 \sigma_2}{j+2} + \frac{\sigma_2 + c_9 \sigma_1}{j+3} - \frac{\sigma_1 + c_9}{j+4} + \frac{1}{j+5}$$

and the "$\times$" indicate certain values. Deduce from (5.34) and $e_{11} \neq 0$ that the most left matrix of (5.34) is singular. This implies that the right-hand matrix of (5.34) is of rank 2 and yields equation (5.25c).

7. Prove that the 8th order method given by (5.20; $s = 12$) does not possess a 6th order embedding with $\widehat{b}_{12} \neq b_{12}$, not even if one adds the numerical result y_1 as 13th stage (FSAL).

II.6 Dense Output, Discontinuities, Derivatives

> ... providing "interpolation" for Runge-Kutta methods. ... this
> capability and the features it makes possible will be the hallmark
> of the next generation of Runge-Kutta codes.
>
> (L.F. Shampine 1986)

The present section is mainly devoted to the construction of dense output formulas for Runge-Kutta methods. This is important for many practical questions such as graphical output, event location or the treatment of discontinuities in differential equations. Further, the numerical computation of derivatives with respect to initial values and parameters is discussed, which is particularly useful for the integration of boundary value problems.

Dense Output

Classical Runge-Kutta methods are inefficient, if the number of output points becomes very large (Shampine, Watts & Davenport 1976). This motivated the construction of dense output formulas (Horn 1983). These are Runge-Kutta methods which provide, in addition to the numerical result y_1, cheap numerical approximations to $y(x_0 + \theta h)$ for the *whole* integration interval $0 \le \theta \le 1$. "Cheap" means without or, at most, with only a few additional function evaluations.

We start from an s-stage Runge-Kutta method with given coefficients c_i, a_{ij} and b_j, eventually add $s^* - s$ new stages, and consider formulas of the form

$$u(\theta) = y_0 + h \sum_{i=1}^{s^*} b_i(\theta) k_i, \tag{6.1}$$

where

$$k_i = f\left(x_0 + c_i h, y_0 + h \sum_{j=1}^{i-1} a_{ij} k_j\right), \qquad i = 1, \dots, s^* \tag{6.2}$$

and $b_i(\theta)$ are polynomials to be determined such that

$$u(\theta) - y(x_0 + \theta h) = \mathcal{O}(h^{p^*+1}). \tag{6.3}$$

Usually $s^* \ge s + 1$ since we include (at least) the first function evaluation of the subsequent step $k_{s+1} = h f(x_0 + h, y_1)$ in the formula with $a_{s+1,j} = b_j$ for all j. A Runge-Kutta method, provided with a formula (6.1), will be called a *continuous* Runge-Kutta method.

Theorem 6.1. *The error of the approximation (6.1) is of order p^* (i.e., the local error satisfies (6.3)), if and only if*

$$\sum_{j=1}^{s^*} b_j(\theta) \Phi_j(t) = \frac{\theta^{\varrho(t)}}{\gamma(t)} \qquad for \qquad \varrho(t) \leq p^* \tag{6.4}$$

with $\Phi_j(t)$, $\varrho(t)$, $\gamma(t)$ given in Section II.2.

Proof. The qth derivative (with respect to h) of the numerical approximation is given by (2.14) with b_j replaced by $b_j(\theta)$; that of the exact solution $y(x_0 + \theta h)$ is $\theta^q y^{(q)}(x_0)$. The statement thus follows as in Theorem 2.13.   □

Corollary 6.2. *Condition (6.4) implies that the derivatives of (6.1) approximate the derivatives of the exact solution as*

$$h^{-k} u^{(k)}(\theta) - y^{(k)}(x_0 + \theta h) = \mathcal{O}(h^{p^* - k + 1}). \tag{6.5}$$

Proof. Comparing the qth derivative (with respect to h) of $u'(\theta)$ with that of $hy'(x_0 + \theta h)$ we find that (6.5) (for $k = 1$) is equivalent to

$$\sum_{j=1}^{s^*} b_j'(\theta) \Phi_j(t) = \frac{\varrho(t) \theta^{\varrho(t)-1}}{\gamma(t)} \qquad for \qquad \varrho(t) \leq p^*.$$

This, however, follows from (6.4) by differentiation. The case $k > 1$ is obtained similarly.   □

We write the polynomials $b_j(\theta)$ as

$$b_j(\theta) = \sum_{q=1}^{p^*} b_{jq} \theta^q, \tag{6.6}$$

so that the equations (6.4) become a system of simultaneous linear equations of the form

$$\underbrace{\begin{pmatrix} 1 & 1 & .. & 1 \\ \Phi_1(t_{21}) & \Phi_2(t_{21}) & .. & \Phi_{s^*}(t_{21}) \\ \Phi_1(t_{31}) & \Phi_2(t_{31}) & .. & \Phi_{s^*}(t_{31}) \\ \vdots & \vdots & & \vdots \end{pmatrix}}_{\Phi} \underbrace{\begin{pmatrix} b_{11} & b_{12} & b_{13} & .. \\ b_{21} & b_{22} & b_{23} & .. \\ \vdots & \vdots & \vdots & \\ b_{s^*1} & b_{s^*2} & b_{s^*3} & .. \end{pmatrix}}_{B} = \underbrace{\begin{pmatrix} 1 & 0 & 0 & .. \\ 0 & \frac{1}{2} & 0 & .. \\ 0 & 0 & \frac{1}{3} & .. \\ 0 & 0 & \frac{1}{6} & .. \\ \vdots & \vdots & \vdots & \end{pmatrix}}_{G} \tag{6.4'}$$

where the $\Phi_j(t)$ are known numbers depending on a_{ij} and c_i. Using standard linear algebra the solution of this system can easily be discussed. It may happen, however, that the order p^* of the dense output is smaller than the order p of the underlying method.

Example. For "the" Runge-Kutta method of Table 1.2 (with $s^* = s = 4$) equations (6.4') with $p^* = 3$ produce a unique solution

$$b_1(\theta) = \theta - \frac{3\theta^2}{2} + \frac{2\theta^3}{3}, \qquad b_2(\theta) = b_3(\theta) = \theta^2 - \frac{2\theta^3}{3}, \qquad b_4(\theta) = -\frac{\theta^2}{2} + \frac{2\theta^3}{3}$$

which constitutes a dense output solution which is globally continuous but not C^1.

Hermite interpolation. A much easier way (than solving (6.4')) and more efficient for low order dense output formulas is the use of Hermite interpolation (Shampine 1985). Whatever the method is, we have two function values y_0, y_1 and two derivatives $f_0 = f(x_0, y_0)$, $f_1 = f(x_0 + h, y_1)$ at our disposal and can thus do cubic polynomial interpolation. The resulting formula is

$$u(\theta) = (1-\theta)y_0 + \theta y_1 + \theta(\theta - 1)\Big((1 - 2\theta)(y_1 - y_0) + (\theta - 1)hf_0 + \theta h f_1\Big). \quad (6.7)$$

Inserting the definition of y_1 into (6.7) shows that Hermite interpolation is a special case of (6.1). Whenever the underlying method is of order $p \geq 3$ we thus obtain a continuous Runge-Kutta method of order 3.

Since the function and derivative values on the right side of the first interval coincide with those on the left side of the second interval, Hermite interpolation leads to a globally C^1 approximation of the solution.

The 4-stage 4th order methods of Section II.1 do not possess a dense output of order 4 without any additional function evaluations (see Exercise 1). Therefore the question arises whether it is really important to have a dense output of the same order. Let us consider an interval far away from the initial value, say $[x_n, x_{n+1}]$, and denote by $z(x)$ the local solution, i.e., the solution of the differential equation which passes through (x_n, y_n). Then the error of the dense output is composed of two terms:

$$u(\theta) - y(x_n + \theta h) = \big(u(\theta) - z(x_n + \theta h)\big) + \big(z(x_n + \theta h) - y(x_n + \theta h)\big).$$

The term to the far right reflects the global error of the method and is of size $\mathcal{O}(h^p)$. In order that both terms be of the same order of magnitude it is thus sufficient to require $p^* = p - 1$.

The situation changes, if we also need accurate values of the derivative $y'(x_n + \theta h)$ (see Section 5 of Enright, Jackson, Nørsett & Thomsen (1986) for a discussion of problems where this is important). We have

$$h^{-1}u'(\theta) - y'(x_n + \theta h) = \big(h^{-1}u'(\theta) - z'(x_n + \theta h)\big) + \big(z'(x_n + \theta h) - y'(x_n + \theta h)\big)$$

and the term to the far right is of size $\mathcal{O}(h^p)$ if $f(x, y)$ satisfies a Lipschitz condition. A comparison with (6.5) shows that we need $p^* = p$ in order that both error terms be of comparable size.

Boot-strapping process (Enright, Jackson, Nørsett & Thomsen 1986). This is a general procedure for increasing iteratively the order of dense output formulas.

Suppose that we already have a 3rd order dense output at our disposal (e.g., from Hermite interpolation). We then fix arbitrarily an $\alpha \in (0,1)$ and denote the 3rd order approximation at $x_0 + \alpha h$ by y_α. The idea is now that $hf(x_0 + \alpha h, y_\alpha)$ is a 4th order approximation to $hy'(x_0 + \alpha h)$. Consequently, the 4th degree polynomial $u(\theta)$ defined by

$$
\begin{aligned}
u(0) &= y_0; & u'(0) &= hf(x_0, y_0) \\
u(1) &= y_1, & u'(1) &= hf(x_0 + h, y_1) \\
& & u'(\alpha) &= hf(x_0 + \alpha h, y_\alpha)
\end{aligned}
\tag{6.8}
$$

(which exists uniquely for $\alpha \neq 1/2$) yields the desired formula. The interpolation error is $\mathcal{O}(h^5)$ and each quantity of (6.8) approximates the corresponding exact solution value with an error of $\mathcal{O}(h^5)$.

The extension to arbitrary order is straightforward. Suppose that a dense output formula $u_0(\theta)$ of order $p^* < p$ is known. We then evaluate this polynomial at $p^* - 2$ distinct points $\alpha_i \in (0,1)$ and compute the values $f\big(x_0 + \alpha_i h, u_0(\alpha_i)\big)$. The interpolation polynomial $u_1(\theta)$ of degree $p^* + 1$, defined by

$$
\begin{aligned}
u_1(0) &= y_0, & u_1'(0) &= hf(x_0, y_0) \\
u_1(1) &= y_1, & u_1'(1) &= hf(x_0 + h, y_1) \\
u_1'(\alpha_i) &= hf\big(x_0 + \alpha_i h, u_0(\alpha_i)\big), & i &= 1, \dots p^* - 2,
\end{aligned}
\tag{6.9}
$$

yields an interpolation formula of order $p^* + 1$. Obviously, the α_i in (6.9) have to be chosen such that the corresponding interpolation problem admits a solution.

Continuous Dormand & Prince Pairs

The method of Dormand & Prince (Table 5.2) is of order 5(4) so that we are mainly interested in dense output formulas with $p^* = 4$ and $p^* = 5$.

Order 4. A continuous formula of order 4 can be obtained without any additional function evaluation. Since the coefficients satisfy (5.7), it follows from the difference of the order conditions for the trees t_{31} and t_{32} (notation of Table 2.2) that

$$
b_2(\theta) = 0
\tag{6.10}
$$

is necessary. This condition together with (5.7) and (5.15) then implies that the order conditions are equivalent for the following pairs of trees: t_{31} and t_{32}, t_{41} and t_{42}, t_{41} and t_{43}. Hence, for order 4, only 5 conditions have to be considered (the four quadrature conditions and $\sum_i b_i(\theta)a_{i2} = 0$). We can arbitrarily choose $b_7(\theta)$ and the coefficients $b_1(\theta), b_3(\theta), \dots, b_6(\theta)$ are then uniquely determined.

As for the choice of $b_7(\theta)$, Shampine (1986) proposed minimizing, for each θ, the error coefficients (Theorem 3.2)

$$e(t) = \theta^5 - \gamma(t) \sum_{j=1}^{7} b_j(\theta) \Phi_j(t) \qquad \text{for} \qquad t \in T_5, \qquad (6.11)$$

weighted by $\alpha(t)$ of Definition 2.5, in the square norm. These expressions can be seen to depend linearly on $b_7(\theta)$,

$$\alpha(t)e(t) = \zeta(t,\theta) - b_7(\theta)\eta(t),$$

thus the minimal value is found for

$$b_7(\theta) = \sum_{t \in T_5} \zeta(t,\theta)\eta(t) \Big/ \sum_{t \in T_5} \eta^2(t).$$

The resulting formula, given by Dormand & Prince (1986), is

$$b_7(\theta) = \theta^2(\theta - 1) + \theta^2(\theta - 1)^2 10 \cdot (7414447 - 8293050\theta)/29380423. \qquad (6.12)$$

The other coefficients, written in a fashion which makes the Hermite-part clearly visible, are then given by

$$b_1(\theta) = \theta^2(3 - 2\theta) \cdot b_1 + \theta(\theta - 1)^2$$
$$- \theta^2(\theta - 1)^2 5 \cdot (2558722523 - 314030160\theta)/11282082432$$
$$b_3(\theta) = \theta^2(3 - 2\theta) \cdot b_3 + \theta^2(\theta - 1)^2 100 \cdot (882725551 - 157015080\theta)/32700410799$$
$$b_4(\theta) = \theta^2(3 - 2\theta) \cdot b_4 - \theta^2(\theta - 1)^2 25 \cdot (443332067 - 314030160\theta)/1880347072$$
$$b_5(\theta) = \theta^2(3 - 2\theta) \cdot b_5 + \theta^2(\theta - 1)^2 32805 \cdot (23143187 - 34892240\theta)/199316789632$$
$$b_6(\theta) = \theta^2(3 - 2\theta) \cdot b_6 - \theta^2(\theta - 1)^2 55 \cdot (29972135 - 70767360\theta)/822651844. $$

$$(6.13)$$

It can be directly verified that the interpolation polynomial $u(\theta)$ defined by (6.10), (6.12) and (6.13) satisfies

$$\begin{aligned} u(0) &= y_0, & u'(0) &= hf(x_0, y_0), \\ u(1) &= y_1, & u'(1) &= hf(x_0 + h, y_1), \end{aligned} \qquad (6.14)$$

so that it produces globally a C^1 approximation of the solution.

Instead of using the above 5th degree polynomial $u(\theta)$, Shampine (1986) suggests evaluating it only at the midpoint, $y_{1/2} = u(1/2)$, and then doing quartic polynomial interpolation with the five values y_0, $hf(x_0, y_0)$, y_1, $hf(x_0 + h, y_1)$, $y_{1/2}$. This dense output is also C^1, is easier to implement and the difference to the above formula "... is not significant" (Dormand & Prince 1986).

We have implemented Shampine's dense output in the code DOPRI5 (see Appendix). The advantages of such a dense output for graphical representations of the solution can already be seen from Fig. 0.1 of the introduction to Chapter II. For a more thorough study we have applied DOPRI5 to the Brusselator (4.15) with initial values $y_1(0) = 1.5$, $y_2(0) = 3$, integration interval $0 \leq x \leq 10$ and error tolerance $Atol = Rtol = 10^{-4}$. The global error of the above 4th order continuous solution is displayed in Fig. 6.1 for both components. The error shows the same quality throughout; the grid points, which are represented by the symbols □ and ○, are by no means outstanding.

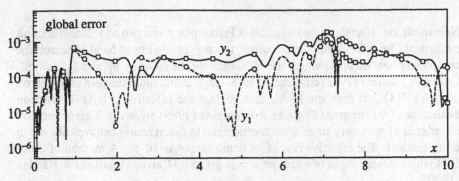

Fig. 6.1. Error of dense output of DOPRI5

Order 5. For a dense output of order $p^* = 5$ for the Dormand & Prince method the linear system (6.4') has no solution since

$$\text{rank}\,(\Phi|G) = 9 \qquad \text{and} \qquad \text{rank}\,(\Phi) = 7 \qquad (6.15)$$

as can be verified by Gaussian elimination. Such a linear system has a solution if and only if the two ranks in (6.15) are *equal*. So we must append additional stages to the method. Each new stage adds a new column to the matrix Φ, thus may increase the rank of Φ by one without changing rank $(\Phi|G)$. Therefore we obtain

Lemma 6.3 (Owren & Zennaro 1991). *Consider a Runge-Kutta method of order p. For the construction of a continuous extension of order $p^* = p$ one has to add at least*

$$\delta := \text{rank}\,(\Phi|G) - \text{rank}\,(\Phi) \qquad (6.16)$$

stages.   □

For the Dormand & Prince method we thus need at least two additional stages. There are several possibilities for constructing such dense output formulas:

a) Shampine (1986) shows that one new function evaluation allows one to compute a 5th order approximation at the midpoint $x_0 + h/2$. If one evaluates anew the function at this point to get an approximation of $y'(x_0 + h/2)$, one can do quintic Hermite interpolation to get a dense output of order 5.

b) Use the 4th order formula constructed above at two different output points and do boot-strapping. This has been done by Calvé & Vaillancourt (1990).

c) Add two arbitrary new stages and solve the order conditions. This leads to methods with 10 free parameters (Calvo, Montijano & Rández 1992) which can then be used to minimize the error terms. This seems to give the best output formulas.

New methods. If anyhow the Dormand & Prince pair needs two additional function evaluations for a 5th order dense output, the suggestion lies at hand to search for completely new methods which use *all* stages for the solution y_1 and $\widehat{y}_1$ as well. Owren & Zennaro (1992) constructed an 8-stage continuous Runge-Kutta method of order 5(4). It uses the FSAL idea so that the effective cost is 7 function evaluations (*fe*) per step. Bogacki & Shampine (1989) present a 7-stage method of order 5(4) with very small error coefficients, so that it nearly behaves like a 6th order method. The effective cost of its dense output is 10 *fe*. A method of order 6(5) with a dense output of order $p^* = 5$ is given by Calvo, Montijano & Rández (1990).

Dense Output for DOP853

We are interested in a continuous extension of the 8th order method of Section II.5 (formula (5.20)). A dense output of order 6 can be obtained for free (add y_1 as 13th stage and solve the linear system (6.19a-c) below with $s^* = s + 1 = 13$). Following Dormand & Prince we shall construct a dense output of order $p^* = 7$. We add three further stages (by Lemma 6.3 this is the minimal number of additional stages). The values for c_{14}, c_{15}, c_{16} are chosen arbitrarily as

$$c_{14} = 0.1, \qquad c_{15} = 0.2, \qquad c_{16} = 7/9 \qquad (6.17)$$

and the coefficients a_{ij} are assumed to satisfy, for $i \in \{14, 15, 16\}$,

$$\sum_{j=1}^{i-1} a_{ij} c_j^{q-1} = c_i^q/q, \qquad q = 1, \dots, 6 \qquad (6.18a)$$

$$a_{i2} = a_{i3} = a_{i4} = a_{i5} = 0 \qquad (6.18b)$$

$$\sum_{j=k+1}^{i-1} a_{ij} a_{jk} = 0, \qquad k = 4, 5. \qquad (6.18c)$$

This system can easily be solved (step 5 of Fig. 5.3). We are still free to set some coefficients equal to 0 (see Fig. 5.3).

We next search for polynomials $b_i(\theta)$ such that the conditions (6.4) are satisfied for all trees of order ≤ 7. We find the following necessary conditions ($s^* = 16$)

$$\sum_{i=1}^{s^*} b_i(\theta) c_i^{q-1} = \theta^q/q, \qquad q = 1, \dots, 7 \qquad (6.19a)$$

$$b_2(\theta) = b_3(\theta) = b_4(\theta) = b_5(\theta) = 0 \qquad (6.19b)$$

$$\sum_{i=j+1}^{s^*} b_i(\theta) a_{ij} = 0, \qquad j = 4, 5 \qquad (6.19c)$$

$$\sum_{i=j+1}^{s^*} b_i(\theta)c_i a_{ij} = 0, \qquad j = 4,5 \qquad\qquad (6.19d)$$

$$\sum_{i,j=1}^{s^*} b_i(\theta)a_{ij} c_j^5 = \theta^7/42. \qquad\qquad (6.19e)$$

Here (6.19a,e) are order conditions for $[\tau, \ldots, \tau]$ and $[[\tau, \tau, \tau, \tau, \tau]]$. The property $b_2(\theta) = 0$ follows from $0 = \sum_i b_i(\theta)(\sum_j a_{ij}c_j - c_i^2/2) = -b_2(\theta)c_2^2/2$ and the other three conditions of (6.19b) are a consequence of the relations $0 = \sum_i b_i(\theta)c_i^{q-1}(\sum_j a_{ij}c_j^3 - c_i^4/4) = 0$ for $q = 1,2,3$. The necessity of the conditions (6.19c,d) is seen similarly.

On the other hand, the conditions (6.19) are also sufficient for the dense output to be of order 7. We first remark that (6.19), (6.18) and (5.20) imply

$$\sum_{i,j=k+1}^{s^*} b_i(\theta)a_{ij}a_{jk} = 0, \qquad k = 4,5 \qquad\qquad (6.20)$$

(see Exercise 3). The verification of the order conditions (6.4) is then possible without difficulty.

System (6.19) consists of 16 linear equations for 16 unknowns which possess a unique solution. An interesting property of the continuous solution (6.1) obtained in this manner is that it yields a global C^1-approximation to the solution, i.e.,

$$u(0) = y_0, \qquad u(1) = y_1, \qquad u'(0) = hf(y_0), \qquad u'(1) = hf(y_1). \qquad (6.21)$$

For the verification of this property we define a polynomial $q(\theta)$ of degree 7 by the relations (6.21) and by $q(\theta_i) = u(\theta_i)$ for 4 distinct values θ_i which are different from 0 and 1. Obviously, $q(\theta)$ is of the form (6.1) and defines a dense output of order 7. Due to the uniqueness of the $b_i(\theta)$ we must have $q(\theta) \equiv u(\theta)$ so that (6.21) is verified.

Event Location

Often the output value x_{end} for which the solutions are wanted is not known in advance, but depends implicitly on the computed solutions. An example of such a situation is the search for periodic solutions and limit cycles discussed in Section I.16, where we wanted to know when the solution reaches the Poincaré-section for the first time.

Such problems are very easily treated when a dense output $u(x)$ is available. Suppose we want to determine x such that

$$g(x, y(x)) = 0. \qquad\qquad (6.22)$$

Algorithm 6.4. Compute the solution step-by-step until a sign change appears between $g(x_i, y_i)$ and $g(x_{i+1}, y_{i+1})$ (this is, however, not completely safe because g may change sign twice in an integration interval; use the dense output at intermediate values if more safety is needed). Then replace $y(x)$ in (6.22) by the approximation $u(x)$ and solve the resulting equation numerically, e.g. by bisection or Newton iterations.

This algorithm can be conveniently done in the subroutine SOLOUT, which is called after every accepted step (see Appendix). If the value of x, satisfying (6.22), has been found, the integration is stopped by setting IRTRN $= -1$.

Whenever the function g of (6.22) also depends on $y'(x)$, it is advisable to use a dense output of order $p^* = p$.

Discontinuous Equations

> If you write some software which is half-way useful, sooner or later someone will use it on discontinuities. You have to scope about ...
> (A.R. Curtis 1986)

In many applications the function defining a differential equation is not analytic or continuous everywhere. A common example is a problem which (at least locally) can be written in the form

$$y' = \begin{cases} f_I(y) & \text{if} \quad g(y) > 0 \\ f_{II}(y) & \text{if} \quad g(y) < 0 \end{cases} \tag{6.23}$$

with sufficiently differentiable functions g, f_I and f_{II}. The derivative of the solution is thus in general discontinuous on the surface

$$S = \{y; \; g(y) = 0\}.$$

The function $g(y)$ is called a *switching function*.

In order to understand the situations which can occur when the solution of (6.23) meets the surface S in a point y_0 (i.e., $g(y_0) = 0$), we consider the scalar products

$$\begin{aligned} a_I &= \langle \operatorname{grad} g(y_0), f_I(y_0) \rangle \\ a_{II} &= -\langle \operatorname{grad} g(y_0), f_{II}(y_0) \rangle \end{aligned} \tag{6.24}$$

which can be approximated numerically by $a_I \approx g(y_0 + \delta f_I(y_0))/\delta$ with small enough δ. Since the vector $\operatorname{grad} g(y_0)$ points towards the domain of f_I, the inequality $a_I < 0$ tells us that the flow for f_I is "pushing" against S, while for $a_I > 0$ the flow is "pulling". The same argument holds for a_{II} and the flow for f_{II}. Therefore, apart from degenerate cases where either a_I or a_{II} vanishes, we can distinguish the following four cases (see Fig. 6.2):

1) $a_I > 0, a_{II} < 0$: the flow traverses S from $g < 0$ to $g > 0$.

2) $a_I < 0, a_{II} > 0$: the flow traverses S from $g > 0$ to $g < 0$.

3) $a_I > 0, a_{II} > 0$: the flow "pulls" on both sides; the solution is not unique; except in the case of an unhappily chosen initial value, this situation would normally not occur.

4) $a_I < 0, a_{II} < 0$: here *both* flows push against S; the solution is trapped in S and the problem no longer has a classical solution.

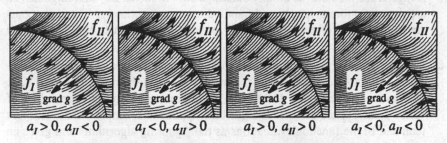

$$a_I > 0, a_{II} < 0 \qquad a_I < 0, a_{II} > 0 \qquad a_I > 0, a_{II} > 0 \qquad a_I < 0, a_{II} < 0$$

Fig. 6.2. Solutions near the surface of discontinuity

Crossing a discontinuity. The *numerical* computation of a solution crossing a discontinuity (cases 1 and 2) can be performed as follows:

a) *Ignoring the discontinuity:* apply a variable step size code with local error control (such as DOPRI5) and hope that the step size mechanism would handle the discontinuity appropriately. Consider the example (which represents the flow of the second picture of Fig. 6.2)

$$y' = \begin{cases} x^2 + 2y^2 & \text{if } (x + 0.05)^2 + (y + 0.15)^2 \le 1 \\ 2x^2 + 3y^2 - 2 & \text{if } (x + 0.05)^2 + (y + 0.15)^2 > 1 \end{cases} \tag{6.25}$$

with initial value $y(0) = 0.3$. The discontinuity for this problem occurs at $x \approx 0.6234$ and the code, applied with $Atol = Rtol = 10^{-5}$, detects the discontinuity fairly well by means of numerous rejected steps (see Fig. 6.3; this figure, however, is much less dramatic than an analogous drawing (see Gear & Østerby 1984) for multistep methods). The numerical solution for $x = 1$ then has an error of $5.9 \cdot 10^{-4}$.

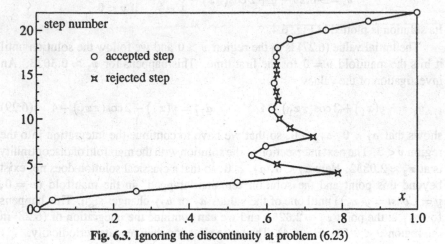

Fig. 6.3. Ignoring the discontinuity at problem (6.23)

b) *Singularity detecting codes.* Concepts have been developed (Gear & Østerby (1984) for multistep methods, Enright, Jackson, Nørsett & Thomsen (1988) for Runge-Kutta methods) to modify existing codes in such a way that singularities are detected more precisely and handled more appropriately. These concepts are mainly based on the behaviour of the local error estimate compared to the step size.

c) *Use the switching function:* stop the computation at the surface of discontinuity using Algorithm 6.4 and restart the integration with the new right-hand side. One has to take care that during one integration step only function values of either f_I or f_{II} are used. This algorithm, applied to Example (6.25), uses less than half of the function evaluations as the "ignoring algorithm" and gives an error of $6.6 \cdot 10^{-6}$ at the point $x = 1$. It is thus not only faster, but also much more reliable.

Example 6.5. Coulomb's law of friction (Coulomb 1785), which states that the force of friction is *independent* of the speed, gives rise to many situations with discontinuous differential equations. Consider the example (see Den Hartog 1930, Reissig 1954, Taubert 1976)

$$y'' + 2Dy' + \mu \operatorname{sign} y' + y = A\cos(\omega x). \tag{6.26}$$

where the Coulomb-force $\mu \operatorname{sign} y'$ is accompanied by a viscosity term Dy'. We fix the parameters as $D = 0.1$, $\mu = 4$, $A = 2$ and $\omega = \pi$, and choose the initial values

$$y(0) = 3, \qquad y'(0) = 4. \tag{6.27}$$

Equation (6.26), written in the form (6.23), is

$$\begin{aligned} y' &= v \\ v' &= -0.2v - y + 2\cos(\pi x) - \begin{cases} 4 & \text{if } v > 0 \\ -4 & \text{if } v < 0. \end{cases} \end{aligned} \tag{6.28}$$

Its solution is plotted in Fig. 6.4.

The initial value (6.27) is in the region $v > 0$ and we follow the solution until it hits the manifold $v = 0$ for the first time. This happens for $x_1 \approx 0.5628$. An investigation of the values

$$a_I = -y(x_1) + 2\cos(\pi x_1) - 4, \qquad a_{II} = y(x_1) - 2\cos(\pi x_1) - 4 \tag{6.29}$$

shows that $a_I < 0$, $a_{II} > 0$, so that we have to continue the integration into the region $v < 0$. The next intersection of the solution with the manifold of discontinuity is at $x_2 \approx 2.0352$. Here $a_I < 0$, $a_{II} < 0$, so that a classical solution does not exist beyond this point and the solution remains "trapped" in the manifold ($v = 0$, $y = Const = y(x_2)$) until one of the values a_I or a_{II} changes sign. This happens for a_{II} at the point $x_3 \approx 2.6281$ and we can continue the integration of (6.28) in the region $v < 0$ (see Fig. 6.4). The same situation then repeats periodically.

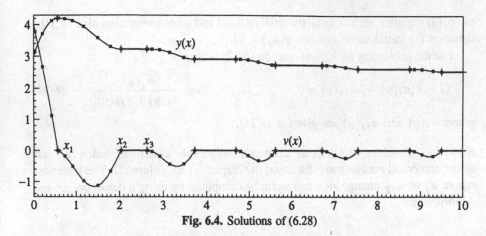

Fig. 6.4. Solutions of (6.28)

Solutions in the manifold. In the case $a_I < 0$, $a_{II} < 0$ the solution of (6.23) can neither be continued along the flow of $y' = f_I(y)$ nor along that of $y' = f_{II}(y)$. However, the physical process, described by the differential equation (6.23), possesses a solution (see Example 6.5). Early papers on this subject studied the convergence of Euler polygons, pushed across the border again and again by the conflicting vector fields (see, e.g., Taubert 1976). Later it became clear that it is much more advantageous to pursue the solution *in* the manifold S, i.e., solve a so-called differential algebraic problem. This approach is advocated by Eich (1992), who attributes the ideas to the thesis of G. Bock, by Eich, Kastner-Maresch & Reich (unpublished manuscript, 1991), and by Stewart (1990). We must decide, however, *which* vector field in S should determine the solution. Several motivations (see Exercises 8 and 9 below) suggest to search this field in the convex hull

$$f(y, \lambda) = (1 - \lambda)f_I(y) + \lambda f_{II}(y), \tag{6.30}$$

of f_I and f_{II}. This coincides, for the special problem (6.23), with Filippov's "generalized solution" (Filippov 1960); but other homotopies may be of interest as well. The value of λ must be chosen in such a way that the solution remains in S. This means that we have to solve the problem

$$y' = f(y, \lambda) \tag{6.31a}$$
$$0 = g(y). \tag{6.31b}$$

Differentiating (6.31b) with respect to time yields

$$0 = \operatorname{grad} g(y)y' = \operatorname{grad} g(y)f(y, \lambda). \tag{6.32}$$

If this relation allows λ to be expressed as a function of y, say as $\lambda = G(y)$, then (6.31a) becomes the ordinary differential equation

$$y' = f(y, G(y)) \tag{6.33}$$

which can be solved by standard integration methods. Obviously, the solution of (6.33) together with $\lambda = G(y)$ satisfy (6.32) and after integration also (6.31b) (because the initial value satisfies $g(y_0) = 0$).

For the homotopy (6.30) the relation (6.32) becomes

$$(1-\lambda)a_I(y) - \lambda a_{II}(y) = 0, \qquad \text{i.e.,} \qquad \lambda = \frac{a_I(y)}{a_I(y) + a_{II}(y)}, \qquad (6.34)$$

where $a_I(y)$ and $a_{II}(y)$ are given in (6.24).

Remark . Problem (6.31) is a "differential-algebraic system of index 2" and direct numerical methods are discussed in Chapter VI of Volume II. The instances where a_I or a_{II} change sign can again be computed by using a dense output and Algorithm 6.4.

Numerical Computation of Derivatives with Respect to Initial Values and Parameters

For the efficient computation of boundary value problems by a shooting technique as explained in Section I.15, we need to compute the derivatives of the solutions with respect to (the missing) initial values. Also, if we want to adjust unknown parameters from given data, say by a nonlinear least squares procedure, we have to compute the derivatives of the solutions with respect to parameters in the differential equation.

We shall restrict our discussion to the problem

$$y' = f(x, y, B), \qquad y(x_0) = y_0(B) \qquad (6.35)$$

where the right-hand side function and the initial values depend on a real parameter B. The generalization to more than one parameter is straightforward. There are several possibilities for computing the derivative $\partial y / \partial B$.

External differentiation. Denote the numerical solution, obtained by a variable step size code with a fixed tolerance, by $y_{Tol}(x_{\text{end}}, x_0, B)$. Then the most simple device is to approximate the derivative by a finite difference

$$\frac{1}{\Delta B}\Big(y_{Tol}(x_{\text{end}}, x_0, B + \Delta B) - y_{Tol}(x_{\text{end}}, x_0, B)\Big). \qquad (6.36)$$

However, due to the error control mechanism with its IF's and THEN's and step rejections, the function $y_{Tol}(x_{\text{end}}, x_0, B)$ is by no means a smooth function of the parameter B. Therefore, the errors of the two numerical results in (6.36) are not correlated, so that the error of (6.36) as an approximation to $\partial y / \partial B(x_{\text{end}}, x_0, B)$ is of size $\mathcal{O}(Tol/\Delta B) + \mathcal{O}(\Delta B)$, the second term coming from the discretization (6.36). This suggests taking for ΔB something like $\sqrt{Tol}$, and the error of (6.36) becomes of size $\mathcal{O}(\sqrt{Tol})$.

Internal differentiation. We know from Section I.14 that $\Psi = \partial y/\partial B$ is the solution of the variational equation

$$\Psi' = \frac{\partial f}{\partial y}(x,y,B)\Psi + \frac{\partial f}{\partial B}(x,y,B), \qquad \Psi(x_0) = \frac{\partial y_0}{\partial B}(B). \qquad (6.37)$$

Here y is the solution of (6.35). Hence, (6.35) and (6.37) together constitute a differential system for y and Ψ, which can be solved simultaneously by any code. If the partial derivatives $\partial f/\partial y$ and $\partial f/\partial B$ are available analytically, then the error of $\partial y/\partial B$, obtained by this procedure, is obviously of size Tol. This algorithm is equivalent to "internal differentiation" as introduced by Bock (1981).

If $\partial f/\partial y$ and $\partial f/\partial B$ are not available one can approximate them by finite differences so that (6.37) becomes

$$\Psi' = \frac{1}{\Delta B}\Big(f(x,y+\Delta B\cdot\Psi,B+\Delta B)-f(x,y,B)\Big). \qquad (6.38)$$

The solution of (6.38), when inserted into (6.37), gives raise to a defect of size $\mathcal{O}(\Delta B)+\mathcal{O}(eps/\Delta B)$, where eps is the precision of the computer (independent of Tol). By Theorem I.10.2, the difference of the solutions of (6.38) and (6.37) is of the same size. Choosing $\Delta B \approx \sqrt{eps}$ the error of the approximation to $\partial y/\partial B$, obtained by solving (6.35), (6.38), will be of order $Tol + \sqrt{eps}$, so that for $Tol \geq \sqrt{eps}$ the result is as precise as that obtained by integration of (6.37). Observe that external differentiation and the numerical solution of (6.35), (6.38) need about the same number of function evaluations.

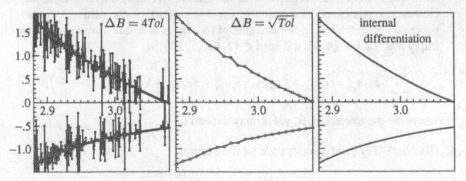

Fig. 6.5. Derivatives of the solution of (6.39) with respect to B

As an example we consider the Brusselator

$$\begin{aligned} y_1' &= 1 + y_1^2 y_2 - (B+1)y_1 & y_1(0) &= 1.3 \\ y_2' &= By_1 - y_1^2 y_2 & y_2(0) &= B \end{aligned} \qquad (6.39)$$

and compute $\partial y/\partial B$ at $x = 20$ for various B ranging from $B = 2.88$ to $B = 3.08$. We applied the code DOPRI5 with $Atol = Rtol = Tol = 10^{-4}$. The numerical result is displayed in Fig. 6.5. External differentiation has been applied, once with $\Delta B = \sqrt{Tol}$ and a second time with $\Delta B = 4Tol$. This numerical example clearly demonstrates that internal differentiation is to be preferred.

Exercises

1. (Owren & Zennaro 1991, Carnicer 1991). The 4-stage 4th order methods of Section II.1 do not possess a dense output of order 4 (also if the numerical solution y_1 is included as 5th stage). Prove this statement.

2. Consider a Runge-Kutta method of order p and use Richardson extrapolation for step size control. Besides the numerical solution y_0, y_1, y_2 we consider the extrapolated values (see Section II.4)

$$\widehat{y}_1 = y_1 + \frac{y_2 - w}{(2^p - 1)2}, \qquad \widehat{y}_2 = y_2 + \frac{y_2 - w}{2^p - 1}$$

and do quintic polynomial interpolation based on y_0, $f(x_0, y_0)$, $\widehat{y}_1$, $f(x_0 + h, y_1)$, $\widehat{y}_2$, $f(x_0 + 2h, \widehat{y}_2)$. Prove that the resulting dense output formula is of order $p^* = \min(5, p + 1)$.

   Remark. It is not necessary to evaluate f at $\widehat{y}_1$.

3. Prove that the conditions (6.19), (6.18) and (5.20) imply (6.20).

   Hint. The system (6.19) together with one relation of (6.20) is overdetermined. However, it possesses the solution b_i for $\theta = 1$. Further, the values $b_i c_i$ also solve this system if the right-hand side of (6.19a) is adapted. These properties imply that for $k \in \{4, 5\}$ and for $i \in \{1, 6, \ldots, 16\}$

$$\sum_{j=k+1}^{i-1} a_{ij} a_{jk} = \alpha a_{i4} + \beta a_{i5} + \gamma c_i a_{i4} + \delta c_i a_{i5} + \varepsilon \left(\sum_{j=1}^{i-1} a_{ij} c_j^5 - \frac{c_i^6}{6} \right),$$

   where the parameters $\alpha, \beta, \gamma, \delta, \varepsilon$ may depend on k.

4. (Butcher). Try your favorite code on the example

$$\begin{aligned} y_1' &= f_1(y_1, y_2), & y_1(0) &= 1 \\ y_2' &= f_2(y_1, y_2), & y_2(0) &= 0 \end{aligned}$$

   where f is defined as follows.

   If $(|y_1| > |y_2|)$ then
   $$f_1 = 0, \quad f_2 = \mathrm{sign}\,(y_1)$$
   Else
   $$f_2 = 0, \quad f_1 = -\mathrm{sign}\,(y_2)$$
   End If.

   Compute $y_1(8), y_2(8)$. Show that the exact solution is periodic.

5. Do numerical computations for the problem $y' = f(y)$, $y(0) = 1$, $y(3) = ?$ where

$$f(y) = \begin{cases} y^2 & \text{if } 0 \le y \le 2 \\ \left.\begin{array}{ll} \text{a)} & 1 \\ \text{b)} & 4 \\ \text{c)} & -4 + 4y \end{array}\right\} & \text{if } 2 < y \end{cases}$$

Remark. The correct answer would be (a) 4.5, (b) 12, (c) $\exp(10) + 1$.

6. Consider an s-stage Runge-Kutta method and denote by $\tilde{s}$ the number of distinct c_i. Prove that the order of any continuous extension is $\le \tilde{s}$.

   Hint. Let $q(x)$ be a polynomial of degree $\tilde{s}$ satisfying $q(c_i) = 0$ (for $i = 1, \dots, s$) and investigate the expression $\sum_i b_i(\theta) q(c_i)$.

7. (Step size freeze). Consider the following algorithm for the computation of $\partial y / \partial B$: first compute numerically the solution of (6.35) and denote it by $y_h(x_{\text{end}}, B)$. At the same time memorize all the selected step sizes. This step size sequence is then used to solve (6.35) with B replaced by $B + \Delta B$. The result is denoted by $y_h(x_{\text{end}}, B + \Delta B)$. Then approximate the derivative $\partial y / \partial B$ by

$$\frac{1}{\Delta B} \Big(y_h(x_{\text{end}}, B + \Delta B) - y_h(x_{\text{end}}, B) \Big).$$

Prove that this algorithm is equivalent to the solution of the system (6.35), (6.38), if only the components of y are considered for error control and step size selection.

Remark. For large systems this algorithm needs less storage requirements than internal differentiation, in particular if the derivative with respect to several parameters is computed.

8. (Taubert 1976). Show that for the discontinuous problem (6.23) the Euler polygons converge to Filippov's solution (6.30), (6.31).

   Hint. The difference quotient of a piece of the Euler polygon lies in the convex hull of points $f_I(y)$ and $f_{II}(y)$.

   Remark. This result can either be interpreted as pleading for myriads of Euler steps, or as a motivation for the homotopy (6.30).

9. Another motivation for formula (6.30): suppose that a small particle of radius ε is transported in a possibly discontinuous flow. Then its movement might be described by the mean of f

$$f_\varepsilon(y) = \int_{B_\varepsilon(y)} f(z)\, dz \Big/ \int_{B_\varepsilon(y)} dz$$

which is continuous in y. Show that the solution of $y_\varepsilon' = f_\varepsilon(y)$ becomes, for $\varepsilon \to 0$, that of (6.33) and (6.34).

II.7 Implicit Runge-Kutta Methods

It has been traditional to consider only explicit processes
(J.C. Butcher 1964a)

The high speed computing machines make it possible to enjoy
the advantage of intricate methods
(P.C. Hammer & J.W. Hollingsworth 1955)

The first *implicit* RK methods were used by Cauchy (1824) for the sake of — you
have guessed correctly — error estimation (Méthodes diverses qui peuvent être
employées au Calcul numérique ...; see Exercise 5). Cauchy inserted the mean
value theorem into the integral studied in Sections I.8 and II.1,

$$y(x_1) = y(x_0) + \int_{x_0}^{x_1} f(x, y(x)) \, dx, \qquad (7.1)$$

to obtain

$$y_1 = y_0 + hf(x_0 + \theta h, y_0 + \Theta(y_1 - y_0)) \qquad (7.2)$$

with $0 \leq \theta, \Theta \leq 1$ (the "θ-method"). The extreme cases are $\theta = \Theta = 0$ (the explicit
Euler method) and $\theta = \Theta = 1$

$$y_1 = y_0 + hf(x_1, y_1), \qquad (7.3)$$

which we call the *implicit* or *backward Euler method*.

For the sake of more efficient numerical processes, we apply, as we did in
Section II.1, the midpoint rule ($\theta = \Theta = 1/2$) and obtain from (7.2) by setting
$k_1 = (y_1 - y_0)/h$:

$$
\begin{aligned}
k_1 &= f\left(x_0 + \frac{h}{2}, y_0 + \frac{h}{2} k_1\right), \\
y_1 &= y_0 + hk_1.
\end{aligned}
\qquad (7.4)
$$

This method is called the *implicit midpoint rule*.

Still another possibility is to approximate (7.1) by the *trapezoidal rule* and to
obtain

$$y_1 = y_0 + \frac{h}{2}\left(f(x_0, y_0) + f(x_1, y_1)\right). \qquad (7.5)$$

Let us also look at the Radau scheme

$$
\begin{aligned}
y(x_1) - y(x_0) &= \int_{x_0}^{x_0+h} f(x, y(x)) \, dx \\
&\approx \frac{h}{4}\left(f(x_0, y_0) + 3f(x_0 + \frac{2}{3}h, y(x_0 + \frac{2}{3}h))\right).
\end{aligned}
$$

Here we need to approximate $y(x_0 + 2h/3)$. One idea would be the use of quadratic interpolation based on y_0, y_0' and $y(x_1)$,

$$y\left(x_0 + \frac{2}{3} h\right) \approx \frac{5}{9} y_0 + \frac{4}{9} y(x_1) + \frac{2}{9} hf(x_0, y_0).$$

The resulting method, given by Hammer & Hollingsworth (1955), is

$$k_1 = f(x_0, y_0)$$
$$k_2 = f\left(x_0 + \frac{2}{3} h, y_0 + \frac{h}{3}(k_1 + k_2)\right) \qquad (7.6)$$
$$y_1 = y_0 + \frac{h}{4}(k_1 + 3k_2).$$

All these schemes are of the form (1.8) if the summations are extended up to "s".

Definition 7.1. Let b_i, a_{ij} $(i, j = 1, \ldots, s)$ be real numbers and let c_i be defined by (1.9). The method

$$k_i = f\left(x_0 + c_i h, y_0 + h \sum_{j=1}^{s} a_{ij} k_j\right) \qquad i = 1, \ldots, s$$
$$\qquad (7.7)$$
$$y_1 = y_0 + h \sum_{i=1}^{s} b_i k_i$$

is called an *s-stage Runge-Kutta method*. When $a_{ij} = 0$ for $i \le j$ we have an explicit (ERK) method. If $a_{ij} = 0$ for $i < j$ and at least one $a_{ii} \ne 0$, we have a *diagonal implicit Runge-Kutta method* (DIRK). If in addition all diagonal elements are identical ($a_{ii} = \gamma$ for $i = 1, \ldots, s$), we speak of a *singly diagonal* implicit (SDIRK) method. In all other cases we speak of an *implicit* Runge-Kutta method (IRK).

The tableau of coefficients used above for ERK-methods is obviously extended to include all the other non-zero a_{ij}'s above the diagonal. For methods (7.3), (7.4) and (7.6) it is given in Table 7.1.

Renewed interest in implicit Runge-Kutta methods arose in connection with *stiff* differential equations (see Volume II).

Table 7.1. Implicit Runge-Kutta methods

1	1		1/2	1/2		0	0	0
	1			1		2/3	1/3	1/3
							1/4	3/4

Implicit Euler   Implicit midpoint rule   Hammer & Hollingsworth

Existence of a Numerical Solution

For implicit methods, the k_i's can no longer be evaluated successively, since (7.7) constitutes a system of implicit equations for the determination of k_i. For DIRK-methods we have a sequence of implicit equations of dimension n for k_1, then for k_2, etc. For fully implicit methods $s \cdot n$ unknowns (k_i, $i = 1, \ldots, s$; each of dimension n) have to be determined simultaneously, which still increases the difficulty. A natural question is therefore (the reason for which the original version of Butcher (1964a) was returned by the editors): do equations (7.7) possess a solution at all?

Theorem 7.2. *Let $f : \mathbb{R} \times \mathbb{R}^n \to \mathbb{R}^n$ be continuous and satisfy a Lipschitz condition with constant L (with respect to y). If*

$$h < \frac{1}{L \, \max_i \sum_j |a_{ij}|} \tag{7.8}$$

there exists a unique solution of (7.7), which can be obtained by iteration. If $f(x, y)$ is p times continuously differentiable, the functions k_i (as functions of h) are also in C^p.

Proof. We prove the existence by iteration ("... on la résoudra facilement par des approximations successives ...", Cauchy 1824)

$$k_i^{(m+1)} = f\left(x_0 + c_i h, y_0 + h \sum_{j=1}^{s} a_{ij} k_j^{(m)}\right).$$

We define $K \in \mathbb{R}^{sn}$ as $K = (k_1, \ldots, k_s)^T$ and use the norm $\|K\| = \max_i(\|k_i\|)$. Then (7.7) can be written as $K = F(K)$ where

$$F_i(K) = f\left(x_0 + c_i h, y_0 + h \sum_{j=1}^{s} a_{ij} k_j\right), \quad i = 1, \ldots, s.$$

The Lipschitz condition and a repeated use of the triangle inequality then show that

$$\|F(K_1) - F(K_2)\| \le hL \max_{i=1,\ldots,s} \sum_{j=1}^{s} |a_{ij}| \cdot \|K_1 - K_2\|$$

which from (7.8) is a contraction. The contraction mapping principle then ensures the existence and uniqueness of the solution and the convergence of the fixed-point iteration.

The differentiability result is ensured by the Implicit Function Theorem of classical analysis: (7.7) is written as $\Phi(h, K) = K - F(K) = 0$. The matrix of partial derivatives $\partial \Phi / \partial K$ for $h = 0$ is the identity matrix and therefore the solution of $\Phi(h, K) = 0$, which for $h = 0$ is $k_i = f(x_0, y_0)$, is continuously differentiable in a neighbourhood of $h = 0$. $\qquad \square$

If the assumptions on f in Theorem 7.2 are only satisfied in a neighbourhood of the initial value, then further restrictions on h are needed in order that the argument of f remains in this neighbourhood. Uniqueness is then only of local nature.

The step size restriction (7.8) becomes useless for stiff problems (L large). We return to this question in Vol. II, Sections IV.8 and IV.14.

The definition of *order* is the same as for explicit methods and the order conditions are derived in precisely the same way as in Section II.2.

Example 7.3. Let us study implicit two-stage methods of order 3: the order conditions become (see Theorem 2.1)

$$b_1 + b_2 = 1, \qquad b_1 c_1 + b_2 c_2 = \frac{1}{2}, \qquad b_1 c_1^2 + b_2 c_2^2 = \frac{1}{3}$$

$$b_1(a_{11}c_1 + a_{12}c_2) + b_2(a_{21}c_1 + a_{22}c_2) = \frac{1}{6}. \tag{7.9}$$

The first three equations imply the following orthogonality relation (from the theory of Gaussian integration):

$$\int_0^1 (x - c_1)(x - c_2)\,dx = 0, \quad \text{i.e.,} \quad c_2 = \frac{2 - 3c_1}{3 - 6c_1} \quad (c_1 \neq 1/2) \tag{7.10}$$

and

$$b_1 = \frac{c_2 - 1/2}{c_2 - c_1}, \qquad b_2 = \frac{c_1 - 1/2}{c_1 - c_2}.$$

In the fourth equation we insert $a_{21} = c_2 - a_{22}$, $a_{11} = c_1 - a_{12}$ and consider a_{12} and c_1 as free parameters. This gives

$$a_{22} = \frac{1/6 - b_1 a_{12}(c_2 - c_1) - c_1/2}{b_2(c_2 - c_1)}. \tag{7.11}$$

For $a_{12} = 0$ we obtain a one-parameter family of DIRK-methods of order 3. An SDIRK-method is obtained if we still require $a_{11} = a_{22}$ (Nørsett 1974b, Crouzeix 1975, see Table 7.2). For order 4 we have 4 additional conditions, with only two free parameters left. Nevertheless there exists a unique solution (see Table 7.3).

Table 7.2. SDIRK method, order 3

γ	γ	0
$1 - \gamma$	$1 - 2\gamma$	γ
	$1/2$	$1/2$

$$\gamma = \frac{3 \pm \sqrt{3}}{6}$$

Table 7.3. Hammer & Hollingsworth, order 4

$\dfrac{1}{2} - \dfrac{\sqrt{3}}{6}$	$\dfrac{1}{4}$	$\dfrac{1}{4} - \dfrac{\sqrt{3}}{6}$
$\dfrac{1}{2} + \dfrac{\sqrt{3}}{6}$	$\dfrac{1}{4} + \dfrac{\sqrt{3}}{6}$	$\dfrac{1}{4}$
	$1/2$	$1/2$

The Methods of Kuntzmann and Butcher of Order 2s

It is clear that formula (7.4) and the method of Table 7.3 extend the one-point and two-point Gaussian quadrature formulas, respectively. Kuntzmann (1961) (see Ceschino & Kuntzmann 1963, p. 106) and Butcher (1964a) then discovered that for all s there exist IRK-methods of order $2s$. The main tools of proof are the following *simplifying assumptions*

$$B(p): \quad \sum_{i=1}^{s} b_i c_i^{q-1} = \frac{1}{q} \qquad q = 1, \ldots, p,$$

$$C(\eta): \quad \sum_{j=1}^{s} a_{ij} c_j^{q-1} = \frac{c_i^q}{q} \qquad i = 1, \ldots, s, \; q = 1, \ldots, \eta,$$

$$D(\zeta): \quad \sum_{i=1}^{s} b_i c_i^{q-1} a_{ij} = \frac{b_j}{q}(1 - c_j^q) \qquad j = 1, \ldots, s, \; q = 1, \ldots, \zeta.$$

Condition $B(p)$ simply means that the quadrature formula (b_i, c_i) is of order p or, equivalently, that the order conditions (2.21) are satisfied for the bushy trees $[\tau, \ldots, \tau]$ up to order p.

The assumption $C(\eta)$ implies that the pairs of trees in Fig. 7.1 give identical order conditions for $q \leq \eta$. In contrast to explicit Runge-Kutta methods (see (5.7) and (5.15)) there is no need to require conditions such as $b_2 = 0$ (see (5.8)), because $\sum_j a_{ij} c_j^{q-1} = c_i^q/q$ is valid for all i.

The assumption $D(\zeta)$ is an extension of (1.12). It means that the order condition of the left-hand tree of Fig. 7.2 is implied by those of the two right-hand trees if $q \leq \zeta$.

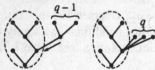

Fig. 7.1. Reduction with $C(q)$

Fig. 7.2. Reduction with $D(q)$

Theorem 7.4 (Butcher 1964a). *If $B(p)$, $C(\eta)$ and $D(\zeta)$ are satisfied with $p \leq 2\eta + 2$ and $p \leq \zeta + \eta + 1$, then the method is of order p.*

Proof. The above reduction by $C(\eta)$ implies that it is sufficient to consider trees $t = [t_1, \ldots, t_m]$ of order $\leq p$, where the subtrees $t_1, \ldots, t_m$ are either equal to τ or of order $\geq \eta + 1$. Since $p \leq 2\eta + 2$ either all subtrees are equal to τ or there is exactly one subtree different from τ. In the second case the number of τ's is $\leq \zeta - 1$ by $p \leq \eta + \zeta + 1$ and the reduction by $D(\zeta)$ can be applied. Therefore, after all these reductions, only the bushy trees are left and they are satisfied by $B(p)$. $\qquad \square$

To obtain the formulas of order $2s$, Butcher assumed $B(2s)$ (i.e., the c_i and b_i are the coefficients of the Gaussian quadrature formula) and $C(s)$. This implies $D(s)$ (see Exercise 7) so that Theorem 7.4 can be applied with $p = 2s$, $\eta = s$ and $\zeta = s$. Hence the method, obtained in this way, is of order $2s$. For $s = 3$ and 4 the coefficients are given in Tables 7.4 and 7.5. They can still be expressed by radicals for $s = 5$ and are given in Butcher (1964a), p. 57.

Impressive numerical results from celestial mechanics for these methods were first reported in the thesis of D. Sommer (see Sommer 1965).

Table 7.4. Kuntzmann & Butcher method, order 6

$$
\begin{array}{c|ccc}
\frac{1}{2} - \frac{\sqrt{15}}{10} & \frac{5}{36} & \frac{2}{9} - \frac{\sqrt{15}}{15} & \frac{5}{36} - \frac{\sqrt{15}}{30} \\[2mm]
\frac{1}{2} & \frac{5}{36} + \frac{\sqrt{15}}{24} & \frac{2}{9} & \frac{5}{36} - \frac{\sqrt{15}}{24} \\[2mm]
\frac{1}{2} + \frac{\sqrt{15}}{10} & \frac{5}{36} + \frac{\sqrt{15}}{30} & \frac{2}{9} + \frac{\sqrt{15}}{15} & \frac{5}{36} \\[2mm]
\hline
 & \frac{5}{18} & \frac{4}{9} & \frac{5}{18}
\end{array}
$$

Table 7.5. Kuntzmann & Butcher method, order 8

$$
\begin{array}{c|cccc}
\frac{1}{2} - \omega_2 & \omega_1 & \omega_1' - \omega_3 + \omega_4' & \omega_1' - \omega_3 - \omega_4' & \omega_1 - \omega_5 \\[2mm]
\frac{1}{2} - \omega_2' & \omega_1 - \omega_3' + \omega_4 & \omega_1' & \omega_1' - \omega_5' & \omega_1 - \omega_3' - \omega_4 \\[2mm]
\frac{1}{2} + \omega_2' & \omega_1 + \omega_3' + \omega_4 & \omega_1' + \omega_5' & \omega_1' & \omega_1 + \omega_3' - \omega_4 \\[2mm]
\frac{1}{2} + \omega_2 & \omega_1 + \omega_5 & \omega_1' + \omega_3 + \omega_4' & \omega_1' + \omega_3 - \omega_4' & \omega_1 \\[2mm]
\hline
 & 2\omega_1 & 2\omega_1' & 2\omega_1' & 2\omega_1
\end{array}
$$

$$\omega_1 = \frac{1}{8} - \frac{\sqrt{30}}{144}, \qquad \omega_1' = \frac{1}{8} + \frac{\sqrt{30}}{144},$$

$$\omega_2 = \frac{1}{2}\sqrt{\frac{15 + 2\sqrt{30}}{35}}, \qquad \omega_2' = \frac{1}{2}\sqrt{\frac{15 - 2\sqrt{30}}{35}},$$

$$\omega_3 = \omega_2\left(\frac{1}{6} + \frac{\sqrt{30}}{24}\right), \qquad \omega_3' = \omega_2'\left(\frac{1}{6} - \frac{\sqrt{30}}{24}\right),$$

$$\omega_4 = \omega_2\left(\frac{1}{21} + \frac{5\sqrt{30}}{168}\right), \qquad \omega_4' = \omega_2'\left(\frac{1}{21} - \frac{5\sqrt{30}}{168}\right),$$

$$\omega_5 = \omega_2 - 2\omega_3, \qquad \omega_5' = \omega_2' - 2\omega_3'.$$

An important interpretation of the assumption $C(\eta)$ is the following:

Lemma 7.5. *The assumption* $C(\eta)$ *implies that the internal stages*

$$g_i = y_0 + h \sum_{j=1}^{s} a_{ij} k_j, \qquad k_j = f(x_0 + c_j h, g_j) \qquad (7.12)$$

satisfy for $i = 1, \ldots, s$

$$g_i - y(x_0 + c_i h) = \mathcal{O}(h^{\eta+1}). \qquad (7.13)$$

Proof. Because of $C(\eta)$ the exact solution satisfies (Taylor expansion)

$$y(x_0 + c_i h) = y_0 + h \sum_{j=1}^{s} a_{ij} y'(x_0 + c_j h) + \mathcal{O}(h^{\eta+1}). \qquad (7.14)$$

Subtracting (7.14) from (7.12) yields

$$g_i - y(x_0 + c_i h) = h \sum_{j=1}^{s} a_{ij} \Big(f(x_0 + c_j h, g_j) - f(x_0 + c_j h, y(x_0 + c_j h)) \Big)$$

$$+ \mathcal{O}(h^{\eta+1})$$

and Lipschitz continuity of f proves (7.13). $\qquad \square$

IRK Methods Based on Lobatto Quadrature

Lobatto quadrature rules (Lobatto 1852, Radau 1880, p. 307) modify the idea of Gaussian quadrature by requiring that the first and the last node coincide with the interval ends, i.e., $c_1 = 0$, $c_s = 1$. These points are easier to handle and, in a step-by-step procedure, can be used twice. The remaining c's are then adjusted optimally, i.e., as the zeros of the Jacobi orthogonal polynomial $P_{s-2}^{(1,1)}(x)$ or of $P_{s-1}'(x)$ (see e.g., Abramowitz & Stegun 1964, 25.4.32 for the interval [-1,1]) and lead to formulas of order $2s - 2$.

J.C. Butcher (1964a, p. 51, 1964c) then found that Lobatto quadrature rules can be extended to IRK-methods whose coefficient matrix is zero in the first line and the last column. The first and the last stage then become *explicit* and the number of implicit stages reduces to $s - 2$. The methods are characterized by $B(2s - 2)$ and $C(s - 1)$. As in Exercise 7 this implies $D(s - 1)$ so that by Theorem 7.4 the method is of order $2s - 2$. For $s = 3$ and 4, the coefficients are given in Table 7.6.

We shall see in Volume II (Section IV.3, Table 3.1) that these methods, although preferable as concerns the relation between order and implicit stages, are not sufficiently stable for stiff differential equations.

Table 7.6. Butcher's Lobatto formulas of orders 4 and 6

0	0	0	0		0	0	0	0	0
$\dfrac{1}{2}$	$\dfrac{1}{4}$	$\dfrac{1}{4}$	0		$\dfrac{5-\sqrt5}{10}$	$\dfrac{5+\sqrt5}{60}$	$\dfrac{1}{6}$	$\dfrac{15-7\sqrt5}{60}$	0
1	0	1	0		$\dfrac{5+\sqrt5}{10}$	$\dfrac{5-\sqrt5}{60}$	$\dfrac{15+7\sqrt5}{60}$	$\dfrac{1}{6}$	0
					1	$\dfrac{1}{6}$	$\dfrac{5-\sqrt5}{12}$	$\dfrac{5+\sqrt5}{12}$	0
	$\dfrac{1}{6}$	$\dfrac{2}{3}$	$\dfrac{1}{6}$			$\dfrac{1}{12}$	$\dfrac{5}{12}$	$\dfrac{5}{12}$	$\dfrac{1}{12}$

Collocation Methods

> Es ist erstaunlich dass die Methode trotz ihrer Primitivität und der geringen Rechenarbeit in vielen Fällen ... sogar gute Ergebnisse liefert.
> (L. Collatz 1951)

> Nous allons montrer l'équivalence de notre définition avec la définition traditionnelle de certaines formules de Runge Kutta implicites.
> (Guillou & Soulé 1969)

The concept of collocation is old and universal in numerical analysis (see e.g., pp. 28,29,32,181,411,453,483,495 of Collatz 1960, Frazer, Jones & Skan 1937). For ordinary differential equations it consists in searching for a polynomial of degree s whose derivative coincides ("co-locates") at s given points with the vector field of the differential equation (Guillou & Soulé 1969, Wright 1970). Still another approach is to combine Galerkin's method with numerical quadrature (see Hulme 1972).

Definition 7.6. For s a positive integer and $c_1,\ldots,c_s$ distinct real numbers (typically between 0 and 1), the corresponding *collocation polynomial* $u(x)$ of degree s is defined by

$$u(x_0) = y_0 \qquad \text{(initial value)} \qquad (7.15a)$$
$$u'(x_0 + c_i h) = f(x_0 + c_i h, u(x_0 + c_i h)), \qquad i = 1,\ldots,s. \qquad (7.15b)$$

The numerical solution is then given by

$$y_1 = u(x_0 + h). \qquad (7.15c)$$

If some of the c_i coincide, the collocation condition (7.15b) will contain higher derivatives and lead to multi-derivative methods (see Section II.13). Accordingly, for the moment, we suppose them all distinct.

Theorem 7.7 (Guillou & Soulé 1969, Wright 1970). *The collocation method (7.15) is equivalent to the s-stage IRK-method (7.7) with coefficients*

$$a_{ij} = \int_0^{c_i} \ell_j(t)\, dt, \qquad b_j = \int_0^1 \ell_j(t)\, dt \qquad i,j = 1,\dots,s, \qquad (7.16)$$

where the $\ell_j(t)$ are the Lagrange polynomials

$$\ell_j(t) = \prod_{k \neq j} \frac{(t - c_k)}{(c_j - c_k)}. \qquad (7.17)$$

Proof. Put $u'(x_0 + c_i h) = k_i$, so that

$$u'(x_0 + th) = \sum_{j=1}^{s} k_j \cdot \ell_j(t) \qquad \text{(Lagrange)}.$$

Then integrate

$$u(x_0 + c_i h) = y_0 + h \int_0^{c_i} u'(x_0 + th)\, dt \qquad (7.18)$$

and insert into (7.15b) together with (7.16). The IRK-method (7.7) then comes out.   □

As a consequence of this result, the existence and uniqueness of the collocation polynomial (for sufficiently small h) follows from Theorem 7.2.

Theorem 7.8. *An implicit Runge-Kutta method with all c_i different and of order at least s is a collocation method iff $C(s)$ is true.*

Proof. $C(s)$ determines the a_{ij} uniquely. We write it as

$$\sum_{j=1}^{s} a_{ij} p(c_j) = \int_0^{c_i} p(t)\, dt \qquad (7.19)$$

for all polynomials p of degree $\leq s - 1$. The a_{ij} given by (7.16) satisfy this relation, because (7.16) inserted into (7.19) is just the Lagrange interpolation formula.   □

Theorem 7.9. *Let $M(t) = \prod_{i=1}^{s}(t - c_i)$ and suppose that M is orthogonal to polynomials of degree $r - 1$,*

$$\int_0^1 M(t) t^{q-1}\, dt = 0, \qquad q = 1,\dots,r, \qquad (7.20)$$

then method (7.15) has order $p = s + r$.

Proof. The following proof uses the Gröbner & Alekseev Formula, which gives nice insight in the background of the result. An alternative proof is indicated in Exercise 7 below. One can also linearize the equation, apply the *linear* variation-of-constants formula and estimate the error (Guillou & Soulé 1969).

The orthogonality condition (7.20) means that the quadrature formula

$$\int_{x_0}^{x_0+h} g(t)\,dt = h \sum_{j=1}^{s} b_j g(x_0 + c_j h) + err(g) \tag{7.21}$$

is of order $s + r = p$, and its error is bounded by

$$|err(g)| \leq Ch^{p+1} \cdot \max |g^{(p)}(x)|. \tag{7.22}$$

The principal idea of the proof is now the following: we consider

$$u'(x) = f(x, u(x)) + (u'(x) - f(x, u(x)))$$

as a perturbation of

$$y'(x) = f(x, y(x))$$

and integrate the Gröbner & Alekseev Formula (I.14.18) with the quadrature formula (7.21). Due to (7.15b), the result is identically zero, since at the collocation points the defect is zero. Thus from (7.21) and (7.22)

$$\|y(x_0 + h) - u(x_0 + h)\| = \|err(g)\| \leq C \cdot h^{p+1} \cdot \max_{x_0 \leq t \leq x_0+h} \|g^{(p)}(t)\|, \tag{7.23}$$

where

$$g(t) = \frac{\partial y}{\partial y_0}(x, t, u(t)) \cdot (u'(t) - f(t, u(t))) ,$$

and we see that the local error behaves like $\mathcal{O}(h^{p+1})$.

There remains, however, a small technical detail: to show that the derivatives of $g(t)$ remain bounded for $h \to 0$. These derivatives contain partial derivatives of $f(t, y)$ and derivatives of $u(t)$. We shall see in the next theorem that these derivatives remain bounded for $h \to 0$.   □

Theorem 7.10. *The collocation polynomial $u(x)$ gives rise to a continuous IRK method of order s, i.e., for all $x_0 \leq x \leq x_0 + h$ we have*

$$\|y(x) - u(x)\| \leq C \cdot h^{s+1}. \tag{7.24}$$

Moreover, for the derivatives of $u(x)$ we have

$$\|y^{(k)}(x) - u^{(k)}(x)\| \leq C \cdot h^{s+1-k} \qquad k = 0, \ldots, s. \tag{7.25}$$

Proof. The exact solution $y(x)$ satisfies the collocation condition everywhere, hence *also* at the points $x_0 + c_i h$. So, in exactly the same way as in the proof of

Theorem 7.7, we apply the Lagrange interpolation formula to $y'(x)$:

$$y'(x_0 + th) = \sum_{j=1}^{s} f(x_0 + c_j h, y(x_0 + c_j h)) \ell_j(t) + h^s R(t, h)$$

where $R(t, h)$ is a smooth function of both variables. Integration and subtraction from (7.18) gives

$$y(x_0 + th) - u(x_0 + th) = h \sum_{j=1}^{s} \Delta f_j \cdot \int_0^t \ell_j(\tau) \, d\tau + h^{s+1} \int_0^t R(\tau, h) \, d\tau, \quad (7.26)$$

where

$$\Delta f_j = f(x_0 + c_j h, y(x_0 + c_j h)) - f(x_0 + c_j h, u(x_0 + c_j h)).$$

The kth derivative of (7.26) with respect to t is

$$h^k \left(y^{(k)}(x_0 + th) - u^{(k)}(x_0 + th) \right) = h \sum_{j=1}^{s} \Delta f_j \cdot \ell_j^{(k-1)}(t) + h^{s+1} \frac{\partial^{k-1} R}{\partial t^{k-1}}(t, h),$$

so that the result follows from the boundedness of the derivatives of $R(t, h)$ and from $\Delta f_j = \mathcal{O}(h^{s+1})$ which is a consequence of Lemma 7.5.   □

Remark. Only *some* IRK methods are collocation methods. An extension of the collocation idea ("Perturbed Collocation", see Nørsett & Wanner 1981) applies to *all* IRK methods.

Exercises

1. Compute the one-point collocation method ($s = 1$) with $c_i = \theta$ and compare with (7.2). Determine its order in dependence of θ.

2. Compute all collocation methods with $s = 2$ of order 2 in dependence of c_1 and c_2.

3. Specify in the method of Exercise 2 $c_1 = 1/3, c_2 = 1$ as well as $c_1 = 0, c_2 = 2/3$. Determine the orders of the obtained methods and explain.

4. Interpret the implicit midpoint rule (7.4) and the explicit Euler method as collocation methods. Is method (7.5) a collocation method? Method (7.6)?

5. (Cauchy 1824). Find from equation (7.2) conditions for the function $f(x, y)$ such that for scalar differential equations

$$y_1(\text{explicit Euler}) \geq y(x_1) \geq y_1(\text{implicit Euler}).$$

Compute five steps with $h = 0.2$ with both methods to obtain upper and lower bounds for $y(1)$, the solution of

$$y' = \cos \frac{x+y}{5}, \qquad y(0) = 0.$$

Cauchy's result: $0.9659 \le y(1) \le 0.9810$. For one single step with $h = 1$ he obtained $0.926 \le y(1) \le 1$.

Compute the exact solution by elementary integration.

6. Determine the orders of the methods of Table 7.7. Generalize to arbitrary s (Ehle 1968).

   Hint. Use Theorems 7.8 and 7.9.

Table 7.7. Methods of Ehle

Radau IIA, order 5

$\frac{4-\sqrt{6}}{10}$	$\frac{88-7\sqrt{6}}{360}$	$\frac{296-169\sqrt{6}}{1800}$	$\frac{-2+3\sqrt{6}}{225}$
$\frac{4+\sqrt{6}}{10}$	$\frac{296+169\sqrt{6}}{1800}$	$\frac{88+7\sqrt{6}}{360}$	$\frac{-2-3\sqrt{6}}{225}$
1	$\frac{16-\sqrt{6}}{36}$	$\frac{16+\sqrt{6}}{36}$	$\frac{1}{9}$
	$\frac{16-\sqrt{6}}{36}$	$\frac{16+\sqrt{6}}{36}$	$\frac{1}{9}$

Lobatto IIIA, order 4

0	0	0	0
$\frac{1}{2}$	$\frac{5}{24}$	$\frac{1}{3}$	$-\frac{1}{24}$
1	$\frac{1}{6}$	$\frac{2}{3}$	$\frac{1}{6}$
	$\frac{1}{6}$	$\frac{2}{3}$	$\frac{1}{6}$

7. (Butcher 1964a). Give an algebraic proof of Theorem 7.9.

   Hint. From Theorem 7.8 we have $C(s)$. Next the condition $B(p)$ with $p = s + r$ (theory of Gaussian quadrature formulas) implies $D(r)$. To see this, multiply the two vectors $u_j = \sum_i b_i c_i^{q-1} a_{ij}$ and $v_j = b_j(1 - c_j^q)/q$ ($j = 1, \ldots, s$) by the Vandermonde matrix

$$V = \begin{pmatrix} 1 & 1 & \cdots & 1 \\ c_1 & c_2 & \cdots & c_s \\ \vdots & \vdots & & \vdots \\ c_1^{s-1} & c_2^{s-1} & \cdots & c_s^{s-1} \end{pmatrix}.$$

Finally apply Theorem 7.4.

II.8 Asymptotic Expansion of the Global Error

Mein Verzicht auf das Restglied war leichtsinnig ...

(W. Romberg 1979)

Our next goal will be to perfect Richardson's extrapolation method (see Section II.4) by doing *repeated* extrapolation and eliminating more and more terms Ch^{p+k} of the error. A sound theoretical basis for this procedure is given by the study of the asymptotic behaviour of the global error. For problems of the type $y' = f(x)$, which lead to integration, the answer is given by the Euler-Maclaurin formula and has been exploited by Romberg (1955) and his successors. The first rigorous treatments for differential equations are due to Henrici (1962) and Gragg (1964) (see also Stetter 1973). We shall follow here the successive elimination of the error terms given by Hairer & Lubich (1984), which also generalizes to multistep methods.

Suppose we have a one-step method which we write, in Henrici's notation, as

$$y_{n+1} = y_n + h\Phi(x_n, y_n, h).\tag{8.1}$$

If the method is of order p, it possesses at each point of the solution $y(x)$ a *local error* of the form

$$y(x+h) - y(x) - h\Phi(x, y(x), h) = \\ d_{p+1}(x)h^{p+1} + \ldots + d_{N+1}(x)h^{N+1} + \mathcal{O}(h^{N+2})\tag{8.2}$$

whenever the differential equation is sufficiently differentiable. For Runge-Kutta methods these error terms were computed in Section II.2 (see also Theorem 3.2).

The Global Error

Let us now set $y_n =: y_h(x)$ for the numerical solution at $x = x_0 + nh$. We then know from Theorem 3.6 that the global error behaves like h^p. We shall search for a function $e_p(x)$ such that

$$y(x) - y_h(x) = e_p(x)h^p + o(h^p).\tag{8.3}$$

The idea is to consider

$$y_h(x) + e_p(x)h^p =: \widehat{y}_h(x)\tag{8.4a}$$

as the numerical solution of a new method

$$\widehat{y}_{n+1} = \widehat{y}_n + h\widehat{\Phi}(x_n, \widehat{y}_n, h). \tag{8.4b}$$

By comparison with (8.1), we see that the increment function for the new method is

$$\widehat{\Phi}(x, \widehat{y}, h) = \Phi\big(x, \widehat{y} - e_p(x)h^p, h\big) + \big(e_p(x+h) - e_p(x)\big)h^{p-1}. \tag{8.5}$$

Our task is to find a function $e_p(x)$, with $e_p(x_0) = 0$, such that the method with increment function $\widehat{\Phi}$ is of order $p+1$.

Expanding the local error of the one-step method $\widehat{\Phi}$ into powers of h we obtain

$$
\begin{aligned}
&y(x+h) - y(x) - h\widehat{\Phi}(x, y(x), h) \\
&= \Big(d_{p+1}(x) + \frac{\partial f}{\partial y}\big(x, y(x)\big)e_p(x) - e_p'(x)\Big)h^{p+1} + \mathcal{O}(h^{p+2})
\end{aligned}
\tag{8.6}
$$

where we have used

$$\frac{\partial \Phi}{\partial y}(x, y, 0) = \frac{\partial f}{\partial y}(x, y). \tag{8.7}$$

The term in h^{p+1} vanishes if $e_p(x)$ is defined as the solution of

$$e_p'(x) = \frac{\partial f}{\partial y}\big(x, y(x)\big)e_p(x) + d_{p+1}(x), \qquad e_p(x_0) = 0. \tag{8.8}$$

By Theorem 3.6, applied to the method $\widehat{\Phi}$, we now have

$$y(x) - y_h(x) = e_p(x)h^p + \mathcal{O}(h^{p+1}) \tag{8.9}$$

and the first term of the desired asymptotic expansion has been determined.

We now repeat the procedure with the method with increment function $\widehat{\Phi}$. It is of order $p+1$ and again satisfies condition (8.7). The final result of this procedure is the following

Theorem 8.1 (Gragg 1964). *Suppose that a given method with sufficiently smooth increment function Φ satisfies the consistency condition $\Phi(x, y, 0) = f(x, y)$ and possesses an expansion (8.2) for the local error. Then the global error has an asymptotic expansion of the form*

$$y(x) - y_h(x) = e_p(x)h^p + \ldots + e_N(x)h^N + E_h(x)h^{N+1} \tag{8.10}$$

where the $e_j(x)$ are solutions of inhomogeneous differential equations of the form (8.8) with $e_j(x_0) = 0$ and $E_h(x)$ is bounded for $x_0 \leq x \leq x_{\text{end}}$ and $0 \leq h \leq h_0$.
□

The differentiability properties of the $e_j(x)$ depend on those of f and Φ (see (8.8) and (8.2)). The expansion (8.10) will be the theoretical basis for all discussions of extrapolation methods.

Examples. 1. For the equation $y' = y$ and Euler's method we have with $h = 1/n$ and $x = 1$, using the binomial theorem,

$$y_h(1) = \left(1 + \frac{1}{n}\right)^n = 1 + 1 + \left(1 - \frac{1}{n}\right)\frac{1}{2!} + \left(1 - \frac{1}{n}\right)\left(1 - \frac{2}{n}\right)\frac{1}{3!} + \dots.$$

By multiplying out, this gives

$$y(1) - y_h(1) = -\sum_{i=1}^{\infty} h^i \sum_{j=1}^{\infty} \frac{S_{i+j}^{(j)}}{(i+j)!} = 1.359h - 1.246h^2 \pm \dots$$

where the $S_i^{(j)}$ are the Stirling numbers of the first kind (1730, see Abramowitz & Stegun 1964, Section 24.1.3). This is, of course, the Taylor series for the function

$$e - (1+h)^{1/h} = e - \exp\left(1 - \frac{h}{2} + \frac{h^2}{3} \pm \dots\right) = e\left(\frac{1}{2}h - \frac{11}{24}h^2 + \frac{7}{16}h^3 \pm \dots\right)$$

with *convergence radius* $r = 1$.

2. For the differential equation $y' = f(x)$ and the trapezoidal rule (7.5), the expansion (8.10) becomes

$$\int_0^1 f(x)\,dx - y_h(1) = -\sum_{k=1}^{N} \frac{h^{2k}}{(2k)!} B_{2k}\left(f^{(2k-1)}(1) - f^{(2k-1)}(0)\right) + \mathcal{O}(h^{2N+1}),$$

the well known Euler-Maclaurin formula (1736). For $N \to \infty$, the series will usually diverge, due to the fast growth of the Bernoulli numbers for large k. It may, however, be useful for small values of N and we call it an *asymptotic expansion* (Poincaré 1893).

Variable h

Theorem 8.1 is not only valid for equal step sizes. A reasonable assumption for the case of variable step sizes is the existence of a function $\tau(x) > 0$ such that the step sizes depend as

$$x_{n+1} - x_n = \tau(x_n)h \tag{8.11}$$

on a parameter h. Then the local error expansion (8.2) becomes

$$y(x + \tau(x)h) - y(x) - h\tau(x)\Phi(x, y(x), \tau(x)h) = d_{p+1}(x)\tau^{p+1}(x)h^{p+1} + \dots$$

and instead of (8.5) we have

$$\widehat{\Phi}(x, \widehat{y}, \tau(x)h) = \Phi(x, \widehat{y} - e_p(x)h^p, \tau(x)h) + \frac{h^p}{h\tau(x)}\left(e_p(x + \tau(x)h) - e_p(x)\right).$$

With this the local error expansion for the new method becomes, instead of (8.6),

$$y(x + \tau(x)h) - y(x) - h\tau(x)\widehat{\Phi}(x, y(x), \tau(x)h)$$

$$= \tau(x)\left(d_{p+1}(x)\tau^p(x) + \frac{\partial f}{\partial y}(x, y(x))e_p(x) - e_p'(x)\right)h^{p+1} + \mathcal{O}(h^{p+2})$$

and the proof of Theorem 8.1 generalizes with slight modifications.

Negative h

The most important extrapolation algorithms will use asymptotic expansions with *even* powers of h. In order to provide a theoretical basis for these methods, we need to explain the meaning of $y_h(x)$ for h *negative*.

Motivation. We write (8.1) as

$$y_h(x+h) = y_h(x) + h\Phi(x, y_h(x), h) \tag{8.1'}$$

and replace h by $-h$ to obtain

$$y_{-h}(x-h) = y_{-h}(x) - h\Phi(x, y_{-h}(x), -h).$$

Next we replace x by $x+h$ which gives

$$y_{-h}(x) = y_{-h}(x+h) - h\Phi(x+h, y_{-h}(x+h), -h). \tag{8.12}$$

This is an implicit equation for $y_{-h}(x+h)$, which possesses a unique solution for sufficiently small h (by the implicit function theorem). We write this solution in the form

$$y_{-h}(x+h) = y_{-h}(x) + h\Phi^*(x, y_{-h}(x), h). \tag{8.13}$$

The comparison of (8.12) and (8.13) (with $A = y_{-h}(x+h)$, $B = y_{-h}(x)$) leads us to the following definition.

Definition 8.2. Let $\Phi(x, y, h)$ be the increment function of a method. Then we define the increment function $\Phi^*(x, y, h)$ of the *adjoint method* by the pair of formulas

$$\begin{aligned} B &= A - h\Phi(x+h, A, -h) \\ A &= B + h\Phi^*(x, B, h). \end{aligned} \tag{8.14}$$

Example. The adjoint method of explicit Euler is implicit Euler.

Theorem 8.3. *Let Φ be the Runge-Kutta method (7.7) with coefficients a_{ij}, b_j, c_i $(i, j = 1, \ldots, s)$. Then the adjoint method Φ^* is equivalent to a Runge-Kutta method with s stages and with coefficients*

$$\begin{aligned} c_i^* &= 1 - c_{s+1-i} \\ a_{ij}^* &= b_{s+1-j} - a_{s+1-i, s+1-j} \\ b_j^* &= b_{s+1-j}. \end{aligned}$$

Proof. The formulas (8.14) indicate that for the definition of the adjoint method we have, starting from (7.7), to exchange $y_0 \leftrightarrow y_1$, $h \leftrightarrow -h$ and replace $x_0 \to x_0 + h$. This then leads to

$$k_i = f\left(x_0 + (1 - c_i)h, y_0 + h\sum_{j=1}^{s}(b_j - a_{ij})k_j\right)$$

$$y_1 = y_0 + h\sum_{j=1}^{s} b_j k_j.$$

In order to preserve the usual natural ordering of $c_1, \ldots, c_s$, we also permute the k_i-values and replace all indices i by $s + 1 - i$.   □

Properties of the Adjoint Method

Theorem 8.4. $\Phi^{**} = \Phi$.

Proof. This property, which is the reason for the name "adjoint", is seen by replacing $h \to -h$ and then $x \to x + h$, $B \to A$, $A \to B$ in (8.14).   □

Theorem 8.5. *The adjoint method has the same order as the original method. Its principal error term is the error term of the first method multiplied by $(-1)^p$.*

Proof. We replace h by $-h$ in (8.2), then $x \to x + h$ and rearrange the terms. This gives (using $d_{p+1}(x + h) = d_{p+1}(x) + \mathcal{O}(h)$)

$$y(x) + d_{p+1}(x)h^{p+1}(-1)^p + \mathcal{O}(h^{p+2})$$
$$= y(x + h) - h\Phi(x + h, y(x + h), -h).$$

Here we let B be the left-hand side of this identity, $A = y(x + h)$, and use (8.14). This leads to

$$y(x + h) = y(x) + d_{p+1}(x)h^{p+1}(-1)^p + h\Phi^*(x, y(x), h) + \mathcal{O}(h^{p+2}),$$

which expresses the statement of the theorem.   □

Theorem 8.6. *The adjoint method has exactly the same asymptotic expansion (8.10) as the original method, with h replaced by $-h$.*

Proof. We repeat the procedure which led to the proof of Theorem 8.1, with h negative. The first separated term corresponding to (8.9) will be

$$y(x) - y_{-h}(x) = e_p(x)(-h)^p + \mathcal{O}(h^{p+1}). \tag{8.9'}$$

This is true because the solution of (8.8) with initial value $e_p(x_0) = 0$ has the same sign change as the inhomogenity $d_{p+1}(x)$. This settles the first term. To continue, we prove that the transformation (8.4b) commutes with the adjunction operation, i.e., that

$$(\widehat{\Phi})^* = (\Phi^*)\widehat{}.\tag{8.15}$$

In order to prove (8.15), we obtain from (8.4a) and the definition of $\widehat{\Phi}$

$$y_h(x+h) + e_p(x+h)h^p = y_h(x) + e_p(x)h^p + h\widehat{\Phi}\big(x, y_h(x) + e_p(x)h^p, h\big).$$

Here again, we substitute $h \to -h$ followed by $x \to x+h$. Finally, we apply (8.14) with $B = y_{-h}(x) + e_p(x)(-h)^p$ and $A = y_{-h}(x+h) + e_p(x+h)(-h)^p$ to obtain

$$
\begin{aligned}
y_{-h}(x+h) &+ e_p(x+h)(-h)^p \\
&= y_{-h}(x) + e_p(x)(-h)^p + h(\widehat{\Phi})^*\big(x, y_{-h}(x) + e_p(x)(-h)^p, h\big).
\end{aligned}
\tag{8.16}
$$

On the other hand, if we perform the transformation (see Theorem 8.5)

$$\widehat{y}_{-h}(x) = y_{-h}(x) + e_p(x)(-h)^p\tag{8.4'}$$

and insert this into (8.13), we obtain (8.16) again, but this time with $(\Phi^*)\widehat{}$ instead of $(\widehat{\Phi})^*$. This proves (8.15).   □

Symmetric Methods

Definition 8.7. A method is *symmetric* if $\Phi = \Phi^*$.

Example. The trapezoidal rule (7.5) and the implicit mid-point rule (7.4) are symmetric: the exchanges $y_1 \leftrightarrow y_0$, $h \leftrightarrow -h$ and $x_0 \leftrightarrow x_0 + h$ leave these methods invariant. The following two theorems (Wanner 1973) characterize symmetric IRK methods.

Theorem 8.8. *If*

$$a_{s+1-i, s+1-j} + a_{ij} = b_{s+1-j} = b_j, \qquad i,j = 1,\ldots,s,\tag{8.17}$$

then the corresponding Runge-Kutta method is symmetric. Moreover, if the b_i are nonzero and the c_i distinct and ordered as $c_1 < c_2 < \ldots < c_s$, then condition (8.17) is also necessary for symmetry.

Proof. The sufficiency of (8.17) follows from Theorem 8.3. The condition $c_i = 1 - c_{s+1-i}$ can be verified by adding up (8.17) for $j = 1,\ldots,s$.

Symmetry implies that the original method (with coefficients c_i, a_{ij}, b_j) and the adjoint method (c_i^*, a_{ij}^*, b_j^*) give identical numerical results. If we apply both methods to $y' = f(x)$ we obtain

$$\sum_{i=1}^{s} b_i f(c_i) = \sum_{i=1}^{s} b_i^* f(c_i^*)$$

for all $f(x)$. Our assumption on b_i and c_i thus yields

$$b_i^* = b_i, \qquad c_i^* = c_i \qquad \text{for all} \quad i.$$

We next apply both methods to $y_1' = f(x)$, $y_2' = x^q y_1$ and obtain

$$\sum_{i,j=1}^{s} b_i c_i^q a_{ij} f(c_j) = \sum_{i,j=1}^{s} b_i^* c_i^{*q} a_{ij}^* f(c_j^*).$$

This implies $\sum_i b_i c_i^q a_{ij} = \sum_i b_i c_i^q a_{ij}^*$ for $q = 0, 1, \ldots$ and hence also $a_{ij}^* = a_{ij}$ for all i, j. $\qquad \square$

Theorem 8.9. *A collocation method based on symmetrically distributed collocation points is symmetric.*

Proof. If $c_i = 1 - c_{s+1-i}$, the Lagrange polynomials satisfy $\ell_i(t) = \ell_{s+1-i}(1-t)$. Condition (8.17) is then an easy consequence of (7.19). $\qquad \square$

The following important property of symmetric methods, known intuitively for many years, now follows from the above results.

Theorem 8.10. *If in addition to the assumptions of Theorem 8.1 the underlying method is symmetric, then the asymptotic expansion (8.10) contains only even powers of h:*

$$y(x) - y_h(x) = e_{2q}(x)h^{2q} + e_{2q+2}(x)h^{2q+2} + \ldots \tag{8.18}$$

with $e_{2j}(x_0) = 0$.

Proof. If $\Phi^* = \Phi$, we have $y_{-h}(x) = y_h(x)$ from (8.13) and the result follows from Theorem 8.6. $\qquad \square$

Exercises

1. Assume the one-step method (8.1) to be of order $p \geq 2$ and in addition to $\Phi(x, y, 0) = f(x, y)$ assume

$$\frac{\partial \Phi}{\partial h}(x, y, 0) = \frac{1}{2}\left(\frac{\partial f}{\partial x}(x, y) + \frac{\partial f}{\partial y}(x, y) \cdot f(x, y)\right). \qquad (8.19)$$

Show that the principal local error term of the method $\widehat{\Phi}$ defined in (8.5) is then given by

$$\widehat{d}_{p+2}(x) = d_{p+2}(x) - \frac{1}{2}\frac{\partial f}{\partial y}(x, y(x))\, d_{p+1}(x) - \frac{1}{2}d'_{p+1}(x).$$

Verify that (8.19) is satisfied for all RK-methods of order ≥ 2.

2. Consider the second order method

$$
\begin{array}{c|cc}
0 & & \\
1 & 1 & \\
\hline
 & 1/2 & 1/2
\end{array}
$$

applied to the problem $y' = y$, $y(0) = 1$. Show that

$$d_3(x) = \frac{1}{6}e^x, \quad d_4(x) = \frac{1}{24}e^x, \quad e_2(x) = \frac{1}{6}xe^x, \quad \widehat{d}_4(x) = -\frac{1}{8}e^x.$$

3. Consider the second order method

$$
\begin{array}{c|ccc}
0 & & & \\
1/2 & 1/2 & & \\
1 & 0 & 1 & \\
\hline
 & 1/4 & 1/2 & 1/4
\end{array}
$$

Show that for this method

$$d_3(x) = \frac{1}{24}\left(F(t_{32})(y(x)) - \frac{1}{2}F(t_{31})(y(x))\right)$$

$$d_4(x) = \frac{1}{24}\left(F(t_{44})(y(x)) + \frac{1}{4}F(t_{43})(y(x)) - \frac{1}{4}F(t_{41})(y(x))\right)$$

in the notation of Table 2.2. Show that this implies

$$\widehat{d}_4(x) = 0 \qquad \text{and} \qquad e_3(x) = 0,$$

so that one step of Richardson extrapolation increases the order of the method by two. Find a connection between this method and the GBS-algorithm of Section II.9.

4. Discuss the symmetry of the IRK methods of Section II.7.

II.9 Extrapolation Methods

> The following method of approximation may or may not be new,
> but as I believe it to be of practical importance...
>
> (S.A. Corey 1906)

> The h^2-extrapolation was discovered by a hint from theory fol-
> lowed by arithmetical experiments, which gave pleasing results.
>
> (L.F. Richardson 1927)

> Extrapolation constitutes a powerful means ...
>
> (R. Bulirsch & J. Stoer 1966)

> Extrapolation does not appear to be a particularly effective way
> ..., our tests raise the question as to whether there is any point to
> pursuing it as a separate method.
>
> (L.F. Shampine & L.S. Baca 1986)

Definition of the Method

Let $y' = f(x, y)$, $y(x_0) = y_0$ be a given differential system and $H > 0$ a basic step size. We choose a sequence of positive integers

$$n_1 < n_2 < n_3 < \ldots \tag{9.1}$$

and define the corresponding step sizes $h_1 > h_2 > h_3 > \ldots$ by $h_i = H/n_i$. We then choose a numerical method of order p and compute the numerical results of our initial value problem by performing n_i steps with step size h_i to obtain

$$y_{h_i}(x_0 + H) =: T_{i,1} \tag{9.2}$$

(the letter "T" stands historically for "trapezoidal rule"). We then eliminate as many terms as possible from the asymptotic expansion (8.10) by computing the interpolation polynomial

$$p(h) = \widehat{y} - e_p h^p - e_{p+1} h^{p+1} - \ldots - e_{p+k-2} h^{p+k-2} \tag{9.3}$$

such that

$$p(h_i) = T_{i,1} \qquad i = j, \ j-1, \ldots, j-k+1. \tag{9.4}$$

Finally we *"extrapolate to the limit"* $h \to 0$ and use

$$p(0) = \widehat{y} =: T_{j,k}$$

as numerical result. Conditions (9.4) consist of k linear equations for the k unknowns $\widehat{y}, e_p, \ldots, e_{p+k-2}$.

Example. For $k = 2$, $n_1 = 1$, $n_2 = 2$ the above definition is identical to Richardson's extrapolation discussed in Section II.4.

Theorem 9.1. *The value $T_{j,k}$ represents a numerical method of order $p + k - 1$.*

Proof. We compare (9.4) and (9.3) with the asymptotic expansion (8.10) which we write in the form (with $N = p + k - 1$)

$$T_{i,1} = y(x_0 + H) - e_p(x_0 + H)h_i^p - \ldots - e_{p+k-2}(x_0 + H)h_i^{p+k-2} - \Delta_i, \quad (9.4')$$

where

$$\Delta_i = e_{p+k-1}(x_0 + H)h_i^{p+k-1} + E_{h_i}(x_0 + H)h_i^{p+k} = \mathcal{O}(H^{p+k})$$

because $e_{p+k-1}(x_0) = 0$ and $h_i \le H$. This is a linear system for the unknowns $y(x_0 + H)$, $H^p e_p(x_0 + H), \ldots, H^{p+k-2} e_{p+k-2}(x_0 + H)$ with the Vandermonde-like matrix

$$A = \begin{pmatrix} 1 & \dfrac{1}{n_j^p} & \cdots & \dfrac{1}{n_j^{p+k-2}} \\ \vdots & \vdots & & \vdots \\ 1 & \dfrac{1}{n_{j-k+1}^p} & \cdots & \dfrac{1}{n_{j-k+1}^{p+k-2}} \end{pmatrix}.$$

It is the same as (9.4), just with the right-hand side perturbed by the $\mathcal{O}(H^{p+k})$-terms Δ_i. The matrix A is invertible (see Exercise 6). Therefore by subtraction we obtain

$$|y(x_0 + H) - \widehat{y}| \le \|A^{-1}\|_\infty \cdot \max |\Delta_i| = \mathcal{O}(H^{p+k}). \qquad \square$$

Remark. The case $p = 1$ (as well as $p = 2$ with expansions in h^2) can also be treated by interpreting the difference $y(x_0 + H) - \widehat{y}$ as an interpolation error (see (9.21)).

A great advantage of the method is that it provides a complete table of numerical results

$$\begin{array}{lllll} T_{11} & & & & \\ T_{21} & T_{22} & & & \\ T_{31} & T_{32} & T_{33} & & \\ T_{41} & T_{42} & T_{43} & T_{44} & \\ \cdots & \cdots & \cdots & \cdots & \cdots \end{array} \qquad (9.5)$$

which form a sequence of embedded methods and allow easy estimates of the local error and strategies for variable order. Several step-number sequences are in use for (9.1):

The *"Romberg sequence"* (Romberg 1955):

$$1, 2, 4, 8, 16, 32, 64, 128, 256, 512, \ldots \qquad (9.6)$$

The *"Bulirsch sequence"* (see also Romberg 1955):

$$1, 2, 3, 4, 6, 8, 12, 16, 24, 32, \ldots \tag{9.7}$$

alternating powers of 2 with 1.5 times 2^k. This sequence needs fewer function evaluations for higher orders than the previous one and became prominent through the success of the "Gragg-Bulirsch-Stoer algorithm" (Bulirsch & Stoer 1966).

The above sequences have the property that for integration problems $y' = f(x)$ many function values can be saved and re-used for smaller h_i. Further, $\liminf(n_{i+1}/n_i)$ remains bounded away from 1 ("Toeplitz condition") which allows convergence proofs for $j = k \to \infty$ (Bauer, Rutishauser & Stiefel 1963). However, if we work with differential equations and with fixed or bounded order, the most economic sequence is the *"harmonic sequence"* (Deuflhard 1983)

$$1, 2, 3, 4, 5, 6, 7, 8, 9, 10, \ldots . \tag{9.8}$$

The Aitken - Neville Algorithm

For the case $p = 1$, (9.3) and (9.4) become a classical interpolation problem and we can compute the values of $T_{j,k}$ economically by the use of classical methods. Since we need only the values of the interpolation polynomials at the point $h = 0$, the most economical algorithm is that of "Aitken - Neville" (Aitken 1932, Neville 1934, based on ideas of Jordan 1928) which leads to

$$T_{j,k+1} = T_{j,k} + \frac{T_{j,k} - T_{j-1,k}}{(n_j/n_{j-k}) - 1}. \tag{9.9}$$

If the basic method used is *symmetric,* we know that the underlying asymptotic expansion is in powers of h^2 (Theorem 8.9), and each extrapolation eliminates *two* powers of h. We may thus simply replace in (9.3) h by h^2 and for $p = 2$ (i.e., $q = 1$ in (8.18)) also use the Aitken - Neville algorithm with this modification. This leads to

$$T_{j,k+1} = T_{j,k} + \frac{T_{j,k} - T_{j-1,k}}{(n_j/n_{j-k})^2 - 1} \tag{9.10}$$

instead of (9.9).

Numerical example. We solve the problem

$$y' = (-y \sin x + 2 \tan x)y, \qquad y(\pi/6) = 2/\sqrt{3} \tag{9.11}$$

with true solution $y(x) = 1/\cos x$ and basic step size $H = 0.2$ by Euler's method. Fig. 9.1 represents, for each of the entries $T_{j,k}$ of the extrapolation tableau, the *numerical work* $(1 + n_j - 1 + n_{j-1} - 1 + \ldots + n_{j-k+1} - 1)$ compared to the *precision* $(|T_{j,k} - y(x_0 + H)|)$ in double logarithmic scale. The first picture is for the Romberg sequence (9.6), the second for the Bulirsch sequence (9.7), and the last

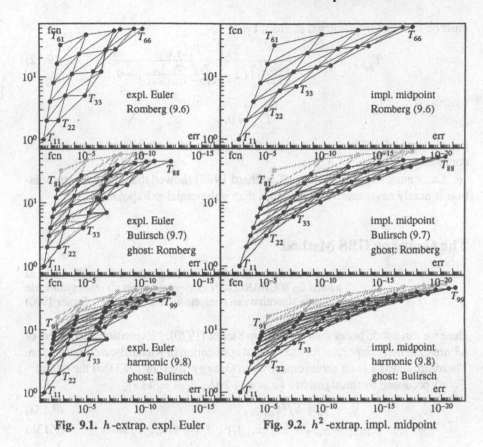

Fig. 9.1. h-extrap. expl. Euler

Fig. 9.2. h^2-extrap. impl. midpoint

for the harmonic sequence (9.8). In pictures 2 and 3 the results of the foregoing graphics are repeated as a shaded "ghost" (... of Canterville) in order to demonstrate how the results are better than those for the predecessor. Nobody is perfect, however. The "best" method in these comparisons, the harmonic sequence, suffers for high orders from a strong influence of rounding errors (see Exercise 5 below; the computations of Fig. 9.1, 9.2 and 9.4 have been made in quadruple precision).

The analogous results for the symmetric implicit mid-point rule (7.4) are presented in Fig. 9.2. Although implicit, this method is easy to implement for this particular example. We again use the same basic step size $H = 0.2$ as above and the same step-number sequences (9.6), (9.7), (9.8). Here, the "numerical work" $(n_j + n_{j-1} + \ldots + n_{j-k+1})$ represents *implicit* stages and therefore can not be compared to the values of the explicit method. The precisions, however, show a drastic improvement.

Rational Extrapolation. Many authors in the sixties claimed that it is better to use rational functions instead of polynomials in (9.3). In this case the formula (9.9) must be replaced by (Bulirsch & Stoer 1964)

$$T_{j,k+1} = T_{j,k} + \frac{T_{j,k} - T_{j-1,k}}{\left(\frac{n_j}{n_{j-k}}\right)\left(1 - \frac{T_{j,k}-T_{j-1,k}}{T_{j,k}-T_{j-1,k-1}}\right) - 1} \tag{9.12}$$

where

$$T_{j,0} = 0.$$

For systems of differential equations the division of vectors is to be understood componentwise.

Later numerical experiments (Deuflhard 1983) showed that rational extrapolation is nearly never more advantageous than polynomial extrapolation.

The Gragg or GBS Method

> Since it is fully explicit GRAGG's algorithm is so ideally suited as a basis for RICHARDSON extrapolation that no other symmetric two-step algorithm can compete with it.   (H.J. Stetter 1970)

Here we can not do better than quote from Stetter (1970): "Expansions in powers of h^2 are extremely important for an efficient application of Richardson extrapolation. Therefore it was a great achievement when Gragg proved in 1963 that the quantity $S_h(x)$ produced by the algorithm ($x = x_0 + 2nh$, $x_i = x_0 + ih$)

$$y_1 = y_0 + hf(x_0, y_0) \tag{9.13a}$$

$$y_{i+1} = y_{i-1} + 2hf(x_i, y_i) \qquad i = 1, 2, \ldots, 2n \tag{9.13b}$$

$$S_h(x) = \frac{1}{4}\left(y_{2n-1} + 2y_{2n} + y_{2n+1}\right) \tag{9.13c}$$

possesses an asymptotic expansion in even powers of h and has satisfactory stability properties. This led to the construction of the very powerful G(ragg)-B(ulirsch)-S(toer)-extrapolation algorithm ...".

Gragg's *proof* of this property was very long and complicated and it was again "a great achievement" that Stetter had the elegant idea of interpreting (9.13b) as a *one-step* algorithm by rewriting (9.13) in terms of odd and even indices: for this purpose we define

$$h^* = 2h, \qquad x_k^* = x_0 + kh^*, \qquad u_0 = v_0 = y_0,$$

$$u_k = y_{2k}, \qquad v_k = y_{2k+1} - hf(x_{2k}, y_{2k}) = \frac{1}{2}\left(y_{2k+1} + y_{2k-1}\right). \tag{9.14}$$

Then the method (9.13) can be rewritten as (see Fig. 9.3)

$$\begin{pmatrix} u_{k+1} \\ v_{k+1} \end{pmatrix} = \begin{pmatrix} u_k \\ v_k \end{pmatrix} + h^* \begin{pmatrix} f\left(x_k^* + \frac{h^*}{2}, \, v_k + \frac{h^*}{2}f(x_k^*, u_k)\right) \\ \frac{1}{2}\left(f(x_k^* + h^*, u_{k+1}) + f(x_k^*, u_k)\right) \end{pmatrix}. \tag{9.15}$$

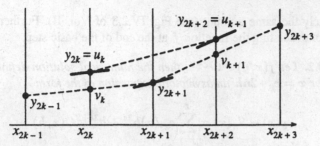

Fig. 9.3. Symmetry of the Gragg method

This method, which maps the pair (u_k, v_k) to (u_{k+1}, v_{k+1}), can be seen from Fig. 9.3 to be *symmetric*. The symmetry can also be checked analytically (see Definition 8.7) by exchanging $u_{k+1} \leftrightarrow u_k$, $v_{k+1} \leftrightarrow v_k$, $h^* \leftrightarrow -h^*$, $x_k^* \leftrightarrow x_k^* + h^*$. A trivial calculation then shows that this leaves formula (9.15) invariant. Method (9.15) is consistent with the differential equation (let $h^* \to 0$ in the increment function)

$$
\begin{aligned}
u' &= f(x, v) & u(x_0) &= y_0 \\
v' &= f(x, u) & v(x_0) &= y_0,
\end{aligned}
\tag{9.16}
$$

whose exact solution is simply $u(x) = v(x) = y(x)$. Therefore, we have from Theorem 8.10 that

$$
y(x) - u_{h^*}(x) = \sum_{j=1}^{\ell} a_{2j}(x)(h^*)^{2j} + (h^*)^{2\ell+2} A(x, h^*)
\tag{9.17a}
$$

$$
y(x) - v_{h^*}(x) = \sum_{j=1}^{\ell} b_{2j}(x)(h^*)^{2j} + (h^*)^{2\ell+2} B(x, h^*)
\tag{9.17b}
$$

and $a_{2j}(x_0) = b_{2j}(x_0) = 0$. We see from (9.14) and (9.17a) that $y_h(x)$ possesses an expansion in even powers of h, provided that the number of steps is even; i.e., for $x = x_0 + 2nh$,

$$
y(x) - y_h(x) = \sum_{j=1}^{\ell} \widehat{a}_{2j}(x) h^{2j} + h^{2\ell+2} \widehat{A}(x, h)
\tag{9.18}
$$

where $\widehat{a}_{2j}(x) = 2^{2j} a_{2j}(x)$ and $\widehat{A}(x, h) = 2^{2\ell+2} A(x, 2h)$.

The so-called *smoothing step*, i.e., formula

$$
S_h(x_0 + 2nh) = \frac{1}{4}(y_{2n-1} + 2y_{2n} + y_{2n+1}) = \frac{1}{2}(u_n + v_n)
$$

(see (9.13c) and (9.14)) had its historical origin in the "weak stability" of the explicit midpoint rule (9.13b) (see also Fig. III.9.2). However, since the method is anyway followed by extrapolation, this step is not of great importance (Shampine & Baca 1983). It is a little more costly and increases the "stability domain" by approximately the same amount (see Fig. IV.2.3 of Vol. II). Further, it has the advantage of evaluating the function f at the end of the basic step.

Theorem 9.2. *Let $f(x,y) \in C^{2\ell+2}$, then the numerical solution defined in (9.13) possesses for $x = x_0 + 2nh$ an asymptotic expansion of the form*

$$y(x) - S_h(x) = \sum_{j=1}^{\ell} e_{2j}(x)h^{2j} + h^{2\ell+2}C(x,h) \qquad (9.19)$$

with $e_{2j}(x_0) = 0$ and $C(x,h)$ bounded for $x_0 \leq x \leq \bar{x}$ and $0 \leq h \leq h_0$.

Proof. By adding (9.17a) and (9.17b) and using $h^* = 2h$ we obtain (9.19) with $e_{2j}(x) = (a_{2j}(x) + b_{2j}(x))2^{2j-1}$. $\qquad \Box$

This method can thus be used for Richardson extrapolation in the same way as symmetric methods above: we choose a step-number sequence, with the condition that the n_j are even, i.e.,

$$2, 4, 8, 16, 32, 64, 128, 256, \ldots \qquad (9.6')$$
$$2, 4, 6, 8, 12, 16, 24, 32, 48, \ldots \qquad (9.7')$$
$$2, 4, 6, 8, 10, 12, 14, 16, 18, \ldots \qquad (9.8')$$

set

$$T_{i,1} := S_{h_i}(x_0 + H)$$

and compute the extrapolated expressions $T_{i,j}$, based on the h^2-expansion, by the Aitken-Neville formula (9.10).

Numerical example. Fig. 9.4 represents the numerical results of this algorithm applied to Example (9.11) with step size $H = 0.2$. The step size sequences are Romberg (9.6') (above), Bulirsch (9.7') (middle), and harmonic (9.8') (below). The algorithm *with* smoothing step (numerical work $= 1 + n_j + n_{j-1} + \ldots + n_{j-k+1}$) is represented *left*, the results *without* smoothing step (numerical work $= 1 + n_j - 1 + n_{j-1} - 1 + \ldots + n_{j-k+1} - 1$) are on the *right*.

The results are nearly identical to those for the implicit midpoint rule (Fig. 9.2), but much more valuable, since here the method is explicit. In the pictures on the left the values for extrapolated Euler (from Fig. 9.1) are repeated as a "ghost" and demonstrate clearly the importance of the h^2-expansion, especially in the diagonal T_{kk} for large values of k. The ghost in the pictures on the right are the values *with* smoothing step from the left; the differences are seen to be tiny.

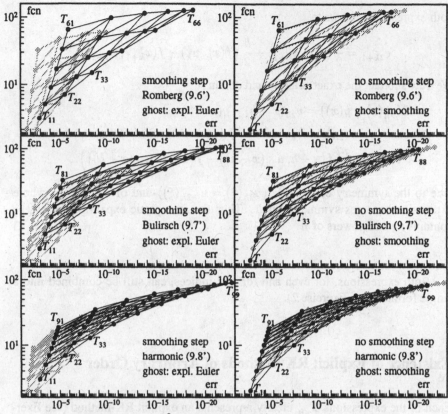

Fig. 9.4. Precision of h^2-extrapolated Gragg method for Example (9.11)

Asymptotic Expansion for Odd Indices

For completeness, we still want to derive the existence of an h^2 expansion for y_{2k+1} from (9.17b), although this is of no practical importance for the numerical algorithm described above.

Theorem 9.3 (Gragg 1964). *For* $x = x_0 + (2k+1)h$ *we have*

$$y(x) - y_h(x) = \sum_{j=1}^{\ell} \widehat{b}_{2j}(x) h^{2j} + h^{2\ell+2} \widehat{B}(x, h) \tag{9.20}$$

where the coefficients $\widehat{b}_{2j}(x)$ *are in general different from those for even indices and* $\widehat{b}_{2j}(x_0) \neq 0$.

Proof. y_{2k+1} can be computed (see Fig. 9.3) either from v_k by a forward step or from v_{k+1} by a backward step. For the sake of symmetry, we take the mean of both expressions and write

$$y_{2k+1} = \frac{1}{2}\left(v_k + v_{k+1}\right) + \frac{h}{2}\left(f(x_k^*, u_k) - f(x_{k+1}^*, u_{k+1})\right).$$

We now subtract the exact solution and obtain

$$2\big(y_h(x) - y(x)\big) = v_{2h}(x-h) - y(x-h)$$
$$+ v_{2h}(x+h) - y(x+h) + y(x-h) - 2y(x) + y(x+h)$$
$$+ h\Big(f\big(x-h, u_{2h}(x-h)\big) - f\big(x+h, u_{2h}(x+h)\big)\Big).$$

Due to the symmetry of $u_{2h}(x)$ $(u_{2h}(\xi) = u_{-2h}(\xi))$ and of $v_{2h}(x)$ the whole expression becomes symmetric in h. Thus the asymptotic expansion for y_{2k+1} contains no odd powers of h. □

Both expressions, for even and for odd indices, can still be combined into a single formula (see Exercise 2).

Existence of Explicit RK Methods of Arbitrary Order

Each of the expressions $T_{j,k}$ clearly represents an explicit RK-method (see Exercise 1). If we apply the well-known error formula for polynomial interpolation (see e.g., Abramowitz & Stegun 1964, formula 25.2.27) to (9.19), we obtain

$$y(x_0 + H) - T_{j,k} = \frac{(-1)^k}{n_j^2 \cdot \ldots \cdot n_{j-k+1}^2} e_{2k}(x_0 + H)H^{2k} + \mathcal{O}(H^{2k+2}). \quad (9.21)$$

Since $e_k(x_0) = 0$, we have

$$y(x_0 + H) - T_{j,k} = \frac{(-1)^k}{n_j^2 \cdot \ldots \cdot n_{j-k+1}^2} e'_{2k}(x_0)H^{2k+1} + \mathcal{O}(H^{2k+2}). \quad (9.22)$$

This shows that $T_{j,k}$ represents an explicit Runge-Kutta method of order $2k$. As an application of this result we have:

Theorem 9.4 (Gragg 1964). *For p even, there exists an explicit RK-method of order p with $s = p^2/4 + 1$ stages.*

Proof. This result is obtained by counting the number of necessary function evaluations of the GBS-algorithm using the harmonic sequence and without the final smoothing step. □

Remark. The extrapolated Euler method leads to explicit Runge-Kutta methods with $s = p(p-1)/2 + 1$ stages. This shows once again the importance of the h^2 expansion.

Order and Step Size Control

Extrapolation methods have the advantage that in addition to the step size also the order (i.e., number of columns) can be changed at each step. Because of this double freedom, the *practical implementation* in an optimal way is more complicated than for fixed-order RK-methods. The first codes were developed by Bulirsch & Stoer (1966) and their students. Very successful extrapolation codes due to P. Deuflhard and his collaborators are described in Deuflhard (code DIFEX1, 1983).

The choice of the *step size* can be performed in exactly the same way as for fixed-order embedded methods (see Section II.4). If the first k lines of the extrapolation tableau are computed, we have T_{kk} as the highest-order approximation (of order $2k$ by (9.22)) and in addition $T_{k,k-1}$ of order $2k-2$. It is therefore natural to use the expression

$$err_k = \|T_{k,k-1} - T_{k,k}\| \tag{9.23}$$

for step size control. The norm is the same as in (4.11). As in (4.12) we get for the optimal step size the formula

$$H_k = H \cdot 0.94 \cdot (0.65/err_k)^{1/(2k-1)} \tag{9.24}$$

where this time we have chosen a safety factor depending partly on the order.

For the choice of an *optimal order* we need a measure of work, which allows us to compare different methods. The work for computing T_{kk} can be measured by the number A_k of function evaluations. For the GBS-algorithm it is given recursively by

$$A_1 = n_1 + 1$$
$$A_k = A_{k-1} + n_k. \tag{9.25}$$

However, a large number of function evaluations can be compensated by a large step size H_k, given by (9.24). We therefore consider

$$W_k = \frac{A_k}{H_k}, \tag{9.26}$$

the *work per unit step,* as a measure of work. The idea is now to choose the order (i.e., the index k) in such a way that W_k is minimized.

Let us describe the *combined order and step size control* in some more detail. We assume that at some point of integration the step size H and the index k $(k > 2)$ are proposed. The step is then realized in the following way: we first compute $k-1$ lines of the extrapolation tableau and also the values $H_{k-2}, W_{k-2}, err_{k-1}, H_{k-1}, W_{k-1}$.

a) *Convergence in line $k-1$.* If $err_{k-1} \leq 1$, we accept $T_{k-1,k-1}$ as numerical solution and continue the integration with the new proposed quantities

$$k_{\text{new}} = \begin{cases} k & \text{if } W_{k-1} < 0.9 \cdot W_{k-2} \\ k-1 & \text{else} \end{cases}$$

$$H_{\text{new}} = \begin{cases} H_{k_{\text{new}}} & \text{if } k_{\text{new}} \leq k-1 \\ H_{k-1}(A_k/A_{k-1}) & \text{if } k_{\text{new}} = k. \end{cases} \tag{9.27}$$

In (9.27), the only non-trivial formula is the choice of the step size H_{new} in the case of an order-increase $k_{\text{new}} = k$. In this case we want to avoid the computation of err_k, so that H_k and W_k are unknown. However, since our k is assumed to be close to the optimal value, we have $W_k \approx W_{k-1}$ which leads to the proposed step size increase.

b) *Convergence monitor.* If $err_{k-1} > 1$, we first decide whether we may expect convergence at least in line $k+1$. It follows from (9.22) that, asymptotically,

$$\|T_{k,k-2} - T_{k,k-1}\| \approx \left(\frac{n_2}{n_k}\right)^2 err_{k-1} \tag{9.28}$$

with err_{k-1} given by (9.23). Unfortunately, err_k cannot be compared with (9.28), since different factors (depending on the differential equation to be solved) are involved in the asymptotic formula (cf. (9.22)). If we nevertheless assume that err_k is $(n_2/n_1)^2$ times smaller than (9.28) we obtain $err_k \approx (n_1/n_k)^2 err_{k-1}$. We therefore already reject the step at this point, if

$$err_{k-1} > \left(\frac{n_{k+1} n_k}{n_1 n_1}\right)^2 \tag{9.29}$$

and restart with $k_{\text{new}} \leq k-1$ and H_{new} according to (9.27). If the contrary of (9.29) holds, we compute the next line of the extrapolation tableau, i.e., $T_{k,k}$, err_k, H_k and W_k.

c) *Convergence in line k.* If $err_k \leq 1$, we accept T_{kk} as numerical solution and continue the integration with the new proposed values

$$k_{\text{new}} = \begin{cases} k-1 & \text{if } W_{k-1} < 0.9 \cdot W_k \\ k+1 & \text{if } W_k < 0.9 \cdot W_{k-1} \\ k & \text{in all other cases} \end{cases} \tag{9.30}$$

$$H_{\text{new}} = \begin{cases} H_{k_{\text{new}}} & \text{if } k_{\text{new}} \leq k \\ H_k(A_{k+1}/A_k) & \text{if } k_{\text{new}} = k+1. \end{cases}$$

d) *Second convergence monitor.* If $err_k > 1$, we check, as in (b), the relation

$$err_k > \left(\frac{n_{k+1}}{n_1}\right)^2. \tag{9.31}$$

If (9.31) is satisfied, the step is rejected and we restart with $k_{\text{new}} \leq k$ and H_{new} of (9.30). Otherwise we continue.

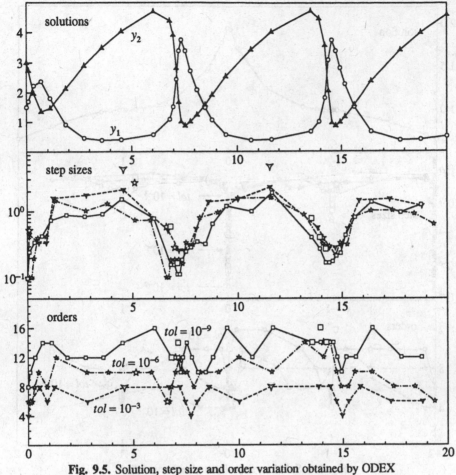

Fig. 9.5. Solution, step size and order variation obtained by ODEX

e) *Hope for convergence in line* $k+1$. We compute err_{k+1}, H_{k+1} and W_{k+1}. If $err_{k+1} \leq 1$, we accept $T_{k+1,k+1}$ as numerical solution and continue the integration with the new proposed order

$$k_{\text{new}} := k$$

$$\text{if } (W_{k-1} < 0.9 \cdot W_k) \qquad k_{\text{new}} := k-1 \qquad (9.32)$$

$$\text{if } (W_{k+1} < 0.9 \cdot W_{k_{\text{new}}}) \qquad k_{\text{new}} := k+1.$$

If $err_{k+1} > 1$ the step is rejected and we restart with $k_{\text{new}} \leq k$ and H_{new} of (9.24).

The following slight modifications of the above algorithm are recommended:

i) Storage considerations lead to a limitation of the number of columns of the extrapolation tableau, say by $k_{\max}$ (e.g., $k_{\max} = 9$). For the proposed index k_{new} we require $2 \leq k_{\text{new}} \leq k_{\max} - 1$. This allows us to activate (e) at each step.

ii) After a step-rejection the step size and the order may not be increased.

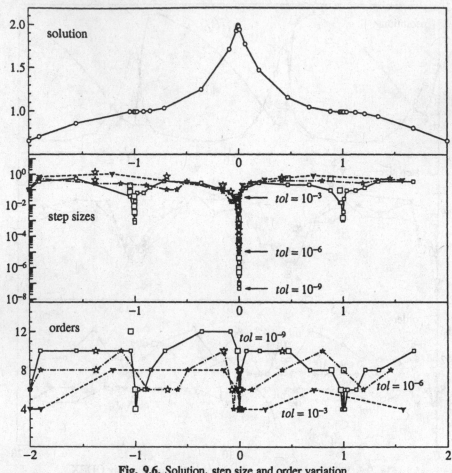

Fig. 9.6. Solution, step size and order variation
obtained by ODEX at the discontinuous example (9.33)

Numerical study of the combined step size and order control. We show in the following examples how the step size and the order vary for the above algorithm. For this purpose we have written the FORTRAN-subroutine ODEX (see Appendix).

As a first example we again take the *Brusselator* (cf. Section II.4). As in Fig. 4.1, the first picture of Fig. 9.5 shows the two components of the solution (obtained with $Atol = Rtol = 10^{-9}$). In the remaining two pictures we have plotted the step sizes and orders for the three tolerances 10^{-3} (broken line), 10^{-6} (dashes and dots) and 10^{-9} (solid line). One can easily observe that the extrapolation code automatically chooses a suitable order (depending essentially on Tol). Step-rejections are indicated by larger symbols.

We next study the behaviour of the order control near discontinuities. In the example

$$y' = -\text{sign}(x)\left|1 - |x|\right| y^2, \qquad y(-2) = 2/3, \qquad -2 \le x \le 2 \qquad (9.33)$$

we have a discontinuity in the first derivative of $y(x)$ at $x = 0$ and two discontinuities in the second derivative (at $x = \pm 1$). The numerical results are shown in Fig. 9.6 for three tolerances. In all cases the error at the endpoint is about $10 \cdot Tol$. The discontinuities at $x = \pm 1$ are not recognized in the computations with $Tol = 10^{-3}$ and $Tol = 10^{-6}$. Whenever a discontinuity is detected, the order drops to 4 (lowest possible) in its neighbourhood, so that these points are passed rather efficiently.

Dense Output for the GBS Method

Extrapolation methods are methods best suited for high precision which typically take very large (basic) step sizes during integration. The reasons for the need of a dense output formula (discussed in Section II.6) are therefore particularly important here. First attempts to provide extrapolation methods with a dense output are due to Lindberg (1972) for the implicit trapezoidal rule, and to Shampine, Baca & Bauer (1983) who constructed a 3rd order dense output for the GBS method. We present here the approach of Hairer & Ostermann (1990) (see also Simonsen 1990).

It turned out that the existence of high order dense output is only possible if the step number sequence satisfies some restrictions such as

$$n_{j+1} - n_j = 0 \,(\,\text{mod}\, 4\,) \qquad \text{for} \quad j = 1, 2, 3, \ldots \qquad (9.34)$$

which, for example, is fulfilled by the sequence

$$\{2, 6, 10, 14, 18, 22, 26, 30, 34, \ldots\}\,. \qquad (9.35)$$

The idea is, once again, to do Hermite interpolation. To begin with, high order approximations are as usual at our disposal for the values y_0, y_0', y_1, y_1' by using y_0, $f(x_0, y_0)$, T_{kk}, $f(x_0 + H, T_{kk})$, where T_{kk} is supposed to be the highest order approximation computed and used for continuation of the solution.

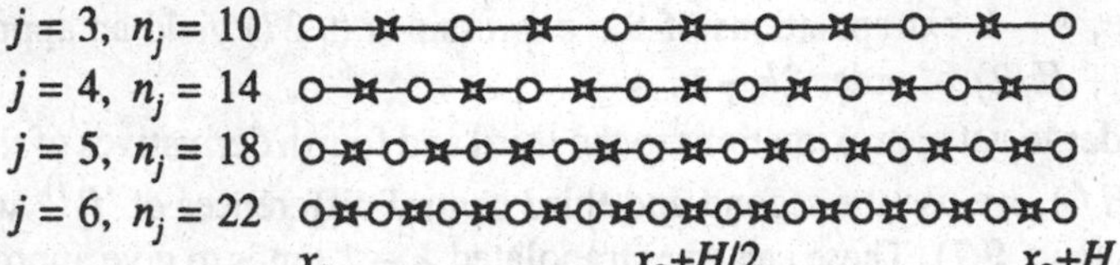

Fig. 9.7. Evaluation points for a GBS step

For more inspiration, we represent in Fig. 9.7 the steps taken by Gragg's midpoint rule for the step number sequence (9.35). The symbols o and × indicate that the even steps and the odd steps possess a *different* asymptotic expansion (see Theorem 9.3) and must not be blended. We see that, owing to condition (9.34), the midpoint values $y_{n_j/2}^{(j)}$, obtained during the computation of T_{j1}, all have the same parity and can therefore also be extrapolated to yield an approximation for $y(x_0 + H/2)$ of order $2k - 1$ (remember that in Theorem 9.3, $\widehat{b}_{2j}(x_0) \neq 0$).

We next insert (9.20) for $x = x_0 + H/2$ into $f(x, y)$

$$f_{n_j/2}^{(j)} := f(x, y_{n_j/2}^{(j)}) = f\left(x, y(x) - h_j^2 \widehat{b}_2(x) - h_j^4 \widehat{b}_4(x) \dots \right)$$

and develop in powers of h_j to obtain

$$y'(x) - f_{n_j/2}^{(j)} = h_j^2 a_{2,1}(x) + h_j^4 a_{4,1}(x) + \dots . \tag{9.36}$$

This shows that the f-values at the midpoint $x_0 + H/2$ (for $j = 1, 2, \dots k$) possess an asymptotic expansion and can be extrapolated $k - 1$ times to yield an approximation to $y'(x_0 + H/2)$ of order $2k - 1$.

But this is not enough. We now consider, similar to an idea which goes back to the papers of Deuflhard & Nowak (1987) and Lubich (1989), the central differences $\delta f_i = f_{i+1} - f_{i-1}$ at the midpoint which, by Fig. 9.7, are available for $j = 1, 2, \dots, k$ and are based on even parity. By using (9.18) and by developing into powers of h_j we obtain

$$\frac{\delta f_{n_j/2}^{(j)}}{2h_j} = \frac{f(x+h_j, y_{n_j/2+1}^{(j)}) - f(x-h_j, y_{n_j/2-1}^{(j)})}{2h_j}$$

$$= \left(f\left(x+h_j, y(x+h_j) - h_j^2 \widehat{a}_2(x+h_j) - h_j^4 \widehat{a}_4(x+h_j) - \dots \right) - \right.$$

$$\left. f\left(x-h_j, y(x-h_j) - h_j^2 \widehat{a}_2(x-h_j) - h_j^4 \widehat{a}_4(x-h_j) - \dots \right) \right) \Big/ 2h_j$$

$$= \frac{y'(x+h_j) - y'(x-h_j)}{2h_j} - h_j^2 c_2(x) - h_j^4 c_4(x) - \dots .$$

Finally we insert the Taylor series for $y'(x+h)$ and $y'(x-h)$ to obtain an expansion

$$y''(x) - \frac{\delta f_{n_j/2}^{(j)}}{2h_j} = h_j^2 a_{2,2}(x) + h_j^4 a_{4,2}(x) + \dots . \tag{9.38}$$

Therefore, $k - 1$ extrapolations of the expressions (9.37) yield an approximation to $y''(x_0 + H/2)$ of order $2k - 1$.

In order to get approximations to the third and fourth derivatives of the solution at $x_0 + H/2$, we use the second and third central differences of $f_i^{(j)}$ which exist for $j \geq 2$ (Fig. 9.7). These can be extrapolated $k - 2$ times to give approximations of order $2k - 3$.

The continuation of this process yields the following algorithm:

Step 1. For each $j \in \{1, \ldots, k\}$, compute approximations to the derivatives of $y(x)$ at $x_0 + H/2$ by:

$$d_j^{(0)} = y_{n_j/2}^{(j)} \,, \qquad d_j^{(\kappa)} = \frac{\delta^{\kappa-1} f_{n_j/2}^{(j)}}{(2h_j)^{\kappa-1}} \quad \text{for } \kappa = 1, \ldots, 2j \,. \tag{9.39}$$

Step 2. Extrapolate $d_j^{(0)}$ $(k-1)$ times and $d_j^{(2\ell-1)}$, $d_j^{(2\ell)}$ $(k-\ell)$ times to obtain improved approximations $d^{(\kappa)}$ to $y^{(\kappa)}(x_0 + H/2)$.

Step 3. For given μ $(-1 \leq \mu \leq 2k)$ define the polynomial $P_\mu(\theta)$ of degree $\mu + 4$ by

$$\begin{aligned}
P_\mu(0) &= y_0 \,, & P_\mu'(0) &= Hf(x_0, y_0) \,, \\
P_\mu(1) &= T_{kk} \,, & P_\mu'(1) &= Hf(x_0 + H, T_{kk}) \\
P_\mu^{(\kappa)}(1/2) &= H^\kappa d^{(\kappa)} & \text{for } &\kappa = 0, \ldots, \mu \,.
\end{aligned} \tag{9.40}$$

This computation of $P_\mu(\theta)$ does not need any further function evaluation since $f(x_0 + H, T_{kk})$ has to be computed anyway for the next step. Further, $P_\mu(\theta)$ gives a global C^1 approximation to the solution.

Theorem 9.5 (Hairer & Ostermann 1990). *If the step number sequence satisfies (9.34), then the error of the dense output polynomial $P_\mu(\theta)$ satisfies*

$$y(x_0 + \theta H) - P_\mu(\theta) = \begin{cases} \mathcal{O}(H^{2k+1}) & \text{if } n_1 = 4 \text{ and } \mu \geq 2k-4 \\ \mathcal{O}(H^{2k}) & \text{if } n_1 = 2 \text{ and } \mu \geq 2k-5. \end{cases} \tag{9.40}$$

Proof. Since $P_\mu(\theta)$ is a polynomial of degree $\mu + 4$ the error due to interpolation is of size $\mathcal{O}(H^{\mu+5})$. This explains the restriction on μ in (9.40). As explained above, the function value and derivative data used for Hermite interpolation have the required precision

$$H^\kappa y^{(\kappa)}(x_0 + H/2) - H^\kappa d^{(\kappa)} = \begin{cases} \mathcal{O}(H^{2k}) & \text{if } \kappa = 0, \\ \mathcal{O}(H^{2k+1}) & \text{if } \kappa \text{ is odd,} \\ \mathcal{O}(H^{2k+2}) & \text{if } \kappa \geq 2 \text{ is even.} \end{cases}$$

In the case $n_1 = 4$ the parity of the central point $x_0 + H/2$ is *even* (in contrary to Fig. 9.7), we therefore apply (9.18) and gain one order because then the functions $a_{i,0}(x)$ vanish at x_0. $\qquad\Box$

Control of the Interpolation Error

> At one time ... every young mathematician was familiar with sn u, cn u, and dn u, and algebraic identities between these functions figured in every examination.
>
> (E.H. Neville, Jacobian Elliptic Functions, 1944)

Numerical example. We apply the above dense output formula with $\mu = 2k - 3$ (as is standard in ODEX) to the differential equations of the Jacobian elliptic functions sn, cn, dn (see Abramowitz & Stegun 1964, 16.16):

$$
\begin{aligned}
y_1' &= y_2 y_3 & y_1(0) &= 0 \\
y_2' &= -y_1 y_3 & y_2(0) &= 1 \\
y_3' &= -0.51 \cdot y_1 y_2 & y_3(0) &= 1
\end{aligned}
\tag{9.41}
$$

with integration interval $0 \le x \le 10$ and error tolerance $Atol = Rtol = 10^{-9}$. The error for the three components of the obtained continuous solution is displayed in Fig. 9.8 (upper picture; the ghosts are the solution curves) and gives a quite disappointing impression when compared with the precision at the grid points. We shall now see that these horrible bumps are nothing else than interpolation errors.

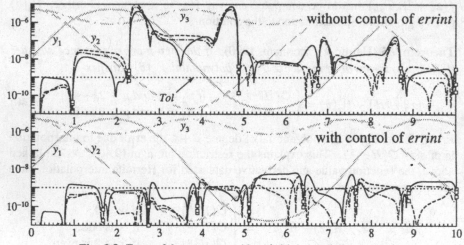

Fig. 9.8. Error of dense output without/with interpolation control

Assume that in the definition of $P_\mu(\theta)$ the basic function and derivative values are replaced by the exact values $y(x_0 + H)$, $y'(x_0 + H)$, and $y^{(\kappa)}(x_0 + H/2)$. Then the error of $P_\mu(\theta)$ is given by

$$
\theta^2 (1 - \theta)^2 \left(\theta - \frac{1}{2} \right)^{\mu+1} \frac{y^{(\mu+5)}(\xi)}{(\mu+5)!} H^{\mu+5}
\tag{9.42}
$$

where $\xi \in (x_0, x_0 + H)$ (possibly different for each component). The function $\theta^2 (1 - \theta)^2 (\theta - 1/2)^{\mu+1}$ has its maximum at

$$\theta_{\mu+1} = \frac{1}{2} \pm \frac{1}{2} \sqrt{\frac{\mu+1}{\mu+5}} \tag{9.43}$$

which, for large μ, are close to the ends of the integration intervals and indicate precisely the locations of the large bumps in Fig. 9.8. This demonstrates the need for a code which not only controls the error at the grid points, but also takes care of the interpolation error. To this end we denote by a_μ the coefficient of $\theta^{\mu+4}$ in the polynomial $P_\mu(\theta)$ and consider (Hairer & Ostermann 1992)

$$P_\mu(\theta) - P_{\mu-1}(\theta) = \theta^2 (1 - \theta)^2 \left(\theta - \frac{1}{2}\right)^\mu a_\mu \tag{9.44}$$

as an approximation for the interpolation error for $P_{\mu-1}(\theta)$ and use

$$errint = \| P_\mu(\theta_\mu) - P_{\mu-1}(\theta_\mu) \| \tag{9.45}$$

as error estimator (the norm is again that of (4.11)). Then, if $errint > 10$ the step is rejected and recomputed with

$$H_{\text{int}} = H \left(1/errint\right)^{1/(\mu+4)}$$

because $errint = \mathcal{O}(H^{\mu+4})$. Otherwise the subsequent step is computed subject to the restriction $H \le H_{\text{int}}$.

This modified step size strategy makes the code, together with its dense output, more robust. The corresponding numerical results for the problem (9.41) are presented in the lower graph of Fig. 9.8.

Exercises

1. Show that the extrapolated Euler methods $T_{3,1}, T_{3,2}, T_{3,3}$ (with step-number sequence (9.8)) are equivalent to the Runge-Kutta methods of Table 9.1. Compute also the Runge-Kutta schemes corresponding to the first elements of the GBS algorithm.

Table 9.1. Extrapolation methods as Runge-Kutta methods

			0				0		
0			1/2	1/2			1/2	1/2	
1/3	1/3		1/3	1/3 0			1/3	1/3 0	
2/3	1/3 1/3		2/3	1/3 0 1/3			2/3	1/3 0 1/3	
	1/3 1/3 1/3			0 −1 1 1				0 −2 3/2 3/2	

$T_{3,1}$ order 1 $T_{3,2}$ order 2 $T_{3,3}$ order 3

2. Combine (9.18) and (9.19) into the formula ($x = x_0 + kh$)

$$y(x) - y_k = \sum_{j=1}^{\ell} \left(\alpha_{2j}(x) + (-1)^k \beta_{2j}(x) \right) h^{2j} + h^{2\ell+2} E(x, h)$$

for the asymptotic expansion of the Gragg method defined by (9.13a,b).

3. (Stetter 1970). Prove that for every real b (generally between 0 and 1) the method

$$y_1 = y_0 + h \Big(bf(x_0, y_0) + (1-b)f(x_1, y_1) \Big)$$

$$y_{i+1} = y_{i-1} + h \Big((1-b)f(x_{i-1}, y_{i-1}) + 2bf(x_i, y_i) + (1-b)f(x_{i+1}, y_{i+1}) \Big)$$

possesses an expansion in powers of h^2. Prove the same property for the smoothing step

$$S_h(x) = \frac{1}{2} \Big(y_{2n} + y_{2n-1} + h(1-b)f(x_{2n-1}, y_{2n-1}) + hbf(x_{2n}, y_{2n}) \Big).$$

4. (Stetter 1970). Is the Euler step (9.13a) essential for an h^2-expansion? Prove that a first order starting procedure

$$y_1 = y_0 + h\Phi(x_0, y_0, h)$$

for (9.13a) produces an h^2-expansion if the quantities $y_{-1} = y_0 - h\Phi(x_0, y_0, -h)$, y_0, and y_1 satisfy (9.13b) for $i = 0$.

5. Study the numerical instability of the extrapolation scheme for the harmonic sequence, i.e., suppose that the entries T_{11}, T_{21}, $T_{31} \ldots$ are disturbed with rounding errors ε, $-\varepsilon$, $\varepsilon, \ldots$ and compute the propagation of these errors into the extrapolation tableau (9.5).

Result. Due to the linearity of the extrapolation scheme, we suppose the T_{ik} equal zero and $\varepsilon = 1$. Then the results for sequence (9.8') are

1.							
−1.	−1.67						
1.	2.60	3.13					
−1.	−3.57	−5.63	−6.21				
1.	4.56	9.13	11.94	12.69			
−1.	−5.55	−13.63	−21.21	−25.35	−26.44		
1.	6.54	19.13	35.01	47.65	54.14	55.82	
−1.	−7.53	−25.63	−54.31	−84.09	−105.64	−116.30	−119.03
1.	8.53	33.13	80.13	140.14	195.34	232.96	251.10	255.73

hence, for order 18, we lose approximately two digits due to roundoff errors.

6. (Laguerre 1883[*]). If $a_1, a_2, \ldots, a_n$ are distinct positive real numbers and $r_1, r_2, \ldots, r_n$ are distinct reals, then

$$A = \begin{pmatrix} a_1^{r_1} & a_1^{r_2} & \cdots & a_1^{r_n} \\ a_2^{r_1} & a_2^{r_2} & \cdots & a_2^{r_n} \\ \vdots & \vdots & & \vdots \\ a_n^{r_1} & a_n^{r_2} & \cdots & a_n^{r_n} \end{pmatrix}$$

is invertible.

Hint (Pólya & Szegö 1925, Vol. II, Abschn. V, Problems 76-77[*]). Show by induction on n that, if the function $g(t) = \sum_{i=1}^{n} \alpha_i t^{r_i}$ has n distinct positive zeros, then $g(t) \equiv 0$. By Rolle's theorem the function

$$\frac{d}{dt}\left(t^{-r_1} g(t)\right) = \sum_{i=2}^{n} \alpha_i (r_i - r_1) t^{r_i - r_1 - 1}$$

has $n - 1$ positive distinct zeros and the induction hypothesis can be applied.

[*] We are grateful to our colleague J. Steinig for these references.

II.10 Numerical Comparisons

The Pleiades seem to be among the first stars mentioned in astronomical literature, appearing in Chinese annals of 2357 B.C. ...
(R.H. Allen, Star names, their love and meaning, 1899, Dover 1963)

If you enjoy fooling around making pictures, instead of typesetting ordinary text, TEX will be a source of endless frustration/amusement for you, ...
(D. Knuth, The TEXbook, p. 389)

Problems

EULR — Euler's equation of rotation of a rigid body ("Diese merkwürdig symmetrischen und eleganten Formeln ...", A. Sommerfeld 1942, vol. I, § 26.1, Euler 1758)

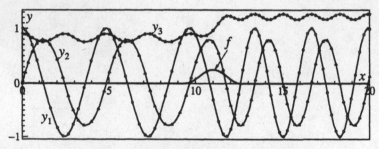

Fig. 10.1. Solutions of Euler's equations (10.1)

$$I_1\, y_1' = (I_2 - I_3)\, y_2 y_3$$
$$I_2\, y_2' = (I_3 - I_1)\, y_3 y_1 \tag{10.1}$$
$$I_3\, y_3' = (I_1 - I_2)\, y_1 y_2 + f(x)$$

where y_1, y_2, y_3 are the coordinates of $\vec{\omega}$, the rotation vector, and I_1, I_2, I_3 are the principal moments of inertia. The third coordinate has an additional exterior force

$$f(x) = \begin{cases} 0.25 \cdot \sin^2 x & \text{if } 3\pi \le x \le 4\pi \\ 0 & \text{otherwise} \end{cases} \tag{10.1'}$$

which is discontinuous in its second derivative. We choose the constants and initial values as

$$I_1 = 0.5, \quad I_2 = 2, \quad I_3 = 3, \quad y_1(0) = 1, \quad y_2(0) = 0, \quad y_3(0) = 0.9$$

(see Fig. 10.1) and check the numerical precision at the output points

$$x_{\text{end}} = 10 \quad \text{and} \quad x_{\text{end}} = 20 .$$

AREN — the Arenstorf orbit (0.1) for the restricted three body problem with initial values (0.2) integrated over one period $0 \leq x \leq x_{\text{end}}$ (see Fig. 0.1). The precision is checked at the endpoint, here the solution is most sensitive to errors of the initial phase.

LRNZ — the solution of the Saltzman-Lorenz equations (I.16.17) displayed in Fig. I.16.8, i.e., with constants and initial values

$$\sigma = 10, \quad r = 28, \quad b = \frac{8}{3}, \quad y_1(0) = -8, \quad y_2(0) = 8, \quad y_3(0) = 27 . \tag{10.2}$$

The solution is, for large values of x, *extremely* sensitive to the errors of the first integration steps (see Fig. I.16.10 and its discussion). For example, at $x = 50$ the numerical solution becomes totally wrong, even if the computations are performed in quadruple precision with $Tol = 10^{-20}$. Hence the numerical results of *all* methods would be equally useless and no comparison makes any sense. Therefore we choose

$$x_{\text{end}} = 16$$

and check the numerical solution at this point. Even here, all computations with $Tol \geq 10^{-7}$, say, fall into a chaotic cloud of meaningless results (see Fig. 10.5).

PLEI — a celestial mechanics problem (which we call "the Pleiades"): seven stars in the plane with coordinates x_i, y_i and masses $m_i = i$ $(i = 1, \ldots, 7)$:

$$\begin{aligned} x_i'' &= \sum_{j \neq i} m_j (x_j - x_i)/r_{ij} \\ y_i'' &= \sum_{j \neq i} m_j (y_j - y_i)/r_{ij} \end{aligned} \tag{10.3}$$

where

$$r_{ij} = \left((x_i - x_j)^2 + (y_i - y_j)^2\right)^{3/2}, \qquad i, j = 1, \ldots, 7.$$

The initial values are

$$\begin{aligned} &x_1(0) = 3, \quad x_2(0) = 3, \quad x_3(0) = -1, \quad x_4(0) = -3, \\ &x_5(0) = 2, \quad x_6(0) = -2, \quad x_7(0) = 2, \\ &y_1(0) = 3, \quad y_2(0) = -3, \quad y_3(0) = 2, \quad y_4(0) = 0, \\ &y_5(0) = 0, \quad y_6(0) = -4, \quad y_7(0) = 4, \\ &x_i'(0) = y_i'(0) = 0, \quad \text{for all } i \text{ with the exception of} \\ &x_6'(0) = 1.75, \quad x_7'(0) = -1.5, \quad y_4'(0) = -1.25, \quad y_5'(0) = 1, \end{aligned} \tag{10.4}$$

and we integrate for $0 \le t \le t_{\text{end}} = 3$. Fig. 10.2a represents the movement of these 7 bodies in phase coordinates. The initial value is marked by an "i", the final value at $t = t_{\text{end}}$ is marked by an "f". Between these points, 19 time-equidistant output points are plotted and connected by a dense output formula. There occur several quasi-collisions which are displayed in Table 10.1.

Table 10.1. Quasi-collisions in the PLEI problem

Body$_1$	1	1	3	1	2	5
Body$_2$	7	3	5	7	6	7
r_{ij}^2	0.0129	0.0193	0.0031	0.0011	0.1005	0.0700
time	1.23	1.46	1.63	1.68	1.94	2.14

The resulting violent shapes of the derivatives $x_i'(t)$, $y_i'(t)$ are displayed in Fig. 10.2b and show that automatic step size control is essential for this example.

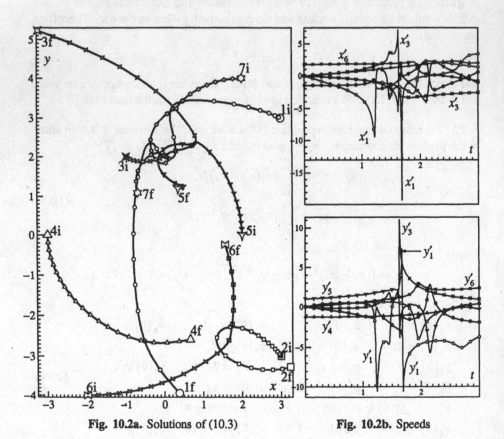

Fig. 10.2a. Solutions of (10.3)   **Fig. 10.2b.** Speeds

ROPE — the movement of a hanging rope (see Fig. 10.3a) of length 1 under gravitation and under the influence of a horizontal force

$$F_y(t) = \left(\frac{1}{\cosh(4t - 2.5)} \right)^4 \tag{10.5a}$$

acting at the point $s = 0.75$ as well as a vertical force

$$F_x(t) = 0.4 \tag{10.5b}$$

acting at the endpoint $s = 1$.

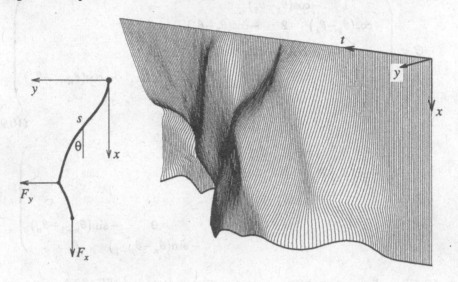

If this problem is discretized, then Lagrange theory (see (I.6.18); see also Exercises IV.1.2 and IV.1.4 of Volume II) leads to the following equations for the unknown angles θ_k:

$$\sum_{k=1}^{n} a_{lk} \ddot{\theta}_k = - \sum_{k=1}^{n} b_{lk} \dot{\theta}_k^2 - n\left(n + \frac{1}{2} - l\right) \sin \theta_l \tag{10.6}$$

$$- n^2 \sin \theta_l \cdot F_x(t) + \begin{cases} n^2 \cos \theta_l \cdot F_y(t) & \text{if } l \leq 3n/4 \\ 0 & \text{if } l > 3n/4, \end{cases} \quad l = 1, \dots, n$$

where

$$a_{lk} = g_{lk} \cos(\theta_l - \theta_k), \qquad b_{lk} = g_{lk} \sin(\theta_l - \theta_k), \qquad g_{lk} = n + \frac{1}{2} - \max(l, k). \tag{10.7}$$

We choose

$$n = 40, \qquad \theta_l(0) = \dot{\theta}_l(0) = 0, \qquad 0 \leq t \leq 3.723. \tag{10.8}$$

The resulting system is of dimension 80. The special structure of G^{-1} (see (IV.1.16–18) of Volume II) allows one to evaluate $\ddot{\theta}_l$ with the following algorithm:

a) Let $v_l = -n\left(n+\tfrac{1}{2}-l\right)\sin\theta_l - n^2 \sin\theta_l \cdot F_x + \begin{cases} n^2 \cos\theta_l \cdot F_y \\ 0 \end{cases}$

b) Compute $w = Dv + \dot{\theta}^2$,

c) Solve the tridiagonal system $Cu = w$,

d) Compute $\ddot{\theta} = Cv + Du$,

where

$$
C = \begin{pmatrix}
1 & -\cos(\theta_1-\theta_2) & & & & \\
-\cos(\theta_2-\theta_1) & 2 & -\cos(\theta_2-\theta_3) & & & \\
& -\cos(\theta_3-\theta_2) & \ddots & & \ddots & \\
& & \ddots & & 2 & -\cos(\theta_{n-1}-\theta_n) \\
& & & & -\cos(\theta_n-\theta_{n-1}) & 3
\end{pmatrix}
$$

(10.9)

$$
D = \begin{pmatrix}
0 & -\sin(\theta_1-\theta_2) & & & & \\
-\sin(\theta_2-\theta_1) & 0 & -\sin(\theta_2-\theta_3) & & & \\
& -\sin(\theta_3-\theta_2) & \ddots & & \ddots & \\
& & \ddots & & 0 & -\sin(\theta_{n-1}-\theta_n) \\
& & & & -\sin(\theta_n-\theta_{n-1}) & 0
\end{pmatrix} .
$$

BRUS — the reaction-diffusion equation (Brusselator with diffusion)

$$
\frac{\partial u}{\partial t} = 1 + u^2 v - 4.4u + \alpha\left(\frac{\partial^2 u}{\partial x^2} + \frac{\partial^2 u}{\partial y^2}\right)
$$
$$
\frac{\partial v}{\partial t} = 3.4u - u^2 v + \alpha\left(\frac{\partial^2 v}{\partial x^2} + \frac{\partial^2 v}{\partial y^2}\right)
$$

(10.10)

for $0 \le x \le 1, 0 \le y \le 1, t \ge 0, \alpha = 2\cdot10^{-3}$ together with the Neumann boundary conditions

$$
\frac{\partial u}{\partial n} = 0, \qquad \frac{\partial v}{\partial n} = 0,
$$

(10.11)

and the initial conditions

$$
u(x,y,0) = 0.5+y, \qquad v(x,y,0) = 1+5x .
$$

(10.12)

By the method of lines (cf. Section I.6) this problem becomes a system of ordinary differential equations. We put

$$
x_i = \frac{i-1}{N-1}, \qquad y_j = \frac{j-1}{N-1}, \qquad i,j = 1,\dots,N
$$

and define

$$U_{ij}(t) = u(x_i, y_j, t), \qquad V_{ij}(t) = v(x_i, y_j, t) . \tag{10.13}$$

Discretizing the derivatives in (10.10) with respect to the space variables we obtain for $i, j = 1, \ldots, N$

$$U'_{ij} = 1 + U^2_{ij}V_{ij} - 4.4U_{ij} + \alpha(N-1)^2 \left(U_{i+1,j} + U_{i-1,j} + U_{i,j+1} + U_{i,j-1} - 4U_{ij} \right)$$

$$V'_{ij} = 3.4U_{ij} - U^2_{ij}V_{ij} + \alpha(N-1)^2 \left(V_{i+1,j} + V_{i-1,j} + V_{i,j+1} + V_{i,j-1} - 4V_{ij} \right), \tag{10.14}$$

an ODE of dimension $2N^2$. Because of the boundary condition (10.11) we have

$$U_{0,j} = U_{2,j} , \quad U_{N+1,j} = U_{N-1,j} , \quad U_{i,0} = U_{i,2} , \quad U_{i,N+1} = U_{i,N-1}$$

and similarly for the V_{ij}-quantities. We choose $N = 21$ so that the system is of dimension 882 and check the numerical solutions at the output point $t_{end} = 7.5$. The solution of (10.14) (in the (x, y)-space) is represented in Fig. 10.4a and Fig. 10.4b for u and v respectively.

Performance of the Codes

Several codes were applied to each of the test problems with $Tol = 10^{-3}$, $Tol = 10^{-3-1/8}$, $Tol = 10^{-3-2/8}$, $Tol = 10^{-3-3/8}$, ... (for the large problems with $Tol = 10^{-3}$, $Tol = 10^{-3-1/4}$, $Tol = 10^{-3-2/4}$, ...) up to, in general, $Tol = 10^{-14}$, then the numerical result at the output points were compared with an "exact solution" (computed very precisely in quadruple precision). Each of these results then corresponds to one point of Fig. 10.5, where this precision is compared (in double logarithmic scale) to the number of function evaluations. The "integer" tolerances 10^{-3}, 10^{-4}, 10^{-5}, ... are distinguishable as enlarged symbols. All codes were applied with complete "standard" parameter settings and were not at all "tuned" to these particular problems.

A comparison of the *computing time* (instead of the number of function evaluations) gave no significant difference. Therefore, only one representative of the small problems (LRNZ) and one large problem (BRUS) are displayed in Fig. 10.6. All computations have been performed in REAL*8 ($Uround = 1.11 \cdot 10^{-16}$) on an Apollo Domain 4000 Workstation.

The codes used are the following:

RKF45 — symbol ⋈ — a product of Shampine and Watts' programming art based on Fehlberg's pair of orders 4 and 5 (Table 5.1). The method is used in the "local extrapolation mode", i.e., the numerical solution is advanced with the 5th order result. The code is usually, except for low precision, the slowest of all, which is explained by its low order. The results of the "time"-picture Fig. 10.6 for this

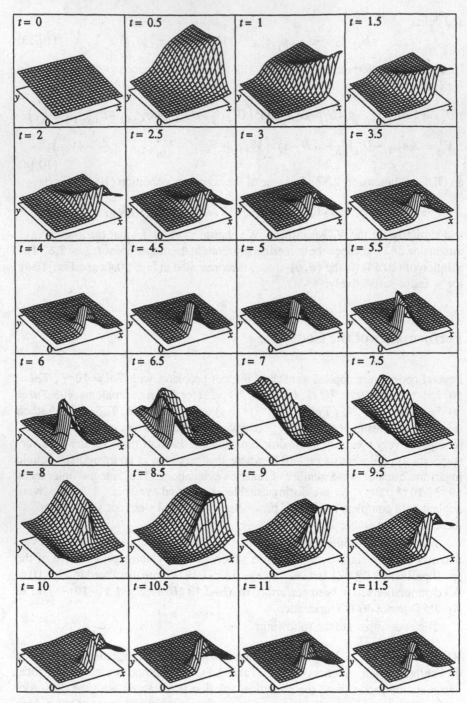

Fig. 10.4a. Solution $u(x, y, t)$ for the BRUS problem

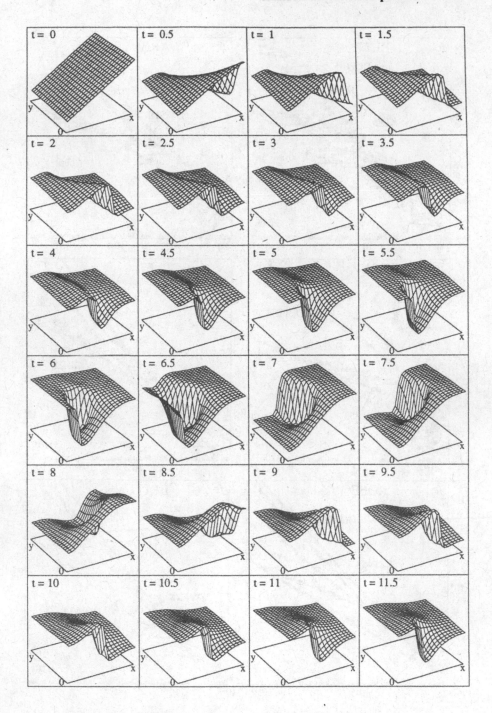

Fig. 10.4b. Solution $v(x, y, t)$ for the BRUS problem

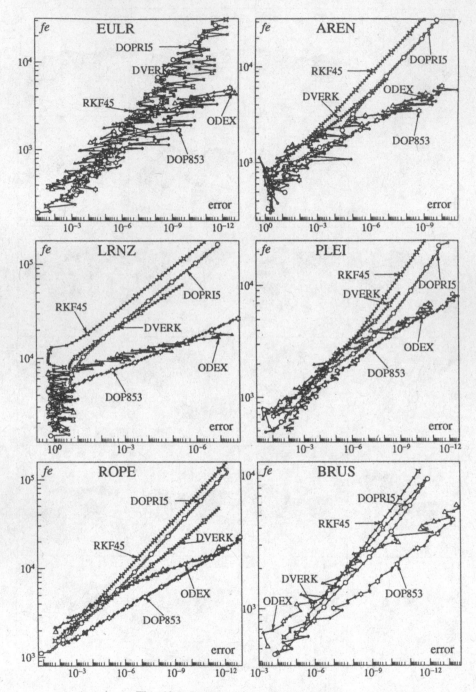

Fig. 10.5. Precision versus function calls

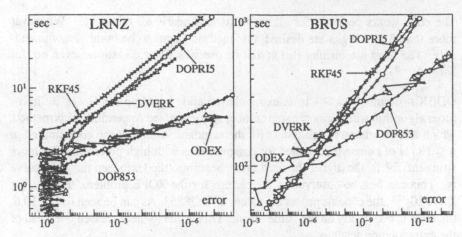

Fig. 10.6. Precision versus computing time

code are relatively better than those on the "function calls" front (Fig. 10.5). This indicates that the code has particularly small overhead.

DOPRI5 — symbol O — the method of Dormand & Prince of order 5 with embedded error estimator of order 4 (see Table 5.2). The code is explained in the Appendix. The method has precisely the same order as that used in RKF45, but the error constants are much more optimized. Therefore the "error curves" in Fig. 10.5 are nicely parallel to those of RKF45, but appear translated to the side of higher precision. One usually gains between a half and one digit of numerical precision for comparable numerical work. The code performs specially well between $Tol = 10^{-3}$ and $Tol = 10^{-8}$ in the AREN problem. This is simply due to an accidental sign change of the error for the most sensitive solution component.

DVERK — symbol ⊠ — this widely known code implements Verner's 6th order method of Table 5.4 and was written by Hull, Enright & Jackson. It has been included in the IMSL library for many years and the source code is available through na-net. The corresponding error curves in Fig. 10.5 appear to be less steep than those of DOPRI5, which illustrates the higher order of the method. However, the error constants seem to be less optimal so that this code surpasses the performance of DOPRI5 only for very stringent tolerances. It is significantly better than DOPRI5 solely in problems EULR and ROPE. The code, as it was, failed at the BRUS problem for $Tol = 10^{-3}$. Therefore these computations were started with $Tol = 10^{-4}$.

DOP853 — symbol ✩— is the method of Dormand & Prince of order 8 explained in Section II.5 (formulas (5.20) – (5.30), see Appendix). The 6th order error estimator (5.29), (5.30) has been replaced by a 5th order estimator with 3rd order correction (see below). This was necessary to make the code robust for the EULR problem.

The code works perfectly for all problems and nearly all tolerances. Whenever more than 3 or 4 digits are desired, this method seems to be highly recommendable. The most astonishing fact is that its use was never disastrous, even not for $Tol = 10^{-3}$.

ODEX — symbol $\triangle$ — is an extrapolation code based on the Gragg-Bulirsch-Stoer algorithm with harmonic step number sequence (see Appendix). This method, which allows arbitrary high orders (in the standard version of the code limited to $p \leq 18$) is of course predestined for computations with high precision. The more stringent Tol is, the higher the used order becomes, the less steep the error curve is. This can best be observed in the picture for the ROPE problem. Finally, for $Tol \approx 10^{-12}$, the code surpasses the values of DOP853. As can be seen in Fig. 10.6, the code loses slightly on the "time"-front. This is due to the increased overhead of the extrapolation scheme.

The numerical results of ODEX behave very similarly to those of DIFEX1 (Deuflhard 1983).

A "Stretched" Error Estimator for DOP853

In preliminary stages of our numerical tests we had written a code "DOPR86" based on the method of order 8 of Dormand & Prince with the 6th order error estimator described in Section II.5. For most problems the results were excellent. However, there are some situations in which the error control of DOPR86 did not work safely:

When applied to the BRUS problem with $Tol = 10^{-3}$ or $Tol = 10^{-4}$ the code stopped with an overflow message. The reason was the following: when the step size is too large, the internal stages are too far away from the solution and their modulus increases at each stage (e.g., by a factor 10^5 between stage 11 and stage 12). Due to the fact that $\widehat{b}_{12} = b_{12}$ (see (5.30) (5.26) and (5.25b)) the difference $\widehat{y}_1 - y_1$ is not influenced by the last stage and is smaller (by a factor of 10^5) than the modulus of y_1. Hence, the error estimator scaled by (4.10) is $\leq 10^{-5}$ and a completely wrong step will be accepted.

The code DOPR86 also had severe difficulties when applied to problems with discontinuities such as EULR. The worst results were obtained for the problem

$$
\begin{aligned}
y_1' &= y_2 y_3 & y_1(0) &= 0 \\
y_2' &= -y_3 y_1 & y_2(0) &= 1 & (10.15) \\
y_3' &= -0.51 \cdot y_1 y_2 + f(x) & y_3(0) &= 1
\end{aligned}
$$

where $f(x)$, given in (10.1'), has a discontinuous second derivative. The results for this problem and the code DOPR86 for very many different Tol values ($Tol = 10^{-3}, 10^{-3-1/24}, 10^{-3-2/24}, \ldots, 10^{-14}$) are displayed in Fig. 10.7. There,

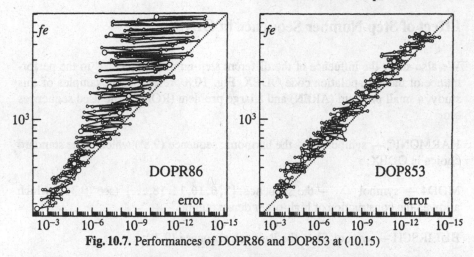

Fig. 10.7. Performances of DOPR86 and DOP853 at (10.15)

the (dotted) diagonal is of exact slope $1/8$ and represents the theoretical convergence speed of the method of order 8. It can be observed that this convergence is well attained by *some* results, but others lose precision of up to 8 digits from the desired tolerance. We explain this disappointing behaviour by the fact that $\hat{b}_{12} = b_{12}$ and that the 12th stage is the only one where the function is evaluated at the endpoint of the step. Whenever the discontinuity of f'' is by accident slightly to the left of a grid point, the error estimator ignores it and the code reports a wrong value.

Unfortunately, the basic 8th order method does not possess a 6th order embedding with $\hat{b}_{12} \neq b_{12}$ (unless additional function evaluations are used). Therefore, we decided to construct a 5th order approximation $\hat{y}_1$. It can be obtained by taking $\hat{b}_6$, $\hat{b}_7$, $\hat{b}_{12}$ as free parameters, e.g.,

$$\hat{b}_6 = b_6/2 + 1, \qquad \hat{b}_7 = b_7/2 + 0.45, \qquad \hat{b}_{12} = b_{12}/2,$$

by putting $\hat{b}_2 = \hat{b}_3 = \hat{b}_4 = \hat{b}_5 = 0$ and by determining the remaining coefficients such that this quadrature formula has order 5. Due to the simplifying assumptions (5.20) all conditions for order 5 are then satisfied. In order to prevent a serious *over*-estimation of the error, we consider a second embedded method $\tilde{y}_1$ of order 3 based on the nodes $c_1 = 0$, c_9 and $c_{12} = 1$ so that two error estimators

$$err_5 = \|\hat{y}_1 - y_1\| = \mathcal{O}(h^6), \qquad err_3 = \|\tilde{y}_1 - y_1\| = \mathcal{O}(h^4) \qquad (10.16)$$

are available. Similarly to a procedure which is common for quadrature formulas (R. Piessens, E. de Doncker-Kapenga, C.W. Überhuber & D.K. Kahaner 1983, Berntsen & Espelid 1991) we consider

$$err = err_5 \cdot \frac{err_5}{\sqrt{err_5^2 + 0.01 \cdot err_3^2}} = \mathcal{O}(h^8) \qquad (10.17)$$

as error estimator. It behaves asymptotically like the global error of the method. The corresponding code DOP853 gives satisfactory results for all the above problems (see right picture in Fig. 10.7).

Effect of Step-Number Sequence in ODEX

We also study the influence of the different step-number sequences to the performance of the extrapolation code ODEX. Fig. 10.8 presents two examples of this study, a small problem (AREN) and a large problem (ROPE). The used sequences are

HARMONIC — symbol O — the harmonic sequence (9.8') which is the standard choice in ODEX;

MOD4 — symbol △ — the sequence $\{2, 6, 10, 14, 18, \ldots\}$ (see (9.35)) which allowed the construction of high-order dense output;

BULIRSCH — symbol □ — the Bulirsch sequence (9.7');

ROMBERG — symbol ◯ — the Romberg sequence (9.6');

DNSECTRL — symbol ✳ — the error control for the MOD4 sequence taking into account the interpolation error of the dense output solution (9.42). This is included only in the small problem, since (complete) dense output on large problems would need too much memory.

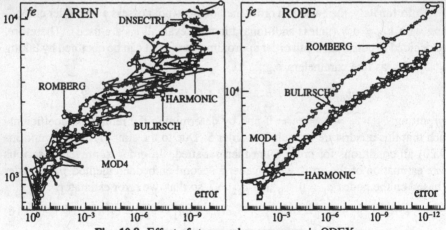

Fig. 10.8. Effect of step-number sequences in ODEX

Discussion. With the exception of the clear inferiority of the Romberg sequence, especially for high precision, and a certain price to be paid for the dense output error control, there is not much difference between the first three sequences. Although the harmonic sequence appears to be slightly superior, the difference is statistically not very significant.

II.11 Parallel Methods

> We suppose that we have a computer with a number of arithmetic
> processors capable of simultaneous operation and seek to devise
> parallel integration algorithms for execution on such a computer.
>
> (W.L. Miranker & W. Liniger 1967)

> "PARALYSING ODES"   (K. Burrage, talk in Helsinki 1990)

Parallel machines are computers with more than one processor and this facility
might help us to speed up the computations in ordinary differential equations.
This is particularly interesting for very large problems, for very costly function
evaluation, or for fast real-time simulations. A second motivation is the desire to
make a code, with the help of parallel computations, not necessarily faster but more
robust and reliable.

Early attempts for finding parallel methods are Nievergelt (1964) and Miranker
& Liniger (1967). See also the survey papers Miranker (1971) and Jackson (1991).

We distinguish today essentially between two types of parallel architectures:

SIMD (single instruction multiple data): all processors execute the same instructions with possibly different input data.

MIMD (multiple instruction multiple data): the different processors can act
independently.

The exploitation of parallelism for an ordinary differential equation

$$y' = f(x, y), \qquad y(x_0) = y_0 \tag{11.1}$$

can be classified into two main categories (Gear 1987, 1988):

Parallelism across the system. Often the problem itself offers more or less trivial
applications for parallelism, e.g.,

> if several solutions are required for various initial or parameter values;

> if the right-hand side of (11.1) is very costly, but structured in such a way that
  the computation of *one* function evaluation can be split efficiently across the
  various processors;

> space discretizations of partial differential equations (such as the Brusselator
  problem (10.14)) whose function evaluation can be done simultaneously for all
  components on an SIMD machine with thousands of processors;

> the solution of boundary value problems with the multiple shooting method
  (see Section I.15) where all computations on the various sub-intervals can be
  done in parallel;

> doing all the high-dimensional linear algebra in the Runge-Kutta method (11.2) in parallel;

> parallelism in the linear algebra for Newton's method for *implicit* Runge-Kutta methods (see Section IV.8).

These types of parallelism, of course, depend strongly on the problem and on the type of the computer.

Parallelism across the method. This is problem-independent and means that, due to a special structure of the method, several function values can be evaluated in parallel within one integration step. This will be discussed in this section in more detail.

Parallel Runge-Kutta Methods

> ... it seems that *explicit* Runge-Kutta methods are not facilitated much by parallelism at the method level.
>
> (Iserles & Nørsett 1990)

Consider an explicit Runge-Kutta method

$$k_i = f\left(x_0 + c_i h, y_0 + h \sum_{j=1}^{i-1} a_{ij} k_j\right), \quad i = 1, \ldots, s$$

$$y_1 = y_0 + h \sum_{i=1}^{s} b_i k_i. \tag{11.2}$$

Suppose, for example, that the coefficients have the zero-pattern indicated in Fig. 11.1.

$$
\begin{array}{c|cccc}
0 & & & & \\
\times & \times & & & \\
\times & \times & 0 & & \\
\times & \times & \times & \times & \\
\hline
 & \times & \times & \times & \times \\
\end{array}
$$

Fig. 11.1. Parallel method

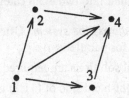

Fig. 11.2. Production graph

Each arrow in the corresponding "production graph" G (Fig. 11.2), pointing from vertex "i" to vertex "j", stands for a non-zero a_{ji}. Here the vertices 2 and 3 are independent and can be evaluated in parallel. We call the number of vertices in the longest chain of successive arrows (here 3) the *number of sequential function evaluations* σ.

In general, if the Runge-Kutta matrix A can be partitioned (possibly after a permutation of the stages) as

$$A = \begin{pmatrix} 0 & & & & \\ A_{21} & 0 & & & \\ A_{31} & A_{32} & 0 & & \\ \vdots & \vdots & & \ddots & \\ A_{\sigma 1} & A_{\sigma 2} & \cdots & A_{\sigma,\sigma-1} & 0 \end{pmatrix}, \tag{11.3}$$

where A_{ij} is a matrix of size $\mu_i \times \mu_j$, then the derivatives $k_1, \ldots, k_{\mu_1}$ as well as $k_{\mu_1+1}, \ldots, k_{\mu_1+\mu_2}$, and so on, can be computed in parallel and one step of the method is executed in σ sequential function evaluations (if $\mu = \max_i \mu_i$ processors are at disposal). The following theorem is a severe restriction on parallel methods. It appeared in hand-written notes by K. Jackson & S. Nørsett around 1986. For a publication see Jackson & Nørsett (1992) and Iserles & Nørsett (1990).

Theorem 11.1. *For an explicit Runge-Kutta method with σ sequential stages the order p satisfies*

$$p \leq \sigma, \tag{11.4}$$

for any number μ of available processors.

Proof. Each non-zero term of the expressions $\Phi_i(t)$ for the "tall" trees t_{21}, t_{32}, t_{44}, $t_{59}, \ldots$ (see Table 2.2 and Definition 2.9) $\sum a_{ij} a_{jk} a_{k\ell} a_{\ell m} \cdots$ corresponds to a connected chain of arrows in the production graph. Since their length is limited by σ, these terms are all zero for $\varrho(t) > \sigma$. □

Methods with $p = \sigma$ will be called *P-optimal methods*. The Runge-Kutta methods of Section II.1 for $p \leq 4$ are all P-optimal. Only for $p > 4$ does the subsequent construction of P-optimal methods allow one to increase the order with the help of parallelism.

Remark. The fact that the "stability function" (see Section IV.2) of an explicit parallel Runge-Kutta method is a polynomial of degree $\leq \sigma$ allows a second proof of Theorem 11.1. Further, P-optimal methods all have the same stability function $1 + z + z^2/2! + \ldots + z^\sigma/\sigma!$.

Parallel Iterated Runge-Kutta Methods

One possibility of constructing P-optimal methods is by fixed point iteration. Consider an arbitrary (explicit or implicit) Runge-Kutta method with coefficients

$$c = (c_1, \ldots, c_s)^T, \qquad A = (a_{ij})_{i,j=1}^s, \qquad b^T = (b_1, \ldots, b_s)$$

and define $\widehat{y}_1$ by

$$k_i^{(0)} = 0$$

$$k_i^{(\ell)} = f\left(x_0 + c_i h, y_0 + h \sum_{j=1}^s a_{ij} k_j^{(\ell-1)}\right), \quad \ell = 1, \ldots, \sigma \tag{11.5}$$

$$\widehat{y}_1 = y_0 + h \sum_{i=1}^s b_i k_i^{(\sigma)}.$$

This algorithm can be interpreted as an explicit Runge-Kutta method with scheme

$$
\begin{array}{c|cccccc}
0 & 0 \\
c & A & 0 \\
c & 0 & A & 0 \\
\vdots & \vdots & & \ddots & \ddots \\
c & 0 & \cdots & 0 & A & 0 \\
\hline
 & 0 & \cdots & 0 & 0 & b^T
\end{array}
\tag{11.6}
$$

It has σ sequential stages if s processors are available. To compute its order we use a Lipschitz condition for $f(x, y)$ and obtain

$$\max_i \| k_i^{(\ell)} - k_i \| \le C h \cdot \max_i \| k_i^{(\ell-1)} - k_i \|$$

where k_i are the stage-vectors of the basic method. Since $k_i^{(0)} - k_i = \mathcal{O}(1)$ this implies $k_i^{(\sigma)} - k_i = \mathcal{O}(h^\sigma)$ and consequently the difference to the solution of the basic method satisfies $\widehat{y}_1 - y_1 = \mathcal{O}(h^{\sigma+1})$.

Theorem 11.2. *The parallel iterated Runge-Kutta method (11.5) is of order*

$$p = \min(p_0, \sigma), \tag{11.7}$$

if p_0 denotes the order of the basic method.

Proof. The statement follows from

$$\widehat{y}_1 - y(x_0 + h) = \widehat{y}_1 - y_1 + y_1 - y(x_0 + h) = \mathcal{O}(h^{\sigma+1}) + \mathcal{O}(h^{p_0+1}). \qquad \square$$

This theorem shows that the choice $\sigma = p_0$ in (11.5) yields P-optimal explicit Runge-Kutta methods (i.e., $\sigma = p$). If we take as basic method the s-stage collocation method based on the Gaussian quadrature ($p_0 = 2s$) then we obtain a method of order $p = 2s$ which is P-optimal on s processors. P.J. van der Houwen

& B.P. Sommeijer (1990) have done extensive numerical experiments with this method.

Extrapolation Methods

It turns out that the GBS-algorithm (Section II.9) without smoothing step is also P-optimal. Indeed, all the values T_{j1} can be computed independently of each other. If we choose the step number sequence $\{2, 4, 6, 8, 10, 12, \ldots\}$ then the computation of T_{k1} requires $2k$ sequential function evaluations. Hence, if k processors are available (one for each T_{j1}), the numerical approximation T_{kk}, which is of order $p = 2k$, can be computed with $\sigma = 2k$ sequential stages. When the processors are of type MIMD we can compute T_{11} and $T_{k-1,1}$ on one processor $(2 + 2(k-1) = 2k$ function evaluations). Similarly, T_{21} and $T_{k-2,1}$ occupy another processor, etc. In this way, the number of necessary processors is reduced by a factor close to 2 without increasing the number of sequential stages.

The order and step size strategy, discussed in Section II.9, should, of course, be adapted for an implementation on parallel computers. The "hope for convergence in line $k + 1$" no longer makes sense because this part of the algorithm is now as costly as the whole step. Similarly, there is no reason to accept already $T_{k-1,k-1}$ as numerical approximation, because T_{kk} is computed on the same time level as $T_{k-1,k-1}$. Moreover, the numbers A_k of (9.25) should be replaced by $A_k = n_k$ which will in general increase the order used by the code.

Increasing Reliability

> ... using parallelism to improve *reliability* and *functionality* rather than efficiency.   (W.H. Enright & D.J. Higham 1991)

For a given Runge-Kutta method parallel computation can be used to give a reliable error estimate or an accurate dense output. This has been advocated by Enright & Higham (1991) and will be the subject of this subsection.

Consider a Runge-Kutta method of order p, choose distinct numbers $0 = \sigma_0 < \sigma_1 < \ldots < \sigma_k = 1$ and apply the Runge-Kutta method in parallel with step sizes $\sigma_1 h, \ldots, \sigma_{k-1} h, \sigma_k h = h$. This gives approximations

$$y_{\sigma_i} \approx y(x_0 + \sigma_i h) . \tag{11.8}$$

Then compute $f(x_0 + \sigma_i h, y_{\sigma_i})$ and do Hermite interpolation with the values

$$y_{\sigma_i}, \quad hf(x_0 + \sigma_i h, y_{\sigma_i}), \qquad i = 0, 1, \ldots, k, \tag{11.9}$$

i.e., compute

$$u(\theta) = \sum_{i=0}^{k} v_i(\theta) y_{\sigma_i} + h \sum_{i=0}^{k} w_i(\theta) f(x_0 + \sigma_i h, y_{\sigma_i}) \tag{11.10}$$

where $v_i(\theta)$ and $w_i(\theta)$ are the scalar polynomials

$$\left.\begin{array}{l} v_i(\theta) = \ell_i^2(\theta) \cdot \left(1 - 2\ell_i'(\sigma_i)(\theta - \sigma_i)\right) \\ w_i(\theta) = \ell_i^2(\theta) \cdot (\theta - \sigma_i) \end{array}\right\} \quad \text{with} \quad \ell_i(\theta) = \prod_{\substack{j=0 \\ j \neq i}}^{k} \frac{(\theta - \sigma_j)}{(\sigma_i - \sigma_j)}. \quad (11.11)$$

The interpolation error, which is $\mathcal{O}(h^{2k+2})$, may be neglected if $2k+2 > p+1$.

As to the choice of σ_i we denote the local error of the method by $le = y_1 - y(x_0 + h)$. It follows from Taylor expansion (see Theorem 3.2) that

$$y_{\sigma_i} - y(x_0 + \sigma_i h) = \sigma_i^{p+1} \cdot le + \mathcal{O}(h^{p+2})$$

and consequently the error of (11.10) satisfies (for $2k+2 > p+1$)

$$u(\theta) - y(x_0 + \theta h) = \left(\sum_{i=1}^{k} \sigma_i^{p+1} v_i(\theta)\right) \cdot le + \mathcal{O}(h^{p+2}). \quad (11.12)$$

The coefficient of le is equal to 1 for $\theta = 1$ and it is natural to search for suitable σ_i such that

$$\left|\sum_{i=1}^{k} \sigma_i^{p+1} v_i(\theta)\right| \leq 1 \quad \text{for all} \quad \theta \in [0,1]. \quad (11.13)$$

Indeed, under the assumption $2k-1 \leq p < 2k+1$, it can be shown that numbers $0 = \sigma_0 < \sigma_1 < \ldots < \sigma_{k-1} < \sigma_k = 1$ exist satisfying (11.13) (see Exercise 1). Selected values of σ_i proposed by Enright & Higham (1991), which satisfy this condition are given in Table 11.1. For such a choice of σ_i the error (11.12) of the dense output is bounded (at least asymptotically) by the local error le at the endpoint of integration. This implementation of a dense output provides a simple way to estimate le. Since $u(\theta)$ is an $\mathcal{O}(h^{p+1})$-approximation of $y(x_0 + \theta h)$, the defect of $u(\theta)$ satisfies

$$u'(\theta) - hf(x_0 + \theta h, u(\theta)) = \left(\sum_{i=1}^{k} \sigma_i^{p+1} v_i'(\theta)\right) \cdot le + \mathcal{O}(h^{p+2}). \quad (11.14)$$

If we take a σ^* different from σ_i such that $\sum_{i=1}^{k} \sigma_i^{p+1} v_i'(\sigma^*) \neq 0$ (see Table 11.1) then only one function evaluation, namely $f(x_0 + \sigma^* h, u(\sigma^*))$, allows the computation of an asymptotically correct approximation of le from (11.14). This error estimate can be used for step size selection and for improving the numerical result (local extrapolation). In the local extrapolation mode one then loses the C^1 continuity of the dense output.

With the use of an additional processor the quantities y_{σ^*} and $f(x_0 + \sigma^* h, y_{\sigma^*})$ can be computed simultaneously with y_{σ_i} and $f(x_0 + \sigma_i h, y_{\sigma_i})$. If the polynomial $u(\theta)$ is required to satisfy $u(\sigma^*) = y_{\sigma^*}$, but not $u'(\sigma^*) = hf(x_0 + \sigma^* h, y_{\sigma^*})$, then the estimate (11.14) of the local error le does not need any further evaluation of f.

Table 11.1. Good values for σ_i

p	k	$\sigma_1, \ldots, \sigma_{k-1}$	σ^*
5	3	0.2, 0.4	0.88
6	3	0.2, 0.4	0.88
7	4	0.2, 0.4, 0.7	0.94
8	4	0.2, 0.4, 0.6	0.93

Exercises

1. Let the positive integers k and p satisfy $2k - 1 \leq p < 2k + 1$. Then show that there exist numbers $0 = \sigma_0 < \sigma_1 < \ldots < \sigma_{k-1} < \sigma_k = 1$ such that (11.13) is true for all $\theta \in [0, 1]$.

   Hint. Put $\sigma_j = j\epsilon$ for $j = 1, \ldots, k - 1$ and show that (11.13) is verified for sufficiently small $\epsilon > 0$. Of course, in a computer program, one should use σ_j which satisfy (11.13) and are well separated in order to avoid roundoff errors.

II.12 Composition of B-Series

At the Dundee Conference in 1969, a paper by J. Butcher was read
which contained a surprising result.   (H.J. Stetter 1971)

We shall now derive a theorem on the composition of what we call B-series (in
honour of J. Butcher). This will have many applications and will lead to a better
understanding of order conditions for all general classes of methods (composition
of methods, multiderivative methods of Section II.13, general linear methods of
Section III.8, Rosenbrock methods in Exercise 2 of Section IV.7).

Composition of Runge-Kutta Methods

There is no five-stage explicit Runge-Kutta method of order 5 (Section II.5). This
led Butcher (1969) to the idea of searching for different five-stage methods such
that a certain *composition* of these methods produces a fifth-order result ("effective
order"). Although not of much practical interest (mainly due to the problem of
changing step size), this was the starting point of a fascinating algebraic theory of
numerical methods.

Suppose we have two methods, say of three stages,

$$
\begin{array}{c|ccc}
0 & & & \\
\widehat{c}_2 & \widehat{a}_{21} & & \\
\widehat{c}_3 & \widehat{a}_{31} & \widehat{a}_{32} & \\
\hline
 & \widehat{b}_1 & \widehat{b}_2 & \widehat{b}_3
\end{array}
\qquad
\begin{array}{c|ccc}
0 & & & \\
\widetilde{c}_2 & \widetilde{a}_{21} & & \\
\widetilde{c}_3 & \widetilde{a}_{31} & \widetilde{a}_{32} & \\
\hline
 & \widetilde{b}_1 & \widetilde{b}_2 & \widetilde{b}_3
\end{array}
\qquad (12.1)
$$

which are applied one after the other to a starting value y_0 with the same step size:

$$
g_i = y_0 + h \sum_j \widehat{a}_{ij} f(g_j), \qquad y_1 = y_0 + h \sum_j \widehat{b}_j f(g_j) \qquad (12.2)
$$

$$
\ell_i = y_1 + h \sum_j \widetilde{a}_{ij} f(\ell_j), \qquad y_2 = y_1 + h \sum_j \widetilde{b}_j f(\ell_j). \qquad (12.3)
$$

If we insert y_1 from (12.2) into (12.3) and group all g_i, ℓ_i together, we see that the composition can be interpreted as a large Runge-Kutta method with coefficients

$$
\begin{array}{c|cccccc}
0 \\
\widehat{c}_2 & \widehat{a}_{21} \\
\widehat{c}_3 & \widehat{a}_{31} & \widehat{a}_{32} \\
\sum \widehat{b}_i & \widehat{b}_1 & \widehat{b}_2 & \widehat{b}_3 \\
\sum \widehat{b}_i + \widetilde{c}_2 & \widehat{b}_1 & \widehat{b}_2 & \widehat{b}_3 & \widetilde{a}_{21} \\
\sum \widehat{b}_i + \widetilde{c}_3 & \widehat{b}_1 & \widehat{b}_2 & \widehat{b}_3 & \widetilde{a}_{31} & \widetilde{a}_{32} \\
\hline
& \widehat{b}_1 & \widehat{b}_2 & \widehat{b}_3 & \widetilde{b}_1 & \widetilde{b}_2 & \widetilde{b}_3
\end{array}
\qquad \equiv \qquad
\begin{array}{c|cccccc}
0 \\
c_2 & a_{21} \\
c_3 & a_{31} & a_{32} \\
c_4 & a_{41} & a_{42} & a_{43} \\
c_5 & a_{51} & a_{52} & a_{53} & a_{54} \\
c_6 & a_{61} & a_{62} & a_{63} & a_{64} & a_{65} \\
\hline
& b_1 & b_2 & b_3 & b_4 & b_5 & b_6
\end{array}
\qquad (12.4)
$$

It is now of interest to study the *order conditions* of the new method. For this, we have to compute the expressions (see Table 2.2)

$$
\sum b_i, \quad 2 \sum b_i c_i, \quad 3 \sum b_i c_i^2, \quad 6 \sum b_i a_{ij} c_j, \quad \text{etc.}
$$

If we insert the values from the left tableau of (12.4), a computation, which for low orders is still not too difficult, shows that these expressions can be written in terms of the corresponding expressions for the two methods (12.1). We shall denote these expressions for the *first* method by $\mathbf{a}(t)$, for the *second* method by $\mathbf{b}(t)$, and for the *composite* method by $\mathbf{ab}(t)$:

$$
\mathbf{a}(\,.\,) = \sum \widehat{b}_i, \quad \mathbf{a}(\,\diagup\,) = 2 \cdot \sum \widehat{b}_i \widehat{c}_i, \quad \mathbf{a}(\vee) = 3 \cdot \sum \widehat{b}_i \widehat{c}_i^2, \quad \ldots \quad (12.5a)
$$

$$
\mathbf{b}(\,.\,) = \sum \widetilde{b}_i, \quad \mathbf{b}(\,\diagup\,) = 2 \cdot \sum \widetilde{b}_i \widetilde{c}_i, \quad \mathbf{b}(\vee) = 3 \cdot \sum \widetilde{b}_i \widetilde{c}_i^2, \quad \ldots \quad (12.5b)
$$

$$
\mathbf{ab}(\,.\,) = \sum b_i, \quad \mathbf{ab}(\,\diagup\,) = 2 \cdot \sum b_i c_i, \quad \mathbf{ab}(\vee) = 3 \cdot \sum b_i c_i^2, \quad \ldots \quad (12.5c)
$$

The above mentioned formulas are then

$$
\begin{aligned}
\mathbf{ab}(\,.\,) &= \mathbf{a}(\,.\,) + \mathbf{b}(\,.\,) \\
\mathbf{ab}(\,\diagup\,) &= \mathbf{a}(\,\diagup\,) + 2\mathbf{b}(\,.\,)\mathbf{a}(\,.\,) + \mathbf{b}(\,\diagup\,) \\
\mathbf{ab}(\vee) &= \mathbf{a}(\vee) + 3\mathbf{b}(\,.\,)\mathbf{a}(\,.\,)^2 + 3\mathbf{b}(\,\diagup\,)\mathbf{a}(\,.\,) + \mathbf{b}(\vee) \\
\mathbf{ab}(\mathord{\succ}) &= \mathbf{a}(\mathord{\succ}) + 3\mathbf{b}(\,.\,)\mathbf{a}(\,\diagup\,) + 3\mathbf{b}(\,\diagup\,)\mathbf{a}(\,.\,) + \mathbf{b}(\mathord{\succ})
\end{aligned}
\qquad (12.6)
$$

etc.

It is now, of course, of interest to have a general understanding of these formulas for arbitrary trees. This, however, is not easy in the above framework ("... a tedious calculation shows that ..."). Further, there are problems of identifying different methods with identical numerical results (see Exercise 1 below). Also, we want the theory to include more general processes than Runge-Kutta methods, for example the exact solution or multi-derivative methods.

B-Series

All these difficulties can be avoided if we consider directly the composition of the series appearing in Section II.2. We define by

$$T = \{\emptyset\} \cup T_1 \cup T_2 \cup \ldots, \qquad LT = \{\emptyset\} \cup LT_1 \cup LT_2 \cup \ldots$$

the sets of all trees and labelled trees, respectively.

Definition 12.1 (Hairer & Wanner 1974). Let $\mathbf{a}(\emptyset)$, $\mathbf{a}(\,.\,)$, $\mathbf{a}(\nearrow)$, $\mathbf{a}(\vee), \ldots$ be a sequence of real coefficients defined for all trees $\mathbf{a} : T \to \mathbf{R}$. Then we call the series (see Theorem 2.11, Definitions 2.2, 2.3)

$$B(\mathbf{a}, y) = \mathbf{a}(\emptyset)y + h\mathbf{a}(\,.\,)f(y) + \frac{h^2}{2!}\,\mathbf{a}(\nearrow)F(\nearrow)(y) + \ldots$$

$$= \sum_{t \in LT} \frac{h^{\varrho(t)}}{\varrho(t)!}\,\mathbf{a}(t)F(t)(y) = \sum_{t \in T} \frac{h^{\varrho(t)}}{\varrho(t)!}\,\alpha(t)\,\mathbf{a}(t)F(t)(y) \qquad (12.7)$$

a *B-series*.

We have seen in Theorems 2.11 and 2.6 that the numerical solution of a Runge-Kutta method as well as the exact solution are B-series. The coefficients of the latter are all equal to 1.

Usually we are only interested in a finite number of terms of these series (only as high as the orders of the methods under consideration, or as far as f is differentiable) and all subsequent results are valid modulo error terms $\mathcal{O}(h^{k+1})$.

Definition 12.2. Let $t \in LT$ be a labelled tree of order $q = \varrho(t)$ and $0 \le i \le q$ be a fixed integer. Then we denote by $s_i(t) = s$ the *subtree* formed by the first i indices and by $d_i(t)$ (the *difference set*) the set of subtrees formed by the remaining indices. In the graphical representation we distinguish the subtree s by fat nodes and doubled lines.

Example 12.3. For the labelled tree $t = {}^{p}\!\!\underset{j}{\overset{l}{\bigvee}}{}^{m}_{k}$ we have:

$$i = 0: \qquad s_0(t) = \emptyset, \qquad d_0(t) = \{\vee\}$$
$$i = 1: \qquad s_1(t) = ., \qquad d_1(t) = \{.,.,\nearrow\}$$
$$i = 2: \qquad s_2(t) = \nearrow, \qquad d_2(t) = \{.,.,.\}$$
$$i = 3: \qquad s_3(t) = \vee, \qquad d_3(t) = \{.,.\}$$
$$i = 4: \qquad s_4(t) = \vee, \qquad d_4(t) = \{.\}$$
$$i = 5: \qquad s_5(t) = t = \vee, \qquad d_5(t) = \emptyset$$

Definition 12.4. Let $\mathbf{a}: T \to \mathbb{R}$ and $\mathbf{b}: T \to \mathbb{R}$ be two sequences of coefficients such that $\mathbf{a}(\emptyset) = 1$. Then for a tree t of order $q = \varrho(t)$ we define the *composition*

$$\mathbf{ab}(t) = \frac{1}{\alpha(t)} \sum \left(\sum_{i=0}^{q} \binom{q}{i} \mathbf{b}(s_i(t)) \prod_{z \in d_i(t)} \mathbf{a}(z) \right) \tag{12.8}$$

where the first summation is over all $\alpha(t)$ different labellings of t (see Definition 2.5).

Example 12.5. It is easily seen that the formulas of (12.6) are special cases of (12.8). The tree t of Example 12.3 possesses 6 different labellings

These lead to

$$\begin{aligned}
\mathbf{ab}(\psi) = {} & \mathbf{b}(\emptyset)\mathbf{a}(\psi) + 5\mathbf{b}(\,.\,)\mathbf{a}(\,.\,)^2\mathbf{a}(\,\nearrow) \\
& + 10\left(\frac{1}{2} \mathbf{b}(\nearrow)\mathbf{a}(\,.\,)\mathbf{a}(\nearrow) + \frac{1}{2} \mathbf{b}(\nearrow)\mathbf{a}(\,.\,)^3 \right) \\
& + 10\left(\frac{1}{6} \mathbf{b}(\vee)\mathbf{a}(\nearrow) + \frac{4}{6} \mathbf{b}(\vee)\mathbf{a}(\,.\,)^2 + \frac{1}{6} \mathbf{b}(\curlyvee)\mathbf{a}(\,.\,)^2 \right) \\
& + 5\left(\frac{1}{2} \mathbf{b}(\psi)\mathbf{a}(\,.\,) + \frac{1}{2} \mathbf{b}(\dot\vee)\mathbf{a}(\,.\,) \right) + \mathbf{b}(\psi).
\end{aligned} \tag{12.9}$$

Here is the main theorem of this section:

Theorem 12.6 (Hairer & Wanner 1974). *As above, let $\mathbf{a}: T \to \mathbb{R}$ and $\mathbf{b}: T \to \mathbb{R}$ be two sequences of coefficients such that $\mathbf{a}(\emptyset) = 1$. Then the composition of the two corresponding B-series is again a B-series*

$$B(\mathbf{b}, B(\mathbf{a}, y)) = B(\mathbf{ab}, y) \tag{12.10}$$

where the "product" $\mathbf{ab}: T \to \mathbb{R}$ is that of Definition 12.4.

Proof. We denote the inner series by

$$B(\mathbf{a}, y) = g(h). \tag{12.11}$$

Then the proof is similar to the development of Section II.2 (see Fig. 2.2), with the difference that, instead of $f(g)$, we now start from

$$B(\mathbf{b}, g) = \sum_{s \in LT} \frac{h^{\varrho(s)}}{\varrho(s)!} \mathbf{b}(s) F(s)(g) \tag{12.12}$$

and have to compute the derivatives of this function: let us select the term $s = \curlyvee$ of this series,

$$\frac{h^3}{3!} \, b(\succ) \sum_{L,M} f_L^K(g) f_M^L(g) f^M(g). \tag{12.13}$$

The q th derivative of this expression, for $h = 0$, is by Leibniz' formula

$$\binom{q}{3} b(\succ) \sum_{L,M} \left(f_L^K(g) f_M^L(g) f^M(g) \right)^{(q-3)} \Big|_{h=0}. \tag{12.14}$$

We now compute, as we did in Lemma 2.8, the derivatives of

$$f_L^K(g) f_M^L(g) f^M(g) \tag{12.15}$$

using the classical rules of differential calculus; this gives for the first derivative

$$\sum_N f_{LN}^K \cdot (g^N)' f_M^L f^M + \sum_N f_L^K f_{MN}^L \cdot (g^N)' f^M + \sum_N f_L^K f_M^L f_N^M \cdot (g^N)'$$

and so on. We again represent this in graphical form in Fig. 12.1.

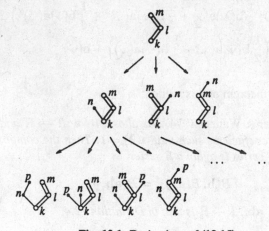

Fig. 12.1. Derivatives of (12.15)

We see that we arrive at trees u of order q such that $s_3(u) = s$ (where $3 = \varrho(s)$) and the elements of $d_3(u)$ have no ramifications. The corresponding expressions are similar to (2.6;q-1) in Lemma 2.8. We finally have to insert the derivatives of g (see (12.11)) and rearrange the terms. Then, as in Fig. 2.4, the tall branches of $d_3(u)$ are replaced by trees z of order δ, multiplied by $a(z)$. Thus the coefficient which we obtain for a given tree t is just given by (12.8).

The factor $1/\alpha(t)$ is due to the fact that in $B(ab, y)$ the term with $ab(t)F(t)$ appears $\alpha(t)$ times. $\square$

Since $hf(y) = B(\mathbf{b}, y)$ is a special B-series with $\mathbf{b}(\,.\,) = 1$ and all other $\mathbf{b}(t) = 0$, we have the following

Corollary 12.7. *If* $\mathbf{a} : T \to \mathbb{R}$ *with* $\mathbf{a}(\emptyset) = 1$, *then*

$$hf(B(\mathbf{a}, y)) = B(\mathbf{a}', y)$$

with

$$\mathbf{a}'(\emptyset) = 0, \quad \mathbf{a}'(\,.\,) = 1$$

$$\mathbf{a}'([t_1, \ldots, t_m]) = \varrho(t)\, \mathbf{a}(t_1) \cdot \ldots \cdot \mathbf{a}(t_m) \tag{12.16}$$

where $t = [t_1, \ldots, t_m]$ *means that* $d_1(t) = \{t_1, t_2, \ldots, t_m\}$ *(Definition 2.12).*

Proof. We obtain (12.16) from (12.8) with $i = 1$, $q = \varrho(t)$ and the fact that the expression in brackets is independent of the labelling of t. □

Order Conditions for Runge-Kutta Methods

As an application of Corollary 12.7, we demonstrate the derivation of order conditions for Runge-Kutta methods: we write method (2.3) as

$$g_i = y_0 + \sum_{j=1}^{s} a_{ij} k_j, \qquad k_i = hf(g_i), \qquad y_1 = y_0 + \sum_{j=1}^{s} b_j k_j. \tag{12.17}$$

If we assume g_i, k_i and y_1 to be B-series, whose coefficients we denote by $\mathbf{g}_i, \mathbf{k}_i, \mathbf{y}_1$

$$g_i = B(\mathbf{g}_i, y_0), \qquad k_i = B(\mathbf{k}_i, y_0), \qquad y_1 = B(\mathbf{y}_1, y_0),$$

then Corollary 12.7 immediately allows us to transcribe formulas (12.17) as

$$\mathbf{g}_i(\emptyset) = 1, \qquad\qquad \mathbf{k}_i(\,.\,) = 1, \qquad\qquad\qquad \mathbf{y}_1(\emptyset) = 1,$$

$$\mathbf{g}_i(t) = \sum_{j=1}^{s} a_{ij} \mathbf{k}_j(t), \quad \mathbf{k}_i(t) = \varrho(t)\, \mathbf{g}_i(t_1) \cdot \ldots \cdot \mathbf{g}_i(t_m), \quad \mathbf{y}_1(t) = \sum_{j=1}^{s} b_j \mathbf{k}_j(t)$$

which leads easily to formulas (2.17), (2.19) and Theorem 2.11.

Also, if we put $y(h) = B(\mathbf{y}, y_0)$ for the *true solution,* and compare the derivative $hy'(h)$ of the series (12.7) with $hf(y(h))$ from Corollary 12.7, we immediately obtain $\mathbf{y}(t) = 1$ for all t, so that Theorem 2.6 drops out. The order conditions are then obtained as in Theorem 2.13 by comparing the coefficients of the B-series $B(\mathbf{y}, y_0)$ and $B(\mathbf{y}_1, y_0)$.

Butcher's "Effective Order"

We search for a 5-stage Runge-Kutta method $\mathbf{a}$ and for a method $\mathbf{d}$, such that $\mathbf{dad}^{-1}$ represents a fifth order method $\mathbf{u}$. This means that we have to satisfy

$$\mathbf{da}(t) = \mathbf{yd}(t) \qquad \text{for} \qquad \varrho(t) \leq 5, \tag{12.18}$$

where $\mathbf{y}(t) = 1$ represents the B-series of the exact solution. Then

$$(\mathbf{dad}^{-1})^k = \mathbf{da}^k \mathbf{d}^{-1} = (\mathbf{da})\mathbf{a}^{k-2}(\mathbf{ad}^{-1}). \tag{12.19}$$

If now two Runge-Kutta methods $\mathbf{b}$ and $\mathbf{c}$ are constructed such that $\mathbf{b} = \mathbf{da}$ and $\mathbf{c} = \mathbf{ad}^{-1}$ up to order 5, then applying one step of $\mathbf{b}$ followed by $k - 2$ steps of $\mathbf{a}$ and a final step of $\mathbf{c}$ is equivalent (up to order 5) to k steps of the 5th order method $\mathbf{dad}^{-1}$ (see Fig. 12.2). A possible set of coefficients, computed by Butcher (1969), is given in Table 12.1 (method $\mathbf{a}$ has classical order 4).

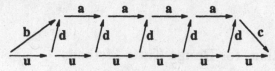

Fig. 12.2. Effective increase of order

Stetter's approach. Soon after the appearance of Butcher's purely algebraic proof, Stetter (1971) gave an elegant analytic explanation. Consider the principal global error term $e_p(x)$ which satisfies the variational equation (8.8). The question is, under which conditions on the local error $d_{p+1}(x)$ (see (8.8)) this equation can be solved, for special initial values, without effort. We write equation (8.8) as

$$e'(x) - \frac{\partial f}{\partial y}(y(x)) \cdot e(x) = d(x) \tag{12.20}$$

and want $e(x)$ to possess an expansion of the form

$$e(x) = \sum_{t \in T_p} \alpha(t) \, e(t) \, F(t)(y(x)) \tag{12.21}$$

with constant coefficients $e(t)$. Simply inserting (12.21) into (12.20) yields

$$d(x) = \sum_{t \in T_p} \alpha(t) \, e(t) \Big(\frac{d}{dx}\big(F(t)(y(x))\big) - f'(y(x)) \cdot F(t)(y(x)) \Big). \tag{12.22}$$

Thus, *(12.21) is the exact solution of the variational equation, if the local error $d(x)$ has the symmetric form (12.22).* Then, if we replace the initial value y_0 by the "starting procedure"

$$\widehat{y}_0 := y_0 - h^p e(x_0) = y_0 - h^p \sum_{t \in T_p} \alpha(t) \, e(t) \, F(t)(y_0) \tag{12.23}$$

Table 12.1. Butcher's method of effective order 5

Method a

0					
$\dfrac{1}{5}$	$\dfrac{1}{5}$				
$\dfrac{2}{5}$	0	$\dfrac{2}{5}$			
$\dfrac{1}{2}$	$\dfrac{3}{16}$	0	$\dfrac{5}{16}$		
1	$\dfrac{1}{4}$	0	$-\dfrac{5}{4}$	2	
	$\dfrac{1}{6}$	0	0	$\dfrac{2}{3}$	$\dfrac{1}{6}$

Method b

0					
$\dfrac{1}{5}$	$\dfrac{1}{5}$				
$\dfrac{2}{5}$	0	$\dfrac{2}{5}$			
$\dfrac{3}{4}$	$\dfrac{75}{64}$	$-\dfrac{9}{4}$	$\dfrac{117}{64}$		
1	$-\dfrac{37}{36}$	$\dfrac{7}{3}$	$-\dfrac{3}{4}$	$\dfrac{4}{9}$	
	$\dfrac{19}{144}$	0	$\dfrac{25}{48}$	$\dfrac{2}{9}$	$\dfrac{1}{8}$

Method c

0					
$\dfrac{1}{5}$	$\dfrac{1}{5}$				
$\dfrac{2}{5}$	0	$\dfrac{2}{5}$			
$\dfrac{3}{4}$	$\dfrac{161}{192}$	$-\dfrac{19}{12}$	$\dfrac{287}{192}$		
1	$-\dfrac{27}{28}$	$\dfrac{19}{7}$	$-\dfrac{291}{196}$	$\dfrac{36}{49}$	
	$\dfrac{7}{48}$	0	$\dfrac{475}{1008}$	$\dfrac{2}{7}$	$\dfrac{7}{72}$

(or by a Runge-Kutta method equivalent to this up to order $p+1$; this would represent "method **d**" in Fig. 12.2), its error satisfies $y(x_0) - \widehat{y}_0 = h^p e(x_0) + \mathcal{O}(h^{p+1})$. By Theorem 8.1 the numerical solution $\widehat{y}_n$ of the Runge-Kutta method applied to $\widehat{y}_0$ satisfies $y(x_n) - \widehat{y}_n = h^p e(x_n) + \mathcal{O}(h^{p+1})$. Therefore the "finishing procedure"

$$y_n := \widehat{y}_n + h^p e(x_n) = \widehat{y}_n + h^p \sum_{t \in T_p} \alpha(t)\, e(t)\, F(t)(\widehat{y}_n) + \mathcal{O}(h^{p+1}) \qquad (12.24)$$

(or some equivalent Runge-Kutta method) gives a $(p+1)$th order approximation to the solution.

Example. Butcher's method **a** of Table 12.1 has the local error

$$d_6(x) = \frac{1}{6!}\left(-\frac{1}{24} F(\curlyvee) - \frac{1}{4} F(\curlyvee) - \frac{1}{8} F(\curlyvee) + \frac{1}{6} F(\curlyvee) + \frac{1}{2} F(\curlyvee) \right). \qquad (12.25)$$

The right-hand side of (12.22) would be (the derivation $\frac{d}{dx} F$ attaches a new twig to each of the nodes, the product $f'(y) \cdot F$ lifts the tree on a stilt)

$$
\begin{aligned}
&\mathbf{e}(\curlyvee)\Big(F(\curlyvee) + 3F(\curlywedge) - F(\Upsilon)\Big) \\
&+3\mathbf{e}(\vartriangle)\Big(F(\curlywedge) + F(\curlyvee) + F(\curlyvee) + F(\curlywedge) - F(\Upsilon)\Big) \\
&+\mathbf{e}(\Upsilon)\Big(F(\curlyvee) + F(\Upsilon) + 2F(\Upsilon) - F(\Upsilon)\Big) \\
&+\mathbf{e}(\vartriangle)\Big(F(\curlywedge) + F(\Upsilon) + F(\Upsilon) + F(\vartriangle) - F(\vartriangle)\Big).
\end{aligned}
\tag{12.26}
$$

Comparison of (12.25) and (12.26) shows that this method does indeed have the desired symmetry if

$$
\mathbf{e}(\curlyvee) = \mathbf{e}(\vartriangle) = -\frac{1}{6!}\cdot\frac{1}{24}, \qquad \mathbf{e}(\Upsilon) = \mathbf{e}(\vartriangle) = \frac{1}{6!}\cdot\frac{1}{8}.
$$

This allows one to construct a Runge-Kutta method as starting procedure corresponding to (12.23) up to the desired order.

Exercises

1. Show that the pairs of methods given in Tables 12.2 - 12.4 produce, at least for h sufficiently small, identical numerical results.

   Result. a) is seen by permutation of the stages, b) by neglecting superfluous stages (Dahlquist & Jeltsch 1979), c) by identifying equal stages (Stetter 1973, Hundsdorfer & Spijker 1981). See also the survey on "The Runge-Kutta space" by Butcher (1984).

2. Extend formulas (12.6) by computing the composition $\mathbf{ab}(t)$ for all trees of order 4 and 5.

3. Verify that the methods given in Table 12.1 satisfy the stated order properties.

4. Prove, using Theorem 12.6, that the set

$$
G = \{\mathbf{a} : T \to \mathbf{R} \mid \mathbf{a}(\emptyset) = 1\}
$$

   together with the composition law of Definition 12.4 is a (non-commutative) group.

5. (Equivalence of Butcher's and Stetter's approach). Let $\mathbf{a} : T \to \mathbf{R}$ represent a Runge-Kutta method of classical order p and effective order $p+1$, i.e., $\mathbf{a}(t) = 1$ for $\varrho(t) \leq p$ and

$$
\mathbf{da}(t) = \mathbf{yd}(t) \qquad \text{for} \qquad \varrho(t) \leq p+1
\tag{12.27}
$$

   for some $\mathbf{d} : T \to \mathbf{R}$ and with $\mathbf{y}(t)$ as in (12.18). Prove that then the local error $h^{p+1}\mathbf{d}(x) + \mathcal{O}(h^{p+2})$ of the method $\mathbf{a}$ has the symmetric form (12.22). This

Table 12.2. Equivalent methods a)

$$
\begin{array}{c|cc}
0 & & \\
1 & 1 & 0 \\
\hline
 & 1/4 & 3/4
\end{array}
\qquad
\begin{array}{c|cc}
1 & 0 & 1 \\
0 & 0 & 0 \\
\hline
 & 3/4 & 1/4
\end{array}
$$

Table 12.3. Equivalent methods b)

$$
\begin{array}{c|cccc}
1 & 2 & 0 & 0 & -1 \\
3 & 0 & 1 & 2 & 0 \\
7 & 0 & 3 & 4 & 0 \\
2 & 1 & 0 & 0 & 1 \\
\hline
 & 1/2 & 0 & 0 & 1/2
\end{array}
\qquad
\begin{array}{c|cc}
1 & 2 & -1 \\
2 & 1 & 1 \\
\hline
 & 1/2 & 1/2
\end{array}
$$

Table 12.4. Equivalent methods c)

$$
\begin{array}{c|cccc}
1 & 1 & 1 & 1 & -2 \\
1 & 2 & 2 & -1 & -2 \\
1 & -1 & -1 & 5 & -2 \\
-1 & -1 & 2 & 1 & -3 \\
\hline
 & 1/4 & 1/4 & 1/4 & 1/4
\end{array}
\qquad
\begin{array}{c|cc}
1 & 3 & -2 \\
-1 & 2 & -3 \\
\hline
 & 3/4 & 1/4
\end{array}
$$

means that, in this situation, Butcher's effective order is equivalent to Stetter's approach.

Hint. Start by expanding condition (12.27) (using (12.8)) for the first trees. Possible simplifications are then best seen if the second sum $\sum_{i=0}^{q}$ (for **yd**) is arranged *downwards* ($i = q, q-1, \ldots, 0$). One then arrives recursively at the result

$$\mathbf{d}(t) = \mathbf{d}(\,.\,)^{\varrho(t)} \qquad \text{for } \varrho(t) \le p-1.$$

Then express the error coefficients $\mathbf{a}(t) - 1$ for $\varrho(t) = p+1$ in terms of $\mathbf{d}(s) - \mathbf{d}(\,.\,)^{\varrho(s)}$ where $\varrho(s) = p$. Formula (12.22) then becomes visible.

6. Prove that for $t = [t_1, \ldots, t_m]$ the coefficient $\alpha(t)$ of Definition 2.5 satisfies the recurrence relation

$$\alpha(t) = \binom{\varrho(t)-1}{\varrho(t_1), \ldots, \varrho(t_m)} \alpha(t_1) \cdot \ldots \cdot \alpha(t_m) \cdot \frac{1}{\mu_1! \mu_2! \ldots}. \qquad (12.28)$$

The integers $\mu_1, \mu_2, \ldots$ count the equal trees among $t_1, \ldots, t_m$.

Hint. The multinomial coefficient in (12.28) counts the possible partitionnings of the labels $2, \ldots, \varrho(t)$ to the m subtrees $t_1, \ldots, t_m$. Equal subtrees lead to equal labellings. Hence the division by $\mu_1! \mu_2! \ldots$.

II.13 Higher Derivative Methods

In Section I.8 we studied the computation of higher derivatives of solutions of

$$(y^J)' = f^J(x, y^1, \ldots, y^n), \qquad J = 1, \ldots, n. \tag{13.1}$$

The chain rule

$$(y^J)'' = \frac{\partial f^J}{\partial x}(x, y) + \frac{\partial f^J}{\partial y^1}(x, y) \cdot f^1(x, y) + \ldots + \frac{\partial f^J}{\partial y^n}(x, y) \cdot f^n(x, y) \tag{13.2}$$

leads to the differential operator D which, when applied to a function $\Psi(x, y)$, is given by

$$(D\Psi)(x, y) = \frac{\partial \Psi}{\partial x}(x, y) + \frac{\partial \Psi}{\partial y^1}(x, y) \cdot f^1(x, y) + \ldots + \frac{\partial \Psi}{\partial y^n}(x, y) \cdot f^n(x, y). \tag{13.2'}$$

Since $Dy^J = f^J$, we see by extending (13.2) that

$$(y^J)^{(\ell)} = (D^\ell y^J)(x, y), \qquad \ell = 0, 1, 2, \ldots. \tag{13.3}$$

This notation allows us to define a new class of methods which combine features of Runge-Kutta methods as well as Taylor series methods:

Definition 13.1. Let $a_{ij}^{(r)}$, $b_j^{(r)}$, $(i, j = 1, \ldots, s,\ r = 1, \ldots, q)$ be real coefficients. Then the method

$$k_i^{(\ell)} = \frac{h^\ell}{\ell!}(D^\ell y)\left(x_0 + c_i h, y_0 + \sum_{r=1}^{q} \sum_{j=1}^{s} a_{ij}^{(r)} k_j^{(r)}\right)$$

$$y_1 = y_0 + \sum_{r=1}^{q} \sum_{j=1}^{s} b_j^{(r)} k_j^{(r)} \tag{13.4}$$

is called an s-stage q-derivative Runge-Kutta method. If $a_{ij}^{(r)} = 0$ for $i \leq j$, the method is explicit, otherwise implicit.

A natural extension of (1.9) is here, because of $Dx = 1$, $D^\ell x = 0$ $(\ell \geq 2)$,

$$c_i = \sum_{j=1}^{s} a_{ij}^{(1)}. \tag{13.5}$$

Definition 13.1 is from Kastlunger & Wanner (1972), but special methods of this type have been considered earlier in the literature. In particular, the very successful methods of Fehlberg (1958, 1964) have this structure.

Collocation Methods

A natural way of obtaining s-stage q-derivative methods is to use the collocation idea with *multiple nodes*, i.e., to replace (7.15b) by

$$u^{(\ell)}(x_0 + c_i h) = (D^\ell y)(x_0 + c_i h, u(x_0 + c_i h)) \quad i = 1, \ldots, s, \quad \ell = 1, \ldots, q_i$$
(13.6)

where $u(x)$ is a polynomial of degree $q_1 + q_2 + \ldots + q_s$ and $q_1, \ldots, q_s$, the "multiplicities" of the nodes $c_1, \ldots, c_s$, are given integers. For example $q_1 = m$, $q_2 = \ldots = q_s = 1$ leads to Fehlberg-type methods.

In order to generalize the results and ideas of Section II.7, we have to replace the Lagrange interpolation of Theorem 7.7 by *Hermite* interpolation (Hermite 1878: "Je me suis proposé de trouver un polynôme ..."). The reason is that (13.6) can be interpreted as an ordinary collocation condition with clusters of q_i nodes "infinitely" close together (Rolle's theorem). We write Hermite's formula as

$$p(t) = \sum_{j=1}^{s} \sum_{r=1}^{q_j} \frac{1}{r!} \ell_{jr}(t) p^{(r-1)}(c_j)$$
(13.7)

for polynomials $p(t)$ of degree $\sum q_j - 1$. Here the "basis" polynomials $\ell_{jr}(t)$ of degree $\sum q_j - 1$ must satisfy

$$l_{jr}^{(k)}(c_i) = \begin{cases} r! & \text{if } i = j \text{ and } k = r - 1 \\ 0 & \text{else} \end{cases}$$
(13.8)

and are best obtained from Newton's interpolation formula (with multiple nodes). We now use this formula, as we did in Section II.7, for $p(t) = h u'(x_0 + th)$:

$$h u'(x_0 + th) = \sum_{j=1}^{s} \sum_{r=1}^{q_j} \ell_{jr}(t) k_j^{(r)},$$
(13.9)

with

$$k_j^{(r)} = \frac{h^r}{r!} u^{(r)}(x_0 + c_j h).$$
(13.10)

If we insert

$$u(x_0 + c_i h) = y_0 + \int_0^{c_i} h u'(x_0 + th)\, dt$$

together with (13.9) into (13.6), we get:

Theorem 13.2. *The collocation method (13.6) is equivalent to an s-stage q-derivative implicit Runge-Kutta method (13.4) with*

$$a_{ij}^{(r)} = \int_0^{c_i} \ell_{jr}(t)\,dt, \qquad b_j^{(r)} = \int_0^1 \ell_{jr}(t)\,dt. \tag{13.11}$$

□

Theorems 7.8, 7.9, and 7.10 now generalize immediately to the case of "confluent" quadrature formulas; i.e., the q-derivative Runge-Kutta method possesses the *same order* as the underlying quadrature formula

$$\int_0^1 p(t)\,dt \approx \sum_{j=1}^{s} \sum_{r=1}^{q_j} b_j^{(r)} p^{(r-1)}(c_j).$$

The "algebraic" proof of this result (extending Exercise 7 of Section II.7) is more complicated and is given, for the case $q_j = q$, in Kastlunger & Wanner (1972b).

The formulas corresponding to condition $C(\eta)$ are given by

$$\sum_{j=1}^{s} \sum_{r=1}^{q_j} a_{ij}^{(r)} \binom{\varrho}{r} c_j^{\varrho-r} = c_i^\varrho, \qquad \varrho = 1, 2, \dots, \sum_{j=1}^{s} q_j. \tag{13.12}$$

These equations uniquely determine the $a_{ij}^{(r)}$, once the c_i have been chosen, by a linear system with a "confluent" Vandermonde matrix (see e.g., Gautschi 1962). Formula (13.12) is obtained by setting $p(t) = t^{\varrho-1}$ in (13.7) and then integrating from 0 to c_i.

Examples of methods. "Gaussian" quadrature formulas with multiple nodes exist for *odd q* (Stroud & Stancu 1965) and extend to q-derivative implicit Runge-Kutta methods (Kastlunger & Wanner 1972b): for $s = 1$ we have, of course, $c_1 = 1/2$ which yields

$$b_1^{(2k)} = 0, \qquad b_1^{(2k+1)} = 2^{-2k}, \qquad a_{11}^{(k)} = (-1)^{k+1} 2^{-k}.$$

We give also the coefficients for the case $s = 2$ and $q_1 = q_2 = 3$. The nodes c_i and the weights $b_i^{(k)}$ are those of Stroud & Stancu. The method has order 8:

$$
\begin{aligned}
c_1 &= 0.185394435825045 & c_2 &= 1 - c_1 \\
b_1^{(1)} &= 0.5 & b_2^{(1)} &= b_1^{(1)} \\
b_1^{(2)}/2! &= 0.0240729420844974 & b_2^{(2)} &= -b_1^{(2)} \\
b_1^{(3)}/3! &= 0.00366264960671727 & b_2^{(3)} &= b_1^{(3)}
\end{aligned}
$$

$$a_{ij}^{(1)} = \begin{pmatrix} 0.201854115831005 & -0.0164596800059598 \\ 0.516459680005959 & 0.298145884168994 \end{pmatrix}$$

$$a_{ij}^{(2)} = \begin{pmatrix} -0.0223466569080541 & 0.00868878773082417 \\ 0.0568346718998190 & -0.0704925410770490 \end{pmatrix}$$

$$a_{ij}^{(3)} = \begin{pmatrix} 0.0116739668400997 & -0.00215351251065784 \\ 0.0241294101509615 & 0.0103019308002039 \end{pmatrix}$$

Hermite-Obreschkoff Methods

We now consider the special case of collocation methods with $s = 2$, $c_1 = 0$, $c_2 = 1$. These methods can be obtained in closed form by repeated partial integration as follows (Darboux 1876, Hermite 1878):

Lemma 13.3. *Let m be a given positive integer and $P(t)$ a polynomial of exact degree m. Then*

$$\sum_{j=0}^{m} h^j (D^j y)(x_1, y_1) P^{(m-j)}(0) = \sum_{j=0}^{m} h^j (D^j y)(x_0, y_0) P^{(m-j)}(1) \qquad (13.13)$$

defines a multiderivative method (13.4) of order m.

Proof. We let $y(x)$ be the exact solution and start from

$$h^{m+1} \int_0^1 y^{(m+1)}(x_0 + ht) P(1 - t)\, dt = \mathcal{O}(h^{m+1}).$$

This integral is now transformed by repeated partial integration until all derivatives of the polynomial $P(1 - t)$ are used up. This leads to

$$\sum_{j=0}^{m} h^j y^{(j)}(x_1) P^{(m-j)}(0) = \sum_{j=0}^{m} h^j y^{(j)}(x_0) P^{(m-j)}(1) + \mathcal{O}(h^{m+1}).$$

If this is subtracted from (13.13) we find the difference of the left-hand sides to be $\mathcal{O}(h^{m+1})$, which shows by the implicit function theorem that (13.13) determines y_1 to this order if $P^{(m)}$, which is a constant, is $\neq 0$. $\square$

The argument $1 - t$ in P (instead of the more natural t) avoids the sign changes in the partial integrations.

A good choice for $P(t)$ is, of course, a polynomial for which most derivatives disappear at $t = 0$ and $t = 1$. Then the method (13.13) is, by keeping the same order m, most economical. We write

$$P(t) = \frac{t^k (t - 1)^\ell}{(k + \ell)!}$$

and obtain

$$y_1 - \frac{\ell}{(k+\ell)}\frac{h}{1!}(Dy)(x_1,y_1) + \frac{\ell(\ell-1)}{(k+\ell)(k+\ell-1)}\frac{h^2}{2!}(D^2y)(x_1,y_1) \pm \dots$$
$$= y_0 + \frac{k}{(k+\ell)}\frac{h}{1!}(Dy)(x_0,y_0) + \frac{k(k-1)}{(k+\ell)(k+\ell-1)}\frac{h^2}{2!}(D^2y)(x_0,y_0) + \dots$$

(13.14)

which is a method of order $m = k + \ell$. After the ℓth term in the first line and the kth term in the second line, the coefficients automatically become zero. Special cases of this method are:

$$k = 1, \quad \ell = 0: \quad \text{explicit Euler}$$
$$k \geq 1, \quad \ell = 0: \quad \text{Taylor series}$$
$$k = 0, \quad \ell = 1: \quad \text{implicit Euler}$$
$$k = 1, \quad \ell = 1: \quad \text{trapezoidal rule.}$$

Darboux and Hermite advocated the use of this formula for the approximations of functions, Obreschkoff (1940) for the computation of integrals, Loscalzo & Schoenberg (1967), Loscalzo (1969) as well as Nørsett (1974a) for the solution of differential equations.

Fehlberg Methods

Another class of multiderivative methods is due to Fehlberg (1958, 1964): the idea is to subtract from the solution of $y' = f(x,y)$, $y(x_0) = y_0$ m terms of the Taylor series (see Section I.8)

$$\widehat{y}(x) := y(x) - \sum_{i=0}^{m} Y_i(x-x_0)^i,$$

(13.15)

and to solve the resulting differential equation $\widehat{y}'(x) = \widehat{f}(x,\widehat{y}(x))$, where

$$\widehat{f}(x,\widehat{y}(x)) = f\left(x, \widehat{y} + \sum_{i=0}^{m} Y_i(x-x_0)^i\right) - \sum_{i=1}^{m} Y_i i(x-x_0)^{i-1},$$

(13.16)

by a Runge-Kutta method. Thus, knowing that the solution of (13.16) and its first m derivatives are zero at the initial value, we can achieve much higher orders.

In order to understand this, we develop the Taylor series of the solution for the non-autonomous case, as we did at the beginning of Section II.1. We thereby omit the hats and suppose the transformation (13.15) already carried out. We then have from (1.6) (see also Exercise 3 of Section II.2)

$$f = 0,$$
$$f_x + f_y f = 0,$$
$$f_{xx} + 2f_{xy}f + f_{yy}f^2 + f_y(f_x + f_y f) = 0, \text{ etc.}$$

These formulas recursively imply that $f = 0$, $f_x = 0$,..., $\partial^{m-1} f / \partial x^{m-1} = 0$. All elementary differentials of order $\leq m$ and most of those of higher orders then become *zero* and the corresponding order conditions can be omitted. The first non-zero terms are

$$\frac{\partial^m f}{\partial x^m} \qquad \text{for order} \quad m+1,$$

$$\frac{\partial^{m+1} f}{\partial x^{m+1}} \quad \text{and} \quad \frac{\partial f}{\partial y} \cdot \frac{\partial^m f}{\partial x^m} \quad \text{for order} \quad m+2,$$

and so on. The corresponding order conditions are then

$$\sum_{i=1}^{s} b_i c_i^m = \frac{1}{m+1}$$

for order $m+1$,

$$\sum_{i=1}^{s} b_i c_i^{m+1} = \frac{1}{m+2} \quad \text{and} \quad \sum_{i,j} b_i a_{ij} c_j^m = \frac{1}{(m+1)(m+2)}$$

for order $m+2$, and so on.

The condition $\sum a_{ij} = c_i$, which usually allows several terms of (1.6) to be grouped together, is not necessary, because all these other terms are zero.

A complete insight is obtained by considering the method as being *partitioned* applied to the *partitioned system* $y' = f(x, y)$, $x' = 1$. This will be explained in Section II.15 (see Fig. 15.4).

Example 13.4. A solution with $s = 3$ stages of the (seven) conditions for order $m + 3$ is given by Fehlberg (1964). The choice $c_1 = c_3 = 1$ minimizes the numerical work for the evaluation of (13.16) and the other coefficients are then uniquely determined (see Table 13.1).

Fehlberg (1964) also derived an embedded method with two additional stages of orders $m + 3$ $(m + 4)$. These methods were widely used in the sixties for scientific computations.

Table 13.1. Fehlberg, order $m + 3$

1				
θ	$\dfrac{\theta^m}{m+3}$			$\theta = \dfrac{m+1}{m+3}$
1	$-\dfrac{1}{m+1}$	$\dfrac{2}{(m+1)\theta^m}$		
	0	$\dfrac{m+3}{2(m+1)(m+2)\theta^m}$	$\dfrac{1}{2(m+2)}$	

General Theory of Order Conditions

For the same reason as in Section II.2 we assume that (13.1) is autonomous. The general form of the order conditions for method (13.4) was derived in the thesis of Kastlunger (see Kastlunger & Wanner 1972). It later became a simple application of the composition theorem for B-series (Hairer & Wanner 1974). The point is that from Theorem 2.6,

$$\frac{h^i}{i!}(D^i y)(y_0) = \sum_{t \in LT, \varrho(t)=i} \frac{h^i}{i!} F(t)(y_0) = B(\mathbf{y}^{(i)}, y_0) \tag{13.17}$$

is a B-series with coefficients

$$\mathbf{y}^{(i)}(t) = \begin{cases} 1 & \text{if } \varrho(t) = i \\ 0 & \text{otherwise.} \end{cases} \tag{13.18}$$

Thus, in extension of Corollary 12.7, we have

$$\frac{h^i}{i!}(D^i y)(B(\mathbf{a}, y_0)) = B(\mathbf{a}^{(i)}, y_0) \tag{13.19}$$

where, from formula (12.8) with $q = \varrho(t)$,

$$\mathbf{a}^{(i)}(t) = (\mathbf{a}\mathbf{y}^{(i)})(t) = \frac{1}{\alpha(t)}\binom{q}{i}\sum \prod_{z \in d_i(t)} \mathbf{a}(z), \tag{13.20}$$

and the sum is over all $\alpha(t)$ different labellings of t. This allows us to compute recursively the coefficients of the B-series which appear in (13.4).

Example 13.5. The tree $t = \diamondsuit$ sketched in Fig. 13.1 possesses three different labellings, two of which produce the same difference set $d_2(t)$, so that formula (13.20) becomes

$$\mathbf{a}''(\diamondsuit) = 2(2(\mathbf{a}(\,.\,))^2 + \mathbf{a}(\,\cdot\,)). \tag{13.21}$$

Fig. 13.1. Different labellings of $\diamondsuit$

For all other trees of order ≤ 4 we have $\alpha(t) = 1$ and (13.20) leads to the following table of second derivatives

$$
\begin{aligned}
&\mathbf{a}''(\,.\,) = 0 &\qquad &\mathbf{a}''(\,\cdot\,) = 1 \\
&\mathbf{a}''(\vee) = 3\mathbf{a}(\,.\,) &\qquad &\mathbf{a}''(\succ) = 3\mathbf{a}(\,.\,) \\
&\mathbf{a}''(\vee\hspace{-0.3em}\cdot) = 6(\mathbf{a}(\,.\,))^2 &\qquad &\mathbf{a}''(\diamondsuit) = 4(\mathbf{a}(\,.\,))^2 + 2\mathbf{a}(\,\cdot\,) \\
&\mathbf{a}''(Y) = 6(\mathbf{a}(\,.\,))^2 &\qquad &\mathbf{a}''(\lessgtr) = 6\mathbf{a}(\,\cdot\,).
\end{aligned} \tag{13.22}
$$

Once these expressions have been established, we write formulas (13.4) in the form

$$k_i^{(\ell)} = \frac{h^\ell}{\ell!}(D^\ell y)(g_i)$$

$$g_i = y_0 + \sum_{r=1}^{q}\sum_{j=1}^{s} a_{ij}^{(r)} k_j^{(r)}, \qquad y_1 = y_0 + \sum_{r=1}^{q}\sum_{j=1}^{s} b_j^{(r)} k_j^{(r)} \qquad (13.23)$$

and suppose the expressions $k_i^{(\ell)}$, g_i, y_1 to be B-series

$$k_i^{(\ell)} = B(\mathbf{k}_i^{(\ell)}, y_0), \qquad g_i = B(\mathbf{g}_i, y_0), \qquad y_1 = B(\mathbf{y}_1, y_0).$$

Then equations (13.23) can be translated into

$$\mathbf{k}_i^{(1)}(t) = \varrho(t)\mathbf{g}_i(t_1)\cdot \ldots \cdot \mathbf{g}_i(t_m), \qquad \mathbf{k}_i^{(1)}(\tau) = 1 \qquad \text{(see (12.16))}$$

$$\mathbf{k}_i^{(2)}(t) = \mathbf{g}_i''(t) \qquad \text{from (13.22)}$$

$$\mathbf{k}_i^{(3)}(t) = \mathbf{g}_i'''(t) \qquad \text{from Exercise 1 or Exercise 2, etc.}$$

$$\mathbf{g}_i(t) = \sum_{r=1}^{q}\sum_{j=1}^{s} a_{ij}^{(r)} \mathbf{k}_j^{(r)}(t), \qquad \mathbf{y}_1(t) = \sum_{r=1}^{q}\sum_{j=1}^{s} b_j^{(r)} \mathbf{k}_j^{(r)}(t).$$

These formulas recursively determine all the coefficients. Method (13.4) (together with (13.5)) is then of order p if, as usual,

$$\mathbf{y}_1(t) = 1 \quad \text{for all } t \text{ with } \varrho(t) \le p. \qquad (13.24)$$

More details and special methods are given in Kastlunger & Wanner (1972); see also Exercise 3.

Exercises

1. Extend Example 13.5 and obtain formulas for $\mathbf{a}^{(3)}(t)$ for all trees of order ≤ 4.

2. (Kastlunger). Prove the following variant form of formula (13.20) which extends (12.16) more directly and can also be used to obtain the formulas of Example 13.5. If $t = [t_1, \ldots, t_m]$ then

$$\mathbf{a}^{(i)}(t) = \frac{\varrho(t)}{i} \sum_{\substack{\lambda_1 + \ldots + \lambda_m = i-1 \\ \lambda_1, \ldots, \lambda_m \ge 0}} \mathbf{a}^{(\lambda_1)}(t_1) \ldots \mathbf{a}^{(\lambda_m)}(t_m)$$

Hint. See Kastlunger & Wanner (1972); Hairer & Wanner (1973), Section 5.

3. Show that the conditions for order 3 of method (13.4) are given by

$$\sum_i b_i^{(1)} = 1$$

$$2 \sum_i b_i^{(1)} c_i + \sum_i b_i^{(2)} = 1$$

$$3 \sum_i b_i^{(1)} c_i^2 + 3 \sum_i b_i^{(2)} c_i = 1$$

$$6 \sum_{i,j} b_i^{(1)} a_{ij}^{(1)} c_j + 3 \sum_i b_i^{(1)} e_i + 3 \sum_i b_i^{(2)} c_i = 1,$$

where $c_i = \sum_j a_{ij}^{(1)}$, $e_i = \sum_j a_{ij}^{(2)}$.

4. (Zurmühl 1952, Albrecht 1955). Differentiate a given first order system of differential equations $y' = f(x, y)$ to obtain

$$y'' = (D^2 y)(x, y), \qquad y(x_0) = y_0, \qquad y'(x_0) = f_0.$$

Apply to this equation a special method for higher order systems (see the following Section II.14) to obtain higher-derivative methods. Show that the following method is of order six

$$k_1 = h^2 g(x_0, y_0)$$

$$k_2 = h^2 g\left(x_0 + \frac{h}{4}, y_0 + \frac{h}{4} f_0 + \frac{1}{32} k_1\right)$$

$$k_3 = h^2 g\left(x_0 + \frac{h}{2}, y_0 + \frac{h}{2} f_0 + \frac{1}{24}\left(-k_1 + 4k_2\right)\right)$$

$$k_4 = h^2 g\left(x_0 + \frac{3h}{4}, y_0 + \frac{3h}{4} f_0 + \frac{1}{32}\left(3k_1 + 4k_2 + 2k_3\right)\right)$$

$$y_1 = y_0 + h f_0 + \frac{1}{90}\left(7k_1 + 24k_2 + 6k_3 + 8k_4\right)$$

where $g(x, y) = (D^2 y)(x, y) = Df(x, y) = f_x(x, y) + f_y(x, y) \cdot f(x, y)$.

II.14 Numerical Methods
for Second Order Differential Equations

Many differential equations which appear in practice are systems of the *second
order*

$$y'' = f(x, y, y').$$ (14.1)

This is mainly due to the fact that the forces are proportional to acceleration, i.e., to
second derivatives. As mentioned in Section I.1, such a system can be transformed
into a first order differential equation of doubled dimension by considering the
vector (y, y') as the new variable:

$$\begin{pmatrix} y \\ y' \end{pmatrix}' = \begin{pmatrix} y' \\ f(x, y, y') \end{pmatrix} \qquad \begin{aligned} y(x_0) &= y_0 \\ y'(x_0) &= y_0'. \end{aligned}$$ (14.2)

In order to solve (14.1) numerically, one can for instance apply a Runge-Kutta
method (explicit or implicit) to (14.2). This yields

$$k_i = y_0' + h \sum_{j=1}^{s} a_{ij} k_j'$$

$$k_i' = f\left(x_0 + c_i h, \ y_0 + h \sum_{j=1}^{s} a_{ij} k_j, \ y_0' + h \sum_{j=1}^{s} a_{ij} k_j'\right)$$ (14.3)

$$y_1 = y_0 + h \sum_{i=1}^{s} b_i k_i, \qquad y_1' = y_0' + h \sum_{i=1}^{s} b_i k_i'.$$

If we insert the first formula of (14.3) into the others we obtain (assuming (1.9) and
an order ≥ 1)

$$k_i' = f\left(x_0 + c_i h, \ y_0 + c_i h y_0' + h^2 \sum_{j=1}^{s} \bar{a}_{ij} k_j', \ y_0' + h \sum_{j=1}^{s} a_{ij} k_j'\right)$$ (14.4)

$$y_1 = y_0 + h y_0' + h^2 \sum_{i=1}^{s} \bar{b}_i k_i', \qquad y_1' = y_0' + h \sum_{i=1}^{s} b_i k_i'$$

where

$$\bar{a}_{ij} = \sum_{k=1}^{s} a_{ik} a_{kj}, \qquad \bar{b}_i = \sum_{j=1}^{s} b_j a_{ji}.$$ (14.5)

For an implementation the representation (14.4) is preferable to (14.3), since about half of the storage can be saved. This may be important, in particular if the dimension of equation (14.1) is large.

Nyström Methods

> R.H. Merson: "... I have not seen the paper by Nyström. Was it in English?"
> J.M. Bennett: "In German actually, not Finnish."
> (From the discussion following a talk of Merson 1957)

E.J. Nyström (1925) was the first to consider methods of the form (14.4) in which the coefficients do not necessarily satisfy (14.5) ("Da bis jetzt die *direkte* Anwendung der Rungeschen Methode auf den wichtigen Fall von Differentialgleichungen zweiter Ordnung nicht behandelt war ..." Nyström, 1925). Such direct methods are called *Nyström methods*.

Definition 14.1. A Nyström method (14.4) has *order p* if for sufficiently smooth problems (14.1)

$$y(x_0 + h) - y_1 = \mathcal{O}(h^{p+1}), \qquad y'(x_0 + h) - y_1' = \mathcal{O}(h^{p+1}). \tag{14.6}$$

An example of an explicit Nyström method where condition (14.5) is violated is given in Table 14.1. Nyström claimed that this method would be simpler to apply than "Runge-Kutta's" and reduce the work by about 25%. This is, of course, not true if the Runge-Kutta method is applied as in (14.4) (see also Exercise 2).

Table 14.1. Nyström, order 4

c_i	$\bar{a}_{ij}$				a_{ij}			
0								
$\frac{1}{2}$	$\frac{1}{8}$				$\frac{1}{2}$			
$\frac{1}{2}$	$\frac{1}{8}$	0			0	$\frac{1}{2}$		
1	0	0	$\frac{1}{2}$		0	0	1	
$\bar{b}_i \rightarrow$	$\frac{1}{6}$	$\frac{1}{6}$	$\frac{1}{6}$	0	$\frac{1}{6}$	$\frac{2}{6}$	$\frac{2}{6}$	$\frac{1}{6} \leftarrow b_i$

A *real* improvement can be achieved in the case where the right-hand side of (14.1) does not depend on y', i.e.,

$$y'' = f(x, y). \tag{14.7}$$

Here the Nyström method becomes

$$k_i' = f(x_0 + c_i h,\ y_0 + c_i h y_0' + h^2 \sum_{j=1}^{s} \bar{a}_{ij} k_j')$$

$$y_1 = y_0 + h y_0' + h^2 \sum_{i=1}^{s} \bar{b}_i k_i', \qquad y_1' = y_0' + h \sum_{i=1}^{s} b_i k_i', \tag{14.8}$$

and the coefficients a_{ij} are no longer needed. Some examples are given in Table 14.2. The fifth-order method of Table 14.2 needs only four evaluations of f. This is a considerable improvement compared to Runge-Kutta methods where at least six evaluations are necessary (cf. Theorem 5.1).

Table 14.2. Methods for $y'' = f(x, y)$

Nyström, order 4

c_i		$\bar{a}_{ij}$	
0			
$\frac{1}{2}$	$\frac{1}{8}$		
1	0	$\frac{1}{2}$	
$\bar{b}_i$	$\frac{1}{6}$	$\frac{1}{3}$	0
b_i	$\frac{1}{6}$	$\frac{4}{6}$	$\frac{1}{6}$

Nyström, order 5

		$\bar{a}_{ij}$		
0				
$\frac{1}{5}$	$\frac{1}{50}$			
$\frac{2}{3}$	$\frac{-1}{27}$	$\frac{7}{27}$		
1	$\frac{3}{10}$	$\frac{-2}{35}$	$\frac{9}{35}$	
$\bar{b}_i$	$\frac{14}{336}$	$\frac{100}{336}$	$\frac{54}{336}$	0
b_i	$\frac{14}{336}$	$\frac{125}{336}$	$\frac{162}{336}$	$\frac{35}{336}$

Global convergence. Introducing the variable $z_n = (y_n, y_n')^T$, a Nyström method (14.4) can be written in the form

$$z_1 = z_0 + h\Phi(x_0, z_0, h) \tag{14.9}$$

where

$$\Phi(x_0, z_0, h) = \begin{pmatrix} y_0' + h \sum_i \bar{b}_i k_i' \\ \sum_i b_i k_i' \end{pmatrix}.$$

(14.9) is just a special one-step method for the differential equation (14.2). For a pth order Nyström method the local error $(y(x_0 + h) - y_1,\ y'(x_0 + h) - y_1')^T$ can be bounded by Ch^{p+1} (Definition 14.1), which is in agreement with formula (3.27). The convergence theorems of Section II.3 and the results on asymptotic expansions of the global error (Section II.8) are also valid here.

Our next aim is to derive the order conditions for Nyström methods. For this purpose we extend the theory of Section II.2 to second order differential equations (Hairer & Wanner 1976).

The Derivatives of the Exact Solution

As for first order equations we may restrict ourselves to systems of autonomous differential equations

$$(y^J)'' = f^J(y^1, \ldots, y^n, y'^1, \ldots, y'^n) \tag{14.10}$$

(if necessary, add $x'' = 0$). The superscript index J denotes the Jth component of the corresponding vector. We now calculate the derivatives of the exact solution of (14.10). The second derivative is given by (14.10):

$$(y^J)^{(2)} = f^J(y, y'). \tag{14.11;2}$$

A repeated differentiation of this equation, using (14.10), leads to

$$(y^J)^{(3)} = \sum_K \frac{\partial f^J}{\partial y^K}(y, y') \cdot y'^K + \sum_K \frac{\partial f^J}{\partial y'^K}(y, y') f^K(y, y') \tag{14.11;3}$$

$$(y^J)^{(4)} = \sum_{K,L} \frac{\partial^2 f^J}{\partial y^K \partial y^L}(y, y') \cdot y'^K \cdot y'^L \tag{14.11;4}$$

$$+ \sum_{K,L} \frac{\partial^2 f^J}{\partial y^K \partial y'^L}(y, y') \cdot y'^K \cdot f^L(y, y') + \sum_K \frac{\partial f^J}{\partial y^K}(y, y') f^K(y, y')$$

$$+ \sum_{K,L} \frac{\partial^2 f^J}{\partial y'^K \partial y^L}(y, y') f^K(y, y') \cdot y'^L$$

$$+ \sum_{K,L} \frac{\partial^2 f^J}{\partial y'^K \partial y'^L}(y, y') f^K(y, y') f^L(y, y')$$

$$+ \sum_{K,L} \frac{\partial f^J}{\partial y'^K}(y, y') \frac{\partial f^K}{\partial y^L}(y, y') y'^L$$

$$+ \sum_{K,L} \frac{\partial f^J}{\partial y'^K}(y, y') \frac{\partial f^K}{\partial y'^L}(y, y') f^L(y, y')$$

The continuation of this process becomes even more complex than for first order differential equations. A graphical representation of the above formulas will therefore be very helpful. In order to distinguish the derivatives with respect to y and y' we need two kinds of vertices: "meagre" and "fat". Fig. 14.1 shows the graphs that correspond to the above formulas.

Definition 14.2. A *labelled N-tree of order q* is a labelled tree (see Definition 2.2)

$$t : A_q \setminus \{j\} \to A_q$$

together with a mapping

$$t' : A_q \to \{\text{"meagre", "fat"}\}$$

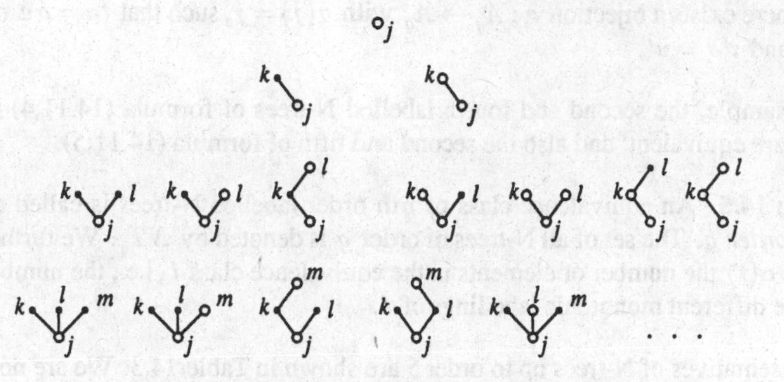

$$\overset{\circ}{}_{j} \tag{14.11;2}$$

$$\overset{k}{\diagdown}_{j} \qquad \overset{k}{\diagdown}_{j} \tag{14.11;3}$$

$$\underset{j}{\overset{k \quad l}{\bigvee}} \quad \underset{j}{\overset{k \quad l}{\bigvee}} \quad \underset{j}{\overset{k \diagup l}{\diagdown}} \qquad \underset{j}{\overset{k \quad l}{\bigvee}} \quad \underset{j}{\overset{k \quad l}{\bigvee}} \quad \underset{j}{\overset{k \diagup l}{\diagdown}} \quad \underset{j}{\overset{k \diagup l}{\diagdown}} \tag{14.11;4}$$

$$\underset{j}{\overset{k \quad l \quad m}{\bigvee}} \quad \underset{j}{\overset{k \quad l \quad m}{\bigvee}} \quad \underset{j}{\overset{k \quad l}{\diagdown}} \quad \underset{j}{\overset{k \quad l}{\diagup}} \quad \underset{j}{\overset{k \quad l \quad m}{\bigvee}} \qquad \cdots \tag{14.11;5}$$

Fig. 14.1. The derivatives of the exact solution

which satisfies:

a) the root of t is always fat; i.e., $t'(j) = $ "fat";

b) a meagre vertex has at most one son and this son has to be fat.

We denote by LNT_q the set of all labelled N-trees of order q.

The reason for condition (b) in Definition 14.2 is that all derivatives of $g(y, y') = y'$ vanish identically with the exception of the first derivative with respect to y'.

In the sequel we use the notation *end-vertex* for a vertex which has no son. If no confusion is possible, we write t instead of (t, t') for a labelled N-tree.

Definition 14.3. For a labelled N-tree t we denote by

$$F^J(t)(y, y')$$

the expression which is a *sum* over the indices of all fat vertices of t (without "j", the index of the root) and over the indices of all meagre end-vertices. The *general term* of this sum is a product of expressions

$$\frac{\partial^r f^K}{\partial y^L \ldots \partial y'^M \ldots}(y, y') \qquad \text{and} \qquad y'^K. \tag{14.12}$$

A factor of the first type appears if the fat vertex k is connected via a meagre son with $l, \ldots$ and directly with a fat son $m, \ldots$; a factor y'^K appears if "k" is the index of a meagre end-vertex. The vector $F(t)(y, y')$ is again called an *elementary differential*.

For some examples see Table 14.3 below. Observe that the indices of the meagre vertices, which are not end-vertices, play no role in the above definition. In analogy to Definition 2.4 we have

Definition 14.4. Two labelled N-trees (t, t') and (u, u') are *equivalent,* if they differ only by a permutation of their indices; i.e., if they have the same order, say

q, and if there exists a bijection $\sigma : A_q \to A_q$ with $\sigma(j) = j$, such that $t\sigma = \sigma u$ on $A_q \setminus \{j\}$ and $t'\sigma = u'$.

For example, the second and fourth labelled N-trees of formula (14.11;4) in Fig. 14.1 are equivalent; and also the second and fifth of formula (14.11;5).

Definition 14.5. An equivalence class of qth order labelled N-trees is called an *N-tree of order q*. The set of all N-trees of order q is denoted by NT_q. We further denote by $\alpha(t)$ the number of elements in the equivalence class t, i.e., the number of possible different monotonic labellings of t.

Representatives of N-trees up to order 5 are shown in Table 14.3. We are now able to give a closed formula for the derivatives of the exact solution of (14.10).

Theorem 14.6. *The exact solution of (14.10) satisfies*

$$y^{(q)} = \sum_{t \in LNT_{q-1}} F(t)(y,y') = \sum_{t \in NT_{q-1}} \alpha(t) F(t)(y,y'). \qquad (14.11;q)$$

Proof. The general formula is obtained by continuing the computation for (14.11;2-4) as in Section II.2.   □

The Derivatives of the Numerical Solution

We first rewrite (14.4) as

$$g_i = y_0 + c_i h y_0' + \sum_{j=1}^{s} \bar{a}_{ij} h^2 f(g_j, g_j'), \qquad g_i' = y_0' + \sum_{j=1}^{s} a_{ij} h f(g_j, g_j')$$

$$y_1 = y_0 + h y_0' + \sum_{i=1}^{s} \bar{b}_i h^2 f(g_i, g_i'), \qquad y_1' = y_0' + \sum_{i=1}^{s} b_i h f(g_i, g_i')$$

$$(14.13)$$

so that the intermediate values g_i, g_i' are treated in the same way as y_1, y_1'. In (14.13) there appear expressions of the form $h^2 \varphi(h)$ and $h\varphi(h)$. Therefore we have to use in addition to (2.4) the formula

$$\left(h^2 \varphi(h) \right)^{(q)} \Big|_{h=0} = q \cdot (q-1) \cdot \left(\varphi(h) \right)^{(q-2)} \Big|_{h=0}. \qquad (14.14)$$

We now compute successively the derivatives of g_i^J and $g_i'^J$ at $h = 0$:

$$\left(g_i^J \right)^{(1)} \Big|_{h=0} = c_i y_0'^J \qquad (14.15;1)$$

$$\left(g_i'^J \right)^{(1)} \Big|_{h=0} = \sum_j a_{ij} f^J \Big|_{y_0, y_0'} \qquad (14.16;1)$$

$$(g_i^J)^{(2)}\big|_{h=0} = 2 \sum_j \bar{a}_{ij} f^J \big|_{y_0, y_0'}. \tag{14.15;2}$$

For a further differentiation we need

$$(f^J(g_j, g_j'))^{(1)} = \sum_K \frac{\partial f^J}{\partial y^K}(g_j, g_j')(g_j^K)^{(1)} + \sum_K \frac{\partial f^J}{\partial y'^K}(g_j, g_j')(g_j'^K)^{(1)}. \tag{14.17}$$

With this formula we then obtain

$$
\begin{aligned}
(g_i'^J)^{(2)}\big|_{h=0} = {}& 2 \sum_j a_{ij} c_j \sum_K \frac{\partial f^J}{\partial y^K} \cdot y'^K \Big|_{y_0, y_0'} \\
& + 2 \sum_{j,k} a_{ij} a_{jk} \sum_K \frac{\partial f^J}{\partial y'^K} \cdot f^K \Big|_{y_0, y_0'}
\end{aligned}
\tag{14.16;2}
$$

$$
\begin{aligned}
(g_i^J)^{(3)}\big|_{h=0} = {}& 3 \cdot 2 \sum_j \bar{a}_{ij} c_j \sum_K \frac{\partial f^J}{\partial y^K} \cdot y'^K \Big|_{y_0, y_0'} \\
& + 3 \cdot 2 \sum_{j,k} \bar{a}_{ij} a_{jk} \sum_K \frac{\partial f^J}{\partial y'^K} \cdot f^K \Big|_{y_0, y_0'}.
\end{aligned}
\tag{14.15;3}
$$

To write down a general formula we need

Definition 14.7. For a labelled N-tree we denote by $\Phi_j(t)$ the expression which is a sum over the indices of all fat vertices of t (without "j", the index of the root). The general term of the sum is a product of

a_{kl} if the fat vertex "k" has a fat son "l";

$\bar{a}_{kl}$ if the fat vertex "k" is connected via a meagre son with "l"; and

c_k^m if the fat vertex "k" is connected with m meagre end-vertices.

Theorem 14.8. *The* g_i, g_i' *of (14.13) satisfy*

$$(g_i)^{(q+1)}\big|_{h=0} = (q+1) \sum_{t \in LNT_q} \gamma(t) \sum_{j=1}^s \bar{a}_{ij} \Phi_j(t) \, F(t)(y_0, y_0') \tag{14.15;q+1}$$

$$(g_i')^{(q)}\big|_{h=0} = \sum_{t \in LNT_q} \gamma(t) \sum_{j=1}^s a_{ij} \Phi_j(t) \, F(t)(y_0, y_0') \tag{14.16;q}$$

where $\gamma(t)$ *is given in Definition 2.10.*

Proof. For small values of q these formulas were obtained above; for general values of q they are proved like Theorem 2.11. System (14.2) is a special case of what will later be treated as a *partitioned system* (see Section II.15). Theorem 14.8 will then appear again in a new light. □

Because of the similarity of the formulas for g_i and y_1, g_i' and y_1' we have

Theorem 14.9. *The numerical solution* y_1, y_1' *of (14.13) satisfies*

$$(y_1)^{(q)}\big|_{h=0} = q \sum_{t \in LNT_{q-1}} \gamma(t) \sum_{i=1}^{s} \bar{b}_i \, \Phi_i(t) \, F(t)(y_0, y_0') \qquad (14.18;q)$$

$$(y_1')^{(q-1)}\big|_{h=0} = \sum_{t \in LNT_{q-1}} \gamma(t) \sum_{i=1}^{s} b_i \, \Phi_i(t) \, F(t)(y_0, y_0') \,. \qquad (14.19;q\text{-}1)$$

$\square$

The Order Conditions

For the study of the order of a Nyström method (Definition 14.1) one has to compare the Taylor series of y_1, y_1' with that of the true solution $y(x_0 + h), y'(x_0 + h)$.

Theorem 14.10. *A Nyström method (14.4) is of order* p *iff*

$$\sum_{i=1}^{s} \bar{b}_i \Phi_i(t) = \frac{1}{(\varrho(t)+1)\cdot\gamma(t)} \quad \text{for N-trees } t \text{ with } \varrho(t) \le p-1, \qquad (14.20)$$

$$\sum_{i=1}^{s} b_i \Phi_i(t) = \frac{1}{\gamma(t)} \quad \text{for N-trees } t \text{ with } \varrho(t) \le p \,. \qquad (14.21)$$

Here $\varrho(t)$ *denotes the order of the N-tree* t, $\Phi_i(t)$ *and* $\gamma(t)$ *are given by Definition 14.7 and formula (2.17).*

Proof. The "if" part is an immediate consequence of Theorems 14.6 and 14.9. The "only if" part can be shown in the same way as for first order equations (cf. Exercise 4 of Section II.2). $\square$

Let us briefly discuss whether the extra freedom in the choice of the parameters of (14.4) (by discarding the assumption (14.5)) can lead to a considerable improvement. Since the order conditions for Runge-Kutta methods (Theorem 2.13) are a subset of (14.21) (see Exercise 3 below), it is impossible to gain order with this extra freedom. Only some (never all) error coefficients can be made smaller. Therefore we shall turn to Nyström methods (14.8) for special second order differential equations (14.7).

For the study of the order conditions for (14.8) we write (14.7) in autonomous form

$$y'' = f(y). \qquad (14.22)$$

This special form implies that those elementary differentials which contain derivatives with respect to y' vanish identically. Consequently, only the following subset of N-trees has to be considered:

Definition 14.11. An N-tree t is called a *special N-tree* or *SN-tree*, if the fat vertices have only meagre sons.

Theorem 14.12. *A Nyström method (14.8) for the special differential equation (14.7) is of order p, iff*

$$\sum_{i=1}^{s} \overline{b}_i \Phi_i(t) = \frac{1}{(\varrho(t)+1) \cdot \gamma(t)} \quad \text{for SN-trees } t \text{ with } \varrho(t) \leq p-1, \quad (14.23)$$

$$\sum_{i=1}^{s} b_i \Phi_i(t) = \frac{1}{\gamma(t)} \quad \text{for SN-trees } t \text{ with } \varrho(t) \leq p. \quad (14.24)$$
□

All SN-trees up to order 5, together with the elementary differentials and the expressions Φ_j, ϱ, α, and γ, which are needed for the order conditions, are given in Table 14.3.

Higher order systems. The extension of the ideas of this section to *higher order systems*

$$y^{(n)} = f(x, y, y', \ldots, y^{(n-1)}) \quad (14.25)$$

is now more or less straightforward. Again, a real improvement is only possible in the case when the right-hand side of (14.25) depends only on x and y. A famous paper on this subject is the work of Zurmühl (1948). Tables of order conditions and methods are given in Hebsacker (1982).

On the Construction of Nyström Methods

The following simplifying assumptions are useful for the construction of Nyström methods.

Lemma 14.13. *Under the assumption*

$$\overline{b}_i = b_i(1 - c_i) \qquad i = 1, \ldots, s \quad (14.26)$$

the condition (14.24) implies (14.23).

Proof. Let t be an SN-tree of order $\leq p-1$ and denote by u the SN-tree of order $\varrho(t)+1$ obtained from t by attaching a new branch with a meagre vertex to the root of t. By Definition 14.7 we have $\Phi_i(u) = c_i \Phi_i(t)$ and from formula (2.17) it

Table 14.3. SN-trees, elementary differentials and order conditions

t	graph	$\varrho(t)$	$\alpha(t)$	$\gamma(t)$	$F^J(t)(y,y')$	$\Phi_j(t)$
t_1		1	1	1	f^J	1
t_2		2	1	2	$\sum_K f_K^J y'^K$	c_j
t_3		3	1	3	$\sum_{K,L} f_{KL}^J y'^K y'^L$	c_j^2
t_4		3	1	6	$\sum_L f_L^J f^L$	$\sum_l \bar{a}_{jl}$
t_5		4	1	4	$\sum_{K,L,M} f_{KLM}^J y'^K y'^L y'^M$	c_j^3
t_6		4	3	8	$\sum_{L,M} f_{LM}^J y'^L f^M$	$\sum_m c_j \bar{a}_{jm}$
t_7		4	1	24	$\sum_{L,M} f_L^J f_M^L y'^M$	$\sum_l \bar{a}_{jl} c_l$
t_8		5	1	5	$\sum_{K,L,M,P} f_{KLMP}^J y'^K y'^L y'^M y'^P$	c_j^4
t_9		5	6	10	$\sum_{L,M,P} f_{LMP}^J y'^L y'^M f^P$	$\sum_p c_j^2 \bar{a}_{jp}$
t_{10}		5	3	20	$\sum_{M,P} f_{MP}^J f^M f^P$	$\sum_{m,p} \bar{a}_{jm} \bar{a}_{jp}$
t_{11}		5	4	30	$\sum_{L,M,P} f_{LP}^J f_M^L y'^M y'^P$	$\sum_l c_j \bar{a}_{jl} c_l$
t_{12}		5	1	60	$\sum_{L,M,P} f_L^J f_{MP}^L y'^M y'^P$	$\sum_l \bar{a}_{jl} c_l^2$
t_{13}		5	1	120	$\sum_{L,P} f_L^J f_P^L f^P$	$\sum_{l,p} \bar{a}_{jl} \bar{a}_{lp}$

follows that $\gamma(u) = (\varrho(t)+1)\gamma(t)/\varrho(t)$. The conclusion now follows since

$$\sum_{i=1}^s \bar{b}_i \Phi_i(t) = \sum_{i=1}^s b_i \Phi_i(t) - \sum_{i=1}^s b_i \Phi_i(u) = \frac{1}{\gamma(t)} - \frac{1}{\gamma(u)} = \frac{1}{(\varrho(t)+1)\gamma(t)}. \qquad \square$$

Lemma 14.14. *Let t and u be two SN-trees as sketched in Fig. 14.2, where the encircled parts are assumed to be identical. Then under the assumption*

$$\sum_{j=1}^s \bar{a}_{ij} = \frac{c_i^2}{2} \qquad i = 1,\ldots,s \tag{14.27}$$

the order conditions for t and u are the same.

Proof. It follows from Definition 14.7 and (14.27) that $\Phi_i(t) = \Phi_i(u)/2$ and from formula (2.17) that $\gamma(t) = 2\gamma(u)$. Both order conditions are thus identical. $\square$

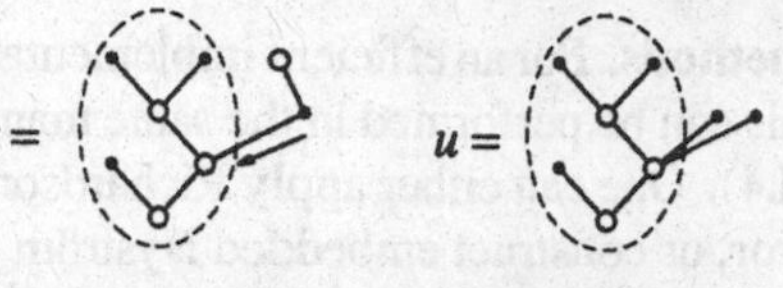

$$t = \qquad\qquad u =$$

Fig. 14.2. Trees of Lemma 14.14

Condition (14.26) allows us to neglect the equations (14.23), while condition (14.27) plays a similar role to that of (1.9) for Runge-Kutta methods. It expresses the fact that the g_i of (14.13) approximate $y(x_0 + c_i h)$ up to $\mathcal{O}(h^3)$. As a consequence of Lemma 14.14, SN-trees which have at least one fat end-vertex can be left out (i.e., t_4, t_6, t_9, t_{10}, t_{13} of Table 14.3).

With the help of (14.26) and (14.27) *explicit Nyström methods* (14.8) *of order 5 with $s = 4$* can now easily be constructed: the order conditions for the trees t_1, t_2, t_3, t_5 and t_8 just indicate that the quadrature formula with nodes $c_1 = 0$, c_2, c_3, c_4 and weights b_1, b_2, b_3, b_4 is of order 5. Thus the nodes c_i have to satisfy the orthogonality relation

$$\int_0^1 x(x - c_2)(x - c_3)(x - c_4)\, dx = 0$$

and we see that two degrees of freedom are still left in the choice of the quadrature formula. The $\bar{a}_{ij}$ are now uniquely determined and can be computed as follows: $\bar{a}_{21}$ is given by (14.27) for $i = 2$. The order conditions for t_7 and t_{11} constitute two linear equations for the unknowns

$$\sum_{j=1}^2 \bar{a}_{3j} c_j \quad \text{and} \quad \sum_{j=1}^3 \bar{a}_{4j} c_j \,.$$

Together with (14.27, $i = 3$) one now obtains $\bar{a}_{31}$ and $\bar{a}_{32}$. Finally, the order condition for t_{12} leads to $\sum_j \bar{a}_{4j} c_j^2$ and the remaining coefficients $\bar{a}_{41}, \bar{a}_{42}, \bar{a}_{43}$ can be computed from a Vandermonde-type linear system. The method of Table 14.2 is obtained in this way.

For still higher order methods it is helpful to use further simplifying assumptions; for example

$$\sum_{j=1}^s \bar{a}_{ij} c_j^q = \frac{c_i^{q+2}}{(q+2)(q+1)} \tag{14.28}$$

which, for $q = 0$, reduces to (14.27), and

$$\sum_{i=1}^s b_i c_i^q \bar{a}_{ij} = b_j \left(\frac{c_j^{q+2}}{(q+2)(q+1)} - \frac{c_j}{q+1} + \frac{1}{q+2} \right) \tag{14.29}$$

which can be considered a generalization of condition $D(\zeta)$ of Section II.7. For more details we refer to Hairer & Wanner (1976) and also to Albrecht (1955), Battin (1976), Beentjes & Gerritsen (1976), Hairer (1977, 1982), where Nyström methods of higher order are presented.

Embedded Nyström methods. For an efficient implementation we need a step size control mechanism. This can be performed in the same manner as for Runge-Kutta methods (see Section II.4). One can either apply Richardson extrapolation in order to estimate the local error, or construct embedded Nyström methods.

A series of embedded Nyström methods has been constructed by Fehlberg (1972). These methods use a $(p+1)$-st order approximation to $y(x_0 + h)$ for step size control. A $(p+1)$-st order approximation to $y'(x_0 + h)$ is not needed, since the lower order approximations are used for step continuation.

As for first order differential equations, local extrapolation — to use the higher order approximations for step continuation — turns out to be superior. Bettis (1973) was apparently the first to use this technique. His proposed method is of order 5(4). A method of order 7(6) has been constructed by Dormand & Prince (1978), methods of order 8(7), 9(8), 10(9) and 11(10) are given by Filippi & Gräf (1986) and further methods of order 8(6) and 12(10) are presented by Dormand, El-Mikkawy & Prince (1987).

In certain situations (see Section II.6) it is important that a Nyström method be equipped with a dense output formula. Such procedures are given by Dormand & Prince (1987) and, for general initial value problems $y'' = f(x, y, y')$, by Fine (1987).

An Extrapolation Method for $y'' = f(x, y)$

> Les calculs originaux, comprenant environ 3.000 pages in-folio avec 358 grandes planches, et encore 3.800 pages de développements mathématiques correspondants, appartiennent maintenant à la collection de manuscrits de la Bibliothèque de l'Université, Christiania.   (Störmer 1921)

If we rewrite the differential equation (14.7) as a first order system

$$\begin{pmatrix} y \\ y' \end{pmatrix}' = \begin{pmatrix} y' \\ f(x,y) \end{pmatrix}, \qquad \begin{pmatrix} y \\ y' \end{pmatrix}(x_0) = \begin{pmatrix} y_0 \\ y_0' \end{pmatrix} \tag{14.30}$$

we can apply the GBS-algorithm (9.13) directly to (14.30); this yields

$$y_1 = y_0 + h y_0' \tag{14.31a}$$
$$y_1' = y_0' + h f(x_0, y_0)$$
$$y_{i+1} = y_{i-1} + 2h y_i' \tag{14.31b}$$
$$y_{i+1}' = y_{i-1}' + 2h f(x_i, y_i) \qquad i = 1, 2, \ldots, 2n$$
$$S_h(x) = (y_{2n-1} + 2y_{2n} + y_{2n+1})/4 \tag{14.31c}$$
$$S_h'(x) = (y_{2n-1}' + 2y_{2n}' + y_{2n+1}')/4.$$

Here, $S_h(x)$ and $S_h'(x)$ are the numerical approximations to $y(x)$ and $y'(x)$ at $x = x_0 + H$, where $H = 2nh$ and $x_i = x_0 + ih$. We now make the following important observation: for the computation of $y_0, y_2, y_4, \ldots, y_{2n}$ (even indices) and of $y_1', y_3', \ldots, y_{2n+1}'$ (odd indices) only the function values $f(x_0, y_0), f(x_2, y_2), \ldots, f(x_{2n}, y_{2n})$ have to be calculated. Furthermore, we know from (9.17) that y_{2n} and $(y_{2n-1}' + y_{2n+1}')/2$ each possess an asymptotic expansion in even powers of h. It is therefore obvious that (14.31c) should be replaced by (Gragg 1965)

$$S_h(x) = y_{2n}$$
$$S_h'(x) = (y_{2n-1}' + y_{2n+1}')/2. \tag{14.31c'}$$

Using this final step, the number of function evaluations is reduced by a factor of two. These numerical approximations can now be used for extrapolation. We take the harmonic sequence (9.8'), put

$$T_{i1} = S_h(x_0 + H), \qquad T_{i1}' = S_h'(x_0 + H)$$

and compute the extrapolated expressions $T_{i,j}$ and $T_{i,j}'$ by the Aitken & Neville formula (9.10).

Remark. Eliminating the y_j'-values in (14.31b) we obtain the equivalent formula

$$y_{i+2} - 2y_i + y_{i-2} = (2h)^2 f(x_i, y_i), \tag{14.32}$$

which is often called *Störmer's rule.* For the implementation the formulation (14.31b) is to be preferred, since it is more stable with respect to round-off errors (see Section III.10).

Dense output. As for the derivation of Section II.9 for the GBS algorithm we shall do Hermite interpolation based on derivatives of the solution at x_0, $x_0 + H$ and $x_0 + H/2$. At the endpoints of the considered interval we have $y_0, y_0', y_0'' = f(x_0, y_0)$ and y_1, y_1', y_1'' at our disposal. The derivatives at the midpoint can be obtained by extrapolation of suitable differences of function values. However, one has to take care of the fact that y_i and $f(x_i, y_i)$ are available only for even indices, whereas y_i' is available for odd indices only. For the same reason as for the GBS method, the step number sequence has to satisfy (9.34). For notational convenience, the following description is restricted to the sequence (9.35).

We suppose that T_{kk} and T_{kk}' are accepted approximations to the solution. Then the construction of a dense output formula can be summarized as follows:

Step 1. For each $j \in \{1, \ldots, k\}$ compute the approximations to the derivatives of $y(x)$ at $x_0 + H/2$ by (δ is the central difference operator):

$$d_j^{(0)} = \frac{1}{2}\left(y_{n_j/2-1} + y_{n_j/2+1}\right), \qquad d_j^{(1)} = y_{n_j/2}',$$

$$d_j^{(\kappa)} = \frac{1}{2} \cdot \frac{1}{(2h_j)^{\kappa-2}}\left(\delta^{\kappa-2} f_{n_j/2-1}^{(j)} + \delta^{\kappa-2} f_{n_j/2+1}^{(j)}\right), \qquad \kappa = 2, 4, \ldots, 2j,$$

$$d_j^{(\kappa)} = \frac{\delta^{\kappa-2} f_{n_j/2}^{(j)}}{(2h_j)^{\kappa-2}}, \qquad \kappa = 3, 5, \ldots, 2j+1. \tag{14.33}$$

Step 2. Extrapolate $d_j^{(0)}$, $d_j^{(1)}$ $(k-1)$ times and $d_j^{(2\ell)}$, $d_j^{(2\ell+1)}$ $(k-\ell)$ times to obtain improved approximations $d^{(\kappa)}$ to $y^{(\kappa)}(x_0 + H/2)$.

Step 3. For given μ $(-1 \leq \mu \leq 2k+1)$ define the polynomial $P_\mu(\theta)$ of degree $\mu + 6$ by

$$
\begin{aligned}
P_\mu(0) &= y_0, & P_\mu'(0) &= y_0', & P_\mu''(0) &= f(x_0, y_0) \\
P_\mu(1) &= T_{kk}, & P_\mu'(1) &= T_{kk}', & P_\mu''(1) &= f(x_0 + H, T_{kk}) \\
P_\mu^{(\kappa)}(1/2) &= H^\kappa d^{(\kappa)} & & \text{for } \kappa = 0, 1, \ldots, \mu.
\end{aligned}
\tag{14.34}
$$

Since T_{kk}, T_{kk}' are the initial values for the next step, the dense output obtained by the above algorithm is a global C^2 approximation to the solution. It satisfies

$$y(x_0 + \theta H) - P_\mu(\theta) = \mathcal{O}(H^{2k}) \qquad \text{if} \quad \mu \geq 2k - 7 \tag{14.35}$$

(compare Theorem 9.5). In the code ODEX2 of the Appendix the value $\mu = 2k - 5$ is suggested as standard choice.

Problems for Numerical Comparisons

PLEI — the celestial mechanics problem (10.3) which is the only problem of Section II.10 already in the special form (14.7).

ARES — the ARERenstorf orbit in Second order form (14.7). This is the restricted three body problem (0.1) with initial values (0.2) integrated over one period $0 \leq x \leq x_{\text{end}}$ (see Fig. 0.1) in a *fixed* coordinate system. Then the equations of motion become

$$
\begin{aligned}
y_1'' &= \mu' \frac{a_1(x) - y_1}{D_1} + \mu \frac{b_1(x) - y_1}{D_2} \\
y_2'' &= \mu' \frac{a_2(x) - y_2}{D_1} + \mu \frac{b_2(x) - y_2}{D_2}
\end{aligned}
\tag{14.36}
$$

where

$$D_1 = \big((y_1 - a_1(x))^2 + (y_2 - a_2(x))^2\big)^{3/2}, \quad D_2 = \big((y_1 - b_1(x))^2 + (y_2 - b_2(x))^2\big)^{3/2}$$

and the movement of sun and moon are described by

$$a_1(x) = -\mu \cos x \quad a_2(x) = -\mu \sin x \quad b_1(x) = \mu' \cos x \quad b_2(x) = \mu' \sin x.$$

The initial values

$$y_1(0) = 0.994, \qquad y_1'(0) = 0, \qquad y_2(0) = 0,$$

$$y_2'(0) = -2.00158510637908252240537862224 + 0.994,$$

$$x_{\text{end}} = 17.0652165601579625588917206249,$$

are those of (0.2) enlarged by the speed of the rotation. The exact solution values are the initial values transformed by the rotation of the coordinate system.

CPEN — the nonlinear Coupled PENdulum (see Fig. 14.3).

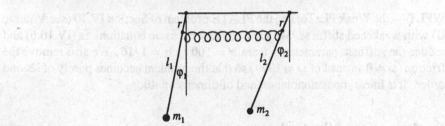

Fig. 14.3. Coupled pendulum

The kinetic as well as potential energies

$$T = \frac{m_1 l_1^2 \dot{\varphi}_1^2}{2} + \frac{m_2 l_2^2 \dot{\varphi}_2^2}{2}$$

$$V = -m_1 l_1 \cos \varphi_1 - m_2 l_2 \cos \varphi_2 + \frac{c_0 r^2 (\sin \varphi_1 - \sin \varphi_2)^2}{2}$$

lead by Lagrange theory (equations (I.6.21)) to

$$\ddot{\varphi}_1 = -\frac{\sin \varphi_1}{l_1} - \frac{c_0 r^2}{m_1 l_1^2}(\sin \varphi_1 - \sin \varphi_2) \cos \varphi_1 + f(t)$$

$$\ddot{\varphi}_2 = -\frac{\sin \varphi_2}{l_2} - \frac{c_0 r^2}{m_2 l_1^2}(\sin \varphi_2 - \sin \varphi_1) \cos \varphi_2.$$

(14.37)

We choose the parameters

$$l_1 = l_2 = 1, \quad m_1 = 1, \quad m_2 = 0.99, \quad r = 0.1, \quad c_0 = 0.01, \quad t_{\text{end}} = 496$$

and all initial values and speeds for $t = 0$ equal to zero. The first pendulum is then pushed into movement by a (somewhat idealized) hammer as

$$f(t) = \begin{cases} \sqrt{1 - (1-t)^2} & \text{if } |t-1| \leq 1; \\ 0 & \text{otherwise.} \end{cases}$$

The resulting solutions are displayed in Fig. 14.4. The nonlinearities in this problem produce quite different sausages (cf. "Mon Oncle" de Jacques Tati 1958) from those people are accustomed to from linear problems (cf. Sommerfeld 1942, §20).

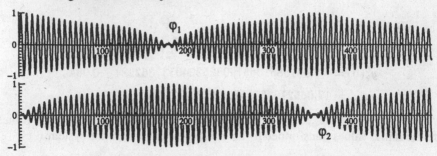

Fig. 14.4. Movement of the coupled pendulum (14.37)

WPLT — the Weak PLaTe, i.e., the PLATE problem of Section IV.10 (see Volume II) with weakened stiffness. We use precisely the same equations as (IV.10.6) and reduce the stiffness parameter σ from $\sigma = 100$ to $\sigma = 1/16$. We also remove the friction ($\omega = 0$ instead of $\omega = 1000$) so that the problem becomes purely of second order. It is linear, nonautonomous, and of dimension 40.

Performance of the Codes

Several codes were applied to each of the above four problems with 89 different tolerances between $Tol = 10^{-3}$ and $Tol = 10^{-14}$ (exactly as in Section II.10). The number of function evaluations (Fig. 14.5) and the computer time on an Apollo Domain 4000 Workstation (Fig. 14.6) are plotted as a function of the global error at the endpoint of the integration interval. The codes used are the following:

RKN6 — symbol ⋈ — is the low order option of the Runge-Kutta-Nyström code presented in Brankin, Gladwell, Dormand, Prince & Seward (1989). It is based on a fixed-order embedded Nyström method of order 6(4), whose coefficients are given in Dormand & Prince (1987). This code is provided with a dense output.

RKN12 — symbol ⋈ — is the high order option of the Runge-Kutta-Nyström code presented in Brankin & al. (1989). It is based on the method of order 12(10), whose coefficients are given in Dormand, El-Mikkawy & Prince (1987). This code is not equipped with a dense output.

ODEX2 — symbol O — is the extrapolation method based on formula (14.31a,b,c') and uses the harmonic step number sequence (see Appendix). It is implemented in the same way as ODEX (the extrapolation code for first order differential equations). In particular, the order and step size strategy is that of Section II.9. A dense output is available. Similar results are obtained by the code DIFEX2 of Deuflhard & Bauer (see Deuflhard 1985).

In order to demonstrate the superiority of the special methods for $y'' = f(x, y)$, we have included the results obtained by DOP853 (symbol ✿) and ODEX (symbol

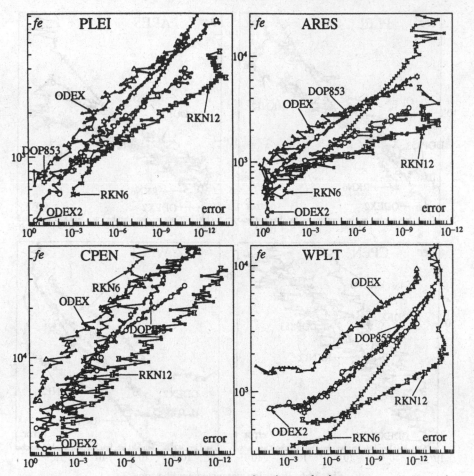

Fig. 14.5. Precision versus function evaluations

$\triangle$) which were already described in Section II.10. For their application we had to rewrite the four problems as a first order system by introducing the first derivatives as new variables. The code ODEX2 is nearly twice as efficient as ODEX which is in agreement with the theoretical considerations. Similarly the Runge-Kutta-Nyström codes RKN6 and RKN12 are a real improvement over DOP853.

A comparison of Fig. 14.5 and 14.6 shows a significant difference. The extrapolation codes ODEX and ODEX2 are relatively better on the "time"-pictures than for the function evaluation counts. With the exception of problem WPLT the performance of the code ODEX2 then becomes comparable to that of RKN12. As can be observed especially at the WPLT problem, the code RKN12 overshoots, for stringent tolerances, significantly the desired precision. It becomes less efficient if *Tol* is chosen too close to *Uround*.

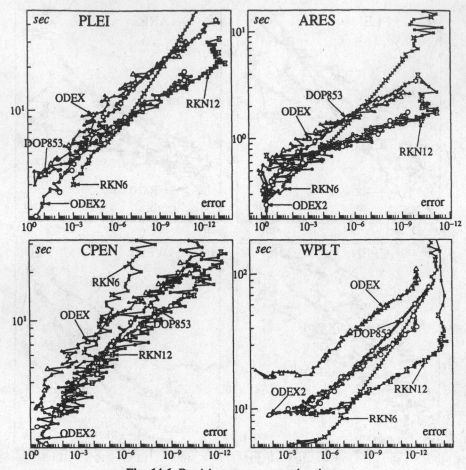

Fig. 14.6. Precision versus computing time

Exercises

1. Verify that the methods of Table 14.2 are of order 4 and 5, respectively.

2. The error coefficients of a pth order Nyström method are defined by

$$
\begin{aligned}
e(t) &= 1 - (\varrho(t) + 1)\gamma(t) \sum_i \overline{b}_i \Phi_i(t) && \text{for } \varrho(t) = p, \\
e'(t) &= 1 - \gamma(t) \sum_i b_i \Phi_i(t) && \text{for } \varrho(t) = p + 1.
\end{aligned}
\tag{14.38}
$$

a) The assumption (14.26) implies that

$$
e(t) = -\varrho(t) e'(u) \qquad \text{for } \varrho(t) = p,
$$

where u is the N-tree obtained from t by adding a branch with a meagre vertex to the root of t.

b) Compute the error coefficients of Nyström's method (Table 14.1) and compare them to those of the classical Runge-Kutta method.

3. Show that the order conditions for Runge-Kutta methods (Theorem 2.13) are a subset of the conditions (14.21). They correspond to the N-trees, all of whose vertices are fat.

4. Sometimes the definition of order of Nyström methods (14.8) is relaxed to

$$y(x_0 + h) - y_1 = \mathcal{O}(h^{p+1})$$
$$y'(x_0 + h) - y_1' = \mathcal{O}(h^p)$$

(14.39)

(see Nyström 1925). Show that the conditions (14.39) are not sufficient to obtain global convergence of order p.

Hint. Investigate the asymptotic expansion of the global error with the help of Theorem 8.1 and formula (8.8).

5. The numerical solutions T_{kk} and T_{kk}' of the extrapolation method of this section are equivalent to a Nyström method of order $p = 2k$ with $s = p^2/8 + p/4 + 1$ stages.

6. A *collocation method* for $y'' = f(x, y, y')$ (or $y'' = f(x, y)$) can be defined as follows: let $u(x)$ be a polynomial of degree $s + 1$ defined by

$$u(x_0) = y_0, \qquad u'(x_0) = y_0'$$

(14.40)

$$u''(x_0 + c_i h) = f(x_0 + c_i h, u(x_0 + c_i h), u'(x_0 + c_i h)), \quad i = 1, \ldots, s,$$

then the numerical solution is given by $y_1 = u(x_0 + h)$, $y_1' = u'(x_0 + h)$.

a) Prove that this collocation method is equivalent to the Nyström method (14.4) where

$$a_{ij} = \int_0^{c_i} \ell_j(t)\, dt, \qquad \bar{a}_{ij} = \int_0^{c_i} (c_i - t)\ell_j(t)\, dt,$$
$$b_i = \int_0^1 \ell_i(t)\, dt, \qquad \bar{b}_i = \int_0^1 (1 - t)\ell_i(t)\, dt,$$

(14.41)

and $\ell_j(t)$ are the Lagrange polynomials of (7.17).

b) The a_{ij} satisfy $C(s)$ (see Theorem 7.8) and the $\bar{a}_{ij}$ satisfy (14.28) for $q = 0, 1, \ldots, s - 1$. These equations uniquely define a_{ij} and $\bar{a}_{ij}$.

c) In general, a_{ij} and $\bar{a}_{ij}$ do not satisfy (14.5).

d) If $M(t) = \prod_{i=1}^s (t - c_i)$ is orthogonal to all polynomials of degree $r - 1$,

$$\int_0^1 M(t)t^{q-1}\, dt = 0, \qquad q = 1, \ldots, r,$$

then the collocation method (14.40) has order $p = s + r$.

e) The polynomial $u(x)$ yields an approximation to the solution $y(x)$ on the whole interval $[x_0, x_0 + h]$. The following estimates hold:

$$y(x) - u(x) = \mathcal{O}(h^{s+2}), \qquad y'(x) - u'(x) = \mathcal{O}(h^{s+1}).$$

II.15 P-Series for Partitioned Differential Equations

Divide ut regnes   (N. Machiavelli 1469-1527)

In the previous section we considered direct methods for second order differential equations $y'' = f(y, y')$. The idea was to write the equation as a partitioned differential system

$$\begin{pmatrix} y \\ y' \end{pmatrix}' = \begin{pmatrix} y' \\ f(y, y') \end{pmatrix} \tag{15.1}$$

and to discretize the two components, y and y', by different formulas. There are many other situations where the problem possesses a natural partitioning. Typical examples are the Hamiltonian equations (I.6.26, I.14.26) and singular perturbation problems (see Chapter VI of Volume II). It may also be of interest to separate linear and nonlinear parts or the "non-stiff" and "stiff" components of a differential equation.

We suppose that the differential system is partitioned as

$$\begin{pmatrix} y_a \\ y_b \end{pmatrix}' = \begin{pmatrix} f_a(y_a, y_b) \\ f_b(y_a, y_b) \end{pmatrix} \tag{15.2}$$

where the solution vector is separated into two components y_a, y_b, each of which may itself be a vector. An extension to more components is straight-forward.

For the numerical solution of (15.2) we consider the *partitioned method*

$$k_i = f_a\left(y_{a0} + h\sum_{j=1}^{s} a_{ij}k_j,\ y_{b0} + h\sum_{j=1}^{s} \widehat{a}_{ij}\ell_j\right)$$

$$\ell_i = f_b\left(y_{a0} + h\sum_{j=1}^{s} a_{ij}k_j,\ y_{b0} + h\sum_{j=1}^{s} \widehat{a}_{ij}\ell_j\right) \tag{15.3}$$

$$y_{a1} = y_{a0} + h\sum_{i=1}^{s} b_i k_i, \qquad y_{b1} = y_{b0} + h\sum_{i=1}^{s} \widehat{b}_i \ell_i$$

where the coefficients a_{ij}, b_i and $\widehat{a}_{ij}, \widehat{b}_i$ represent two different Runge-Kutta schemes. The first methods of this type are due to Hofer (1976) and Griepentrog (1978) who apply an explicit method to the nonstiff part and an implicit method to the stiff part of a differential equation. Later Rentrop (1985) modified this idea by combining explicit Runge-Kutta methods with Rosenbrock-type methods (Section

IV.7). Recent interest for partitioned methods came up when solving Hamiltonian systems (see Section II.16 below).

The subject of this section is the derivation of the order conditions for method (15.3). For order p it is necessary that each of the two Runge-Kutta schemes under consideration be of order p. This can be seen by applying the method to $y'_a = f_a(y_a)$, $y'_b = f_b(y_b)$. But this is not sufficient, the coefficients have to satisfy certain *coupling conditions*. In order to understand this, we first look at the derivatives of the exact solution of (15.2). Then we generalize the theory of B-series (see Section II.12) to the new situation (Hairer 1981) and derive the order conditions in the same way as in II.12 for Runge-Kutta methods.

Derivatives of the Exact Solution, P-Trees

In order to avoid sums and unnecessary indices we assume that y_a and y_b in (15.2) are scalar quantities. All subsequent formulas remain valid for vectors if the derivatives are interpreted as multi-linear mappings. Differentiating (15.2) and inserting (15.2) again for the derivatives we obtain for the first component y_a

$$y_a^{(1)} = f_a \qquad\qquad (15.4;1)$$

$$y_a^{(2)} = \frac{\partial f_a}{\partial y_a} f_a + \frac{\partial f_a}{\partial y_b} f_b \qquad\qquad (15.4;2)$$

$$y_a^{(3)} = \frac{\partial^2 f_a}{\partial y_a^2}(f_a, f_a) + \frac{\partial^2 f_a}{\partial y_b \partial y_a}(f_b, f_a) + \frac{\partial f_a}{\partial y_a}\frac{\partial f_a}{\partial y_a} f_a + \frac{\partial f_a}{\partial y_a}\frac{\partial f_a}{\partial y_b} f_b \qquad (15.4;3)$$

$$+ \frac{\partial^2 f_a}{\partial y_a \partial y_b}(f_a, f_b) + \frac{\partial^2 f_a}{\partial y_b^2}(f_b, f_b) + \frac{\partial f_a}{\partial y_b}\frac{\partial f_b}{\partial y_a} f_a + \frac{\partial f_a}{\partial y_b}\frac{\partial f_b}{\partial y_b} f_b.$$

Similar formulas hold for the derivatives of y_b.

For a graphical representation of these formulas we need two different kinds of vertices. As in Section II.14 we use "meagre" and "fat" vertices, which will correspond to f_a and f_b, respectively. Formulas (15.4) can then be represented as shown in Fig. 15.1.

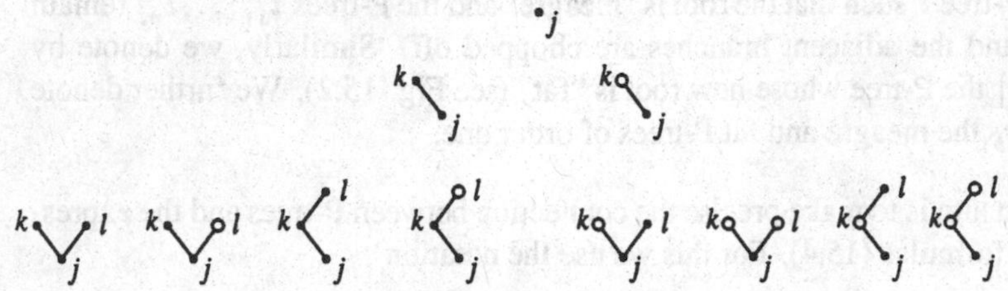

$$(15.4;1)$$

$$(15.4;2)$$

$$(15.4;3)$$

Fig. 15.1. The derivatives of the exact solution y_a

Definition 15.1. A *labelled P-tree* of *order* q is a labelled tree (see Definition 2.2)

$$t : A_q \setminus \{j\} \to A_q$$

together with a mapping

$$t' : A_q \to \{\text{"meagre"}, \text{"fat"}\}.$$

We denote by LTP_q^a the set of those labelled P-trees of order q, whose root is meagre (i.e., $t'(j) = \text{"meagre"}$). Similarly, LTP_q^b is the set of qth order labelled P-trees with a "fat" root.

Due to the symmetry of the second derivative the 2nd and 5th expressions in (15.4;3) are equal. We therefore define:

Definition 15.2. Two labelled P-trees (t, t') and (u, u') are *equivalent*, if they have the same order, say q, and if there exists a bijection $\sigma : A_q \to A_q$ such that $\sigma(j) = j$ and the following diagram commutes:

$$
\begin{array}{ccc}
A_q \setminus \{j\} & \xrightarrow{\ t\ } & A_q \\
\sigma \downarrow & & \sigma \downarrow \\
A_q \setminus \{j\} & \xrightarrow{\ u\ } & A_q
\end{array}
\searrow^{t'} \{\text{"meagre"}, \text{"fat"}\} \nearrow_{u'}
$$

Definition 15.3. An equivalence class of qth order labelled P-trees is called a *P-tree* of *order* q. The set of all P-trees of order q with a meagre root is denoted by TP_q^a, that with a fat root by TP_q^b. For a P-tree t we denote by $\varrho(t)$ the *order* of t, and by $\alpha(t)$ the number of elements in the equivalence class t.

Examples of P-trees together with the numbers $\varrho(t)$ and $\alpha(t)$ are given in Table 15.1 below. We first discuss a recursive representation of P-trees (extension of Definition 2.12), which is fundamental for the following theory.

Definition 15.4. Let $t_1, \ldots, t_m$ be P-trees. We then denote by

$$t = {}_a[t_1, \ldots, t_m] \tag{15.5}$$

the unique P-tree t such that the root is "meagre" and the P-trees $t_1, \ldots, t_m$ remain if the root and the adjacent branches are chopped off. Similarly, we denote by ${}_b[t_1, \ldots, t_m]$ the P-tree whose new root is "fat" (see Fig. 15.2). We further denote by τ_a and τ_b the meagre and fat P-trees of order one.

Our next aim is to make precise the connection between P-trees and the expressions of the formulas (15.4). For this we use the notation

$$w(t) = \begin{cases} a & \text{if the root of } t \text{ is meagre,} \\ b & \text{if the root of } t \text{ is fat.} \end{cases} \tag{15.6}$$

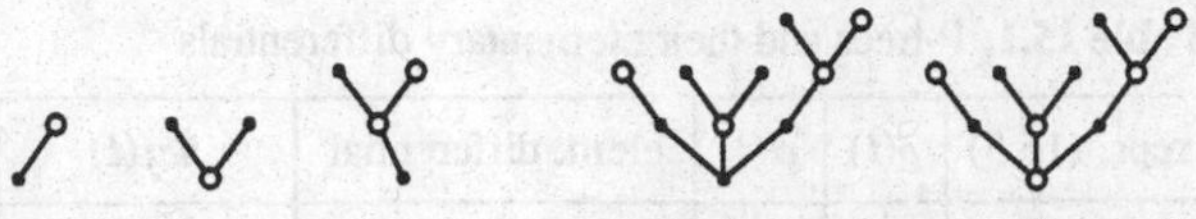

$$t_1 \qquad t_2 \qquad t_3 \qquad t = {}_a\,[t_1,t_2,t_3] \quad t = {}_b\,[t_1,t_2,t_3]$$

Fig. 15.2. Recursive definition of P-trees

Definition 15.5. The *elementary differentials,* corresponding to (15.2), are defined recursively by $(y = (y_a, y_b))$

$$F(\tau_a)(y) = f_a(y), \qquad F(\tau_b)(y) = f_b(y)$$

and

$$F(t)(y) = \frac{\partial^m f_{w(t)}(y)}{\partial y_{w(t_1)} \cdots \partial y_{w(t_m)}} \cdot \left(F(t_1)(y), \ldots, F(t_m)(y)\right)$$

for $t = {}_a[t_1, \ldots, t_m]$ or $t = {}_b[t_1, \ldots, t_m]$.

Elementary differentials for P-trees up to order 3 are given explicitly in Table 15.1.

We now return to the starting-point of this section and continue the differentiation of formulas (15.4). Using the notation of labelled P-trees, one sees that a differentiation of $F(t)(y_a, y_b)$ can be interpreted as an addition of a new branch with a meagre or fat vertex and a new summation letter to each vertex of the labelled P-tree t. In the same way as we proved Theorem 2.6 for non-partitioned differential equations, we arrive at

Theorem 15.6. *The derivatives of the exact solution of (15.2) satisfy*

$$y_a^{(q)} = \sum_{t \in LTP_q^a} F(t)(y_a, y_b) = \sum_{t \in TP_q^a} \alpha(t) F(t)(y_a, y_b) \qquad (15.4;q)$$

$$y_b^{(q)} = \sum_{t \in LTP_q^b} F(t)(y_a, y_b) = \sum_{t \in TP_q^b} \alpha(t) F(t)(y_a, y_b).$$

□

Table 15.1. P-trees and their elementary differentials

P-tree	repr. (15.5)	$\varrho(t)$	$\alpha(t)$	elem. differential	$\Phi_j(t)$
•	τ_a	1	1	f_a	1
	$a[\tau_a]$	2	1	$\frac{\partial f_a}{\partial y_a} f_a$	$\sum_k a_{jk}$
	$a[\tau_b]$	2	1	$\frac{\partial f_a}{\partial y_b} f_b$	$\sum_k \widehat{a}_{jk}$
	$a[\tau_a,\tau_a]$	3	1	$\frac{\partial^2 f_a}{\partial y_a^2}(f_a,f_a)$	$\sum_{k,l} a_{jk}a_{jl}$
	$a[\tau_a,\tau_b]$	3	2	$\frac{\partial^2 f_a}{\partial y_a \partial y_b}(f_a,f_b)$	$\sum_{k,l} a_{jk}\widehat{a}_{jl}$
	$a[\tau_b,\tau_b]$	3	1	$\frac{\partial^2 f_a}{\partial y_b^2}(f_b,f_b)$	$\sum_{k,l} \widehat{a}_{jk}\widehat{a}_{jl}$
	$a[a[\tau_a]]$	3	1	$\frac{\partial f_a}{\partial y_a}\frac{\partial f_a}{\partial y_a} f_a$	$\sum_{k,l} a_{jk}a_{kl}$
	$a[a[\tau_b]]$	3	1	$\frac{\partial f_a}{\partial y_a}\frac{\partial f_a}{\partial y_b} f_b$	$\sum_{k,l} a_{jk}\widehat{a}_{kl}$
	$a[b[\tau_a]]$	3	1	$\frac{\partial f_a}{\partial y_b}\frac{\partial f_b}{\partial y_a} f_a$	$\sum_{k,l} \widehat{a}_{jk}a_{kl}$
	$a[b[\tau_b]]$	3	1	$\frac{\partial f_a}{\partial y_b}\frac{\partial f_b}{\partial y_b} f_b$	$\sum_{k,l} \widehat{a}_{jk}\widehat{a}_{kl}$
...	...	...	...	...	...
o	τ_b	1	1	f_b	1
	$b[\tau_a]$	2	1	$\frac{\partial f_b}{\partial y_a} f_a$	$\sum_k a_{jk}$
	$b[\tau_b]$	2	1	$\frac{\partial f_b}{\partial y_b} f_b$	$\sum_k \widehat{a}_{jk}$
...	...	...	...	...	...

P-Series

In Section II.12 we saw the importance of the key-lemma Corollary 12.7 for the derivation of the order conditions for Runge-Kutta methods. Therefore we extend this result also to partitioned ordinary differential equations.

It is convenient to introduce two new P-trees of order 0, namely $\emptyset_a$ and $\emptyset_b$. The corresponding elementary differentials are $F(\emptyset_a)(y)=y_a$ and $F(\emptyset_b)(y)=y_b$. We further set

$$TP^a = \{\emptyset_a\} \cup TP_1^a \cup TP_2^a \cup \ldots \qquad LTP^a = \{\emptyset_a\} \cup LTP_1^a \cup LTP_2^a \cup \ldots$$
$$TP^b = \{\emptyset_b\} \cup TP_1^b \cup TP_2^b \cup \ldots \qquad LTP^b = \{\emptyset_b\} \cup LTP_1^b \cup LTP_2^b \cup \ldots.$$
$$(15.7)$$

Definition 15.7. Let $c(\emptyset_a)$, $c(\emptyset_b)$, $c(\tau_a)$, $c(\tau_b)$, ... be real coefficients defined for all P-trees, i.e., $c : TP^a \cup TP^b \to \mathbf{R}$. The series

$$P(c, y) = \left(P_a(c, y), P_b(c, y)\right)^T$$

where

$$P_a(c, y) = \sum_{t \in LTP^a} \frac{h^{\varrho(t)}}{\varrho(t)!} c(t) F(t)(y), \qquad P_b(c, y) = \sum_{t \in LTP^b} \frac{h^{\varrho(t)}}{\varrho(t)!} c(t) F(t)(y)$$

is then called a *P-series*.

Theorem 15.6 simply states that the exact solution of (15.2) is a P-series

$$\left(y_a(x_0 + h), y_b(x_0 + h)\right)^T = P(y, (y_a(x_0), y_b(x_0))) \tag{15.8}$$

with $y(t) = 1$ for all P-trees t.

Theorem 15.8. *Let* $c : TP^a \cup TP^b \to \mathbf{R}$ *be a sequence of coefficients such that* $c(\emptyset_a) = c(\emptyset_b) = 1$. *Then*

$$h \begin{pmatrix} f_a\big(P(c, (y_a, y_b))\big) \\ f_b\big(P(c, (y_a, y_b))\big) \end{pmatrix} = P(c', (y_a, y_b)) \tag{15.9}$$

with

$$c'(\emptyset_a) = c'(\emptyset_b) = 0, \qquad c'(\tau_a) = c'(\tau_b) = 1 \tag{15.10}$$
$$c'(t) = \varrho(t)c(t_1)\ldots c(t_m) \qquad if \quad t = {}_a[t_1, \ldots, t_m] \ or \ t = {}_b[t_1, \ldots, t_m].$$

The *proof* is related to that of Theorem 12.6. It is given with more details in Hairer (1981).   □

Order Conditions for Partitioned Runge-Kutta Methods

With the help of Theorem 15.8 the order conditions for method (15.3) can readily be obtained. For this we denote the arguments in (15.3) by

$$g_i = y_{a0} + h \sum_{j=1}^{s} a_{ij} k_j, \qquad \widehat{g}_i = y_{b0} + h \sum_{j=1}^{s} \widehat{a}_{ij} \ell_j, \tag{15.11}$$

and we assume that $G_i = (g_i, \widehat{g}_i)^T$ and $K_i = h(k_i, \ell_i)^T$ are P-series with coefficients $G_i(t)$ and $K_i(t)$, respectively. The formulas (15.11) then yield $G_i(\emptyset_a) = 1$, $G_i(\emptyset_b) = 1$ and

$$G_i(t) = \begin{cases} \sum_{j=1}^{s} a_{ij} K_j(t) & \text{if the root of } t \text{ is meagre,} \\ \sum_{j=1}^{s} \widehat{a}_{ij} K_j(t) & \text{if the root of } t \text{ is fat.} \end{cases} \tag{15.12}$$

Application of Theorem 15.8 to the relations $k_j = f_a(G_j)$, $\ell_j = f_b(G_j)$ shows that $\mathbf{K}_j(t) = \mathbf{G}'_j(t)$ which, together with (15.10) and (15.12), recursively defines the values $\mathbf{K}_j(t)$.

It is usual to write $\mathbf{K}_j(t) = \gamma(t)\Phi_j(t)$ where $\gamma(t)$ is the integer given in Definition 2.10 (see also (2.17)). The coefficient $\Phi_j(t)$ is then obtained in the same way as the corresponding value of standard Runge-Kutta methods (see Definition 2.9) with the exception that a factor a_{ik} has to be replaced by $\hat{a}_{ik}$, if the vertex with label "k" is fat. A comparison of the P-series for the numerical solution $(y_{1a}, y_{1b})^T$ with that for the exact solution (15.8) yields the desired order conditions.

Theorem 15.9. *A partitioned Runge-Kutta method (15.3) is of order p iff*

$$\sum_{j=1}^{s} b_j \Phi_j(t) = \frac{1}{\gamma(t)} \quad \text{and} \quad \sum_{j=1}^{s} \hat{b}_j \Phi_j(t) = \frac{1}{\gamma(t)} \qquad (15.13)$$

for all P-trees of order $\leq p$.   □

Example. A partitioned method (15.3) is of order 2, if and only if each of the two Runge-Kutta schemes has order 2 and if the coupling conditions

$$\sum_{i,j} b_i \hat{a}_{ij} = \frac{1}{2}, \qquad \sum_{i,j} \hat{b}_i a_{ij} = \frac{1}{2},$$

which correspond to trees $_a[\tau_b]$ and $_b[\tau_a]$ of Table 15.1 respectively, are satisfied. This happens if

$$c_i = \hat{c}_i \quad \text{for all } i.$$

This last assumption simplifies the order conditions considerably (the "thickness" of terminating vertices then has no influence). The resulting conditions for order up to 4 have been tabulated by Griepentrog (1978).

Further Applications of P-Series

Runge-Kutta methods violating (1.9). For the non-autonomous differential equation $y' = f(x, y)$ we consider, as in Exercise 6 of Section II.1, the Runge-Kutta method

$$k_i = f\left(x_0 + \hat{c}_i h, \, y_0 + h \sum_{j=1}^{s} a_{ij} k_j\right), \qquad y_1 = y_0 + h \sum_{i=1}^{s} b_i k_i, \qquad (15.14)$$

where $\widehat{c}_i$ is not necessarily equal to $c_i = \sum_j a_{ij}$. Therefore, the x and y components in

$$y' = f(x, y)$$
$$x' = 1. \tag{15.15}$$

are integrated differently. This system is of the form (15.2), if we put $y_a = y$, $y_b = x$, $f_a(y_a, y_b) = f(x, y)$ and $f_b(y_a, y_b) = 1$. Since f_b is constant, all elementary differentials that involve derivatives of f_b vanish identically. Thus, P-trees where at least one fat vertex is not an end-vertex need not be considered. It remains to treat the set

$$T_x = \{t \in TP_a; \text{ all fat vertices are end-vertices}\}. \tag{15.16}$$

Each tree of T_x gives rise to an order condition which is exactly that of Theorem 15.9. It is obtained in the usual way (Section II.2) with the exception that c_k has to be replaced by $\widehat{c}_k$, if the corresponding vertex is a fat one.

Fehlberg methods. The methods of Fehlberg, introduced in Section II.13, are equivalent to (15.14). However, it is known that the exact solution of the differential equation $y' = f(x, y)$ satisfies $y(x_0) = 0$, $y'(x_0) = 0, \ldots, y^{(m)}(x_0) = 0$ at the initial value $x = x_0$. As explained in II.13, this implies that the expressions f, $\partial f/\partial x, \ldots, \partial^{m-1}f/\partial x^{m-1}$ vanish at (x_0, y_0) and consequently also many of the elementary differentials disappear. The elements of T_x which remain to be considered are given in Fig. 15.3.

Fig. 15.3. P-trees for the methods of Fehlberg

Nyström methods. As a last application of Theorem 15.8 we present a new derivation of the order conditions for Nyström methods (Section II.14). The second order differential equation $y'' = f(y, y')$ can be written in partitioned form as

$$\begin{pmatrix} y \\ y' \end{pmatrix}' = \begin{pmatrix} y' \\ f(y, y') \end{pmatrix}. \tag{15.17}$$

In the notation of (15.2) we have $y_a = y$, $y_b = y'$, $f_a(y_a, y_b) = y_b$, $f_b(y_a, y_b) = f(y_a, y_b)$. The special structure of f_a implies that only P-trees which satisfy the condition (see Definition 14.2)

"meagre vertices have at most one son and this son has to be fat"   (15.18) have to be considered. The essential P-trees are thus

$$TN_q^a = \{t \in TP_q^a \; ; \; t \text{ satisfies (15.18)}\}$$
$$TN_q^b = \{t \in TP_q^b \; ; \; t \text{ satisfies (15.18)}\}.$$

It follows that each element of TN_{q+1}^a can be written as $t = {}_a[u]$ with $u \in TN_q^b$. This implies a one-to-one correspondence between TN_{q+1}^a and TN_q^b, leaving the elementary differentials invariant:

$$F({}_a[u])(y_a, y_b) = \frac{\partial y_b}{\partial y_b} \cdot F(u)(y_a, y_b) = F(u)(y_a, y_b).$$

From this property it follows that

$$hP_b(c, (y_a, y_b)) = P_a(c', (y_a, y_b)) \tag{15.19}$$

where $c'(\emptyset_a) = 0$, $c'(\tau_a) = c(\emptyset_b)$ and

$$c'(t) = \varrho(t)c(u) \qquad \text{if } t = {}_a[u]. \tag{15.20}$$

This notation is in agreement with (15.10).

The order conditions of method (14.13) can now be derived as follows: assume g_i, g_i' to be P-series

$$g_i = P_a(c_i, (y_0, y_0')), \qquad g_i' = P_b(c_i, (y_0, y_0')).$$

Theorem 15.8 then implies that

$$hf(g_i, g_i') = P_b(c_i', (y_0, y_0')). \tag{15.21}$$

Multiplying this relation by h it follows from (15.19) that

$$h^2 f(g_i, g_i') = P_a(c_i'', (y_0, y_0')). \tag{15.22}$$

Here $c_i'' = (c_i')'$, i.e.,

$$c_i''(t) = 0 \qquad \text{for } t = \emptyset_a \text{ and } t = \tau_a, \qquad c_i''({}_a[\tau_b]) = 1,$$
$$c_i''(t) = \varrho(t)(\varrho(t) - 1)c_i(t_1)\dots c_i(t_m) \qquad \text{if } t = {}_a[{}_b[t_1, \dots, t_m]].$$

The relations (15.21) and (15.22), when inserted into (14.13), yield

$$c_i(\tau_a) = c_i,$$
$$c_i(t) = \begin{cases} \sum_j \bar{a}_{ij} c_j''(t) & \text{if the root of } t \text{ is meagre}, \\ \sum_j a_{ij} c_j'(t) & \text{if the root of } t \text{ is fat}. \end{cases}$$

Finally, a comparison of the P-series for the exact and numerical solutions gives the order conditions (for order p)

$$\sum_i \bar{b}_i c_i''(t) = 1 \qquad \text{for } t \in TN_q^a, \; q = 2, \dots, p$$

$$\sum_i b_i c_i'(t) = 1 \qquad \text{for } t \in TN_q^b, \; q = 1, \dots, p. \tag{15.23}$$

Exercises

1. Denote the number of elements of TP_q^a (P-trees with meagre root of order q) by α_q (see Table 15.2). Prove that

$$\alpha_1 + \alpha_2 x + \alpha_3 x^2 + \ldots = (1 - x)^{-2\alpha_1}(1 - x^2)^{-2\alpha_2}(1 - x^3)^{-2\alpha_3} \cdots .$$

Compute the first α_q and compare them with the a_q of Table 2.1.

Table 15.2. Number of elements of TP_q^a

q	1	2	3	4	5	6	7	8	9	10
α_q	1	2	7	26	107	458	2058	9498	44947	216598

2. There is no explicit, 4-stage Runge-Kutta method of order 4, which does not satisfy condition (1.9).

   Hint. Use the techniques of the proof of Lemma 1.4.

3. Show that the order conditions (15.23) are the same as those given in Theorem 14.10.

4. Show that the partitioned method of Griepentrog (1978)

$$
\begin{array}{c|ccc}
0 & & a_{ij} & \\
1/2 & 1/2 & & \\
1 & -1 & 2 & \\
\hline
& 1/6 & 2/3 & 1/6
\end{array}
\qquad
\begin{array}{c|ccc}
0 & 0 & & \widehat{a}_{ij} \\
1/2 & -\beta/2 & (1+\beta)/2 & \\
1 & (3+5\beta)/2 & -(1+3\beta) & (1+\beta)/2 \\
\hline
& 1/6 & 2/3 & 1/6
\end{array}
$$

   with $\beta = \sqrt{3}/3$ is of order 3 (the implicit method to the right is A-stable and is provided for the stiff part of the problem).

II.16 Symplectic Integration Methods

> It is natural to look forward to those discrete systems which preserve as much as possible the intrinsic properties of the continuous system.
>
> (Feng Kang 1985)

> Y.V. Rakitskii proposed ... a requirement of the most complete conformity between two dynamical systems: one resulting from the original differential equations and the other resulting from the difference equations of the computational method.
>
> (Y.B. Suris 1989)

Hamiltonian systems, given by

$$\dot{p}_i = -\frac{\partial H}{\partial q_i}(p, q), \qquad \dot{q}_i = \frac{\partial H}{\partial p_i}(p, q), \tag{16.1}$$

have been seen to possess two remarkable properties:

a) the solutions preserve the Hamiltonian $H(p, q)$ (Ex. 5 of Section I.6);

b) the corresponding flow is symplectic, i.e., preserves the differential 2-form

$$\omega^2 = \sum_{i=1}^{n} dp_i \wedge dq_i \tag{16.2}$$

(see Theorem I.14.12). In particular, the flow is volume preserving.

Both properties are usually destroyed by a numerical method applied to (16.1).

After some pioneering papers (de Vogelaere 1956, Ruth 1983, and Feng Kang (冯康 1985) an enormous avalanche of research started around 1988 on the characterization of existing numerical methods which preserve symplecticity or on the construction of new classes of symplectic methods. An excellent overview is presented by Sanz-Serna (1992).

Example 16.1. We consider the harmonic oscillator

$$H(p, q) = \frac{1}{2}\left(p^2 + k^2 q^2\right). \tag{16.3}$$

Here (16.1) becomes

$$\dot{p} = -k^2 q, \qquad \dot{q} = p \tag{16.4}$$

and we study the action of several steps of a numerical method on a well-known set of initial data (p_0, q_0) (see Fig. 16.1):

a) The explicit Euler method (I.7.3)

$$\begin{pmatrix} p_m \\ q_m \end{pmatrix} = \begin{pmatrix} 1 & -hk^2 \\ h & 1 \end{pmatrix} \begin{pmatrix} p_{m-1} \\ q_{m-1} \end{pmatrix}, \qquad h = \frac{\pi}{8k}, \; m = 1, \ldots, 16; \tag{16.5a}$$

b) the implicit (or backward) Euler method (7.3)

$$\begin{pmatrix} p_m \\ q_m \end{pmatrix} = \frac{1}{1+h^2k^2} \begin{pmatrix} 1 & -hk^2 \\ h & 1 \end{pmatrix} \begin{pmatrix} p_{m-1} \\ q_{m-1} \end{pmatrix}, \quad h = \frac{\pi}{8k}, \; m = 1, \ldots, 16;$$

(16.5b)

c) Runge's method (1.4) of order 2

$$\begin{pmatrix} p_m \\ q_m \end{pmatrix} = \begin{pmatrix} 1 - \frac{h^2k^2}{2} & -hk^2 \\ h & 1 - \frac{h^2k^2}{2} \end{pmatrix} \begin{pmatrix} p_{m-1} \\ q_{m-1} \end{pmatrix}, \quad h = \frac{\pi}{4k}, \; m = 1, \ldots, 8;$$

(16.5c)

d) the implicit midpoint rule (7.4) of order 2

$$\begin{pmatrix} p_m \\ q_m \end{pmatrix} = \frac{1}{1+\frac{h^2k^2}{4}} \begin{pmatrix} 1 - \frac{h^2k^2}{4} & -hk^2 \\ h & 1 - \frac{h^2k^2}{4} \end{pmatrix} \begin{pmatrix} p_{m-1} \\ q_{m-1} \end{pmatrix}, \quad h = \frac{\pi}{4k}, \; m = 1, \ldots, 8.$$

(16.5d)

For the exact flow, the last of all these cats would precisely coincide with the first one and all cats would have the same area. Only the last method appears to be area preserving. It also preserves the Hamiltonian in this example.

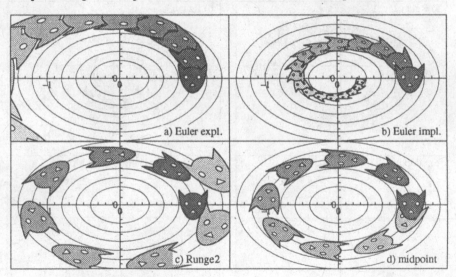

Fig. 16.1. Destruction of symplecticity of a Hamiltonian flow, $k = (\sqrt{5}+1)/2$

Example 16.2. For a nonlinear problem we choose

$$H(p,q) = \frac{p^2}{2} - \cos(q)\left(1 - \frac{p}{6}\right) \tag{16.6}$$

which is similar to the Hamiltonian of the pendulum (I.14.25), but with some of the pendulum's symmetry destroyed. Fig. 16.2 presents 12000 consecutive solution values (p_i, q_i) for

a) Runge's method of order 2 (see (1.4));

b) the implicit Radau method with $s = 2$ and order 3 (see Exercise 6 of Section II.7);

c) the implicit midpoint rule (7.4) of order 2.

The initial values are

$$p_0 = 0, \qquad q_0 = \begin{cases} \arccos(0.5) = \pi/3 & \text{for case (a)} \\ \arccos(-0.8) & \text{for cases (b) and (c).} \end{cases}$$

The computation is done with fixed step sizes

$$h = \begin{cases} 0.15 & \text{for case (a)} \\ 0.3 & \text{for cases (b) and (c).} \end{cases}$$

The solution of method (a) spirals out, that of method (b) spirals in and both by no means preserve the Hamiltonian. Method (c) behaves differently. Although the Hamiltonian is not precisely preserved (see picture (d)), its error remains bounded for long-scale computations.

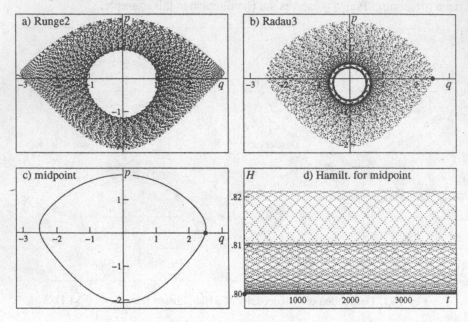

Fig. 16.2. A nonlinear pendulum and behaviour of H
($\bullet \ldots$ indicates the initial position)

Symplectic Runge-Kutta Methods

For a given Hamiltonian system (16.1), for a chosen one-step method (in particular a Runge-Kutta method) and a chosen step size h we denote by

$$\psi_h : \quad \begin{array}{ccc} \mathbf{R}^{2n} & \longrightarrow & \mathbf{R}^{2n} \\ (p_0, q_0) & \longmapsto & (p_1, q_1) \end{array} \tag{16.7}$$

the transformation defined by the method.

Remark. For implicit methods the numerical solution (p_1, q_1) need not exist for all h and all initial values (p_0, q_0) nor need it be uniquely determined (see Exercise 2). Therefore we usually will have to restrict the domain where ψ_h is defined and we will have to select a solution of the nonlinear system such that ψ_h is differentiable on this domain. The subsequent results hold for all possible choices of ψ_h.

Definition 16.4. A one-step method is called *symplectic* if for every smooth Hamiltonian H and for every step size h the mapping ψ_h is symplectic (see Definition I.14.11), i.e., preserves the differential 2-form ω^2 of (16.2).

We start with the easiest result.

Theorem 16.5. *The implicit s-stage Gauss methods of order $2s$ (Kuntzmann & Butcher methods of Section II.7) are symplectic for all s.*

Proof. We simplify the notation by putting $h = 1$ and $t_0 = 0$ and use the fact that the methods under consideration are collocation methods, i.e., the numerical solution after one step is defined by $(u(1), v(1))$ where $(u(t), v(t))$ are polynomials of degree s such that

$$u(0) = p_0, \quad u'(c_i) = -\frac{\partial H}{\partial q}(u(c_i), v(c_i))$$
$$\qquad\qquad\qquad\qquad\qquad\qquad\qquad\qquad i = 1, \ldots, s. \tag{16.8}$$
$$v(0) = q_0, \quad v'(c_i) = \frac{\partial H}{\partial p}(u(c_i), v(c_i))$$

The polynomials $u(t)$ and $v(t)$ are now considered as functions of the initial values. For arbitrary variations ξ_1^0 and ξ_2^0 of the initial point we denote the corresponding variations of u and v as

$$\xi_1^t = \frac{\partial(u(t), v(t))}{\partial(p_0, q_0)} \cdot \xi_1^0, \qquad \xi_2^t = \frac{\partial(u(t), v(t))}{\partial(p_0, q_0)} \cdot \xi_2^0.$$

Symplecticity of the method means that the expression

$$\omega^2(\xi_1^1, \xi_2^1) - \omega^2(\xi_1^0, \xi_2^0) = \int_0^1 \frac{d}{dt} \omega^2(\xi_1^t, \xi_2^t) \, dt \tag{16.9}$$

should vanish. Since ξ_1^t and ξ_2^t are polynomials in t of degree s, the expression $\frac{d}{dt} \omega^2(\xi_1^t, \xi_2^t)$ is a polynomial of degree $2s - 1$. We can thus exactly integrate

(16.9) by the Gaussian quadrature formula and so obtain

$$\omega^2(\xi_1^1, \xi_2^1) - \omega^2(\xi_1^0, \xi_2^0) = \sum_{i=1}^{s} b_i \frac{d}{dt} \omega^2(\xi_1^t, \xi_2^t)\Big|_{t=c_i}. \tag{16.9'}$$

Differentiation of (16.8) with respect to (p_0, q_0) shows that (ξ_1^t, ξ_2^t) satisfies the variational equation (I.14.27) at the collocation points $t = c_i$, $i = 1, \ldots, s$. Therefore, the computations of the proof of Theorem I.14.12 imply that

$$\frac{d}{dt} \omega^2(\xi_1^t, \xi_2^t)\Big|_{t=c_i} = 0 \qquad \text{for} \quad i = 1, \ldots, s. \tag{16.10}$$

This, introduced into (16.9'), completes the proof of symplecticity. $\square$

The following theorem, discovered independently by at least three authors (F. Lasagni 1988, J.M. Sanz-Serna 1988, Y.B. Suris 1989) characterizes the class of all symplectic Runge-Kutta methods:

Theorem 16.6. *If the $s \times s$ matrix M with elements*

$$m_{ij} = b_i a_{ij} + b_j a_{ji} - b_i b_j, \qquad i, j = 1, \ldots, s \tag{16.11}$$

satisfies $M = 0$, then the Runge-Kutta method (7.7) is symplectic.

Proof. The matrix M has been known from nonlinear stability theory for many years (see Theorem IV.12.4). Both theorems have very similar proofs, the one works with the *inner* product, the other with the *exterior* product.

We write method (7.7) applied to problem (16.1) as

$$P_i = p_0 + h \sum_j a_{ij} k_j \qquad Q_i = q_0 + h \sum_j a_{ij} \ell_j \tag{16.12a}$$

$$p_1 = p_0 + h \sum_i b_i k_i \qquad q_1 = q_0 + h \sum_i b_i \ell_i \tag{16.12b}$$

$$k_i = -\frac{\partial H}{\partial q}(P_i, Q_i) \qquad \ell_i = \frac{\partial H}{\partial p}(P_i, Q_i), \tag{16.12c}$$

denote the Jth component of a vector by an upper index J and introduce the linear maps (one-forms)

$$\begin{aligned}
dp_1^J &: \mathbf{R}^{2n} \to \mathbf{R}, & dP_i^J &: \mathbf{R}^{2n} \to \mathbf{R}, \\
\xi &\mapsto \frac{\partial p_1^J}{\partial (p_0, q_0)} \xi & \xi &\mapsto \frac{\partial P_i^J}{\partial (p_0, q_0)} \xi
\end{aligned} \tag{16.13}$$

and similarly also dp_0^J, dk_i^J, dq_0^J, dq_1^J, dQ_i^J, $d\ell_i^J$ (the one-forms dp_0^J and dq_0^J correspond to dp_J and dq_J of Section I.14). Using the notation (16.13), symplecticity of the method is equivalent to

$$\sum_{J=1}^{n} dp_1^J \wedge dq_1^J = \sum_{J=1}^{n} dp_0^J \wedge dq_0^J. \tag{16.14}$$

To check this relation we differentiate (16.12) with respect to the initial values and obtain

$$dP_i^J = dp_0^J + h \sum_j a_{ij} dk_j^J \qquad dQ_i^J = dq_0^J + h \sum_j a_{ij} d\ell_j^J \tag{16.15a}$$

$$dp_1^J = dp_0^J + h \sum_i b_i dk_i^J \qquad dq_1^J = dq_0^J + h \sum_i b_i d\ell_i^J \tag{16.15b}$$

$$dk_i^J = -\sum_{L=1}^{n} \frac{\partial^2 H}{\partial q^J \partial p^L}(P_i, Q_i) \cdot dP_i^L - \sum_{L=1}^{n} \frac{\partial^2 H}{\partial q^J \partial q^L}(P_i, Q_i) \cdot dQ_i^L \tag{16.15c}$$

$$d\ell_i^J = \sum_{L=1}^{n} \frac{\partial^2 H}{\partial p^J \partial p^L}(P_i, Q_i) \cdot dP_i^L + \sum_{L=1}^{n} \frac{\partial^2 H}{\partial p^J \partial q^L}(P_i, Q_i) \cdot dQ_i^L. \tag{16.15d}$$

We now compute

$$dp_1^J \wedge dq_1^J - dp_0^J \wedge dq_0^J \tag{16.16}$$

$$= h \sum_i b_i\, dp_0^J \wedge d\ell_i^J + h \sum_i b_i\, dk_i^J \wedge dq_0^J + h^2 \sum_{i,j} b_i b_j\, dk_i^J \wedge d\ell_j^J$$

by using (16.15b) and the multilinearity of the wedge product. This formula corresponds precisely to (IV.12.6). Exactly as in the proof of Theorem IV.12.5, we now eliminate in (16.16) the quantities dp_0^J and dq_0^J with the help of (16.15a) to obtain

$$dp_1^J \wedge dq_1^J - dp_0^J \wedge dq_0^J \tag{16.17}$$

$$= h \sum_i b_i\, dP_i^J \wedge d\ell_i^J + h \sum_i b_i\, dk_i^J \wedge dQ_i^J - h^2 \sum_{i,j} m_{ij}\, dk_i^J \wedge d\ell_j^J,$$

the formula analogous to (IV.12.7). Equations (16.15c,d) are perfect analogues of the variational equation (I.14.27). Therefore the same computations as in (I.14.39) give

$$\sum_{J=1}^{n} dP_i^J \wedge d\ell_i^J + \sum_{J=1}^{n} dk_i^J \wedge dQ_i^J = 0 \tag{16.18}$$

and the first two terms in (16.17) disappear. The last term vanishes by hypothesis (16.11) and we obtain (16.14).   □

Remark. F. Lasagni (1990) has proved in an unpublished manuscript that for *irreducible* methods (see Definitions IV.12.15 and IV.12.17) the condition $M = 0$ is also *necessary* for symplecticity. For a publication see Abia & Sanz-Serna (1993,

Theorem 5.1), where this proof has been elaborated and adapted to a more general setting.

Remarks. a) Explicit Runge-Kutta methods are never symplectic (Ex. 1).

b) Equations (16.11) imply a substantial simplification of the order conditions (Sanz-Serna & Abia 1991). We shall return to this when treating partitioned methods (see (16.40)).

c) An important tool for the construction of symplectic methods is the W-transformation (see Section IV.5, especially Theorem IV.5.6). As can be seen from formula (IV.12.10), the method under consideration is symplectic if and only if the matrix X is skew-symmetric (with the exception of $x_{11} = 1/2$). Sun Geng (孙耿 1992) constructed several new classes of symplectic Runge-Kutta methods. One of his methods, based on Radau quadrature, is given in Table 16.1.

d) An inspection of Table IV.5.14 shows that all Radau IA, Radau IIA, Lobatto IIIA (in particular the trapezoidal rule), and Lobatto IIIC methods are not symplectic.

Table 16.1. 孙's symplectic Radau method of order 5

$$
\begin{array}{c|ccc}
\dfrac{4-\sqrt{6}}{10} & \dfrac{16-\sqrt{6}}{72} & \dfrac{328-167\sqrt{6}}{1800} & \dfrac{-2+3\sqrt{6}}{450} \\[2ex]
\dfrac{4+\sqrt{6}}{10} & \dfrac{328+167\sqrt{6}}{1800} & \dfrac{16+\sqrt{6}}{72} & \dfrac{-2-3\sqrt{6}}{450} \\[2ex]
1 & \dfrac{85-10\sqrt{6}}{180} & \dfrac{85+10\sqrt{6}}{180} & \dfrac{1}{18} \\[2ex]
\hline
 & \dfrac{16-\sqrt{6}}{36} & \dfrac{16+\sqrt{6}}{36} & \dfrac{1}{9}
\end{array}
$$

Preservation of the Hamiltonian and of first integrals. In Exercise 5 of Section I.6 we have seen that the Hamiltonian $H(p,q)$ is a *first integral* of the system (16.1). This means that every solution $p(t), q(t)$ of (16.1) satisfies $H\big(p(t), q(t)\big) = Const.$ The numerical solution of a symplectic integrator does not share this property in general (see Fig. 16.2). However, we will show that every *quadratic* first integral will be preserved.

Denote $y = (p,q)$ and let G be a symmetric $2n \times 2n$ matrix. We suppose that the quadratic functional

$$\langle y, y \rangle_G := y^T G y$$

is a first integral of the system (16.1). This means that

$$\langle y, J^{-1} \operatorname{grad} H(y) \rangle_G = 0 \qquad \text{with} \qquad J = \begin{pmatrix} 0 & I \\ -I & 0 \end{pmatrix} \qquad (16.19)$$

for all $y \in \mathbf{R}^{2n}$.

Theorem 16.7 (Sanz-Serna 1988). *A symplectic Runge-Kutta method (i.e., a method satisfying (16.11)) leaves all quadratic first integrals of the system (16.1) invariant, i.e., the numerical solution $y_n = (p_n, q_n)$ satisfies*

$$\langle y_1, y_1 \rangle_G = \langle y_0, y_0 \rangle_G \tag{16.20}$$

for all symmetric matrices G satisfying (16.19).

Proof (Cooper 1987). The Runge-Kutta method (7.7) applied to problem (16.1) is given by

$$y_1 = y_0 + \sum_i b_i k_i, \qquad Y_i = y_0 + \sum_j a_{ij} k_j,$$

$$k_i = J^{-1} \operatorname{grad} H(Y_i). \tag{16.21}$$

As in the proof of Theorem 16.6 (see also Theorem IV.12.4) we obtain

$$\langle y_1, y_1 \rangle_G - \langle y_0, y_0 \rangle_G = 2h \sum_i b_i \langle Y_i, k_i \rangle_G - h^2 \sum_{i,j} m_{ij} \langle k_i, k_j \rangle_G.$$

The first term on the right-hand side vanishes by (16.19) and the second one by (16.11). □

An Example from Galactic Dynamics

> Always majestic, usually spectacularly beautiful, galaxies are ...
> (Binney & Tremaine 1987)

While the theoretical meaning of symplecticity of numerical methods is clear, its importance for practical computations is less easy to understand. Numerous numerical experiments have shown that symplectic methods, in a fixed step size mode, show an excellent behaviour for long-scale scientific computations of Hamiltonian systems. We shall demonstrate this on the following example chosen from galactic dynamics and give a theoretical justification later in this section. However, Calvo & Sanz-Serna (1992c) have made the interesting discovery that *variable step size* implementation can *destroy* the advantages of symplectic methods. In order to illustrate this phenomenon we shall include in our computations violent step changes; one with a random number generator and one with the step size changing in function of the solution position.

A galaxy is a set of N stars which are mutually attracted by Newton's law. A relatively easy way to study them is to perform a long-scale computation of the orbit of *one* of its stars in the potential formed by the $N-1$ remaining ones (see Binney & Tremaine 1987, Chapter 3); this potential is assumed to perform a uniform rotation with time, but not to change otherwise. The potential is determined

Fig. 16.3. Galactic orbit

by Poisson's differential equation $\Delta V = 4G\pi\varrho$, where ϱ is the density distribution of the galaxy, and real-life potential-density pairs are difficult to obtain (e.g., de Zeeuw & Pfenniger 1988). A popular issue is to choose a simple formula for V in such a way that the resulting ϱ corresponds to a reasonable galaxy, for example (Binney 1981, Binney & Tremaine 1987, p. 45f, Pfenniger 1990)

$$V = A \ln\left(C + \frac{x^2}{a^2} + \frac{y^2}{b^2} + \frac{z^2}{c^2}\right). \tag{16.22}$$

The Lagrangian for a coordinate system rotating with angular velocity Ω becomes

$$\mathcal{L} = \frac{1}{2}\left((\dot{x} - \Omega y)^2 + (\dot{y} + \Omega x)^2 + \dot{z}^2\right) - V(x, y, z). \tag{16.23}$$

This gives with the coordinates (see (I.6.23))

$$p_1 = \frac{\partial \mathcal{L}}{\partial \dot{x}} = \dot{x} - \Omega y, \qquad p_2 = \frac{\partial \mathcal{L}}{\partial \dot{y}} = \dot{y} + \Omega x, \qquad p_3 = \frac{\partial \mathcal{L}}{\partial \dot{z}} = \dot{z},$$

$$q_1 = x, \qquad\qquad\qquad q_2 = y, \qquad\qquad\qquad q_3 = z,$$

the Hamiltonian

$$H = p_1 \dot{q}_1 + p_2 \dot{q}_2 + p_3 \dot{q}_3 - \mathcal{L} \tag{16.24}$$

$$= \frac{1}{2}\left(p_1^2 + p_2^2 + p_3^2\right) + \Omega\left(p_1 q_2 - p_2 q_1\right) + A \ln\left(C + \frac{q_1^2}{a^2} + \frac{q_2^2}{b^2} + \frac{q_3^2}{c^2}\right).$$

We choose the parameters and initial values as

$$a = 1.25, \quad b = 1, \quad c = 0.75, \quad A = 1, \quad C = 1, \quad \Omega = 0.25,$$

$$q_1(0) = 2.5, \quad q_2(0) = 0, \quad q_3(0) = 0, \quad p_1(0) = 0, \quad p_3(0) = 0.2, \tag{16.25}$$

and take for $p_2(0)$ the larger of the roots for which $H = 2$. Our star then sets out for its voyage through the galaxy, the orbit is represented in Fig. 16.3 for $0 \leq t \leq 15000$. We are interested in its Poincaré sections with the half-plane $q_2 = 0$, $q_1 > 0$, $\dot{q}_2 > 0$ for $0 \leq t \leq 1000000$. These consist, for the exact solution, in 47101 cut points which are presented in Fig. 16.6l. These points were computed with the (non-symplectic) code DOP853 with $Tol = 10^{-17}$ in quadruple precision on a VAX 8700 computer.

Fig. 16.4, Fig. 16.5, and Fig. 16.6 present the obtained numerical results for the methods and step sizes summarized in Table 16.2.

Table 16.2. Methods for numerical experiments

item	method	order	h	points $t < 1000000$	impl.	symplec.	symmet.
a)	Gauss	6	1/5	47093	yes	yes	yes
b)	"	"	2/5	46852	"	"	"
c)	Gauss	6	random	46703	yes	yes	yes
d)	Gauss	6	partially halved	46563	yes	yes	yes
e)	Radau	5	1/10	46597	yes	no	no
f)	"	"	1/5	46266	"	"	"
g)	RK44	4	1/40	47004	no	no	no
h)	"	"	1/10	46192	"	"	"
i)	Lobatto	6	1/5	47091	yes	no	yes
j)	"	"	2/5	46839	"	"	"
k)	Sun Geng	5	1/5	47093	yes	yes	no
l)	exact	–	–	47101	–	–	–

Remarks.

ad a): the Gauss6 method (Kuntzmann & Butcher method based on Gaussian quadrature with $s = 3$ and $p = 6$, see Table 7.4) for $h = 1/5$ is nearly identical to the exact solution;

ad b): Gauss6 for $h = 2/5$ is much better than Gauss6 with random or partially halved step sizes (see item (c) and (d)) where $h \leq 2/5$.

ad c): h was chosen at random uniformly distributed on $(0, 2/5)$;

ad d): h was chosen "partially halved" in the sense that

$$h = \begin{cases} 2/5 & \text{if } q_1 > 0, \\ 1/5 & \text{if } q_1 < 0. \end{cases}$$

This produced the worst result for the 6th order Gauss method. We thus

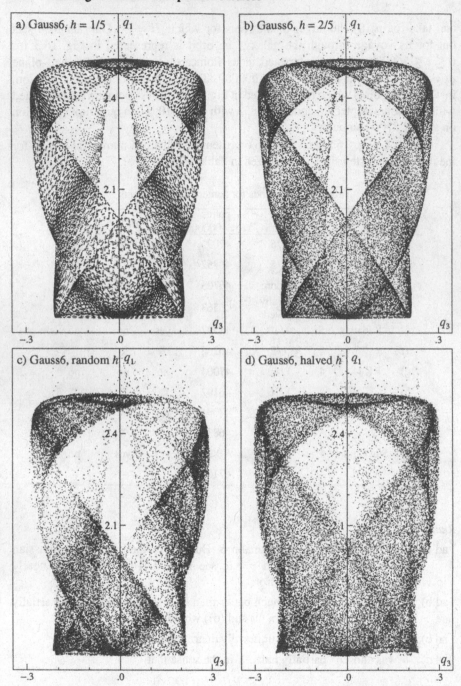

Fig. 16.4. Poincaré cuts for $0 \le t \le 1000000$; methods (a)-(d)

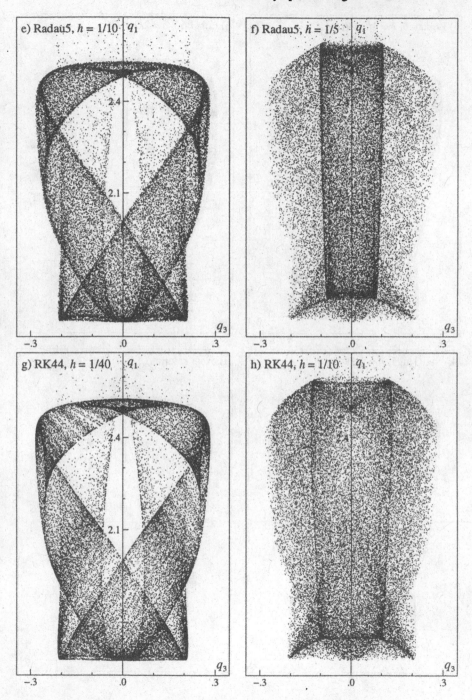

Fig. 16.5. Poincaré cuts for $0 \leq t \leq 1000000$; methods (e)-(h)

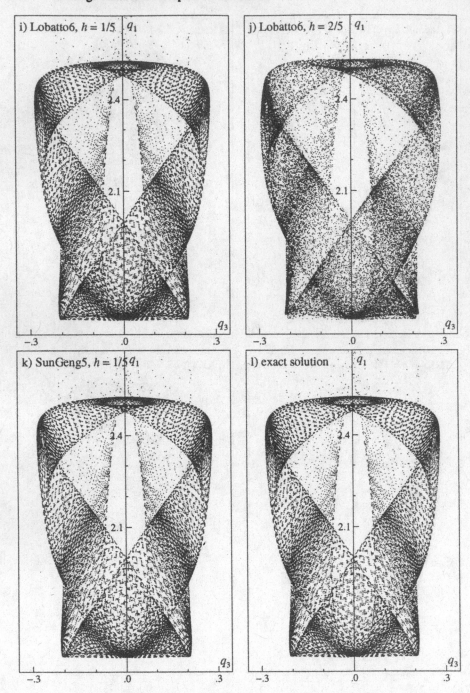

Fig. 16.6. Poincaré cuts for $0 \leq t \leq 1000000$; methods (i)-(l)

see that symplectic and symmetric methods compensate on the way back the errors committed on the outward journey.

ad e), f): Radau5 (method of Ehle based on Radau quadrature with $s = 3$ and $p = 5$, see Table 7.7) is here not at all satisfactory;

ad g): The explicit method RK44 (Runge-Kutta method with $s = p = 4$, see Table 1.2, left) is evidently much faster than the implicit methods, even with a smaller step size;

ad h): With increasing step size RK44 deteriorates drastically;

ad i): this is a non-symplectic but symmetric collocation method based on Lobatto quadrature with $s = 4$ of order 6 (see Table IV.5.8); its good performance on this nonlinear Hamiltonian problem is astonishing;

ad j): with increasing h Lobatto6 is less satisfactory (see also Fig. 16.7);

ad k): this is the symplectic non-symmetric method based on Radau quadrature of order 5 due to Sun Geng 孙耿 (Table 16.1).

The preservation of the Hamiltonian (correct value $H = 2$) during the computation for $0 \le t \le 1000000$ is shown in Fig. 16.7. While the errors for the symplectic and symmetric methods in constant step size mode remain bounded, random h (case c) results in a sort of Brownian motion, and the nonsymplectic methods as well as Gauss6 with partially halved step size result in permanent deterioration.

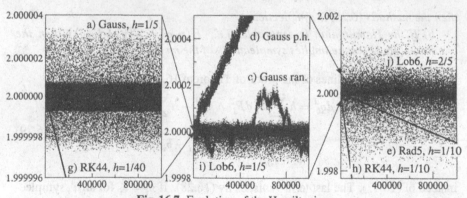

Fig. 16.7. Evolution of the Hamiltonian

Partitioned Runge-Kutta Methods

The fact that the system (16.1) possesses a natural partitioning suggests the use of partitioned Runge-Kutta methods as discussed in Section II.15. The main interest of such methods is for separable Hamiltonians where it is possible to obtain explicit symplectic methods.

A partitioned Runge-Kutta method for system (16.1) is defined by

$$P_i = p_0 + h\sum_j a_{ij}k_j \qquad Q_i = q_0 + h\sum_j \widehat{a}_{ij}\ell_j \qquad (16.26a)$$

$$p_1 = p_0 + h\sum_i b_i k_i \qquad q_1 = q_0 + h\sum_i \widehat{b}_i \ell_i \qquad (16.26b)$$

$$k_i = -\frac{\partial H}{\partial q}(P_i, Q_i) \qquad \ell_i = \frac{\partial H}{\partial p}(P_i, Q_i) \qquad (16.26c)$$

where b_i, a_{ij} and $\widehat{b}_i, \widehat{a}_{ij}$ represent two different Runge-Kutta schemes.

Theorem 16.10 (Sanz-Serna 1992b, Suris 1990). *a) If the coefficients of (16.26) satisfy*

$$b_i = \widehat{b}_i, \qquad i = 1,\ldots,s \qquad (16.27)$$

$$b_i\widehat{a}_{ij} + \widehat{b}_j a_{ji} - b_i\widehat{b}_j = 0, \qquad i,j = 1,\ldots,s \qquad (16.28)$$

then the method (16.26) is symplectic.

b) If the Hamiltonian is separable (i.e., $H(p,q) = T(p) + U(q)$) then the condition (16.28) alone implies symplecticity of the method.

Proof. Following the lines of the proof of Theorem 16.6 we obtain

$$dp_1^J \wedge dq_1^J - dp_0^J \wedge dq_0^J = h\sum_i \widehat{b}_i \, dP_i^J \wedge d\ell_i^J + h\sum_i b_i \, dk_i^J \wedge dQ_i^J$$
$$- h^2 \sum_{i,j} (b_i\widehat{a}_{ij} + \widehat{b}_j a_{ji} - b_i\widehat{b}_j) \, dk_i^J \wedge d\ell_j^J, \qquad (16.29)$$

instead of (16.17). The last term vanishes by (16.28). If $b_i = \widehat{b}_i$ for all i, symplecticity of the method follows from (16.18). If the Hamiltonian is separable (the mixed derivatives $\partial^2 H/\partial q^J \partial p^L$ and $\partial^2 H/\partial p^J \partial q^L$ are not present in (16.15c,d)) then each of the two terms in (16.18) vanishes separately and the method is symplectic without imposing (16.27). □

Remark. If (16.28) is satisfied and if the Hamiltonian is separable, it can be assumed without loss of generality that

$$b_i \neq 0, \quad \widehat{b}_i \neq 0 \qquad \text{for all } i. \qquad (16.30)$$

Indeed, the stage values P_i (for i with $\widehat{b}_i = 0$) and Q_j (for j with $b_j = 0$) don't influence the numerical solution (p_1, q_1) and can be removed from the scheme. Notice however that in the resulting scheme the number of stages P_i may be different from that of Q_j.

Explicit methods for separable Hamiltonians. Let the Hamiltonian be of the form $H(p, q) = T(p) + U(q)$ and consider a partitioned Runge-Kutta method satisfying

$$a_{ij} = 0 \quad \text{for} \quad i < j \quad \text{(diagonally implicit)}$$
$$\widehat{a}_{ij} = 0 \quad \text{for} \quad i \leq j \quad \text{(explicit)}. \tag{16.31}$$

Since $\partial H / \partial q$ depends only on q, the method (16.26) is explicit for such a choice of coefficients. Under the assumption (16.30), the symplecticity condition (16.28) then becomes

$$a_{ij} = b_j \quad \text{for} \quad i \geq j, \qquad \widehat{a}_{ij} = \widehat{b}_j \quad \text{for} \quad i > j, \tag{16.32}$$

so that the method (16.26) is characterized by the two schemes

$$\begin{array}{|llll}
b_1 & & & \\
b_1 & b_2 & & \\
b_1 & b_2 & b_3 & \\
\vdots & \vdots & \ddots & \ddots \\
b_1 & b_2 & \cdots & b_{s-1} \; b_s \\
\hline
b_1 & b_2 & \cdots & b_{s-1} \; b_s
\end{array}
\qquad
\begin{array}{|llll}
0 & & & \\
\widehat{b}_1 & 0 & & \\
\widehat{b}_1 & \widehat{b}_2 & 0 & \\
\vdots & \vdots & \ddots & \ddots \\
\widehat{b}_1 & \widehat{b}_2 & \cdots & \widehat{b}_{s-1} \; 0 \\
\hline
\widehat{b}_1 & \widehat{b}_2 & \cdots & \widehat{b}_{s-1} \; \widehat{b}_s
\end{array} \tag{16.33}$$

If we admit the cases $b_1 = 0$ and/or $\widehat{b}_s = 0$, it can be shown (Exercise 6) that this scheme already represents the most general method (16.26) which is symplectic and explicit. We denote this scheme by

$$\begin{array}{lllll}
b: & b_1 & b_2 & \cdots & b_s \\
\widehat{b}: & \widehat{b}_1 & \widehat{b}_2 & \cdots & \widehat{b}_s.
\end{array} \tag{16.34}$$

This method is particularly easy to implement:

$$\begin{aligned}
& P_0 = p_0, \, Q_1 = q_0 \\
& \text{for } i := 1 \text{ to } s \text{ do} \\
& \qquad P_i = P_{i-1} - h b_i \partial U / \partial q(Q_i) \\
& \qquad Q_{i+1} = Q_i + h \widehat{b}_i \partial T / \partial p(P_i) \\
& p_1 = P_s, \, q_1 = Q_{s+1}
\end{aligned} \tag{16.35}$$

Special case $s = 1$. The combination of the implicit Euler method ($b_1 = 1$) with the explicit Euler method ($\widehat{b}_1 = 1$) gives the following symplectic method of order 1:

$$p_1 = p_0 - h \frac{\partial U}{\partial q}(q_0), \qquad q_1 = q_0 + h \frac{\partial T}{\partial p}(p_1). \tag{16.36a}$$

By interchanging the roles of p and q we obtain the method

$$q_1 = q_0 + h\frac{\partial T}{\partial p}(p_0), \qquad p_1 = p_0 - h\frac{\partial U}{\partial q}(q_1) \qquad (16.36b)$$

which is also symplectic. Methods (16.36a) and (16.36b) are mutually adjoint (see Section II.8).

Construction of higher order methods. The order conditions for general partitioned Runge-Kutta methods applied to general problems (15.2) are derived in Section II.15 (Theorem 15.9). Let us here discuss how these conditions simplify in our special situation.

A) We consider the system (16.1) with separable Hamiltonian. In the notation of Section II.15 this means that $f_a(y_a, y_b)$ depends only on y_b and $f_b(y_a, y_b)$ depends only on y_a. Therefore, many elementary differentials vanish and only P-trees whose meagre and fat vertices alternate in each branch have to be considered. This is a considerable reduction of the order conditions.

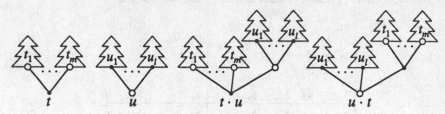

Fig. 16.8. Product of P-trees

B) As observed by Abia & Sanz-Serna (1993) the condition (16.28) acts as a simplifying assumption. Indeed, multiplying (16.28) by $\Phi_i(t) \cdot \Phi_j(u)$ (where $t = {}_a[t_1, \ldots, t_m] \in TP^a$, $u = {}_b[u_1, \ldots, u_l] \in TP^b$) and summing up over all i and j yields

$$\sum_i b_i \Phi_i(t \cdot u) + \sum_j \widehat{b}_j \Phi_j(u \cdot t) - \left(\sum_i b_i \Phi_i(t)\right)\left(\sum_j \widehat{b}_j \Phi_j(u)\right) = 0. \qquad (16.37)$$

Here we have used the notation of Butcher (1987)

$$t \cdot u = {}_a[t_1, \ldots, t_m, u], \qquad u \cdot t = {}_b[u_1, \ldots, u_l, t], \qquad (16.38)$$

illustrated in Fig. 16.8. Since

$$\frac{1}{\gamma(t \cdot u)} + \frac{1}{\gamma(u \cdot t)} - \frac{1}{\gamma(t)} \cdot \frac{1}{\gamma(u)} = 0 \qquad (16.39)$$

(this relation follows from (16.37) by inserting the coefficients of a symplectic Runge-Kutta method of sufficiently high order, e.g., a Gauss method) we obtain the following fact:

let $\varrho(t) + \varrho(u) = p$ *and assume that all order conditions for P-trees of order* $< p$ *are satisfied, then*

$$\sum_i b_i \Phi_i(t \cdot u) = \frac{1}{\gamma(t \cdot u)} \qquad iff \qquad \sum_j \widehat{b}_j \Phi_j(u \cdot t) = \frac{1}{\gamma(u \cdot t)}. \qquad (16.40)$$

From Fig. 16.8 we see that the P-trees $t \cdot u$ and $u \cdot t$ have the same geometrical structure. They differ only in the position of the root. Repeated application of this property implies that of all P-trees with identical geometrical structure only one has to be considered.

A method of order 3 (Ruth 1983). The above reductions leave five order conditions for a method of order 3 which, for $s = 3$, are the following:

$$b_1 + b_2 + b_3 = 1, \qquad \widehat{b}_1 + \widehat{b}_2 + \widehat{b}_3 = 1, \qquad b_2 \widehat{b}_1 + b_3(\widehat{b}_1 + \widehat{b}_2) = 1/2,$$

$$b_2 \widehat{b}_1^2 + b_3(\widehat{b}_1 + \widehat{b}_2)^2 = 1/3, \qquad \widehat{b}_1 b_1^2 + \widehat{b}_2(b_1 + b_2)^2 + \widehat{b}_3(b_1 + b_2 + b_3)^2 = 1/3.$$

This nonlinear system possesses many solutions. A particularly simple solution, proposed by Ruth (1983), is

$$\begin{array}{cccc} b: & 7/24 & 3/4 & -1/24 \\ \widehat{b}: & 2/3 & -2/3 & 1. \end{array} \qquad (16.41)$$

Concatenation of a method with its adjoint. The adjoint method of (16.26) is obtained by replacing h by $-h$ and by exchanging the roles of p_0, q_0 and p_1, q_1 (see Section II.8). This results in a partitioned Runge-Kutta method with coefficients (compare Theorem 8.3)

$$a_{ij}^* = b_{s+1-j} - a_{s+1-i,s+1-j}, \qquad b_i^* = b_{s+1-i},$$

$$\widehat{a}_{ij}^* = \widehat{b}_{s+1-j} - \widehat{a}_{s+1-i,s+1-j}, \qquad \widehat{b}_i^* = \widehat{b}_{s+1-i}.$$

For the adjoint of (16.33) the first method is explicit and the second one is diagonally implicit, but otherwise it has the same structure. Adding dummy stages, it becomes of the form (16.33) with coefficients

$$\begin{array}{cccccc} b^*: & 0 & b_s & b_{s-1} & \cdots & b_1 \\ \widehat{b}^*: & \widehat{b}_s & \widehat{b}_{s-1} & \cdots & \widehat{b}_1 & 0. \end{array} \qquad (16.42)$$

The following idea of Sanz-Serna (1992b) allows one to improve a method of odd order p: one considers the composition of method (16.33) (step size $h/2$) with its adjoint (again with step size $h/2$). The resulting method, which is represented by the coefficients

$$\begin{array}{ccccccccc} b_1/2 & b_2/2 & \cdots & b_{s-1}/2 & b_s/2 & b_s/2 & b_{s-1}/2 & \cdots & b_1/2 \\ \widehat{b}_1/2 & \widehat{b}_2/2 & \cdots & \widehat{b}_{s-1}/2 & \widehat{b}_s & \widehat{b}_{s-1}/2 & \cdots & \widehat{b}_1/2 & 0, \end{array}$$

is symmetric and therefore has an even order which is $\geq p+1$. Concatenating

Ruth's method (16.41) with its adjoint yields the fourth order method

$$
\begin{array}{ccccccc}
b: & 7/48 & 3/8 & -1/48 & -1/48 & 3/8 & 7/48 \\
\widehat{b}: & 1/3 & -1/3 & 1 & & -1/3 & 1/3 & 0.
\end{array}
\tag{16.43}
$$

Symplectic Nyström Methods

A frequent special case of a separable Hamiltonian $H(p,q) = T(p) + U(q)$ is when $T(p)$ is a quadratic functional $T(p) = p^T M p/2$ (with M a constant symmetric matrix). In this situation the Hamiltonian system becomes

$$
\dot{p} = -\frac{\partial U}{\partial q}(q), \qquad \dot{q} = Mp,
$$

which is equivalent to the second order equation

$$
\ddot{q} = -M\frac{\partial U}{\partial q}(q).
\tag{16.44}
$$

It is therefore natural to consider Nyström methods (Section II.14) which for the system (16.44) are given by

$$
Q_i = q_0 + c_i h \dot{q}_0 + h^2 \sum_j \bar{a}_{ij} k_j', \qquad k_j' = -M\frac{\partial U}{\partial q}(Q_j),
$$

$$
q_1 = q_0 + h\dot{q}_0 + h^2 \sum_i \bar{b}_i k_i', \qquad \dot{q}_1 = \dot{q}_0 + h\sum_i b_i k_i'.
$$

Replacing the variable $\dot{q}$ by Mp and k_i' by $M\ell_i$, this method reads

$$
Q_i = q_0 + c_i h M p_0 + h^2 \sum_{j=1}^{s} \bar{a}_{ij} M\ell_j, \qquad \ell_j = -\frac{\partial U}{\partial q}(Q_j),
$$

$$
q_1 = q_0 + hMp_0 + h^2 \sum_{i=1}^{s} \bar{b}_i M\ell_i, \qquad p_1 = p_0 + h\sum_{i=1}^{s} b_i \ell_i.
\tag{16.45}
$$

Theorem 16.11 (Suris 1989). *Consider the system (16.44) where M is a symmetric matrix. Then, the s-stage Nyström method (16.45) is symplectic if the following two conditions are satisfied:*

$$
\bar{b}_i = b_i(1-c_i), \qquad i = 1,\ldots,s
\tag{16.46a}
$$

$$
b_i(\bar{b}_j - \bar{a}_{ij}) = b_j(\bar{b}_i - \bar{a}_{ji}), \qquad i,j = 1,\ldots,s.
\tag{16.46b}
$$

Proof (Okunbor & Skeel 1992). As in the proof of Theorem 16.6 we differentiate the formulas (16.45) and compute

$$
dp_1^J \wedge dq_1^J - dp_0^J \wedge dq_0^J
$$

$$= h \sum_i b_i \, d\ell_i^J \wedge dq_0^J + h \sum_K M_{JK} \, dp_0^J \wedge dp_0^K \qquad (16.47)$$

$$+ h^2 \sum_i b_i \sum_K M_{JK} \, d\ell_i^J \wedge dp_0^K + h^2 \sum_i \bar{b}_i \sum_K M_{JK} \, dp_0^J \wedge d\ell_i^K$$

$$+ h^3 \sum_{i,j} b_i \bar{b}_j \sum_K M_{JK} \, d\ell_i^J \wedge d\ell_j^K .$$

Next we eliminate dq_0^J with the help of the differentiated equation of Q_i, sum over all J and so obtain

$$\sum_{J=1}^n dp_1^J \wedge dq_1^J - \sum_{J=1}^n dp_0^J \wedge dq_0^J$$

$$= h \sum_i b_i \sum_J d\ell_i^J \wedge dQ_i^J + h \sum_{J,K} M_{JK} \, dp_0^J \wedge dp_0^K$$

$$+ h^2 \sum_i (b_i - \bar{b}_i - b_i c_i) \sum_{J,K} M_{JK} \, d\ell_i^J \wedge dp_0^K$$

$$+ h^3 \sum_{i<j} (b_i \bar{b}_j - b_j \bar{b}_i - b_i \bar{a}_{ij} + b_j \bar{a}_{ji}) \sum_{J,K} M_{JK} \, d\ell_i^J \wedge d\ell_j^K .$$

The last two terms disappear by (16.46) whereas the first two terms vanish due to the symmetry of M and of the second derivatives of $U(q)$.   □

We have already encountered condition (16.46a) in Lemma 14.13. There, it was used as a simplifying assumption. It implies that only the order conditions for $\dot{q}_1$ have to be considered.

For Nyström methods satisfying both conditions of (16.46), one can assume without loss of generality that

$$b_i \neq 0 \qquad \text{for} \quad i = 1, \ldots, s. \qquad (16.48)$$

Let $I = \{i \mid b_i = 0\}$, then $\bar{b}_i = 0$ for $i \in I$ and $\bar{a}_{ij} = 0$ for $i \notin I$, $j \in I$. Hence, the stage values Q_i $(i \in I)$ don't influence the numerical result (p_1, q_1) and can be removed from the scheme.

Explicit methods. Our main interest is in methods which satisfy

$$\bar{a}_{ij} = 0 \qquad \text{for} \quad i \leq j. \qquad (16.49)$$

Under the assumption (16.48) the condition (16.46) then implies that the remaining coefficients are given by

$$\bar{a}_{ij} = b_j (c_i - c_j) \qquad \text{for} \quad i > j. \qquad (16.50)$$

In this situation we may also suppose that

$$c_i \neq c_{i-1} \qquad \text{for} \quad i = 2, 3, \ldots, s,$$

because equal consecutive c_i lead (via condition (16.50)) to equal stage values Q_i. Therefore the method is equivalent to one with a smaller number of stages.

The particular form of the coefficients $\bar{a}_{ij}$ allows the following simple implementation (Okunbor & Skeel 1992b)

$$Q_0 = q_0, \quad P_0 = p_0$$
for $i := 1$ to s do
$$Q_i = Q_{i-1} + h(c_i - c_{i-1})MP_{i-1} \qquad \text{(with } c_0 = 0) \qquad (16.51)$$
$$P_i = P_{i-1} - hb_i\partial U/\partial q(Q_i)$$
$$q_1 = Q_s + h(1 - c_s)MP_s, \quad p_1 = P_s.$$

Special case $s = 1$. Putting $b_1 = 1$ (c_1 is a free parameter) yields a symplectic, explicit Nyström method of order 1. For the choice $c_1 = 1/2$ it has order 2.

Special case $s = 3$. To obtain order 3, four order conditions have to be satisfied (see Table 14.3). The first three mean that (b_i, c_i) is a quadrature formula of order 3. They allow us to express b_1, b_2, b_3 in terms of c_1, c_2, c_3. The last condition then becomes (Okunbor & Skeel 1992b)

$$1 + 24\left(c_1 - \frac{1}{2}\right)\left(c_2 - \frac{1}{2}\right) + 24(c_2 - c_1)(c_3 - c_1)(c_3 - c_2) \qquad (16.52)$$
$$+ 144\left(c_1 - \frac{1}{2}\right)\left(c_2 - \frac{1}{2}\right)\left(c_3 - \frac{1}{2}\right)\left(c_1 + c_3 - c_2 - \frac{1}{2}\right) = 0.$$

We thus get a two-parameter family of third order methods. Okunbor & Skeel (1992b) suggest taking

$$c_2 = \frac{1}{2}, \qquad c_1 = 1 - c_3 = \frac{1}{6}\left(2 + \sqrt[3]{2} + \frac{1}{\sqrt[3]{2}}\right) \qquad (16.53)$$

(the real root of $12c_1(2c_1 - 1)^2 = 1$). This method is symmetric and thus of order 4. Another 3-stage method of order 4 has been found by Qin Meng-Zhao & Zhu Wen-jie (1991).

Higher order methods. For the construction of methods of order ≥ 4 it is worthwhile to investigate the effect of the condition (16.46b) on the order conditions. As for partitioned Runge-Kutta methods one can show that SN-trees with the same geometrical structure lead to equivalent order conditions. For details we refer to Calvo & Sanz-Serna (1992). With the notation of Table 14.3, the SN-trees t_6 and t_7 as well as the pairs t_9, t_{12} and t_{10}, t_{13} give rise to equivalent order conditions. Consequently, for order 5, one has to consider 10 conditions. Okunbor & Skeel (1992c) present explicit, symplectic Nyström methods of orders 5 and 6 with 5 and 7 stages, respectively. A 7th order method is given by Calvo & Sanz-Serna (1992b).

Conservation of the Hamiltonian; Backward Analysis

> The differential equation actually solved by the difference scheme will be called the modified equation.
>
> (Warming & Hyett 1974, p. 161)

> The *wrong* solution of the *right* equation; the *right* solution of the *wrong* equation.   (Feng Kang, Beijing Sept. 1, 1992)

We have observed above (Example 16.2 and Fig. 16.6) that for the numerical solution of symplectic methods the Hamiltonian H remained between fixed bounds over any long-term integration, i.e., so-called secular changes of H were absent. Following several authors (Yoshida 1993, Sanz-Serna 1992, Feng Kang 1991b) this phenomenon is explained by interpreting the numerical solution as the *exact* solution of a *perturbed Hamiltonian system*, which is obtained as the formal expansion (16.56) in powers of h. The *exact* conservation of the perturbed Hamiltonian $\widetilde{H}$ then involves the quasi-periodic behaviour of H along the computed points. This resembles Wilkinson's famous idea of backward error analysis in linear algebra and, in the case of differential equations, seems to go back to Warming & Hyett (1974). We demonstrate this idea for the symplectic Euler method (see (16.36b))

$$
\begin{aligned}
p_1 &= p_0 - h H_q(p_0, q_1) \\
q_1 &= q_0 + h H_p(p_0, q_1)
\end{aligned}
\tag{16.54}
$$

which, when expanded around the point (p_0, q_0), gives

$$
\begin{aligned}
p_1 &= p_0 - h H_q - h^2 H_{qq} H_p - \frac{h^3}{2} H_{qqq} H_p H_p - h^3 H_{qq} H_{pq} H_p - \dots \Big|_{p_0, q_0} \\
q_1 &= q_0 + h H_p + h^2 H_{pq} H_p + \frac{h^3}{2} H_{pqq} H_p H_p + h^3 H_{pq} H_{pq} H_p + \dots \Big|_{p_0, q_0}.
\end{aligned}
\tag{16.54'}
$$

In the case of non-scalar equations the p's and q's must here be equipped with various summation indices. We suppress these in the sequel for the sake of simplicity and think of scalar systems only. The exact solution of a perturbed Hamiltonian

$$
\begin{aligned}
\dot{p} &= -\widetilde{H}_q(p, q) \\
\dot{q} &= \widetilde{H}_p(p, q)
\end{aligned}
$$

has a Taylor expansion analogous to Theorem 2.6 as follows

$$
\begin{aligned}
p_1 &= p_0 - h\widetilde{H}_q + \frac{h^2}{2}\left(\widetilde{H}_{qp}\widetilde{H}_q - \widetilde{H}_{qq}\widetilde{H}_p \right) + \dots \\
q_1 &= q_0 + h\widetilde{H}_p + \frac{h^2}{2}\left(-\widetilde{H}_{pp}\widetilde{H}_q + \widetilde{H}_{pq}\widetilde{H}_p \right) + \dots .
\end{aligned}
\tag{16.55}
$$

We now set

$$
\widetilde{H} = H + h H^{(1)} + h^2 H^{(2)} + h^3 H^{(3)} + \dots
\tag{16.56}
$$

with unknown functions $H^{(1)}, H^{(2)}, \ldots$, insert this into (16.55) and compare the resulting formulas with (16.54'). Then the comparison of the h^2 terms gives

$$H_q^{(1)} = \frac{1}{2} H_{qq} H_p + \frac{1}{2} H_{qp} H_q, \qquad H_p^{(1)} = \frac{1}{2} H_{pp} H_q + \frac{1}{2} H_{pq} H_p$$

which by miracle (the "miracle" is in fact a consequence of the symplecticity of method (16.54)) allow the common primitive

$$H^{(1)} = \frac{1}{2} H_p H_q. \tag{16.56;1}$$

The h^3 terms lead to

$$H^{(2)} = \frac{1}{12} \left(H_{pp} H_q^2 + H_{qq} H_p^2 + 4 H_{pq} H_p H_q \right) \tag{16.56;2}$$

and so on.

Connection with the Campbell-Baker-Hausdorff formula. An elegant access to the expansion (16.56), which works for separable Hamiltonians $H(p,q) = T(p) + U(q)$, has been given by Yoshida (1993). We interpret method (16.54) as composition of the two symplectic maps

$$z_0 = \begin{pmatrix} p_0 \\ q_0 \end{pmatrix} \quad \overset{S_T}{\longmapsto} \quad z = \begin{pmatrix} p_0 \\ q_1 \end{pmatrix} \quad \overset{S_U}{\longmapsto} \quad z_1 = \begin{pmatrix} p_1 \\ q_1 \end{pmatrix} \tag{16.57}$$

which consist, respectively, in solving exactly the Hamiltonian systems

$$\begin{aligned} \dot{p} &= 0 \\ \dot{q} &= T_p(p) \end{aligned} \quad \text{and} \quad \begin{aligned} \dot{p} &= -U_q(q) \\ \dot{q} &= 0 \end{aligned} \tag{16.58}$$

and apply some Lie theory. If we introduce for these equations the differential operators given by (13.2')

$$D_T \Psi = \frac{\partial \Psi}{\partial q} T_p(p), \qquad D_U \Psi = -\frac{\partial \Psi}{\partial p} U_q(q), \tag{16.59}$$

the formulas (13.3) allow us to write the Taylor series of the map S_T as

$$z = \sum_{i=0}^{\infty} \frac{h^i}{i!} D_T^i z \Big|_{z=z_0}. \tag{16.60}$$

If now $F(z)$ is an arbitrary function of the solution $z(t) = (p(t), q(t))$ (left equation of (16.58)), we find, as in (13.2), that

$$F(z)' = D_T F, \quad F(z)'' = D_T^2 F, \ldots$$

and (16.60) extends to (Gröbner 1960)

$$F(z) = \sum_{i=0}^{\infty} \frac{h^i}{i!} D_T^i F(z) \Big|_{z=z_0}. \tag{16.60'}$$

We now insert S_U for F and insert for S_U the formula analogous to (16.60) to obtain for the composition (16.57)

$$z_1 = (p_1, q_1) = \sum_{i=0}^{\infty} \frac{h^i}{i!} D_T^i \sum_{j=0}^{\infty} \frac{h^j}{j!} D_U^j z \Big|_{z=z_0}$$

$$= \exp(hD_T) \exp(hD_U)(p, q) \Big|_{p=p_0, q=q_0} .$$

(16.61)

But the product $\exp(hD_T)\exp(hD_U)$ is *not* $\exp(hD_T + hD_U)$, as we have all learned in school, because the operators D_T and D_U do not commute. This is precisely the content of the famous Campbell-Baker-Hausdorff Formula (claimed in 1898 by J.E. Campbell and proved independently by Baker (1905) and in the "kleine Untersuchung" of Hausdorff (1906)) which states, for our problem, that

$$\exp(hD_T) \exp(hD_U) = \exp(h\widetilde{D}) \qquad (16.62)$$

where

$$\widetilde{D} = D_T + D_U + \frac{h}{2}[D_T, D_U] + \frac{h^2}{12}([D_T, [D_T, D_U]] + [D_U, [D_U, D_T]])$$

$$+ \frac{h^3}{24}[D_T, [D_U, [D_U, D_T]]] + \dots \qquad (16.63)$$

and $[D_A, D_B] = D_A D_B - D_B D_A$ is the commutator. Equation (16.62) shows that the map (16.57) is the exact solution of the differential equation corresponding to the differential operator $\widetilde{D}$. A straightforward calculation now shows: If

$$D_A \Psi = -\frac{\partial \Psi}{\partial p} A_q + \frac{\partial \Psi}{\partial q} A_p \qquad \text{and} \qquad D_B \Psi = -\frac{\partial \Psi}{\partial p} B_q + \frac{\partial \Psi}{\partial q} B_p \quad (16.64)$$

are differential operators corresponding to Hamiltonians A and B respectively, then

$$[D_A, D_B]\Psi = D_C \Psi = -\frac{\partial \Psi}{\partial p} C_q + \frac{\partial \Psi}{\partial q} C_p$$

where

$$C = A_p B_q - A_q B_p. \qquad (16.65)$$

A repeated application of (16.65) now allows us to obtain for all brackets in (16.63) a corresponding Hamiltonian which finally leads to

$$\widetilde{H} = T + U + \frac{h}{2}T_p U_q + \frac{h^2}{12}(T_{pp}U_q^2 + U_{qq}T_p^2) + \frac{h^3}{12}T_{pp}U_{qq}T_pU_q + \dots \quad (16.66)$$

which is the specialization of (16.56) to separable Hamiltonians.

Example 16.12 (Yoshida 1993). For the mathematical pendulum

$$H(p, q) = \frac{p^2}{2} - \cos q \qquad (16.67)$$

series (16.66) becomes

$$\widetilde{H} = \frac{p^2}{2} - \cos q + \frac{h}{2} p \sin q + \frac{h^2}{12} (\sin^2 q + p^2 \cos q) + \frac{h^3}{12} p \cos q \sin q + \mathcal{O}(h^4).$$
(16.68)

Fig. 16.9 presents for various step sizes h and for various initial points ($p_0=0$, $q_0=-1.5$; $p_0=0$, $q_0=-2.5$; $p_0=1.5$, $q_0=-\pi$; $p_0=2.5$, $q_0=-\pi$) the numerically computed points for method (16.54) compared to the contour lines of $\widetilde{H} = Const$ given by the terms up to order h^3 in (16.68). The excellent agreement of the results with theory for $h \leq 0.6$ leaves nothing to be desired, while for h beyond 0.9 the dynamics of the numerical method turns rapidly into chaotic behaviour.

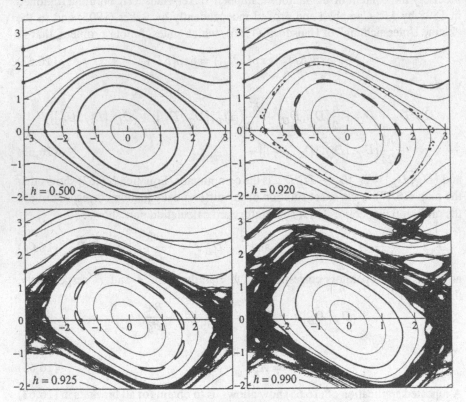

Fig. 16.9. Symplectic method compared to perturbed Hamiltonian
($\bullet \ldots$ indicate the initial positions)

Remark. For much research, especially in the beginning of the "symplectic era", the central role for the construction of canonical difference schemes is played by the Hamilton-Jacobi theory and generating functions. For this, the reader may consult the papers Feng Kang (1986), Feng Kang, Wu Hua-mo, Qin Meng-zhao & Wang Dao-liu (1989), Channell & Scovel (1990) and Miesbach & Pesch (1992). Many additional numerical experiments can be found in Channell & Scovel (1990), Feng Kang (1991), and Pullin & Saffman (1991).

Exercises

1. Show that explicit Runge-Kutta methods are never symplectic.

   Hint. Compute the diagonal elements of M.

2. Study the existence and uniqueness of the numerical solution for the implicit mid-point rule when applied to the Hamiltonian system

   $$\dot{p} = -q^2, \qquad \dot{q} = p.$$

   Show that the method possesses no solution at all for $h^2 q_0 + h^3 p_0/2 < -1$ and two solutions for $h^2 q_0 + h^3 p_0/2 > -1$ ($h \neq 0$). Only one of the solutions tends to (p_0, q_0) for $h \to 0$.

3. A Runge-Kutta method is called *linearly symplectic* if it is symplectic for all linear Hamiltonian systems

   $$\dot{y} = J^{-1} C y$$

   (J is given in (16.19) and C is a symmetric matrix). Prove (Feng Kang 1985) that a Runge-Kutta method is linearly symplectic if and only if its stability function satisfies

   $$R(-z)R(z) = 1 \qquad \text{for all} \quad z \in \mathbf{C}. \tag{16.69}$$

   Hint. For the definition of the stability function see Section IV.2 of Volume II. Then by Theorem 14.13, linear symplecticity is equivalent to

   $$R(hJ^{-1}C)^T J R(hJ^{-1}C) = J.$$

   Furthermore, the matrix $B := J^{-1}C$ is seen to verify $B^T J = -JB$ and hence also $(B^k)^T J = J(-B)^k$ for $k = 0, 1, 2, \ldots$. This implies that

   $$R(hJ^{-1}C)^T J = J R(-hJ^{-1}C).$$

4. Prove that the stability function of a symmetric Runge-Kutta method satisfies (16.69).

5. Compute all quadratic first integrals of the Hamiltonian system (16.4).

6. For a separable Hamiltonian consider the method (16.26) where $a_{ij} = 0$ for $i < j$, $\hat{a}_{ij} = 0$ for $i < j$ and for every i either $a_{ii} = 0$ or $\hat{a}_{ii} = 0$. If the method satisfies (16.28) then it is equivalent to one given by scheme (16.33).

Hint. Remove first all stages which don't influence the numerical result (see the remark after Theorem 16.10). Then deduce from (16.28) relations similar to (16.32). Finally, remove identical stages and add, if necessary, a dummy stage in order that both methods have the same number of stages.

7. (Lasagni 1990). Characterize symplecticity for multi-derivative Runge-Kutta methods. Show that the s-stage q-derivative method of Definition 13.1 is symplectic if its coefficients satisfy

$$b_i^{(r)} b_j^{(m)} - b_i^{(r)} a_{ij}^{(m)} - b_j^{(m)} a_{ji}^{(r)} = \begin{cases} b_i^{(r+m)} & \text{if } i = j \text{ and } r+m \le q, \\ 0 & \text{otherwise.} \end{cases}$$
(16.70)

Hint. Denote $k^{(r)} = D_H^r p$, $\ell^{(r)} = D_H^r q$, where D_H is the differential operator as in (16.59) and (16.64), so that the exact solution of (16.1) is given by

$$p(x_0+h) = p_0 + \sum_{r \ge 1} \frac{h^r}{r!} k^{(r)}(p_0, q_0), \qquad q(x_0+h) = q_0 + \sum_{r \ge 1} \frac{h^r}{r!} \ell^{(r)}(p_0, q_0).$$

Then deduce from the symplecticity of the exact solution that

$$\frac{1}{\varrho!} \left(dp \wedge d\ell^{(\varrho)} + dk^{(\varrho)} \wedge dq \right) + \sum_{r+m=\varrho} \frac{1}{r!} \frac{1}{m!} dk^{(r)} \wedge d\ell^{(m)} = 0. \quad (16.71)$$

This, together with a modification of the proof of Theorem 16.6, allows us to obtain the desired result.

8. (Yoshida 1990, Qin Meng-Zhao & Zhu Wen-Jie 1992). Let $y_1 = \psi_h(y_0)$ denote a symmetric numerical scheme of order $p = 2k$. Prove that the composed method

$$\psi_{c_1 h} \circ \psi_{c_2 h} \circ \psi_{c_1 h}$$

is symmetric and has order $p + 2$ if

$$2c_1 + c_2 = 1, \qquad 2c_1^{2k+1} + c_2^{2k+1} = 0. \quad (16.72)$$

Hence there exist, for separable Hamiltonians, explicit symplectic partitioned methods of arbitrarily high order.

Hint. Proceed as for (4.1)-(4.2) and use Theorem 8.10 (the order of a symmetric method is even).

9. The Hamiltonian function (16.24) for the galactic problem is *not* separable. Nevertheless, the two methods (16.36a) and (16.36b) avoid nonlinear equations. Explain.

II.17 Delay Differential Equations

> Detailed studies of the real world impel us, albeit reluctantly, to take account of the fact that the rate of change of physical systems depends not only on their present state, but also on their past history.
> (Bellman & Cooke 1963)

Delay differential equations are equations with "retarded arguments" or "time lags" such as

$$y'(x) = f\big(x, y(x), y(x - \tau)\big) \tag{17.1}$$

or

$$y'(x) = f\big(x, y(x), y(x - \tau_1), y(x - \tau_2)\big) \tag{17.2}$$

or of even more general form. Here the derivative of the solutions depends also on its values at previous points.

Time lags are present in many models of applied mathematics. They can also be the source of interesting mathematical phenomena such as instabilities, limit cycles, periodic behaviour.

Existence

For equations of the type (17.1) or (17.2), where the delay values $x - \tau$ are bounded away from x by a positive constant, the question of existence is an easy matter: suppose that the solution is known, say

$$y(x) = \varphi(x) \qquad \text{for } x_0 - \tau \le x \le x_0.$$

Then $y(x - \tau)$ is a known function of x for $x_0 \le x \le x_0 + \tau$ and (17.1) becomes an ordinary differential equation, which can be treated by known existence theories. We then know $y(x)$ for $x_0 \le x \le x_0 + \tau$ and can compute the solution for $x_0 + \tau \le x \le x_0 + 2\tau$ and so on. This "method of steps" then yields existence and uniqueness results for all x. For more details we recommend the books of Bellman & Cooke (1963) and Driver (1977, especially Chapter V).

Example 1. We consider the equation

$$y'(x) = -y(x - 1), \qquad y(x) = 1 \quad \text{for } -1 \le x \le 0. \tag{17.3}$$

Proceeding as described above, we obtain

$$y(x) = 1 - x \qquad \text{for } 0 \leq x \leq 1,$$

$$y(x) = 1 - x + \frac{(x-1)^2}{2!} \qquad \text{for } 1 \leq x \leq 2,$$

$$y(x) = 1 - x + \frac{(x-1)^2}{2!} - \frac{(x-2)^3}{3!} \qquad \text{for } 2 \leq x \leq 3, \text{ etc.}$$

The solution is displayed in Fig. 17.1. We observe that despite the fact that the differential equation and the initial function are C^∞, the solution has discontinuities in its derivatives. This results from the fact that the initial function does not satisfy the differential equation. With every time step τ, however, these discontinuities are smoothed out more and more.

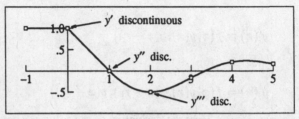

Fig. 17.1. Solution of (17.3)

Example 2. Our next example clearly illustrates the fact that the solutions of a delay equation depend on the entire history between $x_0 - \tau$ and x_0, and not only on the initial value:

$$y'(x) = -1.4 \cdot y(x-1) \tag{17.4}$$

a) $\varphi(x) = 0.8$ for $-1 \leq x \leq 0$,
b) $\varphi(x) = 0.8 + x$ for $-1 \leq x \leq 0$,
c) $\varphi(x) = 0.8 + 2x$ for $-1 \leq x \leq 0$.

The solutions are displayed in Fig. 17.2. An explanation for the oscillatory behaviour of the solutions will be given below.

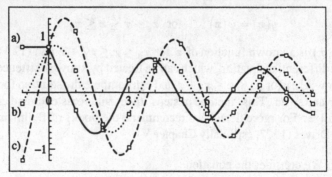

Fig. 17.2. Solutions of (17.4)

Constant Step Size Methods for Constant Delay

If we apply the Runge-Kutta method (1.8) (or (7.7)) to a delay equation (17.1) we obtain

$$g_i^{(n)} = y_n + h \sum_j a_{ij} f\big(x_n + c_j h, g_j^{(n)}, y(x_n + c_j h - \tau)\big)$$

$$y_{n+1} = y_n + h \sum_j b_j f\big(x_n + c_j h, g_j^{(n)}, y(x_n + c_j h - \tau)\big).$$

But which values should we give to $y(x_n + c_j h - \tau)$? If the delay is constant and satisfies $\tau = kh$ for some integer k, the most natural idea is to use the back-values of the old solution

$$g_i^{(n)} = y_n + h \sum_j a_{ij} f\big(x_n + c_j h, g_j^{(n)}, \gamma_j^{(n)}\big) \tag{17.5a}$$

$$y_{n+1} = y_n + h \sum_j b_j f\big(x_n + c_j h, g_j^{(n)}, \gamma_j^{(n)}\big) \tag{17.5b}$$

where

$$\gamma_j^{(n)} = \begin{cases} \varphi(x_n + c_j h - \tau) & \text{if } n < k \\ g_j^{(n-k)} & \text{if } n \geq k. \end{cases} \tag{17.5c}$$

This can be interpreted as solving successively

$$y'(x) = f\big(x, y(x), \varphi(x - \tau)\big) \tag{17.1a}$$

for the interval $[x_0, x_0 + \tau]$, then

$$\begin{aligned} y'(x) &= f\big(x, y(x), z(x)\big) \\ z'(x) &= f\big(x - \tau, z(x), \varphi(x - 2\tau)\big) \end{aligned} \tag{17.1b}$$

for the interval $[x_0 + \tau, x_0 + 2\tau]$, then

$$\begin{aligned} y'(x) &= f\big(x, y(x), z(x)\big) \\ z'(x) &= f\big(x - \tau, z(x), v(x)\big) \\ v'(x) &= f\big(x - 2\tau, v(x), \varphi(x - 3\tau)\big) \end{aligned} \tag{17.1c}$$

for the interval $[x_0 + 2\tau, x_0 + 3\tau]$, and so on. This is the perfect numerical analog of the "method of steps" mentioned above.

Theorem 17.1. *If c_i, a_{ij}, b_j are the coefficients of a p-th order Runge-Kutta method, then (17.5) is convergent of order p.*

Proof. The sequence (17.1a), (17.1b),... are ordinary differential equations normally solved by a pth order Runge-Kutta method. Therefore the result follows immediately from Theorem 3.6. □

Remark. For the collocation method based on Gaussian quadrature formula, Theorem 17.1 yields superconvergence in spite of the use of the low order approximations $\gamma_j^{(n)}$ of (17.5c). Bellen (1984) generalizes this result to the situation where $\tau = \tau(x)$ and $\gamma_j^{(n)}$ is the value of the collocation polynomial at $x_n + c_j h - \tau(x_n + c_j h)$. He proves superconvergence if the grid-points are chosen such that every interval $[x_{n-1}, x_n]$ is mapped, by $x - \tau(x)$, into $[x_{j-1}, x_j]$ for some $j < n$.

Numerical Example. We have integrated the problem

$$y'(x) = \big(1.4 - y(x-1)\big) \cdot y(x)$$

(see (17.12) below) for $0 \leq x \leq 10$ with initial values $y(x) = 0$, $-1 \leq x < 0$, $y(0) = 0.1$, and step sizes $h = 1, 1/2, 1/4, 1/8, \ldots, 1/128$ using Kutta's methods of order 4 (Table 1.2, left). The absolute value of the global errors (and the solution in grey) are presented in Fig. 17.3. The 4th order convergence can clearly be observed. The downward peaks are provoked by sign changes in the error.

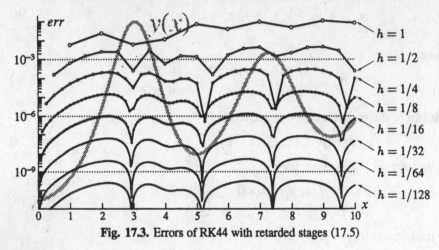

Fig. 17.3. Errors of RK44 with retarded stages (17.5)

Variable Step Size Methods

Although method (17.5) allows efficient and easy to code computations for simple problems with constant delays (such as all the examples of this section), it does not allow to change the step size arbitrarily, and an application to variable delay equations is not straightforward. If complete flexibility is desired, we need a *global* approximation to the solution. Such global approximations are furnished by multistep methods of Adams or BDF type (see Chapter III.1) or the modern Runge-Kutta methods which are constructed together with a dense output. The code RETARD of the appendix is a modification of the code DOPRI5 (method of Dormand &

Prince in Table 5.2 with Shampine's dense output; see (6.12), (6.13) and the subsequent discussion) in such a way that after every successful step of integration the coefficients of the continuous solution are written into memory. Back-values of the solution are then available by calling the function YLAG(I,X,PHI). For example, for problem (17.4) the subroutine FCN would read as

$$F(1) = -1.4D0 * \text{YLAG}(1, X - 1.D0, \text{PHI}).$$

As we have seen, the solutions possess discontinuities in the derivatives at several points, e.g. for (17.1) at $x_0 + \tau$, $x_0 + 2\tau$, $x_0 + 3\tau, \ldots$ etc. Therefore the code RETARD provides a possibility to match given points of discontinuities exactly (specify IWORK(6) and WORK(11),...) which improves precision and computation time.

Earlier Runge-Kutta codes for delay equations have been written by Oppelstrup (1976), Oberle & Pesch (1981) and Bellen & Zennaro (1985). Bock & Schlöder (1981) exploited the natural dense output of multistep methods.

Stability

It can be observed from Fig. 17.1 and Fig. 17.2 that the solutions, after the initial phase, seem to tend to something like $e^{\alpha x} \cos \beta(x - \delta)$. We now try to determine α and β. We study the equation

$$y'(x) = \lambda y(x) + \mu y(x - 1). \tag{17.6}$$

There is no loss of generality in supposing the delay $\tau = 1$, since any delay $\tau \neq 1$ can be reduced to $\tau = 1$ by a coordinate change.

We search for a solution of the form

$$y(x) = e^{\gamma x} \qquad \text{where } \gamma = \alpha + i\beta. \tag{17.7}$$

Introducing this into (17.6) we obtain the following "characteristic equation" for γ

$$\gamma - \lambda - \mu e^{-\gamma} = 0, \tag{17.8}$$

which, for $\mu \neq 0$, possesses an infinity of solutions: in fact, if $|\gamma|$ becomes large, we obtain from (17.8), since λ is fixed, that $\mu e^{-\gamma}$ must be large too and

$$\gamma \approx \mu e^{-\gamma}. \tag{17.8'}$$

This implies that $\gamma = \alpha + i\beta$ is close to the imaginary axis. Hence $|\gamma| \approx |\beta|$ and from (17.8')

$$|\beta| \approx |\mu| e^{-\alpha}.$$

Therefore the roots of (17.8) lie asymptotically on the curves $-\alpha = \log |\beta| - \log |\mu|$. Again from (17.8'), we have a root whenever the argument of $\mu e^{-i\beta}$ is close to $\pi/2$ (for $\beta > 0$), i.e. if

$$\beta \approx \arg \mu - \frac{\pi}{2} + 2k\pi \qquad k = 1, 2, \ldots$$

There are thus two sequences of characteristic values which tend to infinity on logarithmic curves left of the imaginary axis, with 2π as asymptotic distance between two consecutive values.

The "general solution" of (17.6) is thus a Fourier-like superposition of solutions of type (17.7) (Wright 1946, see also Bellman & Cooke 1963, Chapter 4). The larger $-\mathrm{Re}\,\gamma$ is, the faster these solutions "die out" as $x \to \infty$. The dominant solutions are thus (provided that the corresponding coefficients are not zero) those which correspond to the largest real part, i.e., those closest to the origin. For equations (17.3) and (17.4) the characteristic equations are $\gamma + e^{-\gamma} = 0$ and $\gamma + 1.4e^{-\gamma} = 0$ with solutions $\gamma = -0.31813 \pm 1.33724i$ and $\gamma = -0.08170 \pm 1.51699i$ respectively, which explains nicely the behaviour of the asymptotic solutions of Fig. 17.1 and Fig. 17.2.

Remark. For the case of *matrix equations*

$$y'(x) = Ay(x) + By(x-1)$$

where A and B are not simultaneously diagonizable, we set $y(x) = ve^{\gamma x}$ where $v \neq 0$ is a given vector. The equation now leads to

$$\gamma v = Av + Be^{-\gamma}v,$$

which has a nontrivial solution if

$$\det(\gamma I - A - Be^{-\gamma}) = 0, \tag{17.8"}$$

the characteristic equation for the more general case. The shape of the solutions of (17.8") is similar to those of (17.8), there are just $r = \mathrm{rank}(B)$ points in each strip of width 2π instead of one.

All solutions of (17.6) remain *stable* for $x \to \infty$ if all characteristic roots of (17.8) remain in the negative half plane. This result follows either from the above expansion theorem or from the theory of Laplace transforms (e.g., Bellmann & Cooke (1963), Chapter 1), which, in fact, is closely related.

In order to study the boundary of the stability domain, we search for (λ, μ) values for which the first solution γ crosses the imaginary axis, i.e. $\gamma = i\theta$ for θ real. If we insert this into (17.8), we obtain

$$\lambda = -\mu \qquad \text{for } \theta = 0 \ (\gamma \text{ real})$$
$$\lambda = i\theta - \mu e^{-i\theta} \qquad \text{for } \theta \neq 0$$

or, by separating real and imaginary parts,

$$\lambda = \frac{\cos\theta \cdot \theta}{\sin\theta}, \qquad \mu = -\frac{\theta}{\sin\theta}$$

valid for real λ and μ. These paths are sketched in Fig. 17.4 and separate in the (λ, μ)-plane the domains of stability and instability for the solutions of (17.6) (a result of Hayes 1950).

If we put $\theta = \pi/2$, we find that the solutions of $y'(x) = \mu y(x-1)$ remain *stable* for

$$-\frac{\pi}{2} \le \mu \le 0 \qquad (17.9a)$$

and are *unstable* for

$$\mu < -\frac{\pi}{2} \qquad \text{as well as} \qquad \mu > 0. \qquad (17.9b)$$

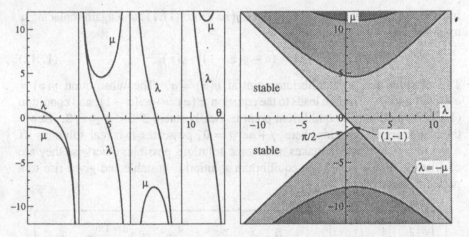

Fig. 17.4. Domain of stability for $y'(x) = \lambda y(x) + \mu y(x-1)$

An Example from Population Dynamics

> Lord Cherwell drew my attention to an equation, equivalent to (8) (here: (17.12)) with $a = \log 2$, which he had encountered in his application of probability methods to the problem of distribution of primes. My thanks are due to him for thus introducing me to an interesting problem.   (E.M. Wright 1945)

We now demonstrate the phenomena discussed above and the power of our programs on a couple of examples drawn from applications. For supplementary applications of delay equations to all sorts of sciences, consult the impressive list in Driver (1977, p. 239-240).

Let $y(x)$ represent the population of a certain species, whose development as a function of time is to be studied. The simple model of infinite exponential growth $y' = \lambda y$ was soon replaced by the hypothesis that the growth rate λ will decrease with increasing population y due to illness and lack of food and space. One then arrives at the model (Verhulst 1845, Pearl & Reed 1922)

$$y'(x) = k \cdot \big(a - y(x)\big) \cdot y(x). \qquad (17.10)$$

"Nous donnerons le nom *logistique* à la courbe caractérisée par l'équation précédente" (Verhulst). It can be solved by elementary functions (Exercise 1). All solutions with initial value $y_0 > 0$ tend asymptotically to a as $x \to \infty$. If we assume the growth rate to depend on the population of the *preceding* generation, (17.10) becomes a delay equation (Cunningham 1954, Wright 1955, Kakutani & Markus 1958)

$$y'(x) = k \cdot \big(a - y(x - \tau)\big) \cdot y(x). \tag{17.11}$$

Introducing the new function $z(x) = k\tau y(\tau x)$ into (17.11) and again replacing z by y and $ka\tau$ by a we obtain

$$y'(x) = \big(a - y(x - 1)\big) \cdot y(x). \tag{17.12}$$

This equation has an equilibrium point at $y(x) = a$. The substitution $y(x) = a + z(x)$ and linearization leads to the equation $z'(x) = -az(x - 1)$, and condition (17.9) shows that this equilibrium point is locally stable if $0 < a \le \pi/2$. Hence the characteristic equation, here $\gamma + ae^{-\gamma} = 0$, possesses two real solutions iff $a < 1/e = 0.368$, which makes monotonic solutions possible; otherwise they are oscillatory. For $a > \pi/2$ the equilibrium solution is unstable and gives rise to a periodic limit cycle.

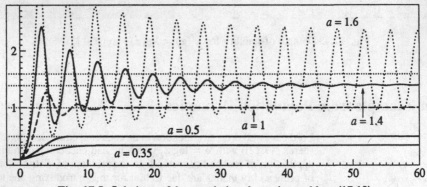

Fig. 17.5. Solutions of the population dynamics problem (17.12)

The solutions in Fig. 17.5 have been computed by the code RETARD of the appendix with subroutine FCN as

$$\mathtt{F(1) = (A - YLAG(1, X - 1.D0, PHI)) * Y(1)}, \quad \mathtt{A} = 0.35, 0.5, 1., 1.4, \text{ and } 1.6.$$

Infectious Disease Modelling

> De tous ceux qui ont traité cette matière, c'est sans contredit M.
> *de la Condamine* qui l'a fait avec plus de succès. Il est déjà venu
> à bout de persuader la meilleure partie du monde raisonnable de
> la grande utilité de l'inoculation: quant aux autres, il serait inutile
> de vouloir employer la raison avec eux: puisqu'ils n'agissent pas
> par principes. Il faut les conduire comme des enfants vers leur
> mieux ...   (Daniel Bernoulli 1760)

Daniel Bernoulli ("Docteur en medecine, Professeur de Physique en l'Université de
Bâle, Associé étranger de l'Academie des Sciences") was the first to use differential
calculus to model infectious diseases in his 1760 paper on smallpox vaccination.
At the beginning of our century, mathematical modelling of epidemics gained new
interest. This finally led to the classical model of Kermack & McKendrick (1927):
let $y_1(x)$ measure the *susceptible* portion of the population, $y_2(x)$ the *infected*,
and $y_3(x)$ the *removed* (e.g. immunized) one. It is then natural to assume that
the number of newly infected people per time unit is proportional to the product
$y_1(x)y_2(x)$, just as in bimolecular chemical reactions (see Section I.16). If we
finally assume the number of newly removed persons to be proportional to the
infected ones, we arrive at the model

$$y_1' = -y_1 y_2, \qquad y_2' = y_1 y_2 - y_2, \qquad y_3' = y_2 \qquad (17.13)$$

where we have taken for simplicity all rate constants equal to one. This system
can be integrated by elementary methods (divide the first two equations and solve
$dy_2/dy_1 = -1 + 1/y_1$). The numerical solution with initial values $y_1(0) = 5$,
$y_2(0) = 0.1$, $y_3(0) = 1$ is painted in gray color in Fig. 17.6: an epidemic breaks
out, everybody finally becomes "removed" and nothing further happens.

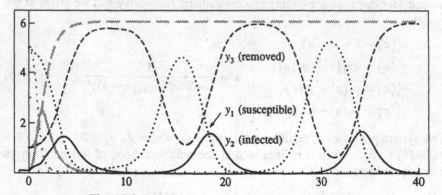

Fig. 17.6. Periodic outbreak of disease, model (17.14)
(in gray: Solution of Kermack - McKendrick model (17.13))

We arrive at a periodic outbreak of the disease, if we assume that immunized
people become susceptible again, say after a fixed time τ ($\tau = 10$). If we also

introduce an incubation period of, say, $\tau_2 = 1$, we arrive at the model

$$y_1'(x) = -y_1(x)y_2(x-1) + y_2(x-10)$$
$$y_2'(x) = y_1(x)y_2(x-1) - y_2(x) \tag{17.14}$$
$$y_3'(x) = y_2(x) - y_2(x-10)$$

instead of (17.13). The solutions of (17.14), for the initial phases $y_1(x) = 5$, $y_2(x) = 0.1$, $y_3(x) = 1$ for $x \le 0$, are shown in Fig. 17.6 and illustrate the periodic outbreak of the disease.

An Example from Enzyme Kinetics

Our next example, more complicated than the preceding ones, is from enzyme kinetics (Okamoto & Hayashi 1984). Consider the following consecutive reactions

$$I \xrightarrow{} Y_1 \xrightarrow{z} Y_2 \xrightarrow{k_2} Y_3 \xrightarrow{k_3} Y_4 \xrightarrow{k_4} \tag{17.15}$$

where I is an exogenous substrate supply which is maintained constant and n molecules of the end product Y_4 inhibit co-operatively the reaction step of $Y_1 \to Y_2$ as

$$z = \frac{k_1}{1 + \alpha(y_4(x))^n}.$$

It is generally expected that the inhibitor molecule must be moved to the position of the regulatory enzyme by forces such as diffusion or active transport. Thus, we consider this time consuming process causing time-delay and we arrive at the model

$$\begin{aligned} y_1'(x) &= I - zy_1(x) \\ y_2'(x) &= zy_1(x) - y_2(x) \\ y_3'(x) &= y_2(x) - y_3(x) \\ y_4'(x) &= y_3(x) - 0.5y_4(x) \end{aligned} \qquad z = \frac{1}{1 + 0.0005(y_4(x-4))^3}. \tag{17.16}$$

This system possesses an equilibrium at $zy_1 = y_2 = y_3 = I$, $y_4 = 2I$, $y_1 = I(1 + 0.004I^3) =: c_1$. When it is linearized in the neighbourhood of this equilibrium point, it becomes

$$\begin{aligned} y_1'(x) &= -c_1 y_1(x) + c_2 y_4(x-4) \\ y_2'(x) &= c_1 y_1(x) - y_2(x) - c_2 y_4(x-4) \\ y_3'(x) &= y_2(x) - y_3(x) \\ y_4'(x) &= y_3(x) - 0.5 y_4(x) \end{aligned} \tag{17.17}$$

where $c_2 = c_1 \cdot I^3 \cdot 0.006$. By setting $y(x) = v \cdot e^{\gamma x}$ we arrive at the characteristic equation (see (17.8")), which becomes after some simplifications

$$(c_1 + \gamma)(1 + \gamma)^2(0.5 + \gamma) + c_2\gamma e^{-4\gamma} = 0. \qquad (17.18)$$

As in the paper of Okamoto & Hayashi, we put $I = 10.5$. Then (17.18) possesses one pair of complex solutions in C^+, namely

$$\gamma = 0.04246 \pm 0.47666i$$

and the equilibrium solution is unstable (see Fig. 17.7). The period of the solution of the linearized equation is thus $T = 2\pi/0.47666 = 13.18$. The solutions then tend to a limit cycle of approximately the same period.

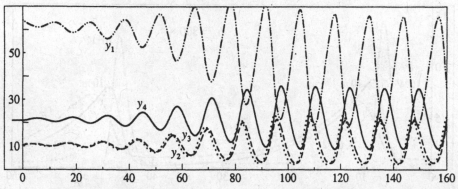

Fig. 17.7. Solutions of the enzyme kinetics problem (17.16), $I = 10.5$.
Initial values close to equilibrium position

A Mathematical Model in Immunology

We conclude our series of examples with Marchuk's model (Marchuk 1975) for the struggle of viruses $V(t)$, antibodies $F(t)$ and plasma cells $C(t)$ in the organism of a person infected by a viral disease. The equations are

$$\frac{dV}{dt} = (h_1 - h_2 F)V$$

$$\frac{dC}{dt} = \xi(m)h_3 F(t-\tau)V(t-\tau) - h_5(C-1) \qquad (17.19)$$

$$\frac{dF}{dt} = h_4(C - F) - h_8 FV.$$

The first is a Volterra - Lotkin like predator-prey equation. The second equation describes the creation of new plasma cells with time lag due to infection, in the absence of which the second term creates an equilibrium at $C = 1$. The third equation models the creation of antibodies from plasma cells $(h_4 C)$ and their

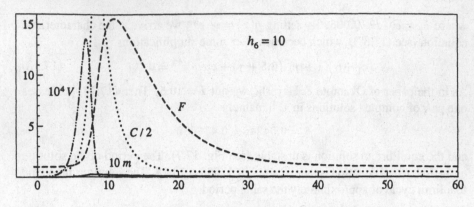

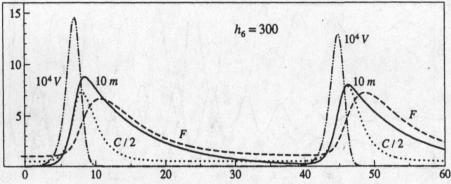

Fig. 17.8. Solutions of the Marchuk immunology model

decrease due to aging $(-h_4 F)$ and binding with antigens $(-h_8 FV)$. The term $\xi(m)$, finally, is defined by

$$\xi(m) = \begin{cases} 1 & \text{if } m \leq 0.1 \\ (1-m)\dfrac{10}{9} & \text{if } 0.1 \leq m \leq 1 \end{cases}$$

and expresses the fact that the creation of plasma cells slows down when the organism is damaged by the viral infection. The relative characteristic $m(t)$ of damaging is given by a fourth equation

$$\frac{dm}{dt} = h_6 V - h_7 m$$

where the first term expresses the damaging and the second recuperation.

This model allows us, by changing the coefficients $h_1, h_2, \ldots, h_8$, to model all sorts of behaviour of stable health, unstable health, acute form of a disease, chronic form etc. See Chapter 2 of Marchuk (1983). In Fig. 17.8 we plot the solutions of this model for $\tau = 0.5$, $h_1 = 2$, $h_2 = 0.8$, $h_3 = 10^4$, $h_4 = 0.17$, $h_5 = 0.5$, $h_7 = 0.12$, $h_8 = 8$ and initial values $V(t) = \max(0, 10^{-6} + t)$ if $t \leq 0$, $C(0) = 1$, $F(t) = 1$ if $t \leq 0$, $m(0) = 0$. In dependence of the value of h_6 ($h_6 = 10$ or

$h_6 = 300$), we then observe either complete recovery (defined by $V(t) < 10^{-16}$), or periodic outbreak of the disease due to damaging ($m(t)$ becomes nearly 1).

Integro-Differential Equations

Often the hypothesis that a system depends on the time lagged solution at a specified fixed value $x - \tau$ is not very realistic, and one should rather suppose this dependence to be stretched out over a longer period of time. Then, instead of (17.1), we would have for example

$$y'(x) = f\left(x, y(x), \int_{x-\tau}^{x} K(x, \xi, y(\xi))\, d\xi\right). \qquad (17.20)$$

The numerical treatment of these problems becomes much more expensive (see Brunner & van der Houwen (1986) for a study of various discretization methods). If $K(x, \xi, y)$ is zero in the neighbourhood of the diagonal $x = \xi$, one can eventually use RETARD and call a quadrature routine for each function evaluation.

Fortunately, many integro-differential equations can be reduced to ordinary or delay differential equations by introducing new variables for the integral function.

Example (Volterra 1934). Consider the equation

$$y'(x) = \left(\varepsilon - \alpha y(x) - \int_0^x k(x - \xi) y(\xi)\, d\xi\right) \cdot y(x) \qquad (17.21)$$

for population dynamics, where the integral term represents a decrease of the reproduction rate due to pollution. If now for example $k(x) = c$, we put

$$\int_0^x y(\xi)\, d\xi = v(x), \qquad y(x) = v'(x)$$

and obtain

$$v''(x) = (\varepsilon - \alpha v'(x) - cv(x)) \cdot v'(x),$$

an ordinary differential equation.

The same method is possible for equations (17.20) with "degenerate kernel"; i.e., where

$$K(x, \xi, y) = \sum_{i=1}^{m} a_i(x) b_i(\xi, y). \qquad (17.22)$$

If we insert this into (17.20) and put

$$v_i(x) = \int_{x-\tau}^{x} b_i(\xi, y(\xi))\, d\xi, \qquad (17.23)$$

we obtain

$$y'(x) = f\left(x, y(x), \sum_{i=1}^{m} a_i(x)v_i(x)\right)$$

$$v_i'(x) = b_i(x, y(x)) - b_i(x - \tau, y(x - \tau)) \qquad i = 1, \ldots, m,$$

(17.20')

a system of delay differential equations.

Exercises

1. Compute the solution of the Verhulst & Pearl equation (17.10).

2. Compute the equilibrium points of Marchuk's equation (17.19) and study their stability.

3. Assume that the kernel $k(x)$ in Volterra's equation (17.21) is given by

$$k(x) = p(x)e^{-\beta x}$$

where $p(x)$ is some polynomial. Show that this problem can be transformed into an ordinary differential equation.

4. Consider the integro-differential equation

$$y'(x) = f\left(x, y(x), \int_0^x K(x, \xi, y(\xi))\, d\xi\right). \tag{17.24}$$

a) For the degenerate kernel (17.22) problem (17.24) becomes equivalent to the ordinary differential equation

$$y'(x) = f\left(x, y(x), \sum_{j=1}^{m} a_j(x)v_j(x)\right)$$

$$v_j'(x) = b_j(x, y(x)).$$

(17.25)

b) Show that an application of an explicit (pth order) Runge-Kutta method to (17.25) yields the formulas (Pouzet 1963)

$$y_{n+1} = y_n + h\sum_{i=1}^{s} b_i f(x_n + c_i h, g_i^{(n)}, u_i^{(n)})$$

$$g_i^{(n)} = y_n + h\sum_{j=1}^{i-1} a_{ij} f(x_n + c_j h, g_j^{(n)}, u_j^{(n)})$$

(17.26)

$$u_i^{(n)} = F_n(x_n + c_i h) + h\sum_{j=1}^{i-1} a_{ij} K(x_n + c_i h, x_n + c_j h, g_j^{(n)})$$

where

$$F_0(x) = 0, \qquad F_{n+1}(x) = F_n(x) + h \sum_{i=1}^{s} b_i K(x, x_n + c_i h, g_i^{(n)}).$$

c) If we apply method (17.26) to problem (17.24), where the kernel does not necessarily satisfy (17.22), we nevertheless have convergence of order p.

Hint. Approximate the kernel by a degenerate one.

5. (Zennaro 1986). For the delay equation (17.1) consider the method (17.5) where (17.5c) is replaced by

$$\gamma_j^{(n)} = \begin{cases} \varphi(x_n + c_j h - \tau) & \text{if } n < k \\ q_{n-k}(c_j) & \text{if } n \geq k. \end{cases} \qquad (17.5c')$$

Here $q_n(\theta)$ is the polynomial given by a continuous Runge-Kutta method (Section II.6)

$$q_n(\theta) = y_n + h \sum_{j=1}^{s} b_j(\theta) f(x_n + c_j h, g_j^{(n)}, \gamma_j^{(n)}).$$

a) Prove that the orthogonality conditions

$$\int_0^1 \theta^{q-1} \left(\gamma(t) \sum_{j=1}^{s} b_j(\theta) \Phi_j(t) - \theta \varrho(t) \right) d\theta = 0 \qquad \text{for } q + \varrho(t) \leq p \tag{17.27}$$

imply convergence of order p, if the underlying Runge-Kutta method is of order p for ordinary differential equations.

Hint. Use the theory of B-series and the Gröbner - Alekseev formula (I.14.18) of Section I.14.

b) If for a given Runge-Kutta method the polynomials $b_j(\theta)$ of degree $\leq [(p+1)/2]$ are such that $b_j(0) = 0$, $b_j(1) = b_j$ and

$$\int_0^1 \theta^{q-1} b_j(\theta) d\theta = \frac{1}{q} b_j(1 - c_j^q), \qquad q = 1, \ldots, [(p-1)/2], \tag{17.28}$$

then (17.27) is satisfied. In addition one has the order conditions

$$\sum_{j=1}^{s} b_j(\theta) \Phi_j(t) = \frac{\theta \varrho(t)}{\gamma(t)} \qquad \text{for } \varrho(t) \leq [(p+1)/2].$$

c) Show that the conditions (17.28) admit unique polynomials $b_j(\theta)$ of degree $[(p+1)/2]$.

6. Solve Volterra's equation (17.21) with $k(x) = c$ and compare the solution with the "pollution free" problem (17.10). Which population lives better, that *with* pollution, or that without?

Chapter III. Multistep Methods and General Linear Methods

This chapter is devoted to the study of multistep and general multivalue methods. After retracing their historical development (Adams, Nyström, Milne, BDF) we study in the subsequent sections the order, stability and convergence properties of these methods. Convergence is most elegantly set in the framework of one-step methods in higher dimensions. Sections III.5 and III.6 are devoted to variable step size and Nordsieck methods. We then discuss the various available codes and compare them on the numerical examples of Section II.10 as well as on some equations of high dimension. Before closing the chapter with a section on special methods for second order equations, we discuss two highly theoretical subjects: one on general linear methods, including Runge-Kutta methods as well as multistep methods and many generalizations, and the other on the asymptotic expansion of the global error of such methods.

III.1 Classical Linear Multistep Formulas

> ..., and my undertaking must have ended here, if I had depended
> upon my own resources. But at this point Professor J.C. Adams
> furnished me with a perfectly satisfactory method of calculating
> by quadratures the exact theoretical forms of drops of fluids from
> the Differential Equation of Laplace,...   (F. Bashforth 1883)

Another improvement of Euler's method was considered even earlier than Runge-Kutta methods — the methods of Adams. These were devised by John Couch Adams in order to solve a problem of F. Bashforth, which occurred in an investigation of capillary action. Both the problem and the numerical integration schemes are published in Bashforth (1883). The actual origin of these methods must date back to at least 1855, since in that year F. Bashforth made an application to the Royal Society for assistance from the Government grant. There he wrote: "...,but I am indebted to Mr Adams for a method of treating the differential equation

$$\frac{\frac{ddz}{du^2}}{\left(1+\frac{dz^2}{du^2}\right)^{3/2}} + \frac{\frac{1}{u}\frac{dz}{du}}{\left(1+\frac{dz^2}{du^2}\right)^{1/2}} - 2\alpha z = \frac{2}{b},$$

when put under the form

$$\frac{b}{\varrho} + \frac{b}{x}\sin\varphi = 2 + 2\alpha b^2\frac{z}{b} = 2 + \beta\frac{z}{b},$$

which gives the theoretical form of the drop with an accuracy exceeding that of the most refined measurements."

In contrast to one-step methods, where the numerical solution is obtained solely from the differential equation and the initial value, the algorithm of Adams consists of two parts: firstly, a *starting procedure* which provides $y_1, \ldots, y_{k-1}$ (approximations to the exact solution at the points $x_0 + h, \ldots, x_0 + (k-1)h$) and, secondly, a *multistep formula* to obtain an approximation to the exact solution $y(x_0 + kh)$. This is then applied recursively, based on the numerical approximations of k successive steps, to compute $y(x_0 + (k+1)h)$, etc.

There are several possibilities for obtaining the missing starting values. J.C. Adams actually computed them using the Taylor series expansion of the exact solution (as described in Section I.8, see also Exercise 2). Another possibility is the use of any one-step method, e.g., a Runge-Kutta method (see Chapter II). It is also usual to start with low-order Adams methods and very small step sizes.

Explicit Adams Methods

We now derive, following Adams, the first explicit multistep formulas. We introduce the notation $x_i = x_0 + ih$ for the grid points and suppose we know the numerical approximations $y_n, y_{n-1}, \ldots, y_{n-k+1}$ to the exact solution $y(x_n), \ldots,$ $y(x_{n-k+1})$ of the differential equation

$$y' = f(x, y), \quad y(x_0) = y_0. \tag{1.1}$$

Adams considers (1.1) in integrated form,

$$y(x_{n+1}) = y(x_n) + \int_{x_n}^{x_{n+1}} f\big(t, y(t)\big)\, dt. \tag{1.2}$$

On the right hand side of (1.2) there appears the unknown solution $y(x)$. But since the approximations $y_{n-k+1}, \ldots, y_n$ are known, the values

$$f_i = f(x_i, y_i) \qquad \text{for} \quad i = n-k+1, \ldots, n \tag{1.3}$$

are also available and it is natural to replace the function $f(t, y(t))$ in (1.2) by the interpolation polynomial through the points $\{(x_i, f_i) \mid i = n-k+1, \ldots, n\}$ (see Fig. 1.1).

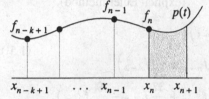
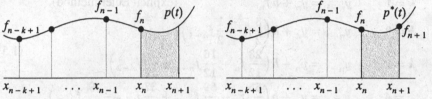

Fig. 1.1. Explicit Adams methods   **Fig. 1.2.** Implicit Adams methods

This polynomial can be expressed in terms of backward differences

$$\nabla^0 f_n = f_n, \qquad \nabla^{j+1} f_n = \nabla^j f_n - \nabla^j f_{n-1}$$

as follows:

$$p(t) = p(x_n + sh) = \sum_{j=0}^{k-1} (-1)^j \binom{-s}{j} \nabla^j f_n \tag{1.4}$$

(Newton's interpolation formula of 1676, published in Newton (1711), see e.g. Henrici (1962), p. 190). The numerical analogue to (1.2) is then given by

$$y_{n+1} = y_n + \int_{x_n}^{x_{n+1}} p(t)\, dt$$

or after insertion of (1.4) by

$$y_{n+1} = y_n + h \sum_{j=0}^{k-1} \gamma_j \nabla^j f_n \tag{1.5}$$

where the coefficients γ_j satisfy

$$\gamma_j = (-1)^j \int_0^1 \binom{-s}{j} ds \qquad (1.6)$$

(see Table 1.1 for their numerical values). A simple recurrence relation for these coefficients will be derived below (formula (1.7)).

Table 1.1. Coefficients for the explicit Adams methods

j	0	1	2	3	4	5	6	7	8
γ_j	1	$\dfrac{1}{2}$	$\dfrac{5}{12}$	$\dfrac{3}{8}$	$\dfrac{251}{720}$	$\dfrac{95}{288}$	$\dfrac{19087}{60480}$	$\dfrac{5257}{17280}$	$\dfrac{1070017}{3628800}$

Special cases of (1.5). For $k = 1, 2, 3, 4$, after expressing the backward differences in terms of f_{n-j}, one obtains the formulas

$$k = 1: \quad y_{n+1} = y_n + h f_n \qquad \text{(explicit Euler method)}$$

$$k = 2: \quad y_{n+1} = y_n + h\left(\frac{3}{2}f_n - \frac{1}{2}f_{n-1}\right)$$

$$k = 3: \quad y_{n+1} = y_n + h\left(\frac{23}{12}f_n - \frac{16}{12}f_{n-1} + \frac{5}{12}f_{n-2}\right) \qquad (1.5')$$

$$k = 4: \quad y_{n+1} = y_n + h\left(\frac{55}{24}f_n - \frac{59}{24}f_{n-1} + \frac{37}{24}f_{n-2} - \frac{9}{24}f_{n-3}\right).$$

Recurrence relation for the coefficients. Using Euler's method of *generating functions* we can deduce a simple recurrence relation for γ_i (see e.g. Henrici 1962). Denote by $G(t)$ the series

$$G(t) = \sum_{j=0}^{\infty} \gamma_j t^j.$$

With the definition of γ_j and the binomial theorem one obtains

$$G(t) = \sum_{j=0}^{\infty}(-t)^j \int_0^1 \binom{-s}{j} ds = \int_0^1 \sum_{j=0}^{\infty}(-t)^j \binom{-s}{j} ds$$

$$= \int_0^1 (1-t)^{-s} ds = -\frac{t}{(1-t)\log(1-t)}.$$

This can be written as

$$-\frac{\log(1-t)}{t} G(t) = \frac{1}{1-t}$$

or as

$$\left(1+\frac{1}{2}t+\frac{1}{3}t^2+\ldots\right)\left(\gamma_0+\gamma_1 t+\gamma_2 t^2+\ldots\right)=\left(1+t+t^2+\ldots\right).$$

Comparing the coefficients of t^m we get the desired recurrence relation

$$\gamma_m+\frac{1}{2}\gamma_{m-1}+\frac{1}{3}\gamma_{m-2}+\ldots+\frac{1}{m+1}\gamma_0=1. \tag{1.7}$$

Implicit Adams Methods

The formulas (1.5) are obtained by integrating the interpolation polynomial (1.4) from x_n to x_{n+1}, i.e., outside the interpolation interval (x_{n-k+1}, x_n). It is well known that an interpolation polynomial is usually a rather poor approximation outside this interval. Adams therefore also investigated methods where (1.4) is replaced by the interpolation polynomial which uses in addition the point (x_{n+1}, f_{n+1}), i.e.,

$$p^*(t)=p^*(x_n+sh)=\sum_{j=0}^{k}(-1)^j\binom{-s+1}{j}\nabla^j f_{n+1} \tag{1.8}$$

(see Fig. 1.2). Inserting this into (1.2) we obtain the following implicit method

$$y_{n+1}=y_n+h\sum_{j=0}^{k}\gamma_j^*\nabla^j f_{n+1} \tag{1.9}$$

where the coefficients γ_j^* satisfy

$$\gamma_j^*=(-1)^j\int_0^1\binom{-s+1}{j}ds \tag{1.10}$$

and are given in Table 1.2 for $j\leq 8$. Again, a simple recurrence relation can be derived for these coefficients (Exercise 3).

Table 1.2. Coefficients for the implicit Adams methods

j	0	1	2	3	4	5	6	7	8
γ_j^*	1	$-\dfrac{1}{2}$	$-\dfrac{1}{12}$	$-\dfrac{1}{24}$	$-\dfrac{19}{720}$	$-\dfrac{3}{160}$	$-\dfrac{863}{60480}$	$-\dfrac{275}{24192}$	$-\dfrac{33953}{3628800}$

The formulas thus obtained are generally of the form

$$y_{n+1}=y_n+h\left(\beta_k f_{n+1}+\ldots+\beta_0 f_{n-k+1}\right). \tag{1.9'}$$

The first examples are as follows

$$k = 0 : \qquad y_{n+1} = y_n + h f_{n+1} = y_n + h f(x_{n+1}, y_{n+1})$$

$$k = 1 : \qquad y_{n+1} = y_n + h \left(\frac{1}{2} f_{n+1} + \frac{1}{2} f_n \right)$$

$$k = 2 : \qquad y_{n+1} = y_n + h \left(\frac{5}{12} f_{n+1} + \frac{8}{12} f_n - \frac{1}{12} f_{n-1} \right) \qquad (1.9'')$$

$$k = 3 : \qquad y_{n+1} = y_n + h \left(\frac{9}{24} f_{n+1} + \frac{19}{24} f_n - \frac{5}{24} f_{n-1} + \frac{1}{24} f_{n-2} \right).$$

The special cases $k = 0$ and $k = 1$ are the implicit Euler method and the trapezoidal rule, respectively. They are actually one-step methods and have already been considered in Chapter II.7.

The methods (1.9) give in general more accurate approximations to the exact solution than (1.5). This will be discussed in detail when the concepts of order and error constant are introduced (Section III.2). The price for this higher accuracy is that y_{n+1} is only defined implicitly by formula (1.9). Therefore, in general a nonlinear equation has to be solved at each step.

Predictor-corrector methods. One possibility for solving this nonlinear equation is to apply fixed point iteration. In practice one proceeds as follows:

P: compute the predictor $\hat{y}_{n+1} = y_n + h \sum_{j=0}^{k-1} \gamma_j \nabla^j f_n$ by the explicit Adams method (1.5); this already yields a reasonable approximation to $y(x_{n+1})$;

E: evaluate the function at this approximation: $\hat{f}_{n+1} = f(x_{n+1}, \hat{y}_{n+1})$;

C: apply the corrector formula

$$y_{n+1} = y_n + h \big(\beta_k \hat{f}_{n+1} + \beta_{k-1} f_n + \ldots + \beta_0 f_{n-k+1} \big) \qquad (1.11)$$

to obtain y_{n+1}.

E: evaluate the function anew, i.e., compute $f_{n+1} = f(x_{n+1}, y_{n+1})$.

This is the most common procedure, denoted by PECE. Other possibilities are: PECECE (two fixed point iterations per step) or PEC (one uses $\hat{f}_{n+1}$ instead of f_{n+1} in the subsequent steps).

This predictor-corrector technique has been used by F.R. Moulton (1926) as well as by W.E. Milne (1926). J.C. Adams actually solved the implicit equation (1.9) by Newton's method, in the same way as is now usual for stiff equations (see Volume II).

Remark. Formula (1.5) is often attributed to Adams-Bashforth. Similarly, the multistep formula (1.9) is usually attributed to Adams-Moulton (Moulton 1926). In fact, both formulas are due to Adams.

Numerical Experiment

We consider the Van der Pol equation (I.16.2) with $\varepsilon = 1$, take as initial values $y_1(0) = A$, $y_2(0) = 0$ on the limit cycle and integrate over one period T (for the values of A and T see Exercise I.16.1). This is exactly the same problem as the one used for the comparison of Runge-Kutta methods (Fig. II.1.1). We have applied the above explicit and implicit Adams methods with several fixed step sizes. The missing starting values were computed with high accuracy by an explicit Runge-Kutta method. Fig. 1.3 shows the errors of both components in dependence of the number of function evaluations. Since we have implemented the implicit method (1.9) in PECE mode it requires 2 function evaluations per step, whereas the explicit method (1.5) needs only one.

This experiment shows that, for the same value of k, the implicit methods usually give a better result (the strange behaviour in the error of the y_2-component for $k \geq 3$ is due to a sign change). Since we have used double logarithmic scales, it is possible to read the "numerical order" from the slope of the corresponding lines. We observe that the global error of the explicit Adams methods behaves like $\mathcal{O}(h^k)$ and that of the implicit methods like $\mathcal{O}(h^{k+1})$. This will be proved in the following sections.

We also remark that the scales used in Fig. 1.3 are exactly the same as those of Fig. II.1.1. This allows a comparison with the Runge-Kutta methods of Section II.1.

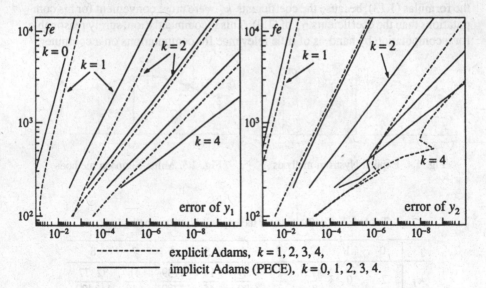

- - - - - - - - - - -   explicit Adams, $k = 1, 2, 3, 4$,
——————   implicit Adams (PECE), $k = 0, 1, 2, 3, 4$.

Fig. 1.3. Global errors versus number of function evaluations

Explicit Nyström Methods

> Die angenäherte Integration hat, besonders in der letzten Zeit, ein ausgedehntes Anwendungsgebiet innerhalb der exakten Wissenschaften und der Technik gefunden.   (E.J. Nyström 1925)

In his review article on the numerical integration of differential equations (which we have already encountered in Section II.14), Nyström (1925) also presents a new class of multistep methods. He considers instead of (1.2) the integral equation

$$y(x_{n+1}) = y(x_{n-1}) + \int_{x_{n-1}}^{x_{n+1}} f(t, y(t)) \, dt. \tag{1.12}$$

In the same way as above he replaces the unknown function $f(t, y(t))$ by the polynomial $p(t)$ of (1.4) and so obtains the formula (see Fig. 1.4)

$$y_{n+1} = y_{n-1} + h \sum_{j=0}^{k-1} \kappa_j \nabla^j f_n \tag{1.13}$$

with the coefficients

$$\kappa_j = (-1)^j \int_{-1}^{1} \binom{-s}{j} \, ds. \tag{1.14}$$

The first of these coefficients are given in Table 1.3. E.J. Nyström recommended the formulas (1.13), because the coefficients κ_j were more convenient for his computations than the coefficients γ_j of (1.6). This recommendation, surely reasonable for a computation by hand, is of little relevance for computations on a computer.

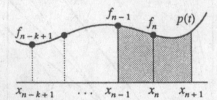

Fig. 1.4. Explicit Nyström methods

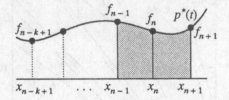

Fig. 1.5. Milne-Simpson methods

Table 1.3. Coefficients for the explicit Nyström methods

j	0	1	2	3	4	5	6	7	8
κ_j	2	0	$\dfrac{1}{3}$	$\dfrac{1}{3}$	$\dfrac{29}{90}$	$\dfrac{14}{45}$	$\dfrac{1139}{3780}$	$\dfrac{41}{140}$	$\dfrac{32377}{113400}$

Special cases. For $k = 1$ the formula

$$y_{n+1} = y_{n-1} + 2hf_n \tag{1.13'}$$

is obtained. It is called the *mid-point rule* and is the simplest two-step method. Its symmetry was extremely useful in the extrapolation schemes of Section II.9. The case $k = 2$ yields nothing new, because $\kappa_1 = 0$. For $k = 3$ one gets

$$y_{n+1} = y_{n-1} + h\left(\frac{7}{3}f_n - \frac{2}{3}f_{n-1} + \frac{1}{3}f_{n-2}\right). \tag{1.13''}$$

Milne–Simpson Methods

We consider again the integral equation (1.12). But now we replace the integrand by the polynomial $p^*(t)$ of (1.8), which in addition to $f_n, \ldots, f_{n-k+1}$ also interpolates the value f_{n+1} (see Fig. 1.5). Proceeding as usual, we get the implicit formulas

$$y_{n+1} = y_{n-1} + h\sum_{j=0}^{k}\kappa_j^*\nabla^j f_{n+1}. \tag{1.15}$$

The coefficients κ_j^* are defined by

$$\kappa_j^* = (-1)^j \int_{-1}^{1} \binom{-s+1}{j} ds, \tag{1.16}$$

and the first of these are given in Table 1.4.

Table 1.4. Coefficients for the Milne-Simpson methods

j	0	1	2	3	4	5	6	7	8
κ_j^*	2	-2	$\frac{1}{3}$	0	$-\frac{1}{90}$	$-\frac{1}{90}$	$-\frac{37}{3780}$	$-\frac{8}{945}$	$-\frac{119}{16200}$

If the backward differences in (1.15) are expressed in terms of f_{n-j}, one obtains the following methods for special values of k:

$$k = 0: \quad y_{n+1} = y_{n-1} + 2hf_{n+1},$$
$$k = 1: \quad y_{n+1} = y_{n-1} + 2hf_n, \tag{1.15'}$$
$$k = 2: \quad y_{n+1} = y_{n-1} + h\left(\frac{1}{3}f_{n+1} + \frac{4}{3}f_n + \frac{1}{3}f_{n-1}\right),$$
$$k = 4: \quad y_{n+1} = y_{n-1} + h\left(\frac{29}{90}f_{n+1} + \frac{124}{90}f_n + \frac{24}{90}f_{n-1} + \frac{4}{90}f_{n-2} - \frac{1}{90}f_{n-3}\right).$$

The special case $k = 0$ is just Euler's implicit method applied with step size $2h$. For $k = 1$ one obtains the previously derived mid-point rule. The particular case

$k = 2$ is an interesting method, known as the *Milne method* (Milne 1926, 1970, p. 66). It is a direct generalization of Simpson's rule.

Many other similar methods have been investigated. They are all based on an integral equation of the form

$$y(x_{n+1}) = y(x_{n-\ell}) + \int_{x_{n-\ell}}^{x_{n+1}} f(t, y(t))\, dt, \qquad (1.17)$$

where $f(t, y(t))$ is replaced either by the interpolating polynomial $p(t)$ (formula (1.4)) or by $p^*(t)$ (formula (1.8)). E.g., for $\ell = 3$ one obtains

$$y_{n+1} = y_{n-3} + h\left(\frac{8}{3}f_n - \frac{4}{3}f_{n-1} + \frac{8}{3}f_{n-2}\right). \qquad (1.18)$$

This particular method has been used by Milne (1926) as a "predictor" for his method: in order to solve the implicit equation (1.15'), Milne uses one or two fixed-point iterations with the numerical value of (1.18) as starting point.

Methods Based on Differentiation (BDF)

> "My name is Gear." — "pardon?"
> "Gear, dshii, ii, ay, are." — "Mr. Jiea?"
> (In a hotel of Paris)

The multistep formulas considered until now are all based on numerical integration, i.e., the integral in (1.17) is approximated numerically using some quadrature formula. The underlying idea of the following multistep formulas is totally different as they are based on the numerical differentiation of a given function.

Assume that the approximations $y_{n-k+1}, \ldots, y_n$ to the exact solution of (1.1) are known. In order to derive a formula for y_{n+1} we consider the polynomial $q(x)$ which interpolates the values $\{(x_i, y_i) \mid i = n-k+1, \ldots, n+1\}$. As in (1.8) this polynomial can be expressed in terms of backward differences, namely

$$q(x) = q(x_n + sh) = \sum_{j=0}^{k} (-1)^j \binom{-s+1}{j} \nabla^j y_{n+1}. \qquad (1.19)$$

The unknown value y_{n+1} will now be determined in such a way that the polynomial $q(x)$ satisfies the differential equation at at least one grid-point, i.e.,

$$q'(x_{n+1-r}) = f(x_{n+1-r}, y_{n+1-r}). \qquad (1.20)$$

For $r = 1$ we obtain *explicit* formulas. For $k = 1$ and $k = 2$, these are equivalent to the explicit Euler method and the mid-point rule, respectively. The case $k = 3$ yields

$$\frac{1}{3}y_{n+1} + \frac{1}{2}y_n - y_{n-1} + \frac{1}{6}y_{n-2} = hf_n. \qquad (1.21)$$

This formula, however, as well as those for $k > 3$, is unstable (see Section III.3) and therefore useless.

Much more interesting are the formulas one obtains when (1.20) is taken for $r = 0$ (see Fig. 1.6).

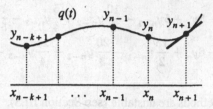

Fig. 1.6. Definition of BDF

In this case one gets the *implicit* formulas

$$\sum_{j=0}^{k} \delta_j^* \nabla^j y_{n+1} = h f_{n+1} \tag{1.22}$$

with the coefficients

$$\delta_j^* = (-1)^j \frac{d}{ds} \binom{-s+1}{j} \bigg|_{s=1}.$$

Using the definition of the binomial coefficient

$$(-1)^j \binom{-s+1}{j} = \frac{1}{j!}(s-1)s(s+1)\dots(s+j-2)$$

the coefficients δ_j^* are obtained by direct differentiation:

$$\delta_0^* = 0, \qquad \delta_j^* = \frac{1}{j} \quad \text{for } j \geq 1. \tag{1.23}$$

Formula (1.22) therefore becomes

$$\sum_{j=1}^{k} \frac{1}{j} \nabla^j y_{n+1} = h f_{n+1}. \tag{1.22'}$$

These multistep formulas, known as *backward differentiation formulas* (or *BDF-methods*), are, since the work of Gear (1971), widely used for the integration of stiff differential equations (see Volume II). They were introduced by Curtiss & Hirschfelder (1952); Mitchell & Craggs (1953) call them "standard step-by-step methods".

For the sake of completeness we give these formulas also in the form which expresses the backward differences in terms of the y_{n-j}.

$$k = 1: \quad y_{n+1} - y_n = h f_{n+1},$$

$$k = 2: \quad \frac{3}{2}y_{n+1} - 2y_n + \frac{1}{2}y_{n-1} = hf_{n+1},$$ (1.22")

$$k = 3: \quad \frac{11}{6}y_{n+1} - 3y_n + \frac{3}{2}y_{n-1} - \frac{1}{3}y_{n-2} = hf_{n+1},$$

$$k = 4: \quad \frac{25}{12}y_{n+1} - 4y_n + 3y_{n-1} - \frac{4}{3}y_{n-2} + \frac{1}{4}y_{n-3} = hf_{n+1},$$

$$k = 5: \quad \frac{137}{60}y_{n+1} - 5y_n + 5y_{n-1} - \frac{10}{3}y_{n-2} + \frac{5}{4}y_{n-3} - \frac{1}{5}y_{n-4} = hf_{n+1},$$

$$k = 6: \quad \frac{147}{60}y_{n+1} - 6y_n + \frac{15}{2}y_{n-1} - \frac{20}{3}y_{n-2} + \frac{15}{4}y_{n-3} - \frac{6}{5}y_{n-4} + \frac{1}{6}y_{n-5}$$
$$= hf_{n+1}.$$

For $k > 6$ the BDF-methods are unstable (see Section III.3).

Exercises

1. Let the differential equation $y' = y^2$, $y(0) = 1$ and the exact starting values $y_i = 1/(1 - x_i)$ for $i = 0, 1, \ldots, k - 1$ be given. Apply the methods of Adams and study the expression $y(x_k) - y_k$ for small step sizes.

2. Consider the differential equation at the beginning of this section. It describes the form of a drop and can be written as (F. Bashforth 1883, page 26; the same problem as Exercise 2 of Section II.1 in a different coordinate system)

$$\frac{dx}{d\varphi} = \varrho \cos\varphi, \qquad \frac{dz}{d\varphi} = \varrho \sin\varphi$$ (1.24)

where

$$\frac{1}{\varrho} + \frac{\sin\varphi}{x} = 2 + \beta z.$$ (1.25)

ϱ may be considered as a function of the coordinates x and z. It can be interpreted as the radius of curvature and φ denotes the angle between the normal to the curve and the z-axis (see Fig. 1.7 for $\beta = 3$). The initial values are given by $x(0) = 0$, $z(0) = 0$, $\varrho(0) = 1$.

Solve the above differential equation along the lines of J.C. Adams:

a) Assuming

$$\varrho = 1 + b_2\varphi^2 + b_4\varphi^4 + \ldots$$

and inserting this expression into (1.24) we obtain after integration the truncated Taylor series of $x(\varphi)$ and $z(\varphi)$ in terms of $b_2, b_4, \ldots$. These parameters can then be calculated from (1.25) by comparing the coefficients of φ^m. In this way one obtains the solution for small values of φ (starting values).

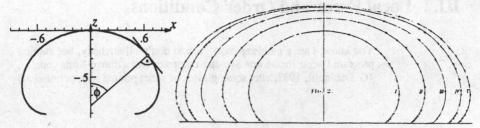

Fig. 1.7. Solution of the differential equation (1.24)
and an illustration from the book of Bashforth

b) Use one of the proposed multistep formulas and calculate the solution for fixed β (say $\beta = 3$) over the interval $[0, \pi]$.

3. Prove that the coefficients γ_j^*, defined by (1.10), satisfy $\gamma_0^* = 1$ and

$$\gamma_m^* + \frac{1}{2}\gamma_{m-1}^* + \frac{1}{3}\gamma_{m-2}^* + \ldots + \frac{1}{m+1}\gamma_0^* = 0 \qquad \text{for} \quad m \geq 1.$$

4. Let $\kappa_j, \kappa_j^*, \gamma_j, \gamma_j^*$ be the coefficients defined by (1.14), (1.16), (1.6), (1.10), respectively. Show that (with $\gamma_{-1} = \gamma_{-1}^* = 0$)

$$\kappa_j = 2\gamma_j - \gamma_{j-1}, \qquad \kappa_j^* = 2\gamma_j^* - \gamma_{j-1}^* \qquad \text{for} \quad j \geq 0.$$

Hint. By splitting the integral in (1.14) one gets $\kappa_j = \gamma_j + \gamma_j^*$. The relation $\gamma_j^* = \gamma_j - \gamma_{j-1}$ is obtained by using a well-known identity for binomial coefficients.

III.2 Local Error and Order Conditions

A general theory of multistep methods was started by the work of Dahlquist
(1956, 1959), and became famous through the classical book of Henrici (1962).
All multistep formulas considered in the previous section have this in common that
the numerical approximations y_i as well as the values f_i appear linearly. We thus
consider the general difference equation

$$\alpha_k y_{n+k} + \alpha_{k-1} y_{n+k-1} + \ldots + \alpha_0 y_n = h(\beta_k f_{n+k} + \ldots + \beta_0 f_n) \qquad (2.1)$$

which includes all considered methods as special cases. In this formula the α_i and
β_i are real parameters, h denotes the step size and

$$f_i = f(x_i, y_i), \qquad x_i = x_0 + ih.$$

Throughout this chapter we shall assume that

$$\alpha_k \neq 0, \qquad |\alpha_0| + |\beta_0| > 0. \qquad (2.2)$$

The first assumption expresses the fact that the implicit equation (2.1) can be solved
with respect to y_{n+k} at least for sufficiently small h. The second relation in (2.2)
can always be achieved by reducing the index k, if necessary.

Formula (2.1) will be called a *linear multistep method* or more precisely a
linear k-step method. We also distinguish between *explicit* ($\beta_k = 0$) and *implicit*
($\beta_k \neq 0$) multistep methods.

Local Error of a Multistep Method

As the numerical solution of a multistep method does not depend only on the initial
value problem (1.1) but also on the choice of the starting values, the definition of
the local error is not as straightforward as for one-step methods (compare Sections
II.2 and II.3).

Definition 2.1. The *local error* of the multistep method (2.1) is defined by

$$y(x_k) - y_k$$

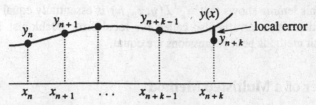

Fig. 2.1. Illustration of the local error

where $y(x)$ is the exact solution of $y' = f(x,y)$, $y(x_0) = y_0$, and y_k is the numerical solution obtained from (2.1) by using the exact starting values $y_i = y(x_i)$ for $i = 0, 1, \ldots, k-1$ (see Fig. 2.1).

In the case $k = 1$ this definition coincides with the definition of the local error for one-step methods. In order to show the connection with other possible definitions of the local error, we associate with (2.1) the linear difference operator L defined by

$$L(y, x, h) = \sum_{i=0}^{k} \Big(\alpha_i y(x + ih) - h\beta_i y'(x + ih) \Big). \tag{2.3}$$

Here $y(x)$ is some differentiable function defined on an interval that contains the values $x + ih$ for $i = 0, 1, \ldots, k$.

Lemma 2.2. *Consider the differential equation (1.1) with $f(x,y)$ continuously differentiable and let $y(x)$ be its solution. For the local error one has*

$$y(x_k) - y_k = \Big(\alpha_k I - h\beta_k \frac{\partial f}{\partial y}(x_k, \eta) \Big)^{-1} L(y, x_0, h).$$

Here η is some value between $y(x_k)$ and y_k, if f is a scalar function. In the case of a vector valued function f, the matrix $\frac{\partial f}{\partial y}(x_k, \eta)$ is the Jacobian whose rows are evaluated at possibly different values lying on the segment joining $y(x_k)$ and y_k.

Proof. By Definition 2.1, y_k is determined implicitly by the equation

$$\sum_{i=0}^{k-1} \Big(\alpha_i y(x_i) - h\beta_i f(x_i, y(x_i)) \Big) + \alpha_k y_k - h\beta_k f(x_k, y_k) = 0.$$

Inserting (2.3) we obtain

$$L(y, x_0, h) = \alpha_k \big(y(x_k) - y_k \big) - h\beta_k \big(f(x_k, y(x_k)) - f(x_k, y_k) \big)$$

and the statement follows from the mean value theorem.   □

This lemma shows that $\alpha_k^{-1} L(y, x_0, h)$ is essentially equal to the local error. Sometimes this term is also called *the* local error (Dahlquist 1956, 1959). For explicit methods both expressions are equal.

Order of a Multistep Method

Once the local error of a multistep method is defined, one can introduce the concept of order in the same way as for one-step methods.

Definition 2.3. The multistep method (2.1) is said to be of *order p*, if one of the following two conditions is satisfied:

i) for all sufficiently regular functions $y(x)$ we have $L(y, x, h) = \mathcal{O}(h^{p+1})$;

ii) the local error of (2.1) is $\mathcal{O}(h^{p+1})$ for all sufficiently regular differential equations (1.1).

Observe that by Lemma 2.2 the above conditions (i) and (ii) are equivalent. Our next aim is to characterize the order of a multistep method in terms of the free parameters α_i and β_i. Dahlquist (1956) was the first to observe the fundamental role of the polynomials

$$\varrho(\zeta) = \alpha_k \zeta^k + \alpha_{k-1} \zeta^{k-1} + \ldots + \alpha_0$$
$$\sigma(\zeta) = \beta_k \zeta^k + \beta_{k-1} \zeta^{k-1} + \ldots + \beta_0. \tag{2.4}$$

They will be called the *generating polynomials* of the multistep method (2.1).

Theorem 2.4. *The multistep method (2.1) is of order p, if and only if one of the following equivalent conditions is satisfied:*

i) $\displaystyle\sum_{i=0}^{k} \alpha_i = 0$ *and* $\displaystyle\sum_{i=0}^{k} \alpha_i i^q = q \sum_{i=0}^{k} \beta_i i^{q-1}$ *for* $q = 1, \ldots, p$;

ii) $\varrho(e^h) - h\sigma(e^h) = \mathcal{O}(h^{p+1})$ *for* $h \to 0$;

iii) $\dfrac{\varrho(\zeta)}{\log \zeta} - \sigma(\zeta) = \mathcal{O}((\zeta - 1)^p)$ *for* $\zeta \to 1$.

Proof. Expanding $y(x + ih)$ and $y'(x + ih)$ into a Taylor series and inserting these series (truncated if necessary) into (2.3) yields

$$L(y, x, h) = \sum_{i=0}^{k} \left(\alpha_i \sum_{q \geq 0} \frac{i^q}{q!} h^q y^{(q)}(x) - h\beta_i \sum_{r \geq 0} \frac{i^r}{r!} h^r y^{(r+1)}(x) \right)$$

$$= y(x) \sum_{i=0}^{k} \alpha_i + \sum_{q \geq 1} \frac{h^q}{q!} y^{(q)}(x) \left(\sum_{i=0}^{k} \alpha_i i^q - q \sum_{i=0}^{k} \beta_i i^{q-1} \right). \tag{2.5}$$

This implies the equivalence of condition (i) with $L(y, x, h) = \mathcal{O}(h^{p+1})$ for all sufficiently regular functions $y(x)$.

It remains to prove that the three conditions of Theorem 2.4 are equivalent. The identity

$$L(\exp, 0, h) = \varrho(e^h) - h\sigma(e^h)$$

where exp denotes the exponential function, together with

$$L(\exp, 0, h) = \sum_{i=0}^{k} \alpha_i + \sum_{q \geq 1} \frac{h^q}{q!} \left(\sum_{i=0}^{k} \alpha_i i^q - q \sum_{i=0}^{k} \beta_i i^{q-1} \right),$$

which follows from (2.5), shows the equivalence of the conditions (i) and (ii).

By use of the transformation $\zeta = e^h$ (or $h = \log \zeta$) condition (ii) can be written in the form

$$\varrho(\zeta) - \log \zeta \cdot \sigma(\zeta) = \mathcal{O}\big((\log \zeta)^{p+1}\big) \qquad \text{for } \zeta \to 1.$$

But this condition is equivalent to (iii), since

$$\log \zeta = (\zeta - 1) + \mathcal{O}\big((\zeta - 1)^2\big) \qquad \text{for } \zeta \to 1. \qquad \square$$

Remark. The conditions for a multistep method to be of order 1, which are usually called *consistency* conditions, can also be written in the form

$$\varrho(1) = 0, \qquad \varrho'(1) = \sigma(1). \tag{2.6}$$

Once the proofs of the above order conditions have been understood, it is not difficult to treat the more general situation of non-equidistant grids (see Section III.5 and the book of Stetter (1973), p. 191).

Example 2.5. *Order of the explicit Adams methods.* Let us first investigate for which differential equations the explicit Adams methods give theoretically the exact solution. This is the case if the polynomial $p(t)$ of (1.4) is equal to $f(t, y(t))$. Suppose now that $f(t, y) = f(t)$ does not depend on y and is a polynomial of degree less than k. Then the explicit Adams methods integrate the differential equations

$$y' = qx^{q-1}, \qquad \text{for } q = 0, 1, \ldots, k$$

exactly. This means that the local error is zero and hence, by Lemma 2.2,

$$0 = L(x^q, 0, h) = h^q \left(\sum_{i=0}^{k} \alpha_i i^q - q \sum_{i=0}^{k} \beta_i i^{q-1} \right) \qquad \text{for } q = 0, \ldots, k.$$

This is just condition (i) of Theorem 2.4 with $p = k$ so that the order of the explicit Adams methods is at least k. In fact it will be shown that the order of these methods is not greater than k (Example 2.7).

Example 2.6. For *implicit Adams methods* the polynomial $p^*(t)$ of (1.8) has degree one higher than that of $p(t)$. Thus the same considerations as in Example 2.5 show that these methods have order at least $k+1$.

All methods of Section III.1 can be treated analogously (see Exercise 3 and Table 2.1).

Table 2.1. Order and error constant of multistep methods

method	formula	order	error constant
explicit Adams	(1.5)	k	γ_k
implicit Adams	(1.9)	$k+1$	γ_{k+1}^*
midpoint rule	(1.13')	2	$1/6$
Nyström, $k>2$	(1.13)	k	$\kappa_k/2$
Milne, $k=2$	(1.15')	4	$-1/180$
Milne-Simpson, $k>3$	(1.15)	$k+1$	$\kappa_{k+1}^*/2$
BDF	(1.22')	k	$-1/(k+1)$

Error Constant

The order of a multistep method indicates how fast the error tends to zero if $h \to 0$. Different methods of the *same* order, however, can have different errors; they are distinguished by the *error constant*. Formula (2.5) shows that the difference operator L, associated with a pth order multistep method, is such that for all sufficiently regular functions $y(x)$

$$L(y,x,h) = C_{p+1}h^{p+1}y^{(p+1)}(x) + \mathcal{O}(h^{p+2}) \qquad (2.7)$$

where the constant C_{p+1} is given by

$$C_{p+1} = \frac{1}{(p+1)!}\left(\sum_{i=0}^{k}\alpha_i i^{p+1} - (p+1)\sum_{i=0}^{k}\beta_i i^p\right). \qquad (2.8)$$

This constant is not suitable as a measure of accuracy, since multiplication of formula (2.1) by a constant can give any value for C_{p+1}, whereas the numerical solution $\{y_n\}$ remains unchanged. A better choice would be the constant $\alpha_k^{-1}C_{p+1}$, since the local error of a multistep method is given by (Lemma 2.2 and formula (2.7))

$$y(x_k) - y_k = \alpha_k^{-1}C_{p+1}h^{p+1}y^{(p+1)}(x_0) + \mathcal{O}(h^{p+2}). \qquad (2.9)$$

For several reasons, however, this is not yet a satisfactory definition, as we shall see from the following motivation: let

$$e_n = \frac{y(x_n) - y_n}{h^p}$$

be the global error scaled by h^p, ans assume for this motivation that $e_n = \mathcal{O}(1)$. Subtracting (2.1) from (2.3) and using (2.7) we have

$$\sum_{i=0}^{k} \alpha_i e_{n+i} = h^{1-p} \sum_{i=0}^{k} \beta_i \Big(f\big(x_{n+i}, y(x_{n+i})\big) - f(x_{n+i}, y_{n+i}) \Big)$$
$$+ C_{p+1} h y^{(p+1)}(x_n) + \mathcal{O}(h^2). \tag{2.10}$$

The point is now to use

$$y^{(p+1)}(x_n) = \frac{1}{\sigma(1)} \sum_{i=0}^{k} \beta_i y^{(p+1)}(x_{n+i}) + \mathcal{O}(h) \tag{2.11}$$

which brings the error term in (2.10) inside the sum with the β_i. We linearize

$$f\big(x_{n+i}, y(x_{n+i})\big) - f(x_{n+i}, y_{n+i}) = \frac{\partial f}{\partial y}\big(x_{n+i}, y(x_{n+i})\big) h^p e_{n+i} + \mathcal{O}(h^{2p})$$

and insert this together with (2.11) into (2.10). Neglecting the $\mathcal{O}(h^2)$ and $\mathcal{O}(h^{2p})$ terms, we can interpret the obtained formula as the multistep method applied to

$$e'(x) = \frac{\partial f}{\partial y}\big(x, y(x)\big)e(x) + C y^{(p+1)}(x), \qquad e(x_0) = 0, \tag{2.12}$$

where

$$C = \frac{C_{p+1}}{\sigma(1)} \tag{2.13}$$

is seen to be a natural measure for the global error and is therefore called *the error constant*.

Another derivation of Definition (2.13) will be given in the section on global convergence (see Exercise 2 of Section III.4). Further, the solution of (2.12) gives the first term of the asymptotic expansion of the global error (see Section III.9).

Example 2.7. *Error constant of the explicit Adams methods.* Consider the differential equation $y' = f(x)$ with $f(x) = (k+1)x^k$, the exact solution of which is $y(x) = x^{k+1}$. As this differential equation is integrated exactly by the $(k+1)$-step explicit Adams method (see Example 2.5), we have

$$y(x_k) - y(x_{k-1}) = h \sum_{j=0}^{k} \gamma_j \nabla^j f_{k-1}.$$

The local error of the k-step explicit Adams method (1.5) is therefore given by

$$y(x_k) - y_k = h \gamma_k \nabla^k f_{k-1} = h^{k+1} \gamma_k f^{(k)}(x_0) = h^{k+1} \gamma_k y^{(k+1)}(x_0).$$

As $\gamma_k \neq 0$, this formula shows that the order of the k-step method is not greater than k (compare Example 2.5). Furthermore, since $\alpha_k = 1$, a comparison with formula (2.9) yields $C_{k+1} = \gamma_k$. Finally, for Adams methods we have $\varrho(\zeta) = \zeta^k - \zeta^{k-1}$ and $\varrho'(1) = 1$, so that by the use of (2.6) the error constant is given by $C = \gamma_k$.

The error constants of all other previously considered multistep methods are summarized in Table 2.1 (observe that $\sigma(1) = 2$ for explicit Nyström and Milne-Simpson methods).

Irreducible Methods

Let $\varrho(\zeta)$ and $\sigma(\zeta)$ of formula (2.4) be the generating polynomials of (2.1) and suppose that they have a common factor $\varphi(\zeta)$. Then the polynomials

$$\varrho^*(\zeta) = \frac{\varrho(\zeta)}{\varphi(\zeta)}, \qquad \sigma^*(\zeta) = \frac{\sigma(\zeta)}{\varphi(\zeta)},$$

are the generating polynomials of a new and simpler multistep method. Using the shift operator E, defined by

$$Ey_n = y_{n+1} \qquad \text{or} \qquad Ey(x) = y(x+h),$$

this multistep method can be written in compact form as

$$\varrho^*(E)y_n = h\sigma^*(E)f_n.$$

Multiplication by $\varphi(E)$ shows that any solution $\{y_n\}$ of this method is also a solution of $\varrho(E)y_n = h\sigma(E)f_n$. The two methods are thus essentially equal. Denote by L^* the difference operator associated with the new reduced method, and by C^*_{p+1} the constant given by (2.7). As

$$L(y,x,h) = \varphi(E)L^*(y,x,h) = C^*_{p+1}h^{p+1}\varphi(E)y^{(p+1)}(x) + \mathcal{O}(h^{p+2})$$
$$= C^*_{p+1}\varphi(1)h^{p+1}y^{(p+1)}(x) + \mathcal{O}(h^{p+2})$$

one immediately obtains $C_{p+1} = \varphi(1)C^*_{p+1}$ and therefore also the relation

$$C_{p+1}/\sigma(1) = C^*_{p+1}/\sigma^*(1)$$

holds. Both methods thus have the same error constant.

The above analysis has shown that multistep methods whose generating polynomials have a common factor are not interesting. We therefore usually assume that

$$\varrho(\zeta) \text{ and } \sigma(\zeta) \text{ have no common factor.} \qquad (2.14)$$

Multistep methods satisfying this property are called *irreducible*.

The Peano Kernel of a Multistep Method

The order and the error constant above do not yet give a complete description of the error, since the subsequent terms of the series for the error may be much larger than C_{p+1}. Several attempts have therefore been made, originally for the error of a quadrature formula, to obtain a complete description of the error. The following discussion is an extension of the ideas of Peano (1913).

Theorem 2.8. *Let the multistep method (2.1) be of order p and let q $(1 \leq q \leq p)$ be an integer. For any $(q+1)$-times continuously differentiable function $y(x)$ we then have*

$$L(y, x, h) = h^{q+1} \int_0^k K_q(s) y^{(q+1)}(x + sh)\, ds, \qquad (2.15)$$

where

$$K_q(s) = \frac{1}{q!} \sum_{i=0}^{k}{}' \alpha_i (i - s)_+^q - \frac{1}{(q-1)!} \sum_{i=0}^{k} \beta_i (i - s)_+^{q-1} \qquad (2.16a)$$

with

$$(i - s)_+^r = \begin{cases} (i - s)^r & \text{for } i - s > 0 \\ 0 & \text{for } i - s \leq 0. \end{cases}$$

$K_q(s)$ *is called the qth Peano kernel of the multistep method (2.1).*

Remark. We see from (2.16a) that $K_q(s)$ is a piecewise polynomial and satisfies

$$K_q(s) = \frac{1}{q!} \sum_{i=j}^{k} \alpha_i (i - s)^q - \frac{1}{(q-1)!} \sum_{i=j}^{k} \beta_i (i - s)^{q-1} \quad \text{for } s \in [j-1, j). \qquad (2.16b)$$

Proof. Taylor's theorem with the integral representation of the remainder yields

$$y(x + ih) = \sum_{r=0}^{q} \frac{i^r}{r!} h^r y^{(r)}(x) + h^{q+1} \int_0^i \frac{(i-s)^q}{q!} y^{(q+1)}(x + sh)\, ds,$$

$$hy'(x + ih) = \sum_{r=1}^{q} \frac{i^{r-1}}{(r-1)!} h^r y^{(r)}(x) + h^{q+1} \int_0^i \frac{(i-s)^{q-1}}{(q-1)!} y^{(q+1)}(x + sh)\, ds.$$

Inserting these two expressions into (2.3), the same considerations as in the proof of Theorem 2.4 show that for $q \leq p$ the polynomials before the integral cancel. The statement then follows from

$$\int_0^i \frac{(i-s)^q}{q!} y^{(q+1)}(x + sh)\, ds = \int_0^k \frac{(i-s)_+^q}{q!} y^{(q+1)}(x + sh)\, ds. \qquad \square$$

Besides the representation (2.16), the Peano kernel $K_q(s)$ has the following properties:

$$K_q(s) = 0 \text{ for } s \in (-\infty, 0) \cup [k, \infty) \text{ and } q = 1, \ldots, p; \qquad (2.17)$$

$K_q(s)$ is $(q-2)$-times continuously differentiable and
$$K_q^i(s) = -K_{q-1}(s) \text{ for } q = 2, \ldots, p \text{ (for } q = 2 \text{ piecewise)}; \qquad (2.18)$$

$K_1(s)$ is a piecewise linear function with discontinuities at $0, 1, \ldots, k$. It has a jump of size β_j at the point j and its slope over the interval $(j-1, j)$ is given by $-(\alpha_j + \alpha_{j+1} + \ldots + \alpha_k)$;   (2.19)

For the constant C_{p+1} of (2.8) we have $C_{p+1} = \int_0^k K_p(s)ds$.   (2.20)

The proofs of Statements (2.17) to (2.20) are as follows: it is an immediate consequence of the definition of the Peano kernel that $K_q(s) = 0$ for $s \geq k$ and $q \leq p$. In order to prove that $K_q(s) = 0$ also for $s < 0$ we consider the polynomial $y(x) = (x - s)^q$ with s as a parameter. Theorem 2.8 then shows that

$$L(y, 0, 1) = \sum_{i=0}^k \alpha_i(i - s)^q - q \sum_{i=0}^k \beta_i(i - s)^{q-1} \equiv 0 \quad \text{for } q \leq p$$

and hence $K_q(s) = 0$ for $s < 0$. This gives (2.17). The relation (2.18) is seen by partial integration of (2.15). As an example, the Peano kernels for the 3-step Nyström method (1.13″) are plotted in Fig. 2.2.

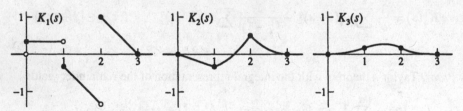

Fig. 2.2. Peano kernels of the 3-step Nyström method

Exercises

1. Construction of multistep methods. Let $\varrho(\zeta)$ be a kth degree polynomial satisfying $\varrho(1) = 0$.

   a) There exists exactly one polynomial $\sigma(\zeta)$ of degree $\leq k$, such that the order of the corresponding multistep method is at least $k+1$.

   b) There exists exactly one polynomial $\sigma(\zeta)$ of degree $< k$, such that the corresponding multistep method, which is then explicit, has order at least k.

   Hint. Use condition (iii) of Theorem 2.4.

2. Find the multistep method of the form

$$y_{n+2} + \alpha_1 y_{n+1} + \alpha_0 y_n = h(\beta_1 f_{n+1} + \beta_0 f_n)$$

   of the highest possible order. Apply this formula to the example $y' = y$, $y(0) = 1$, $h = 0.1$.

3. Verify that the order and the error constant of the BDF-formulas are those of Table 2.1.

4. Show that the Peano kernel $K_p(s)$ does not change sign for the explicit and implicit Adams methods, nor for the BDF-formulas. Deduce from this property that

$$L(y, x, h) = h^{p+1} C_{p+1} y^{(p+1)}(\zeta) \quad \text{with } \zeta \in (x, x + kh)$$

   where the constant C_{p+1} is given by (2.8).

5. Let $y(x)$ be an exact solution of $y' = f(x, y)$ and let $y_i = y(x_i)$, $i = 0, 1, \ldots,$ $k - 1$. Assume that f is continuous and satisfies a Lipschitz condition with respect to y (f not necessarily differentiable). Prove that for consistent multistep methods (i.e., methods with (2.6)) the local error satisfies

$$\|y(x_k) - y_k\| \leq h\omega(h)$$

   where $\omega(h) \to 0$ for $h \to 0$.

III.3 Stability and the First Dahlquist Barrier

... hat der Verfasser seither öfters Verfahren zur numerischen
Integration von Differentialgleichungen beobachtet, die, obschon
zwar mit bestechend kleinem Abbruchfehler behaftet, doch die
grosse Gefahr der numerischen Instabilität in sich bergen.

(H. Rutishauser 1952)

Rutishauser observed in his famous paper that high order and a small local error
are not sufficient for a useful multistep method. The numerical solution can be
"unstable", even though the step size h is taken very small. The same observation
was made by Todd (1950), who applied certain difference methods to second
order differential equations. Our presentation will mainly follow the lines of
Dahlquist (1956), where this effect has been studied systematically. An interesting
presentation of the historical development of numerical stability concepts can be
found in Dahlquist (1985) "33 years of numerical instability, Part I".

Let us start with an example, taken from Dahlquist (1956). Among all explicit
2-step methods we consider the formula with the highest order (see Exercise 2 of
Section III.2). A short calculation using Theorem 2.4 shows that this method of
order 3 is given by

$$y_{n+2} + 4y_{n+1} - 5y_n = h(4f_{n+1} + 2f_n). \tag{3.1}$$

Application to the differential equation

$$y' = y, \qquad y(0) = 1 \tag{3.2}$$

yields the linear difference relation

$$y_{n+2} + 4(1-h)y_{n+1} - (5+2h)y_n = 0. \tag{3.3}$$

As starting values we take $y_0 = 1$ and $y_1 = \exp(h)$, the values on the exact solution.
The numerical solution together with the exact solution $\exp(x)$ is plotted in Fig. 3.1
for the step sizes $h = 1/10$, $h = 1/20$, $h = 1/40$, etc. In spite of the small local
error, the results are very bad and become even worse as the step size decreases.

An explanation for this effect can easily be given. As usual for linear difference
equations (Dan. Bernoulli 1728, Lagrange 1775), we insert $y_j = \zeta^j$ into (3.3). This
leads to the characteristic equation

$$\zeta^2 + 4(1-h)\zeta - (5+2h) = 0. \tag{3.4}$$

The general solution of (3.3) is then given by

$$y_n = A\zeta_1^n(h) + B\zeta_2^n(h) \tag{3.5}$$

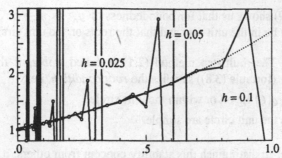

Fig. 3.1. Numerical solution of the unstable method (3.1)

where

$$\zeta_1(h) = 1 + h + \mathcal{O}(h^2), \qquad \zeta_2(h) = -5 + \mathcal{O}(h)$$

are the roots of (3.4) and the coefficients A and B are determined by the starting values y_0 and y_1. Since $\zeta_1(h)$ approximates $\exp(h)$, the first term in (3.5) approximates the exact solution $\exp(x)$ at the point $x = nh$. The second term in (3.5), often called a *parasitic solution,* is the one which causes trouble in our method: since for $h \to 0$ the absolute value of $\zeta_2(h)$ is larger than one, this parasitic solution becomes very large and dominates the solution y_n for increasing n.

We now turn to the stability discussion of the general method (2.1). The essential part is the behaviour of the solution as $n \to \infty$ (or $h \to 0$) with nh fixed. We see from (3.3) that for $h \to 0$ we obtain

$$\alpha_k y_{n+k} + \alpha_{k-1} y_{n+k-1} + \ldots + \alpha_0 y_n = 0. \tag{3.6}$$

This can be interpreted as the numerical solution of the method (2.1) for the differential equation

$$y' = 0. \tag{3.7}$$

We put $y_j = \zeta^j$ in (3.6), divide by ζ^n, and find that ζ must be a root of

$$\varrho(\zeta) = \alpha_k \zeta^k + \alpha_{k-1} \zeta^{k-1} + \ldots + \alpha_0 = 0. \tag{3.8}$$

As in Section I.13, we again have some difficulty when (3.8) possesses a root of *multiplicity* $m > 1$. In this case (Lagrange 1792, see Exercise 1 below) $y_n = n^{j-1}\zeta^n$ ($j = 1, \ldots, m$) are solutions of (3.6) and we obtain by superposition:

Lemma 3.1. *Let* $\zeta_1, \ldots, \zeta_l$ *be the roots of* $\varrho(\zeta)$, *of respective multiplicity* $m_1, \ldots, m_l$. *Then the general solution of (3.6) is given by*

$$y_n = p_1(n)\zeta_1^n + \ldots + p_l(n)\zeta_l^n \tag{3.9}$$

where the $p_j(n)$ *are polynomials of degree* $m_j - 1$. $\qquad\qquad\Box$

Formula (3.9) shows us that for boundedness of y_n, as $n \to \infty$, we need that the roots of (3.8) lie in the unit disc and that the roots on the unit circle be simple.

Definition 3.2. The multistep method (2.1) is called *stable*, if the generating polynomial $\varrho(\zeta)$ (formula (3.8)) satisfies the *root condition*, i.e.,

i) The roots of $\varrho(\zeta)$ lie on or within the unit circle;

ii) The roots on the unit circle are simple.

Remark. In order to distinguish this stability concept from others, it is sometimes called *zero-stability* or, in honour of Dahlquist, also *D-stability*.

Examples. For the explicit and implicit *Adams methods*, $\varrho(\zeta) = \zeta^k - \zeta^{k-1}$. Besides the simple root 1, there is a $(k-1)$-fold root at 0. The Adams methods are therefore stable.

The same is true for the explicit *Nyström* and the *Milne-Simpson methods*, where $\varrho(\zeta) = \zeta^k - \zeta^{k-2}$. Note that here we have a simple root at -1. This root can be dangerous for certain differential equations (see Section III.9 and Section V.1 of Volume II).

Stability of the BDF-Formulas

The investigation of the stability of the BDF-formulas is more difficult. As the characteristic polynomial of $\nabla^j y_{k+n} = 0$ is given by $\zeta^{k-j}(\zeta - 1)^j = 0$ it follows from the representation (1.22') that the generating polynomial $\varrho(\zeta)$ of the BDF-formulas has the form

$$\varrho(\zeta) = \sum_{j=1}^{k} \frac{1}{j} \zeta^{k-j}(\zeta - 1)^j. \tag{3.10}$$

In order to study the zeros of (3.10) it is more convenient to consider the polynomial

$$p(z) = (1-z)^k \varrho\left(\frac{1}{1-z}\right) = \sum_{j=1}^{k} \frac{z^j}{j} \tag{3.11}$$

via the transformation $\zeta = 1/(1-z)$. This polynomial is just the kth partial sum of $-\log(1-z)$. As the roots of $p(z)$ and $\varrho(\zeta)$ are related by the above transformation, we have:

Lemma 3.3. *The k-step BDF-formula (1.22') is stable iff all roots of the polynomial (3.11) are outside the disc $\{z; |z - 1| \leq 1\}$, with simple roots allowed on the boundary.*   □

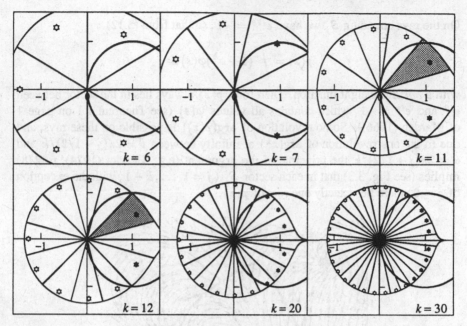

Fig. 3.2. Roots of the polynomial $p(z)$ of (3.11)

The roots of (3.11) are displayed in Fig. 3.2 for different values of k.

Theorem 3.4. *The k-step BDF-formula (1.22') is stable for $k \leq 6$, and unstable for $k \geq 7$.*

Proof. The first assertion can be verified simply by a finite number of numerical calculations (see Fig. 3.2). This was first observed by Mitchell & Craggs (1953). The second statement, however, contains an infinity of cases and is more difficult. The first complete proof was given by Cryer (1971) in a technical report, a condensed version of which is published in Cryer (1972). A second proof is given in Creedon & Miller (1975) (see also Grigorieff (1977), p. 135), based on the Schur-Cohn criterion. This proof is outlined in Exercise 4 below. The following proof, which is given in Hairer & Wanner (1983), is based on the representation

$$ p(z) = \int_0^z \sum_{j=1}^k \zeta^{j-1} d\zeta = \int_0^z \frac{1-\zeta^k}{1-\zeta}\, d\zeta = \int_0^r (1 - e^{ik\theta}s^k)\varphi(s)\, ds \qquad (3.12) $$

with

$$ \zeta = se^{i\theta}, \qquad z = re^{i\theta}, \qquad \varphi(s) = \frac{e^{i\theta}}{1-se^{i\theta}}. $$

We cut the complex plane into k sectors

$$ S_j = \left\{ z\; ; \frac{2\pi}{k}\left(j-\frac{1}{2}\right) < \arg(z) < \frac{2\pi}{k}\left(j+\frac{1}{2}\right)\right\}, \quad j = 0, 1, \ldots, k-1. $$

On the rays bounding S_j we have $e^{ik\theta} = -1$, so that from (3.12)

$$p(z) = \int_0^r (1 + s^k)\varphi(s)\,ds$$

with a *positive* weight function. Therefore, $p(z)$ always lies in the sector between $e^{i\theta}$ and $e^{i\pi} = -1$, which contains all values $\varphi(s)$ (see Theorem 1.1 on page 1 of Marden (1966)). So no revolution of $\arg(p(z))$ is possible on these rays, and due to the one revolution of $\arg(z^k)$ at infinity between $\theta = 2\pi(j - 1/2)/k$ and $\theta = 2\pi(j + 1/2)/k$ the principle of the argument (e.g., Henrici (1974), p. 278) implies (see Fig. 3.3) that in each sector S_j ($j = 1, \ldots, k-1$, with the exception of $j = 0$) there lies exactly one root of $p(z)$.

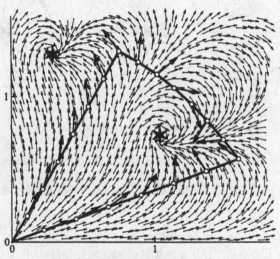

Fig. 3.3. Argument of $p(z)$ of (3.11)

In order to complete the proof, we still have to bound the zeros of $p(z)$ from above: we observe that in (3.12) the term s^k becomes large for $s > 1$. We therefore partition (3.12) into two integrals $p(z) = I_1 - I_2$, where

$$I_1 = \int_0^r \varphi(s)\,ds - \int_0^1 e^{ik\theta}s^k\varphi(s)\,ds, \qquad I_2 = e^{ik\theta}\int_1^r s^k\varphi(s)\,ds.$$

Since $|\varphi(s)| \le B(\theta)$ where

$$B(\theta) = \begin{cases} |\sin\theta|^{-1} & \text{if } 0 < \theta \le \pi/2 \text{ or } 3\pi/2 \le \theta < 2\pi, \\ 1 & \text{otherwise,} \end{cases}$$

we obtain

$$|I_1| \le \left(r + \frac{1}{k+1}\right) B(\theta) < rB(\theta)\frac{k+2}{k+1}, \quad (r > 1). \tag{3.13}$$

Secondly, since s^k is positive,

$$I_2 = e^{ik\theta}\Phi \int_1^r s^k\, ds \quad \text{with} \quad \Phi \in \text{convex hull of } \{\varphi(s); 1 \le s \le r\}.$$

Any element of the above convex hull can be written in the form

$$\Phi = \alpha\varphi(s_1) + (1-\alpha)\varphi(s_2) = \frac{\varphi(s_1)\varphi(s_2)}{\varphi(\widehat{s})}$$

with $\widehat{s} = \alpha s_2 + (1-\alpha)s_1$, $0 \le \alpha \le 1$, $1 \le s_1, s_2 \le r$. Since $|\varphi(s)|$ decreases monotonically for $s \ge 1$, we have $|\Phi| \ge |\varphi(r)|$. Some elementary geometry then leads to $|\Phi| \ge 1/2r$ and we get

$$|I_2| \ge \frac{r^{k+1}-1}{2r(k+1)} > \frac{r(r^{k-1}-1)}{2k+2}, \quad (r>1). \tag{3.14}$$

From (3.13) and (3.14) we see that

$$r \ge R(\theta) = \left((2k+4)B(\theta)+1\right)^{1/(k-1)} \tag{3.15}$$

implies $|I_2| > |I_1|$, so that $p(z)$ cannot be zero. The curve $R(\theta)$ is also plotted in Fig. 3.2 and cuts from the sectors S_j what we call Madame Imhof's cheese pie, each slice of which (with $j \ne 0$) must contain precisely one zero of $p(z)$. A simple analysis shows that for $k = 12$ the cheese pie, cut from S_1, is small enough to ensure the presence of zeros of $p(z)$ inside the disc $\{z; |z-1| \le 1\}$. As $R(\theta)$, for fixed θ, as well as $R(\pi/k)$ are monotonically decreasing in k, the same is true for all $k \ge 12$.

For $6 < k < 12$ numerical calculations show that the method is unstable (see Fig. 3.2 or Exercise 4). □

Highest Attainable Order of Stable Multistep Methods

It is a natural task to investigate the stability of the multistep methods with highest possible order. This has been performed by Dahlquist (1956), resulting in the famous "first Dahlquist-barrier".

Counting the order conditions (Theorem 2.4) shows that for order p the parameters of a linear multistep method have to satisfy $p+1$ linear equations. As $2k+1$ free parameters are involved (without loss of generality one can assume $\alpha_k = 1$), this suggests that $2k$ is the highest attainable order. Indeed, this can be verified (see Exercise 5). However, these methods are of no practical significance, because we shall prove

Theorem 3.5 (The first Dahlquist-barrier). *The order p of a stable linear k-step method satisfies*

$$p \le k+2 \quad \textit{if } k \textit{ is even,}$$
$$p \le k+1 \quad \textit{if } k \textit{ is odd,}$$
$$p \le k \qquad \textit{if } \beta_k/\alpha_k \le 0 \textit{ (in particular if the method is explicit).}$$

We postpone the verification of this theorem and give some notations and lemmas, which will be useful for the proof. First of all we introduce the "Greek-Roman transformation"

$$\zeta = \frac{z+1}{z-1} \quad \text{or} \quad z = \frac{\zeta+1}{\zeta-1}. \tag{3.16}$$

This transformation maps the disk $|\zeta| < 1$ onto the half-plane $\operatorname{Re} z < 0$, the upper half-plane $\operatorname{Im} z > 0$ onto the lower half-plane, the circle $|\zeta| = 1$ to the imaginary axis, the point $\zeta = 1$ to $z = \infty$ and the point $\zeta = -1$ to $z = 0$. We then consider the polynomials

$$R(z) = \left(\frac{z-1}{2}\right)^k \varrho(\zeta) = \sum_{j=0}^{k} a_j z^j,$$

$$S(z) = \left(\frac{z-1}{2}\right)^k \sigma(\zeta) = \sum_{j=0}^{k} b_j z^j. \tag{3.17}$$

Since the zeros of $R(z)$ and of $\varrho(\zeta)$ are connected via the transformation (3.16), the stability condition of a multistep method can be formulated in terms of $R(z)$ as follows: all zeros of $R(z)$ lie in the negative half-plane $\operatorname{Re} z \le 0$ and no multiple zero of $R(z)$ lies on the imaginary axis.

Lemma 3.6. *Suppose the multistep method to be stable and of order at least 0. We then have*

 i) $a_k = 0$ and $a_{k-1} = 2^{1-k} \varrho'(1) \ne 0$;

 ii) *All non-vanishing coefficients of $R(z)$ have the same sign.*

Proof. Dividing formula (3.17) by z^k and putting $z = \infty$, one sees that $a_k = 2^{-k}\varrho(1)$. This expression must vanish, because the method is of order 0. In the same way one gets $a_{k-1} = 2^{1-k}\varrho'(1)$, which is different from zero, since by stability 1 cannot be a multiple root of $\varrho(\zeta)$. The second statement follows from the factorization

$$R(z) = a_{k-1} \prod (z+x_j) \prod ((z+u_j)^2 + v_j^2).$$

where $-x_j$ are the real roots and $-u_j \pm iv_j$ are the conjugate pairs of complex roots. By stability $x_j \ge 0$ and $u_j \ge 0$, implying that all coefficients of $R(z)$ have the same sign. $\qquad\square$

We next express the order conditions of Theorem 2.4 in terms of the polynomials $R(z)$ and $S(z)$.

Lemma 3.7. *The multistep method is of order p if and only if*

$$R(z)\left(\log\frac{z+1}{z-1}\right)^{-1} - S(z) = C_{p+1}\left(\frac{2}{z}\right)^{p-k} + \mathcal{O}\left(\left(\frac{2}{z}\right)^{p-k+1}\right) \quad \text{for } z \to \infty$$

(3.18)

Proof. First, observe that the $\mathcal{O}((\zeta-1)^p)$ term in condition (iii) of Theorem 2.4 is equal to $C_{p+1}(\zeta-1)^p + \mathcal{O}((\zeta-1)^{p+1})$ by formula (2.7). Application of the transformation (3.16) then yields (3.18), because $(\zeta-1) = 2/(z-1) = 2/z + \mathcal{O}((2/z)^2)$ for $z \to \infty$. $\qquad\square$

Lemma 3.8. *The coefficients of the Laurent series*

$$\left(\log\frac{z+1}{z-1}\right)^{-1} = \frac{z}{2} - \mu_1 z^{-1} - \mu_3 z^{-3} - \mu_5 z^{-5} - \ldots$$

(3.19)

satisfy $\mu_{2j+1} > 0$ for all $j \geq 0$.

Proof. We consider the branch of $\log\zeta$ which is analytic in the complex ζ-plane cut along the negative real axis and satisfies $\log 1 = 0$. The transformation (3.16) maps this cut onto the segment from -1 to $+1$ on the real axis. The function $\log((z+1)/(z-1))$ is thus analytic on the complex z-plane cut along this segment (see Fig. 3.4). From the formula

$$\log\frac{z+1}{z-1} = \frac{2}{z}\left(1 + \frac{z^{-2}}{3} + \frac{z^{-4}}{5} + \frac{z^{-6}}{7} + \ldots\right),$$

(3.20)

the existence of (3.19) becomes clear. In order to prove the positivity of the coefficients, we use Cauchy's formula for the coefficients of the function $f(z) = \sum_{n\in\mathbb{Z}} a_n(z-z_0)^n$,

$$a_n = \frac{1}{2\pi i}\int_\gamma \frac{f(z)}{(z-z_0)^{n+1}}\, dz,$$

i.e., in our situation

$$\mu_{2j+1} = -\frac{1}{2\pi i}\int_\gamma z^{2j}\left(\log\frac{z+1}{z-1}\right)^{-1} dz$$

(Cauchy 1831; see also Behnke & Sommer 1962). Here γ is an arbitrary curve enclosing the segment $(-1, 1)$, e.g., the curve plotted in Fig. 3.4.

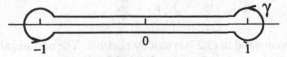

Fig. 3.4. Cut z-plane with curve γ

Observing that $\log((z+1)/(z-1)) = \log((1+x)/(1-x)) - i\pi$ when z approaches the real value $x \in (-1,1)$ from above, and that $\log((z+1)/(z-1)) = \log((1+x)/(1-x)) + i\pi$ when z approaches x from below, we obtain

$$\mu_{2j+1} = -\frac{1}{2\pi i} \int_{-1}^{1} x^{2j} \left[\left(\log \frac{1+x}{1-x} + i\pi \right)^{-1} - \left(\log \frac{1+x}{1-x} - i\pi \right)^{-1} \right] dx$$

$$= \int_{-1}^{1} x^{2j} \left[\left(\log \frac{1+x}{1-x} \right)^2 + \pi^2 \right]^{-1} dx > 0. \qquad \square$$

For another proof of this lemma, which avoids complex analysis, see Exercise 10.

Proof of Theorem 3.5. We insert the series (3.19) into (3.18) and obtain

$$R(z) \left(\log \frac{z+1}{z-1} \right)^{-1} - S(z) = \text{polynomial}(z) + d_1 z^{-1} + d_2 z^{-2} + \mathcal{O}(z^{-3}) \quad (3.21)$$

where

$$\begin{aligned}
d_1 &= -\mu_1 a_0 - \mu_3 a_2 - \mu_5 a_4 - \ldots \\
d_2 &= -\mu_3 a_1 - \mu_5 a_3 - \mu_7 a_5 - \ldots .
\end{aligned} \qquad (3.22)$$

Lemma 3.6 together with the positivity of the μ_j (Lemma 3.8) implies that all summands in the above formulas for d_1 and d_2 have the same sign. Since $a_{k-1} \neq 0$ we therefore have $d_2 \neq 0$ for k even and $d_1 \neq 0$ for k odd. The first two bounds of Theorem 3.5 are now an immediate consequence of formula (3.18).

Finally, we prove that $p \le k$ for $\beta_k/\alpha_k \le 0$: assume, by contradiction, that the order is greater than k. Then by formula (3.18), $S(z)$ is equal to the principal part of $R(z)(\log((z+1)/(z-1)))^{-1}$, and we may write (putting $\mu_j = 0$ for even j)

$$S(z) = R(z) \left(\frac{z}{2} - \sum_{j=1}^{k-1} \mu_j z^{-j} \right) + \sum_{j=1}^{k-1} \left(\sum_{s=j}^{k-1} \mu_s a_{s-j} \right) z^{-j}.$$

Setting $z = 1$ we obtain

$$\frac{S(1)}{R(1)} = \left(\frac{1}{2} - \sum_{j=1}^{k-1} \mu_j \right) + \sum_{j=1}^{k-1} \left(\sum_{s=j}^{k-1} \mu_s a_{s-j} \right) \frac{1}{R(1)}. \qquad (3.23)$$

Since by formula (3.17), $S(1) = \beta_k$ and $R(1) = \alpha_k$, it is sufficient to prove $S(1)/R(1) > 0$. Formula (3.19), for $z \to 1$, gives

$$\sum_{j=1}^{\infty} \mu_j = \frac{1}{2},$$

so that the first summand in (3.23) is strictly positive. The non-negativeness of the second summand is seen from Lemmas 3.6 and 3.8. $\qquad \square$

The stable multistep methods which attain the highest possible order $k+2$ have a very special structure.

Theorem 3.9. *Stable multistep methods of order $k+2$ are symmetric, i.e.,*

$$\alpha_j = -\alpha_{k-j}, \qquad \beta_j = \beta_{k-j} \qquad \text{for all } j. \qquad (3.24)$$

Remark. For symmetric multistep methods we have $\varrho(\zeta) = -\zeta^k \varrho(1/\zeta)$ by definition. Since with ζ_i also $1/\zeta_i$ is a zero of $\varrho(\zeta)$, all roots of stable symmetric multistep methods lie on the unit circle and are simple.

Proof. A comparison of the formulas (3.18) and (3.21) shows that $d_1 = 0$ is necessary for order $k+2$. Since the method is assumed to be stable, Lemma 3.6 implies that all even coefficients of $R(z)$ vanish. Hence, k is even and $R(z)$ satisfies the relation $R(z) = -R(-z)$. By definition of $R(z)$ this relation is equivalent to $\varrho(\zeta) = -\zeta^k \varrho(1/\zeta)$, which implies the first condition of (3.24). Using the above relation for $R(z)$ one obtains from formula (3.18) that $S(z) - S(-z) = \mathcal{O}((2/z)^2)$, implying $S(z) = S(-z)$. If this relation is transformed into an equivalent one for $\sigma(\zeta)$, one gets the second condition of (3.24). $\qquad\square$

Exercises

1. Consider the linear difference equation (3.6) with

$$\varrho(\zeta) = \alpha_k \zeta^k + \alpha_{k-1} \zeta^{k-1} + \ldots + \alpha_0$$

   as characteristic polynomial. Let $\zeta_1, \ldots, \zeta_l$ be the different roots of $\varrho(\zeta)$ and let $m_j \geq 1$ be the multiplicity of the root ζ_j. Show that for $1 \leq j \leq l$ and $0 \leq i \leq m_j - 1$ the sequences

$$\left\{ \binom{n}{i} \zeta_j^{n-i} \right\}_{n \geq 0}$$

   form a system of k linearly independent solutions of (3.6).

2. Show that all roots of the polynomial $p(z)$ of formula (3.11) except the simple root 0 lie in the annulus

$$\frac{k}{k-1} \leq |z| \leq 2.$$

   Hint. Use the following lemma, which can be found in Marden (1966), p.137: if all coefficients of the polynomial $a_k z^k + a_{k-1} z^{k-1} + \ldots + a_0$ are real and positive, then its roots lie in the annulus $\varrho_1 \leq |z| \leq \varrho_2$ with $\varrho_1 = \min(a_j/a_{j+1})$ and $\varrho_2 = \max(a_j/a_{j+1})$.

3. Apply the lemma of the above exercise to $\varrho(\zeta)/(\zeta - 1)$ and show that the BDF-formulas are stable for $k = 1, 2, 3, 4$.

4. Give a different proof of Theorem 3.4 by applying the Schur-Cohn criterion to the polynomial

$$f(z) = z^k \varrho\left(\frac{1}{z}\right) = \sum_{j=1}^{k} \frac{1}{j}(1-z)^j. \tag{3.25}$$

Schur-Cohn criterion (see e.g., Marden (1966), Chapter X). For a given polynomial with real coefficients

$$f(z) = a_0 + a_1 z + \ldots + a_k z^k$$

we consider the coefficients $a_i^{(j)}$ where

$$
\begin{aligned}
a_i^{(0)} &= a_i & i &= 0, 1, \ldots, k \\
a_i^{(j+1)} &= a_0^{(j)} a_i^{(j)} - a_{k-j}^{(j)} a_{k-j-i}^{(j)} & i &= 0, 1, \ldots, k-j-1
\end{aligned} \tag{3.26}
$$

and also the products

$$P_1 = a_0^{(1)}, \qquad P_{j+1} = P_j a_0^{(j+1)} \quad \text{for } j = 1, \ldots, k-1. \tag{3.27}$$

We further denote by n the number of negative elements among the values $P_1, \ldots, P_k$ and by p the number of positive elements. Then $f(z)$ has at least n zeros inside the unit disk and at least p zeros outside it.

a) Prove the following formulas for the coefficients of (3.25):

$$a_0 = \sum_{i=1}^{k} \frac{1}{i}, \qquad a_1 = -k, \qquad a_2 = \frac{k(k-1)}{4},$$

$$a_{k-2} = (-1)^k \frac{k(k-1)}{2(k-2)}, \quad a_{k-1} = (-1)^{k-1} \frac{k}{k-1}, \quad a_k = (-1)^k \frac{1}{k}. \tag{3.28}$$

b) Verify that the coefficients $a_0^{(j)}$ of (3.26) have the sign structure of Table 3.1. For $k < 13$ these tedious calculations can be performed on a computer. The verification of $a_0^{(1)} > 0$ and $a_0^{(2)} > 0$ is easy for all $k > 2$. In order to verify $a_0^{(3)} = (a_0^{(2)})^2 - (a_{k-2}^{(2)})^2 < 0$ for $k \geq 13$ consider the expression

$$
\begin{aligned}
a_0^{(2)} - (-1)^k a_{k-2}^{(2)} = {} & a_0^{(1)}\left(a_0^2 - a_k^2 - a_0|a_{k-2}| + a_2|a_k|\right) \\
& - |a_{k-1}^{(1)}| \cdot (a_0 + |a_k|)(|a_{k-1}| + a_1)
\end{aligned} \tag{3.29}
$$

Table 3.1. Signs of $a_0^{(j)}$.

k	2	3	4	5	6	7	8	9	10	11	12	13	> 13
j=1	+	+	+	+	+	+	+	+	+	+	+	+	+
j=2	0	+	+	+	+	+	+	+	+	+	+	+	+
j=3		0	+	+	+	+	+	+	+	+	+	−	−
j=4			0	+	+	+	−	−	−	−	−	−	
j=5				0	+	−							

which can be written in the form $(a_0 + |a_k|)\varphi(k)$ with

$$\varphi(k) = (a_0 - |a_k|)(a_0^2 - a_k^2 - a_0|a_{k-2}| + a_2|a_k|) - |a_{k-1}^{(1)}|(a_1 + |a_{k-1}|)$$

$$= a_0^3 - a_0^2\left(\frac{k}{2} + \frac{1}{2} + \frac{1}{k-2} + \frac{1}{k}\right)$$

$$+ a_0\left(\frac{5k}{4} + \frac{1}{4} + \frac{1}{2k-4} - \frac{1}{k-1} - \frac{1}{(k-1)^2} - \frac{1}{k^2}\right)$$

$$- \left(k - \frac{3}{4} - \frac{1}{k-1} - \frac{1}{4k} - \frac{1}{k^3}\right).$$

Show that $\varphi(13) < 0$ and that φ is monotonically decreasing for $k \geq 13$ (observe that $a_0 = a_0(k)$ actually depends on k and that $a_0(k+1) = a_0(k) + 1/(k+1)$). Finally, deduce from the negativeness of (3.29) that $a_0^{(3)} < 0$ for $k \geq 13$.

c) Use Table 3.1 and the Schur-Cohn criterion for the verification of Theorem 3.4.

5. (Multistep methods of maximal order). Verify the following statements:

a) there is no k-step method of order $2k + 1$,

b) there is a unique (implicit) k-step method of order $2k$,

c) there is a unique explicit k-step method of order $2k - 1$.

6. Prove that symmetric multistep methods are always of even order. More precisely, if a symmetric multistep method is of order $2s - 1$ then it is also of order $2s$.

7. Show that all stable 4-step methods of order 6 are given by

$$\varrho(\zeta) = (\zeta^2 - 1)(\zeta^2 + 2\mu\zeta + 1), \qquad |\mu| < 1,$$

$$\sigma(\zeta) = \frac{1}{45}(14 - \mu)(\zeta^4 + 1) + \frac{1}{45}(64 + 34\mu)\zeta(\zeta^2 + 1) + \frac{1}{15}(8 + 38\mu)\zeta^2.$$

Compute the error constant and observe that it cannot become arbitrarily small.

Result. $C = -(16 - 5\mu)/(7560(1 + \mu))$.

8. Prove the following bounds for the error constant:
   a) For stable methods of order $k + 2$

$$C \le -2^{-1-k}\mu_{k+1}.$$

   b) For stable methods of order $k + 1$ with odd k we have

$$C \le -2^{-k}\mu_k.$$

   c) For stable explicit methods of order k we have ($\mu_j = 0$ for even j)

$$C \ge 2^{1-k}\left(\frac{1}{2} - \sum_{j=1}^{k-1} \mu_j\right).$$

   Show that all these bounds are optimal.

   Hint. Compare the formulas (3.18) and (3.21) and use the relation $\sigma(1) = 2^{k-1}a_{k-1}$ of Lemma 3.6.

9. The coefficients μ_j of formula (3.19) satisfy the recurrence relation

$$\mu_{2j+1} + \frac{1}{3}\mu_{2j-1} + \ldots + \frac{1}{2j+1}\mu_1 = \frac{1}{4j+6}. \tag{3.30}$$

   The first of these coefficients are given by

$$\mu_1 = \frac{1}{6}, \quad \mu_3 = \frac{2}{45}, \quad \mu_5 = \frac{22}{945}, \quad \mu_7 = \frac{214}{14175}.$$

10. Another proof of Lemma 3.8: multiplying (3.30) by $2j + 3$ and subtracting from it the same formula with j replaced by $j - 1$ yields

$$(2j+3)\mu_{2j+1} + \sum_{i=0}^{j-1} \mu_{2i+1}\left(\frac{2j+3}{2j-2i+1} - \frac{2j+1}{2j-2i-1}\right) = 0.$$

   Show that the expression in brackets is negative and deduce the result of Lemma 3.8 by a simple induction argument.

III.4 Convergence of Multistep Methods

..., ist das Adams'sche Verfahren jedem andern bedeutend überlegen. Wenn es gleichwohl nicht genügend allgemein angewandt wird und, besonders in Deutschland, gegenüber den von Runge, Heun und Kutta entwickelten Methoden zurücktritt, so mag dies daran liegen, dass bisher eine brauchbare Untersuchung der Genauigkeit der Adams'schen Integration gefehlt hat. Diese Lücke soll hier ausgefüllt werden,...

(R. v. Mises 1930)

The convergence of Adams methods was investigated in the influential article of von Mises (1930), which was followed by an avalanche of papers improving the error bounds and applying the ideas to other special multistep methods, e.g., Tollmien (1938), Fricke (1949), Weissinger (1950), Vietoris (1953). A general convergence proof for the method (2.1), however, was first given by Dahlquist (1956), who gave necessary and sufficient conditions for convergence. Great elegance was introduced in the proofs by the ideas of Butcher (1966), where multistep formulas are written as one-step formulas in a higher dimensional space. Furthermore, the resulting presentation can easily be extended to a more general class of integration methods (see Section III.8).

We cannot expect reasonable convergence of numerical methods, if the differential equation problem

$$y' = f(x, y), \qquad y(x_0) = y_0 \tag{4.1}$$

does not possess a unique solution. We therefore make the following assumptions, which were seen in Sections I.7 and I.9 to be natural for our purpose:

$$f \text{ is continuous on } D = \{(x, y) ; x \in [x_0, \hat{x}], \|y(x) - y\| \le b\} \tag{4.2a}$$

where $y(x)$ denotes the exact solution of (4.1) and b is some positive number. We further assume that f satisfies a Lipschitz condition, i.e.,

$$\|f(x, y) - f(x, z)\| \le L\|y - z\| \qquad \text{for } (x, y), (x, z) \in D. \tag{4.2b}$$

If we apply the multistep method (2.1) with step size h to the problem (4.1) we obtain a sequence $\{y_i\}$. For given x and h such that $(x - x_0)/h = n$ is an integer, we introduce the following notation for the numerical solution:

$$y_h(x) = y_n \qquad \text{if } x - x_0 = nh. \tag{4.3}$$

Definition 4.1 (Convergence). i) The linear multistep method (2.1) is called *convergent*, if for all initial value problems (4.1) satisfying (4.2),

$$y(x) - y_h(x) \to 0 \qquad \text{for } h \to 0, \ x \in [x_0, \hat{x}]$$

whenever the starting values satisfy

$$y(x_0 + ih) - y_h(x_0 + ih) \to 0 \qquad \text{for } h \to 0, \ i = 0, 1, \dots, k-1.$$

ii) Method (2.1) is *convergent of order p*, if to any problem (4.1) with f sufficiently differentiable, there exists a positive h_0 such that

$$\|y(x) - y_h(x)\| \leq Ch^p \qquad \text{for } h \leq h_0$$

whenever the starting values satisfy

$$\|y(x_0 + ih) - y_h(x_0 + ih)\| \leq C_0 h^p \qquad \text{for } h \leq h_0, \ i = 0, 1, \dots, k-1.$$

In this definition we clearly assume that a solution of (4.1) exists on $[x_0, \hat{x}]$.

The aim of this section is to prove that stability together with consistency are necessary and sufficient for the convergence of a multistep method. This is expressed in the famous slogan

$$\text{convergence} = \text{stability} + \text{consistency}$$

(compare also Lax & Richtmyer 1956). We begin with the study of necessary conditions for convergence.

Theorem 4.2. *If the multistep method (2.1) is convergent, then it is necessarily*

    i)   stable and
    ii) consistent (i.e. of order 1: $\varrho(1) = 0$, $\varrho'(1) = \sigma(1)$).

Proof. Application of the multistep method (2.1) to the differential equation $y' = 0$, $y(0) = 0$ yields the difference equation (3.6). Suppose, by contradiction, that $\varrho(\zeta)$ has a root ζ_1 with $|\zeta_1| > 1$, or a root ζ_2 on the unit circle whose multiplicity exceeds 1. ζ_1^n and $n\zeta_2^n$ are then divergent solutions of (3.6). Multiplying by $\sqrt{h}$ we achieve that the starting values converges to $y_0 = 0$ for $h \to 0$. Since $y_h(x) = \sqrt{h}\zeta_1^{x/h}$ and $y_h(x) = (x/\sqrt{h})\zeta_2^{x/h}$ remain divergent for every fixed x, we have a contradiction to the assumption of convergence. The method (2.1) must therefore be stable.

We next consider the initial value problem $y' = 0$, $y(0) = 1$ with exact solution $y(x) = 1$. The corresponding difference equation is again that of (3.6), which, in the new notation, can be written as

$$\alpha_k y_h(x + kh) + \alpha_{k-1} y_h(x + (k-1)h) + \dots + \alpha_0 y_h(x) = 0.$$

Letting $h \to 0$, convergence immediately implies that $\varrho(1) = 0$.

Finally we apply method (2.1) to the problem $y' = 1$, $y(0) = 0$. The exact solution is $y(x) = x$. Since we already know that $\varrho(1) = 0$, it is easy to verify that a particular numerical solution is given by $y_n = nhK$ or $y_h(x) = xK$ where $K = \sigma(1)/\varrho'(1)$. By convergence, $K = 1$ is necessary. $\qquad\qquad\square$

Although the statement of Theorem 4.2 was derived from a consideration of almost trivial differential equations, it is remarkable that conditions (i) and (ii) turn out to be not only necessary but also sufficient for convergence.

Formulation as One-Step Method

We are now at the point where it is useful to rewrite a multistep method as a one-step method in a higher dimensional space (see Butcher 1966, Skeel 1976). For this let $\psi = \psi(x_i, y_i, ..., y_{i+k-1}, h)$ be defined implicitly by

$$\psi = \sum_{j=0}^{k-1} \beta'_j f(x_i + jh, y_{i+j}) + \beta'_k f\left(x_i + kh, h\psi - \sum_{j=0}^{k-1} \alpha'_j y_{i+j}\right) \quad (4.4)$$

where $\alpha'_j = \alpha_j/\alpha_k$ and $\beta'_j = \beta_j/\alpha_k$. Multistep formula (2.1) can then be written as

$$y_{i+k} = -\sum_{j=0}^{k-1} \alpha'_j y_{i+j} + h\psi. \quad (4.5)$$

Introducing the $m \cdot k$-dimensional vectors (m is the dimension of the differential equation)

$$Y_i = (y_{i+k-1}, y_{i+k-2}, ..., y_i)^T, \quad i \geq 0 \quad (4.6)$$

and

$$A = \begin{pmatrix} -\alpha'_{k-1} & -\alpha'_{k-2} & \cdots & \cdot & -\alpha'_0 \\ 1 & 0 & \cdots & \cdot & 0 \\ & 1 & & \cdot & 0 \\ & & \ddots & \vdots & \vdots \\ & & & 1 & 0 \end{pmatrix}, \quad e_1 = \begin{pmatrix} 1 \\ 0 \\ 0 \\ \vdots \\ 0 \end{pmatrix}, \quad (4.7)$$

the multistep method (4.5) can be written — after adding some trivial identities — in compact form as

$$Y_{i+1} = (A \otimes I)Y_i + h\Phi(x_i, Y_i, h), \quad i \geq 0 \quad (4.8)$$

with

$$\Phi(x_i, Y_i, h) = (e_1 \otimes I)\psi(x_i, Y_i, h). \quad (4.8a)$$

Here, $A \otimes I$ denotes the Kronecker tensor product, i.e. the $m \cdot k$-dimensional block matrix with (m, m)-blocks $a_{ij}I$. Readers unfamiliar with the notation and properties of this product may assume for simplicity that (4.1) is a scalar equation ($m = 1$) and $A \otimes I = A$.

The following lemmas express the concepts of order and stability in this new notation.

Lemma 4.3. *Let $y(x)$ be the exact solution of (4.1). For $i = 0, 1, 2, \ldots$ we define the vector $\widehat{Y}_{i+1}$ as the numerical solution of one step*

$$\widehat{Y}_{i+1} = (A \otimes I)Y(x_i) + h\Phi(x_i, Y(x_i), h)$$

with correct starting values

$$Y(x_i) = \left(y(x_{i+k-1}), y(x_{i+k-2}), \ldots, y(x_i)\right)^T.$$

i) If the multistep method (2.1) is of order 1 and if f satisfies (4.2), then an $h_0 > 0$ exists such that for $h \leq h_0$,

$$\|Y(x_{i+1}) - \widehat{Y}_{i+1}\| \leq h\omega(h), \qquad 0 \leq i \leq \widehat{x}/h - k$$

where $\omega(h) \to 0$ for $h \to 0$.

ii) If the multistep method (2.1) is of order p and if f is sufficiently differentiable then a constant M exists such that for h small enough,

$$\|Y(x_{i+1}) - \widehat{Y}_{i+1}\| \leq Mh^{p+1}, \qquad 0 \leq i \leq \widehat{x}/h - k.$$

Proof. The first component of $Y(x_{i+1}) - \widehat{Y}_{i+1}$ is the local error as given by Definition 2.1. Since the remaining components all vanish, Exercise 5 of Section III.2 and Definition 2.3 yield the result. □

Lemma 4.4. *Suppose that the multistep method (2.1) is stable. Then there exists a vector norm (on $\mathbf{R}^{mk}$) such that the matrix A of (4.7) satisfies*

$$\|A \otimes I\| \leq 1$$

in the subordinate matrix norm.

Proof. If λ is a root of $\varrho(\zeta)$, then the vector $(\lambda^{k-1}, \lambda^{k-2}, \ldots, 1)$ is an eigenvector of the matrix A with eigenvalue λ. Therefore the eigenvalues of A (which are the roots of $\varrho(\zeta)$) satisfy the root condition by Definition 3.2. A transformation to Jordan canonical form therefore yields (see Section I.12)

$$T^{-1}AT = J = \text{diag}\left\{\lambda_1, \ldots, \lambda_l, \begin{pmatrix} \lambda_{l+1} & \varepsilon_{l+1} & & \\ & \ddots & & \\ & & & \varepsilon_{k-1} \\ & & & \lambda_k \end{pmatrix}\right\} \qquad (4.9)$$

where $\lambda_1, \ldots, \lambda_l$ are the eigenvalues of modulus 1, which must be simple, each ε_j is either 0 or 1. We further find by a suitable multiplication of the columns of T that $|\varepsilon_j| < 1 - |\lambda_j|$ for $j = l+1, \ldots, k-1$. Because of (9.11') of Chapter I we then have $\|J \otimes I\|_\infty \leq 1$. Using the transformation T of (4.9) we define the norm

$$\|x\| := \|(T^{-1} \otimes I)x\|_\infty.$$

This yields

$$\|(A \otimes I)x\| = \|(T^{-1} \otimes I)(A \otimes I)x\|_\infty = \|(J \otimes I)(T^{-1} \otimes I)x\|_\infty$$
$$\leq \|(T^{-1} \otimes I)x\|_\infty = \|x\|$$

and hence also $\|A \otimes I\| \leq 1$. □

Proof of Convergence

The convergence theorem for multistep methods can now be established.

Theorem 4.5. *If the multistep method (2.1) is stable and of order 1 then it is convergent. If method (2.1) is stable and of order p then it is convergent of order p.*

Proof. As in the convergence theorem for one-step methods (Section II.3) we may assume without loss of generality that $f(x, y)$ is defined for all $y \in \mathbb{R}^m$, $x \in [x_0, \widehat{x}]$ and satisfies there a (global) Lipschitz condition. This implies that for sufficiently small h the functions $\psi(x_i, Y_i, h)$ and $\Phi(x_i, Y_i, h)$ satisfy a Lipschitz condition with respect to the second argument (with Lipschitz constant L^*). For the function G, defined by formula (4.8), which maps the vector Y_i onto Y_{i+1} we thus obtain from Lemma 4.4

$$\|G(Y_i) - G(Z_i)\| \leq (1 + hL^*)\|Y_i - Z_i\|. \tag{4.10}$$

The rest of the proof now proceeds in the same way as for one-step methods and is illustrated in Fig. 4.1.

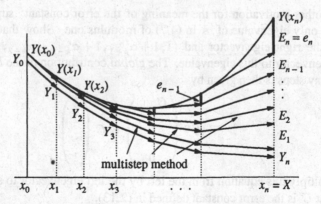

Fig. 4.1. Lady Windermere's Fan for multistep methods

The arrows in Fig. 4.1 indicate the application of G. From Lemma 4.3 we know that $\|Y(x_{i+1}) - G(Y(x_i))\| \leq h\omega(h)$. This together with (4.10) shows that the local error $Y(x_{i+1}) - G(Y(x_i))$ at stage $i+1$ causes an error at stage n, which is at most $h\omega(h)(1 + hL^*)^{n-i+1}$. Thus we have

$$
\begin{aligned}
\|Y(x_n) - Y_n\| &\leq \|Y(x_0) - Y_0\|(1 + hL^*)^n \\
&\quad + h\omega(h)\big((1 + hL^*)^{n-1} + (1 + hL^*)^{n-2} + \ldots + 1\big) \\
&\leq \|Y(x_0) - Y_0\| \exp(nhL^*) + \frac{\omega(h)}{L^*}\big(\exp(nhL^*) - 1\big).
\end{aligned}
\tag{4.11}
$$

Convergence of method (2.1) is now an immediate consequence of formula (4.11). If the multistep method is of order p, the same proof with $\omega(h)$ replaced by Mh^p yields convergence of order p.   □

Exercises

1. Consider the function (for $x \geq 0$)

$$
f(x,y) = \begin{cases}
2x & \text{for } y \leq 0, \\
2x - \dfrac{4y}{x} & \text{for } 0 < y < x^2, \\
-2x & \text{for } y \geq x^2.
\end{cases}
$$

   a) Show that $y(x) = x^2/3$ is the unique solution of $y' = f(x,y)$, $y(0) = 0$, although f does not satisfy a Lipschitz condition near the origin.

   b) Apply the mid-point rule (1.13') with starting values $y_0 = 0$, $y_1 = -h^2$ to the above problem and verify that the numerical solution at $x = nh$ is given by $y_h(x) = (-1)^n x^2$ (Grigorieff 1977).

2. Another motivation for the meaning of the error constant: suppose that 1 is the only eigenvalue of A in (4.7) of modulus one. Show that $(1, 1, \ldots, 1)^T$ is the right eigenvector and $(1, 1 + \alpha'_{k-1}, 1 + \alpha'_{k-1} + \alpha'_{k-2}, \ldots)$ is the left eigenvector to this eigenvalue. The *global* contribution of the *local* error after many steps is then given by

$$
A^\infty \begin{pmatrix} C_{p+1} \\ 0 \\ \vdots \\ 0 \end{pmatrix} = C \begin{pmatrix} 1 \\ 1 \\ \vdots \\ 1 \end{pmatrix}.
\tag{4.12}
$$

   Multiply this equation from the left by the left eigenvector to show with (2.6) that C is the error constant defined in (2.13).

   Remark. For multistep methods with several eigenvalues of modulus 1, formula (4.12) remains valid if A^∞ is replaced by E (see Section III.8).

III.5 Variable Step Size Multistep Methods

It is clear from the considerations of Section II.4 that an efficient integrator must be able to change the step size. However, changing the step size with multistep methods is difficult since the formulas of the preceding sections require the numerical approximations at equidistant points. There are in principle two possibilities for overcoming this difficulty:

i) use polynomial interpolation to reproduce the starting values at the new (equidistant) grid;

ii) construct methods which are adjusted to variable grid points.

This section is devoted to the second approach. We investigate consistency, stability and convergence. The actual implementation (order and step size strategies) will be considered in Section III.7.

Variable Step Size Adams Methods

F. Ceschino (1961) was apparently the first person to propose a "smooth" transition from a step size h to a new step size ωh. C.V.D. Forrington (1961) and later on F.T. Krogh (1969) extended his ideas: we consider an arbitrary grid (x_n) and denote the step sizes by $h_n = x_{n+1} - x_n$. We assume that approximations y_j to $y(x_j)$ are known for $j = n - k + 1, \ldots, n$ and we put $f_j = f(x_j, y_j)$. In the same way as in Section III.1 we denote by $p(t)$ the polynomial which interpolates the values (x_j, f_j) for $j = n - k + 1, \ldots, n$. Using Newton's interpolation formula we have

$$p(t) = \sum_{j=0}^{k-1} \prod_{i=0}^{j-1} (t - x_{n-i}) \, \delta^j f[x_n, x_{n-1}, \ldots, x_{n-j}] \qquad (5.1)$$

where the divided differences $\delta^j f[x_n, \ldots, x_{n-j}]$ are defined recursively by

$$\delta^0 f[x_n] = f_n$$

$$\delta^j f[x_n, \ldots, x_{n-j}] = \frac{\delta^{j-1} f[x_n, \ldots, x_{n-j+1}] - \delta^{j-1} f[x_{n-1}, \ldots, x_{n-j}]}{x_n - x_{n-j}}. \qquad (5.2)$$

For actual computations (see Krogh 1969) it is practical to rewrite (5.1) as

$$p(t) = \sum_{j=0}^{k-1} \prod_{i=0}^{j-1} \frac{t - x_{n-i}}{x_{n+1} - x_{n-i}} \cdot \Phi_j^*(n) \qquad (5.1')$$

where

$$\Phi_j^*(n) = \prod_{i=0}^{j-1} (x_{n+1} - x_{n-i}) \cdot \delta^j f[x_n, \ldots, x_{n-j}]. \qquad (5.3)$$

We now define the approximation to $y(x_{n+1})$ by

$$y_{n+1} = y_n + \int_{x_n}^{x_{n+1}} p(t)\, dt. \qquad (5.4)$$

Inserting formula (5.1') into (5.4) we obtain

$$y_{n+1} = y_n + h_n \sum_{j=0}^{k-1} g_j(n) \Phi_j^*(n) \qquad (5.5)$$

with

$$g_j(n) = \frac{1}{h_n} \int_{x_n}^{x_{n+1}} \prod_{i=0}^{j-1} \frac{t - x_{n-i}}{x_{n+1} - x_{n-i}}\, dt. \qquad (5.6)$$

Formula (5.5) is the extension of the explicit Adams method (1.5) to variable step sizes. Observe that for constant step sizes the above expressions reduce to (Exercise 1)

$$g_j(n) = \gamma_j, \qquad \Phi_j^*(n) = \nabla^j f_n.$$

The variable step size *implicit* Adams methods can be deduced similarly. In analogy to Section III.1 we let $p^*(t)$ be the polynomial of degree k that interpolates (x_j, f_j) for $j = n-k+1, \ldots, n, n+1$ (the value $f_{n+1} = f(x_{n+1}, y_{n+1})$ contains the unknown solution y_{n+1}). Again, using Newton's interpolation formula we obtain

$$p^*(t) = p(t) + \prod_{i=0}^{k-1} (t - x_{n-i}) \cdot \delta^k f[x_{n+1}, x_n, \ldots, x_{n-k+1}].$$

The numerical solution, defined by

$$y_{n+1} = y_n + \int_{x_n}^{x_{n+1}} p^*(t)\, dt,$$

is now given by

$$y_{n+1} = p_{n+1} + h_n g_k(n) \Phi_k(n+1), \qquad (5.7)$$

where p_{n+1} is the numerical approximation obtained by the explicit Adams method

$$p_{n+1} = y_n + h_n \sum_{j=0}^{k-1} g_j(n) \Phi_j^*(n)$$

and where

$$\Phi_k(n+1) = \prod_{i=0}^{k-1}(x_{n+1}-x_{n-i})\cdot\delta^k f[x_{n+1}, x_n, \ldots, x_{n-k+1}]. \qquad (5.8)$$

Recurrence Relations for $g_j(n)$, $\Phi_j(n)$ and $\Phi_j^*(n)$

> The cost of computing integration coefficients is the biggest disadvantage to permitting arbitrary variations in the step size.
>
> (F.T. Krogh 1973)

The values $\Phi_j^*(n)$ $(j = 0, \ldots, k-1)$ and $\Phi_k(n+1)$ can be computed efficiently with the recurrence relations

$$\Phi_0(n) = \Phi_0^*(n) = f_n$$
$$\Phi_{j+1}(n) = \Phi_j(n) - \Phi_j^*(n-1) \qquad (5.9)$$
$$\Phi_j^*(n) = \beta_j(n)\Phi_j(n),$$

which are an immediate consequence of Definitions (5.3) and (5.8). The coefficients

$$\beta_j(n) = \prod_{i=0}^{j-1}\frac{x_{n+1}-x_{n-i}}{x_n - x_{n-i-1}}$$

can be calculated by

$$\beta_0(n) = 1, \qquad \beta_j(n) = \beta_{j-1}(n)\frac{x_{n+1}-x_{n-j+1}}{x_n - x_{n-j}}.$$

The calculation of the coefficients $g_j(n)$ is trickier (F.T. Krogh 1974). We introduce the q-fold integral

$$c_{jq}(x) = \frac{(q-1)!}{h_n^q}\int_{x_n}^x \int_{x_n}^{\xi_{q-1}}\cdots\int_{x_n}^{\xi_1}\prod_{i=0}^{j-1}\frac{\xi_0 - x_{n-i}}{x_{n+1}-x_{n-i}}\,d\xi_0\ldots d\xi_{q-1} \qquad (5.10)$$

and observe that

$$g_j(n) = c_{j1}(x_{n+1}).$$

Lemma 5.1. *We have*

$$c_{0q}(x_{n+1}) = \frac{1}{q}, \qquad c_{1q}(x_{n+1}) = \frac{1}{q(q+1)},$$

$$c_{jq}(x_{n+1}) = c_{j-1,q}(x_{n+1}) - c_{j-1,q+1}(x_{n+1})\frac{h_n}{x_{n+1}-x_{n-j+1}}.$$

Proof. The first two relations follow immediately from (5.10). In order to prove the recurrence relation we denote by $d(x)$ the difference

$$d(x) = c_{jq}(x) - c_{j-1,q}(x)\frac{x - x_{n-j+1}}{x_{n+1} - x_{n-j+1}} + c_{j-1,q+1}(x)\frac{h_n}{x_{n+1} - x_{n-j+1}}.$$

Clearly, $d^{(i)}(x_n) = 0$ for $i = 0, 1, \ldots, q-1$. Moreover, the q-th derivative of $d(x)$ vanishes, since by the Leibniz rule

$$\frac{d^q}{dx^q}\left(c_{j-1,q}(x) \cdot \frac{x - x_{n-j+1}}{x_{n+1} - x_{n-j+1}}\right)$$

$$= c_{j-1,q}^{(q)}(x)\frac{x - x_{n-j+1}}{x_{n+1} - x_{n-j+1}} + qc_{j-1,q}^{(q-1)}(x)\frac{1}{x_{n+1} - x_{n-j+1}}$$

$$= c_{j,q}^{(q)}(x) + c_{j-1,q+1}^{(q)}(x)\frac{h_n}{x_{n+1} - x_{n-j+1}}.$$

Therefore we have $d(x) \equiv 0$ and the statement follows by putting $x = x_{n+1}$.
□

Using the above recurrence relation one can successively compute $c_{2q}(x_{n+1})$ for $q = 1, \ldots, k-1$; $c_{3q}(x_{n+1})$ for $q = 1, \ldots, k-2$; $\ldots$; $c_{kq}(x_{n+1})$ for $q = 1$. This procedure yields in an efficient way the coefficients $g_j(n) = c_{j1}(x_{n+1})$ of the Adams methods.

Variable Step Size BDF

The BDF-formulas (1.22) can also be extended in a natural way to variable step size. Denote by $q(t)$ the polynomial of degree k that interpolates (x_i, y_i) for $i = n+1, n, \ldots, n-k+1$. It can be expressed, using divided differences, by

$$q(t) = \sum_{j=0}^{k} \prod_{i=0}^{j-1}(t - x_{n+1-i}) \cdot \delta^j y[x_{n+1}, x_n, \ldots, x_{n-j+1}]. \tag{5.11}$$

The requirement

$$q'(x_{n+1}) = f(x_{n+1}, y_{n+1})$$

immediately leads to the variable step size BDF-formulas

$$\sum_{j=1}^{k} h_n \prod_{i=1}^{j-1}(x_{n+1} - x_{n+1-i}) \cdot \delta^j y[x_{n+1}, \ldots, x_{n-j+1}] = h_n f(x_{n+1}, y_{n+1}).$$

$$\tag{5.12}$$

The computation of the coefficients is much easier here than for the Adams methods.

General Variable Step Size Methods and Their Orders

For theoretical investigations it is convenient to write the methods in a form where the y_j and f_j values appear linearly. For example, the implicit Adams method (5.7) becomes ($k = 2$)

$$y_{n+1} = y_n + \frac{h_n}{6(1+\omega_n)}\left((3+2\omega_n)f_{n+1} + (3+\omega_n)(1+\omega_n)f_n - \omega_n^2 f_{n-1}\right),$$
(5.13)

where we have introduced the notation $\omega_n = h_n/h_{n-1}$ for the step size ratio. Or, the 2-step BDF-formula (5.12) can be written as

$$y_{n+1} - \frac{(1+\omega_n)^2}{1+2\omega_n}y_n + \frac{\omega_n^2}{1+2\omega_n}y_{n-1} = h_n\frac{1+\omega_n}{1+2\omega_n}f_{n+1}.$$
(5.14)

In order to give a unified theory for all these variable step size multistep methods we consider formulas of the form

$$y_{n+k} + \sum_{j=0}^{k-1}\alpha_{jn}y_{n+j} = h_{n+k-1}\sum_{j=0}^{k}\beta_{jn}f_{n+j}.$$
(5.15)

The coefficients α_{jn} and β_{jn} actually depend on the ratios $\omega_i = h_i/h_{i-1}$ for $i = n+1,\ldots,n+k-1$. In analogy to the constant step size case we give

Definition 5.2. Method (5.15) is *consistent of order* p, if

$$q(x_{n+k}) + \sum_{j=0}^{k-1}\alpha_{jn}q(x_{n+j}) = h_{n+k-1}\sum_{j=0}^{k}\beta_{jn}q'(x_{n+j})$$

holds for all polynomials $q(x)$ of degree $\leq p$ and for all grids (x_j).

By definition, the explicit Adams method (5.5) is of order k, the implicit Adams method (5.7) is of order $k+1$, and the BDF-formula (5.12) is of order k.

The notion of consistency certainly has to be related to the local error. Indeed, if the method is of order p, if the ratios h_j/h_n are bounded for $j = n+1,\ldots,n+k-1$ and if the coefficients satisfy

$$\alpha_{jn}, \beta_{jn} \text{ are bounded},$$
(5.16)

then a Taylor expansion argument implies that

$$y(x_{n+k}) + \sum_{j=0}^{k-1}\alpha_{jn}y(x_{n+j}) - h_{n+k-1}\sum_{j=0}^{k}\beta_{jn}y'(x_{n+j}) = \mathcal{O}(h_n^{p+1}) \quad (5.17)$$

for sufficiently smooth $y(x)$. Interpreting $y(x)$ as the solution of the differential equation, a trivial extension of Lemma 2.2 to variable step sizes shows that the local error at x_{n+k} (cf. Definition 2.1) is also $\mathcal{O}(h_n^{p+1})$.

This motivates the investigation of condition (5.16). The methods (5.13) and (5.14) are seen to satisfy (5.16) whenever the step size ratio h_n/h_{n-1} is bounded from above. In general we have

Lemma 5.3. *For the explicit and implicit Adams methods as well as for the BDF-formulas the coefficients α_{jn} and β_{jn} are bounded whenever for some Ω*

$$h_n/h_{n-1} \leq \Omega.$$

Proof. We prove the statement for the explicit Adams methods only. The proof for the other methods is similar and thus omitted. We see from formula (5.5) that the coefficients α_{jn} do not depend on n and hence are bounded. The β_{jn} are composed of products of $g_j(n)$ with the coefficients of $\Phi_j^*(n)$, when written as a linear combination of $f_n, \ldots, f_{n-j}$. From formula (5.6) we see that $|g_j(n)| \leq 1$. It follows from $(x_{n+1} - x_{n-j+1}) \leq \max(1, \Omega^j)(x_n - x_{n-j})$ and from an induction argument that the coefficients of $\Phi_j^*(n)$ are also bounded. Hence the β_{jn} are bounded, which proves the lemma.   □

The condition $h_n/h_{n-1} \leq \Omega$ is a reasonable assumption which can easily be satisfied by a code.

Stability

                    So geht das einfach ...   (R.D. Grigorieff, Halle 1983)

The study of stability for variable step size methods was begun in the articles of Gear & Tu (1974) and Gear & Watanabe (1974). Further investigations are due to Grigorieff (1983) and Crouzeix & Lisbona (1984).

We have seen in Section III.3 that for equidistant grids stability is equivalent to the boundedness of the numerical solution, when applied to the scalar differential equation $y' = 0$. Let us do the same here for the general case. Method (5.15), applied to $y' = 0$, gives the difference equation with variable coefficients

$$y_{n+k} + \sum_{j=0}^{k-1} \alpha_{jn} y_{n+j} = 0.$$

If we introduce the vector $Y_n = (y_{n+k-1}, \ldots, y_n)^T$, this difference equation is equivalent to

$$Y_{n+1} = A_n Y_n$$

with

$$
A_n = \begin{pmatrix} -\alpha_{k-1,n} & \cdots & \cdots & -\alpha_{1,n} & -\alpha_{0,n} \\ 1 & 0 & \cdots & 0 & 0 \\ & \ddots & \ddots & \vdots & \vdots \\ & & 1 & 0 & 0 \\ & & & 1 & 0 \end{pmatrix}, \tag{5.18}
$$

the companion matrix.

Definition 5.4. Method (5.15) is called *stable*, if

$$
\|A_{n+l}A_{n+l-1}\cdots A_{n+1}A_n\| \le M \tag{5.19}
$$

for all n and $l \ge 0$.

Observe that in general A_n depends on the step ratios $\omega_{n+1},\ldots,\omega_{n+k-1}$. Therefore, condition (5.19) will usually lead to a restriction on these values. For the Adams methods (5.5) and (5.7) the coefficients α_{jn} do not depend on n and hence are stable for any step size sequence.

In the following three theorems we present stability results for general variable step size methods. The first one, taken from Crouzeix & Lisbona (1984), is a sort of perturbation result: the variable step size method is considered as a perturbation of a strongly stable fixed step size method.

Theorem 5.5. *Let the method (5.15) satisfy the following properties:*

a) *it is of order $p \ge 0$, i.e.,* $\displaystyle 1 + \sum_{j=0}^{k-1} \alpha_{jn} = 0$;

b) *the coefficients $\alpha_{jn} = \alpha_j(\omega_{n+1},\ldots,\omega_{n+k-1})$ are continuous in a neighbourhood of $(1,\ldots,1)$;*

c) *the underlying constant step size formula is strongly stable, i.e., all roots of*

$$
\zeta^k + \sum_{j=0}^{k-1} \alpha_j(1,\ldots,1)\zeta^j = 0
$$

lie in the open unit disc $|\zeta| < 1$, with the exception of $\zeta_1 = 1$.
Then there exist real numbers ω, Ω ($\omega < 1 < \Omega$) such that the method is stable if

$$
\omega \le h_n/h_{n-1} \le \Omega \qquad \text{for all } n. \tag{5.20}
$$

Proof. Let A be the companion matrix of the constant step size formula. As in the proof of Lemma 4.4 we transform A to Jordan canonical form and obtain

$$
T^{-1}AT = \begin{pmatrix} \widehat{A} & \begin{matrix} 0 \\ \vdots \\ 0 \\ 1 \end{matrix} \end{pmatrix}
$$

where, by assumption (c), $\|\widehat{A}\|_1 < 1$. Observe that the last column of T, the eigenvector of A corresponding to 1, is given by $t_k = (1, \ldots, 1)^T$. Assumption (a) implies that this vector t_k is also an eigenvector for each A_n. Therefore we have

$$T^{-1} A_n T = \begin{pmatrix} & & & 0 \\ & \widehat{A}_n & & \vdots \\ & & & 0 \\ & & & 1 \end{pmatrix}$$

and, by continuity, $\|\widehat{A}_n\|_1 \leq 1$, if $\omega_{n+1}, \ldots, \omega_{n+k-1}$ are sufficiently close to 1. Stability now follows from the fact that

$$\|T^{-1} A_n T\|_1 = \max(\|\widehat{A}_n\|_1, 1) = 1,$$

which implies that

$$\|A_{n+l} \cdots A_{n+1} A_n\| \leq \|T\| \cdot \|T^{-1}\|. \qquad \square$$

The next result (Grigorieff 1983) is based on a reduction of the dimension of the matrices A_n by one. The idea is to use the transformation

$$T = \begin{pmatrix} 1 & 1 & 1 & .. & 1 \\ & 1 & 1 & .. & 1 \\ & & 1 & .. & 1 \\ 0 & & & \ddots & \vdots \\ & & & & 1 \end{pmatrix}, \qquad T^{-1} = \begin{pmatrix} 1 & -1 & & 0 \\ & 1 & -1 & \\ & & 1 & \ddots \\ 0 & & & \ddots & -1 \\ & & & & 1 \end{pmatrix}.$$

Observe that the last column of T is just t_k of the above proof. A simple calculation shows that

$$T^{-1} A_n T = \begin{pmatrix} A_n^* & 0 \\ e_{k-1}^T & 1 \end{pmatrix}$$

where $e_{k-1}^T = (0, \ldots, 0, 1)$ and

$$A_n^* = \begin{pmatrix} -\alpha_{k-2,n}^* & -\alpha_{k-3,n}^* & \cdots & -\alpha_{1n}^* & -\alpha_{0n}^* \\ 1 & 0 & .. & . & 0 \\ & 1 & .. & . & 0 \\ & & \ddots & \vdots & \vdots \\ & & & 1 & 0 \end{pmatrix} \qquad (5.21)$$

with

$$\alpha_{k-2,n}^* = 1 + \alpha_{k-1,n}, \qquad \alpha_{0n}^* = -\alpha_{0n},$$

$$\alpha_{k-j-1,n}^* - \alpha_{k-j,n}^* = \alpha_{k-j,n} \quad \text{for } j = 2, \ldots, k-1.$$

We remark that the coefficients $\alpha_{j,n}^*$ are just the coefficients of the polynomial defined by

$$(\zeta^k + \alpha_{k-1,n} \zeta^{k-1} + \ldots + \alpha_{1,n} \zeta + \alpha_{0,n})$$
$$= (\zeta - 1)(\zeta^{k-1} + \alpha_{k-2,n}^* \zeta^{k-2} + \ldots + \alpha_{1,n}^* \zeta + \alpha_{0,n}^*).$$

Theorem 5.6. *Let the method (5.15) be of order $p \geq 0$. Then the method is stable if and only if for all n and $l \geq 0$,*

$$a) \qquad \left\| A_{n+l}^* \cdots A_{n+1}^* A_n^* \right\| \leq M_1$$

$$b) \qquad \left\| e_{k-1}^T \sum_{j=n}^{n+l} \prod_{i=n}^{j-1} A_i^* \right\| \leq M_2.$$

Proof. A simple induction argument shows that

$$T^{-1} A_{n+l} \cdots A_n T = \begin{pmatrix} A_{n+l}^* \cdots A_n^* & 0 \\ b_{n,l}^T & 1 \end{pmatrix}$$

with

$$b_{n,l}^T = e_{k-1}^T \sum_{j=n}^{n+l} \prod_{i=n}^{j-1} A_i^*. \qquad \qquad \square$$

Since in this theorem the dimension of the matrices under consideration is reduced by one, it is especially useful for the stability investigation of two-step methods.

Example. Consider the two-step BDF-method (5.14). Here

$$\alpha_{0n} = \frac{\omega_{n+1}^2}{1 + 2\omega_{n+1}}, \qquad \alpha_{1n} = -1 - \alpha_{0n}.$$

The matrix (5.21) becomes in this case

$$A_n^* = (-\alpha_{0n}^*), \qquad -\alpha_{0n}^* = \frac{\omega_{n+1}^2}{1 + 2\omega_{n+1}}.$$

If $|\alpha_{0n}^*| \leq q < 1$ the conditions of Theorem 5.6 are satisfied and imply stability. This is the case, if

$$0 < h_{n+1}/h_n \leq \Omega < 1 + \sqrt{2}.$$

An interesting consequence of the theorem above is the *instability* of the two-step BDF-formula if the step sizes increase at least like $h_{n+1}/h_n \geq 1 + \sqrt{2}$.

The investigation of stability for k-step ($k \geq 3$) methods becomes much more difficult, because several step size ratios $\omega_{n+1}, \omega_{n+2}, \ldots$ are involved. Grigorieff (1983) calculated the bounds (5.20) given in Table 5.1 for the higher order BDF-methods which *ensure* stability. These bounds are surely unrealistic, since all pathological step size variations are admitted.

A less pessimistic result is obtained if the step sizes are supposed to vary more smoothly (Gear & Tu 1974): the local error is known to be of the form $d(x_n)h_n^{p+1} + \mathcal{O}(h_n^{p+2})$, where $d(x)$ is the principal error function. This local error

Table 5.1. Bounds (5.20) for k-step BDF formulas

k	2	3	4	5
ω	0	0.836	0.979	0.997
Ω	2.414	1.127	1.019	1.003

is, by the step size control, kept equal to *Tol*. Hence, if $d(x)$ is bounded away from zero we have

$$h_n = |Tol/d(x_n)|^{1/(p+1)} + \mathcal{O}(h_n)$$

which implies (if $h_{n+1}/h_n \leq \Omega$) that

$$h_{n+1}/h_n = |d(x_n)/d(x_{n+1})|^{1/(p+1)} + \mathcal{O}(h_n).$$

If $d(x)$ is differentiable, we obtain

$$|h_{n+1}/h_n - 1| \leq Ch_n. \tag{5.22}$$

Several stability results of Gear & Tu are based on this hypothesis ("Consequently, we can expect either method to be stable if the fixed step method is stable. ...."). Adding up (5.22) we obtain

$$\sum_{j=n}^{n+l} |h_{j+1}/h_j - 1| \leq C(\widehat{x} - x_0),$$

a condition which contains only step size ratios. This motivates the following theorem:

Theorem 5.7. *Let the coefficients α_{jn} of method (5.15) be continuously differentiable functions of $\omega_{n+1}, \ldots, \omega_{n+k-1}$ in a neighbourhood of the set*

$$\{(\omega_{n+1}, \ldots, \omega_{n+k-1}) \, ; \, \omega \leq \omega_j \leq \Omega\}$$

and assume that the method is stable for constant step sizes (i.e., for $\omega_j = 1$). Then the condition

$$\sum_{j=n}^{n+l} |h_{j+1}/h_j - 1| \leq C \qquad \text{for all } n \text{ and } l \geq 0, \tag{5.23}$$

together with $\omega \leq h_{j+1}/h_j \leq \Omega$, imply the stability condition (5.19).

Proof. As in the proof of Theorem 5.5 we denote by A the companion matrix of the constant step size formula and by T a suitable transformation such that $\|T^{-1}AT\| = 1$. The mean value theorem, applied to $\alpha_j(\omega_{n+1}, \ldots, \omega_{n+k-1}) - \alpha_j(1, \ldots, 1)$, implies that

$$\|T^{-1}A_nT - T^{-1}AT\| \leq K \sum_{j=n+1}^{n+k-1} |\omega_j - 1|.$$

Hence

$$\|T^{-1}A_nT\| \leq 1 + K \sum_{j=n+1}^{n+k-1} |\omega_j - 1| \leq \exp\left(K \sum_{j=n+1}^{n+k-1} |\omega_j - 1|\right).$$

From this inequality we deduce that

$$\|A_{n+l}\cdots A_{n+1}A_n\| \leq \|T\|\cdot\|T^{-1}\|\cdot\exp(K\cdot(k-1)C). \qquad \square$$

Convergence

Convergence for variable step size Adams methods was first studied by Piotrowski (1969). In order to prove convergence for the general case we introduce the vector $Y_n = (y_{n+k-1},\ldots,y_{n+1},y_n)^T$. In analogy to (4.8) the method (5.15) then becomes equivalent to

$$Y_{n+1} = (A_n \otimes I)Y_n + h_{n+k-1}\Phi_n(x_n, Y_n, h_n) \tag{5.24}$$

where A_n is given by (5.18) and

$$\Phi_n(x_n, Y_n, h_n) = (e_1 \otimes I)\Psi_n(x_n, Y_n, h_n).$$

The value $\Psi = \Psi_n(x_n, Y_n, h_n)$ is defined implicitly by

$$\Psi = \sum_{j=0}^{k-1} \beta_{jn} f(x_{n+j}, y_{n+j}) + \beta_{kn} f\left(x_{n+k}, h\Psi - \sum_{j=0}^{k-1} \alpha_{jn} y_{n+j}\right).$$

Let us further denote by

$$Y(x_n) = \left(y(x_{n+k-1}),\ldots,y(x_{n+1}),y(x_n)\right)^T$$

the exact values to be approximated by Y_n. The convergence theorem can now be formulated as follows:

Theorem 5.8. *Assume that*

a) *the method (5.15) is stable, of order p, and has bounded coefficients α_{jn} and β_{jn};*

b) *the starting values satisfy $\|Y(x_0) - Y_0\| = \mathcal{O}(h_0^p)$;*

c) *the step size ratios are bounded $(h_n/h_{n-1} \leq \Omega)$.*

Then the method is convergent of order p, i.e., for each differential equation $y' = f(x,y)$, $y(x_0) = y_0$ with f sufficiently differentiable the global error satisfies

$$\|y(x_n) - y_n\| \leq Ch^p \qquad \text{for } x_n \leq \widehat{x},$$

where $h = \max h_j$.

Proof. Since the method is of order p and the coefficients and step size ratios are bounded, formula (5.17) shows that the local error

$$\delta_{n+1} = Y(x_{n+1}) - (A_n \otimes I)Y(x_n) - h_{n+k-1}\Phi_n(x_n, Y(x_n), h_n) \qquad (5.25)$$

satisfies

$$\delta_{n+1} = \mathcal{O}(h_n^{p+1}). \qquad (5.26)$$

Subtracting (5.24) from (5.25) we obtain

$$\begin{aligned}
Y(x_{n+1}) - Y_{n+1} = {} & (A_n \otimes I)(Y(x_n) - Y_n) \\
& + h_{n+k-1}\big(\Phi_n(x_n, Y(x_n), h_n) - \Phi_n(x_n, Y_n, h_n)\big) + \delta_{n+1}
\end{aligned}$$

and by induction it follows that

$$\begin{aligned}
Y(x_{n+1}) - Y_{n+1} = {} & \big((A_n \dots A_0) \otimes I\big)(Y(x_0) - Y_0) \\
& + \sum_{j=0}^{n} h_{j+k-1}\big((A_n \dots A_{j+1}) \otimes I\big)\big(\Phi_j(x_j, Y(x_j), h_j) - \Phi_j(x_j, Y_j, h_j)\big) \\
& + \sum_{j=0}^{n} \big((A_n \dots A_{j+1}) \otimes I\big)\delta_{j+1}.
\end{aligned}$$

As in the proof of Theorem 4.5 we deduce that the Φ_n satisfy a uniform Lipschitz condition with respect to Y_n. This, together with stability and (5.26), implies that

$$\|Y(x_{n+1}) - Y_{n+1}\| \le \sum_{j=0}^{n} h_{j+k-1}L\|Y(x_j) - Y_j\| + C_1 h^p.$$

In order to solve this inequality we introduce the sequence $\{\varepsilon_n\}$ defined by

$$\varepsilon_0 = \|Y(x_0) - Y_0\|, \qquad \varepsilon_{n+1} = \sum_{j=0}^{n} h_{j+k-1}L\varepsilon_j + C_1 h^p. \qquad (5.27)$$

A simple induction argument shows that

$$\|Y(x_n) - Y_n\| \le \varepsilon_n. \qquad (5.28)$$

From (5.27) we obtain for $n \ge 1$

$$\varepsilon_{n+1} = \varepsilon_n + h_{n+k-1}L\varepsilon_n \le \exp(h_{n+k-1}L)\varepsilon_n$$

so that also

$$\varepsilon_n \le \exp((\hat{x} - x_0)L)\varepsilon_1 = \exp((\hat{x} - x_0)L) \cdot \big(h_{k-1}L\|Y(x_0) - Y_0\| + C_1 h^p\big).$$

This inequality together with (5.28) completes the proof of Theorem 5.8. $\square$

Exercises

1. Prove that for constant step sizes the expressions $g_j(n)$ and $\Phi_j^*(n)$ (formulas (5.3) and (5.6)) reduce to

$$g_j(n) = \gamma_j, \qquad \Phi_j^*(n) = \nabla^j f_n,$$

where γ_j is given by (1.6).

2. (Grigorieff 1983). For the k-step BDF-methods consider grids with constant mesh ratio ω, i.e., $h_n = \omega h_{n-1}$ for all n. In this case the elements of A_n^* (see (5.21)) are independent of n. Show numerically that all eigenvalues of A_n^* are of absolute value less than one for $0 < \omega < R_k$ where

k	2	3	4	5	6
R_k	2.414	1.618	1.280	1.127	1.044

III.6 Nordsieck Methods

> While [the method] is primarily designed to optimize the efficiency of large-scale calculations on automatic computers, its essential procedures also lend themselves well to hand computation.
>
> (A. Nordsieck 1962)

> Two further problems must be dealt with in order to implement the automatic choice and revision of the elementary interval, namely, choosing which quantities to remember in such a way that the interval may be changed rapidly and conveniently ...
>
> (A. Nordsieck 1962)

In an important paper Nordsieck (1962) considered a class of methods for ordinary differential equations which allow a convenient way of changing the step size (see Section III.7). He already remarked that his methods are equivalent to the implicit Adams methods, in a certain sense. Let us begin with his derivation of these methods and then investigate their relation to linear multistep methods.

Nordsieck (1962) remarked "... that all methods of numerical integration are equivalent to finding an approximating polynomial for $y(x)$ ...". His idea was to represent such a polynomial by the 0th to kth derivatives, i.e., by a vector ("the Nordsieck vector")

$$z_n = \left(y_n,\, hy_n',\, \frac{h^2}{2!}y_n'',\, \ldots,\, \frac{h^k}{k!}y_n^{(k)} \right)^T. \tag{6.1}$$

The $y_n^{(j)}$ are meant to be approximations to $y^{(j)}(x_n)$, where $y(x)$ is the exact solution of the differential equation

$$y' = f(x,y). \tag{6.2}$$

In order to define the integration procedure we have to give a rule for determining z_{n+1} when z_n and the differential equation (6.2) are given. By Taylor's expansion, such a rule is (e.g., for $k=3$)

$$
\begin{aligned}
y_{n+1} &= y_n + hy_n' + \tfrac{h^2}{2!}y_n'' + \tfrac{h^3}{3!}y_n''' + \tfrac{h^4}{4!}e \\
hy_{n+1}' &= \phantom{y_n +{}} hy_n' + 2\tfrac{h^2}{2!}y_n'' + 3\tfrac{h^3}{3!}y_n''' + 4\tfrac{h^4}{4!}e \\
\tfrac{h^2}{2!}y_{n+1}'' &= \phantom{y_n + hy_n' +{}} \tfrac{h^2}{2!}y_n'' + 3\tfrac{h^3}{3!}y_n''' + 6\tfrac{h^4}{4!}e \\
\tfrac{h^3}{3!}y_{n+1}''' &= \phantom{y_n + hy_n' + \tfrac{h^2}{2!}y_n'' +{}} \tfrac{h^3}{3!}y_n''' + 4\tfrac{h^4}{4!}e,
\end{aligned}
\tag{6.3}
$$

where the value e is determined in such a way that

$$y_{n+1}' = f(x_{n+1}, y_{n+1}). \tag{6.4}$$

Inserting (6.4) into the second relation of (6.3) yields

$$4\frac{h^4}{4!}e = h\Big(f(x_{n+1}, y_{n+1}) - f_n^p \Big) \tag{6.5}$$

with

$$hf_n^p = hy_n' + 2\frac{h^2}{2!}y_n'' + 3\frac{h^3}{3!}y_n'''.$$

With this relation for e the above method becomes

$$
\begin{aligned}
y_{n+1} &= y_n + hy_n' + \tfrac{h^2}{2!}y_n'' + \tfrac{h^3}{3!}y_n''' + \tfrac{1}{4}h\Big(f(x_{n+1},y_{n+1}) - f_n^p\Big) \\
hy_{n+1}' &= \qquad hy_n' + 2\tfrac{h^2}{2!}y_n'' + 3\tfrac{h^3}{3!}y_n''' + h\Big(f(x_{n+1},y_{n+1}) - f_n^p\Big) \\
\tfrac{h^2}{2!}y_{n+1}'' &= \qquad\qquad \tfrac{h^2}{2!}y_n'' + 3\tfrac{h^3}{3!}y_n''' + \tfrac{3}{2}h\Big(f(x_{n+1},y_{n+1}) - f_n^p\Big) \\
\tfrac{h^3}{3!}y_{n+1}''' &= \qquad\qquad\qquad\quad \tfrac{h^3}{3!}y_n''' + h\Big(f(x_{n+1},y_{n+1}) - f_n^p\Big)
\end{aligned}
\tag{6.6}
$$

The first equation constitutes an implicit formula for y_{n+1}, the others are explicit. Observe that for sufficiently accurate approximations $y_n^{(j)}$ to $y^{(j)}(x_n)$ the value e (formula (6.5)) is an approximation to $y^{(4)}(x_n)$. This seems to be a desirable property from the point of view of accuracy. Unfortunately, method (6.6) is unstable. To see this, we put $f(x,y) = 0$ in (6.6). In this case the method becomes the linear transformation

$$z_{n+1} = Mz_n \tag{6.7}$$

where

$$
M = \begin{pmatrix} 1 & 1 & 1 & 1 \\ 0 & 1 & 2 & 3 \\ 0 & 0 & 1 & 3 \\ 0 & 0 & 0 & 1 \end{pmatrix} - \begin{pmatrix} 1/4 \\ 1 \\ 3/2 \\ 1 \end{pmatrix} (0 \; 1 \; 2 \; 3).
$$

The eigenvalues of M are seen to be $1, 0, -(2+\sqrt{3})$ and $-1/(2+\sqrt{3})$, implying that (6.6) is unstable and therefore of no use. The phenomenon that highly accurate methods are often unstable is, after our experiences in Section III.3, no longer astonishing.

To overcome this difficulty Nordsieck proposed to replace the constants $1/4$, $1, 3/2, 1$ which appear in front of the brackets in (6.6) by arbitrary values (l_0, l_1, l_2, l_3), and to use this extra freedom to achieve stability. In compact form this modification can be written as

$$z_{n+1} = (P \otimes I)z_n + (l \otimes I)\Big(hf(x_{n+1},y_{n+1}) - (e_1^T P \otimes I)z_n\Big). \tag{6.8}$$

Here z_n is given by (6.1), P is the Pascal triangle matrix defined by

$$
p_{ij} = \begin{cases} \dbinom{j}{i} & \text{for } 0 \leq i \leq j \leq k, \\[2mm] 0 & \text{else}, \end{cases}
$$

$l = (l_0, l_1, \ldots, l_k)^T$ and $e_1 = (0, 1, 0, \ldots, 0)^T$. Observe that the indices of vectors and matrices start from zero.

For notational simplicity in the following theorems, we consider from now on scalar differential equations only, so that method (6.8) becomes

$$z_{n+1} = Pz_n + l\left(hf_{n+1} - e_1^T Pz_n\right). \tag{6.8'}$$

All results, of course, remain valid for systems of equations. Condition (6.4), which relates the method to the differential equation, fixes the value of l_1 as

$$l_1 = 1. \tag{6.9}$$

The above stability analysis applied to the general method (6.8) leads to the difference equation (6.7) with

$$M = P - le_1^T P. \tag{6.10}$$

For instance, for $k = 3$ this matrix is given by

$$M = \begin{pmatrix} 1 & 1-l_0 & 1-2l_0 & 1-3l_0 \\ 0 & 0 & 0 & 0 \\ 0 & -l_2 & 1-2l_2 & 3-3l_2 \\ 0 & -l_3 & -2l_3 & 1-3l_3 \end{pmatrix}.$$

One observes that 1 and 0 are two eigenvalues of M and that its characteristic polynomial is independent of l_0. Nordsieck determined $l_2, \dots, l_k$ in such a way that the remaining eigenvalues of M are zero. For $k = 3$ this yields $l_2 = 3/4$ and $l_3 = 1/6$. The coefficient l_0 can be chosen such that the error constant of the method (see Theorem 6.2 below) vanishes. In our situation one gets $l_0 = 3/8$, so that the resulting method is given by

$$l = \left(3/8, 1, 3/4, 1/6\right)^T.$$

It is interesting to note that this method is equivalent to the implicit 3-step Adams method. Indeed, an elimination of the terms $(h^3/3!)y_n'''$ and $(h^2/2!)y_n''$ by using formula (6.8) with reduced indices leads to (cf. formula (1.9"))

$$y_{n+1} = y_n + \frac{h}{24}\left(9y_{n+1}' + 19y_n' - 5y_{n-1}' + y_{n-2}'\right). \tag{6.11}$$

Equivalence with Multistep Methods

More insight into the connection between Nordsieck methods and multistep methods is due to Descloux (1963), Osborne (1966), and Skeel (1979). The following two theorems show that every Nordsieck method is equivalent to a multistep formula and that the order of this method is at least k.

Theorem 6.1. *Consider the Nordsieck method (6.8) where $l_1 = 1$. The first two components of z_n then satisfy the linear multistep formula (for $n \geq 0$)*

$$\sum_{i=0}^{k} \alpha_i y_{n+i} = h \sum_{i=0}^{k} \beta_i f_{n+i} \qquad (6.12)$$

where the generating polynomials are given by

$$\varrho(\zeta) = \det(\zeta I - P) \cdot e_1^T (\zeta I - P)^{-1} l$$
$$\sigma(\zeta) = \det(\zeta I - P) \cdot e_0^T (\zeta I - P)^{-1} l. \qquad (6.13)$$

Proof. The proof of the original papers simplifies considerably, if we work with the generating functions (discrete Laplace transformation)

$$Z(\zeta) = \sum_{n \geq 0} z_n \zeta^n, \quad Y(\zeta) = \sum_{n \geq 0} y_n \zeta^n, \quad F(\zeta) = \sum_{n \geq 0} f_n \zeta^n, \quad \dots$$

Multiplying formula (6.8') by ζ^{n+1} and adding up we obtain

$$Z(\zeta) = \zeta P Z(\zeta) + l\left(hF(\zeta) - e_1^T P \zeta Z(\zeta)\right) + (z_0 - lhf_0). \qquad (6.14)$$

Similarly, the linear multistep method (6.12) can be written as

$$\widehat{\varrho}(\zeta) Y(\zeta) = h\widehat{\sigma}(\zeta) F(\zeta) + p_{k-1}(\zeta), \qquad (6.15)$$

where

$$\widehat{\varrho}(\zeta) = \zeta^k \varrho(1/\zeta), \qquad \widehat{\sigma}(\zeta) = \zeta^k \sigma(1/\zeta) \qquad (6.16)$$

and p_{k-1} is a polynomial of degree $k-1$ depending on the starting values. In order to prove the theorem we have to show that the first two components of $Z(\zeta)$ satisfy a relation of the form (6.15). We first rewrite equation (6.14) in the form

$$Z(\zeta) = (I - \zeta P)^{-1} l\left(hF(\zeta) - e_1^T P \zeta Z(\zeta)\right) + (I - \zeta P)^{-1}(z_0 - lhf_0)$$

so that its first two components become

$$Y(\zeta) = e_0^T (I - \zeta P)^{-1} l\left(hF(\zeta) - e_1^T P \zeta Z(\zeta)\right) + e_0^T (I - \zeta P)^{-1}(z_0 - lhf_0)$$

$$hF(\zeta) = e_1^T (I - \zeta P)^{-1} l\left(hF(\zeta) - e_1^T P \zeta Z(\zeta)\right) + e_1^T (I - \zeta P)^{-1}(z_0 - lhf_0).$$

Eliminating the term in brackets and multiplying by $\det(I - \zeta P)$ we arrive at formula (6.15) with

$$\widehat{\varrho}(\zeta) = \det(I - \zeta P) \cdot e_1^T (I - \zeta P)^{-1} l$$
$$\widehat{\sigma}(\zeta) = \det(I - \zeta P) \cdot e_0^T (I - \zeta P)^{-1} l$$
$$p_{k-1}(\zeta) = \det(I - \zeta P)\left(e_1^T (I - \zeta P)^{-1} l e_0^T (I - \zeta P)^{-1}\right.$$
$$\left. - e_0^T (I - \zeta P)^{-1} l e_1^T (I - \zeta P)^{-1}\right) z_0. \qquad (6.17)$$

With the help of (6.16) we immediately get formulas (6.13). Therefore, it remains to show that p_{k-1}, given by (6.17), is a polynomial of degree $k-1$. Since the dimension of P is $(k+1)$, p_{k-1} behaves like ζ^{k-1} for $|\zeta| \to \infty$. Finally, the relation (6.15) implies that the Laurent series of p_{k-1} cannot contain negative powers. $\qquad\qquad\square$

Putting $(\zeta I - P)^{-1}l = u$ in (6.13) and applying Cramer's rule to the linear system $(\zeta I - P)u = l$ we obtain from (6.13) the elegant expressions

$$\varrho(\zeta) = \det \begin{pmatrix} \zeta-1 & l_0 & -1 & .. & -1 \\ 0 & l_1 & -2 & .. & -k \\ 0 & l_2 & \zeta-1 & .. & . \\ \vdots & \vdots & \vdots & & \vdots \\ 0 & l_k & 0 & .. & \zeta-1 \end{pmatrix} \qquad (6.13a)$$

$$\sigma(\zeta) = \det \begin{pmatrix} l_0 & -1 & -1 & .. & -1 \\ l_1 & \zeta-1 & -2 & .. & -k \\ l_2 & 0 & \zeta-1 & .. & . \\ \vdots & \vdots & \vdots & & \vdots \\ l_k & 0 & 0 & .. & \zeta-1 \end{pmatrix}. \qquad (6.13b)$$

We observe that $\varrho(\zeta)$ does not depend on l_0. Further, $\zeta_0 = 1$ is a simple root of $\varrho(\zeta)$ if and only if $l_k \neq 0$. We have

$$\varrho'(1) = \sigma(1) = k! \, l_k. \qquad (6.18)$$

Condition (6.9) is equivalent to $\alpha_k = 1$.

Theorem 6.2. *Assume that $l_k \neq 0$. The multistep method defined by (6.13) is of order at least k and its error constant (see (2.13)) is given by*

$$C = -\frac{b^T l}{k! \, l_k}.$$

Here the components of

$$b^T = (B_0, B_1, \dots, B_k) = \left(1, -\frac{1}{2}, \frac{1}{6}, 0, -\frac{1}{30}, 0, \frac{1}{42}, \dots\right)$$

are the Bernoulli numbers.

Proof. By Theorem 2.4 we have order k iff

$$\varrho(\zeta) - \log \zeta \cdot \sigma(\zeta) = C_{k+1}(\zeta-1)^{k+1} + \mathcal{O}((\zeta-1)^{k+2}).$$

Since $\det(\zeta I - P) = (\zeta-1)^{k+1}$ this is equivalent to

$$e_1^T(\zeta I - P)^{-1}l - \log \zeta \cdot e_0^T(\zeta I - P)^{-1}l = C_{k+1} + \mathcal{O}((\zeta-1))$$

and, by (6.18), it suffices to show that

$$(\log \zeta \cdot e_0^T - e_1^T)(\zeta I - P)^{-1} = b^T + \mathcal{O}((\zeta - 1)). \tag{6.19}$$

Denoting the left-hand side of (6.19) by $b^T(\zeta)$ we obtain

$$(\zeta I - P)^T b(\zeta) = (\log \zeta \cdot e_0 - e_1). \tag{6.20}$$

The qth component $(q \geq 2)$ of this equation

$$\zeta b_q(\zeta) - \sum_{j=0}^q \binom{q}{j} b_j(\zeta) = 0$$

is equivalent to

$$\frac{\zeta b_q(\zeta)}{q!} - \sum_{j=0}^q \frac{b_j(\zeta)}{j!} \frac{1}{(q-j)!} = 0,$$

which is seen to be a Cauchy product. Hence, formula (6.20) becomes

$$\zeta \sum_{q \geq 0} \frac{t^q}{q!} b_q(\zeta) - e^t \sum_{q \geq 0} \frac{t^q}{q!} b_q(\zeta) = \log \zeta - t$$

which yields

$$\sum_{q \geq 0} \frac{t^q}{q!} b_q(\zeta) = \frac{t - \log \zeta}{e^t - \zeta}.$$

If we set $\zeta = 1$ in this formula we obtain

$$\sum_{q \geq 0} \frac{t^q}{q!} b_q(1) = \frac{t}{e^t - 1},$$

therefore $b_q(1) = B_q$, the qth Bernoulli number (see Abramowitz & Stegun, Chapter 23). $\qquad \square$

We have thus shown that to each Nordsieck method (6.8) there corresponds a linear multistep method of order at least k. Our next aim is to establish a correspondence in the opposite direction.

Theorem 6.3. *Let (ϱ, σ) be the generating polynomials of a k-step method (6.12) of order at least k and assume $\alpha_k = 1$. Then we have:*

a) *There exists a unique vector l such that ϱ and σ are given by (6.13).*

b) *If, in addition, the multistep method is irreducible, then there exists a nonsingular transformation T such that the solution of (6.8') is related to that of (6.12) by*

$$z_n = T^{-1} u_n \tag{6.21}$$

where the j th component of u_n is given by

$$u_j^{(n)} = \begin{cases} \sum_{i=0}^{j}(\alpha_{k-j+i}y_{n+i} - h\beta_{k-j+i}f_{n+i}) & \text{for } 0 \leq j \leq k-1, \\ hf_n & \text{for } j = k. \end{cases} \tag{6.22}$$

Proof. a) For every k th order multistep method the polynomial $\varrho(\zeta)$ is uniquely determined by $\sigma(\zeta)$ (see Theorem 2.4). Expanding the determinant in (6.13b) with respect to the first column we see that

$$\sigma(\zeta) = l_0(\zeta-1)^k + l_1(\zeta-1)^{k-1}r_1(\zeta) + \ldots + l_k r_k(\zeta),$$

where $r_j(\zeta)$ is a polynomial of degree j satisfying $r_j(1) \neq 0$. Hence, l can be computed from $\sigma(\zeta)$.

b) Let $y_0, \ldots, y_{k-1}$ and $f_0, \ldots, f_{k-1}$ be given. Then the polynomial $p_{k-1}(\zeta)$ in (6.15) satisfies

$$p_{k-1}(\zeta) = u_0^{(0)} + u_1^{(0)}\zeta + \ldots + u_{k-1}^{(0)}\zeta^{k-1}.$$

On the other hand, if the starting vector z_0 for the Nordsieck method defined by l of (a) is known, then $p_{k-1}(\zeta)$ is given by (6.17). Equating both expressions we obtain

$$\sum_{j=0}^{k-1} u_j^{(0)}\zeta^j = \left(\widehat{\varrho}(\zeta)e_0^T - \widehat{\sigma}(\zeta)e_1^T\right)(I - \zeta P)^{-1}z_0. \tag{6.23}$$

We now denote by t_j^T $(j = 0, \ldots, k-1)$ the coefficients of the vector polynomial

$$\left(\widehat{\varrho}(\zeta)e_0^T - \widehat{\sigma}(\zeta)e_1^T\right)(I - \zeta P)^{-1} = \sum_{j=0}^{k-1} t_j^T \zeta^j \tag{6.24}$$

and set $t_k^T = e_1^T$. Then let T be the square matrix whose j th row is t_j^T so that $u_0 = Tz_0$ is a consequence of (6.23) and $hf_n = hy_n'$. The same argument applied to $y_n, \ldots, y_{n+k-1}$ and $f_n, \ldots, f_{n+k-1}$ instead of $y_0, \ldots, y_{k-1}$ and $f_0, \ldots, f_{k-1}$ yields $u_n = Tz_n$ for all n.

To complete the proof it remains to verify the non-singularity of T. Let $v = (v_0, v_1, \ldots, v_k)^T$ be a non-zero vector satisfying $Tv = 0$. By definition of t_k^T we have $v_1 = 0$ and from (6.24) it follows (using the transformation (6.16)) that

$$\varrho(\zeta)\tau_0(\zeta) = \sigma(\zeta)\tau_1(\zeta), \tag{6.25}$$

where $\tau_i(\zeta) = \det(\zeta I - P)e_i^T(\zeta I - P)^{-1}v$ are polynomials of degree at most k. Moreover, Cramer's rule shows that the degree of $\tau_1(\zeta)$ is at most $k-1$, since $v_1 = 0$. Hence from (6.25) at least one of the roots of $\varrho(\zeta)$ must be a root of $\sigma(\zeta)$. This is in contradiction with the assumption that the method is irreducible.   □

The vectors l which correspond to the implicit Adams methods and to the BDF-methods are given in Tables 6.1 and 6.2. For these two classes of methods we shall investigate the equivalence in some more detail.

Table 6.1. Coefficients l_j of the k-step implicit Adams methods

	l_0	l_1	l_2	l_3	l_4	l_5	l_6
$k=1$	1/2	1					
$k=2$	5/12	1	1/2				
$k=3$	3/8	1	3/4	1/6			
$k=4$	251/720	1	11/12	1/3	1/24		
$k=5$	95/288	1	25/24	35/72	5/48	1/120	
$k=6$	19087/60480	1	137/120	5/8	17/96	1/40	1/720

Table 6.2. Coefficients l_j of the k-step BDF-methods

	l_0	l_1	l_2	l_3	l_4	l_5	l_6
$k=1$	1	1					
$k=2$	2/3	1	1/3				
$k=3$	6/11	1	6/11	1/11			
$k=4$	12/25	1	7/10	1/5	1/50		
$k=5$	60/137	1	225/274	85/274	15/274	1/274	
$k=6$	20/49	1	58/63	5/12	25/252	1/84	1/1764

Implicit Adams Methods

The following results are due to Byrne & Hindmarsh (1975). Since their "efficient package" EPISODE and the successor VODE are based on the Nordsieck representation of variable step size methods, we extend our considerations to this case. The Adams methods define in a natural way a polynomial which approximates the unknown solution of (6.2). Namely, if y_n and $f_n, \ldots, f_{n-k+1}$ are given, then the k-step Adams method is equivalent to the construction of a polynomial $p_{n+1}(x)$ of degree $k+1$ which satisfies

$$
\begin{aligned}
&p_{n+1}(x_n) = y_n, \qquad p_{n+1}(x_{n+1}) = y_{n+1}, \\
&p'_{n+1}(x_j) = f_j \qquad \text{for } j = n-k+1, \ldots, n+1.
\end{aligned}
\tag{6.26}
$$

Condition (6.26) defines y_{n+1} implicitly. We observe that the difference of two consecutive polynomials, $p_{n+1}(x) - p_n(x)$, vanishes at x_n and that its derivative is zero at $x_{n-k+1}, \ldots, x_n$. Therefore, if we let $e_{n+1} = y_{n+1} - p_n(x_{n+1})$, this difference can be written as

$$p_{n+1}(x) - p_n(x) = \Lambda\left(\frac{x - x_{n+1}}{x_{n+1} - x_n}\right)e_{n+1} \qquad (6.27)$$

where Λ is the unique polynomial of degree $(k+1)$ defined by

$$\Lambda(0) = 1, \qquad \Lambda(-1) = 0$$

$$\Lambda'\left(\frac{x_j - x_{n+1}}{x_{n+1} - x_n}\right) = 0 \qquad \text{for } j = n - k + 1, \ldots, n. \qquad (6.28)$$

The derivative of (6.27) taken at $x = x_{n+1}$ shows that with $h_n = x_{n+1} - x_n$,

$$h_n f_{n+1} - h_n p_n'(x_{n+1}) = \Lambda'(0)e_{n+1}.$$

If we introduce the Nordsieck vector

$$\widetilde{z}_n = \left(p_n(x_n), h_n p_n'(x_n), \ldots, \frac{h_n^{k+1}}{(k+1)!}p_n^{(k+1)}(x_n)\right)^T$$

and the coefficients $\widetilde{l}_j$ by

$$\Lambda(t) = \sum_{j=0}^{k+1} \widetilde{l}_j t^j, \qquad (6.29)$$

then (6.27) becomes equivalent to

$$\widetilde{z}_{n+1} = P\widetilde{z}_n + \widetilde{l}\,\widetilde{l}_1^{-1}\left(hf_{n+1} - e_1^T P\widetilde{z}_n\right) \qquad (6.30)$$

with $\widetilde{l} = (\widetilde{l}_0, \widetilde{l}_1, \ldots, \widetilde{l}_{k+1})^T$. This method is of the form (6.8'). However, it is of dimension $k+2$ and not, as expected by Theorem 6.3, of dimension $k+1$. The reason is the following: let $\widetilde{\varrho}(\zeta)$ and $\widetilde{\sigma}(\zeta)$ be the generating polynomials of the multistep method which corresponds to (6.30). Then the conditions $\Lambda(-1) = 0$ and $\Lambda'(-1) = 0$ imply that $\widetilde{\sigma}(0) = \widetilde{\varrho}(0) = 0$, so that this method is reducible. Nevertheless, method (6.30) is useful, since the last component of $\widetilde{z}_n$ can be used for step size control.

Remark. For $k \geq 2$ the coefficients $\widetilde{l}_j$, defined by (6.29), depend on the step size ratios h_j/h_{j-1} for $j = n - k + 2, \ldots, n$. They can be computed from the formula

$$\Lambda(t) = \frac{\int_{-1}^{t} \prod_{j=1}^{k}(s - t_j)\,ds}{\int_{-1}^{0} \prod_{j=1}^{k}(s - t_j)\,ds}. \qquad (6.31)$$

where $t_j = (x_{n-j+1} - x_{n+1})/(x_{n+1} - x_n)$ (see also Exercise 1).

BDF-Methods

One step of the k-step BDF method consists in constructing a polynomial $q_{n+1}(x)$ of degree k which satisfies

$$q_{n+1}(x_j) = y_j \qquad \text{for } j = n-k+1, \ldots, n+1$$
$$q'_{n+1}(x_{n+1}) = f_{n+1} \tag{6.32}$$

and in computing a value y_{n+1} which makes this possible. As for the Adams methods we have

$$q_{n+1}(x) - q_n(x) = \Lambda\left(\frac{x - x_{n+1}}{x_{n+1} - x_n}\right) \cdot \left(y_{n+1} - q_n(x_{n+1})\right), \tag{6.33}$$

where $\Lambda(t)$ is the polynomial of degree k defined by

$$\Lambda\left(\frac{x_j - x_{n+1}}{x_{n+1} - x_n}\right) = 0 \qquad \text{for } j = n-k+1, \ldots, n,$$
$$\Lambda(0) = 1.$$

With the vector

$$\tilde{z}_n = \left(q_n(x_n), \, h_n q'_n(x_n), \ldots, \frac{h_n^k}{k!} q_n^{(k)}(x_n)\right)^T$$

and the coefficients $\tilde{l}_j$ given by

$$\Lambda(t) = \sum_{j=0}^{k} \tilde{l}_j t^j,$$

equation (6.33) becomes

$$\tilde{z}_{n+1} = P\tilde{z}_n + \tilde{l}\, \tilde{l}_1^{-1}\left(hf_{n+1} - e_1^T P\tilde{z}_n\right). \tag{6.34}$$

The vector $\tilde{l} = (\tilde{l}_0, \tilde{l}_1, \ldots, \tilde{l}_k)^T$ can be computed from the formula

$$\Lambda(t) = \prod_{j=1}^{k}\left(1 + \frac{t}{t_j}\right)$$

where $t_j = (x_{n-j+1} - x_{n+1})/(x_{n+1} - x_n)$. For constant step sizes formula (6.34) corresponds to that of Theorem 6.3 and the coefficients $l_j = \tilde{l}_j/\tilde{l}_1$ coincide with those of Table 6.2.

Exercises

1. Let $l_j^{(k)}(j = 0, \ldots, k)$ be the Nordsieck coefficients of the k-step implicit Adams methods (defined by Theorem 6.3 and given in Table 6.1). Further, denote by $\tilde{l}_j^{(k)}$ $(j = 0, \ldots, k+1)$ the coefficients given by (6.29) and (6.31) for the case of constant step sizes. Show that

$$\frac{\tilde{l}_j^{(k)}}{\tilde{l}_1^{(k)}} = \begin{cases} l_j^{(k)} & \text{for } j = 0 \\ l_j^{(k+1)} & \text{for } j = 1, \ldots, k+1. \end{cases}$$

Use these relations to verify Table 6.1.

2. a) Calculate the matrix T of Theorem 6.3 for the 3-step implicit Adams method.

   Result.

$$T^{-1} = \begin{pmatrix} 1 & 0 & 0 & 3/8 \\ 0 & 0 & 0 & 1 \\ 0 & 6 & 6 & 3/4 \\ 0 & 4 & 12 & 1/6 \end{pmatrix}.$$

   Show that the Nordsieck vector z_n is given by

$$z_n = \left(y_n, \, hf_n, \, (3hf_n - 4hf_{n-1} + hf_{n-2})/4, \, (hf_n - 2hf_{n-1} + hf_{n-2})/6\right)^T.$$

   b) The vector $\tilde{z}_n$ for the 2-step implicit Adams method (6.30) (constant step sizes) also satisfies

$$\tilde{z}_n = \left(y_n, \, hf_n, \, (3hf_n - 4hf_{n-1} + hf_{n-2})/4, \, (hf_n - 2hf_{n-1} + hf_{n-2})/6\right)^T,$$

   but this time y_n is a less accurate approximation to $y(x_n)$.

III.7 Implementation and Numerical Comparisons

There is a great deal of freedom in the implementation of multistep methods (even if we restrict our considerations to the Adams methods). One can either directly use the *variable step size methods* of Section III.5 or one can take a fixed step size method and determine the necessary offgrid values, which are needed for a change of step size, by *interpolation*. Further, it is possible to choose between the *divided difference* formulation (5.7) and the *Nordsieck* representation (6.30).

The historical approach was the use of formula (1.9) together with interpolation (J.C. Adams (1883): "We may, of course, change the value of ω (the step size) whenever the more or less rapid rate of diminution of the successive differences shews that it is expedient to increase or diminish the interval. It is only necessary, by selection from or interpolation between the values already calculated, to find the coordinates for a few values of φ separated from each other by the newly chosen interval."). It is theoretically more satisfactory and more elegant to work with the variable step size method (5.7). For both of these approaches the change of step size is rather expensive whereas the change of order is very simple — one just has to add a further term to the expansion (1.9). If the Nordsieck representation (6.30) is implemented, the situation is the opposite. There, the change of order is not as direct as above, but the step size can be changed simply by multiplying the Nordsieck-vector (6.1) by the diagonal matrix with entries $(1, \omega, \omega^2, \ldots)$ where $\omega = h_{new}/h_{old}$ is the step size ratio. Indeed, this was the main reason for introducing this representation.

Step Size and Order Selection

Much was made of the starting of multistep computations and the need for Runge-Kutta methods in the literature of the 60ies (see e.g., Ralston 1962). Nowadays, codes for multistep methods simply start with order one and very small step sizes and are therefore self-starting. The following step size and order selection is closely related to the description of Shampine & Gordon (1975).

Suppose that the numerical integration has proceeded successfully until x_n and that a further step with step size h_n and order $k+1$ is taken, which yields the approximation y_{n+1} to $y(x_{n+1})$. To decide whether y_{n+1} will be accepted or not, we need an estimate of the local truncation error. Such an estimate is e.g. given by

$$le_{k+1}(n+1) = y_{n+1}^* - y_{n+1}$$

where y_{n+1}^* is the result of the $(k+2)$nd order implicit Adams formula. Subtracting formula (5.7) from the same formula with k replaced by $k+1$, we obtain

$$le_{k+1}(n+1) = h_n\big(g_{k+1}(n) - g_k(n)\big)\Phi_{k+1}(n+1). \tag{7.1}$$

Without changing the leading term in this expression we can replace the expression $\Phi_{k+1}(n+1)$ by

$$\Phi_{k+1}^p(n+1) = \prod_{i=0}^{k}(x_{n+1} - x_{n-i})\,\delta^{k+1}f^p[x_{n+1}, x_n, \ldots, x_{n-k}]. \tag{7.2}$$

The superscript p of f indicates that $f_{n+1} = f(x_{n+1}, y_{n+1})$ is replaced by $f(x_{n+1}, p_{n+1})$ when forming the divided differences. If the implicit equation (5.7) is solved iteratively with p_{n+1} as predictor, then $\Phi_{k+1}^p(n+1)$ has to be calculated anyway. Therefore, the only cost for computing the estimate

$$LE_{k+1}(n+1) = h_n\big(g_{k+1}(n) - g_k(n)\big)\Phi_{k+1}^p(n+1) \tag{7.3}$$

is the computation of $g_{k+1}(n)$. After the expression (7.3) has been calculated, we require (in the norm (4.11) of Section II.4)

$$\|LE_{k+1}(n+1)\| \le 1 \tag{7.4}$$

for the step to be successful.

If the Nordsieck representation (6.30) is considered instead of (5.7), then the estimate of the local error is not as simple, since the $\widetilde{l}$-vectors in (6.30) are totally different for different orders. For a possible error-estimate we refer to the article of Byrne & Hindmarsh (1975).

Suppose now that y_{n+1} is accepted. We next have to choose a new step size and a new order. The idea of the *step size selection* is to find the largest h_{n+1} for which the predicted local error is acceptable, i.e., for which

$$h_{n+1}\cdot\big|g_{k+1}(n+1) - g_k(n+1)\big|\cdot\|\Phi_{k+1}^p(n+2)\| \le 1.$$

However, this procedure is of no practical use, since the expressions $g_j(n+1)$ and $\Phi_{k+1}^p(n+2)$ depend in a complicated manner on the unknown step size h_{n+1}. Also, the coefficients $g_{k+1}(n+1)$ and $g_k(n+1)$ are too expensive to calculate. To overcome this difficulty we assume the grid to be equidistant (this is a doubtful assumption, but leads to a simple formula for the new step size). In this case the local error (for the method of order $k+1$) is of the form $C(x_{n+2})h^{k+2} + \mathcal{O}(h^{k+3})$ with C depending smoothly on x. The local error at x_{n+2} can thus be approximated by that at x_{n+1} and in the same way as for one-step methods (cf. Section II.4 formula

(4.12)) we obtain

$$h_{\text{opt}}^{(k+1)} = h_n \cdot \left(\frac{1}{\|LE_{k+1}(n+1)\|} \right)^{1/(k+2)} \tag{7.5}$$

as optimal step size. The local error $LE_{k+1}(n+1)$ is given by (7.3) or, again under the assumption of an equidistant grid, by

$$LE_{k+1}(n+1) = h_n \gamma_{k+1}^* \Phi_{k+1}^p(n+1) \tag{7.6}$$

with γ_{k+1}^* from Table 1.2 (see Exercise 1 of Section III.5 and Exercise 4 of Section III.1).

We next describe how an *optimal order* can be determined. Since the number of necessary function evaluations is the same for all orders, there are essentially two strategies for selecting the new order. One can choose the order $k+1$ either such that the local error estimate is minimal, or such that the new optimal step size is maximal. Because of the exponent $1/(k+2)$ in formula (7.5), the two strategies are not always equivalent. For more details see the description of the code DEABM below. It should be mentioned that each implementation of the Adams methods — and there are many — contains refinements of the above description and has in addition several ad-hoc devices. One of them is to keep the step size constant if $h_{\text{new}}/h_{\text{old}}$ is near to 1. In this way the computation of the coefficients $g_j(n)$ is simplified.

Some Available Codes

We have chosen the three codes DEABM, VODE and LSODE to illustrate the order- and step size strategies for multistep methods.

DEABM is a modification of the code DE/STEP/INTRP described in the book of Shampine & Gordon (1975). It belongs to the package DEPAC, designed by Shampine & Watts (1979). Our numerical tests use the revised version from February 1984. For European users it is available from the "Rechenzentrum der RWTH Aachen, Seffenter Weg 23, D-5100 Aachen, Germany".

This code implements the variable step size, divided difference representation (5.7) of the Adams formulas. In order to solve the nonlinear equation (5.7) for y_{n+1} the value p_{n+1} is taken as predictor (P), then $f_{n+1}^p = f(x_{n+1}, p_{n+1})$ is calculated (E) and *one* corrector iteration (C) is performed, to obtain y_{n+1}. Finally, in the case of a successful step, $f_{n+1} = f(x_{n+1}, y_{n+1})$ is evaluated (E) for the next step. This PECE implementation needs two function evaluations for each successful step. Let us also outline the order strategy of this code: after performing a step with order $k+1$, one computes $LE_{k-1}(n+1)$, $LE_k(n+1)$ and $LE_{k+1}(n+1)$ using a slight modification of (7.6). Then the order is reduced by one, if

$$\max\left(\|LE_{k-1}(n+1)\|, \|LE_k(n+1)\| \right) \leq \|LE_{k+1}(n+1)\|. \tag{7.7}$$

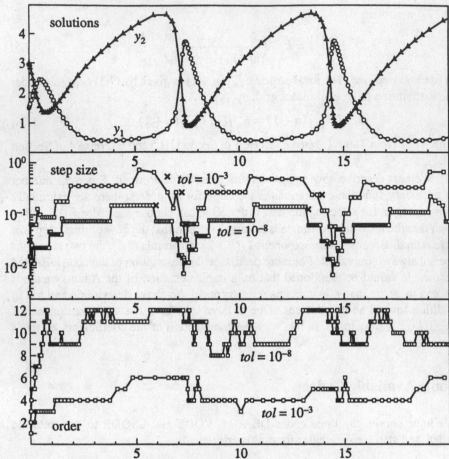

Fig. 7.1. Step size and order variation for the code DEABM

An increase in the order is considered only if the step is successful, (7.7) is violated and a constant step size is used. In this case one computes the estimate

$$LE_{k+2}(n+1) = h_n \gamma^*_{k+2} \Phi_{k+2}(n+1)$$

using the new value $f_{n+1} = f(x_{n+1}, y_{n+1})$ and increases the order by one if

$$\|LE_{k+2}(n+1)\| < \|LE_{k+1}(n+1)\|.$$

In Fig. 7.1 we demonstrate the variation of the step size and order on the example of Section II.4 (see Fig. 4.1 and also Fig. 9.5 of Section II.9). We plot the solution obtained with $Rtol = Atol = 10^{-3}$, the step size and order for the tolerances 10^{-3} and 10^{-8}. We observe that the step size — and not the order — drops significantly at passages where the solution varies more rapidly. Furthermore, constant step sizes are taken over long intervals, and the order is changed rather often (especially for $Tol = 10^{-8}$). This is in agreement with the observation of Shampine & Gordon

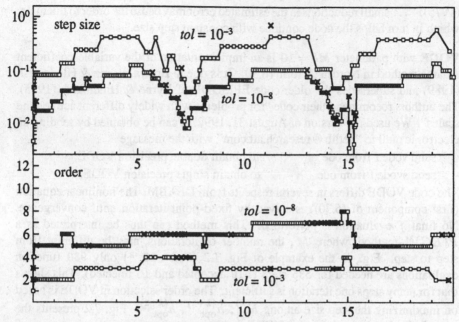

Fig. 7.2. Step size and order variation for the code VODE

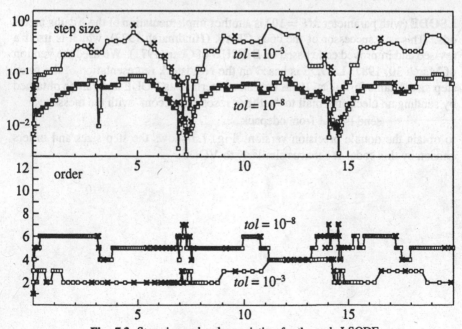

Fig. 7.3. Step size and order variation for the code LSODE

(1975): "... small reductions in the estimated error may cause the order to fluctuate, which in turn helps the code continue with constant step size."

VODE with parameter $MF = 10$ is an implementation of the variable-coefficient Adams method in Nordsieck form (6.30). It is due to Brown, Byrne & Hindmarsh (1989) and supersedes the older code EPISODE of Byrne & Hindmarsh (1975). The authors recommend their code "for problems with widely different active time scales". We used the version of August 31, 1992. It can be obtained by sending an electronic mail to "netlib @ research.att.com" with the message

> send vode.f from ode   to obtain double precision VODE,
> send svode.f from ode   to obtain single precision VODE.

The code VODE differs in several respects from DEABM. The nonlinear equation (first component of (6.30)) is solved by fixed-point iteration until convergence. No final f-evaluation is performed. This method can thus be interpreted as a $P(EC)^M$-method, where M, the number of iterations, may be different from step to step. E.g., in the example of Fig. 7.2 ($Tol = 10^{-8}$) only 930 function evaluations are needed for 535 steps (519 accepted and 16 rejected). This shows that for many steps one iteration is sufficient. The order selection in VODE is based on maximizing the step size among $h_{\mathrm{opt}}^{(k)}$, $h_{\mathrm{opt}}^{(k+1)}$, $h_{\mathrm{opt}}^{(k+2)}$. Fig. 7.2 presents the step size and order variation for VODE for the same example as above: compared to DEABM we observe that much lower orders are taken. Further, the order is constant over long intervals. This is reasonable, since a change in the order is not natural for the Nordsieck representation.

LSODE (with parameter $MF = 10$) is another implementation of the Adams methods. This is a successor of the code GEAR (Hindmarsh 1972), which is itself a revised and improved code based on DIFSUB of Gear (1971). We used the version of March 30, 1987. LSODE is based on the Nordsieck representation of the fixed step size Adams formulas. It has the same interface as VODE and can be obtained by sending an electronic mail to "netlib @ research.att.com" with the message

> send lsode.f from odepack

to obtain the double precision version. Fig. 7.3 shows the step sizes and orders chosen by this code. It behaves similarly to VODE.

Numerical Comparisons

> Of the three families of methods, the fixed order Runge-Kutta is the simplest, in several respects the best understood, and the least efficient.
> (Shampine & Gordon 1975)

It is, of course, interesting to study the numerical performance of the above implementations of the Adams methods:

      DEABM — symbol ⋈

      VODE — symbol ○

      LSODE — symbol △

In order to compare the results with those of a typical one-step Runge-Kutta method we include the results of the code

      DOP853 — symbol ✿

described in Section II.5.

With all these methods we have computed the numerical solution for the six problems EULR, AREN, LRNZ, PLEI, ROPE, BRUS of Section II.10 using many different tolerances between 10^{-3} and 10^{-14} (the "integer" tolerances $10^{-3}, 10^{-4}, \ldots$ are distinguished by enlarged symbols). Fig. 7.4 gives the number of *function evaluations* plotted against the achieved accuracy in double logarithmic scale. Some general tendencies can be distinguished in the crowds of numerical results. LSODE and DEABM require, for equal obtained accuracy, usually less function evaluations, with DEABM becoming champion for higher precision $(Tol \leq 10^{-6})$.

The situation changes dramatically in favour of the Runge-Kutta code DOP853 if *computing time* is measured instead of function evaluations (see Fig. 7.5; the CPU time is that of an Apollo Domain 4000 Workstation). We observe that for problems with cheap function evaluations (EULR, AREN, LRNZ) the Runge-Kutta code needs much less CPU time than the multistep codes, although more function evaluations are necessary in general. For the problems PLEI and ROPE, where the right hand side is rather expensive to evaluate, the discrepancy is not as large. For the last problem (BRUS) the dimension is very high, but the individual components are not too complicated. In this situation, the CPU time of DOP853 is also significantly less than for the multistep codes; this indicates that their overhead also increases with the dimension of the problem.

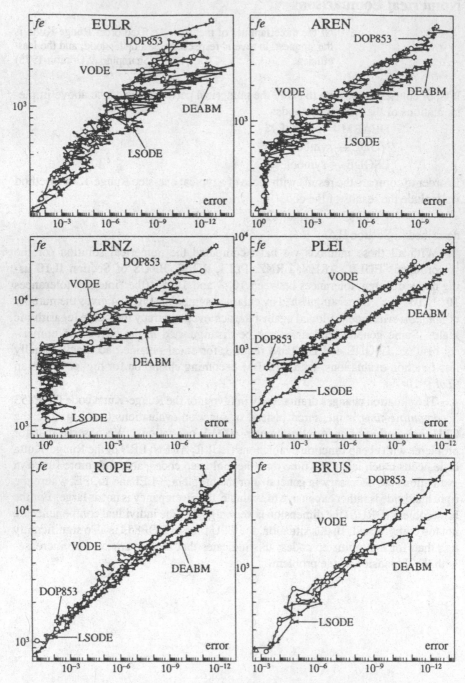

Fig. 7.4. Precision versus function calls for the problems of Section II.10

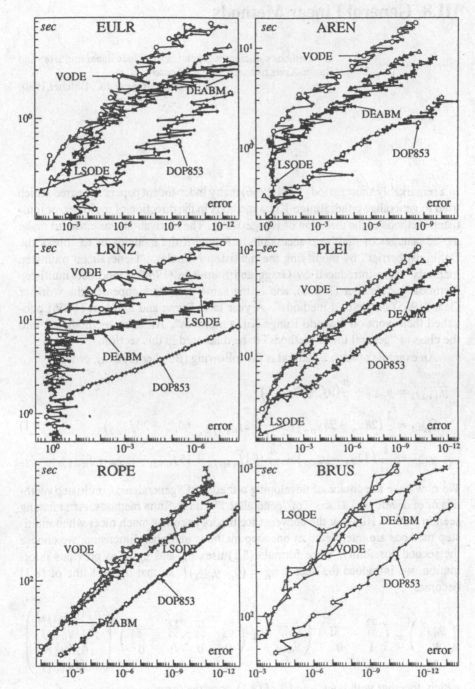

Fig. 7.5. Precision versus computing time for the problems of Section II.10

III.8 General Linear Methods

... methods sufficiently general as to include linear multistep and Runge-Kutta methods as special cases ...

(K. Burrage & J.C. Butcher 1980)

In a remarkably short period (1964-1966) many independent papers appeared which tried to generalize either Runge-Kutta methods in the direction of multistep or multistep methods in the direction of Runge-Kutta. The motivation was either to make the advantages of multistep accessible to Runge-Kutta methods or to "break the Dahlquist barrier" by modifying the multistep formulas. "Generalized multistep methods" were introduced by Gragg and Stetter in (1964), "modified multistep methods" by Butcher (1965a), and in the same year there appeared the work of Gear (1965) on "hybrid methods". A year later Byrne and Lambert (1966) published their work on "pseudo Runge-Kutta methods". All these methods fall into the class of "general linear methods" to be discussed in this section.

An example of such a method is the following (Butcher (1965a), order 5)

$$
\begin{aligned}
\widehat{y}_{n+1/2} &= y_{n-1} + \frac{h}{8}\left(9f_n + 3f_{n-1}\right) \\
\widehat{y}_{n+1} &= \frac{1}{5}\left(28y_n - 23y_{n-1}\right) + \frac{h}{5}\left(32\widehat{f}_{n+1/2} - 60f_n - 26f_{n-1}\right) \\
y_{n+1} &= \frac{1}{31}\left(32y_n - y_{n-1}\right) + \frac{h}{93}\left(64\widehat{f}_{n+1/2} + 15\widehat{f}_{n+1} + 12f_n - f_{n-1}\right).
\end{aligned}
\tag{8.1}
$$

We now have the choice of developing a theory of "generalized" multistep methods or of developing a theory of "generalized" Runge-Kutta methods. After having seen in Section III.4 that the convergence theory becomes much nicer when multistep methods are interpreted as one-step methods in higher dimension, we choose the second possibility: since formula (8.1) uses y_n and y_{n-1} as previous information, we introduce the vector $u_n = (y_n, y_{n-1})^T$ so that the last line of (8.1) becomes

$$
\begin{pmatrix} y_{n+1} \\ y_n \end{pmatrix} = \begin{pmatrix} \frac{32}{31} & -\frac{1}{31} \\ 1 & 0 \end{pmatrix} \begin{pmatrix} y_n \\ y_{n-1} \end{pmatrix} + \begin{pmatrix} \frac{64}{93} & \frac{15}{93} & \frac{12}{93} & -\frac{1}{93} \\ 0 & 0 & 0 & 0 \end{pmatrix} \begin{pmatrix} hf(\widehat{y}_{n+1/2}) \\ hf(\widehat{y}_{n+1}) \\ hf(y_n) \\ hf(y_{n-1}) \end{pmatrix}
$$

which, together with lines 1 and 2 of (8.1), is of the form

$$
u_{n+1} = Su_n + h\Phi(x_n, u_n, h).
\tag{8.2}
$$

Properties of such general methods have been investigated by Butcher (1966), Hairer & Wanner (1973), Skeel (1976), Cooper (1978), Albrecht (1978, 1985) and others. Clearly, nothing prevents us from letting S and Φ be arbitrary, or from allowing also other interpretations of u_n.

A General Integration Procedure

We consider the system

$$y' = f(x, y), \qquad y(x_0) = y_0 \tag{8.3}$$

where f satisfies the regularity condition (4.2). Let m be the dimension of the differential equation (8.3), $q \geq m$ be the dimension of the difference equation (8.2) and $x_n = x_0 + nh$ be the subdivision points of an equidistant grid. The methods under consideration consist of three parts:

i) a *forward step procedure*, i.e., a formula (8.2), where the square matrix S is independent of (8.3).

ii) a *correct value function* $z(x, h)$, which gives an interpretation of the values u_n; $z_n = z(x_n, h)$ is to be approximated by u_n, so that the global error is given by $u_n - z_n$. It is assumed that the exact solution $y(x)$ of (8.3) can be recovered from $z(x, h)$.

iii) a *starting procedure* $\varphi(h)$, which specifies the starting value $u_0 = \varphi(h)$. $\varphi(h)$ approximates $z_0 = z(x_0, h)$.

The discrete problem corresponding to (8.3) is thus given by

$$u_0 = \varphi(h), \tag{8.4a}$$

$$u_{n+1} = S u_n + h\Phi(x_n, u_n, h), \qquad n = 0, 1, 2, \ldots, \tag{8.4b}$$

which yields the numerical solution $u_0, u_1, u_2, \ldots$. We remark that the increment function $\Phi(x, u, h)$, the starting procedure $\varphi(h)$ and the correct value function $z(x, h)$ depend on the differential equation (8.3), although this is not stated explicitly.

Example 8.1. The most simple cases are *one-step methods*. A characteristic feature of these is that the dimensions of the differential and difference equation are equal (i.e., $m = q$) and that S is the identity matrix. Furthermore, $\varphi(h) = y_0$ and $z(x, h) = y(x)$. They have been investigated in Chapter II.

Example 8.2. We have seen in Section III.4 that linear *multistep methods* also fall into the class (8.4). For k-step methods the dimension of the difference equation is $q = km$ and the forward step procedure is given by formula (4.8). A starting procedure yields the vector $\varphi(h) = (y_{k-1}, \ldots, y_1, y_0)^T$ and, finally, the correct value function is given by

$$z(x, h) = \big(y(x + (k-1)h), \ldots, y(x+h), y(x)\big)^T.$$

The most common way of implementing an implicit multistep method is a *predictor-corrector* process (compare (1.11) and Section III.7): an approximation $y_{n+k}^{(0)}$ to y_{n+k} is "predicted" by an explicit multistep method, say

$$\alpha_k^p y_{n+k}^{(0)} + \alpha_{k-1}^p y_{n+k-1} + \ldots + \alpha_0^p y_n = h(\beta_{k-1}^p f_{n+k-1} + \ldots + \beta_0^p f_n) \quad (8.5;P)$$

and is then "corrected" (usually once or twice) by

$$f_{n+k}^{(l-1)} := f(x_{n+k}, y_{n+k}^{(l-1)}) \qquad (8.5;E)$$

$$\alpha_k y_{n+k}^{(l)} + \alpha_{k-1} y_{n+k-1} + \ldots + \alpha_0 y_n = h(\beta_k f_{n+k}^{(l-1)} + \beta_{k-1} f_{n+k-1} + \ldots + \beta_0 f_n).$$
$$(8.5;C)$$

If the iteration (8.5) is carried out until convergence, the process is identical to that of Example 8.2. In practice, however, only a fixed number, say M, of iterations are carried out and the method is theoretically no longer a "pure" multistep method. We distinguish two predictor-corrector (PC) methods, depending on whether it ends with a correction (8.5;C) or not. The first algorithm is symbolized as $P(EC)^M$ and the second possibility, where f_{n+k} is once more updated by (8.5;E) for further use in the subsequent steps, as $P(EC)^M E$. We shall now see how these two procedures can be interpreted as methods of type (8.4).

Example 8.2a. $P(EC)^M E$-methods. The starting procedure and the correct value function are the same as for multistep methods and also $q = km$. Furthermore we have $S = A \otimes I$, where A is given by (4.7) and I is the m-dimensional identity matrix. Observe that S depends only on the corrector-formula and not on the predictor-formula. Here, the increment function is given by

$$\Phi(x, u, h) = (e_1 \otimes I)\psi(x, u, h)$$

with $e_1 = (1, 0, \ldots, 0)^T$. For $u = (u^1, \ldots, u^k)^T$ with $u^j \in \mathbf{R}^m$ the function $\psi(x, u, h)$ is defined by

$$\psi(x, u, h) = \alpha_k^{-1}\Big(\beta_k f(x + kh, y^{(M)})$$
$$+ \beta_{k-1} f(x + (k-1)h, u^1) + \ldots + \beta_0 f(x, u^k)\Big)$$

where the value $y^{(M)}$ is calculated from

$$\alpha_k^p y^{(0)} + \alpha_{k-1}^p u^1 + \ldots + \alpha_0^p u^k$$
$$= h\Big(\beta_{k-1}^p f\big(x + (k-1)h, u^1\big) + \ldots + \beta_0^p f(x, u^k)\Big)$$

$$\alpha_k y^{(l)} + \alpha_{k-1} u^1 + \ldots + \alpha_0 u^k$$
$$= h\Big(\beta_k f\big(x + kh, y^{(l-1)}\big) + \beta_{k-1} f\big(x + (k-1)h, u^1\big) + \ldots + \beta_0 f(x, u^k)\Big)$$

(for $l = 1, \ldots, M$).

Example 8.2b. For $P(EC)^M$-methods, the formulation as a method of type (8.4) becomes more complicated, since the information to be carried over to the next step is determined not only by $y_{n+k-1}, \ldots, y_n$, but also depends on the values $hf_{n+k-1}, \ldots, hf_n$, where $hf_{n+j} = hf(x_{n+j}, y_{n+j}^{(M-1)})$. Therefore the dimension of the difference equation becomes $q = 2km$. A usual starting procedure (as for multistep methods) yields

$$\varphi(h) = \Big(y_{k-1}, \ldots, y_0, hf(x_{k-1}, y_{k-1}), \ldots, hf(x_0, y_0)\Big)^T.$$

If we define the correct value function by

$$z(x, h) = \Big(y(x + (k-1)h), \ldots, y(x), hy'(x + (k-1)h), \ldots, hy'(x)\Big)^T,$$

the forward step procedure is given by

$$S = \begin{pmatrix} A & B \\ 0 & N \end{pmatrix}, \qquad \Phi(x, u, h) = \begin{pmatrix} \beta'_k e_1 \\ e_1 \end{pmatrix} \Psi(x, u, h).$$

Here A is the matrix given by (4.7), $\beta'_j = \beta_j / \alpha_k$ and

$$N = \begin{pmatrix} 0 & 0 & \cdots & 0 & 0 \\ 1 & 0 & \cdots & 0 & 0 \\ \vdots & \vdots & & \vdots & \vdots \\ 0 & 0 & \cdots & 1 & 0 \end{pmatrix}, \qquad B = \begin{pmatrix} \beta'_{k-1} & \cdots & \beta'_0 \\ 0 & \cdots & 0 \\ \vdots & & \vdots \\ 0 & \cdots & 0 \end{pmatrix}, \qquad e_1 = \begin{pmatrix} 1 \\ 0 \\ \vdots \\ 0 \end{pmatrix}.$$

For $u = (u^1, \ldots, u^k, hv^1, \ldots, hv^k)$ the function $\psi(x, u, h) \in \mathbb{R}^q$ is defined by

$$\psi(x, u, h) = f(x + kh, y^{(M-1)})$$

where $y^{(M-1)}$ is given by

$$\alpha_k^p y^{(0)} + \alpha_{k-1}^p u^1 + \ldots + \alpha_0^p u^k = h(\beta_{k-1}^p v^1 + \ldots + \beta_0^p v^k)$$

$$\alpha_k y^{(l)} + \alpha_{k-1} u^1 + \ldots + \alpha_0 u^k = h\big(\beta_k f(x+kh, y^{(l-1)}) + \beta_{k-1} v^1 + \ldots + \beta_0 v^k\big).$$

Again we observe that S depends only on the corrector-formula.

Example 8.3. *Nordsieck methods* are also of the form (8.4). This follows immediately from the representation (6.8). In this case the correct value function

$$z(x, h) = \Big(y(x), hy'(x), \frac{h^2}{2!}y''(x), \ldots, \frac{h^k}{k!}y^{(k)}(x)\Big)^T$$

is composed not only of values of the exact solution, but also contains their derivatives.

Example 8.4. *Cyclic multistep methods.* Donelson & Hansen (1971) have investigated the possibility of basing a discretization scheme on several different k-step methods which are used cyclically. Let S_j and Φ_j represent the forward step procedure of the jth multistep method; then the numerical solution $u_0, u_1, \ldots$ is defined by

$$u_0 = \varphi(h)$$
$$u_{n+1} = S_j u_n + h\Phi_j(x_n, u_n, h) \qquad \text{if } n \equiv (j-1) \bmod m.$$

In order to get a method (8.4) with S independent of the step number, we consider one cycle of the method as one step of a new method

$$u_0^* = \varphi(\frac{h^*}{m})$$
$$u_{n+1}^* = S u_n^* + h^* \Phi(x_n^*, u_n^*, h^*) \tag{8.6}$$

with step size $h^* = mh$. Here $x_n^* = x_0 + nh^*$, $S = S_m \ldots S_2 S_1$ and Φ has to be chosen suitably. E.g., in the case $m = 2$ we have

$$\Phi(x^*, u^*, h^*) = \frac{1}{2} S_2 \Phi_1\left(x^*, u^*, \frac{h^*}{2}\right)$$
$$+ \frac{1}{2} \Phi_2\left(x^* + \frac{h^*}{2}, S_1 u^* + \frac{h^*}{2}\Phi_1(x^*, u^*, \frac{h^*}{2}), \frac{h^*}{2}\right).$$

It is interesting to note that cyclically used k-step methods can lead to convergent methods of order $2k - 1$ (or even $2k$). The "first Dahlquist barrier" (Theorem 3.5) can be broken in this way. For more details see Stetter (1973), Albrecht (1979) and Exercise 2.

Example 8.5. *General linear methods.*

> Following the advice of Aristotle ... (the original Greek can be found in Butcher's paper) ... we look for the greatest good as a mean between extremes.
> (J.C. Butcher 1985a)

Introduced by Burrage & Butcher (1980), these methods are general enough to include all previous examples as special cases, but at the same time the increment function is given explicitly in terms of the differential equation and several free parameters. They are defined by

$$v_i^{(n)} = \sum_{j=1}^{k} \tilde{a}_{ij} u_j^{(n)} + h \sum_{j=1}^{s} \tilde{b}_{ij} f(x_n + c_j h, v_j^{(n)}) \quad i = 1, \ldots, s, \tag{8.7a}$$

$$u_i^{(n+1)} = \sum_{j=1}^{k} a_{ij} u_j^{(n)} + h \sum_{j=1}^{s} b_{ij} f(x_n + c_j h, v_j^{(n)}) \quad i = 1, \ldots, k. \tag{8.7b}$$

The stages $v_i^{(n)}$ $(i = 1, \ldots, s)$ are the *internal stages* and do not leave the "black box" of the current step. The stages $u_i^{(n)}$ $(i = 1, \ldots, k)$ are called the *external stages* since they contain all the necessary information from the previous step used in carrying out the current step. The coefficients a_{ij} in (8.7b) form the matrix S

of (8.4b). Very often, some internal stages are identical to external ones, as for example in method (8.1), where

$$v_n = (\widehat{y}_{n+1/2}, \widehat{y}_{n+1}, y_n, y_{n-1})^T.$$

One-step Runge-Kutta methods are characterized by $k = 1$. At the end of this section we shall discuss the algebraic conditions for general linear methods to be of order p.

Example 8.6. In order to illustrate the fact that the analysis of this section is not only applicable to numerical methods that discretize first order differential equations, we consider the second order initial value problem

$$y'' = g(x, y), \quad y(x_0) = y_0, \quad y'(x_0) = y_0' \qquad (8.8)$$

Replacing $y''(x)$ by a central difference yields

$$y_{n+1} - 2y_n + y_{n-1} = h^2 g(x_n, y_n),$$

and with the additional variables

$$hy_n' = y_{n+1} - y_n$$

this method can be written as

$$\begin{pmatrix} y_{n+1} \\ y_{n+1}' \end{pmatrix} = \begin{pmatrix} 1 & 0 \\ 0 & 1 \end{pmatrix} \begin{pmatrix} y_n \\ y_n' \end{pmatrix} + h \begin{pmatrix} y_n' \\ g(x_{n+1}, y_n + hy_n') \end{pmatrix}.$$

It now has the form of a method (8.4) with the correct value function $z(x, h) = \big(y(x), (y(x + h) - y(x))/h\big)^T$. Here $y(x)$ denotes the exact solution of (8.8).

Clearly, all Nyström methods (Section II.14) fit into this framework, as do multistep methods for second order differential equations. They will be investigated in more detail in Section III.10.

Example 8.7. *Multi-step multi-stage multi-derivative* methods seem to be the most general class of explicitly given linear methods and generalize the methods of Section II.13. In the notation of that section, we can write

$$v_i^{(n)} = \sum_{j=1}^{k} \widetilde{a}_{ij} u_j^{(n)} + \sum_{r=1}^{q} \frac{h^r}{r!} \sum_{j=1}^{s} \widetilde{b}_{ij}^{(r)} D^r y(x_n + c_j h, v_j^{(n)}) \quad i = 1, \dots, s,$$

$$u_i^{(n+1)} = \sum_{j=1}^{k} a_{ij} u_j^{(n)} + \sum_{r=1}^{q} \frac{h^r}{r!} \sum_{j=1}^{s} b_{ij}^{(r)} D^r y(x_n + c_j h, v_j^{(n)}) \quad i = 1, \dots, k.$$

Such methods have been studied in Hairer & Wanner (1973).

Stability and Order

The following study of stability, order and convergence follows mainly the lines of Skeel (1976). Stability of a numerical scheme just requires that for $h \to 0$ the numerical solution remain bounded. This motivates the following definition.

Definition 8.8. Method (8.4) is called *stable* if $\|S^n\|$ is uniformly bounded for all $n \geq 0$.

The local error of method (8.4) is defined in exactly the same way as for one-step methods (Section II.3) and multistep methods (Section III.2).

Definition 8.9. Let $z(x, h)$ be the correct value function for the method (8.4) and let $z_n = z(x_n, h)$. The *local error* is then given by (see Fig. 8.1)

$$
\begin{aligned}
d_0 &= z_0 - \varphi(h) \\
d_{n+1} &= z_{n+1} - Sz_n - h\Phi(x_n, z_n, h), \qquad n = 0, 1, \dots
\end{aligned}
\tag{8.9}
$$

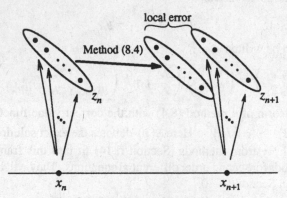

Fig. 8.1. Illustration of the local error

The definition of order is not as straightforward. The requirement that the local error be $\mathcal{O}(h^{p+1})$ (cf. one-step and multistep methods) will turn out to be sufficient but in general not necessary for convergence of order p. For an appropriate definition we need the *spectral decomposition* of the matrix S.

First observe that, whenever the local error (8.9) tends to zero for $h \to 0$ ($nh = x - x_0$ fixed), we get

$$
0 = z(x, 0) - Sz(x, 0), \tag{8.10}
$$

so that 1 is an eigenvalue of S and $z(x, 0)$ a corresponding eigenvector. Furthermore, by stability, no eigenvalue of S can lie outside the unit disc and the eigenvalues of modulus one can not give rise to Jordan chains. Denoting the eigenvalues of modulus one by $\zeta_1(=1), \zeta_2, \dots, \zeta_l$, the Jordan canonical form of S (see

(I.12.14)) is therefore the block diagonal matrix

$$S = T \, \text{diag} \left\{ \begin{pmatrix} 1 & & \\ & \ddots & \\ & & 1 \end{pmatrix}, \begin{pmatrix} \zeta_2 & & \\ & \ddots & \\ & & \zeta_2 \end{pmatrix}, \ldots, \begin{pmatrix} \zeta_l & & \\ & \ddots & \\ & & \zeta_l \end{pmatrix}, \tilde{J} \right\} T^{-1}.$$

If we decompose this matrix into the terms which correspond to the single eigenvalues we obtain

$$S = E + \zeta_2 E_2 + \ldots + \zeta_l E_l + \tilde{E} \tag{8.11}$$

where

$$E = T \, \text{diag} \left\{ I, 0, 0, \ldots \right\} T^{-1}, \tag{8.12}$$

$$E_2 = T \, \text{diag} \left\{ 0, I, 0, \ldots \right\} T^{-1}, \ldots, \quad E_l = T \, \text{diag} \left\{ 0, \ldots, 0, I, 0 \right\} T^{-1},$$

$$\tilde{E} = T \, \text{diag} \left\{ 0, 0, 0, \ldots, \tilde{J} \right\} T^{-1}.$$

We are now prepared to give

Definition 8.10. The method (8.4) is of *order p (consistent of order p)*, if for all problems (8.3) with p times continuously differentiable f, the local error satisfies

$$d_0 = \mathcal{O}(h^p)$$
$$E(d_0 + d_1 + \ldots + d_n) + d_{n+1} = \mathcal{O}(h^p) \qquad \text{for } 0 \le nh \le Const. \tag{8.13}$$

Remark. This property is called *quasi-consistency of order p* by Skeel (1976).

If the right-hand side of the differential equation (8.3) is p-times continuously differentiable then, in general, $\varphi(h)$, $\Phi(x, u, h)$ and $z(x, h)$ are also smooth, so that the local error (8.9) can be expanded into a Taylor series in h:

$$d_0 = \gamma_0 + \gamma_1 h + \ldots + \gamma_{p-1} h^{p-1} + \mathcal{O}(h^p)$$
$$d_{n+1} = \delta_0(x_n) + \delta_1(x_n) h + \ldots + \delta_p(x_n) h^p + \mathcal{O}(h^{p+1}). \tag{8.14}$$

The function $\delta_j(x)$ is then $(p-j+1)$-times continuously differentiable. The following lemma gives a more practical characterization of the order of the methods (8.4).

Lemma 8.11. *Assume that the local error of method (8.4) satisfies (8.14) with continuous $\delta_j(x)$. The method is then of order p, if and only if*

$$d_n = \mathcal{O}(h^p) \quad \text{for } 0 \le nh \le Const, \quad \text{and} \quad E\delta_p(x) = 0. \tag{8.15}$$

Proof. The condition (8.15) is equivalent to

$$d_n = \mathcal{O}(h^p), \qquad E d_{n+1} = \mathcal{O}(h^{p+1}) \qquad \text{for } 0 \le nh \le Const, \tag{8.16}$$

which is clearly sufficient for order p. We now show that (8.15) is also necessary. Since $E^2 = E$ (see (8.12)) order p implies

$$d_n = \mathcal{O}(h^p), \qquad E(d_1 + \ldots + d_n) = \mathcal{O}(h^p) \qquad \text{for } 0 \leq nh \leq Const. \quad (8.17)$$

This is best seen by multiplying (8.13) by E. Consider now pairs (n, h) such that $nh = x - x_0$ for some fixed x. We insert (8.14) (observe that $d_n = \mathcal{O}(h^p)$) into $E(d_1 + \ldots + d_n)$ and approximate the resulting sum by the corresponding Riemann integral

$$E(d_1 + \ldots + d_n) = h^p E \sum_{j=1}^n \delta_p(x_{j-1}) + \mathcal{O}(h^p) = h^{p-1} E \int_{x_0}^x \delta_p(s)\, ds + \mathcal{O}(h^p).$$

It follows from (8.17) that $E \int_{x_0}^x \delta_p(s)\, ds = 0$ and by differentiation that $E\delta_p(x) = 0$. □

Convergence

In addition to the numerical solution given by (8.4) we consider a perturbed numerical solution $(\widehat{u}_n)$ defined by

$$\begin{aligned}
\widehat{u}_0 &= \varphi(h) + r_0 \\
\widehat{u}_{n+1} &= S\widehat{u}_n + h\Phi(x_n, \widehat{u}_n, h) + r_{n+1}, \quad n = 0, 1, \ldots, N-1
\end{aligned} \qquad (8.18)$$

for some perturbation $R = (r_0, r_1, \ldots, r_N)$. For example, the exact solution $z_n = z(x_n, h)$ can be interpreted as a perturbed solution, where the perturbation is just the local error. The following lemma gives the best possible qualitative bound on the difference $u_n - \widehat{u}_n$ in terms of the perturbation R. We have to assume that the increment function $\Phi(x, u, h)$ satisfies a Lipschitz condition with respect to u (on a compact neighbourhood of the solution). This is the case for all reasonable methods.

Lemma 8.12. *Let the method (8.4) be stable and assume the sequences (u_n) and $(\widehat{u}_n)$ be given by (8.4) and (8.18), respectively. Then there exist positive constants c and C such that for any perturbation R and for $hN \leq Const$*

$$c\|R\|_S \leq \max_{0 \leq n \leq N} \|u_n - \widehat{u}_n\| \leq C\|R\|_S$$

with

$$\|R\|_S = \max_{0 \leq n \leq N} \Big\| \sum_{j=0}^n S^{n-j} r_j \Big\|.$$

Remark. $\|R\|_S$ is a norm on $\mathbf{R}^{(N+1)q}$. Its positivity is seen as follows: if $\|R\|_S = 0$ then for $n = 0, 1, 2, \ldots$ one obtains $r_0 = 0, r_1 = 0, \ldots$ recursively.

Proof. Set $\Delta u_n = \widehat{u}_n - u_n$ and $\Delta \Phi_n = \Phi(x_n, \widehat{u}_n, h) - \Phi(x_n, u_n, h)$. Then we have

$$\Delta u_{n+1} = S\Delta u_n + h\Delta\Phi_n + r_{n+1}. \tag{8.19}$$

By assumption there exists a constant L such that $\|\Delta\Phi_n\| \le L\|\Delta u_n\|$. Solving the difference equation (8.19) gives $\Delta u_0 = r_0$ and

$$\Delta u_{n+1} = \sum_{j=0}^{n} S^{n-j} h\Delta\Phi_j + \sum_{j=0}^{n+1} S^{n+1-j} r_j. \tag{8.20}$$

By stability there exists a constant B such that

$$\|S^n\| L \le B \qquad \text{for all } n \ge 0. \tag{8.21}$$

Thus (8.20) becomes

$$\|\Delta u_{n+1}\| \le hB \sum_{j=0}^{n} \|\Delta u_j\| + \|R\|_S.$$

By induction on n it follows that

$$\|\Delta u_n\| \le (1+hB)^n \|R\|_S \le \exp(Const \cdot B) \cdot \|R\|_S,$$

which proves the second inequality in the lemma. From (8.20) and (8.21)

$$\left\| \sum_{j=0}^{n} S^{n-j} r_j \right\| \le (1+nhB) \max_{0 \le n \le N} \|\Delta u_n\|,$$

and we thus obtain for $Nh \le Const$

$$\|R\|_S \le (1 + Const \cdot B) \cdot \max_{0 \le n \le N} \|\widehat{u}_n - u_n\|. \qquad \square$$

Remark. Two-sided error bounds, such as in Lemma 8.12, were first studied, in the case of multi-step methods, by Spijker (1971). This theory has become prominent through the treatment of Stetter (1973, pp. 81-84). Extensions to general linear methods are due to Skeel (1976) and Albrecht (1978).

Using the lemma above we can prove

Theorem 8.13. *Consider a stable method (8.4) and assume that the local error satisfies (8.14) with $\delta_p(x)$ continuously differentiable. The method is then convergent of order p, i.e., the global error $u_n - z_n$ satisfies*

$$u_n - z_n = \mathcal{O}(h^p) \qquad \text{for } 0 \le nh \le Const,$$

if and only if it is consistent of order p.

Proof. The identity

$$E(d_0 + \ldots + d_n) + d_{n+1} = \sum_{j=0}^{n+1} S^{n+1-j} d_j - (S - E) \sum_{j=0}^{n} S^{n-j} d_j,$$

which is a consequence of $ES = E$ (see (8.11) and (8.12)), implies that for $n \leq N - 1$ and $D = (d_0, \ldots, d_N)$,

$$\|E(d_0 + \ldots + d_n) + d_{n+1}\| \leq (1 + \|S - E\|) \cdot \|D\|_S. \tag{8.22}$$

The lower bound of Lemma 8.12, with r_n and $\hat{u}_n$ replaced by d_n and z_n respectively, yields the "only if" part of the theorem.

For the "if" part we use the upper bound of Lemma 8.12. We have to show that consistency of order p implies

$$\max_{0 \leq n \leq N} \left\| \sum_{j=0}^{n} S^{n-j} d_j \right\| = \mathcal{O}(h^p). \tag{8.23}$$

By (8.11) and (8.12) we have

$$S^{n-j} = E + \zeta_2^{n-j} E_2 + \ldots + \zeta_l^{n-j} E_l + \widetilde{E}^{n-j}.$$

This identity together with Lemma 8.11 implies

$$\sum_{j=0}^{n} S^{n-j} d_j = h^p E_2 \sum_{j=1}^{n} \zeta_2^{n-j} \delta_p(x_{j-1}) + \ldots$$

$$+ h^p E_l \sum_{j=1}^{n} \zeta_l^{n-j} \delta_p(x_{j-1}) + \sum_{j=0}^{n} \widetilde{E}^{n-j} d_j + \mathcal{O}(h^p).$$

The last term in this expression is $\mathcal{O}(h^p)$ since in a suitable norm $\|\widetilde{E}\| < 1$ and therefore

$$\left\| \sum_{j=0}^{n} \widetilde{E}^{n-j} d_j \right\| \leq \sum_{j=0}^{n} \|\widetilde{E}\|^{n-j} \|d_j\| \leq \frac{1}{1 - \|\widetilde{E}\|} \cdot \max_{0 \leq n \leq N} \|d_n\|.$$

For the rest we use partial summation (Abel 1826)

$$\sum_{j=1}^{n} \zeta^{n-j} \delta(x_{j-1}) = \frac{1 - \zeta^n}{1 - \zeta} \cdot \delta(x_0) + \sum_{j=1}^{n} \frac{1 - \zeta^{n-j}}{1 - \zeta} \cdot \left(\delta(x_j) - \delta(x_{j-1}) \right) = \mathcal{O}(1),$$

whenever $|\zeta| = 1$, $\zeta \neq 1$ and δ is of bounded variation. $\qquad \square$

Order Conditions for General Linear Methods

For the construction of a pth order general linear method (8.7) the conditions (8.15) are still not very practical. One would like to have instead algebraic conditions in the free parameters, as is the case for Runge-Kutta methods. We shall demonstrate how this can be achieved using the theory of B-series of Section II.12 (see also Burrage & Moss 1980). In order to avoid tensor products we assume in what follows that the differential equation under consideration is a scalar one. All results, however, are also valid for systems. We further assume the differential equation to be autonomous, so that the theory of Section II.12 is directly applicable. This will be justified in Remark 8.17 below.

Suppose now that the components of the correct value function $z(x, h) = (z_1(x, h), \ldots, z_k(x, h))^T$ possess an expansion as a B-series

$$z_i(x, h) = B(z_i, y(x))$$

so that with $z(t) = (z_1(t), \ldots, z_k(t))^T$,

$$z(x, h) = z(\emptyset)y(x) + hz(\tau)f(y(x)) + \ldots . \qquad (8.24)$$

Before deriving the order conditions we observe that (8.7a) makes sense only if $v_j^{(n)} \to y(x_n)$ for $h \to 0$. Otherwise $f(v_j^{(n)})$ need not be defined. Since $u_j^{(n)}$ is an approximation of $z_j(x_n, h)$, this leads to the condition $\sum \tilde{a}_{ij} z_j(\emptyset) = 1$. This together with (8.10) are the so-called *preconsistency conditions*:

$$Az(\emptyset) = z(\emptyset), \qquad \tilde{A}z(\emptyset) = \mathbb{1}. \qquad (8.25)$$

A and $\tilde{A}$ are the matrices with entries a_{ij} and $\tilde{a}_{ij}$, respectively, and $\mathbb{1} = (1, \ldots, 1)^T$. Recall that the local error (8.9) for the general linear method (8.7) is given by

$$d_i^{(n+1)} = z_i(x_n + h, h) - \sum_{j=1}^{k} a_{ij} z_j(x_n, h) - \sum_{j=1}^{s} b_{ij} h f(v_j) \qquad (8.26a)$$

where

$$v_i = \sum_{j=1}^{k} \tilde{a}_{ij} z_j(x_n, h) + \sum_{j=1}^{s} \tilde{b}_{ij} h f(v_j). \qquad (8.26b)$$

For the derivation of the order conditions we write v_i and $d_i^{(n+1)}$ as B-series

$$v_i = B(\mathbf{v}_i, y(x_n)), \qquad d_i^{(n+1)} = B(\mathbf{d}_i, y(x_n)).$$

By the composition theorem for B-series and by formula (12.10) of Section II.12 we have

$$z_i(x_n + h, h) = B(z_i, y(x_n + h)) = B(z_i, B(p, y(x_n))) = B(pz_i, y(x_n)).$$

Inserting all these series into (8.26) and comparing the coefficients we arrive at

$$d_i(t) = (pz_i)(t) - \sum_{j=1}^{k} a_{ij} z_j(t) - \sum_{j=1}^{s} b_{ij} v_j'(t)$$

$$v_i(t) = \sum_{j=1}^{k} \tilde{a}_{ij} z_j(t) + \sum_{j=1}^{s} \tilde{b}_{ij} v_j'(t).$$

(8.27)

An application of Lemma 8.11 now yields

Theorem 8.14. *Let* $d(t) = \big(d_1(t), \ldots, d_k(t)\big)^T$ *with* $d_i(t)$ *be given by (8.27). The general linear method (8.7) is of order* p, *iff*

$$d(t) = 0 \qquad for \ t \in T, \ \varrho(t) \leq p-1,$$

$$Ed(t) = 0 \qquad for \ t \in T, \ \varrho(t) = p,$$

(8.28)

where the matrix E *is defined in (8.12).*   □

Corollary 8.15. *Sufficient conditions for the general linear method to be of order* p *are*

$$d(t) = 0 \qquad for \ t \in T, \ \varrho(t) \leq p.$$

(8.29)
□

Remark 8.16. The expression $(pz_i)(t)$ in (8.27) can be computed using formula (12.8) of Section II.12. Since $p(t) = 1$ for all trees t, we have

$$(pz_i)(t) = \sum_{j=0}^{\varrho(t)} \binom{\varrho(t)}{j} \frac{1}{\alpha(t)} \sum_{\text{all labellings}} z_i(s_j(t)).$$

(8.30)

This rather complicated formula simplifies considerably if we assume that the coefficients $z_i(t)$ of the correct value function depend only on the order of t, i.e., that

$$z_i(t) = z_i(u) \qquad \text{whenever} \ \varrho(t) = \varrho(u) .$$

(8.31)

In this case formula (8.30) becomes

$$(pz_i)(t) = \sum_{j=0}^{\varrho(t)} \binom{\varrho(t)}{j} z_i(\tau^j).$$

(8.32)

Here τ^j represents any tree of order j, e.g.,

$$\tau^j = [\underbrace{\tau, \ldots, \tau}_{j-1}], \qquad \tau^1 = \tau, \qquad \tau^0 = \emptyset.$$

(8.33)

Usually the components of $z(x, h)$ are composed of

$$y(x), \ y(x+jh), \ hy'(x), \ h^2 y''(x), \ldots,$$

in which case assumption (8.31) is satisfied.

Remark 8.17. Non-autonomous systems. For the differential equation $x' = 1$, formula (8.7a) becomes

$$v_n = \widetilde{A} u_n + h\widetilde{B}\mathbb{1}.$$

Assuming that $x' = 1$ is integrated exactly, i.e., $u_n = z(\emptyset)x_n + hz(\tau)$ we obtain $v_n = x_n \mathbb{1} + hc$, where $c = (c_1, \ldots, c_s)^T$ is given by

$$c = \widetilde{A} z(\tau) + \widetilde{B} e. \tag{8.34}$$

This definition of the c_i implies that the numerical results for $y' = f(x, y)$ and for the augmented autonomous differential equation are the same and the above results are also valid in the general case.

Table 8.1 presents the order conditions up to order 3 in addition to the preconsistency conditions (8.25). We assume that (8.31) is satisfied and that c is given by (8.34). Furthermore, c^j denotes the vector $(c_1^j, \ldots, c_s^j)^T$.

Table 8.1. Order conditions for general linear methods

t	$\varrho(t)$	order condition
τ	1	$Az(\tau) + B\mathbb{1} = z(\tau) + z(\emptyset)$
τ^2	2	$Az(\tau^2) + 2Bc = z(\tau^2) + 2z(\tau) + z(\emptyset)$
τ^3	3	$Az(\tau^3) + 3Bc^2 = z(\tau^3) + 3z(\tau^2) + 3z(\tau) + z(\emptyset)$
$[\tau^2]$	3	$Az(\tau^3) + 3Bv(\tau^2) = z(\tau^3) + 3z(\tau^2) + 3z(\tau) + z(\emptyset)$
		with $v(\tau^2) = \widetilde{A}z(\tau^2) + 2\widetilde{B}c$

Construction of General Linear Methods

Let us demonstrate on an example how low order methods can be constructed: we set $k = s = 2$ and fix the correct value function as

$$z(x, h) = \big(y(x), \ y(x - h)\big)^T.$$

This choice satisfies (8.24) and (8.31) with

$$z(\emptyset) = \begin{pmatrix} 1 \\ 1 \end{pmatrix}, \qquad z(\tau) = \begin{pmatrix} 0 \\ -1 \end{pmatrix}, \qquad z(\tau^2) = \begin{pmatrix} 0 \\ 1 \end{pmatrix}, \ldots.$$

Since the second component of $z(x + h, h)$ is equal to the first component of $z(x, h)$, it is natural to look for methods with

$$A = \begin{pmatrix} a_{11} & a_{12} \\ 1 & 0 \end{pmatrix}, \qquad B = \begin{pmatrix} b_{11} & b_{12} \\ 0 & 0 \end{pmatrix}.$$

We further impose

$$\tilde{B} = \begin{pmatrix} 0 & 0 \\ \tilde{b}_{21} & 0 \end{pmatrix}$$

so that the resulting method is explicit.

The preconsistency condition (8.25), formula (8.34) and the order conditions of Table 8.1 yield the following equations to be solved:

$$a_{11} + a_{12} = 1 \tag{8.35a}$$

$$\tilde{a}_{11} + \tilde{a}_{12} = 1, \qquad \tilde{a}_{21} + \tilde{a}_{22} = 1 \tag{8.35b}$$

$$c_1 = -\tilde{a}_{12}, \qquad c_2 = \tilde{b}_{21} - \tilde{a}_{22} \tag{8.35c}$$

$$-a_{12} + b_{11} + b_{12} = 1 \tag{8.35d}$$

$$a_{12} + 2(b_{11}c_1 + b_{12}c_2) = 1 \tag{8.35e}$$

$$-a_{12} + 3(b_{11}c_1^2 + b_{12}c_2^2) = 1 \tag{8.35f}$$

$$-a_{12} + 3(b_{11}\tilde{a}_{12} + b_{12}(\tilde{a}_{22} + 2\tilde{b}_{21}c_1)) = 1. \tag{8.35g}$$

These are 9 equations in 11 unknowns. Letting c_1 and c_2 be free parameters, we obtain the solution in the following way: compute a_{12}, b_{11} and b_{12} from the linear system (8.35d,e,f), then $\tilde{a}_{12}, \tilde{a}_{22}$ and $\tilde{b}_{21}$ from (8.35c,g) and finally $a_{11}, \tilde{a}_{11}$ and $\tilde{a}_{21}$ from (8.35a,b). A particular solution for $c_1 = 1/2, c_2 = -2/5$ is:

$$A = \begin{pmatrix} 16/11 & -5/11 \\ 1 & 0 \end{pmatrix}, \quad B = \begin{pmatrix} 104/99 & -50/99 \\ 0 & 0 \end{pmatrix},$$

$$\tilde{A} = \begin{pmatrix} 3/2 & -1/2 \\ 3/2 & -1/2 \end{pmatrix}, \quad \tilde{B} = \begin{pmatrix} 0 & 0 \\ -9/10 & 0 \end{pmatrix}. \tag{8.36}$$

This method, which represents a stable explicit 2-step, 2-stage method of order 3, is due to Butcher (1984).

The construction of higher order methods soon becomes very complicated, and the use of *simplifying assumptions* will be very helpful:

Theorem 8.18 (Burrage & Moss 1980). *Assume that the correct value function satisfies (8.31). The simplifying assumptions*

$$\tilde{A}z(\tau^j) + j\tilde{B}c^{j-1} = c^j \qquad j = 1, \ldots, p-1 \tag{8.37}$$

together with the preconsistency relations (8.25) and the order conditions for the "bushy trees"

$$\mathbf{d}(\tau^j) = 0 \qquad j = 1, \ldots, p$$

imply that the method (8.7) is of order p.

Proof. An induction argument based on (8.27) implies that

$$\mathbf{v}(t) = \mathbf{v}(\tau^j) \qquad \text{for } \varrho(t) = j, \; j = 1,\ldots,p-1$$

and consequently also that

$$\mathbf{d}(t) = \mathbf{d}(\tau^j) \qquad \text{for } \varrho(t) = j, \; j = 1,\ldots,p. \qquad\qquad \square$$

The simplifying assumptions (8.37) allow an interesting interpretation: they are equivalent to the fact that the internal stages $v_1^{(n)}$ approximate the exact solution at $x_n + c_i h$ up to order $p-1$, i.e., that

$$v_i^{(n)} - y(x_n + c_i h) = \mathcal{O}(h^p).$$

In the case of Runge-Kutta methods (8.37) reduces to the conditions $C(p-1)$ of Section II.7.

For further examples of general linear methods satisfying (8.37) we refer to Burrage & Moss (1980) and Butcher (1981). See also Burrage (1985) and Butcher (1985a).

Exercises

1. Consider the composition of (cf. Example 8.5)
   a) explicit and implicit Euler method;
   b) implicit and explicit Euler method.

   To which methods are they equivalent? What is the order of the composite methods?

2. a) Suppose that each of the m multistep methods (ϱ_i, σ_i) $i = 1,\ldots,m$ is of order p. Prove that the corresponding cyclic method is of order at least p.
   b) Construct a stable, 2-cyclic, 3-step linear multistep method of order 5: find first a one-parameter family of linear 3-step methods of order 5 (which are necessarily unstable).

   Result.

$$\varrho_c(\zeta) = c\zeta^3 + \left(\frac{19}{30} - c\right)\zeta^2 - \left(\frac{8}{30} + c\right)\zeta + \left(c - \frac{11}{30}\right)$$

$$\sigma_c(\zeta) = \left(\frac{1}{9} - \frac{c}{3}\right)\zeta^3 + \left(c + \frac{8}{30}\right)\zeta^2 + \left(\frac{19}{30} - c\right)\zeta + \left(\frac{c}{3} - \frac{1}{90}\right).$$

   Then determine c_1 and c_2, such that the eigenvalues of the matrix S for the composite method become $1, 0, 0$.

3. Prove that the composition of two different general linear methods (with the same correct value function) again gives a general linear method. As a consequence, the cyclic methods of Example 8.4 are general linear methods.

4. Suppose that all eigenvalues of S (except $\zeta_1 = 1$) lie inside the unit circle. Then

$$\|R\|_E = \max_{0 \le n \le N} \left\| r_n + E \sum_{j=0}^{n-1} r_j \right\|$$

is a minimal stability functional.

5. Verify for linear multistep methods that the consistency conditions (2.6) are equivalent to consistency of order 1 in the sense of Lemma 8.11.

6. Write method (8.1) as general linear method (8.7) and determine its order (answer: $p = 5$).

7. Interpret the method of Caira, Costabile & Costabile (1990)

$$k_i^n = h f\left(x_n + c_i h, \, y_n + \sum_{j=1}^{s} \overline{a}_{ij} k_j^{n-1} + \sum_{j=1}^{i-1} a_{ij} k_j^n\right)$$

$$y_{n+1} = y_n + \sum_{i=1}^{s} b_i k_i^n$$

as general linear method. Show that, if

$$\|k_i^{-1} - h y'(x_0 + (c_i - 1)h)\| \le C \cdot h^p,$$

$$\sum_{i=1}^{s} b_i c_i^{q-1} = \frac{1}{q}, \qquad q = 1, \dots, p,$$

$$\sum_{j=1}^{s} \overline{a}_{ij}(c_j - 1)^{q-1} + \sum_{j=1}^{i-1} a_{ij} c_j^{q-1} = \frac{c_i^q}{q}, \qquad q = 1, \dots, p-1,$$

then the method is of order at least p. Find parallels of these conditions with those of Theorem 8.18.

8. Jackiewicz & Zennaro (1992) propose the following two-step Runge-Kutta method

$$Y_i^{n-1} = y_{n-1} + h_{n-1} \sum_{j=1}^{i-1} a_{ij} f(Y_j^{n-1}), \quad Y_i^n = y_n + h_{n-1} \xi \sum_{j=1}^{i-1} a_{ij} f(Y_j^n),$$

$$y_{n+1} = y_n + h_{n-1} \sum_{i=1}^{s} v_i f(Y_i^{n-1}) + h_{n-1} \xi \sum_{i=1}^{s} w_i f(Y_i^n), \qquad (8.38)$$

where $\xi = h_n/h_{n-1}$. The coefficients v_i, w_i may depend on ξ, but the a_{ij} do not. Hence, this method requires s function evaluations per step.

a) Show that the order of method (8.38) is p (according to Definition 8.10) if and only if for all trees t with $1 \le \varrho(t) \le p$

$$\xi^{\varrho(t)} = \sum_{i=1}^{s} v_i(y^{-1}g_i')(t) + \xi^{\varrho(t)} \sum_{i=1}^{s} w_i g_i'(t), \qquad (8.39)$$

where, as for Runge-Kutta methods, $g_i(t) = \sum_{j=1}^{i-1} a_{ij}g_j'(t)$. The coefficients $y^{-1}(t) = (-1)^{\varrho(t)}$ are those of $y(x_n - h) = B(y^{-1}, y(x_n))$.

b) Under the assumption

$$v_i + \xi^p w_i = 0 \qquad \text{for} \quad i = 2, \ldots, s \qquad (8.40)$$

the order conditions (8.39) are equivalent to

$$\xi = \sum_{i=1}^{s} v_i + \xi \sum_{i=1}^{s} w_i, \qquad (8.41a)$$

$$\xi^r = \sum_{j=1}^{r-1} \binom{r}{j}(-1)^{r-j} \sum_{i=1}^{s} v_i c_i^{j-1} + (1 - \xi^{r-p})r \sum_{i=1}^{s} v_i c_i^{r-1}, \quad r = 2, \ldots, p, \qquad (8.41b)$$

$$\sum_{i=1}^{s} v_i \left(g_i'(u) - \varrho(u)c_i^{\varrho(u)-1} \right) = 0 \qquad \text{for } \varrho(u) \le p-1. \qquad (8.41c)$$

c) The conditions (8.41a,b) uniquely define $\sum_i w_i$, $\sum_i v_i c_i^{j-1}$ (for $j = 1, \ldots, p-1$) as functions of $\xi > 0$.

d) For each continuous Runge-Kutta method of order $p-1 \ge 2$ there exists a method (8.38) of order p with the same coefficient matrix (a_{ij}).

Hints. To obtain (8.41c) subtract equation (8.40) from the same equation where t is replaced by the bushy tree of order $\varrho(t)$. Then proceed by induction. The conditions $\sum_i v_i c_i^{j-1} = f_j^p(\xi)$, $j = 1, \ldots, p-1$, obtained from (c), together with (8.41c) have the same structure as the order conditions (order $p-1$) of a continuous Runge-Kutta method (Theorem II.6.1).

III.9 Asymptotic Expansion of the Global Error

The asymptotic expansion of the global error of multistep methods was studied in the famous thesis of Gragg (1964). His proof is very technical and can also be found in a modified version in the book of Stetter (1973), pp. 234-245. The existence of asymptotic expansions for general linear methods was conjectured by Skeel (1976). The proof given below (Hairer & Lubich 1984) is based on the ideas of Section II.8.

An Instructive Example

Let us start with an example in order to understand which kind of asymptotic expansion may be expected. We consider the simple differential equation

$$y' = -y, \qquad y(0) = 1,$$

take a constant step size h and apply the 3-step BDF-formula (1.22') with one of the following three starting procedures:

$$y_0 = 1, \quad y_1 = \exp(-h), \qquad y_2 = \exp(-2h) \quad \text{(exact values)} \quad (9.1a)$$

$$y_0 = 1, \quad y_1 = 1 - h + \frac{h^2}{2} - \frac{h^3}{6}, \qquad y_2 = 1 - 2h + 2h^2 - \frac{4h^3}{3}, \qquad (9.1b)$$

$$y_0 = 1, \quad y_1 = 1 - h + \frac{h^2}{2}, \qquad y_2 = 1 - 2h + 2h^2. \qquad (9.1c)$$

The three pictures on the left of Fig. 9.1 (they correspond to the three starting procedures in the same order) show the global error divided by h^3 for the five step sizes $h = 1/5, 1/10, 1/20, 1/40, 1/80$.

For the first two starting procedures we observe uniform convergence to the function $e_3(x) = xe^{-x}/4$ (cf. formula (2.12)), so that

$$y_n - y(x_n) = e_3(x_n)h^3 + \mathcal{O}(h^4), \qquad (9.2)$$

valid uniformly for $0 \leq nh \leq Const$. In the third case we have convergence to $e_3(x) = (9 + x)e^{-x}/4$ (Exercise 2), but this time the convergence is no longer uniform. Therefore (9.2) only holds for x_n bounded away from x_0, i.e., for $0 < \alpha \leq nh \leq Const$. In the three pictures on the right of Fig. 9.1 the functions

$$(y_n - y(x_n) - e_3(x_n)h^3)/h^4 \qquad (9.3)$$

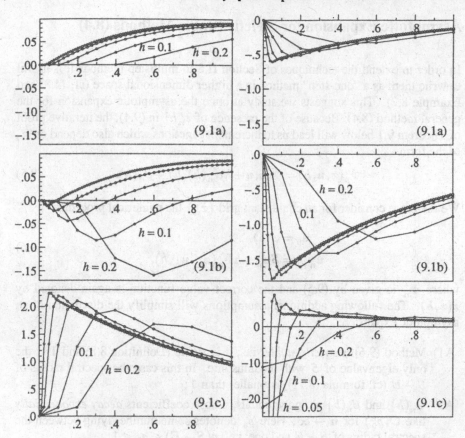

Fig. 9.1. The values $(y_n - y(x_n))/h^3$ (left), $(y_n - y(x_n) - e_3(x_n)h^3)/h^4$ (right)
for the 3-step BDF method and for three different starting procedures

are plotted. Convergence to functions $e_4(x)$ is observed in all cases. Clearly, since
$e_3(x_0) \neq 0$ for the starting procedure (9.1c), the sequence (9.3) diverges at x_0 like
$\mathcal{O}(1/h)$ in this case.

We conclude from this example that for linear multistep methods there is in
general no asymptotic expansion of the form

$$y_n - y(x_n) = e_p(x_n)h^p + e_{p+1}(x_n)h^{p+1} + \dots$$

which holds uniformly for $0 \leq nh \leq Const$. It will be necessary to add perturbation
terms

$$y_n - y(x_n) = \left(e_p(x_n) + \varepsilon_n^p\right)h^p + \left(e_{p+1}(x_n) + \varepsilon_n^{p+1}\right)h^{p+1} + \dots \qquad (9.4)$$

which compensate the irregularity near x_0. If the perturbations ε_n^j decay exponentially (for $n \to \infty$), then they have no influence on the asymptotic expansion for
x_n bounded away from x_0.

Asymptotic Expansion for Strictly Stable Methods (8.4)

In order to extend the techniques of Section II.8 to multistep methods it is useful to write them as a "one-step" method in a higher dimensional space (cf. (4.8) and Example 8.2). This suggests we study at once the asymptotic expansion for the general method (8.4). Because of the presence of $\varepsilon_n^j h^j$ in (9.4), the iterative proof of Theorem 9.1 below will lead us to increment functions which also depend on n, of the form

$$\Phi_n(x, u, h) = \Phi\big(x, u + h\alpha_n(h), h\big) + \beta_n(h). \tag{9.5}$$

We therefore consider for an equidistant grid (x_n) the numerical procedure

$$\begin{aligned} u_0 &= \varphi(h) \\ u_{n+1} &= Su_n + h\Phi_n(x_n, u_n, h), \end{aligned} \tag{9.6}$$

where Φ_n is given by (9.5) and the correct value function is again denoted by $z(x, h)$. The following additional assumptions will simplify the discussion of an asymptotic expansion:

A1) Method (9.6) is *strictly stable;* i.e., it is stable (Definition 8.8) and 1 is the only eigenvalue of S with modulus one. In this case the spectral radius of $S - E$ (cf. formula (8.11)) is smaller than 1;

A2) $\alpha_n(h)$ and $\beta_n(h)$ are polynomials, whose coefficients *decay exponentially* like $\mathcal{O}(\varrho_0^n)$ for $n \to \infty$. Here ϱ_0 denotes some number lying between the spectral radius of $S - E$ and one; i.e. $\varrho(S - E) < \varrho_0 < 1$;

A3) the functions φ, z and Φ are sufficiently differentiable.

Assumption A3 allows us to expand the local error, defined by (8.9), into a Taylor series:

$$\begin{aligned} d_{n+1} &= z(x_n + h, h) - Sz(x_n, h) - h\Phi\big(x_n, z(x_n, h) + h\alpha_n(h), h\big) - h\beta_n(h) \\ &= d_0(x_n) + d_1(x_n)h + \ldots + d_{N+1}(x_n)h^{N+1} \\ &\quad - h^2 \frac{\partial\Phi}{\partial u}\big(x_n, z(x_n, 0), 0\big)\alpha_n(h) - \ldots - h\beta_n(h) + \mathcal{O}(h^{N+1}). \end{aligned}$$

The expressions involving $\alpha_n(h)$ can be simplified further. Indeed, for a smooth function $G(x)$ we have

$$G(x_n)\alpha_n(h) = G(x_0)\alpha_n(h) + hG'(x_0)n\alpha_n(h) + \ldots + h^{N+1}R(n, h).$$

We observe that $n^j\alpha_n(h)$ is again a polynomial in h and that its coefficients decay like $\mathcal{O}(\varrho^n)$ where ϱ satisfies $\varrho_0 < \varrho < 1$. The same argument shows the boundedness of the remainder $R(n, h)$ for $0 \leq nh \leq Const$. As a consequence we can write the local error in the form

$$d_0 = \gamma_0 + \gamma_1 h + \ldots + \gamma_N h^N + \mathcal{O}(h^{N+1})$$

$$d_{n+1} = (d_0(x_n) + \delta_n^0) + \ldots + (d_{N+1}(x_n) + \delta_n^{N+1})h^{N+1} + \mathcal{O}(h^{N+2}) \quad (9.7)$$

$$\text{for } 0 \leq nh \leq Const.$$

The functions $d_j(x)$ are smooth and the perturbations δ_n^j satisfy $\delta_n^j = \mathcal{O}(\varrho^n)$. The expansion (9.7) is unique, because $\delta_n^j \to 0$ for $n \to \infty$.

Method (9.6) is called *consistent of order* p, if the local error (9.7) satisfies (Lemma 8.11)

$$d_n = \mathcal{O}(h^p) \quad \text{for } 0 \leq nh \leq Const, \quad \text{and} \quad Ed_p(x) = 0. \quad (9.8)$$

Observe that by this definition the perturbations δ_n^j have to vanish for $j = 0, \ldots, p - 1$, but no condition is imposed on δ_n^p. The exponential decay of these terms implies that we still have

$$d_{n+1} + E(d_n + \ldots + d_0) = \mathcal{O}(h^p) \quad \text{for } 0 \leq nh \leq Const,$$

in agreement with Definition 8.10. One can now easily verify that Lemma 8.12 (Φ_n satisfies a Lipschitz condition with the same constant as Φ) and the Convergence Theorem 8.13 remain valid for method (9.6). In the following theorem we use, as for one-step methods, the notation $u_h(x) = u_n$ when $x = x_n$.

Theorem 9.1 (Hairer & Lubich 1984). *Let the method (9.6) satisfy A1-A3 and be consistent of order* $p \geq 1$. *Then the global error has an asymptotic expansion of the form*

$$u_h(x) - z(x, h) = e_p(x)h^p + \ldots + e_N(x)h^N + E(x, h)h^{N+1} \quad (9.9)$$

where the $e_j(x)$ *are given in the proof (cf. formula (9.18)) and* $E(x, h)$ *is bounded uniformly in* $h \in [0, h_0]$ *and for* x *in compact intervals not containing* x_0. *More precisely than (9.9), there is an expansion*

$$u_n - z_n = (e_p(x_n) + \varepsilon_n^p)h^p + \ldots + (e_N(x_n) + \varepsilon_n^N)h^N + \widetilde{E}(n, h)h^{N+1} \quad (9.10)$$

where $\varepsilon_n^j = \mathcal{O}(\varrho^n)$ *with* $\varrho(S - E) < \varrho < 1$ *and* $\widetilde{E}(n, h)$ *is bounded for* $0 \leq nh \leq Const.$

Remark. We obtain from (9.10) and (9.9)

$$E(x_n, h) = \widetilde{E}(n, h) + h^{-1}\varepsilon_n^N + h^{-2}\varepsilon_n^{N-1} + \ldots + h^{p-N-1}\varepsilon_n^p,$$

so that the remainder term $E(x, h)$ is in general not uniformly bounded in h for x varying in an interval $[x_0, \bar{x}]$. However, if x is bounded away from x_0, say $x \geq x_0 + \delta$ ($\delta > 0$ fixed), the sequence ε_n^j goes to zero faster than any power of $\delta/n \leq h$.

Proof. a) As for one-step methods (cf. proof of Theorem 8.1, Chapter II) we construct a new method, which has as numerical solution

$$\widehat{u}_n = u_n - \big(e(x_n) + \varepsilon_n\big)h^p \tag{9.11}$$

for a given smooth function $e(x)$ and a given sequence ε_n satisfying $\varepsilon_n = \mathcal{O}(\varrho^n)$. Such a method is given by

$$\widehat{u}_0 = \widehat{\varphi}(h)$$
$$\widehat{u}_{n+1} = S\widehat{u}_n + h\widehat{\Phi}_n(x_n, \widehat{u}_n, h) \tag{9.12}$$

where $\widehat{\varphi}(h) = \varphi(h) - \big(e(x_0) + \varepsilon_0\big)h^p$ and

$$\widehat{\Phi}_n(x, u, h) = \Phi_n\big(x, u + (e(x) + \varepsilon_n)h^p, h\big)$$
$$- \big(e(x+h) - Se(x)\big)h^{p-1} - \big(\varepsilon_{n+1} - S\varepsilon_n\big)h^{p-1}.$$

Since Φ_n is of the form (9.5), $\widehat{\Phi}_n$ is also of this form, so that its local error has an expansion (9.7). We shall now determine $e(x)$ and ε_n in such a way that the method (9.12) is consistent of order $p+1$.

b) The local error $\widehat{d}_n$ of (9.12) can be expanded as

$$\widehat{d}_0 = z_0 - \widehat{u}_0 = \big(\gamma_p + e(x_0) + \varepsilon_0\big)h^p + \mathcal{O}(h^{p+1})$$
$$\widehat{d}_{n+1} = z_{n+1} - Sz_n - h\widehat{\Phi}_n(x_n, z_n, h)$$
$$= d_{n+1} + \big((I - S)e(x_n) + (\varepsilon_{n+1} - S\varepsilon_n)\big)h^p$$
$$+ \big(-G(x_n)(e(x_n) + \varepsilon_n) + e'(x_n)\big)h^{p+1} + \mathcal{O}(h^{p+2}).$$

Here

$$G(x) = \frac{\partial \Phi_n}{\partial u}\big(x, z(x, 0), 0\big)$$

which is independent of n by (9.5). The method (9.12) is consistent of order $p+1$, if (see (9.8))

i) $\varepsilon_0 = -\gamma_p - e(x_0)$,

ii) $d_p(x) + (I - S)e(x) + \delta_n^p + \varepsilon_{n+1} - S\varepsilon_n = 0$ for $x = x_n$,

iii) $Ee'(x) = EG(x)e(x) - Ed_{p+1}(x)$.

We assume for the moment that the system (i)-(iii) can be solved for $e(x)$ and ε_n. This will actually be demonstrated in part (d) of the proof. By the Convergence Theorem 8.13 the method (9.12) is convergent of order $p+1$. Hence

$$\widehat{u}_n - z_n = \mathcal{O}(h^{p+1}) \qquad \text{uniformly for } 0 \le nh \le Const,$$

which yields the statement (9.10) for $N = p$.

c) The method (9.12) satisfies the assumptions of the theorem with p replaced by $p+1$ and ϱ_0 by ϱ. As in Theorem 8.1 (Section II.8) an induction argument yields the result.

d) It remains to find a solution of the system (i)-(iii). Condition (ii) is satisfied if

(iia) $d_p(x) = (S - I)(e(x) + c)$

(iib) $\varepsilon_{n+1} - c = S(\varepsilon_n - c) - \delta_n^p$

hold for some constant c. Using $(I - S + E)^{-1}(I - S) = (I - E)$, which is a consequence of $SE = E^2 = E$ (see (8.11)), formula (iia) is equivalent to

$$(I - S + E)^{-1} d_p(x) = -(I - E)(e(x) + c). \tag{9.13}$$

From (i) we obtain $\varepsilon_0 - c = -\gamma_p - (e(x_0) + c)$, so that by (9.13)

$$(I - E)(\varepsilon_0 - c) = -(I - E)\gamma_p + (I - S + E)^{-1} d_p(x_0).$$

Since $Ed_p(x_0) = 0$, this relation is satisfied in particular if

$$\varepsilon_0 - c = -(I - E)\gamma_p + (I - S + E)^{-1} d_p(x_0). \tag{9.14}$$

The numbers $\varepsilon_n - c$ are now determined by the recurrence relation (iib)

$$\varepsilon_n - c = S^n(\varepsilon_0 - c) - \sum_{j=1}^{n} S^{n-j} \delta_{j-1}^p$$

$$= E(\varepsilon_0 - c) + (S - E)^n(\varepsilon_0 - c) - E\sum_{j=0}^{\infty} \delta_j^p + E\sum_{j=n}^{\infty} \delta_j^p - \sum_{j=1}^{n}(S - E)^{n-j}\delta_{j-1}^p,$$

where we have used $S^n = E + (S - E)^n$. If we put

$$c = E\sum_{j=0}^{\infty} \delta_j^p \tag{9.15}$$

the sequence $\{\varepsilon_n\}$ defined above satisfies $\varepsilon_n = \mathcal{O}(\varrho^n)$, since $E(\varepsilon_0 - c) = 0$ by (9.14) and since $\delta_n^p = \mathcal{O}(\varrho^n)$.

In order to find $e(x)$ we define

$$v(x) = Ee(x).$$

With the help of formulas (9.15) and (9.13) we can recover $e(x)$ from $v(x)$ by

$$e(x) = v(x) - (I - S + E)^{-1} d_p(x). \tag{9.16}$$

Equation (iii) can now be rewritten as the differential equation

$$v'(x) = EG(x)\Big(v(x) - (I - S + E)^{-1} d_p(x)\Big) - Ed_{p+1}(x), \tag{9.17}$$

and condition (i) yields the starting value $v(x_0) = -E(\gamma_p + \varepsilon_0)$. This initial value problem can be solved for $v(x)$ and we obtain $e(x)$ by (9.16). This function and the ε_n defined above represent a solution of (i)-(iii). $\square$

Remarks. a) It follows from (9.15)-(9.17) that the principal error term satisfies

$$e'_p(x) = EG(x)e_p(x) - Ed_{p+1}(x) - (I - S + E)^{-1}d'_p(x)$$

$$e_p(x_0) = -E\gamma_p - E\sum_{j=0}^{\infty} \delta_j^p - (I - S + E)^{-1}d_p(x_0). \tag{9.18}$$

b) Since $e_{p+1}(x)$ is just the principal error term of method (9.12), it satisfies the differential equation (9.18) with d_j replaced by $\hat{d}_{j+1}$. By an induction argument we therefore have for $j \geq p$

$$e'_j(x) = EG(x)e_j(x) + \text{inhomogeneity}(x).$$

Weakly Stable Methods

We next study the asymptotic expansion for stable methods, which are not strictly stable. For example, the explicit mid-point rule (1.13'), treated in connection with the GBS-algorithm (Section II.9), is of this type. As at the beginning of this section, we apply the mid-point rule to the problem $y' = -y$, $y(0) = 1$ and consider the following three starting procedures

$$y_0 = 1, \qquad y_1 = \exp(-h) \tag{9.19a}$$

$$y_0 = 1, \qquad y_1 = 1 - h + \frac{h^2}{2} \tag{9.19b}$$

$$y_0 = 1, \qquad y_1 = 1 - h. \tag{9.19c}$$

The three pictures on the left of Fig. 9.2 show the global error divided by h^2. For the first two starting procedures we have convergence to the function $xe^{-x}/6$, while for (9.19c) the divided error $(y_n - y(x_n))/h^2$ converges to

$$e^{-x}\left(\frac{2x-3}{12}\right) + \frac{e^x}{4} \qquad \text{for } n \text{ even,}$$

$$e^{-x}\left(\frac{2x-3}{12}\right) - \frac{e^x}{4} \qquad \text{for } n \text{ odd.}$$

We then subtract the h^2-term from the global error and divide by h^3 in the case (9.19a) and by h^4 for (b) and (c). The result is plotted in the pictures on the right of Fig. 9.2.

This example nicely illustrates the fact that we no longer have an asymptotic expansion of the form (9.9) or (9.10) but that there exists one expansion for x_n with n even, and a different expansion for x_n with n odd (see also Exercise 2 of Section II.9). Similar results for more general methods will be obtained here.

We say that a method of the form (8.4) is *weakly stable*, if it is stable, but if the matrix S has, besides $\zeta_1 = 1$, further eigenvalues of modulus 1, say $\zeta_2, \ldots, \zeta_l$.

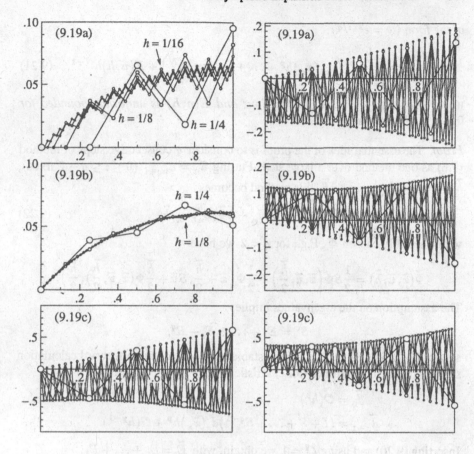

Fig. 9.2. Asymptotic expansion of the mid-point rule
(three different starting procedures)

The matrix S therefore has the representation (cf. (8.11))

$$S = \zeta_1 E_1 + \zeta_2 E_2 + \ldots + \zeta_l E_l + R \tag{9.20}$$

where the E_j are the projectors (corresponding to ζ_j) and the spectral radius of R satisfies $\varrho(R) < 1$.

In what follows we restrict ourselves to the case where all ζ_j $(j = 1, \ldots, l)$ are roots of unity. This allows a simple proof for the existence of an asymptotic expansion and is at the same time by far the most important special case. For the general situation we refer to Hairer & Lubich (1984).

Theorem 9.2. *Let the method (9.6) with Φ_n independent of n be stable, consistent of order p and satisfy A3. If all eigenvalues (of S) of modulus 1 satisfy $\zeta_j^q = 1$ $(j = 1, \ldots, l)$ for some positive integer q, then we have an asymptotic expansion* *of the form* ($\omega = e^{2\pi i/q}$)

$$u_n - z_n = \sum_{s=0}^{q-1} \omega^{ns} \left(e_{ps}(x_n)h^p + \ldots + e_{Ns}(x_n)h^N \right) + E(n,h)h^{N+1} \quad (9.21)$$

where the $e_{j_s}(x)$ *are smooth functions and* $E(n,h)$ *is uniformly bounded for* $0 < \delta \leq nh \leq Const.$

Proof. The essential idea of the proof is to consider q consecutive steps of method (9.6) as one method over a large step. Putting $\tilde{u}_n = u_{nq+i}$ ($0 \leq i \leq q-1$ fixed), $\tilde{h} = qh$ and $\tilde{x}_n = x_i + n\tilde{h}$, this method becomes

$$\tilde{u}_{n+1} = S^q \tilde{u}_n + \tilde{h}\tilde{\Phi}(\tilde{x}_n, \tilde{u}_n, \tilde{h}) \quad (9.22)$$

with a suitably chosen $\tilde{\Phi}$. E.g., for $q = 2$ we have

$$\tilde{\Phi}(\tilde{x}, \tilde{u}, \tilde{h}) = \frac{1}{2}S\Phi\left(\tilde{x}, \tilde{u}, \frac{\tilde{h}}{2}\right) + \frac{1}{2}\Phi\left(\tilde{x} + \frac{\tilde{h}}{2}, S\tilde{u} + \frac{\tilde{h}}{2}\Phi(\tilde{x}, \tilde{u}, \frac{\tilde{h}}{2}), \frac{\tilde{h}}{2}\right).$$

The assumption on the eigenvalues implies

$$S^q = E_1 + \ldots + E_l + R^q$$

so that (9.22) is seen to be a strictly stable method. A straightforward calculation shows that the local error of (9.22) satisfies

$$\tilde{d}_0 = \mathcal{O}(h^p)$$

$$\tilde{d}_{n+1} = (I + S + \ldots + S^{q-1})d_p(\tilde{x}_n)h^p + \mathcal{O}(h^{p+1}).$$

Inserting (9.20) and using $\zeta_j^q = 1$ we obtain, with $\tilde{E} = E_1 + \ldots + E_l$,

$$\tilde{E}(I + S + \ldots + S^{q-1})d_p(x)$$

$$= \tilde{E}\left(I - \tilde{E} + qE_1 + \sum_{j=2}^{l} \frac{1-\zeta_j^q}{1-\zeta_j}E_j + \sum_{j=1}^{q-1} R^j\right)d_p(x) = qE_1 d_p(x),$$

which vanishes by (8.15). Hence, also method (9.22) is consistent of order p. All the assumptions of Theorem 9.1 are thus verified for method (9.22). We therefore obtain

$$u_{nq+i} - z_{nq+i} = \tilde{e}_{pi}(x_{nq+i})h^p + \ldots + \tilde{e}_{Ni}(x_{nq+i})h^N + E_i(n,h)h^{N+1}$$

where $E_i(n,h)$ has the desired boundedness properties. If we define $e_{j_s}(x)$ as a solution of the Vandermonde-type system

$$\sum_{s=0}^{q-1} \omega^{is} e_{j_s}(x) = \tilde{e}_{ji}(x)$$

we obtain (9.21). □

The Adjoint Method

For a method (8.4) the correct value function $z(x,h)$, the starting procedure $\varphi(h)$ and the increment function $\Phi(x,u,h)$ are usually also defined for negative h (see the examples of Section III.8). As for one-step methods (Section II.8) we shall give here a precise meaning to the numerical solution $u_h(x)$ for negative h. This then leads in a natural way to the study of asymptotic expansions in even powers of h.

With the notation $u_h(x) = u_n$ for $x = x_0 + nh$ $(h > 0)$ the method (8.4) becomes

$$u_h(x_0) = \varphi(h)$$
$$u_h(x+h) = Su_h(x) + h\Phi(x, u_h(x), h) \qquad \text{for } x = x_0 + nh. \tag{9.23}$$

We first replace h by $-h$ in (9.23) to obtain

$$u_{-h}(x_0) = \varphi(-h)$$
$$u_{-h}(x-h) = Su_{-h}(x) - h\Phi(x, u_{-h}(x), -h)$$

and then x by $x+h$ which gives

$$u_{-h}(x_0) = \varphi(-h)$$
$$u_{-h}(x) = Su_{-h}(x+h) - h\Phi(x+h, u_{-h}(x+h), -h).$$

For sufficiently small h this equation can be solved for $u_{-h}(x+h)$ (Implicit Function Theorem) and we obtain

$$u_{-h}(x_0) = \varphi(-h),$$
$$u_{-h}(x+h) = S^{-1}u_{-h}(x) + h\Phi^*(x, u_{-h}(x), h). \tag{9.24}$$

The method (9.24), which is again of the form (8.4), is called the *adjoint method* of (9.23). Its correct value function is $z^*(x,h) = z(x,-h)$. Observe that for given S and Φ the new increment function Φ^* is just defined by the pair of formulas

$$v = Su - h\Phi(x+h, u, -h)$$
$$u = S^{-1}v + h\Phi^*(x, v, h). \tag{9.25}$$

Example 9.3. Consider a linear multistep method with generating functions

$$\varrho(\zeta) = \sum_{j=0}^{k} \alpha_j \zeta^j, \qquad \sigma(\zeta) = \sum_{j=0}^{k} \beta_j \zeta^j.$$

Then we have

$$S = \begin{pmatrix} -\alpha_{k-1}/\alpha_k & -\alpha_{k-2}/\alpha_k & \cdots & -\alpha_0/\alpha_k \\ 1 & 0 & \cdots & 0 \\ & 1 & \cdot & 0 \\ & & \vdots & \vdots \\ & & 1 & 0 \end{pmatrix}, \quad \Phi(x,u,h) = \begin{pmatrix} 1 \\ 0 \\ \vdots \\ 0 \end{pmatrix} \psi(x,u,h)$$

where $\psi = \psi(x, u, h)$ is the solution of $(u = (u_{k-1}, \ldots, u_0)^T)$

$$\alpha_k \psi = \sum_{j=0}^{k-1} \beta_j f(x+jh, u_j) + \beta_k f\left(x + kh, h\psi - \sum_{j=0}^{k-1} \frac{\alpha_j}{\alpha_k} u_j\right).$$

A straightforward use of the formulas (9.25) shows that

$$S^{-1} = \begin{pmatrix} 0 & & 1 & & \\ 0 & & 0 & & \\ \vdots & & \vdots & & \\ & & & & 1 \\ -\alpha_k/\alpha_0 & -\alpha_{k-1}/\alpha_0 & \cdots & -\alpha_1/\alpha_0 \end{pmatrix}, \quad \Phi^*(x, v, h) = \begin{pmatrix} 0 \\ \vdots \\ 0 \\ 1 \end{pmatrix} \psi^*(x, v, h)$$

where $\psi^* = \psi^*(x, v, h)$ (with $v = (v_0, \ldots, v_{k-1})^T$) is given by

$$-\alpha_0 \psi^* = \sum_{j=0}^{k-1} \beta_{k-j} f\big(x + (j-k+1)h, v_j\big) + \beta_0 f\left(x+h, h\psi^* - \sum_{j=0}^{k-1} \frac{\alpha_{k-j}}{\alpha_0} v_j\right).$$

This shows that the adjoint method is again a linear multistep method. Its generating polynomials are

$$\varrho^*(\zeta) = -\zeta^k \varrho(\zeta^{-1}), \qquad \sigma^*(\zeta) = \zeta^k \sigma(\zeta^{-1}). \qquad (9.26)$$

Our next aim is to prove that the adjoint method has exactly the same asymptotic expansion as the original method, with h replaced by $-h$. For this it is necessary that S^{-1} also be a stable matrix. Therefore all eigenvalues of S must lie on the unit circle.

Theorem 9.4. *Let the method (9.23) be stable, consistent of order p and assume that all eigenvalues of S satisfy $\zeta_j^q = 1$ for some positive integer q. Then the global error has an asymptotic expansion of the form* $(\omega = e^{2\pi i/q})$

$$u_h(x_n) - z(x_n, h) = \sum_{s=0}^{q-1} \omega^{ns}\left(e_{ps}(x_n)h^p + \ldots + e_{Ns}(x_n)h^N\right) + E(x_n, h)h^{N+1},$$

$$(9.27)$$

valid for positive and negative h. The remainder $E(x, h)$ is uniformly bounded for $|h| \leq h_0$ and $x_0 \leq x \leq \widehat{x}$.

Proof. As in the proof of Theorem 9.2 we consider q consecutive steps of method (9.23) as one new method. The assumption on the eigenvalues implies that $S^q = I =$ identity. Therefore the new method is essentially a one-step method. The only difference is that here the starting procedure and the correct value function may depend on h. A straightforward extension of Theorem 8.5 of Chapter II (Exercise 3) implies the existence of an expansion

$$u_h(x_{nq+i}) - z(x_{nq+i}, h) = \widetilde{e}_{pi}(x_{nq+i})h^p + \ldots + \widetilde{e}_{Ni}(x_{nq+i})h^N$$
$$+ E_i(x_{nq+i}, h)h^{N+1}.$$

This expansion is valid for positive and negative h; the remainder $E_i(x,h)$ is bounded for $|h| \leq h_0$ and $x_0 \leq x \leq \hat{x}$. The same argument as in the proof of Theorem 9.2 now leads to the desired expansion.   □

Symmetric Methods

The definition of symmetry for general linear methods is not as straightforward as for one-step methods. In Example 9.3 we saw that the components of the numerical solution of the adjoint method are in inverse order. Therefore, it is too restrictive to require that $\varphi(h) = \varphi(-h)$, $S = S^{-1}$ and $\Phi = \Phi^*$.

However, for many methods of practical interest the correct value function satisfies a *symmetry relation* of the form

$$z(x,h) = Qz(x+qh, -h) \tag{9.28}$$

where Q is a square matrix and q an integer. This is for instance the case for linear multistep methods, where the correct value function is given by

$$z(x,h) = \big(y(x+(k-1)h), \ldots, y(x)\big)^T.$$

The relation (9.28) holds with

$$Q = \begin{pmatrix} & & 1 \\ & \cdot^{\cdot} & \\ 1 & & \end{pmatrix} \qquad \text{and} \qquad q = k-1. \tag{9.29}$$

Definition 9.5. Suppose that the correct value function satisfies (9.28). Method (9.23) is called *symmetric* (with respect to (9.28)), if the numerical solution satisfies its analogue

$$u_h(x) = Qu_{-h}(x+qh). \tag{9.30}$$

Example 9.6. Consider a linear multistep method and suppose that the generating polynomials of the adjoint method (9.26) satisfy

$$\varrho^*(\zeta) = \varrho(\zeta), \qquad \sigma^*(\zeta) = \sigma(\zeta). \tag{9.31}$$

This is equivalent to the requirement (cf. (3.24))

$$\alpha_{k-j} = -\alpha_j, \qquad \beta_{k-j} = \beta_j.$$

A straightforward calculation (using the formulas of Example 9.3) then shows that the symmetry relation (9.30) holds for all $x = x_0 + nh$ whenever it holds for $x = x_0$. This imposes an additional condition on the starting procedure $\varphi(h)$.

Let us finally demonstrate how Theorem 9.4 can be used to prove *asymptotic expansions in even powers* of h. Denote by $u_h^j(x)$ the j th component of $u_h(x)$.

The symmetry relation (9.30) for multistep methods then implies

$$u^k_{-h}(x) = u^1_h(x - (k-1)h)$$

Furthermore, for any multistep method we have

$$u^k_h(x) = u^1_h(x - (k-1)h)$$

so that

$$u^k_h(x) = u^k_{-h}(x)$$

for symmetric methods. As a consequence of Theorem 9.4 the asymptotic expansion of the global error is in even powers of h, whenever the multistep method is symmetric in the sense of Definition 9.5.

Exercises

1. Consider a strictly stable, pth order, linear multistep method written in the form (9.6) (see Example 9.3) and set

$$G(x) = \frac{\partial \Phi}{\partial u}(x, z(x, 0), 0).$$

   a) Prove that

   $$EG(x)\mathbb{1} = \mathbb{1}\frac{\partial f}{\partial y}(x, y(x))$$

   where E is the matrix given by (8.11) and $\mathbb{1} = (1, \ldots, 1)^T$.

   b) Show that the function $e_p(x)$ in the expansion (9.9) is given by $e_p(x) = \mathbb{1}\widehat{e}_p(x)$, where

   $$\widehat{e}'_p(x) = \frac{\partial f}{\partial y}(x, y(x))\widehat{e}_p(x) - Cy^{(p+1)}(x)$$

   and C is the error constant (cf. (2.13)). Compute also $\widehat{e}_p(x_0)$.

2. For the 3-step BDF-method, applied to $y' = -y$, $y(0) = 1$ with starting procedure (9.1c), compute the function $e_3(x)$ and the perturbations $\{\varepsilon^3_n\}_{n \geq 0}$ in the expansion (9.4). Compare your result with Fig. 9.1.

3. Consider the method

   $$u_0 = \varphi(h), \qquad u_{n+1} = u_n + h\Phi(x_n, u_n, h) \qquad (9.32)$$

   with correct value function $z(x, h)$.

   a) Prove that the global error has an asymptotic expansion of the form

   $$u_n - z_n = e_p(x_n)h^p + \ldots + e_N(x_n)h^N + E(x_n, h)h^{N+1}$$

   where $E(x, h)$ is uniformly bounded for $0 \leq h \leq h_0$ and $x_0 \leq x \leq \widehat{x}$.

   b) Show that Theorem 8.5 of Chapter II remains valid for method (9.32).

III.10 Multistep Methods for Second Order Differential Equations

> En 1904 j'eus besoin d'une pareille méthode pour calculer les trajectoires des corpuscules électrisés dans un champ magnétique, et en essayant diverses méthodes déjà connues, mais sans les trouver assez commodes pour mon but, je fus conduit moi-même à élaborer une méthode assez simple, dont je me suis servi ensuite.
>
> (C. Störmer 1921)

Because of their importance, second order differential equations deserve some additional attention. We already saw in Section II.14 that for special second order differential equations certain direct one-step methods are more efficient than the classical Runge-Kutta methods. We now investigate whether a similar situation also holds for multistep methods.

Consider the second order differential equation

$$y'' = f(x, y, y') \tag{10.1}$$

where y is allowed to be a vector. We rewrite (10.1) in the usual way as a first order system and apply a multistep method

$$\sum_{i=0}^{k} \alpha_i y_{n+i} = h \sum_{i=0}^{k} \beta_i y'_{n+i}$$

$$\sum_{i=0}^{k} \alpha_i y'_{n+i} = h \sum_{i=0}^{k} \beta_i f(x_{n+i}, y_{n+i}, y'_{n+i}). \tag{10.2}$$

If the right hand side of the differential equation does not depend on y',

$$y'' = f(x, y), \tag{10.3}$$

it is natural to look for numerical methods which do not involve the first derivative. An elimination of $\{y'_n\}$ in the equations (10.2) results in

$$\sum_{i=0}^{2k} \widehat{\alpha}_i y_{n+i} = h^2 \sum_{i=0}^{2k} \widehat{\beta}_i f(x_{n+i}, y_{n+i}) \tag{10.4}$$

where the new coefficients $\widehat{\alpha}_i, \widehat{\beta}_i$ are given by

$$\sum_{i=0}^{2k} \widehat{\alpha}_i \zeta^i = \left(\sum_{i=0}^{k} \alpha_i \zeta^i \right)^2, \qquad \sum_{i=0}^{2k} \widehat{\beta}_i \zeta^i = \left(\sum_{i=0}^{k} \beta_i \zeta^i \right)^2. \tag{10.5}$$

In what follows we investigate (10.4) with coefficients that do not necessarily satisfy (10.5). It is hoped to achieve the same order with a smaller step number.

Explicit Störmer Methods

> Sein Vortrag ist übrigens ziemlich trocken und langweilig ...
> (B. Riemann's opinion about Encke, 1847)

> Had the Ast. Ges. Essay been entirely free from numerical blunders, ...
> (P.H. Cowell & A.C.D. Crommelin 1910)

Since most differential equations of celestial mechanics are of the form (10.3) it is not surprising that the first attempts at developing special methods for these equations were made by astronomers.

For his extensive numerical calculations concerning the aurora borealis (see below), C. Störmer (1907) developed an accurate and simple method as follows: by adding the Taylor series for $y(x_n + h)$ and $y(x_n - h)$ we obtain

$$y(x_n + h) - 2y(x_n) + y(x_n - h) = h^2 y''(x_n) + \frac{h^4}{12} y^{(4)}(x_n) + \frac{h^6}{360} y^{(6)}(x_n) + \dots .$$

If we insert $y''(x_n)$ from the differential equation (10.3) and neglect higher terms, we get

$$y_{n+1} - 2y_n + y_{n-1} = h^2 f_n$$

as a first simple method, which is sometimes called Störmer's or Encke's method. For greater precision, we replace the higher derivatives of y by central differences of f

$$h^2 y^{(4)}(x_n) = \Delta^2 f_{n-1} - \frac{1}{12} \Delta^4 f_{n-2} + \dots$$
$$h^4 y^{(6)}(x_n) = \Delta^4 f_{n-2} + \dots$$

and obtain

$$y_{n+1} - 2y_n + y_{n-1} = h^2 \left(f_n + \frac{1}{12} \Delta^2 f_{n-1} - \frac{1}{240} \Delta^4 f_{n-2} + \dots \right). \qquad (10.6)$$

This formula is not yet very practical, since the differences of the right hand side contain the unknown expressions f_{n+1} and f_{n+2}. Neglecting fifth-order differences (i.e., putting $\Delta^4 f_{n-2} \approx \Delta^4 f_{n-4}$ and $\Delta^2 f_{n-1} = \Delta^2 f_{n-2} + \Delta^3 f_{n-3} + \Delta^4 f_{n-3} \approx \Delta^2 f_{n-2} + \Delta^3 f_{n-3} + \Delta^4 f_{n-4}$) one gets

$$y_{n+1} - 2y_n + y_{n-1} = h^2 f_n + \frac{h^2}{12} \left(\Delta^2 f_{n-2} + \Delta^3 f_{n-3} + \frac{19}{20} \Delta^4 f_{n-4} \right) \qquad (10.7)$$

("... formule qui est fondamentale dans notre méthode ...", C. Störmer 1907).

Some years later Cowell & Crommelin (1910) used the same ideas to investigate the motion of Halley's comet. They considered one additional term in the series (10.6), namely

$$\frac{31}{60480} \Delta^6 f_{n-3} \approx \frac{1}{1951} \Delta^6 f_{n-3}.$$

Arbitrary orders. Integrating equation (10.3) twice we obtain

$$y(x+h) = y(x) + hy'(x) + h^2 \int_0^1 (1-s)f(x+sh, y(x+sh)) \, ds. \quad (10.8)$$

In order to eliminate the first derivative of $y(x)$ we write the same formula with h replaced by $-h$ and add the two expressions:

$$y(x+h) - 2y(x) + y(x-h) \quad (10.9)$$

$$= h^2 \int_0^1 (1-s)\Big(f(x+sh, y(x+sh)) + f(x-sh, y(x-sh))\Big) \, ds.$$

As in the derivation of the Adams formulas (Section III.1) we replace the unknown function $f(t, y(t))$ by the interpolation polynomial $p(t)$ of formula (1.4). This yields the *explicit* method

$$y_{n+1} - 2y_n + y_{n-1} = h^2 \sum_{j=0}^{k-1} \sigma_j \nabla^j f_n \quad (10.10)$$

with coefficients σ_j given by

$$\sigma_j = (-1)^j \int_0^1 (1-s)\left(\binom{-s}{j} + \binom{s}{j}\right) ds. \quad (10.11)$$

See Table 10.1 for their numerical values and Exercise 2 for their computation.

Table 10.1. Coefficients of the method (10.10)

j	0	1	2	3	4	5	6	7	8	9
σ_j	1	0	$\dfrac{1}{12}$	$\dfrac{1}{12}$	$\dfrac{19}{240}$	$\dfrac{3}{40}$	$\dfrac{863}{12096}$	$\dfrac{275}{4032}$	$\dfrac{33953}{518400}$	$\dfrac{8183}{129600}$

Special cases of (10.10) are

$$k = 2: \quad y_{n+1} - 2y_n + y_{n-1} = h^2 f_n$$

$$k = 3: \quad y_{n+1} - 2y_n + y_{n-1} = h^2 \left(\frac{13}{12}f_n - \frac{1}{6}f_{n-1} + \frac{1}{12}f_{n-2}\right) \quad (10.10')$$

$$k = 4: \quad y_{n+1} - 2y_n + y_{n-1} = h^2 \left(\frac{7}{6}f_n - \frac{5}{12}f_{n-1} + \frac{1}{3}f_{n-2} - \frac{1}{12}f_{n-3}\right).$$

Method (10.10) with $k = 5$ is formula (10.7), the method used by Störmer (1907, 1921), and for $k = 6$ one obtains the method used by Cowell & Crommelin (1910). The simplest of these methods ($k = 1$ or $k = 2$) has been successfully applied as the basis of an extrapolation method (Section II.14, formula (14.32)).

Implicit Störmer Methods

The first terms of (10.6)

$$y_{n+1} - 2y_n + y_{n-1} = h^2\left(f_n + \frac{1}{12}\Delta^2 f_{n-1}\right)$$

$$= \frac{h^2}{12}(f_{n+1} + 10f_n + f_{n-1}) \tag{10.12}$$

form an implicit equation for y_{n+1}. This can either be used in a predictor-corrector fashion, or, as advocated by B. Numerov (1924, 1927), by solving this implicit nonlinear equation directly for y_{n+1}.

To obtain more accurate formulas, analogous to the implicit Adams methods, we use the interpolation polynomial $p^*(t)$ of (1.8), which passes through the additional point (x_{n+1}, f_{n+1}). This yields the implicit method

$$y_{n+1} - 2y_n + y_{n-1} = h^2 \sum_{j=0}^{k} \sigma_j^* \nabla^j f_{n+1}, \tag{10.13}$$

where the coefficients σ_j^* are defined by

$$\sigma_j^* = (-1)^j \int_0^1 (1-s)\left(\binom{-s+1}{j} + \binom{s+1}{j}\right) ds \tag{10.14}$$

and are given in Table 10.2 for $j \leq 9$.

Table 10.2. Coefficients of the implicit method (10.13)

j	0	1	2	3	4	5	6	7	8	9
σ_j^*	1	-1	$\frac{1}{12}$	0	$\frac{-1}{240}$	$\frac{-1}{240}$	$\frac{-221}{60480}$	$\frac{-19}{6048}$	$\frac{-9829}{3628800}$	$\frac{-407}{172800}$

Further methods can be derived by using the ideas of Nyström and Milne for first order equations. With the substitutions $h \to 2h$, $2s \to s$ and $x \to x - h$ formula (10.9) becomes

$$y(x+h) - 2y(x-h) + y(x-3h) = h^2 \int_0^2 (2-s) \tag{10.15}$$

$$\cdot \Big(f\big(x + (s-1)h, y(x+(s-1)h)\big) + f\big(x - (s+1)h, y(x-(s+1)h)\big)\Big)\, ds.$$

If one replaces $f(t, y(t))$ by the polynomial $p(t)$ (respectively $p^*(t)$) one obtains the new classes of explicit (respectively implicit) methods.

Numerical Example

Nous avons calculé plus de 120 trajectoires différentes, travail immense qui a exigé plus de 4500 heures ... Quand on est suffisamment exercé, on calcule environ trois points (R, z) par heure.

(C. Störmer 1907)

We choose the historical problem treated by Störmer in 1907: Störmer's aim was to confirm numerically the conjecture of Birkeland, who explained in 1896 the aurora borealis as being produced by electrical particles emanating from the sun and dancing in the earth's magnetic field. Suppose that an elementary magnet is situated at the origin with its axis along to the z-axis. The trajectory $(x(s), y(s), z(s))$ of an electrical particle in this magnetic field then satisfies

$$x'' = \frac{1}{r^5}\left(3yzz' - (3z^2 - r^2)y'\right)$$

$$y'' = \frac{1}{r^5}\left((3z^2 - r^2)x' - 3xzz'\right) \qquad (10.16)$$

$$z'' = \frac{1}{r^5}\left(3xzy' - 3yzx'\right)$$

where $r^2 = x^2 + y^2 + z^2$. Introducing the polar coordinates

$$x = R\cos\varphi, \qquad y = R\sin\varphi \qquad (10.17)$$

the system (10.16) becomes equivalent to

$$R'' = \left(\frac{2\gamma}{R} + \frac{R}{r^3}\right)\left(\frac{2\gamma}{R^2} + \frac{3R^2}{r^5} - \frac{1}{r^3}\right) \qquad (10.18a)$$

$$z'' = \left(\frac{2\gamma}{R} + \frac{R}{r^3}\right)\frac{3Rz}{r^5} \qquad (10.18b)$$

$$\varphi' = \left(\frac{2\gamma}{R} + \frac{R}{r^3}\right)\frac{1}{R} \qquad (10.18c)$$

where now $r^2 = R^2 + z^2$ and γ is some constant arising from the integration of φ''. The two equations (10.18a,b) constitute a second order differential equation of type (10.3), which can be solved numerically by the methods of this section. φ is then obtained by simple integration of (10.18c). Störmer found after long calculations that the initial values

$$R_0 = 0.257453, \qquad z_0 = 0.314687, \qquad \gamma = -0.5,$$
$$R_0' = \sqrt{Q_0}\cos u, \qquad z_0' = \sqrt{Q_0}\sin u, \qquad u = 5\pi/4 \qquad (10.18d)$$
$$r_0 = \sqrt{R_0^2 + z_0^2}, \qquad Q_0 = 1 - (2\gamma/R_0 + R_0/r_0^3)^2$$

produce a specially interesting solution curve approaching very closely the North Pole. Fig. 10.1 shows 125 solution curves (in the x, y, z-space) with these and neighbouring initial values to give an impression of how an aurora borealis comes into being.

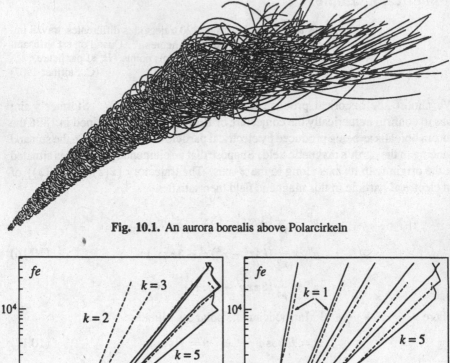

Fig. 10.1. An aurora borealis above Polarcirkeln

Fig. 10.2. Performance of Störmer and Adams methods

Fig. 10.2 compares the performance of the Störmer methods (10.10) and (10.13) (in PECE mode) with that of the Adams methods by integrating subsystem (10.18a,b) with initial values (10.18d) for $0 \leq s \leq 0.3$. The diagrams compare the Euclidean norm in $\mathbf{R}^2$ of the error of the final solution point (R, z) with the number of function evaluations fe. The step numbers used are $\{n = 50 \cdot 2^{0.3 \cdot i}\}_{i=0,1,\ldots,30} = \{50, 61, 75, 93, 114, \ldots, 25600\}$. The starting values were computed very precisely with an explicit Runge-Kutta method and step size $h_{RK} = h/10$. It can be observed that the Störmer methods are substantially more precise due to the smaller error constants (compare Tables 10.1 and 10.2 with Tables 1.1 and 1.2). In addition, they have lower overhead. However, they must be implemented carefully in order to avoid rounding errors (see below).

General Formulation

Our next aim is to study stability, consistency and convergence of general linear multistep methods for (10.3). We write them in the form

$$\sum_{i=0}^{k} \alpha_i y_{n+i} = h^2 \sum_{i=0}^{k} \beta_i f(x_{n+i}, y_{n+i}). \tag{10.19}$$

The generating polynomials of the coefficients α_i and β_i are again denoted by

$$\varrho(\zeta) = \sum_{i=0}^{k} \alpha_i \zeta^i, \qquad \sigma(\zeta) = \sum_{i=0}^{k} \beta_i \zeta^i. \tag{10.20}$$

If we apply method (10.19) to the initial value problem

$$y'' = f(x, y), \qquad y(x_0) = y_0, \qquad y'(x_0) = y_0' \tag{10.21}$$

it is natural to require that the starting values be consistent with both initial values, i.e., that

$$\frac{y_i - y_0 - ihy_0'}{h} \to 0 \qquad \text{for} \quad h \to 0, \qquad i = 0, 1, \dots, k-1. \tag{10.22}$$

For the *stability condition* of method (10.19) we consider the simple problem

$$y'' = 0, \qquad y_0 = 0, \qquad y_0' = 0.$$

Its numerical solution satisfies a linear difference equation with $\varrho(\zeta)$ as characteristic polynomial. The same considerations as in the proof of Theorem 4.2 show that the following stability condition is necessary for convergence.

Definition 10.1. Method (10.19) is called *stable*, if the generating polynomial $\varrho(\zeta)$ satisfies:

i) The roots of $\varrho(\zeta)$ lie on or within the unit circle;

ii) The multiplicity of the roots on the unit circle is at most two.

For the *order conditions* we introduce, similarly to formula (2.3), the linear difference operator

$$L(y, x, h) = \varrho(E)y(x) - h^2 \sigma(E)y''(x)$$
$$= \sum_{i=0}^{k} \Big(\alpha_i y(x+ih) - h^2 \beta_i y''(x+ih)\Big), \tag{10.23}$$

where E is the shift operator. As in Definition 2.3 we now have:

Definition 10.2. Method (10.19) is *consistent of order p* if for all sufficiently smooth functions $y(x)$,

$$L(y, x, h) = \mathcal{O}(h^{p+2}). \tag{10.24}$$

The following theorem is then proved similarly to Theorem 2.4.

Theorem 10.3. *The multistep method (10.19) is of order p if and only if the following equivalent conditions hold:*

i) $\sum_{i=0}^{k} \alpha_i = 0, \qquad \sum_{i=0}^{k} i\alpha_i = 0$

   and $\sum_{i=0}^{k} \alpha_i i^q = q(q-1)\sum_{i=0}^{k} \beta_i i^{q-2}$ *for* $q = 2, \ldots, p+1,$

ii) $\varrho(e^h) - h^2\sigma(e^h) = \mathcal{O}(h^{p+2})$ *for* $h \to 0,$

iii) $\dfrac{\varrho(\zeta)}{(\log \zeta)^2} - \sigma(\zeta) = \mathcal{O}((\zeta-1)^p)$ *for* $\zeta \to 1.$

$\square$

As for Adams methods one easily verifies that the method (10.10) is of order k, and that (10.13) is of order $k+1$.

The following order barriers are similar to those of Theorems 3.5 and 3.9; their proofs are similar too (see, e.g., Dahlquist 1959, Henrici 1962):

Theorem 10.4. *The order p of a stable linear multistep method (10.19) satisfies*

$$p \leq k+2 \quad \textit{if } k \textit{ is even,}$$
$$p \leq k+1 \quad \textit{if } k \textit{ is odd.} \qquad \square$$

Theorem 10.5. *Stable multistep methods (10.19) of order $k+2$ are symmetric, i.e.,*

$$\alpha_j = \alpha_{k-j}, \qquad \beta_j = \beta_{k-j} \quad \textit{for all } j. \qquad \square$$

Convergence

Theorem 10.6. *Suppose that method (10.19) is stable, of order p, and that the starting values satisfy*

$$y(x_j) - y_j = \mathcal{O}(h^{p+1}) \quad \textit{for } j = 0, 1, \ldots, k-1. \tag{10.25}$$

Then we have convergence of order p, i.e.,

$$\|y(x_n) - y_n\| \leq Ch^p \qquad \textit{for } 0 \leq hn \leq Const.$$

Proof. It is possible to develop a theory analogous to that of Sections III.2 - III.4. This is due to Dahlquist (1959) and can also be found in the book of Henrici (1962). We prefer to rewrite (10.19) *in a one-step formulation* of the form (8.4) and to apply directly the results of Section III.8 and III.9 (see Example 8.6). In order to achieve this goal, we could put $u_n = (y_{n+k-1}, \ldots, y_n)^T$, which seems to be a natural choice. But then the corresponding matrix S does not satisfy the stability condition of Definition 8.8 because of the double roots of modulus 1. To overcome this difficulty we separate these roots. We split the characteristic polynomial $\varrho(\zeta)$ into

$$\varrho(\zeta) = \varrho_1(\zeta) \cdot \varrho_2(\zeta) \tag{10.26}$$

such that each polynomial $(l + k = m)$

$$\varrho_1(\zeta) = \sum_{i=0}^{l} \gamma_i \zeta^i, \qquad \varrho_2(\zeta) = \sum_{i=0}^{m} \kappa_i \zeta^i \tag{10.27}$$

has only simple roots of modulus 1. Without loss of generality we assume in the sequel that $m \geq l$ and $\alpha_k = \gamma_l = \kappa_m = 1$. Using the shift operator E, method (10.19) can be written as

$$\varrho(E)y_n = h^2\sigma(E)f_n.$$

The main idea is to introduce $\varrho_2(E)y_n$ as a new variable, say hv_n, so that the multistep formula becomes equivalent to the system

$$\varrho_1(E)v_n = h\sigma(E)f_n, \qquad \varrho_2(E)y_n = hv_n. \tag{10.28}$$

Introducing the vector

$$u_n = (v_{n+l-1}, \ldots, v_n, y_{n+m-1}, \ldots, y_n)^T$$

formula (10.28) can be written as

$$u_{n+1} = Su_n + h\Phi(x_n, u_n, h) \tag{10.29a}$$

where

$$S = \begin{pmatrix} G & 0 \\ 0 & K \end{pmatrix}, \qquad \Phi(x_n, u_n, h) = \begin{pmatrix} e_1\psi(x_n, u_n, h) \\ e_1 v_n \end{pmatrix}. \tag{10.30}$$

The matrices G and K are the companion matrices

$$G = \begin{pmatrix} -\gamma_{l-1} & -\gamma_{l-2} & \cdots & \cdot & -\gamma_0 \\ 1 & 0 & \cdots & \cdot & 0 \\ & 1 & & \cdot & 0 \\ & & \vdots & \vdots & \\ & & & 1 & 0 \end{pmatrix}, \qquad K = \begin{pmatrix} -\kappa_{m-1} & -\kappa_{m-2} & \cdots & \cdot & -\kappa_0 \\ 1 & 0 & \cdots & \cdot & 0 \\ & 1 & & \cdot & 0 \\ & & \vdots & \vdots & \\ & & & 1 & 0 \end{pmatrix},$$

$e_1 = (1, 0, \ldots, 0)^T$, and $\psi = \psi(x_n, u_n, h)$ is implicitly defined by

$$\psi = \sum_{j=0}^{k-1} \beta_j f(x_n + jh, y_{n+j}) + \beta_k f\left(x_n + kh, h^2\psi - \sum_{j=0}^{k-1} \alpha_j y_{n+j}\right). \tag{10.31}$$

In this formula ψ is written as a function of $x_n, (y_{n+k-1}, \ldots, y_n)$ and h. But the second relation of (10.28) shows that each value $y_{n+k-1}, \ldots, y_{n+m}$ can be expressed as a linear combination of the elements of u_n. Therefore ψ is in fact a function of (x_n, u_n, h).

Formula (10.29a) defines our forward step procedure. The corresponding starting procedure is

$$\varphi(h) = (v_{l-1}, \ldots, v_0, y_{m-1}, \ldots, y_0)^T \qquad (10.29b)$$

which, by (10.28), is uniquely determined by $(y_{k-1}, \ldots, y_0)^T$. As correct value function we have

$$z(x, h) = \left(\frac{1}{h} \varrho_2(E) y(x+(l-1)h), \ldots, \frac{1}{h} \varrho_2(E) y(x), \ y(x+(m-1)h, \ldots, y(x) \right)^T.$$
$$(10.29c)$$

By our choice of $\varrho_1(\zeta)$ and $\varrho_2(\zeta)$ (both have only simple roots of modulus 1) the matrices G and K are power bounded. Therefore S is also power bounded and method (10.29) *is stable* in the sense of Definition 8.8.

We now verify the conditions of Definition 8.10 and for this start with the error in the initial values

$$d_0 = z(x_0, h) - \varphi(h).$$

The first l components of this vector are

$$\frac{1}{h} \varrho_2(E) y(x_j) - v_j = \frac{1}{h} \sum_{i=0}^{m} \kappa_i \big(y(x_{i+j}) - y_{i+j} \big), \qquad j = 0, \ldots, l-1$$

and the last m components are just

$$y(x_j) - y_j, \qquad j = 0, \ldots, m-1.$$

Thus hypothesis (10.25) ensures that $d_0 = \mathcal{O}(h^p)$. Consider next the local error at x_n,

$$d_{n+1} = z(x_n + h, h) - S z(x_n, h) - h\Phi\big(x_n, z(x_n, h), h\big).$$

All components of d_{n+1} vanish except the first, which equals

$$d_{n+1}^{(1)} = \frac{1}{h} \varrho(E) y(x_n) - h\psi(x_n, z(x_n, h), h).$$

Using formula (10.31), an application of the mean value theorem yields

$$d_{n+1}^{(1)} = \frac{1}{h} L(y, x_n, h) + h^2 \beta_k f'(x_{n+k}, \eta) \cdot d_{n+1}^{(1)} \qquad (10.32)$$

with η as in Lemma 2.2. We therefore have

$$d_{n+1} = \mathcal{O}(h^{p+1}) \qquad \text{since} \qquad L(y, x_n, h) = \mathcal{O}(h^{p+2}).$$

Finally Theorem 8.13 yields the stated convergence result.   □

Asymptotic Formula for the Global Error

Assume that the method (10.19) is stable and consistent of order p. The local truncation error of (10.29) is then given by

$$d_{n+1} = e_1 h^{p+1} C_{p+2} y^{(p+2)}(x_n) + \mathcal{O}(h^{p+2}) \tag{10.33}$$

with

$$C_{p+2} = \frac{1}{(p+2)!} \sum_{i=0}^{k} \left(\alpha_i i^{p+2} - (p+2)(p+1)\beta_i i^p \right).$$

Formula (10.33) can be verified by developing $L(y, x_n, h)$ into a Taylor series in (10.32). An application of Theorem 9.1 (if 1 is the only root of modulus 1 of $\varrho(\zeta)$) or of Theorem 9.2 shows that the global error of method (10.29) is of the form

$$u_h(x) - z(x, h) = e(x) h^p + \mathcal{O}(h^{p+1})$$

where $e(x)$ is the solution of

$$e'(x) = E \frac{\partial \Phi}{\partial u}(x, z(x,0), 0) e(x) - E e_1 \cdot C_{p+2} y^{(p+2)}(x). \tag{10.34}$$

Here E is the matrix defined in (8.12). Since no h^p-term is present in the local error (10.33), it follows from (9.16) that $e(x) = E e(x)$. Therefore (see Exercise 4a) this function can be written as

$$e(x) = \begin{pmatrix} \gamma(x) \mathbb{1} \\ \kappa(x) \mathbb{1} \end{pmatrix}.$$

A straightforward calculation of $\frac{\partial \Phi}{\partial u}(x, z(x,0), 0)$ and $E e_1$ (for details see Exercise 4) shows that (10.34) becomes equivalent to the system

$$\gamma'(x) = \frac{\sigma(1)}{\varrho_1'(1)} \frac{\partial f}{\partial y}(x, y(x)) \kappa(x) - \frac{C_{p+2}}{\varrho_1'(1)} y^{(p+2)}(x) \tag{10.35a}$$

$$\kappa'(x) = \frac{1}{\varrho_2'(1)} \gamma(x). \tag{10.35b}$$

Differentiating (10.35b) and inserting $\gamma'(x)$ from (10.35a), we finally obtain

$$\kappa''(x) = \frac{\partial f}{\partial y}(x, y(x)) \kappa(x) - C y^{(p+2)}(x) \tag{10.36}$$

with

$$C = \frac{C_{p+2}}{\sigma(1)}. \tag{10.37}$$

Here we have used the relation $\sigma(1) = \varrho_1'(1) \cdot \varrho_2'(1)$, which is an immediate consequence of (10.26), and the assumption that the order of the method is at least 1. The constant C in (10.37) is called the *error constant* of method (10.19). It plays the same role as (2.13) for first order equations.

Since the last component of the vector u_n is y_n we have the desired result

$$y_n - y(x_n) = \kappa(x_n)h^p + \mathcal{O}(h^{p+1})$$

with $\kappa(x)$ satisfying (10.36). Further terms in the asymptotic expansion of the global error can also be obtained by specializing the results of III.9.

Rounding Errors

A *direct* implementation of Störmer's methods, for which (10.19) specializes to

$$y_{n+1} - 2y_n + y_{n-1} = h^2 \sum_{i=0}^{k} \beta_i f_{n+i-k+1}, \tag{10.38}$$

by storing the y-values $y_0, y_1, \ldots, y_{k-1}$ and computing successively the values $y_k, y_{k+1}, \ldots$ with the help of (10.38) leads to numerical instabilities for small h. This instability is caused by the double root of $\varrho(\zeta)$ on the unit circle. It can be observed numerically in Fig. 10.3, where the left picture is a zoom of Fig. 10.2, while the right image contains the results of a code implementing (10.38) directly.

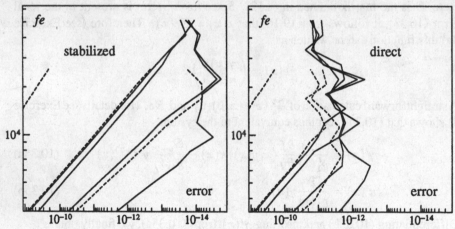

Fig. 10.3. Rounding errors caused by a direct application of (10.38)

In order to obtain the stabilized version of the algorithm, we apply the following two ideas:

a) Split, as in (10.26), the polynomial $\varrho(\zeta)$ as $(\zeta - 1)(\zeta - 1)$. Then (10.28) leads to $hv_n = y_{n+1} - y_n$ and (10.38) becomes the mathematically equivalent formulation

$$v_n - v_{n-1} = h \sum_{i=0}^{k} \beta_i f_{n+i-k+1}, \qquad y_{n+1} - y_n = hv_n. \tag{10.38'}$$

Here the corresponding matrix S of (10.30) is stable.

b) Avoid the use of $v_n = (y_{n+1} - y_n)/h$ for the computation of the starting values $v_0, v_1, \ldots, v_{k-2}$, since the difference is a numerically unstable operation. Instead, add up the increments of the Runge-Kutta method, which you use for the computation of the starting values, directly.

These two ideas together then produce the "stabilized" results in Fig. 10.3 and Fig. 10.2.

Exercises

1. Compute the solution of Störmer's problem (10.18) with one of the methods of this section.

2. a) Show that the generating functions of the coefficients σ_i and σ_j^* (defined in (10.11) and (10.14))

$$S(t) = \sum_{j=0}^{\infty} \sigma_j t^j, \qquad S^*(t) = \sum_{j=0}^{\infty} \sigma_j^* t^j$$

satisfy

$$S(t) = \left(\frac{t}{\log(1-t)}\right)^2 \frac{1}{1-t}, \qquad S^*(t) = \left(\frac{t}{\log(1-t)}\right)^2.$$

b) Compute the coefficients d_j of

$$\sum_{j=0}^{\infty} d_j t^j = \left(\frac{\log(1-t)}{t}\right)^2 = \left(1 + \frac{t}{2} + \frac{t^2}{3} + \frac{t^3}{4} + \ldots\right)^2$$

and derive a recurrence relation for the σ_j and σ_j^*.

c) Prove that $\sigma_j^* = \sigma_j - \sigma_{j-1}$.

3. Let $\varrho(\zeta)$ be a polynomial of degree k which has 1 as root of multiplicity 2. Then there exists a unique $\sigma(\zeta)$ such that the corresponding method is of order $k+1$.

4. Consider the method (10.29) and, for simplicity, assume the differential equation to be a scalar one.

a) For any vector w in $\mathbf{R}^k$ the image vector Ew, with E given by (8.12), satisfies

$$Ew = \begin{pmatrix} \gamma \mathbb{1} \\ \kappa \mathbb{1} \end{pmatrix}$$

where γ, κ are real numbers and $\mathbb{1}$ is the vector with all elements equal to 1. The dimensions of $\gamma \mathbb{1}$ and $\kappa \mathbb{1}$ are l and m, respectively.

b) Verify that for $e_1 = (1, 0, \ldots, 0)^T$,

$$E \cdot \begin{pmatrix} \alpha e_1 \\ \beta e_1 \end{pmatrix} = \begin{pmatrix} (\alpha/\varrho_1'(1))\mathbb{1} \\ (\beta/\varrho_2'(1))\mathbb{1} \end{pmatrix}.$$

c) Show that

$$E \frac{\partial \Phi}{\partial u}(x, z(x, 0), 0) \begin{pmatrix} \gamma\mathbb{1} \\ \kappa\mathbb{1} \end{pmatrix} = \begin{pmatrix} (\sigma(1)/\varrho_1'(1))(\partial f/\partial y)(x, y(x))\kappa\mathbb{1} \\ (1/\varrho_2'(1))\gamma\mathbb{1} \end{pmatrix}.$$

Hint. With $Y_n = (y_{n+k-1}, \ldots, y_n)^T$ the formula (10.31) expresses ψ as a function of (x_n, Y_n, h). The second formula of (10.28) relates Y_n and u_n as

$$KY_n = Lu_n + \mathcal{O}(h) \qquad \text{where} \qquad K\mathbb{1} = L\begin{pmatrix} 0 \\ \mathbb{1} \end{pmatrix}$$

and K is invertible. Use the chain rule for the computation of $\partial\psi/\partial u$. See also Exercise 2 of Section III.4 and Exercise 1 of Section III.9.

5. Compute the error constant (10.37) for the methods (10.10) and (10.13).

Result. σ_k and σ_{k+1}^*, respectively.

Appendix. Fortran Codes

> ... but the software is in various states of development from experimental (a euphemism for badly written) to what we might call
> ...
> (C.W. Gear, in Aiken 1985)

Several Fortran codes have been developed for our numerical computations. Those of the first edition have been improved and several new options have been included, e.g., automatic choice of initial step size, stiffness detection, dense output. We have seen many of the ideas, which are incorporated in these codes, in the programs of P. Deuflhard, A.C. Hindmarsh and L.F. Shampine.

Experiences with all of our codes are welcome. The programs can be obtained from the authors' homepage (http://www.unige.ch/folks/hairer).

Address: Section de Mathématiques, Case postale 240,
CH-1211 Genève 24, Switzerland

E-mail: Ernst.Hairer@math.unige.ch   Gerhard.Wanner@math.unige.ch

Driver for the Code DOPRI5

The driver given here is for the differential equation (II.0.1) with initial values and x_{end} given in (II.0.2). This is the problem AREN of Section II.10. The subroutine FAREN ("F for AREN") computes the right-hand side of this differential equation. The subroutine SOLOUT ("Solution out"), which is called by DOPRI5 after every successful step, and the dense output routine CONTD5 are used to print the solution at equidistant points. The (optional) common block STATD5 gives statistical information after the call to DOPRI5. The common blocks COD5R and COD5I transfer the necessary information to CONTD5.

```
        IMPLICIT REAL*8 (A-H,O-Z)
        PARAMETER (NDGL=4,LWORK=8*NDGL+10,LIWORK=10)
        PARAMETER (NRDENS=2,LRCONT=5*NRDENS+2,LICONT=NRDENS+1)
        DIMENSION Y(NDGL),WORK(LWORK),IWORK(LIWORK)
        COMMON/STATD5/NFCN,NSTEP,NACCPT,NREJCT
        COMMON /COD5R/RCONT(LRCONT)
        COMMON /COD5I/ICONT(LICONT)
        EXTERNAL FAREN,SOLOUT
C --- DIMENSION OF THE SYSTEM
        N=NDGL
C --- OUTPUT ROUTINE (AND DENSE OUTPUT) IS USED DURING INTEGRATION
        IOUT=2
```

```
C --- INITIAL VALUES AND ENDPOINT OF INTEGRATION
      X=0.0D0
      Y(1)=0.994D0
      Y(2)=0.0D0
      Y(3)=0.0D0
      Y(4)=-2.00158510637908252240537862224D0
      XEND=17.0652165601579625588917206249D0
C --- REQUIRED (RELATIVE AND ABSOLUTE) TOLERANCE
      ITOL=0
      RTOL=1.0D-7
      ATOL=RTOL
C --- DEFAULT VALUES FOR PARAMETERS
      DO 10 I=1,10
      IWORK(I)=0
10    WORK(I)=0.D0
C --- DENSE OUTPUT IS USED FOR THE TWO POSITION COORDINATES 1 AND 2
      IWORK(5)=NRDENS
      ICONT(2)=1
      ICONT(3)=2
C --- CALL OF THE SUBROUTINE DOPRI5
      CALL DOPRI5(N,FAREN,X,Y,XEND,
     +                    RTOL,ATOL,ITOL,
     +                    SOLOUT,IOUT,
     +                    WORK,LWORK,IWORK,LIWORK,LRCONT,LICONT,IDID)
C --- PRINT FINAL SOLUTION
      WRITE (6,99) Y(1),Y(2)
99    FORMAT(1X,'X = XEND      Y =',2E18.10)
C --- PRINT STATISTICS
      WRITE (6,91) RTOL,NFCN,NSTEP,NACCPT,NREJCT
91    FORMAT('     tol=',D8.2,'  fcn=',I5,' step=',I4,
     +              ' accpt=',I4,' rejct=',I3)
      STOP
      END
C
      SUBROUTINE SOLOUT (NR,XOLD,X,Y,N,IRTRN)
C --- PRINTS SOLUTION AT EQUIDISTANT OUTPUT-POINTS BY USING "CONTD5"
      IMPLICIT REAL*8 (A-H,O-Z)
      DIMENSION Y(N)
      COMMON /INTERN/XOUT
      IF (NR.EQ.1) THEN
         WRITE (6,99) X,Y(1),Y(2),NR-1
         XOUT=X+2.0D0
      ELSE
10       CONTINUE
         IF (X.GE.XOUT) THEN
            WRITE (6,99) XOUT,CONTD5(1,XOUT),CONTD5(2,XOUT),NR-1
            XOUT=XOUT+2.0D0
            GOTO 10
         END IF
      END IF
99    FORMAT(1X,'X =',F6.2,'    Y =',2E18.10,'    NSTEP =',I4)
      RETURN
      END
C
      SUBROUTINE FAREN(N,X,Y,F)
C --- ARENSTORF ORBIT
      IMPLICIT REAL*8 (A-H,O-Z)
      DIMENSION Y(N),F(N)
      AMU=0.012277471D0
      AMUP=1.D0-AMU
```

```
      F(1)=Y(3)
      F(2)=Y(4)
      R1=(Y(1)+AMU)**2+Y(2)**2
      R1=R1*SQRT(R1)
      R2=(Y(1)-AMUP)**2+Y(2)**2
      R2=R2*SQRT(R2)
      F(3)=Y(1)+2*Y(4)-AMUP*(Y(1)+AMU)/R1-AMU*(Y(1)-AMUP)/R2
      F(4)=Y(2)-2*Y(3)-AMUP*Y(2)/R1-AMU*Y(2)/R2
      RETURN
      END
```

The result, obtained on an Apollo workstation, is the following:

```
X =   0.00    Y =  0.9940000000E+00    0.0000000000E+00    NSTEP =    0
X =   2.00    Y = -0.5798781411E+00    0.6090775251E+00    NSTEP =   60
X =   4.00    Y = -0.1983335270E+00    0.1137638086E+01    NSTEP =   73
X =   6.00    Y = -0.4735743943E+00    0.2239068118E+00    NSTEP =   91
X =   8.00    Y = -0.1174553350E+01   -0.2759466982E+00    NSTEP =  110
X =  10.00    Y = -0.8398073466E+00    0.4468302268E+00    NSTEP =  122
X =  12.00    Y =  0.1314712468E-01   -0.8385751499E+00    NSTEP =  145
X =  14.00    Y = -0.6031129504E+00   -0.9912598031E+00    NSTEP =  159
X =  16.00    Y =  0.2427110999E+00   -0.3899948833E+00    NSTEP =  177
X = XEND     Y =  0.9940021016E+00    0.8911185978E-05
     tol=0.10E-06    fcn= 1442 step= 240 accpt= 216 rejct= 22
```

Subroutine DOPRI5

Explicit Runge-Kutta code based on the method of Dormand & Prince (see Table 5.2 of Section II.5). It is provided with the step control algorithm of Section II.4 and the dense output of Section II.6.

```
      SUBROUTINE DOPRI5(N,FCN,X,Y,XEND,
     +                  RTOL,ATOL,ITOL,
     +                  SOLOUT,IOUT,
     +                  WORK,LWORK,IWORK,LIWORK,LRCONT,LICONT,IDID)
C ------------------------------------------------------------
C     NUMERICAL SOLUTION OF A SYSTEM OF FIRST ORDER
C     ORDINARY DIFFERENTIAL EQUATIONS  Y'=F(X,Y).
C     THIS IS AN EXPLICIT RUNGE-KUTTA METHOD OF ORDER (4)5
C     DUE TO DORMAND & PRINCE (WITH STEPSIZE CONTROL AND
C     DENSE OUTPUT).
C
C     AUTHORS: E. HAIRER AND G. WANNER
C              UNIVERSITE DE GENEVE, DEPT. DE MATHEMATIQUES
C              CH-1211 GENEVE 24, SWITZERLAND
C              E-MAIL:  HAIRER@ UNI2A.UNIGE.CH,  WANNER@ UNI2A.UNIGE.CH
C
C     THIS CODE IS DESCRIBED IN:
C         E. HAIRER, S.P. NORSETT AND G. WANNER, SOLVING ORDINARY
C         DIFFERENTIAL EQUATIONS I. NONSTIFF PROBLEMS. 2ND EDITION.
C         SPRINGER SERIES IN COMPUTATIONAL MATHEMATICS,
C         SPRINGER-VERLAG (1993)
C
```

```
C      VERSION OF OCTOBER 3, 1991
C
C      INPUT PARAMETERS
C      ----------------
C      N          DIMENSION OF THE SYSTEM
C
C      FCN        NAME (EXTERNAL) OF SUBROUTINE COMPUTING THE
C                 VALUE OF F(X,Y):
C                     SUBROUTINE FCN(N,X,Y,F)
C                     REAL*8 X,Y(N),F(N)
C                     F(1)=...    ETC.
C
C      X          INITIAL X-VALUE
C
C      Y(N)       INITIAL VALUES FOR Y
C
C      XEND       FINAL X-VALUE (XEND-X MAY BE POSITIVE OR NEGATIVE)
C
C      RTOL,ATOL  RELATIVE AND ABSOLUTE ERROR TOLERANCES. THEY
C                 CAN BE BOTH SCALARS OR ELSE BOTH VECTORS OF LENGTH N.
C
C      ITOL       SWITCH FOR RTOL AND ATOL:
C                     ITOL=0:  BOTH RTOL AND ATOL ARE SCALARS.
C                     THE CODE KEEPS, ROUGHLY, THE LOCAL ERROR OF
C                     Y(I) BELOW RTOL*ABS(Y(I))+ATOL
C                     ITOL=1:  BOTH RTOL AND ATOL ARE VECTORS.
C                     THE CODE KEEPS THE LOCAL ERROR OF Y(I) BELOW
C                     RTOL(I)*ABS(Y(I))+ATOL(I).
C
C      SOLOUT     NAME (EXTERNAL) OF SUBROUTINE PROVIDING THE
C                 NUMERICAL SOLUTION DURING INTEGRATION.
C                 IF IOUT.GE.1, IT IS CALLED AFTER EVERY SUCCESSFUL STEP.
C                 SUPPLY A DUMMY SUBROUTINE IF IOUT=0.
C                 IT MUST HAVE THE FORM
C                     SUBROUTINE SOLOUT (NR,XOLD,X,Y,N,IRTRN)
C                     REAL*8 X,Y(N)
C                     ....
C                 SOLOUT FURNISHES THE SOLUTION "Y" AT THE NR-TH
C                     GRID-POINT "X" (THEREBY THE INITIAL VALUE IS
C                     THE FIRST GRID-POINT).
C                 "XOLD" IS THE PRECEEDING GRID-POINT.
C                 "IRTRN" SERVES TO INTERRUPT THE INTEGRATION. IF IRTRN
C                     IS SET <0, DOPRI5 WILL RETURN TO THE CALLING PROGRAM.
C
C          -----  CONTINUOUS OUTPUT: -----
C                 DURING CALLS TO "SOLOUT", A CONTINUOUS SOLUTION
C                 FOR THE INTERVAL [XOLD,X] IS AVAILABLE THROUGH
C                 THE FUNCTION
C                     >>>     CONTD5(I,S)    <<<
C                 WHICH PROVIDES AN APPROXIMATION TO THE I-TH
C                 COMPONENT OF THE SOLUTION AT THE POINT S. THE VALUE
C                 S SHOULD LIE IN THE INTERVAL [XOLD,X].
C
C      IOUT       SWITCH FOR CALLING THE SUBROUTINE SOLOUT:
C                     IOUT=0:  SUBROUTINE IS NEVER CALLED
C                     IOUT=1:  SUBROUTINE IS USED FOR OUTPUT.
C                     IOUT=2:  DENSE OUTPUT IS PERFORMED IN SOLOUT
C                              (IN THIS CASE WORK(5) MUST BE SPECIFIED)
C
C      WORK       ARRAY OF WORKING SPACE OF LENGTH "LWORK".
```

```
C                         "LWORK" MUST BE AT LEAST  8*N+10
C
C       LWORK           DECLARED LENGHT OF ARRAY "WORK".
C
C       IWORK           INTEGER WORKING SPACE OF LENGHT "LIWORK".
C                       IWORK(1),...,IWORK(5) SERVE AS PARAMETERS
C                       FOR THE CODE. FOR STANDARD USE, SET THEM
C                       TO ZERO BEFORE CALLING.
C                       "LIWORK" MUST BE AT LEAST 10 .
C
C       LIWORK          DECLARED LENGHT OF ARRAY "IWORK".
C
C       LRCONT          DECLARED LENGTH OF COMMON BLOCK
C                           >>> COMMON /COD5R/RCONT(LRCONT)  <<<
C                       WHICH MUST BE DECLARED IN THE CALLING PROGRAM.
C                       "LRCONT" MUST BE AT LEAST
C                               5 * NRDENS + 2
C                       WHERE NRDENS=IWORK(5) (SEE BELOW).
C
C       LICONT          DECLARED LENGTH OF COMMON BLOCK
C                           >>> COMMON /COD5I/ICONT(LICONT)  <<<
C                       WHICH MUST BE DECLARED IN THE CALLING PROGRAM.
C                       "LICONT" MUST BE AT LEAST
C                               NRDENS + 1
C                       THESE COMMON BLOCKS ARE USED FOR STORING THE COEFFICIENTS
C                       OF THE CONTINUOUS SOLUTION AND MAKES THE CALLING LIST FOR
C                       THE FUNCTION "CONTD5" AS SIMPLE AS POSSIBLE.
C
C-------------------------------------------------------------------------
C
C       SOPHISTICATED SETTING OF PARAMETERS
C       -----------------------------------
C                       SEVERAL PARAMETERS (WORK(1),...,IWORK(1),...)  ALLOW
C                       TO ADAPT THE CODE TO THE PROBLEM AND TO THE NEEDS OF
C                       THE USER. FOR ZERO INPUT, THE CODE CHOOSES DEFAULT VALUES.
C
C       WORK(1)   UROUND, THE ROUNDING UNIT, DEFAULT 2.3D-16.
C
C       WORK(2)   THE SAFETY FACTOR IN STEP SIZE PREDICTION,
C                 DEFAULT 0.9D0.
C
C       WORK(3), WORK(4)   PARAMETERS FOR STEP SIZE SELECTION
C                 THE NEW STEP SIZE IS CHOSEN SUBJECT TO THE RESTRICTION
C                 WORK(3) <= HNEW/HOLD <= WORK(4)
C                 DEFAULT VALUES: WORK(3)=0.2D0, WORK(4)=10.D0
C
C       WORK(5)   IS THE "BETA" FOR STABILIZED STEP SIZE CONTROL
C                 (SEE SECTION IV.2).  LARGER VALUES OF BETA ( <= 0.1 )
C                 MAKE THE STEP SIZE CONTROL MORE STABLE. DOPRI5 NEEDS
C                 A LARGER BETA THAN HIGHAM & HALL. NEGATIVE WORK(5)
C                 PROVOKE BETA=0.
C                 DEFAULT 0.04D0.
C
C       WORK(6)   MAXIMAL STEP SIZE, DEFAULT XEND-X.
C
C       WORK(7)   INITIAL STEP SIZE, FOR WORK(7)=0.D0 AN INITIAL GUESS
C                 IS COMPUTED WITH HELP OF THE FUNCTION HINIT
C
C       IWORK(1)  THIS IS THE MAXIMAL NUMBER OF ALLOWED STEPS.
C                 THE DEFAULT VALUE (FOR IWORK(1)=0) IS 100000.
```

```
C
C     IWORK(2)   SWITCH FOR THE CHOICE OF THE COEFFICIENTS
C                IF IWORK(2).EQ.1  METHOD DOPRI5 OF DORMAND AND PRINCE
C                (TABLE 5.2 OF SECTION II.5).
C                AT THE MOMENT THIS IS THE ONLY POSSIBLE CHOICE.
C                THE DEFAULT VALUE (FOR IWORK(2)=0) IS IWORK(2)=1.
C
C     IWORK(3)   SWITCH FOR PRINTING ERROR MESSAGES
C                IF IWORK(3).LT.0 NO MESSAGES ARE BEING PRINTED
C                IF IWORK(3).GT.0 MESSAGES ARE PRINTED WITH
C                WRITE (IWORK(3),*) ...
C                DEFAULT VALUE (FOR IWORK(3)=0) IS IWORK(3)=6
C
C     IWORK(4)   TEST FOR STIFFNESS IS ACTIVATED AFTER STEP NUMBER
C                J*IWORK(4) (J INTEGER), PROVIDED IWORK(4).GT.0.
C                FOR NEGATIVE IWORK(4) THE STIFFNESS TEST IS
C                NEVER ACTIVATED; DEFAULT VALUE IS IWORK(4)=1000
C
C     IWORK(5)   = NRDENS = NUMBER OF COMPONENTS, FOR WHICH DENSE OUTPUT
C                IS REQUIRED; DEFAULT VALUE IS IWORK(5)=0;
C                FOR   0 < NRDENS < N   THE COMPONENTS (FOR WHICH DENSE
C                OUTPUT IS REQUIRED) HAVE TO BE SPECIFIED IN
C                ICONT(2),...,ICONT(NRDENS+1);
C                FOR  NRDENS=N  THIS IS DONE BY THE CODE.
C
C-----------------------------------------------------------------------
C
C     OUTPUT PARAMETERS
C     -----------------
C
C     X          X-VALUE FOR WHICH THE SOLUTION HAS BEEN COMPUTED
C                (AFTER SUCCESSFUL RETURN X=XEND).
C
C     Y(N)       NUMERICAL SOLUTION AT X
C
C     H          PREDICTED STEP SIZE OF THE LAST ACCEPTED STEP
C
C     IDID       REPORTS ON SUCCESSFULNESS UPON RETURN:
C                  IDID= 1  COMPUTATION SUCCESSFUL,
C                  IDID= 2  COMPUT. SUCCESSFUL (INTERRUPTED BY SOLOUT)
C                  IDID=-1  INPUT IS NOT CONSISTENT,
C                  IDID=-2  LARGER NMAX IS NEEDED,
C                  IDID=-3  STEP SIZE BECOMES TOO SMALL.
C                  IDID=-4  PROBLEM IS PROBABLY STIFF (INTERRUPTED).
C
C-----------------------------------------------------------------------
C *** *** *** *** *** *** *** *** *** *** *** *** ***
C          DECLARATIONS
C *** *** *** *** *** *** *** *** *** *** *** *** ***
      IMPLICIT REAL*8 (A-H,O-Z)
      DIMENSION Y(N),ATOL(1),RTOL(1),WORK(LWORK),IWORK(LIWORK)
      LOGICAL ARRET
      EXTERNAL FCN,SOLOUT
      COMMON/STATD5/NFCN,NSTEP,NACCPT,NREJCT
C --- COMMON STATD5 CAN BE INSPECTED FOR STATISTICAL PURPOSES:
C ---    NFCN     NUMBER OF FUNCTION EVALUATIONS
C ---    NSTEP    NUMBER OF COMPUTED STEPS
C ---    NACCPT   NUMBER OF ACCEPTED STEPS
C ---    NREJCT   NUMBER OF REJECTED STEPS (AFTER AT LEAST ONE STEP
C                 HAS BEEN ACCEPTED)
      .........
```

Subroutine DOP853

Explicit Runge-Kutta code of order 8 based on the method of Dormand & Prince, described in Section II.5. The local error estimation and the step size control is based on embedded formulas or orders 5 and 3 (see Section II.10). This method is provided with a dense output of order 7. In the following description we have omitted the parts which are identical to those for DOPRI5.

```
      SUBROUTINE DOP853(N,FCN,X,Y,XEND,
     +                  RTOL,ATOL,ITOL,
     +                  SOLOUT,IOUT,
     +                  WORK,LWORK,IWORK,LIWORK,LRCONT,LICONT,IDID)
C -----------------------------------------------------------
C     NUMERICAL SOLUTION OF A SYSTEM OF FIRST ORDER
C     ORDINARY DIFFERENTIAL EQUATIONS  Y'=F(X,Y).
C     THIS IS AN EXPLICIT RUNGE-KUTTA METHOD OF ORDER 8(5,3)
C     DUE TO DORMAND & PRINCE (WITH STEPSIZE CONTROL AND
C     DENSE OUTPUT)
C .........
C
C     VERSION OF NOVEMBER 29, 1992
C .........
C       ----- CONTINUOUS OUTPUT: -----
C              DURING CALLS TO "SOLOUT", A CONTINUOUS SOLUTION
C              FOR THE INTERVAL [XOLD,X] IS AVAILABLE THROUGH
C              THE FUNCTION
C                      >>>   CONTD8(I,S)   <<<
C              WHICH PROVIDES AN APPROXIMATION TO THE I-TH
C .........
C
C     WORK       ARRAY OF WORKING SPACE OF LENGTH "LWORK".
C                "LWORK" MUST BE AT LEAST   11*N+10
C .........
C
C     LRCONT     DECLARED LENGTH OF COMMON BLOCK
C                   >>>   COMMON /COD8R/RCONT(LRCONT)   <<<
C                WHICH MUST BE DECLARED IN THE CALLING PROGRAM.
C                "LRCONT" MUST BE AT LEAST
C                            8 * NRDENS + 2
C                WHERE NRDENS=IWORK(5) (SEE BELOW).
C
C     LICONT     DECLARED LENGTH OF COMMON BLOCK
C                   >>>   COMMON /COD8I/ICONT(LICONT)   <<<
C                WHICH MUST BE DECLARED IN THE CALLING PROGRAM.
C                "LICONT" MUST BE AT LEAST
C                            NRDENS + 1
C                THESE COMMON BLOCKS ARE USED FOR STORING THE COEFFICIENTS
C                OF THE CONTINUOUS SOLUTION AND MAKES THE CALLING LIST FOR
C                THE FUNCTION "CONTD8" AS SIMPLE AS POSSIBLE.
C .........
C
C     WORK(3), WORK(4)   PARAMETERS FOR STEP SIZE SELECTION
C                THE NEW STEP SIZE IS CHOSEN SUBJECT TO THE RESTRICTION
C                    WORK(3) <= HNEW/HOLD <= WORK(4)
C                DEFAULT VALUES: WORK(3)=0.333D0, WORK(4)=6.D0
C .........
```

Subroutine ODEX

Extrapolation code for $y' = f(x,y)$, based on the GBS algorithm (Section II.9). It uses variable order and variable step sizes and is provided with a high-order dense output. Again, the missing parts in the description are identical to those of DOPRI5.

```
      SUBROUTINE ODEX(N,FCN,X,Y,XEND,H,
     +                RTOL,ATOL,ITOL,
     +                SOLOUT,IOUT,
     +                WORK,LWORK,IWORK,LIWORK,LRCONT,LICONT,IDID)
C ----------------------------------------------------------------
C     NUMERICAL SOLUTION OF A SYSTEM OF FIRST ORDER
C     ORDINARY DIFFERENTIAL EQUATIONS  Y'=F(X,Y).
C     THIS IS AN EXTRAPOLATION-ALGORITHM (GBS), BASED ON THE
C     EXPLICIT MIDPOINT RULE (WITH STEPSIZE CONTROL,
C     ORDER SELECTION AND DENSE OUTPUT).
C
C     AUTHORS: E. HAIRER AND G. WANNER
C              UNIVERSITE DE GENEVE, DEPT. DE MATHEMATIQUES
C              CH-1211 GENEVE 24, SWITZERLAND
C              E-MAIL:  HAIRER@ UNI2A.UNIGE.CH,  WANNER@ UNI2A.UNIGE.CH
C              DENSE OUTPUT WRITTEN BY E. HAIRER AND A. OSTERMANN
C .........
C
C     VERSION DECEMBER 18, 1991
C .........
C
C     H          INITIAL STEP SIZE GUESS;
C                H=1.D0/(NORM OF F'), USUALLY 1.D-1 OR 1.D-3, IS GOOD.
C                THIS CHOICE IS NOT VERY IMPORTANT, THE CODE QUICKLY
C                ADAPTS ITS STEP SIZE. WHEN YOU ARE NOT SURE, THEN
C                STUDY THE CHOSEN VALUES FOR A FEW
C                STEPS IN SUBROUTINE "SOLOUT".
C                (IF H=0.D0, THE CODE PUTS H=1.D-4).
C .........
C
C              ----- CONTINUOUS OUTPUT (IF IOUT=2):  -----
C                DURING CALLS TO "SOLOUT", A CONTINUOUS SOLUTION
C                FOR THE INTERVAL [XOLD,X] IS AVAILABLE THROUGH
C                THE REAL*8 FUNCTION
C                   >>>   CONTEX(I,S)   <<<
C                WHICH PROVIDES AN APPROXIMATION TO THE I-TH
C                COMPONENT OF THE SOLUTION AT THE POINT S. THE VALUE
C                S SHOULD LIE IN THE INTERVAL [XOLD,X].
C .........
C
C     WORK       ARRAY OF WORKING SPACE OF LENGTH "LWORK".
C                SERVES AS WORKING SPACE FOR ALL VECTORS.
C                "LWORK" MUST BE AT LEAST
C                   N*(KM+5)+5*KM+10+2*KM*(KM+1)*NRDENS
C                WHERE NRDENS=IWORK(8) (SEE BELOW) AND
C                   KM=9               IF IWORK(2)=0
C                   KM=IWORK(2)        IF IWORK(2).GT.0
C                WORK(1),...,WORK(10) SERVE AS PARAMETERS
C                FOR THE CODE. FOR STANDARD USE, SET THESE
C                PARAMETERS TO ZERO BEFORE CALLING.
C .........
```

```
C
C      IWORK         INTEGER WORKING SPACE OF LENGTH "LIWORK".
C                    "LIWORK" MUST BE AT LEAST
C                              2*KM+10+NRDENS
C                    IWORK(1),...,IWORK(9) SERVE AS PARAMETERS
C                    FOR THE CODE. FOR STANDARD USE, SET THESE
C                    PARAMETERS TO ZERO BEFORE CALLING.
C .........
C
C      LRCONT        DECLARED LENGTH OF COMMON BLOCK
C                    >>>   COMMON /CONTR/RCONT(LRCONT)   <<<
C                    WHICH MUST BE DECLARED IN THE CALLING PROGRAM.
C                    "LRCONT" MUST BE AT LEAST
C                              ( 2 * KM + 5 ) * NRDENS + 2
C                    WHERE KM=IWORK(2) AND NRDENS=IWORK(8) (SEE BELOW).
C
C      LICONT        DECLARED LENGTH OF COMMON BLOCK
C                    >>>   COMMON /CONTI/ICONT(LICONT)   <<<
C                    WHICH MUST BE DECLARED IN THE CALLING PROGRAM.
C                    "LICONT" MUST BE AT LEAST
C                              NRDENS + 2
C                    THESE COMMON BLOCKS ARE USED FOR STORING THE COEFFICIENTS
C                    OF THE CONTINUOUS SOLUTION AND MAKES THE CALLING LIST FOR
C                    THE FUNCTION "CONTEX" AS SIMPLE AS POSSIBLE.
C .........
C
C   WORK(2)   MAXIMAL STEP SIZE, DEFAULT XEND-X.
C
C   WORK(3)   STEP SIZE IS REDUCED BY FACTOR WORK(3), IF THE
C             STABILITY CHECK IS NEGATIVE, DEFAULT 0.5.
C
C   WORK(4), WORK(5)   PARAMETERS FOR STEP SIZE SELECTION
C             THE NEW STEP SIZE FOR THE J-TH DIAGONAL ENTRY IS
C             CHOSEN SUBJECT TO THE RESTRICTION
C                 FACMIN/WORK(5) <= HNEW(J)/HOLD <= 1/FACMIN
C             WHERE FACMIN=WORK(4)**(1/(2*J-1))
C             DEFAULT VALUES: WORK(4)=0.02D0, WORK(5)=4.D0
C
C   WORK(6), WORK(7)   PARAMETERS FOR THE ORDER SELECTION
C             STEP SIZE IS DECREASED IF    W(K-1) <= W(K)*WORK(6)
C             STEP SIZE IS INCREASED IF    W(K) <= W(K-1)*WORK(7)
C             DEFAULT VALUES: WORK(6)=0.8D0, WORK(7)=0.9D0
C
C   WORK(8), WORK(9)   SAFETY FACTORS FOR STEP CONTROL ALGORITHM
C             HNEW=H*WORK(9)*(WORK(8)*TOL/ERR)**(1/(J-1))
C             DEFAULT VALUES: WORK(8)=0.65D0,
C                    WORK(9)=0.94D0  IF "HOPE FOR CONVERGENCE"
C                    WORK(9)=0.90D0  IF "NO HOPE FOR CONVERGENCE"
C .........
C
C   IWORK(2)   THE MAXIMUM NUMBER OF COLUMNS IN THE EXTRAPOLATION
C             TABLE. THE DEFAULT VALUE (FOR IWORK(2)=0) IS 9.
C             IF IWORK(2).NE.0 THEN IWORK(2) SHOULD BE .GE.3.
C
C   IWORK(3)   SWITCH FOR THE STEP SIZE SEQUENCE (EVEN NUMBERS ONLY)
C             IF IWORK(3).EQ.1 THEN 2,4,6,8,10,12,14,16,...
C             IF IWORK(3).EQ.2 THEN 2,4,8,12,16,20,24,28,...
C             IF IWORK(3).EQ.3 THEN 2,4,6,8,12,16,24,32,...
C             IF IWORK(3).EQ.4 THEN 2,6,10,14,18,22,26,30,...
C             IF IWORK(3).EQ.5 THEN 4,8,12,16,20,24,28,32,...
```

```
C                     THE DEFAULT VALUE IS IWORK(3)=1 IF IOUT.LE.1;
C                     THE DEFAULT VALUE IS IWORK(3)=4 IF IOUT.GE.2.
C
C     IWORK(4)  STABILITY CHECK IS ACTIVATED AT MOST IWORK(4) TIMES IN
C               ONE LINE OF THE EXTRAP. TABLE, DEFAULT IWORK(4)=1.
C
C     IWORK(5)  STABILITY CHECK IS ACTIVATED ONLY IN THE LINES
C               1 TO IWORK(5) OF THE EXTRAP. TABLE, DEFAULT IWORK(5)=1.
C
C     IWORK(6)  IF  IWORK(6)=0  ERROR ESTIMATOR IN THE DENSE
C               OUTPUT FORMULA IS ACTIVATED. IT CAN BE SUPPRESSED
C               BY PUTTING IWORK(6)=1.
C               DEFAULT IWORK(6)=0  (IF IOUT.GE.2).
C
C     IWORK(7)  DETERMINES THE DEGREE OF INTERPOLATION FORMULA
C               MU = 2 * KAPPA - IWORK(7) + 1
C               IWORK(7) SHOULD LIE BETWEEN 1 AND 6
C               DEFAULT IWORK(7)=4  (IF IWORK(7)=0).
C
C     IWORK(8)  = NRDENS = NUMBER OF COMPONENTS, FOR WHICH DENSE OUTPUT
C               IS REQUIRED
C
C     IWORK(10),...,IWORK(NRDENS+9) INDICATE THE COMPONENTS, FOR WHICH
C               DENSE OUTPUT IS REQUIRED
C .........
C
C     IDID      REPORTS ON SUCCESSFULNESS UPON RETURN:
C                   IDID=1  COMPUTATION SUCCESSFUL,
C                   IDID=-1 COMPUTATION UNSUCCESSFUL.
C .........
```

Subroutine ODEX2

Extrapolation code for second order differential equations $y'' = f(x, y)$ (Section II.14). It uses variable order and variable step sizes and is provided with a high-order dense output. The missing parts of the description are identical to those of ODEX.

```
      SUBROUTINE ODEX2(N,FCN,X,Y,YP,XEND,H,
     +                 RTOL,ATOL,ITOL,
     +                 SOLOUT,IOUT,
     +                 WORK,LWORK,IWORK,LIWORK,LRCONT,LICONT,IDID)
C ------------------------------------------------------------
C     NUMERICAL SOLUTION OF A SYSTEM OF SECOND ORDER
C     ORDINARY DIFFERENTIAL EQUATIONS  Y''=F(X,Y).
C     THIS IS AN EXTRAPOLATION-ALGORITHM, BASED ON
C     THE STOERMER RULE (WITH STEPSIZE CONTROL
C     ORDER SELECTION AND DENSE OUTPUT).
C .........
C
C     VERSION MARCH 30, 1992
C .........
C
C     Y(N)          INITIAL VALUES FOR Y
C
```

```
C     YP(N)        INITIAL VALUES FOR Y'
.........
C
C     ITOL         SWITCH FOR RTOL AND ATOL:
C                     ITOL=0:  BOTH RTOL AND ATOL ARE SCALARS.
C                     THE CODE KEEPS, ROUGHLY, THE LOCAL ERROR OF
C                     Y(I)  BELOW  RTOL*ABS(Y(I))+ATOL
C                     YP(I) BELOW  RTOL*ABS(YP(I))+ATOL
C                     ITOL=1:  BOTH RTOL AND ATOL ARE VECTORS.
C                     THE CODE KEEPS THE LOCAL ERROR OF
C                     Y(I)  BELOW  RTOL(I)*ABS(Y(I))+ATOL(I).
C                     YP(I) BELOW  RTOL(I+N)*ABS(YP(I))+ATOL(I+N).
C
C     SOLOUT       NAME (EXTERNAL) OF SUBROUTINE PROVIDING THE
C                  NUMERICAL SOLUTION DURING INTEGRATION.
C                  IF IOUT>=1, IT IS CALLED AFTER EVERY SUCCESSFUL STEP.
C                  SUPPLY A DUMMY SUBROUTINE IF IOUT=0.
C                  IT MUST HAVE THE FORM
C                     SUBROUTINE SOLOUT (NR,XOLD,X,Y,YP,N,IRTRN)
C                     REAL*8 X,Y(N),YP(N)
C                     ....
C                  SOLOUT FURNISHES THE SOLUTIONS "Y, YP" AT THE NR-TH
C                  GRID-POINT "X" (THEREBY THE INITIAL VALUE IS
C                  THE FIRST GRID-POINT).
C                  "XOLD" IS THE PRECEEDING GRID-POINT.
C                  "IRTRN" SERVES TO INTERRUPT THE INTEGRATION. IF IRTRN
C                     IS SET <0, ODEX2 WILL RETURN TO THE CALLING PROGRAM.
C
C           -----  CONTINUOUS OUTPUT (IF IOUT=2):  -----
C                  DURING CALLS TO "SOLOUT", A CONTINUOUS SOLUTION
C                  FOR THE INTERVAL [XOLD,X] IS AVAILABLE THROUGH
C                  THE REAL*8 FUNCTION
C                         >>>   CONTX2(I,S)   <<<
C                  WHICH PROVIDES AN APPROXIMATION TO THE I-TH
C                  COMPONENT OF THE SOLUTION AT THE POINT S. THE VALUE
C                  S SHOULD LIE IN THE INTERVAL [XOLD,X].
.........
C
C     WORK         ARRAY OF WORKING SPACE OF LENGTH "LWORK".
C                  SERVES AS WORKING SPACE FOR ALL VECTORS.
C                  "LWORK" MUST BE AT LEAST
C                     N*(2*KM+6)+5*KM+10+KM*(2*KM+3)*NRDENS
C                  WHERE NRDENS=IWORK(8) (SEE BELOW) AND
C                     KM=9            IF IWORK(2)=0
C                     KM=IWORK(2)     IF IWORK(2).GT.0
C                  WORK(1),...,WORK(10) SERVE AS PARAMETERS
C                  FOR THE CODE. FOR STANDARD USE, SET THESE
C                  PARAMETERS TO ZERO BEFORE CALLING.
.........
C
C     IWORK        INTEGER WORKING SPACE OF LENGTH "LIWORK".
C                  "LIWORK" MUST BE AT LEAST
C                            KM+9+NRDENS
C                  IWORK(1),...,IWORK(9) SERVE AS PARAMETERS
C                  FOR THE CODE. FOR STANDARD USE, SET THESE
C                  PARAMETERS TO ZERO BEFORE CALLING.
.........
C
C     LRCONT       DECLARED LENGTH OF COMMON BLOCK
C                     >>>  COMMON /CONTR2/RCONT(LRCONT)  <<<
```

```
C                     WHICH MUST BE DECLARED IN THE CALLING PROGRAM.
C                     "LRCONT" MUST BE AT LEAST
C                              ( 2 * KM + 6 ) * NRDENS + 2
C                     WHERE KM=IWORK(2) AND NRDENS=IWORK(8) (SEE BELOW).
C
C      LICONT         DECLARED LENGTH OF COMMON BLOCK
C                       >>>  COMMON /CONTI2/ICONT(LICONT)  <<<
C                     WHICH MUST BE DECLARED IN THE CALLING PROGRAM.
C                     "LICONT" MUST BE AT LEAST
C                                   NRDENS + 2
C                     THESE COMMON BLOCKS ARE USED FOR STORING THE COEFFICIENTS
C                     OF THE CONTINUOUS SOLUTION AND MAKES THE CALLING LIST FOR
C                     THE FUNCTION "CONTX2" AS SIMPLE AS POSSIBLE.
C .........
C
C      WORK(3)        STEP SIZE IS REDUCED BY FACTOR WORK(3), IF DURING THE
C                     COMPUTATION OF THE EXTRAPOLATION TABLEAU DIVERGENCE
C                     IS OBSERVED; DEFAULT 0.5.
C .........
C
C      IWORK(3)       SWITCH FOR THE STEP SIZE SEQUENCE (EVEN NUMBERS ONLY)
C                     IF IWORK(3).EQ.1 THEN 2,4,6,8,10,12,14,16,...
C                     IF IWORK(3).EQ.2 THEN 2,4,8,12,16,20,24,28,...
C                     IF IWORK(3).EQ.3 THEN 2,4,6,8,12,16,24,32,...
C                     IF IWORK(3).EQ.4 THEN 2,6,10,14,18,22,26,30,...
C                     THE DEFAULT VALUE IS IWORK(3)=1 IF IOUT.LE.1;
C                     THE DEFAULT VALUE IS IWORK(3)=4 IF IOUT.GE.2.
C .........
C
C      IWORK(7)       DETERMINES THE DEGREE OF INTERPOLATION FORMULA
C                     MU = 2 * KAPPA - IWORK(7) + 1
C                     IWORK(7) SHOULD LIE BETWEEN 1 AND 8
C                     DEFAULT IWORK(7)=6  (IF IWORK(7)=0).
C .........
```

Driver for the Code RETARD

We consider the delay equation (II.17.14) with initial values and initial functions given there. This is a 3-dimensional problem, but only the second component is used with retarded argument (hence NRDENS=1). We require that the points $1, 2, 3, \ldots, 9, 10, 20$ (points of discontinuity of the derivatives of the solution) are hitten exactly by the integration routine.

```
        IMPLICIT REAL*8 (A-H,O-Z)
        PARAMETER (NDGL=3,NGRID=11,LWORK=8*NDGL+11+NGRID,LIWORK=10)
        PARAMETER (NRDENS=1,LRCONT=500,LICONT=NRDENS+1)
        DIMENSION Y(NDGL),WORK(LWORK),IWORK(LIWORK)
        COMMON/STATRE/NFCN,NSTEP,NACCPT,NREJCT
        COMMON /CORER/RCONT(LRCONT)
        COMMON /COREI/ICONT(LICONT)
        EXTERNAL FCN,SOLOUT
C --- DIMENSION OF THE SYSTEM
        N=NDGL
C --- OUTPUT ROUTINE IS USED DURING INTEGRATION
```

```
        IOUT=1
C --- INITIAL VALUES AND ENDPOINT OF INTEGRATION
        X=0.0D0
        Y(1)=5.0D0
        Y(2)=0.1D0
        Y(3)=1.0D0
        XEND=40.D0
C --- REQUIRED (RELATIVE AND ABSOLUTE) TOLERANCE
        ITOL=0
        RTOL=1.0D-5
        ATOL=RTOL
C --- DEFAULT VALUES FOR PARAMETERS
        DO 10 I=1,10
        IWORK(I)=0
   10   WORK(I)=0.D0
C --- SECOND COMPONENT USES RETARDED ARGUMENT
        IWORK(5)=NRDENS
        ICONT(2)=2
C --- USE AS GRID-POINTS
        IWORK(6)=NGRID
        DO 12 I=1,NGRID-1
   12   WORK(10+I)=I
        WORK(10+NGRID)=20.D0
C --- CALL OF THE SUBROUTINE RETARD
        CALL RETARD(N,FCN,X,Y,XEND,
     +              RTOL,ATOL,ITOL,
     +              SOLOUT,IOUT,
     +              WORK,LWORK,IWORK,LIWORK,LRCONT,LICONT,IDID)
C --- PRINT FINAL SOLUTION
        WRITE (6,99) Y(1),Y(2),Y(3)
   99   FORMAT(1X,'X = XEND    Y =',3E18.10)
C --- PRINT STATISTICS
        WRITE (6,91) RTOL,NFCN,NSTEP,NACCPT,NREJCT
   91   FORMAT('    tol=',D8.2,'   fcn=',I5,' step=',I4,
     +          ' accpt=',I4,' rejct=',I3)
        STOP
        END
C
C
        SUBROUTINE SOLOUT (NR,XOLD,X,Y,N,IRTRN)
C --- PRINTS SOLUTION AT EQUIDISTANT OUTPUT-POINTS
        IMPLICIT REAL*8 (A-H,O-Z)
        DIMENSION Y(N)
        EXTERNAL PHI
        COMMON /INTERN/XOUT
        IF (NR.EQ.1) THEN
           WRITE (6,99) X,Y(1),NR-1
           XOUT=X+5.D0
        ELSE
   10      CONTINUE
           IF (X.GE.XOUT) THEN
              WRITE (6,99) X,Y(1),NR-1
              XOUT=XOUT+5.D0
              GOTO 10
           END IF
        END IF
   99   FORMAT(1X,'X =',F6.2,'    Y =',E18.10,'    NSTEP =',I4)
        RETURN
        END
C
```

```
          SUBROUTINE FCN(N,X,Y,F)
          IMPLICIT REAL*8 (A-H,O-Z)
          DIMENSION Y(N),F(N)
          EXTERNAL PHI
          Y2L1=YLAG(2,X-1.D0,PHI)
          Y2L10=YLAG(2,X-10.D0,PHI)
          F(1)=-Y(1)*Y2L1+Y2L10
          F(2)=Y(1)*Y2L1-Y(2)
          F(3)=Y(2)-Y2L10
          RETURN
          END
  C
          FUNCTION PHI(I,X)
          IMPLICIT REAL*8 (A-H,O-Z)
          IF (I.EQ.2) PHI=0.1D0
          RETURN
          END
```

The result, obtained on an Apollo workstation, is the following:

```
X =   0.00    Y =  0.5000000000E+01    NSTEP =   0
X =   5.00    Y =  0.2533855892E+00    NSTEP =  18
X =  10.00    Y =  0.3328560326E+00    NSTEP =  32
X =  15.29    Y =  0.4539376456E+01    NSTEP =  40
X =  20.00    Y =  0.1706635702E+00    NSTEP =  52
X =  25.22    Y =  0.2524799457E+00    NSTEP =  62
X =  30.48    Y =  0.5134266860E+01    NSTEP =  68
X =  35.10    Y =  0.3610797907E+00    NSTEP =  78
X =  40.00    Y =  0.9125544555E-01    NSTEP =  89
X = XEND     Y =  0.9125544555E-01    0.2029882456E-01   0.5988445730E+01
     tol=0.10E-04    fcn= 586 step= 97 accpt=  89 rejct=  8
```

Subroutine RETARD

Modification of the code DOPRI5 for delay differential equations (see Section II.17). The missing parts of the description are identical to those of DOPRI5.

```
        SUBROUTINE RETARD(N,FCN,X,Y,XEND,
       +                  RTOL,ATOL,ITOL,
       +                  SOLOUT,IOUT,
       +                  WORK,LWORK,IWORK,LIWORK,LRCONT,LICONT,IDID)
C ------------------------------------------------------------
C    NUMERICAL SOLUTION OF A SYSTEM OF FIRST ORDER DELAY
C    ORDINARY DIFFERENTIAL EQUATIONS  Y'(X)=F(X,Y(X),Y(X-A),...).
C    THIS CODE IS BASED ON AN EXPLICIT RUNGE-KUTTA METHOD OF
C    ORDER (4)5 DUE TO DORMAND & PRINCE (WITH STEPSIZE CONTROL
C    AND DENSE OUTPUT).
C .........
C
C    VERSION OF APRIL 24, 1992
C .........
C
C    FCN         NAME (EXTERNAL) OF SUBROUTINE COMPUTING THE RIGHT-
C                HAND-SIDE OF THE DELAY EQUATION, E.G.,
```

```
C                        SUBROUTINE FCN(N,X,Y,F)
C                        REAL*8 X,Y(N),F(N)
C                        EXTERNAL PHI
C                        F(1)=(1.4D0-YLAG(1,X-1.D0,PHI))*Y(1)
C                        F(2)=...      ETC.
C                    FOR AN EXPLICATION OF YLAG SEE BELOW.
C                    DO NOT USE YLAG(I,X-0.D0,PHI) !
C                    THE INITIAL FUNCTION HAS TO BE SUPPLIED BY:
C                        FUNCTION PHI(I,X)
C                        REAL*8 PHI,X
C                    WHERE I IS THE COMPONENT AND X THE ARGUMENT
C .........
C
C       Y(N)        INITIAL VALUES FOR Y (MAY BE DIFFERENT FROM PHI (I,X),
C                   IN THIS CASE IT IS HIGHLY RECOMMENDED TO SET IWORK(6)
C                   AND WORK(11),..., SEE BELOW)
C .........
C
C           -----  CONTINUOUS OUTPUT: -----
C                  DURING CALLS TO "SOLOUT" AS WELL AS TO "FCN", A
C                  CONTINUOUS SOLUTION IS AVAILABLE THROUGH THE FUNCTION
C                          >>>   YLAG(I,S;PHI)   <<<
C                  WHICH PROVIDES AN APPROXIMATION TO THE I-TH
C                  COMPONENT OF THE SOLUTION AT THE POINT S. THE VALUE S
C                  HAS TO LIE IN AN INTERVAL WHERE THE NUMERICAL SOLUTION
C                  IS ALREADY COMPUTED. IT DEPENDS ON THE SIZE OF LRCONT
C                  (SEE BELOW) HOW FAR BACK THE SOLUTION IS AVAILABLE.
C
C       IOUT        SWITCH FOR CALLING THE SUBROUTINE SOLOUT:
C                     IOUT=0:  SUBROUTINE IS NEVER CALLED
C                     IOUT=1:  SUBROUTINE IS USED FOR OUTPUT.
C
C       WORK        ARRAY OF WORKING SPACE OF LENGTH "LWORK".
C                   "LWORK" MUST BE AT LEAST   8*N+11+NGRID
C                   WHERE NGRID=IWORK(6)
C .........
C
C       LRCONT      DECLARED LENGTH OF COMMON BLOCK
C                        >>>   COMMON /CORER/RCONT(LRCONT)   <<<
C                   WHICH MUST BE DECLARED IN THE CALLING PROGRAM.
C                   "LRCONT" MUST BE SUFFICIENTLY LARGE. IF THE DENSE
C                   OUTPUT OF MXST BACK STEPS HAS TO BE STORED, IT MUST
C                   BE AT LEAST
C                            MXST * ( 5 * NRDENS + 2 )
C                   WHERE NRDENS=IWORK(5) (SEE BELOW).
C
C       LICONT      DECLARED LENGTH OF COMMON BLOCK
C                        >>>   COMMON /COREI/ICONT(LICONT)   <<<
C                   WHICH MUST BE DECLARED IN THE CALLING PROGRAM.
C                   "LICONT" MUST BE AT LEAST
C                             NRDENS + 1
C                   THESE COMMON BLOCKS ARE USED FOR STORING THE COEFFICIENTS
C                   OF THE CONTINUOUS SOLUTION AND MAKES THE CALLING LIST FOR
C                   THE FUNCTION "CONTD5" AS SIMPLE AS POSSIBLE.
C .........
C
C   WORK(11),...,WORK(10+NGRID)  PRESCRIBED POINTS, WHICH THE
C              INTEGRATION METHOD HAS TO TAKE AS GRID-POINTS
C              X < WORK(11) < WORK(12) < ...  < WORK(10+NGRID) <= XEND
C .........
```

```
C
C     IWORK(5)   = NRDENS = NUMBER OF COMPONENTS, FOR WHICH DENSE OUTPUT
C                IS REQUIRED (EITHER BY "SOLOUT" OR BY "FCN");
C                DEFAULT VALUE (FOR IWORK(5)=0) IS IWORK(5)=N;
C                FOR   0 < NRDENS < N   THE COMPONENTS (FOR WHICH DENSE
C                OUTPUT IS REQUIRED) HAVE TO BE SPECIFIED IN
C                ICONT(2),...,ICONT(NRDENS+1);
C                FOR  NRDENS=N  THIS IS DONE BY THE CODE.
C
C     IWORK(6)   = NGRID = NUMBER OF PRESCRIBED POINTS IN THE
C                INTEGRATION INTERVAL WHICH HAVE TO BE GRID-POINTS
C                IN THE INTEGRATION. USUALLY, AT THESE POINTS THE
C                SOLUTION OR ONE OF ITS DERIVATIVE HAS A DISCONTINUITY.
C                DEFINE THESE POINTS IN WORK(11),...,WORK(10+NGRID)
C                DEFAULT VALUE:  IWORK(6)=0
C .........
C
C     IDID       REPORTS ON SUCCESSFULNESS UPON RETURN:
C                IDID= 1  COMPUTATION SUCCESSFUL,
C                IDID= 2  COMPUT. SUCCESSFUL (INTERRUPTED BY SOLOUT)
C                IDID=-1  INPUT IS NOT CONSISTENT,
C                IDID=-2  LARGER NMAX IS NEEDED,
C                IDID=-3  STEP SIZE BECOMES TOO SMALL.
C                IDID=-4  PROBLEM IS PROBABLY STIFF (INTERRUPTED).
C                IDID=-5  COMPUT. INTERRUPTED BY YLAG
C .........
```

Bibliography

This bibliography includes the publications referred to in the text. Italic numbers in square brackets following a reference indicate the sections where the reference is cited.

N.H. Abel (1826): *Untersuchungen über die Reihe:*
$$1 + \frac{m}{1}x + \frac{m(m-1)}{1\cdot 2}x^2 + \frac{m(m-1)(m-2)}{1\cdot 2\cdot 3}x^3 + \ldots \; u.s.w.$$
Crelle J. f. d. r. u. angew. Math. (in zwanglosen Heften), Vol.1, p.311-339. *[III.8]*

N.H. Abel (1827): *Ueber einige bestimmte Integrale.* Crelle J. f. d. r. u. angew. Math., Vol.2, p.22-30. *[I.11]*

L. Abia & J.M. Sanz-Serna (1993): *Partitioned Runge-Kutta methods for separable Hamiltonian problems.* Math. Comp., Vol.60, p.617-634. *[II.16]*

L. Abia, see also J.M. Sanz-Serna & L. Abia.

M. Abramowitz & I.A. Stegun (1964): *Handbook of mathematical functions.* Dover, 1000 pages. *[II.7], [II.8], [II.9]*

J.C. Adams (1883): see F.Bashforth (1883).

R.C. Aiken ed. (1985): *Stiff computation.* Oxford, Univ. Press, 462pp. *[Appendix]*

A.C. Aitken (1932): *On interpolation by iteration of proportional parts, without the use of differences.* Proc. Edinburgh Math. Soc. Second ser., Vol.3, p.56-76. *[II.9]*

J. Albrecht (1955): *Beiträge zum Runge-Kutta-Verfahren.* ZAMM, Vol.35, p.100-110. *[II.13], [II.14]*

P. Albrecht (1978): *Explicit, optimal stability functionals and their application to cyclic discretization methods.* Computing, Vol.19, p.233-249. *[III.8]*

P. Albrecht (1979): *Die numerische Behandlung gewöhnlicher Differentialgleichungen.* Akademie Verlag, Berlin; Hanser Verlag, München. *[III.8]*

P. Albrecht (1985): *Numerical treatment of O.D.E.s.: The theory of A-methods.* Numer. Math., Vol.47, p.59-87. *[III.8]*

V.M. Alekseev (1961): *An estimate for the perturbations of the solution of ordinary differential equations (Russian).* Vestn. Mosk. Univ., Ser.I, Math. Meh, 2, p.28-36. *[I.14]*

J.le Rond d'Alembert (1743): *Traité de dynamique, dans lequel les loix de l'équilibre & du mouvement des corps sont réduites au plus petit nombre possible, & démontrées d'une maniére nouvelle, & où l'on donne un principe général pour trouver le mouvement de plusieurs corps qui agissent les uns sur les autres, d'une maniére quelconque.* à Paris, MDCCXLIII, 186p., 70 figs. *[I.6]*

J.le Rond d'Alembert (1747): *Recherches sur la courbe que forme une corde tenduë mise en vibration.* Hist. de l'Acad. Royale de Berlin, Tom.3, Année MDCCXLVII, publ. 1749, p.214-219 et 220-249. *[I.6]*

J.le Rond d'Alembert (1748): *Suite des recherches sur le calcul intégral, quatrìeme partie: Méthodes pour intégrer quelques équations différentielles.* Hist. Acad. Berlin, Tom.IV, p.275-291. *[I.4]*

R.F. Arenstorf (1963): *Periodic solutions of the restricted three body problem representing analytic continuations of Keplerian elliptic motions.* Amer. J. Math., Vol.LXXXV, p.27-35. *[II.0]*

V.I. Arnol'd (1974): *Mathematical methods of classical mechanics.* Nauka, Moscow; French transl. Mir 1976; Engl. transl. Springer-Verlag 1978 (2nd edition 1989). *[I.14]*

C. Arzelà (1895): *Sulle funzioni di linee.* Memorie dell. R. Accad. delle Sc. di Bologna, 5e serie, Vol.V, p.225-244, see also: Vol.V, p.257-270, Vol.VI, (1896), p.131-140. *[I.7]*

U.M. Ascher, R.M.M. Mattheij & R.D. Russel (1988): *Numerical Solution of Boundary Value Problems for Ordinary Differential Equations.* Prentice Hall, Englewood Cliffs. *[I.15]*

L.S. Baca, see L.F. Shampine & L.S. Baca, L.F. Shampine, L.S. Baca & H.-J. Bauer.

H.F. Baker (1905): *Alternants and continuous groups.* Proc. London Math. Soc., Second Ser., Vol.3, p.24-47. *[II.16]*

N. Bakhvalov (1976): *Méthodes numériques.* Editions Mir, Moscou 600pp., russian edition 1973. *[I.9]*

F. Bashforth (1883): *An attempt to test the theories of capillary action by comparing the theoretical and measured forms of drops of fluid. With an explanation of the method of integration employed in constructing the tables which give the theoretical form of such drops, by J.C.Adams.* Cambridge Univ. Press. *[III.1]*

R.H. Battin (1976): *Resolution of Runge-Kutta-Nyström condition equations through eighth order.* AIAA J., Vol.14, p.1012-1021. *[II.14]*

F.L. Bauer, H. Rutishauser & E. Stiefel (1963): *New aspects in numerical quadrature.* Proc. of Symposia in Appl. Math., Vol.15, p.199-218, Am. Math. Soc. *[II.9]*

H.-J. Bauer, see L.F. Shampine, L.S. Baca & H.-J. Bauer.

P.A. Beentjes & W.J. Gerritsen (1976): *Higher order Runge-Kutta methods for the numerical solution of second order differential equations without first derivatives.* Report NW 34/76, Math. Centrum, Amsterdam. *[II.14]*

H. Behnke & F. Sommer (1962): *Theorie der analytischen Funktionen einer komplexen Veränderlichen.* Zweite Auflage. Springer Verlag, Berlin-Göttingen-Heidelberg. *[III.2]*

A. Bellen (1984): *One-step collocation for delay differential equations.* J. Comput. Appl. Math., Vol.10, p.275-283. *[II.17]*

A. Bellen & M. Zennaro (1985): *Numerical solution of delay differential equations by uniform corrections to an implicit Runge-Kutta method.* Numer. Math., Vol.47, p.301-316. *[II.17]*

R. Bellman & K.L. Cooke (1963): *Differential-Difference equations*. Academic Press, 462pp. *[II.17]*

I. Bendixson (1893): *Sur le calcul des intégrales d'un système d'équations différentielles par des approximations successives*. Stock. Akad. Öfversigt Förh., Vol.50, p.599-612. *[I.8]*

I. Bendixson (1901): *Sur les courbes définies par des équations différentielles*. Acta Mathematica, Vol.24, p.1-88. *[I.16]*

I.S. Berezin & N.P Zhidkov (1965): *Computing methods (Metody vychislenii)*. 2 Volumes, Fizmatgiz Moscow, Engl. transl.: Pergamon Press, 464 & 679pp. *[I.1]*

Dan. Bernoulli (1728): *Observationes de seriebus quae formantur ex additione vel substractione quacunque terminorum se mutuo consequentium, ubi praesertim earundem insignis usus pro inveniendis radicum omnium aequationum algebraicarum ostenditur*. Comm. Acad. Sci. Imperialis Petrop., Tom.III, 1728 (1732), p.85-100; Werke, Bd.2, p.49-70. *[III.3]*

Dan. Bernoulli (1732): *Theoremata de oscillationibus corporum filo flexili connexorum et catenae verticaliter suspensae*. Comm. Acad. Sci. Imperialis Petrop., Tom.VI, ad annus MDCCXXXII & MDCCXXXIII, p.108-122. *[I.6]*

Dan. Bernoulli (1760): *Essai d'une nouvelle analyse de la mortalité causée par la petite vérole, et des avantages de l'inoculation pour la prévenir*. Hist. et Mém. de l'Acad. Roy. Sciences Paris, 1760, p.1-45; Werke Bd. 2, p.235-267. *[II.17]*

Jac. Bernoulli (1690): *Analysis problematis ante hac propositi, de inventione lineae descensus a corpore gravi percurrendae uniformiter, sic ut temporibus aequalibus aequales altitudines emetiatur: & alterius cujusdam Problematis Propositio*. Acta Erudit. Lipsiae, Anno MDCLXXXX, p. 217-219. *[I.3]*

Jac. Bernoulli (1695): *Explicationes, Annotationes & Additiones ad ea, quae in Actis sup. anni de Curva Elastica, Isochrona Paracentrica, & Velaria, hinc inde memorata, & partim controversa legundur; ubi de Linea mediarum directionum, aliisque novis*. Acta Erudit. Lipsiae, Anno MDCXCV, p. 537-553. *[I.3]*

Jac. Bernoulli (1697): *Solutio Problematum Fraternorum, Peculiari Programmate Cal. Jan. 1697 Groningae, nec non Actorum Lips. mense Jun. & Dec. 1696, & Febr. 1697 propositorum: una cum Propositione reciproca aliorum*. Acta Erud. Lips. MDCXCVII, p.211-217. *[I.2]*

Joh. Bernoulli (1691): *Solutio problematis funicularii, exhibita à Johanne Bernoulli, Basil. Med. Cand.*. Acta Erud. Lips. MDCXCI, p.274, Opera Omnia, Vol.I, p.48-51, Lausannae & Genevae 1742. *[I.3]*

Joh. Bernoulli (1696): *Problema novum Mathematicis propositorum*. Acta Erud. Lips. MDCXCVI, p.269, Opera Omnia, Vol.I, p.161 and 165, Lausannae & Genevae 1742. *[I.2]*

Joh. Bernoulli (1697): *De Conoidibus et Sphaeroidibus quaedam. Solutio analytica Aequationis in Actis A. 1695, pag. 553 propositae*. Acta Erud. Lips., MDCXCVII, p.113-118. Opera Omnia, Vol.I, p.174-179. *[I.3]*

Joh. Bernoulli (1697b): *Solutioque Problematis a se in Actis 1696, p.269, proposit, de invenienda Linea Brachystochrona*. Acta Erud.Lips. MDCXCVII, p.206, Opera Omnia, Vol.I, p.187-193. *[I.2]*

Joh. Bernoulli (1727): *Meditationes de chordis vibrantibus* .... Comm. Acad. Sci. Imperialis Petrop., Tom.III, p.13; Opera, Vol.III, p.198-210. *[I.6]*

494   Bibliography

J. Berntsen & T.O. Espelid (1991): *Error estimation in automatic quadrature routines.* ACM Trans. on Math. Software, Vol.17, p.233-255. *[II.10]*

D.G. Bettis (1973): *A Runge-Kutta Nyström algorithm.* Celestial Mechanics, Vol.8, p.229-233. *[II.14]*

L. Bieberbach (1923): *Theorie der Differentialgleichungen.* Grundlehren Bd.VI, Springer Verlag. *[II.3]*

L. Bieberbach (1951): *On the remainder of the Runge-Kutta formula in the theory of ordinary differential equations.* ZAMP, Vol.2, p.233-248. *[II.3]*

J. Binney (1981): *Resonant excitation of motion perpendicular to galactic planes.* Mon. Not. R. astr. Soc., Vol.196, p.455-467. *[II.16]*

J. Binney & S. Tremaine (1987): *Galactic dynamics.* Princeton Univ. Press, 733pp. *[II.16]*

J.B. Biot (1804): *Mémoire sur la propagation de la chaleur, lu à la classe des sciences math. et phys. de l'Institut national.* Bibl. britann. Sept 1804, 27, p.310. *[I.6]*

G. Birkhoff & R.S. Varga (1965): *Discretization errors for well-set Cauchy problems I.* Journal of Math. and Physics, Vol.XLIV, p.1-23. *[I.13]*

H.G. Bock (1981): *Numerical treatment of inverse problems in chemical reaction kinetics.* In: Modelling of chemical reaction systems, ed. by K.H. Ebert, P. Deuflhard & W. Jäger, Springer Series in Chem. Phys., Vol.18, p.102-125. *[II.6]*

H.G. Bock & J. Schlöder (1981): *Numerical solution of retarded differential equations with statedependent time lages.* ZAMM, Vol.61, p.269-271. *[II.17]*

P. Bogacki & L.F. Shampine (1989): *An efficient Runge-Kutta (4,5) pair.* SMU Math Rept 89-20. *[II.6]*

N. Bogoliuboff, see N. Kryloff & N. Bogoliuboff.

R.W. Brankin, I. Gladwell, J.R. Dormand, P.J. Prince & W.L. Seward (1989): *Algorithm 670. A Runge-Kutta-Nyström code.* ACM Trans. Math. Softw., Vol.15, p.31-40. *[II.14]*

R.W. Brankin, see also I. Gladwell, L.F. Shampine & R.W. Brankin.

P.N. Brown, G.D. Byrne & A.C. Hindmarsh (1989): *VODE: a variable-coefficient ODE solver.* SIAM J. Sci. Stat. Comput., Vol.10, p.1038-1051. *[III.7]*

H. Brunner & P.J. van der Houwen (1986): *The numerical solution of Volterra equations.* North-Holland, Amsterdam, 588pp. *[II.17]*

R. Bulirsch & J. Stoer (1964): *Fehlerabschätzungen und Extrapolation mit rationalen Funktionen bei Verfahren vom Richardson-Typus.* Num. Math., Vol.6, p.413-427. *[II.9]*

R. Bulirsch & J. Stoer (1966): *Numerical treatment of ordinary differential equations by extrapolation methods.* Num. Math., Vol.8, p.1-13. *[II.9]*

K. Burrage (1985): *Order and stability properties of explicit multivalue methods.* Appl. Numer. Anal., Vol.1, p.363-379. *[III.8]*

K. Burrage & J.C. Butcher (1980): *Non-linear stability of a general class of differential equation methods.* BIT, Vol.20, p.185-203. *[III.8]*

K. Burrage & P. Moss (1980): *Simplifying assumptions for the order of partitioned multivalue methods.* BIT, Vol.20, p.452-465. *[III.8]*

J.C. Butcher (1963): *Coefficients for the study of Runge-Kutta integration processes.* J. Austral. Math. Soc., Vol.3, p.185-201. *[II.2]*

J.C. Butcher (1963a): *On the integration process of A. Huťa.* J. Austral. Math. Soc., Vol.3, p.202-206. *[II.2]*

J.C. Butcher (1964a): *Implicit Runge-Kutta Processes.* Math. Comput., Vol.18, p.50-64. *[II.7], [II.16]*

J.C. Butcher (1964b): *On Runge-Kutta processes of high order.* J.Austral. Math. Soc., Vol.IV, Part2, p.179-194. *[II.1], [II.5]*

J.C. Butcher (1964c): *Integration processes based on Radau quadrature formulas.* Math. Comput., Vol.18, p.233-244. *[II.7]*

J.C. Butcher (1965a): *A modified multistep method for the numerical integration of ordinary differential equations.* J. ACM, Vol.12, p.124-135. *[III.8]*

J.C. Butcher (1965b): *On the attainable order of Runge-Kutta methods.* Math. of Comp., Vol.19, p.408-417. *[II.5]*

J.C. Butcher (1966): *On the convergence of numerical solutions to ordinary differential equations.* Math. Comput., Vol.20, p.1-10. *[III.4], [III.8]*

J.C. Butcher (1969): *The effective order of Runge-Kutta methods.* in: Conference on the numerical solution of differential equations, Lecture Notes in Math., Vol.109, p.133-139. *[II.12]*

J.C. Butcher (1981): *A generalization of singly-implicit methods.* BIT, Vol.21, p.175-189. *[III.8]*

J.C. Butcher (1984): *An application of the Runge-Kutta space.* BIT, Vol.24, p.425-440. *[II.12], [III.8]*

J.C. Butcher (1985a): *General linear method: a survey.* Appl. Num. Math., Vol.1, p.273-284. *[III.8]*

J.C. Butcher (1985b): *The non-existence of ten stage eighth order explicit Runge-Kutta methods.* BIT, Vol.25, p.521-540. *[II.5]*

J.C. Butcher (1987): *The numerical analysis of ordinary differential equations. Runge-Kutta and general linear methods.* John Wiley & Sons, Chichester, 512pp. *[II.16]*

J.C. Butcher, see also K. Burrage & J.C. Butcher.

G.D. Byrne & A.C. Hindmarsh (1975): *A polyalgorithm for the numerical solution of ordinary differential equations.* ACM Trans. on Math. Software, Vol.1, No.1, p.71-96. *[III.6], [III.7]*

G.D. Byrne & R.J. Lambert (1966): *Pseudo-Runge-Kutta methods involving two points.* J. Assoc. Comput. Mach., Vol.13, p.114-123. *[III.8]*

G.D. Byrne, see also P.N. Brown, G.D. Byrne & A.C. Hindmarsh.

R. Caira, C. Costabile & F. Costabile (1990): *A class of pseudo Runge-Kutta methods.* BIT, Vol.30, p.642-649. *[III.8]*

M. Calvé & R. Vaillancourt (1990): *Interpolants for Runge-Kutta pairs of order four and five.* Computing, Vol.45, p.383-388. *[II.6]*

M. Calvo, J.I. Montijano & L. Rández (1990): *A new embedded pair of Runge-Kutta formulas of orders 5 and 6*. Computers Math. Applic., Vol.20, p.15-24. *[II.6]*

M. Calvo, J.I. Montijano & L. Rández (1992): *New continuous extensions for the Dormand and Prince RK method*. In: Computational ordinary differential equations, ed. by J.R. Cash & I. Gladwell, Clarendon Press, Oxford, p.135-164. *[II.6]*

M.P. Calvo & J.M. Sanz-Serna (1992): *Order conditions for canonical Runge-Kutta-Nyström methods*. BIT, Vol.32, p.131-142. *[II.16]*

M.P. Calvo & J.M. Sanz-Serna (1992b): *High order symplectic Runge-Kutta-Nyström methods*. SIAM J. Sci. Stat. Comput., Vol.14 (1993), p.1237-1252. *[II.16]*

M.P. Calvo & J.M. Sanz-Serna (1992c): *Reasons for a failure. The integration of the two-body problem with a symplectic Runge-Kutta method with step changing facilities*. Intern. Conf. on Differential Equations, Vol. 1, 2 (Barcelona, 1991), 93-102, World Sci. Publ., River Edge, NJ, 1993. *[II.16]*

J.M. Carnicer (1991): *A lower bound for the number of stages of an explicit continuous Runge-Kutta method to obtain convergence of given order*. BIT, Vol.31, p.364-368. *[II.6]*

E. Cartan (1899): *Sur certaines expressions différentielles et le problème de Pfaff*. Ann. Ecol. Normale, Vol.16, p.239-332, Oeuvres partie II, p.303-396. *[I.14]*

A.L. Cauchy (1824): *Résumé des Leçons données à l'Ecole Royale Polytechnique. Suite du Calcul Infinitésimal;* published: Equations différentielles ordinaires, ed. Chr. Gilain, Johnson 1981. *[I.2], [I.7], [I.9], [II.3], [II.7]*

A.L. Cauchy (1831): *Sur la mecanique celeste et sur un nouveau calcul appelé calcul des limites*. lu à l'acad. de Turin le 11 oct 1831; also: exerc. d'anal. et de pysique math, 2, Paris 1841; oeuvres (2), 12. *[III.3]*

A.L. Cauchy (1839-42): *Several articles in Comptes Rendus de l'Acad. des Sciences de Paris*. (Aug. 5, Nov. 21, 1839, June 29, Oct. 26, 1840, etc). *[I.8]*

A. Cayley (1857): *On the theory of the analytic forms called trees*. Phil. Magazine, Vol.XIII, p.172-176, Mathematical Papers, Vol.3, Nr.203, p.242-246. *[II.2]*

A. Cayley (1858): *A memoir on the theory of matrices*. Phil. Trans. of Royal Soc. of London, Vol.CXLVIII, p.17-37, Mathematical Papers, Vol.2, Nr.152, p.475.

F. Ceschino (1961): *Modification de la longueur du pas dans l'intégration numérique par les méthodes à pas liés*. Chiffres, Vol.2, p.101-106. *[II.4], [III.5]*

F. Ceschino (1962): *Evaluation de l'erreur par pas dans les problèmes différentiels*. Chiffres, Vol.5, p.223-229. *[II.4]*

F. Ceschino & J. Kuntzmann (1963): *Problèmes différentiels de conditions initiales (méthodes numériques)*. Dunod Paris, 372pp.; english translation: Numerical solutions of initial value problems, Prentice Hall 1966 *[II.5], [II.7]*

P.J. Channell & C. Scovel (1990): *Symplectic integration of Hamiltonian systems*. Nonlinearity, Vol.3, p.231-259. *[II.16]*

A.C. Clairaut (1734): *Solution de plusieurs problèmes où il s'agit de trouver des courbes dont la propriété consiste dans une certaine relation entre leurs branches, exprimée par une Equation donnée*. Mémoires de Math. et de Phys. de l'Acad. Royale des Sciences, Paris, Année MDCCXXXIV, p.196-215. *[I.2]*

L. Collatz (1951): *Numerische Behandlung von Differentialgleichungen.* Grundlehren Band LX, Springer Verlag, 458pp; second edition 1955; third edition and english translation 1960. *[II.7]*

L. Collatz (1967): *Differentialgleichungen. Eine Einführung unter besonderer Berücksichtigung der Anwendungen.* Leitfäden der angewandten Mathematik, Teubner 226pp. English translation: *Differential equations. An introduction with applications,* Wiley, 372pp., (1986). *[I.15]*

P. Collet & J.P. Eckmann (1980): *Iterated maps on the interval as dynamical systems.* Birkhäuser, 248pp. *[I.16]*

K.L. Cooke, see R. Bellman & K.L. Cooke.

G.J. Cooper (1978): *The order of convergence of general linear methods for ordinary differential equations.* SIAM, J. Numer. Anal., Vol.15, p.643-661. *[III.8]*

G.J. Cooper (1987): *Stability of Runge-Kutta methods for trajectory problems.* IMA J. Numer. Anal., Vol.7, p.1-13. *[II.16]*

G.J. Cooper & J.H. Verner (1972): *Some explicit Runge-Kutta methods of high order.* SIAM J.Numer. Anal., Vol.9, p.389-405. *[II.5]*

S.A. Corey (1906): *A method of approximation.* Amer. Math. Monthly, Vol.13, p.137-140. *[II.9]*

C. Costabile, see R. Caira, C. Costabile & F. Costabile.

F. Costabile, see R. Caira, C. Costabile & F. Costabile.

C.A. de Coulomb (1785): *Théorie des machines simples, en ayant égard au frottement de leurs parties, et a la roideur des cordages. Pièce qui a remporté le Prix double de l'Académie des Sciences pour l'année 1781.* Mémoires des Savans Etrangers, tome X, p. 163-332; réimprimé 1809 chez Bachelier, Paris. *[II.6]*

P.H. Cowell & A.C.D. Crommelin (1910): *Investigation of the motion of Halley's comet from 1759 to 1910.* Appendix to Greenwich Observations for 1909, Edinburgh, p.1-84. *[III.10]*

J.W. Craggs, see A.R. Mitchell & J.W. Craggs.

D.M. Creedon & J.J.H. Miller (1975): *The stability properties of q-step backward-difference schemes.* BIT, Vol.15, p.244-249. *[III.3]*

A.C.D. Crommelin, see P.H. Cowell & A.C.D. Crommelin.

M. Crouzeix (1975): *Sur l'approximation des équations différentielles operationnelles linéaires par des méthodes de Runge-Kutta.* Thèse d'état, Univ. Paris 6, 192pp. *[II.2]*, *[II.7]*

M. Crouzeix & F.J. Lisbona (1984): *The convergence of variable-stepsize, variable formula, multistep methods.* SIAM J. Num. Anal., Vol.21, p.512-534. *[III.5]*

C.W. Cryer (1971): *A proof of the instability of backward-difference multistep methods for the numerical integration of ordinary differential equations.* Tech. Rep. No.117, Comp. Sci. Dept., Univ. of Wisconsin, p.1-52. *[III.3]*

C.W. Cryer (1972): *On the instability of high order backward-difference multistep methods.* BIT, Vol.12, p.17-25. *[III.3]*

W.J. Cunningham (1954): *A nonlinear differential-difference equation of growth.* Proc. Mat. Acad. Sci., USA, Vol.40, p.708-713. *[II.17]*

498   Bibliography

A.R. Curtis (1970): *An eighth order Runge-Kutta process with eleven function evaluations per step.* Numer. Math., Vol.16, p.268-277. *[II.5]*

A.R. Curtis (1975): *High-order explicit Runge-Kutta formulae, their uses, and limitations.* J.Inst. Maths Applics, Vol.16, p.35-55. *[II.5]*

C.F. Curtiss & J.O. Hirschfelder (1952): *Integration of stiff equations.* Proc. of the National Academy of Sciences of U.S., Vol.38, p.235-243. *[III.1]*

G. Dahlquist (1956): *Convergence and stability in the numerical integration of ordinary differential equations.* Math. Scand., Vol.4, p.33-53. *[III.2], [III.3], [III.4]*

G. Dahlquist (1959): *Stability and error bounds in the numerical integration of ordinary differential equations.* Trans. of the Royal Inst. of Techn., Stockholm, Sweden, Nr.130, 87pp. *[I.10], [III.2], [III.10]*

G. Dahlquist (1985): *33 years of numerical instability, Part I.* BIT, Vol.25, p.188-204. *[III.3]*

G. Dahlquist & R. Jeltsch (1979): *Generalized disks of contractivity for explicit and implicit Runge-Kutta methods.* Report TRITA-NA-7906, NADA, Roy. Inst. Techn. Stockholm. *[II.12]*

G. Darboux (1876): *Sur les développements en série des fonctions d'une seule variable.* J. des Mathématiques pures et appl., 3ème série, t. II, p.291-312. *[II.13]*

G. H. Darwin (Sir George) (1898): *Periodic orbits.* Acta Mathematica, Vol.21, p.99-242, plates I-IV. *[II.0]*

S.M. Davenport, see L.F. Shampine, H.A. Watts & S.M. Davenport.

F. Debaune (1638): *Letter to Descartes.* lost; answer of Descartes: Feb 20, 1639. *[I.2]*

J.P. Den Hartog (1930): *Forced vibrations with combined viscous and Coulomb damping.* Phil. Mag. Ser.7, Vol.9, p.801-817. *[II.6]*

J. Descloux (1963): *A note on a paper by A. Nordsieck.* Report No.131, Dept. of Comp. Sci., Univ. of Illinois at Urbana-Champaign. *[III.6]*

P. Deuflhard (1980): *Recent advances in multiple shooting techniques.* In: Computational techniques for ordinary differential equations (Gladwell-Sayers, ed.), Section 10, p.217-272, Academic Press. *[I.15]*

P. Deuflhard (1983): *Order and stepsize control in extrapolation methods.* Num. Math., Vol.41, p.399-422. *[II.9], [II.10]*

P. Deuflhard (1985): *Recent progress in extrapolation methods for ordinary differential equations.* SIAM Rev., Vol.27, p.505-535. *[II.14]*

P. Deuflhard & U. Nowak (1987): *Extrapolation integrators for quasilinear implicit ODEs.* In: P. Deuflhard, B. Engquist (eds.), Large-scale scientific computing, Birkhäuser, Boston. *[II.9]*

E. de Doncker-Kapenga, see R. Piessens, E. de Doncker-Kapenga, C.W. Überhuber & D.K. Kahaner.

J. Donelson & E. Hansen (1971): *Cyclic composite multistep predictor-corrector methods.* SIAM, J. Numer. Anal., Vol.8, p.137-157. *[III.8]*

J.R. Dormand, M.E.A. El-Mikkawy & P.J. Prince (1987): *High-order embedded Runge-Kutta-Nystrom formulae.* IMA J. Numer. Anal., Vol.7, p.423-430. *[II.14]*

J.R. Dormand & P.J. Prince (1978): *New Runge-Kutta algorithms for numerical simulation in dynamical astronomy*. Celestial Mechanics, Vol.18, p.223-232. *[II.14]*

J.R. Dormand & P.J. Prince (1980): *A family of embedded Runge-Kutta formulae*. J.Comp. Appl. Math., Vol.6, p.19-26. *[II.5]*

J.R. Dormand & P.J. Prince (1986): *Runge-Kutta triples*. Comp. & Maths. with Applc., Vol.12A, p.1007-1017. *[II.6]*

J.R. Dormand & P.J. Prince (1987): *Runge-Kutta-Nystrom triples*. Comput. Math. Applic., Vol.13(12), p.937-949. *[II.14]*

J.R. Dormand & P.J. Prince (1989): *Practical Runge-Kutta processes*. SIAM J. Sci. Stat. Comput., Vol.10, p.977-989. *[II.5]*

J.R. Dormand, see also P.J. Prince & J.R. Dormand, R.W. Brankin, I. Gladwell, J.R. Dormand, P.J. Prince & W.L. Seward.

R.D. Driver (1977): *Ordinary and delay differential equations*. Applied Math. Sciences 20, Springer Verlag, 501pp. *[II.17]*

J.P. Eckmann, see P. Collet & J.P. Eckmann.

B.L. Ehle (1968): *High order A-stable methods for the numerical solution of systems of D.E.'s*. BIT, Vol.8, p.276-278. *[II.7]*

E. Eich (1992): *Projizierende Mehrschrittverfahren zur numerischen Lösung von Bewegungsgleichungen technischer Mehrkörpersysteme mit Zwangsbedingungen unmd Unstetigkeiten*. Fortschritt-Ber. VDI, Reihe 18, Nr.109, VDI-Verlag Düsseldorf, 188pp. *[II.6]*.

N.F. Eispack (1974): *B.T.Smith, J.M. Boyle, B.S.Garbow, Y.Jkebe, V.C.Klema, C.B.Moler: Matrix Eigensystem Routines*. (Fortran-translations of algorithms published in Reinsch & Wilkinson), Lecture Notes in Computer Science, Vol.6, Springer Verlag. *[I.12], [I.13]*

M.E.A. El-Mikkawy, see J.R. Dormand, M.E.A. El-Mikkawy & P.J. Prince.

H. Eltermann (1955): *Fehlerabschätzung bei näherungsweiser Lösung von Systemen von Differentialgleichungen erster Ordnung*. Math. Zeitschr., Vol.62, p.469-501. *[I.10]*

R. England (1969): *Error estimates for Runge-Kutta type solutions to systems of ordinary differential equations*. The Computer J. Vol.12, p.166-170. *[II.4]*

W.H. Enright & D.J. Higham (1991): *Parallel defect control*. BIT, Vol.31, p.647-663. *[II.11]*

W.H. Enright, K.R. Jackson, S.P. Nørsett & P.G. Thomson (1986): *Interpolants for Runge-Kutta formulas*. ACM Trans. Math. Softw., Vol.12, p.193-218. *[II.6] [II.6]*

W.H. Enright, K.R. Jackson, S.P. Nørsett & P.G. Thomson (1988): *Effective solution of discontinuous IVPs using a Runge-Kutta formula pair with interpolants*. Appl. Math. and Comput., Vol.27, p.313-335. *[II.6]*

T.O. Espelid, see J. Berntsen & T.O. Espelid.

L. Euler (1728): *Nova methodus innumerabiles aequationes differentiales secundi gradus reducendi ad aequationes differentiales primi gradus*. Comm. acad. scient. Petrop., Vol.3, p.124-137; Opera Omnia, Vol.XXII, p.1-14. *[I.3]*

L. Euler (1743): *De integratione aequationum differentialium altiorum graduum*. Miscellanea Berolinensia, Vol.7, p.193-242; Opera Omnia, Vol.XXII, p.108-149. See also: Letter from Euler to Joh. Bernoulli, 15.Sept.1739. *[I.4]*

L. Euler (1744): *Methodus inveniendi lineas curvas maximi minimive proprietate gaudentes* ... Lausannae & Genevae, Opera Omnia (intr. by Caratheodory) Vol.XXIV, p.1-308. *[I.2]*

L. Euler (1747): *Recherches sur le mouvement des corps celestes en général.* Hist. de l'Acad. Royale de Berlin, Tom.3, Année MDCCXLVII, publ. 1749, p.93-143. *[I.6]*

L. Euler (1750): *Methodus aequationes differentiales altiorum graduum integrandi ulterius promota.* Novi Comment. acad. scient. Petrop., Vol.3, p.3-35; Opera Omnia, Vol.XXII, p.181-213. *[I.4]*

L. Euler (1755): *Institutiones calculi differentialis cum eius vsu in analysi finitorum ac doctrina serierum.* Imp. Acad. Imper. Scient. Petropolitanae, Opera Omnia, Vol.X, *[I.6]*

L. Euler (1756): *Elementa calculi variationum.* presented September 16, 1756 at the Acad. of Science, Berlin; printed 1766, Opera Omnia, Vol.XXV,p.141-176. *[I.2]*

L. Euler (1758): *Du mouvement de rotation des corps solides autour d'un axe variable.* Hist. de l'Acad. Royale de Berlin, Tom.14, Année MDCCLVIII, pp.154-193. Opera Omnia Ser.II, Vol.8, p.200-235. *[II.10]*

L. Euler (1768): *Institutionum Calculi Integralis.* Volumen Primum, Opera Omnia, Vol.XI. *[I.7]*, *[I.8]*, *[II.1]*

L. Euler (1769): *Institutionum Calculi Integralis.* Volumen Secundum, Opera Omnia, Vol.XII. *[I.3]*, *[I.5]*

L. Euler (1769b): *De formulis integralibus duplicatis.* Novi Comment. acad. scient. Petrop., Vol.14, I, 1770, p.72-103; Opera Omnia, Vol.XVII, p.289-315. *[I.14]*

L. Euler (1778): *Specimen transformationis singularis serienum.* Nova acta. acad. Petrop., Vol.12 (1794), p.58-70, Opera Omnia, Vol.XVI, Sectio Altera, p.41-55. *[I.5]*

E. Fehlberg (1958): *Eine Methode zur Fehlerverkleinerung beim Runge-Kutta-Verfahren.* ZAMM, Vol.38, p.421-426. *[II.13]*

E. Fehlberg (1964): *New high-order Runge-Kutta formulas with step size control for systems of first and second order differential equations.* ZAMM, Vol.44, Sonderheft T17-T19. *[II.4]*, *[II.13]*

E. Fehlberg (1968): *Classical fifth-, sixth-, seventh-, and eighth order Runge-Kutta formulas with step size control.* NASA Technical Report 287 (1968); extract published in Computing, Vol.4, p.93-106 (1969). *[II.4]*, *[II.5]*

E. Fehlberg (1969): *Low-order classical Runge-Kutta formulas with step size control and their application to some heat transfer problems.* NASA Technical Report 315 (1969), extract published in Computing, Vol.6, p.61-71 (1970). *[II.4]*, *[II.5]*

E. Fehlberg (1972): *Classical eighth- and lower-order Runge-Kutta-Nyström formulas with stepsize control for special second-order differential equations.* NASA Technical Report R-381. *[II.14]*

M. Feigenbaum (1978): *Quantitative universality for a class of nonlinear transformations.* J.Stat. Phys., Vol.19, p.25-52, Vol.21 (1979), p.669-706. *[I.16]*

Feng Kang (冯康) (1985): *On difference schemes and symplectic geometry.* Proceedings of the 5-th Intern. Symposium on differential geometry & differential equations, August 1984, Beijing, p.42-58. *[II.16]*

Feng Kang (1986): *Difference schemes for Hamiltonian formalism and symplectic geometry.* J. Comp. Math., Vol.4, p.279-289. *[II.16]*

Feng Kang (1991): *How to compute properly Newton's equation of motion?* Proceedings of the second conference on numerical methods for partial differential equations, Nankai Univ., Tianjin, China, Eds. Ying Lungan & Guo Benyu, World Scientific, p.15-22. *[II.16]*

Feng Kang (1991b): *Formal power series and numerical algorithms for dynamical systems.* Proceedings of international conference on scientific computation, Hangzhou, China, Eds. Tony Chan & Zhong-Ci Shi, Series on Appl. Math., Vol.1, pp.28-35. *[II.16]*

Feng Kang, Wu Hua-mo, Qin Meng-zhao & Wang Dao-liu (1989): *Construction of canonical difference schemes for Hamiltonian formalism via generating functions.* J. Comp. Math., Vol.11, p.71-96. *[II.16]*

J.R. Field & R.M. Noyes (1974): *Oscillations in chemical systems. IV. Limit cycle behavior in a model of a real chemical reaction.* J. Chem. Physics, Vol.60, p.1877-1884. *[I.16]*

S. Filippi & J. Gräf (1986): *New Runge-Kutta-Nyström formula-pairs of order 8(7), 9(8), 10(9) and 11(10) for differential equations of the form $y'' = f(x, y)$.* J. Comput. and Applied Math., Vol.14, p.361-370. *[II.14]*

A.F. Filippov (1960): *Differential equations with discontinuous right-hand side.* Mat. Sbornik (N.S.) Vol.51(93), p.99-128; Amer. Math. Soc. Transl. Ser.2, Vol.42, p.199-231. *[II.6]*

J.M. Fine (1987): *Interpolants for Runge-Kutta-Nyström methods.* Computing, Vol.39, p.27-42. *[II.14]*

R. Fletcher & D.C. Sorensen (1983): *An algorithmic derivation of the Jordan canonical form.* Amer. Math. Monthly, Vol.90, No.1, p.12-16. *[I.12]*

C.V.D. Forrington (1961-62): *Extensions of the predictor-corrector method for the solution of systems of ordinary differential equations.* Comput. J. 4, p.80-84. *[III.5]*

J.B.J. Fourier (1807): *Sur la propagation de la chaleur.* Unpublished manuscript; published: La théorie analytique de la chaleur, Paris 1822. *[I.6]*

R.A. Frazer, W.P. Jones & S.W. Skan (1937): *Approximations to functions and to the solutions of differential equations.* Reports and Memoranda Nr.1799 (2913), Aeronautical Research Committee. 33pp. *[II.7]*

A. Fricke (1949): *Ueber die Fehlerabschätzung des Adamsschen Verfahrens zur Integration gewöhnlicher Differentialgleichungen erster Ordnung.* ZAMM, Vol.29, p.165-178. *[III.4]*

G. Frobenius (1873): *Ueber die Integration der linearen Differentialgleichungen durch Reihen.* Journal für Math. LXXVI, p.214-235 *[I.5]*

M. Frommer (1934): *Ueber das Auftreten von Wirbeln und Strudeln (geschlossener und spiraliger Integralkurven) inder Umgebung rationaler Unbestimmtheitsstellen.* Math. Ann., Vol.109, p.395-424. *[I.16]*

L. Fuchs (1866, 68): *Zur Theorie der linearen Differentialgleichungen mit veränderlichen Coefficienten.* Crelle J. f. d. r. u. angew. Math., Vol.66, p.121-160 (prepublished in "Programm der städtischen Gewerbeschule zu Berlin, Ostern 1865"). Ergänzung: J. f. Math. LXVIII, p. 354-385. *[I.5], [I.11]*

C.F. Gauss (1812): *Disquisitiones generales circa seriem infinitam*
$$1 + \tfrac{\alpha\beta}{1\cdot\gamma}x + \tfrac{\alpha(\alpha+1)\beta(\beta+1)}{1\cdot2\cdot\gamma(\gamma+1)}xx + \tfrac{\alpha(\alpha+1)(\alpha+2)\beta(\beta+1)(\beta+2)}{1\cdot2\cdot3\cdot\gamma(\gamma+1)(\gamma+2)}x^3 + etc,$$
Werke, Vol.3, p.123-162. *[I.5]*

W. Gautschi (1962): *On inverses of Vandermonde and confluent Vandermonde matrices.* Numer. Math., Vol.4, p.117-123. *[II.13]*

C.W. Gear (1965): *Hybrid methods for initial value problems in ordinary differential equations.* SIAM J. Numer. Anal., ser.B, Vol.2, p.69-86. *[III.8]*

C.W. Gear (1971): *Numerical initial value problems in ordinary differential equations.* Prentice-Hall, 253pp. *[II.2]*, *[III.1]*, *[III.7]*

C.W. Gear (1987): *The potential for parallelism in ordinary differential equations.* In: Computational mathematics II, Proc. 2nd Int. Conf. Numer. Anal. Appl., Benin City/Niger. 1986, Conf. Ser. Boole Press 11, p. 33-48. *[II.11]*

C.W. Gear (1988): *Parallel methods for ordinary differential equations.* Calcolo, Vol.25, No.1/2, p. 1-20. *[II.11]*

C.W. Gear & O. Østerby (1984): *Solving ordinary differential equations with discontinuities.* ACM Trans. Math. Softw., Vol.10, p.23-44. *[II.6]*

C.W. Gear & K.W. Tu (1974): *The effect of variable mesh size on the stability of multistep methods.* SIAM J. Num. Anal., Vol.11, p.1025-1043. *[III.5]*

C.W. Gear & D.S. Watanabe (1974): *Stability and convergence of variable order multistep methods.* SIAM J. Num. Anal., Vol.11, p.1044-1058. *[III.3]*

W.J. Gerritsen, see P.A. Beentjes & W.J. Gerritsen.

A. Gibbons (1960): *A program for the automatic integration of differential equations using the method of Taylor series.* Computer J., Vol.3, p.108-111. *[I.8]*

S. Gill (1951): *A process for the step-by-step integration of differential equations in an automatic digital computing machine.* Proc. Cambridge Philos. Soc., Vol.47, p.95-108. *[II.1]*, *[II.2]*

S. Gill (1956): Discussion in Merson (1957). *[II.2]*

B. Giovannini, L. Weiss-Parmeggiani & B.T. Ulrich (1978): *Phase locking in coupled Josephson weak links.* Helvet. Physica Acta, Vol.51, p.69-74. *[I.16]*

I. Gladwell, L.F. Shampine & R.W. Brankin (1987): *Automatic selection of the initial step size for an ODE solver.* J. Comp. Appl. Math., Vol.18, p.175-192. *[II.4]*

I. Gladwell, see also R.W. Brankin, I. Gladwell, J.R. Dormand, P.J. Prince & W.L. Seward.

G.H. Golub & J.H. Wilkinson (1976): *Ill-conditioned eigensystems and the computation of the Jordan canonical form.* SIAM Review, Vol.18, p.578-619. *[I.12]*

M.K. Gordon, see L.F. Shampine & M.K. Gordon.

J. Gräf, see S. Filippi & J. Gräf.

W.B. Gragg (1964): *Repeated extrapolation to the limit in the numerical solution of ordinary differential equations.* Thesis, Univ. of California; see also SIAM J. Numer. Anal., Vol.2, p.384-403 (1965). *[II.8]*, *[II.9]*

W.B. Gragg (1965): *On extrapolation algorithms for ordinary initial value problems.* SIAM J. Num. Anal., ser.B, Vol.2, p.384-403. *[II.14]*

W.B. Gragg & H.J. Stetter (1964): *Generalized multistep predictor-corrector methods.* J. ACM, Vol.11, p.188-209. *[III.8]*

E. Griepentrog (1978): *Gemischte Runge-Kutta-Verfahren für steife Systeme.* In: Seminar-bericht Nr. 11, Sekt. Math., Humboldt-Univ. Berlin, p.19-29. *[II.15]*

R.D. Grigorieff (1977): *Numerik gewöhnlicher Differentialgleichungen 2.* Teubner Studien-bücher, Stuttgart. *[III.3], [III.4]*

R.D. Grigorieff (1983): *Stability of multistep-methods on variable grids.* Numer. Math. 42, p.359-377. *[III.5]*

W. Gröbner (1960): *Die Liereihen und ihre Anwendungen.* VEB Deutscher Verlag der Wiss., Berlin 1960, 2nd ed. 1967. *[I.14], [II.16]*

T.H. Gronwall (1919): *Note on the derivatives with respect to a parameter of the solutions of a system of differential equations.* Ann. Math., Vol.20, p.292-296. *[I.10], [I.14]*

A. Guillou & J.L. Soulé (1969): *La résolution numérique des problèmes différentiels aux conditions initiales par des méthodes de collocation.* R.I.R.O, No R-3, p.17-44. *[II.7]*

P. Habetsand, see N. Rouche, P. Habetsand & M. Laloy.

H. Hahn (1921): *Theorie der reellen Funktionen.* Springer Verlag Berlin, 600pp. *[I.7]*

W. Hahn (1967): *Stability of motion.* Springer Verlag, 446pp. *[I.13]*

E. Hairer (1977): *Méthodes de Nyström pour l'équation différentielle $y'' = f(x,y)$.* Numer. Math., Vol.27, p.283-300. *[II.14]*

E. Hairer (1978): *A Runge-Kutta method of order 10.* J.Inst. Maths Applics, Vol.21, p.47-59. *[II.5]*

E. Hairer (1981): *Order conditions for numerical methods for partitioned ordinary differential equations.* Numer. Math., Vol.36, p.431-445. *[II.15]*

E. Hairer (1982): *A one-step method of order 10 for $y'' = f(x,y)$.* IMA J. Num. Anal., Vol.2, p.83-94. *[II.14]*

E. Hairer & Ch. Lubich (1984): *Asymptotic expansions of the global error of fixed-stepsize methods.* Numer. Math., Vol.45, p.345-360. *[II.8], [III.9]*

E. Hairer & A. Ostermann (1990): *Dense output for extrapolation methods.* Numer. Math., Vol.58, p.419-439. *[III.9]*

E. Hairer & A. Ostermann (1992): *Dense output for the GBS extrapolation method.* In: Computational ordinary differential equations, ed. by J.R. Cash & I. Gladwell, Clarendon Press, Oxford, p.107-114. *[II.9]*

E. Hairer & G. Wanner (1973): *Multistep-multistage-multiderivative methods for ordinary differential equations.* Computing, Vol.11, p.287-303. *[II.13], [III.8]*

E. Hairer & G. Wanner (1974): *On the Butcher group and general multi-value methods.* Computing, Vol.13, p.1-15. *[II.2], [II.12], [II.13]*

E. Hairer & G. Wanner (1976): *A theory for Nyström methods.* Numer. Math., Vol.25, p.383-400. *[II.14]*

E. Hairer & G. Wanner (1983): *On the instability of the BDF formulas.* SIAM J. Numer. Anal., Vol.20, No.6, p.1206-1209. *[III.3]*

Sir W. R. Hamilton (1833): *On a general method of expressing the paths of light, and of the planets, by the coefficients of a characteristic function.* Dublin Univ. Review, p.795-826; Math. Papers, Vol.I, p.311-332. *[I.6]*

Sir W. R. Hamilton (1834): *On a general method in dynamics; by which the study of the motions of all free systems of attracting or repelling points is reduced to the search and differentiation of one central relation, or characteristic function.* Phil. Trans. Roy. Soc. Part II for 1834, p.247-308; Math. Papers, Vol.II, p.103-161. *[I.6]*

Sir W. R. Hamilton (1835): *Second essay on a general method in dynamics.* Phil. Trans. Roy. Soc. Part I for 1835, p.95-144; Math. Papers, Vol.II, p.162-211. *[I.6]*

P.C. Hammer & J.W. Hollingsworth (1955): *Trapezoidal methods of approximating solutions of differential equations.* MTAC, Vol.9, p.92-96. *[II.7]*

E. Hansen, see J. Donelson & E. Hansen.

F. Hausdorff (1906): *Die symbolische Exponentialformel in der Gruppentheorie.* Berichte ü. d. Verh. Königl. Sächs. Ges. d. Wiss. Leipzig, Math.-Phys. Klasse, Vol.58, p.19-48. *[II.16]*

K. Hayashi, see M. Okamoto & K. Hayashi.

N.D. Hayes (1950): *Roots of the transzendental equation associated with a certain difference-differential equation.* J. of London Math. Soc., Vol.25, p.226-232. *[II.17]*

H.M. Hebsacker (1982): *Conditions for the coefficients of Runge-Kutta methods for systems of n-th order differential equations.* J. Comput. Appl. Math., Vol.8, p.3-14. *[II.14]*

P. Henrici (1962): *Discrete variable methods in ordinary differential equations.* John Wiley & Sons, Inc., New-York-London-Sydney. *[II.2], [II.8], [III.1], [III.2], [III.10]*

P. Henrici (1974): *Applied and computational complex analysis.* Volume 1, John Wiley & Sons, New York, 682pp. *[I.13], [III.3]*

Ch. Hermite (1878): *Extrait d'une lettre de M. Ch. Hermite à M. Borchardt sur la formule d'interpolation de Lagrange.* J. de Crelle, Vol.84, p.70; Oeuvres, tome III, p.432-443. *[II.13]*

K. Heun (1900): *Neue Methode zur approximativen Integration der Differentialgleichungen einer unabhängigen Veränderlichen.* Zeitschr. für Math. u. Phys., Vol.45, p.23-38. *[I.5], [II.1]*

D.J. Higham, see W.H. Enright & D.J. Higham.

A.C. Hindmarsh (1972): *GEAR: ordinary differential equation system solver.* UCID-30001, Rev.2, LLL, Livermore, Calif. *[III.7]*

A.C. Hindmarsh (1980): *LSODE and LSODI, two new initial value ordinary differential equation solvers.* ACM Signum Newsletter 15,4. *[II.4]*

A.C. Hindmarsh, see also P.N. Brown, G.D. Byrne & A.C. Hindmarsh, G.D. Byrne & A.C. Hindmarsh.

J.O. Hirschfelder, see C.F. Curtiss & J.O. Hirschfelder.

E.W. Hobson (1921): *The theory of functions of a real variable.* Vol.I, Cambridge, 670pp. *[I.10]*

E. Hofer (1976): *A partially implicit method for large stiff systems of ODEs with only few equations introducing small time-constants.* SIAM J. Numer. Anal., vol 13, No.5, p.645-663. *[II.15]*

J.W. Hollingsworth, see P.C. Hammer & J.W. Hollingsworth.

G.'t Hooft (1974): *Magnetic monopoles in unified gauge theories.* Nucl. Phys., Vol.B79, p.276-284. *[I.6]*

E. Hopf (1942): *Abzweigung einer periodischen Lösung von einer stationären Lösung eines Differentialsystems.* Ber. math. physik. Kl. Akad. d. Wiss. Leipzig, Bd.XCIV, p.3-22. *[I.16]*

M.K. Horn (1983): *Fourth and fifth-order scaled Runge-Kutta algorithms for treating dense output.* SIAM J.Numer. Anal., Vol.20, p.558-568. *[II.6]*

P.J. van der Houwen (1977): *Construction of integration formulas for initial value problems.* North-Holland Amsterdam, 269pp. *[II.1]*

P.J. van der Houwen & B.P. Sommeijer (1990): *Parallel iteration of high-order Runge-Kutta methods with step size control.* J. Comput. Appl. Math., Vol.29, p.111-127. *[II.11]*

P.J. van der Houwen, see also H. Brunner & P.J. van der Houwen.

T.E. Hull (1967): A search for optimum methods for the numerical integration of ordinary differential equations. SIAM Rev., Vol.9, p.647-654. *[II.1], [II.3]*

T.E. Hull & R.L. Johnston (1964): *Optimum Runge-Kutta methods.* Math. Comput., Vol.18, p.306-310. *[II.3]*

B.L. Hulme (1972): *One-step piecewise polynomial Galerkin methods for initial value problems.* Math. of Comput., Vol.26, p.415-426. *[II.7]*

W.H. Hundsdorfer & M.N. Spijker (1981): *A note on B-stability of Runge-Kutta methods.* Num. Math., Vol.36, p.319-331. *[II.12]*

A. Hurwitz (1895): *Ueber die Bedingungen, unter welchen eine Gleichung nur Wurzeln mit negativen reellen Theilen besitzt.* Math. Ann., Vol.46, p.273-284; Werke, Vol.2, p.533ff. *[I.13]*

A. Huťa (1956): *Une amélioration de la méthode de Runge-Kutta-Nyström pour la résolution numérique des équations différentielles du premier ordre.* Acta Fac. Rerum Natur. Univ. Comenianae (Bratislava) Math., Vol.1, p.201-224. *[II.5]*

B.J. Hyett, see R.F. Warming & B.J. Hyett.

E.L. Ince (1944): *Ordinary differential equations.* Dover Publications, New York, 558pp. *[I.2], [I.3]*

A. Iserles & S.P. Nørsett (1990): On the theory of parallel Runge-Kutta methods. IMA J. Numer. Anal., Vol.10, p.463-488. *[II.11]*

Z. Jackiewicz & M. Zennaro (1992): *Variable stepsize explicit two-step Runge-Kutta methods.* Math. Comput., Vol.59, p.421-438. *[III.8]*

K.R. Jackson (1991): *A survey of parallel numerical methods for initial value problems for ordinary differential equations.* IEEE Trans. on Magnetics, Vol.27, p.3792-3797. *[II.11]*

K.R. Jackson & S.P. Nørsett (1992): *The potential of parallelism in Runge-Kutta methods. Part 1: RK formulas in standard form.* Report. *[II.11]*

K.R. Jackson, see also W.H. Enright, K.R. Jackson, S.P. Nørsett & P.G. Thomson.

C.G.J. Jacobi (1841): *De determantibus functionalibus.* Crelle J. f. d. r. u. angew. Math, Vol.22, p.319-359, Werke, Vol.III, p.393-438. *[I.14]*

C.G.J. Jacobi (1842/43): *Vorlesungen über Dynamik,* gehalten an der Universität zu Königsberg im Wintersemester 1842–1843 und nach einem von C.W. Borchardt ausgearbeiteten Hefte, edited 1866 by A. Clebsch, Werke, Vol. VIII. *[I.6],[I.14]*

C.G.J. Jacobi (1845): *Theoria novi multiplicatoris systemati aequationum differentialum vulgarium applicandi.* Crelle J. f. d. r. u. angew. Math, Vol.29, p.213-279, 333-376. Werke, Vol.IV, p.395-509. *[I.11]*

R. Jeltsch, see G. Dahlquist & R. Jeltsch.

R.L. Johnston, see T.E. Hull & R.L. Johnston.

W.P. Jones, see R.A. Frazer, W.P. Jones & S.W. Skan.

C. Jordan (1870): *Traité des Substitutions et des équations algébriques.* Paris 667pp. *[I.12]*

C. Jordan (1928): *Sur une formule d'interpolation.* Atti Congresso Bologna, vol 6, p.157-177 *[II.9]*

B. Kågström & A. Ruhe (1980): *An algorithm for numerical computation of the Jordan normal form of a complex matrix.* ACM Trans. Math. Software, Vol.6, p.398-419. (Received May 1975, revised Aug. 1977, accepted May 1979). *[I.12]*

D.K. Kahaner, see R. Piessens, E. de Doncker-Kapenga, C.W. Überhuber & D.K. Kahaner.

S. Kakutani & L. Marcus (1958): *On the non-linear difference-differential equation* $y'(t) = [A - By(t - \tau)]y(t)$. In: Contributions to the theory of nonlinear oscillations, ed. by S.Lefschetz, Princeton, Vol.IV, p.1-18. *[II.17]*

E. Kamke (1930): *Ueber die eindeutige Bestimmtheit der Integrale von Differentialgleichungen II.* Sitz. Ber. Heidelberg Akad. Wiss. Math. Naturw. Kl., 17. Abhandl., see also Math. Zeitschr., Vol.32, p.101-107. *[I.10]*

E. Kamke (1942): *Differentialgleichungen, Lösungsmethoden und Lösungen.* Becker & Erler, Leipzig, 642pp. *[I.3]*

K.H. Kastlunger & G. Wanner (1972): *Runge Kutta processes with multiple nodes.* Computing, Vol.9, p.9-24. *[II.13]*

K.H. Kastlunger & G. Wanner (1972b): *On Turan type implicit Runge-Kutta methods.* Computing, Vol.9, p.317-325. *[II.13]*

A.G.Mc. Kendrick, see W.O. Kermack & A.G.Mc. Kendrick.

W.O. Kermack & A.G.Mc. Kendrick (1927): *Contributions to the mathematical theory of epidemics (Part I).* Proc. Roy. Soc., A, Vol.115, p.700-721. *[II.17]*

H. Knapp & G. Wanner (1969): *LIESE II, A program for ordinary differential equations using Lie-series.* MRC Report No.1008, Math. Research Center, Univ. Wisconsin, Madison, Wisc. 53706. *[I.8]*

H. König, see C. Runge & H. König.

F.T. Krogh (1969): *A variable step variable order multistep method for the numerical solution of ordinary differential equations.* Information Processing 68, North-Holland, Amsterdam, p.194-199. *[III.5]*

F.T. Krogh (1973): *Algorithms for changing the step size.* SIAM J. Num. Anal. 10, p.949-965. *[III.5]*

F.T. Krogh (1974): *Changing step size in the integration of differential equations using modified devided differences*. Proceedings of the Conference on the Num. Sol. of ODE, Lecture Notes in Math. No.362, Springer Verlag New York, p.22-71. *[III.5]*

N. Kryloff & N. Bogoliuboff (1947): *Introduction to non-linear Mechanics*. Free translation by S. Lefschetz, Princeton Univ. Press, 105pp. *[I.16]*

E.E. Kummer (1839): *Note sur l'intégration de l'équation* $d^n y/dx^n = x^m y$ *par des intégrales définies*. Crelle J. f. d. r. u. angew. Math., Vol.19, p.286-288. *[I.11]*

J. Kuntzmann (1961): *Neuere Entwickelungen der Methode von Runge-Kutta*. ZAMM, Vol.41, p.28-31. *[II.7]*

J. Kuntzmann, see also F. Ceschino & J. Kuntzmann.

W. Kutta (1901): *Beitrag zur näherungsweisen Integration totaler Differentialgleichungen*. Zeitschr. für Math. u. Phys., Vol.46, p.435-453. *[II.1]*, *[II.2]*, *[II.3]*, *[II.5]*

J.L.de Lagrange (1759): *Recherches sur la nature et la propagation du son*. Miscell. Taurinensia t.I, Oeuvres t.1, p.39-148. *[I.6]*

J.L.de Lagrange (1762): *Solution de différents problèmes de Calcul Intégral*. Miscell. Taurinensa, t.III, Oeuvres t.1, p.471-668. *[I.6]*

J.L.de Lagrange (1774): *Sur les Intégrales particulières des Equations différentielles*. Oeuvres, tom.4, p.5-108. *[I.2]*

J.L.de Lagrange (1775): *Recherche sur les Suites Récurrentes*. Nouveaux Mém. de l'Acad. royale des Sciences et Belles-Lettres, Berlin. Oeuvres, Vol.4, p.159. *[I.4]*, *[III.3]*

J.L.de Lagrange (1788): *Mécanique analytique*. Paris, Oeuvres t.11 et 12. *[I.4]*, *[I.6]*, *[I.12]*

J.L.de Lagrange (1792): *Mémoire sur l'expression du terme géneral des séries récurrentes, lorsque l'équation génératrice a des racines égales*. Nouv. Mém. de l'Acad. royale des Sciences de Berlin, Oeuvres t.5, p.627-641. *[III.3]*

J.L.de Lagrange (1797): *Théorie des fonctions analytiques, contenant les principes du calcul différentiel, dégagés de toute considération d'infiniment petits, d'évanouissants, de limites et de fluxions, et réduits à l'analyse algébrique des quantités finies*. Paris, 1797, nouv. ed. 1813, Oeuvres Tome 9. *[II.3]*

E. Laguerre (1883): *Mémiore sur la théorie des équations numériques*. J. Math. pures appl. (3e série), Vol.9, p.99-146 (also in *Oeuvres* I, p.3-47). *[II.9]*

J.D. Lambert (1987): *Developments in stability theory for ordinary differential equations*. In: The state of the art in numerical analysis, ed. by A. Iserles & M.J.D. Powell, Clarendon Press, Oxford, p.409-431. *[I.13]*

M. Laloy, see N. Rouche, P. Habetsand & M. Laloy.

R.J. Lambert, see G.D. Byrne & R.J. Lambert.

P.S. Laplace (An XIII = 1805): *Supplément au dixième livre du Traité de mécanique céleste sur l'action capillaire*. Paris chez Courcier, 65+78pp. *[II.1]*

F.M. Lasagni (1988): *Canonical Runge-Kutta methods*. ZAMP Vol.39, p.952-953. *[II.16]*

F.M. Lasagni (1990): *Integration methods for Hamiltonian differential equations*. Unpublished manuscript. *[II.16]*

P.D. Lax & R.D. Richtmyer (1956): *Survey of the stability of linear limite difference equations*. Comm. Pure Appl. Math., Vol.9, p.267-293. *[III.4]*

R. Lefever & G. Nicolis (1971): *Chemical Instabilities and sustained oscillations*. J. theor. Biol., Vol.30, p.267-284. *[I.16]*

A.M. Legendre (1787): *Mémoire sur l'intégration de quelques équations aux différences partielles*. Histoire Acad. R. Sciences, Paris, Année MDCCLXXXVII, à Paris MDC-CLXXXIX, p.309-351. *[I.6]*

G.W. Leibniz (1684): *Nova methodus pro maximis et minimis, itemque tangentibus, quae nec fractas, nec irrationales quantitates moratur, & singulare pro illis calculi genus*. Acta Eruditorum, Lipsiae, MDCLXXXIV, p.467-473. *[I.2]*

G.W. Leibniz (1691): *Methodus, qua innummerarum linearum construction ex data proprietate tangentium seu aequatio inter abscissam et ordinatam ex dato valore subtangentialis, exhibetur*. Letter to Huygens, in: C.I. Gerhardt, Leibnizens math. Schriften, 1850, Band II, p.116-121. *[I.3]*

G.W. Leibniz (1693) (Gothofredi Guilielmi Leibnitzii): *Supplementum Geometriae Dimensoriae seugeneralissima omnium tetra gonismorum effectio per motum: Similiterque multiplex constructio linea ex data tangentium conditione*. Acta Eruditorum, Lipsiae, p.385-392; german translation: G. Kowalewski, Leibniz über die Analysis des Unendlichen, Ostwalds Klassiker Nr.162 (1908), p.24-34. *[I.2]*

A.M. Liapunov (1892): *Problème général de la stabilité du mouvement*. Russ., trad. en français 1907 (Annales de la Faculté des Sciences de Toulouse), reprinted 1947 Princeton Univ. Press, 474pp. *[I.13]*

A.M. Liénard (1928): *Etude des oscillations entretenues*. Revue générale de l'Electricité, tome XXIII, p. 901-912 et 946-954. *[I.16]*

B. Lindberg (1972): *A simple interpolation algorithm for improvement of the numerical solution of a differential equation*. SIAM J. Numer. Anal., Vol.9, p.662-668. *[II.9]*

E. Lindelöf (1894): *Sur l'application des méthodes d'approximation successives à l'étude des intégrales réelles des équations différentielles ordinaires*. J. de Math., 4e série, Vol.10, p.117-128. *[I.8]*

W. Liniger, see W.L. Miranker & W. Liniger.

J. Liouville (1836): *Sur le développement des fonctions ou parties de fonctions en séries dont les divers termes sont assujétis à satisfaire à une même équation différentielle du second ordre, contenant un paramètre variable*. Journ. de Math. pures et appl., Vol.1, p.253-265. *[I.8], [I.15]*

J. Liouville (1838): *Sur la Théorie de la variation des constantes arbitraires*. Liouville J. de Math., Vol.3, p.342-349. *[I.8], [I.11]*

J. Liouville (1841): *Remarques nouvelles sur l'équation de Riccati*. J. des Math. pures et appl., Vol.6, p.1-13. *[I.3]*

R. Lipschitz (1876): *Sur la possibilité d'intégrer complètement un système donné d'équations différentielles*. Bulletin des Sciences Math. et Astr., Paris, Vol.10, p.149-159. *[I.7]*

F.J. Lisbona, see M. Crouzeix & F.J. Lisbona.

R. Lobatto (1852): *Lessen over Differentiaal- en Integraal-Rekening*. 2 Vol., La Haye 1851-52. *[II.7]*

E.N. Lorenz (1979): *On the prevalence of aperiodicity in simple systems.* Global Analysis, Calgary 1978, ed. by M.Grmela and J.E.Marsden, Lecture Notes in Mathematics, Vol.755, p.53-75. *[I.16]*

F.R. Loscalzo (1969): *An introduction to the application of spline functions to initial value problems.* In: Theory and Applications of spline functions, ed. T.N.E. Greville, Acad. Press 1969, p.37-64. *[II.13]*

F.R. Loscalzo & I.J. Schoenberg (1967): *On the use of spline functions for the approximation of solutions of ordinary differential equations.* Tech. Summ. Rep. # 723, Math. Res. Center, Univ. Wisconsin, Madison. *[II.13]*

M. Lotkin (1951): *On the accuracy of Runge-Kutta methods.* MTAC Vol.5, p.128-132. *[II.3]*

Ch. Lubich (1989): *Linearly implicit extrapolation methods for differential-algebraic systems.* Numer. Math., Vol.55, p.197-211. *[II.9]*

Ch. Lubich, see also E. Hairer & Ch. Lubich.

G.I. Marchuk (1975): *Prostejshaya matematicheskaya model virusnogo zabolevaniya.* Novosibirsk, VTS SO AN SSSR. Preprint. *[II.17]*

G.I. Marchuk (1983): *Mathematical models in immunology.* Translation series, Optimization Software, New York, Springer Verlag, 351pp. *[II.17]*

L. Marcus, see S. Kakutani & L. Marcus.

M. Marden (1966): *Geometry of polynomials.* American Mathematical Society, Providence, Rhode Island, 2nd edition. *[III.3]*

R.M. May (1976): *Simple mathematical models with very complicated dynamics.* Nature, Vol.261, p.459-467 *[I.16]*

R.M.M. Mattheij, see U.M. Ascher, R.M.M. Mattheij & R.D. Russel.

R.H. Merson (1957): *An operational method for the study of integration processes.* Proc. Symp. Data Processing, Weapons Research Establishment, Salisbury, Australia, p.110-1 to 110-25. *[II.2], [II.4], [II.14]*

S. Miesbach & H.J. Pesch (1992): *Symplectic phase flow approximation for the numerical integration of canonical systems.* Numer.Math., Vol.61, p.501-521. *[II.16]*

J.J.H. Miller, see D.M. Creedon & J.J.H. Miller.

W.E. Milne (1926): *Numerical integration of ordinary differential equations.* Amer. Math. Monthly, Vol.33, p.455-460. *[III.1]*

W.E. Milne (1970): *Numerical solution of differential equations.* Dover Publications, Inc., New York, second edition. *[III.1]*

W.L. Miranker (1971): *A survey of parallelism in numerical analysis.* SIAM Review, Vol.13, p.524-547. *[II.11]*

W.L. Miranker & W. Liniger (1967): *Parallel methods for the numerical integration of ordinary differential equations.* Math. Comput., Vol.21, p. 303-320. *[II.11]*

R. von Mises (1930): *Zur numerischen Integration von Differentialgleichungen.* ZAMM, Vol.10, p.81-92. *[III.4]*

A.R. Mitchell & J.W. Craggs (1953): *Stability of difference relations in the solution of ordinary differential equations*. Math. Tables Aids Comput., Vol.7, p.127-129. *[III.1]*, *[III.3]*

C. Moler & C. Van Loan (1978): *Nineteen dubious ways to compute the exponential of a matrix;* SIAM Review, Vol.20, p.801-836. *[I.12]*

J.I. Montijano, see M. Calvo, J.I. Montijano & L. Rández.

R.E. Moore (1966): *Interval Analysis*. Prentice-Hall, Inc, 145pp. *[I.8]*

R.E. Moore (1979): *Methods and applications of interval analysis*. SIAM studies in Appl. Math., 190pp. *[I.8]*

P. Moss, see K. Burrage & P. Moss.

F.R. Moulton (1926): *New methods in exterior ballistics*. Univ. Chicago Press. *[III.1]*

M. Müller (1926): *Ueber das Fundamentaltheorem in der Theorie der gewöhnlichen Differentialgleichungen*. Math. Zeitschr., Vol.26, p.619-645. (Kap.III). *[I.10]*

F.D. Murnaghan, see A. Wintner & F.D. Murnaghan.

O. Nevanlinna (1989): *Remarks on Picard-Lindelöf iteration*. BIT, Vol.29, p.328-346 and 535-562. *[I.8]*

E.H. Neville (1934): *Iterative interpolation*. Ind. Math. Soc. J. Vol.20, p.87-120. *[II.9]*

I. Newton (1671): *Methodus Fluxionum et Serierum Infinitarum*. edita Londini 1736, Opuscula mathematica, Vol.I, Traduit en français par M.de Buffon, Paris MDCCXL. *[I.2]*

I. Newton (1687): *Philosophiae naturalis principia mathematica*. Imprimatur S. Pepys, Reg. Soc. Praeses, julii 5, 1686, Londini anno MDCLXXXVII. *[I.6]*, *[II.14]*

I. Newton (1711): *Methodus differentialis (Analysis per quantitatum, series, fluxiones, ac differentias: cum enumeratione linearum tertii ordinis)*. London 1711. *[III.1]*

G. Nicolis, see R. Lefever & G. Nicolis.

J. Nievergelt (1964): *Parallel methods for integrating ordinary differential equations*. Comm. ACM, Vol.7, p.731-733. *[II.11]*

S.P. Nørsett (1974a): *One-step methods of Hermite type for numerical integration of stiff systems*. BIT, Vol.14, p.63-77. *[II.13]*

S.P. Nørsett (1974b): *Semi explicit Runge-Kutta methods*. Report No.6/74, ISBN 82-7151-009-6, Dept. Math. Univ. Trondheim, Norway, 68+7pp. *[II.7]*

S.P. Nørsett & G. Wanner (1981): *Perturbed collocation and Runge-Kutta methods*. Numer. Math., Vol.38, p.193-208. *[II.7]*

S.P. Nørsett, see also A. Iserles & S.P. Nørsett, K.R. Jackson & S.P. Nørsett, W.H. Enright, K.R. Jackson, S.P. Nørsett & P.G. Thomson.

A. Nordsieck (1962): *On numerical integration of ordinary differential equations*. Math. Comp., Vol.16, p.22-49. *[III.6]*

U. Nowak, see P. Deuflhard & U. Nowak.

R.M. Noyes, see J.R. Field & R.M. Noyes.

B. Numerov (B.V.Noumerov) (1924): *A method of extrapolation of perturbations*. Monthly notices of the Royal Astronomical Society, Vol.84, p.592-601. *[III.10]*

B. Numerov (1927): *Note on the numerical integration of $d^2x/dt^2 = f(x,t)$.* Astron. Nachrichten, Vol.230, p.359-364. *[III.10]*

E.J. Nyström (1925): *Ueber die numerische Integration von Differentialgleichungen.* Acta Soc. Sci. Fenn., Vol.50, No.13, p.1-54. *[II.2], [II.14], [III.1]*

H.J. Oberle & H.J. Pesch (1981): *Numerical treatment of delay differential equations by Hermite interpolation.* Numer. Math., Vol.37, p.235-255. *[II.17]*

N. Obreschkoff (1940): *Neue Quadraturformeln.* Abh. der Preuss. Akad. der Wiss., Math.-naturwiss. Klasse, Nr.4, Berlin. *[II.13]*

M. Okamoto & K. Hayashi (1984): *Frequency conversion mechanism in enzymatic feedback systems.* J. Theor. Biol., Vol.108, p.529-537. *[II.17]*

D. Okunbor & R.D. Skeel (1992): *An explicit Runge-Kutta-Nyström method is canonical if and only if its adjoint is explicit.* SIAM J. Numer. Anal., Vol.29, p. 521-527. *[II.16]*

D. Okunbor & R.D. Skeel (1992b): *Explicit canonical methods for Hamiltonian systems.* Math. Comput., Vol.59, p.439-455. *[II.16]*

D. Okunbor & R.D. Skeel (1992c): *Canonical Runge-Kutta-Nyström methods of orders 5 and 6,* Working Document 92-1, Dep. Computer Science, Univ. Illinois. *[II.16]*

J. Oliver (1975): *A curiosity of low-order explicit Runge-Kutta methods.* Math. Comp., Vol.29, p.1032-1036. *[II.1]*

J. Oppelstrup (1976): *The RKFHB4 method for delay-differential equations.* Lect. Notes Math., Nr. 631, p.133-146. *[II.17]*

M.R. Osborne (1966): *On Nordsieck's method for the numerical solution of ordinary differential equations.* BIT, Vol.6, p.51-57. *[III.6]*

O. Østerby, see C.W. Gear & O. Østerby.

A. Ostermann, see E. Hairer & A. Ostermann.

B. Owren & M. Zennaro (1991): *Order barriers for continuous explicit Runge-Kutta methods.* Math. Comput., Vol.56, p.645-661. *[II.6]*

B. Owren & M. Zennaro (1992): *Derivation of efficient, continuous, explicit Runge-Kutta methods.* SIAM J. Sci. Stat. Comput., Vol.13, p.1488-1501. *[II.6]*

B.N. Parlett (1976): *A recurrence among the elements of functions of triangular matrices.* Linear Algebra Appl., Vol.14, p.117-121. *[I.12]*

G. Peano (1888): *Intégration par séries des équations différentielles linéaires.* Math. Annalen, Vol.32, p.450-456. *[I.8], [I.9]*

G. Peano (1890): *Démonstration de l'intégrabilité des équations différentielles ordinaires,* Math. Annalen, Vol.37, p.182-228; see also the german translation and commentation: G. Mie, Math. Annalen, Vol.43 (1893), p.553-568. *[I.1], [I.7], [I.9], [I.10]*

G. Peano (1913): *Resto nelle formule di quadratura, espresso con un integrale definito.* Atti Della Reale Accad. Dei Lincei, Rendiconti, Vol.22, N.9, p.562-569, Roma. *[III.2]*

R. Pearl & L.J. Reed (1922): *A further note on the mathematical theory of population growth.* Proceedings of the National Acad. of Sciences, Vol.8, No.12, p.365-368. *[II.17]*

L.M. Perko (1984): *Limit cycles of quadratic systems in the plane.* Rocky Mountain J. of Math., Vol.14, p.619-645. *[I.16]*

512   Bibliography

O. Perron (1915): *Ein neuer Existenzbeweis für die Integrale der Differentialgleichung* $y' = f(x, y)$. Math.Annalen, Vol.76, p.471-484. *[I.10]*

O. Perron (1918, zur Zeit im Felde): *Ein neuer Existenzbeweis für die Integrale eines Systems gewöhnlicher Differentialgleichungen.* Math. Annalen, Vol.78, p.378-384. *[I.7]*

O. Perron (1930): *Ueber ein vermeintliches Stabilitätskriterium.* Nachrichten Göttingen, (1930) p.28-29 (see also Fort.d.Math. 1930 I, p.380.) *[I.13]*

H.J. Pesch, see H.J. Oberle & H.J. Pesch, S. Miesbach & H.J. Pesch.

D. Pfenniger (1990): *Stability of the Lagrangian points in stellar bars.* Astron. Astrophys., Vol.230, p.55-66. *[II.16]*

D. Pfenniger, see also T. de Zeeuw & D. Pfenniger.

E. Picard (1890): *Mémoire sur la théorie des équations aux dérivées partielles et la méthode des approximations successives.* J. de Math. pures et appl., 4e série, Vol.6, p.145-210. *[I.8]*

E. Picard (1891-96): *Traité d'Analyse.* 3 vols. Paris. *[I.7]*, *[I.8]*

R. Piessens, E. de Doncker-Kapenga, C.W. Überhuber & D.K. Kahaner (1983): *QUADPACK. A subroutine package for automatic integration.* Springer Series in Comput. Math., Vol.1, 301pp. *[II.10]*

P. Piotrowsky (1969): *Stability, consistency and convergence of variable k -step methods for numerical integration of large systems of ordinary differential equations.* Lecture Notes in Math., 109, Dundee 1969, p.221-227. *[III.5]*

H. Poincaré (1881,82,85): *Mémoire sur les courbes définies par les équations différentielles.* J. de Math., 3e série, t.7, p.375-422, 3e série, t.8, p.251-296, 4e série, t.1, p.167-244, Oeuvres t.1, p.3-84, 90-161. *[I.12]*, *[I.16]*

H. Poincaré (1892,1893,1899): *Les méthodes nouvelles de la mécanique céleste.* Tome I 385pp., Tome II 480pp., Tome III 414pp., Gauthier-Villars Paris. *[I.6]*, *[I.16]*, *[I.14]*, *[II.8]*

S.D. Poisson (1835): *Théorie mathématique de la chaleur.* Paris, Bachelier, 532pp., Supplément 1837, 72pp. *[I.15]*

B. Van der Pol (1926): *On "Relaxation Oscillations".* Phil. Mag., Vol.2, p.978-992; reproduced in: B. van der Pol, Selected Scientific Papers, Vol.I, North. Holland Publ. Comp. Amsterdam (1960). *[I.16]*

G. Pólya & G. Szegö (1925): *Aufgaben und Lehrsätze aus der Analysis.* Two volumes, Springer Verlag; many later editions and translations. *[II.9]*

P. Pouzet (1963): *Etude en vue de leur traitement numérique des équations intégrales de type Volterra.* Rev. Français Traitement Information (Chiffres), Vol.6, p.79-112. *[II.17]*

P.J. Prince & J.R. Dormand (1981): *High order embedded Runge-Kutta formulae.* J. Comp. Appl. Math., Vol.7, p.67-75. *[II.5]*

P.J. Prince, see also J.R. Dormand, M.E.A. El-Mikkawy & P.J. Prince, J.R. Dormand & P.J. Prince, R.W. Brankin, I. Gladwell, J.R. Dormand, P.J. Prince & W.L. Seward.

H. Prüfer (1926): *Neue Herleitung der Sturm-Liouvillschen Reihenentwicklung stetiger Funktionen.* Math. Annalen, Vol.95, p.499-518. *[I.15]*

D.I. Pullin & P.G. Saffman (1991): *Long-time symplectic integration: the example of four-vortex motion.* Proc. R. Soc. London, A, Vol.432, p.481-494. *[II.16]*

Qin Meng-Zhao & Zhu Wen-Jie (1991): *Canonical Runge-Kutta-Nyström (RKN) methods for second order ordinary differential equations.* Computers Math. Applic., Vol.22, p.85-95. *[II.16]*

Qin Meng-Zhao & Zhu Wen-Jie (1992): *Construction of higher order symplectic schemes by composition.* Computing, Vol.47, p.309-321. *[II.16]*

Qin Meng-zhao, see also Feng Kang, Wu Hua-mo, Qin Meng-zhao & Wang Dao-liu.

R. Radau (1880): *Étude sur les formes d'approximation qui servent à calculer la valeur numérique d'une intégrale définie.* Liouville J. de Mathém. pures et appl., 3eser., tome VI, p.283-336. (Voir p.307). *[II.7]*

A. Ralston (1962): *Runge-Kutta methods with minimum error bounds.* Math. Comput., Vol.16, p.431-437, corr., Vol.17, p.488. *[II.1]*, *[II.3]*, *[III.7]*

L. Rández, see M. Calvo, J.I. Montijano & L. Rández.

Lord Rayleigh (1883): *On maintained vibrations.* Phil. Mag. Ser.5, Vol.15, p.229-235. *[I.16]*

L.J. Reed, see R. Pearl & L.J. Reed.

W.T. Reid (1980): *Sturmian theory for ordinary differential equations.* Springer Verlag, Appl. Math., Serie31, 559pp. *[I.15]*

C. Reinsch, see J.H. Wilkinson & C. Reinsch.

R. Reissig (1954): *Erzwungene Schwingungen mit zäher und trockener Reibung.* Math. Nachrichten, Vol.11, p.345-384; see also p.231. *[II.6]*

P. Rentrop (1985): *Partitioned Runge-Kutta methods with stiffness detection and stepsize control.* Numer. Math., Vol.47, p.545-564. *II.15*

J. Riccati (1712): *Soluzione generale del Problema inverso intorno à raggi osculatori,..., determinar la curva, a cui convenga una tal'espressione.* Giornale de'Letterati d'Italia, Vol.11, p.204-220. *[I.3]*

J. Riccati (1723): *Animadversiones in aequationes differentiales secundi gradus.* Acta Erud. Lips., anno MDCCXXIII, p.502-510. *[I.3]*

L.F. Richardson (1910): *The approximate arithmetical solution by finite differences of physical problems including differential equations, with an application to the stresses in a masonry dam.* Phil. Trans., A, Vol.210, p.307-357. *[II.4]*

L.F. Richardson (1927): *The deferred approach to the limit.* Phil. Trans., A, Vol.226, p.299-349. *[II.4]*, *[II.9]*

R.D. Richtmyer, see P.D. Lax & R.D. Richtmyer.

B. Riemann (1854): *Ueber die Darstellbarkeit einer Function durch eine trigonometrische Reihe.* Von dem Verfasser behufs seiner Habilitation an der Universität zu Göttingen der philosophischen Facultät eingereicht; collected works p. 227-265. *[I.6]*

W. Romberg (1955): *Vereinfachte numerische Integration.* Norske Vid. Selsk. Forhdl, Vol.28, p.30-36. *[II.8]*, *[II.9]*

E. Rothe (1930): *Zweidimensionale parabolische Randwertaufgaben als Grenzfall eindimensionaler Randwertaufgaben.* Math. Annalen, Vol.102, p. 650-670. *[I.1]*

N. Rouche, P. Habetsand & M. Laloy (1977): *Stability theory by Liapunov's direct method.* Appl. Math. Sci. 22, Springer Verlag, 396pp. *[I.13]*

E.J. Routh (1877): *A Treatise on the stability of a given state of motions.* Being the essay to which the Adams prize was adjudged in 1877, in the University of Cambridge. London 108pp. *[I.13]*

E.J. Routh (1884): *A Treatise on the dynamics of a system of rigid bodies, part I and II.* 4th edition (1st ed. 1860, 6th ed. 1897, german translation with remarks of F.Klein 1898). *[I.12]*

D. Ruelle & F. Takens (1971): *On the nature of turbulence.* Commun. Math. Physics, Vol.20, p.167-192. *[I.16]*

A. Ruhe, see B. Kågström & A. Ruhe.

C. Runge (1895): *Ueber die numerische Auflösung von Differentialgleichungen.* Math. Ann., Vol.46, p.167-178. *[II.1], [II.4]*

C. Runge (1905): *Ueber die numerische Auflösung totaler Differentialgleichungen.* Göttinger Nachr., p.252-257. *[II.1], [II.3]*

C. Runge & H. König (1924): *Vorlesungen über numerisches Rechnen.* Grundlehren XI, Springer Verlag, 372pp. *[I.8], [II.1]*

R.D. Russel, see U.M. Ascher, R.M.M. Mattheij & R.D. Russel.

R.D. Ruth (1983): *A canonical integration technique.* IEEE Trans. Nuclear Scince, Vol.NS-30, p.2669-2671. *[II.16]*

H. Rutishauser (1952): *Ueber die Instabilität von Methoden zur Integration gewöhnlicher Differentialgleichungen.* ZAMP, Vol.3, p.65-74. *[III.3]*

H. Rutishauser, see also F.L. Bauer, H. Rutishauser & E. Stiefel.

P.G. Saffman, see D.I. Pullin & P.G. Saffman.

J.M. Sanz-Serna (1988): *Runge-Kutta schemes for Hamiltonian systems.* BIT Vol.28, p.877-883. *[II.16]*

J.M. Sanz-Serna (1992): *Symplectic integrators for Hamiltonian problems: an overview.* Acta Numerica, Vol.1, p.243-286. *[II.16]*

J.M. Sanz-Serna (1992b): *The numerical integration of Hamiltonian systems.* In: Computational ordinary differential equations, ed. by J.R. Cash & I. Gladwell, Clarendon Press, Oxford, p.437-449. *[II.16]*

J.M. Sanz-Serna & L. Abia (1991): *Order conditions for canonical Runge-Kutta schemes.* SIAM J. Numer. Anal., Vol.28, p. 1081-1096. *[II.16]*

J.M. Sanz-Serna, see also M.P. Calvo & J.M. Sanz-Serna, L. Abia & J.M. Sanz-Serna.

D. Sarafyan (1966): *Error estimation for Runge-Kutta methods through pseudo-iterative formulas.* Techn. Rep. No 14, Lousiana State Univ., New Orleans, May 1966. *[II.4]*

L. Scheeffer (1884): *Zur Theorie der stetigen Funktionen einer reellen Veränderlichen.* Acta Mathematica, Vol.5, p.183-194. *[I.10]*

J. Schlöder, see H.G. Bock & J. Schlöder.

I.J. Schoenberg, see F.R. Loscalzo & I.J. Schoenberg.

I. Schur (1909): *Ueber die charakteristischen Wurzeln einer linearen Substitution mit einer Anwendung auf die Theorie der Integralgleichungen.* Math. Ann., Vol.66, p.488-510. *[I.12]*

C. Scovel, see P.J. Channell & C. Scovel.

W.L. Seward, see R.W. Brankin, I. Gladwell, J.R. Dormand, P.J. Prince & W.L. Seward.

L.F. Shampine (1979): *Storage reduction for Runge-Kutta codes.* ACM Trans. Math. Software, Vol.5, p.245-250. *[II.5]*

L.F. Shampine (1985): *Interpolation for Runge-Kutta methods.* SIAM J. Numer. Anal., Vol.22, p.1014-1027. *[II.6]*

L.F. Shampine (1986): *Some practical Runge-Kutta formulas.* Math. Comp., Vol.46, p.135-150. *[II.5]*, *[II.6]*

L.F. Shampine & L.S. Baca (1983): *Smoothing the extrapolated midpoint rule.* Numer. Math., Vol.41, p.165-175. *[II.9]*

L.F. Shampine & L.S. Baca (1986): *Fixed versus variable order Runge-Kutta.* ACM Trans. Math. Softw., Vol.12, p.1-23. *[II.9]*

L.F. Shampine, L.S. Baca & H.-J. Bauer (1983): *Output in extrapolation codes.* Comp. & Maths. with Appls., Vol.9, p.245-255. *[II.9]*

L.F. Shampine & M.K. Gordon (1975): *Computer Solution of Ordinary Differential Equations, The Initial Value Problem.* Freeman and Company, San Francisco, 318pp. *[III.7]*

L.F. Shampine & H.A. Watts (1979): *The art of writing a Runge-Kutta code. II.* Appl. Math. Comput., Vol.5, p.93-121. *[II.4]*, *[III.7]*

L.F. Shampine, H.A. Watts & S.M. Davenport (1976): *Solving nonstiff ordinary differential equations - The state of the art.* SIAM Rev., Vol.18, p.376-410. *[II.6]*

L.F. Shampine, see also I. Gladwell, L.F. Shampine & R.W. Brankin, P. Bogacki & L.F. Shampine.

E.B. Shanks (1966): *Solutions of differential equations by evaluations of functions.* Math. of Comp., Vol.20, p.21-38. *[II.5]*

Shi Songling (1980): *A concrete example of the existence of four limit cycles for plane quadratic systems.* Sci. Sinica, Vol.23, p.153-158. *[I.16]*

G.F. Simmons (1972): *Differential equations with applications and historical notes.* MC Graw-Hill, 465pp. *[I.16]*

H.H. Simonsen (1990): *Extrapolation methods for ODE's: continuous approximations, a parallel approach.* Dr.Ing. Thesis, Norwegian Inst. Tech., Div. of Math. Sciences. *[II.9]*

S.W. Skan, see R.A. Frazer, W.P. Jones & S.W. Skan.

R. Skeel (1976): *Analysis of fixed-stepsize methods.* SIAM J. Numer. Anal., Vol.13, p.664-685. *[III.4]*, *[III.8]*, *[III.9]*

R.D. Skeel (1979): *Equivalent forms of multistep formulas.* Math. Comput., Vol.33, p.1229-1250. *[III.6]*

R.D. Skeel, see also D. Okunbor & R.D. Skeel.

B.P. Sommeijer, see P.J. van der Houwen & B.P. Sommeijer.

516   Bibliography

D. Sommer (1965): *Numerische Anwendung impliziter Runge-Kutta-Formeln.* ZAMM, Vol. 45, Sonderheft, p. T77-T79. *[II.7]*

F. Sommer, see H. Behnke & F. Sommer.

A. Sommerfeld (1942): *Vorlesungen über theoretische Physik.* Bd.1., Mechanik; translated from the 4th german ed.: Acad. Press. *[II.10], [II.14]*

D.C. Sorensen, see R. Fletcher & D.C. Sorensen.

J.L. Soulé, see A. Guillou & J.L. Soulé.

M.N. Spijker (1971): *On the structure of error estimates for finite difference methods.* Numer. Math., Vol.18, pp.73-100. *[III.8]*

M.N. Spijker, see also W.H. Hundsdorfer & M.N. Spijker.

D.D. Stancu, see A.H. Stroud & D.D. Stancu.

J.F. Steffensen (1956): *On the restricted problem of three bodies.* K. danske Vidensk. Selsk., Mat-fys. Medd. 30 Nr.18. *[I.8]*

I.A. Stegun, see M. Abramowitz & I.A. Stegun.

H.J. Stetter (1970): *Symmetric two-step algorithms for ordinary differential equations.* Computing, Vol.5, p.267-280. *[II.9]*

H.J. Stetter (1971): *Local estimation of the global discretization error.* SIAM J. Numer. Anal., Vol.8, p.512-523. *[II.12]*

H.J. Stetter (1973): *Analysis of discretization methods for ordinary differential equations.* Springer Verlag, Berlin-Heidelberg-New York. *[II.8], [II.12], [III.2], [III.8], [III.9]*

H.J. Stetter, see also W.B. Gragg & H.J. Stetter.

D. Stewart (1990): *A high accuracy method for solving ODEs with discontinuous right-hand side.* Numer. Math., Vol.58, p.299-328. *[II.6]*

E. Stiefel, see F.L. Bauer, H. Rutishauser & E. Stiefel.

J. Stoer, see R. Bulirsch & J. Stoer.

C. Störmer (1907): *Sur les trajectoires des corpuscules électrisés.* Arch. sci. phys. nat., Genève, Vol.24, p.5-18, 113-158, 221-247. *[III.10]*

C. Störmer (1921): *Méthodes d'intégration numérique des équations différentielles ordinaires.* C.R. congr. intern. math., Strasbourg, p.243-257. *[II.14], [III.10]*

A.H. Stroud & D.D. Stancu (1965): *Quadrature formulas with multiple Gaussian nodes.* SIAM J. Numer. Anal., ser.B., Vol.2, p.129-143. *[II.13]*

Ch. Sturm (1829): *Bulletin des Sciences de Férussac.* Tome XI, p.419, see also: Algèbre de Choquet et Mayer (1832). *[I.13]*

Ch. Sturm (1836): *Sur les équations différentielles linéaires du second ordre.* Journal de Math. pures et appl. (Liouville), Vol.1, p.106-186 (see also p.253, p.269, p.373 of this volume). *[I.15]*

Sun Geng (孙 耿) (1992): *Construction of high order symplectic Runge-Kutta Methods.* Comput. Math., Vol.11 (1993), p.250-260. *[II.16]*

Y.B. Suris (1989): *The canonicity of mappings generated by Runge-Kutta type methods when integrating the systems* $\ddot{x} = -\partial U/\partial x$. Zh. Vychisl. Mat. i Mat. Fiz., vol 29, p.202-211 (in Russian); same as U.S.S.R. Comput. Maths. Phys., vol 29., p.138-144. *[II.16]*

Y.B. Suris (1990): *Hamiltonian Runge-Kutta type methods and their variational formulation.* Mathematical Simulation, Vol.2, p.78-87 (Russian). *[II.16]*

V. Szebehely (1967): *Theory of orbits. The restricted problem of three bodies.* Acad. Press, New York, 668pp. *[II.0]*

G. Szegö, see G. Pólya & G. Szegö.

P.G.Tait, see W. Thomson (Lord Kelvin) & P.G.Tait.

F. Takens, see D. Ruelle & F. Takens.

K. Taubert (1976): *Differenzenverfahren für Schwingungen mit trockener und zäher Reibung und für Regelungssysteme.* Numer. Math., Vol.26, p.379-395. *[II.6]*

B. Taylor (1715): *Methodus incrementorum directa et inversa.* Londini 1715. *[I.6]*

W. Thomson (Lord Kelvin) & P.G.Tait (1879): *Treatise on natural philosophy (Vol.I., Part I).* Cambridge; New edition 1890, 508pp. *[I.12]*

P.G. Thomson, see W.H. Enright, K.R. Jackson, S.P. Nørsett & P.G. Thomson.

J. Todd (1950): *Notes on modern numerical analysis, I.* Math. Tables Aids Comput., Vol.4, p.39-44. *[III.3]*

W. Tollmien (1938): *Ueber die Fehlerabschätzung beim Adamsschen Verfahren zur Integration gewöhnlicher Differentialgleichungen.* ZAMM, Vol.18, p.83-90. *[III.4]*

S. Tremaine, see J. Binney & S. Tremaine.

K.W. Tu, see C.W. Gear & K.W. Tu.

C.W. Überhuber, see R. Piessens, E. de Doncker-Kapenga, C.W. Überhuber & D.K. Kahaner.

W. Uhlmann (1957): *Fehlerabschätzungen bei Anfangswertaufgaben gewöhnlicher Differentialgleichungssysteme 1. Ordnung.* ZAMM, Vol.37, p.88-99. *[I.10]*

B.T. Ulrich, see B. Giovannini, L. Weiss-Parmeggiani & B.T. Ulrich.

R. Vaillancourt, see M. Calvé & R. Vaillancourt.

C. Van Loan, see C. Moler & C. Van Loan.

R.S. Varga, see G. Birkhoff & R.S. Varga,

P.F. Verhulst (1845): *Recherches mathématiques sur la loi d'accroissement de la population.* Nuov. Mem. Acad. Roy. Bruxelles, Vol.18, p.3-38. *[II.17]*

J.H. Verner (1971): *On deriving explicit Runge-Kutta methods.* Proc. Conf. on Appl. Numer. Analysis, Lecture Notes in Mathematics 228, Springer Verlag, p.340-347. *[II.5]*

J.H. Verner (1978): *Explicit Runge-Kutta methods with estimates of the local truncation error.* SIAM J.Numer. Anal., Vol.15, p.772-790. *[II.5]*

J.H. Verner, see also G.J. Cooper & J.H. Verner.

L. Vietoris (1953): *Der Richtungsfehler einer durch das Adamssche Interpolationsverfahren gewonnenen Näherungslösung einer Gleichung* $y' = f(x, y)$. Oesterr. Akad. Wiss., Math.-naturw. Kl., Abt. IIa, Vol.162, p.157-167 and p.293-299. *[III.4]*

R. de Vogelaere (1956): *Methods of integration which preserve the contact transformation property of the Hamiltonian equations.* Report No. 4, Dept. Mathem., Unive. of Notre Dame, Notre Dame, Ind. *[II.16]*

V. Volterra (1934): *Remarques sur la Note de M. Régnier et Mlle Lambin.* C.R.Acad. Sc. t. CXCIX, p.1682. See also: V.Volterra - U.d'Ancona , Les associations biologiques au point de vue mathématique, Paris 1935. *[II.17]*

W. Walter (1970): *Differential and integral inequalities.* Springer Verlag 352pp., german edition 1964. *[I.10]*

W. Walter (1971): *There is an elementary proof of Peano's existence theorem.* Amer. Math. Monthly, Vol.78, p.170-173. *[I.7]*

Wang Dao-liu, see Feng Kang, Wu Hua-mo, Qin Meng-zhao & Wang Dao-liu.

G. Wanner (1969): *Integration gewöhnlicher Differentialgleichungen, Lie Reihen, Runge-Kutta-Methoden.* BI Mannheim Htb. 831/831a, 182pp. *[I.8]*

G. Wanner (1973): *Runge-Kutta methods with expansions in even powers of h.* Computing, Vol.11, p.81-85. *[II.8]*

G. Wanner (1983): *On Shi's counter example for the 16th Hilbert problem.* Internal Rep. Sect. de Math., Univ. Genève 1982; in german in: Jahrbuch Ueberblicke Mathematik 1983, ed. Chatterji, Fenyö, Kulisch, Laugwitz, Liedl, BI Mannheim, p.9-24. *[I.13], [I.16]*

G. Wanner, see also K.H. Kastlunger & G. Wanner, S.P. Nørsett & G. Wanner, E. Hairer & G. Wanner, H. Knapp & G. Wanner.

R.F. Warming & B.J. Hyett (1974): *The modified equation approach to the stability and accuracy analysis of finite-difference methods.* J. Comp. Phys., Vol.14, p.159-179. *[II.16]*

D.S. Watanabe, see C.W. Gear & D.S. Watanabe.

H.A. Watts (1983): *Starting stepsize for an ODE solver.* J. Comp. Appl. Math., Vol.9, p.177-191. *[II.4]*

H.A. Watts, see also L.F. Shampine & H.A. Watts, L.F. Shampine, H.A. Watts & S.M. Davenport.

K. Weierstrass (1858): *Ueber ein die homogenen Functionen zweiten Grades betreffendes Theorem, nebst Anwendung desselben auf die Theorie der kleinen Schwingungen.* Monats-ber. der Königl. Akad. der Wiss., 4. März 1858, Werke Bd.I, p.233-246. *[I.6]*

L. Weiss-Parmeggiani, see B. Giovannini, L. Weiss-Parmeggiani & B.T. Ulrich.

J. Weissinger (1950): *Eine verschärfte Fehlerabschätzung zum Extrapolationsverfahren von Adams.* ZAMM, Vol.30, p.356-363. *[III.4]*

H. Weyl (1939): *The classical groups.* Princeton, 302pp. *[I.14]*

O. Wilde (1892): *Lady Windermere's Fan, Comedy in four acts.* *[I.7]*

J.H. Wilkinson (1965): *The algebraic eigenvalue problem, Monographs on numerical analysis.* Oxford, 662pp. *[I.9]*

J.H. Wilkinson & C. Reinsch (1970): *Linear Algebra.* Grundlehren Band 186, Springer Verlag, 439pp. *[I.12]*

J.H. Wilkinson, see also G.H. Golub & J.H. Wilkinson.

A. Wintner & F.D. Murnaghan (1931): *A canonical form for real matrices under orthogonal transformations.* Proc. Nat. Acad. Sci. U.S.A., Vol.17, p.417-420. *[I.12]*

E.M. Wright (1945): *On a sequence defined by a non-linear recurrence formula.* J. of London Math. Soc., Vol.20, p.68-73. *[II.17]*

E.M. Wright (1946): *The non-linear difference-differential equation.* Quart. J. of Math., Vol.17, p.245-252. *[II.17]*

E.M. Wright (1955): *A non-linear difference-differential equation.* J.f.d.r.u. angew. Math., Vol.194, p.66-87. *[II.17]*

K. Wright (1970): *Some relationships between implicit Runge-Kutta collocation and Lanczos τ methods, and their stability properties.* BIT Vol.10, p.217-227. *[II.7]*

H. Wronski (1810): *Premier principe des méthodes algorithmiques comme base de la technique algorithmique.* Publication refused by the Acad. de Paris (for more details see: S.Dickstein, Int. Math. Congress 1904, p.515). *[I.11]*

Wu Hua-mo, see Feng Kang, Wu Hua-mo, Qin Meng-zhao & Wang Dao-liu.

H. Yoshida (1990): *Construction of higher order symplectic integrators.* Phys. Lett. A, Vol.150, p.262-268. *[II.16]*

H. Yoshida (1993): *Recent progress in the theory and application of symplectic integrators.* Celestial Mechanics Dynam. Astr., Vol.56, p.27-43. *[II.16]*

T. de Zeeuw & D. Pfenniger (1988): *Potential-density pairs for galaxies.* Mon. Not. R. astr. Soc., Vol.235, p.949-995. *[II.16]*

M. Zennaro (1986): *Natural continuous extensions of Runge-Kutta methods.* Math. Comput., Vol.46, p.119-133. *[II.17]*

M. Zennaro, see also A. Bellen & M. Zennaro, B. Owren & M. Zennaro, Z. Jackiewicz & M. Zennaro.

N.P Zhidkov, see I.S. Berezin & N.P Zhidkov.

Zhu Wen-Jie, see Qin Meng-Zhao & Zhu Wen-Jie.

J.A. Zonneveld (1963): *Automatic integration of ordinary differential equations.* Report R743, Mathematisch Centrum, Postbus 4079, 1009AB Amsterdam. Appeared in book form 1964. *[II.4]*

R. Zurmühl (1948): *Runge-Kutta-Verfahren zur numerischen Integration von Differentialgleichungen n-ter Ordnung.* ZAMM, Vol.28, p.173-182. *[II.14]*

R. Zurmühl (1952): *Runge-Kutta Verfahren unter Verwendung höherer Ableitungen.* ZAMM, Vol.32, p.153-154. *[II.13]*

Symbol Index

Subject Index